ALGEBRA

Lines

Slope of the line through $P_1 = (x_1, y_1)$ and $P_2 = (x_2, y_2)$:

$$m = \frac{y_2 - y_1}{x_2 - x_1}$$

Slope-intercept equation of line with slope m and y-intercept b:

$$y = mx + b$$

Point-slope equation of line through $P_1 = (x_1, y_1)$ with slope m:

$$y - y_1 = m(x - x_1)$$

Point-point equation of line through $P_1 = (x_1, y_1)$ and $P_2 = (x_2, y_2)$:

$$y - y_1 = m(x - x_1) \quad \text{where } m = \frac{y_2 - y_1}{x_2 - x_1}$$

Lines of slope m_1 and m_2 are parallel if and only if $m_1 = m_2$.
Lines of slope m_1 and m_2 are perpendicular if and only if $m_1 = -\frac{1}{m_2}$.

Circles

Equation of the circle with center (a, b) and radius r:

$$(x - a)^2 + (y - b)^2 = r^2$$

Distance and Midpoint Formulas

Distance between $P_1 = (x_1, y_1)$ and $P_2 = (x_2, y_2)$:

$$d = \sqrt{(x_2 - x_1)^2 + (y_2 - y_1)^2}$$

Midpoint of $\overline{P_1 P_2}$: $\left(\dfrac{x_1 + x_2}{2}, \dfrac{y_1 + y_2}{2} \right)$

Laws of Exponents

$$x^m x^n = x^{m+n} \qquad \frac{x^m}{x^n} = x^{m-n} \qquad (x^m)^n = x^{mn}$$

$$x^{-n} = \frac{1}{x^n} \qquad (xy)^n = x^n y^n \qquad \left(\frac{x}{y}\right)^n = \frac{x^n}{y^n}$$

$$x^{1/n} = \sqrt[n]{x} \qquad \sqrt[n]{xy} = \sqrt[n]{x}\,\sqrt[n]{y} \qquad \sqrt[n]{\frac{x}{y}} = \frac{\sqrt[n]{x}}{\sqrt[n]{y}}$$

$$x^{m/n} = \sqrt[n]{x^m} = \left(\sqrt[n]{x}\right)^m$$

Special Factorizations

$$x^2 - y^2 = (x + y)(x - y)$$
$$x^3 + y^3 = (x + y)(x^2 - xy + y^2)$$
$$x^3 - y^3 = (x - y)(x^2 + xy + y^2)$$

Binomial Theorem

$$(x + y)^2 = x^2 + 2xy + y^2$$
$$(x - y)^2 = x^2 - 2xy + y^2$$
$$(x + y)^3 = x^3 + 3x^2 y + 3xy^2 + y^3$$
$$(x - y)^3 = x^3 - 3x^2 y + 3xy^2 - y^3$$
$$(x + y)^n = x^n + nx^{n-1} y + \frac{n(n-1)}{2} x^{n-2} y^2$$
$$+ \cdots + \binom{n}{k} x^{n-k} y^k + \cdots + nxy^{n-1} + y^n$$

where $\dbinom{n}{k} = \dfrac{n(n-1) \cdots (n - k + 1)}{1 \cdot 2 \cdot 3 \cdots \cdot k}$

Quadratic Formula

If $ax^2 + bx + c = 0$, then $x = \dfrac{-b \pm \sqrt{b^2 - 4ac}}{2a}$.

Inequalities and Absolute Value

If $a < b$ and $b < c$, then $a < c$.
If $a < b$, then $a + c < b + c$.
If $a < b$ and $c > 0$, then $ca < cb$.
If $a < b$ and $c < 0$, then $ca > cb$.
$|x| = x \quad$ if $x \geq 0$
$|x| = -x \quad$ if $x \leq 0$

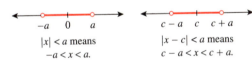

$|x| < a$ means $-a < x < a$.

$|x - c| < a$ means $c - a < x < c + a$.

GEOMETRY

Formulas for area A, circumference C, and volume V

Triangle
$A = \frac{1}{2}bh$
$= \frac{1}{2}ab \sin\theta$

Circle
$A = \pi r^2$
$C = 2\pi r$

Sector of Circle
$A = \frac{1}{2}r^2\theta$
$s = r\theta$
(θ in radians)

Sphere
$V = \frac{4}{3}\pi r^3$
$A = 4\pi r^2$

Cylinder
$V = \pi r^2 h$

Cone
$V = \frac{1}{3}\pi r^2 h$
$A = \pi r \sqrt{r^2 + h^2}$

Cone with arbitrary base
$V = \frac{1}{3}Ah$
where A is the area of the base

Pythagorean Theorem: For a right triangle with hypotenuse of length c and legs of lengths a and b, $c^2 = a^2 + b^2$.

TRIGONOMETRY

Angle Measurement

π radians $= 180°$

$1° = \dfrac{\pi}{180}$ rad $\qquad$ 1 rad $= \dfrac{180°}{\pi}$

$s = r\theta \quad$ (θ in radians)

Right Triangle Definitions

$\sin\theta = \dfrac{\text{opp}}{\text{hyp}} \qquad\qquad \cos\theta = \dfrac{\text{adj}}{\text{hyp}}$

$\tan\theta = \dfrac{\sin\theta}{\cos\theta} = \dfrac{\text{opp}}{\text{adj}} \qquad \cot\theta = \dfrac{\cos\theta}{\sin\theta} = \dfrac{\text{adj}}{\text{opp}}$

$\sec\theta = \dfrac{1}{\cos\theta} = \dfrac{\text{hyp}}{\text{adj}} \qquad \csc\theta = \dfrac{1}{\sin\theta} = \dfrac{\text{hyp}}{\text{opp}}$

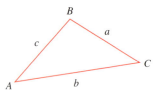

Trigonometric Functions

$\sin\theta = \dfrac{y}{r} \qquad\qquad \csc\theta = \dfrac{r}{y}$

$\cos\theta = \dfrac{x}{r} \qquad\qquad \sec\theta = \dfrac{r}{x}$

$\tan\theta = \dfrac{y}{x} \qquad\qquad \cot\theta = \dfrac{x}{y}$

$\lim\limits_{\theta\to 0} \dfrac{\sin\theta}{\theta} = 1 \qquad \lim\limits_{\theta\to 0} \dfrac{1-\cos\theta}{\theta} = 0$

$P = (r\cos\theta, r\sin\theta)$

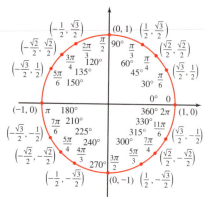

Fundamental Identities

$\sin^2\theta + \cos^2\theta = 1 \qquad\qquad \sin(-\theta) = -\sin\theta$

$1 + \tan^2\theta = \sec^2\theta \qquad\qquad \cos(-\theta) = \cos\theta$

$1 + \cot^2\theta = \csc^2\theta \qquad\qquad \tan(-\theta) = -\tan\theta$

$\sin\left(\dfrac{\pi}{2} - \theta\right) = \cos\theta \qquad\quad \sin(\theta + 2\pi) = \sin\theta$

$\cos\left(\dfrac{\pi}{2} - \theta\right) = \sin\theta \qquad\quad \cos(\theta + 2\pi) = \cos\theta$

$\tan\left(\dfrac{\pi}{2} - \theta\right) = \cot\theta \qquad\quad \tan(\theta + \pi) = \tan\theta$

The Law of Sines

$\dfrac{\sin A}{a} = \dfrac{\sin B}{b} = \dfrac{\sin C}{c}$

The Law of Cosines

$a^2 = b^2 + c^2 - 2bc\cos A$

Addition and Subtraction Formulas

$\sin(x + y) = \sin x \cos y + \cos x \sin y$

$\sin(x - y) = \sin x \cos y - \cos x \sin y$

$\cos(x + y) = \cos x \cos y - \sin x \sin y$

$\cos(x - y) = \cos x \cos y + \sin x \sin y$

$\tan(x + y) = \dfrac{\tan x + \tan y}{1 - \tan x \tan y}$

$\tan(x - y) = \dfrac{\tan x - \tan y}{1 + \tan x \tan y}$

Double-Angle Formulas

$\sin 2x = 2\sin x \cos x$

$\cos 2x = \cos^2 x - \sin^2 x = 2\cos^2 x - 1 = 1 - 2\sin^2 x$

$\tan 2x = \dfrac{2\tan x}{1 - \tan^2 x}$

$\sin^2 x = \dfrac{1 - \cos 2x}{2} \qquad\qquad \cos^2 x = \dfrac{1 + \cos 2x}{2}$

Graphs of Trigonometric Functions

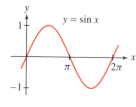

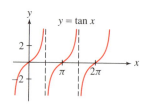

(third top graph)

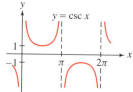

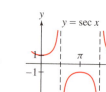

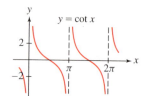

ALGEBRA

Lines

Slope of the line through $P_1 = (x_1, y_1)$ and $P_2 = (x_2, y_2)$:

$$m = \frac{y_2 - y_1}{x_2 - x_1}$$

Slope-intercept equation of line with slope m and y-intercept b:

$$y = mx + b$$

Point-slope equation of line through $P_1 = (x_1, y_1)$ with slope m:

$$y - y_1 = m(x - x_1)$$

Point-point equation of line through $P_1 = (x_1, y_1)$ and $P_2 = (x_2, y_2)$:

$$y - y_1 = m(x - x_1) \quad \text{where } m = \frac{y_2 - y_1}{x_2 - x_1}$$

Lines of slope m_1 and m_2 are parallel if and only if $m_1 = m_2$.
Lines of slope m_1 and m_2 are perpendicular if and only if $m_1 = -\frac{1}{m_2}$.

Circles

Equation of the circle with center (a, b) and radius r:

$$(x - a)^2 + (y - b)^2 = r^2$$

Distance and Midpoint Formulas

Distance between $P_1 = (x_1, y_1)$ and $P_2 = (x_2, y_2)$:

$$d = \sqrt{(x_2 - x_1)^2 + (y_2 - y_1)^2}$$

Midpoint of $\overline{P_1 P_2}$: $\left(\dfrac{x_1 + x_2}{2}, \dfrac{y_1 + y_2}{2} \right)$

Laws of Exponents

$$x^m x^n = x^{m+n} \qquad \frac{x^m}{x^n} = x^{m-n} \qquad (x^m)^n = x^{mn}$$

$$x^{-n} = \frac{1}{x^n} \qquad (xy)^n = x^n y^n \qquad \left(\frac{x}{y}\right)^n = \frac{x^n}{y^n}$$

$$x^{1/n} = \sqrt[n]{x} \qquad \sqrt[n]{xy} = \sqrt[n]{x}\,\sqrt[n]{y} \qquad \sqrt[n]{\frac{x}{y}} = \frac{\sqrt[n]{x}}{\sqrt[n]{y}}$$

$$x^{m/n} = \sqrt[n]{x^m} = \left(\sqrt[n]{x}\right)^m$$

Special Factorizations

$$x^2 - y^2 = (x + y)(x - y)$$
$$x^3 + y^3 = (x + y)(x^2 - xy + y^2)$$
$$x^3 - y^3 = (x - y)(x^2 + xy + y^2)$$

Binomial Theorem

$$(x + y)^2 = x^2 + 2xy + y^2$$
$$(x - y)^2 = x^2 - 2xy + y^2$$
$$(x + y)^3 = x^3 + 3x^2y + 3xy^2 + y^3$$
$$(x - y)^3 = x^3 - 3x^2y + 3xy^2 - y^3$$
$$(x + y)^n = x^n + nx^{n-1}y + \frac{n(n-1)}{2}x^{n-2}y^2$$
$$+ \cdots + \binom{n}{k}x^{n-k}y^k + \cdots + nxy^{n-1} + y^n$$

where $\displaystyle \binom{n}{k} = \frac{n(n-1)\cdots(n-k+1)}{1 \cdot 2 \cdot 3 \cdots k}$

Quadratic Formula

If $ax^2 + bx + c = 0$, then $x = \dfrac{-b \pm \sqrt{b^2 - 4ac}}{2a}$.

Inequalities and Absolute Value

If $a < b$ and $b < c$, then $a < c$.

If $a < b$, then $a + c < b + c$.

If $a < b$ and $c > 0$, then $ca < cb$.

If $a < b$ and $c < 0$, then $ca > cb$.

$|x| = x \quad$ if $x \geq 0$

$|x| = -x \quad$ if $x \leq 0$

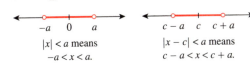

$|x| < a$ means $-a < x < a$.

$|x - c| < a$ means $c - a < x < c + a$.

GEOMETRY

Formulas for area A, circumference C, and volume V

Triangle	Circle	Sector of Circle	Sphere	Cylinder	Cone	Cone with arbitrary base
$A = \frac{1}{2}bh$ $= \frac{1}{2}ab\sin\theta$	$A = \pi r^2$ $C = 2\pi r$	$A = \frac{1}{2}r^2\theta$ $s = r\theta$ (θ in radians)	$V = \frac{4}{3}\pi r^3$ $A = 4\pi r^2$	$V = \pi r^2 h$	$V = \frac{1}{3}\pi r^2 h$ $A = \pi r\sqrt{r^2 + h^2}$	$V = \frac{1}{3}Ah$ where A is the area of the base

Pythagorean Theorem: For a right triangle with hypotenuse of length c and legs of lengths a and b, $c^2 = a^2 + b^2$.

TRIGONOMETRY

Angle Measurement

π radians $= 180°$

$1° = \dfrac{\pi}{180}$ rad $\qquad 1$ rad $= \dfrac{180°}{\pi}$

$s = r\theta \quad$ (θ in radians)

Right Triangle Definitions

$\sin\theta = \dfrac{\text{opp}}{\text{hyp}} \qquad\qquad \cos\theta = \dfrac{\text{adj}}{\text{hyp}}$

$\tan\theta = \dfrac{\sin\theta}{\cos\theta} = \dfrac{\text{opp}}{\text{adj}} \qquad \cot\theta = \dfrac{\cos\theta}{\sin\theta} = \dfrac{\text{adj}}{\text{opp}}$

$\sec\theta = \dfrac{1}{\cos\theta} = \dfrac{\text{hyp}}{\text{adj}} \qquad \csc\theta = \dfrac{1}{\sin\theta} = \dfrac{\text{hyp}}{\text{opp}}$

Trigonometric Functions

$\sin\theta = \dfrac{y}{r} \qquad\qquad \csc\theta = \dfrac{r}{y}$

$\cos\theta = \dfrac{x}{r} \qquad\qquad \sec\theta = \dfrac{r}{x}$

$\tan\theta = \dfrac{y}{x} \qquad\qquad \cot\theta = \dfrac{x}{y}$

$\displaystyle\lim_{\theta\to 0}\dfrac{\sin\theta}{\theta} = 1 \qquad \lim_{\theta\to 0}\dfrac{1-\cos\theta}{\theta} = 0$

$P = (r\cos\theta, r\sin\theta)$

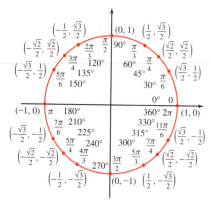

Fundamental Identities

$\sin^2\theta + \cos^2\theta = 1$

$1 + \tan^2\theta = \sec^2\theta$

$1 + \cot^2\theta = \csc^2\theta$

$\sin\left(\dfrac{\pi}{2} - \theta\right) = \cos\theta$

$\cos\left(\dfrac{\pi}{2} - \theta\right) = \sin\theta$

$\tan\left(\dfrac{\pi}{2} - \theta\right) = \cot\theta$

$\sin(-\theta) = -\sin\theta$

$\cos(-\theta) = \cos\theta$

$\tan(-\theta) = -\tan\theta$

$\sin(\theta + 2\pi) = \sin\theta$

$\cos(\theta + 2\pi) = \cos\theta$

$\tan(\theta + \pi) = \tan\theta$

The Law of Sines

$\dfrac{\sin A}{a} = \dfrac{\sin B}{b} = \dfrac{\sin C}{c}$

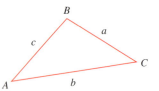

The Law of Cosines

$a^2 = b^2 + c^2 - 2bc\cos A$

Addition and Subtraction Formulas

$\sin(x + y) = \sin x\cos y + \cos x\sin y$

$\sin(x - y) = \sin x\cos y - \cos x\sin y$

$\cos(x + y) = \cos x\cos y - \sin x\sin y$

$\cos(x - y) = \cos x\cos y + \sin x\sin y$

$\tan(x + y) = \dfrac{\tan x + \tan y}{1 - \tan x\tan y}$

$\tan(x - y) = \dfrac{\tan x - \tan y}{1 + \tan x\tan y}$

Double-Angle Formulas

$\sin 2x = 2\sin x\cos x$

$\cos 2x = \cos^2 x - \sin^2 x = 2\cos^2 x - 1 = 1 - 2\sin^2 x$

$\tan 2x = \dfrac{2\tan x}{1 - \tan^2 x}$

$\sin^2 x = \dfrac{1 - \cos 2x}{2} \qquad\qquad \cos^2 x = \dfrac{1 + \cos 2x}{2}$

Graphs of Trigonometric Functions

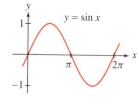

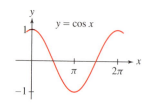

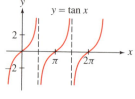

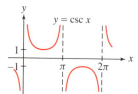

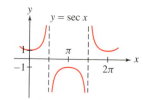

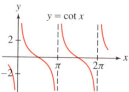

Multivariable

CALCULUS

THIRD EDITION

Early Transcendentals

JON ROGAWSKI
University of California, Los Angeles

COLIN ADAMS
Williams College

W. H. FREEMAN
& COMPANY

A Macmillan Education Imprint

TO JULIE –Jon
TO ALEXA AND COLTON –Colin

Publisher: Terri Ward
Developmental Editors: Tony Palermino, Katrina Wilhelm
Marketing Manager: Cara LeClair
Market Development Manager: Shannon Howard
Executive Media Editor: Laura Judge
Associate Editor: Marie Dripchak
Editorial Assistant: Victoria Garvey
Director of Editing, Design, and Media Production: Tracey Kuehn
Managing Editor: Lisa Kinne
Project Editor: Kerry O'Shaughnessy
Production Manager: Paul Rohloff
Cover and Text Designer: Blake Logan
Illustration Coordinator: Janice Donnola
Illustrations: Network Graphics and Techsetters, Inc.
Photo Editors: Eileen Liang, Christine Buese
Photo Researcher: Eileen Liang
Composition: John Rogosich/Techsetters, Inc.
Printing and Binding: RR Donnelley
Cover photo: ayzek/Shutterstock

Library of Congress Preassigned Control Number: 2014957106

ISBN-13: 978-1-4641-7175-8
ISBN-10: 1-4641-7175-0

Printed in the United States of America
Second printing

W. H. Freeman and Company, 41 Madison Avenue, New York, NY 10010
Houndmills, Basingstoke RG21 6XS, England
www.macmillanhighered.com

ABOUT THE AUTHORS

COLIN ADAMS

Colin Adams is the Thomas T. Read professor of Mathematics at Williams College, where he has taught since 1985. Colin received his undergraduate degree from MIT and his PhD from the University of Wisconsin. His research is in the area of knot theory and low-dimensional topology. He has held various grants to support his research, and written numerous research articles.

Colin is the author or co-author of *The Knot Book*, *How to Ace Calculus: The Streetwise Guide*, *How to Ace the Rest of Calculus: The Streetwise Guide*, *Riot at the Calc Exam and Other Mathematically Bent Stories*, *Why Knot?*, *Introduction to Topology: Pure and Applied*, and *Zombies & Calculus*. He co-wrote and appears in the videos "The Great Pi vs. E Debate" and "Derivative vs. Integral: the Final Smackdown."

He is a recipient of the Haimo National Distinguished Teaching Award from the Mathematical Association of America (MAA) in 1998, an MAA Polya Lecturer for 1998-2000, a Sigma Xi Distinguished Lecturer for 2000-2002, and the recipient of the Robert Foster Cherry Teaching Award in 2003.

Colin has two children and one slightly crazy dog, who is great at providing the entertainment.

JON ROGAWSKI

As a successful teacher for more than 30 years, Jon Rogawski listened and learned much from his own students. These valuable lessons made an impact on his thinking, his writing, and his shaping of a calculus text.

Jon Rogawski received his undergraduate and master's degrees in mathematics simultaneously from Yale University, and he earned his PhD in mathematics from Princeton University, where he studied under Robert Langlands. Before joining the Department of Mathematics at UCLA in 1986, where he was a full professor, he held teaching and visiting positions at the Institute for Advanced Study, the University of Bonn, and the University of Paris at Jussieu and Orsay.

Jon's areas of interest were number theory, automorphic forms, and harmonic analysis on semisimple groups. He published numerous research articles in leading mathematics journals, including the research monograph *Automorphic Representations of Unitary Groups in Three Variables* (Princeton University Press). He was the recipient of a Sloan Fellowship and an editor of the Pacific Journal of Mathematics and the Transactions of the AMS.

Sadly, Jon Rogawski passed away in September 2011. Jon's commitment to presenting the beauty of calculus and the important role it plays in students' understanding of the wider world is the legacy that lives on in each new edition of *Calculus*.

CONTENTS | MULTIVARIABLE CALCULUS

Early Transcendentals

Additional content can be accessed online via **LaunchPad**:

ADDITIONAL PROOFS

- L'Hôpital's Rule
- Error Bounds for Numerical Integration
- Comparison Test for Improper Integrals

ADDITIONAL CONTENT

- Second Order Differential Equations
- Complex Numbers

PREFACE

On Teaching Mathematics

I consider myself very lucky to have a career as a teacher and practitioner of mathematics. When I was young, I decided I wanted to be a writer. I loved telling stories. But I was also good at math, and, once in college, it didn't take me long to become enamored with it. I loved the fact that success in mathematics does not depend on your presentation skills or your interpersonal relationships. You are either right or you are wrong and there is little subjective evaluation involved. And I loved the satisfaction of coming up with a solution. That intensified when I started solving problems that were open research questions that had previously remained unsolved.

So, I became a professor of mathematics. And I soon realized that teaching mathematics is about telling a story. The goal is to explain to students in an intriguing manner, at the right pace, and in as clear a way as possible, how mathematics works and what it can do for you. I find mathematics immensely beautiful. I want students to feel that way, too.

On Writing a Calculus Text

I had always thought I might write a calculus text. But that is a daunting task. These days, calculus books average over a thousand pages. And I would need to convince myself that I had something to offer that was different enough from what already appears in the existing books. Then, I was approached about writing the third edition of Jon Rogawski's calculus book. Here was a book for which I already had great respect. Jon's vision of what a calculus book should be fit very closely with my own. Jon believed that as math teachers, how we say it is as important as what we say. Although he insisted on rigor at all times, he also wanted a book that was written in plain English, a book that could be read and that would entice students to read further and learn more. Moreover, Jon strived to create a text in which exposition, graphics, and layout would work together to enhance all facets of a student's calculus experience.

In writing his book, Jon paid special attention to certain aspects of the text:

1. Clear, accessible exposition that anticipates and addresses student difficulties.

2. Layout and figures that communicate the flow of ideas.

3. Highlighted features that emphasize concepts and mathematical reasoning: Conceptual Insight, Graphical Insight, Assumptions Matter, Reminder, and Historical Perspective.

4. A rich collection of examples and exercises of graduated difficulty that teach basic skills, problem-solving techniques, reinforce conceptual understanding, and motivate calculus through interesting applications. Each section also contains exercises that develop additional insights and challenge students to further develop their skills.

Coming into the project of creating the third edition, I was somewhat apprehensive. Here was an already excellent book that had attained the goals set for it by its author. First and foremost, I wanted to be sure that I did it no harm. On the other hand, I have been teaching calculus now for 30 years, and in that time, I have come to some conclusions about what does and does not work well for students.

As a mathematician, I want to make sure that the theorems, proofs, arguments and development are correct. There is no place in mathematics for sloppiness of any kind. As a teacher, I want the material to be accessible. The book should not be written at the mathematical level of the instructor. Students should be able to use the book to learn the material, with the help of their instructor. Working from the high standard that Jon set, I have tried hard to maintain the level of quality of the previous edition while making the changes that I believe will bring the book to the next level.

Placement of Taylor Polynomials

Taylor polynomials appear in Chapter 8, before infinite series in Chapter 10. The goal here is to present Taylor polynomials as a natural extension of linear approximation. When teaching infinite series, the primary focus is on convergence, a topic that many students find challenging. By the time we have covered the basic convergence tests and studied the convergence of power series, students are ready to tackle the issues involved in representing a function by its Taylor series. They can then rely on their previous work with Taylor polynomials and the error bound from Chapter 8. However, the section on Taylor polynomials is written so that you can cover this topic together with the materials on infinite series if this order is preferred.

Careful, Precise Development

W. H. Freeman is committed to high quality and precise textbooks and supplements. From this project's inception and throughout its development and production, quality and precision have been given significant priority. We have in place unparalleled procedures to ensure the accuracy of the text:

- Exercises and Examples
- Exposition
- Figures
- Editing
- Composition

Together, these procedures far exceed prior industry standards to safeguard the quality and precision of a calculus textbook.

New to the Third Edition

There are a variety of changes that have been implemented in this edition. Following are some of the most important.

MORE FOCUS ON CONCEPTS   The emphasis has been shifted to focus less on the memorization of specific formulas, and more on understanding the underlying concepts. Memorization can never be completely avoided, but it is in no way the crux of calculus. Students will remember how to apply a procedure or technique if they see the logical progression that generates it. And they then understand the underlying concepts rather than seeing the topic as a black box in which you insert numbers. Specific examples include:

- (**Section 1.2**) Removed the general formula for the completion of a square and instead, emphasized the method so students need not memorize the formula.
- (**Section 7.2**) Changed the methods for evaluating trigonometric integrals to focus on techniques to apply rather than formulas to memorize.
- (**Chapter 9**) Discouraged the memorization of solutions of specific types of differential equations and instead, encouraged the use of methods of solution.
- (**Section 12.2**) Decreased number of formulas for parametrizing a line from two to one, as the second can easily be derived from the first.
- (**Section 12.6**) De-emphasized the memorization of the various formulas for quadric surfaces. Instead, moved the focus to slicing with planes to find curves and using those to determine the shape of the surface. These methods will be useful regardless of the type of surface it is.
- (**Section 14.4**) Decreased the number of essential formulas for linear approximation of functions of two variables from four to two, providing the background to derive the others from these.

CHANGES IN NOTATION   There are numerous notational changes. Some were made to bring the notation more into line with standard usage in mathematics and other fields in which mathematics is applied. Some were implemented to make it easier for students to remember the meaning of the notation. Some were made to help make the corresponding concepts that are represented more transparent. Specific examples include:

- **(Section 4.6)** Presented a new notation for graphing that gives the signs of the first and second derivative and then simple symbols (slanted up and down arrows and up and down u's) to help the student keep track of when the graph is increasing or decreasing and concave up or concave down over the given interval.
- **(Section 7.1)** Simplified the notation for integration by parts and provided a visual method for remembering it.
- **(Chapter 10)** Changed names of the various tests for convergence/divergence of infinite series to evoke the usage of the test and thereby make it easier for students to remember them.
- **(Chapters 13–17)** Rather than using $c(t)$ for a path, we consistently switched to the vector-valued function $\mathbf{r}(t)$. This also allowed us to replace $d\mathbf{s}$ with $d\mathbf{r}$ as a differential, which means there is less likely to be confusion with $d\mathbf{s}$, dS and $d\mathbf{S}$.

MORE EXPLANATIONS OF DERIVATIONS   Occasionally, in the previous edition, a result was given and verified, without motivating where the derivation came from. I believe it is important for students to understand how someone might come up with a particular result, thereby helping them to picture how they might themselves one day be able to derive results.

- **(Section 14.4)** Developed the equation of the tangent plane in a manner that makes geometric sense.
- **(Section 14.5)** Included a proof of the fact the gradient of a function f of three variables is orthogonal to the surfaces that are the level sets of f.
- **(Section 14.8)** Gave an intuitive explanation for why the Method of Lagrange Multipliers works.
- **(Section 15.5)** Developed the center of mass formulas by first discussing the one-dimensional case of a seesaw.

REORDERING AND ADDING TOPICS   There were some specific rearrangements among the sections and additions. These include:

- A subsection on piecewise-defined functions has been added to Section 1.3.
- The section on implicit differentiation in Chapter 3 (previously Section 3.10) has been moved up to become Section 3.8 and has absorbed the previous Section 3.8 (inverse functions) so that implicit differentiation can be applied to derive the various derivatives as necessary.
- The section on indefinite integrals (previously Section 4.9) has been moved from Chapter 4 (Applications of the Derivative) to Chapter 5 (The Integral). This is a more natural placement for it.
- A new section on choosing from amongst the various methods of integration has been added to Chapter 7.
- A subsection on choosing the appropriate convergence/divergence test has been added to Section 10.5.
- An explanation of how to find indefinite limits using power series has been added to Section 10.6.
- The definitions of divergence and curl have been moved from Chapter 17 to Section 16.1. This allows us to utilize them at an appropriate earlier point in the text.
- A list all of the different types of integrals that have been introduced in Chapter 16 has been added to Section 16.5.
- A subsection on the Vector Form of Green's Theorem has been added to Section 17.1.

NEW EXAMPLES, FIGURES, AND EXERCISES   Numerous examples and accompanying figures have been added to clarify concepts. A variety of exercises have also been added throughout the text, particularly where new applications are available or further conceptual development is advantageous. Figures marked with a DF icon have been made dynamic and can be accessed via *LaunchPad*. A selection of these figures also includes brief tutorial videos explaining the concepts at work.

SUPPLEMENTS

For Instructors

Instructor's Solutions Manual
Contains worked-out solutions to all exercises in the text.

Test Bank
Computerized (CD-ROM), ISBN:1-3190-0939-5
Includes a comprehensive set of multiple-choice test items.

Instructor's Resource Manual
Provides sample course outlines, suggested class time, key points, lecture material, discussion topics, class activities, worksheets, projects, and questions to accompany the Dynamic Figures.

For Students

Student Solutions Manual
Single Variable ISBN: 1-4641-7188-2
Multivariable ISBN: 1-4641-7189-0
Contains worked-out solutions to all odd-numbered exercises in the text.

Software Manuals
Maple™ and Mathematica® software manuals serve as basic introductions to popular mathematical software options.

ONLINE HOMEWORK OPTIONS

Our new course space, LaunchPad, combines an interactive e-Book with high-quality multimedia content and ready-made assessment options, including LearningCurve adaptive quizzing. Pre-built, curated units are easy to assign or adapt with your own material, such as readings, videos, quizzes, discussion groups, and more. LaunchPad includes a gradebook that provides a clear window on performance for your whole class, for individual students, and for individual assignments. While a streamlined interface helps students focus on what's due next, social commenting tools let them engage, make connections, and learn from each other. Use LaunchPad on its own or integrate it with your school's learning management system so your class is always on the same page. Contact your rep to make sure you have access.

Assets integrated into LaunchPad include:

Interactive e-Book: Every LaunchPad e-Book comes with powerful study tools for students, video and multimedia content, and easy customization for instructors. Students can search, highlight, and bookmark, making it easier to study and access key content. And instructors can make sure their class gets just the book they want to deliver: customize and rearrange chapters, add and share notes and discussions, and link to quizzes, activities, and other resources.

LearningCurve provides students and instructors with powerful adaptive quizzing, a game-like format, direct links to the e-Book, and instant feedback. The quizzing system features questions tailored specifically to the text and adapts to students' responses, providing material at different difficulty levels and topics based on student performance.

Dynamic Figures: Over 250 figures from the text have been recreated in a new interactive format for students and instructors to manipulate and explore, making the visual aspects and dimensions of calculus concepts easier to grasp. Brief tutorial videos accompany selected figures and explain the concepts at work.

CalcClips: These whiteboard tutorials provide animated and narrated step-by-step solutions to exercises that are based on key problems in the text.

SolutionMaster offers an easy-to-use Web-based version of the instructor's solutions, allowing instructors to generate a solution file for any set of homework exercises.

www.webassign.net/freeman.com

WebAssign Premium integrates the book's exercises into the world's most popular and trusted online homework system, making it easy to assign algorithmically generated homework and quizzes. Algorithmic exercises offer the instructor optional algorithmic solutions. WebAssign Premium also offers access to resources, including Dynamic Figures, CalcClips whiteboard tutorials, and a "Show My Work" feature. In addition, WebAssign Premium is available with a fully customizable e-Book option.

webwork.maa.org

W. H. Freeman offers thousands of algorithmically generated questions (with full solutions) through this free, open-source online homework system created at the University of Rochester. Adopters also have access to a shared national library test bank with thousands of additional questions, including 2,500 problem sets matched to the book's table of contents.

FEATURES

Conceptual Insights

encourage students to develop a conceptual understanding of calculus by explaining important ideas clearly but informally.

CONCEPTUAL INSIGHT The parametric equations for the ellipse in Example 5 illustrate a key difference between the path $c(t)$ and its underlying curve C. The curve C is an ellipse in the plane, whereas $c(t)$ describes a particular, counterclockwise motion of a particle along the ellipse. If we let t vary from 0 to 4π, then the particle goes around the ellipse twice.

A key feature of parametrizations is that they are not unique. In fact, every curve can be parametrized in infinitely many different ways. For instance, the parabola $y = x^2$ is parametrized not only by (t, t^2) but also by (t^3, t^6), or (t^5, t^{10}), and so on.

Ch. 11, p. 582

GRAPHICAL INSIGHT Imagine standing at point Q in Figure 2(A). We want to increase the value of f while remaining on the constraint curve. The gradient vector ∇f_Q points in the direction of *maximum* increase, but we cannot move in the gradient direction because that would take us off the constraint curve. However, the gradient points to the right, so we can still increase f somewhat by moving to the right along the constraint curve.

We keep moving to the right until we arrive at the point P, where ∇f_P is orthogonal to the constraint curve [Figure 2(B)]. Once at P, we cannot increase f further by moving either to the right or to the left along the constraint curve. Thus, $f(P)$ is a local maximum subject to the constraint.

Now, the vector ∇g_P is also orthogonal to the constraint curve, so ∇f_P and ∇g_P must point in the same or opposite directions. In other words, $\nabla f_P = \lambda \nabla g_P$ for some scalar λ (called a **Lagrange multiplier**). Graphically, this means that a local max subject to the constraint occurs at points P where the level curves of f and g are tangent.

Graphical Insights enhance students' visual understanding by making the crucial connections between graphical properties and the underlying concepts.

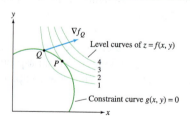

(A) f increases as we move to the right along the constraint curve.

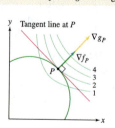

(B) The local maximum of f on the constraint curve occurs where ∇f_P and ∇g_P are parallel.

FIGURE 2

Ch. 14, p. 809

SUPPLEMENTS

For Instructors

Instructor's Solutions Manual
Contains worked-out solutions to all exercises in the text.

Test Bank
Computerized (CD-ROM), ISBN:1-3190-0939-5
Includes a comprehensive set of multiple-choice test items.

Instructor's Resource Manual
Provides sample course outlines, suggested class time, key points, lecture material, discussion topics, class activities, worksheets, projects, and questions to accompany the Dynamic Figures.

For Students

Student Solutions Manual
Single Variable ISBN: 1-4641-7188-2
Multivariable ISBN: 1-4641-7189-0
Contains worked-out solutions to all odd-numbered exercises in the text.

Software Manuals
Maple™ and Mathematica® software manuals serve as basic introductions to popular mathematical software options.

ONLINE HOMEWORK OPTIONS

Our new course space, LaunchPad, combines an interactive e-Book with high-quality multimedia content and ready-made assessment options, including LearningCurve adaptive quizzing. Pre-built, curated units are easy to assign or adapt with your own material, such as readings, videos, quizzes, discussion groups, and more. LaunchPad includes a gradebook that provides a clear window on performance for your whole class, for individual students, and for individual assignments. While a streamlined interface helps students focus on what's due next, social commenting tools let them engage, make connections, and learn from each other. Use LaunchPad on its own or integrate it with your school's learning management system so your class is always on the same page. Contact your rep to make sure you have access.

Assets integrated into LaunchPad include:

Interactive e-Book: Every LaunchPad e-Book comes with powerful study tools for students, video and multimedia content, and easy customization for instructors. Students can search, highlight, and bookmark, making it easier to study and access key content. And instructors can make sure their class gets just the book they want to deliver: customize and rearrange chapters, add and share notes and discussions, and link to quizzes, activities, and other resources.

LearningCurve provides students and instructors with powerful adaptive quizzing, a game-like format, direct links to the e-Book, and instant feedback. The quizzing system features questions tailored specifically to the text and adapts to students' responses, providing material at different difficulty levels and topics based on student performance.

Dynamic Figures: Over 250 figures from the text have been recreated in a new interactive format for students and instructors to manipulate and explore, making the visual aspects and dimensions of calculus concepts easier to grasp. Brief tutorial videos accompany selected figures and explain the concepts at work.

CalcClips: These whiteboard tutorials provide animated and narrated step-by-step solutions to exercises that are based on key problems in the text.

SolutionMaster offers an easy-to-use Web-based version of the instructor's solutions, allowing instructors to generate a solution file for any set of homework exercises.

www.webassign.net/freeman.com

WebAssign Premium integrates the book's exercises into the world's most popular and trusted online homework system, making it easy to assign algorithmically generated homework and quizzes. Algorithmic exercises offer the instructor optional algorithmic solutions. WebAssign Premium also offers access to resources, including Dynamic Figures, CalcClips whiteboard tutorials, and a "Show My Work" feature. In addition, WebAssign Premium is available with a fully customizable e-Book option.

**WeBWorK**

webwork.maa.org

W. H. Freeman offers thousands of algorithmically generated questions (with full solutions) through this free, open-source online homework system created at the University of Rochester. Adopters also have access to a shared national library test bank with thousands of additional questions, including 2,500 problem sets matched to the book's table of contents.

FEATURES

Conceptual Insights encourage students to develop a conceptual understanding of calculus by explaining important ideas clearly but informally.

CONCEPTUAL INSIGHT   The parametric equations for the ellipse in Example 5 illustrate a key difference between the path $c(t)$ and its underlying curve C. The curve C is an ellipse in the plane, whereas $c(t)$ describes a particular, counterclockwise motion of a particle along the ellipse. If we let t vary from 0 to 4π, then the particle goes around the ellipse twice.

A key feature of parametrizations is that they are not unique. In fact, every curve can be parametrized in infinitely many different ways. For instance, the parabola $y = x^2$ is parametrized not only by (t, t^2) but also by (t^3, t^6), or (t^5, t^{10}), and so on.

Ch. 11, p. 582

GRAPHICAL INSIGHT   Imagine standing at point Q in Figure 2(A). We want to increase the value of f while remaining on the constraint curve. The gradient vector ∇f_Q points in the direction of *maximum* increase, but we cannot move in the gradient direction because that would take us off the constraint curve. However, the gradient points to the right, so we can still increase f somewhat by moving to the right along the constraint curve.

We keep moving to the right until we arrive at the point P, where ∇f_P is orthogonal to the constraint curve [Figure 2(B)]. Once at P, we cannot increase f further by moving either to the right or to the left along the constraint curve. Thus, $f(P)$ is a local maximum subject to the constraint.

Now, the vector ∇g_P is also orthogonal to the constraint curve, so ∇f_P and ∇g_P must point in the same or opposite directions. In other words, $\nabla f_P = \lambda \nabla g_P$ for some scalar λ (called a **Lagrange multiplier**). Graphically, this means that a local max subject to the constraint occurs at points P where the level curves of f and g are tangent.

Graphical Insights enhance students' visual understanding by making the crucial connections between graphical properties and the underlying concepts.

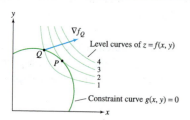

(A) f increases as we move to the right along the constraint curve.

(B) The local maximum of f on the constraint curve occurs where ∇f_P and ∇g_P are parallel.

DF FIGURE 2

Ch. 14, p. 809

Reminders

are margin notes that link the current discussion to important concepts introduced earlier in the text to give students a quick review and make connections with related ideas.

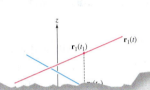

←·· REMINDER *Eq. (8) is similar to the definition of an average value in one variable:*

$$\overline{f} = \frac{1}{b-a}\int_a^b f(x)\,dx = \frac{\int_a^b f(x)\,dx}{\int_a^b 1\,dx}$$

The **average value** (or **mean value**) of a function $f(x, y)$ on a domain $\mathcal{D}$, which we denote by $\overline{f}$, is the quantity

$$\boxed{\overline{f} = \frac{1}{\text{Area}(\mathcal{D})}\iint_{\mathcal{D}} f(x, y)\,dA = \frac{\iint_{\mathcal{D}} f(x, y)\,dA}{\iint_{\mathcal{D}} 1\,dA}}\qquad \boxed{8}$$

Ch. 15, p. 838

Caution Notes

warn students of common pitfalls they may encounter in understanding the material.

CAUTION *We cannot assume in Eq. (8) that the parameter values t_1 and t_2 are equal. The point of intersection may correspond to different parameter values on the two lines.*

$$\mathbf{r}_1(t) = \langle 1, 0, 1\rangle + t\,\langle 3, 3, 5\rangle$$
$$\mathbf{r}_2(t) = \langle 3, 6, 1\rangle + t\,\langle 4, -2, 7\rangle$$

Solution The two lines intersect if there exist parameter values t_1 and t_2 such that $\mathbf{r}_1(t_1) = \mathbf{r}_2(t_2)$—that is, if

$$\langle 1, 0, 1\rangle + t_1\,\langle 3, 3, 5\rangle = \langle 3, 6, 1\rangle + t_2\,\langle 4, -2, 7\rangle \qquad \boxed{8}$$

This is equivalent to three equations for the components:

Ch. 12, p. 641

Historical Perspectives

are brief vignettes that place key discoveries and conceptual advances in their historical context. They give students a glimpse into some of the accomplishments of great mathematicians and an appreciation for their significance.

HISTORICAL PERSPECTIVE

(© SSPL/The Image Works)

James Clerk Maxwell
(1831–1879)

Vector analysis was developed in the nineteenth century, in large part, to express the laws of electricity and magnetism. Electromagnetism was studied intensively in the period 1750–1890, culminating in the famous Maxwell Equations, which provide a unified understanding in terms of two vector fields: the electric field $\mathbf{E}$ and the magnetic field $\mathbf{B}$. In a region of empty space (where there are no charged particles), the Maxwell Equations are

$$\boxed{\begin{aligned} \text{div}(\mathbf{E}) &= 0, & \text{div}(\mathbf{B}) &= 0 \\ \text{curl}(\mathbf{E}) &= -\frac{\partial \mathbf{B}}{\partial t}, & \text{curl}(\mathbf{B}) &= \mu_0\epsilon_0 \frac{\partial \mathbf{E}}{\partial t} \end{aligned}}$$

where μ_0 and ϵ_0 are experimentally determined constants. In SI units,

$$\mu_0 = 4\pi \times 10^{-7} \text{ henries/m}$$
$$\epsilon_0 \approx 8.85 \times 10^{-12} \text{ farads/m}$$

These equations led Maxwell to make two predictions of fundamental importance: (1) that electromagnetic waves exist (this was confirmed by H. Hertz in 1887), and (2) that light is an electromagnetic wave.

How do the Maxwell Equations suggest that electromagnetic waves exist? And why did Maxwell conclude that light is an electromagnetic wave? It was known to mathematicians in the eighteenth century that waves traveling with velocity c may be described by functions $\varphi(x, y, z, t)$ that satisfy the *wave equation*

$$\boxed{\Delta\varphi = \frac{1}{c^2}\frac{\partial^2 \varphi}{\partial t^2}} \qquad \boxed{7}$$

where Δ is the Laplace operator (or "Laplacian")

$$\Delta\varphi = \frac{\partial^2 \varphi}{\partial x^2} + \frac{\partial^2 \varphi}{\partial y^2} + \frac{\partial^2 \varphi}{\partial z^2}$$

We will show that the components of $\mathbf{E}$ satisfy this wave equation. Take the curl of both sides of Maxwell's third equation:

$$\text{curl}(\text{curl}(\mathbf{E})) = \text{curl}\left(-\frac{\partial \mathbf{B}}{\partial t}\right) = -\frac{\partial}{\partial t}\text{curl}(\mathbf{B})$$

Then apply Maxwell's fourth equation to obtain

$$\text{curl}(\text{curl}(\mathbf{E})) = -\frac{\partial}{\partial t}\left(\mu_0\epsilon_0 \frac{\partial \mathbf{E}}{\partial t}\right)$$
$$= -\mu_0\epsilon_0 \frac{\partial^2 \mathbf{E}}{\partial t^2} \qquad \boxed{8}$$

Finally, let us define the Laplacian of a vector field

$$\mathbf{F} = \langle F_1, F_2, F_3\rangle$$

by applying the Laplacian Δ to each component, $\Delta\mathbf{F} = \langle \Delta F_1, \Delta F_2, \Delta F_3\rangle$. Then the following identity holds (see Exercise 34):

$$\text{curl}(\text{curl}(\mathbf{F})) = \nabla(\text{div}(\mathbf{F})) - \Delta\mathbf{F}$$

Applying this identity to $\mathbf{E}$, we obtain $\text{curl}(\text{curl}(\mathbf{E})) = -\Delta\mathbf{E}$ because $\text{div}(\mathbf{E}) = 0$ by Maxwell's first equation. Thus, Eq. (8) yields

$$\boxed{\Delta\mathbf{E} = \mu_0\epsilon_0 \frac{\partial^2 \mathbf{E}}{\partial t^2}}$$

In other words, each component of the electric field satisfies the wave equation (7), with $c = (\mu_0\epsilon_0)^{-1/2}$. This tells us that the E-field (and similarly the B-field) can propagate through space like a wave, giving rise to electromagnetic radiation (Figure 17).

Maxwell computed the velocity c of an electromagnetic wave:

$$c = (\mu_0\epsilon_0)^{-1/2} \approx 3 \times 10^8 \text{ m/s}$$

and observed that the value is suspiciously close to the velocity of light (first measured by Olaf Römer in 1676). This had to be more than a coincidence, as Maxwell wrote in 1862: "We can scarcely avoid the conclusion that light consists in the transverse undulations of the same medium which is the cause of electric and magnetic phenomena." Needless to say, the wireless technologies that drive our modern society rely on the unseen electromagnetic radiation whose existence Maxwell first predicted on mathematical grounds.

Ch. 17, p. 990

Assumptions Matter

uses short explanations and well-chosen counterexamples to help students appreciate why hypotheses are needed in theorems.

Local linearity is used in the next section to prove the Chain Rule for Paths, upon which the fundamental properties of the gradient are based.

Ch. 14, p. 769

Assumptions Matter Local linearity plays a key role, and although most reasonable functions are locally linear, the mere existence of the partial derivatives does not guarantee local linearity. This is in contrast to the one-variable case, where $f(x)$ is automatically locally linear at $x = a$ if $f'(a)$ exists (Exercise 44).

The function $g(x, y)$ in Figure 6(A) shows what can go wrong. The graph contains the x- and y-axes—in other words, $g(x, 0) = 0$ and $g(0, y) = 0$—and therefore, the partial derivatives $g_x(0, 0)$ and $g_y(0, 0)$ are both zero. The tangent plane at the origin $(0, 0)$, if it existed, would have to be the xy-plane. However, Figure 6(B) shows that the graph also contains lines through the origin that do not lie in the xy-plane (in fact, the graph is composed entirely of lines through the origin). As we zoom in on the origin, these lines remain at an angle to the xy-plane, and the surface does not get any flatter. Thus, $g(x, y)$ cannot be locally linear at $(0, 0)$, and the tangent plane does not exist. In particular, $g(x, y)$ cannot satisfy the assumptions of Theorem 1, so the partial derivatives $g_x(x, y)$ and $g_y(x, y)$ cannot be continuous at the origin (see Exercise 45 for details).

Section Summaries summarize a section's key points in a concise and useful way and emphasize for students what is most important in each section.

Section Exercise Sets offer a comprehensive set of exercises closely coordinated with the text. These exercises vary in difficulty from routine, to moderate, to more challenging. Also included are icons indicating problems that require the student to give a written response or require the use of technology T.

Chapter Review Exercises offer a comprehensive set of exercises closely coordinated with the chapter material to provide additional problems for self-study or assignments.

ACKNOWLEDGMENTS

Colin Adams and W. H. Freeman and Company are grateful to the many instructors from across the United States and Canada who have offered comments that assisted in the development and refinement of this book. These contributions included class testing, manuscript reviewing, problems reviewing, and participating in surveys about the book and general course needs.

ALABAMA Tammy Potter, *Gadsden State Community College*; David Dempsey, *Jacksonville State University*; Edwin Smith, *Jacksonville State University*; Jeff Dodd, *Jacksonville State University*; Douglas Bailer, *Northeast Alabama Community College*; Michael Hicks, *Shelton State Community College*; Patricia C. Eiland, *Troy University, Montgomery Campus*; Chadia Affane Aji, *Tuskegee University*; James L. Wang, *The University of Alabama*; Stephen Brick, *University of South Alabama*; Jo-erg Feldvoss, *University of South Alabama* **ALASKA** Mark A. Fitch, *University of Alaska Anchorage*; Kamal Narang, *University of Alaska Anchorage*; Alexei Rybkin, *University of Alaska Fairbanks*; Martin Getz, *University of Alaska Fairbanks* **ARIZONA** Stefania Tracogna, *Arizona State University*; Bruno Welfert, *Arizona State University*; Light Bryant, *Arizona Western College*; Daniel Russow, *Arizona Western College*; Jennifer Jameson, *Coconino College*; George Cole, *Mesa Community College*; David Schultz, *Mesa Community College*; Michael Bezusko, *Pima Community College, Desert Vista Campus*; Garry Carpenter, *Pima Community College, Northwest Campus*; Paul Flasch, *Pima County Community College*; Jessica Knapp, *Pima Community College, Northwest Campus*; Roger Werbylo, *Pima County Community College*; Katie Louchart, *Northern Arizona University*; Janet McShane, *Northern Arizona University*; Donna M. Krawczyk, *The University of Arizona* **ARKANSAS** Deborah Parker, *Arkansas Northeastern College*; J. Michael Hall, *Arkansas State University*; Kevin Cornelius, *Ouachita Baptist University*; Hyungkoo Mark Park, *Southern Arkansas University*; Katherine Pinzon, *University of Arkansas at Fort Smith*; Denise LeGrand, *University of Arkansas at Little Rock*; John Annulis, *University of Arkansas at Monticello*; Erin Haller, *University of Arkansas, Fayetteville*; Shannon Dingman, *University of Arkansas, Fayetteville*; Daniel J. Arrigo, *University of Central Arkansas* **CALIFORNIA** Michael S. Gagliardo, *California Lutheran University*; Harvey Greenwald, *California Polytechnic State University, San Luis Obispo*; Charles Hale, *California Polytechnic State University, San Luis Obispo*; John Hagen, *California Polytechnic State University, San Luis Obispo*; Donald Hartig, *California Polytechnic State University, San Luis Obispo*; Colleen Margarita Kirk, *California Polytechnic State University, San Luis Obispo*; Lawrence Sze, *California Polytechnic State University, San Luis Obispo*; Raymond Terry, *California Polytechnic State University, San Luis Obispo*; James R. McKinney, *California State Polytechnic University, Pomona*; Robin Wilson, *California State Polytechnic University, Pomona*; Charles Lam, *California State University, Bakersfield*; David McKay, *California State University, Long Beach*; Melvin Lax, *California State University, Long Beach*; Wallace A. Etterbeek, *California State University, Sacramento*; Mohamed Allali, *Chapman University*; George Rhys, *College of the Canyons*; Janice Hector, *DeAnza College*; Isabelle Saber, *Glendale Community College*; Peter Stathis, *Glendale Community College*; Douglas B. Lloyd, *Golden West College*; Thomas Scardina, *Golden West College*; Kristin Hartford, *Long Beach City College*; Eduardo Arismendi-Pardi, *Orange Coast College*; Mitchell Alves, *Orange Coast College*; Yenkanh Vu, *Orange Coast*

College; Yan Tian, *Palomar College*; Donna E. Nordstrom, *Pasadena City College*; Don L. Hancock, *Pepperdine University*; Kevin Iga, *Pepperdine University*; Adolfo J. Rumbos, *Pomona College*; Virginia May, *Sacramento City College*; Carlos de la Lama, *San Diego City College*; Matthias Beck, *San Francisco State University*; Arek Goetz, *San Francisco State University*; Nick Bykov, *San Joaquin Delta College*; Eleanor Lang Kendrick, *San Jose City College*; Elizabeth Hodes, *Santa Barbara City College*; William Konya, *Santa Monica College*; John Kennedy, *Santa Monica College*; Peter Lee, *Santa Monica College*; Richard Salome, *Scotts Valley High School*; Norman Feldman, *Sonoma State University*; Elaine McDonald, *Sonoma State University*; John D. Eggers, *University of California, San Diego*; Adam Bowers, *University of California, San Diego*; Bruno Nachtergaele, *University of California, Davis*; Boumediene Hamzi, *University of California, Davis*; Olga Radko, *University of California, Los Angeles*; Richard Leborne, *University of California, San Diego*; Peter Stevenhagen, *University of California, San Diego*; Jeffrey Stopple, *University of California, Santa Barbara*; Guofang Wei, *University of California, Santa Barbara*; Rick A. Simon, *University of La Verne*; Alexander E. Koonce, *University of Redlands*; Mohamad A. Alwash, *West Los Angeles College*; Calder Daenzer, *University of California, Berkeley*; Jude Thaddeus Socrates, *Pasadena City College*; Cheuk Ying Lam, *California State University Bakersfield*; Borislava Gutarts, *California State University, Los Angeles*; Daniel Rogalski, *University of California, San Diego*; Don Hartig, *California Polytechnic State University*; Anne Voth, *Palomar College*; Jay Wiestling, *Palomar College*; Lindsey Bramlett-Smith, *Santa Barbara City College*; Dennis Morrow, *College of the Canyons*; Sydney Shanks, *College of the Canyons*; Bob Tolar, *College of the Canyons*; Gene W. Majors, *Fullerton College*; Robert Diaz, *Fullerton College*; Gregory Nguyen, *Fullerton College*; Paul Sjoberg, *Fullerton College*; Deborah Ritchie, *Moorpark College*; Maya Rahnamaie, *Moorpark College*; Kathy Fink, *Moorpark College*; Christine Cole, *Moorpark College*; K. Di Passero, *Moorpark College*; Sid Kolpas, *Glendale Community College*; Miriam Castrconde, *Irvine Valley College*; Ilkner Erbas-White, *Irvine Valley College*; Corey Manchester, *Grossmont College*; Donald Murray, *Santa Monica College*; Barbara McGee, *Cuesta College*; Marie Larsen, *Cuesta College*; Joe Vasta, *Cuesta College*; Mike Kinter, *Cuesta College*; Mark Turner, *Cuesta College*; G. Lewis, *Cuesta College*; Daniel Kleinfelter, *College of the Desert*; Esmeralda Medrano, *Citrus College*; James Swatzel, *Citrus College*; Mark Littrell, *Rio Hondo College*; Rich Zucker, *Irvine Valley College*; Cindy Torigison, *Palomar College*; Craig Chamberline, *Palomar College*; Lindsey Lang, *Diablo Valley College*; Sam Needham, *Diablo Valley College*; Dan Bach, *Diablo Valley College*; Ted Nirgiotis, *Diablo Valley College*; Monte Collazo, *Diablo Valley College*; Tina Levy, *Diablo Valley College*; Mona Panchal, *East Los Angeles College*; Ron Sandvick, *San Diego Mesa College*; Larry Handa, *West Valley College*; Frederick Utter, *Santa Rose Junior College*; Farshod Mosh, *DeAnza College*; Doli Bambhania, *DeAnza College*; Charles Klein, *DeAnza College*; Tammi Marshall, *Cauyamaca College*; Inwon Leu, *Cauyamaca College*; Michael Moretti, *Bakersfield College*; Janet Tarjan, *Bakersfield College*; Hoat Le, *San Diego City College*; Richard Fielding, *Southwestern College*; Shannon Gracey, *Southwestern College*; Janet Mazzarella, *Southwestern College*; Christina Soderlund, *California Lutheran University*; Rudy Gonzalez, *Citrus College*; Robert Crise, *Crafton Hills College*; Joseph Kazimir, *East Los Angeles College*; Randall Rogers, *Fullerton College*; Peter Bouzar, *Golden West College*; Linda Ternes, *Golden West College*; Hsiao-Ling Liu, *Los Angeles Trade Tech Community College*; Yu-Chung Chang-Hou, *Pasadena City College*; Guillermo Alvarez, *San Diego City College*; Ken Kuniyuki, *San Diego Mesa College*; Laleh Howard, *San Diego Mesa College*; Sharareh Masooman, *Santa Barbara City College*; Jared Hersh, *Santa Barbara City College*; Betty Wong, *Santa Monica College*; Brian Rodas, *Santa Monica College*; Veasna Chiek, *Riverside City College* COLORADO Tony Weathers, *Adams State College*; Erica Johnson, *Arapahoe Community College*; Karen Walters, *Arapahoe Community College*; Joshua D. Laison, *Colorado College*; G. Gustave Greivel, *Colorado School of Mines*; Holly Eklund, *Colorado School of the Mines*; Mike Nicholas, *Colorado School of the Mines*; Jim Thomas, *Colorado State University*; Eleanor Storey, *Front Range Community College*; Larry Johnson, *Metropolitan State College of Denver*; Carol Kuper, *Morgan Community College*; Larry A. Pontaski, *Pueblo Community College*; Terry Chen Reeves, *Red Rocks Community College*; Debra S. Carney, *Colorado School of the Mines*; Louis A. Talman, *Metropolitan State College of Denver*; Mary A. Nelson, *University of Colorado at Boulder*; J. Kyle Pula, *University of Denver*; Jon Von Stroh, *University of Denver*; Sharon Butz, *University of Denver*; Daniel Daly, *University of Denver*; Tracy Lawrence, *Arapahoe Community College*; Shawna Mahan, *University of Colorado Denver*; Adam Norris, *University of Colorado at Boulder*; Anca Radulescu, *University of Colorado at Boulder*; Mike Kawai, *University of Colorado Denver*; Janet Barnett, *Colorado State University–Pueblo*; Byron Hurley, *Colorado State University–Pueblo*; Jonathan Portiz, *Colorado State University–Pueblo*; Bill Emerson, *Metropolitan State College of Denver*; Suzanne Caulk, *Regis University*; Anton Dzhamay, *University of Northern Colorado* CONNECTICUT Jeffrey McGowan, *Central Connecticut State University*; Ivan Gotchev, *Central Connecticut State University*; Charles Waiveris, *Central Connecticut State University*; Christopher Hammond, *Connecticut College*; Anthony Y. Aidoo, *Eastern Connecticut State University*; Kim Ward, *Eastern Connecticut State University*; Joan W. Weiss, *Fairfield University*; Theresa M. Sandifer, *Southern Connecticut State University*; Cristian Rios, *Trinity College*; Melanie Stein, *Trinity College*; Steven Orszag, *Yale University* DELAWARE Patrick F. Mwerinde, *University of Delaware* DISTRICT OF COLUMBIA Jeffrey Hakim, *American University*; Joshua M. Lansky, *American University*; James A. Nickerson, *Gallaudet University* FLORIDA Gregory Spradlin, *Embry-Riddle University at Daytona Beach*; Daniela Popova, *Florida Atlantic University*; Abbas Zadegan, *Florida International University*; Gerardo Aladro, *Florida International University*; Gregory Henderson, *Hillsborough Community College*; Pam Crawford, *Jacksonville University*; Penny Morris, *Polk Community College*; George Schultz, *St. Petersburg College*; Jimmy Chang, *St. Petersburg College*; Carolyn Kistner, *St. Petersburg College*; Aida Kadic-Galeb, *The University of Tampa*; Constance Schober, *University of Central Florida*; S. Roy Choudhury, *University of Central Florida*; Kurt Overhiser, *Valencia Community College*; Jiongmin Yong, *University of Central Florida*; Giray Okten, *The Florida State University*; Frederick Hoffman, *Florida Atlantic University*; Thomas Beatty, *Florida Gulf Coast University*; Witny Librun, *Palm Beach Community College North*; Joe Castillo, *Broward County College*; Joann Lewin, *Edison College*; Donald Ransford, *Edison College*; Scott Berthiaume, *Edison College*; Alexander Ambrioso, *Hillsborough Community College*; Jane Golden, *Hillsborough Community College*; Susan Hiatt, *Polk Community College–Lakeland Campus*; Li Zhou, *Polk Community College–Winter Haven Campus*; Heather Edwards, *Seminole Community College*; Benjamin Landon, *Daytona State College*; Tony Malaret, *Seminole Community College*; Lane Vosbury, *Seminole Community College*; William Rickman, *Seminole Community College*; Cheryl Cantwell, *Seminole Community College*; Michael Schramm, *Indian River State College*; Janette Campbell, *Palm Beach Community College–Lake Worth*; Kwai-Lee Chui, *University of Florida*; Shu-Jen Huang, *University of Florida* GEORGIA Christian Barrientos, *Clayton State University*; Thomas T. Morley, *Georgia Institute of Technology*; Doron Lubinsky, *Georgia Institute of Technology*; Ralph Wildy, *Georgia Military College*; Shahram Nazari, *Georgia Perimeter College*; Alice Eiko Pierce, *Georgia Perimeter College, Clarkson Campus*; Susan Nelson, *Georgia Perimeter College, Clarkson Campus*; Laurene Fausett, *Georgia Southern University*; Scott N. Kersey, *Georgia Southern University*; Jimmy L. Solomon, *Georgia Southern University*; Allen G. Fuller, *Gordon College*; Marwan Zabdawi, *Gordon College*; Carolyn A. Yackel, *Mercer University*; Blane Hollingsworth, *Middle Georgia State College*; Shahryar Heydari, *Piedmont College*; Dan Kannan, *The University of Georgia*; June Jones, *Middle Georgia State College*; Abdelkrim Brania, *Morehouse College*; Ying Wang, *Augusta State University*; James M. Benedict, *Augusta State University*; Kouong Law, *Georgia Perimeter College*; Rob Williams, *Georgia Perimeter College*; Alvina Atkinson, *Georgia Gwinnett Col-*

lege; Amy Erickson, *Georgia Gwinnett College* HAWAII Shuguang Li, *University of Hawaii at Hilo*; Raina B. Ivanova, *University of Hawaii at Hilo* IDAHO Uwe Kaiser, *Boise State University*; Charles Kerr, *Boise State University*; Zach Teitler, *Boise State University*; Otis Kenny, *Boise State University*; Alex Feldman, *Boise State University*; Doug Bullock, *Boise State University*; Brian Dietel, *Lewis-Clark State College*; Ed Korntved, *Northwest Nazarene University*; Cynthia Piez, *University of Idaho* ILLINOIS Chris Morin, *Blackburn College*; Alberto L. Delgado, *Bradley University*; John Haverhals, *Bradley University*; Herbert E. Kasube, *Bradley University*; Marvin Doubet, *Lake Forest College*; Marvin A. Gordon, *Lake Forest Graduate School of Management*; Richard J. Maher, *Loyola University Chicago*; Joseph H. Mayne, *Loyola University Chicago*; Marian Gidea, *Northeastern Illinois University*; John M. Alongi, *Northwestern University*; Miguel Angel Lerma, *Northwestern University*; Mehmet Dik, *Rockford College*; Tammy Voepel, *Southern Illinois University Edwardsville*; Rahim G. Karimpour, *Southern Illinois University*; Thomas Smith, *University of Chicago*; Laura DeMarco, *University of Illinois*; Evangelos Kobotis, *University of Illinois at Chicago*; Jennifer McNeilly, *University of Illinois at Urbana-Champaign*; Timur Oikhberg, *University of Illinois at Urbana-Champaign*; Manouchehr Azad, *Harper College*; Minhua Liu, *Harper College*; Mary Hill, *College of DuPage*; Arthur N. DiVito, *Harold Washington College* INDIANA Vania Mascioni, *Ball State University*; Julie A. Killingbeck, *Ball State University*; Kathie Freed, *Butler University*; Zhixin Wu, *DePauw University*; John P. Boardman, *Franklin College*; Robert N. Talbert, *Franklin College*; Robin Symonds, *Indiana University Kokomo*; Henry L. Wyzinski, *Indiana University Northwest*; Melvin Royer, *Indiana Wesleyan University*; Gail P. Greene, *Indiana Wesleyan University*; David L. Finn, *Rose-Hulman Institute of Technology*; Chong Keat Arthur Lim, *University of Notre Dame* IOWA Nasser Dastrange, *Buena Vista University*; Mark A. Mills, *Central College*; Karen Ernst, *Hawkeye Community College*; Richard Mason, *Indian Hills Community College*; Robert S. Keller, *Loras College*; Eric Robert Westlund, *Luther College*; Weimin Han, *The University of Iowa* KANSAS Timothy W. Flood, *Pittsburg State University*; Sarah Cook, *Washburn University*; Kevin E. Charlwood, *Washburn University*; Conrad Uwe, *Cowley County Community College*; David N. Yetter, *Kansas State University* KENTUCKY Alex M. McAllister, *Center College*; Sandy Spears, *Jefferson Community & Technical College*; Leanne Faulkner, *Kentucky Wesleyan College*; Donald O. Clayton, *Madisonville Community College*; Thomas Riedel, *University of Louisville*; Manabendra Das, *University of Louisville*; Lee Larson, *University of Louisville*; Jens E. Harlander, *Western Kentucky University*; Philip McCartney, *Northern Kentucky University*; Andy Long, *Northern Kentucky University*; Omer Yayenie, *Murray State University*; Donald Krug, *Northern Kentucky University* LOUISIANA William Forrest, *Baton Rouge Community College*; Paul Wayne Britt, *Louisiana State University*; Galen Turner, *Louisiana Tech University*; Randall Wills, *Southeastern Louisiana University*; Kent Neuerburg, *Southeastern Louisiana University*; Guoli Ding, *Louisiana State University*; Julia Ledet, *Louisiana State University*; Brent Strunk, *University of Louisiana at Monroe* MAINE Andrew Knightly, *The University of Maine*; Sergey Lvin, *The University of Maine*; Joel W. Irish, *University of Southern Maine*; Laurie Woodman, *University of Southern Maine*; David M. Bradley, *The University of Maine*; William O. Bray, *The University of Maine* MARYLAND Leonid Stern, *Towson University*; Jacob Kogan, *University of Maryland Baltimore County*; Mark E. Williams, *University of Maryland Eastern Shore*; Austin A. Lobo, *Washington College*; Supawan Lertskrai, *Harford Community College*; Fary Sami, *Harford Community College*; Andrew Bulleri, *Howard Community College* MASSACHUSETTS Sean McGrath, *Algonquin Regional High School*; Norton Starr, *Amherst College*; Renato Mirollo, *Boston College*; Emma Previato, *Boston University*; Laura K Gross, *Bridgewater State University*; Richard H. Stout, *Gordon College*; Matthew P. Leingang, *Harvard University*; Suellen Robinson, *North Shore Community College*; Walter Stone, *North Shore Community College*; Barbara Loud, *Regis College*; Andrew B. Perry, *Springfield College*; Tawanda Gwena, *Tufts University*; Gary Simundza, *Wentworth Institute of Technology*; Mikhail Chkhenkeli, *Western New England College*; David Daniels, *Western New England College*; Alan Gorfin, *Western New England College*; Saeed Ghahramani, *Western New England College*; Julian Fleron, *Westfield State College*; Maria Fung, *Worchester State University*; Brigitte Servatius, *Worcester Polytechnic Institute*; John Goulet, *Worcester Polytechnic Institute*; Alexander Martsinkovsky, *Northeastern University*; Marie Clote, *Boston College*; Alexander Kastner, *Williams College*; Margaret Peard, *Williams College*; Mihai Stoiciu, *Williams College* MICHIGAN Mark E. Bollman, *Albion College*; Jim Chesla, *Grand Rapids Community College*; Jeanne Wald, *Michigan State University*; Allan A. Struthers, *Michigan Technological University*; Debra Pharo, *Northwestern Michigan College*; Anna Maria Spagnuolo, *Oakland University*; Diana Faoro, *Romeo Senior High School*; Andrew Strowe, *University of Michigan–Dearborn*; Daniel Stephen Drucker, *Wayne State University*; Christopher Cartwright, *Lawrence Technological University*; Jay Treiman, *Western Michigan University* MINNESOTA Bruce Bordwell, *Anoka-Ramsey Community College*; Robert Dobrow, *Carleton College*; Jessie K. Lenarz, *Concordia College–Moorhead Minnesota*; Bill Tomhave, *Concordia College*; David L. Frank, *University of Minnesota*; Steven I. Sperber, *University of Minnesota*; Jeffrey T. McLean, *University of St. Thomas*; Chehrzad Shakiban, *University of St. Thomas*; Melissa Loe, *University of St. Thomas*; Nick Christopher Fiala, *St. Cloud State University*; Victor Padron, *Normandale Community College*; Mark Ahrens, *Normandale Community College*; Gerry Naughton, *Century Community College*; Carrie Naughton, *Inver Hills Community College* MISSISSIPPI Vivien G. Miller, *Mississippi State University*; Ted Dobson, *Mississippi State University*; Len Miller, *Mississippi State University*; Tristan Denley, *The University of Mississippi* MISSOURI Robert Robertson, *Drury University*; Gregory A. Mitchell, *Metropolitan Community College–Penn Valley*; Charles N. Curtis, *Missouri Southern State University*; Vivek Narayanan, *Moberly Area Community College*; Russell Blyth, *Saint Louis University*; Julianne Rainbolt, *Saint Louis University*; Blake Thornton, *Saint Louis University*; Kevin W. Hopkins, *Southwest Baptist University*; Joe Howe, *St. Charles Community College*; Wanda Long, *St. Charles Community College*; Andrew Stephan, *St. Charles Community College* MONTANA Kelly Cline, *Carroll College*; Veronica Baker, *Montana State University, Bozeman*; Richard C. Swanson, *Montana State University*; Thomas Hayes-McGoff, *Montana State University*; Nikolaus Vonessen, *The University of Montana* NEBRASKA Edward G. Reinke Jr., *Concordia University*; Judith Downey, *University of Nebraska at Omaha* NEVADA Jennifer Gorman, *College of Southern Nevada*; Jonathan Pearsall, *College of Southern Nevada*; Rohan Dalpatadu, *University of Nevada, Las Vegas*; Paul Aizley, *University of Nevada, Las Vegas* NEW HAMPSHIRE Richard Jardine, *Keene State College*; Michael Cullinane, *Keene State College*; Roberta Kieronski, *University of New Hampshire at Manchester*; Erik Van Erp, *Dartmouth College* NEW JERSEY Paul S. Rossi, *College of Saint Elizabeth*; Mark Galit, *Essex County College*; Katarzyna Potocka, *Ramapo College of New Jersey*; Nora S. Thornber, *Raritan Valley Community College*; Abdulkadir Hassen, *Rowan University*; Olcay Ilicasu, *Rowan University*; Avraham Soffer, *Rutgers, The State University of New Jersey*; Chengwen Wang, *Rutgers, The State University of New Jersey*; Shabnam Beheshti, *Rutgers University, The State University of New Jersey*; Stephen J. Greenfield, *Rutgers, The State University of New Jersey*; John T. Saccoman, *Seton Hall University*; Lawrence E. Levine, *Stevens Institute of Technology*; Jana Gevertz, *The College of New Jersey*; Barry Burd, *Drew University*; Penny Luczak, *Camden County College*; John Climent, *Cecil Community College*; Kristyanna Erickson, *Cecil Community College*; Eric Compton, *Brookdale Community College*; John Atsu-Swanzy, *Atlantic Cape Community College* NEW MEXICO Kevin Leith, *Central New Mexico Community College*; David Blankenbaker, *Central New Mexico Community College*; Joseph Lakey, *New Mexico State University*; Kees Onneweer, *University of New Mexico*; Jurg Bolli, *The University of New Mexico* NEW YORK Robert C. Williams, *Alfred University*; Timmy G. Bremer, *Broome Community College State University of New York*; Joaquin O. Carbonara, *Buffalo State College*; Robin Sue Sanders, *Buffalo State College*; Daniel Cunningham, *Buffalo*

College; Yan Tian, *Palomar College*; Donna E. Nordstrom, *Pasadena City College*; Don L. Hancock, *Pepperdine University*; Kevin Iga, *Pepperdine University*; Adolfo J. Rumbos, *Pomona College*; Virginia May, *Sacramento City College*; Carlos de la Lama, *San Diego City College*; Matthias Beck, *San Francisco State University*; Arek Goetz, *San Francisco State University*; Nick Bykov, *San Joaquin Delta College*; Eleanor Lang Kendrick, *San Jose City College*; Elizabeth Hodes, *Santa Barbara City College*; William Konya, *Santa Monica College*; John Kennedy, *Santa Monica College*; Peter Lee, *Santa Monica College*; Richard Salome, *Scotts Valley High School*; Norman Feldman, *Sonoma State University*; Elaine McDonald, *Sonoma State University*; John D. Eggers, *University of California, San Diego*; Adam Bowers, *University of California, San Diego*; Bruno Nachtergaele, *University of California, Davis*; Boumediene Hamzi, *University of California, Davis*; Olga Radko, *University of California, Los Angeles*; Richard Leborne, *University of California, San Diego*; Peter Stevenhagen, *University of California, San Diego*; Jeffrey Stopple, *University of California, Santa Barbara*; Guofang Wei, *University of California, Santa Barbara*; Rick A. Simon, *University of La Verne*; Alexander E. Koonce, *University of Redlands*; Mohamad A. Alwash, *West Los Angeles College*; Calder Daenzer, *University of California, Berkeley*; Jude Thaddeus Socrates, *Pasadena City College*; Cheuk Ying Lam, *California State University Bakersfield*; Borislava Gutarts, *California State University, Los Angeles*; Daniel Rogalski, *University of California, San Diego*; Don Hartig, *California Polytechnic State University*; Anne Voth, *Palomar College*; Jay Wiestling, *Palomar College*; Lindsey Bramlett-Smith, *Santa Barbara City College*; Dennis Morrow, *College of the Canyons*; Sydney Shanks, *College of the Canyons*; Bob Tolar, *College of the Canyons*; Gene W. Majors, *Fullerton College*; Robert Diaz, *Fullerton College*; Gregory Nguyen, *Fullerton College*; Paul Sjoberg, *Fullerton College*; Deborah Ritchie, *Moorpark College*; Maya Rahnamaie, *Moorpark College*; Kathy Fink, *Moorpark College*; Christine Cole, *Moorpark College*; K. Di Passero, *Moorpark College*; Sid Kolpas, *Glendale Community College*; Miriam Castrconde, *Irvine Valley College*; Ilkner Erbas-White, *Irvine Valley College*; Corey Manchester, *Grossmont College*; Donald Murray, *Santa Monica College*; Barbara McGee, *Cuesta College*; Marie Larsen, *Cuesta College*; Joe Vasta, *Cuesta College*; Mike Kinter, *Cuesta College*; Mark Turner, *Cuesta College*; G. Lewis, *Cuesta College*; Daniel Kleinfelter, *College of the Desert*; Esmeralda Medrano, *Citrus College*; James Swatzel, *Citrus College*; Mark Littrell, *Rio Hondo College*; Rich Zucker, *Irvine Valley College*; Cindy Torigison, *Palomar College*; Craig Chamberline, *Diablo Valley College*; Lindsey Lang, *Diablo Valley College*; Sam Needham, *Diablo Valley College*; Dan Bach, *Diablo Valley College*; Ted Nirgiotis, *Diablo Valley College*; Monte Collazo, *Diablo Valley College*; Tina Levy, *Diablo Valley College*; Mona Panchal, *East Los Angeles College*; Ron Sandvick, *San Diego Mesa College*; Larry Handa, *West Valley College*; Frederick Utter, *Santa Rose Junior College*; Farshod Mosh, *DeAnza College*; Doli Bambhania, *DeAnza College*; Charles Klein, *DeAnza College*; Tammi Marshall, *Cauyamaca College*; Inwon Leu, *Cauyamaca College*; Michael Moretti, *Bakersfield College*; Janet Tarjan, *Bakersfield College*; Hoat Le, *San Diego City College*; Richard Fielding, *Southwestern College*; Shannon Gracey, *Southwestern College*; Janet Mazzarella, *Southwestern College*; Christina Soderlund, *California Lutheran University*; Rudy Gonzalez, *Citrus College*; Robert Crise, *Crafton Hills College*; Joseph Kazimir, *East Los Angeles College*; Randall Rogers, *Fullerton College*; Peter Bouzar, *Golden West College*; Linda Ternes, *Golden West College*; Hsiao-Ling Liu, *Los Angeles Trade Tech Community College*; Yu-Chung Chang-Hou, *Pasadena City College*; Guillermo Alvarez, *San Diego City College*; Ken Kuniyuki, *San Diego Mesa College*; Laleh Howard, *San Diego Mesa College*; Sharareh Masooman, *Santa Barbara City College*; Jared Hersh, *Santa Monica College*; Betty Wong, *Santa Monica College*; Brian Rodas, *Santa Monica College*; Veasna Chiek, *Riverside City College* **COLORADO** Tony Weathers, *Adams State College*; Erica Johnson, *Arapahoe Community College*; Karen Walters, *Arapahoe Community College*; Joshua D. Laison, *Colorado College*; G. Gustave Greivel, *Colorado School of Mines*; Holly Eklund, *Colorado School of the Mines*; Mike Nicholas, *Colorado*

School of the Mines; Jim Thomas, *Colorado State University*; Eleanor Storey, *Front Range Community College*; Larry Johnson, *Metropolitan State College of Denver*; Carol Kuper, *Morgan Community College*; Larry A. Pontaski, *Pueblo Community College*; Terry Chen Reeves, *Red Rocks Community College*; Debra S. Carney, *Colorado School of the Mines*; Louis A. Talman, *Metropolitan State College of Denver*; Mary A. Nelson, *University of Colorado at Boulder*; J. Kyle Pula, *University of Denver*; Jon Von Stroh, *University of Denver*; Sharon Butz, *University of Denver*; Daniel Daly, *University of Denver*; Tracy Lawrence, *Arapahoe Community College*; Shawna Mahan, *University of Colorado Denver*; Adam Norris, *University of Colorado at Boulder*; Anca Radulescu, *University of Colorado at Boulder*; Mike Kawai, *University of Colorado Denver*; Janet Barnett, *Colorado State University–Pueblo*; Byron Hurley, *Colorado State University–Pueblo*; Jonathan Portiz, *Colorado State University–Pueblo*; Bill Emerson, *Metropolitan State College of Denver*; Suzanne Caulk, *Regis University*; Anton Dzhamay, *University of Northern Colorado* **CONNECTICUT** Jeffrey McGowan, *Central Connecticut State University*; Ivan Gotchev, *Central Connecticut State University*; Charles Waiveris, *Central Connecticut State University*; Christopher Hammond, *Connecticut College*; Anthony Y. Aidoo, *Eastern Connecticut State University*; Kim Ward, *Eastern Connecticut State University*; Joan W. Weiss, *Fairfield University*; Theresa M. Sandifer, *Southern Connecticut State University*; Cristian Rios, *Trinity College*; Melanie Stein, *Trinity College*; Steven Orszag, *Yale University* **DELAWARE** Patrick F. Mwerinde, *University of Delaware* **DISTRICT OF COLUMBIA** Jeffrey Hakim, *American University*; Joshua M. Lansky, *American University*; James A. Nickerson, *Gallaudet University* **FLORIDA** Gregory Spradlin, *Embry-Riddle University at Daytona Beach*; Daniela Popova, *Florida Atlantic University*; Abbas Zadegan, *Florida International University*; Gerardo Aladro, *Florida International University*; Gregory Henderson, *Hillsborough Community College*; Pam Crawford, *Jacksonville University*; Penny Morris, *Polk Community College*; George Schultz, *St. Petersburg College*; Jimmy Chang, *St. Petersburg College*; Carolyn Kistner, *St. Petersburg College*; Aida Kadic-Galeb, *The University of Tampa*; Constance Schober, *University of Central Florida*; S. Roy Choudhury, *University of Central Florida*; Kurt Overhiser, *Valencia Community College*; Jiongmin Yong, *University of Central Florida*; Giray Okten, *The Florida State University*; Frederick Hoffman, *Florida Atlantic University*; Thomas Beatty, *Florida Gulf Coast University*; Witny Librun, *Palm Beach Community College North*; Joe Castillo, *Broward County College*; Joann Lewin, *Edison College*; Donald Ransford, *Edison College*; Scott Berthiaume, *Edison College*; Alexander Ambrioso, *Hillsborough Community College*; Jane Golden, *Hillsborough Community College*; Susan Hiatt, *Polk Community College–Lakeland Campus*; Li Zhou, *Polk Community College–Winter Haven Campus*; Heather Edwards, *Seminole Community College*; Benjamin Landon, *Daytona State College*; Tony Malaret, *Seminole Community College*; Lane Vosbury, *Seminole Community College*; William Rickman, *Seminole Community College*; Cheryl Cantwell, *Seminole Community College*; Michael Schramm, *Indian River State College*; Janette Campbell, *Palm Beach Community College–Lake Worth*; Kwai-Lee Chui, *University of Florida*; Shu-Jen Huang, *University of Florida* **GEORGIA** Christian Barrientos, *Clayton State University*; Thomas T. Morley, *Georgia Institute of Technology*; Doron Lubinsky, *Georgia Institute of Technology*; Ralph Wildy, *Georgia Military College*; Shahram Nazari, *Georgia Perimeter College*; Alice Eiko Pierce, *Georgia Perimeter College, Clarkson Campus*; Susan Nelson, *Georgia Perimeter College, Clarkson Campus*; Laurene Fausett, *Georgia Southern University*; Scott N. Kersey, *Georgia Southern University*; Jimmy L. Solomon, *Georgia Southern University*; Allen G. Fuller, *Gordon College*; Marwan Zabdawi, *Gordon College*; Carolyn A. Yackel, *Mercer University*; Blane Hollingsworth, *Middle Georgia State College*; Shahryar Heydari, *Piedmont College*; Dan Kannan, *The University of Georgia*; June Jones, *Middle Georgia State College*; Abdelkrim Brania, *Morehouse College*; Ying Wang, *Augusta State University*; James M. Benedict, *Augusta State University*; Kouong Law, *Georgia Perimeter College*; Rob Williams, *Georgia Perimeter College*; Alvina Atkinson, *Georgia Gwinnett Col-*

lege; Amy Erickson, *Georgia Gwinnett College* **HAWAII** Shuguang Li, *University of Hawaii at Hilo*; Raina B. Ivanova, *University of Hawaii at Hilo* **IDAHO** Uwe Kaiser, *Boise State University*; Charles Kerr, *Boise State University*; Zach Teitler, *Boise State University*; Otis Kenny, *Boise State University*; Alex Feldman, *Boise State University*; Doug Bullock, *Boise State University*; Brian Dietel, *Lewis-Clark State College*; Ed Korntved, *Northwest Nazarene University*; Cynthia Piez, *University of Idaho* **ILLINOIS** Chris Morin, *Blackburn College*; Alberto L. Delgado, *Bradley University*; John Haverhals, *Bradley University*; Herbert E. Kasube, *Bradley University*; Marvin Doubet, *Lake Forest College*; Marvin A. Gordon, *Lake Forest Graduate School of Management*; Richard J. Maher, *Loyola University Chicago*; Joseph H. Mayne, *Loyola University Chicago*; Marian Gidea, *Northeastern Illinois University*; John M. Alongi, *Northwestern University*; Miguel Angel Lerma, *Northwestern University*; Mehmet Dik, *Rockford College*; Tammy Voepel, *Southern Illinois University Edwardsville*; Rahim G. Karimpour, *Southern Illinois University*; Thomas Smith, *University of Chicago*; Laura DeMarco, *University of Illinois*; Evangelos Kobotis, *University of Illinois at Chicago*; Jennifer McNeilly, *University of Illinois at Urbana-Champaign*; Timur Oikhberg, *University of Illinois at Urbana-Champaign*; Manouchehr Azad, *Harper College*; Minhua Liu, *Harper College*; Mary Hill, *College of DuPage*; Arthur N. DiVito, *Harold Washington College* **INDIANA** Vania Mascioni, *Ball State University*; Julie A. Killingbeck, *Ball State University*; Kathie Freed, *Butler University*; Zhixin Wu, *DePauw University*; John P. Boardman, *Franklin College*; Robert N. Talbert, *Franklin College*; Robin Symonds, *Indiana University Kokomo*; Henry L. Wyzinski, *Indiana University Northwest*; Melvin Royer, *Indiana Wesleyan University*; Gail P. Greene, *Indiana Wesleyan University*; David L. Finn, *Rose-Hulman Institute of Technology*; Chong Keat Arthur Lim, *University of Notre Dame* **IOWA** Nasser Dastrange, *Buena Vista University*; Mark A. Mills, *Central College*; Karen Ernst, *Hawkeye Community College*; Richard Mason, *Indian Hills Community College*; Robert S. Keller, *Loras College*; Eric Robert Westlund, *Luther College*; Weimin Han, *The University of Iowa* **KANSAS** Timothy W. Flood, *Pittsburg State University*; Sarah Cook, *Washburn University*; Kevin E. Charlwood, *Washburn University*; Conrad Uwe, *Cowley County Community College*; David N. Yetter, *Kansas State University* **KENTUCKY** Alex M. McAllister, *Center College*; Sandy Spears, *Jefferson Community & Technical College*; Leanne Faulkner, *Kentucky Wesleyan College*; Donald O. Clayton, *Madisonville Community College*; Thomas Riedel, *University of Louisville*; Manabendra Das, *University of Louisville*; Lee Larson, *University of Louisville*; Jens E. Harlander, *Western Kentucky University*; Philip McCartney, *Northern Kentucky University*; Andy Long, *Northern Kentucky University*; Omer Yayenie, *Murray State University*; Donald Krug, *Northern Kentucky University* **LOUISIANA** William Forrest, *Baton Rouge Community College*; Paul Wayne Britt, *Louisiana State University*; Galen Turner, *Louisiana Tech University*; Randall Wills, *Southeastern Louisiana University*; Kent Neuerburg, *Southeastern Louisiana University*; Guoli Ding, *Louisiana State University*; Julia Ledet, *Louisiana State University*; Brent Strunk, *University of Louisiana at Monroe* **MAINE** Andrew Knightly, *The University of Maine*; Sergey Lvin, *The University of Maine*; Joel W. Irish, *University of Southern Maine*; Laurie Woodman, *University of Southern Maine*; David M. Bradley, *The University of Maine*; William O. Bray, *The University of Maine* **MARYLAND** Leonid Stern, *Towson University*; Jacob Kogan, *University of Maryland Baltimore County*; Mark E. Williams, *University of Maryland Eastern Shore*; Austin A. Lobo, *Washington College*; Supawan Lertskrai, *Harford Community College*; Fary Sami, *Harford Community College*; Andrew Bulleri, *Howard Community College* **MASSACHUSETTS** Sean McGrath, *Algonquin Regional High School*; Norton Starr, *Amherst College*; Renato Mirollo, *Boston College*; Emma Previato, *Boston University*; Laura K Gross, *Bridgewater State University*; Richard H. Stout, *Gordon College*; Matthew P. Leingang, *Harvard University*; Suellen Robinson, *North Shore Community College*; Walter Stone, *North Shore Community College*; Barbara Loud, *Regis College*; Andrew B. Perry, *Springfield College*; Tawanda Gwena, *Tufts University*; Gary Simundza, *Wentworth Institute of*

Technology; Mikhail Chkhenkeli, *Western New England College*; David Daniels, *Western New England College*; Alan Gorfin, *Western New England College*; Saeed Ghahramani, *Western New England College*; Julian Fleron, *Westfield State College*; Maria Fung, *Worchester State University*; Brigitte Servatius, *Worcester Polytechnic Institute*; John Goulet, *Worcester Polytechnic Institute*; Alexander Martsinkovsky, *Northeastern University*; Marie Clote, *Boston College*; Alexander Kastner, *Williams College*; Margaret Peard, *Williams College*; Mihai Stoiciu, *Williams College* **MICHIGAN** Mark E. Bollman, *Albion College*; Jim Chesla, *Grand Rapids Community College*; Jeanne Wald, *Michigan State University*; Allan A. Struthers, *Michigan Technological University*; Debra Pharo, *Northwestern Michigan College*; Anna Maria Spagnuolo, *Oakland University*; Diana Faoro, *Romeo Senior High School*; Andrew Strowe, *University of Michigan–Dearborn*; Daniel Stephen Drucker, *Wayne State University*; Christopher Cartwright, *Lawrence Technological University*; Jay Treiman, *Western Michigan University* **MINNESOTA** Bruce Bordwell, *Anoka-Ramsey Community College*; Robert Dobrow, *Carleton College*; Jessie K. Lenarz, *Concordia College–Moorhead Minnesota*; Bill Tomhave, *Concordia College*; David L. Frank, *University of Minnesota*; Steven I. Sperber, *University of Minnesota*; Jeffrey T. McLean, *University of St. Thomas*; Chehrzad Shakiban, *University of St. Thomas*; Melissa Loe, *University of St. Thomas*; Nick Christopher Fiala, *St. Cloud State University*; Victor Padron, *Normandale Community College*; Mark Ahrens, *Normandale Community College*; Gerry Naughton, *Century Community College*; Carrie Naughton, *Inver Hills Community College* **MISSISSIPPI** Vivien G. Miller, *Mississippi State University*; Ted Dobson, *Mississippi State University*; Len Miller, *Mississippi State University*; Tristan Denley, *The University of Mississippi* **MISSOURI** Robert Robertson, *Drury University*; Gregory A. Mitchell, *Metropolitan Community College–Penn Valley*; Charles N. Curtis, *Missouri Southern State University*; Vivek Narayanan, *Moberly Area Community College*; Russell Blyth, *Saint Louis University*; Julianne Rainbolt, *Saint Louis University*; Blake Thornton, *Saint Louis University*; Kevin W. Hopkins, *Southwest Baptist University*; Joe Howe, *St. Charles Community College*; Wanda Long, *St. Charles Community College*; Andrew Stephan, *St. Charles Community College* **MONTANA** Kelly Cline, *Carroll College*; Veronica Baker, *Montana State University, Bozeman*; Richard C. Swanson, *Montana State University*; Thomas Hayes-McGoff, *Montana State University*; Nikolaus Vonessen, *The University of Montana* **NEBRASKA** Edward G. Reinke Jr., *Concordia University*; Judith Downey, *University of Nebraska at Omaha* **NEVADA** Jennifer Gorman, *College of Southern Nevada*; Jonathan Pearsall, *College of Southern Nevada*; Rohan Dalpatadu, *University of Nevada, Las Vegas*; Paul Aizley, *University of Nevada, Las Vegas* **NEW HAMPSHIRE** Richard Jardine, *Keene State College*; Michael Cullinane, *Keene State College*; Roberta Kieronski, *University of New Hampshire at Manchester*; Erik Van Erp, *Dartmouth College* **NEW JERSEY** Paul S. Rossi, *College of Saint Elizabeth*; Mark Galit, *Essex County College*; Katarzyna Potocka, *Ramapo College of New Jersey*; Nora S. Thornber, *Raritan Valley Community College*; Abdulkadir Hassen, *Rowan University*; Olcay Ilicasu, *Rowan University*; Avraham Soffer, *Rutgers, The State University of New Jersey*; Chengwen Wang, *Rutgers, The State University of New Jersey*; Shabnam Beheshti, *Rutgers University, The State University of New Jersey*; Stephen J. Greenfield, *Rutgers, The State University of New Jersey*; John T. Saccoman, *Seton Hall University*; Lawrence E. Levine, *Stevens Institute of Technology*; Jana Gevertz, *The College of New Jersey*; Barry Burd, *Drew University*; Penny Luczak, *Camden County College*; John Climent, *Cecil Community College*; Kristyanna Erickson, *Cecil Community College*; Eric Compton, *Brookdale Community College*; John Atsu-Swanzy, *Atlantic Cape Community College* **NEW MEXICO** Kevin Leith, *Central New Mexico Community College*; David Blankenbaker, *Central New Mexico Community College*; Joseph Lakey, *New Mexico State University*; Kees Onneweer, *University of New Mexico*; Jurg Bolli, *The University of New Mexico* **NEW YORK** Robert C. Williams, *Alfred University*; Timmy G. Bremer, *Broome Community College State University of New York*; Joaquin O. Carbonara, *Buffalo State College*; Robin Sue Sanders, *Buffalo State College*; Daniel Cunningham, *Buffalo*

State College; Rose Marie Castner, *Canisius College*; Sharon L. Sullivan, *Catawba College*; Fabio Nironi, *Columbia University*; Camil Muscalu, *Cornell University*; Maria S. Terrell, *Cornell University*; Margaret Mulligan, *Dominican College of Blauvelt*; Robert Andersen, *Farmingdale State University of New York*; Leonard Nissim, *Fordham University*; Jennifer Roche, *Hobart and William Smith Colleges*; James E. Carpenter, *Iona College*; Peter Shenkin, *John Jay College of Criminal Justice/CUNY*; Gordon Crandall, *LaGuardia Community College/CUNY*; Gilbert Traub, *Maritime College, State University of New York*; Paul E. Seeburger, *Monroe Community College Brighton Campus*; Abraham S. Mantell, *Nassau Community College*; Daniel D. Birmajer, *Nazareth College*; Sybil G. Shaver, *Pace University*; Margaret Kiehl, *Rensselaer Polytechnic Institute*; Carl V. Lutzer, *Rochester Institute of Technology*; Michael A. Radin, *Rochester Institute of Technology*; Hossein Shahmohamad, *Rochester Institute of Technology*; Thomas Rousseau, *Siena College*; Jason Hofstein, *Siena College*; Leon E. Gerber, *St. Johns University*; Christopher Bishop, *Stony Brook University*; James Fulton, *Suffolk County Community College*; John G. Michaels, *SUNY Brockport*; Howard J. Skogman, *SUNY Brockport*; Cristina Bacuta, *SUNY Cortland*; Jean Harper, *SUNY Fredonia*; David Hobby, *SUNY New Paltz*; Kelly Black, *Union College*; Thomas W. Cusick, *University at Buffalo/The State University of New York*; Gino Biondini, *University at Buffalo/The State University of New York*; Robert Koehler, *University at Buffalo/The State University of New York*; Donald Larson, *University of Rochester*; Robert Thompson, *Hunter College*; Ed Grossman, *The City College of New York* NORTH CAROLINA Jeffrey Clark, *Elon University*; William L. Burgin, *Gaston College*; Manouchehr H. Misaghian, *Johnson C. Smith University*; Legunchim L. Emmanwori, *North Carolina A&T State University*; Drew Pasteur, *North Carolina State University*; Demetrio Labate, *North Carolina State University*; Mohammad Kazemi, *The University of North Carolina at Charlotte*; Richard Carmichael, *Wake Forest University*; Gretchen Wilke Whipple, *Warren Wilson College*; John Russell Taylor, *University of North Carolina at Charlotte*; Mark Ellis, *Piedmont Community College* NORTH DAKOTA Jim Coykendall, *North Dakota State University*; Anthony J. Bevelacqua, *The University of North Dakota*; Richard P. Millspaugh, *The University of North Dakota*; Thomas Gilsdorf, *The University of North Dakota*; Michele Iiams, *The University of North Dakota*; Mohammad Khavanin, *University of North Dakota* OHIO Christopher Butler, *Case Western Reserve University*; Pamela Pierce, *The College of Wooster*; Barbara H. Margolius, *Cleveland State University*; Tzu-Yi Alan Yang, *Columbus State Community College*; Greg S. Goodhart, *Columbus State Community College*; Kelly C. Stady, *Cuyahoga Community College*; Brian T. Van Pelt, *Cuyahoga Community College*; David Robert Ericson, *Miami University*; Frederick S. Gass, *Miami University*; Thomas Stacklin, *Ohio Dominican University*; Vitaly Bergelson, *The Ohio State University*; Robert Knight, *Ohio University*; John R. Pather, *Ohio University, Eastern Campus*; Teresa Contenza, *Otterbein College*; Ali Hajjafar, *The University of Akron*; Jianping Zhu, *The University of Akron*; Ian Clough, *University of Cincinnati Clermont College*; Atif Abueida, *University of Dayton*; Judith McCrory, *The University at Findlay*; Thomas Smotzer, *Youngstown State University*; Angela Spalsbury, *Youngstown State University*; James Osterburg, *The University of Cincinnati*; Mihaela A. Poplicher, *University of Cincinnati*; Frederick Thulin, *University of Illinois at Chicago*; Weimin Han, *The Ohio State University*; Crichton Ogle, *The Ohio State University*; Jackie Miller, *The Ohio State University*; Walter Mackey, *Owens Community College*; Jonathan Baker, *Columbus State Community College* OKLAHOMA Christopher Francisco, *Oklahoma State University*; Michael McClendon, *University of Central Oklahoma*; Teri Jo Murphy, *The University of Oklahoma*; Kimberly Adams, *University of Tulsa*; Shirley Pomeranz, *University of Tulsa* OREGON Lorna TenEyck, *Chemeketa Community College*; Angela Martinek, *Linn-Benton Community College*; Filix Maisch, *Oregon State University*; Tevian Dray, *Oregon State University*; Mark Ferguson, *Chemekata Community College*; Andrew Flight, *Portland State University*; Austina Fong, *Portland State University*; Jeanette R. Palmiter, *Portland State University* PENNSYLVANIA John B. Polhill, *Bloomsburg University of Pennsylvania*; Russell C. Walker, *Carnegie Mellon University*; Jon A. Beal, *Clarion University of Pennsylvania*; Kathleen Kane, *Community College of Allegheny County*; David A. Santos, *Community College of Philadelphia*; David S. Richeson, *Dickinson College*; Christine Marie Cedzo, *Gannon University*; Monica Pierri-Galvao, *Gannon University*; John H. Ellison, *Grove City College*; Gary L. Thompson, *Grove City College*; Dale McIntyre, *Grove City College*; Dennis Benchoff, *Harrisburg Area Community College*; William A. Drumin, *King's College*; Denise Reboli, *King's College*; Chawne Kimber, *Lafayette College*; Elizabeth McMahon, *Lafayette College*; Lorenzo Traldi, *Lafayette College*; David L. Johnson, *Lehigh University*; Matthew Hyatt, *Lehigh University*; Zia Uddin, *Lock Haven University of Pennsylvania*; Donna A. Dietz, *Mansfield University of Pennsylvania*; Samuel Wilcock, *Messiah College*; Richard R. Kern, *Montgomery County Community College*; Michael Fraboni, *Moravian College*; Neena T. Chopra, *The Pennsylvania State University*; Boris A. Datskovsky, *Temple University*; Dennis M. DeTurck, *University of Pennsylvania*; Jacob Burbea, *University of Pittsburgh*; Mohammed Yahdi, *Ursinus College*; Timothy Feeman, *Villanova University*; Douglas Norton, *Villanova University*; Robert Styer, *Villanova University*; Michael J. Fisher, *West Chester University of Pennsylvania*; Peter Brooksbank, *Bucknell University*; Emily Dryden, *Bucknell University*; Larry Friesen, *Butler County Community College*; Lisa Angelo, *Bucks County College*; Elaine Fitt, *Bucks County College*; Pauline Chow, *Harrisburg Area Community College*; Diane Benner, *Harrisburg Area Community College*; Emily B. Dryden, *Bucknell University*; Erica Chauvet, *Waynesburg University* RHODE ISLAND Thomas F. Banchoff, *Brown University*; Yajni Warnapala-Yehiya, *Roger Williams University*; Carol Gibbons, *Salve Regina University*; Joe Allen, *Community College of Rhode Island*; Michael Latina, *Community College of Rhode Island* SOUTH CAROLINA Stanley O. Perrine, *Charleston Southern University*; Joan Hoffacker, *Clemson University*; Constance C. Edwards, *Coastal Carolina University*; Thomas L. Fitzkee, *Francis Marion University*; Richard West, *Francis Marion University*; John Harris, *Furman University*; Douglas B. Meade, *University of South Carolina*; George Androulakis, *University of South Carolina*; Art Mark, *University of South Carolina Aiken*; Sherry Biggers, *Clemson University*; Mary Zachary Krohn, *Clemson University*; Andrew Incognito, *Coastal Carolina University*; Deanna Caveny, *College of Charleston* SOUTH DAKOTA Dan Kemp, *South Dakota State University* TENNESSEE Andrew Miller, *Belmont University*; Arthur A. Yanushka, *Christian Brothers University*; Laurie Plunk Dishman, *Cumberland University*; Maria Siopsis, *Maryville College*; Beth Long, *Pellissippi State Technical Community College*; Judith Fethe, *Pellissippi State Technical Community College*; Andrzej Gutek, *Tennessee Technological University*; Sabine Le Borne, *Tennessee Technological University*; Richard Le Borne, *Tennessee Technological University*; Maria F. Bothelho, *University of Memphis*; Roberto Triggiani, *University of Memphis*; Jim Conant, *The University of Tennessee*; Pavlos Tzermias, *The University of Tennessee*; Luis Renato Abib Finotti, *University of Tennessee, Knoxville*; Jennifer Fowler, *University of Tennessee, Knoxville*; Jo Ann W. Staples, *Vanderbilt University*; Dave Vinson, *Pellissippi State Community College*; Jonathan Lamb, *Pellissippi State Community College* TEXAS Sally Haas, *Angelina College*; Karl Havlak, *Angelo State University*; Michael Huff, *Austin Community College*; John M. Davis, *Baylor University*; Scott Wilde, *Baylor University and The University of Texas at Arlington*; Rob Eby, *Blinn College*; Tim Sever, *Houston Community College–Central*; Ernest Lowery, *Houston Community College–Northwest*; Brian Loft, *Sam Houston State University*; Jianzhong Wang, *Sam Houston State University*; Shirley Davis, *South Plains College*; Todd M. Steckler, *South Texas College*; Mary E. Wagner-Krankel, *St. Mary's University*; Elise Z. Price, *Tarrant County College, Southeast Campus*; David Price, *Tarrant County College, Southeast Campus*; Runchang Lin, *Texas A&M University*; Michael Stecher, *Texas A&M University*; Philip B. Yasskin, *Texas A&M University*; Brock Williams, *Texas Tech University*; I. Wayne Lewis, *Texas Tech University*; Robert E. Byerly, *Texas Tech University*; Ellina Grigorieva, *Texas Woman's University*; Abraham Haje, *Tomball College*; Scott Chapman, *Trinity University*; Elias Y. Deeba, *University of Houston Downtown*; Jianping Zhu, *The Uni-*

versity of Texas at Arlington; Tuncay Aktosun, *The University of Texas at Arlington*; John E. Gilbert, *The University of Texas at Austin*; Jorge R. Viramontes-Olivias, *The University of Texas at El Paso*; Fengxin Chen, *University of Texas at San Antonio*; Melanie Ledwig, *The Victoria College*; Gary L. Walls, *West Texas A&M University*; William Heierman, *Wharton County Junior College*; Lisa Rezac, *University of St. Thomas*; Raymond J. Cannon, *Baylor University*; Kathryn Flores, *McMurry University*; Jacqueline A. Jensen, *Sam Houston State University*; James Galloway, *Collin County College*; Raja Khoury, *Collin County College*; Annette Benbow, *Tarrant County College–Northwest*; Greta Harland, *Tarrant County College–Northeast*; Doug Smith, *Tarrant County College–Northeast*; Marcus McGuff, *Austin Community College*; Clarence McGuff, *Austin Community College*; Steve Rodi, *Austin Community College*; Vicki Payne, *Austin Community College*; Anne Pradera, *Austin Community College*; Christy Babu, *Laredo Community College*; Deborah Hewitt, *McLennan Community College*; W. Duncan, *McLennan Community College*; Hugh Griffith, *Mt. San Antonio College* UTAH Ruth Trygstad, *Salt Lake City Community College* VIRGINIA Verne E. Leininger, *Bridgewater College*; Brian Bradie, *Christopher Newport University*; Hongwei Chen, *Christopher Newport University*; John J. Avioli, *Christopher Newport University*; James H. Martin, *Christopher Newport University*; David Walnut, *George Mason University*; Mike Shirazi, *Germanna Community College*; Julie Clark, *Hollins University*; Ramon A. Mata-Toledo, *James Madison University*; Adrian Riskin, *Mary Baldwin College*; Josephine Letts, *Ocean Lakes High School*; Przemyslaw Bogacki, *Old Dominion University*; Deborah Denvir, *Randolph-Macon Woman's College*; Linda Powers, *Virginia Tech*; Gregory Dresden, *Washington and Lee University*; Jacob A. Siehler, *Washington and Lee University*; Yuan-Jen Chiang, *University of Mary Washington*; Nicholas Hamblet, *University of Virginia*; Bernard Fulgham, *University of Virginia*; Manouchehr "Mike" Mohajeri, *University of Virginia*; Lester Frank Caudill, *University of Richmond* VERMONT David Dorman, *Middlebury College*; Rachel Repstad, *Vermont Technical College* WASHINGTON Jennifer Laveglia, *Bellevue Community College*; David Whittaker, *Cascadia Community College*; Sharon Saxton, *Cascadia Community College*; Aaron Montgomery, *Central Washington University*; Patrick Averbeck, *Edmonds Community College*; Tana Knudson, *Heritage University*; Kelly Brooks, *Pierce College*; Shana P. Calaway, *Shoreline Community College*; Abel Gage, *Skagit Valley College*; Scott MacDonald, *Tacoma Community College*; Jason Preszler, *University of Puget Sound*; Martha A. Gady, *Whitworth College*; Wayne L. Neidhardt, *Edmonds Community College*; Simrat Ghuman, *Bellevue College*; Jeff Eldridge, *Edmonds Community College*; Kris Kissel, *Green River Community College*; Laura Moore-Mueller, *Green River Community College*; David Stacy, *Bellevue College*; Eric Schultz, *Walla Walla Community College*; Julianne Sachs, *Walla Walla Community College* WEST VIRGINIA David Cusick, *Marshall University*; Ralph Oberste-Vorth, *Marshall University*; Suda Kunyosying, *Shepard University*; Nicholas Martin, *Shepherd University*; Rajeev Rajaram, *Shepherd University*; Xiaohong Zhang, *West Virginia State University*; Sam B. Nadler, *West Virginia University* WYOMING Claudia Stewart, *Casper College*; Pete Wildman, *Casper College*; Charles Newberg, *Western Wyoming Community College*; Lynne Ipina, *University of Wyoming*; John Spitler, *University of Wyoming* WISCONSIN Erik R. Tou, *Carthage College*; Paul Bankston, *Marquette University*; Jane Nichols, *Milwaukee School of Engineering*; Yvonne Yaz, *Milwaukee School of Engineering*; Simei Tong, *University of Wisconsin–Eau Claire*; Terry Nyman, *University of Wisconsin–Fox Valley*; Robert L. Wilson, *University of Wisconsin–Madison*; Dietrich A. Uhlenbrock, *University of Wisconsin–Madison*; Paul Milewski, *University of Wisconsin–Madison*; Donald Solomon, *University of Wisconsin–Milwaukee*; Kandasamy Muthuvel, *University of Wisconsin–Oshkosh*; Sheryl Wills, *University of Wisconsin–Platteville*; Kathy A. Tomlinson, *University of Wisconsin–River Falls*; Cynthia L. McCabe, *University of Wisconsin–Stevens Point*; Matthew Welz, *University of Wisconsin–Stevens Point*; Joy Becker, *University of Wisconsin-Stout*; Jeganathan Sriskandarajah , *Madison Area Tech College*; Wayne Sigelko, *Madison Area Tech College* CANADA Don St. Jean, *George Brown College*; Robert Dawson, *St. Mary's University*; Len Bos, *University of Calgary*; Tony Ware, *University of Calgary*; Peter David Papez, *University of Calgary*; John O'Conner, *Grant MacEwan University*; Michael P. Lamoureux, *University of Calgary*; Yousry Elsabrouty, *University of Calgary*; Darja Kalajdzievska, *University of Manitoba*; Andrew Skelton, *University of Guelph*; Douglas Farenick, *University of Regina*

The creation of this third edition could not have happened without the help of many people. First, I want to thank the individuals whom I have worked with at W. H. Freeman. Terri Ward and Ruth Baruth convinced me that I should take on this project, and I am grateful to them for their support and their confidence in my ability to tackle it. Throughout this process, Terri has been a huge help. I can always count on her to keep this train on track. Katrina Wilhelm has also been an amazing resource. She brings calm competence and organizational skills that constantly impress me. Tony Palermino has provided expert editorial help throughout the process. He is incredibly knowledgeable about all aspects of mathematics textbooks and has an eye for the details that make a book work. Kerry O'Shaughnessy kept the production process moving forward in a timely manner without ever resorting to threats. John Rogosich was the superb compositor. Patti Brecht handled the copyediting in an expert manner. My thanks are also due to W. H. Freeman's superb production team: Janice Donnola, Eileen Liang, Blake Logan, Paul Rohloff, and to Ron Weickart at Network Graphics for his skilled and creative execution of the art program.

Many faculty gave critical feedback on the second edition, and their names appear above. I am deeply grateful to them. I do want to particularly thank all of the advisory board members who gave me feedback month after month. Maria Shea Terrell continually sent me excellent unsolicited feedback until I asked to have her on the board. Then it became solicited. The accuracy reviewers, John Alongi, CK Cheung, Kwai-Lee Chui, John Davis, John Eggers, Stephen Greenfield, Roger Lipsett, Vivek Narayanan, and Olga Radko, helped to bring the final version into the form in which it now appears. You think you have found the errors, but you have not.

I also want to thank my colleagues in the Mathematics and Statistics Department at Williams College. I am incredibly lucky to be a member of this department. There are so many interesting projects and clever pedagogical ideas coming out of the department that it motivates me just because I am trying to keep up.

I would further like to thank my students. Their enthusiasm is what makes teaching fun. I enjoy coming to work every day, and they are what make it such a pleasure.

Finally, I want to thank my two children, Alexa and Colton. They are the ones who keep me grounded, who remind me what works and what doesn't in the real world. This book is dedicated to them.

Colin Adams

10 INFINITE SERIES

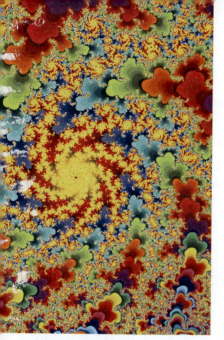

Infinite series allow us to make sense of a process that iterates infinitely, as occurs with fractals, generating pictures such as this one.

(Gregory Sams/Science Source)

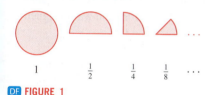

FIGURE 1

The theory of infinite series is a third branch of calculus, in addition to differential and integral calculus. Infinite series yield a new perspective on functions and on many interesting numbers. Two examples are the infinite series for the exponential function

$$e^x = 1 + x + \frac{x^2}{2!} + \frac{x^3}{3!} + \frac{x^4}{4!} + \cdots$$

and the Gregory–Leibniz series (see Exercise 63 in Section 10.2)

$$\frac{\pi}{4} = 1 - \frac{1}{3} + \frac{1}{5} - \frac{1}{7} + \frac{1}{9} - \cdots$$

The first shows that e^x can be expressed as an "infinite polynomial," and the second reveals that π is related to the reciprocals of the odd integers in an unexpected way. To make sense of infinite series, we need to define precisely what it means to add up infinitely many terms. Limits play a key role here, just as they do in differential and integral calculus.

10.1 Sequences

Sequences of numbers appear in diverse situations. If you divide a cake in half, and then divide the remaining half in half, and continue dividing in half indefinitely (Figure 1), then the fraction of cake remaining at each step forms the sequence

$$1, \quad \frac{1}{2}, \quad \frac{1}{4}, \quad \frac{1}{8}, \quad \cdots$$

This is the sequence of values of the function $f(n) = \dfrac{1}{2^n}$ for $n = 0, 1, 2, \ldots$.

> **DEFINITION** **Sequence** A **sequence** $\{a_n\}$ is an ordered collection of numbers defined by a function f on a set of sequential integers. The values $a_n = f(n)$ are called the **terms** of the sequence, and n is called the **index**. Informally, we think of a sequence $\{a_n\}$ as a list of terms:
>
> $$a_1, \quad a_2, \quad a_3, \quad a_4, \quad \ldots$$
>
> The sequence does not have to start at $n = 1$. It can start at $n = 0$, $n = 2$, or any other integer.

The sequence b_n is the Balmer series of absorption wavelengths of the hydrogen atom in nanometers. It plays a key role in spectroscopy.

General term	Domain	Sequence
$a_n = 1 - \dfrac{1}{n}$	$n \geq 1$	$0, \ \dfrac{1}{2}, \ \dfrac{2}{3}, \ \dfrac{3}{4}, \ \dfrac{4}{5}, \ \ldots$
$a_n = (-1)^n n$	$n \geq 0$	$0, \ -1, \ 2, \ -3, \ 4, \ \ldots$
$b_n = \dfrac{364.5 n^2}{n^2 - 4}$	$n \geq 3$	$656.1, \ 486, \ 433.9, \ 410.1, \ 396.9, \ \ldots$

Not all sequences are generated by a formula. For instance, here is a sequence:

$$3, \ 1, \ 4, \ 1, \ 5, \ 9, \ 2, \ 6, \ldots$$

This sequence is simply the digits of π, and there is no formula for the nth digit of π. When a_n is given by a formula, we refer to a_n as the **general term**.

The sequence in the next example is defined *recursively*. For such a sequence, the first one or more terms may be given, and then the nth term is computed in terms of the preceding terms using some formula.

■ **EXAMPLE 1** **The Fibonacci Sequence** We define the sequence by taking $F_1 = 1$, $F_2 = 1$ and $F_n = F_{n-1} + F_{n-2}$ for all integers $n > 2$. In other words, each subsequent term is obtained by adding together the two preceding terms. From this, every term can easily be determined. Calculating out terms, we find the sequence to be

$$1, 1, 2, 3, 5, 8, 13, 21, 34 \ldots$$

This particular sequence appears in a surprisingly wide variety of situations, particularly in nature. For instance, the number of spiral arms in a sunflower almost always turns out to be a number from the Fibonacci sequence, as in Figure 2.

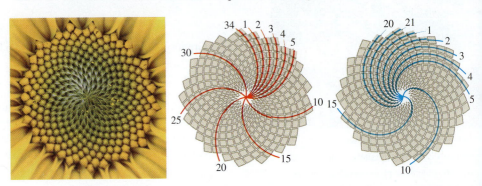

FIGURE 2 The number of spiral arms in a sunflower when counted at different angles are Fibonacci numbers. In the first case, we get 34, and in the second, we get 21. (*Eiji Ueda/Shutterstock*)

If we take the limit of the $n + 1$st term divided by the nth term, we obtain $\phi = \frac{1+\sqrt{5}}{2} \approx 1.618$, which is known as the Golden Mean. According to the ancient Greeks, this quantity gives the ideal ratio of the sides of a rectangle. The face of the Parthenon was designed to reflect this ratio. ■

■ **EXAMPLE 2** **Recursive Sequence** Compute a_2, a_3, a_4 for the sequence defined recursively by

$$a_1 = 1, \qquad a_n = \frac{1}{2}\left(a_{n-1} + \frac{2}{a_{n-1}}\right)$$

Solution

$$a_2 = \frac{1}{2}\left(a_1 + \frac{2}{a_1}\right) = \frac{1}{2}\left(1 + \frac{2}{1}\right) = \frac{3}{2} = 1.5$$

$$a_3 = \frac{1}{2}\left(a_2 + \frac{2}{a_2}\right) = \frac{1}{2}\left(\frac{3}{2} + \frac{2}{3/2}\right) = \frac{17}{12} \approx 1.4167$$

$$a_4 = \frac{1}{2}\left(a_3 + \frac{2}{a_3}\right) = \frac{1}{2}\left(\frac{17}{12} + \frac{2}{17/12}\right) = \frac{577}{408} \approx 1.414216 \quad ■$$

You may recognize the sequence in Example 2 as the sequence of approximations to $\sqrt{2} \approx 1.4142136$ produced by Newton's Method with a starting value $a_1 = 1$ (see Section 4.8). As n tends to infinity, a_n approaches $\sqrt{2}$.

Our main goal is to study convergence of sequences. A sequence $\{a_n\}$ converges to a limit L if $|a_n - L|$ becomes arbitrary small when n is sufficiently large. Here is the formal definition.

DEFINITION **Limit of a Sequence** We say that $\{a_n\}$ **converges to a limit** L, and we write

$$\lim_{n \to \infty} a_n = L \qquad \text{or} \qquad a_n \to L$$

if, for every $\epsilon > 0$, there is a number M such that $|a_n - L| < \epsilon$ for all $n > M$.

- If no limit exists, we say that $\{a_n\}$ **diverges**.
- If the terms increase without bound, we say that $\{a_n\}$ **diverges to infinity**.

If $\{a_n\}$ converges, then its limit L is unique. A good way to visualize the limit is to plot the points $(1, a_1)$, $(2, a_2)$, $(3, a_3)$, . . ., as in Figure 3. The sequence converges to L if, for every $\epsilon > 0$, the plotted points eventually remain within an ϵ-band around the horizontal line $y = L$. Figure 4 shows the plot of a sequence converging to $L = 1$. On the other hand, we can show that the sequence $a_n = \cos n$ in Figure 5 has no limit.

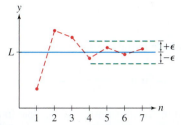

DF **FIGURE 3** Plot of a sequence with limit L. For any ϵ, the dots eventually remain within an ϵ-band around L.

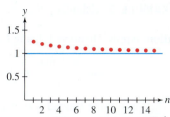

DF **FIGURE 4** The sequence $a_n = \dfrac{n+4}{n+1}$.

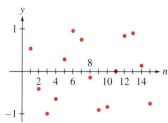

DF **FIGURE 5** The sequence $a_n = \cos n$ has no limit.

■ **EXAMPLE 3** **Proving Convergence** Let $a_n = \dfrac{n+4}{n+1}$. Prove formally $\lim\limits_{n\to\infty} a_n = 1$.

Solution The definition requires us to find, for every $\epsilon > 0$, a number M such that

$$|a_n - 1| < \epsilon \qquad \text{for all } n > M \qquad \boxed{1}$$

We have

$$|a_n - 1| = \left| \frac{n+4}{n+1} - 1 \right| = \frac{3}{n+1}$$

Therefore, $|a_n - 1| < \epsilon$ if

$$\frac{3}{n+1} < \epsilon \qquad \text{or} \qquad n > \frac{3}{\epsilon} - 1$$

In other words, Eq. (1) is valid with $M = \frac{3}{\epsilon} - 1$. This proves that $\lim\limits_{n\to\infty} a_n = 1$. ■

Note the following two facts about sequences:

- The limit does not change if we change or drop finitely many terms of the sequence.
- If C is a constant and $a_n = C$ for all n greater than some fixed value N, then $\lim\limits_{n\to\infty} a_n = C$.

Many of the sequences we consider are defined by functions; that is, $a_n = f(n)$ for some function f. For example,

$$a_n = \frac{n-1}{n} \qquad \text{is defined by} \qquad f(x) = \frac{x-1}{x}$$

We will often use the fact that if $f(x)$ approaches a limit L as $x \to \infty$, then the sequence $a_n = f(n)$ approaches the same limit L (Figure 6). Indeed, for all $\epsilon > 0$, we can find a positive real number M so that $|f(x) - L| < \epsilon$ for all $x > M$. It follows automatically that $|f(n) - L| < \epsilon$ for all integers $n > M$.

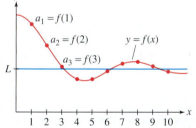

FIGURE 6 If $f(x)$ converges to L, then the sequence $a_n = f(n)$ also converges to L.

> **THEOREM 1** **Sequence Defined by a Function** If $\lim\limits_{x\to\infty} f(x)$ exists, then the sequence $a_n = f(n)$ converges to the same limit:
>
> $$\lim_{n\to\infty} a_n = \lim_{x\to\infty} f(x)$$

■ **EXAMPLE 4** Find the limit of the sequence

$$\frac{2^2 - 2}{2^2}, \quad \frac{3^2 - 2}{3^2}, \quad \frac{4^2 - 2}{4^2}, \quad \frac{5^2 - 2}{5^2}, \quad \cdots$$

Solution This is the sequence with general term

$$a_n = \frac{n^2 - 2}{n^2} = 1 - \frac{2}{n}$$

Therefore, we apply Theorem 1 with $f(x) = 1 - \frac{2}{x}$:

$$\lim_{n\to\infty} a_n = \lim_{x\to\infty} \left(1 - \frac{2}{x} \right) = 1 - \lim_{x\to\infty} \frac{2}{x} = 1 - 0 = 1 \qquad ■$$

■ **EXAMPLE 5** Calculate $\lim\limits_{n\to\infty} \dfrac{n + \ln n}{n^2}$.

Solution Apply Theorem 1, using L'Hôpital's Rule in the second step:

$$\lim_{n\to\infty} \frac{n + \ln n}{n^2} = \lim_{x\to\infty} \frac{x + \ln x}{x^2} = \lim_{x\to\infty} \frac{1 + (1/x)}{2x} = 0$$

The limit of the Balmer wavelengths b_n in the next example plays a role in physics and chemistry because it determines the ionization energy of the hydrogen atom. Table 1 suggests that b_n approaches 364.5 nanometers (nm). Figure 7 gives the graph, and in Figure 8, the wavelengths are shown "crowding in" toward their limiting value.

TABLE 1
Balmer Wavelengths (nm)

n	b_n
3	656.1
4	486
5	433.9
6	410.1
7	396.9
10	379.7
20	368.2
40	365.4
60	364.9
80	364.7
100	364.6

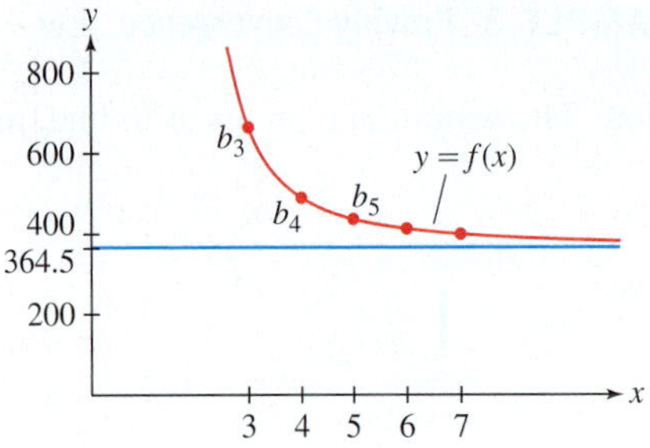

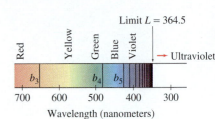

FIGURE 7 The sequence and the function approach the same limit.

FIGURE 8

■ **EXAMPLE 6** **Balmer Wavelengths** Calculate the limit of the Balmer wavelengths $b_n = \dfrac{364.5 n^2}{n^2 - 4}$ in nanometers, where $n \geq 3$.

Solution Apply Theorem 1 with $f(x) = \dfrac{364.5 x^2}{x^2 - 4}$:

$$\lim_{n\to\infty} b_n = \lim_{x\to\infty} \frac{364.5 x^2}{x^2 - 4} = \lim_{x\to\infty} \frac{364.5 x^2 \frac{1}{x^2}}{(x^2 - 4)\frac{1}{x^2}}$$

$$= \lim_{x\to\infty} \frac{364.5}{1 - 4/x^2} = \frac{364.5}{\lim\limits_{x\to\infty}(1 - 4/x^2)} = 364.5 \text{ nm}$$

A **geometric sequence** is a sequence $a_n = cr^n$, where c and r are nonzero constants. Each term is r times the previous term; that is, $a_n/a_{n-1} = r$. The number r is called the **common ratio**. For instance, if $r = 3$ and $c = 2$, we obtain the sequence (starting at $n = 0$)

$$2, \quad 2 \cdot 3, \quad 2 \cdot 3^2, \quad 2 \cdot 3^3, \quad 2 \cdot 3^4, \quad 2 \cdot 3^5, \quad \ldots$$

In the next example, we determine when a geometric series converges. Recall that $\{a_n\}$ **diverges to** ∞ if the terms a_n increase beyond all bounds (Figure 9); that is,

$$\lim_{n\to\infty} a_n = \infty \quad \text{if, for every number } N, a_n > N \text{ for all sufficiently large } n$$

We define $\lim\limits_{n\to\infty} a_n = -\infty$ similarly.

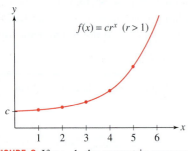

FIGURE 9 If $r > 1$, the geometric sequence $a_n = r^n$ diverges to ∞.

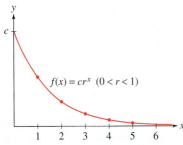

FIGURE 10 If $0 < r < 1$, the geometric sequence $a_n = r^n$ converges to 0.

■ **EXAMPLE 7** **Geometric Sequences with $r \geq 0$** Prove that for $r \geq 0$ and $c > 0$,

$$\lim_{n\to\infty} cr^n = \begin{cases} 0 & \text{if} & 0 \leq r < 1 \\ c & \text{if} & r = 1 \\ \infty & \text{if} & r > 1 \end{cases}$$

Solution Set $f(r) = cr^x$. If $0 \leq r < 1$, then (Figure 10)

$$\lim_{n\to\infty} cr^n = \lim_{x\to\infty} f(x) = c \lim_{x\to\infty} r^x = 0$$

If $r > 1$, then both $f(x)$ and the sequence $\{cr^n\}$ diverge to ∞ (because $c > 0$) (Figure 9). If $r = 1$, then $cr^n = c$ for all n, and the limit is c. ■

This last example will prove extremely useful when we consider geometric series in Section 10.2.

The limit laws we have used for functions also apply to sequences and are proved in a similar fashion.

THEOREM 2 Limit Laws for Sequences Assume that $\{a_n\}$ and $\{b_n\}$ are convergent sequences with

$$\lim_{n \to \infty} a_n = L, \qquad \lim_{n \to \infty} b_n = M$$

Then:

(i) $\displaystyle \lim_{n \to \infty} (a_n \pm b_n) = \lim_{n \to \infty} a_n \pm \lim_{n \to \infty} b_n = L \pm M$

(ii) $\displaystyle \lim_{n \to \infty} a_n b_n = \left(\lim_{n \to \infty} a_n \right) \left(\lim_{n \to \infty} b_n \right) = LM$

(iii) $\displaystyle \lim_{n \to \infty} \frac{a_n}{b_n} = \frac{\lim_{n \to \infty} a_n}{\lim_{n \to \infty} b_n} = \frac{L}{M}$ if $M \neq 0$

(iv) $\displaystyle \lim_{n \to \infty} c a_n = c \lim_{n \to \infty} a_n = cL$ for any constant c

---

THEOREM 3 Squeeze Theorem for Sequences Let $\{a_n\}$, $\{b_n\}$, $\{c_n\}$ be sequences such that for some number M,

$$b_n \leq a_n \leq c_n \quad \text{for } n > M \qquad \text{and} \qquad \lim_{n \to \infty} b_n = \lim_{n \to \infty} c_n = L$$

Then $\displaystyle \lim_{n \to \infty} a_n = L$.

◄··· **REMINDER** $n!$ *(n-factorial) is the number*

$$n! = n(n-1)(n-2) \cdots 2 \cdot 1$$

For example, $4! = 4 \cdot 3 \cdot 2 \cdot 1 = 24$. By definition, $0! = 1$.

■ **EXAMPLE 8** Show that if $\displaystyle \lim_{n \to \infty} |a_n| = 0$, then $\displaystyle \lim_{n \to \infty} a_n = 0$.

Solution We have

$$-|a_n| \leq a_n \leq |a_n|$$

By hypothesis, $\displaystyle \lim_{n \to \infty} |a_n| = 0$, and thus also $\displaystyle \lim_{n \to \infty} -|a_n| = -\lim_{n \to \infty} |a_n| = 0$. Therefore, we can apply the Squeeze Theorem to conclude that $\displaystyle \lim_{n \to \infty} a_n = 0$. ■

■ **EXAMPLE 9** **Geometric Sequences with $r < 0$** Prove that for $c \neq 0$,

$$\lim_{n \to \infty} c r^n = \begin{cases} 0 & \text{if} & -1 < r < 0 \\ \text{diverges} & \text{if} & r \leq -1 \end{cases}$$

Solution If $-1 < r < 0$, then $0 < |r| < 1$ and $\displaystyle \lim_{n \to \infty} |c r^n| = 0$ by Example 7. Thus, $\displaystyle \lim_{n \to \infty} c r^n = 0$ by Example 8. If $r = -1$, then the sequence $c r^n = (-1)^n c$ alternates in sign and does not approach a limit. The sequence also diverges if $r < -1$ because $c r^n$ alternates in sign and $|c r^n|$ grows arbitrarily large. ■

As another application of the Squeeze Theorem, consider the sequence

$$a_n = \frac{5^n}{n!}$$

FIGURE 11 Graph of $a_n = \dfrac{5^n}{n!}$.

TABLE 2

n	$a_n = \dfrac{5^n}{n!}$
1	5
2	12.5
3	20.83
4	26.04
10	2.69
15	0.023
20	0.000039
50	2.92×10^{-30}

Both the numerator and the denominator grow without bound, so it is not clear in advance whether $\{a_n\}$ converges. Figure 11 and Table 2 suggest that a_n increases initially and then tends to zero. In the next example, we verify that $a_n = R^n/n!$ converges to zero for all R. This fact is used in the discussion of Taylor series in Section 10.7.

■ **EXAMPLE 10** Prove that $\lim\limits_{n \to \infty} \dfrac{R^n}{n!} = 0$ for all R.

Solution Assume first that $R > 0$ and let M be the nonnegative integer such that

$$M \le R < M + 1$$

For $n > M$, we write $R^n/n!$ as a product of n factors:

$$\frac{R^n}{n!} = \underbrace{\left(\frac{R}{1} \frac{R}{2} \cdots \frac{R}{M} \right)}_{\text{Call this constant } C} \underbrace{\left(\frac{R}{M+1} \right) \left(\frac{R}{M+2} \right) \cdots \left(\frac{R}{n} \right)}_{\text{Each factor is less than 1}} \le C \left(\frac{R}{n} \right) \qquad \boxed{2}$$

The first M factors are greater than or equal to 1 and the last $n - M$ factors are less than 1. If we lump together the first M factors and call the product C, and drop all the remaining factors except the last factor R/n, we see that

$$0 \le \frac{R^n}{n!} \le \frac{CR}{n}$$

Since $CR/n \to 0$, the Squeeze Theorem gives us $\lim\limits_{n \to \infty} R^n/n! = 0$ as claimed. If $R < 0$, the limit is also zero by Example 8 because $\left| R^n/n! \right|$ tends to zero. ■

Given a sequence $\{a_n\}$ and a function f, we can form the new sequence $f(a_n)$. It is useful to know that if f is continuous and $a_n \to L$, then $f(a_n) \to f(L)$. A proof is given in Appendix D.

THEOREM 4 If f is continuous and $\lim\limits_{n \to \infty} a_n = L$, then

$$\lim_{n \to \infty} f(a_n) = f \left(\lim_{n \to \infty} a_n \right) = f(L)$$

In other words, we may "bring a limit of a sequence inside a continuous function."

■ **EXAMPLE 11** Apply Theorem 4 to the sequence $a_n = \dfrac{3n}{n+1}$ and to the functions

(a) $f(x) = e^x$ and **(b)** $g(x) = x^2$.

Solution Observe first that

$$L = \lim_{n \to \infty} a_n = \lim_{n \to \infty} \frac{3n}{n+1} = \lim_{n \to \infty} \frac{3}{1 + n^{-1}} = 3$$

(a) With $f(x) = e^x$, we have $f(a_n) = e^{a_n} = e^{\frac{3n}{n+1}}$. According to Theorem 4,

$$\lim_{n \to \infty} f(a_n) = f \left(\lim_{n \to \infty} a_n \right) = e^{\lim_{n \to \infty} \frac{3n}{n+1}} = e^3$$

(b) With $g(x) = x^2$, we have $g(a_n) = a_n^2$. According to Theorem 4,

$$\lim_{n \to \infty} g(a_n) = g \left(\lim_{n \to \infty} a_n \right) = \left(\lim_{n \to \infty} \frac{3n}{n+1} \right)^2 = 3^2 = 9 \qquad ■$$

Of great importance for understanding convergence are the concepts of a bounded sequence and a monotonic sequence.

DEFINITION Bounded Sequences A sequence $\{a_n\}$ is:

- **Bounded from above** if there is a number M such that $a_n \le M$ for all n. The number M is called an *upper bound*.
- **Bounded from below** if there is a number m such that $a_n \ge m$ for all n. The number m is called a *lower bound*.

The sequence $\{a_n\}$ is called **bounded** if it is bounded from above and below. A sequence that is not bounded is called an **unbounded sequence**.

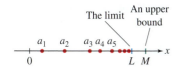

FIGURE 12 A convergent sequence is bounded.

Thus, for instance, the sequence given by $a_n = 3 - \frac{1}{n}$ is clearly bounded above by 3. It is also bounded below by 0, since all the terms are positive. Hence, this sequence is bounded.

Upper and lower bounds are not unique. If M is an upper bound, then any larger number is also an upper bound, and if m is a lower bound, then any smaller number is also a lower bound (Figure 12).

As we might expect, a convergent sequence $\{a_n\}$ is necessarily bounded because the terms a_n get closer and closer to the limit. This fact is recorded in the next theorem.

> **THEOREM 5 Convergent Sequences Are Bounded** If $\{a_n\}$ converges, then $\{a_n\}$ is bounded.

Proof Let $L = \lim\limits_{n \to \infty} a_n$. Then there exists $N > 0$ such that $|a_n - L| < 1$ for $n > N$. In other words,

$$L - 1 < a_n < L + 1 \qquad \text{for } n > N$$

If M is any number larger than $L + 1$ and also larger than the numbers $a_1, a_2, \ldots, a_N$, then $a_n < M$ for all n. Thus, M is an upper bound. Similarly, any number m smaller than $L - 1$ and also smaller than the numbers $a_1, a_2, \ldots, a_N$ is a lower bound. ∎

There are two ways that a sequence $\{a_n\}$ can diverge. One way is by being unbounded. For example, the unbounded sequence $a_n = n$ diverges:

$$1, \quad 2, \quad 3, \quad 4, \quad 5, \quad 6, \quad \ldots$$

However, a sequence can diverge even if it is bounded. This is the case with $a_n = (-1)^{n+1}$, whose terms a_n bounce back and forth but never settle down to approach a limit:

$$1, \quad -1, \quad 1, \quad -1, \quad 1, \quad -1, \quad \ldots$$

There is no surefire method for determining whether a sequence $\{a_n\}$ converges, unless the sequence happens to be both bounded and **monotonic**. By definition, $\{a_n\}$ is monotonic if it is either increasing or decreasing:

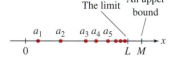

FIGURE 13 An increasing sequence with upper bound M approaches a limit L.

- $\{a_n\}$ is *increasing* if $a_n < a_{n+1}$ for all n.
- $\{a_n\}$ is *decreasing* if $a_n > a_{n+1}$ for all n.

Intuitively, if $\{a_n\}$ is increasing and bounded above by M, then the terms must bunch up near some limiting value L that is not greater than M (Figure 13). See Appendix B for a proof of the next theorem.

> **THEOREM 6 Bounded Monotonic Sequences Converge**
> - If $\{a_n\}$ is increasing and $a_n \le M$, then $\{a_n\}$ converges and $\lim\limits_{n \to \infty} a_n \le M$.
> - If $\{a_n\}$ is decreasing and $a_n \ge m$, then $\{a_n\}$ converges and $\lim\limits_{n \to \infty} a_n \ge m$.

■ **EXAMPLE 12** Verify that $a_n = \sqrt{n+1} - \sqrt{n}$ is decreasing and bounded below. Does $\lim\limits_{n \to \infty} a_n$ exist?

TABLE 3

$a_n = \sqrt{n+1} - \sqrt{n}$
$a_1 \approx 0.4142$
$a_2 \approx 0.3178$
$a_3 \approx 0.2679$
$a_4 \approx 0.2361$
$a_5 \approx 0.2134$
$a_6 \approx 0.1963$
$a_7 \approx 0.1827$
$a_8 \approx 0.1716$

Solution The function $f(x) = \sqrt{x+1} - \sqrt{x}$ is decreasing because its derivative is negative:

$$f'(x) = \frac{1}{2\sqrt{x+1}} - \frac{1}{2\sqrt{x}} < 0 \qquad \text{for } x > 0$$

It follows that $a_n = f(n)$ is decreasing (see Table 3). Furthermore, $a_n > 0$ for all n, so the sequence has lower bound $m = 0$. Theorem 6 guarantees that $L = \lim\limits_{n \to \infty} a_n$ exists and $L \ge 0$. In fact, we can show that $L = 0$ by noting that $f(x)$ can be rewritten as $f(x) = \dfrac{1}{\sqrt{x+1} + \sqrt{x}}$. Hence, $\lim\limits_{x \to \infty} f(x) = 0$. ■

■ **EXAMPLE 13** Show that the following sequence is bounded and increasing:

$$a_1 = \sqrt{2}, \quad a_2 = \sqrt{2\sqrt{2}}, \quad a_3 = \sqrt{2\sqrt{2\sqrt{2}}}, \quad \ldots$$

Then prove that $L = \lim_{n\to\infty} a_n$ exists and compute its value.

Solution

Step 1. **Show that $\{a_n\}$ is bounded above.**
We claim that $M = 2$ is an upper bound. We certainly have $a_1 < 2$ because $a_1 = \sqrt{2} \approx 1.414$. On the other hand,

$$\text{if} \quad a_n < 2, \qquad \text{then} \quad a_{n+1} < 2 \qquad\qquad \boxed{3}$$

is true because $a_{n+1} = \sqrt{2a_n} < \sqrt{2 \cdot 2} = 2$. Now, since $a_1 < 2$, we can apply (3) to conclude that $a_2 < 2$. Similarly, $a_2 < 2$ implies $a_3 < 2$, etc. for all n. (Formally speaking, this is a proof by induction.)

Step 2. **Show that $\{a_n\}$ is increasing.**
Since a_n is positive and $a_n < 2$, we have

$$a_{n+1} = \sqrt{2a_n} > \sqrt{a_n \cdot a_n} = a_n$$

This shows that $\{a_n\}$ is increasing. Since the sequence is bounded above and increasing, we conclude that the limit L exists.

Now that we know the limit L exists, we can find its value as follows. The idea is that L "contains a copy" of itself under the square root sign:

$$L = \sqrt{2\sqrt{2\sqrt{2\sqrt{2}\cdots}}} = \sqrt{2\left(\sqrt{2\sqrt{2\sqrt{2}\cdots}}\right)} = \sqrt{2L}$$

Thus, $L^2 = 2L$, which implies that $L = 2$ or $L = 0$. We eliminate $L = 0$ because the terms a_n are positive and increasing (as shown below), so we must have $L = 2$ (see Table 4).

This argument is phrased more formally by noting that the sequence is defined recursively by

$$a_1 = \sqrt{2}, \qquad a_{n+1} = \sqrt{2a_n}$$

If a_n converges to L, then the sequence $b_n = a_{n+1}$ also converges to L (because it is the same sequence, with terms shifted one to the left). Then, applying Theorem 4 to $f(x) = \sqrt{x}$, we have

$$L = \lim_{n\to\infty} a_{n+1} = \lim_{n\to\infty} \sqrt{2a_n} = \sqrt{2 \lim_{n\to\infty} a_n} = \sqrt{2L}$$

Hence, $L = 2$. ■

TABLE 4	Recursive Sequence $a_{n+1} = \sqrt{2a_n}$
a_1	1.4142
a_2	1.6818
a_3	1.8340
a_4	1.9152
a_5	1.9571
a_6	1.9785
a_7	1.9892
a_8	1.9946

10.1 SUMMARY

- A sequence $\{a_n\}$ *converges* to a limit L if, for every $\epsilon > 0$, there is a number M such that

$$|a_n - L| < \epsilon \qquad \text{for all } n > M$$

We write $\lim_{n\to\infty} a_n = L$ or $a_n \to L$.
- If no limit exists, we say that $\{a_n\}$ *diverges*.
- In particular, if the terms increase without bound, we say that $\{a_n\}$ diverges to infinity.
- If $a_n = f(n)$ and $\lim_{x\to\infty} f(x) = L$, then $\lim_{n\to\infty} a_n = L$.
- A *geometric sequence* is a sequence $a_n = cr^n$, where c and r are nonzero. It converges to 0 for $-1 < r < 1$, converges to c for $r = 1$, and diverges otherwise.
- The Basic Limit Laws and the Squeeze Theorem apply to sequences.

• If f is continuous and $\lim\limits_{n\to\infty} a_n = L$, then $\lim\limits_{n\to\infty} f(a_n) = f(L)$.
• A sequence $\{a_n\}$ is

  – *bounded above* by M if $a_n \leq M$ for all n.
  – *bounded below* by m if $a_n \geq m$ for all n.

  If $\{a_n\}$ is bounded above and below, $\{a_n\}$ is called *bounded*.
• A sequence $\{a_n\}$ is *monotonic* if it is increasing ($a_n < a_{n+1}$) or decreasing ($a_n > a_{n+1}$).
• Bounded monotonic sequences converge (Theorem 6).

10.1 EXERCISES

Preliminary Questions

1. What is a_4 for the sequence $a_n = n^2 - n$?

2. Which of the following sequences converge to zero?

(a) $\dfrac{n^2}{n^2 + 1}$ **(b)** 2^n **(c)** $\left(\dfrac{-1}{2}\right)^n$

3. Let a_n be the nth decimal approximation to $\sqrt{2}$. That is, $a_1 = 1$, $a_2 = 1.4$, $a_3 = 1.41$, etc. What is $\lim\limits_{n\to\infty} a_n$?

4. Which of the following sequences is defined recursively?

(a) $a_n = \sqrt{4 + n}$ **(b)** $b_n = \sqrt{4 + b_{n-1}}$

5. Theorem 5 says that every convergent sequence is bounded. Determine if the following statements are true or false, and if false, give a counterexample:

(a) If $\{a_n\}$ is bounded, then it converges.

(b) If $\{a_n\}$ is not bounded, then it diverges.

(c) If $\{a_n\}$ diverges, then it is not bounded.

Exercises

1. Match each sequence with its general term:

$a_1, a_2, a_3, a_4, \ldots$	General term
(a) $\frac{1}{2}, \frac{2}{3}, \frac{3}{4}, \frac{4}{5}, \ldots$	**(i)** $\cos \pi n$
(b) $-1, 1, -1, 1, \ldots$	**(ii)** $\dfrac{n!}{2^n}$
(c) $1, -1, 1, -1, \ldots$	**(iii)** $(-1)^{n+1}$
(d) $\frac{1}{2}, \frac{2}{4}, \frac{6}{8}, \frac{24}{16} \ldots$	**(iv)** $\dfrac{n}{n+1}$

2. Let $a_n = \dfrac{1}{2n - 1}$ for $n = 1, 2, 3, \ldots$. Write out the first three terms of the following sequences.

(a) $b_n = a_{n+1}$ **(b)** $c_n = a_{n+3}$

(c) $d_n = a_n^2$ **(d)** $e_n = 2a_n - a_{n+1}$

In Exercises 3–12, calculate the first four terms of the sequence, starting with $n = 1$.

3. $c_n = \dfrac{3^n}{n!}$

4. $b_n = \dfrac{(2n - 1)!}{n!}$

5. $a_1 = 2$, $a_{n+1} = 2a_n^2 - 3$

6. $b_1 = 1$, $b_n = b_{n-1} + \dfrac{1}{b_{n-1}}$

7. $b_n = 5 + \cos \pi n$

8. $c_n = (-1)^{2n+1}$

9. $c_n = 1 + \dfrac{1}{2} + \dfrac{1}{3} + \cdots + \dfrac{1}{n}$

10. $a_n = n + (n + 1) + (n + 2) + \cdots + (2n)$

11. $b_1 = 2$, $b_2 = 3$, $b_n = 2b_{n-1} + b_{n-2}$

12. $c_n = n$-place decimal approximation to e

13. Find a formula for the nth term of each sequence.

(a) $\dfrac{1}{1}, \dfrac{-1}{8}, \dfrac{1}{27}, \ldots$ **(b)** $\dfrac{2}{6}, \dfrac{3}{7}, \dfrac{4}{8}, \ldots$

14. Suppose that $\lim\limits_{n\to\infty} a_n = 4$ and $\lim\limits_{n\to\infty} b_n = 7$. Determine:

(a) $\lim\limits_{n\to\infty} (a_n + b_n)$ **(b)** $\lim\limits_{n\to\infty} a_n^3$

(c) $\lim\limits_{n\to\infty} \cos(\pi b_n)$ **(d)** $\lim\limits_{n\to\infty} (a_n^2 - 2a_n b_n)$

In Exercises 15–26, use Theorem 1 to determine the limit of the sequence or state that the sequence diverges.

15. $a_n = 12$

16. $a_n = 20 - \dfrac{4}{n^2}$

17. $b_n = \dfrac{5n - 1}{12n + 9}$

18. $a_n = \dfrac{4 + n - 3n^2}{4n^2 + 1}$

19. $c_n = -2^{-n}$

20. $z_n = \left(\dfrac{1}{3}\right)^n$

21. $c_n = 9^n$

22. $z_n = 10^{-1/n}$

23. $a_n = \dfrac{n}{\sqrt{n^2 + 1}}$

24. $a_n = \dfrac{n}{\sqrt{n^3 + 1}}$

25. $a_n = \ln\left(\dfrac{12n + 2}{-9 + 4n}\right)$

26. $r_n = \ln n - \ln(n^2 + 1)$

In Exercises 27–30, use Theorem 4 to determine the limit of the sequence.

27. $a_n = \sqrt{4 + \dfrac{1}{n}}$

28. $a_n = e^{4n/(3n+9)}$

29. $a_n = \cos^{-1}\left(\dfrac{n^3}{2n^3 + 1}\right)$

30. $a_n = \tan^{-1}(e^{-n})$

31. Let $a_n = \dfrac{n}{n + 1}$. Find a number M such that:

(a) $|a_n - 1| \leq 0.001$ for $n \geq M$.

(b) $|a_n - 1| \leq 0.00001$ for $n \geq M$.

Then use the limit definition to prove that $\lim\limits_{n\to\infty} a_n = 1$.

32. Let $b_n = \left(\frac{1}{3}\right)^n$.

(a) Find a value of M such that $|b_n| \leq 10^{-5}$ for $n \geq M$.

(b) Use the limit definition to prove that $\lim_{n\to\infty} b_n = 0$.

33. Use the limit definition to prove that $\lim_{n\to\infty} n^{-2} = 0$.

34. Use the limit definition to prove that $\lim_{n\to\infty} \frac{n}{n+n^{-1}} = 1$.

In Exercises 35–62, use the appropriate limit laws and theorems to determine the limit of the sequence or show that it diverges.

35. $a_n = 10 + \left(-\frac{1}{9}\right)^n$

36. $d_n = \sqrt{n+3} - \sqrt{n}$

37. $c_n = 1.01^n$

38. $b_n = e^{1-n^2}$

39. $a_n = 2^{1/n}$

40. $b_n = n^{1/n}$

41. $c_n = \frac{9^n}{n!}$

42. $a_n = \frac{8^{2n}}{n!}$

43. $a_n = \frac{3n^2+n+2}{2n^2-3}$

44. $a_n = \frac{\sqrt{n}}{\sqrt{n}+4}$

45. $a_n = \frac{\cos n}{n}$

46. $c_n = \frac{(-1)^n}{\sqrt{n}}$

47. $d_n = \ln 5^n - \ln n!$

48. $d_n = \ln(n^2+4) - \ln(n^2-1)$

49. $a_n = \left(2+\frac{4}{n^2}\right)^{1/3}$

50. $b_n = \tan^{-1}\left(1-\frac{2}{n}\right)$

51. $c_n = \ln\left(\frac{2n+1}{3n+4}\right)$

52. $c_n = \frac{n}{n+n^{1/n}}$

53. $y_n = \frac{e^n}{2^n}$

54. $a_n = \frac{n}{2^n}$

55. $y_n = \frac{e^n+(-3)^n}{5^n}$

56. $b_n = \frac{(-1)^n n^3 + 2^{-n}}{3n^3 + 4^{-n}}$

57. $a_n = n\sin\frac{\pi}{n}$

58. $b_n = \frac{n!}{\pi^n}$

59. $b_n = \frac{3-4^n}{2+7\cdot 4^n}$

60. $a_n = \frac{3-4^n}{2+7\cdot 3^n}$

61. $a_n = \left(1+\frac{1}{n}\right)^n$

62. $a_n = \left(1+\frac{1}{n^2}\right)^n$

In Exercises 63–66, find the limit of the sequence using L'Hôpital's Rule.

63. $a_n = \frac{(\ln n)^2}{n}$

64. $b_n = \sqrt{n}\ln\left(1+\frac{1}{n}\right)$

65. $c_n = n\left(\sqrt{n^2+1}-n\right)$

66. $d_n = n^2\left(\sqrt[3]{n^3+1}-n\right)$

In Exercises 67–70, use the Squeeze Theorem to evaluate $\lim_{n\to\infty} a_n$ by verifying the given inequality.

67. $a_n = \frac{1}{\sqrt{n^4+n^8}}$, $\quad \frac{1}{\sqrt{2}n^4} \leq a_n \leq \frac{1}{\sqrt{2}n^2}$

68. $c_n = \frac{1}{\sqrt{n^2+1}} + \frac{1}{\sqrt{n^2+2}} + \cdots + \frac{1}{\sqrt{n^2+n}}$,

$\frac{n}{\sqrt{n^2+n}} \leq c_n \leq \frac{n}{\sqrt{n^2+1}}$

69. $a_n = (2^n+3^n)^{1/n}$, $\quad 3 \leq a_n \leq (2\cdot 3^n)^{1/n} = 2^{1/n}\cdot 3$

70. $a_n = (n+10^n)^{1/n}$, $\quad 10 \leq a_n \leq (2\cdot 10^n)^{1/n}$

71. Which of the following statements is equivalent to the assertion $\lim_{n\to\infty} a_n = L$? Explain.

(a) For every $\epsilon > 0$, the interval $(L-\epsilon, L+\epsilon)$ contains at least one element of the sequence $\{a_n\}$.

(b) For every $\epsilon > 0$, the interval $(L-\epsilon, L+\epsilon)$ contains all but at most finitely many elements of the sequence $\{a_n\}$.

72. Show that $a_n = \frac{1}{2n+1}$ is decreasing.

73. Show that $a_n = \frac{3n^2}{n^2+2}$ is increasing. Find an upper bound.

74. Show that $a_n = \sqrt[3]{n+1}-n$ is decreasing.

75. Give an example of a divergent sequence $\{a_n\}$ such that $\lim_{n\to\infty} |a_n|$ converges.

76. Give an example of divergent sequences $\{a_n\}$ and $\{b_n\}$ such that $\{a_n+b_n\}$ converges.

77. Using the limit definition, prove that if $\{a_n\}$ converges and $\{b_n\}$ diverges, then $\{a_n+b_n\}$ diverges.

78. Use the limit definition to prove that if $\{a_n\}$ is a convergent sequence of integers with limit L, then there exists a number M such that $a_n = L$ for all $n \geq M$.

79. Theorem 1 states that if $\lim_{x\to\infty} f(x) = L$, then the sequence $a_n = f(n)$ converges and $\lim_{n\to\infty} a_n = L$. Show that the *converse* is false. In other words, find a function f such that $a_n = f(n)$ converges but $\lim_{x\to\infty} f(x)$ does not exist.

80. Use the limit definition to prove that the limit does not change if a finite number of terms are added or removed from a convergent sequence.

81. Let $b_n = a_{n+1}$. Use the limit definition to prove that if $\{a_n\}$ converges, then $\{b_n\}$ also converges and $\lim_{n\to\infty} a_n = \lim_{n\to\infty} b_n$.

82. Let $\{a_n\}$ be a sequence such that $\lim_{n\to\infty} |a_n|$ exists and is nonzero. Show that $\lim_{n\to\infty} a_n$ exists if and only if there exists an integer M such that the sign of a_n does not change for $n > M$.

83. Proceed as in Example 13 to show that the sequence $\sqrt{3}, \sqrt{3\sqrt{3}}, \sqrt{3\sqrt{3\sqrt{3}}}, \ldots$ is increasing and bounded above by $M = 3$. Then prove that the limit exists and find its value.

84. Let $\{a_n\}$ be the sequence defined recursively by

$$a_0 = 0, \qquad a_{n+1} = \sqrt{2+a_n}$$

Thus, $a_1 = \sqrt{2}$, $a_2 = \sqrt{2+\sqrt{2}}$, $a_3 = \sqrt{2+\sqrt{2+\sqrt{2}}}, \ldots$.

(a) Show that if $a_n < 2$, then $a_{n+1} < 2$. Conclude by induction that $a_n < 2$ for all n.

(b) Show that if $a_n < 2$, then $a_n \leq a_{n+1}$. Conclude by induction that $\{a_n\}$ is increasing.

(c) Use (a) and (b) to conclude that $L = \lim_{n\to\infty} a_n$ exists. Then compute L by showing that $L = \sqrt{2+L}$.

Further Insights and Challenges

85. Show that $\lim\limits_{n\to\infty} \sqrt[n]{n!} = \infty$. *Hint:* Verify that $n! \geq (n/2)^{n/2}$ by observing that half of the factors of $n!$ are greater than or equal to $n/2$.

86. Let $b_n = \dfrac{\sqrt[n]{n!}}{n}$.

(a) Show that $\ln b_n = \dfrac{1}{n}\sum\limits_{k=1}^{n} \ln \dfrac{k}{n}$.

(b) Show that $\ln b_n$ converges to $\displaystyle\int_0^1 \ln x \, dx$, and conclude that $b_n \to e^{-1}$.

87. Given positive numbers $a_1 < b_1$, define two sequences recursively by
$$a_{n+1} = \sqrt{a_n b_n}, \qquad b_{n+1} = \frac{a_n + b_n}{2}$$

(a) Show that $a_n \leq b_n$ for all n (Figure 14).

(b) Show that $\{a_n\}$ is increasing and $\{b_n\}$ is decreasing.

(c) Show that $b_{n+1} - a_{n+1} \leq \dfrac{b_n - a_n}{2}$.

(d) Prove that both $\{a_n\}$ and $\{b_n\}$ converge and have the same limit. This limit, denoted $\text{AGM}(a_1, b_1)$, is called the **arithmetic-geometric mean** of a_1 and b_1.

(e) Estimate $\text{AGM}(1, \sqrt{2})$ to three decimal places.

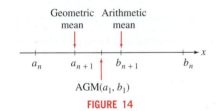

Geometric   Arithmetic
mean   mean

$\text{AGM}(a_1, b_1)$

FIGURE 14

88. Let $c_n = \dfrac{1}{n} + \dfrac{1}{n+1} + \dfrac{1}{n+2} + \cdots + \dfrac{1}{2n}$.

(a) Calculate c_1, c_2, c_3, c_4.

(b) Use a comparison of rectangles with the area under $y = x^{-1}$ over the interval $[n, 2n]$ to prove that
$$\int_n^{2n} \frac{dx}{x} + \frac{1}{2n} \leq c_n \leq \int_n^{2n} \frac{dx}{x} + \frac{1}{n}$$

(c) Use the Squeeze Theorem to determine $\lim\limits_{n\to\infty} c_n$.

89. Let $a_n = H_n - \ln n$, where H_n is the nth harmonic number:
$$H_n = 1 + \frac{1}{2} + \frac{1}{3} + \cdots + \frac{1}{n}$$

(a) Show that $a_n \geq 0$ for $n \geq 1$. *Hint:* Show that $H_n \geq \displaystyle\int_1^{n+1} \frac{dx}{x}$.

(b) Show that $\{a_n\}$ is decreasing by interpreting $a_n - a_{n+1}$ as an area.

(c) Prove that $\lim\limits_{n\to\infty} a_n$ exists.

This limit, denoted γ, is known as *Euler's Constant*. It appears in many areas of mathematics, including analysis and number theory, and has been calculated to more than 100 million decimal places, but it is still not known whether γ is an irrational number. The first 10 digits are $\gamma \approx 0.5772156649$.

10.2 Summing an Infinite Series

Many quantities that arise in applications cannot be computed exactly. We cannot write down an exact decimal expression for the number π or for values of the sine function such as $\sin 1$. However, sometimes these quantities can be represented as infinite sums. For example, using Taylor series (Section 10.7), we can show that

$$\sin 1 = 1 - \frac{1}{3!} + \frac{1}{5!} - \frac{1}{7!} + \frac{1}{9!} - \frac{1}{11!} + \cdots \qquad \boxed{1}$$

Infinite sums of this type are called **infinite series**. We think of them as having been obtained by adding up all of the terms in an infinite sequence.

But what precisely does Eq. (1) mean? It is impossible to add up infinitely many numbers, but what we can do is compute the **partial sums** S_N, defined as the finite sum of the terms up to and including the Nth term. Here are the first five partial sums of the infinite series for $\sin 1$:

$$S_1 = 1$$
$$S_2 = 1 - \frac{1}{3!} = 1 - \frac{1}{6} \qquad\qquad \approx 0.833$$
$$S_3 = 1 - \frac{1}{3!} + \frac{1}{5!} = 1 - \frac{1}{6} + \frac{1}{120} \qquad \approx 0.841667$$
$$S_4 = 1 - \frac{1}{6} + \frac{1}{120} - \frac{1}{5040} \qquad\qquad \approx 0.841468$$
$$S_5 = 1 - \frac{1}{6} + \frac{1}{120} - \frac{1}{5040} + \frac{1}{362{,}880} \approx \mathbf{0.8414709846}$$

Compare these values with the value obtained from a calculator:

$$\sin 1 \approx \mathbf{0.8414709848}079 \qquad \text{(calculator value)}$$

We see that S_5 differs from $\sin 1$ by less than 10^{-9}. This suggests that the partial sums converge to $\sin 1$, and in fact, we will prove that

$$\sin 1 = \lim_{N \to \infty} S_N$$

(see Example 2 in Section 10.7). So although we cannot add up infinitely many numbers, it makes sense to *define* the sum of an infinite series as a limit of partial sums.

In general, an infinite series is an expression of the form

$$\sum_{n=1}^{\infty} a_n = a_1 + a_2 + a_3 + a_4 + \cdots$$

where $\{a_n\}$ is any sequence. For example,

> • *Infinite series may begin with any value for the index. For example,*
>
> $$\sum_{n=3}^{\infty} \frac{1}{n} = \frac{1}{3} + \frac{1}{4} + \frac{1}{5} + \cdots$$
>
> *When it is not necessary to specify the starting point, we write simply $\sum a_n$.*
> • *Any letter may be used for the index. Thus, we may write a_m, a_k, a_i, etc.*

Sequence	General term	Infinite series
$\dfrac{1}{3}, \dfrac{1}{9}, \dfrac{1}{27}, \dots$	$a_n = \dfrac{1}{3^n}$	$\displaystyle\sum_{n=1}^{\infty} \frac{1}{3^n} = \frac{1}{3} + \frac{1}{9} + \frac{1}{27} + \frac{1}{81} + \cdots$
$\dfrac{1}{1}, \dfrac{1}{4}, \dfrac{1}{9}, \dfrac{1}{16}, \dots$	$a_n = \dfrac{1}{n^2}$	$\displaystyle\sum_{n=1}^{\infty} \frac{1}{n^2} = \frac{1}{1} + \frac{1}{4} + \frac{1}{9} + \frac{1}{16} + \cdots$

The Nth partial sum S_N is the finite sum of the terms up to and including a_N:

$$S_N = \sum_{n=1}^{N} a_n = a_1 + a_2 + a_3 + \cdots + a_N$$

If the series begins at k, then $S_N = a_k + a_{k+1} + \cdots + a_N$.

DEFINITION Convergence of an Infinite Series An infinite series $\displaystyle\sum_{n=k}^{\infty} a_n$ *converges* to the sum S if the sequence of its partial sums $\{S_N\}$ converge to S:

$$\lim_{N \to \infty} S_N = S$$

In this case, we write $S = \displaystyle\sum_{n=k}^{\infty} a_n$.

- If the limit does not exist, we say that the infinite series *diverges*.
- If the limit is infinite, we say that the infinite series *diverges to infinity*.

We can investigate series numerically by computing several partial sums S_N. If the sequence of partial sums shows a trend of convergence to some number S, then we have evidence (but not proof) that the series converges to S. The next example treats a **telescoping series**, where the partial sums are particularly easy to evaluate.

■ **EXAMPLE 1** **Telescoping Series** Investigate numerically:

$$S = \sum_{n=1}^{\infty} \frac{1}{n(n+1)} = \frac{1}{1(2)} + \frac{1}{2(3)} + \frac{1}{3(4)} + \frac{1}{4(5)} + \cdots$$

Then compute the sum S using the identity:

$$\frac{1}{n(n+1)} = \frac{1}{n} - \frac{1}{n+1}$$

Solution The values of the partial sums listed in Table 1 suggest convergence to $S = 1$. To prove this, we observe that because of the identity, each partial sum collapses down to just two terms:

$$S_1 = \frac{1}{1(2)} = \frac{1}{1} - \frac{1}{2}$$

$$S_2 = \frac{1}{1(2)} + \frac{1}{2(3)} = \left(\frac{1}{1} - \frac{1}{2}\right) + \left(\frac{1}{2} - \frac{1}{3}\right) = 1 - \frac{1}{3}$$

$$S_3 = \frac{1}{1(2)} + \frac{1}{2(3)} + \frac{1}{3(4)} = \left(\frac{1}{1} - \frac{1}{2}\right) + \left(\frac{1}{2} - \frac{1}{3}\right) + \left(\frac{1}{3} - \frac{1}{4}\right) = 1 - \frac{1}{4}$$

In general,

$$S_N = \left(\frac{1}{1} - \frac{1}{2}\right) + \left(\frac{1}{2} - \frac{1}{3}\right) + \cdots + \left(\frac{1}{N-1} - \frac{1}{N}\right) + \left(\frac{1}{N} - \frac{1}{N+1}\right)$$

$$= 1 - \frac{1}{N+1}$$

2

The sum S is the limit of the sequence of partial sums:

$$S = \lim_{N \to \infty} S_N = \lim_{N \to \infty} \left(1 - \frac{1}{N+1}\right) = 1$$

■

It is important to keep in mind the difference between a sequence $\{a_n\}$ and an infinite series $\sum_{n=1}^{\infty} a_n$.

■ **EXAMPLE 2 Sequences Versus Series** Discuss the difference between $\{a_n\}$ and $\sum_{n=1}^{\infty} a_n$, where $a_n = \frac{1}{n(n+1)}$.

Solution The sequence is the list of numbers $\frac{1}{1(2)}, \frac{1}{2(3)}, \frac{1}{3(4)}, \dots$. This sequence converges to zero:

$$\lim_{n \to \infty} a_n = \lim_{n \to \infty} \frac{1}{n(n+1)} = 0$$

The infinite series is the *sum* of the numbers a_n, defined formally as the limit of the sequence of partial sums. This sum is not zero. In fact, the sum is equal to 1 by Example 1:

$$\sum_{n=1}^{\infty} a_n = \sum_{n=1}^{\infty} \frac{1}{n(n+1)} = \frac{1}{1(2)} + \frac{1}{2(3)} + \frac{1}{3(4)} + \cdots = 1$$

■

The next theorem shows that infinite series may be added or subtracted like ordinary sums, *provided that the series converge.*

> **THEOREM 1 Linearity of Infinite Series** If $\sum a_n$ and $\sum b_n$ converge, then $\sum (a_n \pm b_n)$ and $\sum c a_n$ also converge (c any constant), and
>
> $$\sum a_n + \sum b_n = \sum (a_n + b_n)$$
>
> $$\sum a_n - \sum b_n = \sum (a_n - b_n)$$
>
> $$\sum c a_n = c \sum a_n \qquad (c \text{ any constant})$$

DF TABLE 1 Partial Sums for $\displaystyle\sum_{n=1}^{\infty} \frac{1}{n(n+1)}$

N	S_N
10	0.90909
50	0.98039
100	0.990099
200	0.995025
300	0.996678

In most cases (apart from telescoping series and the geometric series introduced below), there is no simple formula like Eq. (2) for the partial sum S_N. Therefore, we shall develop techniques that do not rely on formulas for S_N.

Make sure you understand the difference between sequences and series.

- *With a sequence, we consider the limit of the individual terms a_n.*
- *With a series, we are interested in the sum of the terms*

$$a_1 + a_2 + a_3 + \cdots$$

which is defined as the limit of the sequence of partial sums.

Proof These rules follow from the corresponding linearity rules for limits. For example,

$$\sum_{n=1}^{\infty}(a_n + b_n) = \lim_{N\to\infty}\sum_{n=1}^{N}(a_n + b_n) = \lim_{N\to\infty}\left(\sum_{n=1}^{N}a_n + \sum_{n=1}^{N}b_n\right)$$

$$= \lim_{N\to\infty}\sum_{n=1}^{N}a_n + \lim_{N\to\infty}\sum_{n=1}^{\infty}b_n = \sum_{n=1}^{\infty}a_n + \sum_{n=1}^{\infty}b_n \qquad \blacksquare$$

A main goal in this chapter is to develop techniques for determining whether a series converges or diverges. It is easy to give examples of series that diverge:

- $S = \displaystyle\sum_{n=1}^{\infty} 1$ diverges to infinity (the partial sums increase without bound):

$$S_1 = 1, \quad S_2 = 1 + 1 = 2, \quad S_3 = 1 + 1 + 1 = 3, \quad S_4 = 1 + 1 + 1 + 1 = 4, \quad \ldots$$

- $\displaystyle\sum_{n=1}^{\infty}(-1)^{n-1}$ diverges (the partial sums jump between 1 and 0):

$$S_1 = 1, \quad S_2 = 1 - 1 = 0, \quad S_3 = 1 - 1 + 1 = 1, \quad S_4 = 1 - 1 + 1 - 1 = 0, \quad \ldots$$

Next, we study geometric series, which converge or diverge depending on the common ratio r.

A **geometric series** with common ratio $r \neq 0$ is a series defined by a geometric sequence cr^n, where $c \neq 0$. If the series begins at $n = 0$, then

$$S = \sum_{n=0}^{\infty} cr^n = c + cr + cr^2 + cr^3 + cr^4 + cr^5 + \cdots$$

For $r = \frac{1}{2}$ and $c = 1$, we can visualize the geometric series starting at $n = 1$ (Figure 1):

$$S = \sum_{n=1}^{\infty} \frac{1}{2^n} = \frac{1}{2} + \frac{1}{4} + \frac{1}{8} + \frac{1}{16} + \cdots = 1$$

Adding up the terms corresponds to moving stepwise from 0 to 1, where each step is a move to the right by half of the remaining distance. Thus, $S = 1$.

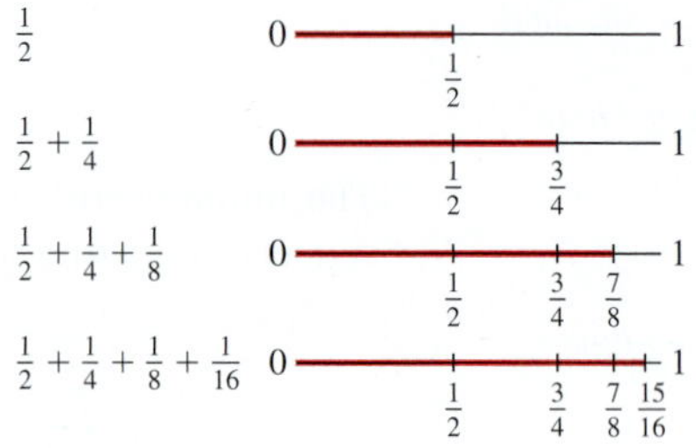

FIGURE 1 Partial sums of $\displaystyle\sum_{n=1}^{\infty} \frac{1}{2^n}$.

There is a simple device for computing the partial sums of a geometric series:

$$S_N = c + cr + cr^2 + cr^3 + \cdots + cr^N$$

$$rS_N = \qquad cr + cr^2 + cr^3 + \cdots + cr^N + cr^{N+1}$$

$$S_N - rS_N = c - cr^{N+1}$$

$$S_N(1 - r) = c(1 - r^{N+1})$$

If $r \neq 1$, we may divide by $(1 - r)$ to obtain

$$\boxed{S_N = c + cr + cr^2 + cr^3 + \cdots + cr^N = \frac{c(1 - r^{N+1})}{1 - r}} \qquad \boxed{3}$$

This formula enables us to sum the geometric series.

THEOREM 2 **Sum of a Geometric Series**   Let $c \neq 0$. If $|r| < 1$, then

$$\sum_{n=0}^{\infty} cr^n = c + cr + cr^2 + cr^3 + \cdots = \frac{c}{1-r} \qquad \boxed{4}$$

$$\sum_{n=M}^{\infty} cr^n = cr^M + cr^{M+1} + cr^{M+2} + cr^{M+3} + \cdots = \frac{cr^M}{1-r} \qquad \boxed{5}$$

If $|r| \geq 1$, then the geometric series diverges.

Proof If $r = 1$, then the series certainly diverges because the partial sums $S_N = Nc$ grow arbitrarily large. If $r \neq 1$, then Eq. (3) yields

$$S = \lim_{N \to \infty} S_N = \lim_{N \to \infty} \frac{c(1 - r^{N+1})}{1-r} = \frac{c}{1-r} - \frac{c}{1-r} \lim_{N \to \infty} r^{N+1}$$

If $|r| < 1$, then $\lim_{N \to \infty} r^{N+1} = 0$ and we obtain Eq. (4). If $|r| \geq 1$ and $r \neq 1$, then $\lim_{N \to \infty} r^{N+1}$ does not exist and the geometric series diverges. Finally, if the series starts with cr^M rather than cr^0, then

$$S = cr^M + cr^{M+1} + cr^{M+2} + cr^{M+3} + \cdots = r^M \sum_{n=0}^{\infty} cr^n = \frac{cr^M}{1-r} \qquad \blacksquare$$

■ **EXAMPLE 3** Evaluate $\displaystyle\sum_{n=0}^{\infty} 5^{-n}$.

Solution This is a geometric series with $r = 5^{-1}$. By Eq. (4),

$$\sum_{n=0}^{\infty} 5^{-n} = 1 + \frac{1}{5} + \frac{1}{5^2} + \frac{1}{5^3} + \cdots = \frac{1}{1 - 5^{-1}} = \frac{5}{4} \qquad ■$$

■ **EXAMPLE 4** Evaluate $\displaystyle\sum_{n=3}^{\infty} 7\left(-\frac{3}{4}\right)^n = 7\left(-\frac{3}{4}\right)^3 + 7\left(-\frac{3}{4}\right)^4 + 7\left(-\frac{3}{4}\right)^5 + \cdots$.

Solution This is a geometric series with $r = -\frac{3}{4}$ and $c = 7$, starting at $n = 3$. By Eq. (5),

$$\sum_{n=3}^{\infty} 7\left(-\frac{3}{4}\right)^n = \frac{7\left(-\frac{3}{4}\right)^3}{1 - \left(-\frac{3}{4}\right)} = -\frac{27}{16} \qquad ■$$

■ **EXAMPLE 5** Find a fraction that has repeated decimal expansion $0.212121\ldots$.

Solution We can write this decimal as the series $\frac{21}{100} + \frac{21}{100^2} + \frac{21}{100^3} + \cdots$. This is a geometric series with $a = \frac{21}{100}$ and $r = \frac{1}{100}$. Thus, it converges to

$$\frac{a}{1-r} = \frac{\frac{21}{100}}{1 - \frac{1}{100}} = \frac{21}{99} = \frac{7}{33} \qquad ■$$

■ **EXAMPLE 6** Evaluate $S = \displaystyle\sum_{n=0}^{\infty} \frac{2 + 3^n}{5^n}$.

Solution Write S as a sum of two geometric series. This is valid by Theorem 1 because both geometric series converge:

$$\overbrace{\phantom{\sum_{n=0}^{\infty} \frac{1}{5^n} + \sum_{n=0}^{\infty} \left(\frac{3}{5}\right)^n}}^{\text{Both geometric series converge}}$$

$$\sum_{n=0}^{\infty} \frac{2 + 3^n}{5^n} = \sum_{n=0}^{\infty} \frac{2}{5^n} + \sum_{n=0}^{\infty} \frac{3^n}{5^n} = 2 \sum_{n=0}^{\infty} \frac{1}{5^n} + \sum_{n=0}^{\infty} \left(\frac{3}{5}\right)^n = 2 \cdot \frac{1}{1 - \frac{1}{5}} + \frac{1}{1 - \frac{3}{5}} = 5 \quad ■$$

CONCEPTUAL INSIGHT Sometimes, the following *incorrect argument* is given for summing a geometric series:

$$S = \frac{1}{2} + \frac{1}{4} + \frac{1}{8} + \cdots$$

$$2S = 1 + \frac{1}{2} + \frac{1}{4} + \frac{1}{8} + \cdots = 1 + S$$

Thus, $2S = 1 + S$, or $S = 1$. The answer is correct, so why is the argument wrong? It is wrong because we do not know in advance that the series converges. Observe what happens when this argument is applied to a divergent series:

$$S = 1 + 2 + 4 + 8 + 16 + \cdots$$

$$2S = 2 + 4 + 8 + 16 + \cdots = S - 1$$

This would yield $2S = S - 1$, or $S = -1$, which is absurd because S diverges. We avoid such erroneous conclusions by carefully defining the sum of an infinite series as the limit of its sequence of partial sums.

The infinite series $\sum_{k=1}^{\infty} 1$ diverges because the Nth partial sum $S_N = N$ diverges to infinity. It is less clear whether the following series converges or diverges:

$$\sum_{n=1}^{\infty} (-1)^{n+1} \frac{n}{n+1} = \frac{1}{2} - \frac{2}{3} + \frac{3}{4} - \frac{4}{5} + \frac{5}{6} - \cdots$$

We now introduce a useful test that allows us to conclude that this series diverges. The idea is that if the terms are not shrinking to 0 in size, then the series will not converge. This is typically the first test one applies when attempting to determine whether a series diverges.

The *n***th Term Divergence Test** (also known as the **Divergence Test**) is often stated as follows:

If $\sum_{n=1}^{\infty} a_n$ converges, then $\lim_{n \to \infty} a_n = 0$.

In practice, we use it to prove that a given series diverges. It is important to note that it does NOT say that if $\lim_{n \to \infty} a_n = 0$, then

$\sum_{n=1}^{\infty} a_n$ necessarily converges. We will see

that even though $\lim_{n \to \infty} \frac{1}{n} = 0$, the series

$\sum_{n=1}^{\infty} \frac{1}{n}$ diverges.

THEOREM 3 *n***th Term Divergence Test** If $\lim_{n \to \infty} a_n \neq 0$, then the series $\sum_{n=1}^{\infty} a_n$ diverges.

Proof First, note that $a_n = S_n - S_{n-1}$ because

$$S_n = (a_1 + a_2 + \cdots + a_{n-1}) + a_n = S_{n-1} + a_n$$

If $\sum_{n=1}^{\infty} a_n$ converges with sum S, then

$$\lim_{n \to \infty} a_n = \lim_{n \to \infty} (S_n - S_{n-1}) = \lim_{n \to \infty} S_n - \lim_{n \to \infty} S_{n-1} = S - S = 0$$

Therefore, if a_n does not converge to zero, $\sum_{n=1}^{\infty} a_n$ cannot converge. ∎

■ **EXAMPLE 7** Prove the divergence of $S = \sum_{n=1}^{\infty} \frac{n}{4n+1}$.

Solution We have

$$\lim_{n \to \infty} a_n = \lim_{n \to \infty} \frac{n}{4n+1} = \lim_{n \to \infty} \frac{1}{4 + 1/n} = \frac{1}{4}$$

The nth term a_n does not converge to zero, so the series diverges by the nth Term Divergence Test (Theorem 3). ∎

■ **EXAMPLE 8** Determine the convergence or divergence of

$$S = \sum_{n=1}^{\infty} (-1)^{n-1} \frac{n}{n+1} = \frac{1}{2} - \frac{2}{3} + \frac{3}{4} - \frac{4}{5} + \cdots$$

Solution The general term $a_n = (-1)^{n-1} \dfrac{n}{n+1}$ does not approach a limit. Indeed, $\dfrac{n}{n+1}$ tends to 1, so the odd terms a_{2n+1} tend to 1, and the even terms a_{2n} tend to -1. Because $\lim_{n \to \infty} a_n$ does not exist, the series S diverges by the nth Term Divergence Test. ■

The nth Term Divergence Test tells only part of the story. If a_n does not tend to zero, then $\sum a_n$ certainly diverges. But what if a_n does tend to zero? In this case, the series may converge or it may diverge. In other words, $\lim_{n \to \infty} a_n = 0$ is a *necessary* condition of convergence, but it is *not sufficient*. As we show in the next example, it is possible for a series to diverge even though its terms tend to zero.

■ **EXAMPLE 9** **Sequence Tends to Zero, Yet the Series Diverges** Prove the divergence of

$$\sum_{n=1}^{\infty} \frac{1}{\sqrt{n}} = \frac{1}{\sqrt{1}} + \frac{1}{\sqrt{2}} + \frac{1}{\sqrt{3}} + \cdots$$

Solution The general term $1/\sqrt{n}$ tends to zero. However, because each term in the partial sum S_N is greater than or equal to $1/\sqrt{N}$, we have

$$S_N = \overbrace{\frac{1}{\sqrt{1}} + \frac{1}{\sqrt{2}} + \cdots + \frac{1}{\sqrt{N}}}^{N \text{ terms}}$$

$$\geq \frac{1}{\sqrt{N}} + \frac{1}{\sqrt{N}} + \cdots + \frac{1}{\sqrt{N}}$$

$$= N \left(\frac{1}{\sqrt{N}} \right) = \sqrt{N}$$

This shows that $S_N \geq \sqrt{N}$. But $\sqrt{N}$ increases without bound (Figure 2). Therefore, S_N also increases without bound. This proves that the series diverges. ■

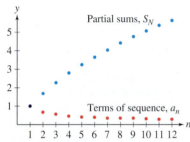

FIGURE 2 The partial sums of

$$\sum_{n=1}^{\infty} \frac{1}{\sqrt{n}}$$

diverge even though the terms $a_n = 1/\sqrt{n}$ tend to zero.

10.2 SUMMARY

- An *infinite series* is an expression

$$\sum_{n=1}^{\infty} a_n = a_1 + a_2 + a_3 + a_4 + \cdots$$

We call a_n the *general term* of the series. An infinite series can begin at $n = k$ for any integer k.
- The Nth *partial sum* is the finite sum of the terms up to and including the Nth term:

$$S_N = \sum_{n=1}^{N} a_n = a_1 + a_2 + a_3 + \cdots + a_N$$

- By definition, the sum of an infinite series is the limit $S = \lim_{N \to \infty} S_N$. If the limit exists, we say that the infinite series is *convergent* or *converges* to the sum S. If the limit does not exist, we say that the infinite series *diverges*.
- If the sequence of partial sums $\{S_N\}$ increase without bound, we say that S diverges to infinity.
- nth Term Divergence Test: If $\lim_{n \to \infty} a_n \neq 0$, then $\sum_{n=1}^{\infty} a_n$ diverges. However, a series may diverge even if its general term $\{a_n\}$ tends to zero.

- Partial sum of a geometric series:

$$c + cr + cr^2 + cr^3 + \cdots + cr^N = \frac{c(1 - r^{N+1})}{1 - r}$$

- Geometric series: If $|r| < 1$, then

$$\sum_{n=0}^{\infty} r^n = 1 + r + r^2 + r^3 + \cdots = \frac{1}{1 - r}$$

$$\sum_{n=M}^{\infty} cr^n = cr^M + cr^{M+1} + cr^{M+2} + \cdots = \frac{cr^M}{1 - r}$$

The geometric series diverges if $|r| \geq 1$.

Archimedes (287 BCE–212 BCE), who discovered the law of the lever, said, "Give me a place to stand on, and I can move the Earth" (quoted by Pappus of Alexandria c. 340 CE).

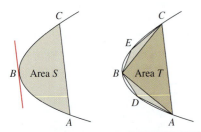

FIGURE 3 Archimedes showed that the area S of the parabolic segment is $\frac{4}{3}T$, where T is the area of $\triangle ABC$.

HISTORICAL PERSPECTIVE

(Mechanics Magazine London, 1824)

Geometric series were used as early as the third century BCE by Archimedes in a brilliant argument for determining the area S of a "parabolic segment" (shaded region in Figure 3). Given two points A and C on a parabola, there is a point B between A and C where the tangent line is parallel to $\overline{AC}$ (apparently, Archimedes was aware of the Mean Value Theorem more than 2000 years before the invention of calculus). Let T be the area of triangle $\triangle ABC$. Archimedes proved that if D is chosen in a similar fashion relative to $\overline{AB}$ and E is chosen relative to $\overline{BC}$, then

$$\frac{1}{4}T = \text{Area}(\triangle ADB) + \text{Area}(\triangle BEC) \quad \boxed{6}$$

This construction of triangles can be continued. The next step would be to construct the four triangles on the segments $\overline{AD}, \overline{DB}, \overline{BE}, \overline{EC}$, of total area $\frac{1}{4}^2 T$. Then construct eight triangles of total area $\frac{1}{4}^3 T$, etc. In this way, we obtain infinitely many triangles that completely fill up the parabolic segment. By the formula for the sum of a geometric series, we get

$$S = T + \frac{1}{4}T + \frac{1}{16}T + \cdots = T \sum_{n=0}^{\infty} \frac{1}{4^n} = \frac{4}{3}T$$

For this and many other achievements, Archi-

medes is ranked together with Newton and Gauss as one of the greatest scientists of all time.

The modern study of infinite series began in the seventeenth century with Newton, Leibniz, and their contemporaries. The divergence of $\sum_{n=1}^{\infty} 1/n$ (called the **harmonic series**) was known to the medieval scholar Nicole d'Oresme (1323–1382), but his proof was lost for centuries, and the result was rediscovered on more than one occasion. It was also known that the sum of the reciprocal squares $\sum_{n=1}^{\infty} 1/n^2$ converges, and in the 1640s, the Italian Pietro Mengoli put forward the challenge of finding its sum. Despite the efforts of the best mathematicians of the day, including Leibniz and the Bernoulli brothers Jakob and Johann, the problem resisted solution for nearly a century. In 1735 the great master Leonhard Euler (at the time, 28 years old) astonished his contemporaries by proving that

$$\frac{1}{1^2} + \frac{1}{2^2} + \frac{1}{3^2} + \frac{1}{4^2} + \frac{1}{5^2} + \frac{1}{6^2} + \cdots = \frac{\pi^2}{6}$$

This formula, surprising in itself, plays a role in a variety of mathematical fields. A theorem from number theory states that two whole numbers, chosen randomly, have no common factor with probability $6/\pi^2 \approx 0.6$ (the reciprocal of Euler's result). On the other hand, Euler's result and its generalizations appear in the field of statistical mechanics.

10.2 EXERCISES

Preliminary Questions

1. What role do partial sums play in defining the sum of an infinite series?

2. What is the sum of the following infinite series?

$$\frac{1}{4} + \frac{1}{8} + \frac{1}{16} + \frac{1}{32} + \frac{1}{64} + \cdots$$

3. What happens if you apply the formula for the sum of a geometric series to the following series? Is the formula valid?

$$1 + 3 + 3^2 + 3^3 + 3^4 + \cdots$$

4. A student asserts that $\sum_{n=1}^{\infty} \frac{1}{n^2} = 0$ because $\frac{1}{n^2}$ tends to zero. Is this valid reasoning?

5. A student claims that $\sum_{n=1}^{\infty} \frac{1}{\sqrt{n}}$ converges because

$$\lim_{n\to\infty} \frac{1}{\sqrt{n}} = 0$$

Is this valid reasoning?

6. Find an N such that $S_N > 25$ for the series $\sum_{n=1}^{\infty} 2$.

7. Does there exist an N such that $S_N > 25$ for the series $\sum_{n=1}^{\infty} 2^{-n}$? Explain.

8. Give an example of a divergent infinite series whose general term tends to zero.

Exercises

1. Find a formula for the general term a_n (not the partial sum) of the infinite series.

(a) $\frac{1}{3} + \frac{1}{9} + \frac{1}{27} + \frac{1}{81} + \cdots$
(b) $\frac{1}{1} + \frac{5}{2} + \frac{25}{4} + \frac{125}{8} + \cdots$

(c) $\frac{1}{1} - \frac{2^2}{2\cdot1} + \frac{3^3}{3\cdot2\cdot1} - \frac{4^4}{4\cdot3\cdot2\cdot1} + \cdots$

(d) $\frac{2}{1^2+1} + \frac{1}{2^2+1} + \frac{2}{3^2+1} + \frac{1}{4^2+1} + \cdots$

2. Write in summation notation:

(a) $1 + \frac{1}{4} + \frac{1}{9} + \frac{1}{16} + \cdots$
(b) $\frac{1}{9} + \frac{1}{16} + \frac{1}{25} + \frac{1}{36} + \cdots$

(c) $1 - \frac{1}{3} + \frac{1}{5} - \frac{1}{7} + \cdots$

(d) $\frac{125}{9} + \frac{625}{16} + \frac{3125}{25} + \frac{15{,}625}{36} + \cdots$

In Exercises 3–6, compute the partial sums S_2, S_4, and S_6.

3. $1 + \frac{1}{2^2} + \frac{1}{3^2} + \frac{1}{4^2} + \cdots$
4. $\sum_{k=1}^{\infty} (-1)^k k^{-1}$

5. $\frac{1}{1\cdot2} + \frac{1}{2\cdot3} + \frac{1}{3\cdot4} + \cdots$
6. $\sum_{j=1}^{\infty} \frac{1}{j!}$

7. The series $S = 1 + \left(\frac{1}{5}\right) + \left(\frac{1}{5}\right)^2 + \left(\frac{1}{5}\right)^3 + \cdots$ converges to $\frac{5}{4}$. Calculate S_N for $N = 1, 2, \ldots$ until you find an S_N that approximates $\frac{5}{4}$ with an error less than 0.0001.

8. The series $S = \frac{1}{0!} - \frac{1}{1!} + \frac{1}{2!} - \frac{1}{3!} + \cdots$ is known to converge to e^{-1} (recall that $0! = 1$). Calculate S_N for $N = 1, 2, \ldots$ until you find an S_N that approximates e^{-1} with an error less than 0.001.

In Exercises 9 and 10, use a computer algebra system to compute S_{10}, S_{100}, S_{500}, and S_{1000} for the series. Do these values suggest convergence to the given value?

9. CAS
$$\frac{\pi - 3}{4} = \frac{1}{2\cdot3\cdot4} - \frac{1}{4\cdot5\cdot6} + \frac{1}{6\cdot7\cdot8} - \frac{1}{8\cdot9\cdot10} + \cdots$$

10. CAS
$$\frac{\pi^4}{90} = 1 + \frac{1}{2^4} + \frac{1}{3^4} + \frac{1}{4^4} + \cdots$$

11. Calculate S_3, S_4, and S_5 and then find the sum of the telescoping series
$$S = \sum_{n=1}^{\infty} \left(\frac{1}{n+1} - \frac{1}{n+2}\right)$$

12. Write $\sum_{n=3}^{\infty} \frac{1}{n(n-1)}$ as a telescoping series and find its sum.

13. Calculate S_3, S_4, and S_5 and then find the sum $S = \sum_{n=1}^{\infty} \frac{1}{4n^2 - 1}$ using the identity
$$\frac{1}{4n^2 - 1} = \frac{1}{2}\left(\frac{1}{2n-1} - \frac{1}{2n+1}\right)$$

14. Use partial fractions to rewrite $\sum_{n=1}^{\infty} \frac{1}{n(n+3)}$ as a telescoping series and find its sum.

15. Find the sum of $\frac{1}{1\cdot3} + \frac{1}{3\cdot5} + \frac{1}{5\cdot7} + \cdots$.

16. Find a formula for the partial sum S_N of $\sum_{n=1}^{\infty} (-1)^{n-1}$ and show that the series diverges.

In Exercises 17–22, the nth Term Divergence Test (Theorem 3) to prove that the following series diverge.

17. $\sum_{n=1}^{\infty} \frac{n}{10n + 12}$
18. $\sum_{n=1}^{\infty} \frac{n}{\sqrt{n^2 + 1}}$

19. $\frac{0}{1} - \frac{1}{2} + \frac{2}{3} - \frac{3}{4} + \cdots$
20. $\sum_{n=1}^{\infty} (-1)^n n^2$

21. $\cos\frac{1}{2} + \cos\frac{1}{3} + \cos\frac{1}{4} + \cdots$
22. $\sum_{n=0}^{\infty} \left(\sqrt{4n^2+1} - n\right)$

In Exercises 23–36, use the formula for the sum of a geometric series to find the sum or state that the series diverges.

23. $\frac{1}{1} + \frac{1}{8} + \frac{1}{8^2} + \cdots$
24. $\frac{4^3}{5^3} + \frac{4^4}{5^4} + \frac{4^5}{5^5} + \cdots$

25. $\sum_{n=3}^{\infty} \left(\frac{3}{11}\right)^{-n}$
26. $\sum_{n=2}^{\infty} \frac{7\cdot(-3)^n}{5^n}$

27. $\sum_{n=-4}^{\infty} \left(-\frac{4}{9}\right)^n$
28. $\sum_{n=0}^{\infty} \left(\frac{\pi}{e}\right)^n$

29. $\sum_{n=1}^{\infty} e^{-n}$
30. $\sum_{n=2}^{\infty} e^{3-2n}$

31. $\sum_{n=0}^{\infty} \frac{8 + 2^n}{5^n}$
32. $\sum_{n=0}^{\infty} \frac{3(-2)^n - 5^n}{8^n}$

33. $5 - \frac{5}{4} + \frac{5}{4^2} - \frac{5}{4^3} + \cdots$

34. $\frac{2^3}{7} + \frac{2^4}{7^2} + \frac{2^5}{7^3} + \frac{2^6}{7^4} + \cdots$

35. $\dfrac{7}{8} - \dfrac{49}{64} + \dfrac{343}{512} - \dfrac{2401}{4096} + \cdots$

36. $\dfrac{25}{9} + \dfrac{5}{3} + 1 + \dfrac{3}{5} + \dfrac{9}{25} + \dfrac{27}{125} + \cdots$

In Exercises 37–40, determine a reduced fraction that has this repeating decimal.

37. $0.222\ldots$ **38.** $0.454545\ldots$

39. $0.313131\ldots$ **40.** $0.217217217\ldots$

41. Determine which reduced fractions have a repeating decimal of the form $0.aaa\ldots$, where $a = 0, 1, 2, \ldots, 9$.

42. Determine which denominators are realized for reduced fractions that have a repeating decimal of the form $0.abcabcabc\ldots$, where each of a, b, and c is a one-digit number from 0 to 9.

43. Which of the following are *not* geometric series?

(a) $\displaystyle\sum_{n=0}^{\infty} \frac{7^n}{29^n}$ **(b)** $\displaystyle\sum_{n=3}^{\infty} \frac{1}{n^4}$

(c) $\displaystyle\sum_{n=0}^{\infty} \frac{n^2}{2^n}$ **(d)** $\displaystyle\sum_{n=5}^{\infty} \pi^{-n}$

44. Use the method of Example 9 to show that $\displaystyle\sum_{k=1}^{\infty} \frac{1}{k^{1/3}}$ diverges.

45. Prove that if $\displaystyle\sum_{n=1}^{\infty} a_n$ converges and $\displaystyle\sum_{n=1}^{\infty} b_n$ diverges, then $\displaystyle\sum_{n=1}^{\infty} (a_n + b_n)$ diverges. *Hint:* If not, derive a contradiction by writing

$$\sum_{n=1}^{\infty} b_n = \sum_{n=1}^{\infty} (a_n + b_n) - \sum_{n=1}^{\infty} a_n$$

46. Prove the divergence of $\displaystyle\sum_{n=0}^{\infty} \frac{9^n + 2^n}{5^n}$.

47. 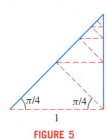   Give a counterexample to show that each of the following statements is false.

(a) If the general term a_n tends to zero, then $\displaystyle\sum_{n=1}^{\infty} a_n = 0$.

(b) The Nth partial sum of the infinite series defined by $\{a_n\}$ is a_N.

(c) If a_n tends to zero, then $\displaystyle\sum_{n=1}^{\infty} a_n$ converges.

(d) If a_n tends to L, then $\displaystyle\sum_{n=1}^{\infty} a_n = L$.

48. Suppose that $S = \displaystyle\sum_{n=1}^{\infty} a_n$ is an infinite series with partial sum $S_N = 5 - \dfrac{2}{N^2}$.

(a) What are the values of $\displaystyle\sum_{n=1}^{10} a_n$ and $\displaystyle\sum_{n=5}^{16} a_n$?

(b) What is the value of a_3?

(c) Find a general formula for a_n.

(d) Find the sum $\displaystyle\sum_{n=1}^{\infty} a_n$.

49. Compute the total area of the (infinitely many) triangles in Figure 4.

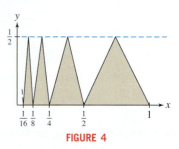

FIGURE 4

50. The winner of a lottery receives m dollars at the end of each year for N years. The present value (PV) of this prize in today's dollars is $\text{PV} = \displaystyle\sum_{i=1}^{N} m(1 + r)^{-i}$, where r is the interest rate. Calculate PV if $m = \$50{,}000$, $r = 0.06$ (corresponding to 6%), and $N = 20$. What is PV if $N = \infty$?

51. If a patient takes a dose of D units of a particular drug, the amount of the dosage that remains in the patient's bloodstream after t days is De^{-kt}, where k is a positive constant depending on the particular drug.

(a) Show that if the patient takes a dose D every day for an extended period, the amount of drug in the bloodstream approaches $R = \dfrac{De^{-k}}{1 - e^{-k}}$.

(b) Show that if the patient takes a dose D once every t days for an extended period, the amount of drug in the bloodstream approaches $R = \dfrac{De^{-kt}}{1 - e^{-kt}}$.

(c) Suppose that a dosage of more than S units is considered dangerous. What is the minimal time between doses that is safe? (*Hint:* $D + R \le S$.)

52. In economics, the multiplier effect refers to the fact that when there is an injection of money to consumers, the consumers spend a certain percentage of it. That amount recirculates through the economy and adds additional income, which comes back to the consumers and of which they spend the same percentage. This process repeats indefinitely, circulating additional money through the economy. Suppose that in order to stimulate the economy, the government institutes a tax cut of \$10 billion. If taxpayers are known to save 10% of any additional money they receive, and to spend 90%, how much total money will be circulated through the economy by that single \$10 billion tax cut?

53. Find the total length of the infinite zigzag path in Figure 5 (each zag occurs at an angle of $\frac{\pi}{4}$).

FIGURE 5

54. Evaluate $\displaystyle\sum_{n=1}^{\infty} \frac{1}{n(n+1)(n+2)}$. *Hint:* Find constants A, B, and C such that

$$\frac{1}{n(n+1)(n+2)} = \frac{A}{n} + \frac{B}{n+1} + \frac{C}{n+2}$$

55. Show that if a is a positive integer, then

$$\sum_{n=1}^{\infty} \frac{1}{n(n+a)} = \frac{1}{a}\left(1 + \frac{1}{2} + \cdots + \frac{1}{a}\right)$$

56. A ball dropped from a height of 10 ft begins to bounce vertically. Each time it strikes the ground, it returns to two-thirds of its previous height. What is the total vertical distance traveled by the ball if it bounces infinitely many times?

57. A ball is dropped from a height of 6 ft and begins to bounce vertically. Each time it strikes the ground, it returns to three-quarters of its previous height. What is the total vertical distance traveled by the ball if it bounces infinitely many times?

58. A unit square is cut into nine equal regions as in Figure 6(A). The central subsquare is painted red. Each of the unpainted squares is then cut into nine equal subsquares and the central square of each is painted red as in Figure 6(B). This procedure is repeated for each of the resulting unpainted squares. After continuing this process an infinite number of times, what fraction of the total area of the original square is painted?

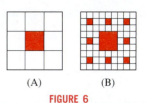

(A)   (B)

FIGURE 6

59. Let $\{b_n\}$ be a sequence and let $a_n = b_n - b_{n-1}$. Show that $\sum_{n=1}^{\infty} a_n$ converges if and only if $\lim_{n\to\infty} b_n$ exists.

60. Assumptions Matter Show, by giving counterexamples, that the assertions of Theorem 1 are not valid if the series $\sum_{n=0}^{\infty} a_n$ and $\sum_{n=0}^{\infty} b_n$ are not convergent.

Further Insights and Challenges

Exercises 61–63 use the formula

$$1 + r + r^2 + \cdots + r^{N-1} = \frac{1 - r^N}{1 - r} \qquad \boxed{7}$$

61. Professor George Andrews of Pennsylvania State University observed that we can use Eq. (7) to calculate the derivative of $f(x) = x^N$ (for $N \geq 0$). Assume that $a \neq 0$ and let $x = ra$. Show that

$$f'(a) = \lim_{x\to a} \frac{x^N - a^N}{x - a} = a^{N-1} \lim_{r\to 1} \frac{r^N - 1}{r - 1}$$

and evaluate the limit.

62. Pierre de Fermat used geometric series to compute the area under the graph of $f(x) = x^N$ over $[0, A]$. For $0 < r < 1$, let $F(r)$ be the sum of the areas of the infinitely many right-endpoint rectangles with endpoints Ar^n, as in Figure 7. As r tends to 1, the rectangles become narrower and $F(r)$ tends to the area under the graph.

(a) Show that $F(r) = A^{N+1} \dfrac{1-r}{1-r^{N+1}}$.

(b) Use Eq. (7) to evaluate $\displaystyle\int_0^A x^N \, dx = \lim_{r\to 1} F(r)$.

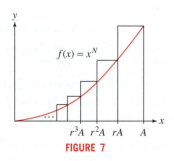

FIGURE 7

63. Verify the Gregory–Leibniz formula as follows.

(a) Set $r = -x^2$ in Eq. (7) and rearrange to show that

$$\frac{1}{1+x^2} = 1 - x^2 + x^4 - \cdots + (-1)^{N-1}x^{2N-2} + \frac{(-1)^N x^{2N}}{1 + x^2}$$

(b) Show, by integrating over $[0, 1]$, that

$$\frac{\pi}{4} = 1 - \frac{1}{3} + \frac{1}{5} - \frac{1}{7} + \cdots + \frac{(-1)^{N-1}}{2N-1} + (-1)^N \int_0^1 \frac{x^{2N} \, dx}{1 + x^2}$$

(c) Use the Comparison Theorem for integrals to prove that

$$0 \leq \int_0^1 \frac{x^{2N} \, dx}{1+x^2} \leq \frac{1}{2N+1}$$

Hint: Observe that the integrand is $\leq x^{2N}$.

(d) Prove that

$$\frac{\pi}{4} = 1 - \frac{1}{3} + \frac{1}{5} - \frac{1}{7} + \frac{1}{9} - \cdots$$

Hint: Use (b) and (c) to show that the partial sums S_N satisfy $\left|S_N - \frac{\pi}{4}\right| \leq \frac{1}{2N+1}$, and thereby conclude that $\lim_{N\to\infty} S_N = \frac{\pi}{4}$.

64. Cantor's Disappearing Table (following Larry Knop of Hamilton College) Take a table of length L (Figure 8). At Stage 1, remove the section of length $L/4$ centered at the midpoint. Two sections remain, each with length less than $L/2$. At Stage 2, remove sections of length $L/4^2$ from each of these two sections (this stage removes $L/8$ of the table). Now four sections remain, each of length less than $L/4$. At Stage 3, remove the four central sections of length $L/4^3$, etc.

(a) Show that at the Nth stage, each remaining section has length less than $L/2^N$ and that the total amount of table removed is

$$L\left(\frac{1}{4} + \frac{1}{8} + \frac{1}{16} + \cdots + \frac{1}{2^{N+1}}\right)$$

(b) Show that in the limit as $N \to \infty$, precisely one-half of the table remains.

This result is intriguing, because there are no nonzero intervals of table left (at each stage, the remaining sections have a length less than $L/2^N$). So, the table has "disappeared." However, we can place any object longer than $L/4$ on the table. The object will not fall through because it will not fit through any of the removed sections.

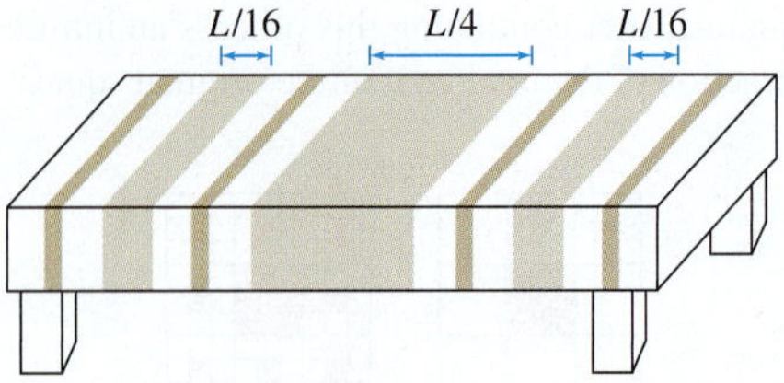

L/16   L/4   L/16

FIGURE 8

65. The **Koch snowflake** (described in 1904 by Swedish mathematician Helge von Koch) is an infinitely jagged "fractal" curve obtained as a limit of polygonal curves (it is continuous but has no tangent line at any point). Begin with an equilateral triangle (Stage 0) and produce Stage 1 by replacing each edge with four edges of one-third the length, arranged as in Figure 9. Continue the process: At the nth stage, replace each edge with four edges of one-third the length of the edge from the $(n-1)$st stage.

(a) Show that the perimeter P_n of the polygon at the nth stage satisfies $P_n = \frac{4}{3}P_{n-1}$. Prove that $\lim_{n\to\infty} P_n = \infty$. The snowflake has infinite length.

(b) Let A_0 be the area of the original equilateral triangle. Show that $(3)4^{n-1}$ new triangles are added at the nth stage, each with area $A_0/9^n$ (for $n \geq 1$). Show that the total area of the Koch snowflake is $\frac{8}{5}A_0$.

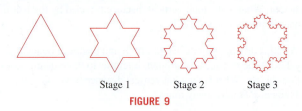

Stage 1   Stage 2   Stage 3

FIGURE 9

10.3 Convergence of Series with Positive Terms

The next three sections develop techniques for determining whether an infinite series converges or diverges. This is easier than finding the sum of an infinite series, which is possible only in special cases.

In this section, we consider **positive series** $\sum a_n$, where $a_n > 0$ for all n. We can visualize the terms of a positive series as rectangles of width 1 and height a_n (Figure 1). The partial sum

$$S_N = a_1 + a_2 + \cdots + a_N$$

is equal to the area of the first N rectangles.

The key feature of positive series is that their partial sums form an increasing sequence:

$$S_N < S_{N+1}$$

for all N. This is because S_{N+1} is obtained from S_N by adding a positive number:

$$S_{N+1} = \left(a_1 + a_2 + \cdots + a_N\right) + a_{N+1} = S_N + \underbrace{a_{N+1}}_{\text{Positive}}$$

Recall that an increasing sequence converges if it is bounded above. Otherwise, it diverges (Theorem 6, Section 10.1). It follows that a positive series behaves in one of two ways.

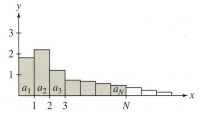

FIGURE 1 The partial sum S_N is the sum of the areas of the N shaded rectangles.

THEOREM 1 Partial Sum Theorem for Positive Series If $S = \displaystyle\sum_{n=1}^{\infty} a_n$ is a positive series, then either:

(i) The partial sums S_N are bounded above. In this case, S converges. Or,

(ii) The partial sums S_N are not bounded above. In this case, S diverges.

- Theorem 1 remains true if $a_n \geq 0$. It is not necessary to assume that $a_n > 0$.
- It also remains true if $a_n > 0$ for all $n \geq M$ for some M, because the convergence or divergence of a series is not affected by the first M terms.

Assumptions Matter   The theorem does not hold for nonpositive series. Consider

$$S = \sum_{n=1}^{\infty}(-1)^{n-1} = 1 - 1 + 1 - 1 + 1 - 1 + \cdots$$

The partial sums are bounded (because $S_N = 1$ or 0), but S diverges.

Our first application of Theorem 1 is the following Integral Test. It is extremely useful because integrals are easier to evaluate than series in most cases.

The Integral Test is valid for any series
$$\sum_{n=k}^{\infty} f(n),$$ provided that for some $M > 0$,
f is a positive, decreasing, and continuous function of x for $x \geq M$. The convergence of the series is determined by the convergence of
$$\int_M^{\infty} f(x)\,dx$$

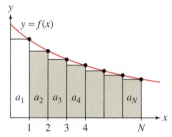

FIGURE 2

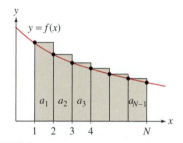

FIGURE 3

The infinite series
$$\sum_{n=1}^{\infty} \frac{1}{n}$$
is called the "harmonic series."

> **THEOREM 2 Integral Test** Let $a_n = f(n)$, where f is a positive, decreasing, and continuous function of x for $x \geq 1$.
>
> **(i)** If $\displaystyle\int_1^{\infty} f(x)\,dx$ converges, then $\displaystyle\sum_{n=1}^{\infty} a_n$ converges.
>
> **(ii)** If $\displaystyle\int_1^{\infty} f(x)\,dx$ diverges, then $\displaystyle\sum_{n=1}^{\infty} a_n$ diverges.

Proof Because f is decreasing, the shaded rectangles in Figure 2 lie below the graph of f, and therefore for all N

$$\underbrace{a_2 + \cdots + a_N}_{\text{Area of shaded rectangles in Figure 2}} \leq \int_1^N f(x)\,dx \leq \int_1^{\infty} f(x)\,dx$$

If the improper integral on the right converges, then the sums $a_2 + \cdots + a_N$ remain bounded. In this case, S_N also remains bounded, and the infinite series converges by the Partial Sum Theorem for Positive Series (Theorem 1). This proves (i).

On the other hand, the rectangles in Figure 3 lie above the graph of $f(x)$, so

$$\int_1^N f(x)\,dx \leq \underbrace{a_1 + a_2 + \cdots + a_{N-1}}_{\text{Area of shaded rectangles in Figure 3}} \qquad \boxed{1}$$

If $\int_1^{\infty} f(x)\,dx$ diverges, then $\int_1^N f(x)\,dx$ tends to ∞, and Eq. (1) shows that S_N also tends to ∞. This proves (ii). ∎

■ **EXAMPLE 1** **The Harmonic Series Diverges** Show that $\displaystyle\sum_{n=1}^{\infty} \frac{1}{n}$ diverges.

Solution Let $f(x) = \frac{1}{x}$. Then $f(n) = \frac{1}{n}$, and the Integral Test applies because f is positive, decreasing, and continuous for $x \geq 1$. The integral diverges:

$$\int_1^{\infty} \frac{dx}{x} = \lim_{R \to \infty} \int_1^R \frac{dx}{x} = \lim_{R \to \infty} \ln R = \infty$$

Therefore, the series $\displaystyle\sum_{n=1}^{\infty} \frac{1}{n}$ diverges. ■

■ **EXAMPLE 2** Does $\displaystyle\sum_{n=1}^{\infty} \frac{n}{(n^2+1)^2} = \frac{1}{2^2} + \frac{2}{5^2} + \frac{3}{10^2} + \cdots$ converge?

Solution The function $f(x) = \dfrac{x}{(x^2+1)^2}$ is positive and continuous for $x \geq 1$. It is decreasing because $f'(x)$ is negative:

$$f'(x) = \frac{1 - 3x^2}{(x^2+1)^3} < 0 \qquad \text{for } x \geq 1$$

Therefore, the Integral Test applies. Using the substitution $u = x^2 + 1$, $du = 2x\,dx$, we have

$$\int_1^{\infty} \frac{x}{(x^2+1)^2}\,dx = \lim_{R \to \infty} \int_1^R \frac{x}{(x^2+1)^2}\,dx = \lim_{R \to \infty} \frac{1}{2} \int_2^{R^2+1} \frac{du}{u^2}$$

$$= \lim_{R \to \infty} \frac{-1}{2u}\Big|_2^{R^2+1} = \lim_{R \to \infty} \left(\frac{1}{4} - \frac{1}{2(R^2+1)} \right) = \frac{1}{4}$$

Thus, the integral converges, and therefore, $\displaystyle\sum_{n=1}^{\infty} \frac{n}{(n^2+1)^2}$ also converges by the Integral Test. ■

The sum of the reciprocal powers n^{-p} is called a **p-series**.

> **THEOREM 3 Convergence of p-Series** The infinite series $\sum_{n=1}^{\infty} \dfrac{1}{n^p}$ converges if $p > 1$ and diverges otherwise.

Proof If $p \leq 0$, then the general term n^{-p} does not tend to zero, so the series diverges by the nth Term Divergence Test. If $p > 0$, then $f(x) = x^{-p}$ is positive and decreasing for $x \geq 1$, so the Integral Test applies. According to Theorem 1 in Section 7.7,

$$\int_1^{\infty} \frac{1}{x^p}\,dx = \begin{cases} \dfrac{1}{p-1} & \text{if } p > 1 \\ \infty & \text{if } p \leq 1 \end{cases}$$

Therefore, $\sum_{n=1}^{\infty} \dfrac{1}{n^p}$ converges for $p > 1$ and diverges for $p \leq 1$. ■

Here are two examples of p-series:

$$p = \frac{1}{3}: \quad \sum_{n=1}^{\infty} \frac{1}{\sqrt[3]{n}} = \frac{1}{\sqrt[3]{1}} + \frac{1}{\sqrt[3]{2}} + \frac{1}{\sqrt[3]{3}} + \frac{1}{\sqrt[3]{4}} + \cdots = \infty \quad \text{diverges}$$

$$p = 2: \quad \sum_{n=1}^{\infty} \frac{1}{n^2} = \frac{1}{1} + \frac{1}{2^2} + \frac{1}{3^2} + \frac{1}{4^2} + \cdots \quad \text{converges}$$

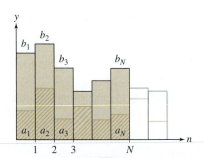

FIGURE 4 The series $\sum a_n$ is dominated by the series $\sum b_n$.

Another powerful method for determining convergence of positive series occurs via comparison with other series. Suppose that $0 \leq a_n \leq b_n$. Figure 4 suggests that if the larger sum $\sum b_n$ *converges*, then the smaller sum $\sum a_n$ also converges. Similarly, if the smaller sum *diverges*, then the larger sum also diverges.

> **THEOREM 4 Direct Comparison Test**
> Assume that there exists $M > 0$ such that $0 \leq a_n \leq b_n$ for $n \geq M$.
> **(i)** If $\sum_{n=1}^{\infty} b_n$ converges, then $\sum_{n=1}^{\infty} a_n$ also converges.
> **(ii)** If $\sum_{n=1}^{\infty} a_n$ diverges, then $\sum_{n=1}^{\infty} b_n$ also diverges.

Proof We can assume, without loss of generality, that $M = 1$. If $S = \sum_{n=1}^{\infty} b_n$ converges, then the partial sums of $\sum_{n=1}^{\infty} a_n$ are bounded above by S because

$$a_1 + a_2 + \cdots + a_N \leq b_1 + b_2 + \cdots + b_N \leq \sum_{n=1}^{\infty} b_n = S \qquad \boxed{2}$$

Therefore, $\sum_{n=1}^{\infty} a_n$ converges by the Partial Sum Theorem for Positive Series (Theorem 1). This proves (i). On the other hand, if $\sum_{n=1}^{\infty} a_n$ diverges, then $\sum_{n=1}^{\infty} b_n$ must also diverge. Otherwise, we would have a contradiction to (i). ■

■ **EXAMPLE 3** Does $\sum_{n=1}^{\infty} \dfrac{1}{\sqrt{n}\,3^n}$ converge?

A good analogy for the Direct Comparison Test, as in Figure 5, is one balloon containing the terms for a_n, inside a bigger balloon containing the terms for b_n. As we blow air into the balloons in amounts corresponding to the subsequent terms a_n and b_n, the balloon containing the b_n terms will always be bigger than the balloon containing the a_n terms since $a_n \leq b_n$. If the bigger balloon does not contain enough air to make it pop, then the smaller balloon does not pop either. Thus, if the bigger series converges, so does the smaller series. On the other hand, if the smaller balloon contains enough air to make it pop, then the bigger balloon must also pop, implying that if the smaller series diverges, the bigger series diverges as well. But in the cases that the bigger balloon pops or the smaller balloon does not pop, it says nothing about the remaining balloon.

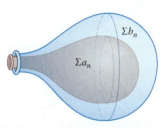

FIGURE 5 The smaller series is contained in the smaller balloon.

Solution For $n \geq 1$, we have

$$\frac{1}{\sqrt{n}\, 3^n} \leq \frac{1}{3^n}$$

The larger series $\sum\limits_{n=1}^{\infty} \frac{1}{3^n}$ converges because it is a geometric series with $r = \frac{1}{3} < 1$. By the Direct Comparison Test, the smaller series $\sum\limits_{n=1}^{\infty} \frac{1}{\sqrt{n}\, 3^n}$ also converges. ■

■ **EXAMPLE 4** Does $S = \sum\limits_{n=2}^{\infty} \frac{1}{(n^2+3)^{1/3}}$ converge?

Solution Let us show that

$$\frac{1}{n} \leq \frac{1}{(n^2+3)^{1/3}} \qquad \text{for } n \geq 2$$

This inequality is equivalent to $(n^2 + 3) \leq n^3$, so we must show that

$$f(x) = x^3 - (x^2 + 3) \geq 0 \qquad \text{for } x \geq 2$$

The function f is increasing because its derivative $f'(x) = 3x\left(x - \frac{2}{3}\right)$ is positive for $x \geq 2$. Since $f(2) = 1$, it follows that $f(x) \geq 1$ for $x \geq 2$, and our original inequality follows. We know that the smaller harmonic series $\sum\limits_{n=2}^{\infty} \frac{1}{n}$ diverges. Therefore, the larger series $\sum\limits_{n=2}^{\infty} \frac{1}{(n^2+1)^{1/3}}$ also diverges. ■

■ **EXAMPLE 5** **Using the Comparison Correctly** Determine the convergence of

$$\sum_{n=2}^{\infty} \frac{1}{n(\ln n)^2}$$

Solution We might be tempted to compare $\sum\limits_{n=2}^{\infty} \frac{1}{n(\ln n)^2}$ to the harmonic series $\sum\limits_{n=2}^{\infty} \frac{1}{n}$ using the inequality (valid for $n \geq 3$ since $\ln 3 > 1$)

$$\frac{1}{n(\ln n)^2} \leq \frac{1}{n}$$

However, $\sum\limits_{n=2}^{\infty} \frac{1}{n}$ diverges, and this says nothing about the *smaller* series $\sum \frac{1}{n(\ln n)^2}$. Fortunately, the Integral Test can be used. The substitution $u = \ln x$ yields

$$\int_{2}^{\infty} \frac{dx}{x(\ln x)^2} = \int_{\ln 2}^{\infty} \frac{du}{u^2} = \lim_{R \to \infty} \left(\frac{1}{\ln 2} - \frac{1}{R} \right) = \frac{1}{\ln 2} < \infty$$

The Integral Test shows that $\sum\limits_{n=2}^{\infty} \frac{1}{n(\ln n)^2}$ converges. ■

Suppose we wish to study the convergence of

$$S = \sum_{n=2}^{\infty} \frac{n^2}{n^4 - n - 1}$$

For large n, the general term is very close to $1/n^2$:

$$\frac{n^2}{n^4 - n - 1} = \frac{1}{n^2 - n^{-1} - n^{-2}} \approx \frac{1}{n^2}$$

Thus, we might try to compare S with $\sum_{n=2}^{\infty} \frac{1}{n^2}$. Unfortunately, however, the inequality goes in the wrong direction:

$$\frac{n^2}{n^4 - n - 1} > \frac{n^2}{n^4} = \frac{1}{n^2}$$

Although the smaller series $\sum_{n=2}^{\infty} \frac{1}{n^2}$ converges, we cannot use the Direct Comparison Test to say anything about our larger series. In this situation, the following variation of the Direct Comparison Test can be used.

THEOREM 5 **Limit Comparison Test** Let $\{a_n\}$ and $\{b_n\}$ be *positive* sequences. Assume that the following limit exists:

$$L = \lim_{n \to \infty} \frac{a_n}{b_n}$$

- If $L > 0$, then $\sum a_n$ converges if and only if $\sum b_n$ converges.
- If $L = \infty$ and $\sum a_n$ converges, then $\sum b_n$ converges.
- If $L = 0$ and $\sum b_n$ converges, then $\sum a_n$ converges.

CAUTION *The Limit Comparison Test is not valid if the series are not positive. See Exercise 44 in Section 10.4.*

Proof Assume first that L is finite (possibly zero) and that $\sum b_n$ converges. Choose a positive number $R > L$. Then $0 \le a_n/b_n \le R$ for all n sufficiently large because a_n/b_n approaches L. Therefore, $a_n \le R b_n$. The series $\sum R b_n$ converges because it is a multiple of the convergent series $\sum b_n$. Thus, $\sum a_n$ converges by the Direct Comparison Test.

Next, suppose that L is nonzero (positive or infinite) and that $\sum a_n$ converges. Let $L^{-1} = \lim_{n \to \infty} b_n/a_n$. Then L^{-1} is finite and we can apply the result of the previous paragraph with the roles of $\{a_n\}$ and $\{b_n\}$ reversed to conclude that $\sum b_n$ converges. ∎

CONCEPTUAL INSIGHT To remember the different cases of the Limit Comparison Test, you can think of it this way: If $L > 0$, then $a_n \approx L b_n$ for large n. In other words, the series $\sum a_n$ and $\sum b_n$ are *roughly* multiples of each other, so one converges if and only if the other converges. If $L = \infty$, then a_n is much larger than b_n (for large n), so if $\sum a_n$ converges, $\sum b_n$ certainly converges. Finally, if $L = 0$, then b_n is much larger than a_n and the convergence of $\sum b_n$ yields the convergence of $\sum a_n$.

■ **EXAMPLE 6** Show that $\sum_{n=2}^{\infty} \frac{n^2}{n^4 - n - 1}$ converges.

Solution Let

$$a_n = \frac{n^2}{n^4 - n - 1} \quad \text{and} \quad b_n = \frac{1}{n^2}$$

We observed above that $a_n \approx b_n$ for large n. To apply the Limit Comparison Test, we observe that the limit L exists and $L > 0$:

$$L = \lim_{n \to \infty} \frac{a_n}{b_n} = \lim_{n \to \infty} \frac{n^2}{n^4 - n - 1} \cdot \frac{n^2}{1} = \lim_{n \to \infty} \frac{1}{1 - n^{-3} - n^{-4}} = 1$$

Since $\sum_{n=2}^{\infty} \frac{1}{n^2}$ converges, our series $\sum_{n=2}^{\infty} \frac{n^2}{n^4 - n - 1}$ also converges by Theorem 5. ∎

■ **EXAMPLE 7** Determine whether $\displaystyle\sum_{n=3}^{\infty} \frac{1}{\sqrt{n^2+4}}$ converges.

Solution Apply the Limit Comparison Test with $a_n = \dfrac{1}{\sqrt{n^2+4}}$ and $b_n = \dfrac{1}{n}$. Then

$$L = \lim_{n\to\infty} \frac{a_n}{b_n} = \lim_{n\to\infty} \frac{n}{\sqrt{n^2+4}} = \lim_{n\to\infty} \frac{1}{\sqrt{1+4/n^2}} = 1$$

Since $\displaystyle\sum_{n=3}^{\infty} \frac{1}{n}$ diverges and $L > 0$, the series $\displaystyle\sum_{n=3}^{\infty} \frac{1}{\sqrt{n^2+4}}$ also diverges. ■

In the Limit Comparison Test, when attempting to find an appropriate b_n to compare with a_n, we typically keep only the largest power of n in the numerator and denominator of a_n, as we did in each of the previous examples.

10.3 SUMMARY

- The partial sums S_N of a positive series $S = \sum a_n$ form an increasing sequence.
- Partial Sum Theorem for Positive Series: A positive series S converges if its partial sums S_N remain bounded. Otherwise, it diverges.
- Integral Test: Assume that f is positive, decreasing, and continuous for $x > M$. Set $a_n = f(n)$. If $\displaystyle\int_M^\infty f(x)\,dx$ converges, then $S = \sum a_n$ converges, and if $\displaystyle\int_M^\infty f(x)\,dx$ diverges, then $S = \sum a_n$ diverges.
- p-Series: The series $\displaystyle\sum_{n=1}^{\infty} \frac{1}{n^p}$ converges if $p > 1$ and diverges if $p \le 1$.
- Direct Comparison Test: Assume there exists $M > 0$ such that $0 \le a_n \le b_n$ for all $n \ge M$. If $\sum b_n$ converges, then $\sum a_n$ converges, and if $\sum a_n$ diverges, then $\sum b_n$ diverges.
- Limit Comparison Test: Assume that $\{a_n\}$ and $\{b_n\}$ are positive and that the following limit exists:

$$L = \lim_{n\to\infty} \frac{a_n}{b_n}$$

  - If $L > 0$, then $\sum a_n$ converges if and only if $\sum b_n$ converges.
  - If $L = \infty$ and $\sum a_n$ converges, then $\sum b_n$ converges.
  - If $L = 0$ and $\sum b_n$ converges, then $\sum a_n$ converges.

10.3 EXERCISES

Preliminary Questions

1. Let $S = \displaystyle\sum_{n=1}^{\infty} a_n$. If the partial sums S_N are increasing, then (choose the correct conclusion):

(a) $\{a_n\}$ is an increasing sequence.

(b) $\{a_n\}$ is a positive sequence.

2. What are the hypotheses of the Integral Test?

3. Which test would you use to determine whether $\displaystyle\sum_{n=1}^{\infty} n^{-3.2}$ converges?

4. Which test would you use to determine whether $\displaystyle\sum_{n=1}^{\infty} \frac{1}{2^n + \sqrt{n}}$ converges?

5. Ralph hopes to investigate the convergence of $\displaystyle\sum_{n=1}^{\infty} \frac{e^{-n}}{n}$ by comparing it with $\displaystyle\sum_{n=1}^{\infty} \frac{1}{n}$. Is Ralph on the right track?

Exercises

In Exercises 1–14, use the Integral Test to determine whether the infinite series is convergent.

1. $\displaystyle\sum_{n=1}^{\infty} \frac{1}{n^4}$

2. $\displaystyle\sum_{n=1}^{\infty} \frac{1}{n+3}$

3. $\displaystyle\sum_{n=1}^{\infty} n^{-1/3}$

4. $\displaystyle\sum_{n=5}^{\infty} \frac{1}{\sqrt{n-4}}$

5. $\displaystyle\sum_{n=25}^{\infty} \frac{n^2}{(n^3+9)^{5/2}}$

6. $\displaystyle\sum_{n=1}^{\infty} \frac{n}{(n^2+1)^{3/5}}$

7. $\displaystyle\sum_{n=1}^{\infty} \frac{1}{n^2+1}$

8. $\displaystyle\sum_{n=4}^{\infty} \frac{1}{n^2-1}$

9. $\displaystyle\sum_{n=1}^{\infty} \frac{1}{n(n+1)}$

10. $\displaystyle\sum_{n=1}^{\infty} ne^{-n^2}$

11. $\displaystyle\sum_{n=2}^{\infty} \frac{1}{n(\ln n)^2}$

12. $\displaystyle\sum_{n=1}^{\infty} \frac{\ln n}{n^2}$

13. $\displaystyle\sum_{n=1}^{\infty} \frac{1}{2^{\ln n}}$

14. $\displaystyle\sum_{n=1}^{\infty} \frac{1}{3^{\ln n}}$

15. Show that $\displaystyle\sum_{n=1}^{\infty} \frac{1}{n^3+8n}$ converges by using the Direct Comparison Test with $\displaystyle\sum_{n=1}^{\infty} n^{-3}$.

16. Show that $\displaystyle\sum_{n=2}^{\infty} \frac{1}{\sqrt{n^2-3}}$ diverges by comparing with $\displaystyle\sum_{n=2}^{\infty} n^{-1}$.

17. Let $S = \displaystyle\sum_{n=1}^{\infty} \frac{1}{n+\sqrt{n}}$. Verify that for $n \geq 1$,

$$\frac{1}{n+\sqrt{n}} \leq \frac{1}{n}, \qquad \frac{1}{n+\sqrt{n}} \leq \frac{1}{\sqrt{n}}$$

Can either inequality be used to show that S diverges? Show that $\dfrac{1}{n+\sqrt{n}} \geq \dfrac{1}{2n}$ for $n \geq 1$ and conclude that S diverges.

18. Which of the following inequalities can be used to study the convergence of $\displaystyle\sum_{n=2}^{\infty} \frac{1}{n^2+\sqrt{n}}$? Explain.

$$\frac{1}{n^2+\sqrt{n}} \leq \frac{1}{\sqrt{n}}, \qquad \frac{1}{n^2+\sqrt{n}} \leq \frac{1}{n^2}$$

In Exercises 19–30, use the Direct Comparison Test to determine whether the infinite series is convergent.

19. $\displaystyle\sum_{n=1}^{\infty} \frac{1}{n2^n}$

20. $\displaystyle\sum_{n=1}^{\infty} \frac{n^3}{n^5+4n+1}$

21. $\displaystyle\sum_{n=1}^{\infty} \frac{1}{n^{1/3}+2^n}$

22. $\displaystyle\sum_{n=1}^{\infty} \frac{1}{\sqrt{n^3+2n-1}}$

23. $\displaystyle\sum_{m=1}^{\infty} \frac{4}{m!+4^m}$

24. $\displaystyle\sum_{n=4}^{\infty} \frac{\sqrt{n}}{n-3}$

25. $\displaystyle\sum_{k=1}^{\infty} \frac{\sin^2 k}{k^2}$

26. $\displaystyle\sum_{k=2}^{\infty} \frac{k^{1/3}}{k^{5/4}-k}$

27. $\displaystyle\sum_{n=1}^{\infty} \frac{2}{3^n+3^{-n}}$

28. $\displaystyle\sum_{k=1}^{\infty} 2^{-k^2}$

29. $\displaystyle\sum_{n=1}^{\infty} \frac{1}{(n+1)!}$

30. $\displaystyle\sum_{n=1}^{\infty} \frac{n!}{n^3}$

Exercise 31–36: For all $a > 0$ and $b > 1$, the inequalities

$$\ln n \leq n^a, \qquad n^a < b^n$$

are true for n sufficiently large (this can be proved using L'Hôpital's Rule). Use this, together with the Direct Comparison Test, to determine whether the series converges or diverges.

31. $\displaystyle\sum_{n=1}^{\infty} \frac{\ln n}{n^3}$

32. $\displaystyle\sum_{m=2}^{\infty} \frac{1}{\ln m}$

33. $\displaystyle\sum_{n=1}^{\infty} \frac{(\ln n)^{100}}{n^{1.1}}$

34. $\displaystyle\sum_{n=1}^{\infty} \frac{1}{(\ln n)^{10}}$

35. $\displaystyle\sum_{n=1}^{\infty} \frac{n}{3^n}$

36. $\displaystyle\sum_{n=1}^{\infty} \frac{n^5}{2^n}$

37. Show that $\displaystyle\sum_{n=1}^{\infty} \sin \frac{1}{n^2}$ converges. *Hint:* Use $\sin x \leq x$ for $x \geq 0$.

38. Does $\displaystyle\sum_{n=2}^{\infty} \frac{\sin(1/n)}{\ln n}$ converge? *Hint:* By Theorem 3 in Section 2.6, $\sin(1/n) > (\cos(1/n))/n$. Thus, $\sin(1/n) > 1/(2n)$ for $n > 2$ [because $\cos(1/n) > \frac{1}{2}$].

In Exercises 39–48, use the Limit Comparison Test to prove convergence or divergence of the infinite series.

39. $\displaystyle\sum_{n=2}^{\infty} \frac{n^2}{n^4-1}$

40. $\displaystyle\sum_{n=2}^{\infty} \frac{1}{n^2-\sqrt{n}}$

41. $\displaystyle\sum_{n=2}^{\infty} \frac{n}{\sqrt{n^3+1}}$

42. $\displaystyle\sum_{n=2}^{\infty} \frac{n^3}{\sqrt{n^7+2n^2+1}}$

43. $\displaystyle\sum_{n=3}^{\infty} \frac{3n+5}{n(n-1)(n-2)}$

44. $\displaystyle\sum_{n=1}^{\infty} \frac{e^n+n}{e^{2n}-n^2}$

45. $\displaystyle\sum_{n=1}^{\infty} \frac{1}{\sqrt{n} + \ln n}$

46. $\displaystyle\sum_{n=1}^{\infty} \frac{\ln(n+4)}{n^{5/2}}$

47. $\displaystyle\sum_{n=1}^{\infty} \left(1 - \cos\frac{1}{n}\right)$ *Hint: Compare with* $\displaystyle\sum_{n=1}^{\infty} n^{-2}$.

48. $\displaystyle\sum_{n=1}^{\infty} (1 - 2^{-1/n})$ *Hint: Compare with the harmonic series.*

In Exercises 49–78, determine convergence or divergence using any method covered so far.

49. $\displaystyle\sum_{n=4}^{\infty} \frac{1}{n^2 - 9}$

50. $\displaystyle\sum_{n=1}^{\infty} \frac{\cos^2 n}{n^2}$

51. $\displaystyle\sum_{n=1}^{\infty} \frac{\sqrt{n}}{4n+9}$

52. $\displaystyle\sum_{n=1}^{\infty} \frac{n - \cos n}{n^3}$

53. $\displaystyle\sum_{n=1}^{\infty} \frac{n^2 - n}{n^5 + n}$

54. $\displaystyle\sum_{n=1}^{\infty} \frac{1}{n^2 + \sin n}$

55. $\displaystyle\sum_{n=5}^{\infty} (4/5)^{-n}$

56. $\displaystyle\sum_{n=1}^{\infty} \frac{1}{3^{n^2}}$

57. $\displaystyle\sum_{n=2}^{\infty} \frac{1}{n^{3/2} \ln n}$

58. $\displaystyle\sum_{n=2}^{\infty} \frac{(\ln n)^{12}}{n^{9/8}}$

59. $\displaystyle\sum_{k=1}^{\infty} 4^{1/k}$

60. $\displaystyle\sum_{n=1}^{\infty} \frac{4^n}{5^n - 2n}$

61. $\displaystyle\sum_{n=2}^{\infty} \frac{1}{(\ln n)^4}$

62. $\displaystyle\sum_{n=1}^{\infty} \frac{2^n}{3^n - n}$

63. $\displaystyle\sum_{n=3}^{\infty} \frac{1}{n \ln n - n}$

64. $\displaystyle\sum_{n=3}^{\infty} \frac{1}{n(\ln n)^2 - n}$

65. $\displaystyle\sum_{n=1}^{\infty} \frac{1}{n^n}$

66. $\displaystyle\sum_{n=1}^{\infty} \frac{n^2 - 4n^{3/2}}{n^3}$

67. $\displaystyle\sum_{n=1}^{\infty} \frac{1 + (-1)^n}{n}$

68. $\displaystyle\sum_{n=1}^{\infty} \frac{2 + (-1)^n}{n^{3/2}}$

69. $\displaystyle\sum_{n=1}^{\infty} \sin\frac{1}{n}$

70. $\displaystyle\sum_{n=1}^{\infty} \frac{\sin(1/n)}{\sqrt{n}}$

71. $\displaystyle\sum_{n=1}^{\infty} \frac{2n+1}{4^n}$

72. $\displaystyle\sum_{n=3}^{\infty} \frac{1}{e^{\sqrt{n}}}$

73. $\displaystyle\sum_{n=4}^{\infty} \frac{\ln n}{n^2 - 3n}$

74. $\displaystyle\sum_{n=1}^{\infty} \frac{1}{3^{\ln n}}$

75. $\displaystyle\sum_{n=2}^{\infty} \frac{1}{n^{1/2} \ln n}$

76. $\displaystyle\sum_{n=1}^{\infty} \frac{1}{n^{3/2} - (\ln n)^4}$

77. $\displaystyle\sum_{n=2}^{\infty} \frac{4n^2 + 15n}{3n^4 - 5n^2 - 17}$

78. $\displaystyle\sum_{n=1}^{\infty} \frac{n}{4^{-n} + 5^{-n}}$

79. For which a does $\displaystyle\sum_{n=2}^{\infty} \frac{1}{n(\ln n)^a}$ converge?

80. For which a does $\displaystyle\sum_{n=2}^{\infty} \frac{1}{n^a \ln n}$ converge?

81. For which values of p does $\displaystyle\sum_{n=1}^{\infty} \frac{n^2}{(n^3 + 1)^p}$ converge?

82. For which values of p does $\displaystyle\sum_{n=1}^{\infty} \frac{e^x}{(1 + e^{2x})^p}$ converge?

Approximating Infinite Sums *In Exercises 83–85, let $a_n = f(n)$, where f is a continuous, decreasing function such that $f(x) \geq 0$ and $\int_1^{\infty} f(x)\,dx$ converges.*

83. Show that

$$\int_1^{\infty} f(x)\,dx \leq \sum_{n=1}^{\infty} a_n \leq a_1 + \int_1^{\infty} f(x)\,dx \qquad \boxed{3}$$

84. *CAS* Using Eq. (3), show that

$$5 \leq \sum_{n=1}^{\infty} \frac{1}{n^{1.2}} \leq 6$$

This series converges slowly. Use a computer algebra system to verify that $S_N < 5$ for $N \leq 43{,}128$ and $S_{43{,}129} \approx 5.00000021$.

85. Let $S = \displaystyle\sum_{n=1}^{\infty} a_n$. Arguing as in Exercise 83, show that

$$\sum_{n=1}^{M} a_n + \int_{M+1}^{\infty} f(x)\,dx \leq S \leq \sum_{n=1}^{M+1} a_n + \int_{M+1}^{\infty} f(x)\,dx \qquad \boxed{4}$$

Conclude that

$$0 \leq S - \left(\sum_{n=1}^{M} a_n + \int_{M+1}^{\infty} f(x)\,dx\right) \leq a_{M+1} \qquad \boxed{5}$$

This provides a method for approximating S with an error of at most a_{M+1}.

86. *CAS* Use Eq. (4) from Exercise 85 with $M = 43{,}129$ to prove that

$$5.5915810 \leq \sum_{n=1}^{\infty} \frac{1}{n^{1.2}} \leq 5.5915839$$

87. *CAS* Apply Eq. (4) from Exercise 85 with $M = 40{,}000$ to show that

$$1.644934066 \leq \sum_{n=1}^{\infty} \frac{1}{n^2} \leq 1.644934068$$

Is this consistent with Euler's result, according to which this infinite series has sum $\pi^2/6$?

88. *CAS* Using a CAS and Eq. (5) from Exercise 85, determine the value of $\displaystyle\sum_{n=1}^{\infty} n^{-6}$ to within an error less than 10^{-4}. Check that your result is consistent with that of Euler, who proved that the sum is equal to $\pi^6/945$.

89. *CAS* Using a CAS and Eq. (5) from Exercise 85, determine the value of $\displaystyle\sum_{n=1}^{\infty} n^{-5}$ to within an error less than 10^{-4}.

90. How far can a stack of identical books (of mass m and unit length) extend without tipping over? The stack will not tip over if the $(n + 1)$st book is placed at the bottom of the stack with its right edge located at or before the center of mass of the first n books (Figure 6). Let c_n be the center of mass of the first n books, measured along the x-axis, where we take the positive x-axis to the left of the origin as in Figure 7. Recall that if an object of mass m_1 has center of mass at x_1 and a second object of m_2 has center of mass x_2, then the center of mass of the system has x-coordinate

$$\frac{m_1 x_1 + m_2 x_2}{m_1 + m_2}$$

(a) Show that if the $(n + 1)$st book is placed with its right edge at c_n, then its center of mass is located at $c_n + \frac{1}{2}$.

(b) Consider the first n books as a single object of mass nm with center of mass at c_n and the $(n + 1)$st book as a second object of mass m. Show that if the $(n + 1)$st book is placed with its right edge at c_n, then

$$c_{n+1} = c_n + \frac{1}{2(n + 1)}.$$

(c) Prove that $\lim_{n \to \infty} c_n = \infty$. Thus, by using enough books, the stack can be extended as far as desired without tipping over.

91. The following argument proves the divergence of the harmonic series $S = \sum_{n=1}^{\infty} 1/n$ without using the Integral Test. Let

$$S_1 = 1 + \frac{1}{3} + \frac{1}{5} + \cdots , \qquad S_2 = \frac{1}{2} + \frac{1}{4} + \frac{1}{6} + \cdots$$

Show that if S converges, then

(a) S_1 and S_2 also converge and $S = S_1 + S_2$.

(b) $S_1 > S_2$ and $S_2 = \frac{1}{2}S$.

Observe that (b) contradicts (a), and conclude that S diverges.

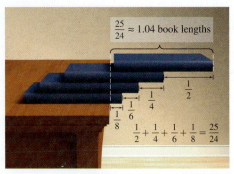

FIGURE 6

$\frac{25}{24} \approx 1.04$ book lengths

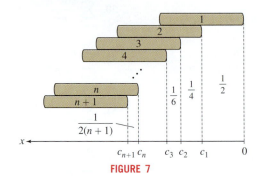

FIGURE 7

Further Insights and Challenges

92. Let $S = \sum_{n=2}^{\infty} a_n$, where $a_n = (\ln(\ln n))^{-\ln n}$.

(a) Show, by taking logarithms, that $a_n = n^{-\ln(\ln(\ln n))}$.

(b) Show that $\ln(\ln(\ln n)) \geq 2$ if $n > C$, where $C = e^{e^{e^2}}$.

(c) Show that S converges.

93. Kummer's Acceleration Method   Suppose we wish to approximate $S = \sum_{n=1}^{\infty} 1/n^2$. There is a similar telescoping series whose value can be computed exactly (Example 2 in Section 10.2):

$$\sum_{n=1}^{\infty} \frac{1}{n(n + 1)} = 1$$

(a) Verify that

$$S = \sum_{n=1}^{\infty} \frac{1}{n(n + 1)} + \sum_{n=1}^{\infty} \left(\frac{1}{n^2} - \frac{1}{n(n + 1)} \right)$$

Thus for M large,

$$S \approx 1 + \sum_{n=1}^{M} \frac{1}{n^2(n + 1)} \qquad \boxed{6}$$

(b) Explain what has been gained. Why is Eq. (6) a better approximation to S than $\sum_{n=1}^{M} 1/n^2$?

(c) **CAS**   Compute

$$\sum_{n=1}^{1000} \frac{1}{n^2}, \qquad 1 + \sum_{n=1}^{100} \frac{1}{n^2(n + 1)}$$

Which is a better approximation to S, whose exact value is $\pi^2/6$?

94. **CAS**   The series $S = \sum_{k=1}^{\infty} k^{-3}$ has been computed to more than 100 million digits. The first 30 digits are

$$S = 1.202056903159594285399738161511$$

Approximate S using Kummer's Acceleration Method of Exercise 93 with $M = 100$ and auxiliary series $R = \sum_{n=1}^{\infty} (n(n + 1)(n + 2))^{-1}$. According to Exercise 54 in Section 10.2, R is a telescoping series with the sum $R = \frac{1}{4}$.

10.4 Absolute and Conditional Convergence

In the previous section, we studied positive series, but we still lack the tools to analyze series with both positive and negative terms. One of the keys to understanding such series is the concept of absolute convergence.

DEFINITION Absolute Convergence The series $\sum a_n$ **converges absolutely** if $\sum |a_n|$ converges.

■ **EXAMPLE 1** Verify that the series

$$\sum_{n=1}^{\infty} \frac{(-1)^{n-1}}{n^2} = \frac{1}{1^2} - \frac{1}{2^2} + \frac{1}{3^2} - \frac{1}{4^2} + \cdots$$

converges absolutely.

Solution This series converges absolutely because the positive series (with absolute values) is a p-series with $p = 2 > 1$:

$$\sum_{n=1}^{\infty} \left| \frac{(-1)^{n-1}}{n^2} \right| = \frac{1}{1^2} + \frac{1}{2^2} + \frac{1}{3^2} + \frac{1}{4^2} + \cdots \qquad \text{(convergent } p\text{-series)} \qquad ■$$

The next theorem tells us that if the series of absolute values converges, then the original series also converges.

THEOREM 1 Absolute Convergence Implies Convergence If $\sum |a_n|$ converges, then $\sum a_n$ also converges.

Proof We have $-|a_n| \leq a_n \leq |a_n|$. By adding $|a_n|$ to all parts of the inequality, we get $0 \leq |a_n| + a_n \leq 2|a_n|$. If $\sum |a_n|$ converges, then $\sum 2|a_n|$ also converges, and therefore, $\sum (a_n + |a_n|)$ converges by the Comparison Test. Our original series converges because it is the difference of two convergent series:

$$\sum a_n = \sum (a_n + |a_n|) - \sum |a_n| \qquad ■$$

■ **EXAMPLE 2** Verify that $S = \sum_{n=1}^{\infty} \frac{(-1)^{n-1}}{n^2}$ converges.

Solution We showed that S converges absolutely in Example 1. By Theorem 1, S itself converges. ■

■ **EXAMPLE 3** Does $S = \sum_{n=1}^{\infty} \frac{(-1)^{n-1}}{\sqrt{n}} = \frac{1}{\sqrt{1}} - \frac{1}{\sqrt{2}} + \frac{1}{\sqrt{3}} - \cdots$ converge absolutely?

Solution The positive series $\sum_{n=1}^{\infty} \frac{1}{\sqrt{n}}$ is a p-series with $p = \frac{1}{2}$. It diverges because $p < 1$. Therefore, S does not converge absolutely. ■

The series in the previous example does not converge *absolutely*, but we still do not know whether or not it converges. A series $\sum a_n$ may converge without converging absolutely. In this case, we say that $\sum a_n$ is conditionally convergent.

> **DEFINITION Conditional Convergence** An infinite series $\sum a_n$ **converges conditionally** if $\sum a_n$ converges but $\sum |a_n|$ diverges.

If a series is not absolutely convergent, how can we determine whether it is conditionally convergent? This is often a more difficult question, because we cannot use the Integral Test or the Comparison Test (they apply only to positive series). However, convergence is guaranteed in the particular case of an **alternating series**

$$S = \sum_{n=1}^{\infty} (-1)^{n-1} b_n = b_1 - b_2 + b_3 - b_4 + \cdots$$

where the terms b_n are positive and decrease to zero (Figure 1).

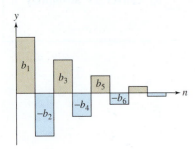

FIGURE 1 An alternating series with decreasing terms. The sum is the signed area, which is at most b_1.

> **THEOREM 2 Alternating Series Test** Assume that $\{b_n\}$ is a positive sequence that is decreasing and converges to 0:
>
> $$b_1 > b_2 > b_3 > b_4 > \cdots > 0, \qquad \lim_{n \to \infty} b_n = 0$$
>
> Then the following alternating series converges:
>
> $$S = \sum_{n=1}^{\infty} (-1)^{n-1} b_n = b_1 - b_2 + b_3 - b_4 + \cdots$$
>
> Furthermore,
>
> $$0 < S < b_1 \quad \text{and} \quad S_{2N} < S < S_{2N+1}, \qquad N \geq 1$$

| **Assumptions Matter** The Alternating Series Test is not valid if we drop the assumption that b_n is decreasing (see Exercise 35).

As we will see, this last fact allows the estimation of such a series to any level of accuracy needed.

Notice that under the same conditions, the series

$$\sum_{n=1}^{\infty} (-1)^{n} b_n = -b_1 + b_2 - b_3 + b_4 - \cdots$$

also converges since it is just -1 times the series appearing in the theorem.

Proof We will prove that the partial sums zigzag above and below the sum S as in Figure 2. Note first that the even partial sums are increasing. Indeed, the odd-numbered terms occur with a plus sign and thus, for example,

$$S_4 + b_5 - b_6 = S_6$$

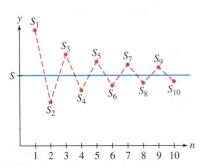

DF **FIGURE 2** The partial sums of an alternating series zigzag above and below the limit. The odd partial sums decrease and the even partial sums increase.

But $b_5 - b_6 > 0$ because b_n is decreasing, and therefore $S_4 < S_6$. In general,

$$S_{2N} + (b_{2N+1} - b_{2N+2}) = S_{2N+2}$$

where $b_{2n+1} - b_{2N+2} > 0$. Thus, $S_{2N} < S_{2N+2}$ and

$$0 < S_2 < S_4 < S_6 < \cdots$$

Similarly,

$$S_{2N-1} - (b_{2N} - b_{2N+1}) = S_{2N+1}$$

Therefore, $S_{2N+1} < S_{2N-1}$, and the sequence of odd partial sums is decreasing:

$$\cdots < S_7 < S_5 < S_3 < S_1$$

Finally, $S_{2N} < S_{2N} + b_{2N+1} = S_{2N+1}$. The picture is as follows:

$$0 < S_2 < S_4 < S_6 < \quad \cdots \quad < S_7 < S_5 < S_3 < S_1$$

Now, because bounded monotonic sequences converge (Theorem 6 of Section 10.1), the even and odd partial sums approach limits that are sandwiched in the middle:

$$0 < S_2 < S_4 < \cdots < \lim_{N\to\infty} S_{2N} \le \lim_{N\to\infty} S_{2N+1} < \cdots < S_5 < S_3 < S_1 \qquad \boxed{1}$$

These two limits must have a common value L because

$$\lim_{N\to\infty} S_{2N+1} - \lim_{N\to\infty} S_{2N} = \lim_{N\to\infty} (S_{2N+1} - S_{2N}) = \lim_{N\to\infty} b_{2N+1} = 0$$

Therefore, $\lim_{N\to\infty} S_N = L$ and the infinite series converges to $S = L$. From Eq. (1) we also see that $0 < S < S_1 = b_1$ and $S_{2N} < S < S_{2N+1}$ for all N as claimed. ∎

The Alternating Series Test is the only test for conditional convergence developed in this text. Other tests, such as Abel's Criterion and the Dirichlet Test, are discussed in textbooks on analysis.

■ **EXAMPLE 4** Show that $S = \displaystyle\sum_{n=1}^{\infty} \frac{(-1)^{n-1}}{\sqrt{n}} = \frac{1}{\sqrt{1}} - \frac{1}{\sqrt{2}} + \frac{1}{\sqrt{3}} - \cdots$ converges conditionally and that $0 \le S \le 1$.

Solution The terms $b_n = 1/\sqrt{n}$ are positive and decreasing, and $\lim_{n\to\infty} b_n = 0$. Therefore, S converges by the Alternating Series Test. Furthermore, $0 \le S \le 1$ because $b_1 = 1$. However, the positive series $\displaystyle\sum_{n=1}^{\infty} 1/\sqrt{n}$ diverges because it is a p-series with $p = \frac{1}{2} < 1$. Thus, S is conditionally convergent but not absolutely convergent (Figure 3). ∎

The inequality $S_{2N} < S < S_{2N+1}$ in Theorem 2 gives us important information about the error; it tells us that $|S_N - S|$ is less than $|S_N - S_{N+1}| = b_{N+1}$ for all N.

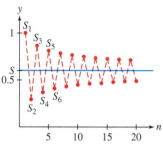

(A) Partial sums of $S = \displaystyle\sum_{n=1}^{\infty} (-1)^{n-1} \frac{1}{\sqrt{n}}$

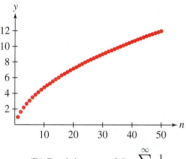

(B) Partial sums of $S = \displaystyle\sum_{n=1}^{\infty} \frac{1}{\sqrt{n}}$

DF **FIGURE 3**

> **THEOREM 3** Let $S = \displaystyle\sum_{n=1}^{\infty} (-1)^{n-1} b_n$, where $\{b_n\}$ is a positive decreasing sequence that converges to 0. Then
>
> $$\boxed{|S - S_N| < b_{N+1}} \qquad \boxed{2}$$
>
> In other words, *when we approximate S by S_N, the error is less than the size of the first omitted term b_{N+1}.*

■ **EXAMPLE 5** **Alternating Harmonic Series** Show that $S = \displaystyle\sum_{n=1}^{\infty} \frac{(-1)^{n-1}}{n}$ converges conditionally. Then:

(a) Show that $|S - S_6| < \frac{1}{7}$.

(b) Find an N such that S_N approximates S with an error less than 10^{-3}.

Solution The terms $b_n = 1/n$ are positive and decreasing, and $\lim_{n\to\infty} b_n = 0$. Therefore, S converges by the Alternating Series Test. The harmonic series $\displaystyle\sum_{n=1}^{\infty} 1/n$ diverges, so S converges conditionally but not absolutely. Now, applying Eq. (2), we have

$$|S - S_N| < b_{N+1} = \frac{1}{N+1}$$

For $N = 6$, we obtain $|S - S_6| < b_7 = \frac{1}{7}$. We can make the error less than 10^{-3} by choosing N so that

$$\frac{1}{N+1} \le 10^{-3} \quad \Rightarrow \quad N + 1 \ge 10^3 \quad \Rightarrow \quad N \ge 999$$

Using a computer algebra system, we find that $S_{999} \approx 0.69365$. In Exercise 86 of Section 10.7, we will prove that $S = \ln 2 \approx 0.69314$, and thus we can verify that

$$|S - S_{999}| \approx |\ln 2 - 0.69365| \approx 0.0005 < 10^{-3} \qquad ■$$

CONCEPTUAL INSIGHT The convergence of an infinite series $\sum a_n$ depends on two factors: (1) how quickly a_n tends to zero, and (2) how much cancellation takes place among the terms. Consider

Harmonic series (diverges): $\qquad 1 + \dfrac{1}{2} + \dfrac{1}{3} + \dfrac{1}{4} + \dfrac{1}{5} + \cdots$

p-Series with $p = 2$ (converges): $\qquad 1 + \dfrac{1}{2^2} + \dfrac{1}{3^2} + \dfrac{1}{4^2} + \dfrac{1}{5^2} + \cdots$

Alternating harmonic series (converges): $1 - \dfrac{1}{2} + \dfrac{1}{3} - \dfrac{1}{4} + \dfrac{1}{5} - \cdots$

The harmonic series diverges because reciprocals $1/n$ do not tend to zero quickly enough. By contrast, the reciprocal squares $1/n^2$ tend to zero quickly enough for the p-series with $p = 2$ to converge. The alternating harmonic series converges, but only due to the cancelation among the terms.

10.4 SUMMARY

- $\sum a_n$ converges *absolutely* if the positive series $\sum |a_n|$ converges.
- Absolute convergence implies convergence: If $\sum |a_n|$ converges, then $\sum a_n$ also converges.
- $\sum a_n$ converges *conditionally* if $\sum a_n$ converges but $\sum |a_n|$ diverges.
- Alternating Series Test: If $\{b_n\}$ is positive and decreasing and $\lim\limits_{n\to\infty} b_n = 0$, then the alternating series

$$S = \sum_{n=1}^{\infty} (-1)^{n-1} b_n = b_1 - b_2 + b_3 - b_4 + b_5 - \cdots$$

converges. Furthermore, $|S - S_N| < b_{N+1}$.
- We have developed two ways to handle nonpositive series: Show absolute convergence if possible, or use the Alternating Series Test if applicable.

10.4 EXERCISES

Preliminary Questions

1. Give an example of a series such that $\sum a_n$ converges but $\sum |a_n|$ diverges.

2. Which of the following statements is equivalent to Theorem 1?

(a) If $\sum\limits_{n=0}^{\infty} |a_n|$ diverges, then $\sum\limits_{n=0}^{\infty} a_n$ also diverges.

(b) If $\sum\limits_{n=0}^{\infty} a_n$ diverges, then $\sum\limits_{n=0}^{\infty} |a_n|$ also diverges.

(c) If $\sum\limits_{n=0}^{\infty} a_n$ converges, then $\sum\limits_{n=0}^{\infty} |a_n|$ also converges.

3. A student argues that $\sum\limits_{n=1}^{\infty} (-1)^n \sqrt{n}$ is an alternating series and therefore converges. Is the student right?

4. Suppose that b_n is positive, decreasing, and tends to 0, and let $S = \sum\limits_{n=1}^{\infty} (-1)^{n-1} b_n$. What can we say about $|S - S_{100}|$ if $a_{101} = 10^{-3}$? Is S larger or smaller than S_{100}?

Exercises

1. Show that

$$\sum_{n=0}^{\infty} \frac{(-1)^n}{2^n}$$

converges absolutely.

2. Show that the following series converges conditionally:

$$\sum_{n=1}^{\infty} (-1)^{n-1} \frac{1}{n^{2/3}} = \frac{1}{1^{2/3}} - \frac{1}{2^{2/3}} + \frac{1}{3^{2/3}} - \frac{1}{4^{2/3}} + \cdots$$

In Exercises 3–10, determine whether the series converges absolutely, conditionally, or not at all.

3. $\displaystyle\sum_{n=1}^{\infty} \frac{(-1)^{n-1}}{n^{1/3}}$

4. $\displaystyle\sum_{n=1}^{\infty} \frac{(-1)^n n^4}{n^3 + 1}$

5. $\displaystyle\sum_{n=0}^{\infty} \frac{(-1)^{n-1}}{(1.1)^n}$

6. $\displaystyle\sum_{n=1}^{\infty} \frac{\sin(\frac{\pi n}{4})}{n^2}$

7. $\displaystyle\sum_{n=2}^{\infty} \frac{(-1)^n}{n \ln n}$

8. $\displaystyle\sum_{n=1}^{\infty} \frac{(-1)^n}{1 + \frac{1}{n}}$

9. $\displaystyle\sum_{n=2}^{\infty} \frac{\cos n\pi}{(\ln n)^2}$

10. $\displaystyle\sum_{n=1}^{\infty} \frac{\cos n}{2^n}$

11. Let $S = \displaystyle\sum_{n=1}^{\infty} (-1)^{n+1} \frac{1}{n^3}$.

(a) Calculate S_n for $1 \le n \le 10$.

(b) Use Eq. (2) to show that $0.9 \le S \le 0.902$.

12. Use Eq. (2) to approximate

$$\sum_{n=1}^{\infty} \frac{(-1)^{n+1}}{n!}$$

to four decimal places.

13. Approximate $\displaystyle\sum_{n=1}^{\infty} \frac{(-1)^{n+1}}{n^4}$ to three decimal places.

14. *CAS* Let

$$S = \sum_{n=1}^{\infty} (-1)^{n-1} \frac{n}{n^2 + 1}$$

Use a computer algebra system to calculate and plot the partial sums S_n for $1 \le n \le 100$. Observe that the partial sums zigzag above and below the limit.

In Exercises 15–16, find a value of N such that S_N approximates the series with an error of at most 10^{-5}. If you have a CAS, compute this value of S_N.

15. $\displaystyle\sum_{n=1}^{\infty} \frac{(-1)^{n+1}}{n(n+2)(n+3)}$

16. $\displaystyle\sum_{n=1}^{\infty} \frac{(-1)^{n+1} \ln n}{n!}$

In Exercises 17–32, determine convergence or divergence by any method.

17. $\displaystyle\sum_{n=0}^{\infty} 7^{-n}$

18. $\displaystyle\sum_{n=1}^{\infty} \frac{1}{n^{7.5}}$

19. $\displaystyle\sum_{n=1}^{\infty} \frac{1}{5^n - 3^n}$

20. $\displaystyle\sum_{n=2}^{\infty} \frac{n}{n^2 - n}$

21. $\displaystyle\sum_{n=1}^{\infty} \frac{1}{3n^4 + 12n}$

22. $\displaystyle\sum_{n=1}^{\infty} \frac{(-1)^n}{\sqrt{n^2 + 1}}$

23. $\displaystyle\sum_{n=1}^{\infty} \frac{1}{\sqrt{n^2 + 1}}$

24. $\displaystyle\sum_{n=0}^{\infty} \frac{(-1)^n n}{\sqrt{n^2 + 1}}$

25. $\displaystyle\sum_{n=1}^{\infty} \frac{3^n + (-2)^n}{5^n}$

26. $\displaystyle\sum_{n=1}^{\infty} \frac{(-1)^{n+1}}{(2n + 1)!}$

27. $\displaystyle\sum_{n=1}^{\infty} (-1)^n n^2 e^{-n^3/3}$

28. $\displaystyle\sum_{n=1}^{\infty} n e^{-n^3/3}$

29. $\displaystyle\sum_{n=2}^{\infty} \frac{(-1)^n}{n^{1/2} (\ln n)^2}$

30. $\displaystyle\sum_{n=2}^{\infty} \frac{1}{n(\ln n)^{1/4}}$

31. $\displaystyle\sum_{n=1}^{\infty} \frac{\ln n}{n^{1.05}}$

32. $\displaystyle\sum_{n=2}^{\infty} \frac{1}{(\ln n)^2}$

33. Show that

$$S = \frac{1}{2} - \frac{1}{2} + \frac{1}{3} - \frac{1}{3} + \frac{1}{4} - \frac{1}{4} + \cdots$$

converges by computing the partial sums. Does it converge absolutely?

34. The Alternating Series Test cannot be applied to

$$\frac{1}{2} - \frac{1}{3} + \frac{1}{2^2} - \frac{1}{3^2} + \frac{1}{2^3} - \frac{1}{3^3} + \cdots$$

Why not? Show that it converges by another method.

35. ✏️ **Assumptions Matter** Show by counterexample that the Alternating Series Test does not remain true if the sequence a_n tends to zero but is not assumed to be decreasing. *Hint:* Consider

$$R = \frac{1}{2} - \frac{1}{4} + \frac{1}{3} - \frac{1}{8} + \frac{1}{4} - \frac{1}{16} + \cdots + \left(\frac{1}{n} - \frac{1}{2^n}\right) + \cdots$$

36. Determine whether the following series converges conditionally:

$$1 - \frac{1}{3} + \frac{1}{2} - \frac{1}{5} + \frac{1}{3} - \frac{1}{7} + \frac{1}{4} - \frac{1}{9} + \frac{1}{5} - \frac{1}{11} + \cdots$$

37. Prove that if $\sum a_n$ converges absolutely, then $\sum a_n^2$ also converges. Give an example where $\sum a_n$ is only conditionally convergent and $\sum a_n^2$ diverges.

Further Insights and Challenges

38. Prove the following variant of the Alternating Series Test: If $\{b_n\}$ is a positive, decreasing sequence with $\displaystyle\lim_{n\to\infty} b_n = 0$, then the series

$$b_1 + b_2 - 2b_3 + a_4 + b_5 - 2a_6 + \cdots$$

converges. *Hint:* Show that S_{3N} is increasing and bounded by $a_1 + a_2$, and continue as in the proof of the Alternating Series Test.

39. Use Exercise 38 to show that the following series converges:

$$S = \frac{1}{\ln 2} + \frac{1}{\ln 3} - \frac{2}{\ln 4} + \frac{1}{\ln 5} + \frac{1}{\ln 6} - \frac{2}{\ln 7} + \cdots$$

40. Prove the conditional convergence of

$$R = 1 + \frac{1}{2} + \frac{1}{3} - \frac{3}{4} + \frac{1}{5} + \frac{1}{6} + \frac{1}{7} - \frac{3}{8} + \cdots$$

41. Show that the following series diverges:

$$S = 1 + \frac{1}{2} + \frac{1}{3} - \frac{2}{4} + \frac{1}{5} + \frac{1}{6} + \frac{1}{7} - \frac{2}{8} + \cdots$$

Hint: Use the result of Exercise 40 to write S as the sum of a convergent series and a divergent series.

42. Prove that

$$\sum_{n=1}^{\infty} (-1)^{n+1} \frac{(\ln n)^a}{n}$$

converges for all exponents a. *Hint:* Show that $f(x) = (\ln x)^a / x$ is decreasing for x sufficiently large.

43. We say that $\{b_n\}$ is a rearrangement of $\{a_n\}$ if $\{b_n\}$ has the same terms as $\{a_n\}$ but occurring in a different order. Show that if $\{b_n\}$ is a rearrangement of $\{a_n\}$ and $S = \sum_{n=1}^{\infty} a_n$ converges absolutely, then $T = \sum_{n=1}^{\infty} b_n$ also converges absolutely. (This result does not hold if S is only conditionally convergent.) *Hint:* Prove that the partial sums $\sum_{n=1}^{N} |b_n|$ are bounded. It can be shown further that $S = T$.

44. Assumptions Matter   In 1829 Lejeune Dirichlet pointed out that the great French mathematician Augustin Louis Cauchy made a mistake in a published paper by improperly assuming the Limit Comparison Test to be valid for nonpositive series. Here are Dirichlet's two series:

$$\sum_{n=1}^{\infty} \frac{(-1)^n}{\sqrt{n}}, \qquad \sum_{n=1}^{\infty} \frac{(-1)^n}{\sqrt{n}} \left(1 + \frac{(-1)^n}{\sqrt{n}}\right)$$

Explain how they provide a counterexample to the Limit Comparison Test when the series are not assumed to be positive.

10.5 The Ratio and Root Tests and Strategies for Choosing Tests

Series such as

$$S = 1 + \frac{2}{1!} + \frac{2^2}{2!} + \frac{2^3}{3!} + \frac{2^4}{4!} + \cdots$$

arise in applications, but the convergence tests developed so far cannot be applied easily. Fortunately, the Ratio Test can be used for this and many other series.

THEOREM 1 Ratio Test   Assume that the following limit exists:

$$\rho = \lim_{n \to \infty} \left| \frac{a_{n+1}}{a_n} \right|$$

The symbol ρ is a lowercase rho, the seventeenth letter of the Greek alphabet.

(i) If $\rho < 1$, then $\sum a_n$ converges absolutely.

(ii) If $\rho > 1$, then $\sum a_n$ diverges.

(iii) If $\rho = 1$, the test is inconclusive (the series may converge or diverge).

Proof   The idea is to compare with a geometric series. If $\rho < 1$, we may choose a number r such that $\rho < r < 1$. Since $|a_{n+1}/a_n|$ converges to ρ, there exists a number M such that $|a_{n+1}/a_n| < r$ for all $n \geq M$. Therefore,

$$|a_{M+1}| < r|a_M|$$

$$|a_{M+2}| < r|a_{M+1}| < r(r|a_M|) = r^2|a_M|$$

$$|a_{M+3}| < r|a_{M+2}| < r^3|a_M|$$

In general, $|a_{M+n}| < r^n|a_M|$, and thus,

$$\sum_{n=M}^{\infty} |a_n| = \sum_{n=0}^{\infty} |a_{M+n}| \leq \sum_{n=0}^{\infty} |a_M| r^n = |a_M| \sum_{n=0}^{\infty} r^n$$

The geometric series on the right converges because $0 < r < 1$, so $\sum_{n=M}^{\infty} |a_n|$ converges by the Direct Comparison Test and thus $\sum a_n$ converges absolutely.

If $\rho > 1$, choose r such that $1 < r < \rho$. Then there exists a number M such that $|a_{n+1}/a_n| > r$ for all $n \geq M$. Arguing as before with the inequalities reversed, we find that $|a_{M+n}| \geq r^n|a_M|$. Since r^n tends to ∞, the terms a_{M+n} do not tend to zero, and consequently, $\sum a_n$ diverges. Finally, Example 4 below shows that both convergence and divergence are possible when $\rho = 1$, so the test is inconclusive in this case. ∎

■ **EXAMPLE 1** Prove that $\displaystyle\sum_{n=1}^{\infty} \frac{2^n}{n!}$ converges.

Solution Compute the ratio and its limit with $a_n = \dfrac{2^n}{n!}$. Note that $(n+1)! = (n+1)n!$. Thus,

$$\left| \frac{a_{n+1}}{a_n} \right| = \frac{2^{n+1}}{(n+1)!} \frac{n!}{2^n} = \frac{2^{n+1}}{2^n} \frac{n!}{(n+1)!} = \frac{2}{n+1}$$

We obtain

$$\rho = \lim_{n\to\infty} \left| \frac{a_{n+1}}{a_n} \right| = \lim_{n\to\infty} \frac{2}{n+1} = 0$$

Since $\rho < 1$, the series $\displaystyle\sum_{n=1}^{\infty} \frac{2^n}{n!}$ converges by the Ratio Test. ■

■ **EXAMPLE 2** Does $\displaystyle\sum_{n=1}^{\infty} \frac{n^2}{2^n}$ converge?

Solution Apply the Ratio Test with $a_n = \dfrac{n^2}{2^n}$:

$$\left| \frac{a_{n+1}}{a_n} \right| = \frac{(n+1)^2}{2^{n+1}} \frac{2^n}{n^2} = \frac{1}{2}\left(\frac{n^2 + 2n + 1}{n^2} \right) = \frac{1}{2}\left(1 + \frac{2}{n} + \frac{1}{n^2} \right)$$

We obtain

$$\rho = \lim_{n\to\infty} \left| \frac{a_{n+1}}{a_n} \right| = \frac{1}{2} \lim_{n\to\infty} \left(1 + \frac{2}{n} + \frac{1}{n^2} \right) = \frac{1}{2}$$

Since $\rho < 1$, the series converges by the Ratio Test. ■

■ **EXAMPLE 3** Does $\displaystyle\sum_{n=0}^{\infty} (-1)^n \frac{n!}{1000^n}$ converge?

Solution This series diverges by the Ratio Test because $\rho > 1$:

$$\rho = \lim_{n\to\infty} \left| \frac{a_{n+1}}{a_n} \right| = \lim_{n\to\infty} \frac{(n+1)!}{1000^{n+1}} \frac{1000^n}{n!} = \lim_{n\to\infty} \frac{n+1}{1000} = \infty$$ ■

■ **EXAMPLE 4** **Ratio Test Inconclusive** Show that both convergence and divergence are possible when $\rho = 1$ by considering $\displaystyle\sum_{n=1}^{\infty} n^2$ and $\displaystyle\sum_{n=1}^{\infty} n^{-2}$.

Solution For $a_n = n^2$, we have

$$\rho = \lim_{n\to\infty} \left| \frac{a_{n+1}}{a_n} \right| = \lim_{n\to\infty} \frac{(n+1)^2}{n^2} = \lim_{n\to\infty} \frac{n^2 + 2n + 1}{n^2} = \lim_{n\to\infty} \left(1 + \frac{2}{n} + \frac{1}{n^2} \right) = 1$$

On the other hand, for $b_n = n^{-2}$,

$$\rho = \lim_{n\to\infty} \left| \frac{b_{n+1}}{b_n} \right| = \lim_{n\to\infty} \left| \frac{a_n}{a_{n+1}} \right| = \frac{1}{\displaystyle\lim_{n\to\infty} \left| \frac{a_{n+1}}{a_n} \right|} = 1$$

Thus, $\rho = 1$ in both cases, but, in fact, $\displaystyle\sum_{n=1}^{\infty} n^2$ diverges by the nth Term Divergence Test since $\displaystyle\lim_{n\to\infty} n^2 = \infty$, and $\displaystyle\sum_{n=1}^{\infty} n^{-2}$ converges since it is a p-series with $p = 2 > 1$. This shows that both convergence and divergence are possible when $\rho = 1$. ■

Our next test is based on the limit of the nth roots $\sqrt[n]{|a_n|}$ rather than the ratios a_{n+1}/a_n. Its proof, like that of the Ratio Test, is based on a comparison with a geometric series (see Exercise 63).

THEOREM 2 Root Test   Assume that the following limit exists:

$$L = \lim_{n\to\infty} \sqrt[n]{|a_n|}$$

(i) If $L < 1$, then $\sum a_n$ converges absolutely.

(ii) If $L > 1$, then $\sum a_n$ diverges.

(iii) If $L = 1$, the test is inconclusive (the series may converge or diverge).

■ **EXAMPLE 5** Does $\displaystyle\sum_{n=1}^{\infty} \left(\frac{n}{2n+3}\right)^n$ converge?

Solution We have $L = \lim_{n\to\infty} \sqrt[n]{a_n} = \lim_{n\to\infty} \dfrac{n}{2n+3} = \dfrac{1}{2}$. Since $L < 1$, the series converges by the Root Test. ■

Determining Which Test to Apply

We end this section with a brief review of all of the tests we have introduced for determining convergence so far and how one decides which test to apply.

Let $\displaystyle\sum_{n=1}^{\infty} a_n$ be given. Keep in mind that the series for which convergence or divergence is known include the geometric series $\displaystyle\sum_{n=0}^{\infty} ar^n$, which converge for $|r| < 1$, and the p-series $\displaystyle\sum_{n=0}^{\infty} \frac{1}{n^p}$, which converge for $p > 1$.

1. The nth Term Divergence Test Always check this test first. If $\lim_{n\to\infty} a_n \neq 0$, then the series diverges. But if $\lim_{n\to\infty} a_n = 0$, we do not know whether the series converges or diverges, and hence we move on to the next step.

2. Positive Series If all terms in the series are positive, try one of the following tests:

(a) The Direct Comparison Test Consider whether dropping terms in the numerator or denominator gives a series that we know either converges or diverges. If a larger series converges or a smaller series diverges, then the original series does the same. For example, $\displaystyle\sum_{n=1}^{\infty} \frac{1}{n^2 + \sqrt{n}}$ converges because $\dfrac{1}{n^2 + \sqrt{n}} < \dfrac{1}{n^2}$, and $\displaystyle\sum_{n=1}^{\infty} \frac{1}{n^2}$ is a p-series with $p = 2 > 1$ so it converges. On the other hand, this does not work for $\displaystyle\sum_{n=2}^{\infty} \frac{1}{n^2 - \sqrt{n}}$ since then the comparison series $\displaystyle\sum_{n=1}^{\infty} \frac{1}{n^2}$, while still converging, is smaller than the original series, so the Direct Comparison Test does not apply. In this case, we can often apply the Limit Comparison Test as follows.

(b) The Limit Comparison Test Consider the dominant term in the numerator and denominator, and compare the original series to the ratio of those terms. For example, for $\displaystyle\sum_{n=2}^{\infty} \frac{1}{n^2 - \sqrt{n}}$, n^2 is dominant over $\sqrt{n}$ as it grows faster as n increases. So, we let $b_n = \dfrac{1}{n^2}$. Then

$$\lim_{n\to\infty} \frac{a_n}{b_n} = \lim_{n\to\infty} \frac{\frac{1}{n^2 - \sqrt{n}}}{\frac{1}{n^2}} = \lim_{n\to\infty} \frac{n^2}{n^2 - \sqrt{n}} = 1$$

The limit is a positive number, so the Limit Comparison Test applies. Since $\displaystyle\sum_{n=1}^{\infty} \frac{1}{n^2}$ converges, so does the original series.

(c) The Ratio Test The Ratio Test is often effective in the presence of a factorial such as $n!$ since in the ratio, the factorial disappears after cancellation. It is also effective when there are constants to the power n, such as 2^n, since in the ratio, the power n disappears after cancelation. For example, if the series is $\displaystyle\sum_{n=1}^{\infty} \frac{3^n}{n!}$, then applying the Ratio Test yields

$$\lim_{n\to\infty} \left| \frac{a_{n+1}}{a_n} \right| = \lim_{n\to\infty} \frac{\frac{3^{n+1}}{(n+1)!}}{\frac{3^n}{(n)!}} = \lim_{n\to\infty} \frac{3}{n+1} = 0 < 1$$

Therefore, the series converges.

(d) The Root Test The Root Test is often effective when there is a term of the form $f(n)^{g(n)}$. For example, $\displaystyle\sum_{n=1}^{\infty} \frac{2^n}{n^{2n}}$ is a good example since applying the Root Test yields

$$\lim_{n\to\infty} |a_n|^{1/n} = \lim_{n\to\infty} \left(\frac{2^n}{n^{2n}} \right)^{1/n} = \lim_{n\to\infty} \frac{2}{n^2} = 0 < 1$$

Thus, the series converges.

(e) The Integral Test When the other tests fail on a positive series, consider the Integral Test. If $a_n = f(n)$ is a decreasing function, then the series converges if and only if the improper integral $\displaystyle\int_1^{\infty} f(x)\,dx$ converges. For example, the other tests do not easily apply to $\displaystyle\sum_{n=2}^{\infty} \frac{1}{n \ln n}$. However $f(x) = \dfrac{1}{n \ln n}$ is a decreasing function and

$$\int_2^{\infty} \frac{1}{x \ln x}\,dx = \ln(\ln x)\Big|_2^{\infty} = \infty.$$ Thus, the integral and, hence, the series diverges.

3. Series That Are Not Positive Series

(a) Alternating Series Test If the series is alternating of the form $\displaystyle\sum_{n=1}^{\infty} (-1)^{n-1} b_n$, show that $b_{n+1} < b_n$ and $\displaystyle\lim_{n\to\infty} b_n = 0$. Then the Alternating Series Test shows the series converges.

(b) Absolute Convergence If the series is not alternating, then see if its absolute value, which is a positive series, converges using the tests for positive series. If so, the original series converges as well.

10.5 SUMMARY

- Ratio Test: Assume that $\rho = \displaystyle\lim_{n\to\infty} \left| \frac{a_{n+1}}{a_n} \right|$ exists. Then $\displaystyle\sum a_n$

    – Converges absolutely if $\rho < 1$.
    – Diverges if $\rho > 1$.
    – Inconclusive if $\rho = 1$.

- Root Test: Assume that $L = \displaystyle\lim_{n\to\infty} \sqrt[n]{|a_n|}$ exists. Then $\displaystyle\sum a_n$

    – Converges absolutely if $L < 1$.
    – Diverges if $L > 1$.
    – Inconclusive if $L = 1$.

10.5 EXERCISES

Preliminary Questions

1. In the Ratio Test, is ρ equal to $\lim\limits_{n\to\infty}\left|\dfrac{a_{n+1}}{a_n}\right|$ or $\lim\limits_{n\to\infty}\left|\dfrac{a_n}{a_{n+1}}\right|$?

2. Is the Ratio Test conclusive for $\sum\limits_{n=1}^{\infty}\dfrac{1}{2^n}$? Is it conclusive for $\sum\limits_{n=1}^{\infty}\dfrac{1}{n}$?

3. Can the Ratio Test be used to show convergence if the series is only conditionally convergent?

Exercises

In Exercises 1–20, apply the Ratio Test to determine convergence or divergence, or state that the Ratio Test is inconclusive.

1. $\sum\limits_{n=1}^{\infty}\dfrac{1}{5^n}$

2. $\sum\limits_{n=1}^{\infty}\dfrac{(-1)^{n-1}n}{5^n}$

3. $\sum\limits_{n=1}^{\infty}\dfrac{1}{n^n}$

4. $\sum\limits_{n=0}^{\infty}\dfrac{3n+2}{5n^3+1}$

5. $\sum\limits_{n=1}^{\infty}\dfrac{n}{n^2+1}$

6. $\sum\limits_{n=1}^{\infty}\dfrac{2^n}{n}$

7. $\sum\limits_{n=1}^{\infty}\dfrac{2^n}{n^{100}}$

8. $\sum\limits_{n=1}^{\infty}\dfrac{n^3}{3^{n^2}}$

9. $\sum\limits_{n=1}^{\infty}\dfrac{10^n}{2^{n^2}}$

10. $\sum\limits_{n=1}^{\infty}\dfrac{e^n}{n!}$

11. $\sum\limits_{n=1}^{\infty}\dfrac{e^n}{n^n}$

12. $\sum\limits_{n=1}^{\infty}\dfrac{n^{40}}{n!}$

13. $\sum\limits_{n=0}^{\infty}\dfrac{n!}{6^n}$

14. $\sum\limits_{n=1}^{\infty}\dfrac{n!}{n^9}$

15. $\sum\limits_{n=2}^{\infty}\dfrac{1}{n\ln n}$

16. $\sum\limits_{n=1}^{\infty}\dfrac{1}{(2n)!}$

17. $\sum\limits_{n=1}^{\infty}\dfrac{n^2}{(2n+1)!}$

18. $\sum\limits_{n=1}^{\infty}\dfrac{(n!)^3}{(3n)!}$

19. $\sum\limits_{n=2}^{\infty}\dfrac{1}{2^n+1}$

20. $\sum\limits_{n=2}^{\infty}\dfrac{1}{\ln n}$

21. Show that $\sum\limits_{n=1}^{\infty}n^k\,3^{-n}$ converges for all exponents k.

22. Show that $\sum\limits_{n=1}^{\infty}n^2x^n$ converges if $|x|<1$.

23. Show that $\sum\limits_{n=1}^{\infty}2^nx^n$ converges if $|x|<\frac{1}{2}$.

24. Show that $\sum\limits_{n=1}^{\infty}\dfrac{r^n}{n!}$ converges for all r.

25. Show that $\sum\limits_{n=1}^{\infty}\dfrac{r^n}{n}$ converges if $|r|<1$.

26. Is there any value of k such that $\sum\limits_{n=1}^{\infty}\dfrac{2^n}{n^k}$ converges?

27. Show that $\sum\limits_{n=1}^{\infty}\dfrac{n!}{n^n}$ converges. *Hint:* Use $\lim\limits_{n\to\infty}\left(1+\dfrac{1}{n}\right)^n=e$.

In Exercises 28–33, assume that $|a_{n+1}/a_n|$ converges to $\rho=\frac{1}{3}$. What can you say about the convergence of the given series?

28. $\sum\limits_{n=1}^{\infty}na_n$

29. $\sum\limits_{n=1}^{\infty}n^3a_n$

30. $\sum\limits_{n=1}^{\infty}2^na_n$

31. $\sum\limits_{n=1}^{\infty}3^na_n$

32. $\sum\limits_{n=1}^{\infty}4^na_n$

33. $\sum\limits_{n=1}^{\infty}a_n^2$

34. Assume that $|a_{n+1}/a_n|$ converges to $\rho=4$. Does $\sum\limits_{n=1}^{\infty}a_n^{-1}$ converge (assume that $a_n\neq0$ for all n)?

35. Is the Ratio Test conclusive for the p-series $\sum\limits_{n=1}^{\infty}\dfrac{1}{n^p}$?

In Exercises 36–41, use the Root Test to determine convergence or divergence (or state that the test is inconclusive).

36. $\sum\limits_{n=0}^{\infty}\dfrac{1}{10^n}$

37. $\sum\limits_{n=1}^{\infty}\dfrac{1}{n^n}$

38. $\sum\limits_{k=0}^{\infty}\left(\dfrac{k}{k+10}\right)^k$

39. $\sum\limits_{k=0}^{\infty}\left(\dfrac{k}{3k+1}\right)^k$

40. $\sum\limits_{n=1}^{\infty}\left(1+\dfrac{1}{n}\right)^{-n}$

41. $\sum\limits_{n=4}^{\infty}\left(1+\dfrac{1}{n}\right)^{-n^2}$

42. Prove that $\sum\limits_{n=1}^{\infty}\dfrac{2^{n^2}}{n!}$ diverges. *Hint:* Use $2^{n^2}=(2^n)^n$ and $n!\leq n^n$.

In Exercises 43–62, determine convergence or divergence using any method covered in the text so far.

43. $\sum\limits_{n=1}^{\infty}\dfrac{2^n+4^n}{7^n}$

44. $\sum\limits_{n=1}^{\infty}\dfrac{n^3}{n!}$

45. $\sum\limits_{n=1}^{\infty}\dfrac{n}{2n+1}$

46. $\sum\limits_{n=1}^{\infty}2^{1/n}$

47. $\sum\limits_{n=1}^{\infty}\dfrac{\sin n}{n^2}$

48. $\sum\limits_{n=1}^{\infty}\dfrac{n!}{(2n)!}$

49. $\sum\limits_{n=1}^{\infty}\dfrac{1}{n+\sqrt{n}}$

50. $\sum\limits_{n=2}^{\infty}\dfrac{1}{n(\ln n)^3}$

51. $\sum\limits_{n=1}^{\infty}\dfrac{n^3}{5^n}$

52. $\sum\limits_{n=2}^{\infty}\dfrac{1}{n(\ln n)^3}$

53. $\sum\limits_{n=2}^{\infty}\dfrac{1}{\sqrt{n^3-n^2}}$

54. $\sum\limits_{n=1}^{\infty}\dfrac{n^2+4n}{3n^4+9}$

55. $\sum\limits_{n=1}^{\infty}n^{-0.8}$

56. $\sum\limits_{n=1}^{\infty}(0.8)^{-n}n^{-0.8}$

57. $\sum\limits_{n=1}^{\infty}4^{-2n+1}$

58. $\sum\limits_{n=1}^{\infty}\dfrac{(-1)^{n-1}}{\sqrt{n}}$

59. $\sum\limits_{n=1}^{\infty}\sin\dfrac{1}{n^2}$

60. $\sum\limits_{n=1}^{\infty}(-1)^n\cos\dfrac{1}{n}$

61. $\sum\limits_{n=1}^{\infty}\dfrac{(-2)^n}{\sqrt{n}}$

62. $\sum\limits_{n=1}^{\infty}\left(\dfrac{n}{n+12}\right)^n$

Further Insights and Challenges

63. ![icon] **Proof of the Root Test** Let $S = \sum\limits_{n=0}^{\infty} a_n$ be a positive series, and assume that $L = \lim\limits_{n\to\infty} \sqrt[n]{a_n}$ exists.

(a) Show that S converges if $L < 1$. *Hint:* Choose R with $L < R < 1$ and show that $a_n \leq R^n$ for n sufficiently large. Then compare with the geometric series $\sum R^n$.

(b) Show that S diverges if $L > 1$.

64. Show that the Ratio Test does not apply, but verify convergence using the Direct Comparison Test for the series

$$\frac{1}{2} + \frac{1}{3^2} + \frac{1}{2^3} + \frac{1}{3^4} + \frac{1}{2^5} + \cdots$$

65. Let $S = \sum\limits_{n=1}^{\infty} \frac{c^n n!}{n^n}$, where c is a constant.

(a) Prove that S converges absolutely if $|c| < e$ and diverges if $|c| > e$.

(b) It is known that $\lim\limits_{n\to\infty} \frac{e^n n!}{n^{n+1/2}} = \sqrt{2\pi}$. Verify this numerically.

(c) Use the Limit Comparison Test to prove that S diverges for $c = e$.

10.6 Power Series

A **power series** with center c is an infinite series

$$F(x) = \sum_{n=0}^{\infty} a_n(x - c)^n = a_0 + a_1(x - c) + a_2(x - c)^2 + a_3(x - c)^3 + \cdots$$

where x is a variable. For example,

$$F(x) = 1 + (x - 2) + 2(x - 2)^2 + 3(x - 2)^3 + \cdots \qquad \boxed{1}$$

is a power series with center $c = 2$.

Many functions that arise in applications can be represented as power series. This includes not only the familiar trigonometric, exponential, logarithm, and root functions, but also the host of "special functions" of physics and engineering such as Bessel functions and elliptic functions.

A power series $F(x) = \sum\limits_{n=0}^{\infty} a_n(x - c)^n$ converges for some values of x and may diverge for others. For example, if we set $x = \frac{9}{4}$ in the power series of Eq. (1), we obtain the infinite series

$$F\left(\frac{9}{4}\right) = 1 + \left(\frac{9}{4} - 2\right) + 2\left(\frac{9}{4} - 2\right)^2 + 3\left(\frac{9}{4} - 2\right)^3 + \cdots$$

$$= 1 + \left(\frac{1}{4}\right) + 2\left(\frac{1}{4}\right)^2 + 3\left(\frac{1}{4}\right)^3 + \cdots + n\left(\frac{1}{4}\right)^n + \cdots$$

This converges by the Ratio Test:

$$\lim_{n\to\infty} \left|\frac{a_{n+1}}{a_n}\right| = \lim_{n\to\infty} \left|\frac{\frac{n+1}{4^{n+1}}}{\frac{n}{4^n}}\right| = \lim_{n\to\infty} \frac{1}{4}\left(\frac{n+1}{n}\right)\frac{1/n}{1/n} = \lim_{n\to\infty} \frac{1}{4}\left(\frac{1+1/n}{1}\right) = \frac{1}{4}$$

On the other hand, the power series in Eq. (1) diverges for $x = 3$ by the nth Term Test:

$$F(3) = 1 + (3 - 2) + 2(3 - 2)^2 + 3(3 - 2)^3 + \cdots$$

$$= 1 + 1 + 2 + 3 + \cdots$$

There is a surprisingly simple way to describe the set of values x at which a power series $F(x)$ converges. According to our next theorem, either $F(x)$ converges absolutely for all values of x or there is a radius of convergence R such that

> $F(x)$ converges absolutely when $|x - c| < R$ and diverges when $|x - c| > R$.

This means that $F(x)$ converges for x in an **interval of convergence** consisting of the open interval $(c - R, c + R)$ and possibly one or both of the endpoints $c - R$ and $c + R$ (Figure 1). Note that $F(x)$ automatically converges at $x = c$ because

$$F(c) = a_0 + a_1(c - c) + a_2(c - c)^2 + a_3(c - c)^3 + \cdots = a_0$$

We set $R = 0$ if $F(x)$ converges only for $x = c$, and we set $R = \infty$ if $F(x)$ converges for all values of x.

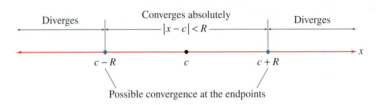

FIGURE 1 Interval of convergence of a power series.

THEOREM 1 Radius of Convergence Every power series

$$F(x) = \sum_{n=0}^{\infty} a_n (x - c)^n$$

has a radius of convergence R, which is either a nonnegative number ($R \geq 0$) or infinity ($R = \infty$). If R is finite, $F(x)$ converges absolutely when $|x - c| < R$ and diverges when $|x - c| > R$. If $R = \infty$, then $F(x)$ converges absolutely for all x.

Proof We assume that $c = 0$ to simplify the notation. If $F(x)$ converges only at $x = 0$, then $R = 0$. Otherwise, $F(x)$ converges for some nonzero value $x = B$. We claim that $F(x)$ must then converge absolutely for all $|x| < |B|$. To prove this, note that because

$$F(B) = \sum_{n=0}^{\infty} a_n B^n$$ converges, the general term $a_n B^n$ tends to zero. In particular, there exists $M > 0$ such that $|a_n B^n| < M$ for all n. Therefore,

$$\sum_{n=0}^{\infty} |a_n x^n| = \sum_{n=0}^{\infty} |a_n B^n| \left| \frac{x}{B} \right|^n < M \sum_{n=0}^{\infty} \left| \frac{x}{B} \right|^n$$

If $|x| < |B|$, then $|x/B| < 1$ and the series on the right is a convergent geometric series. By the Direct Comparison Test, the series on the left also converges. This proves that $F(x)$ converges absolutely if $|x| < |B|$.

Now let S be the set of numbers x such that $F(x)$ converges. Then S contains 0, and we have shown that if S contains a number $B \neq 0$, then S contains the open interval $(-|B|, |B|)$. If S is bounded, then S has a least upper bound $L > 0$ (see marginal note). In this case, there exist numbers $B \in S$ smaller than but arbitrarily close to L, and thus, S contains $(-B, B)$ for all $0 < B < L$. It follows that S contains the open interval $(-L, L)$. The set S cannot contain any number x with $|x| > L$, but S may contain one or both of the endpoints $x = \pm L$. So in this case, F has radius of convergence $R = L$. If S is not bounded, then S contains intervals $(-B, B)$ for B arbitrarily large. In this case, S is the entire real line **R**, and the radius of convergence is $R = \infty$. ■

Least Upper Bound Property: If S is a set of real numbers with an upper bound M (i.e., $x \leq M$ for all $x \in S$), then S has a least upper bound L. See Appendix B.

From Theorem 1, we see that there are two steps in determining the interval of convergence of F:

Step 1. Find the radius of convergence R (using the Ratio Test, in most cases).

Step 2. Check convergence at the endpoints (if $R \neq 0$ or ∞).

■ **EXAMPLE 1** **Using the Ratio Test** Where does $F(x) = \sum_{n=0}^{\infty} \frac{x^n}{2^n}$ converge?

Solution

Step 1. **Find the radius of convergence.**

Let $a_n = \frac{x^n}{2^n}$ and compute ρ from the Ratio Test:

$$\rho = \lim_{n\to\infty} \left| \frac{a_{n+1}}{a_n} \right| = \lim_{n\to\infty} \left| \frac{x^{n+1}}{2^{n+1}} \right| \cdot \left| \frac{2^n}{x^n} \right| = \lim_{n\to\infty} \frac{1}{2}|x| = \frac{1}{2}|x|$$

We find that

$$\rho < 1 \quad \text{if} \quad \frac{1}{2}|x| < 1, \quad \text{that is, if} \quad |x| < 2$$

Thus, $F(x)$ converges if $|x| < 2$. Similarly, $\rho > 1$ if $\frac{1}{2}|x| > 1$, or $|x| > 2$. So, $F(x)$ diverges if $|x| > 2$. Therefore, the radius of convergence is $R = 2$.

Step 2. **Check the endpoints.**

The Ratio Test is inconclusive for $x = \pm 2$, so we must check these cases directly:

$$F(2) = \sum_{n=0}^{\infty} \frac{2^n}{2^n} = 1 + 1 + 1 + 1 + 1 + 1 \cdots$$

$$F(-2) = \sum_{n=0}^{\infty} \frac{(-2)^n}{2^n} = 1 - 1 + 1 - 1 + 1 - 1 \cdots$$

Both series diverge. We conclude that $F(x)$ converges only for $|x| < 2$ (Figure 2). ■

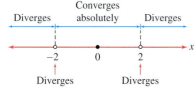

Converges
Diverges   absolutely   Diverges

-2 0 2

Diverges   Diverges

DF FIGURE 2 The power series

$$\sum_{n=0}^{\infty} \frac{x^n}{2^n}$$

has an interval of convergence $(-2, 2)$.

■ **EXAMPLE 2** Where does $F(x) = \sum_{n=1}^{\infty} \frac{(-1)^n}{4^n n}(x-5)^n$ converge?

Solution We compute ρ with $a_n = \frac{(-1)^n}{4^n n}(x-5)^n$:

$$\rho = \lim_{n\to\infty} \left| \frac{a_{n+1}}{a_n} \right| = \lim_{n\to\infty} \left| \frac{(x-5)^{n+1}}{4^{n+1}(n+1)} \frac{4^n n}{(x-5)^n} \right|$$

$$= |x-5| \lim_{n\to\infty} \left| \frac{n}{4(n+1)} \right|$$

$$= \frac{1}{4}|x-5|$$

We find that

$$\rho < 1 \quad \text{if} \quad \frac{1}{4}|x-5| < 1, \quad \text{that is, if} \quad |x-5| < 4$$

Thus, $F(x)$ converges absolutely on the open interval $(1, 9)$ of radius 4 with center $c = 5$. In other words, the radius of convergence is $R = 4$. Next, we check the endpoints:

$$x = 9: \quad \sum_{n=1}^{\infty} \frac{(-1)^n}{4^n n}(9-5)^n = \sum_{n=1}^{\infty} \frac{(-1)^n}{n} \quad \text{converges (Alternating Series Test)}$$

$$x = 1: \quad \sum_{n=1}^{\infty} \frac{(-1)^n}{4^n n}(-4)^n = \sum_{n=1}^{\infty} \frac{1}{n} \quad \text{diverges (harmonic series)}$$

We conclude that $F(x)$ converges for x in the half-open interval $(1, 9]$ shown in Figure 3.

■

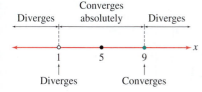

Converges
Diverges   absolutely   Diverges

1 5 9

Diverges   Converges

DF FIGURE 3 The power series

$$\sum_{n=1}^{\infty} \frac{(-1)^n}{4^n n}(x-5)^n$$

has an interval of convergence $(1, 9]$.

Some power series contain only even powers or only odd powers of x. The Ratio Test can still be used to find the radius of convergence.

■ **EXAMPLE 3** **An Even Power Series** Where does $\displaystyle\sum_{n=0}^{\infty} \frac{x^{2n}}{(2n)!}$ converge?

Solution Although this power series has only even powers of x, we can still apply the Ratio Test with $a_n = x^{2n}/(2n)!$. We have

$$a_{n+1} = \frac{x^{2(n+1)}}{(2(n+1))!} = \frac{x^{2n+2}}{(2n+2)!}$$

Furthermore, $(2n+2)! = (2n+2)(2n+1)(2n)!$, so

$$\rho = \lim_{n\to\infty}\left|\frac{a_{n+1}}{a_n}\right| = \lim_{n\to\infty}\frac{x^{2n+2}}{(2n+2)!}\frac{(2n)!}{x^{2n}} = |x|^2 \lim_{n\to\infty}\frac{1}{(2n+2)(2n+1)} = 0$$

Thus, $\rho = 0$ for all x, and $F(x)$ converges for all x. The radius of convergence is $R = \infty$.   ■

| When a function f is represented by a power series on an interval I, we refer to it as the power series expansion of f on I.

Geometric series are important examples of power series. Recall the formula $\displaystyle\sum_{n=0}^{\infty} r^n = 1/(1-r)$, valid for $|r| < 1$. Writing x in place of r, we obtain a power series expansion with radius of convergence $R = 1$:

$$\boxed{\frac{1}{1-x} = \sum_{n=0}^{\infty} x^n \qquad \text{for } |x| < 1}$$ **2**

The next two examples show that we can modify this formula to find the power series expansions of other functions.

■ **EXAMPLE 4** **Geometric Series** Prove that

$$\frac{1}{1-2x} = \sum_{n=0}^{\infty} 2^n x^n \qquad \text{for } |x| < \frac{1}{2}$$

Solution Substitute $2x$ for x in Eq. (2):

$$\frac{1}{1-2x} = \sum_{n=0}^{\infty}(2x)^n = \sum_{n=0}^{\infty} 2^n x^n$$ **3**

Expansion (2) is valid for $|x| < 1$, so Eq. (3) is valid for $|2x| < 1$, or $|x| < \frac{1}{2}$.   ■

■ **EXAMPLE 5** Find a power series expansion with center $c = 0$ for

$$f(x) = \frac{1}{2+x^2}$$

and find the interval of convergence.

Solution We need to rewrite $f(x)$ so we can use Eq. (2). We have

$$\frac{1}{2+x^2} = \frac{1}{2}\left(\frac{1}{1+\frac{1}{2}x^2}\right) = \frac{1}{2}\left(\frac{1}{1-\left(-\frac{1}{2}x^2\right)}\right) = \frac{1}{2}\left(\frac{1}{1-u}\right)$$

where $u = -\frac{1}{2}x^2$. Now substitute $u = -\frac{1}{2}x^2$ for x in Eq. (2) to obtain

$$f(x) = \frac{1}{2+x^2} = \frac{1}{2}\sum_{n=0}^{\infty}\left(-\frac{x^2}{2}\right)^n$$

$$= \sum_{n=0}^{\infty}\frac{(-1)^n x^{2n}}{2^{n+1}}$$

This expansion is valid if $|-x^2/2| < 1$, or $|x| < \sqrt{2}$. The interval of convergence is $(-\sqrt{2}, \sqrt{2})$.   ■

Our next theorem tells us that within the interval of convergence, we can treat a power series as though it were a polynomial; that is, we can differentiate and integrate term by term.

> **THEOREM 2** **Term-by-Term Differentiation and Integration** Assume that
>
> $$F(x) = \sum_{n=0}^{\infty} a_n (x - c)^n$$
>
> has radius of convergence $R > 0$. Then F is differentiable on $(c - R, c + R)$ (or for all x if $R = \infty$). Furthermore, we can integrate and differentiate term by term. For $x \in (c - R, c + R)$,
>
> $$F'(x) = \sum_{n=1}^{\infty} n a_n (x - c)^{n-1}$$
>
> $$\int F(x) \, dx = A + \sum_{n=0}^{\infty} \frac{a_n}{n + 1} (x - c)^{n+1} \qquad (A \text{ any constant})$$
>
> These series have the same radius of convergence R.

The proof of Theorem 2 is somewhat technical and is omitted. See Exercise 66 for a proof that F is continuous.

■ **EXAMPLE 6** **Differentiating a Power Series** Prove that for $-1 < x < 1$,

$$\frac{1}{(1 - x)^2} = 1 + 2x + 3x^2 + 4x^3 + 5x^4 + \cdots$$

Solution The geometric series has radius of convergence $R = 1$:

$$\frac{1}{1 - x} = 1 + x + x^2 + x^3 + x^4 + \cdots$$

By Theorem 2, we can differentiate term by term for $|x| < 1$ to obtain

$$\frac{d}{dx} \left(\frac{1}{1 - x} \right) = \frac{d}{dx} (1 + x + x^2 + x^3 + x^4 + \cdots)$$

$$\frac{1}{(1 - x)^2} = 1 + 2x + 3x^2 + 4x^3 + 5x^4 + \cdots \qquad ■$$

Theorem 2 is a powerful tool in the study of power series.

■ **EXAMPLE 7** **Power Series for Arctangent** Prove that for $-1 < x < 1$,

$$\tan^{-1} x = \sum_{n=0}^{\infty} \frac{(-1)^n x^{2n+1}}{2n + 1} = x - \frac{x^3}{3} + \frac{x^5}{5} - \frac{x^7}{7} + \cdots \qquad \boxed{4}$$

Solution Recall that $\tan^{-1} x$ is an antiderivative of $(1 + x^2)^{-1}$. We obtain a power series expansion of this antiderivative by substituting $-x^2$ for x in the geometric series of Eq. (2):

$$\frac{1}{1 + x^2} = 1 - x^2 + x^4 - x^6 + \cdots$$

This expansion is valid for $|x^2| < 1$—that is, for $|x| < 1$. By Theorem 2, we can integrate series term by term. The resulting expansion is also valid for $|x| < 1$:

$$\tan^{-1} x = \int \frac{dx}{1 + x^2} = \int (1 - x^2 + x^4 - x^6 + \cdots) \, dx$$

$$= A + x - \frac{x^3}{3} + \frac{x^5}{5} - \frac{x^7}{7} + \cdots$$

Setting $x = 0$, we obtain $A = \tan^{-1} 0 = 0$. Thus, Eq. (4) is valid for $-1 < x < 1$. ■

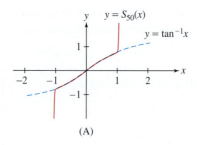

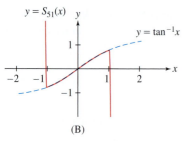

FIGURE 4 $y = S_{50}(x)$ and $y = S_{51}(x)$ are nearly indistinguishable from $y = \tan^{-1} x$ on $(-1, 1)$.

GRAPHICAL INSIGHT Let's examine the expansion of the previous example graphically. The partial sums of the power series for $f(x) = \tan^{-1} x$ are

$$S_N(x) = x - \frac{x^3}{3} + \frac{x^5}{5} - \frac{x^7}{7} + \cdots + (-1)^N \frac{x^{2N+1}}{2N + 1}$$

For large N, we can expect $S_N(x)$ to provide a good approximation to $f(x) = \tan^{-1} x$ on the interval $(-1, 1)$, where the power series expansion is valid. Figure 4 confirms this expectation: The graphs of $y = S_{50}(x)$ and $y = S_{51}(x)$ are nearly indistinguishable from the graph of $y = \tan^{-1} x$ on $(-1, 1)$. Thus, we may use the partial sums to approximate the arctangent. For example, $\tan^{-1}(0.3)$ is approximated by

$$S_4(0.3) = 0.3 - \frac{(0.3)^3}{3} + \frac{(0.3)^5}{5} - \frac{(0.3)^7}{7} + \frac{(0.3)^9}{9} \approx 0.2914569$$

Since the power series is an alternating series, the error is less than the first omitted term:

$$|\tan^{-1}(0.3) - S_4(0.3)| < \frac{(0.3)^{11}}{11} \approx 1.61 \times 10^{-7}$$

The situation changes drastically in the region $|x| > 1$, where the power series diverges and the partial sums $S_N(x)$ deviate sharply from $\tan^{-1} x$.

Power Series Solutions of Differential Equations

Power series are a basic tool in the study of differential equations. To illustrate, consider the differential equation with initial condition

$$y' = y, \qquad y(0) = 1$$

We know that $f(x) = e^x$ is the unique solution, but let's try to find a power series that satisfies this Initial Value Problem. We have

$$F(x) = \sum_{n=0}^{\infty} a_n x^n = a_0 + a_1 x + a_2 x^2 + a_3 x^3 + \cdots$$

$$F'(x) = \sum_{n=0}^{\infty} n a_n x^{n-1} = a_1 + 2a_2 x + 3a_3 x^2 + 4a_4 x^3 + \cdots$$

Therefore, $F'(x) = F(x)$ if

$$a_0 = a_1, \quad a_1 = 2a_2, \quad a_2 = 3a_3, \quad a_3 = 4a_4, \quad \ldots$$

In other words, $F'(x) = F(x)$ if $a_{n-1} = n a_n$, or

$$\boxed{a_n = \frac{a_{n-1}}{n}}$$

An equation of this type is called a *recursion relation*. It enables us to determine all of the coefficients a_n successively from the first coefficient a_0, which may be chosen arbitrarily. For example,

$$n = 1: \qquad a_1 = \frac{a_0}{1}$$

$$n = 2: \qquad a_2 = \frac{a_1}{2} = \frac{a_0}{2 \cdot 1} = \frac{a_0}{2!}$$

$$n = 3: \qquad a_3 = \frac{a_2}{3} = \frac{a_1}{3 \cdot 2} = \frac{a_0}{3 \cdot 2 \cdot 1} = \frac{a_0}{3!}$$

To obtain a general formula for a_n, apply the recursion relation n times:

$$a_n = \frac{a_{n-1}}{n} = \frac{a_{n-2}}{n(n-1)} = \frac{a_{n-3}}{n(n-1)(n-2)} = \cdots = \frac{a_0}{n!}$$

We conclude that

$$F(x) = a_0 \sum_{n=0}^{\infty} \frac{x^n}{n!}$$

In Example 3, we showed that this power series has radius of convergence $R = \infty$, so $y = F(x)$ satisfies $y' = y$ for all x. Moreover, $F(0) = a_0$, so the initial condition $y(0) = 1$ is satisfied with $a_0 = 1$.

What we have shown is that $f(x) = e^x$ and $F(x)$ with $a_0 = 1$ are both solutions of the Initial Value Problem. They must be equal because the solution is unique. This proves that for all x,

$$e^x = \sum_{n=0}^{\infty} \frac{x^n}{n!} = 1 + x + \frac{x^2}{2!} + \frac{x^3}{3!} + \frac{x^4}{4!} + \cdots$$

In this example, we knew in advance that $y = e^x$ is a solution of $y' = y$, but suppose we are given a differential equation whose solution is unknown. We can try to find a solution in the form of a power series $F(x) = \sum_{n=0}^{\infty} a_n x^n$. In favorable cases, the differential equation leads to a recursion relation that enables us to determine the coefficients a_n.

The solution in Example 8 is called the "Bessel function of order 1." The Bessel function of order n is a solution of

$$x^2 y'' + xy' + (x^2 - n^2) y = 0$$

These functions have applications in many areas of physics and engineering.

■ **EXAMPLE 8** Find a power series solution to the Initial Value Problem:

$$x^2 y'' + xy' + (x^2 - 1) y = 0, \qquad y'(0) = 1 \qquad \boxed{5}$$

Solution Assume that Eq. (5) has a power series solution $F(x) = \sum_{n=0}^{\infty} a_n x^n$. Then

$$y' = F'(x) = \sum_{n=0}^{\infty} n a_n x^{n-1} = a_1 + 2a_2 x + 3a_3 x^2 + \cdots$$

$$y'' = F''(x) = \sum_{n=0}^{\infty} n(n-1) a_n x^{n-2} = 2a_2 + 6a_3 x + 12a_4 x^2 + \cdots$$

Now substitute the series for y, y', and y'' into the differential equation (5) to determine the recursion relation satisfied by the coefficients a_n:

$$x^2 y'' + xy' + (x^2 - 1) y$$

In Eq. (6), we combine the first three series into a single series using

$$n(n-1) + n - 1 = n^2 - 1$$

and we shift the fourth series to begin at $n = 2$ rather than $n = 0$.

$$= x^2 \sum_{n=0}^{\infty} n(n-1) a_n x^{n-2} + x \sum_{n=0}^{\infty} n a_n x^{n-1} + (x^2 - 1) \sum_{n=0}^{\infty} a_n x^n$$

$$= \sum_{n=0}^{\infty} n(n-1) a_n x^n + \sum_{n=0}^{\infty} n a_n x^n - \sum_{n=0}^{\infty} a_n x^n + \sum_{n=0}^{\infty} a_n x^{n+2} \qquad \boxed{6}$$

$$= \sum_{n=0}^{\infty} (n^2 - 1) a_n x^n + \sum_{n=2}^{\infty} a_{n-2} x^n = 0$$

The differential equation is satisfied if

$$\sum_{n=0}^{\infty} (n^2 - 1) a_n x^n = - \sum_{n=2}^{\infty} a_{n-2} x^n$$

The first few terms on each side of this equation are

$$-a_0 + 0 \cdot x + 3a_2 x^2 + 8a_3 x^3 + 15a_4 x^4 + \cdots = 0 + 0 \cdot x - a_0 x^2 - a_1 x^3 - a_2 x^4 - \cdots$$

Matching up the coefficients of x^n, we find that

$$-a_0 = 0, \qquad 3a_2 = -a_0, \qquad 8a_3 = -a_1, \qquad 15a_4 = -a_2 \qquad \boxed{7}$$

In general, $(n^2 - 1)a_n = -a_{n-2}$, and this yields the recursion relation

$$\boxed{a_n = -\frac{a_{n-2}}{n^2 - 1} \qquad \text{for } n \geq 2} \qquad \boxed{8}$$

Note that $a_0 = 0$ by Eq. (7). The recursion relation forces all of the even coefficients a_2, $a_4, a_6, \dots$ to be zero:

$$a_2 = \frac{a_0}{2^2 - 1} \text{ so } a_2 = 0, \qquad \text{and then} \qquad a_4 = \frac{a_2}{4^2 - 1} = 0 \text{ so } a_4 = 0, \qquad \text{etc.}$$

As for the odd coefficients, a_1 may be chosen arbitrarily. Because $F'(0) = a_1$, we set $a_1 = 1$ to obtain a solution $y = F(x)$ satisfying $F'(0) = 1$. Now apply Eq. (8):

$$n = 3: \qquad a_3 = -\frac{a_1}{3^2 - 1} = -\frac{1}{3^2 - 1}$$

$$n = 5: \qquad a_5 = -\frac{a_3}{5^2 - 1} = \frac{1}{(5^2 - 1)(3^2 - 1)}$$

$$n = 7: \qquad a_7 = -\frac{a_5}{7^2 - 1} = -\frac{1}{(7^2 - 1)(3^2 - 1)(5^2 - 1)}$$

This shows the general pattern of coefficients. To express the coefficients in a compact form, let $n = 2k + 1$. Then the denominator in the recursion relation (8) can be written

$$n^2 - 1 = (2k + 1)^2 - 1 = 4k^2 + 4k = 4k(k + 1)$$

and

$$a_{2k+1} = -\frac{a_{2k-1}}{4k(k + 1)}$$

Applying this recursion relation k times, we obtain the closed formula

$$a_{2k+1} = (-1)^k \left(\frac{1}{4k(k + 1)} \right) \left(\frac{1}{4(k - 1)k} \right) \cdots \left(\frac{1}{4(1)(2)} \right) = \frac{(-1)^k}{4^k \, k! \, (k + 1)!}$$

Thus, we obtain a power series representation of our solution:

$$F(x) = \sum_{k=0}^{\infty} \frac{(-1)^k}{4^k k!(k + 1)!} x^{2k+1}$$

A straightforward application of the Ratio Test shows that F has an infinite radius of convergence. Therefore, $F(x)$ is a solution of the Initial Value Problem for all x. ∎

10.6 SUMMARY

- A *power series* is an infinite series of the form

$$F(x) = \sum_{n=0}^{\infty} a_n (x - c)^n$$

The constant c is called the *center* of $F(x)$.

- Every power series $F(x)$ has a *radius of convergence* R (Figure 5) such that

  − $F(x)$ converges absolutely for $|x - c| < R$ and diverges for $|x - c| > R$.

  − $F(x)$ may converge or diverge at the endpoints $c - R$ and $c + R$.

We set $R = 0$ if $F(x)$ converges only for $x = c$ and $R = \infty$ if $F(x)$ converges for all x.

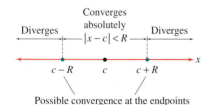

FIGURE 5 Interval of convergence of a power series.

- The *interval of convergence* of F consists of the open interval $(c - R, c + R)$ and possibly one or both endpoints $c - R$ and $c + R$.
- In many cases, the Ratio Test can be used to find the radius of convergence R. It is necessary to check convergence at the endpoints separately.
- If $R > 0$, then F is differentiable on $(c - R, c + R)$ and

$$F'(x) = \sum_{n=1}^{\infty} n a_n (x - c)^{n-1}, \qquad \int F(x)\, dx = A + \sum_{n=0}^{\infty} \frac{a_n}{n + 1} (x - c)^{n+1}$$

(A is any constant). These two power series have the same radius of convergence R.

- The expansion $\dfrac{1}{1 - x} = \sum_{n=0}^{\infty} x^n$ is valid for $|x| < 1$. It can be used to derive expansions of other related functions by substitution, integration, or differentiation.

10.6 EXERCISES

Preliminary Questions

1. Suppose that $\sum a_n x^n$ converges for $x = 5$. Must it also converge for $x = 4$? What about $x = -3$?

2. Suppose that $\sum a_n (x - 6)^n$ converges for $x = 10$. At which of the points (a)–(d) must it also converge?

(a) $x = 8$ (b) $x = 11$ (c) $x = 3$ (d) $x = 0$

3. What is the radius of convergence of $F(3x)$ if $F(x)$ is a power series with radius of convergence $R = 12$?

4. The power series $F(x) = \sum_{n=1}^{\infty} n x^n$ has radius of convergence $R = 1$. What is the power series expansion of $F'(x)$ and what is its radius of convergence?

Exercises

1. Use the Ratio Test to determine the radius of convergence R of $\sum_{n=0}^{\infty} \dfrac{x^n}{2^n}$. Does it converge at the endpoints $x = \pm R$?

2. Use the Ratio Test to show that $\sum_{n=1}^{\infty} \dfrac{x^n}{\sqrt{n} 2^n}$ has radius of convergence $R = 2$. Then determine whether it converges at the endpoints $R = \pm 2$.

3. Show that the power series (a)–(c) have the same radius of convergence. Then show that (a) diverges at both endpoints, (b) converges at one endpoint but diverges at the other, and (c) converges at both endpoints.

(a) $\sum_{n=1}^{\infty} \dfrac{x^n}{3^n}$ (b) $\sum_{n=1}^{\infty} \dfrac{x^n}{n 3^n}$ (c) $\sum_{n=1}^{\infty} \dfrac{x^n}{n^2 3^n}$

4. Repeat Exercise 3 for the following series:

(a) $\sum_{n=1}^{\infty} \dfrac{(x - 5)^n}{9^n}$ (b) $\sum_{n=1}^{\infty} \dfrac{(x - 5)^n}{n 9^n}$ (c) $\sum_{n=1}^{\infty} \dfrac{(x - 5)^n}{n^2 9^n}$

5. Show that $\sum_{n=0}^{\infty} n^n x^n$ diverges for all $x \neq 0$.

6. For which values of x does $\sum_{n=0}^{\infty} n! x^n$ converge?

7. Use the Ratio Test to show that $\sum_{n=0}^{\infty} \dfrac{x^{2n}}{3^n}$ has radius of convergence $R = \sqrt{3}$.

8. Show that $\sum_{n=0}^{\infty} \dfrac{x^{3n+1}}{64^n}$ has radius of convergence $R = 4$.

In Exercises 9–34, find the interval of convergence.

9. $\sum_{n=0}^{\infty} n x^n$

10. $\sum_{n=1}^{\infty} \dfrac{2^n}{n} x^n$

11. $\sum_{n=1}^{\infty} (-1)^n \dfrac{x^{2n+1}}{2^n n}$

12. $\sum_{n=0}^{\infty} (-1)^n \dfrac{n}{4^n} x^{2n}$

13. $\sum_{n=4}^{\infty} \dfrac{x^n}{n^5}$

14. $\sum_{n=8}^{\infty} n^7 x^n$

15. $\sum_{n=0}^{\infty} \dfrac{x^n}{(n!)^2}$

16. $\sum_{n=0}^{\infty} \dfrac{8^n}{n!} x^n$

17. $\sum_{n=0}^{\infty} \dfrac{(2n)!}{(n!)^3} x^n$

18. $\sum_{n=0}^{\infty} \dfrac{4^n}{(2n + 1)!} x^{2n-1}$

19. $\sum_{n=0}^{\infty} \dfrac{(-1)^n x^n}{\sqrt{n^2 + 1}}$

20. $\sum_{n=0}^{\infty} \dfrac{x^n}{n^4 + 2}$

21. $\sum_{n=15}^{\infty} \dfrac{x^{2n+1}}{3n + 1}$

22. $\sum_{n=9}^{\infty} \dfrac{x^n}{n - 4 \ln n}$

23. $\sum_{n=2}^{\infty} \dfrac{x^n}{\ln n}$

24. $\sum_{n=2}^{\infty} \dfrac{x^{3n+2}}{\ln n}$

25. $\sum_{n=1}^{\infty} n(x - 3)^n$

26. $\sum_{n=1}^{\infty} \dfrac{(-5)^n (x - 3)^n}{n^2}$

27. $\displaystyle\sum_{n=1}^{\infty}(-1)^n n^5 (x-7)^n$

28. $\displaystyle\sum_{n=0}^{\infty}27^n (x-1)^{3n+2}$

29. $\displaystyle\sum_{n=1}^{\infty}\frac{2^n}{3n}(x+3)^n$

30. $\displaystyle\sum_{n=0}^{\infty}\frac{(x-4)^n}{n!}$

31. $\displaystyle\sum_{n=0}^{\infty}\frac{(-5)^n}{n!}(x+10)^n$

32. $\displaystyle\sum_{n=10}^{\infty}n!\,(x+5)^n$

33. $\displaystyle\sum_{n=12}^{\infty}e^n (x-2)^n$

34. $\displaystyle\sum_{n=2}^{\infty}\frac{(x+4)^n}{(n\ln n)^2}$

In Exercises 35–40, use Eq. (2) to expand the function in a power series with center $c=0$ and determine the interval of convergence.

35. $f(x)=\dfrac{1}{1-3x}$

36. $f(x)=\dfrac{1}{1+3x}$

37. $f(x)=\dfrac{1}{3-x}$

38. $f(x)=\dfrac{1}{4+3x}$

39. $f(x)=\dfrac{1}{1+x^2}$

40. $f(x)=\dfrac{1}{16+2x^3}$

41. Use the equalities

$$\frac{1}{1-x}=\frac{1}{-3-(x-4)}=\frac{-\frac{1}{3}}{1+\left(\frac{x-4}{3}\right)}$$

to show that for $|x-4|<3$,

$$\frac{1}{1-x}=\sum_{n=0}^{\infty}(-1)^{n+1}\frac{(x-4)^n}{3^{n+1}}$$

42. Use the method of Exercise 41 to expand $1/(1-x)$ in power series with centers $c=2$ and $c=-2$. Determine the interval of convergence.

43. Use the method of Exercise 41 to expand $1/(4-x)$ in a power series with center $c=5$. Determine the interval of convergence.

44. Find a power series that converges only for x in $[2,6)$.

45. Apply integration to the expansion

$$\frac{1}{1+x}=\sum_{n=0}^{\infty}(-1)^n x^n=1-x+x^2-x^3+\cdots$$

to prove that for $-1<x<1$,

$$\ln(1+x)=\sum_{n=1}^{\infty}\frac{(-1)^{n-1}x^n}{n}=x-\frac{x^2}{2}+\frac{x^3}{3}-\frac{x^4}{4}+\cdots$$

46. Use the result of Exercise 45 to prove that

$$\ln\frac{3}{2}=\frac{1}{2}-\frac{1}{2\cdot 2^2}+\frac{1}{3\cdot 2^3}-\frac{1}{4\cdot 2^4}+\cdots$$

Use your knowledge of alternating series to find an N such that the partial sum S_N approximates $\ln\frac{3}{2}$ to within an error of at most 10^{-3}. Confirm using a calculator to compute both S_N and $\ln\frac{3}{2}$.

47. Let $F(x)=(x+1)\ln(1+x)-x$.

(a) Apply integration to the result of Exercise 45 to prove that for $-1<x<1$,

$$F(x)=\sum_{n=1}^{\infty}(-1)^{n+1}\frac{x^{n+1}}{n(n+1)}$$

(b) Evaluate at $x=\frac{1}{2}$ to prove

$$\frac{3}{2}\ln\frac{3}{2}-\frac{1}{2}=\frac{1}{1\cdot 2\cdot 2^2}-\frac{1}{2\cdot 3\cdot 2^3}+\frac{1}{3\cdot 4\cdot 2^4}-\frac{1}{4\cdot 5\cdot 2^5}+\cdots$$

(c) Use a calculator to verify that the partial sum S_4 approximates the left-hand side with an error no greater than the term a_5 of the series.

48. Prove that for $|x|<1$,

$$\int\frac{dx}{x^4+1}=A+x-\frac{x^5}{5}+\frac{x^9}{9}-\cdots$$

Use the first two terms to approximate $\int_0^{1/2}dx/(x^4+1)$ numerically. Use the fact that you have an alternating series to show that the error in this approximation is at most 0.00022.

49. Use the result of Example 7 to show that

$$F(x)=\frac{x^2}{1\cdot 2}-\frac{x^4}{3\cdot 4}+\frac{x^6}{5\cdot 6}-\frac{x^8}{7\cdot 8}+\cdots$$

is an antiderivative of $f(x)=\tan^{-1}x$ satisfying $F(0)=0$. What is the radius of convergence of this power series?

50. Verify that function $F(x)=x\tan^{-1}x-\frac{1}{2}\log(x^2+1)$ is an antiderivative of $f(x)=\tan^{-1}x$ satisfying $F(0)=0$. Then use the result of Exercise 49 with $x=\frac{1}{\sqrt{3}}$ to show that

$$\frac{\pi}{6\sqrt{3}}-\frac{1}{2}\ln\frac{4}{3}=\frac{1}{1\cdot 2(3)}-\frac{1}{3\cdot 4(3^2)}+\frac{1}{5\cdot 6(3^3)}-\frac{1}{7\cdot 8(3^4)}+\cdots$$

Use a calculator to compare the value of the left-hand side with the partial sum S_4 of the series on the right.

51. Evaluate $\displaystyle\sum_{n=1}^{\infty}\frac{n}{2^n}$. *Hint:* Use differentiation to show that

$$(1-x)^{-2}=\sum_{n=1}^{\infty}nx^{n-1}\quad(\text{for }|x|<1)$$

52. Use the power series for $(1+x^2)^{-1}$ and differentiation to prove that for $|x|<1$,

$$\frac{2x}{(x^2+1)^2}=\sum_{n=1}^{\infty}(-1)^{n-1}(2n)x^{2n-1}$$

53. Show that the following series converges absolutely for $|x|<1$ and compute its sum:

$$F(x)=1-x-x^2+x^3-x^4-x^5+x^6-x^7-x^8+\cdots$$

Hint: Write $F(x)$ as a sum of three geometric series with common ratio x^3.

54. Show that for $|x|<1$,

$$\frac{1+2x}{1+x+x^2}=1+x-2x^2+x^3+x^4-2x^5+x^6+x^7-2x^8+\cdots$$

Hint: Use the hint from Exercise 53.

55. Find all values of x such that $\displaystyle\sum_{n=1}^{\infty}\frac{x^{n^2}}{n!}$ converges.

56. Find all values of x such that the following series converges:

$$F(x)=1+3x+x^2+27x^3+x^4+243x^5+\cdots$$

57. Find a power series $P(x)=\displaystyle\sum_{n=0}^{\infty}a_n x^n$ satisfying the differential equation $y'=-y$ with initial condition $y(0)=1$. Then use Theorem 1 of Section 5.9 to conclude that $P(x)=e^{-x}$.

58. Let $C(x)=1-\dfrac{x^2}{2!}+\dfrac{x^4}{4!}-\dfrac{x^6}{6!}+\cdots$.

(a) Show that $C(x)$ has an infinite radius of convergence.

(b) Prove that $C(x)$ and $f(x)=\cos x$ are both solutions of $y''=-y$ with initial conditions $y(0)=1$, $y'(0)=0$. This Initial Value Problem has a unique solution, so we have $C(x)=\cos x$ for all x.

59. Use the power series for $y = e^x$ to show that

$$\frac{1}{e} = \frac{1}{2!} - \frac{1}{3!} + \frac{1}{4!} - \cdots$$

Use your knowledge of alternating series to find an N such that the partial sum S_N approximates e^{-1} to within an error of at most 10^{-3}. Confirm this using a calculator to compute both S_N and e^{-1}.

60. Let $P(x) = \sum_{n=0}^{\infty} a_n x^n$ be a power series solution to $y' = 2xy$ with initial condition $y(0) = 1$.

(a) Show that the odd coefficients a_{2k+1} are all zero.

(b) Prove that $a_{2k} = a_{2k-2}/k$ and use this result to determine the coefficients a_{2k}.

61. Find a power series $P(x)$ satisfying the differential equation

$$y'' - xy' + y = 0 \qquad \boxed{9}$$

with initial condition $y(0) = 1$, $y'(0) = 0$. What is the radius of convergence of the power series?

62. Find a power series satisfying Eq. (9) with initial condition $y(0) = 0$, $y'(0) = 1$.

63. Prove that

$$J_2(x) = \sum_{k=0}^{\infty} \frac{(-1)^k}{2^{2k+2}\, k!\, (k+3)!} x^{2k+2}$$

is a solution of the Bessel differential equation of order 2:

$$x^2 y'' + xy' + (x^2 - 4)y = 0$$

64. [icon] Why is it impossible to expand $f(x) = |x|$ as a power series that converges in an interval around $x = 0$? Explain using Theorem 2.

Further Insights and Challenges

65. Suppose that the coefficients of $F(x) = \sum_{n=0}^{\infty} a_n x^n$ are *periodic*; that is, for some whole number $M > 0$, we have $a_{M+n} = a_n$. Prove that $F(x)$ converges absolutely for $|x| < 1$ and that

$$F(x) = \frac{a_0 + a_1 x + \cdots + a_{M-1} x^{M-1}}{1 - x^M}$$

Hint: Use the hint for Exercise 53.

66. Continuity of Power Series Let $F(x) = \sum_{n=0}^{\infty} a_n x^n$ be a power series with radius of convergence $R > 0$.

(a) Prove the inequality

$$|x^n - y^n| \le n|x - y|(|x|^{n-1} + |y|^{n-1}) \qquad \boxed{10}$$

Hint: $x^n - y^n = (x - y)(x^{n-1} + x^{n-2}y + \cdots + y^{n-1})$.

(b) Choose R_1 with $0 < R_1 < R$. Show that the infinite series

$$M = \sum_{n=0}^{\infty} 2n|a_n|R_1^n \text{ converges. } \textit{Hint: Show that } n|a_n|R_1^n < |a_n||x^n| \text{ for}$$

all n sufficiently large if $R_1 < x < R$.

(c) Use Eq. (10) to show that if $|x| < R_1$ and $|y| < R_1$, then $|F(x) - F(y)| \le M|x - y|$.

(d) Prove that if $|x| < R$, then F is continuous at x. *Hint:* Choose R_1 such that $|x| < R_1 < R$. Show that if $\epsilon > 0$ is given, then $|F(x) - F(y)| \le \epsilon$ for all y such that $|x - y| < \delta$, where δ is any positive number that is less than ϵ/M and $R_1 - |x|$ (see Figure 6).

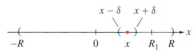

FIGURE 6 If $x > 0$, choose $\delta > 0$ less than ϵ/M and $R_1 - x$.

10.7 Taylor Series

In this section, we develop general methods for finding power series representations. Ultimately, we will see that these power series are, in fact, the extended versions of the Taylor polynomials we discussed in Section 8.4, which we used to approximate values of functions. Suppose that $f(x)$ is represented by a power series centered at $x = c$ on an interval $(c - R, c + R)$ with $R > 0$:

$$f(x) = \sum_{n=0}^{\infty} a_n(x - c)^n = a_0 + a_1(x - c) + a_2(x - c)^2 + \cdots$$

According to Theorem 2 in Section 10.6, we can compute the derivatives of f by differentiating the series expansion term by term:

$$f(x) = \quad a_0 + \quad\quad a_1(x - c) + \quad\quad a_2(x - c)^2 + \quad\quad a_3(x - c)^3 + \cdots$$

$$f'(x) = \quad a_1 + \quad\quad 2a_2(x - c) + \quad\quad 3a_3(x - c)^2 + \quad\quad 4a_4(x - c)^3 + \cdots$$

$$f''(x) = \quad 2a_2 + \quad 2 \cdot 3a_3(x - c) + \quad 3 \cdot 4a_4(x - c)^2 + 4 \cdot 5a_5(x - c)^3 + \cdots$$

$$f'''(x) = 2 \cdot 3a_3 + 2 \cdot 3 \cdot 4a_4(x - 2) + 3 \cdot 4 \cdot 5a_5(x - 2)^2 + \quad\quad\quad \cdots$$

In general,

$$f^{(k)}(x) = k!a_k + \left(2 \cdot 3 \cdots (k+1)\right)a_{k+1}(x-c) + \cdots$$

Setting $x = c$ in each of these series, we find that

$$f(c) = a_0, \quad f'(c) = a_1, \quad f''(c) = 2a_2, \quad f'''(c) = 2 \cdot 3a_2, \quad \ldots, \quad f^{(k)}(c) = k!a_k, \quad \ldots$$

We see that a_k is the kth coefficient of the Taylor polynomial studied in Section 8.4:

$$\boxed{a_k = \frac{f^{(k)}(c)}{k!}} \qquad \boxed{1}$$

Therefore, $f(x) = T(x)$, where $T(x)$ is the **Taylor series** of $f(x)$ centered at $x = c$:

$$T(x) = f(c) + f'(c)(x-c) + \frac{f''(c)}{2!}(x-c)^2 + \frac{f'''(c)}{3!}(x-c)^3 + \cdots$$

This proves the next theorem.

THEOREM 1 Taylor Series Expansion If $f(x)$ is represented by a power series centered at c in an interval $|x - c| < R$ with $R > 0$, then that power series is the Taylor series

$$T(x) = \sum_{n=0}^{\infty} \frac{f^{(n)}(c)}{n!}(x-c)^n$$

In the special case $c = 0$, $T(x)$ is also called the **Maclaurin series**:

$$f(x) = \sum_{n=0}^{\infty} \frac{f^{(n)}(0)}{n!}x^n = f(0) + f'(0)x + \frac{f''(0)}{2!}x^2 + \frac{f'''(0)}{3!}x^3 + \frac{f^{(4)}(0)}{4!}x^4 + \cdots$$

TABLE 1 Finding the Coefficients for the Taylor Series.

n	$f^{(n)}(x)$	$\frac{f^{(n)}(x)}{n!}$	$\frac{f^{(n)}(1)}{n!}$
0	x^{-3}	x^{-3}	1
1	$-3x^{-4}$	$-3x^{-4}$	-3
2	$12x^{-5}$	$6x^{-5}$	6
3	$-60x^{-6}$	$-10x^{-6}$	-10
4	$360x^{-7}$	$15x^{-7}$	15

■ **EXAMPLE 1** Find the Taylor series for $f(x) = x^{-3}$ centered at $c = 1$.

Solution It often helps to create a table, as in Table 1, to see the pattern. The derivatives of $f(x)$ are $f'(x) = -3x^{-4}$, $f''(x) = (-3)(-4)x^{-5}$, and in general,

$$f^{(n)}(x) = (-1)^n (3)(4) \cdots (n+2)x^{-3-n}$$

Note that $(3)(4) \cdots (n+2) = \frac{1}{2}(n+2)!$. Therefore,

$$f^{(n)}(1) = (-1)^n \frac{1}{2}(n+2)!$$

Noting that $(n+2)! = (n+2)(n+1)n!$, we write the coefficients of the Taylor series as

$$a_n = \frac{f^{(n)}(1)}{n!} = \frac{(-1)^n \frac{1}{2}(n+2)!}{n!} = (-1)^n \frac{(n+2)(n+1)}{2}$$

The Taylor series for $f(x) = x^{-3}$ centered at $c = 1$ is

$$T(x) = 1 - 3(x-1) + 6(x-1)^2 - 10(x-1)^3 + \cdots$$

$$= \sum_{n=0}^{\infty} (-1)^n \frac{(n+2)(n+1)}{2}(x-1)^n \qquad ■$$

Theorem 1 tells us that if we want to represent a function f by a power series centered at c, then the only candidate for the job is the Taylor series:

$$T(x) = \sum_{n=0}^{\infty} \frac{f^{(n)}(c)}{n!}(x - c)^n$$

See Exercise 94 for an example where a Taylor series $T(x)$ converges but does not converge to $f(x)$.

However, *there is no guarantee that $T(x)$ converges to $f(x)$, even if $T(x)$ converges.* To study convergence, we consider the kth partial sum, which is the Taylor polynomial of degree k:

$$T_k(x) = f(c) + f'(c)(x - c) + \frac{f''(c)}{2!}(x - c)^2 + \cdots + \frac{f^{(k)}(c)}{k!}(x - c)^k$$

In Section 8.4, we defined the remainder

$$R_k(x) = f(x) - T_k(x)$$

Since $T(x)$ is the limit of the partial sums $T_k(x)$, we see that

The Taylor series converges to $f(x)$ if and only if $\lim_{k \to \infty} R_k(x) = 0$.

There is no general method for determining whether $R_k(x)$ tends to zero, but the following theorem can be applied in some important cases.

*◄·· **REMINDER** $f(x)$ is called "infinitely differentiable" if $f^{(n)}(x)$ exists for all n.*

THEOREM 2 Let $I = (c - R, c + R)$, where $R > 0$. Suppose there exists $K > 0$ such that all derivatives of f are bounded by K on I:

$$|f^{(k)}(x)| \leq K \qquad \text{for all} \quad k \geq 0 \quad \text{and} \quad x \in I$$

Then f is represented by its Taylor series in I:

$$f(x) = \sum_{n=0}^{\infty} \frac{f^{(n)}(c)}{n!}(x - c)^n \qquad \text{for all} \quad x \in I$$

Proof According to the Error Bound for Taylor polynomials (Theorem 2 in Section 8.4),

$$|R_k(x)| = |f(x) - T_k(x)| \leq K \frac{|x - c|^{k+1}}{(k + 1)!}$$

If $x \in I$, then $|x - c| < R$ and

$$|R_k(x)| \leq K \frac{R^{k+1}}{(k + 1)!}$$

We showed in Example 9 of Section 10.1 that $R^k/k!$ tends to zero as $k \to \infty$. Therefore, $\lim_{k \to \infty} R_k(x) = 0$ for all $x \in (c - R, c + R)$, as required. ■

Taylor expansions were studied throughout the seventeenth and eighteenth centuries by Gregory, Leibniz, Newton, Maclaurin, Taylor, Euler, and others. These developments were anticipated by the great Hindu mathematician Madhava (c. 1340–1425), who discovered the expansions of sine and cosine and many other results two centuries earlier.

■ **EXAMPLE 2** **Expansions of Sine and Cosine** Show that the following Maclaurin expansions are valid for all x:

$$\sin x = \sum_{n=0}^{\infty} (-1)^n \frac{x^{2n+1}}{(2n + 1)!} = x - \frac{x^3}{3!} + \frac{x^5}{5!} - \frac{x^7}{7!} + \cdots$$

$$\cos x = \sum_{n=0}^{\infty} (-1)^n \frac{x^{2n}}{(2n)!} = 1 - \frac{x^2}{2!} + \frac{x^4}{4!} - \frac{x^6}{6!} + \cdots$$

Solution Recall that the derivatives of $f(x) = \sin x$ and their values at $x = 0$ form a repeating pattern of period 4:

$f(x)$	$f'(x)$	$f''(x)$	$f'''(x)$	$f^{(4)}(x)$	$\cdots$
$\sin x$	$\cos x$	$-\sin x$	$-\cos x$	$\sin x$	$\cdots$
0	1	0	-1	0	$\cdots$

In other words, the even derivatives are zero and the odd derivatives alternate in sign: $f^{(2n+1)}(0) = (-1)^n$. Therefore, the nonzero Taylor coefficients for $\sin x$ are

$$a_{2n+1} = \frac{(-1)^n}{(2n+1)!}$$

For $f(x) = \cos x$, the situation is reversed. The odd derivatives are zero and the even derivatives alternate in sign: $f^{(2n)}(0) = (-1)^n \cos 0 = (-1)^n$. Therefore, the nonzero Taylor coefficients for $\cos x$ are $a_{2n} = (-1)^n/(2n)!$.

We can apply Theorem 2 with $K = 1$ and any value of R because both sine and cosine satisfy $|f^{(n)}(x)| \le 1$ for all x and n. The conclusion is that the Taylor series converges to $f(x)$ for $|x| < R$. Since R is arbitrary, the Taylor expansions hold for all x. ■

■ **EXAMPLE 3** **Taylor Expansion of $f(x) = e^x$ at $x = c$** Find the Taylor series $T(x)$ of $f(x) = e^x$ at $x = c$.

Solution We have $f^{(n)}(c) = e^c$ for all x. Thus,

$$T(x) = \sum_{n=0}^{\infty} \frac{e^c}{n!}(x - c)^n$$

Because $f(x) = e^x$ is increasing for all $R > 0$, $|f^{(k)}(x)| \le e^{c+R}$ for $x \in (c - R, c + R)$. Applying Theorem 2 with $K = e^{c+R}$, we conclude that $T(x)$ converges to $f(x)$ for all $x \in (c - R, c + R)$. Since R is arbitrary, the Taylor expansion holds for all x. For $c = 0$, we obtain the standard Maclaurin series

$$\boxed{e^x = 1 + x + \frac{x^2}{2!} + \frac{x^3}{3!} + \cdots}$$

■

Shortcuts to Finding Taylor Series

There are several methods for generating new Taylor series from known ones. First of all, we can differentiate and integrate Taylor series term by term within its interval of convergence, by Theorem 2 of Section 10.6. We can also multiply two Taylor series or substitute one Taylor series into another (we omit the proofs of these facts).

In Example 4, we can also write the Maclaurin series as

$$\sum_{n=0}^{\infty} \frac{x^{n+2}}{n!}$$

■ **EXAMPLE 4** Find the Maclaurin series for $f(x) = x^2 e^x$.

Solution Multiply the known Maclaurin series for e^x by x^2:

$$x^2 e^x = x^2 \left(1 + x + \frac{x^2}{2!} + \frac{x^3}{3!} + \frac{x^4}{4!} + \frac{x^5}{5!} + \cdots\right)$$

$$= x^2 + x^3 + \frac{x^4}{2!} + \frac{x^5}{3!} + \frac{x^6}{4!} + \frac{x^7}{5!} + \cdots = \sum_{n=2}^{\infty} \frac{x^n}{(n-2)!}$$

■

■ **EXAMPLE 5** **Substitution** Find the Maclaurin series for $f(x) = e^{-x^2}$.

Solution Substitute $-x^2$ in the Maclaurin series for e^x:

$$e^{-x^2} = \sum_{n=0}^{\infty} \frac{(-x^2)^n}{n!} = \sum_{n=0}^{\infty} \frac{(-1)^n x^{2n}}{n!} = 1 - x^2 + \frac{x^4}{2!} - \frac{x^6}{3!} + \frac{x^8}{4!} - \cdots \qquad \boxed{2}$$

The Taylor expansion of e^x is valid for all x, so this expansion is also valid for all x. ■

■ **EXAMPLE 6** Integration Find the Maclaurin series for $f(x) = \ln(1+x)$.

Solution We integrate the geometric series with common ratio $-x$ (valid for $|x| < 1$):

$$\frac{1}{1+x} = 1 - x + x^2 - x^3 + \cdots$$

$$\ln(1+x) = \int \frac{dx}{1+x} = A + x - \frac{x^2}{2} + \frac{x^3}{3} - \frac{x^4}{4} + \cdots = A + \sum_{n=1}^{\infty} (-1)^{n-1} \frac{x^n}{n}$$

The constant of integration A on the right is zero because $\ln(1+x) = 0$ for $x = 0$, so

$$\ln(1+x) = \sum_{n=1}^{\infty} (-1)^{n-1} \frac{x^n}{n}$$

This expansion is valid for $|x| < 1$. It also holds for $x = 1$ (see Exercise 86). ■

In many cases, there is no convenient general formula for the Taylor coefficients, but we can still compute as many coefficients as desired.

■ **EXAMPLE 7** Multiplying Taylor Series Write out the terms up to degree 5 in the Maclaurin series for $f(x) = e^x \cos x$.

Solution We multiply the fifth-order Taylor polynomials of e^x and $\cos x$ together, dropping the terms of degree greater than 5:

$$\left(1 + x + \frac{x^2}{2} + \frac{x^3}{6} + \frac{x^4}{24} + \frac{x^5}{120}\right)\left(1 - \frac{x^2}{2} + \frac{x^4}{24}\right)$$

Distributing the term on the left (and ignoring terms of degree greater than 5), we obtain

$$\left(1 + x + \frac{x^2}{2} + \frac{x^3}{6} + \frac{x^4}{24} + \frac{x^5}{120}\right) - \left(1 + x + \frac{x^2}{2} + \frac{x^3}{6}\right)\left(\frac{x^2}{2}\right) + (1 + x)\left(\frac{x^4}{24}\right)$$

$$= \underbrace{1 + x - \frac{x^3}{3} - \frac{x^4}{6} - \frac{x^5}{30}}_{\text{Retain terms of degree } \leq 5}$$

We conclude that the fifth Maclaurin polynomial for $f(x) = e^x \cos x$ is

$$T_5(x) = 1 + x - \frac{x^3}{3} - \frac{x^4}{6} - \frac{x^5}{30}$$ ■

In the next example, we express the definite integral of $\sin(x^2)$ as an infinite series. This is useful because the integral cannot be evaluated explicitly. Figure 1 shows the graph of the Taylor polynomial $y = T_{12}(x)$ of the Taylor series expansion of the antiderivative.

FIGURE 1 Graph of $y = T_{12}(x)$ for the power series expansion of the antiderivative

$$F(x) = \int_0^x \sin(t^2)\, dt$$

■ **EXAMPLE 8** Let $J = \int_0^1 \sin(x^2)\, dx$.

(a) Express J as an infinite series.

(b) Determine J to within an error less than 10^{-4}.

Solution

(a) The Maclaurin expansion for $f(x) = \sin x$ is valid for all x, so we have

$$\sin x = \sum_{n=0}^{\infty} \frac{(-1)^n}{(2n+1)!} x^{2n+1} \quad \Rightarrow \quad \sin(x^2) = \sum_{n=0}^{\infty} \frac{(-1)^n}{(2n+1)!} x^{4n+2}$$

We obtain an infinite series for J by integration:

$$J = \int_0^1 \sin(x^2)\, dx = \sum_{n=0}^{\infty} \frac{(-1)^n}{(2n+1)!} \int_0^1 x^{4n+2} dx = \sum_{n=0}^{\infty} \frac{(-1)^n}{(2n+1)!} \left(\frac{1}{4n+3}\right)$$

$$= \frac{1}{3} - \frac{1}{42} + \frac{1}{1320} - \frac{1}{75,600} + \cdots$$

3

(b) The infinite series for J is an alternating series with decreasing terms, so the sum of the first N terms is accurate to within an error that is less than the $(N+1)$st term. The absolute value of the fourth term $1/75{,}600$ is smaller than 10^{-4}, so we obtain the desired accuracy using the first three terms of the series for J:

$$J \approx \frac{1}{3} - \frac{1}{42} + \frac{1}{1320} \approx 0.31028$$

The error satisfies

$$\left| J - \left(\frac{1}{3} - \frac{1}{42} + \frac{1}{1320} \right) \right| < \frac{1}{75{,}600} \approx 1.3 \times 10^{-5}$$

The percentage error is less than 0.005% with just three terms. ∎

Power series have additional applications.

■ **EXAMPLE 9** Determine $\displaystyle\lim_{x \to 0} \frac{x - \sin x}{x^3 \cos x}$.

Solution This limit is of indeterminate form $\frac{0}{0}$, so we could use L'Hôpital's Rule repeatedly. However, instead, we will work with the Maclaurin series. We have

$$\sin x = x - \frac{x^3}{3!} + \frac{x^5}{5!} - \cdots$$

$$\cos x = 1 - \frac{x^2}{2!} + \frac{x^4}{4!} - \cdots$$

Hence, the limit becomes

$$\lim_{x \to 0} \frac{x - \sin x}{x^3 \cos x} = \lim_{x \to 0} \frac{x - \left(x - \frac{x^3}{3!} + \frac{x^5}{5!} - \cdots \right)}{x^3 \left(1 - \frac{x^2}{2!} + \frac{x^4}{4!} - \cdots \right)}$$

$$= \lim_{x \to 0} \frac{\frac{x^3}{3!} - \frac{x^5}{5!} + \cdots}{x^3 \left(1 - \frac{x^2}{2!} + \frac{x^4}{4!} - \cdots \right)}$$

$$= \lim_{x \to 0} \frac{x^3 \left(\frac{1}{3!} - \frac{x^2}{5!} + \cdots \right)}{x^3 \left(1 - \frac{x^2}{2!} + \frac{x^4}{4!} - \cdots \right)}$$

$$= \lim_{x \to 0} \frac{\frac{1}{3!} - \frac{x^2}{5!} + \cdots}{1 - \frac{x^2}{2!} + \frac{x^4}{4!} - \cdots}$$

$$= \frac{1}{3!} = \frac{1}{6}$$

∎

Binomial Series

Isaac Newton discovered an important generalization of the Binomial Theorem around 1665. For any number a (integer or not) and integer $n \geq 0$, we define the **binomial coefficient**:

$$\binom{a}{n} = \frac{a(a-1)(a-2) \cdots (a-n+1)}{n!}, \qquad \binom{a}{0} = 1$$

For example,

$$\binom{6}{3} = \frac{6 \cdot 5 \cdot 4}{3 \cdot 2 \cdot 1} = 20, \qquad \binom{\frac{4}{3}}{3} = \frac{\frac{4}{3} \cdot \frac{1}{3} \cdot \left(-\frac{2}{3} \right)}{3 \cdot 2 \cdot 1} = -\frac{4}{81}$$

Let

$$f(x) = (1 + x)^a$$

The **Binomial Theorem** of algebra (see Appendix C) states that for any whole number a,

$$(r + s)^a = r^a + \binom{a}{1} r^{a-1} s + \binom{a}{2} r^{a-2} s^2 + \cdots + \binom{a}{a-1} r s^{a-1} + s^a$$

Setting $r = 1$ and $s = x$, we obtain the expansion of $f(x)$:

$$(1 + x)^a = 1 + \binom{a}{1}x + \binom{a}{2}x^2 + \cdots + \binom{a}{a-1}x^{a-1} + x^a$$

We derive Newton's generalization by computing the Maclaurin series of $f(x)$ without assuming that a is a whole number. Observe that the derivatives follow a pattern:

$$f(x) = (1 + x)^a \qquad\qquad f(0) = 1$$
$$f'(x) = a(1 + x)^{a-1} \qquad\qquad f'(0) = a$$
$$f''(x) = a(a - 1)(1 + x)^{a-2} \qquad\qquad f''(0) = a(a - 1)$$
$$f'''(x) = a(a - 1)(a - 2)(1 + x)^{a-3} \qquad f'''(0) = a(a - 1)(a - 2)$$

In general, $f^{(n)}(0) = a(a - 1)(a - 2) \cdots (a - n + 1)$ and

$$\frac{f^{(n)}(0)}{n!} = \frac{a(a - 1)(a - 2) \cdots (a - n + 1)}{n!} = \binom{a}{n}$$

When a is a positive whole number, $\binom{a}{n}$ is zero for $n > a$, and in this case, the binomial series breaks off at degree n. The binomial series is an infinite series when a is not a positive whole number.

Hence, the Maclaurin series for $f(x) = (1 + x)^a$ is the binomial series

$$\sum_{n=0}^{\infty} \binom{a}{n}x^n = 1 + ax + \frac{a(a - 1)}{2!}x^2 + \frac{a(a - 1)(a - 2)}{3!}x^3 + \cdots + \binom{a}{n}x^n + \cdots$$

The Ratio Test shows that this series has radius of convergence $R = 1$ (Exercise 88), and an additional argument (developed in Exercise 89) shows that it converges to $(1 + x)^a$ for $|x| < 1$.

THEOREM 3 **The Binomial Series** For any exponent a and for $|x| < 1$,

$$(1 + x)^a = 1 + \frac{a}{1!}x + \frac{a(a - 1)}{2!}x^2 + \frac{a(a - 1)(a - 2)}{3!}x^3 + \cdots + \binom{a}{n}x^n + \cdots$$

■ **EXAMPLE 10** Find the terms through degree 4 in the Maclaurin expansion of

$$f(x) = (1 + x)^{4/3}$$

Solution The binomial coefficients $\binom{a}{n}$ for $a = \frac{4}{3}$ for $0 < n < 4$ are

$$1, \qquad \frac{\frac{4}{3}}{1!} = \frac{4}{3}, \qquad \frac{\frac{4}{3}\left(\frac{1}{3}\right)}{2!} = \frac{2}{9}, \qquad \frac{\frac{4}{3}\left(\frac{1}{3}\right)\left(-\frac{2}{3}\right)}{3!} = -\frac{4}{81}, \qquad \frac{\frac{4}{3}\left(\frac{1}{3}\right)\left(-\frac{2}{3}\right)\left(-\frac{5}{3}\right)}{4!} = \frac{5}{243}$$

Therefore, $(1 + x)^{4/3} \approx 1 + \frac{4}{3}x + \frac{2}{9}x^2 - \frac{4}{81}x^3 + \frac{5}{243}x^4 + \cdots$. ■

■ **EXAMPLE 11** Find the Maclaurin series for

$$f(x) = \frac{1}{\sqrt{1 - x^2}}$$

Solution First, let's find the coefficients in the binomial series for $(1 + x)^{-1/2}$:

$$1, \qquad \frac{-\frac{1}{2}}{1!} = -\frac{1}{2}, \qquad \frac{-\frac{1}{2}\left(-\frac{3}{2}\right)}{1 \cdot 2} = \frac{1 \cdot 3}{2 \cdot 4}, \qquad \frac{-\frac{1}{2}\left(-\frac{3}{2}\right)\left(-\frac{5}{2}\right)}{1 \cdot 2 \cdot 3} = \frac{1 \cdot 3 \cdot 5}{2 \cdot 4 \cdot 6}$$

The general pattern is

$$\binom{-\frac{1}{2}}{n} = \frac{-\frac{1}{2}\left(-\frac{3}{2}\right)\left(-\frac{5}{2}\right) \cdots \left(-\frac{2n-1}{2}\right)}{1 \cdot 2 \cdot 3 \cdots n} = (-1)^n \frac{1 \cdot 3 \cdot 5 \cdots (2n - 1)}{2 \cdot 4 \cdot 6 \cdot 2n}$$

Thus, the following binomial expansion is valid for $|x| < 1$:

$$\frac{1}{\sqrt{1 + x}} = 1 + \sum_{n=1}^{\infty}(-1)^n \frac{1 \cdot 3 \cdot 5 \cdots (2n - 1)}{2 \cdot 4 \cdot 6 \cdots (2n)}x^n = 1 - \frac{1}{2}x + \frac{1 \cdot 3}{2 \cdot 4}x^2 - \cdots$$

If $|x| < 1$, then $|x|^2 < 1$, and we can substitute $-x^2$ for x to obtain

$$\frac{1}{\sqrt{1-x^2}} = 1 + \sum_{n=1}^{\infty} \frac{1 \cdot 3 \cdot 5 \cdots (2n-1)}{2 \cdot 4 \cdot 6 \cdots 2n} x^{2n} = 1 + \frac{1}{2}x^2 + \frac{1 \cdot 3}{2 \cdot 4}x^4 + \cdots \qquad \boxed{4}$$

Taylor series are particularly useful for studying the so-called *special functions* (such as Bessel and hypergeometric functions) that appear in a wide range of physics and engineering applications. One example is the following **elliptic integral of the first kind**, defined for $|k| < 1$:

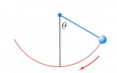

FIGURE 2 Pendulum released at an angle θ.

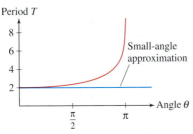

FIGURE 3 The period T of a 1-meter pendulum as a function of the angle θ at which it is released.

$$E(k) = \int_0^{\pi/2} \frac{dt}{\sqrt{1 - k^2 \sin^2 t}}$$

This function is used in physics to compute the period T of pendulum of length L released from an angle θ (Figure 2). We can use the "small-angle approximation" $T \approx 2\pi\sqrt{L/g}$ when θ is small (where $g = 9.8$ m/s^2), but this approximation breaks down for large angles (Figure 3). The exact value of the period is $T = 4\sqrt{L/g}\,E(k)$, where $k = \sin\frac{1}{2}\theta$.

■ **EXAMPLE 12 Elliptic Function** Find the Maclaurin series for $E(k)$ and estimate $E(k)$ for $k = \sin\frac{\pi}{6}$.

Solution Substitute $x = k \sin t$ in the Taylor expansion (4):

$$\frac{1}{\sqrt{1 - k^2 \sin^2 t}} = 1 + \frac{1}{2}k^2 \sin^2 t + \frac{1 \cdot 3}{2 \cdot 4}k^4 \sin^4 t + \frac{1 \cdot 3 \cdot 5}{2 \cdot 4 \cdot 6}k^6 \sin^6 t + \cdots$$

This expansion is valid because $|k| < 1$ and hence $|x| = |k \sin t| < 1$. Thus, $E(k)$ is equal to

$$\int_0^{\pi/2} \frac{dt}{\sqrt{1 - k^2 \sin^2 t}} = \int_0^{\pi/2} dt + \sum_{n=1}^{\infty} \frac{1 \cdot 3 \cdots (2n-1)}{2 \cdot 4 \cdot (2n)} \left(\int_0^{\pi/2} \sin^{2n} t \, dt \right) k^{2n}$$

According to Exercise 80 in Section 7.2,

$$\int_0^{\pi/2} \sin^{2n} t \, dt = \left(\frac{1 \cdot 3 \cdots (2n-1)}{2 \cdot 4 \cdot (2n)} \right) \frac{\pi}{2}$$

This yields

$$E(k) = \frac{\pi}{2} + \frac{\pi}{2} \sum_{n=1}^{\infty} \left(\frac{1 \cdot 3 \cdots (2n-1)^2}{2 \cdot 4 \cdots (2n)} \right)^2 k^{2n}$$

We approximate $E(k)$ for $k = \sin\left(\frac{\pi}{6}\right) = \frac{1}{2}$ using the first five terms:

$$E\left(\frac{1}{2}\right) \approx \frac{\pi}{2}\left(1 + \left(\frac{1}{2}\right)^2 \left(\frac{1}{2}\right)^2 + \left(\frac{1 \cdot 3}{2 \cdot 4}\right)^2 \left(\frac{1}{2}\right)^4 \right.$$

$$\left. + \left(\frac{1 \cdot 3 \cdot 5}{2 \cdot 4 \cdot 6}\right)^2 \left(\frac{1}{2}\right)^6 + \left(\frac{1 \cdot 3 \cdot 5 \cdot 7}{2 \cdot 4 \cdot 6 \cdot 8}\right)^2 \left(\frac{1}{2}\right)^8 \right)$$

$$\approx 1.68517$$

The value given by a computer algebra system to seven places is $E\left(\frac{1}{2}\right) \approx 1.6856325$. ■

In Table 2, we provide a list of useful Taylor series centered at $x = 0$ (also called a Maclaurin series) and the values of x for which they converge.

TABLE 2

$f(x)$	Maclaurin series	Converges to $f(x)$ for
e^x	$\displaystyle\sum_{n=0}^{\infty} \frac{x^n}{n!} = 1 + x + \frac{x^2}{2!} + \frac{x^3}{3!} + \frac{x^4}{4!} + \cdots$	All x
$\sin x$	$\displaystyle\sum_{n=0}^{\infty} \frac{(-1)^n x^{2n+1}}{(2n+1)!} = x - \frac{x^3}{3!} + \frac{x^5}{5!} - \frac{x^7}{7!} + \cdots$	All x
$\cos x$	$\displaystyle\sum_{n=0}^{\infty} \frac{(-1)^n x^{2n}}{(2n)!} = 1 - \frac{x^2}{2!} + \frac{x^4}{4!} - \frac{x^6}{6!} + \cdots$	All x
$\dfrac{1}{1-x}$	$\displaystyle\sum_{n=0}^{\infty} x^n = 1 + x + x^2 + x^3 + x^4 + \cdots$	$|x| < 1$
$\dfrac{1}{1+x}$	$\displaystyle\sum_{n=0}^{\infty} (-1)^n x^n = 1 - x + x^2 - x^3 + x^4 - \cdots$	$|x| < 1$
$\ln(1+x)$	$\displaystyle\sum_{n=1}^{\infty} \frac{(-1)^{n-1} x^n}{n} = x - \frac{x^2}{2} + \frac{x^3}{3} - \frac{x^4}{4} + \cdots$	$|x| < 1$ and $x = 1$
$\tan^{-1} x$	$\displaystyle\sum_{n=0}^{\infty} \frac{(-1)^n x^{2n+1}}{2n+1} = x - \frac{x^3}{3} + \frac{x^5}{5} - \frac{x^7}{7} + \cdots$	$|x| \leq 1$
$(1+x)^a$	$\displaystyle\sum_{n=0}^{\infty} \binom{a}{n} x^n = 1 + ax + \frac{a(a-1)}{2!}x^2 + \frac{a(a-1)(a-2)}{3!}x^3 + \cdots$	$|x| < 1$

10.7 SUMMARY

- *Taylor series of $f(x)$ centered at $x = c$:*

$$T(x) = \sum_{n=0}^{\infty} \frac{f^{(n)}(c)}{n!}(x - c)^n$$

The partial sum $T_k(x)$ is the kth Taylor polynomial.
- *Maclaurin series ($c = 0$):*

$$T(x) = \sum_{n=0}^{\infty} \frac{f^{(n)}(0)}{n!} x^n$$

- If $f(x)$ is represented by a power series $\displaystyle\sum_{n=0}^{\infty} a_n(x - c)^n$ for $|x - c| < R$ with $R > 0$, then this power series is necessarily the Taylor series centered at $x = c$.
- A function f is represented by its Taylor series $T(x)$ if and only if the remainder $R_k(x) = f(x) - T_k(x)$ tends to zero as $k \to \infty$.
- Let $I = (c - R, c + R)$ with $R > 0$. Suppose that there exists $K > 0$ such that $|f^{(k)}(x)| < K$ for all $x \in I$ and all k. Then f is represented by its Taylor series on I; that is, $f(x) = T(x)$ for $x \in I$.
- A good way to find the Taylor series of a function is to start with known Taylor series and apply one of the following operations: differentiation, integration, multiplication, or substitution.
- For any exponent a, the binomial expansion is valid for $|x| < 1$:

$$(1+x)^a = 1 + ax + \frac{a(a-1)}{2!}x^2 + \frac{a(a-1)(a-2)}{3!}x^3 + \cdots + \binom{a}{n}x^n + \cdots$$

10.7 EXERCISES

Preliminary Questions

1. Determine $f(0)$ and $f'''(0)$ for a function f with Maclaurin series

$$T(x) = 3 + 2x + 12x^2 + 5x^3 + \cdots$$

2. Determine $f(-2)$ and $f^{(4)}(-2)$ for a function with Taylor series

$$T(x) = 3(x+2) + (x+2)^2 - 4(x+2)^3 + 2(x+2)^4 + \cdots$$

3. What is the easiest way to find the Maclaurin series for the function $f(x) = \sin(x^2)$?

4. Find the Taylor series for f centered at $c = 3$ if $f(3) = 4$ and $f'(x)$ has a Taylor expansion

$$f'(x) = \sum_{n=1}^{\infty} \frac{(x-3)^n}{n}$$

5. Let $T(x)$ be the Maclaurin series of $f(x)$. Which of the following guarantees that $f(2) = T(2)$?
(a) $T(x)$ converges for $x = 2$.
(b) The remainder $R_k(2)$ approaches a limit as $k \to \infty$.
(c) The remainder $R_k(2)$ approaches zero as $k \to \infty$.

Exercises

1. Write out the first four terms of the Maclaurin series of $f(x)$ if

$$f(0) = 2, \quad f'(0) = 3, \quad f''(0) = 4, \quad f'''(0) = 12$$

2. Write out the first four terms of the Taylor series of $f(x)$ centered at $c = 3$ if

$$f(3) = 1, \quad f'(3) = 2, \quad f''(3) = 12, \quad f'''(3) = 3$$

In Exercises 3–18, find the Maclaurin series and find the interval on which the expansion is valid.

3. $f(x) = \dfrac{1}{1-2x}$

4. $f(x) = \dfrac{x}{1-x^4}$

5. $f(x) = \cos 3x$

6. $f(x) = \sin(2x)$

7. $f(x) = \sin(x^2)$

8. $f(x) = e^{4x}$

9. $f(x) = \ln(1-x^2)$

10. $f(x) = (1-x)^{-1/2}$

11. $f(x) = \tan^{-1}(x^2)$

12. $f(x) = x^2 e^{x^2}$

13. $f(x) = e^{x-2}$

14. $f(x) = \dfrac{1-\cos x}{x}$

15. $f(x) = \ln(1-5x)$

16. $f(x) = (x^2+2x)e^x$

17. $f(x) = \sinh x$

18. $f(x) = \cosh x$

In Exercises 19–28, find the terms through degree 4 of the Maclaurin series of $f(x)$. Use multiplication and substitution as necessary.

19. $f(x) = e^x \sin x$

20. $f(x) = e^x \ln(1-x)$

21. $f(x) = \dfrac{\sin x}{1-x}$

22. $f(x) = \dfrac{1}{1+\sin x}$

23. $f(x) = (1+x)^{1/4}$

24. $f(x) = (1+x)^{-3/2}$

25. $f(x) = e^x \tan^{-1} x$

26. $f(x) = \sin(x^3 - x)$

27. $f(x) = e^{\sin x}$

28. $f(x) = e^{(e^x)}$

In Exercises 29–38, find the Taylor series centered at c and the interval on which the expansion is valid.

29. $f(x) = \dfrac{1}{x}, \quad c = 1$

30. $f(x) = e^{3x}, \quad c = -1$

31. $f(x) = \dfrac{1}{1-x}, \quad c = 5$

32. $f(x) = \sin x, \quad c = \dfrac{\pi}{2}$

33. $f(x) = x^4 + 3x - 1, \quad c = 2$

34. $f(x) = x^4 + 3x - 1, \quad c = 0$

35. $f(x) = \dfrac{1}{x^2}, \quad c = 4$

36. $f(x) = \sqrt{x}, \quad c = 4$

37. $f(x) = \dfrac{1}{1-x^2}, \quad c = 3$

38. $f(x) = \dfrac{1}{3x-2}, \quad c = -1$

39. Use the identity $\cos^2 x = \frac{1}{2}(1 + \cos 2x)$ to find the Maclaurin series for $f(x) = \cos^2 x$.

40. Show that for $|x| < 1$,

$$\tanh^{-1} x = x + \frac{x^3}{3} + \frac{x^5}{5} + \cdots$$

Hint: Recall that $\dfrac{d}{dx} \tanh^{-1} x = \dfrac{1}{1-x^2}$.

41. Use the Maclaurin series for $\ln(1+x)$ and $\ln(1-x)$ to show that

$$\frac{1}{2} \ln\left(\frac{1+x}{1-x}\right) = x + \frac{x^3}{3} + \frac{x^5}{5} + \cdots$$

for $|x| < 1$. What can you conclude by comparing this result with that of Exercise 40?

42. Differentiate the Maclaurin series for $\dfrac{1}{1-x}$ twice to find the Maclaurin series of $\dfrac{1}{(1-x)^3}$.

43. Show, by integrating the Maclaurin series for $f(x) = \dfrac{1}{\sqrt{1-x^2}}$, that for $|x| < 1$,

$$\sin^{-1} x = x + \sum_{n=1}^{\infty} \frac{1 \cdot 3 \cdot 5 \cdots (2n-1)}{2 \cdot 4 \cdot 6 \cdots (2n)} \frac{x^{2n+1}}{2n+1}$$

44. Use the first five terms of the Maclaurin series in Exercise 43 to approximate $\sin^{-1} \frac{1}{2}$. Compare the result with the calculator value.

45. How many terms of the Maclaurin series of $f(x) = \ln(1+x)$ are needed to compute $\ln 1.2$ to within an error of at most 0.0001? Make the computation and compare the result with the calculator value.

46. Show that

$$\pi - \frac{\pi^3}{3!} + \frac{\pi^5}{5!} - \frac{\pi^7}{7!} + \cdots$$

converges to zero. How many terms must be computed to get within 0.01 of zero?

47. Use the Maclaurin expansion for e^{-t^2} to express the function $F(x) = \int_0^x e^{-t^2}\, dt$ as an alternating power series in x (Figure 4).
(a) How many terms of the Maclaurin series are needed to approximate the integral for $x = 1$ to within an error of at most 0.001?

(b) $\mathcal{CAS}$ Carry out the computation and check your answer using a computer algebra system.

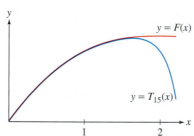

FIGURE 4 The Maclaurin polynomial $T_{15}(x)$ for $F(t) = \int_0^x e^{-t^2} dt$.

48. Let $F(x) = \int_0^x \dfrac{\sin t \, dt}{t}$. Show that

$$F(x) = x - \frac{x^3}{3 \cdot 3!} + \frac{x^5}{5 \cdot 5!} - \frac{x^7}{7 \cdot 7!} + \cdots$$

Evaluate $F(1)$ to three decimal places.

In Exercises 49–52, express the definite integral as an infinite series and find its value to within an error of at most 10^{-4}.

49. $\displaystyle\int_0^1 \cos(x^2) \, dx$ **50.** $\displaystyle\int_0^1 \tan^{-1}(x^2) \, dx$

51. $\displaystyle\int_0^1 e^{-x^3} \, dx$ **52.** $\displaystyle\int_0^1 \frac{dx}{\sqrt{x^4 + 1}}$

In Exercises 53–56, express the integral as an infinite series.

53. $\displaystyle\int_0^x \frac{1 - \cos t}{t} \, dt$, for all x **54.** $\displaystyle\int_0^x \frac{t - \sin t}{t} \, dt$, for all x

55. $\displaystyle\int_0^x \ln(1 + t^2) \, dt$, for $|x| < 1$ **56.** $\displaystyle\int_0^x \frac{dt}{\sqrt{1 - t^4}}$, for $|x| < 1$

57. Which function has Maclaurin series $\displaystyle\sum_{n=0}^{\infty} (-1)^n 2^n x^n$?

58. Which function has Maclaurin series

$$\sum_{k=0}^{\infty} \frac{(-1)^k}{3^{k+1}} (x - 3)^k?$$

For which values of x is the expansion valid?

59. Using Maclaurin series, determine to exactly what value the following series converges:

$$\sum_{n=0}^{\infty} (-1)^n \frac{(\pi)^{2n}}{(2n)!}$$

60. Using Maclaurin series, determine to exactly what value the following series converges:

$$\sum_{n=0}^{\infty} \frac{(\ln 5)^n}{n!}$$

In Exercises 61–64, use Theorem 2 to prove that the $f(x)$ is represented by its Maclaurin series for all x.

61. $f(x) = \sin(x/2) + \cos(x/3)$ **62.** $f(x) = e^{-x}$

63. $f(x) = \sinh x$ **64.** $f(x) = (1 + x)^{100}$

In Exercises 65–68, find the functions with the following Maclaurin series (refer to Table 2 on page 571).

65. $1 + x^3 + \dfrac{x^6}{2!} + \dfrac{x^9}{3!} + \dfrac{x^{12}}{4!} + \cdots$

66. $1 - 4x + 4^2 x^2 - 4^3 x^3 + 4^4 x^4 - 4^5 x^5 + \cdots$

67. $1 - \dfrac{5^3 x^3}{3!} + \dfrac{5^5 x^5}{5!} - \dfrac{5^7 x^7}{7!} + \cdots$

68. $x^4 - \dfrac{x^{12}}{3} + \dfrac{x^{20}}{5} - \dfrac{x^{28}}{7} + \cdots$

In Exercises 69 and 70, let

$$f(x) = \frac{1}{(1 - x)(1 - 2x)}$$

69. Find the Maclaurin series of $f(x)$ using the identity

$$f(x) = \frac{2}{1 - 2x} - \frac{1}{1 - x}$$

70. Find the Taylor series for $f(x)$ at $c = 2$. *Hint:* Rewrite the identity of Exercise 69 as

$$f(x) = \frac{2}{-3 - 2(x - 2)} - \frac{1}{-1 - (x - 2)}$$

71. When a voltage V is applied to a series circuit consisting of a resistor R and an inductor L, the current at time t is

$$I(t) = \left(\frac{V}{R}\right)\left(1 - e^{-Rt/L}\right)$$

Expand $I(t)$ in a Maclaurin series. Show that $I(t) \approx \dfrac{Vt}{L}$ for small t.

72. Use the result of Exercise 71 and your knowledge of alternating series to show that

$$\frac{Vt}{L}\left(1 - \frac{R}{2L}t\right) \le I(t) \le \frac{Vt}{L} \qquad \text{(for all } t\text{)}$$

73. Find the Maclaurin series for $f(x) = \cos(x^3)$ and use it to determine $f^{(6)}(0)$.

74. Find $f^{(7)}(0)$ and $f^{(8)}(0)$ for $f(x) = \tan^{-1} x$ using the Maclaurin series.

75. Use substitution to find the first three terms of the Maclaurin series for $f(x) = e^{x^{20}}$. How does the result show that $f^{(k)}(0) = 0$ for $1 \le k \le 19$?

76. Use the binomial series to find $f^{(8)}(0)$ for $f(x) = \sqrt{1 - x^2}$.

77. Does the Maclaurin series for $f(x) = (1 + x)^{3/4}$ converge to $f(x)$ at $x = 2$? Give numerical evidence to support your answer.

78. Explain the steps required to verify that the Maclaurin series for $f(x) = e^x$ converges to $f(x)$ for all x.

79. GU Let $f(x) = \sqrt{1 + x}$.
(a) Use a graphing calculator to compare the graph of f with the graphs of the first five Taylor polynomials for f. What do they suggest about the interval of convergence of the Taylor series?
(b) Investigate numerically whether or not the Taylor expansion for f is valid for $x = 1$ and $x = -1$.

80. Use the first five terms of the Maclaurin series for the elliptic integral $E(k)$ to estimate the period T of a 1-meter pendulum released at an angle $\theta = \frac{\pi}{4}$ (see Example 12).

81. Use Example 12 and the approximation $\sin x \approx x$ to show that the period T of a pendulum released at an angle θ has the following second-order approximation:

$$T \approx 2\pi \sqrt{\frac{L}{g}} \left(1 + \frac{\theta^2}{16} \right)$$

In Exercises 82–85, find the Maclaurin series of the function and use it to calculate the limit.

82. $\displaystyle \lim_{x \to 0} \frac{\cos x - 1 + \frac{x^2}{2}}{x^4}$

83. $\displaystyle \lim_{x \to 0} \frac{\sin x - x + \frac{x^3}{6}}{x^5}$

84. $\displaystyle \lim_{x \to 0} \frac{\tan^{-1} x - x \cos x - \frac{1}{6}x^3}{x^5}$

85. $\displaystyle \lim_{x \to 0} \left(\frac{\sin(x^2)}{x^4} - \frac{\cos x}{x^2} \right)$

Further Insights and Challenges

86. In this exercise, we show that the Maclaurin expansion of $f(x) = \ln(1 + x)$ is valid for $x = 1$.

(a) Show that for all $x \ne -1$,

$$\frac{1}{1+x} = \sum_{n=0}^{N} (-1)^n x^n + \frac{(-1)^{N+1} x^{N+1}}{1+x}$$

(b) Integrate from 0 to 1 to obtain

$$\ln 2 = \sum_{n=1}^{N} \frac{(-1)^{n-1}}{n} + (-1)^{N+1} \int_0^1 \frac{x^{N+1} \, dx}{1+x}$$

(c) Verify that the integral on the right tends to zero as $N \to \infty$ by showing that it is smaller than $\int_0^1 x^{N+1} dx$.

(d) Prove the formula

$$\ln 2 = 1 - \frac{1}{2} + \frac{1}{3} - \frac{1}{4} + \cdots$$

87. Let $g(t) = \dfrac{1}{1+t^2} - \dfrac{t}{1+t^2}$.

(a) Show that $\displaystyle \int_0^1 g(t) \, dt = \frac{\pi}{4} - \frac{1}{2} \ln 2$.

(b) Show that $g(t) = 1 - t - t^2 + t^3 + t^4 - t^5 - t^6 + \cdots$.

(c) Evaluate $S = 1 - \frac{1}{2} - \frac{1}{3} + \frac{1}{4} + \frac{1}{5} - \frac{1}{6} - \frac{1}{7} + \cdots$.

In Exercises 88 and 89, we investigate the convergence of the binomial series

$$T_a(x) = \sum_{n=0}^{\infty} \binom{a}{n} x^n$$

88. Prove that $T_a(x)$ has radius of convergence $R = 1$ if a is not a whole number. What is the radius of convergence if a is a whole number?

89. By Exercise 88, $T_a(x)$ converges for $|x| < 1$, but we do not yet know whether $T_a(x) = (1 + x)^a$.

(a) Verify the identity

$$a \binom{a}{n} = n \binom{a}{n} + (n+1) \binom{a}{n+1}$$

(b) Use (a) to show that $y = T_a(x)$ satisfies the differential equation $(1 + x)y' = ay$ with initial condition $y(0) = 1$.

(c) Prove $T_a(x) = (1 + x)^a$ for $|x| < 1$ by showing that the derivative of the ratio $\dfrac{T_a(x)}{(1+x)^a}$ is zero.

90. The function $G(k) = \displaystyle \int_0^{\pi/2} \sqrt{1 - k^2 \sin^2 t} \, dt$ is called an **elliptic integral of the second kind**. Prove that for $|k| < 1$,

$$G(k) = \frac{\pi}{2} - \frac{\pi}{2} \sum_{n=1}^{\infty} \left(\frac{1 \cdot 3 \cdots (2n-1)}{2 \cdots 4 \cdot (2n)} \right)^2 \frac{k^{2n}}{2n-1}$$

91. Assume that $a < b$ and let L be the arc length (circumference) of the ellipse $\left(\frac{x}{a}\right)^2 + \left(\frac{y}{b}\right)^2 = 1$ shown in Figure 5. There is no explicit formula for L, but it is known that $L = 4bG(k)$, with $G(k)$ as in Exercise 90 and $k = \sqrt{1 - a^2/b^2}$. Use the first three terms of the expansion of Exercise 90 to estimate L when $a = 4$ and $b = 5$.

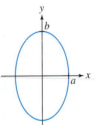

FIGURE 5 The ellipse $\left(\dfrac{x}{a}\right)^2 + \left(\dfrac{y}{b}\right)^2 = 1$.

92. Use Exercise 90 to prove that if $a < b$ and a/b is near 1 (a nearly circular ellipse), then

$$L \approx \frac{\pi}{2} \left(3b + \frac{a^2}{b} \right)$$

Hint: Use the first two terms of the series for $G(k)$.

93. Irrationality of e   Prove that e is an irrational number using the following argument by contradiction. Suppose that $e = M/N$, where M, N are nonzero integers.

(a) Show that $M! \, e^{-1}$ is a whole number.

(b) Use the power series for $f(x) = e^x$ at $x = -1$ to show that there is an integer B such that $M! \, e^{-1}$ equals

$$B + (-1)^{M+1} \left(\frac{1}{M+1} - \frac{1}{(M+1)(M+2)} + \cdots \right)$$

(c) Use your knowledge of alternating series with decreasing terms to conclude that $0 < |M! \, e^{-1} - B| < 1$ and observe that this contradicts (a). Hence, e is not equal to M/N.

94. Use the result of Exercise 75 in Section 4.5 to show that the Maclaurin series of the function

$$f(x) = \begin{cases} e^{-1/x^2} & \text{for } x \ne 0 \\ 0 & \text{for } x = 0 \end{cases}$$

is $T(x) = 0$. This provides an example of a function f whose Maclaurin series converges but does not converge to $f(x)$ (except at $x = 0$).

CHAPTER REVIEW EXERCISES

1. Let $a_n = \dfrac{n-3}{n!}$ and $b_n = a_{n+3}$. Calculate the first three terms in each sequence.

(a) a_n^2

(b) b_n

(c) $a_n b_n$

(d) $2a_{n+1} - 3a_n$

2. Prove that $\lim\limits_{n\to\infty} \dfrac{2n-1}{3n+2} = \dfrac{2}{3}$ using the limit definition.

In Exercises 3–8, compute the limit (or state that it does not exist) assuming that $\lim\limits_{n\to\infty} a_n = 2$.

3. $\lim\limits_{n\to\infty} (5a_n - 2a_n^2)$

4. $\lim\limits_{n\to\infty} \dfrac{1}{a_n}$

5. $\lim\limits_{n\to\infty} e^{a_n}$

6. $\lim\limits_{n\to\infty} \cos(\pi a_n)$

7. $\lim\limits_{n\to\infty} (-1)^n a_n$

8. $\lim\limits_{n\to\infty} \dfrac{a_n + n}{a_n + n^2}$

In Exercises 9–22, determine the limit of the sequence or show that the sequence diverges.

9. $a_n = \sqrt{n+5} - \sqrt{n+2}$

10. $a_n = \dfrac{3n^3 - n}{1 - 2n^3}$

11. $a_n = 2^{1/n^2}$

12. $a_n = \dfrac{10^n}{n!}$

13. $b_m = 1 + (-1)^m$

14. $b_m = \dfrac{1 + (-1)^m}{m}$

15. $b_n = \tan^{-1}\left(\dfrac{n+2}{n+5}\right)$

16. $a_n = \dfrac{100^n}{n!} - \dfrac{3 + \pi^n}{5^n}$

17. $b_n = \sqrt{n^2 + n} - \sqrt{n^2 + 1}$

18. $c_n = \sqrt{n^2 + n} - \sqrt{n^2 - n}$

19. $b_m = \left(1 + \dfrac{1}{m}\right)^{3m}$

20. $c_n = \left(1 + \dfrac{3}{n}\right)^n$

21. $b_n = n\big(\ln(n+1) - \ln n\big)$

22. $c_n = \dfrac{\ln(n^2 + 1)}{\ln(n^3 + 1)}$

23. Use the Squeeze Theorem to show that $\lim\limits_{n\to\infty} \dfrac{\arctan(n^2)}{\sqrt{n}} = 0$.

24. Give an example of a divergent sequence $\{a_n\}$ such that $\{\sin a_n\}$ is convergent.

25. Calculate $\lim\limits_{n\to\infty} \dfrac{a_{n+1}}{a_n}$, where $a_n = \dfrac{1}{2}3^n - \dfrac{1}{3}2^n$.

26. Define $a_{n+1} = \sqrt{a_n + 6}$ with $a_1 = 2$.

(a) Compute a_n for $n = 2, 3, 4, 5$.

(b) Show that $\{a_n\}$ is increasing and is bounded by 3.

(c) Prove that $\lim\limits_{n\to\infty} a_n$ exists and find its value.

27. Calculate the partial sums S_4 and S_7 of the series $\sum\limits_{n=1}^{\infty} \dfrac{n-2}{n^2 + 2n}$.

28. Find the sum $1 - \dfrac{1}{4} + \dfrac{1}{4^2} - \dfrac{1}{4^3} + \cdots$.

29. Find the sum $\dfrac{4}{9} + \dfrac{8}{27} + \dfrac{16}{81} + \dfrac{32}{243} + \cdots$.

30. Use series to determine a reduced fraction that has decimal expansion $0.121212\cdots$.

31. Use series to determine a reduced fraction that has decimal expansion $0.108108108\cdots$.

32. Find the sum $\sum\limits_{n=2}^{\infty} \left(\dfrac{2}{e}\right)^n$.

33. Find the sum $\sum\limits_{n=-1}^{\infty} \dfrac{2^{n+3}}{3^n}$.

34. Show that $\sum\limits_{n=1}^{\infty} \left(b - \tan^{-1} n^2\right)$ diverges if $b \neq \dfrac{\pi}{2}$.

35. Give an example of divergent series $\sum\limits_{n=1}^{\infty} a_n$ and $\sum\limits_{n=1}^{\infty} b_n$ such that $\sum\limits_{n=1}^{\infty} (a_n + b_n) = 1$.

36. Let $S = \sum\limits_{n=1}^{\infty} \left(\dfrac{1}{n} - \dfrac{1}{n+2}\right)$. Compute S_N for $N = 1, 2, 3, 4$. Find S by showing that

$$S_N = \dfrac{3}{2} - \dfrac{1}{N+1} - \dfrac{1}{N+2}$$

37. Evaluate $S = \sum\limits_{n=3}^{\infty} \dfrac{1}{n(n+3)}$.

38. Find the total area of the infinitely many circles on the interval $[0, 1]$ in Figure 1.

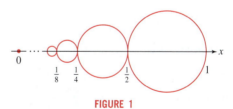

FIGURE 1

In Exercises 39–42, use the Integral Test to determine whether the infinite series converges.

39. $\sum\limits_{n=1}^{\infty} \dfrac{n^2}{n^3 + 1}$

40. $\sum\limits_{n=1}^{\infty} \dfrac{n^2}{(n^3 + 1)^{1.01}}$

41. $\sum\limits_{n=1}^{\infty} \dfrac{1}{(n+2)(\ln(n+2))^3}$

42. $\sum\limits_{n=1}^{\infty} \dfrac{n^3}{e^{n^4}}$

In Exercises 43–50, use the Direct Comparison or Limit Comparison Test to determine whether the infinite series converges.

43. $\sum\limits_{n=1}^{\infty} \dfrac{1}{(n+1)^2}$

44. $\sum\limits_{n=1}^{\infty} \dfrac{1}{\sqrt{n} + n}$

45. $\sum\limits_{n=2}^{\infty} \dfrac{n^2 + 1}{n^{3.5} - 2}$

46. $\sum\limits_{n=1}^{\infty} \dfrac{1}{n - \ln n}$

47. $\sum\limits_{n=2}^{\infty} \dfrac{n}{\sqrt{n^5 + 5}}$

48. $\sum\limits_{n=1}^{\infty} \dfrac{1}{3^n - 2^n}$

49. $\displaystyle\sum_{n=1}^{\infty} \frac{n^{10} + 10^n}{n^{11} + 11^n}$

50. $\displaystyle\sum_{n=1}^{\infty} \frac{n^{20} + 21^n}{n^{21} + 20^n}$

51. Determine the convergence of $\displaystyle\sum_{n=1}^{\infty} \frac{2^n + n}{3^n - 2}$ using the Limit Comparison Test with $b_n = \left(\dfrac{2}{3}\right)^n$.

52. Determine the convergence of $\displaystyle\sum_{n=1}^{\infty} \frac{\ln n}{1.5^n}$ using the Limit Comparison Test with $b_n = \dfrac{1}{1.4^n}$.

53. Let $a_n = 1 - \sqrt{1 - \frac{1}{n}}$. Show that $\displaystyle\lim_{n\to\infty} a_n = 0$ and that $\displaystyle\sum_{n=1}^{\infty} a_n$ diverges. *Hint:* Show that $a_n \geq \dfrac{1}{2n}$.

54. Determine whether $\displaystyle\sum_{n=2}^{\infty} \left(1 - \sqrt{1 - \frac{1}{n^2}}\right)$ converges.

55. Let $S = \displaystyle\sum_{n=1}^{\infty} \frac{n}{(n^2 + 1)^2}$.

(a) Show that S converges.

(b) *CAS* Use Eq. (4) in Exercise 85 of Section 10.3 with $M = 99$ to approximate S. What is the maximum size of the error?

In Exercises 56–59, determine whether the series converges absolutely. If it does not, determine whether it converges conditionally.

56. $\displaystyle\sum_{n=1}^{\infty} \frac{(-1)^n}{\sqrt[3]{n} + 2n}$

57. $\displaystyle\sum_{n=1}^{\infty} \frac{(-1)^n}{n^{1.1} \ln(n + 1)}$

58. $\displaystyle\sum_{n=1}^{\infty} \frac{\cos\left(\frac{\pi}{4} + \pi n\right)}{\sqrt{n}}$

59. $\displaystyle\sum_{n=1}^{\infty} \frac{\cos\left(\frac{\pi}{4} + 2\pi n\right)}{\sqrt{n}}$

60. *CAS* Use a computer algebra system to approximate $\displaystyle\sum_{n=1}^{\infty} \frac{(-1)^n}{n^3 + \sqrt{n}}$ to within an error of at most 10^{-5}.

61. Catalan's constant is defined by $K = \displaystyle\sum_{k=0}^{\infty} \frac{(-1)^k}{(2k + 1)^2}$.

(a) How many terms of the series are needed to calculate K with an error of less than 10^{-6}?

(b) *CAS* Carry out the calculation.

62. Give an example of conditionally convergent series $\displaystyle\sum_{n=1}^{\infty} a_n$ and $\displaystyle\sum_{n=1}^{\infty} b_n$ such that $\displaystyle\sum_{n=1}^{\infty}(a_n + b_n)$ converges absolutely.

63. Let $\displaystyle\sum_{n=1}^{\infty} a_n$ be an absolutely convergent series. Determine whether the following series are convergent or divergent:

(a) $\displaystyle\sum_{n=1}^{\infty} \left(a_n + \frac{1}{n^2}\right)$

(b) $\displaystyle\sum_{n=1}^{\infty} (-1)^n a_n$

(c) $\displaystyle\sum_{n=1}^{\infty} \frac{1}{1 + a_n^2}$

(d) $\displaystyle\sum_{n=1}^{\infty} \frac{|a_n|}{n}$

64. Let $\{a_n\}$ be a positive sequence such that $\displaystyle\lim_{n\to\infty} \sqrt[n]{a_n} = \frac{1}{2}$. Determine whether the following series converge or diverge:

(a) $\displaystyle\sum_{n=1}^{\infty} 2a_n$

(b) $\displaystyle\sum_{n=1}^{\infty} 3^n a_n$

(c) $\displaystyle\sum_{n=1}^{\infty} \sqrt{a_n}$

In Exercises 65–72, apply the Ratio Test to determine convergence or divergence, or state that the Ratio Test is inconclusive.

65. $\displaystyle\sum_{n=1}^{\infty} \frac{n^5}{5^n}$

66. $\displaystyle\sum_{n=1}^{\infty} \frac{\sqrt{n + 1}}{n^8}$

67. $\displaystyle\sum_{n=1}^{\infty} \frac{1}{n2^n + n^3}$

68. $\displaystyle\sum_{n=1}^{\infty} \frac{n^4}{n!}$

69. $\displaystyle\sum_{n=1}^{\infty} \frac{2^{n^2}}{n!}$

70. $\displaystyle\sum_{n=4}^{\infty} \frac{\ln n}{n^{3/2}}$

71. $\displaystyle\sum_{n=1}^{\infty} \left(\frac{n}{2}\right)^n \frac{1}{n!}$

72. $\displaystyle\sum_{n=1}^{\infty} \left(\frac{n}{4}\right)^n \frac{1}{n!}$

In Exercises 73–76, apply the Root Test to determine convergence or divergence, or state that the Root Test is inconclusive.

73. $\displaystyle\sum_{n=1}^{\infty} \frac{1}{4^n}$

74. $\displaystyle\sum_{n=1}^{\infty} \left(\frac{2}{n}\right)^n$

75. $\displaystyle\sum_{n=1}^{\infty} \left(\frac{3}{4n}\right)^n$

76. $\displaystyle\sum_{n=1}^{\infty} \left(\cos \frac{1}{n}\right)^{n^3}$

In Exercises 77–100, determine convergence or divergence using any method covered in the text.

77. $\displaystyle\sum_{n=1}^{\infty} \left(\frac{2}{3}\right)^n$

78. $\displaystyle\sum_{n=1}^{\infty} \frac{\pi^{7n}}{e^{8n}}$

79. $\displaystyle\sum_{n=1}^{\infty} e^{-0.02n}$

80. $\displaystyle\sum_{n=1}^{\infty} ne^{-0.02n}$

81. $\displaystyle\sum_{n=1}^{\infty} \frac{(-1)^{n-1}}{\sqrt{n} + \sqrt{n + 1}}$

82. $\displaystyle\sum_{n=10}^{\infty} \frac{1}{n(\ln n)^{3/2}}$

83. $\displaystyle\sum_{n=2}^{\infty} \frac{(-1)^n}{\ln n}$

84. $\displaystyle\sum_{n=1}^{\infty} \frac{n!}{(2n)!}$

85. $\displaystyle\sum_{n=2}^{\infty} \frac{n}{1 + 100n}$

86. $\displaystyle\sum_{n=2}^{\infty} \frac{n^3 - 2n^2 + n - 4}{2n^4 + 3n^3 - 4n^2 - 1}$

87. $\displaystyle\sum_{n=1}^{\infty} \frac{\cos n}{n^{3/2}}$

88. $\displaystyle\sum_{n=1}^{\infty} \frac{n}{\sqrt{n^{3/2} + 1}}$

89. $\displaystyle\sum_{n=1}^{\infty} \left(\frac{n}{5n + 2}\right)^n$

90. $\displaystyle\sum_{n=1}^{\infty} \frac{e^n}{n!}$

91. $\displaystyle\sum_{n=1}^{\infty} \frac{1}{n\sqrt{n} + \ln n}$

92. $\displaystyle\sum_{n=1}^{\infty} \frac{1}{\sqrt[3]{n}(1 + \sqrt{n})}$

93. $\displaystyle\sum_{n=1}^{\infty} \left(\frac{1}{\sqrt{n}} - \frac{1}{\sqrt{n + 1}}\right)$

94. $\displaystyle\sum_{n=1}^{\infty} \left(\ln n - \ln(n + 1)\right)$

95. $\displaystyle\sum_{n=1}^{\infty} \frac{1}{n + \sqrt{n}}$

96. $\displaystyle\sum_{n=2}^{\infty} \frac{\cos(\pi n)}{n^{2/3}}$

97. $\displaystyle\sum_{n=2}^{\infty} \frac{1}{n^{\ln n}}$

98. $\displaystyle\sum_{n=2}^{\infty} \frac{1}{\ln^3 n}$

99. $\displaystyle\sum_{n=1}^{\infty} \sin^2 \frac{\pi}{n}$

100. $\displaystyle\sum_{n=0}^{\infty} \frac{2^{2n}}{n!}$

In Exercises 101–106, find the interval of convergence of the power series.

101. $\displaystyle\sum_{n=0}^{\infty} \frac{2^n x^n}{n!}$

102. $\displaystyle\sum_{n=0}^{\infty} \frac{x^n}{n + 1}$

103. $\displaystyle\sum_{n=0}^{\infty} \frac{n^6}{n^8 + 1}(x - 3)^n$

104. $\displaystyle\sum_{n=0}^{\infty} n x^n$

105. $\displaystyle\sum_{n=0}^{\infty} (nx)^n$

106. $\displaystyle\sum_{n=2}^{\infty} \frac{(2x - 3)^n}{n \ln n}$

107. Expand $f(x) = \dfrac{2}{4 - 3x}$ as a power series centered at $c = 0$. Determine the values of x for which the series converges.

108. Prove that

$$\sum_{n=0}^{\infty} n e^{-nx} = \frac{e^{-x}}{(1 - e^{-x})^2}$$

Hint: Express the left-hand side as the derivative of a geometric series.

109. Let $F(x) = \displaystyle\sum_{k=0}^{\infty} \frac{x^{2k}}{2^k \cdot k!}$.

(a) Show that $F(x)$ has infinite radius of convergence.

(b) Show that $y = F(x)$ is a solution of

$$y'' = xy' + y, \qquad y(0) = 1, \qquad y'(0) = 0$$

(c) $\mathcal{CRS}$ Plot the partial sums S_N for $N = 1, 3, 5, 7$ on the same set of axes.

110. Find a power series $P(x) = \displaystyle\sum_{n=0}^{\infty} a_n x^n$ that satisfies the Laguerre differential equation

$$xy'' + (1 - x)y' - y = 0$$

with initial condition satisfying $P(0) = 1$.

111. Use power series to evaluate $\displaystyle\lim_{x \to 0} \frac{x^2 e^x}{\cos x - 1}$.

112. Use power series to evaluate $\displaystyle\lim_{x \to 0} \frac{x^2(1 - \ln(x + 1))}{\sin x - x}$.

In Exercises 113–122, find the Taylor series centered at c.

113. $f(x) = e^{4x}, \quad c = 0$

114. $f(x) = e^{2x}, \quad c = -1$

115. $f(x) = x^4, \quad c = 2$

116. $f(x) = x^3 - x, \quad c = -2$

117. $f(x) = \sin x, \quad c = \pi$

118. $f(x) = e^{x-1}, \quad c = -1$

119. $f(x) = \dfrac{1}{1 - 2x}, \quad c = -2$

120. $f(x) = \dfrac{1}{(1 - 2x)^2}, \quad c = -2$

121. $f(x) = \ln \dfrac{x}{2}, \quad c = 2$

122. $f(x) = x \ln\left(1 + \dfrac{x}{2}\right), \quad c = 0$

In Exercises 123–126, find the first three terms of the Maclaurin series of $f(x)$ and use it to calculate $f^{(3)}(0)$.

123. $f(x) = (x^2 - x)e^{x^2}$

124. $f(x) = \tan^{-1}(x^2 - x)$

125. $f(x) = \dfrac{1}{1 + \tan x}$

126. $f(x) = (\sin x)\sqrt{1 + x}$

127. Calculate $\dfrac{\pi}{2} - \dfrac{\pi^3}{2^3 3!} + \dfrac{\pi^5}{2^5 5!} - \dfrac{\pi^7}{2^7 7!} + \cdots$.

128. Find the Maclaurin series of the function $F(x) = \displaystyle\int_0^x \frac{e^t - 1}{t}\,dt$.

11 PARAMETRIC EQUATIONS, POLAR COORDINATES, AND CONIC SECTIONS

This chapter introduces two important new tools. First, we consider parametric equations, which describe curves in a form that is particularly useful for analyzing motion and is indispensable in fields such as computer graphics and computer-aided design. We then study polar coordinates, an alternative to rectangular coordinates that simplifies computations in many applications. The chapter closes with a discussion of the conic sections (ellipses, hyperbolas, and parabolas).

Ellipses, which are a type of conic section, and parametric equations are used to track satellites as they orbit Earth. (*Chad Baker/Getty Images*)

11.1 Parametric Equations

Imagine a particle moving along a curve $\mathcal{C}$ in the plane as in Figure 1. We would like to be able to describe the particle's motion; however, the curve $\mathcal{C}$ is not the graph of a function $y = h(x)$ since it fails the vertical line test. Instead, we can describe the particle's motion by specifying its coordinates as functions of time t:

We use the term "particle" when we treat an object as a moving point, ignoring its internal structure.

$$x = f(t), \qquad y = g(t) \qquad \boxed{1}$$

In other words, at time t, the particle is located at the point

$$c(t) = (f(t), g(t))$$

The equations (1) are called **parametric equations**, and $\mathcal{C}$ is called a **parametric curve**. We refer to $c(t)$ as a **parametrization** with **parameter** t.

Because x and y are functions of t, we often write $c(t) = (x(t), y(t))$ instead of $(f(t), g(t))$. Of course, we are free to use any variable for the parameter (such as s or θ). In plots of parametric curves, the direction of motion is often indicated by an arrow as in Figure 1.

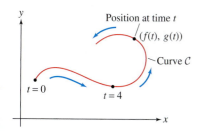

DF **FIGURE 1** Particle moving along a curve $\mathcal{C}$ in the plane.

■ **EXAMPLE 1** Sketch the curve with parametric equations

$$x = 2t - 4, \qquad y = 3 + t^2 \qquad \boxed{2}$$

Solution First compute the x- and y-coordinates for several values of t as in Table 1, and plot the corresponding points (x, y) as in Figure 2. Then join the points by a smooth curve, indicating the direction of motion (direction of increasing t) with an arrow. ■

TABLE 1		
t	$x = 2t - 4$	$y = 3 + t^2$
-2	-8	7
0	-4	3
2	0	7
4	4	19

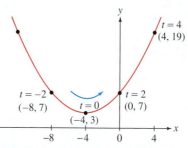

DF **FIGURE 2** The parametric curve $x = 2t - 4$, $y = 3 + t^2$.

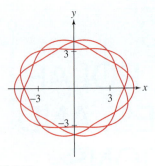

FIGURE 3 The parametric curve
$x = 5\cos(3t)\cos\left(\frac{2}{3}\sin(5t)\right)$,
$y = 4\sin(3t)\cos\left(\frac{2}{3}\sin(5t)\right)$.

CONCEPTUAL INSIGHT The graph of $y = x^2$ can be parametrized in a simple way. We take $x = t$, and we take $y = t^2$. Then, clearly this generates the right curve, since eliminating the variable t returns us to the original equation $y = x^2$. Thus, the parabola $y = x^2$ is parametrized by $c(t) = (t, t^2)$. More generally, we can parametrize the graph of $y = f(x)$ by taking $x = t$ and $y = f(t)$. Therefore, $c(t) = (t, f(t))$ parametrizes the graph. For another example, the graph of $y = e^x$ is parametrized by $c(t) = (t, e^t)$. An advantage of parametric equations is that they enable us to describe curves that are not graphs of functions. For example, the curve in Figure 3 is not of the form $y = f(x)$ but it can be expressed parametrically.

As we have just noted, a parametric curve $c(t)$ need not be the graph of a function. If it is, however, it may be possible to find the function f by "eliminating the parameter" as in the next example.

■ **EXAMPLE 2** **Eliminating the Parameter** Describe the parametric curve

$$c(t) = (2t - 4, 3 + t^2)$$

of the previous example in the form $y = f(x)$.

Solution We "eliminate the parameter" by solving for y as a function of x. First, express t in terms of x: Since $x = 2t - 4$, we have $t = \frac{1}{2}x + 2$. Then substitute

$$y = 3 + t^2 = 3 + \left(\frac{1}{2}x + 2\right)^2 = 7 + 2x + \frac{1}{4}x^2$$

Thus, $c(t)$ traces out the graph of $f(x) = 7 + 2x + \frac{1}{4}x^2$ shown in Figure 2. ■

■ **EXAMPLE 3** A model rocket follows the trajectory

$$c(t) = (80t, 200t - 4.9t^2)$$

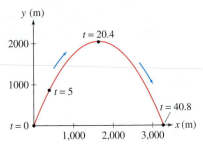

DF **FIGURE 4** Trajectory of rocket.

until it hits the ground, with t in seconds and distance in meters (Figure 4). Find:

(a) The rocket's height at $t = 5$ s.   **(b)** Its maximum height.

Solution The height of the rocket at time t is $y(t) = 200t - 4.9t^2$.

(a) The height at $t = 5$ s is

$$y(5) = 200(5) - 4.9(5^2) = 877.5 \text{ m}$$

CAUTION *The graph of height versus time for an object tossed in the air is a parabola (by Galileo's formula). But keep in mind that Figure 4 is **not** a graph of height versus time. It shows the actual path of the rocket (which has both a vertical and a horizontal displacement).*

(b) The maximum height occurs at the critical point of $y(t)$:

$$y'(t) = \frac{d}{dt}(200t - 4.9t^2) = 200 - 9.8t = 0 \quad \Rightarrow \quad t = \frac{200}{9.8} \approx 20.4 \text{ s}$$

The rocket's maximum height is $y(20.4) = 200(20.4) - 4.9(20.4)^2 \approx 2041$ m.   ■

We now discuss parametrizations of lines and circles. They will appear frequently in later chapters.

THEOREM 1 **Parametrization of a Line**

(a) The line through $P = (a, b)$ of slope m is parametrized by

$$\boxed{x = a + rt, \qquad y = b + st \qquad -\infty < t < \infty}$$ **3**

for any r and s (with $r \neq 0$) such that $m = s/r$.

(b) The line through $P = (a, b)$ and $Q = (c, d)$ has parametrization

$$\boxed{x = a + t(c - a), \qquad y = b + t(d - b) \qquad -\infty < t < \infty}$$ **4**

The segment from P to Q corresponds to $0 \le t \le 1$.

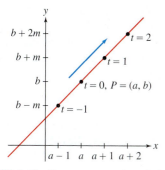

FIGURE 5 The line $y - a = m(x - b)$ has parametrization $c(t) = (a + t, b + mt)$. This corresponds to $r = 1, s = m$ in Eq. (3).

Proof **(a)** Solve $x = a + rt$ for t in terms of x to obtain $t = (x - a)/r$. Then

$$y = b + st = b + s\left(\frac{x - a}{r}\right) = b + m(x - a) \qquad \text{or} \qquad y - b = m(x - a)$$

This is the equation of the line through $P = (a, b)$ of slope m. The choice $r = 1$ and $s = m$ yields the parametrization in Figure 5.

The parametrization in **(b)** defines a line that satisfies $(x(0), y(0)) = (a, b)$ and $(x(1), y(1)) = (c, d)$. Thus, it parametrizes the line through P and Q and traces the segment from P to Q as t varies from 0 to 1. ∎

■ **EXAMPLE 4** **Parametrization of a Line** Find parametric equations for the line through $P = (3, -1)$ of slope $m = 4$.

Solution We can parametrize the line by taking $r = 1$ and $s = 4$ in Eq. (3):

$$x = 3 + t, \qquad y = -1 + 4t$$

This is also written as $c(t) = (3 + t, -1 + 4t)$. Another parametrization of the line is $c(t) = (3 + 5t, -1 + 20t)$, corresponding to $r = 5$ and $s = 20$ in Eq. (3). ■

The circle of radius R centered at the origin has the parametrization

$$x = R\cos\theta, \qquad y = R\sin\theta$$

The parameter θ represents the angle corresponding to the point (x, y) on the circle (Figure 6). The circle is traversed once in the counterclockwise direction as θ varies over a half-open interval of length 2π such as $[0, 2\pi)$ or $[-\pi, \pi)$.

More generally, the circle of radius R with center (a, b) has parametrization (Figure 6)

$$x = a + R\cos\theta, \qquad y = b + R\sin\theta \qquad \boxed{5}$$

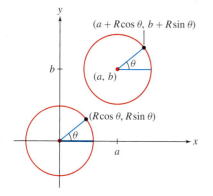

FIGURE 6 Parametrization of a circle of radius R with center (a, b).

As a check, let's verify that a point (x, y) given by Eq. (5) satisfies the equation of the circle of radius R centered at (a, b):

$$(x - a)^2 + (y - b)^2 = (a + R\cos\theta - a)^2 + (b + R\sin\theta - b)^2$$

$$= R^2\cos^2\theta + R^2\sin^2\theta = R^2$$

In general, to **translate** (meaning "to move") a parametric curve horizontally a units and vertically b units, replace $c(t) = (x(t), y(t))$ by $c(t) = (a + x(t), b + y(t))$.

Suppose we have a parametrization $c(t) = (x(t), y(t))$, where $x(t)$ is an even function and $y(t)$ is an odd function, that is, $x(-t) = x(t)$ and $y(-t) = -y(t)$. In this case, $c(-t)$ is the *reflection* of $c(t)$ across the x-axis:

$$c(-t) = (x(-t), y(-t)) = (x(t), -y(t))$$

The curve, therefore, is *symmetric* with respect to the x-axis. We apply this remark in the next example and in Example 7 below.

NOTATION *We use $x(t)$ and $y(t)$ to represent functions of t, where $x(t)$ and $y(t)$ correspond to the x-coordinate and y-coordinate of a parametric curve.*

■ **EXAMPLE 5** **Parametrization of an Ellipse** Verify that the ellipse with equation $\left(\frac{x}{a}\right)^2 + \left(\frac{y}{b}\right)^2 = 1$ is parametrized by

$$c(t) = (a\cos t, \ b\sin t) \qquad (\text{for } -\pi \le t < \pi)$$

Plot the case $a = 4, b = 2$.

Solution To verify that $c(t)$ parametrizes the ellipse, show that the equation of the ellipse is satisfied with $x = a\cos t, y = b\sin t$:

$$\left(\frac{x}{a}\right)^2 + \left(\frac{y}{b}\right)^2 = \left(\frac{a\cos t}{a}\right)^2 + \left(\frac{b\sin t}{b}\right)^2 = \cos^2 t + \sin^2 t = 1$$

TABLE 2

t	$x(t) = 4\cos t$	$y(t) = 2\sin t$
0	4	0
$\dfrac{\pi}{6}$	$2\sqrt{3}$	1
$\dfrac{\pi}{3}$	2	$\sqrt{3}$
$\dfrac{\pi}{2}$	0	2
$\dfrac{2\pi}{3}$	-2	$\sqrt{3}$
$\dfrac{5\pi}{6}$	$-2\sqrt{3}$	1
π	-4	0

To plot the case $a = 4$, $b = 2$, we connect the points for the t-values in Table 2 [see Figure 7(A)]. This gives us the top half of the ellipse for $0 \le t \le \pi$. Then we observe that $x(t) = 4\cos t$ is even and $y(t) = 2\sin t$ is odd. As noted above, this tells us that the bottom half of the ellipse is obtained by symmetry with respect to the x-axis, as in Figure 7(B). Alternatively, we could also evaluate $x(t)$ and $y(t)$ for negative values of t between $-\pi$ and 0 to determine the bottom portion of the ellipse.   ■

A parametric curve $c(t)$ is also called a **path**. This term emphasizes that $c(t)$ describes not just an underlying curve $\mathcal{C}$, but a particular way of moving along the curve.

CONCEPTUAL INSIGHT   The parametric equations for the ellipse in Example 5 illustrate a key difference between the path $c(t)$ and its underlying curve $\mathcal{C}$. The curve $\mathcal{C}$ is an ellipse in the plane, whereas $c(t)$ describes a particular, counterclockwise motion of a particle along the ellipse. If we let t vary from 0 to 4π, then the particle goes around the ellipse twice.

A key feature of parametrizations is that they are not unique. In fact, every curve can be parametrized in infinitely many different ways. For instance, the parabola $y = x^2$ is parametrized not only by (t, t^2) but also by (t^3, t^6), or (t^5, t^{10}), and so on.

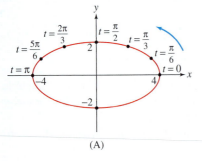

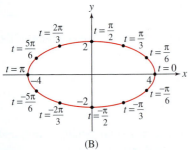

DF **FIGURE 7** Ellipse with parametric equations $x = 4\cos t$, $y = 2\sin t$.

FIGURE 8 Three parametrizations of portions of the parabola.

■ **EXAMPLE 6   Different Parametrizations of the Same Curve**   Describe the motion of a particle moving along each of the following paths:

(a) $c_1(t) = (t^3, t^6)$ **(b)** $c_2(t) = (t^2, t^4)$ **(c)** $c_3(t) = (\cos t, \cos^2 t)$

Solution   Each of these parametrizations satisfies $y = x^2$, so all three parametrize portions of the parabola $y = x^2$.

(a) As t varies from $-\infty$ to ∞, t^3 also varies from $-\infty$ to ∞. Therefore, $c_1(t) = (t^3, t^6)$ traces the entire parabola $y = x^2$, moving from left to right and passing through each point once [Figure 8(A)].

(b) Since $x = t^2 \ge 0$, the path $c_2(t) = (t^2, t^4)$ traces only the right half of the parabola. The particle comes in toward the origin as t varies from $-\infty$ to 0, and it goes back out to the right as t varies from 0 to ∞ [Figure 8(B)].

(c) As t varies from $-\infty$ and ∞, $\cos t$ oscillates between 1 and -1. Thus, a particle following the path $c_3(t) = (\cos t, \cos^2 t)$ oscillates back and forth between the points $(1, 1)$ and $(-1, 1)$ on the parabola [Figure 8(C)].   ■

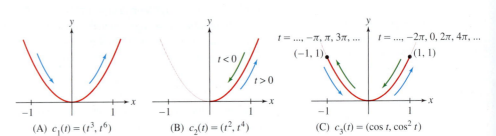

■ **EXAMPLE 7   Using Symmetry to Sketch a Loop**   Sketch the parametric curve

$$c(t) = (t^2 + 1, t^3 - 4t)$$

Label the points corresponding to $t = 0, \pm 1, \pm 2, \pm 2.5$.

Solution

Step 1. **Use symmetry.**
Observe that $x(t) = t^2 + 1$ is an even function and that $y(t) = t^3 - 4t$ is an odd function. As noted before Example 5, this tells us that $c(t)$ is symmetric with respect to the x-axis. Therefore, we will plot the curve for $t \ge 0$ and reflect across the x-axis to obtain the part for $t \le 0$.

***Step 2.* Analyze $x(t)$, $y(t)$ as functions of t.**

We have $x(t) = t^2 + 1$ and $y(t) = t^3 - 4t$. The x-coordinate $x(t) = t^2 + 1$ increases to ∞ as $t \to \infty$. To analyze the y-coordinate, we graph $y(t) = t^3 - 4t = t(t-2)(t+2)$ as a function of t (*not* as a function of x). Since $y(t)$ is the height above the x-axis, Figure 9(A) shows that

$$y(t) < 0 \quad \text{for} \quad 0 < t < 2, \quad \Rightarrow \quad \text{curve below } x\text{-axis}$$

$$y(t) > 0 \quad \text{for} \quad t > 2, \quad \Rightarrow \quad \text{curve above } x\text{-axis}$$

So, the curve starts at $c(0) = (1, 0)$, dips below the x-axis, and returns to the x-axis at $t = 2$. Both $x(t)$ and $y(t)$ tend to ∞ as $t \to \infty$. The curve is concave up because $y(t)$ increases more rapidly than $x(t)$.

***Step 3.* Plot points and join by an arc.**

The points $c(0)$, $c(1)$, $c(2)$, $c(2.5)$ tabulated in Table 3 are plotted and joined by an arc to create the sketch for $t \geq 0$ as in Figure 9(B). The sketch is completed by reflecting across the x-axis as in Figure 9(C). ∎

	TABLE 3	
t	$x = t^2 + 1$	$y = t^3 - 4t$
0	1	0
1	2	−3
2	5	0
2.5	7.25	5.625

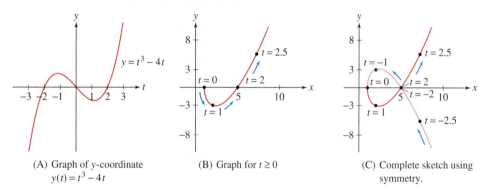

(A) Graph of y-coordinate $y(t) = t^3 - 4t$

(B) Graph for $t \geq 0$

(C) Complete sketch using symmetry.

FIGURE 9 The parametric curve $c(t) = (t^2 + 1, t^3 - 4t)$.

A **cycloid** is a curve traced by a point on the circumference of a rolling wheel as in Figure 10. Cycloids are famous for their "brachistochrone property" (see the marginal note below).

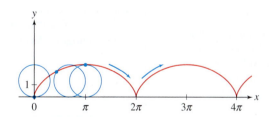

FIGURE 10 A cycloid.

A stellar cast of mathematicians (including Galileo, Pascal, Newton, Leibniz, Huygens, and Bernoulli) studied the cycloid and discovered many of its remarkable properties. A slide designed so that an object sliding down (without friction) reaches the bottom in the least time must have the shape of an inverted cycloid. This is the brachistochrone property, a term derived from the Greek brachistos, "shortest," and chronos, "time."

■ **EXAMPLE 8 Parametrizing the Cycloid** Find parametric equations for the cycloid generated by a point P on the unit circle.

Solution The point P is located at the origin at $t = 0$. At time t, the circle has rolled t radians along the x-axis and the center C of the circle then has coordinates $(t, 1)$ as in Figure 11(A). Figure 11(B) shows that we get from C to P by moving down $\cos t$ units and to the left $\sin t$ units, giving us the parametric equations

$$\boxed{x(t) = t - \sin t, \qquad y(t) = 1 - \cos t} \qquad \boxed{6}$$

∎

The argument in Example 8 shows in a similar fashion that the cycloid generated by a circle of radius R has parametric equations

$$\boxed{x = Rt - R\sin t, \qquad y = R - R\cos t} \qquad \boxed{7}$$

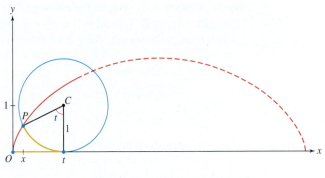

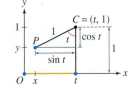

(A) Position of P at time t

(B) P has coordinates
$x = t - \sin t$, $y = 1 - \cos t$

DF FIGURE 11

Tangent Lines to Parametric Curves

Just as we use tangent lines to the graph of $y = f(x)$ to determine the rate of change of the function f, we would like to be able to determine how y changes with x when the curve is described by parametric equations. The slope of the tangent line is the derivative dy/dx, but we have to use the Chain Rule to compute it because y is not given explicitly as a function of x. Write $x = f(t)$, $y = g(t)$. Then, by the Chain Rule,

$$g'(t) = \frac{dy}{dt} = \frac{dy}{dx}\frac{dx}{dt} = \frac{dy}{dx}f'(t)$$

> **NOTATION** *In this section, we write* $f'(t)$, $x'(t)$, $y'(t)$, *and so on to denote the derivative with respect to* t.

If $f'(t) \neq 0$, we can divide by $f'(t)$ to obtain

$$\frac{dy}{dx} = \frac{g'(t)}{f'(t)}$$

This calculation is valid if $f(t)$ and $g(t)$ are differentiable, $f'(t)$ is continuous, and $f'(t) \neq 0$. In this case, the inverse $t = f^{-1}(x)$ exists, and the composite $y = g(f^{-1}(x))$ is a differentiable function of x.

> **CAUTION** *Do not confuse* dy/dx *with the derivatives* dx/dt *and* dy/dt, *which are derivatives with respect to the parameter* t. *Only* dy/dx *is the slope of the tangent line.*

> **THEOREM 2 Slope of the Tangent Line** Let $c(t) = (x(t), y(t))$, where $x(t)$ and $y(t)$ are differentiable. Assume that $x'(t)$ is continuous and $x'(t) \neq 0$. Then
>
> $$\frac{dy}{dx} = \frac{dy/dt}{dx/dt} = \frac{y'(t)}{x'(t)}$$
> **8**

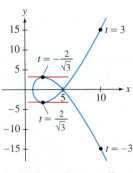

FIGURE 12 Horizontal tangent lines on $c(t) = (t^2 + 1, t^3 - 4t)$.

■ **EXAMPLE 9** Let $c(t) = (t^2 + 1, t^3 - 4t)$. Find:

(a) An equation of the tangent line at $t = 3$

(b) The points where the tangent is horizontal (Figure 12).

Solution We have

$$\frac{dy}{dx} = \frac{y'(t)}{x'(t)} = \frac{(t^3 - 4t)'}{(t^2 + 1)'} = \frac{3t^2 - 4}{2t}$$

(a) The slope at $t = 3$ is

$$\frac{dy}{dx} = \frac{3t^2 - 4}{2t}\bigg|_{t=3} = \frac{3(3)^2 - 4}{2(3)} = \frac{23}{6}$$

Since $c(3) = (10, 15)$, the equation of the tangent line in point-slope form is

$$y - 15 = \frac{23}{6}(x - 10)$$

(b) The slope dy/dx is zero if $y'(t) = 0$ and $x'(t) \neq 0$. We have $y'(t) = 3t^2 - 4 = 0$ for $t = \pm 2/\sqrt{3}$ (and $x'(t) = 2t \neq 0$ for these values of t). Therefore, the tangent line is horizontal at the points

$$c\left(-\frac{2}{\sqrt{3}}\right) = \left(\frac{7}{3}, \frac{16}{3\sqrt{3}}\right), \qquad c\left(\frac{2}{\sqrt{3}}\right) = \left(\frac{7}{3}, -\frac{16}{3\sqrt{3}}\right) \qquad \blacksquare$$

Bézier curves were invented in the 1960s by the French engineer Pierre Bézier (1910–1999), who worked for the Renault car company. They are based on the properties of Bernstein polynomials, introduced 50 years earlier by the Russian mathematician Sergei Bernstein to study the approximation of continuous functions by polynomials. Today, Bézier curves are used in standard graphics programs, such as Adobe Illustrator™ and Corel Draw™, and in the construction and storage of computer fonts such as TrueType™ and PostScript™ fonts.

Parametric curves are widely used in the field of computer graphics. A particularly important class of curves are **Bézier curves**, which we discuss here briefly in the cubic case. Given four "control points" (Figure 13):

$$P_0 = (a_0, b_0), \qquad P_1 = (a_1, b_1), \qquad P_2 = (a_2, b_2), \qquad P_3 = (a_3, b_3)$$

the Bézier curve $c(t) = (x(t), y(t))$ is defined for $0 \leq t \leq 1$ by

$$x(t) = a_0(1-t)^3 + 3a_1 t(1-t)^2 + 3a_2 t^2(1-t) + a_3 t^3 \qquad \boxed{9}$$

$$y(t) = b_0(1-t)^3 + 3b_1 t(1-t)^2 + 3b_2 t^2(1-t) + b_3 t^3 \qquad \boxed{10}$$

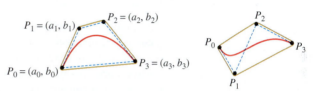

DF FIGURE 13 Cubic Bézier curves specified by four control points.

Note that $c(0) = (a_0, b_0)$ and $c(1) = (a_3, b_3)$, so the Bézier curve begins at P_0 and ends at P_3 (Figure 13). It can also be shown that the Bézier curve is contained within the quadrilateral (shown in blue) with vertices P_0, P_1, P_2, P_3. However, $c(t)$ does not pass through P_1 and P_2. Instead, these intermediate control points determine the slopes of the tangent lines at P_0 and P_3, as we show in the next example (also, see Exercises 67–70).

■ **EXAMPLE 10** Show that the Bézier curve is tangent to segment $\overline{P_0 P_1}$ at P_0.

Solution The Bézier curve passes through P_0 at $t = 0$, so we must show that the slope of the tangent line at $t = 0$ is equal to the slope of $\overline{P_0 P_1}$. To find the slope, we compute the derivatives:

$$x'(t) = -3a_0(1-t)^2 + 3a_1(1 - 4t + 3t^2) + a_2(2t - 3t^2) + 3a_3 t^2$$

$$y'(t) = -3b_0(1-t)^2 + 3b_1(1 - 4t + 3t^2) + b_2(2t - 3t^2) + 3b_3 t^2$$

Evaluating at $t = 0$, we obtain $x'(0) = 3(a_1 - a_0)$, $y'(0) = 3(b_1 - b_0)$, and

$$\left.\frac{dy}{dx}\right|_{t=0} = \frac{y'(0)}{x'(0)} = \frac{3(b_1 - b_0)}{3(a_1 - a_0)} = \frac{b_1 - b_0}{a_1 - a_0}$$

This is equal to the slope of the line through $P_0 = (a_0, b_0)$ and $P_1 = (a_1, b_1)$ as claimed (provided that $a_1 \neq a_0$). ■

Area Under a Parametric Curve

As we know, the area under a curve given by $y = h(x)$ when $h(x) \geq 0$ for $a \leq x \leq b$ is given by

$$A = \int_a^b h(x)\, dx$$

When the curve $y = h(x)$ is traced once by a parametric curve $c(t) = (x(t), y(t))$ as in Figure 14, where $x(t_0) = a$ and $x(t_1) = b$, then we can substitute, replacing $y = h(x)$ by $y(t)$ and dx by $x'(t)dt$, yielding a formula for the area A under the curve:

$$A = \int_{t_0}^{t_1} y(t) x'(t)\, dt \qquad \boxed{11}$$

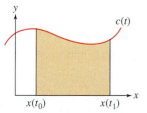

FIGURE 14 Finding area under a parametric curve $c(t)$.

■ **EXAMPLE 11** Find the area inside the ellipse of Example 5.

Solution The ellipse was found to be parametrized by $c(t) = (a \cos t, b \sin t)$ for $-\pi \le t \le \pi$. The top half of the ellipse corresponds to $0 \le t \le \pi$, and since this yields the graph of a function that is nonnegative, we can find the area under the curve using formula (10). Notice, however, that $t_0 = \pi$ and $t_1 = 0$, so $x(t_0) = -a$ and $x(t_1) = a$. Thus, the area of the ellipse is given by

$$A = 2 \int_{\pi}^{0} \underbrace{b \sin t}_{y(t)} \underbrace{(-a \sin t)}_{x'(t)} \, dt = 2ab \int_{0}^{\pi} \sin^2 t \, dt$$

$$= 2ab \int_{0}^{\pi} \frac{1 - \cos 2t}{2} \, dt = 2ab \left(\frac{t}{2} - \frac{\sin 2t}{4} \right) \Big|_{0}^{\pi} = \pi ab$$ ■

11.1 SUMMARY

- A parametric curve $c(t) = (f(t), g(t))$ describes the path of a particle moving along a curve as a function of the parameter t.
- Parametrizations are not unique: Every curve $\mathcal{C}$ can be parametrized in infinitely many ways. Furthermore, the path $c(t)$ may traverse all or part of $\mathcal{C}$ more than once.
- Slope of the tangent line at $c(t)$:

$$\frac{dy}{dx} = \frac{dy/dt}{dx/dt} = \frac{y'(t)}{x'(t)} \qquad \text{[valid if } x'(t) \ne 0 \text{]}$$

- Do not confuse the slope of the tangent line dy/dx with the derivatives dy/dt and dx/dt, with respect to t.
- Standard parametrizations:

  - Line of slope $m = s/r$ through $P = (a, b)$: $c(t) = (a + rt, b + st)$
  - Circle of radius R centered at $P = (a, b)$: $c(t) = (a + R \cos t, b + R \sin t)$
  - Cycloid generated by a circle of radius R: $c(t) = (R(t - \sin t), R(1 - \cos t))$

- Area under a parametric curve $c(t) = (x(t), y(t))$ that does not dip below the x-axis and that traces once the graph of a function is given by $A = \displaystyle\int_{t_0}^{t_1} y(t)x'(t) \, dt$.

11.1 EXERCISES

Preliminary Questions

1. Describe the shape of the curve $x = 3 \cos t$, $y = 3 \sin t$.

2. How does $x = 4 + 3 \cos t$, $y = 5 + 3 \sin t$ differ from the curve in the previous question?

3. What is the maximum height of a particle whose path has parametric equations $x = t^9$, $y = 4 - t^2$?

4. Can the parametric curve $(t, \sin t)$ be represented as a graph $y = f(x)$? What about $(\sin t, t)$?

5. **(a)** Describe the path of an ant that is crawling along the plane according to $c_1(t) = (f(t), f(t))$, where $f(t)$ is an increasing function.

(b) Compare that path to the path of a second ant crawling according to $c_2(t) = (f(2t), f(2t))$.

6. Find three different parametrizations of the graph of $y = x^3$.

7. Match the derivatives with a verbal description:

(a) $\dfrac{dx}{dt}$ **(b)** $\dfrac{dy}{dt}$ **(c)** $\dfrac{dy}{dx}$

(i) Slope of the tangent line to the curve

(ii) Vertical rate of change with respect to time

(iii) Horizontal rate of change with respect to time

Exercises

1. Find the coordinates at times $t = 0, 2, 4$ of a particle following the path $x = 1 + t^3$, $y = 9 - 3t^2$.

2. Find the coordinates at $t = 0, \frac{\pi}{4}, \pi$ of a particle moving along the path $c(t) = (\cos 2t, \sin^2 t)$.

3. Show that the path traced by the bullet in Example 3 is a parabola by eliminating the parameter.

4. Use the table of values to sketch the parametric curve $(x(t), y(t))$, indicating the direction of motion.

t	-3	-2	-1	0	1	2	3
x	-15	0	3	0	-3	0	15
y	5	0	-3	-4	-3	0	5

5. Graph the parametric curves. Include arrows indicating the direction of motion.

(a) (t, t), $-\infty < t < \infty$

(b) $(\sin t, \sin t)$, $0 \le t \le 2\pi$

(c) (e^t, e^t), $-\infty < t < \infty$

(d) (t^3, t^3), $-1 \le t \le 1$

6. Give two different parametrizations of the line through $(4, 1)$ with slope 2.

In Exercises 7–14, express in the form $y = f(x)$ by eliminating the parameter.

7. $x = t + 3$, $y = 4t$

8. $x = t^{-1}$, $y = t^{-2}$

9. $x = t$, $y = \tan^{-1}(t^3 + e^t)$

10. $x = t^2$, $y = t^3 + 1$

11. $x = e^{-2t}$, $y = 6e^{4t}$

12. $x = 1 + t^{-1}$, $y = t^2$

13. $x = \ln t$, $y = 2 - t$

14. $x = \cos t$, $y = \tan t$

In Exercises 15–18, graph the curve and draw an arrow specifying the direction corresponding to motion.

15. $x = \frac{1}{2}t$, $y = 2t^2$

16. $x = 2 + 4t$, $y = 3 + 2t$

17. $x = \pi t$, $y = \sin t$

18. $x = t^2$, $y = t^3$

19. Match the parametrizations (a)–(d) below with their plots in Figure 15, and draw an arrow indicating the direction of motion.

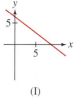
(I)

(II)

(III)

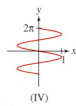
(IV)

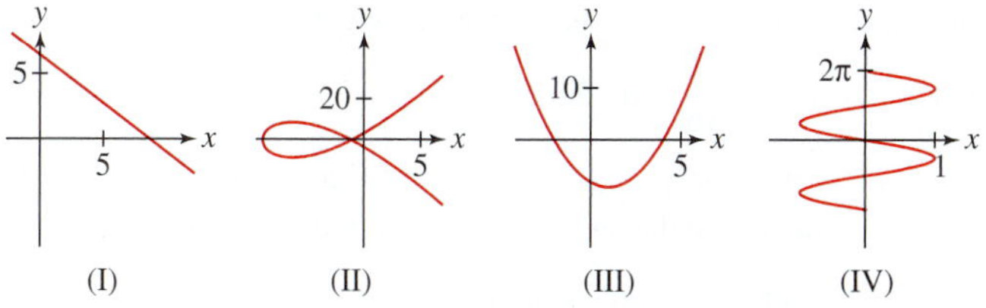

FIGURE 15

(a) $c(t) = (\sin t, -t)$

(b) $c(t) = (t^2 - 9, 8t - t^3)$

(c) $c(t) = (1 - t, t^2 - 9)$

(d) $c(t) = (4t + 2, 5 - 3t)$

20. Find an interval of t-values such that $c(t) = (\cos t, \sin t)$ traces the lower half of the unit circle.

21. A particle follows the trajectory

$$x(t) = \frac{1}{4}t^3 + 2t, \qquad y(t) = 20t - t^2$$

with t in seconds and distance in centimeters.

(a) What is the particle's maximum height?

(b) When does the particle hit the ground and how far from the origin does it land?

22. Find an interval of t-values such that $c(t) = (2t + 1, 4t - 5)$ parametrizes the segment from $(0, -7)$ to $(7, 7)$.

In Exercises 23–38, find parametric equations for the given curve.

23. $y = 9 - 4x$

24. $y = 8x^2 - 3x$

25. $4x - y^2 = 5$

26. $x^2 + y^2 = 49$

27. $(x + 9)^2 + (y - 4)^2 = 49$

28. $\left(\dfrac{x}{5}\right)^2 + \left(\dfrac{y}{12}\right)^2 = 1$

29. Line of slope 8 through $(-4, 9)$

30. Line through $(2, 5)$ perpendicular to $y = 3x$

31. Line through $(3, 1)$ and $(-5, 4)$

32. Line through $\left(\frac{1}{3}, \frac{1}{6}\right)$ and $\left(-\frac{7}{6}, \frac{5}{3}\right)$

33. Segment joining $(1, 1)$ and $(2, 3)$

34. Segment joining $(-3, 0)$ and $(0, 4)$

35. Circle of radius 4 with center $(3, 9)$

36. Ellipse of Exercise 28, with its center translated to $(7, 4)$

37. $y = x^2$, translated so that the minimum occurs at $(-4, -8)$

38. $y = \cos x$, translated so that a maximum occurs at $(3, 5)$

In Exercises 39–42, find a parametrization $c(t)$ of the curve satisfying the given condition.

39. $y = 3x - 4$, $c(0) = (2, 2)$

40. $y = 3x - 4$, $c(3) = (2, 2)$

41. $y = x^2$, $c(0) = (3, 9)$

42. $x^2 + y^2 = 4$, $c(0) = (1, \sqrt{3})$

43. Describe $c(t) = (\sec t, \tan t)$ for $0 \le t < \frac{\pi}{2}$ in the form $y = f(x)$. Specify the domain of x.

44. Find a parametrization of the right branch $(x > 0)$ of the hyperbola

$$\left(\frac{x}{a}\right)^2 - \left(\frac{y}{b}\right)^2 = 1$$

using $\cosh t$ and $\sinh t$. How can you parametrize the branch $x < 0$?

45. The graphs of $x(t)$ and $y(t)$ as functions of t are shown in Figure 16(A). Which of (I)–(III) is the plot of $c(t) = (x(t), y(t))$? Explain.

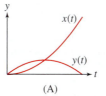
(A)

(I)

(II)

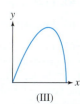
(III)

FIGURE 16

46. Which graph, (I) or (II), is the graph of $x(t)$ and which is the graph of $y(t)$ for the parametric curve in Figure 17(A)?

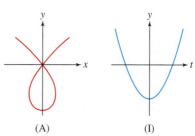
(A)   (I)   (II)

FIGURE 17

47. Sketch $c(t) = (t^3 - 4t, t^2)$ following the steps in Example 7.

48. Sketch $c(t) = (t^2 - 4t, 9 - t^2)$ for $-4 \le t \le 10$.

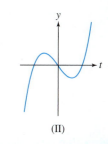

In Exercises 49–54, use Eq. (8) to find dy/dx at the given point.

49. $(t^3, t^2 - 1), \quad t = -4$

50. $(2t + 9, 7t - 9), \quad t = 1$

51. $(s^{-1} - 3s, s^3), \quad s = -1$

52. $(\sin 2\theta, \cos 3\theta), \quad \theta = \frac{\pi}{6}$

53. $(\sin^3 \theta, \cos \theta), \quad \theta = \frac{\pi}{4}$

54. $(e^t, t^2), \quad t = 1$

In Exercises 55–58, find an equation $y = f(x)$ for the parametric curve and compute dy/dx in two ways: using Eq. (8) and by differentiating $f(x)$.

55. $c(t) = (2t + 1, 1 - 9t)$

56. $c(t) = \left(\frac{1}{2}t, \frac{1}{4}t^2 - t\right)$

57. $x = s^3, \quad y = s^6 + s^{-3}$

58. $x = \cos \theta, \quad y = \cos \theta + \sin^2 \theta$

59. Find the points on the parametric curve $c(t) = (3t^2 - 2t, t^3 - 6t)$ where the tangent line has slope 3.

60. Find the equation of the tangent line to the cycloid generated by a circle of radius 4 at $t = \frac{\pi}{2}$.

In Exercises 61–64, let $c(t) = (t^2 - 9, t^2 - 8t)$ (see Figure 18).

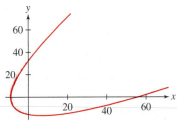

FIGURE 18 Plot of $c(t) = (t^2 - 9, t^2 - 8t)$.

61. Draw an arrow indicating the direction of motion, and determine the interval of t-values corresponding to the portion of the curve in each of the four quadrants.

62. Find the equation of the tangent line at $t = 4$.

63. Find the points where the tangent has slope $\frac{1}{2}$.

64. Find the points where the tangent is horizontal or vertical.

65. Let A and B be the points where the ray of angle θ intersects the two concentric circles of radii $r < R$ centered at the origin (Figure 19). Let P be the point of intersection of the horizontal line through A and the vertical line through B. Express the coordinates of P as a function of θ and describe the curve traced by P for $0 \leq \theta \leq 2\pi$.

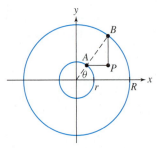

FIGURE 19

66. A 10-ft ladder slides down a wall as its bottom B is pulled away from the wall (Figure 20). Using the angle θ as a parameter, find the parametric equations for the path followed by (a) the top of the ladder A, (b) the bottom of the ladder B, and (c) the point P located 4 ft from the top of the ladder. Show that P describes an ellipse.

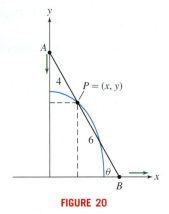

FIGURE 20

In Exercises 67–70, refer to the Bézier curve defined by Eqs. (9) and (10).

67. Show that the Bézier curve with control points

$$P_0 = (1, 4), \quad P_1 = (3, 12), \quad P_2 = (6, 15), \quad P_3 = (7, 4)$$

has parametrization

$$c(t) = (1 + 6t + 3t^2 - 3t^3, 4 + 24t - 15t^2 - 9t^3)$$

Verify that the slope at $t = 0$ is equal to the slope of the segment $\overline{P_0 P_1}$.

68. Find an equation of the tangent line to the Bézier curve in Exercise 67 at $t = \frac{1}{3}$.

69. **CAS** Find and plot the Bézier curve $c(t)$ with control points

$$P_0 = (3, 2), \quad P_1 = (0, 2), \quad P_2 = (5, 4), \quad P_3 = (2, 4)$$

70. Show that a cubic Bézier curve is tangent to the segment $\overline{P_2 P_3}$ at P_3.

71. A bullet fired from a gun follows the trajectory

$$x = at, \quad y = bt - 16t^2 \quad (a, b > 0)$$

Show that the bullet leaves the gun at an angle $\theta = \tan^{-1}\left(\frac{b}{a}\right)$ and lands at a distance $\frac{ab}{16}$ from the origin.

72. **CAS** Plot $c(t) = (t^3 - 4t, t^4 - 12t^2 + 48)$ for $-3 \leq t \leq 3$. Find the points where the tangent line is horizontal or vertical.

73. **CAS** Plot the astroid $x = \cos^3 \theta, y = \sin^3 \theta$ and find the equation of the tangent line at $\theta = \frac{\pi}{3}$.

74. Find the equation of the tangent line at $t = \frac{\pi}{4}$ to the cycloid generated by the unit circle with parametric equation (6).

75. Find the points with a horizontal tangent line on the cycloid with parametric equation (6).

76. Property of the Cycloid Prove that the tangent line at a point P on the cycloid always passes through the top point on the rolling circle as indicated in Figure 21. Assume the generating circle of the cycloid has radius 1.

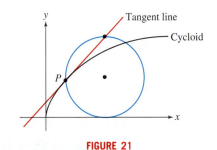

FIGURE 21

77. A *curtate cycloid* (Figure 22) is the curve traced by a point at a distance h from the center of a circle of radius R rolling along the x-axis where $h < R$. Show that this curve has parametric equations $x = Rt - h \sin t$, $y = R - h \cos t$.

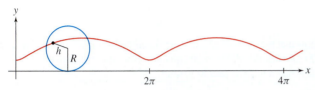

FIGURE 22 Curtate cycloid.

78. *CAS* Use a computer algebra system to explore what happens when $h > R$ in the parametric equations of Exercise 77. Describe the result.

79. ✎ Show that the line of slope t through $(-1, 0)$ intersects the unit circle in the point with coordinates

$$x = \frac{1 - t^2}{t^2 + 1}, \qquad y = \frac{2t}{t^2 + 1} \qquad \boxed{12}$$

Conclude that these equations parametrize the unit circle with the point $(-1, 0)$ excluded (Figure 23). Show further that $t = y/(x + 1)$.

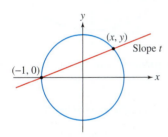

FIGURE 23 Unit circle.

80. The **folium of Descartes** is the curve with equation $x^3 + y^3 = 3axy$, where $a \neq 0$ is a constant (Figure 24).

(a) Show that the line $y = tx$ intersects the folium at the origin and at one other point P for all $t \neq -1, 0$. Express the coordinates of P in terms of t to obtain a parametrization of the folium. Indicate the direction of the parametrization on the graph.

(b) Describe the interval of t-values parametrizing the parts of the curve in quadrants I, II, and IV. Note that $t = -1$ is a point of discontinuity of the parametrization.

(c) Calculate dy/dx as a function of t and find the points with horizontal or vertical tangent.

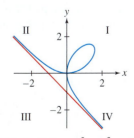

FIGURE 24 Folium $x^3 + y^3 = 3axy$.

81. Use the results of Exercise 80 to show that the asymptote of the folium is the line $x + y = -a$. *Hint:* Show that $\lim_{t \to -1} (x + y) = -a$.

82. Find a parametrization of $x^{2n+1} + y^{2n+1} = ax^n y^n$, where a and n are constants.

83. Second Derivative for a Parametrized Curve Given a parametrized curve $c(t) = (x(t), y(t))$, show that

$$\frac{d}{dt}\left(\frac{dy}{dx}\right) = \frac{x'(t)y''(t) - y'(t)x''(t)}{x'(t)^2}$$

Use this to prove the formula

$$\boxed{\frac{d^2 y}{dx^2} = \frac{x'(t)y''(t) - y'(t)x''(t)}{x'(t)^3}} \qquad \boxed{13}$$

84. The second derivative of $y = x^2$ is $dy^2/d^2x = 2$. Verify that Eq. (13) applied to $c(t) = (t, t^2)$ yields $dy^2/d^2x = 2$. In fact, any parametrization may be used. Check that $c(t) = (t^3, t^6)$ and $c(t) = (\tan t, \tan^2 t)$ also yield $dy^2/d^2x = 2$.

In Exercises 85–88, use Eq. (13) to find d^2y/dx^2.

85. $x = t^3 + t^2$, $\quad y = 7t^2 - 4$, $\quad t = 2$

86. $x = s^{-1} + s$, $\quad y = 4 - s^{-2}$, $\quad s = 1$

87. $x = 8t + 9$, $\quad y = 1 - 4t$, $\quad t = -3$

88. $x = \cos\theta$, $\quad y = \sin\theta$, $\quad \theta = \frac{\pi}{4}$

89. Use Eq. (13) to find the t-intervals on which $c(t) = (t^2, t^3 - 4t)$ is concave up.

90. Use Eq. (13) to find the t-intervals on which $c(t) = (t^2, t^4 - 4t)$ is concave up.

91. Calculate the area under $y = x^2$ over $[0, 1]$ using Eq. (11) with the parametrizations (t^3, t^6) and (t^2, t^4).

92. What does Eq. (11) say if $c(t) = (t, f(t))$?

93. Consider the curve $c(t) = (t^2, t^3)$ for $0 \leq t \leq 1$.
(a) Find the area under the curve using Eq. (11).
(b) Find the area under the curve by expressing y as a function of x and finding the area using the standard method.

94. Compute the area under the parametrized curve $c(t) = (e^t, t)$ for $0 \leq t \leq 1$ using Eq. (11).

95. Compute the area under the parametrized curve given by $c(t) = (\sin t, \cos^2 t)$ for $0 \leq t \leq \pi/2$ using Eq. (11).

96. Sketch the graph of $c(t) = (\ln t, 2 - t)$ for $1 \leq t \leq 2$ and compute the area under the graph using Eq. (11).

97. Galileo tried unsuccessfully to find the area under a cycloid. Around 1630, Gilles de Roberval proved that the area under one arch of the cycloid $c(t) = (Rt - R \sin t, R - R \cos t)$ generated by a circle of radius R is equal to three times the area of the circle (Figure 25). Verify Roberval's result using Eq. (11).

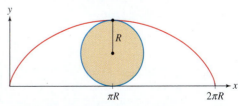

FIGURE 25 The area of one arch of the cycloid equals three times the area of the generating circle.

Further Insights and Challenges

98. Prove the following generalization of Exercise 97: For all $t > 0$, the area of the cycloidal sector OPC is equal to three times the area of the circular segment cut by the chord PC in Figure 26.

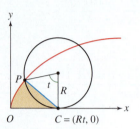

(A) Cycloidal sector OPC (B) Circular segment cut by the chord PC

FIGURE 26

99. ✏️ Derive the formula for the slope of the tangent line to a parametric curve $c(t) = (x(t), y(t))$ using a method different from that presented in the text. Assume that $x'(t_0)$ and $y'(t_0)$ exist and $x'(t_0) \neq 0$. Show that

$$\lim_{h \to 0} \frac{y(t_0 + h) - y(t_0)}{x(t_0 + h) - x(t_0)} = \frac{y'(t_0)}{x'(t_0)}$$

Then explain why this limit is equal to the slope dy/dx. Draw a diagram showing that the ratio in the limit is the slope of a secant line.

100. Verify that the **tractrix** curve ($\ell > 0$)

$$c(t) = \left(t - \ell \tanh \frac{t}{\ell}, \ell \operatorname{sech} \frac{t}{\ell} \right)$$

has the following property: For all t, the segment from $c(t)$ to $(t, 0)$ is tangent to the curve and has length ℓ (Figure 27).

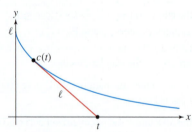

FIGURE 27 The tractrix $c(t) = \left(t - \ell \tanh \frac{t}{\ell}, \ell \operatorname{sech} \frac{t}{\ell} \right)$.

101. In Exercise 59 of Section 9.1, we described the tractrix by the differential equation

$$\frac{dy}{dx} = -\frac{y}{\sqrt{\ell^2 - y^2}}$$

Show that the parametric curve $c(t)$ identified as the tractrix in Exercise 100 satisfies this differential equation. Note that the derivative on the left is taken with respect to x, not t.

In Exercises 102 and 103, refer to Figure 28.

102. In the parametrization $c(t) = (a \cos t, b \sin t)$ of an ellipse, t is *not* an angular parameter unless $a = b$ (in which case, the ellipse is a circle). However, t can be interpreted in terms of area: Show that if $c(t) = (x, y)$, then $t = (2/ab)A$, where A is the area of the shaded region in Figure 28. *Hint:* Use Eq. (11).

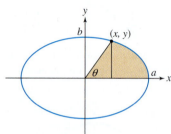

FIGURE 28 The parameter θ on the ellipse $\left(\frac{x}{a}\right)^2 + \left(\frac{y}{b}\right)^2 = 1$.

103. Show that the parametrization of the ellipse by the angle θ is

$$x = \frac{ab \cos \theta}{\sqrt{a^2 \sin^2 \theta + b^2 \cos^2 \theta}}$$

$$y = \frac{ab \sin \theta}{\sqrt{a^2 \sin^2 \theta + b^2 \cos^2 \theta}}$$

11.2 Arc Length and Speed

We now derive a formula for the arc length of a curve in parametric form. Recall that in Section 8.1, arc length was defined as the limit of the lengths of polygonal approximations (Figure 1).

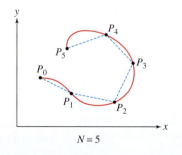

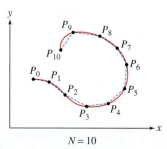

DF FIGURE 1 Polygonal approximations for $N = 5$ and $N = 10$.

Given a parametrization $c(t) = (x(t), y(t))$ for $a \leq t \leq b$, we construct a polygonal approximation L consisting of the N segments by joining points

$$P_0 = c(t_0), \quad P_1 = c(t_1), \quad \ldots, \quad P_N = c(t_N)$$

corresponding to a choice of values $t_0 = a < t_1 < t_2 < \cdots < t_N = b$. By the distance formula,

$$P_{i-1}P_i = \sqrt{\left(x(t_i) - x(t_{i-1})\right)^2 + \left(y(t_i) - y(t_{i-1})\right)^2} \qquad \boxed{1}$$

Now assume that $x(t)$ and $y(t)$ are differentiable. According to the Mean Value Theorem, there are values t_i^* and t_i^{**} in the interval $[t_{i-1}, t_i]$ such that

$$x(t_i) - x(t_{i-1}) = x'(t_i^*)\Delta t_i, \qquad y(t_i) - y(t_{i-1}) = y'(t_i^{**})\Delta t_i$$

where $\Delta t_i = t_i - t_{i-1}$, and therefore,

$$P_{i-1}P_i = \sqrt{x'(t_i^*)^2 \Delta t_i^2 + y'(t_i^{**})^2 \Delta t_i^2} = \sqrt{x'(t_i^*)^2 + y'(t_i^{**})^2}\,\Delta t_i$$

The length of the polygonal approximation L is equal to the sum

$$\sum_{i=1}^{N} P_{i-1}P_i = \sum_{i=1}^{N} \sqrt{x'(t_i^*)^2 + y'(t_i^{**})^2}\,\Delta t_i \qquad \boxed{2}$$

This is *nearly* a Riemann sum for the function $\sqrt{x'(t)^2 + y'(t)^2}$. It would be a true Riemann sum if the intermediate values t_i^* and t_i^{**} were equal. Although they are not necessarily equal, it can be shown (and we will take for granted) that if $x'(t)$ and $y'(t)$ are continuous, then the sum in Eq. (2) still approaches the integral as the widths Δt_i tend to 0. Thus,

$$s = \lim_{\Delta t_i \to 0} \sum_{i=1}^{N} P_{i-1}P_i = \int_a^b \sqrt{x'(t)^2 + y'(t)^2}\,dt$$

Because of the square root, the arc length integral cannot be evaluated explicitly except in special cases, but we can always approximate it numerically.

> **THEOREM 1 Arc Length** Let $c(t) = (x(t), y(t))$, where $x'(t)$ and $y'(t)$ exist and are continuous. Then the arc length s of $c(t)$ for $a \leq t \leq b$ is equal to
>
> $$s = \int_a^b \sqrt{x'(t)^2 + y'(t)^2}\,dt \qquad \boxed{3}$$

The graph of a function $y = f(x)$ has parametrization $c(t) = (t, f(t))$. In this case,

$$\sqrt{x'(t)^2 + y'(t)^2} = \sqrt{1 + f'(t)^2}$$

and Eq. (3) reduces to the arc length formula derived in Section 8.1.

As mentioned above, the arc length integral can be evaluated explicitly only in special cases. The circle and the cycloid are two such cases.

■ **EXAMPLE 1** Use Eq. (3) to calculate the arc length of a circle of radius R.

Solution With the parametrization $x = R\cos\theta$, $y = R\sin\theta$,

$$x'(\theta)^2 + y'(\theta)^2 = (-R\sin\theta)^2 + (R\cos\theta)^2 = R^2(\sin^2\theta + \cos^2\theta) = R^2$$

We obtain the expected result:

$$s = \int_0^{2\pi} \sqrt{x'(\theta)^2 + y'(\theta)^2}\,d\theta = \int_0^{2\pi} R\,d\theta = 2\pi R \qquad ■$$

■ **EXAMPLE 2** Find the arc length of the curve given in parametric form by $c(t) = (t^2, t^3)$ for $0 \le t \le 1$.

Solution The arc length of this curve is given by

$$s = \int_0^1 \sqrt{x'(t)^2 + y'(t)^2}\, dt = \int_0^1 \sqrt{(2t)^2 + (3t^2)^2}\, dt$$

$$= \int_0^1 t\sqrt{4 + 9t^2}\, dt$$

Letting $u = 4 + 9t^2$, and therefore $du = 18t\, dt$, we obtain

$$s = \frac{1}{18}\int_4^{13} \sqrt{u}\, du = \frac{2}{3}\frac{u^{\frac{3}{2}}}{18}\Big|_4^{13} = \frac{1}{27}(13^{\frac{3}{2}} - (4)^{\frac{3}{2}}) \approx 1.4397$$ ■

■ **EXAMPLE 3** Length of the Cycloid Calculate the length s of one arch of the cycloid generated by a circle of radius $R = 2$ (Figure 2).

Solution We use the parametrization of the cycloid in Eq. (7) of Section 11.1:

$$x(t) = 2(t - \sin t), \qquad y(t) = 2(1 - \cos t)$$

$$x'(t) = 2(1 - \cos t), \qquad y'(t) = 2\sin t$$

Thus,

$$x'(t)^2 + y'(t)^2 = 2^2(1 - \cos t)^2 + 2^2 \sin^2 t$$

$$= 4 - 8\cos t + 4\cos^2 t + 4\sin^2 t$$

$$= 8 - 8\cos t$$

$$= 16\sin^2 \frac{t}{2} \qquad \text{(Use the identity recalled in the margin.)}$$

One arch of the cycloid is traced as t varies from 0 to 2π, so

$$s = \int_0^{2\pi} \sqrt{x'(t)^2 + y'(t)^2}\, dt = \int_0^{2\pi} 4\sin\frac{t}{2}\, dt = -8\cos\frac{t}{2}\Big|_0^{2\pi} = -8(-1) + 8 = 16$$

Note that because $\sin\frac{t}{2} \ge 0$ for $0 \le t \le 2\pi$, we did not need an absolute value when taking the square root of $16\sin^2\frac{t}{2}$. ■

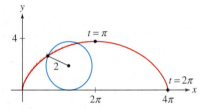

FIGURE 2 One arch of the cycloid generated by a circle of radius 2.

← **REMINDER**

$$\frac{1 - \cos t}{2} = \sin^2\frac{t}{2}$$

In Chapter 13, we will discuss not just the speed but also the velocity of a particle moving along a curved path. Velocity is "speed plus direction" and is represented by a vector.

Now consider a particle moving along a path $c(t)$. The distance traveled by the particle over the time interval $[t_0, t]$ is given by the arc length integral:

$$s(t) = \int_{t_0}^t \sqrt{x'(u)^2 + y'(u)^2}\, du$$

On the other hand, speed is defined as the rate of change of distance traveled with respect to time, so by the Fundamental Theorem of Calculus,

$$\text{Speed} = \frac{ds}{dt} = \frac{d}{dt}\int_{t_0}^t \sqrt{x'(u)^2 + y'(u)^2}\, du = \sqrt{x'(t)^2 + y'(t)^2}$$

THEOREM 2 Speed Along a Parametrized Path   The speed of $c(t) = (x(t), y(t))$ is

$$\boxed{\text{Speed} = \frac{ds}{dt} = \sqrt{x'(t)^2 + y'(t)^2}}$$

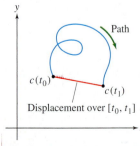

FIGURE 3 The distance along the path is greater than or equal to the displacement.

The next example illustrates the difference between distance traveled along a path and **displacement** (also called the net change in position). The displacement along a path is the distance between the initial point $c(t_0)$ and the endpoint $c(t_1)$. The distance traveled is greater than the displacement unless the particle happens to move in a straight line (Figure 3).

■ **EXAMPLE 4** A particle travels along the path $c(t) = (2t, 1 + t^{3/2})$. Find:

(a) The particle's speed at $t = 1$ (assume units of meters and minutes).

(b) The distance traveled s and displacement d during the interval $0 \le t \le 4$.

Solution We have

$$x'(t) = 2, \qquad y'(t) = \frac{3}{2}t^{1/2}$$

The speed at time t is

$$s'(t) = \sqrt{x'(t)^2 + y'(t)^2} = \sqrt{4 + \frac{9}{4}t} \quad \text{m/min}$$

(a) The particle's speed at $t = 1$ is $s'(1) = \sqrt{4 + \frac{9}{4}} = 2.5$ m/min.

(b) The distance traveled in the first 4 min is

$$s = \int_0^4 \sqrt{4 + \frac{9}{4}t} \, dt = \frac{8}{27}\left(4 + \frac{9}{4}t\right)^{3/2}\bigg|_0^4 = \frac{8}{27}\left(13^{3/2} - 8\right) \approx 11.52 \text{ m}$$

The displacement d is the distance from the initial point $c(0) = (0, 1)$ to the endpoint $c(4) = (8, 1 + 4^{3/2}) = (8, 9)$ (see Figure 4):

$$d = \sqrt{(8 - 0)^2 + (9 - 1)^2} = 8\sqrt{2} \approx 11.31 \text{ m} \qquad ■$$

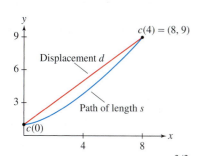

FIGURE 4 The path $c(t) = (2t, 1 + t^{3/2})$.

In physics, we often describe the path of a particle moving with constant speed along a circle of radius R in terms of a constant ω (lowercase Greek omega) as follows:

$$c(t) = (R \cos \omega t, R \sin \omega t)$$

The constant ω, called the *angular velocity*, is the rate of change with respect to time of the particle's angle θ (Figure 5).

■ **EXAMPLE 5** **Angular Velocity** Calculate the speed of the circular path of radius R and angular velocity ω. What is the speed if $R = 3$ m and $\omega = 4$ radians per second (rad/s)?

Solution We have $x = R \cos \omega t$ and $y = R \sin \omega t$, and

$$x'(t) = -\omega R \sin \omega t, \qquad y'(t) = \omega R \cos \omega t$$

The particle's speed is

$$\frac{ds}{dt} = \sqrt{x'(t)^2 + y'(t)^2} = \sqrt{(-\omega R \sin \omega t)^2 + (\omega R \cos \omega t)^2}$$

$$= \sqrt{\omega^2 R^2 (\sin^2 \omega t + \cos^2 \omega t)} = |\omega| R$$

Thus, the speed is constant with value $|\omega| R$. If $R = 3$ m and $\omega = 4$ rad/s, then the speed is $|\omega| R = 3(4) = 12$ m/s. ■

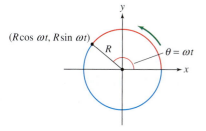

FIGURE 5 A particle moving on a circle of radius R with angular velocity ω has speed $|\omega| R$.

Consider the surface obtained by rotating a parametric curve $c(t) = (x(t), y(t))$ about the x-axis. The surface area is given by Eq. (4) in the next theorem. It can be derived in much the same way as the formula for a surface of revolution of a graph $y = f(x)$ in Section 8.1. In this theorem, we assume that $y(t) \ge 0$ so the parametric curve $c(t)$ lies above the x-axis, and that $x(t)$ is increasing so the curve does not reverse direction.

THEOREM 3 **Surface Area** Let $c(t) = (x(t), y(t))$, where $y(t) \ge 0$, $x(t)$ is increasing, and $x'(t)$ and $y'(t)$ are continuous. Then the surface obtained by rotating $c(t)$ about the x-axis for $a \le t \le b$ has surface area

$$S = 2\pi \int_a^b y(t)\sqrt{x'(t)^2 + y'(t)^2} \, dt \qquad \boxed{4}$$

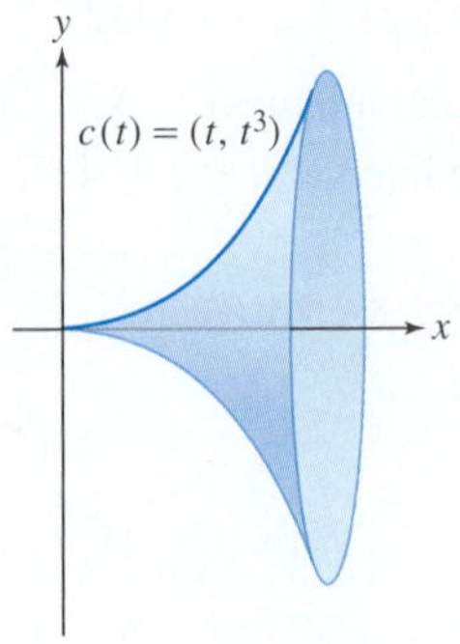

FIGURE 6 Surface generated by revolving the curve about the x-axis.

■ **EXAMPLE 6** Calculate the surface area of the surface obtained by rotating the parametric curve $c(t) = (t, t^3)$ about the x-axis for $0 \le t \le 1$. The surface appears as in Figure 6.

Solution We have $x'(t) = 1$ and $y'(t) = 3t^2$.
Therefore,

$$S = 2\pi \int_0^1 t^3 \sqrt{1 + (3t^2)^2}\, dt = 2\pi \int_0^1 t^3 \sqrt{1 + 9t^4}\, dt$$

With the substitution $u = 1 + 9t^4$ and $du = 36t^3\, dt$, we obtain

$$S = 2\pi \frac{1}{36} \int_1^{10} \sqrt{u}\, du = \frac{\pi}{18} \left(\frac{2}{3}\right) u^{\frac{3}{2}} \Big|_1^{10} = \frac{\pi}{27}(10^{\frac{3}{2}} - 1) \approx 3.5631 \qquad ■$$

11.2 SUMMARY

- Arc length of $c(t) = (x(t), y(t))$ for $a \le t \le b$:

$$s = \text{arc length} = \int_a^b \sqrt{x'(t)^2 + y'(t)^2}\, dt$$

- The arc length is the distance along the path $c(t)$. The *displacement* is the distance from the starting point $c(a)$ to the endpoint $c(b)$.
- Arc length integral as a function of t:

$$s(t) = \int_{t_0}^t \sqrt{x'(u)^2 + y'(u)^2}\, du$$

- Speed at time t:

$$\frac{ds}{dt} = \sqrt{x'(t)^2 + y'(t)^2}$$

- Surface area of the surface obtained by rotating $c(t) = (x(t), y(t))$ about the x-axis for $a \le t \le b$:

$$S = 2\pi \int_a^b y(t)\sqrt{x'(t)^2 + y'(t)^2}\, dt$$

11.2 EXERCISES

Preliminary Questions

1. What is the definition of arc length?

2. Can the distance traveled by a particle ever be less than its displacement? When are they equal?

3. What is the interpretation of $\sqrt{x'(t)^2 + y'(t)^2}$ for a particle following the trajectory $(x(t), y(t))$?

4. A particle travels along a path from $(0, 0)$ to $(3, 4)$. What is the displacement? Can the distance traveled be determined from the information given?

5. A particle traverses the parabola $y = x^2$ with constant speed 3 cm/s. What is the distance traveled during the first minute? *Hint:* Only simple computation is necessary.

6. If the straight line segment given by $c(t) = (3, t)$ for $0 \le t \le 2$ is rotated around the x-axis, what surface area results? *Hint:* Only simple computation is necessary.

Exercises

In Exercises 1–10, use Eq. (3) to find the length of the path over the given interval.

1. $(3t + 1, 9 - 4t)$, $0 \le t \le 2$

2. $(1 + 2t, 2 + 4t)$, $1 \le t \le 4$

3. $(2t^2, 3t^2 - 1)$, $0 \le t \le 4$

4. $(3t, 4t^{3/2})$, $0 \le t \le 1$

5. $(3t^2, 4t^3)$, $1 \le t \le 4$

6. $(t^3 + 1, t^2 - 3)$, $0 \le t \le 1$

7. $(\sin 3t, \cos 3t)$, $0 \le t \le \pi$

8. $(\sin \theta - \theta \cos \theta, \cos \theta + \theta \sin \theta)$, $0 \le \theta \le 2$

In Exercises 9 and 10, use the identity

$$\frac{1 - \cos t}{2} = \sin^2 \frac{t}{2}$$

9. $(2 \cos t - \cos 2t, 2 \sin t - \sin 2t), \quad 0 \leq t \leq \frac{\pi}{2}$

10. $(5(\theta - \sin \theta), 5(1 - \cos \theta)), \quad 0 \leq \theta \leq 2\pi$

11. Show that one arch of a cycloid generated by a circle of radius R has length $8R$.

12. Find the length of the spiral $c(t) = (t \cos t, t \sin t)$ for $0 \leq t \leq 2\pi$ to three decimal places (Figure 7). *Hint:* Use the formula

$$\int \sqrt{1 + t^2}\, dt = \frac{1}{2} t \sqrt{1 + t^2} + \frac{1}{2} \ln\left(t + \sqrt{1 + t^2}\right)$$

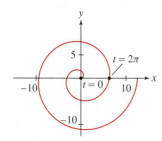

FIGURE 7 The spiral $c(t) = (t \cos t, t \sin t)$.

13. Find the length of the parabola given by $c(t) = (t, t^2)$ for $0 \leq t \leq 1$. See the hint for Exercise 12.

14. *CAS* Find a numerical approximation to the length of $c(t) = (\cos 5t, \sin 3t)$ for $0 \leq t \leq 2\pi$ (Figure 8).

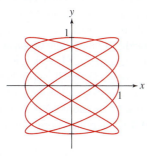

FIGURE 8

In Exercises 15–20, determine the speed $\frac{ds}{dt}$ at time t (assume units of meters and seconds).

15. $(t^3, t^2), \quad t = 2$

16. $(3 \sin 5t, 8 \cos 5t), \quad t = \frac{\pi}{4}$

17. $(5t + 1, 4t - 3), \quad t = 9$

18. $(\ln(t^2 + 1), t^3), \quad t = 1$

19. $t^2, e^t), \quad t = 0$

20. $(\sin^{-1} t, \tan^{-1} t), \quad t = 0$

21. Find the minimum speed of a particle with trajectory $c(t) = (t^3 - 4t, t^2 + 1)$ for $t \geq 0$. *Hint:* It is easier to find the minimum of the square of the speed.

22. Find the minimum speed of a particle with trajectory $c(t) = (t^3, t^{-2})$ for $t \geq 0.5$.

23. Find the speed of the cycloid $c(t) = (4t - 4 \sin t, 4 - 4 \cos t)$ at points where the tangent line is horizontal.

24. Calculate the arc length integral $s(t)$ for the *logarithmic spiral* $c(t) = (e^t \cos t, e^t \sin t)$.

CAS *In Exercises 25–28, plot the curve and use the Midpoint Rule with $N = 10, 20, 30$, and 50 to approximate its length.*

25. $c(t) = (\cos t, e^{\sin t})$ for $0 \leq t \leq 2\pi$

26. $c(t) = (t - \sin 2t, 1 - \cos 2t)$ for $0 \leq t \leq 2\pi$

27. The ellipse $\left(\frac{x}{5}\right)^2 + \left(\frac{y}{3}\right)^2 = 1$

28. $x = \sin 2t, \quad y = \sin 3t$ for $0 \leq t \leq 2\pi$

29. If you unwind thread from a stationary circular spool, keeping the thread taut at all times, then the endpoint traces a curve $\mathcal{C}$ called the **involute** of the circle (Figure 9). Observe that $\overline{PQ}$ has length $R\theta$. Show that $\mathcal{C}$ is parametrized by

$$c(\theta) = \left(R(\cos \theta + \theta \sin \theta), \quad R(\sin \theta - \theta \cos \theta)\right)$$

Then find the length of the involute for $0 \leq \theta \leq 2\pi$.

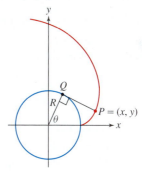

FIGURE 9 Involute of a circle.

30. Let $a > b$ and set

$$k = \sqrt{1 - \frac{b^2}{a^2}}$$

Use a parametric representation to show that the ellipse $\left(\frac{x}{a}\right)^2 + \left(\frac{y}{b}\right)^2 = 1$ has length $L = 4aG\left(\frac{\pi}{2}, k\right)$, where

$$G(\theta, k) = \int_0^\theta \sqrt{1 - k^2 \sin^2 t}\, dt$$

is the *elliptic integral of the second kind*.

In Exercises 31–38, use Eq. (4) to compute the surface area of the given surface.

31. The cone generated by revolving $c(t) = (t, mt)$ about the x-axis for $0 \leq t \leq A$.

32. A sphere of radius R.

33. The surface generated by revolving the curve $c(t) = (t^2, t)$ about the x-axis for $0 \leq t \leq 1$.

34. The surface generated by revolving the curve $c(t) = (t, e^t)$ about the x-axis for $0 \leq t \leq 1$.

35. The surface generated by revolving the curve $c(t) = (\sin^2 t, \cos^2 t)$ about the x-axis for $0 \leq t \leq \frac{\pi}{2}$.

36. The surface generated by revolving the curve $c(t) = (t, \sin t)$ about the x-axis for $0 \leq t \leq 2\pi$.

37. The surface generated by revolving one arch of the cycloid $c(t) = (t - \sin t, 1 - \cos t)$ about the x-axis

38. The surface generated by revolving the astroid $c(t) = (\cos^3 t, \sin^3 t)$ about the x-axis for $0 \leq t \leq \frac{\pi}{2}$

Further Insights and Challenges

39. $\mathcal{CAS}$ Let $b(t)$ be the "Butterfly Curve":

$$x(t) = \sin t \left(e^{\cos t} - 2 \cos 4t - \sin \left(\frac{t}{12} \right)^5 \right)$$

$$y(t) = \cos t \left(e^{\cos t} - 2 \cos 4t - \sin \left(\frac{t}{12} \right)^5 \right)$$

(a) Use a computer algebra system to plot $b(t)$ and the speed $s'(t)$ for $0 \leq t \leq 12\pi$.

(b) Approximate the length $b(t)$ for $0 \leq t \leq 10\pi$.

40. $\mathcal{CAS}$ Let $a \geq b > 0$ and set $k = \dfrac{2\sqrt{ab}}{a - b}$. Show that the **trochoid**

$$x = at - b \sin t, \qquad y = a - b \cos t, \quad 0 \leq t \leq T$$

has length $2(a - b)G\left(\frac{T}{2}, k\right)$, with $G(\theta, k)$ as in Exercise 30.

41. A satellite orbiting at a distance R from the center of the earth follows the circular path $x(t) = R \cos \omega t$, $y(t) = R \sin \omega t$.

(a) Show that the period T (the time of one revolution) is $T = 2\pi/\omega$.

(b) According to Newton's Laws of Motion and Gravity,

$$x''(t) = -Gm_e \frac{x}{R^3}, \qquad y''(t) = -Gm_e \frac{y}{R^3}$$

where G is the universal gravitational constant and m_e is the mass of the earth. Prove that $R^3/T^2 = Gm_e/4\pi^2$. Thus, R^3/T^2 has the same value for all orbits (a special case of Kepler's Third Law).

42. The acceleration due to gravity on the surface of the earth is

$$g = \frac{Gm_e}{R_e^2} = 9.8 \text{ m/s}^2, \quad \text{where } R_e = 6378 \text{ km}$$

Use Exercise 41(b) to show that a satellite orbiting at the earth's surface would have period $T_e = 2\pi \sqrt{R_e/g} \approx 84.5$ min. Then estimate the distance R_m from the moon to the center of the earth. Assume that the period of the moon (sidereal month) is $T_m \approx 27.43$ days.

Polar coordinates are appropriate when distance from the origin or angle plays a role. For example, the gravitational force exerted on a planet by the sun depends only on the distance r from the sun and is conveniently described in polar coordinates.

11.3 Polar Coordinates

The rectangular coordinates that we have utilized up to now provide a useful way to represent points in the plane. However, there are a variety of situations where a different coordinate system is more natural. In polar coordinates, we label a point P by coordinates (r, θ), where r is the distance to the origin O and θ is the angle between $\overline{OP}$ and the positive x-axis (Figure 1). By convention, an angle is positive if the corresponding rotation is counterclockwise. We call r the **radial coordinate** and θ the **angular coordinate**.

The point P in Figure 2 has polar coordinates $(r, \theta) = \left(4, \frac{2\pi}{3}\right)$. It is located at distance $r = 4$ from the origin (so it lies on the circle of radius 4), and it lies on the ray of angle $\theta = \frac{2\pi}{3}$. Notice that it can also be described by $(r, \theta) = \left(4, \frac{-4\pi}{3}\right)$. Unlike Cartesian coordinates, polar coordinates are not unique, as we will discuss in more detail shortly.

Figure 3 shows the two families of **grid lines** in polar coordinates:

$$\text{Circle centered at } O \quad \longleftrightarrow \quad r = \text{constant}$$

$$\text{Ray starting at } O \quad \longleftrightarrow \quad \theta = \text{constant}$$

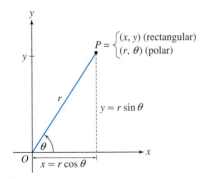

DF FIGURE 1

Every point in the plane other than the origin lies at the intersection of the two grid lines and these two grid lines determine its polar coordinates. For example, point Q in Figure 3 lies on the circle $r = 3$ and the ray $\theta = \frac{5\pi}{6}$, so $Q = \left(3, \frac{5\pi}{6}\right)$ in polar coordinates.

Figure 1 shows that polar and rectangular coordinates are related by the equations $x = r \cos \theta$ and $y = r \sin \theta$. On the other hand, $r^2 = x^2 + y^2$ by the distance formula, and $\tan \theta = y/x$ if $x \neq 0$. This yields the conversion formulas:

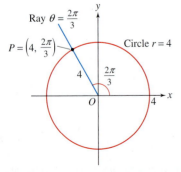

FIGURE 2

Polar to Rectangular	Rectangular to Polar
$x = r \cos \theta$	$r = \sqrt{x^2 + y^2}$
$y = r \sin \theta$	$\tan \theta = \dfrac{y}{x} \quad (x \neq 0)$

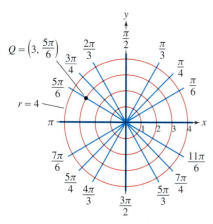

FIGURE 3 Grid lines in polar coordinates.

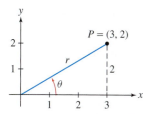

FIGURE 4 The polar coordinates of P satisfy $r = \sqrt{3^2 + 2^2}$ and $\tan \theta = \frac{2}{3}$.

By definition,

$$-\frac{\pi}{2} < \tan^{-1} x < \frac{\pi}{2}$$

If $r > 0$, a coordinate θ of $P = (x, y)$ is

$$\theta = \begin{cases} \tan^{-1} \dfrac{y}{x} & \text{if } x > 0 \\ \tan^{-1} \dfrac{y}{x} + \pi & \text{if } x < 0 \\ \pm \dfrac{\pi}{2} & \text{if } x = 0 \end{cases}$$

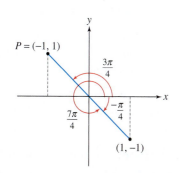

FIGURE 7

■ **EXAMPLE 1** **From Polar to Rectangular Coordinates** Find the rectangular coordinates of point Q in Figure 3.

Solution The point $Q = (r, \theta) = \left(3, \frac{5\pi}{6}\right)$ has rectangular coordinates:

$$x = r \cos \theta = 3 \cos \left(\frac{5\pi}{6}\right) = 3 \left(-\frac{\sqrt{3}}{2}\right) = -\frac{3\sqrt{3}}{2}$$

$$y = r \sin \theta = 3 \sin \left(\frac{5\pi}{6}\right) = 3 \left(\frac{1}{2}\right) = \frac{3}{2}$$

■ **EXAMPLE 2** **From Rectangular to Polar Coordinates** Find the polar coordinates of point P in Figure 4.

Solution Since $P = (x, y) = (3, 2)$,

$$r = \sqrt{x^2 + y^2} = \sqrt{3^2 + 2^2} = \sqrt{13} \approx 3.6$$

$$\tan \theta = \frac{y}{x} = \frac{2}{3}$$

and because P lies in the first quadrant,

$$\theta = \tan^{-1} \frac{y}{x} = \tan^{-1} \frac{2}{3} \approx 0.588$$

Thus, P has polar coordinates $(r, \theta) \approx (3.6, 0.588)$.

A few remarks are in order before proceeding:

- The angular coordinate is not unique because (r, θ) and $(r, \theta + 2\pi n)$ *label the same point* for any integer n. For instance, point P in Figure 5 has radial coordinate $r = 2$, but its angular coordinate can be any one of $\frac{\pi}{2}, \frac{5\pi}{2}, \ldots$ or $-\frac{3\pi}{2}, -\frac{7\pi}{2}, \ldots$.
- The origin O has no well-defined angular coordinate, so we assign to O the polar coordinates $(0, \theta)$ for any angle θ.

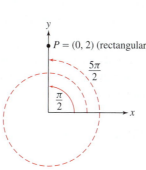

FIGURE 5 The angular coordinate of $P = (0, 2)$ is $\frac{\pi}{2}$ or any angle $\frac{\pi}{2} + 2\pi n$, where n is an integer.

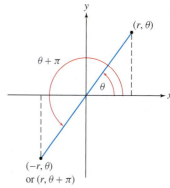

FIGURE 6 Relation between (r, θ) and $(-r, \theta)$.

- By convention, we allow *negative* radial coordinates. By definition, $(-r, \theta)$ is the reflection of (r, θ) through the origin (Figure 6). With this convention, $(-r, \theta)$ and $(r, \theta + \pi)$ represent the same point.
- We may specify unique polar coordinates for points other than the origin by placing restrictions on r and θ. We commonly choose $r > 0$ and $0 \le \theta < 2\pi$, but other choices are sometimes made.

When determining the angular coordinate of a point $P = (x, y)$, remember that there are two angles between 0 and 2π satisfying $\tan \theta = y/x$. You must choose θ so that (r, θ) lies in the quadrant containing P if you use $r > 0$ and in the opposite quadrant if you use $r < 0$ (Figure 7).

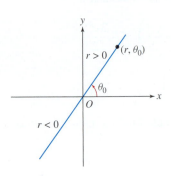

FIGURE 8 Lines through O with polar equation $\theta = \theta_0$.

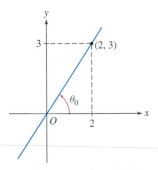

FIGURE 9 Line of slope $\frac{3}{2}$ through the origin.

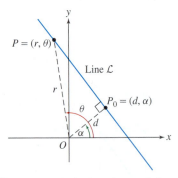

DF **FIGURE 10** P_0 is the point on $\mathcal{L}$ closest to the origin.

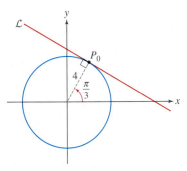

FIGURE 11 The tangent line has equation $r = 4 \sec\left(\theta - \frac{\pi}{3}\right)$.

■ **EXAMPLE 3 Choosing θ Correctly** Find two polar representations of $P = (-1, 1)$, one with $r > 0$ and one with $r < 0$.

Solution The point $P = (x, y) = (-1, 1)$ has polar coordinates (r, θ), where

$$r = \sqrt{(-1)^2 + 1^2} = \sqrt{2}, \qquad \tan\theta = \tan\frac{y}{x} = -1$$

However, θ is not given by

$$\tan^{-1}\frac{y}{x} = \tan^{-1}\left(\frac{1}{-1}\right) = -\frac{\pi}{4}$$

Because $\theta = -\frac{\pi}{4}$, this would place P in the fourth quadrant (Figure 7). Since P is in the second quadrant, the correct angle is

$$\theta = \tan^{-1}\frac{y}{x} + \pi = -\frac{\pi}{4} + \pi = \frac{3\pi}{4}$$

If we wish to use the negative radial coordinate $r = -\sqrt{2}$, then the angle becomes $\theta = -\frac{\pi}{4}$ or $\frac{7\pi}{4}$. Thus,

$$P = \left(\sqrt{2}, \frac{3\pi}{4}\right) \qquad \text{or} \qquad \left(-\sqrt{2}, \frac{7\pi}{4}\right) \qquad ■$$

A curve is described in polar coordinates by an equation involving r and θ, which we call a **polar equation**. By convention, we allow solutions with $r < 0$.

A line through the origin O has the simple equation $\theta = \theta_0$, where θ_0 is the angle between the line and the x-axis (Figure 8). Indeed, the points with $\theta = \theta_0$ are (r, θ_0), where r is arbitrary (positive, negative, or zero).

■ **EXAMPLE 4 Line Through the Origin** Find the polar equation of the line through the origin of slope $\frac{3}{2}$ (Figure 9).

Solution A line of slope m makes an angle θ_0 with the x-axis, where $m = \tan\theta_0$. In our case, $\theta_0 = \tan^{-1}\frac{3}{2} \approx 0.98$. The equation of the line is $\theta = \tan^{-1}\frac{3}{2}$ or $\theta \approx 0.98$. ■

To describe lines that do not pass through the origin, we note that any such line has a unique point P_0 that is *closest* to the origin. The next example shows how to write down the polar equation of the line in terms of P_0 (Figure 10).

■ **EXAMPLE 5 Line Not Passing Through the Origin** Show that

$$\boxed{r = d\sec(\theta - \alpha)} \qquad \boxed{1}$$

is the polar equation of the line $\mathcal{L}$ whose point closest to the origin is $P_0 = (d, \alpha)$.

Solution The point P_0 is obtained by dropping a perpendicular from the origin to $\mathcal{L}$ (Figure 10), and if $P = (r, \theta)$ is any point on $\mathcal{L}$ other than P_0, then $\triangle OPP_0$ is a right triangle. Therefore, $d/r = \cos(\theta - \alpha)$, or $r = d\sec(\theta - \alpha)$, as claimed. ■

■ **EXAMPLE 6** Find the polar equation of the line $\mathcal{L}$ tangent to the circle $r = 4$ at the point with polar coordinates $P_0 = \left(4, \frac{\pi}{3}\right)$.

Solution The point on $\mathcal{L}$ closest to the origin is P_0 itself (Figure 11). Therefore, we take $(d, \alpha) = \left(4, \frac{\pi}{3}\right)$ in Eq. (1) to obtain the equation $r = 4\sec\left(\theta - \frac{\pi}{3}\right)$. ■

■ **EXAMPLE 7** Sketch the curve corresponding to $r = 1 + \sin\theta$.

Solution If we let θ vary from 0 to 2π, we see all possible values of the function, and then it will repeat. So, we consider values in the range from 0 to 2π.

		A	B	C	D	E	F	G	H
θ	0	$\dfrac{\pi}{4}$	$\dfrac{\pi}{2}$	$\dfrac{3\pi}{4}$	π	$\dfrac{5\pi}{4}$	$\dfrac{3\pi}{2}$	$\dfrac{7\pi}{4}$
$r = 1 + \sin\theta$	1	1.707	2	1.707	1	0.293	0	0.293

For each of the given angles, we plot the point as in Figure 12, and then we connect the points with a smooth curve. The resulting curve is called a *cardioid*, which is Greek for the "heart" that it resembles. ■

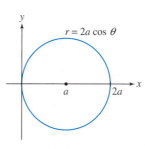

Often, it is hard to guess the shape of a graph of a polar equation. In some cases, it is helpful to rewrite the equation in rectangular coordinates.

■ **EXAMPLE 8** **Converting to Rectangular Coordinates** Identify the curve with polar equation $r = 2a\cos\theta$ (a a positive constant).

Solution Multiply the equation by r to obtain $r^2 = 2ar\cos\theta$. Because $r^2 = x^2 + y^2$ and $x = r\cos\theta$, this equation becomes

$$x^2 + y^2 = 2ax \qquad \text{or} \qquad x^2 - 2ax + y^2 = 0$$

Then complete the square to obtain $(x - a)^2 + y^2 = a^2$. This is the equation of the circle of radius a and center $(a, 0)$ (Figure 13). ■

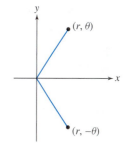

FIGURE 13

FIGURE 14 The points (r, θ) and $(r, -\theta)$ are symmetric with respect to the x-axis.

A similar calculation shows that the circle $x^2 + (y - a)^2 = a^2$ of radius a and center $(0, a)$ has polar equation $r = 2a\sin\theta$. In the next example, we make use of symmetry. Note that the points (r, θ) and $(r, -\theta)$ are symmetric with respect to the x-axis (Figure 14).

■ **EXAMPLE 9** **Symmetry About the x-Axis** Sketch the *limaçon* curve $r = 2\cos\theta - 1$.

Solution Since $f(\theta) = \cos\theta$ is periodic, it suffices to plot points for $-\pi \le \theta \le \pi$.

Step 1. **Plot points.**

To get started, we plot points A–G on a grid and join them by a smooth curve (Figure 15).

		A	B	C	D	E	F	G
θ	0	$\dfrac{\pi}{6}$	$\dfrac{\pi}{3}$	$\dfrac{\pi}{2}$	$\dfrac{2\pi}{3}$	$\dfrac{5\pi}{6}$	π
$r = 2\cos\theta - 1$	1	0.73	0	-1	-2	-2.73	-3

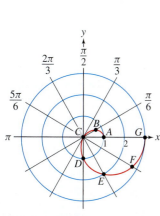

FIGURE 12 The cardioid given by $r = 1 + \sin\theta$.

DF **FIGURE 15** Plotting $r = 2\cos\theta - 1$ using a grid.

Step 2. Analyze r as a function of θ.

For a better understanding, it is helpful to graph r as a function of θ in rectangular coordinates. Figure 16(A) shows that

As θ varies from 0 to $\frac{\pi}{3}$, r varies from 1 to 0.

As θ varies from $\frac{\pi}{3}$ to π, r is *negative* and varies from 0 to -3.

We conclude:

- The graph begins at point A in Figure 16(B) and moves in toward the origin as θ varies from 0 to $\frac{\pi}{3}$.
- Since r is negative for $\frac{\pi}{3} \le \theta \le \pi$, the curve continues into the third and fourth quadrants (rather than into the first and second quadrants), moving toward the point $G = (-3, \pi)$ in Figure 16(C).

Step 3. Use symmetry.

Since $r(\theta) = r(-\theta)$, the curve is symmetric with respect to the x-axis. So, the part of the curve with $-\pi \le \theta \le 0$ is obtained by reflection through the x-axis as in Figure 16(D). ∎

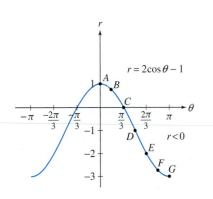

(A) Variation of r as a function of θ.

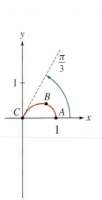

(B) θ varies from 0 to $\pi/3$; r varies from 1 to 0.

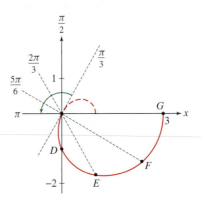

(C) θ varies from $\pi/3$ to π, but r is negative and varies from 0 to -3.

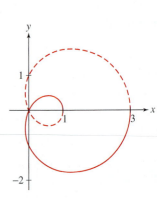

(D) The entire limaçon.

DF **FIGURE 16** The curve $r = 2\cos\theta - 1$ is called the *limaçon*, from the Latin word for "snail." It was first described in 1525 by the German artist Albrecht Dürer.

11.3 SUMMARY

- A point $P = (x, y)$ has polar coordinates (r, θ), where r is the distance to the origin and θ is the angle between the positive x-axis and the segment $\overline{OP}$, measured in the counterclockwise direction:

$$x = r\cos\theta, \qquad r = \sqrt{x^2 + y^2}$$

$$y = r\sin\theta, \qquad \tan\theta = \frac{y}{x} \quad (x \neq 0)$$

- The angular coordinate θ must be chosen so that (r, θ) lies in the proper quadrant. If $r > 0$, then

$$\theta = \begin{cases} \tan^{-1}\dfrac{y}{x} & \text{if } x > 0 \\[2mm] \tan^{-1}\dfrac{y}{x} + \pi & \text{if } x < 0 \\[2mm] \pm\dfrac{\pi}{2} & \text{if } x = 0 \end{cases}$$

- Nonuniqueness: (r, θ) and $(r, \theta + 2n\pi)$ represent the same point for all integers n. The origin O has polar coordinates $(0, \theta)$ for any θ.

- Negative radial coordinates: $(-r, \theta)$ and $(r, \theta + \pi)$ represent the same point.
- Polar equations:

Curve	Polar equation
Circle of radius R, center at the origin	$r = R$
Line through origin of slope $m = \tan \theta_0$	$\theta = \theta_0$
Line on which $P_0 = (d, \alpha)$ is the point closest to the origin	$r = d \sec(\theta - \alpha)$
Circle of radius a, center at $(a, 0)$ $(x - a)^2 + y^2 = a^2$	$r = 2a \cos \theta$
Circle of radius a, center at $(0, a)$ $x^2 + (y - a)^2 = a^2$	$r = 2a \sin \theta$

11.3 EXERCISES

Preliminary Questions

1. Points P and Q with the same radial coordinate (choose the correct answer):

(a) lie on the same circle with the center at the origin.

(b) lie on the same ray based at the origin.

2. Give two polar representations for the point $(x, y) = (0, 1)$, one with negative r and one with positive r.

3. Describe each of the following curves:

(a) $r = 2$ **(b)** $r^2 = 2$ **(c)** $r \cos \theta = 2$

4. If $f(-\theta) = f(\theta)$, then the curve $r = f(\theta)$ is symmetric with respect to the (choose the correct answer):

(a) x-axis.   **(b)** y-axis.   **(c)** origin.

Exercises

1. Find polar coordinates for each of the seven points plotted in Figure 17. (Choose $r \geq 0$ and θ in $[0, 2\pi)$.)

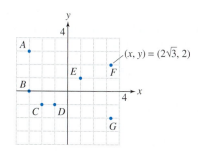

FIGURE 17

2. Plot the points with polar coordinates:

(a) $\left(2, \frac{\pi}{6}\right)$ **(b)** $\left(4, \frac{3\pi}{4}\right)$ **(c)** $\left(3, -\frac{\pi}{2}\right)$ **(d)** $\left(0, \frac{\pi}{6}\right)$

3. Convert from rectangular to polar coordinates:

(a) $(1, 0)$ **(b)** $(3, \sqrt{3})$ **(c)** $(-2, 2)$ **(d)** $(-1, \sqrt{3})$

4. Convert from rectangular to polar coordinates using a calculator (make sure your choice of θ gives the correct quadrant):

(a) $(2, 3)$ **(b)** $(4, -7)$ **(c)** $(-3, -8)$ **(d)** $(-5, 2)$

5. Convert from polar to rectangular coordinates:

(a) $\left(3, \frac{\pi}{6}\right)$ **(b)** $\left(6, \frac{3\pi}{4}\right)$ **(c)** $\left(0, \frac{\pi}{5}\right)$ **(d)** $\left(5, -\frac{\pi}{2}\right)$

6. Which of the following are possible polar coordinates for the point P with rectangular coordinates $(0, -2)$?

(a) $\left(2, \frac{\pi}{2}\right)$ **(b)** $\left(2, \frac{7\pi}{2}\right)$

(c) $\left(-2, -\frac{3\pi}{2}\right)$ **(d)** $\left(-2, \frac{7\pi}{2}\right)$

(e) $\left(-2, -\frac{\pi}{2}\right)$ **(f)** $\left(2, -\frac{7\pi}{2}\right)$

7. Describe each shaded sector in Figure 18 by inequalities in r and θ.

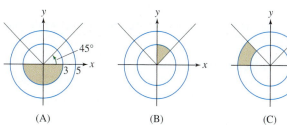

(A)   (B)   (C)

FIGURE 18

8. Find the equation in polar coordinates of the line through the origin with slope $\frac{1}{2}$.

9. What is the slope of the line $\theta = \frac{3\pi}{5}$?

10. Which of $r = 2 \sec \theta$ and $r = 2 \csc \theta$ defines a horizontal line?

In Exercises 11–16, convert to an equation in rectangular coordinates.

11. $r = 7$ **12.** $r = \sin \theta$

13. $r = 2 \sin \theta$ **14.** $r = 2 \csc \theta$

15. $r = \dfrac{1}{\cos \theta - \sin \theta}$ **16.** $r = \dfrac{1}{2 - \cos \theta}$

In Exercises 17–22, convert to an equation in polar coordinates of the form $r = f(\theta)$.

17. $x^2 + y^2 = 5$

18. $x = 5$

19. $y = x^2$

20. $xy = 1$

21. $e^{\sqrt{x^2 + y^2}} = 1$

22. $\ln x = 1$

23. Match each equation with its description:

(a) $r = 2$ (i) Vertical line
(b) $\theta = 2$ (ii) Horizontal line
(c) $r = 2\sec\theta$ (iii) Circle
(d) $r = 2\csc\theta$ (iv) Line through origin

24. Suppose that $P = (x, y)$ has polar coordinates (r, θ). Find the polar coordinates for the points:

(a) $(x, -y)$ (b) $(-x, -y)$ (c) $(-x, y)$ (d) (y, x)

25. Find the values of θ in the plot of $r = 4\cos\theta$ corresponding to points A, B, C, D in Figure 19. Then indicate the portion of the graph traced out as θ varies in the following intervals:

(a) $0 \leq \theta \leq \frac{\pi}{2}$ (b) $\frac{\pi}{2} \leq \theta \leq \pi$ (c) $\pi \leq \theta \leq \frac{3\pi}{2}$

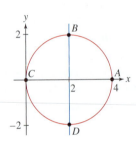

FIGURE 19 Plot of $r = 4\cos\theta$.

26. Match each equation in rectangular coordinates with its equation in polar coordinates:

(a) $x^2 + y^2 = 4$ (i) $r^2(1 - 2\sin^2\theta) = 4$
(b) $x^2 + (y - 1)^2 = 1$ (ii) $r(\cos\theta + \sin\theta) = 4$
(c) $x^2 - y^2 = 4$ (iii) $r = 2\sin\theta$
(d) $x + y = 4$ (iv) $r = 2$

27. What are the polar equations of the lines parallel to the line $r\cos\left(\theta - \frac{\pi}{3}\right) = 1$?

28. Show that the circle with its center at $\left(\frac{1}{2}, \frac{1}{2}\right)$ in Figure 20 has polar equation $r = \sin\theta + \cos\theta$ and find the values of θ between 0 and π corresponding to points A, B, C, and D.

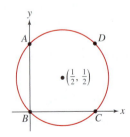

FIGURE 20 Plot of $r = \sin\theta + \cos\theta$.

29. Sketch the curve $r = \frac{1}{2}\theta$ (the spiral of Archimedes) for θ between 0 and 2π by plotting the points for $\theta = 0, \frac{\pi}{4}, \frac{\pi}{2}, \ldots, 2\pi$.

30. Sketch $r = 3\cos\theta - 1$ (see Example 9).

31. Sketch the cardioid curve $r = 1 + \cos\theta$.

32. Show that the cardioid of Exercise 31 has equation

$$(x^2 + y^2 - x)^2 = x^2 + y^2$$

in rectangular coordinates.

33. Figure 21 displays the graphs of $r = \sin 2\theta$ in rectangular coordinates and in polar coordinates, where it is a "rose with four petals." Identify:

(a) The points in (B) corresponding to points A–I in (A).

(b) The parts of the curve in (B) corresponding to the angle intervals $\left[0, \frac{\pi}{2}\right]$, $\left[\frac{\pi}{2}, \pi\right]$, $\left[\pi, \frac{3\pi}{2}\right]$, and $\left[\frac{3\pi}{2}, 2\pi\right]$.

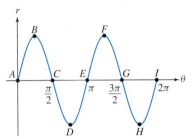 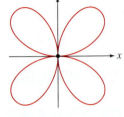

(A) Graph of r as a function of θ, where $r = \sin 2\theta$

(B) Graph of $r = \sin 2\theta$ in polar coordinates

FIGURE 21

34. Sketch the curve $r = \sin 3\theta$. First fill in the table of r-values below and plot the corresponding points of the curve. Notice that the three petals of the curve correspond to the angle intervals $\left[0, \frac{\pi}{3}\right]$, $\left[\frac{\pi}{3}, \frac{2\pi}{3}\right]$, and $\left[\frac{\pi}{3}, \pi\right]$. Then plot $r = \sin 3\theta$ in rectangular coordinates and label the points on this graph corresponding to (r, θ) in the table.

θ	0	$\frac{\pi}{12}$	$\frac{\pi}{6}$	$\frac{\pi}{4}$	$\frac{\pi}{3}$	$\frac{5\pi}{12}$	$\cdots$	$\frac{11\pi}{12}$	π
r									

35. **CAS** Plot the **cissoid** $r = 2\sin\theta\tan\theta$ and show that its equation in rectangular coordinates is

$$y^2 = \frac{x^3}{2 - x}$$

36. Prove that $r = 2a\cos\theta$ is the equation of the circle in Figure 22 using only the fact that a triangle inscribed in a circle with one side a diameter is a right triangle.

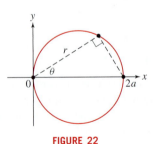

FIGURE 22

37. Show that

$$r = a\cos\theta + b\sin\theta$$

is the equation of a circle passing through the origin. Express the radius and center (in rectangular coordinates) in terms of a and b and write down the equation in rectangular coordinates.

38. Use the previous exercise to write the equation of the circle of radius 5 and center $(3, 4)$ in the form $r = a\cos\theta + b\sin\theta$.

39. Use the identity $\cos 2\theta = \cos^2\theta - \sin^2\theta$ to find a polar equation of the hyperbola $x^2 - y^2 = 1$.

40. Find an equation in rectangular coordinates for the curve $r^2 = \cos 2\theta$.

41. Show that $\cos 3\theta = \cos^3\theta - 3\cos\theta\sin^2\theta$ and use this identity to find an equation in rectangular coordinates for the curve $r = \cos 3\theta$.

42. Use the addition formula for the cosine to show that the line $\mathcal{L}$ with polar equation $r\cos(\theta - \alpha) = d$ has the equation in rectangular coordinates $(\cos\alpha)x + (\sin\alpha)y = d$. Show that $\mathcal{L}$ has slope $m = -\cot\alpha$ and y-intercept $d/\sin\alpha$.

In Exercises 43–46, find an equation in polar coordinates of the line $\mathcal{L}$ with the given description.

43. The point on $\mathcal{L}$ closest to the origin has polar coordinates $\left(2, \frac{\pi}{9}\right)$.

44. The point on $\mathcal{L}$ closest to the origin has rectangular coordinates $(-2, 2)$.

45. $\mathcal{L}$ is tangent to the circle $r = 2\sqrt{10}$ at the point with rectangular coordinates $(-2, -6)$.

46. $\mathcal{L}$ has slope 3 and is tangent to the unit circle in the fourth quadrant.

47. Show that every line that does not pass through the origin has a polar equation of the form

$$r = \frac{b}{\sin\theta - a\cos\theta}$$

where $b \neq 0$.

48. By the Law of Cosines, the distance d between two points (Figure 23) with polar coordinates (r, θ) and (r_0, θ_0) is

$$d^2 = r^2 + r_0^2 - 2rr_0\cos(\theta - \theta_0)$$

Use this distance formula to show that

$$r^2 - 10r\cos\left(\theta - \frac{\pi}{4}\right) = 56$$

is the equation of the circle of radius 9 whose center has polar coordinates $\left(5, \frac{\pi}{4}\right)$.

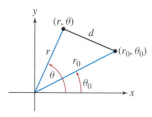

FIGURE 23

49. For $a > 0$, a **lemniscate curve** is the set of points P such that the product of the distances from P to $(a, 0)$ and $(-a, 0)$ is a^2. Show that the equation of the lemniscate is

$$(x^2 + y^2)^2 = 2a^2(x^2 - y^2)$$

Then find the equation in polar coordinates. To obtain the simplest form of the equation, use the identity $\cos 2\theta = \cos^2\theta - \sin^2\theta$. Plot the lemniscate for $a = 2$ if you have a computer algebra system.

50. ◻ Let c be a fixed constant. Explain the relationship between the graphs of:

(a) $y = f(x + c)$ and $y = f(x)$ (rectangular).

(b) $r = f(\theta + c)$ and $r = f(\theta)$ (polar).

(c) $y = f(x) + c$ and $y = f(x)$ (rectangular).

(d) $r = f(\theta) + c$ and $r = f(\theta)$ (polar).

51. The Derivative in Polar Coordinates Show that a polar curve $r = f(\theta)$ has parametric equations

$$x = f(\theta)\cos\theta, \qquad y = f(\theta)\sin\theta$$

Then apply Theorem 2 of Section 11.1 to prove

$$\frac{dy}{dx} = \frac{f(\theta)\cos\theta + f'(\theta)\sin\theta}{-f(\theta)\sin\theta + f'(\theta)\cos\theta} \qquad \boxed{2}$$

where $f'(\theta) = df/d\theta$.

52. Use Eq. (2) to find the slope of the tangent line to $r = \sin\theta$ at $\theta = \frac{\pi}{3}$.

53. Use Eq. (2) to find the slope of the tangent line to $r = \theta$ at $\theta = \frac{\pi}{2}$ and $\theta = \pi$.

54. Find the equation in rectangular coordinates of the tangent line to $r = 4\cos 3\theta$ at $\theta = \frac{\pi}{6}$.

55. Find the polar coordinates of the points on the lemniscate $r^2 = \cos 2\theta$ in Figure 24 where the tangent line is horizontal.

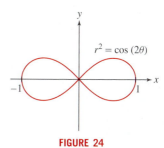

FIGURE 24

56. Find the polar coordinates of the points on the cardioid $r = 1 + \cos\theta$ where the tangent line is horizontal (see Figure 25(A)).

57. Use Eq. (2) to show that for $r = \sin\theta + \cos\theta$,

$$\frac{dy}{dx} = \frac{\cos 2\theta + \sin 2\theta}{\cos 2\theta - \sin 2\theta}$$

Then calculate the slopes of the tangent lines at points A, B, C in Figure 20.

Further Insights and Challenges

58. ◻ Let $y = f(x)$ be a periodic function of period 2π—that is, $f(x) = f(x + 2\pi)$. Explain how this periodicity is reflected in the graph of:

(a) $y = f(x)$ in rectangular coordinates.

(b) $r = f(\theta)$ in polar coordinates.

59. [GU] Use a graphing utility to convince yourself that the polar equations $r = f_1(\theta) = 2\cos\theta - 1$ and $r = f_2(\theta) = 2\cos\theta + 1$ have the same graph. Then explain why. *Hint:* Show that the points $(f_1(\theta + \pi), \theta + \pi)$ and $(f_2(\theta), \theta)$ coincide.

60. [CAS] We investigate how the shape of the limaçon curve $r = b + \cos\theta$ depends on the constant b (see Figure 25).

(a) Argue as in Exercise 59 to show that the constants b and $-b$ yield the same curve.

(b) Plot the limaçon for $b = 0, 0.2, 0.5, 0.8, 1$ and describe how the curve changes.

(c) Plot the limaçon for $b = 1.2, 1.5, 1.8, 2, 2.4$ and describe how the curve changes.

(d) Use Eq. (2) to show that

$$\frac{dy}{dx} = -\left(\frac{b\cos\theta + \cos 2\theta}{b + 2\cos\theta}\right)\csc\theta$$

(e) Find the points where the tangent line is vertical. Note that there are three cases: $0 \le b < 2$, $b = 2$, and $b > 2$. Do the plots constructed in (b) and (c) reflect your results?

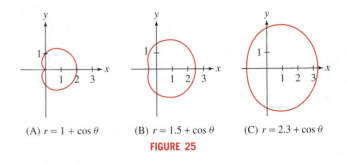

(A) $r = 1 + \cos\theta$ (B) $r = 1.5 + \cos\theta$ (C) $r = 2.3 + \cos\theta$

FIGURE 25

11.4 Area and Arc Length in Polar Coordinates

Integration in polar coordinates involves finding not the area *underneath* a curve, but rather the area of a sector bounded by a curve as in Figure 1(A). Consider the region bounded by the curve $r = f(\theta)$ where $f(\theta) \ge 0$ and the two rays $\theta = \alpha$ and $\theta = \beta$ with $\alpha < \beta$. To derive a formula for the area, divide the region into N narrow sectors of angle $\Delta\theta = (\beta - \alpha)/N$ corresponding to a partition of the interval $[\alpha, \beta]$ as in Figure 1(B):

$$\theta_0 = \alpha < \theta_1 < \theta_2 < \cdots < \theta_N = \beta$$

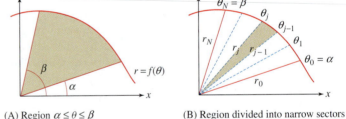

[DF] **FIGURE 1** Area bounded by the curve $r = f(\theta)$ and the two rays $\theta = \alpha$ and $\theta = \beta$.

(A) Region $\alpha \le \theta \le \beta$ (B) Region divided into narrow sectors

Recall that a circular sector of angle $\Delta\theta$ and radius r has area $\frac{1}{2}r^2\Delta\theta$ (Figure 2). If $\Delta\theta$ is small, the jth narrow sector (Figure 3) is nearly a circular sector of radius $r_j = f(\theta_j)$, so its area is *approximately* $\frac{1}{2}r_j^2\Delta\theta$. The total area is approximated by the sum:

$$\text{Area of region} \approx \sum_{j=1}^{N} \frac{1}{2}r_j^2\Delta\theta = \frac{1}{2}\sum_{j=1}^{N} f(\theta_j)^2\Delta\theta \qquad \boxed{1}$$

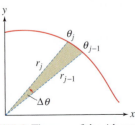

FIGURE 2 The area of a circular sector is the fraction of the total area of the circle given by the angle divided by 2π, that is, $\frac{\Delta\theta}{2\pi}\pi r^2 = \frac{1}{2}r^2\Delta\theta$.

This is a Riemann sum for the integral $\frac{1}{2}\int_{\alpha}^{\beta} f(\theta)^2\, d\theta$. If f is continuous, then the sum approaches the integral as $N \to \infty$, and we obtain the following formula.

THEOREM 1 Area in Polar Coordinates If f is a continuous function, then the area bounded by a curve in polar form $r = f(\theta)$ and the rays $\theta = \alpha$ and $\theta = \beta$ (with $\alpha < \beta$) is equal to

$$\frac{1}{2}\int_{\alpha}^{\beta} r^2\, d\theta = \frac{1}{2}\int_{\alpha}^{\beta} f(\theta)^2\, d\theta \qquad \boxed{2}$$

[DF] **FIGURE 3** The area of the jth sector is approximately $\frac{1}{2}r_j^2\Delta\theta$.

We know that $r = R$ defines a circle of radius R. By Eq. (2), the area is equal to

$$\frac{1}{2}\int_{0}^{2\pi} R^2\, d\theta = \frac{1}{2}R^2(2\pi) = \pi R^2, \text{ as expected.}$$

CAUTION Keep in mind that the integral $\frac{1}{2}\int_\alpha^\beta r^2\,d\theta$ does **not** compute the area **under** a curve as in Figure 4(B), but rather computes the area "swept out" by a radial segment as θ varies from α to β, as in Figure 4(A).

■ **EXAMPLE 1** Use Theorem 1 to compute the area of the right semicircle with equation $r = 4\sin\theta$.

Solution The equation $r = 4\sin\theta$ defines a circle of radius 2 tangent to the x-axis at the origin. The right semicircle is "swept out" as θ varies from 0 to $\frac{\pi}{2}$ as in Figure 4(A). By Eq. (2), the area of the right semicircle is

$$\frac{1}{2}\int_0^{\pi/2} r^2\,d\theta = \frac{1}{2}\int_0^{\pi/2} (4\sin\theta)^2\,d\theta = 8\int_0^{\pi/2}\sin^2\theta\,d\theta \qquad \boxed{3}$$

$$= 8\int_0^{\pi/2}\frac{1}{2}(1-\cos 2\theta)\,d\theta$$

$$= (4\theta - 2\sin 2\theta)\Big|_0^{\pi/2} = 4\left(\frac{\pi}{2}\right) - 0 = 2\pi \qquad ■$$

⤺··· REMINDER In Eq. (3), we use the identity

$$\sin^2\theta = \frac{1}{2}(1-\cos 2\theta) \qquad \boxed{4}$$

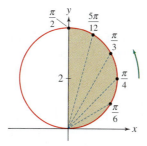

(A) The polar integral computes the area swept out by a radial segment.

(B) The ordinary integral in rectangular coordinates computes the area underneath a curve.

FIGURE 4

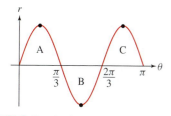

FIGURE 5 Graph of $r = \sin 3\theta$ as a function of θ.

■ **EXAMPLE 2** Sketch $r = \sin 3\theta$ and compute the area of one "petal."

Solution To sketch the curve, we first graph $r = \sin 3\theta$ in rectangular coordinates. Figure 5 shows that the radius r varies from 0 to 1 and back to 0 as θ varies from 0 to $\frac{\pi}{3}$. This gives petal A in Figure 6. Petal B is traced as θ varies from $\frac{\pi}{3}$ to $\frac{2\pi}{3}$ (with $r \le 0$), and petal C is traced for $\frac{2\pi}{3} \le \theta \le \pi$. We find that the area of petal A [using Eq. (4) in the margin to evaluate the integral] is equal to

$$\frac{1}{2}\int_0^{\pi/3}(\sin 3\theta)^2\,d\theta = \frac{1}{2}\int_0^{\pi/3}\left(\frac{1-\cos 6\theta}{2}\right)d\theta = \left(\frac{1}{4}\theta - \frac{1}{24}\sin 6\theta\right)\Big|_0^{\pi/3} = \frac{\pi}{12} \quad ■$$

The region sandwiched between two polar curves $r = f_1(\theta)$ and $r = f_2(\theta)$ with $f_2(\theta) \ge f_1(\theta) > 0$, for $\alpha \le \theta \le \beta$, is called a **radially simple region**. It has the property that every radial line through the origin that intersects the region does so in either a single point or a single line segment that begins on the curve $r = f_1(\theta)$ and ends on the curve $r = f_2(\theta)$ (Figure 7). For such a radially simple region, we have the following formula for its area:

$$\boxed{\text{Area between two curves} = \frac{1}{2}\int_\alpha^\beta \left(f_2(\theta)^2 - f_1(\theta)^2\right)d\theta} \qquad \boxed{5}$$

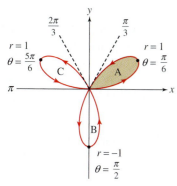

FIGURE 6 Graph of polar curve $r = \sin 3\theta$, a "rose with three petals."

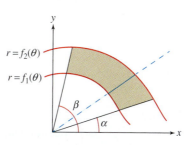

FIGURE 7 A radially simple region.

■ **EXAMPLE 3 Area Between Two Curves** Find the area of the region inside the circle $r = 2\cos\theta$ but outside the circle $r = 1$ [Figure 8(A)].

Solution The two circles intersect at the points where $(r, 2\cos\theta) = (r, 1)$ or, in other words, when $2\cos\theta = 1$. This yields $\cos\theta = \frac{1}{2}$, which has solutions $\theta = \pm\frac{\pi}{3}$.

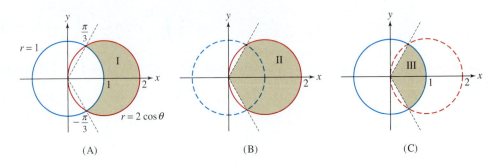

FIGURE 8 Region I is the difference of regions II and III.

We see in Figure 8 that region I is the difference of regions II and III in Figures 8(B) and (C). Therefore,

$$\text{Area of I} = \text{Area of II} - \text{Area of III}$$

◀·· **REMINDER** In Eq. (6), we use the identity
$$\cos^2\theta = \frac{1}{2}(1 + \cos 2\theta)$$

$$= \frac{1}{2}\int_{-\pi/3}^{\pi/3} (2\cos\theta)^2\,d\theta - \frac{1}{2}\int_{-\pi/3}^{\pi/3} (1)^2\,d\theta$$

$$= \frac{1}{2}\int_{-\pi/3}^{\pi/3} (4\cos^2\theta - 1)\,d\theta = \frac{1}{2}\int_{-\pi/3}^{\pi/3} (2\cos 2\theta + 1)\,d\theta \qquad \boxed{6}$$

$$= \frac{1}{2}(\sin 2\theta + \theta)\Big|_{-\pi/3}^{\pi/3} = \frac{\sqrt{3}}{2} + \frac{\pi}{3} \approx 1.91 \qquad ■$$

We close this section by deriving a formula for arc length in polar coordinates. Observe that a polar curve $r = f(\theta)$ has a parametrization with θ as a parameter:

$$x = r\cos\theta = f(\theta)\cos\theta, \qquad y = r\sin\theta = f(\theta)\sin\theta$$

Using a prime to denote the derivative with respect to θ, we have

$$x'(\theta) = \frac{dx}{d\theta} = -f(\theta)\sin\theta + f'(\theta)\cos\theta$$

$$y'(\theta) = \frac{dy}{d\theta} = f(\theta)\cos\theta + f'(\theta)\sin\theta$$

Recall from Section 11.2 that arc length is obtained by integrating $\sqrt{x'(\theta)^2 + y'(\theta)^2}$. Straightforward algebra shows that $x'(\theta)^2 + y'(\theta)^2 = f(\theta)^2 + f'(\theta)^2$; thus,

$$\boxed{\text{Arc length } s = \int_\alpha^\beta \sqrt{f(\theta)^2 + f'(\theta)^2}\,d\theta} \qquad \boxed{7}$$

■ **EXAMPLE 4** Find the total length of the circle $r = 2a\cos\theta$ for $a > 0$.

Solution In this case, $f(\theta) = 2a\cos\theta$ and

$$f(\theta)^2 + f'(\theta)^2 = 4a^2\cos^2\theta + 4a^2\sin^2\theta = 4a^2$$

The total length of this circle of radius a has the expected value:

$$\int_0^\pi \sqrt{f(\theta)^2 + f'(\theta)^2}\,d\theta = \int_0^\pi (2a)\,d\theta = 2\pi a$$

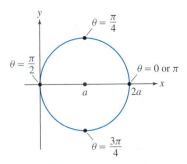

FIGURE 9 Graph of $r = 2a\cos\theta$.

Note that the upper limit of integration is π rather than 2π because the entire circle is traced out as θ varies from 0 to π (see Figure 9).   ■

11.4 SUMMARY

- Area of the sector bounded by a polar curve $r = f(\theta)$ and two rays $\theta = \alpha$ and $\theta = \beta$ (Figure 10):

$$\text{Area} = \frac{1}{2} \int_\alpha^\beta f(\theta)^2 \, d\theta$$

- Area of a radially simple region, which is between $r = f_1(\theta)$ and $r = f_2(\theta)$, where $f_2(\theta) \geq f_1(\theta)$ (Figure 11):

$$\text{Area} = \frac{1}{2} \int_\alpha^\beta \left(f_2(\theta)^2 - f_1(\theta)^2 \right) d\theta$$

- Arc length of the polar curve $r = f(\theta)$ for $\alpha \leq \theta \leq \beta$:

$$\text{Arc length} = \int_\alpha^\beta \sqrt{f(\theta)^2 + f'(\theta)^2} \, d\theta$$

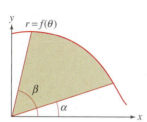

FIGURE 10 Region bounded by the polar curve $r = f(\theta)$ and the rays $\theta = \alpha$, $\theta = \beta$.

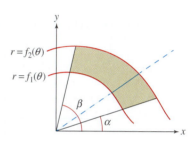

FIGURE 11 Region between two polar curves.

11.4 EXERCISES

Preliminary Questions

1. Polar coordinates are suited to finding the area (choose one):

(a) under a curve between $x = a$ and $x = b$.

(b) bounded by a curve and two rays through the origin.

2. Is the formula for area in polar coordinates valid if $f(\theta)$ takes negative values?

3. The horizontal line $y = 1$ has polar equation $r = \csc \theta$. Which area is represented by the integral $\dfrac{1}{2} \displaystyle\int_{\pi/6}^{\pi/2} \csc^2 \theta \, d\theta$ (Figure 12)?

(a) $\square ABCD$ **(b)** $\triangle ABC$ **(c)** $\triangle ACD$

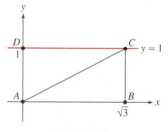

FIGURE 12

Exercises

1. Sketch the area bounded by the circle $r = 5$ and the rays $\theta = \frac{\pi}{2}$ and $\theta = \pi$, and compute its area as an integral in polar coordinates.

2. Sketch the region bounded by the line $r = \sec \theta$ and the rays $\theta = 0$ and $\theta = \frac{\pi}{3}$. Compute its area in two ways: as an integral in polar coordinates and using geometry.

3. Calculate the area of the circle $r = 4 \sin \theta$ as an integral in polar coordinates (see Figure 4). Be careful to choose the correct limits of integration.

4. Find the area of the shaded triangle in Figure 13 as an integral in polar coordinates. Then find the rectangular coordinates of P and Q, and compute the area via geometry.

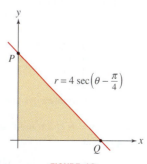

FIGURE 13

5. Find the area of the shaded region in Figure 14. Note that θ varies from 0 to $\frac{\pi}{2}$.

6. Which interval of θ-values corresponds to the the the shaded region in Figure 15? Find the area of the region.

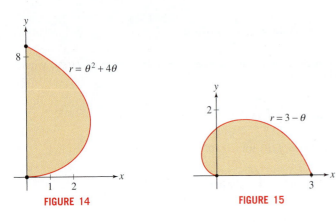

FIGURE 14

FIGURE 15

7. Find the total area enclosed by the cardioid in Figure 16.

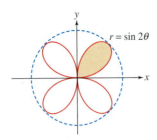

FIGURE 16 The cardioid $r = 1 - \cos\theta$.

8. Find the area of the shaded region in Figure 16.

9. Find the area of one leaf of the "four-petaled rose" $r = \sin 2\theta$ (Figure 17). Then prove that the total area of the rose is equal to one-half the area of the circumscribed circle.

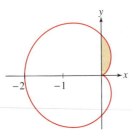

FIGURE 17 Four-petaled rose $r = \sin 2\theta$.

10. Find the area enclosed by one loop of the lemniscate with equation $r^2 = \cos 2\theta$ (Figure 18). Choose your limits of integration carefully.

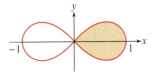

FIGURE 18 The lemniscate $r^2 = \cos 2\theta$.

11. Sketch the spiral $r = \theta$ for $0 \le \theta \le 2\pi$ and find the area bounded by the curve and the first quadrant.

12. Find the area of the intersection of the circles $r = \sin\theta$ and $r = \cos\theta$.

13. Find the area of region A in Figure 19.

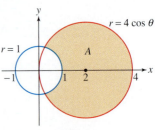

FIGURE 19

14. Find the area of the shaded region in Figure 20 enclosed by the circle $r = \frac{1}{2}$ and a petal of the curve $r = \cos 3\theta$. *Hint:* Compute the area of both the petal and the region inside the petal and outside the circle.

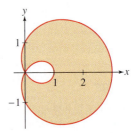

FIGURE 20

15. Find the area of the inner loop of the limaçon with polar equation $r = 2\cos\theta - 1$ (Figure 21).

16. Find the area of the shaded region in Figure 21 between the inner and outer loop of the limaçon $r = 2\cos\theta - 1$.

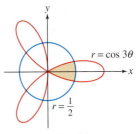

FIGURE 21 The limaçon $r = 2\cos\theta - 1$.

17. Find the area of the part of the circle $r = \sin\theta + \cos\theta$ in the fourth quadrant (see Exercise 28 in Section 11.3).

18. Find the area of the region inside the circle $r = 2\sin\left(\theta + \frac{\pi}{4}\right)$ and above the line $r = \sec\left(\theta - \frac{\pi}{4}\right)$.

19. Find the area between the two curves in Figure 22(A).

20. Find the area between the two curves in Figure 22(B).

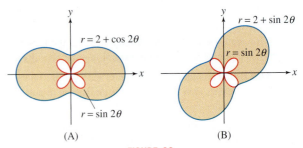

FIGURE 22

21. Find the area inside both curves in Figure 23.

22. Find the area of the region that lies inside one but not both of the curves in Figure 23.

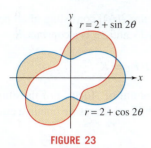

FIGURE 23

23. Calculate the total length of the circle $r = 4\sin\theta$ as an integral in polar coordinates.

24. Sketch the segment $r = \sec\theta$ for $0 \le \theta \le A$. Then compute its length in two ways: as an integral in polar coordinates and using trigonometry.

In Exercises 25–32, compute the length of the polar curve.

25. The length of $r = \theta^2$ for $0 \le \theta \le \pi$

26. The spiral $r = \theta$ for $0 \le \theta \le A$

27. The curve $r = \sin\theta$ for $0 \le \theta \le \pi$

28. The equiangular spiral $r = e^{\theta}$ for $0 \le \theta \le 2\pi$

29. $r = \sqrt{1 + \sin 2\theta}$ for $0 \le \theta \le \pi/4$

30. The cardioid $r = 1 - \cos\theta$ in Figure 16

31. $r = \cos^2\theta$

32. $r = 1 + \theta$ for $0 \le \theta \le \pi/2$

In Exercises 33–36, express the length of the curve as an integral but do not evaluate it.

33. $r = e^{\theta} + 1$, $0 \le \theta \le \pi/2$.

34. $r = (2 - \cos\theta)^{-1}$, $0 \le \theta \le 2\pi$

35. $r = \sin^3\theta$, $0 \le \theta \le 2\pi$

36. $r = \sin\theta\cos\theta$, $0 \le \theta \le \pi$

In Exercises 37–40, use a computer algebra system to calculate the total length to two decimal places.

37. ⌂⌂⌂ The three-petal rose $r = \cos 3\theta$ in Figure 20

38. ⌂⌂⌂ The curve $r = 2 + \sin 2\theta$ in Figure 23

39. ⌂⌂⌂ The curve $r = \theta\sin\theta$ in Figure 24 for $0 \le \theta \le 4\pi$

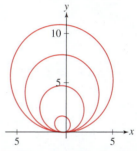

FIGURE 24 $r = \theta\sin\theta$ for $0 \le \theta \le 4\pi$.

40. ⌂⌂⌂ $r = \sqrt{\theta}$, $0 \le \theta \le 4\pi$

Further Insights and Challenges

41. Suppose that the polar coordinates of a moving particle at time t are $(r(t), \theta(t))$. Prove that the particle's speed is equal to
$$\sqrt{(dr/dt)^2 + r^2(d\theta/dt)^2}.$$

42. Compute the speed at time $t = 1$ of a particle whose polar coordinates at time t are $r = t$, $\theta = t$ (use Exercise 41). What would the speed be if the particle's rectangular coordinates were $x = t, y = t$? Why is the speed increasing in one case and constant in the other?

11.5 Conic Sections

The conics were first studied by the ancient Greek mathematicians, beginning possibly with Menaechmus (c. 380–320 BCE) and including Archimedes (287–212 BCE) and Apollonius (c. 262–190 BCE).

Three familiar families of curves—ellipses, hyperbolas, and parabolas—appear throughout mathematics and its applications. They are called **conic sections** because they are obtained as the intersection of a cone with a suitable plane (Figure 1). Our goal in this section is to derive equations for the conic sections from their geometric definitions as curves in the plane.

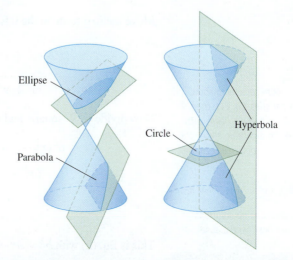

DF **FIGURE 1** The conic sections are obtained by intersecting a plane and a cone.

An **ellipse** is an oval-shaped curve [Figure 2(A)] consisting of all points P such that the sum of the distances to two fixed points F_1 and F_2 is a constant $K > 0$:

$$PF_1 + PF_2 = K \qquad \boxed{1}$$

We assume always that K is greater than the distance F_1F_2 between the foci, because the ellipse reduces to the line segment $\overline{F_1F_2}$ if $K = F_1F_2$, and it has no points at all if $K < F_1F_2$.

The points F_1 and F_2 are called the **foci** (plural of "focus") of the ellipse. Note that if the foci coincide, then Eq. (1) reduces to $2PF_1 = K$ and we obtain a circle of radius $\frac{1}{2}K$ centered at F_1.

We use the following terminology:

- The midpoint of $\overline{F_1F_2}$ is the **center** of the ellipse.
- The line through the foci is the **focal axis**.
- The line through the center perpendicular to the focal axis is the **conjugate axis**.

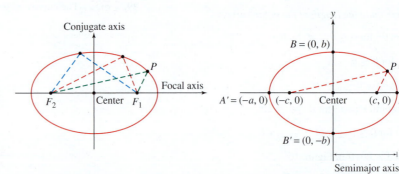

(A) The ellipse consists of all points P such that $PF_1 + PF_2 = K$.

(B) Ellipse in standard position:
$$\left(\frac{x}{a}\right)^2 + \left(\frac{y}{b}\right)^2 = 1$$

FIGURE 2

The ellipse is said to be in **standard position** if the focal and conjugate axes are the x- and y-axes, as shown in Figure 2(B). In this case, the foci have coordinates $F_1 = (c, 0)$ and $F_2 = (-c, 0)$ for some $c > 0$. Let us prove that the equation of this ellipse has the particularly simple form

$$\left(\frac{x}{a}\right)^2 + \left(\frac{y}{b}\right)^2 = 1 \qquad \boxed{2}$$

where $a = K/2$ and $b = \sqrt{a^2 - c^2}$.

By the distance formula, $P = (x, y)$ lies on the ellipse in Figure 2(B) if

$$PF_1 + PF_2 = \sqrt{(x-c)^2 + y^2} + \sqrt{(x+c)^2 + y^2} = 2a \qquad \boxed{3}$$

Move the first term on the left over to the right and square both sides:

$$(x+c)^2 + y^2 = 4a^2 - 4a\sqrt{(x-c)^2 + y^2} + (x-c)^2 + y^2$$

$$4a\sqrt{(x-c)^2 + y^2} = 4a^2 + (x-c)^2 - (x+c)^2 = 4a^2 - 4cx$$

Strictly speaking, it is necessary to show that if $P = (x, y)$ satisfies Eq. (4), then it also satisfies Eq. (3). When we begin with Eq. (4) and reverse the algebraic steps, the process of taking square roots leads to the relation

$$\sqrt{(x-c)^2 + y^2} \pm \sqrt{(x+c)^2 + y^2} = \pm 2a$$

However, this equation has no solutions unless both signs are positive because $a > c$.

Now divide by 4, square, and simplify:

$$a^2(x^2 - 2cx + c^2 + y^2) = a^4 - 2a^2cx + c^2x^2$$

$$(a^2 - c^2)x^2 + a^2y^2 = a^4 - a^2c^2 = a^2(a^2 - c^2)$$

$$\frac{x^2}{a^2} + \frac{y^2}{a^2 - c^2} = 1 \qquad \boxed{4}$$

This is Eq. (2) with $b^2 = a^2 - c^2$ as claimed.

The ellipse intersects the axes in four points A, A', B, B' called **vertices**. Vertices A and A' along the focal axis are called the **focal vertices**. Following common usage, the numbers a and b are referred to as the **semimajor axis** and the **semiminor axis** (even though they are numbers rather than axes).

THEOREM 1 **Ellipse in Standard Position** Let $a > b > 0$, and set $c = \sqrt{a^2 - b^2}$. The ellipse $PF_1 + PF_2 = 2a$ with foci $F_1 = (c, 0)$ and $F_2 = (-c, 0)$ has equation

$$\left(\frac{x}{a}\right)^2 + \left(\frac{y}{b}\right)^2 = 1 \qquad \boxed{5}$$

Furthermore, the ellipse has:

- semimajor axis a, semiminor axis b.
- focal vertices $(\pm a, 0)$, minor vertices $(0, \pm b)$.

If $b > a > 0$, then Eq. (5) defines an ellipse with foci $(0, \pm c)$, where $c = \sqrt{b^2 - a^2}$.

■ **EXAMPLE 1** Find the equation of the ellipse with foci $(\pm\sqrt{11}, 0)$ and semimajor axis $a = 6$. Then find the semiminor axis and sketch the graph.

Solution The foci are $(\pm c, 0)$ with $c = \sqrt{11}$, and the semimajor axis is $a = 6$, so we can use the relation $c = \sqrt{a^2 - b^2}$ to find b:

$$b^2 = a^2 - c^2 = 6^2 - (\sqrt{11})^2 = 25 \quad \Rightarrow \quad b = 5$$

Thus, the semiminor axis is $b = 5$ and the ellipse has equation $\left(\frac{x}{6}\right)^2 + \left(\frac{y}{5}\right)^2 = 1$. To sketch this ellipse, plot the vertices $(\pm 6, 0)$ and $(0, \pm 5)$ and connect them as in Figure 3. ■

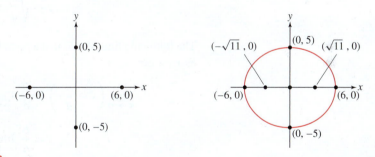

FIGURE 3

To write down the equation of an ellipse with axes parallel to the x- and y-axes and center translated to the point $C = (h, k)$, replace x by $x - h$ and y by $y - k$ in the equation (Figure 4):

$$\left(\frac{x - h}{a}\right)^2 + \left(\frac{y - k}{b}\right)^2 = 1$$

■ **EXAMPLE 2** **Translating an Ellipse** Find an equation of the ellipse with center $C = (6, 7)$, vertical focal axis, semimajor axis 5, and semiminor axis 3. Where are the foci located?

Solution Since the focal axis is vertical, we have $a = 3$ and $b = 5$, so that $a < b$ (Figure 4). The ellipse centered at the origin would have equation $\left(\frac{x}{3}\right)^2 + \left(\frac{y}{5}\right)^2 = 1$. When the center is translated to $(h, k) = (6, 7)$, the equation becomes

$$\left(\frac{x - 6}{3}\right)^2 + \left(\frac{y - 7}{5}\right)^2 = 1$$

Furthermore, $c = \sqrt{b^2 - a^2} = \sqrt{5^2 - 3^2} = 4$, so the foci are located ± 4 vertical units above and below the center—that is, $F_1 = (6, 11)$ and $F_2 = (6, 3)$. ■

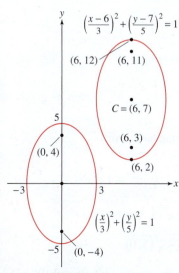

FIGURE 4 An ellipse with vertical major axis and its translate with center $C = (6, 7)$.

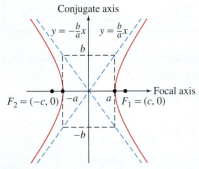

FIGURE 5 A hyperbola with center $(0, 0)$.

A **hyperbola** is the set of all points P such that the difference of the distances from P to two foci F_1 and F_2 is $\pm K$:

$$PF_1 - PF_2 = \pm K \qquad \boxed{6}$$

We assume that K is less than the distance $F_1 F_2$ between the foci (the hyperbola has no points if $K > F_1 F_2$). Note that a hyperbola consists of two branches corresponding to the choices of sign $\pm$ (Figure 5).

As before, the midpoint of $\overline{F_1 F_2}$ is the **center** of the hyperbola, the line through F_1 and F_2 is called the **focal axis**, and the line through the center perpendicular to the focal axis is called the **conjugate axis**. The **vertices** are the points where the focal axis intersects the hyperbola; they are labeled A and A' in Figure 5. The hyperbola is said to be in standard position when the focal and conjugate axes are the x- and y-axes, respectively, as in Figure 6. The next theorem can be verified in much the same way as Theorem 1.

> **THEOREM 2 Hyperbola in Standard Position** Let $a > 0$ and $b > 0$, and set $c = \sqrt{a^2 + b^2}$. The hyperbola $PF_1 - PF_2 = \pm 2a$ with foci $F_1 = (c, 0)$ and $F_2 = (-c, 0)$ has equation
>
> $$\left(\frac{x}{a}\right)^2 - \left(\frac{y}{b}\right)^2 = 1 \qquad \boxed{7}$$

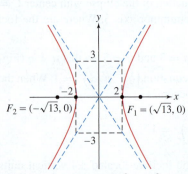

DF **FIGURE 6** Hyperbola in standard position.

A hyperbola has two **asymptotes** $y = \pm\frac{b}{a}x$ that are, we claim, diagonals of the rectangle whose sides pass through $(\pm a, 0)$ and $(0, \pm b)$ as in Figure 6. To prove this, consider a point (x, y) on the hyperbola in the first quadrant. By Eq. (7),

$$y = \sqrt{\frac{b^2}{a^2}x^2 - b^2} = \frac{b}{a}\sqrt{x^2 - a^2}$$

The following limit shows that a point (x, y) on the hyperbola approaches the line $y = \frac{b}{a}x$ as $x \to \infty$:

$$\lim_{x \to \infty}\left(y - \frac{b}{a}x\right) = \frac{b}{a}\lim_{x \to \infty}\left(\sqrt{x^2 - a^2} - x\right)$$

$$= \frac{b}{a}\lim_{x \to \infty}\left(\sqrt{x^2 - a^2} - x\right)\left(\frac{\sqrt{x^2 - a^2} + x}{\sqrt{x^2 - a^2} + x}\right)$$

$$= \frac{b}{a}\lim_{x \to \infty}\left(\frac{-a^2}{\sqrt{x^2 - a^2} + x}\right) = 0$$

The asymptotic behavior in the remaining quadrants is similar.

■ **EXAMPLE 3** Find the foci of the hyperbola $9x^2 - 4y^2 = 36$. Sketch its graph and asymptotes.

Solution First divide by 36 to write the equation in standard form:

$$\frac{x^2}{4} - \frac{y^2}{9} = 1 \qquad \text{or} \qquad \left(\frac{x}{2}\right)^2 - \left(\frac{y}{3}\right)^2 = 1$$

Thus, $a = 2$, $b = 3$, and $c = \sqrt{a^2 + b^2} = \sqrt{4 + 9} = \sqrt{13}$. The foci are

$$F_1 = (\sqrt{13}, 0), \qquad F_2 = (-\sqrt{13}, 0)$$

To sketch the graph, we draw the rectangle through the points $(\pm 2, 0)$ and $(0, \pm 3)$ as in Figure 7. The diagonals of the rectangle are the asymptotes $y = \pm\frac{3}{2}x$. The hyperbola passes through the vertices $(\pm 2, 0)$ and approaches the asymptotes. ■

FIGURE 7 The hyperbola $9x^2 - 4y^2 = 36$.

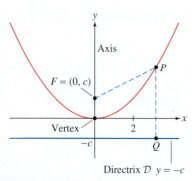

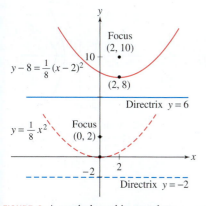

Unlike the ellipse and hyperbola, which are defined in terms of two foci, a **parabola** is the set of points P equidistant from a focus F and a line $\mathcal{D}$ called the **directrix**:

$$PF = PD \qquad \boxed{8}$$

Here, when we speak of the *distance* from a point P to a line $\mathcal{D}$, we mean the distance from P to the point Q on $\mathcal{D}$ closest to P, obtained by dropping a perpendicular from P to $\mathcal{D}$ (Figure 8). We denote this distance by PD.

The line through the focus F perpendicular to $\mathcal{D}$ is called the **axis** of the parabola. The **vertex** is the point where the parabola intersects its axis. We say that the parabola is in standard position if, for some c, the focus is $F = (0, c)$ and the directrix is $y = -c$, as shown in Figure 8. We verify in Exercise 73 that the vertex is then located at the origin and the equation of the parabola is $y = x^2/4c$. If $c < 0$, then the parabola opens downward.

THEOREM 3 Parabola in Standard Position Let $c \neq 0$. The parabola with focus $F = (0, c)$ and directrix $y = -c$ has equation

$$y = \frac{1}{4c}x^2 \qquad \boxed{9}$$

The vertex is located at the origin. The parabola opens upward if $c > 0$ and downward if $c < 0$.

■ **EXAMPLE 4** The standard parabola with directrix $y = -2$ is translated so that its vertex is located at $(2, 8)$. Find its equation, directrix, and focus.

Solution By Eq. (9) with $c = 2$, the standard parabola with directrix $y = -2$ has equation $y = \frac{1}{8}x^2$ (Figure 9). The focus of this standard parabola is $(0, c) = (0, 2)$, which is 2 units above the vertex $(0, 0)$.

To obtain the equation when the parabola is translated with vertex at $(2, 8)$, we replace x by $x - 2$ and y by $y - 8$:

$$y - 8 = \frac{1}{8}(x - 2)^2 \qquad \text{or} \qquad y = \frac{1}{8}x^2 - \frac{1}{2}x + \frac{17}{2}$$

The vertex has moved up 8 units, so the directrix also moves up 8 units to become $y = 6$. The new focus is 2 units above the new vertex $(2, 8)$, so the new focus is $(2, 10)$. ■

Eccentricity

Some ellipses are flatter than others, just as some hyperbolas are steeper. The "shape" of a conic section is measured by a number e called the **eccentricity**. For an ellipse or hyperbola,

$$e = \frac{\text{distance betweeen foci}}{\text{distance between vertices on focal axis}}$$

A parabola is defined to have eccentricity $e = 1$.

← REMINDER

Standard ellipse:

$$\left(\frac{x}{a}\right)^2 + \left(\frac{y}{b}\right)^2 = 1, \quad c = \sqrt{a^2 - b^2}$$

Standard hyperbola:

$$\left(\frac{x}{a}\right)^2 - \left(\frac{y}{b}\right)^2 = 1, \quad c = \sqrt{a^2 + b^2}$$

THEOREM 4 For ellipses and hyperbolas in standard position,

$$e = \frac{c}{a}$$

1. An ellipse has eccentricity $0 \leq e < 1$.
2. A hyperbola has eccentricity $e > 1$.

Proof The foci are located at $(\pm c, 0)$ and the vertices are on the focal axis at $(\pm a, 0)$. Therefore,

$$e = \frac{\text{distance between foci}}{\text{distance between vertices on focal axis}} = \frac{2c}{2a} = \frac{c}{a}$$

For an ellipse, $c = \sqrt{a^2 - b^2}$ and so $e = c/a < 1$. For a hyperbola, $c = \sqrt{a^2 + b^2}$ and thus $e = c/a > 1$. ∎

How does eccentricity determine the shape of a conic [Figure 10(A)]? Consider the ratio b/a of the semiminor axis to the semimajor axis of an ellipse. The ellipse is nearly circular if b/a is close to 1, whereas it is elongated and flat if b/a is small. Now

$$\frac{b}{a} = \frac{\sqrt{a^2 - c^2}}{a} = \sqrt{1 - \frac{c^2}{a^2}} = \sqrt{1 - e^2}$$

This shows that b/a gets smaller (and the ellipse gets flatter) as $e \to 1$ [Figure 10(B)]. The "roundest" ellipse is the circle, with $e = 0$.

Similarly, for a hyperbola,

$$\frac{b}{a} = \sqrt{1 + e^2}$$

The ratios $\pm b/a$ are the slopes of the asymptotes, so the asymptotes get steeper as $e \to \infty$ [Figure 10(C)].

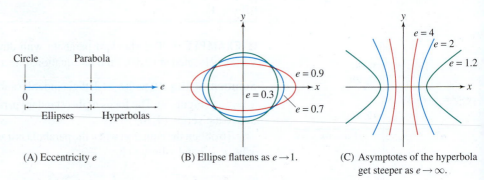

(A) Eccentricity e

(B) Ellipse flattens as $e \to 1$.

(C) Asymptotes of the hyperbola get steeper as $e \to \infty$.

DF FIGURE 10

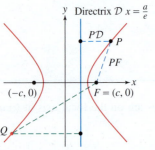

DF FIGURE 11 The ellipse consists of points P such that $PF = ePD$.

FIGURE 12 The hyperbola consists of points P such that $PF = ePD$.

CONCEPTUAL INSIGHT There is a more precise way to explain how eccentricity determines the shape of a conic. We can prove that if two conics $\mathcal{C}_1$ and $\mathcal{C}_2$ have the same eccentricity e, then there is a change of scale that makes $\mathcal{C}_1$ *congruent* to $\mathcal{C}_2$. Changing the scale means changing the units along the x- and y-axes by a common positive factor. A curve scaled by a factor of 10 has the same shape but is ten times as large. This corresponds, for example, to changing units from centimeters to millimeters (smaller units make for a larger figure). By "congruent" we mean that after scaling, it is possible to move $\mathcal{C}_1$ by a rigid motion (involving rotation and translation, but no stretching or bending) so that it lies directly on top of $\mathcal{C}_2$.

All circles ($e = 0$) have the same shape because scaling by a factor $r > 0$ transforms a circle of radius R into a circle of radius rR. Similarly, any two parabolas ($e = 1$) become congruent after suitable scaling. However, an ellipse of eccentricity $e = 0.5$ cannot be made congruent to an ellipse of eccentricity $e = 0.8$ by scaling (see Exercise 74).

Eccentricity can be used to give a unified focus-directrix definition of the conic sections. Given a point F (the focus), a line $\mathcal{D}$ (the directrix), and a number $e > 0$, we consider the set of all points P such that

$$\boxed{PF = eP\mathcal{D}} \qquad \boxed{10}$$

For $e = 1$, this is our definition of a parabola. According to the next theorem, Eq. (10) defines a conic section of eccentricity e for all $e > 0$ (Figures 11 and 12). Note, however, that there is no focus-directrix definition for circles ($e = 0$).

> **THEOREM 5 Focus-Directrix Definition** For all $e > 0$, the set of points satisfying Eq. (10) is a conic section of eccentricity e. Furthermore,
>
> - **Ellipse:** Let $a > b > 0$ and $c = \sqrt{a^2 - b^2}$. The ellipse
>
> $$\left(\frac{x}{a}\right)^2 + \left(\frac{y}{b}\right)^2 = 1$$
>
> satisfies Eq. (10) with $F = (c, 0)$, $e = \frac{c}{a}$, and vertical directrix $x = \frac{a}{e}$.
>
> - **Hyperbola:** Let $a, b > 0$ and $c = \sqrt{a^2 + b^2}$. The hyperbola
>
> $$\left(\frac{x}{a}\right)^2 - \left(\frac{y}{b}\right)^2 = 1$$
>
> satisfies Eq. (10) with $F = (c, 0)$, $e = \frac{c}{a}$, and vertical directrix $x = \frac{a}{e}$.

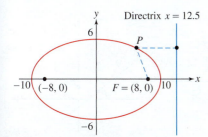

Directrix $\mathcal{D}$ $x = \frac{a}{e}$

$F = (c, 0)$

d

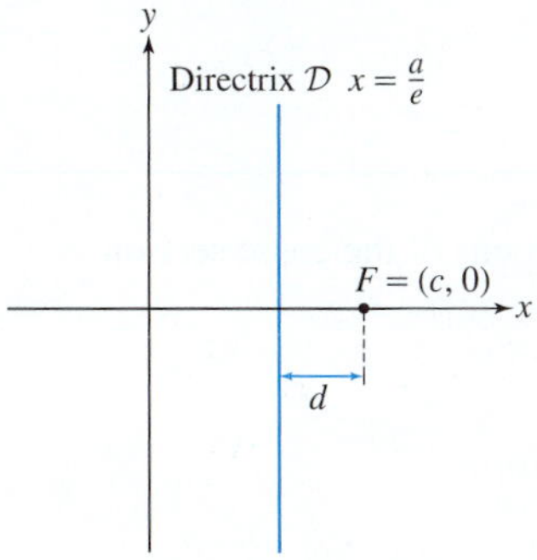

FIGURE 13

Proof Assume that $e > 1$ (the case $e < 1$ is similar, see Exercise 66). We may choose our coordinate axes so that the focus F lies on the x-axis and the directrix is vertical, lying to the left of F, as in Figure 13. Anticipating the final result, we let d be the distance from the focus F to the directrix $\mathcal{D}$ and set

$$c = \frac{d}{1 - e^{-2}}, \qquad a = \frac{c}{e}, \qquad b = \sqrt{c^2 - a^2}$$

Since we are free to shift the y-axis, let us choose the y-axis so that the focus has coordinates $F = (c, 0)$. Then the directrix is the line

$$x = c - d = c - c(1 - e^{-2})$$

$$= c e^{-2} = \frac{a}{e}$$

Now, the equation $PF = ePD$ for a point $P = (x, y)$ may be written

$$\underbrace{\sqrt{(x - c)^2 + y^2}}_{PF} = e \underbrace{\sqrt{\left(x - (a/e)\right)^2}}_{P\mathcal{D}}$$

Algebraic manipulation yields

$$(x - c)^2 + y^2 = e^2 \left(x - (a/e)\right)^2 \qquad \text{(square)}$$

$$x^2 - 2cx + c^2 + y^2 = e^2 x^2 - 2aex + a^2$$

$$x^2 - 2\cancel{aex} + a^2 e^2 + y^2 = e^2 x^2 - 2\cancel{aex} + a^2 \qquad \text{(use } c = ae\text{)}$$

$$(e^2 - 1)x^2 - y^2 = a^2(e^2 - 1) \qquad \text{(rearrange)}$$

$$\frac{x^2}{a^2} - \frac{y^2}{a^2(e^2 - 1)} = 1 \qquad \text{(divide)}$$

This is the desired equation because $a^2(e^2 - 1) = c^2 - a^2 = b^2$. ■

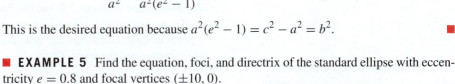

Directrix $x = 12.5$

P

$F = (8, 0)$

$(-8, 0)$

FIGURE 14 Ellipse of eccentricity $e = 0.8$ with focus at $(8, 0)$.

■ **EXAMPLE 5** Find the equation, foci, and directrix of the standard ellipse with eccentricity $e = 0.8$ and focal vertices $(\pm 10, 0)$.

Solution The vertices are $(\pm a, 0)$ with $a = 10$ (Figure 14). By Theorem 5,

$$c = ae = 10(0.8) = 8, \qquad b = \sqrt{a^2 - c^2} = \sqrt{10^2 - 8^2} = 6$$

Thus, our ellipse has equation

$$\left(\frac{x}{10}\right)^2 + \left(\frac{y}{6}\right)^2 = 1$$

The foci are $(\pm c, 0) = (\pm 8, 0)$ and the directrix is $x = \frac{a}{e} = \frac{10}{0.8} = 12.5$. ■

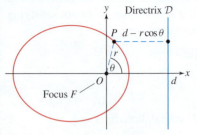

FIGURE 15 Focus-directrix definition of the ellipse in polar coordinates.

In Section 13.6, we discuss the famous law of Johannes Kepler stating that the orbit of a planet around the sun is an ellipse with one focus at the sun. In this discussion, we will need to write the equation of an ellipse in polar coordinates. To derive the polar equations of the conic sections, it is convenient to use the focus-directrix definition with focus F at the origin O and vertical line $x = d$ as directrix $\mathcal{D}$ (Figure 15). Note from the figure that if $P = (r, \theta)$, then

$$PF = r, \qquad P\mathcal{D} = d - r\cos\theta$$

Thus, the focus-directrix equation of the ellipse $PF = eP\mathcal{D}$ becomes $r = e(d - r\cos\theta)$, or $r(1 + e\cos\theta) = ed$. This proves the following result, which is also valid for the hyperbola and parabola (see Exercise 67).

THEOREM 6 Polar Equation of a Conic Section The conic section of eccentricity $e > 0$ with focus at the origin and directrix $x = d$ has polar equation

$$r = \frac{ed}{1 + e\cos\theta} \qquad \boxed{11}$$

■ **EXAMPLE 6** Find the eccentricity, directrix, and focus of the conic section

$$r = \frac{24}{4 + 3\cos\theta}$$

Solution First, we write the equation in the standard form

$$r = \frac{24}{4 + 3\cos\theta} = \frac{6}{1 + \frac{3}{4}\cos\theta}$$

Comparing with Eq. (11), we see that $e = \frac{3}{4}$ and $ed = 6$. Therefore, $d = 8$. Since $e < 1$, the conic is an ellipse. By Theorem 6, the directrix is the line $x = 8$ and the focus is the origin. ■

Reflective Properties of Conic Sections

The conic sections have numerous geometric properties. Especially important are the *reflective properties*, which are used in optics and communications (e.g., in antenna and telescope design; Figure 16). We describe these properties here briefly without proof (but see Exercises 68–70 and Exercise 71 for proofs of the reflective property of ellipses).

- **Ellipse:** The segments F_1P and F_2P make equal angles with the tangent line at a point P on the ellipse. Therefore, a beam of light originating at focus F_1 is reflected off the ellipse toward the second focus F_2 [Figure 17(A)]. See also Figure 18.
- **Hyperbola:** The tangent line at a point P on the hyperbola bisects the angle formed by the segments F_1P and F_2P. Therefore, a beam of light directed toward F_2 is reflected off the hyperbola toward the second focus F_1 [Figure 17(B)].
- **Parabola:** The segment FP and the line through P parallel to the axis make equal angles with the tangent line at a point P on the parabola [Figure 17(C)]. Therefore, a beam of light approaching P from above in the axial direction is reflected off the parabola toward the focus F.

FIGURE 16 The paraboloid shape of this radio telescope directs the incoming signal to the focus. *(Stockbyte)*

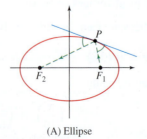

(A) Ellipse

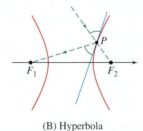

(B) Hyperbola

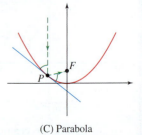

(C) Parabola

DF **FIGURE 17**

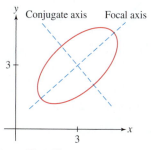

FIGURE 19 The ellipse with equation $6x^2 - 8xy + 8y^2 - 12x - 24y + 38 = 0$.

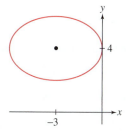

FIGURE 20 The ellipse with equation $4x^2 + 9y^2 + 24x - 72y + 144 = 0$.

If $(\tilde{x}, \tilde{y})$ are coordinates relative to axes rotated by an angle θ as in Figure 21, then

$$x = \tilde{x}\cos\theta - \tilde{y}\sin\theta \qquad \boxed{13}$$

$$y = \tilde{x}\sin\theta + \tilde{y}\cos\theta \qquad \boxed{14}$$

See Exercise 75. In Exercise 76, we show that the cross term disappears when Eq. (12) is rewritten in terms of $\tilde{x}$ and $\tilde{y}$ for the angle

$$\theta = \frac{1}{2}\cot^{-1}\frac{a-c}{b} \qquad \boxed{15}$$

General Equations of Degree 2

The equations of the standard conic sections are special cases of the general equation of degree 2 in x and y:

$$ax^2 + bxy + cy^2 + dx + ey + f = 0 \qquad \boxed{12}$$

Here, a, b, c, d, e, f are constants with a, b, c not all zero. It turns out that this general equation of degree 2 does not give rise to any new types of curves. Apart from certain "degenerate cases," Eq. (12) defines a conic section that is not necessarily in standard position: It need not be centered at the origin, and its focal and conjugate axes may be rotated relative to the coordinate axes. For example, the equation

$$6x^2 - 8xy + 8y^2 - 12x - 24y + 38 = 0$$

defines an ellipse with its center at $(3, 3)$ whose axes are rotated (Figure 19).

We say that Eq. (12) is **degenerate** if the set of solutions is a pair of intersecting lines, a pair of parallel lines, a single line, a point, or the empty set. For example:

- $x^2 - y^2 = 0$ defines a pair of intersecting lines $y = x$ and $y = -x$.
- $x^2 - x = 0$ defines a pair of parallel lines $x = 0$ and $x = 1$.
- $x^2 = 0$ defines a single line (the y-axis).
- $x^2 + y^2 = 0$ has just one solution $(0, 0)$.
- $x^2 + y^2 = -1$ has no solutions.

Now assume that Eq. (12) is nondegenerate. The term bxy is called the *cross term*. When the cross term is zero (i.e., when $b = 0$), we can "complete the square" to show that Eq. (12) defines a translate of the conic in standard position. In other words, the axes of the conic are parallel to the coordinate axes. This is illustrated in the next example.

■ **EXAMPLE 7** **Completing the Square** Show that

$$4x^2 + 9y^2 + 24x - 72y + 144 = 0$$

defines a translate of a conic section in standard position (Figure 20).

Solution Since there is no cross term, we may complete the square of the terms involving x and y separately:

$$4x^2 + 9y^2 + 24x - 72y + 144 = 0$$

$$4(x^2 + 6x + 9 - 9) + 9(y^2 - 8y + 16 - 16) + 144 = 0$$

$$4(x + 3)^2 - 4(9) + 9(y - 4)^2 - 9(16) + 144 = 0$$

$$4(x + 3)^2 + 9(y - 4)^2 = 36$$

Therefore, this quadratic equation can be rewritten in the form

$$\left(\frac{x+3}{3}\right)^2 + \left(\frac{y-4}{2}\right)^2 = 1 \qquad ■$$

When the cross term bxy is nonzero, Eq. (12) defines a conic whose axes are rotated relative to the coordinate axes. The marginal note describes how this may be verified in general. We illustrate with the following example.

■ **EXAMPLE 8** Show that $2xy = 1$ defines a conic section whose focal and conjugate axes are rotated relative to the coordinate axes.

Solution Figure 22(A) shows axes labeled $\tilde{x}$ and $\tilde{y}$ that are rotated by $45°$ relative to the standard coordinate axes. A point P with coordinates (x, y) may also be described by

FIGURE 21

coordinates $(\tilde{x}, \tilde{y})$ relative to these rotated axes. Applying Eqs. (13) and (14) with $\theta = \frac{\pi}{4}$, we find that (x, y) and $(\tilde{x}, \tilde{y})$ are related by the formulas

$$x = \frac{\tilde{x} - \tilde{y}}{\sqrt{2}}, \qquad y = \frac{\tilde{x} + \tilde{y}}{\sqrt{2}}$$

Therefore, if $P = (x, y)$ lies on the hyperbola—that is, if $2xy = 1$—then

$$2xy = 2\left(\frac{\tilde{x} - \tilde{y}}{\sqrt{2}}\right)\left(\frac{\tilde{x} + \tilde{y}}{\sqrt{2}}\right) = \tilde{x}^2 - \tilde{y}^2 = 1$$

Thus, the coordinates $(\tilde{x}, \tilde{y})$ satisfy the equation of the standard hyperbola $\tilde{x}^2 - \tilde{y}^2 = 1$ whose focal and conjugate axes are the $\tilde{x}$- and $\tilde{y}$-axes, respectively. ■

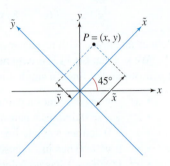

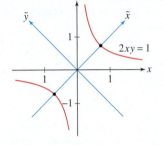

(A) The point $P = (x, y)$ may also be described by coordinates $(\tilde{x}, \tilde{y})$ relative to the rotated axis.

(B) The hyperbola $2xy = 1$ has the standard form $\tilde{x}^2 - \tilde{y}^2 = 1$ relative to the $\tilde{x}, \tilde{y}$ axes.

FIGURE 22 The $\tilde{x}$- and $\tilde{y}$-axes are rotated at a 45° angle relative to the x- and y-axes.

We conclude our discussion of conics by stating the Discriminant Test. Suppose that the equation

$$ax^2 + bxy + cy^2 + dx + ey + f = 0$$

is nondegenerate and thus defines a conic section. According to the Discriminant Test, the type of conic is determined by the **discriminant** D:

$$\boxed{D = b^2 - 4ac}$$

We have the following cases:

- $D < 0$: Ellipse or circle
- $D > 0$: Hyperbola
- $D = 0$: Parabola

For example, the discriminant of the equation $2xy = 1$ is

$$D = b^2 - 4ac = 2^2 - 0 = 4 > 0$$

According to the Discriminant Test, $2xy = 1$ defines a hyperbola. This agrees with our conclusion in Example 8.

11.5 SUMMARY

- An *ellipse* with foci F_1 and F_2 is the set of points P such that $PF_1 + PF_2 = K$, where K is a constant such that $K > F_1F_2$. The equation in standard position is

$$\left(\frac{x}{a}\right)^2 + \left(\frac{y}{b}\right)^2 = 1$$

The vertices of the ellipse are $(\pm a, 0)$ and $(0, \pm b)$.

	Focal axis	Foci	Focal vertices
$a > b$	x-axis	$(\pm c, 0)$ with $c = \sqrt{a^2 - b^2}$	$(\pm a, 0)$
$a < b$	y-axis	$(0, \pm c)$ with $c = \sqrt{b^2 - a^2}$	$(0, \pm b)$

Eccentricity: $e = \frac{c}{a}$ $(0 \le e < 1)$. Directrix: $x = \frac{a}{e}$ (if $a > b$).

• A *hyperbola* with foci F_1 and F_2 is the set of points P such that

$$PF_1 - PF_2 = \pm K$$

where K is a constant such that $0 < K < F_1F_2$. The equation in standard position is

$$\left(\frac{x}{a}\right)^2 - \left(\frac{y}{b}\right)^2 = 1$$

Focal axis	Foci	Vertices	Asymptotes
x-axis	$(\pm c, 0)$ with $c = \sqrt{a^2 + b^2}$	$(\pm a, 0)$	$y = \pm\frac{b}{a}x$

Eccentricity: $e = \frac{c}{a}$ $(e > 1)$. Directrix: $x = \frac{a}{e}$.

• A *parabola* with focus F and directrix $\mathcal{D}$ is the set of points P such that $PF = PD$. The equation in standard position is

$$y = \frac{1}{4c}x^2$$

Focus $F = (0, c)$, directrix $y = -c$, and vertex at the origin $(0, 0)$.
• *Focus-directrix definition* of conic with focus F and directrix $\mathcal{D}$: $PF = ePD$.
• To translate a conic section h units horizontally and k units vertically, replace x by $x - h$ and y by $y - k$ in the equation.
• Polar equation of conic of eccentricity $e > 0$, focus at the origin, directrix $x = d$:

$$r = \frac{ed}{1 + e\cos\theta}$$

11.5 EXERCISES

Preliminary Questions

1. Decide if the equation defines an ellipse, a hyperbola, a parabola, or no conic section at all.

(a) $4x^2 - 9y^2 = 12$ 　　　　**(b)** $-4x + 9y^2 = 0$

(c) $4y^2 + 9x^2 = 12$ 　　　　**(d)** $4x^3 + 9y^3 = 12$

2. For which conic sections do the vertices lie between the foci?

3. What are the foci of

$$\left(\frac{x}{a}\right)^2 + \left(\frac{y}{b}\right)^2 = 1 \quad \text{if } a < b?$$

4. What is the geometric interpretation of b/a in the equation of a hyperbola in standard position?

Exercises

In Exercises 1–6, find the vertices and foci of the conic section.

1. $\left(\frac{x}{9}\right)^2 + \left(\frac{y}{4}\right)^2 = 1$ 　　　　**2.** $\frac{x^2}{9} + \frac{y^2}{4} = 1$

3. $\left(\frac{x}{4}\right)^2 - \left(\frac{y}{9}\right)^2 = 1$ 　　　　**4.** $\frac{x^2}{4} - \frac{y^2}{9} = 36$

5. $\left(\frac{x - 3}{7}\right)^2 - \left(\frac{y + 1}{4}\right)^2 = 1$

6. $\left(\frac{x - 3}{4}\right)^2 + \left(\frac{y + 1}{7}\right)^2 = 1$

In Exercises 7–10, find the equation of the ellipse obtained by translating (as indicated) the ellipse

$$\left(\frac{x - 8}{6}\right)^2 + \left(\frac{y + 4}{3}\right)^2 = 1$$

7. Translated with center at the origin

8. Translated with center at $(-2, -12)$

9. Translated to the right 6 units

10. Translated down 4 units

In Exercises 11–14, find the equation of the given ellipse.

11. Vertices $(\pm 5, 0)$ and $(0, \pm 7)$

12. Foci $(\pm 6, 0)$ and focal vertices $(\pm 10, 0)$

13. Foci $(0, \pm 10)$ and eccentricity $e = \frac{3}{5}$

14. Vertices $(4, 0)$, $(28, 0)$ and eccentricity $e = \frac{2}{3}$

In Exercises 15–20, find the equation of the given hyperbola.

15. Vertices $(\pm 3, 0)$ and foci $(\pm 5, 0)$

16. Vertices $(\pm 3, 0)$ and asymptotes $y = \pm \frac{1}{2} x$

17. Foci $(\pm 4, 0)$ and eccentricity $e = 2$

18. Vertices $(0, \pm 6)$ and eccentricity $e = 3$

19. Vertices $(-3, 0)$, $(7, 0)$ and eccentricity $e = 3$

20. Vertices $(0, -6)$, $(0, 4)$ and foci $(0, -9)$, $(0, 7)$

In Exercises 21–28, find the equation of the parabola with the given properties.

21. Vertex $(0, 0)$, focus $\left(\frac{1}{12}, 0\right)$

22. Vertex $(0, 0)$, focus $(0, 2)$

23. Vertex $(0, 0)$, directrix $y = -5$

24. Vertex $(3, 4)$, directrix $y = -2$

25. Focus $(0, 4)$, directrix $y = -4$

26. Focus $(0, -4)$, directrix $y = 4$

27. Focus $(2, 0)$, directrix $x = -2$

28. Focus $(-2, 0)$, vertex $(2, 0)$

In Exercises 29–38, find the vertices, foci, center (if an ellipse or a hyperbola), and asymptotes (if a hyperbola).

29. $x^2 + 4y^2 = 16$　　　　**30.** $4x^2 + y^2 = 16$

31. $\left(\dfrac{x-3}{4}\right)^2 - \left(\dfrac{y+5}{7}\right)^2 = 1$　　**32.** $3x^2 - 27y^2 = 12$

33. $4x^2 - 3y^2 + 8x + 30y = 215$

34. $y = 4x^2$　　　　　**35.** $y = 4(x - 4)^2$

36. $8y^2 + 6x^2 - 36x - 64y + 134 = 0$

37. $4x^2 + 25y^2 - 8x - 10y = 20$

38. $16x^2 + 25y^2 - 64x - 200y + 64 = 0$

In Exercises 39–42, use the Discriminant Test to determine the type of the conic section (in each case, the equation is nondegenerate). Plot the curve if you have a computer algebra system.

39. $4x^2 + 5xy + 7y^2 = 24$

40. $x^2 - 2xy + y^2 + 24x - 8 = 0$

41. $2x^2 - 8xy + 3y^2 - 4 = 0$

42. $2x^2 - 3xy + 5y^2 - 4 = 0$

43. Show that the "conic" $x^2 + 3y^2 - 6x + 12y + 23 = 0$ has no points.

44. For which values of a does the conic $3x^2 + 2y^2 - 16y + 12x = a$ have at least one point?

45. Show that $\dfrac{b}{a} = \sqrt{1 - e^2}$ for a standard ellipse of eccentricity e.

46. Show that the eccentricity of a hyperbola in standard position is $e = \sqrt{1 + m^2}$, where $\pm m$ are the slopes of the asymptotes.

47. Explain why the dots in Figure 23 lie on a parabola. Where are the focus and directrix located?

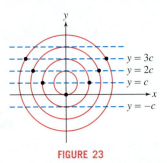

FIGURE 23

48. Find the equation of the ellipse consisting of points P such that $PF_1 + PF_2 = 12$, where $F_1 = (4, 0)$ and $F_2 = (-2, 0)$.

49. A **latus rectum** of a conic section is a chord through a focus parallel to the directrix. Find the area bounded by the parabola $y = x^2/(4c)$ and its latus rectum (refer to Figure 8).

50. Show that the tangent line at a point $P = (x_0, y_0)$ on the hyperbola $\left(\dfrac{x}{a}\right)^2 - \left(\dfrac{y}{b}\right)^2 = 1$ has equation

$$Ax - By = 1$$

where $A = \dfrac{x_0}{a^2}$ and $B = \dfrac{y_0}{b^2}$.

In Exercises 51–54, find the polar equation of the conic with the given eccentricity and directrix, and focus at the origin.

51. $e = \frac{1}{2}$,　$x = 3$　　　**52.** $e = \frac{1}{2}$,　$x = -3$

53. $e = 1$,　$x = 4$　　　　**54.** $e = \frac{3}{2}$,　$x = -4$

In Exercises 55–58, identify the type of conic, the eccentricity, and the equation of the directrix.

55. $r = \dfrac{8}{1 + 4\cos\theta}$　　　**56.** $r = \dfrac{8}{4 + \cos\theta}$

57. $r = \dfrac{8}{4 + 3\cos\theta}$　　　**58.** $r = \dfrac{12}{4 + 3\cos\theta}$

59. Find a polar equation for the hyperbola with focus at the origin, directrix $x = -2$, and eccentricity $e = 1.2$.

60. Let $\mathcal{C}$ be the ellipse $r = de/(1 + e\cos\theta)$, where $e < 1$. Show that the x-coordinates of the points in Figure 24 are as follows:

Point	A	C	F_2	A'
x-coordinate	$\dfrac{de}{e+1}$	$-\dfrac{de^2}{1-e^2}$	$-\dfrac{2de^2}{1-e^2}$	$\dfrac{de}{1-e}$

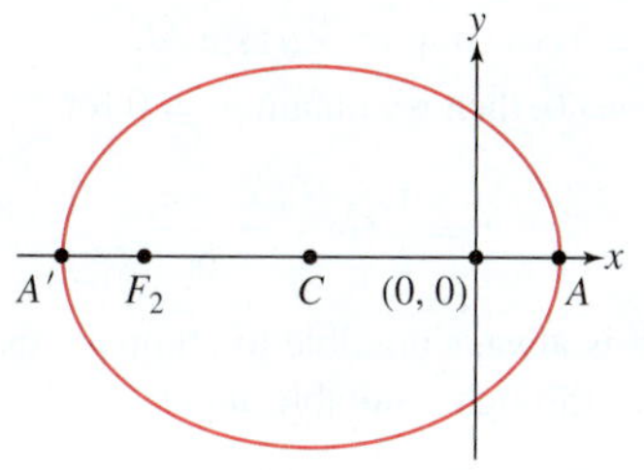

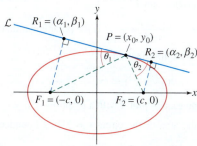

FIGURE 24

61. Find an equation in rectangular coordinates of the conic

$$r = \frac{16}{5 + 3\cos\theta}$$

Hint: Use the results of Exercise 60.

62. Let $e > 1$. Show that the vertices of the hyperbola $r = \dfrac{de}{1 + e\cos\theta}$ have x-coordinates $\dfrac{ed}{e+1}$ and $\dfrac{ed}{e-1}$.

63. Kepler's First Law states that planetary orbits are ellipses with the sun at one focus. The orbit of Pluto has eccentricity $e \approx 0.25$. Its **perihelion** (closest distance to the sun) is approximately 2.7 billion miles. Find the **aphelion** (farthest distance from the sun).

64. Kepler's Third Law states that the ratio $T/a^{3/2}$ is equal to a constant C for all planetary orbits around the sun, where T is the period (time for a complete orbit) and a is the semimajor axis.

(a) Compute C in units of days and kilometers, given that the semimajor axis of the earth's orbit is 150×10^6 km.

(b) Compute the period of Saturn's orbit, given that its semimajor axis is approximately 1.43×10^9 km.

(c) Saturn's orbit has eccentricity $e = 0.056$. Find the perihelion and aphelion of Saturn (see Exercise 63).

Further Insights and Challenges

65. Verify Theorem 2.

66. Verify Theorem 5 in the case $0 < e < 1$. *Hint:* Repeat the proof of Theorem 5, but set $c = d/(e^{-2} - 1)$.

67. Verify that if $e > 1$, then Eq. (11) defines a hyperbola of eccentricity e, with its focus at the origin and directrix at $x = d$.

Reflective Property of the Ellipse *In Exercises 68–70, we prove that the focal radii at a point on an ellipse make equal angles with the tangent line* $\mathcal{L}$. *Let* $P = (x_0, y_0)$ *be a point on the ellipse in Figure 25 with foci* $F_1 = (-c, 0)$ *and* $F_2 = (c, 0)$, *and eccentricity* $e = c/a$.

68. Show that the equation of the tangent line at P is $Ax + By = 1$, where $A = \dfrac{x_0}{a^2}$ and $B = \dfrac{y_0}{b^2}$.

69. Points R_1 and R_2 in Figure 25 are defined so that $\overline{F_1 R_1}$ and $\overline{F_2 R_2}$ are perpendicular to the tangent line.

70. **(a)** Prove that $PF_1 = a + x_0 e$ and $PF_2 = a - x_0 e$. *Hint:* Show that $PF_1{}^2 - PF_2{}^2 = 4x_0 c$. Then use the defining property $PF_1 + PF_2 = 2a$ and the relation $e = c/a$.

(b) Verify that $\dfrac{F_1 R_1}{PF_1} = \dfrac{F_2 R_2}{PF_2}$.

(c) Show that $\sin\theta_1 = \sin\theta_2$. Conclude that $\theta_1 = \theta_2$.

71. Here is another proof of the Reflective Property.

(a) Figure 25 suggests that $\mathcal{L}$ is the unique line that intersects the ellipse only in the point P. Assuming this, prove that $QF_1 + QF_2 > PF_1 + PF_2$ for all points Q on the tangent line other than P.

(b) Use the Principle of Least Distance (Example 6 in Section 4.7) to prove that $\theta_1 = \theta_2$.

72. Show that the length QR in Figure 26 is independent of the point P.

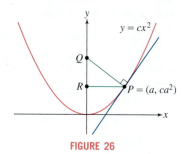

FIGURE 26

73. Show that $y = x^2/4c$ is the equation of a parabola with directrix $y = -c$, focus $(0, c)$, and the vertex at the origin, as stated in Theorem 3.

74. Consider two ellipses in standard position:

$$E_1 : \quad \left(\frac{x}{a_1}\right)^2 + \left(\frac{y}{b_1}\right)^2 = 1$$

$$E_2 : \quad \left(\frac{x}{a_2}\right)^2 + \left(\frac{y}{b_2}\right)^2 = 1$$

We say that E_1 is similar to E_2 under scaling if there exists a factor $r > 0$ such that for all (x, y) on E_1, the point (rx, ry) lies on E_2. Show that E_1 and E_2 are similar under scaling if and only if they have the same eccentricity. Show that any two circles are similar under scaling.

FIGURE 25 The ellipse $\left(\dfrac{x}{a}\right)^2 + \left(\dfrac{y}{b}\right)^2 = 1$.

(a) Show, with A and B as in Exercise 68, that

$$\frac{\alpha_1 + c}{\beta_1} = \frac{\alpha_2 - c}{\beta_2} = \frac{A}{B}$$

(b) Use (a) and the distance formula to show that

$$\frac{F_1 R_1}{F_2 R_2} = \frac{\beta_1}{\beta_2}$$

(c) Use (a) and the equation of the tangent line in Exercise 68 to show that

$$\beta_1 = \frac{B(1 + Ac)}{A^2 + B^2}, \qquad \beta_2 = \frac{B(1 - Ac)}{A^2 + B^2}$$

75. Derive Equations (13) and (14) in the text as follows. Write the coordinates of P with respect to the rotated axes in Figure 21 in polar form $\tilde{x} = r \cos \alpha$, $\tilde{y} = r \sin \alpha$. Explain why P has polar coordinates $(r, \alpha + \theta)$ with respect to the standard x- and y-axes, and derive (13) and (14) using the addition formulas for cosine and sine.

76. If we rewrite the general equation of degree 2 (Eq. 12) in terms of variables $\tilde{x}$ and $\tilde{y}$ that are related to x and y by Eqs. (13) and (14), we obtain a new equation of degree 2 in $\tilde{x}$ and $\tilde{y}$ of the same form but with different coefficients:

$$a'\tilde{x}^2 + b'\tilde{x}\tilde{y} + c'\tilde{y}^2 + d'\tilde{x} + e'\tilde{y} + f' = 0$$

(a) Show that $b' = b \cos 2\theta + (c - a) \sin 2\theta$.

(b) Show that if $b \neq 0$, then we obtain $b' = 0$ for

$$\theta = \frac{1}{2} \cot^{-1} \frac{a - c}{b}$$

This proves that it is always possible to eliminate the cross term bxy by rotating the axes through a suitable angle.

CHAPTER REVIEW EXERCISES

1. Which of the following curves pass through the point $(1, 4)$?

(a) $c(t) = (t^2, t + 3)$ **(b)** $c(t) = (t^2, t - 3)$

(c) $c(t) = (t^2, 3 - t)$ **(d)** $c(t) = (t - 3, t^2)$

2. Find parametric equations for the line through $P = (2, 5)$ perpendicular to the line $y = 4x - 3$.

3. Find parametric equations for the circle of radius 2 with center $(1, 1)$. Use the equations to find the points of intersection of the circle with the x- and y-axes.

4. Find a parametrization $c(t)$ of the line $y = 5 - 2x$ such that $c(0) = (2, 1)$.

5. Find a parametrization $c(\theta)$ of the unit circle such that $c(0) = (-1, 0)$.

6. Find a path $c(t)$ that traces the parabolic arc $y = x^2$ from $(0, 0)$ to $(3, 9)$ for $0 \leq t \leq 1$.

7. Find a path $c(t)$ that traces the line $y = 2x + 1$ from $(1, 3)$ to $(3, 7)$ for $0 \leq t \leq 1$.

8. Sketch the graph $c(t) = (1 + \cos t, \sin 2t)$ for $0 \leq t \leq 2\pi$ and draw arrows specifying the direction of motion.

In Exercises 9–12, express the parametric curve in the form $y = f(x)$.

9. $c(t) = (4t - 3, 10 - t)$ **10.** $c(t) = (t^3 + 1, t^2 - 4)$

11. $c(t) = \left(3 - \frac{2}{t}, t^3 + \frac{1}{t}\right)$ **12.** $x = \tan t$, $y = \sec t$

In Exercises 13–16, calculate dy/dx at the point indicated.

13. $c(t) = (t^3 + t, t^2 - 1)$, $t = 3$

14. $c(\theta) = (\tan^2 \theta, \cos \theta)$, $\theta = \frac{\pi}{4}$

15. $c(t) = (e^t - 1, \sin t)$, $t = 20$

16. $c(t) = (\ln t, 3t^2 - t)$, $P = (0, 2)$

17. *CAS* Find the point on the cycloid $c(t) = (t - \sin t, 1 - \cos t)$ where the tangent line has slope $\frac{1}{2}$.

18. Find the points on $(t + \sin t, t - 2 \sin t)$ where the tangent is vertical or horizontal.

19. Find the equation of the Bézier curve with control points

$$P_0 = (-1, -1), \quad P_1 = (-1, 1), \quad P_2 = (1, 1), \quad P_3(1, -1)$$

20. Find the speed at $t = \frac{\pi}{4}$ of a particle whose position at time t seconds is $c(t) = (\sin 4t, \cos 3t)$.

21. Find the speed (as a function of t) of a particle whose position at time t seconds is $c(t) = (\sin t + t, \cos t + t)$. What is the particle's maximal speed?

22. Find the length of $(3e^t - 3, 4e^t + 7)$ for $0 \leq t \leq 1$.

In Exercises 23 and 24, let $c(t) = (e^{-t} \cos t, e^{-t} \sin t)$.

23. Show that $c(t)$ for $0 \leq t < \infty$ has finite length and calculate its value.

24. Find the first positive value of t_0 such that the tangent line to $c(t_0)$ is vertical, and calculate the speed at $t = t_0$.

25. *CAS* Plot $c(t) = (\sin 2t, 2 \cos t)$ for $0 \leq t \leq \pi$. Express the length of the curve as a definite integral, and approximate it using a computer algebra system.

26. Convert the points $(x, y) = (1, -3)$, $(3, -1)$ from rectangular to polar coordinates.

27. Convert the points $(r, \theta) = \left(1, \frac{\pi}{6}\right)$, $\left(3, \frac{5\pi}{4}\right)$ from polar to rectangular coordinates.

28. Write $(x + y)^2 = xy + 6$ as an equation in polar coordinates.

29. Write $r = \dfrac{2 \cos \theta}{\cos \theta - \sin \theta}$ as an equation in rectangular coordinates.

30. Show that $r = \dfrac{4}{7 \cos \theta - \sin \theta}$ is the polar equation of a line.

31. GU Convert the equation

$$9(x^2 + y^2) = (x^2 + y^2 - 2y)^2$$

to polar coordinates, and plot it with a graphing utility.

32. Calculate the area of the circle $r = 3 \sin \theta$ bounded by the rays $\theta = \frac{\pi}{3}$ and $\theta = \frac{2\pi}{3}$.

33. Calculate the area of one petal of $r = \sin 4\theta$ (see Figure 1).

34. The equation $r = \sin(n\theta)$, where $n \geq 2$ is even, is a "rose" of $2n$ petals (Figure 1). Compute the total area of the flower, and show that it does not depend on n.

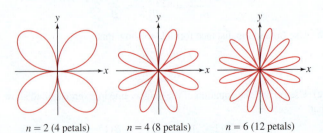

$n = 2$ (4 petals)　　$n = 4$ (8 petals)　　$n = 6$ (12 petals)

FIGURE 1 Plot of $r = \sin(n\theta)$.

35. Calculate the total area enclosed by the curve $r^2 = \cos\theta e^{\sin\theta}$ (Figure 2).

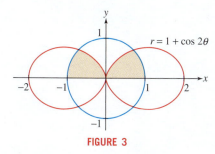

FIGURE 2 Graph of $r^2 = \cos\theta e^{\sin\theta}$.

36. Find the shaded area in Figure 3.

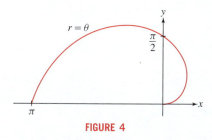

$r = 1 + \cos 2\theta$

FIGURE 3

37. Find the area enclosed by the cardioid $r = a(1 + \cos\theta)$, where $a > 0$.

38. Calculate the length of the curve with polar equation $r = \theta$ in Figure 4.

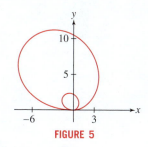

$r = \theta$

FIGURE 4

39. ⌐*CAS* Figure 5 shows the graph of $r = e^{0.5\theta}\sin\theta$ for $0 \le \theta \le 2\pi$. Use a computer algebra system to approximate the difference in length between the outer and inner loops.

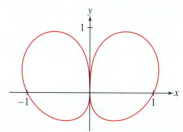

FIGURE 5

40. ✎ Show that $r = f_1(\theta)$ and $r = f_2(\theta)$ define the same curves in polar coordinates if $f_1(\theta) = -f_2(\theta + \pi)$. Use this to show that the following define the same conic section:

$$r = \frac{de}{1 - e\cos\theta}, \qquad r = \frac{-de}{1 + e\cos\theta}$$

In Exercises 41–44, identify the conic section. Find the vertices and foci.

41. $\left(\dfrac{x}{3}\right)^2 + \left(\dfrac{y}{2}\right)^2 = 1$

42. $x^2 - 2y^2 = 4$

43. $\left(2x + \dfrac{1}{2}y\right)^2 = 4 - (x - y)^2$

44. $(y - 3)^2 = 2x^2 - 1$

In Exercises 45–50, find the equation of the conic section indicated.

45. Ellipse with vertices $(\pm 8, 0)$, foci $(\pm\sqrt{3}, 0)$

46. Ellipse with foci $(\pm 8, 0)$, eccentricity $\frac{1}{8}$

47. Hyperbola with vertices $(\pm 8, 0)$, asymptotes $y = \pm\frac{3}{4}x$

48. Hyperbola with foci $(2, 0)$ and $(10, 0)$, eccentricity $e = 4$

49. Parabola with focus $(8, 0)$, directrix $x = -8$

50. Parabola with vertex $(4, -1)$, directrix $x = 15$

51. Find the asymptotes of the hyperbola $3x^2 + 6x - y^2 - 10y = 1$.

52. Show that the "conic section" with equation $x^2 - 4x + y^2 + 5 = 0$ has no points.

53. Show that the relation $\dfrac{dy}{dx} = (e^2 - 1)\dfrac{x}{y}$ holds on a standard ellipse or hyperbola of eccentricity e.

54. The orbit of Jupiter is an ellipse with the sun at a focus. Find the eccentricity of the orbit if the perihelion (closest distance to the sun) equals 740×10^6 km and the aphelion (farthest distance from the sun) equals 816×10^6 km.

55. Refer to Figure 25 in Section 11.5. Prove that the product of the perpendicular distances $F_1 R_1$ and $F_2 R_2$ from the foci to a tangent line of an ellipse is equal to the square b^2 of the semiminor axes.

Ferrofluids contain suspended magnetic nanoparticles. When placed in a magnetic field, they form peaks and valleys to minimize the total energy of the system. Vectors and their cross products play a key role in understanding the behavior of ferrofluids as the magnetic field varies. *(Oliver Hoffmann/Shutterstock)*

12 VECTOR GEOMETRY

Vectors play a role in nearly all areas of mathematics and its applications. In physical settings, they are used to represent quantities that have both magnitude and direction, such as velocity and force. Newtonian mechanics, quantum physics, and special and general relativity all depend fundamentally on vectors. We could not understand electricity and magnetism without vectors, which are used in constructing the theoretical framework.

Vectors also play a critical role in computer graphics, describing how light should be depicted, and providing a means to change the viewpoint of the screen appropriately. Fields such as economics and statistics use vectors to encapsulate information in a manner that provides for efficient manipulation. Vectors are an indispensable tool across a wide array of disciplines.

This chapter develops the basic geometric and algebraic properties of vectors. Although no calculus is required, the concepts developed will be used throughout the remainder of the text.

12.1 Vectors in the Plane

A two-dimensional **vector v** is determined by two points in the plane: an initial point P (also called the "tail" or basepoint) and a terminal point Q (also called the "head"). We write

$$\mathbf{v} = \overrightarrow{PQ}$$

and we draw **v** as an arrow pointing from P to Q. This vector is said to be based at P. Figure 1(A) shows the vector with initial point $P = (2, 2)$ and terminal point $Q = (7, 5)$. The **length** or **magnitude** of **v**, denoted $\|\mathbf{v}\|$, is the distance from P to Q.

The vector $\mathbf{v} = \overrightarrow{OR}$ pointing from the origin to a point R is called the **position vector** of R. Figure 1(B) shows the position vector of the point $R = (3, 5)$.

NOTATION *In this text, vectors are represented by boldface lowercase letters such as* **v**, **w**, **a**, **b**, *etc.*

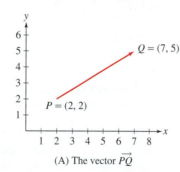
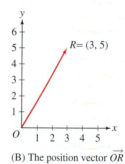

(A) The vector $\overrightarrow{PQ}$ (B) The position vector $\overrightarrow{OR}$

DF FIGURE 1

We now introduce some vector terminology.

- Two vectors **v** and **w** of nonzero length are called **parallel** if the lines through **v** and **w** are parallel. Parallel vectors point either in the same or in opposite directions [Figure 2(A)].
- A vector **v** is said to undergo a **translation** when it is moved to begin at a new point without changing its length or direction. The resulting vector **w** is called a translate of **v** [Figure 2(B)]. Translates have the same length and direction but different basepoints.

(A) Vectors parallel to **v**

(B) **w** is a translate of **v**.

FIGURE 2

In almost all situations, it is convenient to treat vectors with the same length and direction as equivalent, even if they have different basepoints. With this in mind, we say that

- **v** and **w** are **equivalent** if **w** is a translate of **v** [Figure 3(A)].

Every vector can be translated so that its tail is at the origin [Figure 3(C)]. Therefore,

Every vector **v** *is equivalent to a unique vector* **v**$_0$ *based at the origin.*

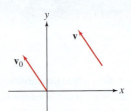

(A) Vectors equivalent   (B) Inequivalent vectors   (C) **v**$_0$ is the unique vector based
to **v** (translates of **v**)   at the origin and equivalent to **v**.

FIGURE 3

To work algebraically, we define the components of a vector (Figure 4).

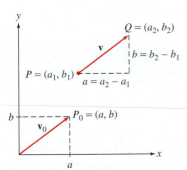

FIGURE 4 Both of the vectors **v** and **v**$_0$ have components $\langle a, b \rangle$.

> **DEFINITION Components of a Vector**   The components of $\mathbf{v} = \overrightarrow{PQ}$, where $P = (a_1, b_1)$ and $Q = (a_2, b_2)$, are the quantities
>
> $$a = a_2 - a_1 \quad (x\text{-component}), \qquad b = b_2 - b_1 \quad (y\text{-component})$$
>
> The pair of components is denoted $\langle a, b \rangle$.

- When the basepoint $P = (0, 0)$, the components of **v** are just the coordinates of its endpoint Q.
- The length of a vector in terms of its components (by the distance formula, see Figure 4) is

$$\|\mathbf{v}\| = \|\overrightarrow{PQ}\| = \sqrt{a^2 + b^2}$$

- The **zero vector** (whose head and tail coincide) is the vector $\mathbf{0} = \langle 0, 0 \rangle$ of length zero. It is the only vector that lacks a direction.

The components $\langle a, b \rangle$ determine the length and direction of **v**, but not its basepoint. Therefore, *two vectors are equivalent if and only if they have the same components.* Nevertheless, the standard practice is to describe a vector by its components, and thus we write

$$\mathbf{v} = \langle a, b \rangle$$

- *In this text, "angle brackets" are used to distinguish between the vector* $\mathbf{v} = \langle a, b \rangle$ *and the point* $P = (a, b)$. *Some textbooks denote both* **v** *and* P *by* (a, b).
- *When referring to vectors, we use the terms "length" and "magnitude" interchangeably. The term "norm" is also commonly used.*

Equivalent vectors have the same length and point in the same direction. But, they can start at any basepoint. When not specified otherwise, we assume that the vector under consideration is based at the origin.

■ **EXAMPLE 1**   Determine whether $\mathbf{v}_1 = \overrightarrow{P_1 Q_1}$ and $\mathbf{v}_2 = \overrightarrow{P_2 Q_2}$ are equivalent, where

$$P_1 = (3, 7), \quad Q_1 = (6, 5) \quad \text{and} \quad P_2 = (-1, 4), \quad Q_2 = (2, 1)$$

What is the magnitude of $\mathbf{v}_1$?

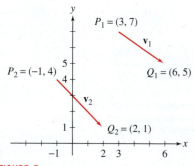

FIGURE 5

Solution We can test for equivalence by computing the components (Figure 5):

$$\mathbf{v}_1 = \langle 6 - 3, 5 - 7 \rangle = \langle 3, -2 \rangle, \qquad \mathbf{v}_2 = \langle 2 - (-1), 1 - 4 \rangle = \langle 3, -3 \rangle$$

The components of $\mathbf{v}_1$ and $\mathbf{v}_2$ are not the same, so $\mathbf{v}_1$ and $\mathbf{v}_2$ are not equivalent. Since $\mathbf{v}_1 = \langle 3, -2 \rangle$, its magnitude is

$$\|\mathbf{v}_1\| = \sqrt{3^2 + (-2)^2} = \sqrt{13}$$ ∎

■ **EXAMPLE 2** Sketch the vector $\mathbf{v} = \langle 2, -3 \rangle$ based at $P = (1, 4)$ and the vector $\mathbf{v}_0$ equivalent to $\mathbf{v}$ based at the origin.

Solution The vector $\mathbf{v} = \langle 2, -3 \rangle$, which is based at $P = (1, 4)$, has the terminal point $Q = (1 + 2, 4 - 3) = (3, 1)$, located 2 units to the right and 3 units down from P as shown in Figure 6. The vector $\mathbf{v}_0$ equivalent to $\mathbf{v}$ based at O has terminal point $(2, -3)$. ∎

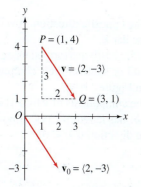

FIGURE 6 The vectors $\mathbf{v}$ and $\mathbf{v}_0$ have the same components but different basepoints.

Vector Algebra

We now define two basic vector operations: vector addition and scalar multiplication.

The vector sum $\mathbf{v} + \mathbf{w}$ is defined when $\mathbf{v}$ and $\mathbf{w}$ have the same basepoint: Translate $\mathbf{w}$ to the equivalent vector $\mathbf{w}'$ whose tail coincides with the head of $\mathbf{v}$. The sum $\mathbf{v} + \mathbf{w}$ is the vector pointing from the tail of $\mathbf{v}$ to the head of $\mathbf{w}'$ [Figure 7(A)]. Alternatively, we can use the **Parallelogram Law**: $\mathbf{v} + \mathbf{w}$ is the vector pointing from the basepoint to the opposite vertex of the parallelogram formed by $\mathbf{v}$ and $\mathbf{w}$ [Figure 7(B)].

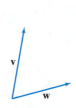 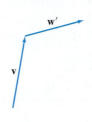 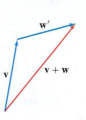 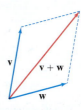

(A) The vector sum $\mathbf{v} + \mathbf{w}$ (B) Addition via the Parallelogram Law

FIGURE 7

To add several vectors $\mathbf{v}_1, \mathbf{v}_2, \ldots, \mathbf{v}_n$, translate the vectors to $\mathbf{v}_1 = \mathbf{v}_1', \mathbf{v}_2', \ldots, \mathbf{v}_n'$ so that they lie head to tail as in Figure 8. The vector sum $\mathbf{v} = \mathbf{v}_1 + \mathbf{v}_2 + \cdots + \mathbf{v}_n$ is the vector whose terminal point is the terminal point of $\mathbf{v}_n'$.

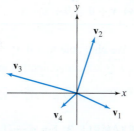 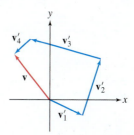

DF **FIGURE 8** The sum $\mathbf{v} = \mathbf{v}_1 + \mathbf{v}_2 + \mathbf{v}_3 + \mathbf{v}_4$.

CAUTION *Remember that the vector* $\mathbf{v} - \mathbf{w}$ *points in the direction from the tip of* $\mathbf{w}$ *to the tip of* $\mathbf{v}$ *(not from the tip of* $\mathbf{v}$ *to the tip of* $\mathbf{w}$*).*

Vector subtraction $\mathbf{v} - \mathbf{w}$ is carried out by adding $-\mathbf{w}$ to $\mathbf{v}$ as in Figure 9(A). Or, more simply, draw the vector pointing from $\mathbf{w}$ to $\mathbf{v}$ as in Figure 9(B), and translate it back to the basepoint to obtain $\mathbf{v} - \mathbf{w}$.

The term **scalar** is another word for "real number," and we often speak of scalar versus vector quantities. Thus, the number 8 is a scalar, while $\langle 8, 2 \rangle$ is a vector. If λ is a scalar and $\mathbf{v}$ is a nonzero vector, the **scalar multiple** $\lambda \mathbf{v}$ is defined as follows (Figure 10):

NOTATION λ *(pronounced "lambda") is the eleventh letter in the Greek alphabet. We use the symbol* λ *often (but not exclusively) to denote a scalar.*

- $\lambda \mathbf{v}$ has length $|\lambda|\,\|\mathbf{v}\|$.
- It points in the same direction as $\mathbf{v}$ if $\lambda > 0$.
- It points in the opposite direction if $\lambda < 0$.

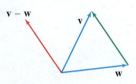

(A) **v** − **w** equals v plus (−**w**).

(B) More simply, **v** − **w** is the translate of the vector pointing from the tip of **w** to the tip of **v**.

FIGURE 9 Vector subtraction.

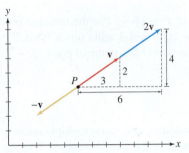

FIGURE 10 Vectors **v** and 2**v** are based at P but 2**v** is twice as long. Vectors **v** and −**v** have the same length but opposite directions.

Note that $0\mathbf{v} = \mathbf{0}$ for all **v**, and

$$\|\lambda\mathbf{v}\| = |\lambda|\,\|\mathbf{v}\|$$

In particular, −**v** has the same length as **v** but points in the opposite direction. A vector **w** is parallel to **v** if and only if $\mathbf{w} = \lambda\mathbf{v}$ for some nonzero scalar λ.

Vector addition and scalar multiplication operations are easily performed using components. To add or subtract two vectors **v** and **w**, we add or subtract their components. This follows from the Parallelogram Law as indicated in Figure 11(A).

Similarly, to multiply **v** by a scalar λ, we multiply the components of **v** by λ [Figures 11(B) and (C)]. Indeed, if $\mathbf{v} = \langle a, b \rangle$ is nonzero, $\langle \lambda a, \lambda b \rangle$ has length $|\lambda|\,\|\mathbf{v}\|$. It points in the same direction as $\langle a, b \rangle$ if $\lambda > 0$, and in the opposite direction if $\lambda < 0$.

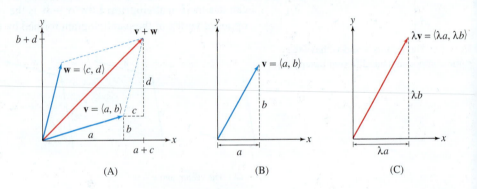

(A)   (B)   (C)

FIGURE 11 Vector operations using components.

Vector Operations Using Components If $\mathbf{v} = \langle a, b \rangle$ and $\mathbf{w} = \langle c, d \rangle$, then:

(i) $\mathbf{v} + \mathbf{w} = \langle a + c, b + d \rangle$

(ii) $\mathbf{v} - \mathbf{w} = \langle a - c, b - d \rangle$

(iii) $\lambda\mathbf{v} = \langle \lambda a, \lambda b \rangle$

(iv) $\mathbf{v} + \mathbf{0} = \mathbf{0} + \mathbf{v} = \mathbf{v}$

We also note that if $P = (a_1, b_1)$ and $Q = (a_2, b_2)$, then components of the vector $\mathbf{v} = \overrightarrow{PQ}$ are conveniently computed as the difference

$$\overrightarrow{PQ} = \overrightarrow{OQ} - \overrightarrow{OP} = \langle a_2, b_2 \rangle - \langle a_1, b_1 \rangle = \langle a_2 - a_1, b_2 - b_1 \rangle$$

■ **EXAMPLE 3** For $\mathbf{v} = \langle 1, 4 \rangle$, $\mathbf{w} = \langle 3, 2 \rangle$, calculate:

(a) $\mathbf{v} + \mathbf{w}$

(b) $5\mathbf{v}$

Solution

$$\mathbf{v} + \mathbf{w} = \langle 1, 4 \rangle + \langle 3, 2 \rangle = \langle 1 + 3, 4 + 2 \rangle = \langle 4, 6 \rangle$$

$$5\mathbf{v} = 5\langle 1, 4 \rangle = \langle 5, 20 \rangle$$

The vector sum is illustrated in Figure 12. ■

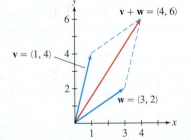

DF FIGURE 12

Vector operations obey the usual laws of algebra.

THEOREM 1 **Basic Properties of Vector Algebra** For all vectors **u**, **v**, **w** and for all scalars λ,

Commutative Law:	$\mathbf{v} + \mathbf{w} = \mathbf{w} + \mathbf{v}$
Associative Law:	$\mathbf{u} + (\mathbf{v} + \mathbf{w}) = (\mathbf{u} + \mathbf{v}) + \mathbf{w}$
Distributive Law for Scalars:	$\lambda(\mathbf{v} + \mathbf{w}) = \lambda\mathbf{v} + \lambda\mathbf{w}$

These properties are verified easily using components. For example, we can check that vector addition is commutative:

$$\langle v_1, v_2 \rangle + \langle w_1, w_2 \rangle = \underbrace{\langle v_1 + w_1, v_2 + w_2 \rangle = \langle w_1 + v_1, w_2 + v_2 \rangle}_{\text{Commutativity of ordinary addition}} = \langle w_1, w_2 \rangle + \langle v_1, v_2 \rangle$$

A **linear combination** of vectors **v** and **w** is a vector

$$r\mathbf{v} + s\mathbf{w}$$

where r and s are scalars. If **v** and **w** are not parallel, then every vector **u** in the plane can be expressed as a linear combination $\mathbf{u} = r\mathbf{v} + s\mathbf{w}$ [Figure 13(A)]. The parallelogram $\mathcal{P}$ whose vertices are the origin and the terminal points of **v**, **w** and $\mathbf{v} + \mathbf{w}$ is called the **parallelogram spanned** by **v** and **w** [Figure 13(B)]. It consists of the linear combinations $r\mathbf{v} + s\mathbf{w}$ with $0 \le r \le 1$ and $0 \le s \le 1$.

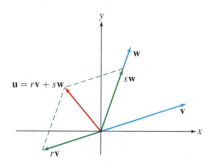

(A) The vector **u** can be expressed as a linear combination $\mathbf{u} = r\mathbf{v} + s\mathbf{w}$. In this figure, $r < 0$.

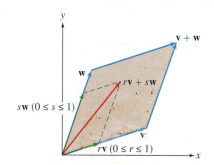

(B) The parallelogram $\mathcal{P}$ spanned by **v** and **w** consists of all linear combinations $r\mathbf{v} + s\mathbf{w}$ with $0 \le r, s \le 1$.

FIGURE 13

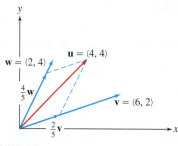

FIGURE 14

■ **EXAMPLE 4** **Linear Combinations** Express the vector $\mathbf{u} = \langle 4, 4 \rangle$ in Figure 14 as a linear combination of $\mathbf{v} = \langle 6, 2 \rangle$ and $\mathbf{w} = \langle 2, 4 \rangle$.

Solution We must find r and s such that $r\mathbf{v} + s\mathbf{w} = \langle 4, 4 \rangle$, or

$$r \langle 6, 2 \rangle + s \langle 2, 4 \rangle = \langle 6r + 2s, 2r + 4s \rangle = \langle 4, 4 \rangle$$

The components must be equal, so we have a system of two linear equations:

$$6r + 2s = 4$$

$$2r + 4s = 4$$

Subtracting the equations, we obtain $4r - 2s = 0$ or $s = 2r$. Setting $s = 2r$ in the first equation yields $6r + 4r = 4$ or $r = \frac{2}{5}$, and then $s = 2r = \frac{4}{5}$. Therefore,

$$\mathbf{u} = \langle 4, 4 \rangle = \frac{2}{5} \langle 6, 2 \rangle + \frac{4}{5} \langle 2, 4 \rangle$$

■

CONCEPTUAL INSIGHT In general, to write a vector $\mathbf{u} = \langle u_1, u_2 \rangle$ as a linear combination of two other vectors $\mathbf{v} = \langle v_1, v_2 \rangle$ and $\mathbf{w} = \langle w_1, w_2 \rangle$, we have to solve a system of two linear equations in two unknowns r and s:

$$r\mathbf{v} + s\mathbf{w} = \mathbf{u} \quad \Leftrightarrow \quad r \langle v_1, v_2 \rangle + s \langle w_1, w_2 \rangle = \langle w_1, w_2 \rangle \quad \Leftrightarrow \quad \begin{cases} rv_1 + sw_1 = u_1 \\ rv_2 + sw_2 = u_2 \end{cases}$$

On the other hand, vectors give us a way of visualizing the system of equations geometrically. The solution is represented by a parallelogram as in Figure 14. This relation between vectors and systems of linear equations extends to any number of variables and is the starting point for the important subject of linear algebra.

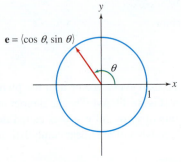

$\mathbf{e} = \langle \cos \theta, \sin \theta \rangle$

FIGURE 15 The head of a unit vector lies on the unit circle.

A vector of length 1 is called a **unit vector**. Unit vectors are often used to indicate direction, when it is not necessary to specify length. The head of a unit vector $\mathbf{e}$ based at the origin lies on the unit circle and can be given as

$$\mathbf{e} = \langle \cos \theta, \sin \theta \rangle$$

where θ is the angle between $\mathbf{e}$ and the positive x-axis (Figure 15). The fact that its length is 1 when represented in this way follows immediately from the trigonometric identity $\sin^2 \theta + \cos^2 \theta = 1$.

We can always scale a nonzero vector $\mathbf{v} = \langle v_1, v_2 \rangle$ to obtain a unit vector pointing in the same direction (Figure 16):

$$\boxed{\mathbf{e_v} = \frac{1}{\|\mathbf{v}\|} \mathbf{v}}$$

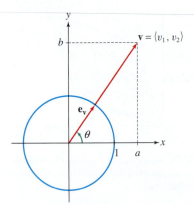

FIGURE 16 Unit vector in the direction of **v**.

Indeed, we can check that $\mathbf{e_v}$ is a unit vector as follows:

$$\|\mathbf{e_v}\| = \left\| \frac{1}{\|\mathbf{v}\|} \mathbf{v} \right\| = \frac{1}{\|\mathbf{v}\|} \|\mathbf{v}\| = 1$$

If $\mathbf{v} = \langle v_1, v_2 \rangle$ makes an angle θ with the positive x-axis, then

$$\boxed{\mathbf{v} = \langle v_1, v_2 \rangle = \|\mathbf{v}\| \mathbf{e_v} = \|\mathbf{v}\| \langle \cos \theta, \sin \theta \rangle} \qquad \boxed{1}$$

■ **EXAMPLE 5** Find the unit vector in the direction of $\mathbf{v} = \langle 3, 5 \rangle$.

Solution $\|\mathbf{v}\| = \sqrt{3^2 + 5^2} = \sqrt{34}$, and thus, $\mathbf{e_v} = \frac{1}{\sqrt{34}} \mathbf{v} = \left\langle \frac{3}{\sqrt{34}}, \frac{5}{\sqrt{34}} \right\rangle$. ■

It is customary to introduce a special notation for the unit vectors in the direction of the positive x- and y-axes (Figure 17):

$$\boxed{\mathbf{i} = \langle 1, 0 \rangle, \qquad \mathbf{j} = \langle 0, 1 \rangle}$$

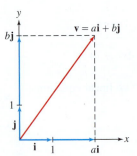

FIGURE 17

The vectors $\mathbf{i}$ and $\mathbf{j}$ are called the **standard basis vectors**. Every vector in the plane is a linear combination of $\mathbf{i}$ and $\mathbf{j}$ (Figure 17):

$$\mathbf{v} = \langle a, b \rangle = a\mathbf{i} + b\mathbf{j}$$

For example, $\langle 4, -2 \rangle = 4\mathbf{i} - 2\mathbf{j}$. Vector addition is performed by adding the $\mathbf{i}$ and $\mathbf{j}$ coefficients. For example,

$$(4\mathbf{i} - 2\mathbf{j}) + (5\mathbf{i} + 7\mathbf{j}) = (4 + 5)\mathbf{i} + (-2 + 7)\mathbf{j} = 9\mathbf{i} + 5\mathbf{j}$$

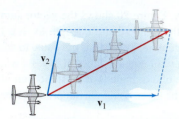

FIGURE 18 When an airplane traveling with velocity $\mathbf{v}_1$ encounters a wind of velocity $\mathbf{v}_2$, its resultant velocity is the vector sum $\mathbf{v}_1 + \mathbf{v}_2$.

CONCEPTUAL INSIGHT It is often said that quantities such as force and velocity are vectors because they have both magnitude and direction, but there is more to this statement than meets the eye. A vector quantity must obey the Law of Vector Addition (Figure 18), so if we say that force is a vector, we are really claiming that forces add according to the Parallelogram Law. In other words, if forces $\mathbf{F}_1$ and $\mathbf{F}_2$ act on an object, then the resultant force is the **vector sum** $\mathbf{F}_1 + \mathbf{F}_2$. This is a physical fact that must be verified experimentally. It was well known to scientists and engineers long before the vector concept was introduced formally in the 1800s.

■ **EXAMPLE 6** Find the forces on cables 1 and 2 in Figure 19(A).

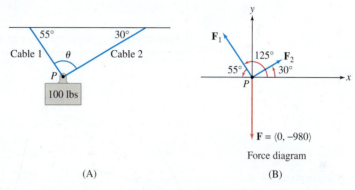

(A)　　　　　　　　　　　(B)

FIGURE 19

Solution Three forces act on the point P in Figure 19(A): the force $\mathbf{F}_g$ due to gravity of $100g = 980$ newtons or N ($g = 9.8$ m/s^2) acting vertically downward, and two unknown forces $\mathbf{F}_1$ and $\mathbf{F}_2$ acting through cables 1 and 2, as indicated in Figure 19(B).

Since the point P is not in motion, the net force on P is zero:

$$\mathbf{F}_1 + \mathbf{F}_2 + \mathbf{F}_g = \mathbf{0}$$

We use this fact to determine $\mathbf{F}_1$ and $\mathbf{F}_2$.

Let $f_1 = \|\mathbf{F}_1\|$ and $f_2 = \|\mathbf{F}_2\|$. Because $\mathbf{F}_1$ makes an angle of $125°$ (the supplement of $55°$) with the positive x-axis, and $\mathbf{F}_2$ makes an angle of $30°$, we can use Eq. (1) and the table in the margin to write these vectors in component form:

$$\mathbf{F}_1 = f_1 \langle \cos 125°, \sin 125° \rangle \approx f_1 \langle -0.573, 0.819 \rangle$$

$$\mathbf{F}_2 = f_2 \langle \cos 30°, \sin 30° \rangle \approx f_2 \langle 0.866, 0.5 \rangle$$

$$\mathbf{F}_g = \langle 0, -980 \rangle$$

θ	$\cos\theta$	$\sin\theta$
$125°$	-0.573	0.819
$30°$	0.866	0.5

$$f_1 \langle -0.573, 0.819 \rangle + f_2 \langle 0.866, 0.5 \rangle + \langle 0, -980 \rangle = \langle 0, 0 \rangle$$

This gives us two equations in two unknowns:

$$-0.573 f_1 + 0.866 f_2 = 0, \qquad 0.819 f_1 + 0.5 f_2 - 980 = 0$$

By the first equation, $f_2 = \left(\frac{0.573}{0.866}\right) f_1$. Substitution in the second equation yields

$$0.819 f_1 + 0.5 \left(\frac{0.573}{0.866}\right) f_1 - 980 \approx 1.15 f_1 - 980 = 0$$

Therefore, the forces in newtons are

$$f_1 \approx \frac{980}{1.15} \approx 852 \text{ newtons} \qquad \text{and} \qquad f_2 \approx \left(\frac{0.573}{0.866}\right) 852 \approx 564 \text{ newtons}$$

Hence,

$$\mathbf{F}_1 \approx 852 \langle -0.573, 0.819 \rangle \approx \langle -488.196, 697.788 \rangle$$

and

$$\mathbf{F}_2 \approx 564 \langle 0.866, 0.5 \rangle \approx \langle 488.424, 282 \rangle$$

■

We close this section with the Triangle Inequality. Figure 20 shows the vector sum $\mathbf{v} + \mathbf{w}$ for three different vectors $\mathbf{w}$ of the same length. Notice that the length $\|\mathbf{v} + \mathbf{w}\|$ varies depending on the angle between $\mathbf{v}$ and $\mathbf{w}$. So in general, $\|\mathbf{v} + \mathbf{w}\|$ is not equal to the sum $\|\mathbf{v}\| + \|\mathbf{w}\|$. What we can say is that $\|\mathbf{v} + \mathbf{w}\|$ is *at most* equal to the sum $\|\mathbf{v}\| + \|\mathbf{w}\|$. This corresponds to the fact that the length of one side of a triangle is at most the sum of the lengths of the other two sides. A formal proof may be given using the dot product (see Exercise 92 in Section 12.3).

THEOREM 2 Triangle Inequality   For any two vectors $\mathbf{v}$ and $\mathbf{w}$,

$$\|\mathbf{v} + \mathbf{w}\| \leq \|\mathbf{v}\| + \|\mathbf{w}\|$$

Equality holds only if $\mathbf{v} = \mathbf{0}$ or $\mathbf{w} = \mathbf{0}$, or if $\mathbf{w} = \lambda\mathbf{v}$, where $\lambda \geq 0$.

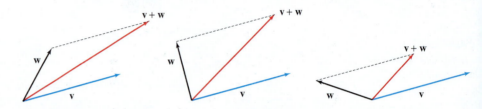

DF **FIGURE 20** The length of $\mathbf{v} + \mathbf{w}$ depends on the angle between $\mathbf{v}$ and $\mathbf{w}$.

12.1 SUMMARY

- A *vector* $\mathbf{v} = \overrightarrow{PQ}$ is determined by a basepoint P (the "tail") and a terminal point Q (the "head").
- Components of $\mathbf{v} = \overrightarrow{PQ}$, where $P = (a_1, b_1)$ and $Q = (a_2, b_2)$:

$$\mathbf{v} = \langle a, b \rangle$$

  with $a = a_2 - a_1$, $b = b_2 - b_1$.
- Length or magnitude: $\|\mathbf{v}\| = \sqrt{a^2 + b^2}$.
- The *length* $\|\mathbf{v}\|$ is the distance from P to Q.
- The *position vector* of $P_0 = (a, b)$ is the vector $\mathbf{v} = \langle a, b \rangle$ pointing from the origin O to P_0.
- Vectors $\mathbf{v}$ and $\mathbf{w}$ are *equivalent* if they have the same magnitude and direction. Two vectors are equivalent if and only if they have the same components.
- The *zero vector* is the vector $\mathbf{0} = \langle 0, 0 \rangle$ of length 0.
- *Vector addition* is defined geometrically by the *Parallelogram Law*. In components,

$$\langle v_1, v_2 \rangle + \langle w_1, w_2 \rangle = \langle v_1 + w_1, v_2 + w_2 \rangle$$

- Scalar multiplication: $\lambda\mathbf{v}$ is the vector of length $|\lambda|\,\|\mathbf{v}\|$ in the same direction as $\mathbf{v}$ if $\lambda > 0$, and in the opposite direction if $\lambda < 0$. In components,

$$\lambda\langle v_1, v_2 \rangle = \langle \lambda v_1, \lambda v_2 \rangle$$

- Nonzero vectors $\mathbf{v}$ and $\mathbf{w}$ are *parallel* if $\mathbf{w} = \lambda\mathbf{v}$ for some scalar λ.
- Unit vector making an angle θ with the positive x-axis: $\mathbf{e} = \langle \cos\theta, \sin\theta \rangle$.
- Unit vector in the direction of $\mathbf{v} \neq \mathbf{0}$: $\mathbf{e_v} = \dfrac{1}{\|\mathbf{v}\|}\mathbf{v}$.
- If $\mathbf{v} = \langle v_1, v_2 \rangle$ makes an angle θ with the positive x-axis, then

$$v_1 = \|\mathbf{v}\|\cos\theta, \qquad v_2 = \|\mathbf{v}\|\sin\theta, \qquad \mathbf{e_v} = \langle \cos\theta, \sin\theta \rangle$$

- Standard basis vectors: $\mathbf{i} = \langle 1, 0 \rangle$ and $\mathbf{j} = \langle 0, 1 \rangle$.
- Every vector $\mathbf{v} = \langle a, b \rangle$ is a linear combination $\mathbf{v} = a\mathbf{i} + b\mathbf{j}$.
- Triangle Inequality: $\|\mathbf{v} + \mathbf{w}\| \leq \|\mathbf{v}\| + \|\mathbf{w}\|$.

12.1 EXERCISES

Preliminary Questions

1. Answer true or false. Every nonzero vector is:

(a) equivalent to a vector based at the origin.

(b) equivalent to a unit vector based at the origin.

(c) parallel to a vector based at the origin.

(d) parallel to a unit vector based at the origin.

2. What is the length of $-3\mathbf{a}$ if $\|\mathbf{a}\| = 5$?

3. Suppose that $\mathbf{v}$ has components $\langle 3, 1 \rangle$. How, if at all, do the components change if you translate $\mathbf{v}$ horizontally 2 units to the left?

4. What are the components of the zero vector based at $P = (3, 5)$?

5. True or false?

(a) The vectors $\mathbf{v}$ and $-2\mathbf{v}$ are parallel.

(b) The vectors $\mathbf{v}$ and $-2\mathbf{v}$ point in the same direction.

6. Explain the commutativity of vector addition in terms of the Parallelogram Law.

Exercises

1. Sketch the vectors $\mathbf{v}_1, \mathbf{v}_2, \mathbf{v}_3, \mathbf{v}_4$ with tail P and head Q, and compute their lengths. Are any two of these vectors equivalent?

	$\mathbf{v}_1$	$\mathbf{v}_2$	$\mathbf{v}_3$	$\mathbf{v}_4$
P	$(2, 4)$	$(-1, 3)$	$(-1, 3)$	$(4, 1)$
Q	$(4, 4)$	$(1, 3)$	$(2, 4)$	$(6, 3)$

2. Sketch the vector $\mathbf{b} = \langle 3, 4 \rangle$ based at $P = (-2, -1)$.

3. What is the terminal point of the vector $\mathbf{a} = \langle 1, 3 \rangle$ based at $P = (2, 2)$? Sketch $\mathbf{a}$ and the vector $\mathbf{a}_0$ based at the origin and equivalent to $\mathbf{a}$.

4. Let $\mathbf{v} = \overrightarrow{PQ}$, where $P = (1, 1)$ and $Q = (2, 2)$. What is the head of the vector $\mathbf{v}'$ equivalent to $\mathbf{v}$ based at $(2, 4)$? What is the head of the vector $\mathbf{v}_0$ equivalent to $\mathbf{v}$ based at the origin? Sketch $\mathbf{v}$, $\mathbf{v}_0$, and $\mathbf{v}'$.

In Exercises 5–8, refer to Figure 21.

5. Find the components of $\mathbf{u}$.

6. Find the components of $\mathbf{v}$.

7. Find the components of $\mathbf{w}$.

8. Find the components of $\mathbf{q}$.

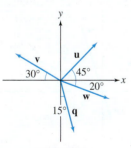

FIGURE 21

In Exercises 9–12, find the components of $\overrightarrow{PQ}$.

9. $P = (3, 2), \quad Q = (2, 7)$

10. $P = (1, -4), \quad Q = (3, 5)$

11. $P = (3, 5), \quad Q = (1, -4)$

12. $P = (0, 2), \quad Q = (5, 0)$

In Exercises 13–18, calculate.

13. $\langle 2, 1 \rangle + \langle 3, 4 \rangle$

14. $\langle -4, 6 \rangle - \langle 3, -2 \rangle$

15. $5\langle 6, 2 \rangle$

16. $4(\langle 1, 1 \rangle + \langle 3, 2 \rangle)$

17. $\left\langle -\frac{1}{2}, \frac{5}{3} \right\rangle + \left\langle 3, \frac{10}{3} \right\rangle$

18. $\langle \ln 2, e \rangle + \langle \ln 3, \pi \rangle$

19. Which of the vectors (A)–(C) in Figure 22 is equivalent to $\mathbf{v} - \mathbf{w}$?

(A)   (B)   (C)

FIGURE 22

20. Sketch $\mathbf{v} + \mathbf{w}$ and $\mathbf{v} - \mathbf{w}$ for the vectors in Figure 23.

FIGURE 23

21. Sketch $2\mathbf{v}, -\mathbf{w}, \mathbf{v} + \mathbf{w}$, and $2\mathbf{v} - \mathbf{w}$ for the vectors in Figure 24.

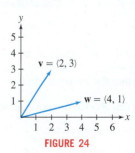

FIGURE 24

22. Sketch $\mathbf{v} = \langle 1, 3 \rangle$, $\mathbf{w} = \langle 2, -2 \rangle$, $\mathbf{v} + \mathbf{w}$, $\mathbf{v} - \mathbf{w}$.

23. Sketch $\mathbf{v} = \langle 0, 2 \rangle$, $\mathbf{w} = \langle -2, 4 \rangle$, $3\mathbf{v} + \mathbf{w}$, $2\mathbf{v} - 2\mathbf{w}$.

24. Sketch $\mathbf{v} = \langle -2, 1 \rangle$, $\mathbf{w} = \langle 2, 2 \rangle$, $\mathbf{v} + 2\mathbf{w}$, $\mathbf{v} - 2\mathbf{w}$.

25. Sketch the vector $\mathbf{v}$ such that $\mathbf{v} + \mathbf{v}_1 + \mathbf{v}_2 = \mathbf{0}$ for $\mathbf{v}_1$ and $\mathbf{v}_2$ in Figure 25(A).

26. Sketch the vector sum $\mathbf{v} = \mathbf{v}_1 + \mathbf{v}_2 + \mathbf{v}_3 + \mathbf{v}_4$ in Figure 25(B).

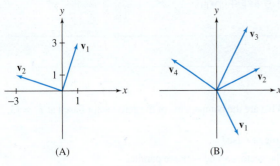

(A)   (B)

FIGURE 25

27. Let $\mathbf{v} = \overrightarrow{PQ}$, where $P = (-2, 5)$, $Q = (1, -2)$. Which of the following vectors with the given tails and heads are equivalent to $\mathbf{v}$?

(a) $(-3, 3)$, $(0, 4)$ **(b)** $(0, 0)$, $(3, -7)$

(c) $(-1, 2)$, $(2, -5)$ **(d)** $(4, -5)$, $(1, 4)$

28. Which of the following vectors are parallel to $\mathbf{v} = \langle 6, 9 \rangle$ and which point in the same direction?

(a) $\langle 12, 18 \rangle$ **(b)** $\langle 3, 2 \rangle$ **(c)** $\langle 2, 3 \rangle$

(d) $\langle -6, -9 \rangle$ **(e)** $\langle -24, -27 \rangle$ **(f)** $\langle -24, -36 \rangle$

In Exercises 29–32, sketch the vectors $\overrightarrow{AB}$ and $\overrightarrow{PQ}$, and determine whether they are equivalent.

29. $A = (1, 1)$, $B = (3, 7)$, $P = (4, -1)$, $Q = (6, 5)$

30. $A = (1, 4)$, $B = (-6, 3)$, $P = (1, 4)$, $Q = (6, 3)$

31. $A = (-3, 2)$, $B = (0, 0)$, $P = (0, 0)$, $Q = (3, -2)$

32. $A = (5, 8)$, $B = (1, 8)$, $P = (1, 8)$, $Q = (-3, 8)$

In Exercises 33–36, are $\overrightarrow{AB}$ and $\overrightarrow{PQ}$ parallel? And if so, do they point in the same direction?

33. $A = (1, 1)$, $B = (3, 4)$, $P = (1, 1)$, $Q = (7, 10)$

34. $A = (-3, 2)$, $B = (0, 0)$, $P = (0, 0)$, $Q = (3, 2)$

35. $A = (2, 2)$, $B = (-6, 3)$, $P = (9, 5)$, $Q = (17, 4)$

36. $A = (5, 8)$, $B = (2, 2)$, $P = (2, 2)$, $Q = (-3, 8)$

In Exercises 37–40, let $R = (-2, 7)$. Calculate the following:

37. The length of $\overrightarrow{OR}$

38. The components of $\mathbf{u} = \overrightarrow{PR}$, where $P = (1, 2)$

39. The point P such that $\overrightarrow{PR}$ has components $\langle -2, 7 \rangle$

40. The point Q such that $\overrightarrow{RQ}$ has components $\langle 8, -3 \rangle$

In Exercises 41–48, find the given vector.

41. Unit vector $\mathbf{e_v}$ where $\mathbf{v} = \langle 3, 4 \rangle$

42. Unit vector $\mathbf{e_w}$ where $\mathbf{w} = \langle 24, 7 \rangle$

43. Vector of length 4 in the direction of $\mathbf{u} = \langle -1, -1 \rangle$

44. Vector of length 3 in the direction of $\mathbf{v} = 4\mathbf{i} + 3\mathbf{j}$

45. Vector of length 2 in the direction opposite to $\mathbf{v} = \mathbf{i} - \mathbf{j}$

46. Unit vector in the direction opposite to $\mathbf{v} = \langle -2, 4 \rangle$

47. Unit vector $\mathbf{e}$ making an angle of $\frac{4\pi}{7}$ with the x-axis

48. Vector $\mathbf{v}$ of length 2 making an angle of $30°$ with the x-axis

49. Find all scalars λ such that $\lambda \langle 2, 3 \rangle$ has length 1.

50. Find a vector $\mathbf{v}$ satisfying $3\mathbf{v} + \langle 5, 20 \rangle = \langle 11, 17 \rangle$.

51. What are the coordinates of the point P in the parallelogram in Figure 26(A)?

52. What are the coordinates a and b in the parallelogram in Figure 26(B)?

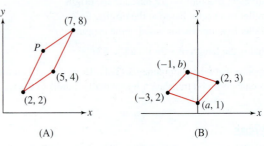

(A)   (B)

FIGURE 26

53. Let $\mathbf{v} = \overrightarrow{AB}$ and $\mathbf{w} = \overrightarrow{AC}$, where A, B, C are three distinct points in the plane. Match (a)–(d) with (i)–(iv). (*Hint:* Draw a picture.)

(a) $-\mathbf{w}$ **(b)** $-\mathbf{v}$ **(c)** $\mathbf{w} - \mathbf{v}$ **(d)** $\mathbf{v} - \mathbf{w}$

(i) $\overrightarrow{CB}$ **(ii)** $\overrightarrow{CA}$ **(iii)** $\overrightarrow{BC}$ **(iv)** $\overrightarrow{BA}$

54. Find the components and length of the following vectors:

(a) $4\mathbf{i} + 3\mathbf{j}$ **(b)** $2\mathbf{i} - 3\mathbf{j}$ **(c)** $\mathbf{i} + \mathbf{j}$ **(d)** $\mathbf{i} - 3\mathbf{j}$

In Exercises 55–58, calculate the linear combination.

55. $3\mathbf{j} + (9\mathbf{i} + 4\mathbf{j})$ **56.** $-\frac{3}{2}\mathbf{i} + 5\left(\frac{1}{2}\mathbf{j} - \frac{1}{2}\mathbf{i}\right)$

57. $(3\mathbf{i} + \mathbf{j}) - 6\mathbf{j} + 2(\mathbf{j} - 4\mathbf{i})$ **58.** $3(3\mathbf{i} - 4\mathbf{j}) + 5(\mathbf{i} + 4\mathbf{j})$

59. For each of the position vectors $\mathbf{u}$ with endpoints A, B, and C in Figure 27, indicate with a diagram the multiples $r\mathbf{v}$ and $s\mathbf{w}$ such that $\mathbf{u} = r\mathbf{v} + s\mathbf{w}$. A sample is shown for $\mathbf{u} = \overrightarrow{OQ}$.

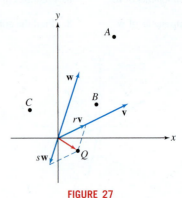

FIGURE 27

60. Sketch the parallelogram spanned by $\mathbf{v} = \langle 1, 4 \rangle$ and $\mathbf{w} = \langle 5, 2 \rangle$. Add the vector $\mathbf{u} = \langle 2, 3 \rangle$ to the sketch and express $\mathbf{u}$ as a linear combination of $\mathbf{v}$ and $\mathbf{w}$.

In Exercises 61 and 62, express $\mathbf{u}$ as a linear combination $\mathbf{u} = r\mathbf{v} + s\mathbf{w}$. Then sketch $\mathbf{u}$, $\mathbf{v}$, $\mathbf{w}$, and the parallelogram formed by $r\mathbf{v}$ and $s\mathbf{w}$.

61. $\mathbf{u} = \langle 3, -1 \rangle$; $\mathbf{v} = \langle 2, 1 \rangle$, $\mathbf{w} = \langle 1, 3 \rangle$

62. $\mathbf{u} = \langle 6, -2 \rangle$; $\mathbf{v} = \langle 1, 1 \rangle$, $\mathbf{w} = \langle 1, -1 \rangle$

63. Calculate the magnitude of the force on cables 1 and 2 in Figure 28.

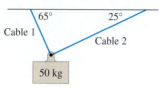

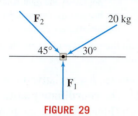

FIGURE 28

64. Determine the magnitude of the forces $\mathbf{F}_1$ and $\mathbf{F}_2$ in Figure 29, assuming that there is no net force on the object.

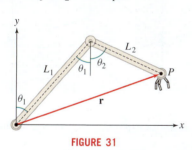

FIGURE 29

65. A plane flying due east at 200 km/h encounters a 40-km/h wind blowing in the northeast direction. The resultant velocity of the plane is the vector sum $\mathbf{v} = \mathbf{v}_1 + \mathbf{v}_2$, where $\mathbf{v}_1$ is the velocity vector of the plane and $\mathbf{v}_2$ is the velocity vector of the wind (Figure 30). The angle between $\mathbf{v}_1$ and $\mathbf{v}_2$ is $\frac{\pi}{4}$. Determine the resultant *speed* of the plane (the length of the vector $\mathbf{v}$).

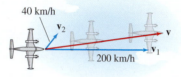

FIGURE 30

Further Insights and Challenges

In Exercises 66–68, refer to Figure 31, which shows a robotic arm consisting of two segments of lengths L_1 and L_2.

66. Find the components of the vector $\mathbf{r} = \overrightarrow{OP}$ in terms of θ_1 and θ_2.

67. Let $L_1 = 5$ and $L_2 = 3$. Find $\mathbf{r}$ for $\theta_1 = \frac{\pi}{3}, \theta_2 = \frac{\pi}{4}$.

68. Let $L_1 = 5$ and $L_2 = 3$. Show that the set of points reachable by the robotic arm with $\theta_1 = \theta_2$ is an ellipse.

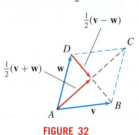

FIGURE 31

69. Use vectors to prove that the diagonals $\overline{AC}$ and $\overline{BD}$ of a parallelogram bisect each other (Figure 32). *Hint:* Observe that the midpoint of $\overline{BD}$ is the terminal point of $\mathbf{w} + \frac{1}{2}(\mathbf{v} - \mathbf{w})$.

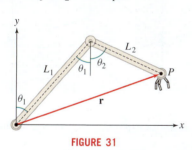

FIGURE 32

70. Use vectors to prove that the segments joining the midpoints of opposite sides of a quadrilateral bisect each other (Figure 33). *Hint:* Show that the midpoints of these segments are the terminal points of

$$\frac{1}{4}(2\mathbf{u} + \mathbf{v} + \mathbf{z}) \quad \text{and} \quad \frac{1}{4}(2\mathbf{v} + \mathbf{w} + \mathbf{u})$$

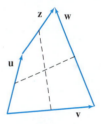

FIGURE 33

71. Prove that two vectors $\mathbf{v} = \langle a, b \rangle$ and $\mathbf{w} = \langle c, d \rangle$ are perpendicular if and only if

$$ac + bd = 0$$

12.2 Vectors in Three Dimensions

This section extends the vector concepts introduced in the previous section to three-dimensional space. We begin with some introductory remarks about the three-dimensional coordinate system.

By convention, we label the axes as in Figure 1(A), where the positive sides of the axes are labeled x, y, and z. This labeling satisfies the **right-hand rule**, which means that when you position your right hand so that your fingers curl from the positive x-axis toward the positive y-axis, your thumb points in the positive z-direction. The axes in Figure 1(B) are not labeled according to the right-hand rule.

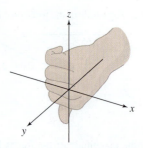

(A) Standard coordinate system
(satisfies the right-hand rule).

(B) This coordinate system does
not satisfy the right-hand rule
because the thumb points in
the negative z-direction.

FIGURE 1 The fingers of the right hand curl from the positive x-axis to the positive y-axis.

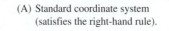

FIGURE 2

Each point in space has unique coordinates (a, b, c) relative to the axes (Figure 2). We denote the set of all triples (a, b, c) by $\mathbf{R}^3$. The **coordinate planes** in $\mathbf{R}^3$ are defined by setting one of the coordinates equal to zero (Figure 3). The xy-plane consists of the points $(a, b, 0)$ and is defined by the equation $z = 0$. Similarly, $x = 0$ defines the yz-plane consisting of the points $(0, b, c)$, and $y = 0$ defines the xz-plane consisting of the points $(a, 0, c)$. The coordinate planes divide $\mathbf{R}^3$ into eight **octants** (analogous to the four quadrants in the plane). Each octant corresponds to a possible combination of signs of the coordinates. The set of points (a, b, c) with $a, b, c > 0$ is called the **first octant**.

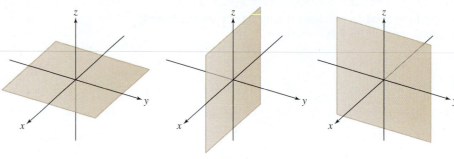

$z = 0$ defines the xy-plane. $y = 0$ defines the xz-plane. $x = 0$ defines the yz-plane.

FIGURE 3

As in two dimensions, we derive the distance formula in $\mathbf{R}^3$ from the Pythagorean Theorem.

THEOREM 1 Distance Formula in $\mathbf{R}^3$ The distance $|P - Q|$ between the points $P = (a_1, b_1, c_1)$ and $Q = (a_2, b_2, c_2)$ is

$$|P - Q| = \sqrt{(a_2 - a_1)^2 + (b_2 - b_1)^2 + (c_2 - c_1)^2} \qquad \boxed{1}$$

Proof First apply the distance formula in the plane to the points P and R (Figure 4):

$$|P - R|^2 = (a_2 - a_1)^2 + (b_2 - b_1)^2$$

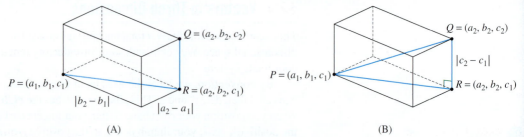

(A)   (B)

FIGURE 4 Compute $|P - Q|$ using the right triangle $\triangle PRQ$.

Then observe that $\triangle PRQ$ is a right triangle [Figure 4(B)] and use the Pythagorean Theorem:

$$|P - Q|^2 = |P - R|^2 + |R - Q|^2 = (a_2 - a_1)^2 + (b_2 - b_1)^2 + (c_2 - c_1)^2 \quad \blacksquare$$

The sphere of radius R with center $Q = (a, b, c)$ consists of all points $P = (x, y, z)$ located a distance R from Q (Figure 5). By the distance formula, the coordinates of $P = (x, y, z)$ must satisfy

$$\sqrt{(x - a)^2 + (y - b)^2 + (z - c)^2} = R$$

On squaring both sides, we obtain the standard equation of the sphere [Eq. (3) below].
  Now consider the equation

$$(x - a)^2 + (y - b)^2 = R^2 \qquad \boxed{2}$$

In the xy-plane, Eq. (2) defines the circle of radius R with center (a, b). However, as an equation in $\mathbf{R}^3$, it defines the right circular cylinder of radius R whose central axis is the vertical line through $(a, b, 0)$ (Figure 6). Indeed, a point (x, y, z) satisfies Eq. (2) for any value of z if (x, y) lies on the circle. It is usually clear from the context which of the following is intended:

$$\text{Circle} = \{(x, y) : (x - a)^2 + (y - b)^2 = R^2\}$$

$$\text{Right circular cylinder} = \{(x, y, z) : (x - a)^2 + (y - b)^2 = R^2\}$$

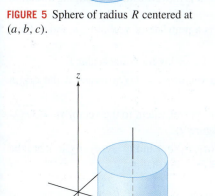

FIGURE 5 Sphere of radius R centered at (a, b, c).

FIGURE 6 Right circular cylinder of radius R centered at $(a, b, 0)$.

Equations of Spheres and Cylinders An equation of the sphere in $\mathbf{R}^3$ of radius R centered at $Q = (a, b, c)$ is

$$(x - a)^2 + (y - b)^2 + (z - c)^2 = R^2 \qquad \boxed{3}$$

An equation of the right circular cylinder in $\mathbf{R}^3$ of radius R whose central axis is the vertical line through $(a, b, 0)$ is

$$(x - a)^2 + (y - b)^2 = R^2 \qquad \boxed{4}$$

■ **EXAMPLE 1** Describe the sets of points defined by the following conditions:

(a) $x^2 + y^2 + z^2 = 4, \quad y \geq 0$
(b) $(x - 3)^2 + (y - 2)^2 = 1, \quad z \geq -1$

Solution

(a) The equation $x^2 + y^2 + z^2 = 4$ defines a sphere of radius 2 centered at the origin. The inequality $y \geq 0$ holds for points lying on the positive side of the xz-plane. We obtain the right hemisphere of radius 2 illustrated in Figure 7(A).
(b) The equation $(x - 3)^2 + (y - 2)^2 = 1$ defines a cylinder of radius 1 whose central axis is the vertical line through $(3, 2, 0)$. The part of the cylinder where $z \geq -1$ is illustrated in Figure 7(B). ■

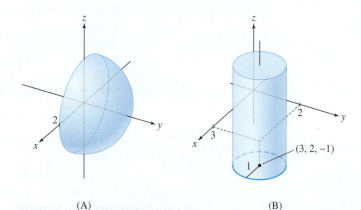

DF FIGURE 7 Hemisphere and upper cylinder.

(A)　　　　　　　(B)

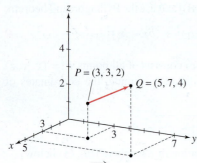

FIGURE 8 A vector $\overrightarrow{PQ}$ in 3-space.

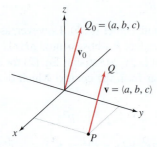

FIGURE 9 A vector **v** and its translate based at the origin.

Our basepoint convention remains in force: Vectors are assumed to be based at the origin unless otherwise indicated.

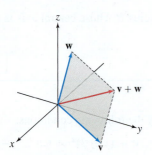

FIGURE 10 Vector addition is defined by the Parallelogram Law.

Vector Concepts

As in the plane, a vector $\mathbf{v} = \overrightarrow{PQ}$ in $\mathbf{R}^3$ is determined by an initial point P and a terminal point Q (Figure 8). If $P = (a_1, b_1, c_1)$ and $Q = (a_2, b_2, c_2)$, then the **length** or **magnitude** of $\mathbf{v} = \overrightarrow{PQ}$, denoted $\|\mathbf{v}\|$, is the distance from P to Q:

$$\|\mathbf{v}\| = \|\overrightarrow{PQ}\| = \sqrt{(a_2 - a_1)^2 + (b_2 - b_1)^2 + (c_2 - c_1)^2}$$

The terminology and basic properties discussed in the previous section carry over to $\mathbf{R}^3$ with little change.

- A vector **v** is said to undergo a **translation** if it is moved without changing direction or magnitude.
- Two vectors **v** and **w** are **equivalent** if **w** is a translate of **v**; that is, **v** and **w** have the same length and direction.
- Two nonzero vectors **v** and **w** are **parallel** if $\mathbf{v} = \lambda\mathbf{w}$ for some scalar λ.
- The **position vector** of a point Q_0 is the vector $\mathbf{v}_0 = \overrightarrow{OQ_0}$ based at the origin (Figure 9).
- A vector $\mathbf{v} = \overrightarrow{PQ}$ with components $\langle a, b, c \rangle$ is equivalent to the vector $\mathbf{v}_0 = \overrightarrow{OQ_0}$ based at the origin with $Q_0 = (a, b, c)$ (Figure 9).
- The **components** of $\mathbf{v} = \overrightarrow{PQ}$, where $P = (a_1, b_1, c_1)$ and $Q = (a_2, b_2, c_2)$, are the differences $a = a_2 - a_1$, $b = b_2 - b_1$, $c = c_2 - c_1$; that is,

$$\mathbf{v} = \overrightarrow{PQ} = \overrightarrow{OQ} - \overrightarrow{OP} = \langle a_2, b_2, c_2 \rangle - \langle a_1, b_1, c_1 \rangle$$

For example, if $P = (3, -4, -4)$ and $Q = (2, 5, -1)$, then

$$\mathbf{v} = \overrightarrow{PQ} = \langle 2, 5, -1 \rangle - \langle 3, -4, -4 \rangle = \langle -1, 9, 3 \rangle$$

- Two vectors are equivalent if and only if they have the same components.
- Vector addition and scalar multiplication are defined as in the two-dimensional case. Vector addition is defined by the Parallelogram Law (Figure 10).
- In terms of components, if $\mathbf{v} = \langle v_1, v_2, v_3 \rangle$ and $\mathbf{w} = \langle w_1, w_2, w_3 \rangle$, then

$$\lambda\mathbf{v} = \lambda \langle v_1, v_2, v_3 \rangle = \langle \lambda v_1, \lambda v_2, \lambda v_3 \rangle$$

$$\mathbf{v} + \mathbf{w} = \langle v_1, v_2, v_3 \rangle + \langle w_1, w_2, w_3 \rangle = \langle v_1 + w_1, v_2 + w_2, v_3 + w_3 \rangle$$

- Vector addition is commutative, is associative, and satisfies the distributive property with respect to scalar multiplication (Theorem 1 in Section 12.1).

■ **EXAMPLE 2** **Vector Calculations** Calculate $\|\mathbf{v}\|$ and $6\mathbf{v} - \frac{1}{2}\mathbf{w}$, where $\mathbf{v} = \langle 3, -1, 2 \rangle$ and $\mathbf{w} = \langle 4, 6, -8 \rangle$. Then determine whether **v** and **w** are parallel.

Solution

$$\|\mathbf{v}\| = \sqrt{3^2 + (-1)^2 + 2^2} = \sqrt{14}$$

$$6\mathbf{v} - \frac{1}{2}\mathbf{w} = 6 \langle 3, -1, 2 \rangle - \frac{1}{2} \langle 4, 6, -8 \rangle$$

$$= \langle 18, -6, 12 \rangle - \langle 2, 3, -4 \rangle$$

$$= \langle 16, -9, 16 \rangle$$

If the vectors **v** and **w** are parallel, then there must be a scalar λ such that $\mathbf{v} = \lambda\mathbf{w}$. Thus,

$$\langle 3, -1, 2 \rangle = \lambda \langle 4, 6, -8 \rangle$$

Considering components, this means that

$$3 = \lambda 4, \quad -1 = \lambda 6, \quad 2 = \lambda(-8)$$

But then $\lambda = \frac{3}{4}$, the value of which does not satisfy either of the other two equations. So, the two vectors are not parallel. ■

The **standard basis vectors** in $\mathbf{R}^3$ are

$$\mathbf{i} = \langle 1, 0, 0 \rangle, \qquad \mathbf{j} = \langle 0, 1, 0 \rangle, \qquad \mathbf{k} = \langle 0, 0, 1 \rangle$$

Every vector is a **linear combination** of the standard basis vectors (Figure 11):

$$\langle a, b, c \rangle = a \langle 1, 0, 0 \rangle + b \langle 0, 1, 0 \rangle + c \langle 0, 0, 1 \rangle = a\mathbf{i} + b\mathbf{j} + c\mathbf{k}$$

For example, $\langle -9, -4, 17 \rangle = -9\mathbf{i} - 4\mathbf{j} + 17\mathbf{k}$.

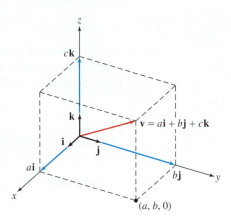

FIGURE 11 Writing $\mathbf{v} = \langle a, b, c \rangle$ as the sum $a\mathbf{i} + b\mathbf{j} + c\mathbf{k}$.

■ **EXAMPLE 3** Find the unit vector $\mathbf{e_v}$ in the direction of $\mathbf{v} = 3\mathbf{i} + 2\mathbf{j} - 4\mathbf{k}$.

Solution Since $\|\mathbf{v}\| = \sqrt{3^2 + 2^2 + (-4)^2} = \sqrt{29}$,

$$\mathbf{e_v} = \frac{1}{\|\mathbf{v}\|}\mathbf{v} = \frac{1}{\sqrt{29}}(3\mathbf{i} + 2\mathbf{j} - 4\mathbf{k}) = \left\langle \frac{3}{\sqrt{29}}, \frac{2}{\sqrt{29}}, \frac{-4}{\sqrt{29}} \right\rangle$$

■

Parametric Equations of a Line

Although the basic vector concepts in two and three dimensions are essentially the same, there is an important difference in the way lines are described. A line in $\mathbf{R}^2$ is defined by a single linear equation such as $y = mx + b$. In $\mathbf{R}^3$, a single linear equation defines a plane rather than a line. Therefore, we describe lines in $\mathbf{R}^3$ in parametric form.

We note first that a line $\mathcal{L}_0$ through the origin consists of the multiples of a nonzero vector $\mathbf{v} = \langle a, b, c \rangle$, as in Figure 12(A). More precisely,

$$t\mathbf{v} = \langle ta, tb, tc \rangle \qquad (-\infty < t < \infty)$$

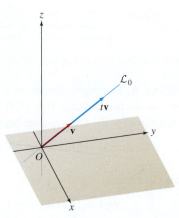

(A) Line through the origin (multiples of **v**)

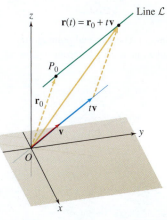

(B) Line through P_0 in the direction of **v**

DF **FIGURE 12**

Then the line $\mathcal{L}_0$ consists of the terminal points (ta, tb, tc) of the vectors $t\mathbf{v}$ as t varies from $-\infty$ to ∞. The coordinates (x, y, z) of the points on the line are given by the parametric equations

$$x = at, \quad y = bt, \quad z = ct$$

Suppose, more generally, that we would like to parametrize the line $\mathcal{L}$ parallel to $\mathbf{v}$ but passing through a point $P_0 = (x_0, y_0, z_0)$ as in Figure 12(B). We must translate the line $t\mathbf{v}$ so that it passes through P_0. To do this, we add the position vector $\mathbf{r}_0 = \overrightarrow{OP_0}$ to the multiples $t\mathbf{v}$:

$$\mathbf{r}(t) = \mathbf{r}_0 + t\mathbf{v} = \langle x_0, y_0, z_0 \rangle + t \langle a, b, c \rangle$$

The terminal point of $\mathbf{r}(t)$ traces out $\mathcal{L}$ as t varies from $-\infty$ to ∞. The vector $\mathbf{v}$ is called a **direction vector** for $\mathcal{L}$, and coordinates (x, y, z) of the points on the line $\mathcal{L}$ are given by the parametric equations

$$x = x_0 + at, \quad y = y_0 + bt, \quad z = z_0 + ct$$

Equation of a Line (Point-Direction Form) The line $\mathcal{L}$ through $P_0 = (x_0, y_0, z_0)$ in the direction of $\mathbf{v} = \langle a, b, c \rangle$ is described by

Vector parametrization:

$$\mathbf{r}(t) = \mathbf{r}_0 + t\mathbf{v} = \langle x_0, y_0, z_0 \rangle + t \langle a, b, c \rangle \qquad \boxed{5}$$

where $\mathbf{r}_0 = \overrightarrow{OP_0}$.

Parametric equations:

$$x = x_0 + at, \quad y = y_0 + bt, \quad z = z_0 + ct \qquad \boxed{6}$$

The parameter t takes on values $-\infty < t < \infty$.

Parametric equations specify the x-, y-, and z-coordinates of a point on the line as a function of the parameter t. These are familiar from our discussion of parametric curves in the plane in Section 11.1. Note that each of the three parametric equations is itself a linear function in the parameter t. What is new here is the notion of a vector parametrization, the idea that $\mathbf{r}(t)$ describes a vector whose terminal point traces out a line as t varies from $-\infty$ to ∞ (Figure 13).

■ **EXAMPLE 4** Find a vector parametrization and parametric equations for the line through $P_0 = (3, -1, 4)$ with direction vector $\mathbf{v} = \langle 2, 1, 7 \rangle$.

Solution By Eq. (5), the following is a vector parametrization:

$$\mathbf{r}(t) = \underbrace{\langle 3, -1, 4 \rangle}_{\text{Coordinates of } P_0} + \ t\underbrace{\langle 2, 1, 7 \rangle}_{\text{Direction vector}} = \langle 3 + 2t, -1 + t, 4 + 7t \rangle$$

The corresponding parametric equations are $x = 3 + 2t$, $y = -1 + t$, $z = 4 + 7t$. ■

■ **EXAMPLE 5** **Parametric Equations of the Line Through Two Points** Find parametric equations for the line through $P = (1, 0, 2)$ and $Q = (2, 5, -1)$. Use them to find parametric equations for the line segment between P and Q.

Solution We can take our direction vector to be $\mathbf{v} = \overrightarrow{PQ} = \langle 2 - 1, 5 - 0, -1 - 2 \rangle = \langle 1, 5, -3 \rangle$.

Hence, we obtain

$$\mathbf{r}(t) = \langle 1, 0, 2 \rangle + t \langle 1, 5, -3 \rangle = \langle 1 + t, 5t, 2 - 3t \rangle$$

Thus, the parametric equations for the line are $x = 1 + t$, $y = 5t$, $z = 2 - 3t$, where $-\infty < t < \infty$.

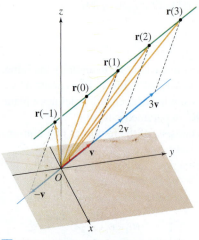

DF FIGURE 13 The terminal point of $\mathbf{r}(t)$ traces out a line as t varies from $-\infty$ to ∞.

To obtain parametric equations for the line segment between P and Q, we note that $\mathbf{r}(0) = \langle 1, 0, 2 \rangle = \overrightarrow{OP}$ and $\mathbf{r}(1) = \langle 2, 5, -1 \rangle = \overrightarrow{OQ}$. Therefore, if we use the same parametric equations, but restrict t so that $0 \le t \le 1$, we obtain parametric equations for the line segment. ∎

The parametrization of a line $\mathcal{L}$ is not unique. We are free to choose any point P_0 on $\mathcal{L}$ and we may replace the direction vector $\mathbf{v}$ by any nonzero scalar multiple $\lambda \mathbf{v}$. However, two lines in $\mathbf{R}^3$ coincide if they are parallel and pass through a common point, so we can always check whether two parametrizations describe the same line.

■ **EXAMPLE 6** Different Parametrizations of the Same Line Show that

$$\mathbf{r}_1(t) = \langle 1, 1, 0 \rangle + t \langle -2, 1, 3 \rangle \quad \text{and} \quad \mathbf{r}_2(t) = \langle -3, 3, 6 \rangle + t \langle 4, -2, -6 \rangle$$

parametrize the same line.

Solution The line $\mathbf{r}_1(t)$ has direction vector $\mathbf{v} = \langle -2, 1, 3 \rangle$, whereas $\mathbf{r}_2(t)$ has direction vector $\mathbf{w} = \langle 4, -2, -6 \rangle$. These vectors are parallel because $\mathbf{w} = -2\mathbf{v}$. Therefore, the lines described by $\mathbf{r}_1(t)$ and $\mathbf{r}_2(t)$ are parallel. We must check that they have a point in common. Choose any point on $\mathbf{r}_1(t)$, say, $P = (1, 1, 0)$ [corresponding to $t = 0$]. This point lies on $\mathbf{r}_2(t)$ if there is a value of t such that

$$\langle 1, 1, 0 \rangle = \langle -3, 3, 6 \rangle + t \langle 4, -2, -6 \rangle \qquad \boxed{7}$$

This yields three equations

$$1 = -3 + 4t, \qquad 1 = 3 - 2t, \qquad 0 = 6 - 6t$$

All three are satisfied with $t = 1$. Therefore, P also lies on $\mathbf{r}_2(t)$. We conclude that $\mathbf{r}_1(t)$ and $\mathbf{r}_2(t)$ parametrize the same line. If Eq. (7) had no solution, we would conclude that $\mathbf{r}_1(t)$ and $\mathbf{r}_2(t)$ are parallel but do not coincide. ∎

■ **EXAMPLE 7** Intersection of Two Lines Determine whether the following two lines intersect:

$$\mathbf{r}_1(t) = \langle 1, 0, 1 \rangle + t \langle 3, 3, 5 \rangle$$
$$\mathbf{r}_2(t) = \langle 3, 6, 1 \rangle + t \langle 4, -2, 7 \rangle$$

CAUTION *We cannot assume in Eq. (8) that the parameter values t_1 and t_2 are equal. The point of intersection may correspond to different parameter values on the two lines.*

Solution The two lines intersect if there exist parameter values t_1 and t_2 such that $\mathbf{r}_1(t_1) = \mathbf{r}_2(t_2)$—that is, if

$$\langle 1, 0, 1 \rangle + t_1 \langle 3, 3, 5 \rangle = \langle 3, 6, 1 \rangle + t_2 \langle 4, -2, 7 \rangle \qquad \boxed{8}$$

This is equivalent to three equations for the components:

$$x = 1 + 3t_1 = 3 + 4t_2, \qquad y = 3t_1 = 6 - 2t_2, \qquad z = 1 + 5t_1 = 1 + 7t_2 \qquad \boxed{9}$$

Let's solve the first two equations for t_1 and t_2. Subtracting the second equation from the first, we get $1 = 6t_2 - 3$ or $t_2 = \frac{2}{3}$. Using this value in the second equation, we get $t_1 = 2 - \frac{2}{3}t_2 = \frac{14}{9}$. The values $t_1 = \frac{14}{9}$ and $t_2 = \frac{2}{3}$ satisfy the first two equations, and thus, $\mathbf{r}_1(t_1)$ and $\mathbf{r}_2(t_2)$ have the same x- and y-coordinates (Figure 14). However, they do not have the same z-coordinates because t_1 and t_2 do not satisfy the third equation in (9):

$$1 + 5\left(\frac{14}{9}\right) \ne 1 + 7\left(\frac{2}{3}\right)$$

Therefore, Eq. (8) has no solution and the lines do not intersect. ∎

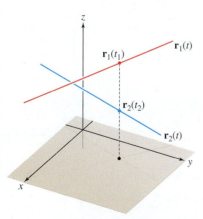

FIGURE 14 The lines $\mathbf{r}_1(t)$ and $\mathbf{r}_2(t)$ do not intersect, but the particular points $\mathbf{r}_1(t_1)$ and $\mathbf{r}_2(t_2)$ have the same x- and y-coordinates.

12.2 SUMMARY

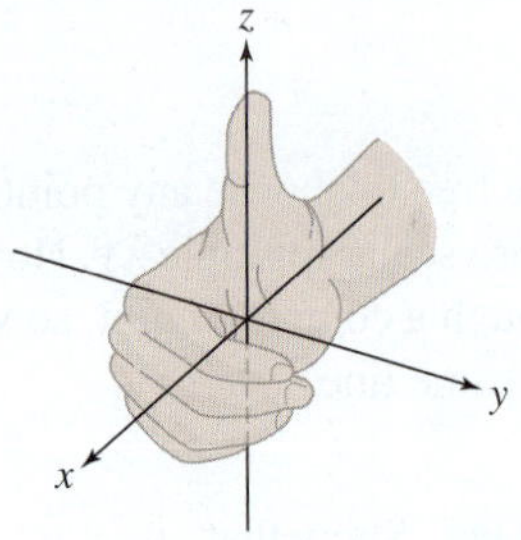

- The axes in $\mathbf{R}^3$ are labeled so that they satisfy the *right-hand rule*: When the fingers of your right hand curl from the positive x-axis toward the positive y-axis, your thumb points in the positive z-direction (Figure 15).

Sphere of radius R and center (a, b, c)	$(x - a)^2 + (y - b)^2 + (z - c)^2 = R^2$
Cylinder of radius R with vertical axis through $(a, b, 0)$	$(x - a)^2 + (y - b)^2 = R^2$

FIGURE 15

- The notation and terminology for vectors in the plane carry over to vectors in $\mathbf{R}^3$.
- The length (or magnitude) of $\mathbf{v} = \overrightarrow{PQ}$, where $P = (a_1, b_1, c_1)$ and $Q = (a_2, b_2, c_2)$, is

$$\|\mathbf{v}\| = \|\overrightarrow{PQ}\| = \sqrt{(a_2 - a_1)^2 + (b_2 - b_1)^2 + (c_2 - c_1)^2}$$

- Equations for the line through $P_0 = (x_0, y_0, z_0)$ with direction vector $\mathbf{v} = \langle a, b, c \rangle$:

$$\text{Vector parametrization:} \quad \mathbf{r}(t) = \overrightarrow{OP_0} + t\mathbf{v} = \langle x_0, y_0, z_0 \rangle + t \langle a, b, c \rangle$$

$$\text{Parametric equations:} \quad x = x_0 + at, \quad y = y_0 + bt, \quad z = z_0 + ct$$

- To obtain the line through $P = (a_1, b_1, c_1)$ and $Q = (a_2, b_2, c_2)$, take direction vector $\mathbf{v} = \overrightarrow{PQ} = \langle a_2 - a_1, b_2 - b_1, c_2 - c_1 \rangle$, and use the parametrization above. The segment $\overline{PQ}$ is parametrized by $\mathbf{r}(t)$ for $0 \le t \le 1$.

12.2 EXERCISES

Preliminary Questions

1. What is the terminal point of the vector $\mathbf{v} = \langle 3, 2, 1 \rangle$ based at the point $P = (1, 1, 1)$?

2. What are the components of the vector $\mathbf{v} = \langle 3, 2, 1 \rangle$ based at the point $P = (1, 1, 1)$?

3. If $\mathbf{v} = -3\mathbf{w}$, then (choose the correct answer):

(a) $\mathbf{v}$ and $\mathbf{w}$ are parallel.

(b) $\mathbf{v}$ and $\mathbf{w}$ point in the same direction.

4. Which of the following is a direction vector for the line through $P = (3, 2, 1)$ and $Q = (1, 1, 1)$?

(a) $\langle 3, 2, 1 \rangle$ **(b)** $\langle 1, 1, 1 \rangle$ **(c)** $\langle 2, 1, 0 \rangle$

5. How many different direction vectors does a line have?

6. True or false? If $\mathbf{v}$ is a direction vector for a line $\mathcal{L}$, then $-\mathbf{v}$ is also a direction vector for $\mathcal{L}$.

Exercises

1. Sketch the vector $\mathbf{v} = \langle 1, 3, 2 \rangle$ and compute its length.

2. Let $\mathbf{v} = \overrightarrow{P_0 Q_0}$, where $P_0 = (1, -2, 5)$ and $Q_0 = (0, 1, -4)$. Which of the following vectors (with tail P and head Q) are equivalent to $\mathbf{v}$?

	$\mathbf{v}_1$	$\mathbf{v}_2$	$\mathbf{v}_3$	$\mathbf{v}_4$
P	$(1, 2, 4)$	$(1, 5, 4)$	$(0, 0, 0)$	$(2, 4, 5)$
Q	$(0, 5, -5)$	$(0, -8, 13)$	$(-1, 3, -9)$	$(1, 7, 4)$

3. Sketch the vector $\mathbf{v} = \langle 1, 1, 0 \rangle$ based at $P = (0, 1, 1)$. Describe this vector in the form $\overrightarrow{PQ}$ for some point Q, and sketch the vector $\mathbf{v}_0$ based at the origin equivalent to $\mathbf{v}$.

4. Determine whether the coordinate systems (A)–(C) in Figure 16 satisfy the right-hand rule.

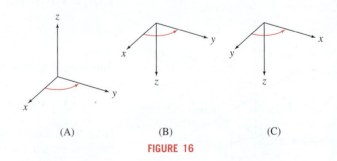

(A)   **(B)**   **(C)**

FIGURE 16

In Exercises 5–8, find the components of the vector $\overrightarrow{PQ}$.

5. $P = (1, 0, 1)$, $Q = (2, 1, 0)$

6. $P = (-3, -4, 2)$, $Q = (1, -4, 3)$

7. $P = (4, 6, 0)$, $Q = \left(-\frac{1}{2}, \frac{9}{2}, 1\right)$

8. $P = \left(-\frac{1}{2}, \frac{9}{2}, 1\right)$, $Q = (4, 6, 0)$

In Exercises 9–12, let R = (1, 4, 3).

9. Calculate the length of $\overrightarrow{OR}$.

10. Find the point Q such that $\mathbf{v} = \overrightarrow{RQ}$ has components $\langle 4, 1, 1 \rangle$, and sketch $\mathbf{v}$.

11. Find the point P such that $\mathbf{w} = \overrightarrow{PR}$ has components $\langle 3, -2, 3 \rangle$, and sketch $\mathbf{w}$.

12. Find the components of $\mathbf{u} = \overrightarrow{PR}$, where $P = (1, 2, 2)$.

13. Let $\mathbf{v} = \langle 4, 8, 12 \rangle$. Which of the following vectors is parallel to $\mathbf{v}$? Which point in the same direction?

(a) $\langle 2, 4, 6 \rangle$ (b) $\langle -1, -2, 3 \rangle$
(c) $\langle -7, -14, -21 \rangle$ (d) $\langle 6, 10, 14 \rangle$

In Exercises 14–17, determine whether $\overrightarrow{AB}$ is equivalent to $\overrightarrow{PQ}$.

14. $\begin{array}{ll} A = (1, 1, 1) & B = (3, 3, 3) \\ P = (1, 4, 5) & Q = (3, 6, 7) \end{array}$

15. $\begin{array}{ll} A = (1, 4, 1) & B = (-2, 2, 0) \\ P = (2, 5, 7) & Q = (-3, 2, 1) \end{array}$

16. $\begin{array}{ll} A = (0, 0, 0) & B = (-4, 2, 3) \\ P = (4, -2, -3) & Q = (0, 0, 0) \end{array}$

17. $\begin{array}{ll} A = (1, 1, 0) & B = (3, 3, 5) \\ P = (2, -9, 7) & Q = (4, -7, 13) \end{array}$

In Exercises 18–23, calculate the linear combinations.

18. $5 \langle 2, 2, -3 \rangle + 3 \langle 1, 7, 2 \rangle$ **19.** $-2 \langle 8, 11, 3 \rangle + 4 \langle 2, 1, 1 \rangle$

20. $6(4\mathbf{j} + 2\mathbf{k}) - 3(2\mathbf{i} + 7\mathbf{k})$ **21.** $\frac{1}{2} \langle 4, -2, 8 \rangle - \frac{1}{3} \langle 12, 3, 3 \rangle$

22. $5(\mathbf{i} + 2\mathbf{j}) - 3(2\mathbf{j} + \mathbf{k}) + 7(2\mathbf{k} - \mathbf{i})$

23. $4 \langle 6, -1, 1 \rangle - 2 \langle 1, 0, -1 \rangle + 3 \langle -2, 1, 1 \rangle$

In Exercises 24–27, determine whether or not the two vectors are parallel.

24. $\mathbf{u} = \langle 1, -2, 5 \rangle , \mathbf{v} = \langle -2, 4, -10 \rangle$

25. $\mathbf{u} = \langle 4, 2, -6 \rangle , \mathbf{v} = \langle 2, -1, 3 \rangle$

26. $\mathbf{u} = \langle 4, 2, -6 \rangle , \mathbf{v} = \langle 2, 1, 3 \rangle$

27. $\mathbf{u} = \langle -3, 1, 4 \rangle , \mathbf{v} = \langle 6, -2, 8 \rangle$

In Exercises 28–31, find the given vector.

28. $\mathbf{e}_{\mathbf{v}}$, where $\mathbf{v} = \langle 1, 1, 2 \rangle$

29. $\mathbf{e}_{\mathbf{w}}$, where $\mathbf{w} = \langle 4, -2, -1 \rangle$

30. Unit vector in the direction of $\mathbf{u} = \langle 1, 0, 7 \rangle$

31. Unit vector in the direction opposite to $\mathbf{v} = \langle -4, 4, 2 \rangle$

32. Sketch the following vectors, and find their components and lengths:

(a) $4\mathbf{i} + 3\mathbf{j} - 2\mathbf{k}$ (b) $\mathbf{i} + \mathbf{j} + \mathbf{k}$
(c) $4\mathbf{j} + 3\mathbf{k}$ (d) $12\mathbf{i} + 8\mathbf{j} - \mathbf{k}$

In Exercises 33–40, find a vector parametrization for the line with the given description.

33. Passes through $P = (1, 2, -8)$, direction vector $\mathbf{v} = \langle 2, 1, 3 \rangle$

34. Passes through $P = (4, 0, 8)$, direction vector $\mathbf{v} = \langle 1, 0, 1 \rangle$

35. Passes through $P = (4, 0, 8)$, direction vector $\mathbf{v} = 7\mathbf{i} + 4\mathbf{k}$

36. Passes through O, direction vector $\mathbf{v} = \langle 3, -1, -4 \rangle$

37. Passes through $(1, 1, 1)$ and $(3, -5, 2)$

38. Passes through $(-2, 0, -2)$ and $(4, 3, 7)$

39. Passes through O and $(4, 1, 1)$

40. Passes through $(1, 1, 1)$ parallel to the line through $(2, 0, -1)$ and $(4, 1, 3)$

In Exercises 41–44, find parametric equations for the lines with the given description.

41. Perpendicular to the xy-plane, passes through the origin

42. Perpendicular to the yz-plane, passes through $(0, 0, 2)$

43. Parallel to the line through $(1, 1, 0)$ and $(0, -1, -2)$, passes through $(0, 0, 4)$

44. Passes through $(1, -1, 0)$ and $(0, -1, 2)$

45. Which of the following is a parametrization of the line through $P = (4, 9, 8)$ perpendicular to the xz-plane (Figure 17)?

(a) $\mathbf{r}(t) = \langle 4, 9, 8 \rangle + t \langle 1, 0, 1 \rangle$ (b) $\mathbf{r}(t) = \langle 4, 9, 8 \rangle + t \langle 0, 0, 1 \rangle$
(c) $\mathbf{r}(t) = \langle 4, 9, 8 \rangle + t \langle 0, 1, 0 \rangle$ (d) $\mathbf{r}(t) = \langle 4, 9, 8 \rangle + t \langle 1, 1, 0 \rangle$

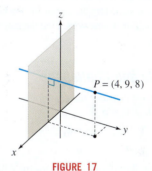

FIGURE 17

46. Find a parametrization of the line through $P = (4, 9, 8)$ perpendicular to the yz-plane.

47. Show that $\mathbf{r}_1(t)$ and $\mathbf{r}_2(t)$ define the same line, where

$$\mathbf{r}_1(t) = \langle 3, -1, 4 \rangle + t \langle 8, 12, -6 \rangle$$

$$\mathbf{r}_2(t) = \langle 11, 11, -2 \rangle + t \langle 4, 6, -3 \rangle$$

Hint: Show that $\mathbf{r}_2(t)$ passes through $(3, -1, 4)$ and that the direction vectors for $\mathbf{r}_1(t)$ and $\mathbf{r}_2(t)$ are parallel.

48. Show that $\mathbf{r}_1(t)$ and $\mathbf{r}_2(t)$ define the same line, where

$$\mathbf{r}_1(t) = t \langle 2, 1, 3 \rangle , \qquad \mathbf{r}_2(t) = \langle -6, -3, -9 \rangle + t \langle 8, 4, 12 \rangle$$

49. Find two different vector parametrizations of the line through $P = (5, 5, 2)$ with direction vector $\mathbf{v} = \langle 0, -2, 1 \rangle$.

50. Find the point of intersection of the lines $\mathbf{r}(t) = \langle 1, 0, 0 \rangle + t \langle -3, 1, 0 \rangle$ and $\mathbf{s}(t) = \langle 0, 1, 1 \rangle + t \langle 2, 0, 1 \rangle$.

51. Show that the lines $\mathbf{r}_1(t) = \langle -1, 2, 2 \rangle + t \langle 4, -2, 1 \rangle$ and $\mathbf{r}_2(t) = \langle 0, 1, 1 \rangle + t \langle 2, 0, 1 \rangle$ do not intersect.

52. Determine whether the lines $\mathbf{r}_1(t) = \langle 2, 1, 1 \rangle + t \langle -4, 0, 1 \rangle$ and $\mathbf{r}_2(s) = \langle -4, 1, 5 \rangle + s \langle 2, 1, -2 \rangle$ intersect, and if so, find the point of intersection.

53. Determine whether the lines $\mathbf{r}_1(t) = \langle 0, 1, 1 \rangle + t \langle 1, 1, 2 \rangle$ and $\mathbf{r}_2(s) = \langle 2, 0, 3 \rangle + s \langle 1, 4, 4 \rangle$ intersect, and if so, find the point of intersection.

54. Find the intersection of the lines $\mathbf{r}_1(t) = \langle -1, 1 \rangle + t \langle 2, 4 \rangle$ and $\mathbf{r}_2(s) = \langle 2, 1 \rangle + s \langle -1, 6 \rangle$ in $\mathbf{R}^2$.

55. A meteor follows a trajectory $\mathbf{r}(t) = \langle 2, 1, 4 \rangle + t \langle 3, 2, -1 \rangle$ km. with t in minutes, near the surface of the earth, which is represented by the xy-plane. Determine at what time the meteor hits the ground.

56. A laser's beam shines along the ray given by $\mathbf{r}_1(t) = \langle 1, 2, 4 \rangle + t \langle 2, 1, -1 \rangle$ for $t \geq 0$. A second laser's beam shines along the ray given by $\mathbf{r}_2(s) = \langle 6, 3, -1 \rangle + s \langle -5, 2, c \rangle$ for $s \geq 0$, where the value of c allows for the adjustment of the z-coordinate of its direction vector. Find the value of c that will make the two beams intersect.

57. Find the components of the vector $\mathbf{v}$ whose tail and head are the midpoints of segments $\overline{AC}$ and $\overline{BC}$ in Figure 18. [Note that the midpoint of (a_1, a_2, a_3) and (b_1, b_2, b_3) is $\left(\frac{a_1+b_1}{2}, \frac{a_2+b_2}{2}, \frac{a_3+b_3}{2} \right)$.]

58. Find the components of the vector $\mathbf{w}$ whose tail is C and head is the midpoint of $\overline{AB}$ in Figure 18.

59. A box that weighs 1000 kg is hanging from a crane at the dock. The crane has a square 20 m by 20 m framework as in Figure 19, with four cables, each of the same length, supporting the box. The box hangs 10 m below the level of the framework. Find the magnitude of the force acting on each cable.

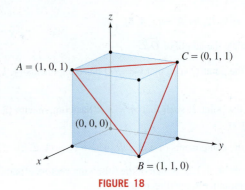

FIGURE 18

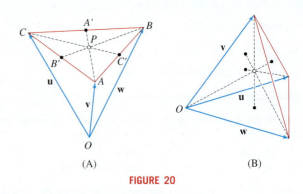

FIGURE 19

Further Insights and Challenges

*In Exercises 60–66, we consider the equations of a line in **symmetric form**, when $a \neq 0$, $b \neq 0$, $c \neq 0$.*

$$\frac{x - x_0}{a} = \frac{y - y_0}{b} = \frac{z - z_0}{c} \qquad \boxed{10}$$

60. Let $\mathcal{L}$ be the line through $P_0 = (x_0, y_0, z_0)$ with direction vector $\mathbf{v} = \langle a, b, c \rangle$. Show that $\mathcal{L}$ is defined by the symmetric equations (10). *Hint:* Use the vector parametrization to show that every point on $\mathcal{L}$ satisfies (10).

61. Find the symmetric equations of the line through $P_0 = (-2, 3, 3)$ with direction vector $\mathbf{v} = \langle 2, 4, 3 \rangle$.

62. Find the symmetric equations of the line through $P = (1, 1, 2)$ and $Q = (-2, 4, 0)$.

63. Find the symmetric equations of the line

$$x = 3 + 2t, \quad y = 4 - 9t, \quad z = 12t$$

64. Find a vector parametrization for the line

$$\frac{x - 5}{9} = \frac{y + 3}{7} = z - 10$$

65. Find a vector parametrization for the line $\frac{x}{2} = \frac{y}{7} = \frac{z}{8}$.

66. Show that the line in the plane through (x_0, y_0) of slope m has symmetric equations

$$x - x_0 = \frac{y - y_0}{m}$$

67. A median of a triangle is a segment joining a vertex to the midpoint of the opposite side. Referring to Figure 20(A), prove that three medians of triangle ABC intersect at the terminal point P of the vector $\frac{1}{3}(\mathbf{u} + \mathbf{v} + \mathbf{w})$. The point P is the *centroid* of the triangle. *Hint:* Show, by parametrizing the segment $\overline{AA'}$, that P lies two-thirds of the way from A to A'. It will follow similarly that P lies on the other two medians.

68. A median of a tetrahedron is a segment joining a vertex to the centroid of the opposite face. The tetrahedron in Figure 20(B) has vertices at the origin and at the terminal points of vectors $\mathbf{u}$, $\mathbf{v}$, and $\mathbf{w}$. Show that the medians intersect at the terminal point of $\frac{1}{4}(\mathbf{u} + \mathbf{v} + \mathbf{w})$.

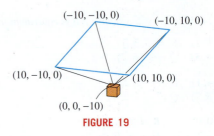

FIGURE 20

12.3 Dot Product and the Angle Between Two Vectors

The dot product is one of the most important vector operations. It plays a role in nearly all aspects of multivariable calculus.

DEFINITION Dot Product   The **dot product** $\mathbf{v} \cdot \mathbf{w}$ of two vectors

$$\mathbf{v} = \langle v_1, v_2, v_3 \rangle, \qquad \mathbf{w} = \langle w_1, w_2, w_3 \rangle$$

is the scalar defined by

$$\boxed{\mathbf{v} \cdot \mathbf{w} = v_1 w_1 + v_2 w_2 + v_3 w_3}$$

Important concepts in mathematics often have multiple names or notations either for historical reasons or because they arise in more than one context. The dot product *is also called the "scalar product" or "inner product," and in many texts,* $\mathbf{v} \cdot \mathbf{w}$ *is denoted* $(\mathbf{v}, \mathbf{w})$ *or* $\langle \mathbf{v}, \mathbf{w} \rangle$.

In words, to compute the dot product, *multiply the corresponding components and add.* For example,

$$\langle 2, 3, 1 \rangle \cdot \langle -4, 2, 5 \rangle = 2(-4) + 3(2) + 1(5) = -8 + 6 + 5 = 3$$

The dot product of vectors $\mathbf{v} = \langle v_1, v_2 \rangle$ and $\mathbf{w} = \langle w_1, w_2 \rangle$ in $\mathbf{R}^2$ is defined similarly:

$$\mathbf{v} \cdot \mathbf{w} = v_1 w_1 + v_2 w_2$$

We will see in a moment that the dot product is closely related to the angle between $\mathbf{v}$ and $\mathbf{w}$. Before getting to this, we describe some elementary properties of dot products.

First, the dot product is *commutative*: $\mathbf{v} \cdot \mathbf{w} = \mathbf{w} \cdot \mathbf{v}$, because the components can be multiplied in either order. Second, the dot product of a vector with itself is the square of the length: If $\mathbf{v} = \langle v_1, v_2, v_3 \rangle$, then

$$\mathbf{v} \cdot \mathbf{v} = v_1 \cdot v_1 + v_2 \cdot v_2 + v_3 \cdot v_3 = v_1^2 + v_2^2 + v_3^2 = \|\mathbf{v}\|^2$$

The dot product appears in a very wide range of applications. To rank how closely a Web document matches a search input at Google,

We take the dot product of the vector of count-weights with the vector of type-weights to compute an IR score for the document.

From "The Anatomy of a Large-Scale Hypertextual Web Search Engine" by Google founders Sergey Brin and Lawrence Page.

The dot product also satisfies the Distributive Law and a scalar property as summarized in the next theorem (see Exercises 88 and 89).

THEOREM 1 Properties of the Dot Product

(i) $\mathbf{0} \cdot \mathbf{v} = \mathbf{v} \cdot \mathbf{0} = 0$

(ii) **Commutativity:** $\mathbf{v} \cdot \mathbf{w} = \mathbf{w} \cdot \mathbf{v}$

(iii) **Pulling out scalars:** $(\lambda \mathbf{v}) \cdot \mathbf{w} = \mathbf{v} \cdot (\lambda \mathbf{w}) = \lambda(\mathbf{v} \cdot \mathbf{w})$

(iv) **Distributive Law:** $\mathbf{u} \cdot (\mathbf{v} + \mathbf{w}) = \mathbf{u} \cdot \mathbf{v} + \mathbf{u} \cdot \mathbf{w}$

$$(\mathbf{v} + \mathbf{w}) \cdot \mathbf{u} = \mathbf{v} \cdot \mathbf{u} + \mathbf{w} \cdot \mathbf{u}$$

(v) **Relation with length:** $\mathbf{v} \cdot \mathbf{v} = \|\mathbf{v}\|^2$

■ **EXAMPLE 1**   Verify the Distributive Law $\mathbf{u} \cdot (\mathbf{v} + \mathbf{w}) = \mathbf{u} \cdot \mathbf{v} + \mathbf{u} \cdot \mathbf{w}$ for

$$\mathbf{u} = \langle 4, 3, 3 \rangle, \qquad \mathbf{v} = \langle 1, 2, 2 \rangle, \qquad \mathbf{w} = \langle 3, -2, 5 \rangle$$

Solution   We compute both sides and check that they are equal:

$$\mathbf{u} \cdot (\mathbf{v} + \mathbf{w}) = \langle 4, 3, 3 \rangle \cdot \big(\langle 1, 2, 2 \rangle + \langle 3, -2, 5 \rangle \big)$$

$$= \langle 4, 3, 3 \rangle \cdot \langle 4, 0, 7 \rangle = 4(4) + 3(0) + 3(7) = 37$$

$$\mathbf{u} \cdot \mathbf{v} + \mathbf{u} \cdot \mathbf{w} = \langle 4, 3, 3 \rangle \cdot \langle 1, 2, 2 \rangle + \langle 4, 3, 3 \rangle \cdot \langle 3, -2, 5 \rangle$$

$$= \big(4(1) + 3(2) + 3(2) \big) + \big(4(3) + 3(-2) + 3(5) \big)$$

$$= 16 + 21 = 37 \qquad \blacksquare$$

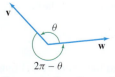

FIGURE 1 By convention, the angle θ between two vectors is chosen so that $0 \leq \theta \leq \pi$.

As mentioned above, the dot product $\mathbf{v} \cdot \mathbf{w}$ is related to the angle θ between $\mathbf{v}$ and $\mathbf{w}$. This angle θ is not uniquely defined because, as we see in Figure 1, both θ and $2\pi - \theta$ can serve as an angle between $\mathbf{v}$ and $\mathbf{w}$. Furthermore, any multiple of 2π may be added

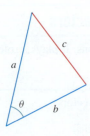

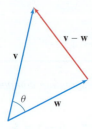

FIGURE 2

to θ. All of these angles have the same cosine, so it does not matter which angle we use in the next theorem. However, we shall adopt the following convention:

> *The angle between two vectors is chosen to satisfy* $0 \leq \theta \leq \pi$.

THEOREM 2 Dot Product and the Angle Let θ be the angle between two nonzero vectors $\mathbf{v}$ and $\mathbf{w}$. Then

$$\mathbf{v} \cdot \mathbf{w} = \|\mathbf{v}\| \, \|\mathbf{w}\| \cos\theta \quad \text{or} \quad \cos\theta = \frac{\mathbf{v} \cdot \mathbf{w}}{\|\mathbf{v}\| \, \|\mathbf{w}\|} \qquad \boxed{1}$$

Proof According to the Law of Cosines, the three sides of a triangle satisfy (Figure 2)

$$c^2 = a^2 + b^2 - 2ab\cos\theta$$

If two sides of the triangle are $\mathbf{v}$ and $\mathbf{w}$, then the third side is $\mathbf{v} - \mathbf{w}$, as in the figure, and the Law of Cosines gives

$$\|\mathbf{v} - \mathbf{w}\|^2 = \|\mathbf{v}\|^2 + \|\mathbf{w}\|^2 - 2\cos\theta \, \|\mathbf{v}\| \, \|\mathbf{w}\| \qquad \boxed{2}$$

Now, by property (v) of Theorem 1 and the Distributive Law,

$$\|\mathbf{v} - \mathbf{w}\|^2 = (\mathbf{v} - \mathbf{w}) \cdot (\mathbf{v} - \mathbf{w}) = \mathbf{v} \cdot \mathbf{v} - 2\mathbf{v} \cdot \mathbf{w} + \mathbf{w} \cdot \mathbf{w}$$

$$= \|\mathbf{v}\|^2 + \|\mathbf{w}\|^2 - 2\mathbf{v} \cdot \mathbf{w} \qquad \boxed{3}$$

Comparing Eq. (2) and Eq. (3), we obtain $-2\cos\theta \, \|\mathbf{v}\| \, \|\mathbf{w}\| = -2\mathbf{v} \cdot \mathbf{w}$, and Eq. (1) follows. ■

By definition of the arccosine, the angle $\theta = \cos^{-1} x$ is the angle in the interval $[0, \pi]$ satisfying $\cos\theta = x$. Thus, for nonzero vectors $\mathbf{v}$ and $\mathbf{w}$, we have

$$\cos\theta = \frac{\mathbf{v} \cdot \mathbf{w}}{\|\mathbf{v}\| \, \|\mathbf{w}\|} \quad \text{or} \quad \theta = \cos^{-1}\left(\frac{\mathbf{v} \cdot \mathbf{w}}{\|\mathbf{v}\| \, \|\mathbf{w}\|}\right)$$

■ **EXAMPLE 2** Find the angle θ between $\mathbf{v} = \langle 3, 6, 2 \rangle$ and $\mathbf{w} = \langle 4, 2, 4 \rangle$.

Solution Compute $\cos\theta$ using the dot product:

$$\|\mathbf{v}\| = \sqrt{3^2 + 6^2 + 2^2} = \sqrt{49} = 7, \qquad \|\mathbf{w}\| = \sqrt{4^2 + 2^2 + 4^2} = \sqrt{36} = 6$$

$$\cos\theta = \frac{\mathbf{v} \cdot \mathbf{w}}{\|\mathbf{v}\| \|\mathbf{w}\|} = \frac{\langle 3, 6, 2 \rangle \cdot \langle 4, 2, 4 \rangle}{7 \cdot 6} = \frac{3 \cdot 4 + 6 \cdot 2 + 2 \cdot 4}{42} = \frac{32}{42} = \frac{16}{21}$$

The angle itself is $\theta = \cos^{-1}\left(\frac{16}{21}\right) \approx 0.705$ radians or rad (Figure 3). ■

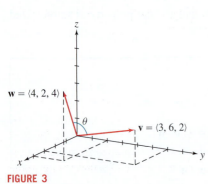

FIGURE 3

The terms "orthogonal" and "perpendicular" are synonymous and are used interchangeably, although we usually use "orthogonal" when dealing with vectors.

Two nonzero vectors $\mathbf{v}$ and $\mathbf{w}$ are called **perpendicular** or **orthogonal** if the angle between them is $\frac{\pi}{2}$. In this case, we write $\mathbf{v} \perp \mathbf{w}$.

We can use the dot product to test whether $\mathbf{v}$ and $\mathbf{w}$ are orthogonal. Because an angle θ between 0 and π satisfies $\cos\theta = 0$ if and only if $\theta = \frac{\pi}{2}$, we see that

$$\mathbf{v} \cdot \mathbf{w} = \|\mathbf{v}\| \, \|\mathbf{w}\| \cos\theta = 0 \quad \Leftrightarrow \quad \theta = \frac{\pi}{2}$$

Defining the zero vector to be orthogonal to all other vectors, we then have

$$\boxed{\mathbf{v} \perp \mathbf{w} \quad \text{if and only if} \quad \mathbf{v} \cdot \mathbf{w} = 0}$$

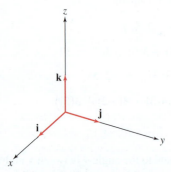

FIGURE 4 The standard basis vectors are mutually orthogonal and have length 1.

The standard basis vectors are mutually orthogonal and have length 1 (Figure 4). In terms of dot products, because $\mathbf{i} = \langle 1, 0, 0 \rangle$, $\mathbf{j} = \langle 0, 1, 0 \rangle$, and $\mathbf{k} = \langle 0, 0, 1 \rangle$,

$$\mathbf{i} \cdot \mathbf{j} = \mathbf{i} \cdot \mathbf{k} = \mathbf{j} \cdot \mathbf{k} = 0, \qquad \mathbf{i} \cdot \mathbf{i} = \mathbf{j} \cdot \mathbf{j} = \mathbf{k} \cdot \mathbf{k} = 1$$

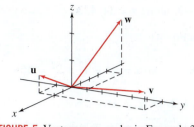

FIGURE 5 Vectors **v**, **w**, and **u** in Example 3.

■ **EXAMPLE 3** **Testing for Orthogonality** Determine whether **v** = ⟨2, 6, 1⟩ is orthogonal to **u** = ⟨2, −1, 1⟩ or **w** = ⟨−4, 1, 2⟩.

Solution We test for orthogonality by computing the dot products (Figure 5):

$$\mathbf{v} \cdot \mathbf{u} = \langle 2, 6, 1 \rangle \cdot \langle 2, -1, 1 \rangle = 2(2) + 6(-1) + 1(1) = -1 \quad \text{(not orthogonal)}$$

$$\mathbf{v} \cdot \mathbf{w} = \langle 2, 6, 1 \rangle \cdot \langle -4, 1, 2 \rangle = 2(-4) + 6(1) + 1(2) = 0 \quad \text{(orthogonal)} \quad ■$$

■ **EXAMPLE 4** **Testing for Obtuseness** Determine whether the angles between the vector **v** = ⟨3, 1, −2⟩ and the vectors **u** = ⟨$\frac{1}{2}, \frac{1}{2}, 5$⟩ and **w** = ⟨4, −3, 0⟩ are obtuse.

Solution By definition, the angle θ between **v** and **u** is obtuse if $\frac{\pi}{2} < \theta \leq \pi$, and this is the case if $\cos \theta < 0$. Since $\mathbf{v} \cdot \mathbf{u} = \|\mathbf{v}\| \|\mathbf{u}\| \cos \theta$ and the lengths $\|\mathbf{v}\|$ and $\|\mathbf{u}\|$ are positive, we see that $\cos \theta$ is negative if and only if $\mathbf{v} \cdot \mathbf{u}$ is negative. Thus,

> The angle θ between **v** and **u** is obtuse if $\mathbf{v} \cdot \mathbf{u} < 0$.

We have

$$\mathbf{v} \cdot \mathbf{u} = \langle 3, 1, -2 \rangle \cdot \left\langle \frac{1}{2}, \frac{1}{2}, 5 \right\rangle = \frac{3}{2} + \frac{1}{2} - 10 = -8 < 0 \quad \text{(angle is obtuse)}$$

$$\mathbf{v} \cdot \mathbf{w} = \langle 3, 1, -2 \rangle \cdot \langle 4, -3, 0 \rangle = 12 - 3 + 0 = 9 > 0 \quad \text{(angle is acute)} \quad ■$$

■ **EXAMPLE 5** **Finding an Orthogonal Vector** Find a vector orthogonal to **w** = ⟨1, 3, 2⟩.

Solution Let **v** = ⟨a, b, c⟩. Then if **v** is orthogonal to **w**, it must be that

$$\mathbf{w} \cdot \mathbf{v} = a + 3b + 2c = 0.$$

So, any a, b, c satisfying this equation such that not all are 0 will do the trick. For instance, we could take $a = 1$ and $b = 1$. Then we have $1 + 3(1) + 2c = 0$ and hence $c = -2$. Thus, **v** = ⟨1, 1, −2⟩ is such an orthogonal vector. Note that the collection of all such orthogonal vectors, when chosen with their basepoint at the origin, together forms a plane through the origin. That plane has equation $x + 3y + 2z = 0$ and it is the unique plane through the origin that is perpendicular to the original vector **w**. We will discuss planes in 3-space further in Section 12.5. ■

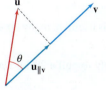

DF **FIGURE 6** The projection **u**$_{\|\mathbf{v}}$ of **u** along **v** has length $\|\mathbf{u}\| \cos \theta$.

Another important use of the dot product is in finding the **projection** **u**$_{\|\mathbf{v}}$ of a vector **u** along a nonzero vector **v**. By definition, **u**$_{\|\mathbf{v}}$ is the vector parallel to **v** obtained by dropping a perpendicular from **u** to the line through **v** as in Figures 6 and 7. We think of **u**$_{\|\mathbf{v}}$ as that part of **u** that is parallel to **v**. It is the shadow vector obtained from **u** when we shine a light down on **v**. In the next theorem, recall that **e**$_\mathbf{v}$ represents a unit vector in the direction of **v**.

> **THEOREM 3** **Projection** Assume $\mathbf{v} \neq \mathbf{0}$. The **projection** of **u** along **v** is the vector
>
> $$\mathbf{u}_{\|\mathbf{v}} = \left(\frac{\mathbf{u} \cdot \mathbf{v}}{\mathbf{v} \cdot \mathbf{v}} \right) \mathbf{v} = \left(\frac{\mathbf{u} \cdot \mathbf{v}}{\|\mathbf{v}\|^2} \right) \mathbf{v} = \left(\frac{\mathbf{u} \cdot \mathbf{v}}{\|\mathbf{v}\|} \right) \mathbf{e}_\mathbf{v} \qquad \boxed{4}$$
>
> This is sometimes denoted proj$_\mathbf{v}\mathbf{u}$.
> The scalar $\dfrac{\mathbf{u} \cdot \mathbf{v}}{\|\mathbf{v}\|}$ is called the **component** or the **scalar component** of **u** along **v** and is sometimes denoted comp$_\mathbf{v}\mathbf{u}$.

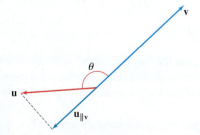

DF **FIGURE 7** When θ is obtuse, **u**$_{\|\mathbf{v}}$ and **v** point in opposite directions.

Proof Referring to Figures 6 and 7, we see by trigonometry that **u**$_{\|\mathbf{v}}$ has length $\|\mathbf{u}\| |\cos \theta|$. If θ is acute, then **u**$_{\|\mathbf{v}}$ is a positive multiple of **v**, and thus, **u**$_{\|\mathbf{v}}$ = $(\|\mathbf{u}\| \cos \theta) \mathbf{e}_\mathbf{v}$ since $\cos \theta > 0$. Similarly, if θ is obtuse, then **u**$_{\|\mathbf{v}}$ is a negative multiple of **e**$_\mathbf{v}$ and again **u**$_{\|\mathbf{v}}$ = $(\|\mathbf{u}\| \cos \theta) \mathbf{e}_\mathbf{v}$ since $\cos \theta < 0$. The first formula for **u**$_{\|\mathbf{v}}$ now follows because

$$\mathbf{u}_{\|\mathbf{v}} = (\|\mathbf{u}\| \cos \theta) \mathbf{e}_\mathbf{v} = \|\mathbf{u}\| \cos \theta \, \frac{\mathbf{v}}{\|\mathbf{v}\|} = \frac{\|\mathbf{u}\| \|\mathbf{v}\| \cos \theta}{\|\mathbf{v}\|^2} \mathbf{v} = \left(\frac{\mathbf{u} \cdot \mathbf{v}}{\mathbf{v} \cdot \mathbf{v}} \right) \mathbf{v} \qquad ■$$

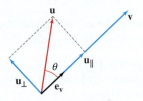

DF FIGURE 8 Decomposition of **u** as a sum $\mathbf{u} = \mathbf{u}_{||\mathbf{v}} + \mathbf{u}_{\perp\mathbf{v}}$ of vectors parallel and orthogonal to **v**.

■ **EXAMPLE 6** Find the projection of $\mathbf{u} = \langle 5, 1, -3 \rangle$ along $\mathbf{v} = \langle 4, 4, 2 \rangle$.

Solution It is convenient to use the first formula in Eq. (4):

$$\mathbf{u} \cdot \mathbf{v} = \langle 5, 1, -3 \rangle \cdot \langle 4, 4, 2 \rangle = 20 + 4 - 6 = 18, \qquad \mathbf{v} \cdot \mathbf{v} = 4^2 + 4^2 + 2^2 = 36$$

$$\mathbf{u}_{||\mathbf{v}} = \left(\frac{\mathbf{u} \cdot \mathbf{v}}{\mathbf{v} \cdot \mathbf{v}} \right) \mathbf{v} = \left(\frac{18}{36} \right) \langle 4, 4, 2 \rangle = \langle 2, 2, 1 \rangle \qquad ■$$

We show now that if $\mathbf{v} \neq \mathbf{0}$, then every vector **u** can be written as the sum of the projection $\mathbf{u}_{||\mathbf{v}}$ and a vector $\mathbf{u}_{\perp\mathbf{v}}$ that is orthogonal to **v** (see Figure 8). In fact, if we set

$$\mathbf{u}_{\perp\mathbf{v}} = \mathbf{u} - \mathbf{u}_{||\mathbf{v}}$$

then we have

$$\boxed{\mathbf{u} = \mathbf{u}_{||\mathbf{v}} + \mathbf{u}_{\perp\mathbf{v}}} \qquad \boxed{5}$$

Eq. (5) is called the **decomposition** of **u** with respect to **v**. It expresses **u** as a sum of vectors, one parallel to **v** and one perpendicular to **v**. We must verify, however, that $\mathbf{u}_{\perp\mathbf{v}}$ is perpendicular to **v**. We do this by showing that the dot product is zero:

$$\mathbf{u}_{\perp\mathbf{v}} \cdot \mathbf{v} = (\mathbf{u} - \mathbf{u}_{||\mathbf{v}}) \cdot \mathbf{v} = \left(\mathbf{u} - \left(\frac{\mathbf{u} \cdot \mathbf{v}}{\mathbf{v} \cdot \mathbf{v}} \right) \mathbf{v} \right) \cdot \mathbf{v} = \mathbf{u} \cdot \mathbf{v} - \left(\frac{\mathbf{u} \cdot \mathbf{v}}{\mathbf{v} \cdot \mathbf{v}} \right) (\mathbf{v} \cdot \mathbf{v}) = 0$$

■ **EXAMPLE 7** Find the decomposition of $\mathbf{u} = \langle 5, 1, -3 \rangle$ with respect to $\mathbf{v} = \langle 4, 4, 2 \rangle$.

Solution In Example 6, we showed that $\mathbf{u}_{||\mathbf{v}} = \langle 2, 2, 1 \rangle$. The orthogonal vector is

$$\mathbf{u}_{\perp\mathbf{v}} = \mathbf{u} - \mathbf{u}_{||\mathbf{v}} = \langle 5, 1, -3 \rangle - \langle 2, 2, 1 \rangle = \langle 3, -1, -4 \rangle$$

The decomposition of **u** with respect to **v** is

$$\mathbf{u} = \langle 5, 1, -3 \rangle = \mathbf{u}_{||\mathbf{v}} + \mathbf{u}_{\perp\mathbf{v}} = \underbrace{\langle 2, 2, 1 \rangle}_{\text{Projection along } \mathbf{v}} + \underbrace{\langle 3, -1, -4 \rangle}_{\text{Orthogonal to } \mathbf{v}} \qquad ■$$

The decomposition into parallel and orthogonal vectors is useful in many applications.

■ **EXAMPLE 8** What is the minimum force you must apply to pull a 20-kg wagon up a frictionless ramp inclined at an angle $\theta = 15°$?

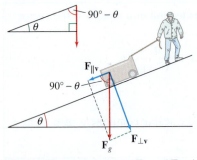

FIGURE 9 The angle between $\mathbf{F}_g$ and $\mathbf{F}_{||\mathbf{v}}$ is $90° - \theta$.

Solution Let **v** be a vector in the direction of the ramp, and let $\mathbf{F}_g$ be the force on the wagon due to gravity. It has magnitude $20g$ newtons or N with $g = 9.8$. Referring to Figure 9, we decompose $\mathbf{F}_g$ as a sum

$$\mathbf{F}_g = \mathbf{F}_{||\mathbf{v}} + \mathbf{F}_{\perp\mathbf{v}}$$

where $\mathbf{F}_{||\mathbf{v}}$ is the projection along the ramp and $\mathbf{F}_{\perp\mathbf{v}}$ is the "normal force" orthogonal to the ramp. The normal force $\mathbf{F}_{\perp\mathbf{v}}$ is canceled by the ramp pushing back against the wagon in the normal direction, and thus (because there is no friction), you need only pull against $\mathbf{F}_{||\mathbf{v}}$.

Notice that the angle between $\mathbf{F}_g$ and the ramp is the complementary angle $90° - \theta$. Since $\mathbf{F}_{||\mathbf{v}}$ is parallel to the ramp, the angle between $\mathbf{F}_g$ and $\mathbf{F}_{||\mathbf{v}}$ is also $90° - \theta$, or $75°$, and

$$\|\mathbf{F}_{||\mathbf{v}}\| = \|\mathbf{F}_g\| \cos(75°) \approx 20(9.8)(0.26) \approx 51 \text{ newtons}$$

Since gravity pulls the wagon down the ramp with a 51-newton force, it takes a minimum force of 51 newtons to pull the wagon up the ramp. ■

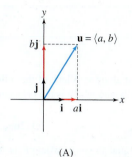

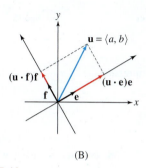

(A)

(B)

FIGURE 10

GRAPHICAL INSIGHT It seems that we are using the term "component" in two ways. We say that a vector $\mathbf{u} = \langle a, b \rangle$ has components a and b. On the other hand, $\mathbf{u} \cdot \mathbf{e}$ is called the component of $\mathbf{u}$ along the unit vector $\mathbf{e}$.

In fact, these two notions of component are the same. The components a and b are the dot products of $\mathbf{u}$ with the standard unit vectors:

$$\mathbf{u} \cdot \mathbf{i} = \langle a, b \rangle \cdot \langle 1, 0 \rangle = a$$

$$\mathbf{u} \cdot \mathbf{j} = \langle a, b \rangle \cdot \langle 0, 1 \rangle = b$$

and we have the decomposition [Figure 10(A)]

$$\mathbf{u} = a\mathbf{i} + b\mathbf{j}$$

But any two orthogonal unit vectors $\mathbf{e}$ and $\mathbf{f}$ give rise to a rotated coordinate system, and we see in Figure 10(B) that

$$\mathbf{u} = (\mathbf{u} \cdot \mathbf{e})\mathbf{e} + (\mathbf{u} \cdot \mathbf{f})\mathbf{f}$$

In other words, $\mathbf{u} \cdot \mathbf{e}$ and $\mathbf{u} \cdot \mathbf{f}$ really are the components when we express $\mathbf{u}$ relative to the rotated system.

12.3 SUMMARY

- The *dot product* of $\mathbf{v} = \langle a_1, b_1, c_1 \rangle$ and $\mathbf{w} = \langle a_2, b_2, c_2 \rangle$ is

$$\mathbf{v} \cdot \mathbf{w} = a_1 a_2 + b_1 b_2 + c_1 c_2$$

- Basic Properties:

  - Commutativity: $\mathbf{v} \cdot \mathbf{w} = \mathbf{w} \cdot \mathbf{v}$
  - Pulling out scalars: $(\lambda \mathbf{v}) \cdot \mathbf{w} = \mathbf{v} \cdot (\lambda \mathbf{w}) = \lambda(\mathbf{v} \cdot \mathbf{w})$
  - Distributive Law: $\mathbf{u} \cdot (\mathbf{v} + \mathbf{w}) = \mathbf{u} \cdot \mathbf{v} + \mathbf{u} \cdot \mathbf{w}$
    $$(\mathbf{v} + \mathbf{w}) \cdot \mathbf{u} = \mathbf{v} \cdot \mathbf{u} + \mathbf{w} \cdot \mathbf{u}$$

  - $\boxed{\mathbf{v} \cdot \mathbf{v} = \|\mathbf{v}\|^2}$

  - $\boxed{\mathbf{v} \cdot \mathbf{w} = \|\mathbf{v}\| \|\mathbf{w}\| \cos\theta}$,   where θ is the angle between $\mathbf{v}$ and $\mathbf{w}$

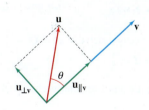

FIGURE 11

- By convention, the angle θ is chosen to satisfy $0 \le \theta \le \pi$.
- Test for orthogonality: $\mathbf{v} \perp \mathbf{w}$ if and only if $\mathbf{v} \cdot \mathbf{w} = 0$.
- The angle between $\mathbf{v}$ and $\mathbf{w}$ is acute if $\mathbf{v} \cdot \mathbf{w} > 0$ and obtuse if $\mathbf{v} \cdot \mathbf{w} < 0$.
- Assume $\mathbf{v} \ne \mathbf{0}$. Every vector $\mathbf{u}$ has a decomposition $\mathbf{u} = \mathbf{u}_{\|\mathbf{v}} + \mathbf{u}_{\perp\mathbf{v}}$, where $\mathbf{u}_{\|\mathbf{v}}$ is parallel to $\mathbf{v}$, and $\mathbf{u}_{\perp\mathbf{v}}$ is orthogonal to $\mathbf{v}$ (see Figure 11). The vector $\mathbf{u}_{\|\mathbf{v}}$ is called the *projection* of $\mathbf{u}$ along $\mathbf{v}$.
- Let $\mathbf{e}_\mathbf{v} = \dfrac{\mathbf{v}}{\|\mathbf{v}\|}$. Then

$$\boxed{\mathbf{u}_{\|\mathbf{v}} = \left(\frac{\mathbf{u} \cdot \mathbf{v}}{\mathbf{v} \cdot \mathbf{v}}\right)\mathbf{v} = \left(\frac{\mathbf{u} \cdot \mathbf{v}}{\|\mathbf{v}\|^2}\right)\mathbf{v} = \left(\frac{\mathbf{u} \cdot \mathbf{v}}{\|\mathbf{v}\|}\right)\mathbf{e}_\mathbf{v}, \qquad \mathbf{u}_{\perp\mathbf{v}} = \mathbf{u} - \mathbf{u}_{\|\mathbf{v}}}$$

- The coefficient $\dfrac{\mathbf{u} \cdot \mathbf{v}}{\|\mathbf{v}\|}$ is called the *component* of $\mathbf{u}$ along $\mathbf{v}$:

$$\boxed{\text{Component of } \mathbf{u} \text{ along } \mathbf{v} = \frac{\mathbf{u} \cdot \mathbf{v}}{\|\mathbf{v}\|} = \|\mathbf{u}\| \cos\theta}$$

12.3 EXERCISES

Preliminary Questions

1. Is the dot product of two vectors a scalar or a vector?

2. What can you say about the angle between **a** and **b** if $\mathbf{a} \cdot \mathbf{b} < 0$?

3. Which property of dot products allows us to conclude that if **v** is orthogonal to both **u** and **w**, then **v** is orthogonal to $\mathbf{u} + \mathbf{w}$?

4. Which is the projection of **v** along **v**: (a) **v** or (b) $\mathbf{e_v}$?

5. Let $\mathbf{u}_{||\mathbf{v}}$ be the projection of **u** along **v**. Which of the following is the projection **u** along the vector $2\mathbf{v}$ and which is the projection of $2\mathbf{u}$ along **v**?

(a) $\frac{1}{2}\mathbf{u}_{||\mathbf{v}}$ (b) $\mathbf{u}_{||\mathbf{v}}$ (c) $2\mathbf{u}_{||\mathbf{v}}$

6. Which of the following is equal to $\cos\theta$, where θ is the angle between **u** and **v**?

(a) $\mathbf{u} \cdot \mathbf{v}$ (b) $\mathbf{u} \cdot \mathbf{e_v}$ (c) $\mathbf{e_u} \cdot \mathbf{e_v}$

Exercises

In Exercises 1–12, compute the dot product.

1. $\langle 1, 2, 1 \rangle \cdot \langle 4, 3, 5 \rangle$

2. $\langle 3, -2, 2 \rangle \cdot \langle 1, 0, 1 \rangle$

3. $\langle 0, 1, 0 \rangle \cdot \langle 7, 41, -3 \rangle$

4. $\langle 1, 1, 1 \rangle \cdot \langle 6, 4, 2 \rangle$

5. $\langle 3, 1 \rangle \cdot \langle 4, -7 \rangle$

6. $\left\langle \frac{1}{6}, \frac{1}{2} \right\rangle \cdot \left\langle 3, \frac{1}{2} \right\rangle$

7. $\mathbf{k} \cdot \mathbf{j}$

8. $\mathbf{k} \cdot \mathbf{k}$

9. $(\mathbf{i} + \mathbf{j}) \cdot (\mathbf{j} + \mathbf{k})$

10. $(3\mathbf{j} + 2\mathbf{k}) \cdot (\mathbf{i} - 4\mathbf{k})$

11. $(\mathbf{i} + \mathbf{j} + \mathbf{k}) \cdot (3\mathbf{i} + 2\mathbf{j} - 5\mathbf{k})$

12. $(-\mathbf{k}) \cdot (\mathbf{i} - 2\mathbf{j} + 7\mathbf{k})$

In Exercises 13–18, determine whether the two vectors are orthogonal and, if not, whether the angle between them is acute or obtuse.

13. $\langle 1, 1, 1 \rangle$, $\langle 1, -2, -2 \rangle$

14. $\langle 0, 2, 4 \rangle$, $\langle -5, 0, 0 \rangle$

15. $\langle 1, 2, 1 \rangle$, $\langle 7, -3, -1 \rangle$

16. $\langle 0, 2, 4 \rangle$, $\langle 3, 1, 0 \rangle$

17. $\left\langle \frac{12}{5}, -\frac{4}{5} \right\rangle$, $\left\langle \frac{1}{2}, -\frac{7}{4} \right\rangle$

18. $\langle 12, 6 \rangle$, $\langle 2, -4 \rangle$

In Exercises 19–22, find the cosine of the angle between the vectors.

19. $\langle 0, 3, 1 \rangle$, $\langle 4, 0, 0 \rangle$

20. $\langle 1, 1, 1 \rangle$, $\langle 2, -1, 2 \rangle$

21. $\mathbf{i} + \mathbf{j}$, $\mathbf{j} + 2\mathbf{k}$

22. $3\mathbf{i} + \mathbf{k}$, $\mathbf{i} + \mathbf{j} + \mathbf{k}$

In Exercises 23–28, find the angle between the vectors. Use a calculator if necessary.

23. $\left\langle 2, \sqrt{2} \right\rangle$, $\left\langle 1 + \sqrt{2}, 1 - \sqrt{2} \right\rangle$

24. $\left\langle 5, \sqrt{3} \right\rangle$, $\left\langle \sqrt{3}, 2 \right\rangle$

25. $\langle 1, 1, 1 \rangle$, $\langle 1, 0, 1 \rangle$

26. $\langle 3, 1, 1 \rangle$, $\langle 2, -4, 2 \rangle$

27. $\langle 0, 1, 1 \rangle$, $\langle 1, -1, 0 \rangle$

28. $\langle 1, 1, -1 \rangle$, $\langle 1, -2, -1 \rangle$

29. Find all values of b for which the vectors are orthogonal.

(a) $\langle b, 3, 2 \rangle$, $\langle 1, b, 1 \rangle$ (b) $\langle 4, -2, 7 \rangle$, $\left\langle b^2, b, 0 \right\rangle$

30. Find a vector that is orthogonal to $\langle -1, 2, 2 \rangle$.

31. Find two vectors that are not multiples of each other and are both orthogonal to $\langle 2, 0, -3 \rangle$.

32. Find a vector that is orthogonal to $\mathbf{v} = \langle 1, 2, 1 \rangle$ but not to $\mathbf{w} = \langle 1, 0, -1 \rangle$.

33. Find $\mathbf{v} \cdot \mathbf{e}$ where $\|\mathbf{v}\| = 3$, **e** is a unit vector, and the angle between **e** and **v** is $\frac{2\pi}{3}$.

34. Assume that **v** lies in the yz-plane. Which of the following dot products is equal to zero for all choices of **v**?

(a) $\mathbf{v} \cdot \langle 0, 2, 1 \rangle$ (b) $\mathbf{v} \cdot \mathbf{k}$

(c) $\mathbf{v} \cdot \langle -3, 0, 0 \rangle$ (d) $\mathbf{v} \cdot \mathbf{j}$

In Exercises 35–38, simplify the expression.

35. $(\mathbf{v} - \mathbf{w}) \cdot \mathbf{v} + \mathbf{v} \cdot \mathbf{w}$

36. $(\mathbf{v} + \mathbf{w}) \cdot (\mathbf{v} + \mathbf{w}) - 2\mathbf{v} \cdot \mathbf{w}$

37. $(\mathbf{v} + \mathbf{w}) \cdot \mathbf{v} - (\mathbf{v} + \mathbf{w}) \cdot \mathbf{w}$

38. $(\mathbf{v} + \mathbf{w}) \cdot \mathbf{v} - (\mathbf{v} - \mathbf{w}) \cdot \mathbf{w}$

In Exercises 39–42, use the properties of the dot product to evaluate the expression, assuming that $\mathbf{u} \cdot \mathbf{v} = 2$, $\|\mathbf{u}\| = 1$, and $\|\mathbf{v}\| = 3$.

39. $\mathbf{u} \cdot (4\mathbf{v})$

40. $(\mathbf{u} + \mathbf{v}) \cdot \mathbf{v}$

41. $2\mathbf{u} \cdot (3\mathbf{u} - \mathbf{v})$

42. $(\mathbf{u} + \mathbf{v}) \cdot (\mathbf{u} - \mathbf{v})$

43. Find the angle between **v** and **w** if $\mathbf{v} \cdot \mathbf{w} = -\|\mathbf{v}\| \, \|\mathbf{w}\|$.

44. Find the angle between **v** and **w** if $\mathbf{v} \cdot \mathbf{w} = \frac{1}{2}\|\mathbf{v}\| \, \|\mathbf{w}\|$.

45. Assume that $\|\mathbf{v}\| = 3$, $\|\mathbf{w}\| = 5$, and the angle between **v** and **w** is $\theta = \frac{\pi}{3}$.

(a) Use the relation $\|\mathbf{v} + \mathbf{w}\|^2 = (\mathbf{v} + \mathbf{w}) \cdot (\mathbf{v} + \mathbf{w})$ to show that $\|\mathbf{v} + \mathbf{w}\|^2 = 3^2 + 5^2 + 2\mathbf{v} \cdot \mathbf{w}$.

(b) Find $\|\mathbf{v} + \mathbf{w}\|$.

46. Assume that $\|\mathbf{v}\| = 2$, $\|\mathbf{w}\| = 3$, and the angle between **v** and **w** is $120°$. Determine:

(a) $\mathbf{v} \cdot \mathbf{w}$ (b) $\|2\mathbf{v} + \mathbf{w}\|$ (c) $\|2\mathbf{v} - 3\mathbf{w}\|$

47. Show that if **e** and **f** are unit vectors such that $\|\mathbf{e} + \mathbf{f}\| = \frac{3}{2}$, then $\|\mathbf{e} - \mathbf{f}\| = \frac{\sqrt{7}}{2}$. *Hint:* Show that $\mathbf{e} \cdot \mathbf{f} = \frac{1}{8}$.

48. Find $\|2\mathbf{e} - 3\mathbf{f}\|$, assuming that **e** and **f** are unit vectors such that $\|\mathbf{e} + \mathbf{f}\| = \sqrt{3/2}$.

49. Find the angle θ in the triangle in Figure 12.

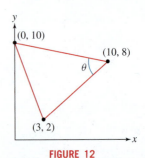

FIGURE 12

50. Find all three angles in the triangle in Figure 13.

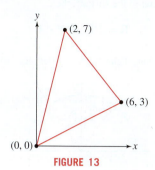

FIGURE 13

51. (a) Draw $\mathbf{u}_{\|\mathbf{v}}$ and $\mathbf{v}_{\|\mathbf{u}}$ for the vectors appearing as in Figure 14.
(b) Which of $\mathbf{u}_{\|\mathbf{v}}$ and $\mathbf{v}_{\|\mathbf{u}}$ has the greater magnitude?

FIGURE 14

52. Let $\mathbf{u}$ and $\mathbf{v}$ be two nonzero vectors.
(a) Is it possible for the component of $\mathbf{u}$ along $\mathbf{v}$ to have the opposite sign from the component of $\mathbf{v}$ along $\mathbf{u}$? Why or why not?
(b) What must be true of the vectors if either of these two components is 0?

In Exercises 53–60, find the projection of $\mathbf{u}$ along $\mathbf{v}$.

53. $\mathbf{u} = \langle 2, 5 \rangle, \quad \mathbf{v} = \langle 1, 1 \rangle$ **54.** $\mathbf{u} = \langle 2, -3 \rangle, \quad \mathbf{v} = \langle 1, 2 \rangle$

55. $\mathbf{u} = \langle -1, 2, 0 \rangle, \quad \mathbf{v} = \langle 2, 0, 1 \rangle$

56. $\mathbf{u} = \langle 1, 1, 1 \rangle, \quad \mathbf{v} = \langle 1, 1, 0 \rangle$

57. $\mathbf{u} = 5\mathbf{i} + 7\mathbf{j} - 4\mathbf{k}, \quad \mathbf{v} = \mathbf{k}$ **58.** $\mathbf{u} = \mathbf{i} + 29\mathbf{k}, \quad \mathbf{v} = \mathbf{j}$

59. $\mathbf{u} = \langle a, b, c \rangle, \quad \mathbf{v} = \mathbf{i}$ **60.** $\mathbf{u} = \langle a, a, b \rangle, \quad \mathbf{v} = \mathbf{i} - \mathbf{j}$

In Exercises 61 and 62, compute the component of $\mathbf{u}$ along $\mathbf{v}$.

61. $\mathbf{u} = \langle 3, 2, 1 \rangle, \quad \mathbf{v} = \langle 1, 0, 1 \rangle$

62. $\mathbf{u} = \langle 3, 0, 9 \rangle, \quad \mathbf{v} = \langle 1, 2, 2 \rangle$

63. Find the length of $\overline{OP}$ in Figure 15.

64. Find $\|\mathbf{u}_{\perp \mathbf{v}}\|$ in Figure 15.

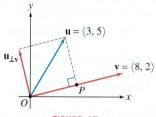

FIGURE 15

In Exercises 65–70, find the decomposition $\mathbf{a} = \mathbf{a}_{\|\mathbf{b}} + \mathbf{a}_{\perp \mathbf{b}}$ with respect to $\mathbf{b}$.

65. $\mathbf{a} = \langle 1, 0 \rangle, \quad \mathbf{b} = \langle 1, 1 \rangle$ **66.** $\mathbf{a} = \langle 2, -3 \rangle, \quad \mathbf{b} = \langle 5, 0 \rangle$

67. $\mathbf{a} = \langle 4, -1, 0 \rangle, \quad \mathbf{b} = \langle 0, 1, 1 \rangle$

68. $\mathbf{a} = \langle 4, -1, 5 \rangle, \quad \mathbf{b} = \langle 2, 1, 1 \rangle$

69. $\mathbf{a} = \langle x, y \rangle, \quad \mathbf{b} = \langle 1, -1 \rangle$

70. $\mathbf{a} = \langle x, y, z \rangle, \quad \mathbf{b} = \langle 1, 1, 1 \rangle$

71. Let $\mathbf{e}_\theta = \langle \cos\theta, \sin\theta \rangle$. Show that $\mathbf{e}_\theta \cdot \mathbf{e}_\psi = \cos(\theta - \psi)$ for any two angles θ and ψ.

72.   Let $\mathbf{v}$ and $\mathbf{w}$ be vectors in the plane.
(a) Use Theorem 2 to explain why the dot product $\mathbf{v} \cdot \mathbf{w}$ does not change if both $\mathbf{v}$ and $\mathbf{w}$ are rotated by the same angle θ.
(b) Sketch the vectors $\mathbf{e}_1 = \langle 1, 0 \rangle$ and $\mathbf{e}_2 = \left\langle \frac{\sqrt{2}}{2}, \frac{\sqrt{2}}{2} \right\rangle$, and determine the vectors $\mathbf{e}_1'$, $\mathbf{e}_2'$ obtained by rotating $\mathbf{e}_1$, $\mathbf{e}_2$ through an angle $\frac{\pi}{4}$. Verify that $\mathbf{e}_1 \cdot \mathbf{e}_2 = \mathbf{e}_1' \cdot \mathbf{e}_2'$.

In Exercises 73–76, refer to Figure 16.

73. Find the angle between $\overline{AB}$ and $\overline{AC}$.

74. Find the angle between $\overline{AB}$ and $\overline{AD}$.

75. Calculate the projection of $\overrightarrow{AC}$ along $\overrightarrow{AD}$.

76. Calculate the projection of $\overrightarrow{AD}$ along $\overrightarrow{AB}$.

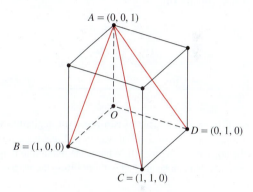

FIGURE 16 Unit cube in $\mathbf{R}^3$.

77. The methane molecule CH_4 consists of a carbon molecule bonded to four hydrogen molecules that are spaced as far apart from each other as possible. The hydrogen atoms then sit at the vertices of a tetrahedron, with the carbon atom at its center, as in Figure 17. We can model this with the carbon atom at the point $(\frac{1}{2}, \frac{1}{2}, \frac{1}{2})$ and the hydrogen atoms at $(0, 0, 0)$, $(1, 1, 0)$, $(1, 0, 1)$, and $(0, 1, 1)$. Use the dot product to find the bond angle α formed between any two of the line segments from the carbon atom to the hydrogen atoms.

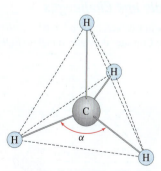

FIGURE 17 A methane molecule.

78. Iron forms a crystal lattice where each central atom appears at the center of a cube, the corners of which correspond to additional iron atoms, as in Figure 18. Use the dot product to find the angle β between the line segments from the central atom to two adjacent outer atoms. *Hint: Take the central atom to be situated at the origin and the corner atoms to occur at $(\pm 1, \pm 1, \pm 1)$.*

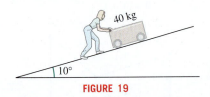

FIGURE 18 An iron crystal.

79. [✎]   Let **v** and **w** be nonzero vectors and set $\mathbf{u} = \mathbf{e_v} + \mathbf{e_w}$. Use the dot product to show that the angle between **u** and **v** is equal to the angle between **u** and **w**. Explain this result geometrically with a diagram.

80. [✎]   Let **v**, **w**, and **a** be nonzero vectors such that $\mathbf{v} \cdot \mathbf{a} = \mathbf{w} \cdot \mathbf{a}$. Is it true that $\mathbf{v} = \mathbf{w}$? Either prove this or give a counterexample.

81. Calculate the force (in newtons) required to push a 40-kg wagon up a 10° incline (Figure 19).

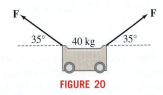

FIGURE 19

82. A force **F** is applied to each of two ropes (of negligible weight) attached to opposite ends of a 40-kg wagon and making an angle of 35° with the horizontal (Figure 20). What is the maximum magnitude of **F** (in newtons) that can be applied without lifting the wagon off the ground?

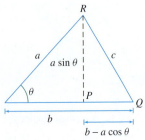

FIGURE 20

83. A light beam travels along the ray determined by a unit vector **L**, strikes a flat surface at point P, and is reflected along the ray determined

by a unit vector **R**, where $\theta_1 = \theta_2$ (Figure 21). Show that if **N** is the unit vector orthogonal to the surface, then

$$\mathbf{R} = 2(\mathbf{L} \cdot \mathbf{N})\mathbf{N} - \mathbf{L}$$

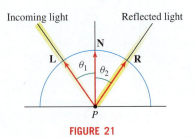

FIGURE 21

84. Let P and Q be antipodal (opposite) points on a sphere of radius r centered at the origin and let R be a third point on the sphere (Figure 22). Prove that $\overline{PR}$ and $\overline{QR}$ are orthogonal.

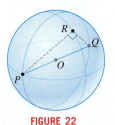

FIGURE 22

85. Prove that $\|\mathbf{v} + \mathbf{w}\|^2 - \|\mathbf{v} - \mathbf{w}\|^2 = 4\mathbf{v} \cdot \mathbf{w}$.

86. Use Exercise 85 to show that **v** and **w** are orthogonal if and only if $\|\mathbf{v} - \mathbf{w}\| = \|\mathbf{v} + \mathbf{w}\|$.

87. Show that the two diagonals of a parallelogram are perpendicular if and only if its sides have equal length. *Hint:* Use Exercise 86 to show that $\mathbf{v} - \mathbf{w}$ and $\mathbf{v} + \mathbf{w}$ are orthogonal if and only if $\|\mathbf{v}\| = \|\mathbf{w}\|$.

88. Verify the Distributive Law:

$$\mathbf{u} \cdot (\mathbf{v} + \mathbf{w}) = \mathbf{u} \cdot \mathbf{v} + \mathbf{u} \cdot \mathbf{w}$$

89. Verify that $(\lambda\mathbf{v}) \cdot \mathbf{w} = \lambda(\mathbf{v} \cdot \mathbf{w})$ for any scalar λ.

Further Insights and Challenges

90. Prove the Law of Cosines, $c^2 = a^2 + b^2 - 2ab\cos\theta$, by referring to Figure 23. *Hint:* Consider the right triangle $\triangle PQR$.

FIGURE 23

91. In this exercise, we prove the Cauchy–Schwarz inequality: If **v** and **w** are any two vectors, then

$$|\mathbf{v} \cdot \mathbf{w}| \leq \|\mathbf{v}\| \, \|\mathbf{w}\| \qquad \boxed{6}$$

(a) Let $f(x) = \|x\mathbf{v} + \mathbf{w}\|^2$ for x a scalar. Show that $f(x) = ax^2 + bx + c$, where $a = \|\mathbf{v}\|^2$, $b = 2\mathbf{v} \cdot \mathbf{w}$, and $c = \|\mathbf{w}\|^2$.

(b) Conclude that $b^2 - 4ac \leq 0$. *Hint:* Observe that $f(x) \geq 0$ for all x.

92. Use (6) to prove the Triangle Inequality:

$$\|\mathbf{v} + \mathbf{w}\| \leq \|\mathbf{v}\| + \|\mathbf{w}\|$$

Hint: First use the Triangle Inequality for numbers to prove

$$|(\mathbf{v} + \mathbf{w}) \cdot (\mathbf{v} + \mathbf{w})| \leq |(\mathbf{v} + \mathbf{w}) \cdot \mathbf{v}| + |(\mathbf{v} + \mathbf{w}) \cdot \mathbf{w}|$$

93. This exercise gives another proof of the relation between the dot product and the angle θ between two vectors $\mathbf{v} = \langle a_1, b_1 \rangle$ and $\mathbf{w} = \langle a_2, b_2 \rangle$ in the plane. Observe that $\mathbf{v} = \|\mathbf{v}\| \langle \cos\theta_1, \sin\theta_1 \rangle$ and $\mathbf{w} = \|\mathbf{w}\| \langle \cos\theta_2, \sin\theta_2 \rangle$, with θ_1 and θ_2 as in Figure 24. Then use the addition formula for the cosine to show that

$$\mathbf{v} \cdot \mathbf{w} = \|\mathbf{v}\| \, \|\mathbf{w}\| \cos\theta$$

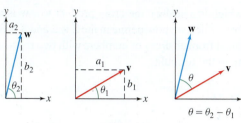

FIGURE 24

94. Let $\mathbf{v} = \langle x, y \rangle$ and

$$\mathbf{v}_\theta = \langle x \cos\theta + y \sin\theta, -x \sin\theta + y \cos\theta \rangle$$

Prove that the angle between $\mathbf{v}$ and $\mathbf{v}_\theta$ is θ.

95. Let $\mathbf{v}$ be a nonzero vector. The angles α, β, γ between $\mathbf{v}$ and the unit vectors $\mathbf{i}, \mathbf{j}, \mathbf{k}$ are called the direction angles of $\mathbf{v}$ (Figure 25). The cosines of these angles are called the **direction cosines** of $\mathbf{v}$. Prove that

$$\cos^2\alpha + \cos^2\beta + \cos^2\gamma = 1$$

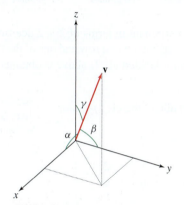

FIGURE 25 Direction angles of $\mathbf{v}$.

96. Find the direction cosines of $\mathbf{v} = \langle 3, 6, -2 \rangle$.

97. The set of all points $X = (x, y, z)$ equidistant from two points P, Q in $\mathbf{R}^3$ is a plane (Figure 26). Show that X lies on this plane if

$$\overrightarrow{PQ} \cdot \overrightarrow{OX} = \frac{1}{2}\left(\|\overrightarrow{OQ}\|^2 - \|\overrightarrow{OP}\|^2\right) \qquad \boxed{7}$$

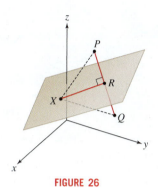

FIGURE 26

Hint: If R is the midpoint of $\overline{PQ}$, then X is equidistant from P and Q if and only if $\overrightarrow{XR}$ is orthogonal to $\overrightarrow{PQ}$.

98. Sketch the plane consisting of all points $X = (x, y, z)$ equidistant from the points $P = (0, 1, 0)$ and $Q = (0, 0, 1)$. Use Eq. (7) to show that X lies on this plane if and only if $y = z$.

99. Use Eq. (7) to find the equation of the plane consisting of all points $X = (x, y, z)$ equidistant from $P = (2, 1, 1)$ and $Q = (1, 0, 2)$.

12.4 The Cross Product

This section introduces the **cross product** $\mathbf{v} \times \mathbf{w}$ of two 3-dimensional vectors $\mathbf{v}$ and $\mathbf{w}$. The cross product (sometimes called the **vector product**) is used in physics and engineering to describe quantities involving rotation, such as torque and angular momentum. In electromagnetic theory, magnetic forces are described using cross products (Figures 1 and 2).

FIGURE 1 The spiral paths of charged particles in a bubble chamber in the presence of a magnetic field are described using cross products. (*Science Source*)

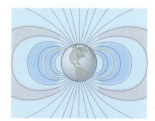

FIGURE 2 The Van Allen radiation belts, located thousands of miles above the earth's surface, are made up of streams of protons and electrons that oscillate back and forth in helical paths between two "magnetic mirrors" set up by the earth's magnetic field. This helical motion is explained by the "cross-product" nature of magnetic forces.

Unlike the dot product $\mathbf{v} \cdot \mathbf{w}$ (which is a scalar), the cross product $\mathbf{v} \times \mathbf{w}$ is again a vector. It is defined using determinants, which we now define in the 2×2 and 3×3 cases. A 2×2 determinant is a number formed from an array of numbers with two rows and two columns (called a **matrix**) according to the formula

$$\begin{vmatrix} a & b \\ c & d \end{vmatrix} = ad - bc \qquad \boxed{1}$$

Note that the determinant is the difference of the diagonal products. For example,

$$\begin{vmatrix} 3 & 2 \\ \frac{1}{2} & 4 \end{vmatrix} = \begin{vmatrix} 3 & 2 \\ \frac{1}{2} & 4 \end{vmatrix} - \begin{vmatrix} 3 & 2 \\ \frac{1}{2} & 4 \end{vmatrix} = 3 \cdot 4 - 2 \cdot \frac{1}{2} = 11$$

The determinant of a 3×3 matrix is defined by the formula

$$\begin{vmatrix} a_1 & b_1 & c_1 \\ a_2 & b_2 & c_2 \\ a_3 & b_3 & c_3 \end{vmatrix} = a_1 \underbrace{\begin{vmatrix} b_2 & c_2 \\ b_3 & c_3 \end{vmatrix}}_{(1,1)\text{-minor}} - b_1 \underbrace{\begin{vmatrix} a_2 & c_2 \\ a_3 & c_3 \end{vmatrix}}_{(1,2)\text{-minor}} + c_1 \underbrace{\begin{vmatrix} a_2 & b_2 \\ a_3 & b_3 \end{vmatrix}}_{(1,3)\text{-minor}} \qquad \boxed{2}$$

This formula expresses the 3×3 determinant in terms of 2×2 determinants called **minors**. The minors are obtained by crossing out the first row and one of the three columns of the 3×3 matrix. For example, the minor labeled $(1, 2)$ above is obtained as follows:

$$\begin{vmatrix} a_1 & b_1 & c_1 \\ a_2 & b_2 & c_2 \\ a_3 & b_3 & c_3 \end{vmatrix} \qquad \text{to obtain the } (1, 2)\text{-minor} \qquad \begin{vmatrix} a_2 & c_2 \\ a_3 & c_3 \end{vmatrix}$$

Cross out row 1 and column 2 $(1,2)$-minor

The theory of matrices and determinants is part of linear algebra, a subject of great importance throughout mathematics. In this section, we discuss just a few basic definitions and facts needed for our treatment of multivariable calculus.

■ **EXAMPLE 1   A 3×3 Determinant**   Calculate $\begin{vmatrix} 2 & 4 & 3 \\ 0 & 1 & -7 \\ -1 & 5 & 3 \end{vmatrix}$.

Solution

$$\begin{vmatrix} 2 & 4 & 3 \\ 0 & 1 & -7 \\ -1 & 5 & 3 \end{vmatrix} = 2 \begin{vmatrix} 1 & -7 \\ 5 & 3 \end{vmatrix} - 4 \begin{vmatrix} 0 & -7 \\ -1 & 3 \end{vmatrix} + 3 \begin{vmatrix} 0 & 1 \\ -1 & 5 \end{vmatrix}$$

$$= 2(38) - 4(-7) + 3(1) = 107 \qquad ■$$

Later in this section we will see how determinants are related to area and volume. First, we introduce the cross product, which is defined as a "symbolic" determinant whose first row has the vector entries $\mathbf{i}, \mathbf{j}, \mathbf{k}$.

CAUTION Note in Eq. (3) that the middle term comes with a minus sign.

DEFINITION   The Cross Product   The cross product of vectors $\mathbf{v} = \langle v_1, v_2, v_3 \rangle$ and $\mathbf{w} = \langle w_1, w_2, w_3 \rangle$ is the vector

$$\mathbf{v} \times \mathbf{w} = \begin{vmatrix} \mathbf{i} & \mathbf{j} & \mathbf{k} \\ v_1 & v_2 & v_3 \\ w_1 & w_2 & w_3 \end{vmatrix} = \begin{vmatrix} v_2 & v_3 \\ w_2 & w_3 \end{vmatrix} \mathbf{i} - \begin{vmatrix} v_1 & v_3 \\ w_1 & w_3 \end{vmatrix} \mathbf{j} + \begin{vmatrix} v_1 & v_2 \\ w_1 & w_2 \end{vmatrix} \mathbf{k} \qquad \boxed{3}$$

The cross product differs fundamentally from the dot product in that $\mathbf{u} \times \mathbf{v}$ is a vector, whereas $\mathbf{u} \cdot \mathbf{v}$ is a number.

■ **EXAMPLE 2**   Calculate $\mathbf{v} \times \mathbf{w}$, where $\mathbf{v} = \langle -2, 1, 4 \rangle$ and $\mathbf{w} = \langle 3, 2, 5 \rangle$.

Solution

$$\mathbf{v} \times \mathbf{w} = \begin{vmatrix} \mathbf{i} & \mathbf{j} & \mathbf{k} \\ -2 & 1 & 4 \\ 3 & 2 & 5 \end{vmatrix} = \begin{vmatrix} 1 & 4 \\ 2 & 5 \end{vmatrix} \mathbf{i} - \begin{vmatrix} -2 & 4 \\ 3 & 5 \end{vmatrix} \mathbf{j} + \begin{vmatrix} -2 & 1 \\ 3 & 2 \end{vmatrix} \mathbf{k}$$

$$= (-3)\mathbf{i} - (-22)\mathbf{j} + (-7)\mathbf{k} = \langle -3, 22, -7 \rangle \qquad ■$$

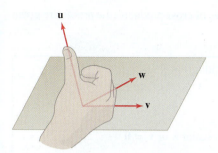

FIGURE 3 $\{\mathbf{v}, \mathbf{w}, \mathbf{u}\}$ forms a right-handed system.

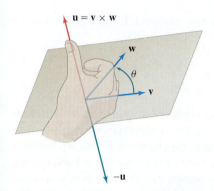

FIGURE 4 There are two vectors orthogonal to $\mathbf{v}$ and $\mathbf{w}$ with length $\|\mathbf{v}\| \, \|\mathbf{w}\| \sin \theta$. The right-hand rule determines which is $\mathbf{v} \times \mathbf{w}$.

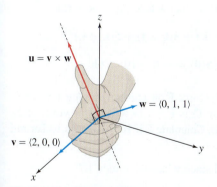

FIGURE 5 The direction of $\mathbf{u} = \mathbf{v} \times \mathbf{w}$ is determined by the right-hand rule. Thus, $\mathbf{u}$ has a positive z-component.

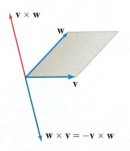

FIGURE 6

Formula (3) gives no hint of the geometric meaning of the cross product. However, there is a simple way to visualize the vector $\mathbf{v} \times \mathbf{w}$ using the **right-hand rule**. Suppose that $\mathbf{v}$, $\mathbf{w}$, and $\mathbf{u}$ are nonzero vectors that do not all lie in a plane. We say that $\{\mathbf{v}, \mathbf{w}, \mathbf{u}\}$ forms a **right-handed system** if the direction of $\mathbf{u}$ is determined by the right-hand rule: *When the fingers of your right hand curl from $\mathbf{v}$ to $\mathbf{w}$, your thumb points to the same side of the plane spanned by $\mathbf{v}$ and $\mathbf{w}$ as $\mathbf{u}$* (Figure 3). The following theorem is proved at the end of this section.

THEOREM 1 **Geometric Description of the Cross Product**   Given two nonzero non-parallel vectors $\mathbf{v}$ and $\mathbf{w}$, the cross product $\mathbf{v} \times \mathbf{w}$ is the unique vector with the following three properties:

   (i) $\mathbf{v} \times \mathbf{w}$ is orthogonal to $\mathbf{v}$ and $\mathbf{w}$.
   (ii) $\mathbf{v} \times \mathbf{w}$ has length $\|\mathbf{v}\| \, \|\mathbf{w}\| \sin \theta$ (θ is the angle between $\mathbf{v}$ and $\mathbf{w}$, $0 \le \theta \le \pi$).
   (iii) $\{\mathbf{v}, \mathbf{w}, \mathbf{v} \times \mathbf{w}\}$ forms a right-handed system.

How do the three properties in Theorem 1 determine $\mathbf{v} \times \mathbf{w}$? By property (i), $\mathbf{v} \times \mathbf{w}$ lies on the line orthogonal to $\mathbf{v}$ and $\mathbf{w}$. By property (ii), $\mathbf{v} \times \mathbf{w}$ is one of the two vectors on this line of length $\|\mathbf{v}\| \, \|\mathbf{w}\| \sin \theta$. Finally, property (iii) tells us which of these two vectors is $\mathbf{v} \times \mathbf{w}$—namely, the vector for which $\{\mathbf{v}, \mathbf{w}, \mathbf{u}\}$ is right-handed (Figure 4).

■ **EXAMPLE 3** Let $\mathbf{v} = \langle 2, 0, 0 \rangle$ and $\mathbf{w} = \langle 0, 1, 1 \rangle$. Determine $\mathbf{u} = \mathbf{v} \times \mathbf{w}$ using the geometric properties of the cross product rather than Eq. (3).

Solution We use Theorem 1. First, by property (i), $\mathbf{u} = \mathbf{v} \times \mathbf{w}$ is orthogonal to $\mathbf{v}$ and $\mathbf{w}$. Since $\mathbf{v}$ lies along the x-axis, $\mathbf{u}$ must lie in the yz-plane (Figure 5). In other words, $\mathbf{u} = \langle 0, b, c \rangle$. But $\mathbf{u}$ is also orthogonal to $\mathbf{w} = \langle 0, 1, 1 \rangle$, so $\mathbf{u} \cdot \mathbf{w} = b + c = 0$ and thus $\mathbf{u} = \langle 0, b, -b \rangle$.

Next, direct computation shows that $\|\mathbf{v}\| = 2$ and $\|\mathbf{w}\| = \sqrt{2}$. Furthermore, the angle between $\mathbf{v}$ and $\mathbf{w}$ is $\theta = \frac{\pi}{2}$ since $\mathbf{v} \cdot \mathbf{w} = 0$. By property (ii),

$$\|\mathbf{u}\| = \sqrt{b^2 + (-b)^2} = |b|\sqrt{2} \quad \text{is equal to} \quad \|\mathbf{v}\| \, \|\mathbf{w}\| \sin \frac{\pi}{2} = 2\sqrt{2}.$$

Therefore, $|b| = 2$ and $b = \pm 2$. Finally, property (iii) tells us that $\mathbf{u}$ points in the positive z-direction (Figure 5). Thus, $b = -2$ and $\mathbf{u} = \langle 0, -2, 2 \rangle$. You can verify that the formula for the cross product yields the same answer. ■

One of the most striking properties of the cross product is that it is *anticommutative*. Reversing the order changes the sign:

$$\boxed{\mathbf{w} \times \mathbf{v} = -\mathbf{v} \times \mathbf{w}} \qquad \boxed{4}$$

We verify this using Eq. (3): When we interchange $\mathbf{v}$ and $\mathbf{w}$, each of the 2×2 determinants changes sign. For example,

$$\begin{vmatrix} v_1 & v_2 \\ w_1 & w_2 \end{vmatrix} = v_1 w_2 - v_2 w_1$$

$$= -(v_2 w_1 - v_1 w_2) = -\begin{vmatrix} w_1 & w_2 \\ v_1 & v_2 \end{vmatrix}$$

Anticommutativity also follows from the geometric description of the cross product. By properties (i) and (ii) in Theorem 1, $\mathbf{v} \times \mathbf{w}$ and $\mathbf{w} \times \mathbf{v}$ are both orthogonal to $\mathbf{v}$ and $\mathbf{w}$ and have the same length. However, $\mathbf{v} \times \mathbf{w}$ and $\mathbf{w} \times \mathbf{v}$ point in opposite directions by the right-hand rule, and thus $\mathbf{v} \times \mathbf{w} = -\mathbf{w} \times \mathbf{v}$ (Figure 6). In particular, $\mathbf{v} \times \mathbf{v} = -\mathbf{v} \times \mathbf{v}$ and hence $\mathbf{v} \times \mathbf{v} = \mathbf{0}$.

The next theorem lists some further properties of cross products (the proofs are given as Exercises 49–52).

Note an important distinction between the dot product and cross product of a vector with itself:

$$\mathbf{v} \times \mathbf{v} = \mathbf{0}$$

$$\mathbf{v} \cdot \mathbf{v} = \|\mathbf{v}\|^2$$

THEOREM 2   Basic Properties of the Cross Product

(i) $\mathbf{w} \times \mathbf{v} = -\mathbf{v} \times \mathbf{w}$

(ii) $\mathbf{v} \times \mathbf{v} = \mathbf{0}$

(iii) $\mathbf{v} \times \mathbf{w} = \mathbf{0}$ if and only if $\mathbf{w} = \lambda\mathbf{v}$ for some scalar λ or $\mathbf{v} = \mathbf{0}$

(iv) $(\lambda\mathbf{v}) \times \mathbf{w} = \mathbf{v} \times (\lambda\mathbf{w}) = \lambda(\mathbf{v} \times \mathbf{w})$

(v) $(\mathbf{u} + \mathbf{v}) \times \mathbf{w} = \mathbf{u} \times \mathbf{w} + \mathbf{v} \times \mathbf{w}$

$\quad\ \mathbf{u} \times (\mathbf{v} + \mathbf{w}) = \mathbf{u} \times \mathbf{v} + \mathbf{u} \times \mathbf{w}$

The cross product of any two of the standard basis vectors $\mathbf{i}$, $\mathbf{j}$, and $\mathbf{k}$ is equal to the third, possibly with a minus sign. More precisely (see Exercise 53),

$$\mathbf{i} \times \mathbf{j} = \mathbf{k}, \qquad \mathbf{j} \times \mathbf{k} = \mathbf{i}, \qquad \mathbf{k} \times \mathbf{i} = \mathbf{j} \qquad \boxed{5}$$

$$\mathbf{i} \times \mathbf{i} = \mathbf{j} \times \mathbf{j} = \mathbf{k} \times \mathbf{k} = \mathbf{0}$$

Since the cross product is anticommutative, minus signs occur when the cross products are taken in the opposite order. An easy way to remember these relations is to draw $\mathbf{i}$, $\mathbf{j}$, and $\mathbf{k}$ in a circle as in Figure 7. Go around the circle in the clockwise direction (starting at any point) and you obtain one of the relations (5). For example, starting at $\mathbf{i}$ and moving clockwise yield $\mathbf{i} \times \mathbf{j} = \mathbf{k}$. If you go around in the counterclockwise direction, you obtain the relations with a minus sign. Thus, starting at $\mathbf{k}$ and going counterclockwise give the relation $\mathbf{k} \times \mathbf{j} = -\mathbf{i}$.

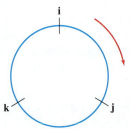

FIGURE 7 Circle for computing the cross products of the basis vectors.

■ **EXAMPLE 4   Using the ijk Relations**   Compute $(2\mathbf{i} + \mathbf{k}) \times (3\mathbf{j} + 5\mathbf{k})$.

Solution   We use the Distributive Law for cross products:

$$(2\mathbf{i} + \mathbf{k}) \times (3\mathbf{j} + 5\mathbf{k}) = (2\mathbf{i}) \times (3\mathbf{j}) + (2\mathbf{i}) \times (5\mathbf{k}) + \mathbf{k} \times (3\mathbf{j}) + \mathbf{k} \times (5\mathbf{k})$$

$$= 6(\mathbf{i} \times \mathbf{j}) + 10(\mathbf{i} \times \mathbf{k}) + 3(\mathbf{k} \times \mathbf{j}) + 5(\mathbf{k} \times \mathbf{k})$$

$$= 6\mathbf{k} - 10\mathbf{j} - 3\mathbf{i} + 5(\mathbf{0}) = -3\mathbf{i} - 10\mathbf{j} + 6\mathbf{k} \qquad ■$$

■ **EXAMPLE 5   Velocity in a Magnetic Field**   The force $\mathbf{F}$ on a proton moving at velocity $\mathbf{v}$ m/s in a uniform magnetic field $\mathbf{B}$ (in teslas or T) is $\mathbf{F} = q(\mathbf{v} \times \mathbf{B})$ in newtons or N, where $q = 1.6 \times 10^{-19}$ coulombs or C (Figure 8). Calculate $\mathbf{F}$ if $\mathbf{B} = 0.0004\mathbf{k}$ teslas and $\mathbf{v}$ has magnitude 10^6 m/s in the direction $-\mathbf{j} + \mathbf{k}$.

Solution   The vector $-\mathbf{j} + \mathbf{k}$ has length $\sqrt{2}$, and since $\mathbf{v}$ has magnitude 10^6,

$$\mathbf{v} = 10^6 \left(\frac{-\mathbf{j} + \mathbf{k}}{\sqrt{2}} \right)$$

Therefore, the force (in newtons) is

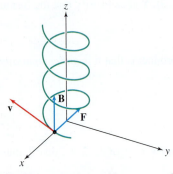

FIGURE 8 A proton in a uniform magnetic field travels in a helical path.

$$\mathbf{F} = q(\mathbf{v} \times \mathbf{B}) = 10^6 q \left(\frac{-\mathbf{j} + \mathbf{k}}{\sqrt{2}} \right) \times (0.0004\mathbf{k}) = \frac{400q}{\sqrt{2}} \left((-\mathbf{j} + \mathbf{k}) \times \mathbf{k} \right)$$

$$= -\frac{400q}{\sqrt{2}} \mathbf{i} = \frac{-400(1.6 \times 10^{-19})}{\sqrt{2}} \mathbf{i} \approx -(4.5 \times 10^{-17})\mathbf{i} \qquad ■$$

Cross Products, Area, and Volume

Cross products and determinants are closely related to area and volume. Consider the parallelogram $\mathcal{P}$ spanned by nonzero vectors $\mathbf{v}$ and $\mathbf{w}$ with a common basepoint. In Figure 9(A), we see that $\mathcal{P}$ has base $b = \|\mathbf{v}\|$ and height $h = \|\mathbf{w}\| \sin \theta$, where θ is the angle between $\mathbf{v}$ and $\mathbf{w}$. Therefore, $\mathcal{P}$ has area $A = bh = \|\mathbf{v}\| \|\mathbf{w}\| \sin \theta = \|\mathbf{v} \times \mathbf{w}\|$.

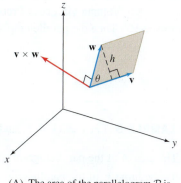

(A) The area of the parallelogram $\mathcal{P}$ is
$\|\mathbf{v} \times \mathbf{w}\| = \|\mathbf{v}\| \, \|\mathbf{w}\| \sin\theta$.

(B) The area of the triangle T is $\|\mathbf{v} \times \mathbf{w}\|/2$

DF **FIGURE 9**

Notice also, as in Figure 9(B), that we also know the area of the triangle $\mathcal{T}$ spanned by nonzero vectors $\mathbf{v}$ and $\mathbf{w}$ is exactly half the area of the parallelogram. Thus, we have the following:

Areas   If $\mathcal{P}$ is the parallelogram spanned by $\mathbf{v}$ and $\mathbf{w}$, and $\mathcal{T}$ is the triangle spanned by $\mathbf{v}$ and $\mathbf{w}$, then

$$\text{Area}(\mathcal{P}) = \|\mathbf{v} \times \mathbf{w}\| \qquad \boxed{6}$$

and

$$\text{Area}(\mathcal{T}) = \frac{\|\mathbf{v} \times \mathbf{w}\|}{2} \qquad \boxed{7}$$

A "parallelepiped" is the solid spanned by three vectors. Each face is a parallelogram.

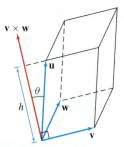

FIGURE 10 The volume of the parallelepiped is $|\mathbf{u} \cdot (\mathbf{v} \times \mathbf{w})|$.

We use the following notation for the determinant of the matrix whose rows are the vectors $\mathbf{v}$, $\mathbf{w}$, $\mathbf{u}$:

$$\det\begin{pmatrix} \mathbf{u} \\ \mathbf{v} \\ \mathbf{w} \end{pmatrix} = \begin{vmatrix} u_1 & u_2 & u_3 \\ v_1 & v_2 & v_3 \\ w_1 & w_2 & w_3 \end{vmatrix}$$

It is awkward to write the absolute value of a determinant in the notation on the right, but we may denote it as

$$\left| \det\begin{pmatrix} \mathbf{u} \\ \mathbf{v} \\ \mathbf{w} \end{pmatrix} \right|$$

Next, consider the **parallelepiped P** spanned by three nonzero vectors $\mathbf{u}$, $\mathbf{v}$, $\mathbf{w}$ in $\mathbf{R}^3$ (the three-dimensional prism in Figure 10). The base of $\mathbf{P}$ is the parallelogram spanned by $\mathbf{v}$ and $\mathbf{w}$, so the area of the base is $\|\mathbf{v} \times \mathbf{w}\|$. The height of $\mathbf{P}$ is $h = \|\mathbf{u}\| \cdot |\cos\theta|$, where θ is the angle between $\mathbf{u}$ and $\mathbf{v} \times \mathbf{w}$. Therefore,

$$\text{Volume of } \mathbf{P} = (\text{area of base})(\text{height}) = \|\mathbf{v} \times \mathbf{w}\| \cdot \|\mathbf{u}\| \cdot |\cos\theta|$$

But, $\|\mathbf{v} \times \mathbf{w}\| \, \|\mathbf{u}\| \cos\theta$ is equal to the dot product of $\mathbf{v} \times \mathbf{w}$ and $\mathbf{u}$. This proves the formula

$$\text{Volume of } \mathbf{P} = |\mathbf{u} \cdot (\mathbf{v} \times \mathbf{w})|$$

The quantity $\mathbf{u} \cdot (\mathbf{v} \times \mathbf{w})$, called the **vector triple product**, can be expressed as a determinant. Let

$$\mathbf{u} = \langle u_1, u_2, u_3 \rangle, \qquad \mathbf{v} = \langle v_1, v_2, v_3 \rangle, \qquad \mathbf{w} = \langle w_1, w_2, w_3 \rangle$$

Then

$$\mathbf{u} \cdot (\mathbf{v} \times \mathbf{w}) = \mathbf{u} \cdot \left(\begin{vmatrix} v_2 & v_3 \\ w_2 & w_3 \end{vmatrix} \mathbf{i} - \begin{vmatrix} v_1 & v_3 \\ w_1 & w_3 \end{vmatrix} \mathbf{j} + \begin{vmatrix} v_1 & v_2 \\ w_1 & w_2 \end{vmatrix} \mathbf{k} \right)$$

$$= u_1 \begin{vmatrix} v_2 & v_3 \\ w_2 & w_3 \end{vmatrix} - u_2 \begin{vmatrix} v_1 & v_3 \\ w_1 & w_3 \end{vmatrix} + u_3 \begin{vmatrix} v_1 & v_2 \\ w_1 & w_2 \end{vmatrix}$$

$$= \begin{vmatrix} u_1 & u_2 & u_3 \\ v_1 & v_2 & v_3 \\ w_1 & w_2 & w_3 \end{vmatrix} = \det\begin{pmatrix} \mathbf{u} \\ \mathbf{v} \\ \mathbf{w} \end{pmatrix} \qquad \boxed{8}$$

We obtain the following formulas for area and volume.

THEOREM 3 Volume via Cross Products and Determinants Let $\mathbf{u}, \mathbf{v}, \mathbf{w}$ be nonzero vectors in $\mathbf{R}^3$. Then the parallelepiped $\mathbf{P}$ spanned by $\mathbf{u}$, $\mathbf{v}$, and $\mathbf{w}$ has volume

$$V = |\mathbf{u} \cdot (\mathbf{v} \times \mathbf{w})| = \left| \det \begin{pmatrix} \mathbf{u} \\ \mathbf{v} \\ \mathbf{w} \end{pmatrix} \right| \qquad \boxed{9}$$

■ **EXAMPLE 6** Let $\mathbf{v} = \langle 1, 4, 5 \rangle$ and $\mathbf{w} = \langle -2, -1, 2 \rangle$. Calculate:

(a) The area A of the parallelogram spanned by $\mathbf{v}$ and $\mathbf{w}$

(b) The volume V of the parallelepiped in Figure 11

Solution We compute the cross product and apply Theorem 3:

$$\mathbf{v} \times \mathbf{w} = \begin{vmatrix} 4 & 5 \\ -1 & 2 \end{vmatrix} \mathbf{i} - \begin{vmatrix} 1 & 5 \\ -2 & 2 \end{vmatrix} \mathbf{j} + \begin{vmatrix} 1 & 4 \\ -2 & -1 \end{vmatrix} \mathbf{k} = \langle 13, -12, 7 \rangle$$

(a) The area of the parallelogram spanned by $\mathbf{v}$ and $\mathbf{w}$ is

$$A = \|\mathbf{v} \times \mathbf{w}\| = \sqrt{13^2 + (-12)^2 + 7^2} = \sqrt{362} \approx 19$$

(b) The vertical leg of the parallelepiped is the vector $6\mathbf{k}$, so by Eq. (9),

$$V = |(6\mathbf{k}) \cdot (\mathbf{v} \times \mathbf{w})| = |\langle 0, 0, 6 \rangle \cdot \langle 13, -12, 7 \rangle| = 6(7) = 42 \qquad ■$$

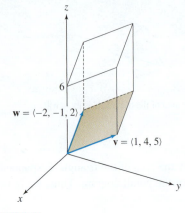

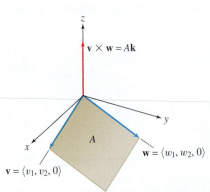

FIGURE 11

We can compute the area A of the parallelogram spanned by vectors $\mathbf{v} = \langle v_1, v_2 \rangle$ and $\mathbf{w} = \langle w_1, w_2 \rangle$ by regarding $\mathbf{v}$ and $\mathbf{w}$ as vectors in $\mathbf{R}^3$ with a zero component in the z-direction (Figure 12). Thus, we write $\mathbf{v} = \langle v_1, v_2, 0 \rangle$ and $\mathbf{w} = \langle w_1, w_2, 0 \rangle$. The cross product $\mathbf{v} \times \mathbf{w}$ is a vector pointing in the z-direction:

$$\mathbf{v} \times \mathbf{w} = \begin{vmatrix} \mathbf{i} & \mathbf{j} & \mathbf{k} \\ v_1 & v_2 & 0 \\ w_1 & w_2 & 0 \end{vmatrix} = \begin{vmatrix} v_2 & 0 \\ w_2 & 0 \end{vmatrix} \mathbf{i} - \begin{vmatrix} v_1 & 0 \\ w_1 & 0 \end{vmatrix} \mathbf{j} + \begin{vmatrix} v_1 & v_2 \\ w_1 & w_2 \end{vmatrix} \mathbf{k} = \begin{vmatrix} v_1 & v_2 \\ w_1 & w_2 \end{vmatrix} \mathbf{k}$$

FIGURE 12 Parallelogram spanned by $\mathbf{v}$ and $\mathbf{w}$ in the xy-plane.

By Theorem 3, the parallelogram spanned by $\mathbf{v}$ and $\mathbf{w}$ has area $A = \|\mathbf{v} \times \mathbf{w}\|$, and thus,

$$A = \left| \det \begin{pmatrix} \mathbf{v} \\ \mathbf{w} \end{pmatrix} \right| = \left| \det \begin{pmatrix} v_1 & v_2 \\ w_1 & w_2 \end{pmatrix} \right| = |v_1 w_2 - v_2 w_1| \qquad \boxed{10}$$

■ **EXAMPLE 7** Compute the area A of the parallelogram in Figure 13.

Solution We have $\begin{vmatrix} \mathbf{v} \\ \mathbf{w} \end{vmatrix} = \begin{vmatrix} 1 & 4 \\ 3 & 2 \end{vmatrix} = 1 \cdot 2 - 3 \cdot 4 = -10$. The area is the absolute value $A = |-10| = 10$. ■

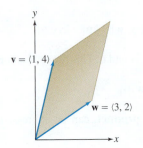

FIGURE 13

Proofs of Cross-Product Properties

We now derive the properties of the cross product listed in Theorem 1. Let

$$\mathbf{v} = \langle v_1, v_2, v_3 \rangle, \qquad \mathbf{w} = \langle w_1, w_2, w_3 \rangle$$

We prove that $\mathbf{v} \times \mathbf{w}$ is orthogonal to $\mathbf{v}$ by showing that $\mathbf{v} \cdot (\mathbf{v} \times \mathbf{w}) = 0$. By Eq. (8),

$$\mathbf{v} \cdot (\mathbf{v} \times \mathbf{w}) = \det \begin{pmatrix} \mathbf{v} \\ \mathbf{v} \\ \mathbf{w} \end{pmatrix} = v_1 \begin{vmatrix} v_2 & v_3 \\ w_2 & w_3 \end{vmatrix} - v_2 \begin{vmatrix} v_1 & v_3 \\ w_1 & w_3 \end{vmatrix} + v_3 \begin{vmatrix} v_1 & v_2 \\ w_1 & w_2 \end{vmatrix} \qquad \boxed{11}$$

Straightforward algebra (left to the reader) shows that the right-hand side of Eq. (11) is equal to zero. This shows that $\mathbf{v} \cdot (\mathbf{v} \times \mathbf{w}) = 0$ and thus $\mathbf{v} \times \mathbf{w}$ is orthogonal to $\mathbf{v}$ as claimed. Interchanging the roles of $\mathbf{v}$ and $\mathbf{w}$, we conclude also that $\mathbf{w} \times \mathbf{v}$ is orthogonal to **w**, and since $\mathbf{v} \times \mathbf{w} = -\mathbf{w} \times \mathbf{v}$, it follows that $\mathbf{v} \times \mathbf{w}$ is orthogonal to **w**. This proves part (i) of Theorem 1. To prove (ii), we shall use the following identity:

$$\|\mathbf{v} \times \mathbf{w}\|^2 = \|\mathbf{v}\|^2 \|\mathbf{w}\|^2 - (\mathbf{v} \cdot \mathbf{w})^2$$

12

To verify this identity, we compute $\|\mathbf{v} \times \mathbf{w}\|^2$ as the sum of the squares of the components of $\mathbf{v} \times \mathbf{w}$:

$$\|\mathbf{v} \times \mathbf{w}\|^2 = \begin{vmatrix} v_2 & v_3 \\ w_2 & w_3 \end{vmatrix}^2 + \begin{vmatrix} v_1 & v_3 \\ w_1 & w_3 \end{vmatrix}^2 + \begin{vmatrix} v_1 & v_2 \\ w_1 & w_2 \end{vmatrix}^2$$

$$= (v_2 w_3 - v_3 w_2)^2 + (v_1 w_3 - v_3 w_1)^2 + (v_1 w_2 - v_2 w_1)^2$$

13

On the other hand,

$$\|\mathbf{v}\|^2 \|\mathbf{w}\|^2 - (\mathbf{v} \cdot \mathbf{w})^2 = (v_1^2 + v_2^2 + v_3^2)(w_1^2 + w_2^2 + w_3^2) - (v_1 w_1 + v_2 w_2 + v_3 w_3)^2$$

14

Again, algebra (left to the reader) shows that Eq. (13) is equal to Eq. (14).

Now let θ be the angle between **v** and **w**. By Eq. (12),

$$\|\mathbf{v} \times \mathbf{w}\|^2 = \|\mathbf{v}\|^2 \|\mathbf{w}\|^2 - (\mathbf{v} \cdot \mathbf{w})^2 = \|\mathbf{v}\|^2 \|\mathbf{w}\|^2 - \|\mathbf{v}\|^2 \|\mathbf{w}\|^2 \cos^2 \theta$$

$$= \|\mathbf{v}\|^2 \|\mathbf{w}\|^2 (1 - \cos^2 \theta) = \|\mathbf{v}\|^2 \|\mathbf{w}\|^2 \sin^2 \theta$$

Therefore, $\|\mathbf{v} \times \mathbf{w}\| = \|\mathbf{v}\| \|\mathbf{w}\| \sin \theta$. Note that $\sin \theta \geq 0$ since, by convention, θ lies between 0 and π. This proves (ii).

Part (iii) of Theorem 1 asserts that $\{\mathbf{v}, \mathbf{w}, \mathbf{v} \times \mathbf{w}\}$ is a right-handed system. This is a more subtle property that cannot be verified by algebra alone. We must rely on the following relation between right-handedness and the sign of the determinant, which can be established using the continuity of determinants:

$$\det \begin{pmatrix} \mathbf{u} \\ \mathbf{v} \\ \mathbf{w} \end{pmatrix} > 0 \quad \text{if and only if } \{\mathbf{u}, \mathbf{v}, \mathbf{w}\} \text{ is a right-handed system}$$

Furthermore, it can be checked directly from Eq. (2) that the determinant does not change when we replace $\{\mathbf{u}, \mathbf{v}, \mathbf{w}\}$ by $\{\mathbf{v}, \mathbf{w}, \mathbf{u}\}$ (or $\{\mathbf{w}, \mathbf{u}, \mathbf{v}\}$). Granting this and using Eq. (8), we obtain

$$\det \begin{pmatrix} \mathbf{v} \\ \mathbf{w} \\ \mathbf{v} \times \mathbf{w} \end{pmatrix} = \det \begin{pmatrix} \mathbf{v} \times \mathbf{w} \\ \mathbf{v} \\ \mathbf{w} \end{pmatrix} = (\mathbf{v} \times \mathbf{w}) \cdot (\mathbf{v} \times \mathbf{w}) = \|\mathbf{v} \times \mathbf{w}\|^2 > 0$$

Therefore $\{\mathbf{v}, \mathbf{w}, \mathbf{v} \times \mathbf{w}\}$ is right-handed as claimed (Figure 14).

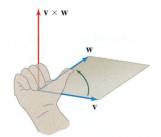

FIGURE 14 Both $\{\mathbf{v} \times \mathbf{w}, \mathbf{v}, \mathbf{w}\}$ and $\{\mathbf{v}, \mathbf{w}, \mathbf{v} \times \mathbf{w}\}$ are right-handed.

12.4 SUMMARY

- Determinants of sizes 2×2 and 3×3:

$$\begin{vmatrix} a_{11} & a_{12} \\ a_{21} & a_{22} \end{vmatrix} = a_{11} a_{22} - a_{12} a_{21}$$

$$\begin{vmatrix} a_{11} & a_{12} & a_{13} \\ a_{21} & a_{22} & a_{23} \\ a_{31} & a_{32} & a_{33} \end{vmatrix} = a_{11} \begin{vmatrix} a_{22} & a_{23} \\ a_{32} & a_{33} \end{vmatrix} - a_{12} \begin{vmatrix} a_{21} & a_{23} \\ a_{31} & a_{33} \end{vmatrix} + a_{13} \begin{vmatrix} a_{21} & a_{22} \\ a_{31} & a_{32} \end{vmatrix}$$

- The *cross product* of $\mathbf{v} = \langle v_1, v_2, v_3 \rangle$ and $\mathbf{w} = \langle w_1, w_2, w_3 \rangle$ is the symbolic determinant

$$\mathbf{v} \times \mathbf{w} = \begin{vmatrix} \mathbf{i} & \mathbf{j} & \mathbf{k} \\ v_1 & v_2 & v_3 \\ w_1 & w_2 & w_3 \end{vmatrix} = \begin{vmatrix} v_2 & v_3 \\ w_2 & w_3 \end{vmatrix} \mathbf{i} - \begin{vmatrix} v_1 & v_3 \\ w_1 & w_3 \end{vmatrix} \mathbf{j} + \begin{vmatrix} v_1 & v_2 \\ w_1 & w_2 \end{vmatrix} \mathbf{k}$$

- The cross product $\mathbf{v} \times \mathbf{w}$ is the unique vector with the following three properties:

  (i) $\mathbf{v} \times \mathbf{w}$ is orthogonal to $\mathbf{v}$ and $\mathbf{w}$.

  (ii) $\mathbf{v} \times \mathbf{w}$ has length $\|\mathbf{v}\| \|\mathbf{w}\| \sin \theta$ (θ is the angle between $\mathbf{v}$ and $\mathbf{w}$, $0 \le \theta \le \pi$).

  (iii) $\{\mathbf{v}, \mathbf{w}, \mathbf{v} \times \mathbf{w}\}$ is a right-handed system.

- Properties of the cross product:

  (i) $\mathbf{w} \times \mathbf{v} = -\mathbf{v} \times \mathbf{w}$

  (ii) $\mathbf{v} \times \mathbf{w} = \mathbf{0}$ if and only if $\mathbf{w} = \lambda \mathbf{v}$ for some scalar or $\mathbf{v} = \mathbf{0}$

  (iii) $(\lambda \mathbf{v}) \times \mathbf{w} = \mathbf{v} \times (\lambda \mathbf{w}) = \lambda (\mathbf{v} \times \mathbf{w})$

  (iv) $(\mathbf{u} + \mathbf{v}) \times \mathbf{w} = \mathbf{u} \times \mathbf{w} + \mathbf{v} \times \mathbf{w}$

  $\qquad \mathbf{v} \times (\mathbf{u} + \mathbf{w}) = \mathbf{v} \times \mathbf{u} + \mathbf{v} \times \mathbf{w}$

- Cross products of standard basis vectors (Figure 15):

$$\mathbf{i} \times \mathbf{j} = \mathbf{k}, \qquad \mathbf{j} \times \mathbf{k} = \mathbf{i}, \qquad \mathbf{k} \times \mathbf{i} = \mathbf{j}$$

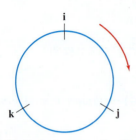

FIGURE 15 Circle for computing the cross products of the basis vectors.

- The parallelogram spanned by $\mathbf{v}$ and $\mathbf{w}$ has area $\|\mathbf{v} \times \mathbf{w}\|$.
- The triangle spanned by $\mathbf{v}$ and $\mathbf{w}$ has area $\dfrac{\|\mathbf{v} \times \mathbf{w}\|}{2}$.
- Cross-product identity: $\|\mathbf{v} \times \mathbf{w}\|^2 = \|\mathbf{v}\|^2 \|\mathbf{w}\|^2 - (\mathbf{v} \cdot \mathbf{w})^2$.
- The *vector triple product* is defined by $\mathbf{u} \cdot (\mathbf{v} \times \mathbf{w})$. We have

$$\mathbf{u} \cdot (\mathbf{v} \times \mathbf{w}) = \det \begin{pmatrix} \mathbf{u} \\ \mathbf{v} \\ \mathbf{w} \end{pmatrix}$$

- The parallelepiped spanned by $\mathbf{u}, \mathbf{v},$ and $\mathbf{w}$ has volume $|\mathbf{u} \cdot (\mathbf{v} \times \mathbf{w})|$.

12.4 EXERCISES

Preliminary Questions

1. What is the $(1, 3)$ minor of the matrix $\begin{vmatrix} 3 & 4 & 2 \\ -5 & -1 & 1 \\ 4 & 0 & 3 \end{vmatrix}$?

2. The angle between two unit vectors $\mathbf{e}$ and $\mathbf{f}$ is $\frac{\pi}{6}$. What is the length of $\mathbf{e} \times \mathbf{f}$?

3. What is $\mathbf{u} \times \mathbf{w}$, assuming that $\mathbf{w} \times \mathbf{u} = \langle 2, 2, 1 \rangle$?

4. Find the cross product without using the formula:

(a) $\langle 4, 8, 2 \rangle \times \langle 4, 8, 2 \rangle$ **(b)** $\langle 4, 8, 2 \rangle \times \langle 2, 4, 1 \rangle$

5. What are $\mathbf{i} \times \mathbf{j}$ and $\mathbf{i} \times \mathbf{k}$?

6. When is the cross product $\mathbf{v} \times \mathbf{w}$ equal to zero?

7. Which of the following are meaningful and which are not? Explain.

(a) $(\mathbf{u} \cdot \mathbf{v}) \times \mathbf{w}$

(b) $(\mathbf{u} \times \mathbf{v}) \cdot \mathbf{w}$

(c) $\|\mathbf{w}\|(\mathbf{u} \cdot \mathbf{v})$

(d) $\|\mathbf{w}\|(\mathbf{u} \times \mathbf{v})$

8. Which of the following vectors are equal to $\mathbf{j} \times \mathbf{i}$?

(a) $\mathbf{i} \times \mathbf{k}$

(b) $-\mathbf{k}$

(c) $\mathbf{i} \times \mathbf{j}$

Exercises

In Exercises 1–4, calculate the 2×2 determinant.

1. $\begin{vmatrix} 1 & 2 \\ 4 & 3 \end{vmatrix}$

2. $\begin{vmatrix} \frac{2}{3} & \frac{1}{6} \\ -5 & 2 \end{vmatrix}$

3. $\begin{vmatrix} -6 & 9 \\ 1 & 1 \end{vmatrix}$

4. $\begin{vmatrix} 9 & 25 \\ 5 & 14 \end{vmatrix}$

In Exercises 5–8, calculate the 3×3 determinant.

5. $\begin{vmatrix} 1 & 2 & 1 \\ 4 & -3 & 0 \\ 1 & 0 & 1 \end{vmatrix}$

6. $\begin{vmatrix} 1 & 0 & 1 \\ -2 & 0 & 3 \\ 1 & 3 & -1 \end{vmatrix}$

7. $\begin{vmatrix} 1 & 2 & 3 \\ 2 & 4 & 6 \\ -3 & -4 & 2 \end{vmatrix}$

8. $\begin{vmatrix} 1 & 0 & 0 \\ 0 & 0 & -1 \\ 0 & 1 & 0 \end{vmatrix}$

In Exercises 9–12, calculate $\mathbf{v} \times \mathbf{w}$.

9. $\mathbf{v} = \langle 1, 2, 1 \rangle, \quad \mathbf{w} = \langle 3, 1, 1 \rangle$

10. $\mathbf{v} = \langle 2, 0, 0 \rangle, \quad \mathbf{w} = \langle -1, 0, 1 \rangle$

11. $\mathbf{v} = \langle \frac{2}{3}, 1, \frac{1}{2} \rangle, \quad \mathbf{w} = \langle 4, -6, 3 \rangle$

12. $\mathbf{v} = \langle 1, 1, 0 \rangle, \quad \mathbf{w} = \langle 0, 1, 1 \rangle$

In Exercises 13–16, use the relations in Eq. (5) to calculate the cross product.

13. $(\mathbf{i} + \mathbf{j}) \times \mathbf{k}$

14. $(\mathbf{j} - \mathbf{k}) \times (\mathbf{j} + \mathbf{k})$

15. $(\mathbf{i} - 3\mathbf{j} + 2\mathbf{k}) \times (\mathbf{j} - \mathbf{k})$

16. $(2\mathbf{i} - 3\mathbf{j} + 4\mathbf{k}) \times (\mathbf{i} + \mathbf{j} - 7\mathbf{k})$

In Exercises 17–22, calculate the cross product assuming that

$$\mathbf{u} \times \mathbf{v} = \langle 1, 1, 0 \rangle, \quad \mathbf{u} \times \mathbf{w} = \langle 0, 3, 1 \rangle, \quad \mathbf{v} \times \mathbf{w} = \langle 2, -1, 1 \rangle$$

17. $\mathbf{v} \times \mathbf{u}$

18. $\mathbf{v} \times (\mathbf{u} + \mathbf{v})$

19. $\mathbf{w} \times (\mathbf{u} + \mathbf{v})$

20. $(3\mathbf{u} + 4\mathbf{w}) \times \mathbf{w}$

21. $(\mathbf{u} - 2\mathbf{v}) \times (\mathbf{u} + 2\mathbf{v})$

22. $(\mathbf{v} + \mathbf{w}) \times (3\mathbf{u} + 2\mathbf{v})$

23. Let $\mathbf{v} = \langle a, b, c \rangle$. Calculate $\mathbf{v} \times \mathbf{i}$, $\mathbf{v} \times \mathbf{j}$, and $\mathbf{v} \times \mathbf{k}$.

24. Find $\mathbf{v} \times \mathbf{w}$, where $\mathbf{v}$ and $\mathbf{w}$ are vectors of length 3 in the xz-plane, oriented as in Figure 16, and $\theta = \frac{\pi}{6}$.

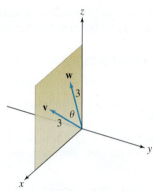

FIGURE 16

In Exercises 25 and 26, refer to Figure 17.

25. Which of $\mathbf{u}$ and $-\mathbf{u}$ is equal to $\mathbf{v} \times \mathbf{w}$?

26. Which of the following form a right-handed system?

(a) $\{\mathbf{v}, \mathbf{w}, \mathbf{u}\}$ **(b)** $\{\mathbf{w}, \mathbf{v}, \mathbf{u}\}$ **(c)** $\{\mathbf{v}, \mathbf{u}, \mathbf{w}\}$

(d) $\{\mathbf{u}, \mathbf{v}, \mathbf{w}\}$ **(e)** $\{\mathbf{w}, \mathbf{v}, -\mathbf{u}\}$ **(f)** $\{\mathbf{v}, -\mathbf{u}, \mathbf{w}\}$

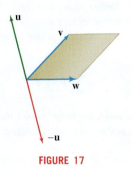

FIGURE 17

27. Let $\mathbf{v} = \langle 3, 0, 0 \rangle$ and $\mathbf{w} = \langle 0, 1, -1 \rangle$. Determine $\mathbf{u} = \mathbf{v} \times \mathbf{w}$ using the geometric properties of the cross product rather than the formula.

28. What are the possible angles θ between two unit vectors $\mathbf{e}$ and $\mathbf{f}$ if $\|\mathbf{e} \times \mathbf{f}\| = \frac{1}{2}$?

29. Show that if $\mathbf{v}$ and $\mathbf{w}$ lie in the yz-plane, then $\mathbf{v} \times \mathbf{w}$ is a multiple of $\mathbf{i}$.

30. Find the two unit vectors orthogonal to both $\mathbf{a} = \langle 3, 1, 1 \rangle$ and $\mathbf{b} = \langle -1, 2, 1 \rangle$.

31. Let $\mathbf{e}$ and $\mathbf{e}'$ be unit vectors in $\mathbf{R}^3$ such that $\mathbf{e} \perp \mathbf{e}'$. Use the geometric properties of the cross product to compute $\mathbf{e} \times (\mathbf{e}' \times \mathbf{e})$.

32. Calculate the force $\mathbf{F}$ on an electron (charge $q = -1.6 \times 10^{-19}$ C) moving with velocity 10^5 m/s in the direction $\mathbf{i}$ in a uniform magnetic field $\mathbf{B}$, where $\mathbf{B} = 0.0004\mathbf{i} + 0.0001\mathbf{j}$ teslas (see Example 5).

33. An electron moving with velocity $\mathbf{v}$ in the plane experiences a force $\mathbf{F} = q(\mathbf{v} \times \mathbf{B})$, where q is the charge on the electron and $\mathbf{B}$ is a uniform magnetic field pointing directly out of the page. Which of the two vectors $\mathbf{F}_1$ or $\mathbf{F}_2$ in Figure 18 represents the force on the electron? Remember that q is negative.

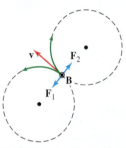

FIGURE 18 The magnetic field vector $\mathbf{B}$ points directly out of the page.

34. Calculate the scalar triple product $\mathbf{u} \cdot (\mathbf{v} \times \mathbf{w})$, where $\mathbf{u} = \langle 1, 1, 0 \rangle$, $\mathbf{v} = \langle 3, -2, 2 \rangle$, and $\mathbf{w} = \langle 4, -1, 2 \rangle$.

35. Verify identity (12) for vectors $\mathbf{v} = \langle 3, -2, 2 \rangle$ and $\mathbf{w} = \langle 4, -1, 2 \rangle$.

36. Find the volume of the parallelepiped spanned by $\mathbf{u}$, $\mathbf{v}$, and $\mathbf{w}$ in Figure 19.

37. Find the area of the parallelogram spanned by $\mathbf{v}$ and $\mathbf{w}$ in Figure 19.

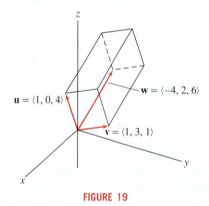

FIGURE 19

38. Calculate the volume of the parallelepiped spanned by

$$\mathbf{u} = \langle 2, 2, 1 \rangle, \quad \mathbf{v} = \langle 1, 0, 3 \rangle, \quad \mathbf{w} = \langle 0, -4, 0 \rangle$$

39. Sketch and compute the volume of the parallelepiped spanned by

$$\mathbf{u} = \langle 1, 0, 0 \rangle, \quad \mathbf{v} = \langle 0, 2, 0 \rangle, \quad \mathbf{w} = \langle 1, 1, 2 \rangle$$

40. Sketch the parallelogram spanned by $\mathbf{u} = \langle 1, 1, 1 \rangle$ and $\mathbf{v} = \langle 0, 0, 4 \rangle$, and compute its area.

41. Calculate the area of the parallelogram spanned by $\mathbf{u} = \langle 1, 0, 3 \rangle$ and $\mathbf{v} = \langle 2, 1, 1 \rangle$.

42. Find the area of the parallelogram determined by the vectors $\langle a, 0, 0 \rangle$ and $\langle 0, b, c \rangle$.

43. Sketch the triangle with vertices at the origin O, $P = (3, 3, 0)$, and $Q = (0, 3, 3)$, and compute its area using cross products.

44. Use the cross product to find the area of the triangle with vertices $P = (1, 1, 5)$, $Q = (3, 4, 3)$, and $R = (1, 5, 7)$ (Figure 20).

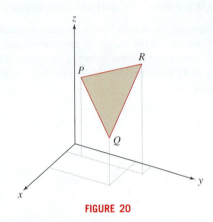

FIGURE 20

45. Use cross products to find the area of the triangle in the xy-plane defined by $(1, 2)$, $(3, 4)$, and $(-2, 2)$.

46. Use cross products to find the area of the quadrilateral in the xy-plane defined by $(0, 0)$, $(1, -1)$, $(3, 1)$, and $(2, 4)$.

47. Check that the four points $P(2, 4, 4)$, $Q(3, 1, 6)$, $R(2, 8, 0)$, and $S(7, 3, 1)$ all lie in a plane. Then use vectors to find the area of the quadrilateral they define.

48. Find three nonzero vectors $\mathbf{a}$, $\mathbf{b}$, and $\mathbf{c}$ such that $\mathbf{a} \times \mathbf{b} = \mathbf{a} \times \mathbf{c} \neq \mathbf{0}$ but $\mathbf{b} \neq \mathbf{c}$.

In Exercises 49–51, verify the identity using the formula for the cross product.

49. $\mathbf{v} \times \mathbf{w} = -\mathbf{w} \times \mathbf{v}$

50. $(\lambda \mathbf{v}) \times \mathbf{w} = \lambda (\mathbf{v} \times \mathbf{w})$ (λ a scalar)

51. $(\mathbf{u} + \mathbf{v}) \times \mathbf{w} = \mathbf{u} \times \mathbf{w} + \mathbf{v} \times \mathbf{w}$

52. Use the geometric description in Theorem 1 to prove Theorem 2 (iii): $\mathbf{v} \times \mathbf{w} = \mathbf{0}$ if and only if $\mathbf{w} = \lambda \mathbf{v}$ for some scalar λ or $\mathbf{v} = \mathbf{0}$.

53. Verify the relations (5).

54. Show that

$$(\mathbf{i} \times \mathbf{j}) \times \mathbf{j} \neq \mathbf{i} \times (\mathbf{j} \times \mathbf{j})$$

Conclude that the Associative Law does not hold for cross products.

55. The components of the cross product have a geometric interpretation. Show that the absolute value of the **k**-component of $\mathbf{v} \times \mathbf{w}$ is equal to the area of the parallelogram spanned by the projections $\mathbf{v}_0$ and $\mathbf{w}_0$ onto the xy-plane (Figure 21).

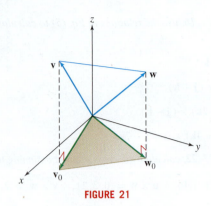

FIGURE 21

56. Formulate and prove analogs of the result in Exercise 55 for the **i**- and **j**-components of $\mathbf{v} \times \mathbf{w}$.

57. Show that three points P, Q, R are collinear (lie on a line) if and only if $\overrightarrow{PQ} \times \overrightarrow{PR} = \mathbf{0}$.

58. Use the result of Exercise 57 to determine whether the points P, Q, and R are collinear, and if not, find a vector normal to the plane containing them.

(a) $P = (2, 1, 0)$, $Q = (1, 5, 2)$, $R = (-1, 13, 6)$

(b) $P = (2, 1, 0)$, $Q = (-3, 21, 10)$, $R = (5, -2, 9)$

(c) $P = (1, 1, 0)$, $Q = (1, -2, -1)$, $R = (3, 2, -4)$

59. Solve the equation $\langle 1, 1, 1 \rangle \times \mathbf{X} = \langle 1, -1, 0 \rangle$, where $\mathbf{X} = \langle x, y, z \rangle$. *Note:* There are infinitely many solutions.

60. Explain geometrically why $\langle 1, 1, 1 \rangle \times \mathbf{X} = \langle 1, 0, 0 \rangle$ has no solution, where $\mathbf{X} = \langle x, y, z \rangle$.

61. Let $\mathbf{X} = \langle x, y, z \rangle$. Show that $\mathbf{i} \times \mathbf{X} = \mathbf{v}$ has a solution if and only if $\mathbf{v}$ is contained in the yz-plane (the **i**-component is zero).

62. Suppose that vectors $\mathbf{u}$, $\mathbf{v}$, and $\mathbf{w}$ are mutually orthogonal—that is, $\mathbf{u} \perp \mathbf{v}$, $\mathbf{u} \perp \mathbf{w}$, and $\mathbf{v} \perp \mathbf{w}$. Prove that $(\mathbf{u} \times \mathbf{v}) \times \mathbf{w} = \mathbf{0}$ and $\mathbf{u} \times (\mathbf{v} \times \mathbf{w}) = \mathbf{0}$.

*In Exercises 63–66, the **torque** about the origin O due to a force $\mathbf{F}$ acting on an object with position vector $\mathbf{r}$ is the vector quantity $\boldsymbol{\tau} = \mathbf{r} \times \mathbf{F}$. If several forces $\mathbf{F}_j$ act at positions $\mathbf{r}_j$, then the net torque (units: N-m or lb-ft) is the sum*

$$\boldsymbol{\tau} = \sum \mathbf{r}_j \times \mathbf{F}_j$$

Torque measures how much the force causes the object to rotate. By Newton's Laws, $\boldsymbol{\tau}$ is equal to the rate of change of angular momentum.

63. Calculate the torque $\boldsymbol{\tau}$ about O acting at the point P on the mechanical arm in Figure 22(A), assuming that a 25-newton force acts as indicated. Ignore the weight of the arm itself.

64. Calculate the net torque about O at P, assuming that a 30-kg mass is attached at P [Figure 22(B)]. The force $\mathbf{F}_g$ due to gravity on a mass m has magnitude $9.8m$ m/s^2 in the downward direction.

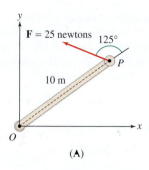

(A)

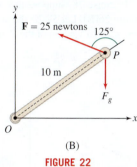

(B)

FIGURE 22

Show that the position vectors of the masses m_1, m_2, and m_3 are

$$\mathbf{r}_1 = \frac{1}{2}L_1(\sin\theta_1\mathbf{i} + \cos\theta_1\mathbf{j})$$

$$\mathbf{r}_2 = L_1(\sin\theta_1\mathbf{i} + \cos\theta_1\mathbf{j}) + \frac{1}{2}L_2(\sin\theta_2\mathbf{i} - \cos\theta_2\mathbf{j})$$

$$\mathbf{r}_3 = L_1(\sin\theta_1\mathbf{i} + \cos\theta_1\mathbf{j}) + L_2(\sin\theta_2\mathbf{i} - \cos\theta_2\mathbf{j})$$

Then show that

$$\tau = -g\left(L_1\left(\frac{1}{2}m_1 + m_2 + m_3\right)\sin\theta_1 + L_2\left(\frac{1}{2}m_2 + m_3\right)\sin\theta_2\right)\mathbf{k}$$

where $g = 9.8$ m/s^2. To simplify the computation, note that all three gravitational forces act in the $-\mathbf{j}$ direction, so the $\mathbf{j}$-components of the position vectors $\mathbf{r}_i$ do not contribute to the torque.

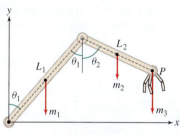

FIGURE 23

65. Let τ be the net torque about O acting on the robotic arm of Figure 23. Assume that the two segments of the arms have mass m_1 and m_2 (in kilograms) and that a weight of m_3 kg is located at the endpoint P. In calculating the torque, we may assume that the entire mass of each arm segment lies at the midpoint of the arm (its center of mass).

66. Continuing with Exercise 65, suppose that $L_1 = 3$ m, $L_2 = 2$ m, $m_1 = 15$ kg, $m_2 = 20$ kg, and $m_3 = 18$ kg. If the angles θ_1, θ_2 are equal (say, to θ), what is the maximum allowable value of θ if we assume that the robotic arm can sustain a maximum torque of 1200 N-m?

Further Insights and Challenges

67. Show that 3×3 determinants can be computed using the **diagonal rule**: Repeat the first two columns of the matrix and form the products of the numbers along the six diagonals indicated. Then add the products for the diagonals that slant from left to right and subtract the products for the diagonals that slant from right to left.

$$\det(A) = \begin{vmatrix} a_{11} & a_{12} & a_{13} & a_{11} & a_{12} \\ a_{21} & a_{22} & a_{23} & a_{21} & a_{22} \\ a_{31} & a_{32} & a_{33} & a_{31} & a_{32} \end{vmatrix}$$

$$= a_{11}a_{22}a_{33} + a_{12}a_{23}a_{31} + a_{13}a_{21}a_{32}$$

$$- a_{13}a_{22}a_{31} - a_{11}a_{23}a_{32} - a_{12}a_{21}a_{33}$$

68. Use the diagonal rule to calculate $\begin{vmatrix} 2 & 4 & 3 \\ 0 & 1 & -7 \\ -1 & 5 & 3 \end{vmatrix}$.

69. Prove that $\mathbf{v} \times \mathbf{w} = \mathbf{v} \times \mathbf{u}$ if and only if $\mathbf{u} = \mathbf{w} + \lambda\mathbf{v}$ for some scalar λ. Assume that $\mathbf{v} \neq \mathbf{0}$.

70. Use Eq. (12) to prove the Cauchy–Schwarz inequality:

$$|\mathbf{v} \cdot \mathbf{w}| \le \|\mathbf{v}\| \, \|\mathbf{w}\|$$

Show that equality holds if and only if $\mathbf{w}$ is a multiple of $\mathbf{v}$ or at least one of $\mathbf{v}$ and $\mathbf{w}$ is zero.

71. Show that if $\mathbf{u}$, $\mathbf{v}$, and $\mathbf{w}$ are nonzero vectors and $(\mathbf{u} \times \mathbf{v}) \times \mathbf{w} = \mathbf{0}$, then either (i) $\mathbf{u}$ and $\mathbf{v}$ are parallel, or (ii) $\mathbf{w}$ is orthogonal to $\mathbf{u}$ and $\mathbf{v}$.

72. Suppose that $\mathbf{u}$, $\mathbf{v}$, $\mathbf{w}$ are nonzero and

$$(\mathbf{u} \times \mathbf{v}) \times \mathbf{w} = \mathbf{u} \times (\mathbf{v} \times \mathbf{w}) = \mathbf{0}$$

Show that $\mathbf{u}$, $\mathbf{v}$, and $\mathbf{w}$ are either mutually parallel or mutually perpendicular. *Hint: Use Exercise 71.*

73. Let $\mathbf{a}$, $\mathbf{b}$, $\mathbf{c}$ be nonzero vectors. Assume that $\mathbf{b}$ and $\mathbf{c}$ are not parallel, and set

$$\mathbf{v} = \mathbf{a} \times (\mathbf{b} \times \mathbf{c}), \qquad \mathbf{w} = (\mathbf{a} \cdot \mathbf{c})\mathbf{b} - (\mathbf{a} \cdot \mathbf{b})\mathbf{c}$$

(a) Prove that:

(i) $\mathbf{v}$ lies in the plane spanned by $\mathbf{b}$ and $\mathbf{c}$.

(ii) $\mathbf{v}$ is orthogonal to $\mathbf{a}$.

(b) Prove that $\mathbf{w}$ also satisfies (i) and (ii). Conclude that $\mathbf{v}$ and $\mathbf{w}$ are parallel.

(c) Show algebraically that $\mathbf{v} = \mathbf{w}$ (Figure 24).

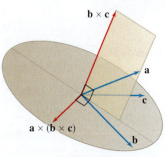

FIGURE 24

74. Use Exercise 73 to prove the identity

$$(\mathbf{a} \times \mathbf{b}) \times \mathbf{c} - \mathbf{a} \times (\mathbf{b} \times \mathbf{c}) = (\mathbf{a} \cdot \mathbf{b})\mathbf{c} - (\mathbf{b} \cdot \mathbf{c})\mathbf{a}$$

75. Show that if $\mathbf{a}, \mathbf{b}$ are nonzero vectors such that $\mathbf{a} \perp \mathbf{b}$, then there exists a vector $\mathbf{X}$ such that

$$\mathbf{a} \times \mathbf{X} = \mathbf{b} \qquad \boxed{15}$$

Hint: Show that if $\mathbf{X}$ is orthogonal to $\mathbf{b}$ and is not a multiple of $\mathbf{a}$, then $\mathbf{a} \times \mathbf{X}$ is a multiple of $\mathbf{b}$.

76. Show that if $\mathbf{a}, \mathbf{b}$ are nonzero vectors such that $\mathbf{a} \perp \mathbf{b}$, then the set of all solutions of Eq. (15) is a line with $\mathbf{a}$ as direction vector. *Hint:* Let $\mathbf{X}_0$ be any solution (which exists by Exercise 75), and show that every other solution is of the form $\mathbf{X}_0 + \lambda\mathbf{a}$ for some scalar λ.

77. Assume that $\mathbf{v}$ and $\mathbf{w}$ lie in the first quadrant in $\mathbf{R}^2$ as in Figure 25. Use geometry to prove that the area of the parallelogram is equal to $\det \begin{pmatrix} \mathbf{v} \\ \mathbf{w} \end{pmatrix}$.

78. Consider the tetrahedron spanned by vectors $\mathbf{a}, \mathbf{b}$, and $\mathbf{c}$ as in Figure 26(A). Let A, B, C be the faces containing the origin O, and let D be the fourth face opposite O. For each face F, let $\mathbf{v}_F$ be the vector normal to the face, pointing outside the tetrahedron, of magnitude equal to twice the area of F. Prove the relations

$$\mathbf{v}_A + \mathbf{v}_B + \mathbf{v}_C = \mathbf{a} \times \mathbf{b} + \mathbf{b} \times \mathbf{c} + \mathbf{c} \times \mathbf{a}$$

$$\mathbf{v}_A + \mathbf{v}_B + \mathbf{v}_C + \mathbf{v}_D = 0$$

Hint: Show that $\mathbf{v}_D = (\mathbf{c} - \mathbf{b}) \times (\mathbf{b} - \mathbf{a})$.

79. In the notation of Exercise 78, suppose that $\mathbf{a}, \mathbf{b}, \mathbf{c}$ are mutually perpendicular as in Figure 26(B). Let S_F be the area of face F. Prove the following three-dimensional version of the Pythagorean Theorem:

$$S_A^2 + S_B^2 + S_C^2 = S_D^2$$

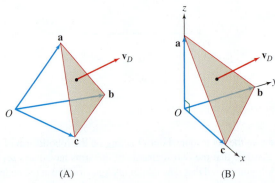

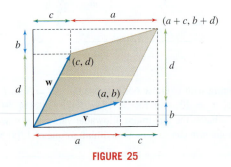

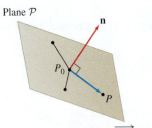

FIGURE 25

FIGURE 26 The vector $\mathbf{v}_D$ is perpendicular to the face.

12.5 Planes in 3-Space

A linear equation $ax + by = c$ in two variables defines a line in $\mathbf{R}^2$. In this section, we show that a linear equation $ax + by + cz = d$ in three variables defines a plane in $\mathbf{R}^3$.

Consider a plane $\mathcal{P}$ that passes through a point $P_0 = (x_0, y_0, z_0)$. We can determine $\mathcal{P}$ completely by specifying a nonzero vector $\mathbf{n} = \langle a, b, c \rangle$ that is orthogonal to $\mathcal{P}$. Such a vector is called a **normal vector**. Basing $\mathbf{n}$ at P_0 as in Figure 1, we see that a point $P = (x, y, z)$ lies on $\mathcal{P}$ precisely when $\overrightarrow{P_0P}$ is orthogonal to $\mathbf{n}$. Therefore, P lies on the plane if

$$\mathbf{n} \cdot \overrightarrow{P_0P} = 0 \qquad \boxed{1}$$

> The term "normal" is another word for "orthogonal" or "perpendicular."

In components, $\overrightarrow{P_0P} = \langle x - x_0, y - y_0, z - z_0 \rangle$, so Eq. (1) reads

$$\langle a, b, c \rangle \cdot \langle x - x_0, y - y_0, z - z_0 \rangle = 0$$

This gives us the following equation for the plane:

$$a(x - x_0) + b(y - y_0) + c(z - z_0) = 0$$

This can also be written

$$ax + by + cz = ax_0 + by_0 + cz_0 \qquad \text{or} \qquad \mathbf{n} \cdot \overrightarrow{OP} = \mathbf{n} \cdot \overrightarrow{OP_0} \qquad \boxed{2}$$

When we set $d = ax_0 + by_0 + cz_0$, Eq. (2) becomes $\mathbf{n} \cdot \langle x, y, z \rangle = d$, or

$$ax + by + cz = d$$

FIGURE 1 A point P lies on $\mathcal{P}$ if $\overrightarrow{P_0P} \perp \mathbf{n}$.

> **THEOREM 1 Equation of a Plane** Plane through $P_0 = (x_0, y_0, z_0)$ with normal vector $\mathbf{n} = \langle a, b, c \rangle$:
>
> | | | |
> |---|---|---|
> | **Vector form:** | $\mathbf{n} \cdot \langle x, y, z \rangle = d$ | **3** |
> | **Scalar forms:** | $a(x - x_0) + b(y - y_0) + c(z - z_0) = 0$ | **4** |
> | | $ax + by + cz = d$ | **5** |
>
> where $d = ax_0 + by_0 + cz_0$.

Note that the equation $ax + by + cz = d$ of a plane in 3-space is the direct generalization of the equation $ax + by = c$ of a line in 2-space. In this sense, planes generalize lines.

To show how the plane equation works in a simple case, consider the plane $\mathcal{P}$ through $P_0 = (1, 2, 0)$ with normal vector $\mathbf{n} = \langle 0, 0, 3 \rangle$ (Figure 2). Because $\mathbf{n}$ points in the z-direction, $\mathcal{P}$ must be parallel to the xy-plane. On the other hand, P_0 lies on the xy-plane, so $\mathcal{P}$ must be the xy-plane itself. This is precisely what Eq. (3) gives us:

$$\mathbf{n} \cdot \langle x, y, z \rangle = \mathbf{n} \cdot \langle 1, 2, 0 \rangle$$

$$\langle 0, 0, 3 \rangle \cdot \langle x, y, z \rangle = \langle 0, 0, 3 \rangle \cdot \langle 1, 2, 0 \rangle$$

$$3z = 0 \quad \text{or} \quad z = 0$$

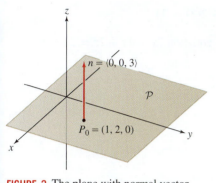

FIGURE 2 The plane with normal vector $\mathbf{n} = \langle 0, 0, 3 \rangle$ passing through $P_0 = (1, 2, 0)$ is the xy-plane.

In other words, $\mathcal{P}$ has equation $z = 0$, so $\mathcal{P}$ is the xy-plane.

■ **EXAMPLE 1** Find an equation of the plane through $P_0 = (3, 1, 0)$ with normal vector $\mathbf{n} = \langle 3, 2, -5 \rangle$.

Solution Using Eq. (4), we obtain

$$3(x - 3) + 2(y - 1) - 5z = 0$$

Alternatively, we can compute

$$d = \mathbf{n} \cdot \overrightarrow{OP_0} = \langle 3, 2, -5 \rangle \cdot \langle 3, 1, 0 \rangle = 11$$

and write the equation as $\langle 3, 2, -5 \rangle \cdot \langle x, y, z \rangle = 11$, or $3x + 2y - 5z = 11$. ■

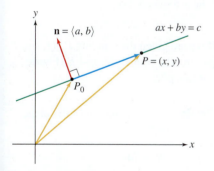

FIGURE 3 A line with normal vector $\mathbf{n}$.

> **CONCEPTUAL INSIGHT** Keep in mind that the components of a normal vector are "lurking" inside the equation $ax + by + cz = d$, because $\mathbf{n} = \langle a, b, c \rangle$. The same is true for lines in $\mathbf{R}^2$. The line $ax + by = c$ in Figure 3 has normal vector $\mathbf{n} = \langle a, b \rangle$ because the line has slope $-a/b$ and the vector $\mathbf{n}$ has slope b/a (lines are orthogonal if the product of their slopes is -1).

Note that if $\mathbf{n}$ is normal to a plane $\mathcal{P}$, then so is every nonzero scalar multiple $\lambda\mathbf{n}$. When we use $\lambda\mathbf{n}$ instead of $\mathbf{n}$, the resulting equation for $\mathcal{P}$ changes by a factor of λ. For example, the following two equations define the same plane:

$$x + y + z = 1, \qquad 4x + 4y + 4z = 4$$

The first equation uses the normal $\langle 1, 1, 1 \rangle$, and the second uses the normal $\langle 4, 4, 4 \rangle$.

On the other hand, two planes $\mathcal{P}$ and $\mathcal{P}'$ are parallel if they have a common normal vector. The following planes are parallel because each is normal to $\mathbf{n} = \langle 1, 1, 1 \rangle$:

$$x + y + z = 1, \qquad x + y + z = 2, \qquad 4x + 4y + 4z = 7$$

In general, a family of parallel planes is obtained by choosing a normal vector $\mathbf{n} = \langle a, b, c \rangle$ and varying the constant d in the equation

$$ax + by + cz = d$$

The unique plane in this family through the origin has equation $ax + by + cz = 0$.

■ **EXAMPLE 2 Parallel Planes** Let $\mathcal{P}$ have equation $7x - 4y + 2z = -10$. Find an equation of the plane parallel to $\mathcal{P}$ passing through:

(a) The origin.   **(b)** $Q = (2, -1, 3)$.

Solution The planes parallel to $\mathcal{P}$ have an equation of the form (Figure 4)

$$7x - 4y + 2z = d \qquad \boxed{6}$$

(a) For $d = 0$, we get the plane through the origin: $7x - 4y + 2z = 0$.

(b) The point $Q = (2, -1, 3)$ satisfies Eq. (6) with

$$d = 7(2) - 4(-1) + 2(3) = 24$$

Therefore, the plane parallel to $\mathcal{P}$ through Q has equation $7x - 4y + 2z = 24$. ■

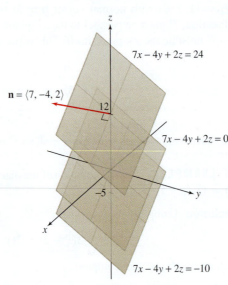

DF FIGURE 4 Parallel planes with normal vector $\mathbf{n} = \langle 7, -4, 2 \rangle$.

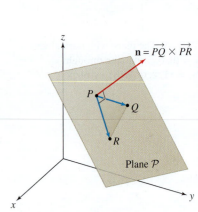

FIGURE 5 Three points P, Q, and R determine a plane (assuming they do not lie in a straight line).

Points that lie on a line are called **collinear**. If we are given three points P, Q, and R that are not collinear, then there is just one plane passing through P, Q, and R (Figure 5). The next example shows how to find an equation of this plane.

■ **EXAMPLE 3 The Plane Determined by Three Points** Find an equation of the plane $\mathcal{P}$ determined by the points

$$P = (1, 0, -1), \qquad Q = (2, 2, 1), \qquad R = (4, 1, 2)$$

Solution

Step 1. **Find a normal vector.**

The vectors $\overrightarrow{PQ}$ and $\overrightarrow{PR}$ lie in the plane $\mathcal{P}$, so their cross product is normal to $\mathcal{P}$:

$$\overrightarrow{PQ} = \langle 2, 2, 1 \rangle - \langle 1, 0, -1 \rangle = \langle 1, 2, 2 \rangle$$

$$\overrightarrow{PR} = \langle 4, 1, 2 \rangle - \langle 1, 0, -1 \rangle = \langle 3, 1, 3 \rangle$$

In Example 3, we could just as well have used the vectors $\overrightarrow{QP}$ and $\overrightarrow{QR}$ (or $\overrightarrow{RP}$ and $\overrightarrow{RQ}$) to find a normal vector $\mathbf{n}$.

$$\mathbf{n} = \overrightarrow{PQ} \times \overrightarrow{PR} = \begin{vmatrix} \mathbf{i} & \mathbf{j} & \mathbf{k} \\ 1 & 2 & 2 \\ 3 & 1 & 3 \end{vmatrix} = 4\mathbf{i} + 3\mathbf{j} - 5\mathbf{k} = \langle 4, 3, -5 \rangle$$

By Eq. (5), $\mathcal{P}$ has equation $4x + 3y - 5z = d$ for some d.

Step 2. **Choose a point on the plane and compute *d*.**

Now choose any one of the three points—say, $P = (1, 0, -1)$—and compute

$$d = \mathbf{n} \cdot \overrightarrow{OP} = \langle 4, 3, -5 \rangle \cdot \langle 1, 0, -1 \rangle = 9$$

We conclude that $\mathcal{P}$ has equation $4x + 3y - 5z = 9$. ∎

■ **EXAMPLE 4** **Intersection of a Plane and a Line** Find the point P where the plane $3x - 9y + 2z = 7$ and the line $\mathbf{r}(t) = \langle 1, 2, 1 \rangle + t \langle -2, 0, 1 \rangle$ intersect.

Solution The line has parametric equations

$$x = 1 - 2t, \qquad y = 2, \qquad z = 1 + t$$

Substitute in the equation of the plane and solve for t:

$$3x - 9y + 2z = 3(1 - 2t) - 9(2) + 2(1 + t) = 7$$

Simplification yields $-4t - 13 = 7$ or $t = -5$. Therefore, P has coordinates

$$x = 1 - 2(-5) = 11, \qquad y = 2, \qquad z = 1 + (-5) = -4$$

The plane and line intersect at the point $P = (11, 2, -4)$. ∎

The intersection of a plane $\mathcal{P}$ with a coordinate plane or a plane parallel to a coordinate plane is called a **trace**. The trace is a line unless $\mathcal{P}$ is parallel to the coordinate plane (in which case, the trace is empty or is $\mathcal{P}$ itself).

■ **EXAMPLE 5** **Traces of the Plane** Graph the plane $-2x + 3y + z = 6$ and then find its traces in the coordinate planes.

Solution To draw the plane, we determine its intersections with the coordinate axes. To find where it intersects the x-axis, we set $y = z = 0$ and obtain

$$-2x = 6 \qquad \text{so} \qquad x = -3$$

It intersects the y-axis when $x = z = 0$, giving

$$3y = 6 \qquad \text{so} \qquad y = 2$$

It intersects the z-axis when $x = y = 0$, giving

$$z = 6$$

Thus, the plane appears as in Figure 6.

We obtain the trace in the xy-plane by setting $z = 0$ in the equation of the plane. Therefore, the trace is the line $-2x + 3y = 6$ in the xy-plane (Figure 6).

Similarly, the trace in the xz-plane is obtained by setting $y = 0$, which gives the line $-2x + z = 6$ in the xz-plane. Finally, the trace in the yz-plane is $3y + z = 6$. ∎

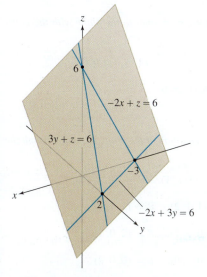

FIGURE 6 The three blue lines are the traces of the plane $-2x + 3y + z = 6$ in the coordinate planes.

To draw a plane, it works well to find its intersections with the coordinate axes, unless the plane is parallel to one of the axes and therefore does not intersect it, or the plane passes through the origin, and therefore intersects all three axes at the same point. In these cases, the normal vector can be used to see how the plane sits in space.

12.5 SUMMARY

- Equation of plane through $P_0 = (x_0, y_0, z_0)$ with normal vector $\mathbf{n} = \langle a, b, c \rangle$:

Vector form: $\mathbf{n} \cdot \langle x, y, z \rangle = d$

Scalar forms: $a(x - x_0) + b(y - y_0) + c(z - z_0) = 0$

$$ax + by + cz = d$$

where $d = \mathbf{n} \cdot \langle x_0, y_0, z_0 \rangle = ax_0 + by_0 + cz_0$.

- The family of parallel planes with given normal vector $\mathbf{n} = \langle a, b, c \rangle$ consists of all planes with equation $ax + by + cz = d$ for some d.
- The plane through three points P, Q, R that are not collinear:

$$- \quad \mathbf{n} = \overrightarrow{PQ} \times \overrightarrow{PR}$$
$$- \quad d = \mathbf{n} \cdot \langle x_0, y_0, z_0 \rangle, \text{ where } P = (x_0, y_0, z_0)$$

- The intersection of a plane $\mathcal{P}$ with a coordinate plane or a plane parallel to a coordinate plane is called a *trace*. The trace in the yz-plane is obtained by setting $x = 0$ in the equation of the plane (and similarly for the traces in the xz- and xy-planes).

12.5 EXERCISES

Preliminary Questions

1. What is the equation of the plane parallel to $3x + 4y - z = 5$ passing through the origin?

2. The vector $\mathbf{k}$ is normal to which of the following planes?

(a) $x = 1$ **(b)** $y = 1$ **(c)** $z = 1$

3. Which of the following planes is not parallel to the plane $x + y + z = 1$?

(a) $2x + 2y + 2z = 1$ **(b)** $x + y + z = 3$

(c) $x - y + z = 0$

4. To which coordinate plane is the plane $y = 1$ parallel?

5. Which of the following planes contains the z-axis?

(a) $z = 1$ **(b)** $x + y = 1$ **(c)** $x + y = 0$

6. Suppose that a plane $\mathcal{P}$ with normal vector $\mathbf{n}$ and a line $\mathcal{L}$ with direction vector $\mathbf{v}$ both pass through the origin and that $\mathbf{n} \cdot \mathbf{v} = 0$. Which of the following statements is correct?

(a) $\mathcal{L}$ is contained in $\mathcal{P}$.

(b) $\mathcal{L}$ is orthogonal to $\mathcal{P}$.

Exercises

In Exercises 1–8, write the equation of the plane with normal vector $\mathbf{n}$ passing through the given point in the scalar form $ax + by + cz = d$.

1. $\mathbf{n} = \langle 1, 3, 2 \rangle$, $(4, -1, 1)$ **2.** $\mathbf{n} = \langle -1, 2, 1 \rangle$, $(3, 1, 9)$

3. $\mathbf{n} = \langle -1, 2, 1 \rangle$, $(4, 1, 5)$ **4.** $\mathbf{n} = \langle 2, -4, 1 \rangle$, $\left(\frac{1}{3}, \frac{2}{3}, 1 \right)$

5. $\mathbf{n} = \mathbf{i}$, $(3, 1, -9)$ **6.** $\mathbf{n} = \mathbf{j}$, $\left(-5, \frac{1}{2}, \frac{1}{2} \right)$

7. $\mathbf{n} = \mathbf{k}$, $(6, 7, 2)$ **8.** $\mathbf{n} = \mathbf{i} - \mathbf{k}$, $(4, 2, -8)$

9. Write down the equation of any plane through the origin.

10. Write down the equations of any two distinct planes with normal vector $\mathbf{n} = \langle 3, 2, 1 \rangle$ that do not pass through the origin.

11. Which of the following statements are true of a plane that is parallel to the yz-plane?

(a) $\mathbf{n} = \langle 0, 0, 1 \rangle$ is a normal vector.

(b) $\mathbf{n} = \langle 1, 0, 0 \rangle$ is a normal vector.

(c) The equation has the form $ay + bz = d$

(d) The equation has the form $x = d$

12. Find a normal vector $\mathbf{n}$ and an equation for the planes in Figures 7(A)–(C).

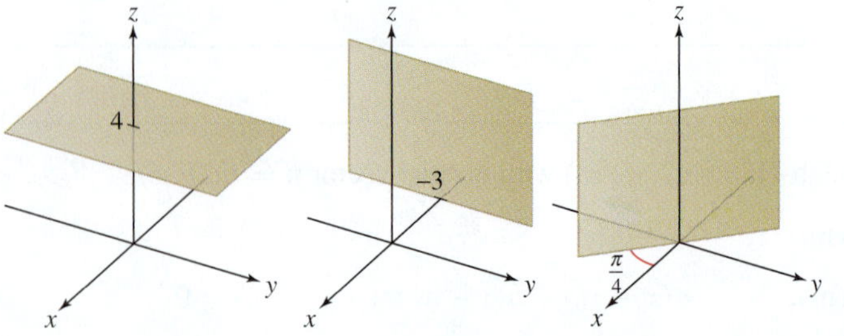

(A)   (B)   (C)

FIGURE 7

In Exercises 13–16, find a vector normal to the plane with the given equation.

13. $9x - 4y - 11z = 2$ **14.** $x - z = 0$

15. $3(x - 4) - 8(y - 1) + 11z = 0$

16. $x = 1$

In Exercises 17–20, find an equation of the plane passing through the three points given.

17. $P = (2, -1, 4)$, $Q = (1, 1, 1)$, $R = (3, 1, -2)$

18. $P = (5, 1, 1)$, $Q = (1, 1, 2)$, $R = (2, 1, 1)$

19. $P = (1, 0, 0)$, $Q = (0, 1, 1)$, $R = (2, 0, 1)$

20. $P = (2, 0, 0)$, $Q = (0, 4, 0)$, $R = (0, 0, 2)$

In Exercises 21–28, find the equation of the plane with the given description.

21. Passes through O and is parallel to $4x - 9y + z = 3$

22. Passes through $(4, 1, 9)$ and is parallel to $x + y + z = 3$

23. Passes through $(4, 1, 9)$ and is parallel to $x = 3$

24. Passes through $P = (3, 5, -9)$ and is parallel to the xz-plane

25. Passes through $(-2, -3, 5)$ and has normal vector $\mathbf{i} + \mathbf{k}$

26. Contains the lines $\mathbf{r}_1(t) = \langle t, 2t, 3t \rangle$ and $\mathbf{r}_2(t) = \langle 3t, t, 8t \rangle$

27. Contains the lines $\mathbf{r}_1(t) = \langle 2, 1, 0 \rangle + \langle t, 2t, 3t \rangle$ and $\mathbf{r}_2(t) = \langle 2, 1, 0 \rangle + \langle 3t, t, 8t \rangle$

28. Contains $P = (-1, 0, 1)$ and $\mathbf{r}(t) = \langle t + 1, 2t, 3t - 1 \rangle$

29. Are the planes $\frac{1}{2}x + 2y - z = 5$ and $3x + 12y - 6z = 1$ parallel?

30. Are the planes $2x - 4y - z = 3$ and $-6x + 12y + 3z = 1$ parallel?

In Exercises 31–35, draw the plane given by the equation.

31. $x + y + z = 4$

32. $3x + 2y - 6z = 12$

33. $12x - 6y + 4z = 6$

34. $x + 2y = 6$

35. $x + y + z = 0$

36. Let a, b, c be constants. Which two of the following equations define the plane passing through $(a, 0, 0)$, $(0, b, 0)$, $(0, 0, c)$?

(a) $ax + by + cz = 1$

(b) $bcx + acy + abz = abc$

(c) $bx + cy + az = 1$

(d) $\dfrac{x}{a} + \dfrac{y}{b} + \dfrac{z}{c} = 1$

37. Find an equation of the plane $\mathcal{P}$ in Figure 8.

FIGURE 8

38. Verify that the plane $x - y + 5z = 10$ and the line $\mathbf{r}(t) = \langle 1, 0, 1 \rangle + t \langle -2, 1, 1 \rangle$ intersect at $P = (-3, 2, 3)$.

In Exercises 39–42, find the intersection of the line and the plane.

39. $x + y + z = 14$, $\mathbf{r}(t) = \langle 1, 1, 0 \rangle + t \langle 0, 2, 4 \rangle$

40. $2x + y = 3$, $\mathbf{r}(t) = \langle 2, -1, -1 \rangle + t \langle 1, 2, -4 \rangle$

41. $z = 12$, $\mathbf{r}(t) = t \langle -6, 9, 36 \rangle$

42. $x - z = 6$, $\mathbf{r}(t) = \langle 1, 0, -1 \rangle + t \langle 4, 9, 2 \rangle$

In Exercises 43–48, find the trace of the plane in the given coordinate plane.

43. $3x - 9y + 4z = 5$, yz

44. $3x - 9y + 4z = 5$, xz

45. $3x + 4z = -2$, xy

46. $3x + 4z = -2$, xz

47. $-x + y = 4$, xz

48. $-x + y = 4$, yz

49. Does the plane $x = 5$ have a trace in the yz-plane? Explain.

50. Give equations for two distinct planes whose trace in the xy-plane has equation $4x + 3y = 8$.

51. Give equations for two distinct planes whose trace in the yz-plane has equation $y = 4z$.

52. Find parametric equations for the line through $P_0 = (3, -1, 1)$ perpendicular to the plane $3x + 5y - 7z = 29$.

53. Find all planes in $\mathbf{R}^3$ whose intersection with the xz-plane is the line with equation $3x + 2z = 5$.

54. Find all planes in $\mathbf{R}^3$ whose intersection with the xy-plane is the line $\mathbf{r}(t) = t \langle 2, 1, 0 \rangle$.

In Exercises 55–60, compute the angle between the two planes, defined as the angle θ (between 0 and π) between their normal vectors (Figure 9).

55. Planes with normals $\mathbf{n}_1 = \langle 1, 0, 1 \rangle$, $\mathbf{n}_2 = \langle -1, 1, 1 \rangle$

56. Planes with normals $\mathbf{n}_1 = \langle 1, 2, 1 \rangle$, $\mathbf{n}_2 = \langle 4, 1, 3 \rangle$

57. $2x + 3y + 7z = 2$ and $4x - 2y + 2z = 4$

58. $x - 3y + z = 3$ and $2x - 3z = 4$

59. $3(x - 1) - 5y + 2(z - 12) = 0$ and the plane with normal $\mathbf{n} = \langle 1, 0, 1 \rangle$

60. The plane through $(1, 0, 0)$, $(0, 1, 0)$, and $(0, 0, 1)$ and the yz-plane

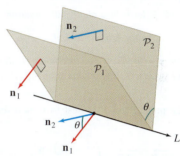

FIGURE 9 By definition, the angle between two planes is the angle between their normal vectors.

61. Find an equation of a plane making an angle of $\frac{\pi}{2}$ with the plane $3x + y - 4z = 2$.

62. Let $\mathcal{P}_1$ and $\mathcal{P}_2$ be planes with normal vectors $\mathbf{n}_1$ and $\mathbf{n}_2$. Assume that the planes are not parallel, and let $\mathcal{L}$ be their intersection (a line). Show that $\mathbf{n}_1 \times \mathbf{n}_2$ is a direction vector for $\mathcal{L}$.

63. Find a plane that is perpendicular to the two planes $x + y = 3$ and $x + 2y - z = 4$.

64. Let $\mathcal{L}$ be the intersection of the planes $x + y + z = 1$ and $x + 2y + 3z = 1$. Use Exercise 62 to find a direction vector for $\mathcal{L}$. Then find a point P on $\mathcal{L}$ by *inspection*, and write down the parametric equations for $\mathcal{L}$.

65. Let $\mathcal{L}$ denote the intersection of the planes $x - y - z = 1$ and $2x + 3y + z = 2$. Find parametric equations for the line $\mathcal{L}$. *Hint:* To find a point on $\mathcal{L}$, substitute an arbitrary value for z (say, $z = 2$) and then solve the resulting pair of equations for x and y.

66. Find parametric equations for the intersection of the planes $2x + y - 3z = 0$ and $x + y = 1$.

67. Two vectors $\mathbf{v}$ and $\mathbf{w}$, each of length 12, lie in the plane $x + 2y - 2z = 0$. The angle between $\mathbf{v}$ and $\mathbf{w}$ is $\pi/6$. This information determines $\mathbf{v} \times \mathbf{w}$ up to a sign ± 1. What are the two possible values of $\mathbf{v} \times \mathbf{w}$?

68. The plane

$$\frac{x}{2} + \frac{y}{4} + \frac{z}{3} = 1$$

intersects the x-, y-, and z-axes in points P, Q, and R. Find the area of the triangle $\triangle PQR$.

69. 📖 In this exercise, we show that the orthogonal distance D from the plane $\mathcal{P}$ with equation $ax + by + cz = d$ to the origin O is equal to (Figure 10)

$$D = \frac{|d|}{\sqrt{a^2 + b^2 + c^2}}$$

Let $\mathbf{n} = \langle a, b, c \rangle$, and let P be the point where the line through $\mathbf{n}$ intersects $\mathcal{P}$. By definition, the orthogonal distance from $\mathcal{P}$ to O is the distance from P to O.

(a) Show that P is the terminal point of $\mathbf{v} = \left(\dfrac{d}{\mathbf{n} \cdot \mathbf{n}} \right) \mathbf{n}$.

(b) Show that the distance from P to O is D.

70. Use Exercise 69 to compute the orthogonal distance from the plane $x + 2y + 3z = 5$ to the origin.

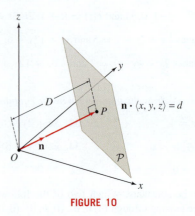

FIGURE 10

Further Insights and Challenges

In Exercises 71 and 72, let $\mathcal{P}$ be a plane with equation

$$ax + by + cz = d$$

and normal vector $\mathbf{n} = \langle a, b, c \rangle$. For any point Q, there is a unique point P on $\mathcal{P}$ that is closest to Q, and is such that $\overrightarrow{PQ}$ is orthogonal to $\mathcal{P}$ (Figure 11).

71. Show that the point P on $\mathcal{P}$ closest to Q is determined by the equation

$$\overrightarrow{OP} = \overrightarrow{OQ} + \left(\frac{d - \overrightarrow{OQ} \cdot \mathbf{n}}{\mathbf{n} \cdot \mathbf{n}} \right) \mathbf{n} \qquad \boxed{7}$$

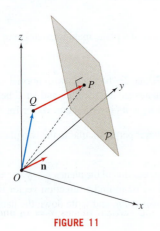

FIGURE 11

72. By definition, the distance from $Q = (x_1, y_1, z_1)$ to the plane $\mathcal{P}$ is the distance to the point P on $\mathcal{P}$ closest to Q. Prove

$$\text{Distance from } Q \text{ to } \mathcal{P} = \frac{|ax_1 + by_1 + cz_1 - d|}{\|\mathbf{n}\|} \qquad \boxed{8}$$

73. Use Eq. (7) to find the point P nearest to $Q = (2, 1, 2)$ on the plane $x + y + z = 1$.

74. Find the point P nearest to $Q = (-1, 3, -1)$ on the plane

$$x - 4z = 2$$

75. Use Eq. (8) to find the distance from $Q = (1, 1, 1)$ to the plane $2x + y + 5z = 2$.

76. Find the distance from $Q = (1, 2, 2)$ to the plane $\mathbf{n} \cdot \langle x, y, z \rangle = 3$, where $\mathbf{n} = \left\langle \frac{3}{5}, \frac{4}{5}, 0 \right\rangle$.

77. What is the distance from $Q = (a, b, c)$ to the plane $x = 0$? Visualize your answer geometrically and explain without computation. Then verify that Eq. (8) yields the same answer.

78. The equation of a plane $\mathbf{n} \cdot \langle x, y, z \rangle = d$ is said to be in **normal form** if $\mathbf{n}$ is a unit vector. Show that in this case, $|d|$ is the distance from the plane to the origin. Write the equation of the plane $4x - 2y + 4z = 24$ in normal form.

12.6 A Survey of Quadric Surfaces

Quadric surfaces are the surface analogs of conic sections. Recall that a conic section is a curve in $\mathbf{R}^2$ defined by a quadratic equation in two variables. A quadric surface is defined by a quadratic equation in *three* variables:

$$Ax^2 + By^2 + Cz^2 + Dxy + Eyz + Fzx + ax + by + cz + d = 0 \qquad \boxed{1}$$

To ensure that Eq. (1) is genuinely quadratic, we assume that the degree-2 coefficients A, B, C, D, E, F are not all zero.

Like conic sections, quadric surfaces are classified into a small number of types. When the coordinate axes are chosen to coincide with the axes of the quadric, the equation of the quadric has a simple form. The quadric is then said to be in **standard position**. In standard position, the coefficients D, E, F are all zero and the linear part ($ax + by + cz + d$) also reduces to just the constant term d. In this short survey of quadric surfaces, we restrict our attention to quadrics in standard position. The idea here is not to memorize the formulas for the various quadric surfaces, but rather to be able to recognize and graph them using cross sections obtained by slicing the surface with certain planes, as we describe below.

The surface analogs of ellipses are the egg-shaped **ellipsoids** (Figure 1). In standard form, an ellipsoid has the equation

$$\boxed{\textbf{Ellipsoid} \quad \left(\frac{x}{a}\right)^2 + \left(\frac{y}{b}\right)^2 + \left(\frac{z}{c}\right)^2 = 1}$$

For $a = b = c$, this equation is equivalent to $x^2 + y^2 + z^2 = a^2$ and the ellipsoid is a sphere of radius a.

Surfaces are often represented graphically by a mesh of curves called **traces**, obtained by intersecting the surface with planes parallel to one of the coordinate planes (Figure 2), yielding certain cross sections of the surface. Algebraically, this corresponds to **freezing** one of the three variables (holding it constant). For example, the intersection of the horizontal plane $z = z_0$ with the surface is a horizontal trace curve.

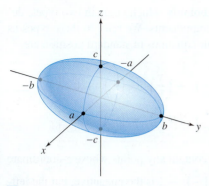

DF FIGURE 1 Ellipsoid with equation $\left(\frac{x}{a}\right)^2 + \left(\frac{y}{b}\right)^2 + \left(\frac{z}{c}\right)^2 = 1$.

■ **EXAMPLE 1** **The Traces of an Ellipsoid** Describe the traces of the ellipsoid

$$\left(\frac{x}{5}\right)^2 + \left(\frac{y}{7}\right)^2 + \left(\frac{z}{9}\right)^2 = 1$$

Solution First we observe that the traces in the coordinate planes are ellipses (Figure 3A):

xy-trace (set $z = 0$, blue in figure): $\left(\frac{x}{5}\right)^2 + \left(\frac{y}{7}\right)^2 = 1$

yz-trace (set $x = 0$, green in figure): $\left(\frac{y}{7}\right)^2 + \left(\frac{z}{9}\right)^2 = 1$

xz-trace (set $y = 0$, red in figure): $\left(\frac{x}{5}\right)^2 + \left(\frac{z}{9}\right)^2 = 1$

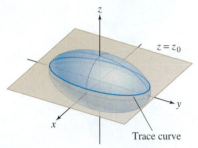

DF FIGURE 2 The intersection of the plane $z = z_0$ with an ellipsoid is an ellipse.

In fact, all the traces of an ellipsoid are ellipses (or just single points). For example, the horizontal trace defined by setting $z = z_0$ is the ellipse [Figure 3(B)]

$$\text{Trace at height } z_0: \quad \left(\frac{x}{5}\right)^2 + \left(\frac{y}{7}\right)^2 + \left(\frac{z_0}{9}\right)^2 = 1 \quad \text{or} \quad \frac{x^2}{25} + \frac{y^2}{49} = \underbrace{1 - \frac{z_0^2}{81}}_{\text{A constant}}$$

The trace at height $z_0 = 9$ is the single point $(0, 0, 9)$ because $x^2/25 + y^2/49 = 0$ has only one solution: $x = 0$, $y = 0$. Similarly, for $z_0 = -9$, the trace is the point $(0, 0, -9)$. If $|z_0| > 9$, then $1 - z_0^2/81 < 0$ and the plane lies above or below the ellipsoid. The trace has no points in this case. The traces in the vertical planes $x = x_0$ and $y = y_0$ have a similar description [Figure 3(C)]. ■

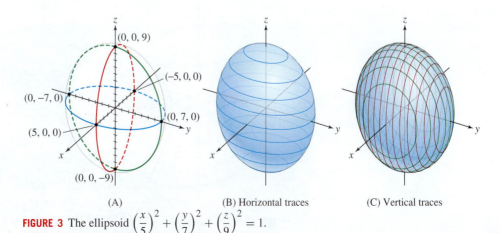

(A)　　　　　(B) Horizontal traces　　　　　(C) Vertical traces

FIGURE 3 The ellipsoid $\left(\frac{x}{5}\right)^2 + \left(\frac{y}{7}\right)^2 + \left(\frac{z}{9}\right)^2 = 1$.

The analogs of the hyperbolas are the **hyperboloids**, which come in two types, depending on whether the surface has one or two components. We refer to these types as hyperboloids of one or two sheets (Figure 4). Their equations in standard position are

Hyperboloids   One Sheet: $\left(\dfrac{x}{a}\right)^2 + \left(\dfrac{y}{b}\right)^2 = \left(\dfrac{z}{c}\right)^2 + 1$

Two Sheets: $\left(\dfrac{x}{a}\right)^2 + \left(\dfrac{y}{b}\right)^2 = \left(\dfrac{z}{c}\right)^2 - 1$

2

Notice that a hyperboloid of two sheets does not contain any points whose z-coordinate satisfies $-c < z < c$ because the right-hand side $\left(\dfrac{z}{c}\right)^2 - 1$ is then negative, but the left-hand side of the equation is greater than or equal to zero.

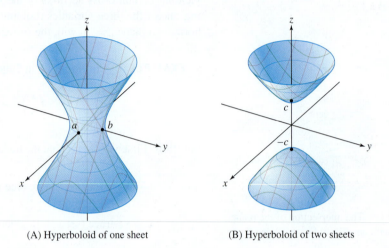

DF **FIGURE 4** Hyperboloids of one and two sheets.

(A) Hyperboloid of one sheet   (B) Hyperboloid of two sheets

■ **EXAMPLE 2   The Traces of a Hyperboloid of One Sheet**   Determine the traces of the hyperboloid $\left(\dfrac{x}{2}\right)^2 + \left(\dfrac{y}{3}\right)^2 = \left(\dfrac{z}{4}\right)^2 + 1$.

Solution The horizontal traces are ellipses and the vertical traces (parallel to either the yz-plane or the xz-plane) are hyperbolas or pairs of crossed lines (Figure 5):

Trace $z = z_0$ (ellipse, blue in figure): $\left(\dfrac{x}{2}\right)^2 + \left(\dfrac{y}{3}\right)^2 = \left(\dfrac{z_0}{4}\right)^2 + 1$

Trace $x = x_0$ (hyperbola, green in figure): $\left(\dfrac{y}{3}\right)^2 - \left(\dfrac{z}{4}\right)^2 = 1 - \left(\dfrac{x_0}{2}\right)^2$

Trace $y = y_0$ (hyperbola, red in figure): $\left(\dfrac{x}{2}\right)^2 - \left(\dfrac{z}{4}\right)^2 = 1 - \left(\dfrac{y_0}{3}\right)^2$ ■

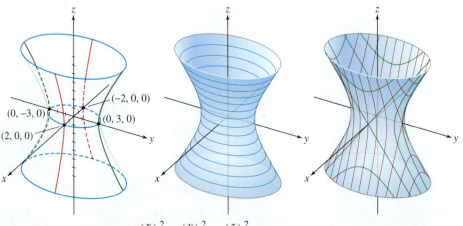

DF **FIGURE 5** The hyperboloid $\left(\dfrac{x}{2}\right)^2 + \left(\dfrac{y}{3}\right)^2 = \left(\dfrac{z}{4}\right)^2 + 1$.

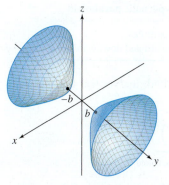

Hyperboloid of two sheets

DF **FIGURE 6** The two-sheeted hyperboloid $\left(\frac{x}{a}\right)^2 + \left(\frac{z}{c}\right)^2 = \left(\frac{y}{b}\right)^2 - 1$.

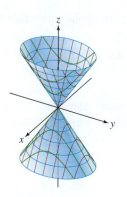

DF **FIGURE 7** Elliptic cone $\left(\frac{x}{a}\right)^2 + \left(\frac{y}{b}\right)^2 = \left(\frac{z}{c}\right)^2$.

Paraboloids play an important role in the optimization of functions of two variables. The elliptic paraboloid in Figure 8 has a local minimum at the origin. The hyperbolic paraboloid is a "saddle shape" at the origin, which is an analog for surfaces of a point of inflection.

■ **EXAMPLE 3** Hyperboloid of Two Sheets Symmetric About the *y*-axis Show that $\left(\frac{x}{a}\right)^2 + \left(\frac{z}{c}\right)^2 = \left(\frac{y}{b}\right)^2 - 1$ has no points for $-b < y < b$.

Solution This equation does not have the same form as Eq. (3) because the variables *y* and *z* have been interchanged. This hyperboloid is symmetric about the *y*-axis rather than the *z*-axis (Figure 6). The left-hand side of the equation is always ≥ 0. Thus, there are no solutions with $|y| < b$ because the right-hand side is $\left(\frac{y}{b}\right)^2 - 1 < 0$. Therefore, the hyperboloid has two sheets, corresponding to $y \geq b$ and $y \leq -b$. ■

The following equation defines an **elliptic cone** (Figure 7):

> **Elliptic Cone:** $\left(\frac{x}{a}\right)^2 + \left(\frac{y}{b}\right)^2 = \left(\frac{z}{c}\right)^2$

An elliptic cone may be thought of as a limiting case of a hyperboloid of one sheet in which we "pinch the waist" down to a point. Elliptic cones can be difficult to graph. Notice what happens when we intersect the elliptic cone with the coordinate planes. When $z = 0$, we get $\left(\frac{x}{a}\right)^2 + \left(\frac{y}{b}\right)^2 = 0$, which forces $x = y = 0$. Thus, the only intersection with the *xy*-plane is the origin. When $y = 0$, we get $\left(\frac{x}{a}\right)^2 = \left(\frac{z}{c}\right)^2$, which is the pair of diagonal lines in the *xz*-plane given by $z = \pm \left(\frac{c}{a}\right) x$. When $x = 0$, we have $\left(\frac{y}{b}\right)^2 = \left(\frac{z}{c}\right)^2$, which is the pair of diagonal lines in the *yz*-plane given by $z = \pm \left(\frac{c}{b}\right) y$. But given these diagonal lines, how could we see they are generating an elliptic cone? The trick is to slice the surface with planes parallel to the *xy*-plane. For instance, if we set $z = 1$, then the trace in the plane $z = 1$ is given by $\left(\frac{x}{a}\right)^2 + \left(\frac{y}{b}\right)^2 = \left(\frac{1}{c}\right)^2$. This is the equation of an ellipse. If we slice with the plane $z = 2$, we obtain a larger ellipse. We obtain similar ellipses when we slice with the planes $z = -1$ and $z = -2$. Hence, we can recognize that the resulting surface is the elliptic cone appearing in Figure 7.

The third main family of quadric surfaces are the **paraboloids**. There are two types—elliptic and hyperbolic. In standard position, their equations are

> **Paraboloids** Elliptic: $z = \left(\frac{x}{a}\right)^2 + \left(\frac{y}{b}\right)^2$
>
> Hyperbolic: $z = \left(\frac{x}{a}\right)^2 - \left(\frac{y}{b}\right)^2$

3

Let's compare their traces (Figure 8):

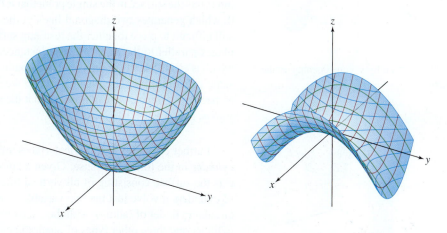

(A) Elliptic paraboloid
$z = \left(\frac{x}{2}\right)^2 + \left(\frac{y}{3}\right)^2$

(B) Hyperbolic paraboloid
$z = \left(\frac{x}{2}\right)^2 - \left(\frac{y}{3}\right)^2$

FIGURE 8

	Elliptic paraboloid	Hyperbolic paraboloid
Horizontal traces	ellipses	hyperbolas
Vertical traces	upward parabolas	upward and downward parabolas

Notice, for example, that for the hyperbolic paraboloid, the vertical traces $x = x_0$ are downward parabolas (green in the figure)

$$\underbrace{z = -\left(\frac{y}{b}\right)^2 + \left(\frac{x_0}{a}\right)^2}_{\text{Trace } x = x_0 \text{ of hyperbolic paraboloid}}$$

whereas the vertical traces $y = y_0$ are upward parabolas (red in the figure)

$$\underbrace{z = \left(\frac{x}{a}\right)^2 - \left(\frac{y_0}{b}\right)^2}_{\text{Trace } y = y_0 \text{ of hyperbolic paraboloid}}$$

It is somewhat surprising to realize that if we slice the hyperbolic paraboloid with the plane $z = 0$, we obtain $0 = \left(\frac{x}{a}\right)^2 - \left(\frac{y}{b}\right)^2$, which yields the pair of diagonal lines given by $y = \left(\frac{b}{a}\right)x$ as the trace in the xy-plane. It is not obvious that this surface would contain these two diagonal lines, but in fact it does.

■ **EXAMPLE 4** **Alternative Form of a Hyperbolic Paraboloid** Show that $z = 4xy$ is a hyperbolic paraboloid by writing the equation in terms of the variables $u = x + y$ and $v = x - y$.

Solution Note that $u + v = 2x$ and $u - v = 2y$. Therefore,

$$4xy = (u + v)(u - v) = u^2 - v^2$$

and thus the equation takes the form $z = u^2 - v^2$ in the coordinates $\{u, v, z\}$. The coordinates $\{u, v, z\}$ are obtained by rotating the coordinates $\{x, y, z\}$ by $45°$ about the z-axis (Figure 9). ■

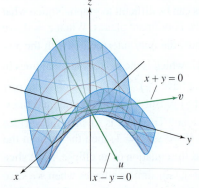

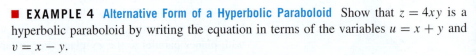

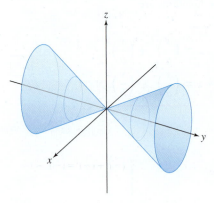

DF FIGURE 9 The hyperbolic paraboloid is defined by $z = 4xy$ or $z = u^2 - v^2$.

■ **EXAMPLE 5** Without referring back to the formulas, use traces to determine and graph the quadric surface given by $x^2 + 2z^2 - y^2 = 0$.

Solution We first slice with the coordinate planes. When $x = 0$, we obtain $2z^2 - y^2 = 0$, and therefore the trace in the yz-plane is the pair of diagonal lines $z = \pm\frac{y}{\sqrt{2}}$. When $y = 0$, we have $x^2 + 2z^2 = 0$, which has the solution $x = z = 0$. Hence, the xz-plane only intersects the surface in the single point that is the origin. When $z = 0$, we obtain $x^2 - y^2 = 0$, which generates two diagonal lines in the xy-plane given by $y = \pm x$. However, it is still difficult to piece together the resultant surface from these traces. So, we will slice with planes parallel to the xz-plane. For instance, if we set $y = 1$, we obtain $x^2 + 2z^2 = 1$, which is an ellipse. If we set $y = 2$, we obtain $x^2 + 2z^2 = 4$, which is a larger ellipse. The same pair of ellipses are obtained when we slice with the planes $y = -1$ and $y = -2$, respectively. Hence, we can now see that the surface is an elliptic cone opening along the y-axis, as in Figure 10. ■

FIGURE 10 An elliptic cone opening on the y-axis.

Further examples of quadric surfaces are the **quadratic cylinders**. We use the term *cylinder* in the following sense: Given a curve C in the xy-plane, the cylinder with base C is the surface consisting of all vertical lines passing through C (Figure 11). Equations of cylinders involve just the two variables x and y. The equation $x^2 + y^2 = r^2$ defines a circular cylinder of radius r with the z-axis as the central axis. Figure 12 shows a circular cylinder and three other types of quadratic cylinders.

The ellipsoids, hyperboloids, paraboloids, and quadratic cylinders are called **nondegenerate** quadric surfaces. There are also a certain number of "degenerate" quadric surfaces. For example, $x^2 + y^2 + z^2 = 0$ is a quadric that reduces to a single point $(0, 0, 0)$, and $(x + y + z)^2 = 1$ reduces to the union of the two planes $x + y + z = \pm 1$.

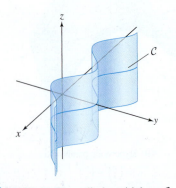

DF FIGURE 11 The cylinder with base C.

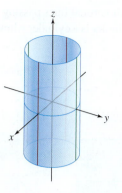

$x^2 + y^2 = r^2$

Right-circular cylinder of radius r

DF **FIGURE 12**

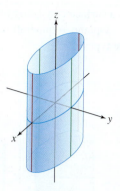

$\left(\dfrac{x}{a}\right)^2 + \left(\dfrac{y}{b}\right)^2 = 1$

Elliptic cylinder

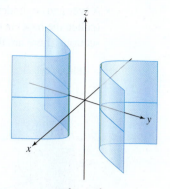

$\left(\dfrac{x}{a}\right)^2 - \left(\dfrac{y}{b}\right)^2 = 1$

Hyperbolic cylinder

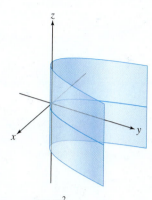

$y = ax^2$

Parabolic cylinder

12.6 SUMMARY

- A *quadric surface* is a surface defined by a quadratic equation in three variables in which the coefficients $A–F$ are not all zero:

$$Ax^2 + By^2 + Cz^2 + Dxy + Eyz + Fzx + ax + by + cz + d = 0$$

- Quadric surfaces in standard position:

Ellipsoid

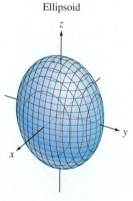

$$\left(\frac{x}{a}\right)^2 + \left(\frac{y}{b}\right)^2 + \left(\frac{z}{c}\right)^2 = 1$$

Hyperboloid (one sheet)

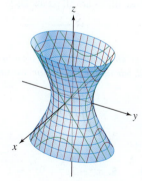

$$\left(\frac{x}{a}\right)^2 + \left(\frac{y}{b}\right)^2 = \left(\frac{z}{c}\right)^2 + 1$$

Hyperboloid (two sheets)

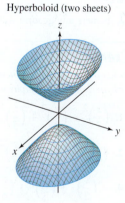

$$\left(\frac{x}{a}\right)^2 + \left(\frac{y}{b}\right)^2 = \left(\frac{z}{c}\right)^2 - 1$$

Paraboloid (elliptic)

$$z = \left(\frac{x}{a}\right)^2 + \left(\frac{y}{b}\right)^2$$

Paraboloid (hyperbolic)

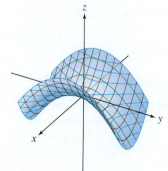

$$z = \left(\frac{x}{a}\right)^2 - \left(\frac{y}{b}\right)^2$$

Cone (elliptic)

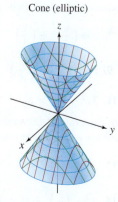

$$\left(\frac{x}{a}\right)^2 + \left(\frac{y}{b}\right)^2 = \left(\frac{z}{c}\right)^2$$

• A (vertical) cylinder is a surface consisting of all vertical lines passing through a curve (called the base) in the xy-plane. A quadratic cylinder is a cylinder whose base is a conic section. There are three types:

Elliptic cylinder

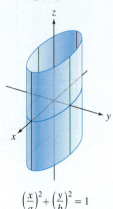

$$\left(\frac{x}{a}\right)^2 + \left(\frac{y}{b}\right)^2 = 1$$

Hyperbolic cylinder

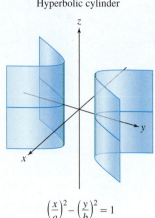

$$\left(\frac{x}{a}\right)^2 - \left(\frac{y}{b}\right)^2 = 1$$

Parabolic cylinder

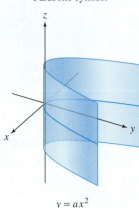

$$y = ax^2$$

12.6 EXERCISES

Preliminary Questions

1. True or false? All traces of an ellipsoid are ellipses.

2. True or false? All traces of a hyperboloid are hyperbolas.

3. Which quadric surfaces have both hyperbolas and parabolas as traces?

4. Is there any quadric surface whose traces are all parabolas?

5. A surface is called **bounded** if there exists $M > 0$ such that every point on the surface lies at a distance of at most M from the origin. Which of the quadric surfaces are bounded?

6. What is the definition of a parabolic cylinder?

Exercises

In Exercises 1–6, state whether the given equation defines an ellipsoid or hyperboloid, and if a hyperboloid, whether it is of one or two sheets.

1. $\left(\frac{x}{2}\right)^2 + \left(\frac{y}{3}\right)^2 + \left(\frac{z}{5}\right)^2 = 1$

2. $\left(\frac{x}{5}\right)^2 + \left(\frac{y}{5}\right)^2 - \left(\frac{z}{7}\right)^2 = 1$ **3.** $x^2 + 3y^2 + 9z^2 = 1$

4. $-\left(\frac{x}{2}\right)^2 - \left(\frac{y}{3}\right)^2 + \left(\frac{z}{5}\right)^2 = 1$ **5.** $x^2 - 3y^2 + 9z^2 = 1$

6. $x^2 - 3y^2 - 9z^2 = 1$

In Exercises 7–12, state whether the given equation defines an elliptic paraboloid, a hyperbolic paraboloid, or an elliptic cone.

7. $z = \left(\frac{x}{4}\right)^2 + \left(\frac{y}{3}\right)^2$ **8.** $z^2 = \left(\frac{x}{4}\right)^2 + \left(\frac{y}{3}\right)^2$

9. $z = \left(\frac{x}{9}\right)^2 - \left(\frac{y}{12}\right)^2$ **10.** $4z = 9x^2 + 5y^2$

11. $3x^2 - 7y^2 = z$ **12.** $3x^2 + 7y^2 = 14z^2$

In Exercises 13–20, state the type of the quadric surface and describe the trace obtained by intersecting with the given plane.

13. $x^2 + \left(\frac{y}{4}\right)^2 + z^2 = 1$, $y = 0$

14. $x^2 + \left(\frac{y}{4}\right)^2 + z^2 = 1$, $y = 5$

15. $x^2 + \left(\frac{y}{4}\right)^2 + z^2 = 1$, $z = \frac{1}{4}$

16. $\left(\frac{x}{2}\right)^2 + \left(\frac{y}{5}\right)^2 - 5z^2 = 1$, $x = 0$

17. $\left(\frac{x}{3}\right)^2 + \left(\frac{y}{5}\right)^2 - 5z^2 = 1$, $y = 1$

18. $4x^2 + \left(\frac{y}{3}\right)^2 - 2z^2 = -1$, $z = 1$

19. $y = 3x^2$, $z = 27$ **20.** $y = 3x^2$, $y = 27$

21. Match each of the ellipsoids in Figure 13 with the correct equation:
(a) $x^2 + 4y^2 + 4z^2 = 16$ (b) $4x^2 + y^2 + 4z^2 = 16$
(c) $4x^2 + 4y^2 + z^2 = 16$

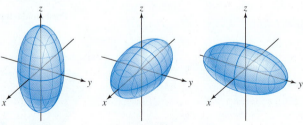

FIGURE 13

22. Describe the surface that is obtained when, in the equation $\pm 8x^2 \pm 3y^2 \pm z^2 = 1$, we choose (a) all plus signs, (b) one minus sign, and (c) two minus signs.

23. What is the equation of the surface obtained when the elliptic paraboloid $z = \left(\frac{x}{2}\right)^2 + \left(\frac{y}{4}\right)^2$ is rotated about the x-axis by $90°$? Refer to Figure 14.

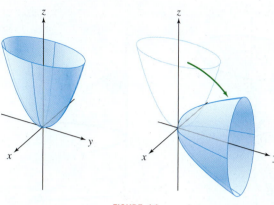

FIGURE 14

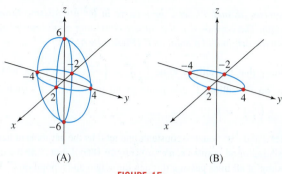

FIGURE 15

24. Describe the intersection of the horizontal plane $z = h$ and the hyperboloid $-x^2 - 4y^2 + 4z^2 = 1$. For which values of h is the intersection empty?

In Exercises 25–38, sketch the given surface.

25. $x^2 + y^2 - z^2 = 1$

26. $\left(\dfrac{x}{4}\right)^2 + \left(\dfrac{y}{8}\right)^2 + \left(\dfrac{z}{12}\right)^2 = 1$

27. $z = \left(\dfrac{x}{4}\right)^2 + \left(\dfrac{y}{8}\right)^2$

28. $z = \left(\dfrac{x}{4}\right)^2 - \left(\dfrac{y}{8}\right)^2$

29. $z^2 = \left(\dfrac{x}{4}\right)^2 + \left(\dfrac{y}{8}\right)^2$

30. $z = -x^2$

31. $x^2 - y^2 + 9z^2 = 9$

32. $y^2 + z^2 = 1$

33. $x = \sin y$

34. $x = 2y^2 - z^2$

35. $x = 1 + y^2 + z^2$

36. $x^2 - 4y^2 = z$

37. $x^2 + 9y^2 + 4z^2 = 36$

38. $y^2 - 4x^2 - z^2 = 4$

39. Find the equation of the ellipsoid passing through the points marked in Figure 15(A).

40. Find the equation of the elliptic cylinder passing through the points marked in Figure 15(B).

41. Find the equation of the hyperboloid shown in Figure 16(A).

42. Find the equation of the quadric surface shown in Figure 16(B).

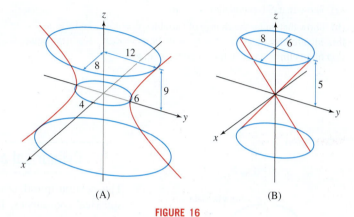

FIGURE 16

43. Determine the vertical traces of elliptic and parabolic cylinders in standard form.

44. What is the equation of a hyperboloid of one or two sheets in standard form if every horizontal trace is a circle?

45. Let C be an ellipse in a horizonal plane lying above the xy-plane. Which type of quadric surface is made up of all lines passing through the origin and a point on C?

46. The eccentricity of a conic section is defined in Section 11.5. Show that the horizontal traces of the ellipsoid

$$\left(\frac{x}{a}\right)^2 + \left(\frac{y}{b}\right)^2 + \left(\frac{z}{c}\right)^2 = 1$$

are ellipses of the same eccentricity (apart from the traces at height $h = \pm c$, which reduce to a single point). Find the eccentricity.

Further Insights and Challenges

47. Let S be the hyperboloid $x^2 + y^2 = z^2 + 1$ and let $P = (\alpha, \beta, 0)$ be a point on S in the (x, y)-plane. Show that there are precisely two lines through P entirely contained in S (Figure 17). *Hint:* Consider the line $\mathbf{r}(t) = \langle \alpha + at, \beta + bt, t \rangle$ through P. Show that $\mathbf{r}(t)$ is contained in S if (a, b) is one of the two points on the unit circle obtained by rotating (α, β) through $\pm\frac{\pi}{2}$. This proves that a hyperboloid of one sheet is a **doubly ruled surface**, which means that it can be swept out by moving a line in space in two different ways.

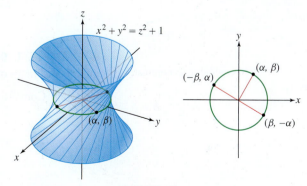

FIGURE 17

In Exercises 48 and 49, let C be a curve in $\mathbf{R}^3$ not passing through the origin. The cone on C is the surface consisting of all lines passing through the origin and a point on C [Figure 18(A)].

48. Show that the elliptic cone $\left(\dfrac{z}{c}\right)^2 = \left(\dfrac{x}{a}\right)^2 + \left(\dfrac{y}{b}\right)^2$ is, in fact, a cone on the ellipse C consisting of all points (x, y, c) such that $\left(\dfrac{x}{a}\right)^2 + \left(\dfrac{y}{b}\right)^2 = 1$.

49. Let a and c be nonzero constants and let C be the parabola at height c consisting of all points (x, ax^2, c) [Figure 18(B)]. Let S be the cone consisting of all lines passing through the origin and a point on C. This exercise shows that S is also an elliptic cone.

(a) Show that S has equation $yz = acx^2$.

(b) Show that under the change of variables $y = u + v$ and $z = u - v$, this equation becomes $acx^2 = u^2 - v^2$ or $u^2 = acx^2 + v^2$ (the equation of an elliptic cone in the variables x, v, u).

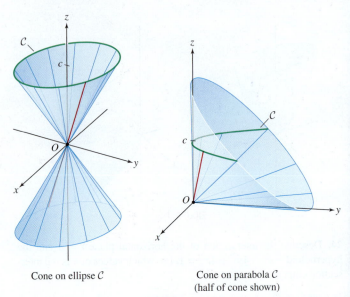

Cone on ellipse C

Cone on parabola C
(half of cone shown)

FIGURE 18

12.7 Cylindrical and Spherical Coordinates

This section introduces two generalizations of polar coordinates to $\mathbf{R}^3$: cylindrical and spherical coordinates. These coordinate systems are commonly used in problems having symmetry about an axis or rotational symmetry. For example, the magnetic field generated by a current flowing in a long, straight wire is conveniently expressed in cylindrical coordinates (Figure 1). We will also see the benefits of cylindrical and spherical coordinates when we study change of variables for multiple integrals.

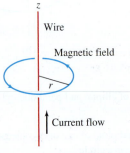

FIGURE 1 The magnetic field generated by a current flowing in a long, straight wire is conveniently expressed in cylindrical coordinates.

Cylindrical Coordinates

In cylindrical coordinates, we replace the x- and y-coordinates of a point $P = (x, y, z)$ by polar coordinates. Thus, the **cylindrical coordinates** of P are (r, θ, z), where (r, θ) are polar coordinates of the projection $Q = (x, y, 0)$ of P onto the xy-plane (Figure 2). Note that the points at fixed distance r from the z-axis make up a cylinder; hence the name cylindrical coordinates.

We convert between rectangular and cylindrical coordinates using the rectangular-polar formulas of Section 11.3. In cylindrical coordinates, we usually assume $r \geq 0$.

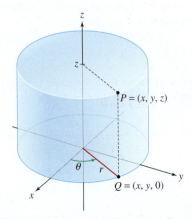

DF **FIGURE 2** P has cylindrical coordinates (r, θ, z).

Cylindrical to rectangular	Rectangular to cylindrical
$x = r \cos \theta$	$r = \sqrt{x^2 + y^2}$
$y = r \sin \theta$	$\tan \theta = \dfrac{y}{x}$
$z = z$	$z = z$

■ **EXAMPLE 1** **Converting from Cylindrical to Rectangular Coordinates** Find the rectangular coordinates of the point P with cylindrical coordinates $(r, \theta, z) = \left(2, \frac{3\pi}{4}, 5\right)$.

Solution Converting to rectangular coordinates is straightforward (Figure 3):

$$x = r \cos \theta = 2 \cos \frac{3\pi}{4} = 2\left(-\frac{\sqrt{2}}{2}\right) = -\sqrt{2}$$

$$y = r \sin \theta = 2 \sin \frac{3\pi}{4} = 2\left(\frac{\sqrt{2}}{2}\right) = \sqrt{2}$$

The z-coordinate is unchanged, so $(x, y, z) = (-\sqrt{2}, \sqrt{2}, 5)$. ■

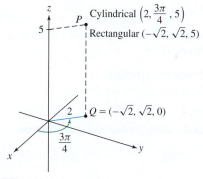

Cylindrical $\left(2, \frac{3\pi}{4}, 5\right)$

Rectangular $(-\sqrt{2}, \sqrt{2}, 5)$

$Q = (-\sqrt{2}, \sqrt{2}, 0)$

DF FIGURE 3

■ **EXAMPLE 2** **Converting from Rectangular to Cylindrical Coordinates** Find cylindrical coordinates for the point with rectangular coordinates $(x, y, z) = (-3\sqrt{3}, -3, 5)$.

Solution We have $r = \sqrt{x^2 + y^2} = \sqrt{(-3\sqrt{3})^2 + (-3)^2} = 6$. The angle θ satisfies

$$\tan \theta = \frac{y}{x} = \frac{-3}{-3\sqrt{3}} = \frac{1}{\sqrt{3}} \quad \Rightarrow \quad \theta = \frac{\pi}{6} \text{ or } \frac{7\pi}{6}$$

The correct choice is $\theta = \frac{7\pi}{6}$ because the projection $Q = (-3\sqrt{3}, -3, 0)$ lies in the third quadrant (Figure 4). The cylindrical coordinates are $(r, \theta, z) = \left(6, \frac{7\pi}{6}, 5\right)$. ■

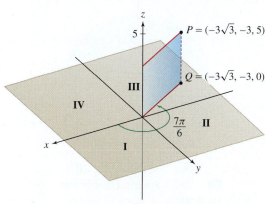

FIGURE 4 The projection Q lies in the third quadrant. Therefore, $\theta = \frac{7\pi}{6}$.

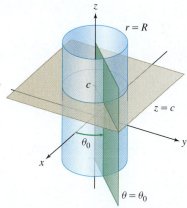

DF FIGURE 5 Level surfaces in cylindrical coordinates.

Level Surfaces in Cylindrical Coordinates:

$r = R$ *Cylinder of radius R with the z-axis as axis of symmetry*

$\theta = \theta_0$ *Half-plane through the z-axis making an angle θ_0 with the xz-plane*

$z = c$ *Horizontal plane at height c*

The **level surfaces** of a coordinate system are the surfaces obtained by setting one of the coordinates equal to a constant. In rectangular coordinates, the level surfaces are the planes $x = x_0$, $y = y_0$, and $z = z_0$. In cylindrical coordinates, the level surfaces come in three types (Figure 5). The surface $r = R$ is the cylinder of radius R consisting of all points located a distance R from the z-axis. The equation $\theta = \theta_0$ defines the half-plane of all points that project onto the ray $\theta = \theta_0$ in the (x, y)-plane. Finally, $z = c$ is the horizontal plane at height c.

■ **EXAMPLE 3** **Equations in Cylindrical Coordinates** Find an equation of the form $z = f(r, \theta)$ for the surfaces:

(a) $x^2 + y^2 + z^2 = 9$ **(b)** $x + y + z = 1$

Solution We use the formulas

$$x^2 + y^2 = r^2, \qquad x = r\cos\theta, \qquad y = r\sin\theta$$

(a) The equation $x^2 + y^2 + z^2 = 9$ becomes $r^2 + z^2 = 9$, or $z = \pm\sqrt{9 - r^2}$. This is a sphere of radius 3.

(b) The plane $x + y + z = 1$ becomes

$$z = 1 - x - y = 1 - r\cos\theta - r\sin\theta \qquad \text{or} \qquad z = 1 - r(\cos\theta + \sin\theta) \quad ■$$

■ **EXAMPLE 4** **Graphing Equations in Cylindrical Coordinates** Graph the surface corresponding to the equation in cylindrical coordinates given by $z = r^2$.

Solution We could convert the equation to rectangular coordinates and then see that the equation was a quadric surface and use the techniques developed for them to graph it. However, graphing it without converting will be easier. We first graph its intersection with the yz-plane. Note that for a point in the yz-plane, $|y|$ is the distance back to the z-axis. However, r is always the distance back to the z-axis. So in the yz-plane, y and r behave similarly. In this example, we can therefore replace $z = r^2$ with $z = y^2$ in the yz-plane. Once we have graphed this parabola, since there is no constraint on θ, we rotate the parabola around the z-axis to obtain the circular paraboloid shown in Figure 6. ■

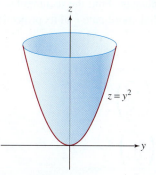

$z = y^2$

FIGURE 6 Rotating the curve $z = y^2$ about the z-axis yields the surface $z = r^2$.

Spherical Coordinates

Spherical coordinates make use of the fact that a point P on a sphere of radius ρ is determined by two angular coordinates θ and ϕ (Figure 7):

- θ is the polar angle of the projection Q of P onto the xy-plane.
- ϕ is the **angle of declination**, which measures how much the ray through P declines from the vertical.

Thus, P is determined by the triple (ρ, θ, ϕ), which are called **spherical coordinates**.

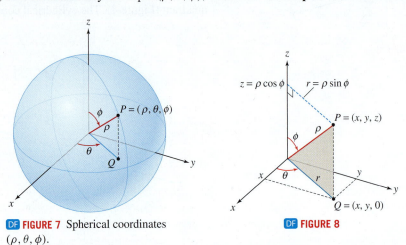

DF **FIGURE 7** Spherical coordinates (ρ, θ, ϕ).

DF **FIGURE 8**

- The symbol ϕ (usually pronounced "fee," but sometimes pronounced "fie") is the twenty-first letter of the Greek alphabet.
- We use ρ (written out as "rho" and pronounced "row") for the radial coordinate, although r is also used to denote distance from the origin in other contexts.

Suppose that $P = (x, y, z)$ in rectangular coordinates. Since ρ is the distance from P to the origin,

$$\rho = \sqrt{x^2 + y^2 + z^2}$$

On the other hand, we see in Figure 8 that

$$\tan \theta = \frac{y}{x}, \qquad \cos \phi = \frac{z}{\rho}$$

The radial coordinate r of $Q = (x, y, 0)$ is $r = \rho \sin \phi$, and therefore,

$$x = r \cos \theta = \rho \sin \phi \cos \theta, \qquad y = r \sin \theta = \rho \sin \phi \sin \theta, \qquad z = \rho \cos \phi$$

Spherical Coordinates:

ρ = distance from origin

θ = polar angle in the xy-plane

ϕ = angle of declination from the vertical

In some textbooks, θ is referred to as the azimuthal angle and ϕ as the polar angle.

Spherical to rectangular	Rectangular to spherical
$x = \rho \sin \phi \cos \theta$	$\rho = \sqrt{x^2 + y^2 + z^2}$
$y = \rho \sin \phi \sin \theta$	$\tan \theta = \dfrac{y}{x}$
$z = \rho \cos \phi$	$\cos \phi = \dfrac{z}{\rho}$

■ **EXAMPLE 5** **From Spherical to Rectangular Coordinates** Find the rectangular coordinates of $P = (\rho, \theta, \phi) = \left(3, \frac{\pi}{3}, \frac{\pi}{4}\right)$, and find the radial coordinate r of its projection Q onto the xy-plane.

Solution By the formulas above,

$$x = \rho \sin \phi \cos \theta = 3 \sin \frac{\pi}{4} \cos \frac{\pi}{3} = 3 \left(\frac{\sqrt{2}}{2}\right) \frac{1}{2} = \frac{3\sqrt{2}}{4}$$

$$y = \rho \sin \phi \sin \theta = 3 \sin \frac{\pi}{4} \sin \frac{\pi}{3} = 3 \left(\frac{\sqrt{2}}{2}\right) \frac{\sqrt{3}}{2} = \frac{3\sqrt{6}}{4}$$

$$z = \rho \cos \phi = 3 \cos \frac{\pi}{4} = 3 \frac{\sqrt{2}}{2} = \frac{3\sqrt{2}}{2}$$

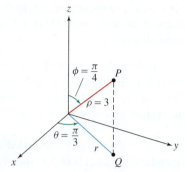

FIGURE 9 Point with spherical coordinates $\left(3, \frac{\pi}{3}, \frac{\pi}{4}\right)$.

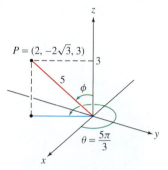

DF **FIGURE 10** Point with rectangular coordinates $(2, -2\sqrt{3}, 3)$.

Now consider the projection $Q = (x, y, 0) = \left(\frac{3\sqrt{2}}{4}, \frac{3\sqrt{6}}{4}, 0\right)$ (Figure 9). The radial coordinate r of Q satisfies

$$r^2 = x^2 + y^2 = \left(\frac{3\sqrt{2}}{4}\right)^2 + \left(\frac{3\sqrt{6}}{4}\right)^2 = \frac{9}{2}$$

Therefore, $r = 3/\sqrt{2}$. ∎

■ **EXAMPLE 6** **From Rectangular to Spherical Coordinates** Find the spherical coordinates of the point $P = (x, y, z) = (2, -2\sqrt{3}, 3)$.

Solution The radial coordinate is $\rho = \sqrt{2^2 + (-2\sqrt{3})^2 + 3^2} = \sqrt{25} = 5$. The angular coordinate θ satisfies

$$\tan \theta = \frac{y}{x} = \frac{-2\sqrt{3}}{2} = -\sqrt{3} \quad \Rightarrow \quad \theta = \frac{2\pi}{3} \ \text{or} \ \frac{5\pi}{3}$$

Since the point $(x, y) = (2, -2\sqrt{3})$ lies in the fourth quadrant, the correct choice is $\theta = \frac{5\pi}{3}$ (Figure 10). Finally, $\cos \phi = \frac{z}{\rho} = \frac{3}{5}$ and so $\phi = \cos^{-1} \frac{3}{5} \approx 0.93$. Therefore, P has spherical coordinates $\left(5, \frac{5\pi}{3}, 0.93\right)$. ∎

Figure 11 shows the three types of level surfaces in spherical coordinates. Notice that if $\phi \neq 0, \frac{\pi}{2}$ or π, then the level surface $\phi = \phi_0$ is the right circular cone consisting of points P such that $\overline{OP}$ makes an angle ϕ_0 with the z-axis. There are three exceptional cases: $\phi = \frac{\pi}{2}$ defines the xy-plane, $\phi = 0$ is the positive z-axis, and $\phi = \pi$ is the negative z-axis.

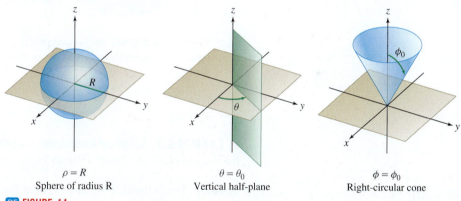

$\rho = R$	$\theta = \theta_0$	$\phi = \phi_0$
Sphere of radius R	Vertical half-plane	Right-circular cone

DF **FIGURE 11**

■ **EXAMPLE 7** **Finding an Equation in Spherical Coordinates** Find an equation of the form $\rho = f(\theta, \phi)$ for the following surfaces:

(a) $x^2 + y^2 + z^2 = 9$ **(b)** $z = x^2 - y^2$

Solution

(a) The equation $x^2 + y^2 + z^2 = 9$ defines the sphere of radius 3 centered at the origin. Since $\rho^2 = x^2 + y^2 + z^2$, the equation in spherical coordinates is $\rho = 3$.

(b) To convert $z = x^2 - y^2$ to spherical coordinates, we substitute the formulas for x, y, and z in terms of ρ, θ, and ϕ:

$$\underbrace{\rho \cos \phi}_{z} = \underbrace{(\rho \sin \phi \cos \theta)^2}_{x^2} - \underbrace{(\rho \sin \phi \sin \theta)^2}_{y^2}$$

$$\cos \phi = \rho \sin^2 \phi (\cos^2 \theta - \sin^2 \theta) \qquad \text{(divide by } \rho \text{ and factor)}$$

$$\cos \phi = \rho \sin^2 \phi \cos 2\theta \qquad \text{(since } \cos^2 \theta - \sin^2 \theta = \cos 2\theta\text{)}$$

Solving for ρ, we obtain $\rho = \dfrac{\cos \phi}{\sin^2 \phi \cos 2\theta}$, which is valid for $\phi \neq 0, \pi$, and when $\theta \neq \pi/4, 3\pi/4, 5\pi/4, 7\pi/4$. ∎

The angular coordinates (θ, ϕ) on a sphere of fixed radius are closely related to the longitude-latitude system used to identify points on the surface of the earth (Figure 12). By convention, in this system we use degrees rather than radians.

- A **longitude** is a half-circle stretching from the North to the South Pole (Figure 13). The axes are chosen so that $\theta = 0$ passes through Greenwich, England (this longitude is called the *prime meridian*). We designate the longitude by an angle between 0 and 180° together with a label E or W, according to whether it lies to the east or west of the prime meridian.
- The set of points on the sphere satisfying $\phi = \phi_0$ is a horizontal circle called a **latitude**. We measure latitudes from the equator and use the label N or S to specify the Northern or Southern Hemisphere. Thus, in the upper hemisphere $0 \le \phi_0 \le 90°$, a spherical coordinate ϕ_0 corresponds to the latitude $(90° - \phi_0)$ N. In the lower hemisphere $90° \le \phi_0 \le 180°$, ϕ_0 corresponds to the latitude $(\phi_0 - 90°)$ S.

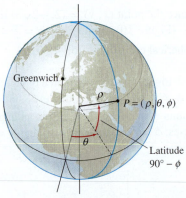

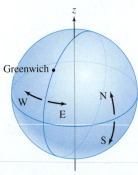

FIGURE 12 Longitude and latitude provide spherical coordinates on the surface of the earth.

FIGURE 13 Latitude is measured from the equator and is labeled N (north) in the upper hemisphere, and S (south) in the lower hemisphere.

■ **EXAMPLE 8** Spherical Coordinates via Longitude and Latitude Find the angles (θ, ϕ) for Nairobi (1.17° S, 36.48° E) and Ottawa (45.27° N, 75.42° W).

Solution For Nairobi, $\theta = 36.48°$ since the longitude lies to the east of Greenwich. Nairobi's latitude is south of the equator, so $1.17 = \phi_0 - 90$ and $\phi_0 = 91.17°$.

For Ottawa, we have $\theta = 360 - 75.42 = 284.58°$ because 75.42° W refers to 75.42° in the negative θ direction. Since the latitude of Ottawa is north of the equator, $45.27 = 90 - \phi_0$ and $\phi_0 = 44.73°$. ■

■ **EXAMPLE 9** Graphing Equations in Spherical Coordinates Graph the surface corresponding to the equation in spherical coordinates given by $\rho = \sec \phi$.

Solution We could plug in values for ϕ and obtain the corresponding values for ρ and then plot points. but that will not be entirely accurate. Instead, notice that we can rewrite the equation:

$$\rho = \frac{1}{\cos \phi}$$

$$\rho \cos \phi = 1$$

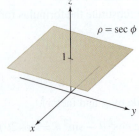

FIGURE 14 This plane is the graph of $\rho = \sec \phi$.

From our conversion equations, we see that this is $z = 1$. Hence, our surface is simply the horizontal plane at height $z = 1$, as in Figure 14. ■

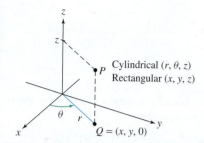

FIGURE 15 Cylindrical coordinates (r, θ, z).

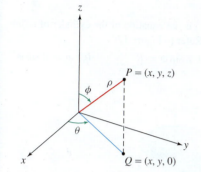

FIGURE 16 Spherical coordinates (ρ, θ, ϕ).

12.7 SUMMARY

- Conversion from rectangular to cylindrical (Figure 15) and spherical coordinates (Figure 16):

	Cylindrical	Spherical
	$r = \sqrt{x^2 + y^2}$	$\rho = \sqrt{x^2 + y^2 + z^2}$
	$\tan\theta = \dfrac{y}{x}$	$\tan\theta = \dfrac{y}{x}$
	$z = z$	$\cos\phi = \dfrac{z}{\rho}$

The angles are chosen so that

$$0 \le \theta < 2\pi \quad \text{(cylindrical or spherical)}, \qquad 0 \le \phi \le \pi \quad \text{(spherical)}$$

- Conversion to rectangular coordinates:

Cylindrical (r, θ, z)	Spherical (ρ, θ, ϕ)
$x = r\cos\theta$	$x = \rho\sin\phi\cos\theta$
$y = r\sin\theta$	$y = \rho\sin\phi\sin\theta$
$z = z$	$z = \rho\cos\phi$

- Level surfaces:

Cylindrical		Spherical	
$r = R$:	Cylinder of radius R	$\rho = R$:	Sphere of radius R
$\theta = \theta_0$:	Vertical half-plane	$\theta = \theta_0$:	Vertical half-plane
$z = c$:	Horizontal plane	$\phi = \phi_0$:	Right circular cone

12.7 EXERCISES

Preliminary Questions

1. Describe the surfaces $r = R$ in cylindrical coordinates and $\rho = R$ in spherical coordinates.

2. Which statement about cylindrical coordinates is correct?
(a) If $\theta = 0$, then P lies on the z-axis.
(b) If $\theta = 0$, then P lies in the xz-plane.

3. Which statement about spherical coordinates is correct?

(a) If $\phi = 0$, then P lies on the z-axis.
(b) If $\phi = 0$, then P lies in the xy-plane.

4. The level surface $\phi = \phi_0$ in spherical coordinates, usually a cone, reduces to a half-line for two values of ϕ_0. Which two values?

5. For which value of ϕ_0 is $\phi = \phi_0$ a plane? Which plane?

Exercises

In Exercises 1–4, convert from cylindrical to rectangular coordinates.

1. $(4, \pi, 4)$

2. $\left(2, \dfrac{\pi}{3}, -8\right)$

3. $\left(0, \dfrac{\pi}{5}, \dfrac{1}{2}\right)$

4. $\left(1, \dfrac{\pi}{2}, -2\right)$

In Exercises 5–10, convert from rectangular to cylindrical coordinates.

5. $(1, -1, 1)$

6. $(2, 2, 1)$

7. $(1, \sqrt{3}, 7)$

8. $\left(\dfrac{3}{2}, \dfrac{3\sqrt{3}}{2}, 9\right)$

9. $\left(\dfrac{5}{\sqrt{2}}, \dfrac{5}{\sqrt{2}}, 2\right)$

10. $(3, 3\sqrt{3}, 2)$

In Exercises 11–16, describe the set in cylindrical coordinates.

11. $x^2 + y^2 \le 1$

12. $x^2 + y^2 + z^2 \le 1$

13. $y^2 + z^2 \le 4, \quad x = 0$

14. $x^2 + y^2 + z^2 = 4, \quad x \ge 0, \quad y \ge 0, \quad z \ge 0$

15. $x^2 + y^2 \le 9, \quad x \ge y$

16. $y^2 + z^2 \le 9, \quad x \ge y$

In Exercises 17–26, sketch the set (described in cylindrical coordinates).

17. $r = 4$

18. $\theta = \dfrac{\pi}{3}$

19. $z = -2$

20. $r = 2, \quad z = 3$

21. $1 \leq r \leq 3, \quad 0 \leq z \leq 4$

22. $z = r$

23. $r = \sin\theta$ (*Hint:* Convert to rectangular.)

24. $1 \leq r \leq 3, \quad 0 \leq \theta \leq \frac{\pi}{2}, \quad 0 \leq z \leq 4$

25. $z^2 + r^2 \leq 4$

26. $r \leq 3, \quad \pi \leq \theta \leq \frac{3\pi}{2}, \quad z = 4$

In Exercises 27–32, find an equation of the form $r = f(\theta, z)$ in cylindrical coordinates for the following surfaces.

27. $z = x + y$

28. $x^2 + y^2 + z^2 = 4$

29. $\frac{x^2}{yz} = 1$

30. $x^2 - y^2 = 4$

31. $x^2 + y^2 = 4$

32. $z = 3xy$

In Exercises 33–38, convert from spherical to rectangular coordinates.

33. $\left(3, 0, \frac{\pi}{2}\right)$

34. $\left(2, \frac{\pi}{4}, \frac{\pi}{3}\right)$

35. $(3, \pi, 0)$

36. $\left(5, \frac{3\pi}{4}, \frac{\pi}{4}\right)$

37. $\left(6, \frac{\pi}{6}, \frac{5\pi}{6}\right)$

38. $(0.5, 3.7, 2)$

In Exercises 39–44, convert from rectangular to spherical coordinates.

39. $(\sqrt{3}, 0, 1)$

40. $\left(\frac{\sqrt{3}}{2}, \frac{3}{2}, 1\right)$

41. $(1, 1, 1)$

42. $(1, -1, 1)$

43. $\left(\frac{1}{2}, \frac{\sqrt{3}}{2}, \sqrt{3}\right)$

44. $\left(\frac{\sqrt{2}}{2}, \frac{\sqrt{2}}{2}, \sqrt{3}\right)$

In Exercises 45 and 46, convert from cylindrical to spherical coordinates.

45. $(2, 0, 2)$

46. $(3, \pi, \sqrt{3})$

In Exercises 47 and 48, convert from spherical to cylindrical coordinates.

47. $\left(4, 0, \frac{\pi}{4}\right)$

48. $\left(2, \frac{\pi}{3}, \frac{\pi}{6}\right)$

In Exercises 49–54, describe the given set in spherical coordinates.

49. $x^2 + y^2 + z^2 \leq 1$

50. $x^2 + y^2 + z^2 = 1, \quad z \geq 0$

51. $x^2 + y^2 + z^2 = 1, \quad x \geq 0, \quad y \geq 0, \quad z \geq 0$

52. $x^2 + y^2 + z^2 \leq 1, \quad x = y, \quad x \geq 0, \quad y \geq 0$

53. $y^2 + z^2 \leq 4, \quad x = 0$

54. $x^2 + y^2 = 3z^2$

In Exercises 55–64, sketch the set of points (described in spherical coordinates).

55. $\rho = 4$

56. $\phi = \frac{\pi}{4}$

57. $\rho = 2, \quad \theta = \frac{\pi}{4}$

58. $\rho = 2, \quad \phi = \frac{\pi}{4}$

59. $\rho = 2, \quad 0 \leq \phi \leq \frac{\pi}{2}$

60. $\theta = \frac{\pi}{2}, \quad \phi = \frac{\pi}{4}, \quad \rho \geq 1$

61. $\rho \leq 2, \quad 0 \leq \theta \leq \frac{\pi}{2}, \quad \frac{\pi}{2} \leq \phi \leq \pi$

62. $\rho = 1, \quad \frac{\pi}{3} \leq \phi \leq \frac{2\pi}{3}$

63. $\rho = \csc\phi$

64. $\rho = \csc\phi\cot\phi$

In Exercises 65–70, find an equation of the form $\rho = f(\theta, \phi)$ in spherical coordinates for the following surfaces.

65. $z = 2$ **66.** $z^2 = 3(x^2 + y^2)$ **67.** $x = z^2$

68. $z = x^2 + y^2$ **69.** $x^2 - y^2 = 4$ **70.** $xy = z$

71. Which of (a)–(c) is the equation of the cylinder of radius R in spherical coordinates? Refer to Figure 17.

(a) $R\rho = \sin\phi$ **(b)** $\rho\sin\phi = R$ **(c)** $\rho = R\sin\phi$

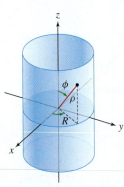

FIGURE 17

72. Let $P_1 = (1, -\sqrt{3}, 5)$ and $P_2 = (-1, \sqrt{3}, 5)$ in rectangular coordinates. In which quadrants do the projections of P_1 and P_2 onto the xy-plane lie? Find the polar angle θ of each point.

73. Find the spherical angles (θ, ϕ) for Helsinki, Finland (60.1° N, 25.0° E), and São Paulo, Brazil (23.52° S, 46.52° W).

74. Find the longitude and latitude for the points on the globe with angular coordinates $(\theta, \phi) = (\pi/8, 7\pi/12)$ and $(4, 2)$.

75. Consider a rectangular coordinate system with its origin at the center of the earth, z-axis through the North Pole, and x-axis through the prime meridian. Find the rectangular coordinates of Sydney, Australia (34° S, 151° E), and Bogotá, Colombia (4° 32′ N, 74° 15′ W). A minute is $1/60°$. Assume that the earth is a sphere of radius $R = 6370$ km.

76. Find the equation in rectangular coordinates of the quadric surface consisting of the two cones $\phi = \frac{\pi}{4}$ and $\phi = \frac{3\pi}{4}$.

77. Find an equation of the form $z = f(r, \theta)$ in cylindrical coordinates for $z^2 = x^2 - y^2$.

78. Show that $\rho = 2\cos\phi$ is the equation of a sphere with its center on the z-axis. Find its radius and center.

79. An apple modeled by taking all the points in and on a sphere of radius 2 in. is cored with a vertical cylinder of radius 1 in. Use inequalities in cylindrical coordinates to describe the set of all points that remain in the apple once the core is removed.

80. Repeat Problem 79 using inequalities in spherical coordinates.

81. 📖 Explain the following statement: If the equation of a surface in cylindrical or spherical coordinates does not involve the coordinate θ, then the surface is rotationally symmetric with respect to the z-axis.

82. *CAS* Plot the surface $\rho = 1 - \cos\phi$. Then plot the trace of S in the xz-plane and explain why S is obtained by rotating this trace.

83. Find equations $r = g(\theta, z)$ (cylindrical) and $\rho = f(\theta, \phi)$ (spherical) for the hyperboloid $x^2 + y^2 = z^2 + 1$ (Figure 18). Do there exist points on the hyperboloid with $\phi = 0$ or π? Which values of ϕ occur for points on the hyperboloid?

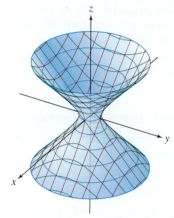

FIGURE 18 The hyperboloid $x^2 + y^2 = z^2 + 1$.

Further Insights and Challenges

*In Exercises 84–88, a **great circle** on a sphere S with center O and radius R is a circle obtained by intersecting S with a plane that passes through O (Figure 19). If P and Q are not antipodal (on opposite sides), there is a unique great circle through P and Q on S (intersect S with the plane through O, P, and Q). The geodesic distance from P to Q is defined as the length of the smaller of the two circular arcs of this great circle.*

84. Show that the geodesic distance from P to Q is equal to $R\psi$, where ψ is the *central angle* between P and Q (the angle between the vectors $\mathbf{v} = \overrightarrow{OP}$ and $\mathbf{u} = \overrightarrow{OQ}$).

85. Show that the geodesic distance from $Q = (a, b, c)$ to the North Pole $P = (0, 0, R)$ is equal to $R\cos^{-1}\left(\dfrac{c}{R}\right)$.

86. The coordinates of Los Angeles are 34° N and 118° W. Find the geodesic distance from the North Pole to Los Angeles, assuming that the earth is a sphere of radius $R = 6370$ km.

87. Show that the central angle ψ between points P and Q on a sphere (of any radius) with angular coordinates (θ, ϕ) and (θ', ϕ') is equal to

$$\psi = \cos^{-1}\left(\sin\phi\sin\phi'\cos(\theta - \theta') + \cos\phi\cos\phi'\right)$$

Hint: Compute the dot product of $\overrightarrow{OP}$ and $\overrightarrow{OQ}$. Check this formula by computing the geodesic distance between the North and South Poles.

88. Use Exercise 87 to find the geodesic distance between Los Angeles (34° N, 118° W) and Bombay (19° N, 72.8° E).

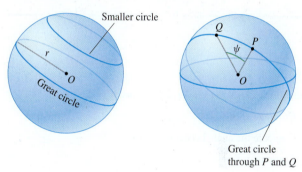

FIGURE 19

CHAPTER REVIEW EXERCISES

In Exercises 1–6, let $\mathbf{v} = \langle -2, 5\rangle$ and $\mathbf{w} = \langle 3, -2\rangle$.

1. Calculate $5\mathbf{w} - 3\mathbf{v}$ and $5\mathbf{v} - 3\mathbf{w}$.

2. Sketch $\mathbf{v}$, $\mathbf{w}$, and $2\mathbf{v} - 3\mathbf{w}$.

3. Find the unit vector in the direction of $\mathbf{v}$.

4. Find the length of $\mathbf{v} + \mathbf{w}$.

5. Express $\mathbf{i}$ as a linear combination $r\mathbf{v} + s\mathbf{w}$.

6. Find a scalar α such that $\|\mathbf{v} + \alpha\mathbf{w}\| = 6$.

7. If $P = (1, 4)$ and $Q = (-3, 5)$, what are the components of $\overrightarrow{PQ}$? What is the length of $\overrightarrow{PQ}$?

8. Let $A = (2, -1)$, $B = (1, 4)$, and $P = (2, 3)$. Find the point Q such that $\overrightarrow{PQ}$ is equivalent to $\overrightarrow{AB}$. Sketch $\overrightarrow{PQ}$ and $\overrightarrow{AB}$.

9. Find the vector with length 3 making an angle of $\frac{7\pi}{4}$ with the positive x-axis.

10. Calculate $3(\mathbf{i} - 2\mathbf{j}) - 6(\mathbf{i} + 6\mathbf{j})$.

11. Find the value of β for which $\mathbf{w} = \langle -2, \beta\rangle$ is parallel to $\mathbf{v} = \langle 4, -3\rangle$.

12. Let $P = (1, 4, -3)$.

(a) Find the point Q such that $\overrightarrow{PQ}$ is equivalent to $\langle 3, -1, 5\rangle$.

(b) Find a unit vector $\mathbf{e}$ equivalent to $\overrightarrow{PQ}$.

13. Let $\mathbf{w} = \langle 2, -2, 1\rangle$ and $\mathbf{v} = \langle 4, 5, -4\rangle$. Solve for $\mathbf{u}$ if $\mathbf{v} + 5\mathbf{u} = 3\mathbf{w} - \mathbf{u}$.

14. Let $\mathbf{v} = 3\mathbf{i} - \mathbf{j} + 4\mathbf{k}$. Find the length of $\mathbf{v}$ and the vector $2\mathbf{v} + 3(4\mathbf{i} - \mathbf{k})$.

15. Find a parametrization $\mathbf{r}_1(t)$ of the line passing through $(1, 4, 5)$ and $(-2, 3, -1)$. Then find a parametrization $\mathbf{r}_2(t)$ of the line parallel to $\mathbf{r}_1$ passing through $(1, 0, 0)$.

16. Let $\mathbf{r}_1(t) = \mathbf{v}_1 + t\mathbf{w}_1$ and $\mathbf{r}_2(t) = \mathbf{v}_2 + t\mathbf{w}_2$ be parametrizations of lines $\mathcal{L}_1$ and $\mathcal{L}_2$. For each statement (a)–(e), provide a proof if the statement is true and a counterexample if it is false.

(a) If $\mathcal{L}_1 = \mathcal{L}_2$, then $\mathbf{v}_1 = \mathbf{v}_2$ and $\mathbf{w}_1 = \mathbf{w}_2$.
(b) If $\mathcal{L}_1 = \mathcal{L}_2$ and $\mathbf{v}_1 = \mathbf{v}_2$, then $\mathbf{w}_1 = \mathbf{w}_2$.
(c) If $\mathcal{L}_1 = \mathcal{L}_2$ and $\mathbf{w}_1 = \mathbf{w}_2$, then $\mathbf{v}_1 = \mathbf{v}_2$.
(d) If $\mathcal{L}_1$ is parallel to $\mathcal{L}_2$, then $\mathbf{w}_1 = \mathbf{w}_2$.
(e) If $\mathcal{L}_1$ is parallel to $\mathcal{L}_2$, then $\mathbf{w}_1 = \lambda\mathbf{w}_2$ for some scalar λ.

17. Find a and b such that the lines $\mathbf{r}_1 = \langle 1, 2, 1 \rangle + t \langle 1, -1, 1 \rangle$ and $\mathbf{r}_2 = \langle 3, -1, 1 \rangle + t \langle a, b, -2 \rangle$ are parallel.

18. Find a such that the lines $\mathbf{r}_1 = \langle 1, 2, 1 \rangle + t \langle 1, -1, 1 \rangle$ and $\mathbf{r}_2 = \langle 3, -1, 1 \rangle + t \langle a, 4, -2 \rangle$ intersect.

19. Sketch the vector sum $\mathbf{v} = \mathbf{v}_1 - \mathbf{v}_2 + \mathbf{v}_3$ for the vectors in Figure 1(A).

20. Sketch the sums $\mathbf{v}_1 + \mathbf{v}_2 + \mathbf{v}_3$, $\mathbf{v}_1 + 2\mathbf{v}_2$, and $\mathbf{v}_2 - \mathbf{v}_3$ for the vectors in Figure 1(B).

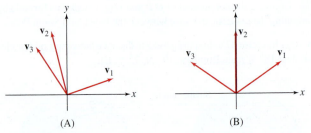

(A)   (B)

FIGURE 1

In Exercises 21–26, let $\mathbf{v} = \langle 1, 3, -2 \rangle$ and $\mathbf{w} = \langle 2, -1, 4 \rangle$.

21. Compute $\mathbf{v} \cdot \mathbf{w}$.

22. Compute the angle between $\mathbf{v}$ and $\mathbf{w}$.

23. Compute $\mathbf{v} \times \mathbf{w}$.

24. Find the area of the parallelogram spanned by $\mathbf{v}$ and $\mathbf{w}$.

25. Find the volume of the parallelepiped spanned by $\mathbf{v}$, $\mathbf{w}$, and $\mathbf{u} = \langle 1, 2, 6 \rangle$.

26. Find all the vectors orthogonal to both $\mathbf{v}$ and $\mathbf{w}$.

27. Use vectors to prove that the line connecting the midpoints of two sides of a triangle is parallel to the third side.

28. Let $\mathbf{v} = \langle 1, -1, 3 \rangle$ and $\mathbf{w} = \langle 4, -2, 1 \rangle$.
(a) Find the decomposition $\mathbf{v} = \mathbf{v}_{\|\mathbf{w}} + \mathbf{v}_{\perp\mathbf{w}}$ with respect to $\mathbf{w}$.
(b) Find the decomposition $\mathbf{w} = \mathbf{w}_{\|\mathbf{v}} + \mathbf{w}_{\perp\mathbf{v}}$ with respect to $\mathbf{v}$.

29. Calculate the component of $\mathbf{v} = \left\langle -2, \frac{1}{2}, 3 \right\rangle$ along $\mathbf{w} = \langle 1, 2, 2 \rangle$.

30. Calculate the magnitude of the forces on the two ropes in Figure 2.

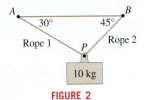

FIGURE 2

31. A 50-kg wagon is pulled to the right by a force $\mathbf{F}_1$ making an angle of $30°$ with the ground. At the same time, the wagon is pulled to the left by a horizontal force $\mathbf{F}_2$.

(a) Find the magnitude of $\mathbf{F}_1$ in terms of the magnitude of $\mathbf{F}_2$ if the wagon does not move.

(b) What is the maximal magnitude of $\mathbf{F}_1$ that can be applied to the wagon without lifting it?

32. Let $\mathbf{v}$, $\mathbf{w}$, and $\mathbf{u}$ be the vectors in $\mathbf{R}^3$. Which of the following is a scalar?

(a) $\mathbf{v} \times (\mathbf{u} + \mathbf{w})$
(b) $(\mathbf{u} + \mathbf{w}) \cdot (\mathbf{v} \times \mathbf{w})$
(c) $(\mathbf{u} \times \mathbf{w}) + (\mathbf{w} - \mathbf{v})$

In Exercises 33–36, let $\mathbf{v} = \langle 1, 2, 4 \rangle$, $\mathbf{u} = \langle 6, -1, 2 \rangle$, and $\mathbf{w} = \langle 1, 0, -3 \rangle$. Calculate the given quantity.

33. $\mathbf{v} \times \mathbf{w}$ **34.** $\mathbf{w} \times \mathbf{u}$

35. $\det \begin{pmatrix} \mathbf{u} \\ \mathbf{v} \\ \mathbf{w} \end{pmatrix}$ **36.** $\mathbf{v} \cdot (\mathbf{u} \times \mathbf{w})$

37. Use the cross product to find the area of the triangle whose vertices are $(1, 3, -1)$, $(2, -1, 3)$, and $(4, 1, 1)$.

38. Calculate $\|\mathbf{v} \times \mathbf{w}\|$ if $\|\mathbf{v}\| = 2$, $\mathbf{v} \cdot \mathbf{w} = 3$, and the angle between $\mathbf{v}$ and $\mathbf{w}$ is $\frac{\pi}{6}$.

39. Show that if the vectors $\mathbf{v}$, $\mathbf{w}$ are orthogonal, then $\|\mathbf{v} + \mathbf{w}\|^2 = \|\mathbf{v}\|^2 + \|\mathbf{w}\|^2$.

40. Find the angle between $\mathbf{v}$ and $\mathbf{w}$ if $\|\mathbf{v} + \mathbf{w}\| = \|\mathbf{v}\| = \|\mathbf{w}\|$.

41. Find $\|\mathbf{e} - 4\mathbf{f}\|$, assuming that $\mathbf{e}$ and $\mathbf{f}$ are unit vectors such that $\|\mathbf{e} + \mathbf{f}\| = \sqrt{3}$.

42. Find the area of the parallelogram spanned by vectors $\mathbf{v}$ and $\mathbf{w}$ such that $\|\mathbf{v}\| = \|\mathbf{w}\| = 2$ and $\mathbf{v} \cdot \mathbf{w} = 1$.

43. Show that the equation $\langle 1, 2, 3 \rangle \times \mathbf{v} = \langle -1, 2, a \rangle$ has no solution for $a \neq -1$.

44. Prove with a diagram the following: If $\mathbf{e}$ is a unit vector orthogonal to $\mathbf{v}$, then $\mathbf{e} \times (\mathbf{v} \times \mathbf{e}) = (\mathbf{e} \times \mathbf{v}) \times \mathbf{e} = \mathbf{v}$.

45. Use the identity

$$\mathbf{u} \times (\mathbf{v} \times \mathbf{w}) = (\mathbf{u} \cdot \mathbf{w}) \mathbf{v} - (\mathbf{u} \cdot \mathbf{v}) \mathbf{w}$$

to prove that

$$\mathbf{u} \times (\mathbf{v} \times \mathbf{w}) + \mathbf{v} \times (\mathbf{w} \times \mathbf{u}) + \mathbf{w} \times (\mathbf{u} \times \mathbf{v}) = \mathbf{0}$$

46. Find an equation of the plane through $(1, -3, 5)$ with normal vector $\mathbf{n} = \langle 2, 1, -4 \rangle$.

47. Write the equation of the plane $\mathcal{P}$ with vector equation

$$\langle 1, 4, -3 \rangle \cdot \langle x, y, z \rangle = 7$$

in the form

$$a(x - x_0) + b(y - y_0) + c(z - z_0) = 0$$

Hint: You must find a point $P = (x_0, y_0, z_0)$ on $\mathcal{P}$.

48. Find all the planes parallel to the plane passing through the points $(1, 2, 3)$, $(1, 2, 7)$, and $(1, 1, -3)$.

49. Find the plane through $P = (4, -1, 9)$ containing the line $\mathbf{r}(t) = \langle 1, 4, -3 \rangle + t \langle 2, 1, 1 \rangle$.

50. Find the intersection of the line $\mathbf{r}(t) = \langle 3t + 2, 1, -7t \rangle$ and the plane $2x - 3y + z = 5$.

51. Find the trace of the plane $3x - 2y + 5z = 4$ in the xy-plane.

52. Find the intersection of the planes $x + y + z = 1$ and $3x - 2y + z = 5$.

In Exercises 53–58, determine the type of the quadric surface.

53. $\left(\dfrac{x}{3}\right)^2 + \left(\dfrac{y}{4}\right)^2 + 2z^2 = 1$

54. $\left(\dfrac{x}{3}\right)^2 - \left(\dfrac{y}{4}\right)^2 + 2z^2 = 1$

55. $\left(\dfrac{x}{3}\right)^2 + \left(\dfrac{y}{4}\right)^2 - 2z = 0$

56. $\left(\dfrac{x}{3}\right)^2 - \left(\dfrac{y}{4}\right)^2 - 2z = 0$

57. $\left(\dfrac{x}{3}\right)^2 - \left(\dfrac{y}{4}\right)^2 - 2z^2 = 0$

58. $\left(\dfrac{x}{3}\right)^2 - \left(\dfrac{y}{4}\right)^2 - 2z^2 = 1$

59. Determine the type of the quadric surface $ax^2 + by^2 - z^2 = 1$ if:

(a) $a < 0, \quad b < 0$

(b) $a > 0, \quad b > 0$

(c) $a > 0, \quad b < 0$

60. Describe the traces of the surface

$$\left(\frac{x}{2}\right)^2 - y^2 + \left(\frac{z}{2}\right)^2 = 1$$

in the three coordinate planes.

61. Convert $(x, y, z) = (3, 4, -1)$ from rectangular to cylindrical and spherical coordinates.

62. Convert $(r, \theta, z) = \left(3, \frac{\pi}{6}, 4\right)$ from cylindrical to spherical coordinates.

63. Convert the point $(\rho, \theta, \phi) = \left(3, \frac{\pi}{6}, \frac{\pi}{3}\right)$ from spherical to cylindrical coordinates.

64. Describe the set of all points $P = (x, y, z)$ satisfying $x^2 + y^2 \leq 4$ in both cylindrical and spherical coordinates.

65. Sketch the graph of the cylindrical equation $z = 2r \cos \theta$ and write the equation in rectangular coordinates.

66. Write the surface $x^2 + y^2 - z^2 = 2(x + y)$ as an equation $r = f(\theta, z)$ in cylindrical coordinates.

67. Show that the cylindrical equation

$$r^2(1 - 2\sin^2 \theta) + z^2 = 1$$

is a hyperboloid of one sheet.

68. Sketch the graph of the spherical equation $\rho = 2\cos\theta \sin\phi$ and write the equation in rectangular coordinates.

69. Describe how the surface with spherical equation

$$\rho^2(1 + A\cos^2 \phi) = 1$$

depends on the constant A.

70. Show that the spherical equation $\cot \phi = 2\cos\theta + \sin\theta$ defines a plane through the origin (with the origin excluded). Find a normal vector to this plane.

71. Let c be a scalar, $\mathbf{a}$ and $\mathbf{b}$ be vectors, and $\mathbf{X} = \langle x, y, z \rangle$. Show that the equation $(\mathbf{X} - \mathbf{a}) \cdot (\mathbf{X} - \mathbf{b}) = c^2$ defines a sphere with center $\mathbf{m} = \frac{1}{2}(\mathbf{a} + \mathbf{b})$ and radius R, where $R^2 = c^2 + \left\|\frac{1}{2}(\mathbf{a} - \mathbf{b})\right\|^2$.

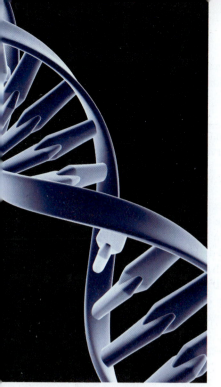

DNA polymers form helical curves whose spatial orientation influences their biochemical properties. (*Alfred Pasieka/Science Source*)

13 CALCULUS OF VECTOR-VALUED FUNCTIONS

I n this chapter, we study vector-valued functions and their derivatives, and we use them to analyze curves and motion in 3-space. Although many techniques from single-variable calculus carry over to the vector setting, there is an important new aspect to the derivative. A real-valued function f can change in just one of two ways: It can increase or decrease. By contrast, a vector-valued function can change not just in magnitude but also in direction, and the rate of change is not a single number but is itself a vector. To develop these new concepts, we begin with an introduction to vector-valued functions.

13.1 Vector-Valued Functions

Consider a particle moving in $\mathbf{R}^3$ whose coordinates at time t are $(x(t), y(t), z(t))$. It is convenient to represent the particle's path by the **vector-valued function**

Functions f (with real number values) are often called scalar-valued to distinguish them from vector-valued functions.

$$\boxed{\mathbf{r}(t) = \langle x(t), y(t), z(t)\rangle = x(t)\mathbf{i} + y(t)\mathbf{j} + z(t)\mathbf{k}} \qquad \boxed{1}$$

Think of $\mathbf{r}(t)$ as a moving vector that points from the origin to the position of the particle at time t (Figure 1).

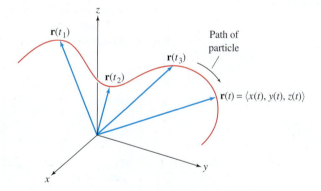

DF **FIGURE 1**

The parameter is often called t (for time), but we are free to use any other variable such as s or θ. It is best to avoid writing $\mathbf{r}(x)$ or $\mathbf{r}(y)$ to prevent confusion with the x- and y-components of $\mathbf{r}$.

More generally, a vector-valued function is any function $\mathbf{r}(t)$ of the form in Eq. (1) whose domain $\mathcal{D}$ is a set of real numbers and whose range is a set of position vectors. The variable t is called a **parameter**, and the functions $x(t), y(t), z(t)$ are called the **components** or **coordinate functions**. We usually take as domain the set of all values of t for which $\mathbf{r}(t)$ is defined—that is, all values of t that belong to the domains of all three coordinate functions $x(t), y(t), z(t)$. For example,

$$\mathbf{r}(t) = \langle t^2, e^t, 4 - 7t\rangle, \qquad \text{domain } \mathcal{D} = \mathbf{R}$$
$$\mathbf{r}(s) = \langle \sqrt{s}, e^s, s^{-1}\rangle, \qquad \text{domain } \mathcal{D} = \{s \in \mathbf{R} : s > 0\}$$

The terminal point of a vector-valued function $\mathbf{r}(t)$ traces a path in $\mathbf{R}^3$ as t varies. We refer to $\mathbf{r}(t)$ either as a path or as a **vector parametrization** of a path. We shall assume throughout this chapter that the components of $\mathbf{r}(t)$ have continuous derivatives.

We have already studied special cases of vector parametrizations. In Chapter 12, we described lines in $\mathbf{R}^3$ using vector parametrizations. Recall that

$$\mathbf{r}(t) = \langle x_0, y_0, z_0 \rangle + t\mathbf{v} = \langle x_0 + ta, y_0 + tb, z_0 + tc \rangle$$

parametrizes the line through $P = (x_0, y_0, z_0)$ in the direction of the vector $\mathbf{v} = \langle a, b, c \rangle$.

In Chapter 11, we studied parametrized curves in the plane $\mathbf{R}^2$ in the form

$$c(t) = (x(t), y(t))$$

Such a curve is described equally well by the vector-valued function $\mathbf{r}(t) = \langle x(t), y(t) \rangle$. The difference lies only in whether we visualize the path as traced by a "moving point" $c(t)$ or a "moving vector" $\mathbf{r}(t)$. The vector form is used in this chapter because it leads most naturally to the definition of vector-valued derivatives.

It is important to distinguish between the path parametrized by $\mathbf{r}(t)$ and the underlying curve $\mathcal{C}$ traced by $\mathbf{r}(t)$. The curve $\mathcal{C}$ is the set of all points $(x(t), y(t), z(t))$ as t ranges over the domain of $\mathbf{r}(t)$. The path is a particular way of traversing the curve; it may traverse the curve several times, reverse direction, or move back and forth, etc.

■ **EXAMPLE 1**   **The Path Versus the Curve**   Describe the path

$$\mathbf{r}(t) = \langle \cos t, \sin t, 1 \rangle, \qquad -\infty < t < \infty$$

How are the path and the curve $\mathcal{C}$ traced by $\mathbf{r}(t)$ different?

Solution   As t varies from $-\infty$ to ∞, the endpoint of the vector $\mathbf{r}(t)$ moves around a unit circle at height $z = 1$ infinitely many times in the counterclockwise direction when viewed from above (Figure 2). The underlying curve $\mathcal{C}$ traced by $\mathbf{r}(t)$ is the circle itself.   ■

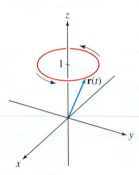

FIGURE 2 Plot of $\mathbf{r}(t) = \langle \cos t, \sin t, 1 \rangle$.

A curve in $\mathbf{R}^3$ is also referred to as a **space curve** (as opposed to a curve in $\mathbf{R}^2$, which is called a **plane curve**). Space curves can be quite complicated and difficult to sketch by hand. The most effective way to visualize a space curve is to plot it from different viewpoints using a computer (Figure 3). As an aid to visualization, we plot a "thickened" curve as in Figures 3 and 5, but keep in mind that space curves are one-dimensional and have no thickness.

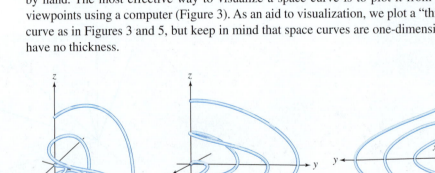

FIGURE 3 The curve $\mathbf{r}(t) = \langle t \sin 2t \cos t, t \sin^2 t, t \cos t \rangle$ for $0 \le t \le 4\pi$, seen from three different viewpoints.

The projections onto the coordinate planes are another aid in visualizing space curves. The projection of a path $\mathbf{r}(t) = \langle x(t), y(t), z(t) \rangle$ onto the xy-plane is the path $\mathbf{p}(t) = \langle x(t), y(t), 0 \rangle$ (Figure 4). Similarly, the projections onto the yz- and xz-planes are the paths $\langle 0, y(t), z(t) \rangle$ and $\langle x(t), 0, z(t) \rangle$, respectively.

■ **EXAMPLE 2**   **Helix**   Describe the curve traced by $\mathbf{r}(t) = \langle -\sin t, \cos t, t \rangle$ for $t \ge 0$ in terms of its projections onto the coordinate planes.

Solution   The projections are as follows (Figure 4):

- xy-plane (set $z = 0$): the path $\mathbf{p}(t) = \langle -\sin t, \cos t, 0 \rangle$, which describes a point moving counterclockwise around the unit circle starting at $\mathbf{p}(0) = (0, 1, 0)$

- xz-plane (set $y = 0$): the path $\langle -\sin t, 0, t \rangle$, which is a wave in the z-direction
- yz-plane (set $x = 0$): the path $\langle 0, \cos t, t \rangle$, which is a wave in the z-direction

The function $\mathbf{r}(t)$ describes a point moving above the unit circle in the xy-plane, while its height $z = t$ increases linearly, resulting in the helix of Figure 4. ∎

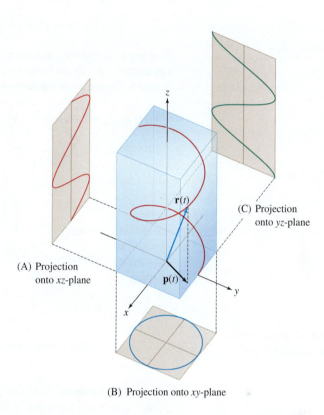

(C) Projection onto yz-plane

(A) Projection onto xz-plane

(B) Projection onto xy-plane

FIGURE 4 Projections of the helix $\mathbf{r}(t) = \langle -\sin t, \cos t, t \rangle$.

Every curve can be parametrized in infinitely many ways (because there are infinitely many ways that a particle can traverse a curve as a function of time). The next example describes two very different parametrizations of the same curve.

■ **EXAMPLE 3** **Parametrizing the Intersection of Surfaces** Parametrize the curve $\mathcal{C}$ obtained as the intersection of the surfaces $x^2 - y^2 = z - 1$ and $x^2 + y^2 = 4$ (Figure 5).

Solution We have to express the coordinates (x, y, z) of a point on the curve as functions of a parameter t. Here are two ways of doing this.

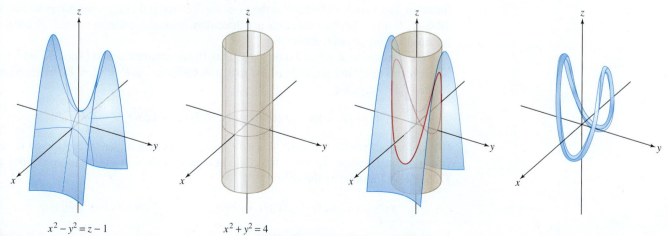

$x^2 - y^2 = z - 1$ $x^2 + y^2 = 4$

FIGURE 5 Intersection of surfaces $x^2 - y^2 = z - 1$ and $x^2 + y^2 = 4$.

First method: Solve the given equations for y and z in terms of x. First, solve for y:

$$x^2 + y^2 = 4 \quad \Rightarrow \quad y^2 = 4 - x^2 \quad \Rightarrow \quad y = \pm\sqrt{4 - x^2}$$

The equation $x^2 - y^2 = z - 1$ can be written $z = x^2 - y^2 + 1$. Thus, we can substitute $y^2 = 4 - x^2$ to solve for z:

$$z = x^2 - y^2 + 1 = x^2 - (4 - x^2) + 1 = 2x^2 - 3$$

Now use $t = x$ as the parameter. Then $y = \pm\sqrt{4 - t^2}$, $z = 2t^2 - 3$. The two signs of the square root correspond to the two halves of the curve where $y > 0$ and $y < 0$, as shown in Figure 6. Therefore, we need two vector-valued functions to parametrize the entire curve:

$$\mathbf{r}_1(t) = \left\langle t, \sqrt{4 - t^2}, 2t^2 - 3 \right\rangle, \quad \mathbf{r}_2(t) = \left\langle t, -\sqrt{4 - t^2}, 2t^2 - 3 \right\rangle, \quad -2 \le t \le 2$$

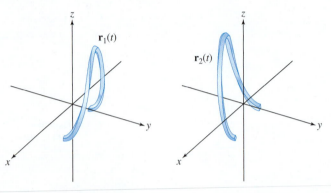

Part of curve where $y > 0$ Part of curve where $y < 0$

FIGURE 6 Two halves of the curve of intersection in Example 3.

Second method: Note that $x^2 + y^2 = 4$ has a trigonometric parametrization: $x = 2\cos t$, $y = 2\sin t$ for $0 \le t < 2\pi$. The equation $x^2 - y^2 = z - 1$ gives us

$$z = x^2 - y^2 + 1 = 4\cos^2 t - 4\sin^2 t + 1 = 4\cos 2t + 1$$

Thus, we may parametrize the entire curve by a single vector-valued function:

$$\mathbf{r}(t) = \langle 2\cos t, 2\sin t, 4\cos 2t + 1 \rangle, \qquad 0 \le t < 2\pi \qquad ■$$

■ **EXAMPLE 4** Parametrize the circle of radius 3 with its center $P = (2, 6, 8)$ located in a plane:

(a) parallel to the xy-plane.   **(b)** parallel to the xz-plane.

Solution **(a)** A circle of radius R in the xy-plane centered at the origin has parametrization $\langle R\cos t, R\sin t \rangle$. To place the circle in a three-dimensional coordinate system, we use the parametrization $\langle R\cos t, R\sin t, 0 \rangle$.

Thus, the circle of radius 3 centered at $(0, 0, 0)$ has parametrization $\langle 3\cos t, 3\sin t, 0 \rangle$. To move this circle in a parallel fashion so that its center lies at $P = (2, 6, 8)$, we translate by the vector $\langle 2, 6, 8 \rangle$:

$$\mathbf{r}_1(t) = \langle 2, 6, 8 \rangle + \langle 3\cos t, 3\sin t, 0 \rangle = \langle 2 + 3\cos t, 6 + 3\sin t, 8 \rangle$$

(b) The parametrization $\langle 3\cos t, 0, 3\sin t \rangle$ gives us a circle of radius 3 centered at the origin in the xz-plane. To move the circle in a parallel fashion so that its center lies at $(2, 6, 8)$, we translate by the vector $\langle 2, 6, 8 \rangle$:

$$\mathbf{r}_2(t) = \langle 2, 6, 8 \rangle + \langle 3\cos t, 0, 3\sin t \rangle = \langle 2 + 3\cos t, 6, 8 + 3\sin t \rangle$$

These two circles are shown in Figure 7.   ■

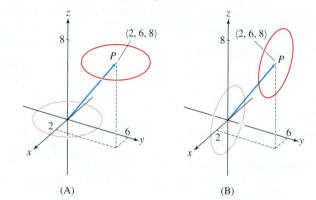

FIGURE 7 Horizontal and vertical circles of radius 3 and center $P = (2, 6, 8)$ obtained by translation.

(A)   (B)

13.1 SUMMARY

- A *vector-valued function* is a function of the form

$$\mathbf{r}(t) = \langle x(t), y(t), z(t) \rangle = x(t)\mathbf{i} + y(t)\mathbf{j} + z(t)\mathbf{k}$$

- We often think of t as time and $\mathbf{r}(t)$ as a "moving vector" whose terminal point traces out a path as a function of time. We refer to $\mathbf{r}(t)$ as a *vector parametrization* of the path, or simply as a "path."
- The underlying curve $\mathcal{C}$ traced by $\mathbf{r}(t)$ is the set of all points $(x(t), y(t), z(t))$ in $\mathbf{R}^3$ for t in the domain of $\mathbf{r}(t)$. A curve in $\mathbf{R}^3$ is also called a *space curve*.
- Every curve $\mathcal{C}$ can be parametrized in infinitely many ways.
- The projection of $\mathbf{r}(t)$ onto the xy-plane is the curve traced by $\langle x(t), y(t), 0 \rangle$. The projection onto the xz-plane is $\langle x(t), 0, z(t) \rangle$, and the projection onto the yz-plane is $\langle 0, y(t), z(t) \rangle$.

13.1 EXERCISES

Preliminary Questions

1. Which one of the following does *not* parametrize a line?

(a) $\mathbf{r}_1(t) = \langle 8 - t, 2t, 3t \rangle$

(b) $\mathbf{r}_2(t) = t^3\mathbf{i} - 7t^3\mathbf{j} + t^3\mathbf{k}$

(c) $\mathbf{r}_3(t) = \langle 8 - 4t^3, 2 + 5t^2, 9t^3 \rangle$

2. What is the projection of $\mathbf{r}(t) = t\mathbf{i} + t^4\mathbf{j} + e^t\mathbf{k}$ onto the xz-plane?

3. Which projection of $\langle \cos t, \cos 2t, \sin t \rangle$ is a circle?

4. What is the center of the circle with the following parametrization?

$$\mathbf{r}(t) = (-2 + \cos t)\mathbf{i} + 2\mathbf{j} + (3 - \sin t)\mathbf{k}$$

5. How do the paths $\mathbf{r}_1(t) = \langle \cos t, \sin t \rangle$ and $\mathbf{r}_2(t) = \langle \sin t, \cos t \rangle$ around the unit circle differ?

6. Which three of the following vector-valued functions parametrize the same space curve?

(a) $(-2 + \cos t)\mathbf{i} + 9\mathbf{j} + (3 - \sin t)\mathbf{k}$

(b) $(2 + \cos t)\mathbf{i} - 9\mathbf{j} + (-3 - \sin t)\mathbf{k}$

(c) $(-2 + \cos 3t)\mathbf{i} + 9\mathbf{j} + (3 - \sin 3t)\mathbf{k}$

(d) $(-2 - \cos t)\mathbf{i} + 9\mathbf{j} + (3 + \sin t)\mathbf{k}$

(e) $(2 + \cos t)\mathbf{i} + 9\mathbf{j} + (3 + \sin t)\mathbf{k}$

Exercises

1. What is the domain of $\mathbf{r}(t) = e^t\mathbf{i} + \dfrac{1}{t}\mathbf{j} + (t + 1)^{-3}\mathbf{k}$?

2. What is the domain of $\mathbf{r}(s) = e^s\mathbf{i} + \sqrt{s}\mathbf{j} + \cos s\mathbf{k}$?

3. Evaluate $\mathbf{r}(2)$ and $\mathbf{r}(-1)$ for $\mathbf{r}(t) = \left\langle \sin \frac{\pi}{2}t, t^2, (t^2 + 1)^{-1} \right\rangle$.

4. Does either of $P = (4, 11, 20)$ or $Q = (-1, 6, 16)$ lie on the path $\mathbf{r}(t) = \langle 1 + t, 2 + t^2, t^4 \rangle$?

5. Find a vector parametrization of the line through $P = (3, -5, 7)$ in the direction $\mathbf{v} = \langle 3, 0, 1 \rangle$.

6. Find a direction vector for the line with parametrization $\mathbf{r}(t) = (4 - t)\mathbf{i} + (2 + 5t)\mathbf{j} + \frac{1}{2}t\mathbf{k}$.

7. Determine whether the space curve given by $\mathbf{r}(t) = \langle \sin t, \cos t/2, t \rangle$ intersects the z-axis, and if it does, determine where.

8. Determine whether the space curve given by $\mathbf{r}(t) = \langle t^2, t^2 - 2t - 3, t - 3 \rangle$ intersects the x-axis, and if it does, determine where.

9. Determine whether the space curve given by $\mathbf{r}(t) = \langle t, t^3, t^2 + 1 \rangle$ intersects the xy-plane, and if it does, determine where.

10. Determine whether the space curves given by $\mathbf{r}_1(t) = \langle t, t^2, t + 1 \rangle$ and $\mathbf{r}_2(s) = \langle \sqrt{s}, s, s - 1 \rangle$ intersect, and if they do, determine where.

11. Match the space curves in Figure 8 with their projections onto the xy-plane in Figure 9.

12. Match the space curves in Figure 8 with the following vector-valued functions:

(a) $\mathbf{r}_1(t) = \langle \cos 2t, \cos t, \sin t \rangle$ **(b)** $\mathbf{r}_2(t) = \langle t, \cos 2t, \sin 2t \rangle$
(c) $\mathbf{r}_3(t) = \langle 1, t, t \rangle$

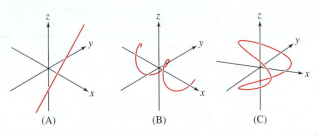

(A)   (B)   (C)

FIGURE 8

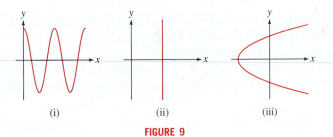

(i)   (ii)   (iii)

FIGURE 9

13. Match the vector-valued functions (a)–(f) with the space curves (i)–(vi) in Figure 10.

(a) $\mathbf{r}(t) = \langle t + 15, e^{0.08t} \cos t, e^{0.08t} \sin t \rangle$

(b) $\mathbf{r}(t) = \langle \cos t, \sin t, \sin 12t \rangle$ **(c)** $\mathbf{r}(t) = \left\langle t, t, \dfrac{25t}{1+t^2} \right\rangle$

(d) $\mathbf{r}(t) = \langle \cos^3 t, \sin^3 t, \sin 2t \rangle$ **(e)** $\mathbf{r}(t) = \langle t, t^2, 2t \rangle$
(f) $\mathbf{r}(t) = \langle \cos t, \sin t, \cos t \sin 12t \rangle$

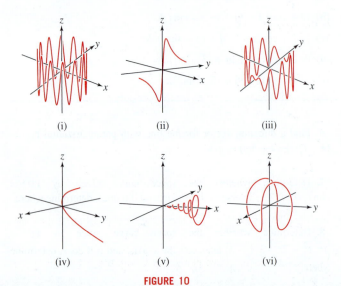

(i)   (ii)   (iii)

(iv)   (v)   (vi)

FIGURE 10

14. Which of the following curves have the same projection onto the xy-plane?

(a) $\mathbf{r}_1(t) = \langle t, t^2, e^t \rangle$ **(b)** $\mathbf{r}_2(t) = \langle e^t, t^2, t \rangle$
(c) $\mathbf{r}_3(t) = \langle t, t^2, \cos t \rangle$

15. Match the space curves (A)–(C) in Figure 11 with their projections (i)–(iii) onto the xy-plane.

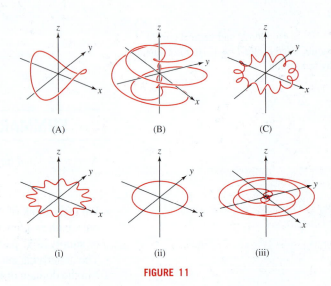

(A)   (B)   (C)

(i)   (ii)   (iii)

FIGURE 11

16. Describe the projections of the circle $\mathbf{r}(t) = \langle \sin t, 0, 4 + \cos t \rangle$ onto the coordinate planes.

In Exercises 17–20, the function $\mathbf{r}(t)$ traces a circle. Determine the radius, center, and plane containing the circle.

17. $\mathbf{r}(t) = (9 \cos t)\mathbf{i} + (9 \sin t)\mathbf{j}$

18. $\mathbf{r}(t) = 7\mathbf{i} + (12 \cos t)\mathbf{j} + (12 \sin t)\mathbf{k}$

19. $\mathbf{r}(t) = \langle \sin t, 0, 4 + \cos t \rangle$

20. $\mathbf{r}(t) = \langle 6 + 3 \sin t, 9, 4 + 3 \cos t \rangle$

21. Let $\mathcal{C}$ be the curve given by $\mathbf{r}(t) = \langle t \cos t, t \sin t, t \rangle$.

(a) Show that $\mathcal{C}$ lies on the cone $x^2 + y^2 = z^2$.

(b) Sketch the cone and make a rough sketch of $\mathcal{C}$ on the cone.

22. **CAS** Use a computer algebra system to plot the projections onto the xy- and xz-planes of the curve $\mathbf{r}(t) = \langle t \cos t, t \sin t, t \rangle$ in Exercise 21.

In Exercises 23 and 24, let

$$\mathbf{r}(t) = \langle \sin t, \cos t, \sin t \cos 2t \rangle$$

be the parametrization of the curve shown in Figure 12.

23. Find the points where $\mathbf{r}(t)$ intersects the xy-plane.

24. Show that the projection of $\mathbf{r}(t)$ onto the xz-plane is the curve

$$z = x - 2x^3 \quad \text{for} \quad -1 \le x \le 1$$

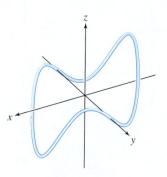

FIGURE 12

25. Parametrize the intersection of the surfaces

$$y^2 - z^2 = x - 2, \qquad y^2 + z^2 = 9$$

using $t = y$ as the parameter (two vector functions are needed as in Example 3).

26. Find a parametrization of the curve in Exercise 25 using trigonometric functions.

27. Viviani's Curve C is the intersection of the surfaces (Figure 13)

$$x^2 + y^2 = z^2, \qquad y = z^2$$

(a) Parametrize each of the two parts of C corresponding to $x \geq 0$ and $x \leq 0$, taking $t = z$ as the parameter.

(b) Describe the projection of C onto the xy-plane.

(c) Show that C lies on the sphere of radius 1 with its center $(0, 1, 0)$. This curve looks like a figure eight lying on a sphere [Figure 13(B)].

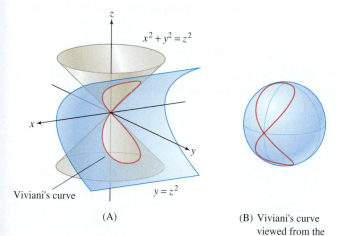

Viviani's curve

(A)

(B) Viviani's curve
viewed from the
negative y-axis

FIGURE 13 Viviani's curve is the intersection of the surfaces $x^2 + y^2 = z^2$ and $y = z^2$.

28. (a) Show that any point on $x^2 + y^2 = z^2$ can be written in the form $(z \cos\theta, z \sin\theta, z)$ for some θ.

(b) Use this to find a parametrization of Viviani's curve (Exercise 27) with θ as the parameter.

29. Use sine and cosine to parametrize the intersection of the cylinders $x^2 + y^2 = 1$ and $x^2 + z^2 = 1$ (use two vector-valued functions). Then describe the projections of this curve onto the three coordinate planes.

30. Use hyperbolic functions to parametrize the intersection of the surfaces $x^2 - y^2 = 4$ and $z = xy$.

31. Use sine and cosine to parametrize the intersection of the surfaces $x^2 + y^2 = 1$ and $z = 4x^2$ (Figure 14).

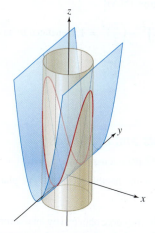

FIGURE 14 Intersection of the surfaces $x^2 + y^2 = 1$ and $z = 4x^2$.

In Exercises 32–34, two paths $\mathbf{r}_1(t)$ and $\mathbf{r}_2(t)$ intersect if there is a point P lying on both curves. We say that $\mathbf{r}_1(t)$ and $\mathbf{r}_2(t)$ collide if $\mathbf{r}_1(t_0) = \mathbf{r}_2(t_0)$ at some time t_0.

32. Which of the following statements are true?

(a) If $\mathbf{r}_1(t)$ and $\mathbf{r}_2(t)$ intersect, then they collide.

(b) If $\mathbf{r}_1(t)$ and $\mathbf{r}_2(t)$ collide, then they intersect.

(c) Intersection depends only on the underlying curves traced by $\mathbf{r}_1$ and $\mathbf{r}_2$, but collision depends on the actual parametrizations.

33. Determine whether $\mathbf{r}_1(t)$ and $\mathbf{r}_2(t)$ collide or intersect, giving the coordinates of the corresponding points if they exist:

$$\mathbf{r}_1(t) = \langle t^2 + 3, t + 1, 6t^{-1} \rangle$$

$$\mathbf{r}_2(t) = \langle 4t, 2t - 2, t^2 - 7 \rangle$$

34. Determine whether $\mathbf{r}_1(t)$ and $\mathbf{r}_2(t)$ collide or intersect, giving the coordinates of the corresponding points if they exist:

$$\mathbf{r}_1(t) = \langle t, t^2, t^3 \rangle, \qquad \mathbf{r}_2(t) = \langle 4t + 6, 4t^2, 7 - t \rangle$$

In Exercises 35–44, find a parametrization of the curve.

35. The vertical line passing through the point $(3, 2, 0)$

36. The line passing through $(1, 0, 4)$ and $(4, 1, 2)$

37. The line through the origin whose projection on the xy-plane is a line of slope 3 and whose projection on the yz-plane is a line of slope 5 (i.e., $\Delta z/\Delta y = 5$)

38. The horizontal circle of radius 1 with center $(2, -1, 4)$

39. The circle of radius 2 with center $(1, 2, 5)$ in a plane parallel to the yz-plane

40. The ellipse $\left(\dfrac{x}{2}\right)^2 + \left(\dfrac{y}{3}\right)^2 = 1$ in the xy-plane, translated to have center $(9, -4, 0)$

41. The intersection of the plane $y = \frac{1}{2}$ with the sphere $x^2 + y^2 + z^2 = 1$

42. The intersection of the surfaces

$$z = x^2 - y^2 \qquad \text{and} \qquad z = x^2 + xy - 1$$

43. The ellipse $\left(\dfrac{x}{2}\right)^2 + \left(\dfrac{z}{3}\right)^2 = 1$ in the xz-plane, translated to have center $(3, 1, 5)$ [Figure 15(A)]

44. The ellipse $\left(\dfrac{y}{2}\right)^2 + \left(\dfrac{z}{3}\right)^2 = 1$, translated to have center $(3, 1, 5)$ [Figure 15(B)]

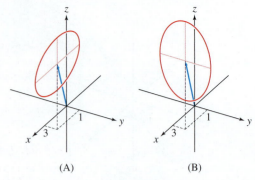

(A) (B)

FIGURE 15 The ellipses described in Exercises 43 and 44.

Further Insights and Challenges

45. Sketch the curve parametrized by $\mathbf{r}(t) = \langle |t| + t, |t| - t \rangle$.

46. Find the maximum height above the xy-plane of a point on $\mathbf{r}(t) = \langle e^t, \sin t, t(4 - t) \rangle$.

47. ▱ Let C be the curve obtained by intersecting a cylinder of radius r and a plane. Insert two spheres of radius r into the cylinder above and below the plane, and let F_1 and F_2 be the points where the plane is tangent to the spheres [Figure 16(A)]. Let K be the vertical distance between the equators of the two spheres. Rediscover Archimedes's proof that C is an ellipse by showing that every point P on C satisfies

$$PF_1 + PF_2 = K \qquad \boxed{2}$$

Hint: If two lines through a point P are tangent to a sphere and intersect the sphere at Q_1 and Q_2 as in Figure 16(B), then the segments $\overline{PQ_1}$ and $\overline{PQ_2}$ have equal length. Use this to show that $PF_1 = PR_1$ and $PF_2 = PR_2$.

48. Assume that the cylinder in Figure 16 has equation $x^2 + y^2 = r^2$ and the plane has equation $z = ax + by$. Find a vector parametrization $\mathbf{r}(t)$ of the curve of intersection using the trigonometric functions $y = \cos t$ and $y = \sin t$.

49. ⌂⌂⌂ Now reprove the result of Exercise 47 using vector geometry. Assume that the cylinder has equation $x^2 + y^2 = r^2$ and the plane has equation $z = ax + by$.

(a) Show that the upper and lower spheres in Figure 16 have centers

$$C_1 = \left(0, 0, r\sqrt{a^2 + b^2 + 1}\right)$$
$$C_2 = \left(0, 0, -r\sqrt{a^2 + b^2 + 1}\right)$$

(b) Show that the points where the plane is tangent to the sphere are

$$F_1 = \frac{r}{\sqrt{a^2 + b^2 + 1}} \left(a, b, a^2 + b^2\right)$$

$$F_2 = \frac{-r}{\sqrt{a^2 + b^2 + 1}} \left(a, b, a^2 + b^2\right)$$

Hint: Show that $\overline{C_1 F_1}$ and $\overline{C_2 F_2}$ have length r and are orthogonal to the plane.

(c) Verify, with the aid of a computer algebra system, that Eq. (2) holds with

$$K = 2r\sqrt{a^2 + b^2 + 1}$$

To simplify the algebra, observe that since a and b are arbitrary, it suffices to verify Eq. (2) for the point $P = (r, 0, ar)$.

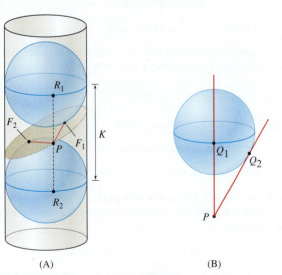

(A) (B)

FIGURE 16

13.2 Calculus of Vector-Valued Functions

In this section, we extend differentiation and integration to vector-valued functions. This is straightforward because the techniques of single-variable calculus carry over with little change. What is new and important, however, is the geometric interpretation of the derivative as a tangent vector. We describe this later in the section.

The first step is to define the limits of vector-valued functions.

> **DEFINITION Limit of a Vector-Valued Function** A vector-valued function $\mathbf{r}(t)$ approaches the limit $\mathbf{u}$ (a vector) as t approaches t_0 if $\lim_{t \to t_0} \|\mathbf{r}(t) - \mathbf{u}\| = 0$. In this case, we write
>
> $$\lim_{t \to t_0} \mathbf{r}(t) = \mathbf{u}$$

We can visualize the limit of a vector-valued function as a vector $\mathbf{r}(t)$ "moving" toward the limit vector $\mathbf{u}$ (Figure 1). According to the next theorem, vector limits may be computed componentwise.

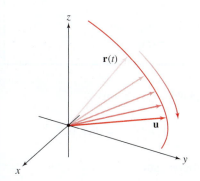

DF **FIGURE 1** The vector-valued function $\mathbf{r}(t)$ approaches the vector $\mathbf{u}$ as $t \to t_0$.

The Limit Laws of scalar functions remain valid in the vector-valued case. They are verified by applying the Limit Laws to the components.

> **THEOREM 1 Vector-Valued Limits Are Computed Componentwise** A vector-valued function $\mathbf{r}(t) = \langle x(t), y(t), z(t) \rangle$ approaches a limit as $t \to t_0$ if and only if each component approaches a limit, and in this case,
>
> $$\lim_{t \to t_0} \mathbf{r}(t) = \left\langle \lim_{t \to t_0} x(t), \lim_{t \to t_0} y(t), \lim_{t \to t_0} z(t) \right\rangle \qquad \boxed{1}$$

Proof Let $\mathbf{u} = \langle a, b, c \rangle$ and consider the square of the length

$$\|\mathbf{r}(t) - \mathbf{u}\|^2 = (x(t) - a)^2 + (y(t) - b)^2 + (z(t) - c)^2 \qquad \boxed{2}$$

The term on the left approaches zero if and only if each term on the right approaches zero (because these terms are nonnegative). It follows that $\|\mathbf{r}(t) - \mathbf{u}\|$ approaches zero if and only if $|x(t) - a|$, $|y(t) - b|$, and $|z(t) - c|$ tend to zero. Therefore, $\mathbf{r}(t)$ approaches a limit $\mathbf{u}$ as $t \to t_0$ if and only if $x(t)$, $y(t)$, and $z(t)$ converge to the components a, b, and c. ■

■ **EXAMPLE 1** Calculate $\lim_{t \to 3} \mathbf{r}(t)$, where $\mathbf{r}(t) = \langle t^2, 1 - t, t^{-1} \rangle$.

Solution By Theorem 1,

$$\lim_{t \to 3} \mathbf{r}(t) = \lim_{t \to 3} \langle t^2, 1 - t, t^{-1} \rangle = \left\langle \lim_{t \to 3} t^2, \lim_{t \to 3} (1 - t), \lim_{t \to 3} t^{-1} \right\rangle = \left\langle 9, -2, \frac{1}{3} \right\rangle \qquad ■$$

Continuity of vector-valued functions is defined in the same way as in the scalar case. A vector-valued function $\mathbf{r}(t) = \langle x(t), y(t), z(t) \rangle$ is **continuous** at t_0 if

$$\lim_{t \to t_0} \mathbf{r}(t) = \mathbf{r}(t_0)$$

By Theorem 1, $\mathbf{r}(t)$ is continuous at t_0 if and only if the components $x(t)$, $y(t)$, $z(t)$ are continuous at t_0.

We define the derivative of $\mathbf{r}(t)$ as the limit of the difference quotient:

$$\mathbf{r}'(t) = \frac{d}{dt}\mathbf{r}(t) = \lim_{h \to 0} \frac{\mathbf{r}(t + h) - \mathbf{r}(t)}{h} \qquad \boxed{3}$$

In Leibniz notation, the derivative is written $d\mathbf{r}/dt$.

We say that $\mathbf{r}(t)$ is *differentiable* at t if the limit in Eq. (3) exists. Notice that the components of the difference quotient are themselves difference quotients:

$$\lim_{h \to 0} \frac{\mathbf{r}(t+h) - \mathbf{r}(t)}{h} = \lim_{h \to 0} \left\langle \frac{x(t+h) - x(t)}{h}, \frac{y(t+h) - y(t)}{h}, \frac{z(t+h) - z(t)}{h} \right\rangle$$

and by Theorem 1, $\mathbf{r}(t)$ is differentiable if and only if the components are differentiable. In this case, $\mathbf{r}'(t)$ is equal to the vector of derivatives $\langle x'(t), y'(t), z'(t) \rangle$.

By Theorems 1 and 2, vector-valued limits and derivatives are computed "componentwise," so they are no more difficult to compute than ordinary limits and derivatives.

> **THEOREM 2 Vector-Valued Derivatives Are Computed Componentwise** A vector-valued function $\mathbf{r}(t) = \langle x(t), y(t), z(t) \rangle$ is differentiable if and only if each component is differentiable. In this case,
>
> $$\mathbf{r}'(t) = \frac{d}{dt} \mathbf{r}(t) = \langle x'(t), y'(t), z'(t) \rangle$$

Here are some vector-valued derivatives, computed componentwise:

$$\frac{d}{dt} \langle t^2, t^3, \sin t \rangle = \langle 2t, 3t^2, \cos t \rangle, \qquad \frac{d}{dt} \langle \cos t, -1, e^{2t} \rangle = \langle -\sin t, 0, 2e^{2t} \rangle$$

Higher order derivatives are defined by repeated differentiation:

$$\mathbf{r}''(t) = \frac{d}{dt} \mathbf{r}'(t), \quad \mathbf{r}'''(t) = \frac{d}{dt} \mathbf{r}''(t), \quad \dots$$

■ **EXAMPLE 2** Calculate $\mathbf{r}''(3)$, where $\mathbf{r}(t) = \langle \ln t, t, t^2 \rangle$.

Solution We perform the differentiation componentwise:

$$\mathbf{r}'(t) = \frac{d}{dt} \langle \ln t, t, t^2 \rangle = \langle t^{-1}, 1, 2t \rangle$$

$$\mathbf{r}''(t) = \frac{d}{dt} \langle t^{-1}, 1, 2t \rangle = \langle -t^{-2}, 0, 2 \rangle$$

Therefore, $\mathbf{r}''(3) = \langle -\frac{1}{9}, 0, 2 \rangle$. ■

The differentiation rules of single-variable calculus carry over to the vector setting.

> **Differentiation Rules** Assume that $\mathbf{r}(t)$, $\mathbf{r}_1(t)$, and $\mathbf{r}_2(t)$ are differentiable. Then
>
> - **Sum Rule:** $(\mathbf{r}_1(t) + \mathbf{r}_2(t))' = \mathbf{r}_1'(t) + \mathbf{r}_2'(t)$
> - **Constant Multiple Rule:** For any constant c, $(c\,\mathbf{r}(t))' = c\,\mathbf{r}'(t)$.
> - **Product Rule:** For any differentiable scalar-valued function f,
>
> $$\frac{d}{dt}\big(f(t)\mathbf{r}(t)\big) = f'(t)\mathbf{r}(t) + f(t)\mathbf{r}'(t)$$
>
> - **Chain Rule:** For any differentiable scalar-valued function g,
>
> $$\frac{d}{dt}\mathbf{r}(g(t)) = \mathbf{r}'(g(t))g'(t)$$

Proof Each rule is proved by applying the differentiation rules to the components. For example, to prove the Product Rule (we consider vector-valued functions in the plane, to keep the notation simple), we write

$$f(t)\mathbf{r}(t) = f(t)\langle x(t), y(t) \rangle = \langle f(t)x(t), f(t)y(t) \rangle$$

Now apply the Product Rule to each component:

$$\frac{d}{dt} f(t)\mathbf{r}(t) = \left\langle \frac{d}{dt} f(t)x(t), \frac{d}{dt} f(t)y(t) \right\rangle$$

$$= \langle f'(t)x(t) + f(t)x'(t), f'(t)y(t) + f(t)y'(t) \rangle$$

$$= \langle f'(t)x(t), f'(t)y(t) \rangle + \langle f(t)x'(t), f(t)y'(t) \rangle$$

$$= f'(t)\langle x(t), y(t) \rangle + f(t)\langle x'(t), y'(t) \rangle = f'(t)\mathbf{r}(t) + f(t)\mathbf{r}'(t)$$

The remaining proofs are left as exercises (Exercises 69–70). ■

■ **EXAMPLE 3** Let $\mathbf{r}(t) = \langle t^2, 5t, 1 \rangle$ and $f(t) = e^{3t}$. Calculate:

(a) $\dfrac{d}{dt} f(t)\mathbf{r}(t)$ **(b)** $\dfrac{d}{dt}\mathbf{r}(f(t))$

Solution We have $\mathbf{r}'(t) = \langle 2t, 5, 0 \rangle$ and $f'(t) = 3e^{3t}$.

(a) By the Product Rule,

$$\frac{d}{dt} f(t)\mathbf{r}(t) = f'(t)\mathbf{r}(t) + f(t)\mathbf{r}'(t) = 3e^{3t}\langle t^2, 5t, 1 \rangle + e^{3t}\langle 2t, 5, 0 \rangle$$

$$= \langle (3t^2 + 2t)e^{3t}, (15t + 5)e^{3t}, 3e^{3t} \rangle$$

Note that we could have first found $f(t)\mathbf{r}(t) = e^{3t}\langle t^2, 5t, 1 \rangle = \langle e^{3t}t^2, e^{3t}5t, e^{3t} \rangle$, and then differentiated to obtain the same answer but our goal here was to demonstrate how to apply the Product Rule for a scalar-valued function f.

(b) By the Chain Rule,

$$\frac{d}{dt}\mathbf{r}(f(t)) = \mathbf{r}'(f(t))f'(t) = \mathbf{r}'(e^{3t})3e^{3t} = \langle 2e^{3t}, 5, 0 \rangle 3e^{3t} = \langle 6e^{6t}, 15e^{3t}, 0 \rangle$$ ■

There are three different Product Rules for vector-valued functions. In addition to the rule for the product of a scalar function f and a vector-valued function $\mathbf{r}$ stated above, there are Product Rules for the dot and cross products. These rules are very important in applications, as we will see.

CAUTION *Order is important in the Product Rule for cross products. The first term in Eq. (5) must be written as*

$$\mathbf{r}_1'(t) \times \mathbf{r}_2(t)$$

not $\mathbf{r}_2(t) \times \mathbf{r}_1'(t)$. Similarly, the second term is $\mathbf{r}_1(t) \times \mathbf{r}_2'(t)$. Why is order not a concern for dot products?

THEOREM 3 Product Rule for Dot and Cross Products Assume that $\mathbf{r}_1(t)$ and $\mathbf{r}_2(t)$ are differentiable. Then

Dot Products: $\dfrac{d}{dt}(\mathbf{r}_1(t) \cdot \mathbf{r}_2(t)) = \mathbf{r}_1'(t) \cdot \mathbf{r}_2(t) + \mathbf{r}_1(t) \cdot \mathbf{r}_2'(t)$ $\boxed{4}$

Cross Products: $\dfrac{d}{dt}(\mathbf{r}_1(t) \times \mathbf{r}_2(t)) = [\mathbf{r}_1'(t) \times \mathbf{r}_2(t)] + [\mathbf{r}_1(t) \times \mathbf{r}_2'(t)]$ $\boxed{5}$

Proof We verify Eq. (4) for vector-valued functions in the plane. If $\mathbf{r}_1(t) = \langle x_1(t), y_1(t) \rangle$ and $\mathbf{r}_2(t) = \langle x_2(t), y_2(t) \rangle$, then

$$\frac{d}{dt}(\mathbf{r}_1(t) \cdot \mathbf{r}_2(t)) = \frac{d}{dt}(x_1(t)x_2(t) + y_1(t)y_2(t))$$

$$= x_1'(t)x_2(t) + x_1(t)x_2'(t) + y_1'(t)y_2(t) + y_1(t)y_2'(t)$$

$$= (x_1'(t)x_2(t) + y_1'(t)y_2(t)) + (x_1(t)x_2'(t) + y_1(t)y_2'(t))$$

$$= \mathbf{r}_1'(t) \cdot \mathbf{r}_2(t) + \mathbf{r}_1(t) \cdot \mathbf{r}_2'(t)$$

The proof of Eq. (5) is left as an exercise (Exercise 71). ■

In the next example and throughout this chapter, *all vector-valued functions are assumed differentiable, unless otherwise stated.*

■ **EXAMPLE 4** Prove the formula $\dfrac{d}{dt}\big(\mathbf{r}(t) \times \mathbf{r}'(t)\big) = \mathbf{r}(t) \times \mathbf{r}''(t)$.

Solution By the Product Formula for cross products,

$$\frac{d}{dt}\big(\mathbf{r}(t) \times \mathbf{r}'(t)\big) = \underbrace{\mathbf{r}'(t) \times \mathbf{r}'(t)}_{\text{Equals } \mathbf{0}} + \mathbf{r}(t) \times \mathbf{r}''(t) = \mathbf{r}(t) \times \mathbf{r}''(t)$$

Here, $\mathbf{r}'(t) \times \mathbf{r}'(t) = \mathbf{0}$ because the cross product of a vector with itself is zero.   ■

The Derivative as a Tangent Vector

The derivative vector $\mathbf{r}'(t_0)$ has an important geometric property: It points in the direction tangent to the path traced by $\mathbf{r}(t)$ at $t = t_0$.

To understand why, consider the difference quotient, where $\Delta\mathbf{r} = \mathbf{r}(t_0 + h) - \mathbf{r}(t_0)$ and $\Delta t = h$ with $h \neq 0$:

$$\frac{\Delta\mathbf{r}}{\Delta t} = \frac{\mathbf{r}(t_0 + h) - \mathbf{r}(t_0)}{h} \qquad \boxed{6}$$

The vector $\Delta\mathbf{r}$ points from the head of $\mathbf{r}(t_0)$ to the head of $\mathbf{r}(t_0 + h)$ as in Figure 2(A). The difference quotient $\Delta\mathbf{r}/\Delta t$ is a scalar multiple of $\Delta\mathbf{r}$ and therefore points in the same direction [Figure 2(B)].

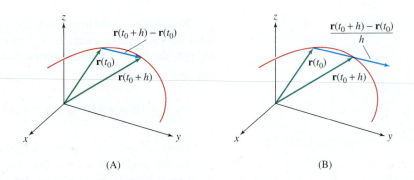

DF **FIGURE 2** The difference quotient points in the direction of $\Delta\mathbf{r} = \mathbf{r}(t_0 + h) - \mathbf{r}(t_0)$.

As $h = \Delta t$ tends to zero, $\Delta\mathbf{r}$ also tends to zero but the quotient $\Delta\mathbf{r}/\Delta t$ approaches a vector $\mathbf{r}'(t_0)$ (assuming it exists), which, if nonzero, points in the direction tangent to the curve. Figure 3 illustrates the limiting process. We refer to $\mathbf{r}'(t_0)$ as the **tangent vector** or the **velocity vector** at $\mathbf{r}(t_0)$.

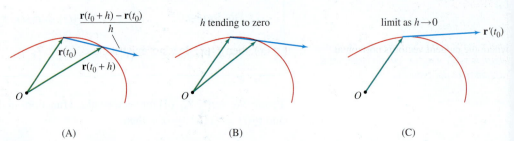

DF **FIGURE 3** The difference quotient converges to a vector $\mathbf{r}'(t_0)$, tangent to the curve.

Although it has been our convention to regard all vectors as based at the origin, the tangent vector $\mathbf{r}'(t)$ is an exception; we visualize it as a vector based at the terminal point of $\mathbf{r}(t)$. This makes sense because $\mathbf{r}'(t)$ then appears as a vector tangent to the curve (Figure 3).

The tangent vector $\mathbf{r}'(t_0)$ (if it exists and is nonzero) is a direction vector for the tangent line to the curve. Therefore, the tangent line has vector parametrization:

$$\boxed{\text{Tangent line at } \mathbf{r}(t_0): \qquad \mathbf{L}(t) = \mathbf{r}(t_0) + t\,\mathbf{r}'(t_0)} \qquad \boxed{7}$$

■ **EXAMPLE 5** **Plotting Tangent Vectors** *CAS* Plot $\mathbf{r}(t) = \langle \cos t, \sin t, 4\cos^2 t \rangle$ together with its tangent vectors at $t = \frac{\pi}{4}$ and $\frac{3\pi}{2}$. Find a parametrization of the tangent line at $t = \frac{\pi}{4}$.

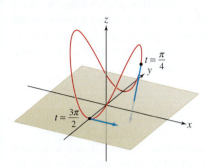

FIGURE 4 Tangent vectors to

$$\mathbf{r}(t) = \langle \cos t, \sin t, 4 \cos^2 t \rangle$$

at $t = \frac{\pi}{4}$ and $\frac{3\pi}{2}$.

Solution The derivative is $\mathbf{r}'(t) = \langle -\sin t, \cos t, -8 \cos t \sin t \rangle$, and thus the tangent vectors at $t = \frac{\pi}{4}$ and $\frac{3\pi}{2}$ are

$$\mathbf{r}'\left(\frac{\pi}{4}\right) = \left\langle -\frac{\sqrt{2}}{2}, \frac{\sqrt{2}}{2}, -4 \right\rangle, \qquad \mathbf{r}'\left(\frac{3\pi}{2}\right) = \langle 1, 0, 0 \rangle$$

Figure 4 shows a plot of $\mathbf{r}(t)$ with $\mathbf{r}'\left(\frac{\pi}{4}\right)$ based at $\mathbf{r}\left(\frac{\pi}{4}\right)$ and $\mathbf{r}'\left(\frac{3\pi}{2}\right)$ based at $\mathbf{r}\left(\frac{3\pi}{2}\right)$. At $t = \frac{\pi}{4}$, $\mathbf{r}\left(\frac{\pi}{4}\right) = \left\langle \frac{\sqrt{2}}{2}, \frac{\sqrt{2}}{2}, 2 \right\rangle$, and thus the tangent line is parametrized by

$$\mathbf{L}(t) = \mathbf{r}\left(\frac{\pi}{4}\right) + t\,\mathbf{r}'\left(\frac{\pi}{4}\right) = \left\langle \frac{\sqrt{2}}{2}, \frac{\sqrt{2}}{2}, 2 \right\rangle + t \left\langle -\frac{\sqrt{2}}{2}, \frac{\sqrt{2}}{2}, -4 \right\rangle \qquad \blacksquare$$

There are some important differences between vector- and scalar-valued derivatives. The tangent line to a plane curve $y = f(x)$ is horizontal at x_0 exactly when $f'(x_0) = 0$. But in a vector parametrization, the tangent vector $\mathbf{r}'(t_0) = \langle x'(t_0), y'(t_0) \rangle$ is horizontal and nonzero if $y'(t_0) = 0$ but $x'(t_0) \neq 0$.

Imagining the case where a vector-valued function $\mathbf{r}(t)$ describes a particle moving along a curve, when it occurs that $\mathbf{r}'(t_0) = 0$, the particle has momentarily stopped its forward motion. It could subsequently continue to move forward, or move in any other direction from that point, including reversing itself and returning along the path upon which it arrived. We will see this phenomenon in the next example.

■ **EXAMPLE 6** **Horizontal Tangent Vectors on the Cycloid** The function

$$\mathbf{r}(t) = \langle t - \sin t, 1 - \cos t \rangle$$

traces a cycloid. Find the points where:

(a) $\mathbf{r}'(t)$ is horizontal and nonzero. **(b)** $\mathbf{r}'(t)$ is the zero vector.

Solution The tangent vector is $\mathbf{r}'(t) = \langle 1 - \cos t, \sin t \rangle$. The y-component of $\mathbf{r}'(t)$ is zero if $\sin t = 0$—that is, if $t = 0, \pi, 2\pi, \dots$. We have

$$\mathbf{r}(0) = \langle 0, 0 \rangle, \quad \mathbf{r}'(0) = \langle 1 - \cos 0, \sin 0 \rangle = \langle 0, 0 \rangle \quad \text{(zero vector)}$$

$$\mathbf{r}(\pi) = \langle \pi, 2 \rangle, \quad \mathbf{r}'(\pi) = \langle 1 - \cos \pi, \sin \pi \rangle = \langle 2, 0 \rangle \quad \text{(horizontal)}$$

By periodicity, we conclude that $\mathbf{r}'(t)$ is nonzero and horizontal for $t = \pi, 3\pi, 5\pi, \dots$ and $\mathbf{r}'(t) = \mathbf{0}$ for $t = 0, 2\pi, 4\pi, \dots$ (Figure 5). ■

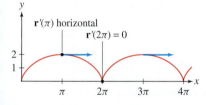

DF **FIGURE 5** Points on the cycloid

$$r(t) = \langle t - \sin t, 1 - \cos t \rangle$$

where the tangent vector is horizontal.

CONCEPTUAL INSIGHT The cycloid in Figure 5 has sharp points called **cusps** at points where $x = 0, 2\pi, 4\pi, \dots$. If we represent the cycloid as the graph of a function $y = f(x)$, then $f'(x)$ does not exist at these points. By contrast, the vector derivative $\mathbf{r}'(t) = \langle 1 - \cos t, \sin t \rangle$ exists *for all t*, but $\mathbf{r}'(t) = \mathbf{0}$ at the cusps. In general, $\mathbf{r}'(t)$ is a direction vector for the tangent line whenever it exists, but we get no information about the tangent line (which may or may not exist) at points where $\mathbf{r}'(t) = \mathbf{0}$.

The next example establishes an important property of vector-valued functions that will be used in Sections 13.4 through 13.6.

■ **EXAMPLE 7** **Orthogonality of r and r′ When r Has Constant Length** Prove that if $\mathbf{r}(t)$ has constant length, then $\mathbf{r}(t)$ is orthogonal to $\mathbf{r}'(t)$.

Solution By the Product Rule for Dot Products,

$$\frac{d}{dt} \|\mathbf{r}(t)\|^2 = \frac{d}{dt}\left(\mathbf{r}(t) \cdot \mathbf{r}(t)\right) = \mathbf{r}'(t) \cdot \mathbf{r}(t) + \mathbf{r}(t) \cdot \mathbf{r}'(t) = 2\mathbf{r}'(t) \cdot \mathbf{r}(t)$$

This derivative is zero because $\|\mathbf{r}(t)\|$ is constant. Therefore, $\mathbf{r}'(t) \cdot \mathbf{r}(t) = 0$, and $\mathbf{r}(t)$ is orthogonal to $\mathbf{r}'(t)$ [or $\mathbf{r}'(t) = \mathbf{0}$]. ■

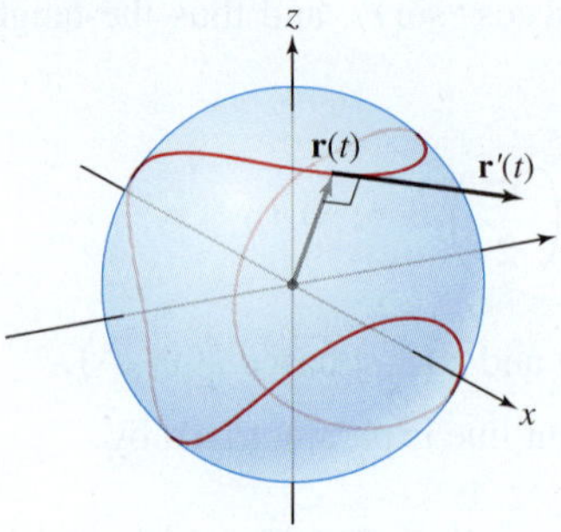

FIGURE 6 $\mathbf{r}'(t)$ is orthogonal to $\mathbf{r}(t)$ if $\mathbf{r}(t)$ has fixed length.

GRAPHICAL INSIGHT   The result of Example 7 has a geometric explanation. A vector parametrization $\mathbf{r}(t)$ consisting of vectors of constant length R traces a curve on the surface of a sphere of radius R with its center at the origin (Figure 6). Thus, $\mathbf{r}'(t)$ is tangent to this sphere. But any line that is tangent to a sphere at a point P is orthogonal to the radial vector through P, and thus $\mathbf{r}(t)$ is orthogonal to $\mathbf{r}'(t)$.

Vector-Valued Integration

The integral of a vector-valued function can be defined in terms of Riemann sums as in Chapter 5. We will define it more simply via componentwise integration (the two definitions are equivalent). In other words,

$$\int_a^b \mathbf{r}(t)\,dt = \left\langle \int_a^b x(t)\,dt, \int_a^b y(t)\,dt, \int_a^b z(t)\,dt \right\rangle$$

The integral exists if each of the components $x(t)$, $y(t)$, $z(t)$ is integrable. For example,

$$\int_0^\pi \langle 1, t, \sin t \rangle\,dt = \left\langle \int_0^\pi 1\,dt, \int_0^\pi t\,dt, \int_0^\pi \sin t\,dt \right\rangle = \left\langle \pi, \frac{1}{2}\pi^2, 2 \right\rangle$$

Vector-valued integrals obey the same linearity rules as scalar-valued integrals (see Exercise 72).

An **antiderivative** of $\mathbf{r}(t)$ is a vector-valued function $\mathbf{R}(t)$ such that $\mathbf{R}'(t) = \mathbf{r}(t)$. In the single-variable case, two functions f_1 and f_2 with the same derivative differ by a constant. Similarly, two vector-valued functions with the same derivative differ by a *constant vector* (i.e., a vector that does not depend on t). This is proved by applying the scalar result to each component of $\mathbf{r}(t)$.

THEOREM 4   If $\mathbf{R}_1(t)$ and $\mathbf{R}_2(t)$ are differentiable and $\mathbf{R}_1'(t) = \mathbf{R}_2'(t)$, then

$$\mathbf{R}_1(t) = \mathbf{R}_2(t) + \mathbf{c}$$

for some constant vector $\mathbf{c}$.

The general antiderivative of $\mathbf{r}(t)$ is written

$$\int \mathbf{r}(t)\,dt = \mathbf{R}(t) + \mathbf{c}$$

where $\mathbf{c} = \langle c_1, c_2, c_3 \rangle$ is an arbitrary constant vector. For example,

$$\int \langle 1, t, \sin t \rangle\,dt = \left\langle t, \frac{1}{2}t^2, -\cos t \right\rangle + \mathbf{c} = \left\langle t + c_1, \frac{1}{2}t^2 + c_2, -\cos t + c_3 \right\rangle$$

Fundamental Theorem of Calculus for Vector-Valued Functions   If $\mathbf{r}(t)$ is continuous on $[a, b]$, and $\mathbf{R}(t)$ is an antiderivative of $\mathbf{r}(t)$, then

$$\int_a^b \mathbf{r}(t)\,dt = \mathbf{R}(b) - \mathbf{R}(a)$$

■ **EXAMPLE 8   Finding Position via Vector-Valued Differential Equations**   The path of a particle satisfies

$$\frac{d\mathbf{r}}{dt} = \left\langle 1 - 6\sin 3t, \frac{1}{5}t \right\rangle$$

Find the particle's location at $t = 4$ if $\mathbf{r}(0) = \langle 4, 1 \rangle$.

Solution   The general solution is obtained by integration:

$$\mathbf{r}(t) = \int \left\langle 1 - 6\sin 3t, \frac{1}{5}t \right\rangle\,dt = \left\langle t + 2\cos 3t, \frac{1}{10}t^2 \right\rangle + \mathbf{c}$$

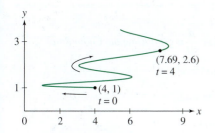

FIGURE 7 Particle path

$$\mathbf{r}(t) = \left\langle t + 2\cos 3t + 2, \tfrac{1}{10}t^2 + 1\right\rangle$$

The initial condition $\mathbf{r}(0) = \langle 4, 1 \rangle$ gives us

$$\mathbf{r}(0) = \langle 2, 0 \rangle + \mathbf{c} = \langle 4, 1 \rangle$$

Therefore, $\mathbf{c} = \langle 2, 1 \rangle$ and (Figure 7)

$$\mathbf{r}(t) = \left\langle t + 2\cos 3t, \frac{1}{10}t^2\right\rangle + \langle 2, 1 \rangle = \left\langle t + 2\cos 3t + 2, \frac{1}{10}t^2 + 1\right\rangle$$

The particle's position at $t = 4$ is

$$\mathbf{r}(4) = \left\langle 4 + 2\cos 12 + 2, \frac{1}{10}(4^2) + 1\right\rangle \approx \langle 7.69, 2.6 \rangle \qquad \blacksquare$$

13.2 SUMMARY

- Limits, differentiation, and integration of vector-valued functions are performed componentwise.
- Differentation rules:
  - Sum Rule: $(\mathbf{r}_1(t) + \mathbf{r}_2(t))' = \mathbf{r}_1'(t) + \mathbf{r}_2'(t)$
  - Constant Multiple Rule: $(c\,\mathbf{r}(t))' = c\,\mathbf{r}'(t)$
  - Chain Rule: $\dfrac{d}{dt}\,\mathbf{r}(g(t)) = g'(t)\mathbf{r}'(g(t))$
- Product Rules:

  Scalar times vector: $\qquad \dfrac{d}{dt}\big(f(t)\mathbf{r}(t)\big) = f'(t)\mathbf{r}(t) + f(t)\mathbf{r}'(t)$

  Dot product: $\qquad \dfrac{d}{dt}\big(\mathbf{r}_1(t) \cdot \mathbf{r}_2(t)\big) = \mathbf{r}_1'(t) \cdot \mathbf{r}_2(t) + \mathbf{r}_1(t) \cdot \mathbf{r}_2'(t)$

  Cross product: $\qquad \dfrac{d}{dt}\big(\mathbf{r}_1(t) \times \mathbf{r}_2(t)\big) = \big[\mathbf{r}_1'(t) \times \mathbf{r}_2(t)\big] + \big[\mathbf{r}_1(t) \times \mathbf{r}_2'(t)\big]$

- The derivative $\mathbf{r}'(t_0)$ is called the *tangent vector* or *velocity vector*.
- If $\mathbf{r}'(t_0)$ is nonzero, then it points in the direction tangent to the curve at $\mathbf{r}(t_0)$. The tangent line at $\mathbf{r}(t_0)$ has vector parametrization

$$\mathbf{L}(t) = \mathbf{r}(t_0) + t\mathbf{r}'(t_0)$$

- If $\mathbf{R}_1'(t) = \mathbf{R}_2'(t)$, then $\mathbf{R}_1(t) = \mathbf{R}_2(t) + \mathbf{c}$ for some constant vector $\mathbf{c}$.
- The Fundamental Theorem for vector-valued functions: If $\mathbf{r}(t)$ is continuous and $\mathbf{R}(t)$ is an antiderivative of $\mathbf{r}(t)$, then

$$\int_a^b \mathbf{r}(t)\,dt = \mathbf{R}(b) - \mathbf{R}(a)$$

13.2 EXERCISES

Preliminary Questions

1. State the three forms of the Product Rule for vector-valued functions.

In Questions 2–6, indicate whether the statement is true or false, and if it is false, provide a correct statement.

2. The derivative of a vector-valued function is defined as the limit of the difference quotient, just as in the scalar-valued case.

3. There are two Chain Rules for vector-valued functions: one for the composite of two vector-valued functions and one for the composite of a vector-valued and a scalar-valued function.

4. The terms "velocity vector" and "tangent vector" for a path $\mathbf{r}(t)$ mean one and the same thing.

5. The derivative of a vector-valued function is the slope of the tangent line, just as in the scalar case.

6. The derivative of the cross product is the cross product of the derivatives.

7. State whether the following derivatives of vector-valued functions $\mathbf{r}_1(t)$ and $\mathbf{r}_2(t)$ are scalars or vectors:

(a) $\dfrac{d}{dt}\mathbf{r}_1(t)$ **(b)** $\dfrac{d}{dt}\big(\mathbf{r}_1(t) \cdot \mathbf{r}_2(t)\big)$ **(c)** $\dfrac{d}{dt}\big(\mathbf{r}_1(t) \times \mathbf{r}_2(t)\big)$

Exercises

In Exercises 1–6, evaluate the limit.

1. $\lim_{t \to 3} \left\langle t^2, 4t, \frac{1}{t} \right\rangle$

2. $\lim_{t \to \pi} \sin 2t\mathbf{i} + \cos t\mathbf{j} + \tan 4t\mathbf{k}$

3. $\lim_{t \to 0} e^{2t}\mathbf{i} + \ln(t+1)\mathbf{j} + 4\mathbf{k}$

4. $\lim_{t \to 0} \left\langle \frac{1}{t+1}, \frac{e^t - 1}{t}, 4t \right\rangle$

5. Evaluate $\lim_{h \to 0} \frac{\mathbf{r}(t+h) - \mathbf{r}(t)}{h}$ for $\mathbf{r}(t) = \left\langle t^{-1}, \sin t, 4 \right\rangle$.

6. Evaluate $\lim_{t \to 0} \frac{\mathbf{r}(t)}{t}$ for $\mathbf{r}(t) = \langle \sin t, 1 - \cos t, -2t \rangle$.

In Exercises 7–12, compute the derivative.

7. $\mathbf{r}(t) = \left\langle t, t^2, t^3 \right\rangle$

8. $\mathbf{r}(t) = \left\langle 7 - t, 4\sqrt{t}, 8 \right\rangle$

9. $\mathbf{r}(s) = \left\langle e^{3s}, e^{-s}, s^4 \right\rangle$

10. $\mathbf{b}(t) = \left\langle e^{3t-4}, e^{6-t}, (t+1)^{-1} \right\rangle$ **11.** $\mathbf{c}(t) = t^{-1}\mathbf{i} - e^{2t}\mathbf{k}$

12. $\mathbf{a}(\theta) = (\cos 3\theta)\mathbf{i} + (\sin^2 \theta)\mathbf{j} + (\tan \theta)\mathbf{k}$

13. Calculate $\mathbf{r}'(t)$ and $\mathbf{r}''(t)$ for $\mathbf{r}(t) = \left\langle t, t^2, t^3 \right\rangle$.

14. Sketch the curve parametrized by $\mathbf{r}(t) = \left\langle 1 - t^2, t \right\rangle$ for $-1 \le t \le 1$. Compute the tangent vector at $t = 1$ and add it to the sketch.

15. Sketch the curve parametrized by $\mathbf{r}_1(t) = \left\langle t, t^2 \right\rangle$ together with its tangent vector at $t = 1$. Then do the same for $\mathbf{r}_2(t) = \left\langle t^3, t^6 \right\rangle$.

16. Sketch the cycloid $\mathbf{r}(t) = \langle t - \sin t, 1 - \cos t \rangle$ together with its tangent vectors at $t = \frac{\pi}{3}$ and $\frac{3\pi}{4}$.

In Exercises 17–20, evaluate the derivative by using the appropriate Product Rule, where

$$\mathbf{r}_1(t) = \left\langle t^2, t^3, t \right\rangle, \qquad \mathbf{r}_2(t) = \left\langle e^{3t}, e^{2t}, e^t \right\rangle$$

17. $\dfrac{d}{dt}\left(\mathbf{r}_1(t) \cdot \mathbf{r}_2(t)\right)$

18. $\dfrac{d}{dt}\left(t^4 \mathbf{r}_1(t)\right)$

19. $\dfrac{d}{dt}\left(\mathbf{r}_1(t) \times \mathbf{r}_2(t)\right)$

20. $\dfrac{d}{dt}\left(\mathbf{r}(t) \cdot \mathbf{r}_1(t)\right)\Big|_{t=2}$, assuming that

$$\mathbf{r}(2) = \langle 2, 1, 0 \rangle, \qquad \mathbf{r}'(2) = \langle 1, 4, 3 \rangle$$

In Exercises 21 and 22, let

$$\mathbf{r}_1(t) = \left\langle t^2, 1, 2t \right\rangle, \qquad \mathbf{r}_2(t) = \left\langle 1, 2, e^t \right\rangle$$

21. Compute $\dfrac{d}{dt}\mathbf{r}_1(t) \cdot \mathbf{r}_2(t)\Big|_{t=1}$ in two ways:

(a) Calculate $\mathbf{r}_1(t) \cdot \mathbf{r}_2(t)$ and differentiate.

(b) Use the Product Rule.

22. Compute $\dfrac{d}{dt}\mathbf{r}_1(t) \times \mathbf{r}_2(t)\Big|_{t=1}$ in two ways:

(a) Calculate $\mathbf{r}_1(t) \times \mathbf{r}_2(t)$ and differentiate.

(b) Use the Product Rule.

In Exercises 23–26, evaluate $\dfrac{d}{dt}\mathbf{r}(g(t))$ using the Chain Rule.

23. $\mathbf{r}(t) = \left\langle t^2, 1 - t \right\rangle, \quad g(t) = e^t$

24. $\mathbf{r}(t) = \left\langle t^2, t^3 \right\rangle, \quad g(t) = \sin t$

25. $\mathbf{r}(t) = \left\langle e^t, e^{2t}, 4 \right\rangle, \quad g(t) = 4t + 9$

26. $\mathbf{r}(t) = \langle 4 \sin 2t, 6 \cos 2t \rangle, \quad g(t) = t^2$

27. Let $\mathbf{r}(t) = \left\langle t^2, 1 - t, 4t \right\rangle$. Calculate the derivative of $\mathbf{r}(t) \cdot \mathbf{a}(t)$ at $t = 2$, assuming that $\mathbf{a}(2) = \langle 1, 3, 3 \rangle$ and $\mathbf{a}'(2) = \langle -1, 4, 1 \rangle$.

28. Let $\mathbf{v}(s) = s^2\mathbf{i} + 2s\mathbf{j} + 9s^{-2}\mathbf{k}$. Evaluate $\dfrac{d}{ds}\mathbf{v}(g(s))$ at $s = 4$, assuming that $g(4) = 3$ and $g'(4) = -9$.

In Exercises 29–34, find a parametrization of the tangent line at the point indicated.

29. $\mathbf{r}(t) = \left\langle t^2, t^4 \right\rangle, \quad t = -2$

30. $\mathbf{r}(t) = \langle \cos 2t, \sin 3t \rangle, \quad t = \frac{\pi}{4}$

31. $\mathbf{r}(t) = \left\langle 1 - t^2, 5t, 2t^3 \right\rangle, \quad t = 2$

32. $\mathbf{r}(t) = \langle 4t, 5t, 9t \rangle, \quad t = -4$

33. $\mathbf{r}(s) = 4s^{-1}\mathbf{i} - \frac{8}{3}s^{-3}\mathbf{k}, \quad s = 2$

34. $\mathbf{r}(s) = (\ln s)\mathbf{i} + s^{-1}\mathbf{j} + 9s\mathbf{k}, \quad s = 1$

35. Use Example 4 to calculate $\dfrac{d}{dt}(\mathbf{r} \times \mathbf{r}')$, where $\mathbf{r}(t) = \left\langle t, t^2, e^t \right\rangle$.

36. Let $\mathbf{r}(t) = \langle 3 \cos t, 5 \sin t, 4 \cos t \rangle$. Show that $\|\mathbf{r}(t)\|$ is constant and conclude, using Example 7, that $\mathbf{r}(t)$ and $\mathbf{r}'(t)$ are orthogonal. Then compute $\mathbf{r}'(t)$ and verify directly that $\mathbf{r}'(t)$ is orthogonal to $\mathbf{r}(t)$.

37. Show that the *derivative of the norm* is not equal to the *norm of the derivative* by verifying that $\|\mathbf{r}(t)\|' \ne \|\mathbf{r}(t)'\|$ for $\mathbf{r}(t) = \langle t, 1, 1 \rangle$.

38. Show that $\dfrac{d}{dt}(\mathbf{a} \times \mathbf{r}) = \mathbf{a} \times \mathbf{r}'$ for any constant vector $\mathbf{a}$.

In Exercises 39–46, evaluate the integrals.

39. $\displaystyle\int_{-1}^{3} \left\langle 8t^2 - t, 6t^3 + t \right\rangle dt$

40. $\displaystyle\int_{0}^{1} \left\langle \frac{1}{1+s^2}, \frac{s}{1+s^2} \right\rangle ds$

41. $\displaystyle\int_{-2}^{2} \left(u^3\mathbf{i} + u^5\mathbf{j} \right) du$

42. $\displaystyle\int_{0}^{1} \left(te^{-t^2}\mathbf{i} + t \ln(t^2 + 1)\mathbf{j} \right) dt$

43. $\displaystyle\int_{0}^{1} \langle 2t, 4t, -\cos 3t \rangle \, dt$

44. $\displaystyle\int_{1/2}^{1} \left\langle \frac{1}{u^2}, \frac{1}{u^4}, \frac{1}{u^5} \right\rangle du$

45. $\displaystyle\int_{1}^{4} \left(t^{-1}\mathbf{i} + 4\sqrt{t}\,\mathbf{j} - 8t^{3/2}\mathbf{k} \right) dt$ **46.** $\displaystyle\int_{0}^{t} \left(3s\mathbf{i} + 6s^2\mathbf{j} + 9\mathbf{k} \right) ds$

In Exercises 47–54, find both the general solution of the differential equation and the solution with the given initial condition.

47. $\dfrac{d\mathbf{r}}{dt} = \langle 1 - 2t, 4t \rangle, \quad \mathbf{r}(0) = \langle 3, 1 \rangle$

48. $\mathbf{r}'(t) = \mathbf{i} - \mathbf{j}, \quad \mathbf{r}(0) = 2\mathbf{i} + 3\mathbf{k}$

49. $\mathbf{r}'(t) = t^2\mathbf{i} + 5t\mathbf{j} + \mathbf{k}, \quad \mathbf{r}(1) = \mathbf{j} + 2\mathbf{k}$

50. $\mathbf{r}'(t) = \langle \sin 3t, \sin 3t, t \rangle$, $\mathbf{r}\left(\frac{\pi}{2}\right) = \left\langle 2, 4, \frac{\pi^2}{4} \right\rangle$

51. $\mathbf{r}''(t) = 16\mathbf{k}$, $\mathbf{r}(0) = \langle 1, 0, 0 \rangle$, $\mathbf{r}'(0) = \langle 0, 1, 0 \rangle$

52. $\mathbf{r}''(t) = \left\langle e^{2t-2}, t^2 - 1, 1 \right\rangle$, $\mathbf{r}(1) = \langle 0, 0, 1 \rangle$, $\mathbf{r}'(1) = \langle 2, 0, 0 \rangle$

53. $\mathbf{r}''(t) = \langle 0, 2, 0 \rangle$, $\mathbf{r}(3) = \langle 1, 1, 0 \rangle$, $\mathbf{r}'(3) = \langle 0, 0, 1 \rangle$

54. $\mathbf{r}''(t) = \left\langle e^t, \sin t, \cos t \right\rangle$, $\mathbf{r}(0) = \langle 1, 0, 1 \rangle$, $\mathbf{r}'(0) = \langle 0, 2, 2 \rangle$

55. Find the location at $t = 3$ of a particle whose path (Figure 8) satisfies

$$\frac{d\mathbf{r}}{dt} = \left\langle 2t - \frac{1}{(t+1)^2}, 2t - 4 \right\rangle, \qquad \mathbf{r}(0) = \langle 3, 8 \rangle$$

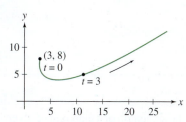

FIGURE 8 Particle path.

56. Find the location and velocity at $t = 4$ of a particle whose path satisfies

$$\frac{d\mathbf{r}}{dt} = \left\langle 2t^{-1/2}, 6, 8t \right\rangle, \qquad \mathbf{r}(1) = \langle 4, 9, 2 \rangle$$

57. A fighter plane, which can shoot a laser beam straight ahead, travels along the path $\mathbf{r}(t) = \langle 5 - t, 21 - t^2, 3 - t^3/27 \rangle$. Show that there is precisely one time t at which the pilot can hit a target located at the origin.

58. The fighter plane of Exercise 57 travels along the path $\mathbf{r}(t) = \langle t - t^3, 12 - t^2, 3 - t \rangle$. Show that the pilot cannot hit any target on the x-axis.

59. Find all solutions to $\mathbf{r}'(t) = \mathbf{v}$ with initial condition $\mathbf{r}(1) = \mathbf{w}$, where $\mathbf{v}$ and $\mathbf{w}$ are constant vectors in $\mathbf{R}^3$.

60. Let $\mathbf{u}$ be a constant vector in $\mathbf{R}^3$. Find the solution of the equation $\mathbf{r}'(t) = (\sin t)\mathbf{u}$ satisfying $\mathbf{r}'(0) = \mathbf{0}$.

61. Find all solutions to $\mathbf{r}'(t) = 2\mathbf{r}(t)$, where $\mathbf{r}(t)$ is a vector-valued function in 3-space.

62. Show that $\mathbf{w}(t) = \langle \sin(3t + 4), \sin(3t - 2), \cos 3t \rangle$ satisfies the differential equation $\mathbf{w}''(t) = -9\mathbf{w}(t)$.

63. Prove that the **Bernoulli spiral** (Figure 9) with parametrization $\mathbf{r}(t) = \langle e^t \cos 4t, e^t \sin 4t \rangle$ has the property that the angle ψ between the position vector and the tangent vector is constant. Find the angle ψ in degrees.

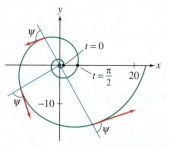

FIGURE 9 Bernoulli spiral.

64. A curve in polar form $r = f(\theta)$ has parametrization

$$\mathbf{r}(\theta) = f(\theta) \langle \cos \theta, \sin \theta \rangle$$

Let ψ be the angle between the radial and tangent vectors (Figure 10). Prove that

$$\tan \psi = \frac{r}{dr/d\theta} = \frac{f(\theta)}{f'(\theta)}$$

Hint: Compute $\mathbf{r}(\theta) \times \mathbf{r}'(\theta)$ and $\mathbf{r}(\theta) \cdot \mathbf{r}'(\theta)$.

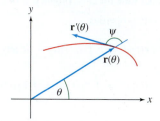

FIGURE 10 Curve with polar parametrization $\mathbf{r}(\theta) = f(\theta) \langle \cos \theta, \sin \theta \rangle$.

65. Prove that if $\|\mathbf{r}(t)\|$ takes on a local minimum or maximum value at t_0, then $\mathbf{r}(t_0)$ is orthogonal to $\mathbf{r}'(t_0)$. Explain how this result is related to Figure 11. *Hint:* Observe that if $\|\mathbf{r}(t_0)\|$ is a minimum, then $\mathbf{r}(t)$ is tangent at t_0 to the sphere of radius $\|\mathbf{r}(t_0)\|$ centered at the origin.

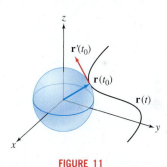

FIGURE 11

66. Newton's Second Law of Motion in vector form states that $\mathbf{F} = \dfrac{d\mathbf{p}}{dt}$, where $\mathbf{F}$ is the force acting on an object of mass m and $\mathbf{p} = m\mathbf{r}'(t)$ is the object's momentum. The analogs of force and momentum for rotational motion are the **torque** $\boldsymbol{\tau} = \mathbf{r} \times \mathbf{F}$ and **angular momentum**

$$\mathbf{J} = \mathbf{r}(t) \times \mathbf{p}(t)$$

Use the Second Law to prove that $\boldsymbol{\tau} = \dfrac{d\mathbf{J}}{dt}$.

Further Insights and Challenges

67. Let $\mathbf{r}(t) = \langle x(t), y(t) \rangle$ trace a plane curve $\mathcal{C}$. Assume that $x'(t_0) \neq 0$. Show that the slope of the tangent vector $\mathbf{r}'(t_0)$ is equal to the slope dy/dx of the curve at $\mathbf{r}(t_0)$.

68. Prove that $\dfrac{d}{dt}(\mathbf{r} \cdot (\mathbf{r}' \times \mathbf{r}'')) = \mathbf{r} \cdot (\mathbf{r}' \times \mathbf{r}''')$.

69. Verify the Sum and Product Rules for derivatives of vector-valued functions.

70. Verify the Chain Rule for vector-valued functions.

71. Verify the Product Rule for cross products [Eq. (5)].

72. Verify the linearity properties

$$\int c\mathbf{r}(t)\,dt = c\int \mathbf{r}(t)\,dt \qquad (c \text{ any constant})$$

$$\int (\mathbf{r}_1(t) + \mathbf{r}_2(t))\,dt = \int \mathbf{r}_1(t)\,dt + \int \mathbf{r}_2(t)\,dt$$

73. Prove the Substitution Rule [where g is a differentiable scalar function]:

$$\int_a^b \mathbf{r}(g(t))g'(t)\,dt = \int_{g^{-1}(a)}^{g^{-1}(b)} \mathbf{r}(u)\,du$$

74. Prove that if $\|\mathbf{r}(t)\| \leq K$ for $t \in [a, b]$, then

$$\left\| \int_a^b \mathbf{r}(t)\,dt \right\| \leq K(b - a)$$

13.3 Arc Length and Speed

In Section 11.2, we derived a formula for the arc length of a plane curve given in parametric form. This discussion applies to paths in 3-space with only minor changes.

←·· REMINDER The length of a path or curve is referred to as the arc length.

Recall that arc length is defined as the limit of the lengths of polygonal approximations. To produce a polygonal approximation to a path

$$\mathbf{r}(t) = \langle x(t), y(t), z(t) \rangle, \qquad a \leq t \leq b$$

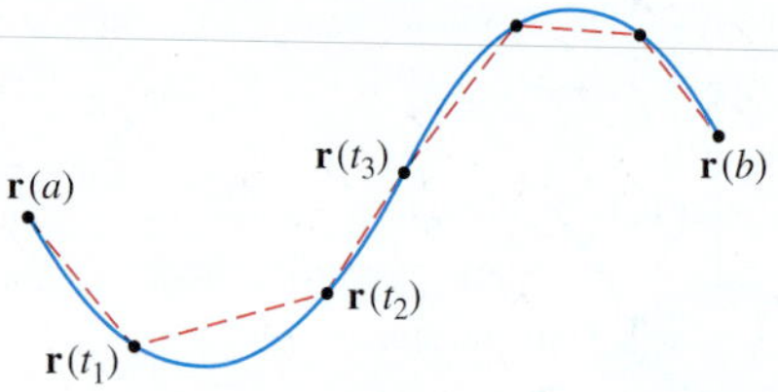

we choose a partition $a = t_0 < t_1 < t_2 < \cdots < t_N = b$ and join the terminal points of the vectors $\mathbf{r}(t_j)$ by segments, as in Figure 1. As in Section 11.2, we find that if $\mathbf{r}'(t)$ exists and is continuous on the interval $[a, b]$, then the lengths of the polygonal approximations approach a limit L as the maximum of the widths $|t_j - t_{j-1}|$ tends to zero. This limit is the length s of the path that is computed by the integral in the next theorem.

FIGURE 1 Polygonal approximation to the arc $\mathbf{r}(t)$ for $a \leq t \leq b$.

Keep in mind that the length s in Eq. (1) is the distance traveled by a particle following the path $\mathbf{r}(t)$. The path length s is not equal to the length of the underlying curve unless $\mathbf{r}(t)$ traverses the curve only once without reversing direction.

THEOREM 1 Length of a Path Assume that $\mathbf{r}(t)$ is differentiable and that $\mathbf{r}'(t)$ is continuous on $[a, b]$. Then the length s of the path $\mathbf{r}(t)$ for $a \leq t \leq b$ is equal to

$$s = \int_a^b \|\mathbf{r}'(t)\|\,dt = \int_a^b \sqrt{x'(t)^2 + y'(t)^2 + z'(t)^2}\,dt \qquad \boxed{1}$$

■ **EXAMPLE 1** Find the arc length s of the helix given by $\mathbf{r}(t) = \langle \cos 3t, \sin 3t, 3t \rangle$ for $0 \leq t \leq 2\pi$.

Solution The derivative is $\mathbf{r}'(t) = \langle -3\sin 3t, 3\cos 3t, 3 \rangle$, and

$$\|\mathbf{r}'(t)\|^2 = 9\sin^2 3t + 9\cos^2 3t + 9 = 9(\sin^2 3t + \cos^2 3t) + 9 = 18$$

Therefore, $s = \displaystyle\int_0^{2\pi} \|\mathbf{r}'(t)\|\,dt = \int_0^{2\pi} \sqrt{18}\,dt = 6\sqrt{2}\pi$. ■

Speed, by definition, is the rate of change of distance traveled with respect to time t. To calculate the speed, we define the **arc length function**:

$$s(t) = \int_a^t \|\mathbf{r}'(u)\|\,du$$

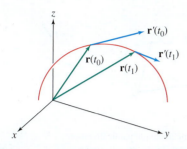

DF FIGURE 2 The velocity vector is longer at t_0 than at t_1, indicating that the particle is moving faster at t_0.

Thus, $s(t)$ is the distance traveled during the time interval $[a, t]$. By the Fundamental Theorem of Calculus (Part II),

$$\text{Speed at time } t = \frac{ds}{dt} = \|\mathbf{r}'(t)\|$$

Now we can see why $\mathbf{r}'(t)$ is known as the velocity vector (and also as the tangent vector). It points in the direction of motion, and its magnitude is the speed (Figure 2). We often denote the velocity vector by $\mathbf{v}(t)$ and the speed by $v(t)$:

$$\mathbf{v}(t) = \mathbf{r}'(t), \qquad v(t) = \|\mathbf{v}(t)\|$$

■ **EXAMPLE 2** Find the speed at time $t = 2$ seconds (s) of a particle whose position vector is

$$\mathbf{r}(t) = t^3\mathbf{i} - e^t\mathbf{j} + 4t\mathbf{k}$$

Solution The velocity vector is $\mathbf{v}(t) = \mathbf{r}'(t) = 3t^2\mathbf{i} - e^t\mathbf{j} + 4\mathbf{k}$, and at $t = 2$,

$$\mathbf{v}(2) = 12\mathbf{i} - e^2\mathbf{j} + 4\mathbf{k}$$

The particle's speed is $v(2) = \|\mathbf{v}(2)\| = \sqrt{12^2 + (-e^2)^2 + 4^2} \approx 14.65$ ft/s. ■

Arc Length Parametrization

Keep in mind that a parametrization $\mathbf{r}(t)$ *describes not just a curve, but also how a particle traverses the curve, possibly speeding up, slowing down, or reversing direction along the way. Changing the parametrization amounts to describing a different way of traversing the same underlying curve.*

We have seen that parametrizations are not unique. For example, $\mathbf{r}_1(t) = \langle t, t^2 \rangle$ and $\mathbf{r}_2(u) = \langle u^3, u^6 \rangle$ both parametrize the parabola $y = x^2$. Notice in this case that $\mathbf{r}_2(u)$ is obtained by substituting $t = u^3$ in $\mathbf{r}_1(t)$.

In general, we obtain a new parametrization by making a substitution $t = g(u)$—that is, by replacing $\mathbf{r}(t)$ with $\mathbf{r}_1(u) = \mathbf{r}(g(u))$ (Figure 3). If $t = g(u)$ increases from a to b as u varies from c to d, then the path $\mathbf{r}(t)$ for $a \leq t \leq b$ is also parametrized by $\mathbf{r}_1(u)$ for $c \leq u \leq d$.

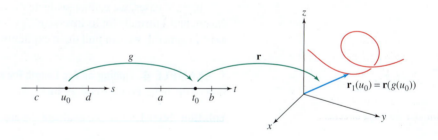

FIGURE 3 The path is parametrized by $\mathbf{r}(t)$ and by $\mathbf{r}_1(u) = \mathbf{r}(g(u))$.

■ **EXAMPLE 3** Parametrize the path $\mathbf{r}(t) = (t^2, \sin t, t)$ for $3 \leq t \leq 9$ using the parameter u, where $t = g(u) = e^u$.

Solution Substituting $t = e^u$ in $\mathbf{r}(t)$, we obtain the parametrization

$$\mathbf{r}_1(u) = \mathbf{r}(g(u)) = \langle e^{2u}, \sin e^u, e^u \rangle$$

Because $u = \ln t$, the parameter t varies from 3 to 9 as u varies from $\ln 3$ to $\ln 9$. Therefore, the path is parametrized by $\mathbf{r}_1(u)$ for $\ln 3 \leq u \leq \ln 9$. ■

One way of parametrizing a path is to choose a starting point and "walk along the path" at unit speed (say, 1 m/s). A parametrization of this type is called an **arc length parametrization** [Figure 4(A)]. It is defined by the property that the speed has constant value 1:

$$\|\mathbf{r}'(t)\| = 1 \qquad \text{for all } t$$

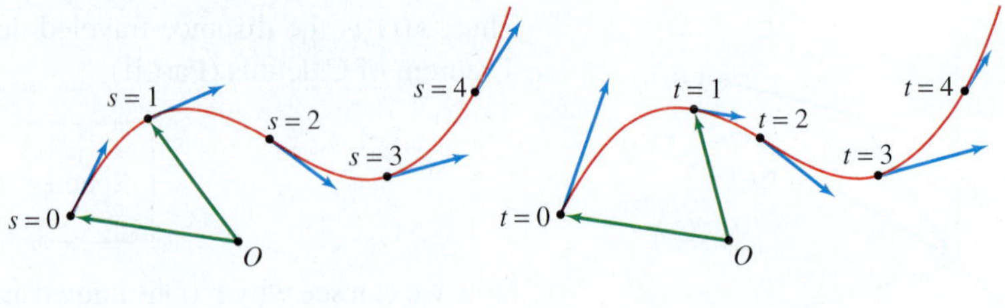

(A) An arc length parametrization:
All tangent vectors have length 1,
so speed is 1.

(B) Not an arc length parametrization:
Lengths of tangent vectors vary,
so the speed is changing.

FIGURE 4

Arc length parametrizations are also called **unit speed parametrizations.** *We will use arc length parametrizations to define curvature in Section 13.4.*

In an arc length parametrization, the distance traveled over any time interval $[a, b]$ is equal to the length of the interval:

$$\text{Distance traveled over } [a, b] = \int_a^b \|\mathbf{r}'(t)\| \, dt = \int_a^b 1 \, dt = b - a$$

The letter "s" is often used as the parameter in an arc length parametrization.

Finding an arc length parametrization: Start with any parametrization $\mathbf{r}(t)$ such that $\mathbf{r}'(t) \neq \mathbf{0}$ for all t.

Step 1. Form the arc length integral

$$s = g(t) = \int_0^t \|\mathbf{r}'(u)\| \, du$$

Step 2. Because $\|\mathbf{r}'(t)\| \neq 0$, $s = g(t)$ is an increasing function and therefore it has an inverse $t = g^{-1}(s)$.

Step 3. Take the new parametrization

$$\mathbf{r}_1(s) = \mathbf{r}(g^{-1}(s))$$

This is our arc length parametrization.

In most cases, we cannot evaluate the arc length integral $s = g(t)$ explicitly, and we cannot find a formula for its inverse $g^{-1}(s)$ either. So although arc length parametrizations exist in general, we can find them explicitly only in special cases.

■ **EXAMPLE 4   Finding an Arc Length Parametrization**   Find the arc length parametrization of the helix $\mathbf{r}(t) = \langle \cos 4t, \sin 4t, 3t \rangle$.

Solution   Step 1. First, we evaluate the arc length function

$$\|\mathbf{r}'(t)\| = \|\langle -4 \sin 4t, 4 \cos 4t, 3 \rangle\| = \sqrt{16 \sin^2 4t + 16 \cos^2 4t + 3^2} = 5$$

$$s = g(t) = \int_0^t \|\mathbf{r}'(t)\| \, dt = \int_0^t 5 \, dt = 5t$$

Step 2. Then we observe that the inverse of $s = 5t$ is $t = s/5$; that is, $g^{-1}(s) = s/5$.
Step 3. Substituting $\frac{s}{5}$ for each appearance of t in the original parametrization, we obtain the arc length parametrization

$$\mathbf{r}_1(s) = \mathbf{r}(g^{-1}(s)) = \mathbf{r}\left(\frac{s}{5}\right) = \left\langle \cos \frac{4s}{5}, \sin \frac{4s}{5}, \frac{3s}{5} \right\rangle$$

As a check, let's verify that $\mathbf{r}_1(s)$ has unit speed:

$$\|\mathbf{r}_1'(s)\| = \left\| \left\langle -\frac{4}{5} \sin \frac{4s}{5}, \frac{4}{5} \cos \frac{4s}{5}, \frac{3}{5} \right\rangle \right\| = \sqrt{\frac{16}{25} \sin^2 \frac{4s}{5} + \frac{16}{25} \cos^2 \frac{4s}{5} + \frac{9}{25}} = 1 \quad \blacksquare$$

13.3 SUMMARY

- The length s of a path $\mathbf{r}(t) = \langle x(t), y(t), z(t) \rangle$ for $a \leq t \leq b$ is

$$s = \int_a^b \|\mathbf{r}'(t)\| \, dt = \int_a^b \sqrt{x'(t)^2 + y'(t)^2 + z'(t)^2} \, dt$$

- Arc length function: $s(t) = \int_a^t \|\mathbf{r}'(u)\| \, du$
- Speed is the derivative of distance traveled with respect to time:

$$v(t) = \frac{ds}{dt} = \|\mathbf{r}'(t)\|$$

- The velocity vector $\mathbf{v}(t) = \mathbf{r}'(t)$ points in the direction of motion [provided that $\mathbf{r}'(t) \neq \mathbf{0}$] and its magnitude $v(t) = \|\mathbf{r}'(t)\|$ is the object's speed.
- We say that $\mathbf{r}(s)$ is an *arc length parametrization* if $\|\mathbf{r}'(s)\| = 1$ for all s. In this case, the length of the path for $a \leq s \leq b$ is $b - a$.
- If $\mathbf{r}(t)$ is any parametrization such that $\mathbf{r}'(t) \neq \mathbf{0}$ for all t, then

$$\mathbf{r}_1(s) = \mathbf{r}(g^{-1}(s))$$

is an arc length parametrization, where $t = g^{-1}(s)$ is the inverse of the arc length function $s = g(t)$.

13.3 EXERCISES

Preliminary Questions

1. At a given instant, a car on a roller coaster has velocity vector $\mathbf{r}' = \langle 25, -35, 10 \rangle$ (in miles per hour). What would the velocity vector be if the speed were doubled? What would it be if the car's direction were reversed but its speed remained unchanged?

2. Two cars travel in the same direction along the same roller coaster (at different times). Which of the following statements about their velocity vectors at a given point P on the roller coaster are true?

(a) The velocity vectors are identical.

(b) The velocity vectors point in the same direction but may have different lengths.

(c) The velocity vectors may point in opposite directions.

3. A mosquito flies along a parabola with speed $v(t) = t^2$. Let $L(t)$ be the total distance traveled at time t.

(a) How fast is $L(t)$ changing at $t = 2$?

(b) Is $L(t)$ equal to the mosquito's distance from the origin?

4. What is the length of the path traced by $\mathbf{r}(t)$ for $4 \leq t \leq 10$ if $\mathbf{r}(t)$ is an arc length parametrization?

Exercises

In Exercises 1–8, compute the length of the curve over the given interval.

1. $\mathbf{r}(t) = \langle 3t, 4t - 3, 6t + 1 \rangle$, $\quad 0 \leq t \leq 3$

2. $\mathbf{r}(t) = 2t\mathbf{i} - 3t\mathbf{k}$, $\quad 11 \leq t \leq 15$

3. $\mathbf{r}(t) = \langle 2t, \ln t, t^2 \rangle$, $\quad 1 \leq t \leq 4$

4. $\mathbf{r}(t) = \langle \cos t, \sin t, t^{3/2} \rangle$, $\quad 0 \leq t \leq 2\pi$

5. $\mathbf{r}(t) = \langle t, 4t^{3/2}, 2t^{3/2} \rangle$, $\quad 0 \leq t \leq 3$

6. $\mathbf{r}(t) = \langle 2t^2 + 1, 2t^2 - 1, t^3 \rangle$, $\quad 0 \leq t \leq 2$

7. $\mathbf{r}(t) = \langle t \cos t, t \sin t, 3t \rangle$, $\quad 0 \leq t \leq 2\pi$ *Hint:*

$$\int \sqrt{t^2 + a^2} \, dt = \frac{1}{2} t \sqrt{t^2 + a^2} + \frac{1}{2} a^2 \ln\left(t + \sqrt{t^2 + a^2}\right)$$

8. $\mathbf{r}(t) = t\mathbf{i} + 2t\mathbf{j} + (t^2 - 3)\mathbf{k}$, $\quad 0 \leq t \leq 2$ (See the hint above.)

In Exercises 9 and 10, compute the arc length function

$$s(t) = \int_a^t \|\mathbf{r}'(u)\| \, du \text{ for the given value of } a.$$

9. $\mathbf{r}(t) = \langle t^2, 2t^2, t^3 \rangle$, $\quad a = 0$

10. $\mathbf{r}(t) = \langle 4t^{1/2}, \ln t, 2t \rangle$, $\quad a = 1$

In Exercises 11–16, find the speed at the given value of t.

11. $\mathbf{r}(t) = \langle 2t + 3, 4t - 3, 5 - t \rangle$, $\quad t = 4$

12. $\mathbf{r}(t) = \langle t, t^2, t^3 \rangle$, $\quad t = 1$

13. $\mathbf{r}(t) = \langle t, \ln t, (\ln t)^2 \rangle$, $\quad t = 1$

14. $\mathbf{r}(t) = \langle e^{t-3}, 12, 3t^{-1} \rangle$, $\quad t = 3$

15. $\mathbf{r}(t) = \langle \sin 3t, \cos 4t, \cos 5t \rangle$, $\quad t = \frac{\pi}{2}$

16. $\mathbf{r}(t) = \langle \cosh t, \sinh t, t \rangle$, $\quad t = 0$

17. At an air show, a jet has a trajectory following the curve $y = x^2$. If when the jet is at the point $(1, 1)$, it has a speed of 500 km/h, determine its tangent vector at this point.

18. What is the velocity vector of a particle traveling to the right along the hyperbola $y = x^{-1}$ with constant speed 5 cm/s when the particle's location is $\left(2, \frac{1}{2}\right)$?

19. A bee with velocity vector $\mathbf{r}'(t)$ starts out at the origin at $t = 0$ and flies around for T seconds. Where is the bee located at time T if $\int_0^T \mathbf{r}'(u)\, du = 0$? What does the quantity $\int_0^T \|\mathbf{r}'(u)\|\, du$ represent?

20. The DNA molecule comes in the form of a double helix, meaning two helices that wrap around one another. Suppose a single one of the helices has a radius of 10Å (1 angstrom Å $= 10^{-8}$ cm) and one full turn of the helix has a height of 34Å.

(a) Show that the helix can be parametrized by $\mathbf{r}(t) = \left\langle 10\cos t,\ 10\sin t,\ \frac{34t}{2\pi} \right\rangle$.

(b) Find the arc length of one full turn of the helix.

21. Let

$$\mathbf{r}(t) = \left\langle R\cos\left(\frac{2\pi N t}{h}\right),\ R\sin\left(\frac{2\pi N t}{h}\right),\ t \right\rangle, \qquad 0 \le t \le h$$

(a) Show that $\mathbf{r}(t)$ parametrizes a helix of radius R and height h making N complete turns.

(b) Guess which of the two springs in Figure 5 uses more wire.

(c) Compute the lengths of the two springs and compare.

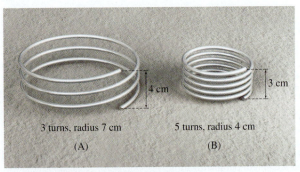

3 turns, radius 7 cm   5 turns, radius 4 cm

(A)   (B)

FIGURE 5 Which spring uses more wire?

22. Use Exercise 21 to find a general formula for the length of a helix of radius R and height h that makes N complete turns.

23. The cycloid generated by the unit circle has parametrization

$$\mathbf{r}(t) = \langle t - \sin t,\ 1 - \cos t \rangle$$

(a) Find the value of t in $[0, 2\pi]$ where the speed is at a maximum.

(b) Show that one arch of the cycloid has length 8. Recall the identity $\sin^2(t/2) = (1 - \cos t)/2$.

24. Which of the following is an arc length parametrization of a circle of radius 4 centered at the origin?

(a) $\mathbf{r}_1(t) = \langle 4\sin t,\ 4\cos t \rangle$

(b) $\mathbf{r}_2(t) = \langle 4\sin 4t,\ 4\cos 4t \rangle$

(c) $\mathbf{r}_3(t) = \left\langle 4\sin \frac{t}{4},\ 4\cos \frac{t}{4} \right\rangle$

25. Let $\mathbf{r}(t) = \langle 3t + 1,\ 4t - 5,\ 2t \rangle$.

(a) Evaluate the arc length integral $s(t) = \int_0^t \|\mathbf{r}'(u)\|\, du$.

(b) Find the inverse $g(s)$ of $s(t)$.

(c) Verify that $\mathbf{r}_1(s) = \mathbf{r}(g(s))$ is an arc length parametrization.

26. Find an arc length parametrization of the line $y = 4x + 9$.

27. Let $\mathbf{r}(t) = \mathbf{w} + t\mathbf{v}$ be the parametrization of a line.

(a) Show that the arc length function $s(t) = \int_0^t \|\mathbf{r}'(u)\|\, du$ is given by $s(t) = t\|\mathbf{v}\|$. This shows that $\mathbf{r}(t)$ is an arc length parametrizaton if and only if $\mathbf{v}$ is a unit vector.

(b) Find an arc length parametrization of the line with $\mathbf{w} = \langle 1, 2, 3 \rangle$ and $\mathbf{v} = \langle 3, 4, 5 \rangle$.

28. Find an arc length parametrization of the circle in the plane $z = 9$ with radius 4 and center $(1, 4, 9)$.

29. Find a path that traces the circle in the plane $y = 10$ with radius 4 and center $(2, 10, -3)$ with constant speed 8.

30. Find an arc length parametrization of the curve $\mathbf{r}(t) = \left\langle t,\ \frac{2}{3}t^{3/2},\ \frac{2}{\sqrt{3}}t^{3/2} \right\rangle$, with the parameter s measuring from $(0, 0, 0)$.

31. Find an arc length parametrization of the curve $\mathbf{r}(t) = \left\langle \cos t,\ \sin t,\ \frac{2}{3}t^{3/2} \right\rangle$, with the parameter s measuring from $(1, 0, 0)$.

32. Find an arc length parametrization of $\mathbf{r}(t) = \langle e^t \sin t,\ e^t \cos t,\ e^t \rangle$.

33. Find an arc length parametrization of $\mathbf{r}(t) = \langle t^2,\ t^3 \rangle$.

34. Find an arc length parametrization of the cycloid with parametrization $\mathbf{r}(t) = \langle t - \sin t,\ 1 - \cos t \rangle$.

35. Find an arc length parametrization of the line $y = mx$ for an arbitrary slope m.

36. Express the arc length L of $y = x^3$ for $0 \le x \le 8$ as an integral in two ways, using the parametrizations $\mathbf{r}_1(t) = \langle t, t^3 \rangle$ and $\mathbf{r}_2(t) = \langle t^3, t^9 \rangle$. Do not evaluate the integrals, but use substitution to show that they yield the same result.

37. The curve known as the **Bernoulli spiral** (Figure 6) has parametrization $\mathbf{r}(t) = \langle e^t \cos 4t,\ e^t \sin 4t \rangle$.

(a) Evaluate $s(t) = \int_{-\infty}^t \|\mathbf{r}'(u)\|\, du$. It is convenient to take lower limit $-\infty$ because $\mathbf{r}(-\infty) = \langle 0, 0 \rangle$.

(b) Use (a) to find an arc length parametrization of $\mathbf{r}(t)$.

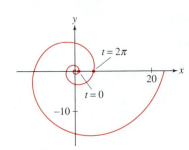

FIGURE 6 Bernoulli spiral.

Further Insights and Challenges

38. Prove that the length of a curve as computed using the arc length integral does not depend on its parametrization. More precisely, let $\mathcal{C}$ be the curve traced by $\mathbf{r}(t)$ for $a \leq t \leq b$. Let $f(s)$ be a differentiable function such that $f'(s) > 0$ and $f(c) = a$ and $f(d) = b$. Then $\mathbf{r}_1(s) = \mathbf{r}(f(s))$ parametrizes $\mathcal{C}$ for $c \leq s \leq d$. Verify that

$$\int_a^b \|\mathbf{r}'(t)\| \, dt = \int_c^d \|\mathbf{r}_1'(s)\| \, ds$$

39. The unit circle with the point $(-1, 0)$ removed has parametrization (see Exercise 79 in Section 11.1)

$$\mathbf{r}(t) = \left\langle \frac{1 - t^2}{1 + t^2}, \frac{2t}{1 + t^2} \right\rangle, \qquad -\infty < t < \infty$$

Use this parametrization to compute the length of the unit circle as an improper integral. *Hint:* The expression for $\|\mathbf{r}'(t)\|$ simplifies.

40. The involute of a circle (Figure 7), traced by a point at the end of a thread unwinding from a circular spool of radius R, has parametrization (see Exercise 29 in Section 11.2)

$$\mathbf{r}(\theta) = \langle R(\cos \theta + \theta \sin \theta), R(\sin \theta - \theta \cos \theta) \rangle$$

Find an arc length parametrization of the involute.

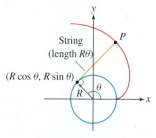

FIGURE 7 The involute of a circle.

41. The curve $\mathbf{r}(t) = \langle t - \tanh t, \operatorname{sech} t \rangle$ is called a **tractrix** (see Exercise 100 in Section 11.1).

(a) Show that $s(t) = \displaystyle\int_0^t \|\mathbf{r}'(u)\| \, du$ is equal to $s(t) = \ln(\cosh t)$.

(b) Show that $t = g(s) = \ln(e^s + \sqrt{e^{2s} - 1})$ is an inverse of $s(t)$ and verify that

$$\mathbf{r}_1(s) = \left\langle \tanh^{-1}\left(\sqrt{1 - e^{-2s}}\right) - \sqrt{1 - e^{-2s}}, e^{-s} \right\rangle$$

is an arc length parametrization of the tractrix.

13.4 Curvature

Curvature is a measure of how much a curve bends. It is used to study geometric properties of curves and motion along curves, and has applications in diverse areas such as roller coaster design (Figure 1), optics, eye surgery (see Exercise 68), and biochemistry (Figure 2).

FIGURE 1 Curvature is a key ingredient in roller coaster design. *(Robin Smith/Getty Images)*

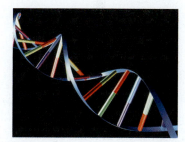

FIGURE 2 Biochemists study the effect of the curvature of DNA strands on biological processes. *(Alfred Pasieka/Science Source)*

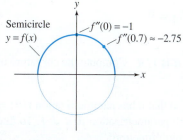

FIGURE 3 The second derivative of $f(x) = \sqrt{1 - x^2}$ does not capture the curvature of the circle, which by symmetry should be the same at all points.

In Chapter 4, we used the value of second derivative $f''(x)$ to measure the bending or concavity of the graph of $y = f(x)$, so it might seem natural to take $f''(x)$ as our definition of curvature. However, there are two reasons why this proposed definition will not work. First, $f''(x)$ makes sense only for a graph $y = f(x)$ in the plane, and our goal is to define curvature for curves in 3-space. A more serious problem is that $f''(x)$ does not truly capture the intrinsic curvature of a curve. A circle, for example, is symmetric, so its curvature ought to be the same at every point (Figure 3). But the upper semicircle is the graph of $f(x) = (1 - x^2)^{1/2}$ and the second derivative $f''(x) = -(1 - x^2)^{-3/2}$ does not have the same value at each point of the semicircle. We must look for a definition that depends only on the curve itself and not how it is oriented relative to the axes.

Consider a path with parametrization $\mathbf{r}(t) = \langle x(t), y(t), z(t) \rangle$. We assume that $\mathbf{r}'(t) \neq \mathbf{0}$ for all t in the domain of $\mathbf{r}(t)$. A parametrization with this property is called **regular**. At every point P along the path, there is a **unit tangent vector** $\mathbf{T} = \mathbf{T}_P$ that points in the direction of motion of the parametrization. We write $\mathbf{T}(t)$ for the unit tangent vector at the terminal point of $\mathbf{r}(t)$:

$$\text{Unit tangent vector} = \mathbf{T}(t) = \frac{\mathbf{r}'(t)}{\|\mathbf{r}'(t)\|}$$

For example, if $\mathbf{r}(t) = \langle t, t^2, t^3 \rangle$, then $\mathbf{r}'(t) = \langle 1, 2t, 3t^2 \rangle$, and the unit tangent vector at $P = (1, 1, 1)$, which is the terminal point of $\mathbf{r}(1) = \langle 1, 1, 1 \rangle$, is

$$\mathbf{T}_P = \frac{\langle 1, 2, 3 \rangle}{\|\langle 1, 2, 3 \rangle\|} = \frac{\langle 1, 2, 3 \rangle}{\sqrt{1^2 + 2^2 + 3^2}} = \left\langle \frac{1}{\sqrt{14}}, \frac{2}{\sqrt{14}}, \frac{3}{\sqrt{14}} \right\rangle$$

If we choose another parametrization, say, $\mathbf{r}_1(s)$, then we can also view $\mathbf{T}$ as function of s: $\mathbf{T}(s)$ is the unit tangent vector at the terminal point of $\mathbf{r}_1(s)$.

Now imagine walking along a path and observing how the unit tangent vector $\mathbf{T}$ changes direction (Figure 4). A change in $\mathbf{T}$ indicates that the path is bending, and the more rapidly $\mathbf{T}$ changes, the more the path bends. Thus, $\left\| \dfrac{d\mathbf{T}}{dt} \right\|$ would seem to be a good measure of curvature. However, $\left\| \dfrac{d\mathbf{T}}{dt} \right\|$ depends on how fast you walk (when you walk faster, the unit tangent vector changes more quickly). Therefore, we assume that you walk at unit speed. In other words, curvature is the magnitude $\kappa(s) = \left\| \dfrac{d\mathbf{T}}{ds} \right\|$, where s is the parameter of an arc length parametrization. Recall that $\mathbf{r}(s)$ is an arc length parametrization if $\|\mathbf{r}'(s)\| = 1$ for all s.

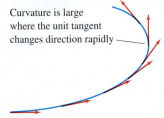

Curvature is large where the unit tangent changes direction rapidly

DF **FIGURE 4** The unit tangent vector varies in direction but not in length.

DEFINITION Curvature Let $\mathbf{r}(s)$ be an arc length parametrization and $\mathbf{T}$ the unit tangent vector. The **curvature** at $\mathbf{r}(s)$ is the quantity (denoted by a lowercase Greek letter "kappa")

$$\kappa(s) = \left\| \frac{d\mathbf{T}}{ds} \right\| \qquad \boxed{1}$$

Our first two examples illustrate curvature in the case of lines and circles.

■ **EXAMPLE 1** **A Line Has Zero Curvature** Compute the curvature at each point on the line $\mathbf{r}(t) = \langle x_0, y_0, z_0 \rangle + t\mathbf{u}$, where $\|\mathbf{u}\| = 1$.

Solution First, we note that because $\mathbf{u}$ is a unit vector, $\mathbf{r}(t)$ is an arc length parametrization. Indeed, $\mathbf{r}'(t) = \mathbf{u}$ and thus $\|\mathbf{r}'(t)\| = \|\mathbf{u}\| = 1$. Therefore, we have $\mathbf{T}(t) = \mathbf{r}'(t)/\|\mathbf{r}'(t)\| = \mathbf{r}'(t)$ and hence $\mathbf{T}'(t) = \mathbf{r}''(t) = \mathbf{0}$ [because $\mathbf{r}'(t) = \mathbf{u}$ is constant]. As expected, the curvature is zero at all points on a line:

$$\kappa(t) = \left\| \frac{d\mathbf{T}}{dt} \right\| = \|\mathbf{r}''(t)\| = 0 \qquad ■$$

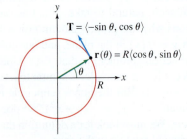

$\mathbf{T} = \langle -\sin\theta, \cos\theta \rangle$

$\mathbf{r}(\theta) = R\langle \cos\theta, \sin\theta \rangle$

FIGURE 5 The unit tangent vector at a point on a circle of radius R.

■ **EXAMPLE 2** **The Curvature of a Circle of Radius R Is $1/R$** Compute the curvature of a circle of radius R.

Solution Assume the circle is centered at the origin, so that it has parametrization $\mathbf{r}(\theta) = \langle R\cos\theta, R\sin\theta \rangle$ (Figure 5). This is not an arc length parametrization if $R \neq 1$. To find an arc length parametrization, we compute the arc length function:

$$s(\theta) = \int_0^\theta \|\mathbf{r}'(u)\| \, du = \int_0^\theta R \, du = R\theta$$

Example 2 shows that a circle of large radius R has small curvature 1/R. This makes sense because your direction of motion changes slowly when you walk at unit speed along a circle of large radius.

Thus, $s = R\theta$, and the inverse of the arc length function $s = g(\theta)$ is $\theta = g^{-1}(s) = s/R$. In Section 13.3, we showed that $\mathbf{r}_1(s) = \mathbf{r}(g^{-1}(s))$ is an arc length parametrization. In our case, we obtain

$$\mathbf{r}_1(s) = \mathbf{r}(g(s)) = \mathbf{r}\left(\frac{s}{R}\right) = \left\langle R\cos\frac{s}{R}, R\sin\frac{s}{R}\right\rangle$$

The unit tangent vector and its derivative are

$$\mathbf{T}(s) = \frac{d\mathbf{r}_1}{ds} = \frac{d}{ds}\left\langle R\cos\frac{s}{R}, R\sin\frac{s}{R}\right\rangle = \left\langle -\sin\frac{s}{R}, \cos\frac{s}{R}\right\rangle$$

$$\frac{d\mathbf{T}}{ds} = -\frac{1}{R}\left\langle \cos\frac{s}{R}, \sin\frac{s}{R}\right\rangle$$

By definition of curvature,

$$\kappa(s) = \left\|\frac{d\mathbf{T}}{ds}\right\| = \frac{1}{R}\left\|\left\langle\cos\frac{s}{R}, \sin\frac{s}{R}\right\rangle\right\| = \frac{1}{R}$$

This shows that the curvature is $1/R$ at all points on the circle. ■

In practice, it is often impossible to find an arc length parametrization explicitly. Fortunately, we can compute curvature using any regular parametrization $\mathbf{r}(t)$.

Since arc length s is a function of time t, the derivatives of $\mathbf{T}$ with respect to t and s are related by the Chain Rule. Denoting the derivative with respect to t by a prime, we have

$$\mathbf{T}'(t) = \frac{d\mathbf{T}}{dt} = \frac{d\mathbf{T}}{ds}\frac{ds}{dt} = v(t)\frac{d\mathbf{T}}{ds}$$

where $v(t) = \dfrac{ds}{dt} = \|\mathbf{r}'(t)\|$ is the speed of $\mathbf{r}(t)$. Since curvature is the magnitude $\left\|\dfrac{d\mathbf{T}}{ds}\right\|$, we obtain

$$\|\mathbf{T}'(t)\| = v(t)\kappa(t)$$

This yields an alternative formula for the curvature in the case where we do not have an arc length parametrization:

$$\kappa(t) = \frac{1}{v(t)}\|\mathbf{T}'(t)\| \qquad \boxed{2}$$

We can directly apply this formula to find curvature, but we can also use it to derive another option for calculations.

To apply Eq. (3) to plane curves, replace $\mathbf{r}(t) = \langle x(t), y(t)\rangle$ by $\mathbf{r}(t) = \langle x(t), y(t), 0\rangle$ and compute the cross product.

THEOREM 1 **Formula for Curvature** If $\mathbf{r}(t)$ is a regular parametrization, then the curvature at $\mathbf{r}(t)$ is

$$\kappa(t) = \frac{\|\mathbf{r}'(t) \times \mathbf{r}''(t)\|}{\|\mathbf{r}'(t)\|^3} \qquad \boxed{3}$$

Proof Since $v(t) = \|\mathbf{r}'(t)\|$, we have $\mathbf{r}'(t) = v(t)\mathbf{T}(t)$. By the Product Rule,

$$\mathbf{r}''(t) = v'(t)\mathbf{T}(t) + v(t)\mathbf{T}'(t)$$

Now compute the following cross product, using the fact that $\mathbf{T}(t) \times \mathbf{T}(t) = \mathbf{0}$:

$$\mathbf{r}'(t) \times \mathbf{r}''(t) = v(t)\mathbf{T}(t) \times \big(v'(t)\mathbf{T}(t) + v(t)\mathbf{T}'(t)\big)$$

$$= v(t)^2\mathbf{T}(t) \times \mathbf{T}'(t) \qquad \boxed{4}$$

←·· REMINDER To prove that $\mathbf{T}(t)$ and $\mathbf{T}'(t)$
are orthogonal (see the first marginal note), we have (by the second

←·· **REMINDER** To prove that $\mathbf{T}(t)$ and $\mathbf{T}'(t)$
are orthogonal, note that $\mathbf{T}(t)$ is a unit
vector, so $\mathbf{T}(t) \cdot \mathbf{T}(t) = 1$. Differentiate
using the Product Rule for Dot Products:

$$\frac{d}{dt}(\mathbf{T}(t) \cdot \mathbf{T}(t)) = 2\mathbf{T}'(t) \cdot \mathbf{T}(t) = 0$$

This shows that $\mathbf{T}'(t) \cdot \mathbf{T}(t) = 0$ (This is
the same argument that we used in
Example 7 of Section 13.2.)

←·· **REMINDER** By Theorem 1 in Section
12.4,

$$\|\mathbf{v} \times \mathbf{w}\| = \|\mathbf{v}\| \, \|\mathbf{w}\| \sin \theta$$

where θ is the angle between $\mathbf{v}$ and $\mathbf{w}$.

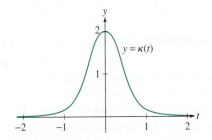

FIGURE 6 Graph of the curvature $\kappa(t)$ of
the twisted cubic $\mathbf{r}(t) = \langle t, t^2, t^3 \rangle$.

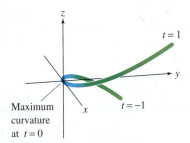

DF **FIGURE 7** Graph of the twisted cubic
$\mathbf{r}(t) = \langle t, t^2, t^3 \rangle$ colored by curvature.

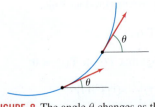

DF **FIGURE 8** The angle θ changes as the
curve bends.

Because $\mathbf{T}(t)$ and $\mathbf{T}'(t)$ are orthogonal (see the first marginal note), we have (by the second
marginal note)

$$\|\mathbf{T}(t) \times \mathbf{T}'(t)\| = \|\mathbf{T}(t)\| \, \|\mathbf{T}'(t)\| \sin \frac{\pi}{2} = \|\mathbf{T}'(t)\|$$

Eq. (4) yields $\|\mathbf{r}'(t) \times \mathbf{r}''(t)\| = v(t)^2 \|\mathbf{T}'(t)\|$. Using Eq. (2), we obtain

$$\|\mathbf{r}'(t) \times \mathbf{r}''(t)\| = v(t)^2 \|\mathbf{T}'(t)\| = v(t)^3 \kappa(t) = \|\mathbf{r}'(t)\|^3 \kappa(t)$$

This yields the desired formula. ∎

■ **EXAMPLE 3** **Twisted Cubic Curve** *CAS* Calculate the curvature $\kappa(t)$ of the twisted
cubic $\mathbf{r}(t) = \langle t, t^2, t^3 \rangle$. Then plot the graph of $\kappa(t)$ and determine where the curvature is
largest.

Solution The derivatives are

$$\mathbf{r}'(t) = \langle 1, 2t, 3t^2 \rangle, \qquad \mathbf{r}''(t) = \langle 0, 2, 6t \rangle$$

The parametrization is regular because $\mathbf{r}'(t) \neq \mathbf{0}$ for all t, so we may use Eq. (3):

$$\mathbf{r}'(t) \times \mathbf{r}''(t) = \begin{vmatrix} \mathbf{i} & \mathbf{j} & \mathbf{k} \\ 1 & 2t & 3t^2 \\ 0 & 2 & 6t \end{vmatrix} = 6t^2 \mathbf{i} - 6t \mathbf{j} + 2\mathbf{k}$$

$$\kappa(t) = \frac{\|\mathbf{r}'(t) \times \mathbf{r}''(t)\|}{\|\mathbf{r}'(t)\|^3} = \frac{\sqrt{36t^4 + 36t^2 + 4}}{(1 + 4t^2 + 9t^4)^{3/2}}$$

The graph of $\kappa(t)$ in Figure 6 shows that the curvature is largest at $t = 0$. The curve $\mathbf{r}(t)$ is
illustrated in Figure 7. The plot is colored by curvature, with large curvature represented
in blue, small curvature in green. ∎

In the second paragraph of this section, we pointed out that the curvature of a graph
$y = f(x)$ must involve more than just the value of the second derivative $f''(x)$. We now
show that the curvature can be expressed in terms of both $f''(x)$ and $f'(x)$.

> **THEOREM 2** **Curvature of a Graph in the Plane** The curvature at the point $(x, f(x))$
> on the graph of $y = f(x)$ is equal to
>
> $$\kappa(x) = \frac{|f''(x)|}{\left(1 + f'(x)^2\right)^{3/2}}$$
> 5

Proof The curve $y = f(x)$ has parametrization $\mathbf{r}(x) = \langle x, f(x) \rangle$. Therefore, $\mathbf{r}'(x) =
\langle 1, f'(x) \rangle$ and $\mathbf{r}''(x) = \langle 0, f''(x) \rangle$. To apply Theorem 1, we treat $\mathbf{r}'(x)$ and $\mathbf{r}''(x)$ as vectors
in $\mathbf{R}^3$ with the z-component equal to zero. Then

$$\mathbf{r}'(x) \times \mathbf{r}''(x) = \begin{vmatrix} \mathbf{i} & \mathbf{j} & \mathbf{k} \\ 1 & f'(x) & 0 \\ 0 & f''(x) & 0 \end{vmatrix} = f''(x)\mathbf{k}$$

Since $\|\mathbf{r}'(x)\| = \|\langle 1, f'(x) \rangle\| = (1 + f'(x)^2)^{1/2}$, Eq. (3) yields

$$\kappa(x) = \frac{\|\mathbf{r}'(x) \times \mathbf{r}''(x)\|}{\|\mathbf{r}'(x)\|^3} = \frac{|f''(x)|}{\left(1 + f'(x)^2\right)^{3/2}}$$ ∎

CONCEPTUAL INSIGHT Curvature for plane curves has a geometric interpretation in
terms of the angle of inclination, defined as the angle θ between the tangent vector
and the horizontal (Figure 8). The angle θ changes as the curve bends, and we can show
that the curvature κ is the rate of change of θ as you walk along the curve at unit speed
(see Exercise 69).

■ **EXAMPLE 4** Compute the curvature of $f(x) = x^3 - 3x^2 + 4$ at $x = 0, 1, 2, 3$.

Solution We apply Eq. (5):

$$f'(x) = 3x^2 - 6x = 3x(x-2), \qquad f''(x) = 6x - 6$$

$$\kappa(x) = \frac{|f''(x)|}{\left(1 + f'(x)^2\right)^{3/2}} = \frac{|6x - 6|}{\left(1 + 9x^2(x-2)^2\right)^{3/2}}$$

We obtain the following values:

$$\kappa(0) = \frac{6}{(1+0)^{3/2}} = 6, \qquad \kappa(1) = \frac{0}{(1+9)^{3/2}} = 0$$

$$\kappa(2) = \frac{6}{(1+0)^{3/2}} = 6, \qquad \kappa(3) = \frac{12}{82^{3/2}} \approx 0.016$$

Figure 9 shows that the graph bends more where the curvature is large. ■

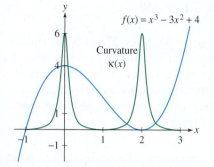

DF **FIGURE 9** Graph of $f(x) = x^3 - 3x^2 + 4$ and the curvature $\kappa(x)$.

The Normal Vector

We noted above that $\mathbf{T}'(t)$ and $\mathbf{T}(t)$ are orthogonal. The unit vector in the direction of $\mathbf{T}'(t)$, assuming it is nonzero, is called the **normal vector** and denoted $\mathbf{N}(t)$ or simply $\mathbf{N}$:

$$\text{Normal vector:} \qquad \mathbf{N}(t) = \frac{\mathbf{T}'(t)}{\|\mathbf{T}'(t)\|} \qquad \boxed{6}$$

This pair $\mathbf{T}$ and $\mathbf{N}$ of orthogonal unit vectors play a critical role in understanding a given space curve. Since $\|\mathbf{T}'(t)\| = v(t)\kappa(t)$ by Eq. (2), we have

$$\mathbf{T}'(t) = v(t)\kappa(t)\mathbf{N}(t) \qquad \boxed{7}$$

Intuitively, $\mathbf{N}$ points in the direction in which the curve is turning (Figure 10). This is particularly clear for a plane curve. In this case, there are two unit vectors orthogonal to $\mathbf{T}$ (Figure 11), and of these two, $\mathbf{N}$ is the vector that points to the "inside" of the curve.

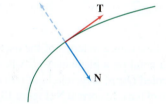

FIGURE 10 For a plane curve, the unit normal vector points in the direction of bending.

■ **EXAMPLE 5** **Normal Vector to a Helix** Find the unit normal vector at $t = \frac{\pi}{4}$ to the helix $\mathbf{r}(t) = \langle \cos t, \sin t, t \rangle$.

Solution The tangent vector $\mathbf{r}'(t) = \langle -\sin t, \cos t, 1 \rangle$ has length $\|\mathbf{r}'(t)\| = \sqrt{(-\sin t)^2 + (\cos t)^2 + 1} = \sqrt{2}$, so

$$\mathbf{T}(t) = \frac{\mathbf{r}'(t)}{\|\mathbf{r}'(t)\|} = \frac{1}{\sqrt{2}} \langle -\sin t, \cos t, 1 \rangle$$

$$\mathbf{T}'(t) = \frac{1}{\sqrt{2}} \langle -\cos t, -\sin t, 0 \rangle$$

$$\|\mathbf{T}'(t)\| = \frac{1}{\sqrt{2}} \sqrt{(-\cos t)^2 + (-\sin t)^2 + 0} = \frac{1}{\sqrt{2}}$$

$$\mathbf{N}(t) = \frac{\mathbf{T}'(t)}{\|\mathbf{T}'(t)\|} = \langle -\cos t, -\sin t, 0 \rangle$$

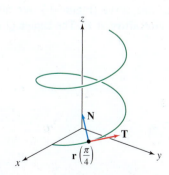

DF **FIGURE 11** Unit tangent and unit normal vectors at $t = \frac{\pi}{4}$ on the helix in Example 5.

Hence, $\mathbf{N}\left(\frac{\pi}{4}\right) = \left\langle -\frac{\sqrt{2}}{2}, -\frac{\sqrt{2}}{2}, 0 \right\rangle$ (Figure 11).

Note that at every point on the helix, the normal vector is a horizontal vector that points directly back at the z-axis. This reflects the fact that the curve is always turning back toward the z-axis. ■

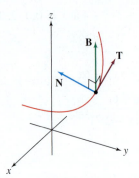

FIGURE 12 The Frenet frame at a point on a curve.

The Frenet frame, also known as the Frenet–Serret frame or the TNB frame, depends only on the intrinsic properties of the space curve, and needs no outside reference coordinate system. As such, it is very useful in analyzing the path of a spacecraft, satellite, or asteroid.

The Frenet Frame

Given a point on a curve corresponding to a parameter value t, we have now introduced two unit vectors at that point: the unit tangent vector $\mathbf{T}(t)$ that points in the direction of motion and the unit normal vector $\mathbf{N}(t)$ that points in the direction of turning. There is a third vector that is related to the first two, and that together with those two, provides a frame of reference at the point. The **binormal vector**, denoted $\mathbf{B}(t)$, is given by

$$\text{Binormal vector:} \qquad \mathbf{B}(t) = \mathbf{T}(t) \times \mathbf{N}(t) \qquad \boxed{8}$$

The fact $\mathbf{B}$ is a unit vector is verified in Exercise 83. Since $\mathbf{T}$ and $\mathbf{N}$ are already perpendicular to one another, and since, by its definition in terms of a cross product, $\mathbf{B}$ is perpendicular to each of them, these three vectors form a set $(\mathbf{T}, \mathbf{N}, \mathbf{B})$ of mutually perpendicular unit vectors, which we call the **Frenet frame**, after the French geometer Jean Frenet (1816–1900). As we move along a space curve, this frame, always centered at the point, moves and twists along with us, as in Figure 12.

■ **EXAMPLE 6** Continue Example 5 to find a formula for the binormal vector for a point on the helix $\mathbf{r}(t) = \langle \cos t, \sin t, t \rangle$.

Solution We have already calculated

$$\mathbf{T}(t) = \frac{1}{\sqrt{2}} \langle -\sin t, \cos t, 1 \rangle$$

$$\mathbf{N}(t) = \langle -\cos t, -\sin t, 0 \rangle$$

Hence,

$$\mathbf{B}(t) = \mathbf{T}(t) \times \mathbf{N}(t) = \begin{vmatrix} \mathbf{i} & \mathbf{j} & \mathbf{k} \\ \dfrac{-\sin t}{\sqrt{2}} & \dfrac{\cos t}{\sqrt{2}} & \dfrac{1}{\sqrt{2}} \\ -\cos t & -\sin t & 0 \end{vmatrix} = \frac{\sin t}{\sqrt{2}}\mathbf{i} + \frac{-\cos t}{\sqrt{2}}\mathbf{j} + \frac{1}{\sqrt{2}}\mathbf{k} \qquad ■$$

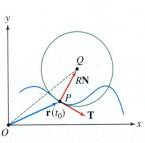

FIGURE 13 The center Q of the osculating circle at P lies a distance $R = \kappa_P^{-1}$ from P in the normal direction.

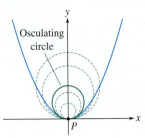

FIGURE 14 Among all circles tangent to the curve at P, the osculating circle is the "best fit" to the curve.

We conclude by describing another interpretation of curvature in terms of the osculating or "best-fitting circle" circle. Suppose that P is a point on a plane curve $\mathcal{C}$ where the curvature κ_P is nonzero. The **osculating circle**, denoted Osc_P, is the circle of radius $R = 1/\kappa_P$ through P whose center Q lies in the direction of the unit normal $\mathbf{N}$ (Figure 13). In other words, the center Q is determined by

$$\overrightarrow{OQ} = \mathbf{r}(t_0) + \frac{1}{\kappa_P}\mathbf{N} = \mathbf{r}(t_0) + R\mathbf{N} \qquad \boxed{9}$$

Among all circles tangent to $\mathcal{C}$ at P, Osc_P "best fits" the curve (Figure 14; see also Exercise 79). We refer to $R = 1/\kappa_P$ as the **radius of curvature** at P. The center Q of Osc_P is called the **center of curvature** at P.

■ **EXAMPLE 7** Parametrize the osculating circle to $y = x^2$ at $x = \frac{1}{2}$.

Solution Let $f(x) = x^2$. We use the parametrization

$$\mathbf{r}(x) = \langle x, f(x) \rangle = \langle x, x^2 \rangle$$

and proceed by the following steps.

Step 1. **Find the radius.**

Apply Eq. (5) to $f(x) = x^2$ to compute the curvature:

$$\kappa(x) = \frac{|f''(x)|}{\left(1 + f'(x)^2\right)^{3/2}} = \frac{2}{\left(1 + 4x^2\right)^{3/2}}, \qquad \kappa\left(\frac{1}{2}\right) = \frac{2}{2^{3/2}} = \frac{1}{\sqrt{2}}$$

The osculating circle has radius $R = 1/\kappa\left(\frac{1}{2}\right) = \sqrt{2}$.

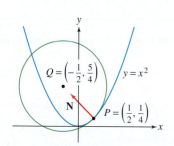

FIGURE 15 The osculating circle to $y = x^2$ at $x = \frac{1}{2}$ has center Q and radius $R = \sqrt{2}$.

Step 2. Find N at $t = \frac{1}{2}$.

For a plane curve, there is an easy way to find **N** without computing **T**′. The tangent vector is $\mathbf{r}'(x) = \langle 1, 2x \rangle$, and we know that $\langle 2x, -1 \rangle$ is orthogonal to $\mathbf{r}'(x)$ (because their dot product is zero). Therefore, $\mathbf{N}(x)$ is the unit vector in one of the two directions $\pm \langle 2x, -1 \rangle$. Figure 15 shows that the normal vector points in the positive y-direction (the direction of bending). Therefore,

$$\mathbf{N}(x) = \frac{\langle -2x, 1 \rangle}{\| \langle -2x, 1 \rangle \|} = \frac{\langle -2x, 1 \rangle}{\sqrt{1 + 4x^2}}, \qquad \mathbf{N}\left(\frac{1}{2}\right) = \frac{1}{\sqrt{2}} \langle -1, 1 \rangle$$

Step 3. Find the center Q.

Apply Eq. (9) with $t_0 = \frac{1}{2}$:

$$\overrightarrow{OQ} = \mathbf{r}\left(\frac{1}{2}\right) + \frac{1}{\kappa(1/2)} \mathbf{N}\left(\frac{1}{2}\right) = \left\langle \frac{1}{2}, \frac{1}{4} \right\rangle + \sqrt{2}\left(\frac{\langle -1, 1 \rangle}{\sqrt{2}}\right) = \left\langle -\frac{1}{2}, \frac{5}{4} \right\rangle$$

Step 4. Parametrize the osculating circle.

The osculating circle has radius $R = \sqrt{2}$, so it has parametrization

$$\mathbf{r}(t) = \underbrace{\left\langle -\frac{1}{2}, \frac{5}{4} \right\rangle}_{\text{Center}} + \sqrt{2} \langle \cos t, \sin t \rangle$$

If a curve C lies in a plane, then this plane is the osculating plane. For a general curve in 3-space, the osculating plane varies from point to point.

To define the osculating circle at a point P on a space curve C, we must first specify the plane in which the circle lies. The **osculating plane** is the plane through P determined by the unit tangent $\mathbf{T}_P$ and the unit normal $\mathbf{N}_P$ at P (we assume that $\mathbf{T}' \neq 0$, so $\mathbf{N}$ is defined). Intuitively, the osculating plane is the plane that "most nearly" contains the curve C near P (see Figure 16). Notice that because it contains the vectors $\mathbf{T}_P$ and $\mathbf{N}_P$, the vector $\mathbf{B}_P$ is a normal vector for the plane. The osculating circle is the circle in the osculating plane through P of radius $R = 1/\kappa_P$ whose center is located in the normal direction N_P from P. Equation (9) remains valid for space curves.

FIGURE 16 Osculating circles to $\mathbf{r}(t) = \langle \cos t, \sin t, \sin 2t \rangle$.

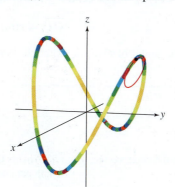

(A) Osculating circle at $t = \frac{\pi}{4}$
Curvature is $\kappa = 4.12$.

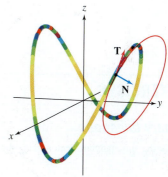

(B) Osculating circle at $t = \frac{\pi}{8}$
Curvature is $\kappa = 0.64$.

EXAMPLE 8 Continuing Examples 5 and 6, find the osculating plane for the helix $\mathbf{r}(t) = \langle \cos t, \sin t, t \rangle$ at the point on the helix corresponding to $t = \frac{\pi}{4}$.

Solution The point on the helix is given by $\mathbf{r}(\frac{\pi}{4}) = \left\langle \frac{\sqrt{2}}{2}, \frac{\sqrt{2}}{2}, \frac{\pi}{4} \right\rangle$.

In Example 6, we found $\mathbf{B}(t) = \left\langle \frac{\sin t}{\sqrt{2}}, \frac{-\cos t}{\sqrt{2}}, \frac{1}{\sqrt{2}} \right\rangle$. Therefore, $\mathbf{B}(\frac{\pi}{4}) = \left\langle \frac{1}{2}, \frac{-1}{2}, \frac{1}{\sqrt{2}} \right\rangle$.

Hence, the equation of the osculating plane, which contains this point and has its normal vector given by **B**, is

$$\frac{1}{2}\left(x - \frac{\sqrt{2}}{2}\right) - \frac{1}{2}\left(y - \frac{\sqrt{2}}{2}\right) + \frac{1}{\sqrt{2}}\left(z - \frac{\pi}{4}\right) = 0$$

or

$$x - y + \sqrt{2}z = \frac{\sqrt{2}\pi}{4}$$

13.4 SUMMARY

- A parametrization $\mathbf{r}(t)$ is called *regular* if $\mathbf{r}'(t) \neq \mathbf{0}$ for all t. If $\mathbf{r}(t)$ is regular, we define the *unit tangent vector* $\mathbf{T}(t) = \dfrac{\mathbf{r}'(t)}{\|\mathbf{r}'(t)\|}$.

- *Curvature* is defined by $\kappa(s) = \left\| \dfrac{d\mathbf{T}}{ds} \right\|$, where $\mathbf{r}(s)$ is an arc length parametrization or

  $\kappa(s) = \dfrac{1}{v(t)} \left\| \dfrac{d\mathbf{T}}{dt} \right\|$ if $\mathbf{r}(t)$ is not an arc length parametrization

- In practice, we compute curvature using the following formula, which is valid for arbitrary regular parametrizations:

$$\kappa(t) = \frac{\|\mathbf{r}'(t) \times \mathbf{r}''(t)\|}{\|\mathbf{r}'(t)\|^3}$$

- The curvature at a point on a graph $y = f(x)$ in the plane is

$$\kappa(x) = \frac{|f''(x)|}{\left(1 + f'(x)^2\right)^{3/2}}$$

- If $\|\mathbf{T}'(t)\| \neq 0$, we define the *unit normal vector* $\mathbf{N}(t) = \dfrac{\mathbf{T}'(t)}{\|\mathbf{T}'(t)\|}$.

- $\mathbf{T}'(t) = \kappa(t)v(t)\mathbf{N}(t)$
- The *binormal vector* is defined by $\mathbf{B} = \mathbf{T} \times \mathbf{N}$.
- The *osculating plane* at a point P on a curve $\mathcal{C}$ is the plane through P determined by the vectors $\mathbf{T}_P$ and $\mathbf{N}_P$. It has normal vector $\mathbf{B}_P$. It is defined only if the curvature κ_P at P is nonzero.
- The *osculating circle* Osc_P is the circle in the osculating plane through P of radius $R = 1/\kappa_P$ whose center Q lies in the normal direction $\mathbf{N}_P$:

$$\overrightarrow{OQ} = \mathbf{r}(t_0) + \kappa_P^{-1}\mathbf{N}_P = \mathbf{r}(t_0) + R\mathbf{N}_P$$

  The center of Osc_P is called the *center of curvature* and R the *radius of curvature*.

13.4 EXERCISES

Preliminary Questions

1. What is the unit tangent vector of a line with direction vector $\mathbf{v} = \langle 2, 1, -2 \rangle$?

2. What is the curvature of a circle of radius 4?

3. Which has larger curvature, a circle of radius 2 or a circle of radius 4?

4. What is the curvature of $\mathbf{r}(t) = \langle 2 + 3t, 7t, 5 - t \rangle$?

5. What is the curvature at a point where $\mathbf{T}'(s) = \langle 1, 2, 3 \rangle$ in an arc length parametrization $\mathbf{r}(s)$?

6. What is the radius of curvature of a circle of radius 4?

7. What is the radius of curvature at P if $\kappa_P = 9$?

Exercises

In Exercises 1–6, calculate $\mathbf{r}'(t)$ and $\mathbf{T}(t)$, and evaluate $\mathbf{T}(1)$.

1. $\mathbf{r}(t) = \langle 4t^2, 9t \rangle$

2. $\mathbf{r}(t) = \langle e^t, t^2 \rangle$

3. $\mathbf{r}(t) = \langle 3 + 4t, 3 - 5t, 9t \rangle$

4. $\mathbf{r}(t) = \langle 1 + 2t, t^2, 3 - t^2 \rangle$

5. $\mathbf{r}(t) = \langle \cos \pi t, \sin \pi t, t \rangle$

6. $\mathbf{r}(t) = \langle e^t, e^{-t}, t^2 \rangle$

In Exercises 7–10, use Eq. (3) to calculate the curvature function $\kappa(t)$.

7. $\mathbf{r}(t) = \langle 1, e^t, t \rangle$

8. $\mathbf{r}(t) = \langle 4 \cos t, t, 4 \sin t \rangle$

9. $\mathbf{r}(t) = \langle 4t + 1, 4t - 3, 2t \rangle$

10. $\mathbf{r}(t) = \langle t^{-1}, 1, t \rangle$

In Exercises 11–14, use Eq. (3) to evaluate the curvature at the given point.

11. $\mathbf{r}(t) = \langle 1/t, 1/t^2, t^2 \rangle$, $t = -1$

12. $\mathbf{r}(t) = \langle 3 - t, e^{t-4}, 8t - t^2 \rangle$, $t = 4$

13. $\mathbf{r}(t) = \langle \cos t, \sin t, t^2 \rangle$, $t = \frac{\pi}{2}$

14. $\mathbf{r}(t) = \langle \cosh t, \sinh t, t \rangle$, $t = 0$

In Exercises 15–18, find the curvature of the plane curve at the point indicated.

15. $y = e^t$, $t = 3$

16. $y = \cos x$, $x = 0$

17. $y = t^4, \quad t = 2$

18. $y = t^n, \quad t = 1$

19. Find the curvature of $\mathbf{r}(t) = \langle 2 \sin t, \cos 3t, t \rangle$ at $t = \frac{\pi}{3}$ and $t = \frac{\pi}{2}$ (Figure 17).

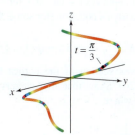

FIGURE 17 The curve $\mathbf{r}(t) = \langle 2 \sin t, \cos 3t, t \rangle$.

20. GU Find the curvature function $\kappa(x)$ for $y = \sin x$. Use a computer algebra system to plot $\kappa(x)$ for $0 \le x \le 2\pi$. Prove that the curvature takes its maximum at $x = \frac{\pi}{2}$ and $\frac{3\pi}{2}$. *Hint:* As a shortcut to finding the max, observe that the maximum of the numerator and the minimum of the denominator of $\kappa(x)$ occur at the same points.

21. Show that the tractrix $\mathbf{r}(t) = \langle t - \tanh t, \operatorname{sech} t \rangle$ has the curvature function $\kappa(t) = \operatorname{sech} t$.

22. Show that curvature at an inflection point of a plane curve $y = f(x)$ is zero.

23. Find the value(s) of α such that the curvature of $y = e^{\alpha x}$ at $x = 0$ is as large as possible.

24. Find the point of maximum curvature on $y = e^x$.

25. Show that the curvature function of the parametrization $\mathbf{r}(t) = \langle a \cos t, b \sin t \rangle$ of the ellipse $\left(\frac{x}{a}\right)^2 + \left(\frac{y}{b}\right)^2 = 1$ is

$$\kappa(t) = \frac{ab}{(b^2 \cos^2 t + a^2 \sin^2 t)^{3/2}} \qquad \boxed{10}$$

26. Use a sketch to predict where the points of minimal and maximal curvature occur on an ellipse. Then use Eq. (10) to confirm or refute your prediction.

27. In the notation of Exercise 25, assume that $a \ge b$. Show that $b/a^2 \le \kappa(t) \le a/b^2$ for all t.

28. Use Eq. (3) to prove that for a plane curve $\mathbf{r}(t) = \langle x(t), y(t) \rangle$,

$$\kappa(t) = \frac{|x'(t)y''(t) - x''(t)y'(t)|}{\left(x'(t)^2 + y'(t)^2\right)^{3/2}} \qquad \boxed{11}$$

In Exercises 29–32, use Eq. (11) to compute the curvature at the given point.

29. $\langle t^2, t^3 \rangle, \quad t = 2$

30. $\langle \cosh s, s \rangle, \quad s = 0$

31. $\langle t \cos t, \sin t \rangle, \quad t = \pi$

32. $\langle \sin 3s, 2 \sin 4s \rangle, \quad s = \frac{\pi}{2}$

33. Let $s(t) = \int_{-\infty}^{t} \|\mathbf{r}'(u)\| \, du$ for the Bernoulli spiral $\mathbf{r}(t) = \langle e^t \cos 4t, e^t \sin 4t \rangle$ (see Exercise 37 in Section 13.3). Show that the radius of curvature is proportional to $s(t)$.

34. The **Cornu spiral** is the plane curve $\mathbf{r}(t) = \langle x(t), y(t) \rangle$, where

$$x(t) = \int_0^t \sin \frac{u^2}{2} \, du, \qquad y(t) = \int_0^t \cos \frac{u^2}{2} \, du$$

Verify that $\kappa(t) = |t|$. Since the curvature increases linearly, the Cornu spiral is used in highway design to create transitions between straight and curved road segments (Figure 18).

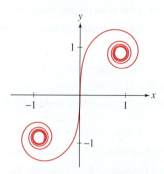

FIGURE 18 Cornu spiral.

35. CAS Plot the **clothoid** $\mathbf{r}(t) = \langle x(t), y(t) \rangle$, and compute its curvature $\kappa(t)$ where

$$x(t) = \int_0^t \sin \frac{u^3}{3} \, du, \qquad y(t) = \int_0^t \cos \frac{u^3}{3} \, du$$

36. Find the normal vector $\mathbf{N}(\theta)$ to $\mathbf{r}(\theta) = R \langle \cos \theta, \sin \theta \rangle$, the circle of radius R. Does $\mathbf{N}(\theta)$ point inside or outside the circle? Draw $\mathbf{N}(\theta)$ at $\theta = \frac{\pi}{4}$ with $R = 4$.

37. Find the normal vector $\mathbf{N}(t)$ to $\mathbf{r}(t) = \langle 4, \sin 2t, \cos 2t \rangle$.

38. Sketch the graph of $\mathbf{r}(t) = \langle t, t^3 \rangle$. Since $\mathbf{r}'(t) = \langle 1, 3t^2 \rangle$, the unit normal $\mathbf{N}(t)$ points in one of the two directions $\pm \langle -3t^2, 1 \rangle$. Which sign is correct at $t = 1$? Which is correct at $t = -1$?

39. Find the normal vectors to $\mathbf{r}(t) = \langle t, \cos t \rangle$ at $t = \frac{\pi}{4}$ and $t = \frac{3\pi}{4}$.

40. Find the normal vector to the Cornu spiral (Exercise 34) at $t = \sqrt{\pi}$.

In Exercises 41–44, find $\mathbf{T}$, $\mathbf{N}$, and $\mathbf{B}$ for the curve at the indicated point. Hint: After finding $\mathbf{T}'$, plug in the specific value for t before computing $\mathbf{N}$ and $\mathbf{B}$.

41. $\mathbf{r}(t) = \left\langle 0, t, t^2 \right\rangle$ at $(0, 1, 1)$. In this case, draw the curve and the three resultant vectors in 3-space.

42. $\mathbf{r}(t) = \langle \cos t, \sin t, 2 \rangle$ at $(1, 0, 0)$. In this case, draw the curve and the three resultant vectors in 3-space.

43. $\mathbf{r}(t) = \left\langle t, t^2, \frac{2}{3} t^3 \right\rangle$ at $\left(1, 1, \frac{2}{3}\right)$.

44. $\mathbf{r}(t) = \langle t, t, e^t \rangle$ at $(0, 0, 1)$.

45. Find the normal vector to the clothoid (Exercise 35) at $t = \pi^{1/3}$.

46. Method for Computing N Let $v(t) = \|\mathbf{r}'(t)\|$. Show that

$$\mathbf{N}(t) = \frac{v(t)\mathbf{r}''(t) - v'(t)\mathbf{r}'(t)}{\|v(t)\mathbf{r}''(t) - v'(t)\mathbf{r}'(t)\|} \qquad \boxed{12}$$

Hint: $\mathbf{N}$ is the unit vector in the direction $\mathbf{T}'(t)$. Differentiate $\mathbf{T}(t) = \mathbf{r}'(t)/v(t)$ to show that $v(t)\mathbf{r}''(t) - v'(t)\mathbf{r}'(t)$ is a positive multiple of $\mathbf{T}'(t)$.

In Exercises 47–52, use Eq. (12) to find $\mathbf{N}$ at the point indicated.

47. $\langle t^2, t^3 \rangle, \quad t = 1$

48. $\langle t - \sin t, 1 - \cos t \rangle, \quad t = \pi$

49. $\langle t^2/2, t^3/3, t \rangle, \quad t = 1$

50. $\langle t^{-1}, t, t^2 \rangle, \quad t = -1$

51. $\langle t, e^t, t \rangle$, $t = 0$

52. $\langle \cosh t, \sinh t, t^2 \rangle$, $t = 0$

53. Let $\mathbf{r}(t) = \left\langle t, \frac{4}{3}t^{3/2}, t^2 \right\rangle$.

(a) Find $\mathbf{T}$, $\mathbf{N}$, and $\mathbf{B}$ at the point corresponding to $t = 1$.

(b) Find the equation of the osculating plane at the point corresponding to $t = 1$.

54. Let $\mathbf{r}(t) = \langle \cos t, \sin t, \ln(\cos t) \rangle$.

(a) Find $\mathbf{T}$, $\mathbf{N}$, and $\mathbf{B}$ at $(1, 0, 0)$.

(b) Find the equation of the osculating plane at $(1, 0, 0)$.

55. Let $\mathbf{r}(t) = \left\langle t, 1 - t, t^2 \right\rangle$.

(a) Find the general formulas for $\mathbf{T}$ and $\mathbf{N}$ as functions of t.

(b) Find the general formula for $\mathbf{B}$ as a function of t.

(c) What can you conclude about the osculating planes of the curve based on your answer to *b*?

56. (a) What does it mean for a space curve to have a constant unit tangent vector $\mathbf{T}$?

(b) What does it mean for a space curve to have a constant normal vector $\mathbf{N}$?

(c) What does it mean for a space curve to have a constant binormal vector $\mathbf{B}$?

57. Let $f(x) = x^2$. Show that the center of the osculating circle at (x_0, x_0^2) is given by $\left(-4x_0^3, \frac{1}{2} + 3x_0^2 \right)$.

58. Use Eq. (9) to find the center of curvature to $\mathbf{r}(t) = \langle t^2, t^3 \rangle$ at $t = 1$.

In Exercises 59–66, find a parametrization of the osculating circle at the point indicated.

59. $\mathbf{r}(t) = \langle \cos t, \sin t \rangle$, $t = \frac{\pi}{4}$

60. $\mathbf{r}(t) = \langle \sin t, \cos t \rangle$, $t = 0$

61. $y = x^2$, $x = 1$

62. $y = \sin x$, $x = \frac{\pi}{2}$

63. $\langle t - \sin t, 1 - \cos t \rangle$, $t = \pi$

64. $\mathbf{r}(t) = \langle t^2/2, t^3/3, t \rangle$, $t = 0$

65. $\mathbf{r}(t) = \langle \cos t, \sin t, t \rangle$, $t = 0$

66. $\mathbf{r}(t) = \langle \cosh t, \sinh t, t \rangle$, $t = 0$

67. Figure 19 shows the graph of the half-ellipse $y = \pm\sqrt{2rx - px^2}$, where r and p are positive constants. Show that the radius of curvature at the origin is equal to r. *Hint:* One way of proceeding is to write the ellipse in the form of Exercise 25 and apply Eq. (10).

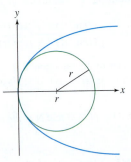

FIGURE 19 The curve $y = \pm\sqrt{2rx - px^2}$ and the osculating circle at the origin.

68. In a recent study of laser eye surgery by Gatinel, Hoang-Xuan, and Azar, a vertical cross section of the cornea is modeled by the half-ellipse of Exercise 67. Show that the half-ellipse can be written in the form $x = f(y)$, where $f(y) = p^{-1}\left(r - \sqrt{r^2 - py^2}\right)$. During surgery, tissue is removed to a depth $t(y)$ at height y for $-S \leq y \leq S$, where $t(y)$ is given by Munnerlyn's equation (for some $R > r$):

$$t(y) = \sqrt{R^2 - S^2} - \sqrt{R^2 - y^2} - \sqrt{r^2 - S^2} + \sqrt{r^2 - y^2}$$

After surgery, the cross section of the cornea has the shape $x = f(y) + t(y)$ (Figure 20). Show that after surgery, the radius of curvature at the point P (where $y = 0$) is R.

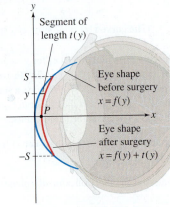

FIGURE 20 Contour of cornea before and after surgery.

69. The **angle of inclination** at a point P on a plane curve is the angle θ between the unit tangent vector $\mathbf{T}$ and the x-axis (Figure 21). Assume that $\mathbf{r}(s)$ is a arc length parametrization, and let $\theta = \theta(s)$ be the angle of inclination at $\mathbf{r}(s)$. Prove that

$$\kappa(s) = \left| \frac{d\theta}{ds} \right| \qquad \boxed{13}$$

Hint: Observe that $\mathbf{T}(s) = \langle \cos\theta(s), \sin\theta(s) \rangle$.

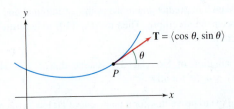

FIGURE 21 The curvature at P is the quantity $|d\theta/ds|$.

70. A particle moves along the path $y = x^3$ with unit speed. How fast is the tangent turning (i.e., how fast is the angle of inclination changing) when the particle passes through the point $(2, 8)$?

71. Let $\theta(x)$ be the angle of inclination at a point on the graph $y = f(x)$ (see Exercise 69).

(a) Use the relation $f'(x) = \tan\theta$ to prove that $\dfrac{d\theta}{dx} = \dfrac{f''(x)}{(1 + f'(x)^2)}$.

(b) Use the arc length integral to show that $\dfrac{ds}{dx} = \sqrt{1 + f'(x)^2}$.

(c) Now give a proof of Eq. (5) using Eq. (13).

72. Use the parametrization $\mathbf{r}(\theta) = \langle f(\theta)\cos\theta, f(\theta)\sin\theta \rangle$ to show that a curve $r = f(\theta)$ in polar coordinates has curvature

$$\kappa(\theta) = \frac{|f(\theta)^2 + 2f'(\theta)^2 - f(\theta)f''(\theta)|}{\left(f(\theta)^2 + f'(\theta)^2\right)^{3/2}} \qquad \boxed{14}$$

In Exercises 73–75, use Eq. (14) to find the curvature of the curve given in polar form.

73. $f(\theta) = 2\cos\theta$ **74.** $f(\theta) = \theta$ **75.** $f(\theta) = e^\theta$

76. Use Eq. (14) to find the curvature of the general Bernoulli spiral $r = ae^{b\theta}$ in polar form (a and b are constants).

77. Show that both $\mathbf{r}'(t)$ and $\mathbf{r}''(t)$ lie in the osculating plane for a vector function $\mathbf{r}(t)$. *Hint:* Differentiate $\mathbf{r}'(t) = v(t)\mathbf{T}(t)$.

78. Show that

$$\gamma(s) = \mathbf{r}(t_0) + \frac{1}{\kappa}\mathbf{N} + \frac{1}{\kappa}\left((\sin\kappa s)\mathbf{T} - (\cos\kappa s)\mathbf{N}\right)$$

is an arc length parametrization of the osculating circle at $\mathbf{r}(t_0)$.

79. Two vector-valued functions $\mathbf{r}_1(s)$ and $\mathbf{r}_2(s)$ are said to *agree to order 2* at s_0 if

$$\mathbf{r}_1(s_0) = \mathbf{r}_2(s_0), \quad \mathbf{r}_1'(s_0) = \mathbf{r}_2'(s_0), \quad \mathbf{r}_1''(s_0) = \mathbf{r}_2''(s_0)$$

Let $\mathbf{r}(s)$ be an arc length parametrization of a curve $\mathcal{C}$, and let P be the terminal point of $\mathbf{r}(0)$. Let $\gamma(s)$ be the arc length parametrization of the osculating circle given in Exercise 78. Show that $\mathbf{r}(s)$ and $\gamma(s)$ agree to order 2 at $s = 0$ (in fact, the osculating circle is the unique circle that approximates $\mathcal{C}$ to order 2 at P).

80. Let $\mathbf{r}(t) = \langle x(t), y(t), z(t) \rangle$ be a path with curvature $\kappa(t)$ and define the scaled path $\mathbf{r}_1(t) = \langle \lambda x(t), \lambda y(t), \lambda z(t) \rangle$, where $\lambda \neq 0$ is a constant. Prove that curvature varies inversely with the scale factor. That is, prove that the curvature $\kappa_1(t)$ of $\mathbf{r}_1(t)$ is $\kappa_1(t) = \lambda^{-1}\kappa(t)$. This explains why the curvature of a circle of radius R is proportional to $1/R$ (in fact, it is equal to $1/R$). *Hint:* Use Eq. (3).

Further Insights and Challenges

81. Show that the curvature of Viviani's curve, given by $\mathbf{r}(t) = \langle 1 + \cos t, \sin t, 2\sin(t/2) \rangle$, is

$$\kappa(t) = \frac{\sqrt{13 + 3\cos t}}{(3 + \cos t)^{3/2}}$$

82. Let $\mathbf{r}(s)$ be an arc length parametrization of a closed curve $\mathcal{C}$ of length L. We call $\mathcal{C}$ an **oval** if $d\theta/ds > 0$ (see Exercise 69). Observe that $-\mathbf{N}$ points to the *outside* of $\mathcal{C}$. For $k > 0$, the curve $\mathcal{C}_1$ defined by $\mathbf{r}_1(s) = \mathbf{r}(s) - k\mathbf{N}$ is called the expansion of $c(s)$ in the normal direction.

(a) Show that $\|\mathbf{r}_1'(s)\| = \|\mathbf{r}'(s)\| + k\kappa(s)$.

(b) As P moves around the oval counterclockwise, θ increases by 2π [Figure 22(A)]. Use this and a change of variables to prove that $\int_0^L \kappa(s)\,ds = 2\pi$.

(c) Show that $\mathcal{C}_1$ has length $L + 2\pi k$.

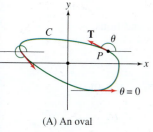

(A) An oval

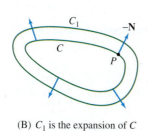

(B) C_1 is the expansion of C in the normal direction.

FIGURE 22 As P moves around the oval, θ increases by 2π.

In Exercises 83–91, we investigate the binormal vector further.

83. Show that $\mathbf{B}$ is a unit vector.

84. Follow steps (a)–(c) to prove that there is a number τ (lowercase Greek "tau") called the **torsion** such that

$$\frac{d\mathbf{B}}{ds} = -\tau\mathbf{N} \qquad \boxed{15}$$

(a) Show that $\dfrac{d\mathbf{B}}{ds} = \mathbf{T} \times \dfrac{d\mathbf{N}}{ds}$ and conclude that $d\mathbf{B}/ds$ is orthogonal to $\mathbf{T}$.

(b) Differentiate $\mathbf{B} \cdot \mathbf{B} = 1$ with respect to s to show that $d\mathbf{B}/ds$ is orthogonal to $\mathbf{B}$.

(c) Conclude that $d\mathbf{B}/ds$ is a multiple of $\mathbf{N}$.

85. Show that if $\mathcal{C}$ is contained in a plane $\mathcal{P}$, then $\mathbf{B}$ is a unit vector normal to $\mathcal{P}$. Conclude that $\tau = 0$ for a plane curve.

86. Torsion means "twisting." Is this an appropriate name for τ? Explain by interpreting τ geometrically.

87. Use the identity

$$\mathbf{a} \times (\mathbf{b} \times \mathbf{c}) = (\mathbf{a} \cdot \mathbf{c})\mathbf{b} - (\mathbf{a} \cdot \mathbf{b})\mathbf{c}$$

to prove

$$\mathbf{N} \times \mathbf{B} = \mathbf{T}, \qquad \mathbf{B} \times \mathbf{T} = \mathbf{N} \qquad \boxed{16}$$

88. Follow steps (a)–(b) to prove

$$\frac{d\mathbf{N}}{ds} = -\kappa\mathbf{T} + \tau\mathbf{B} \qquad \boxed{17}$$

(a) Show that $d\mathbf{N}/ds$ is orthogonal to $\mathbf{N}$. Conclude that $d\mathbf{N}/ds$ lies in the plane spanned by $\mathbf{T}$ and $\mathbf{B}$, and hence, $d\mathbf{N}/ds = a\mathbf{T} + b\mathbf{B}$ for some scalars a, b.

(b) Use $\mathbf{N} \cdot \mathbf{T} = 0$ to show that $\mathbf{T} \cdot \dfrac{d\mathbf{N}}{ds} = -\mathbf{N} \cdot \dfrac{d\mathbf{T}}{ds}$ and compute a. Compute b similarly. Equations (15) and (17) together with $d\mathbf{T}/dt = \kappa\mathbf{N}$ are called the **Frenet formulas**.

89. Show that $\mathbf{r}' \times \mathbf{r}''$ is a multiple of $\mathbf{B}$. Conclude that

$$\mathbf{B} = \frac{\mathbf{r}' \times \mathbf{r}''}{\|\mathbf{r}' \times \mathbf{r}''\|} \qquad \boxed{18}$$

90. Use the formula from the preceding problem to find $\mathbf{B}$ for the space curve given by $\mathbf{r}(t) = \langle \sin t, -\cos t, \sin t \rangle$. Conclude that the space curve lies in a plane.

91. The vector $\mathbf{N}$ can be computed using $\mathbf{N} = \mathbf{B} \times \mathbf{T}$ [Eq. (16)] with $\mathbf{B}$, as in Eq. (18). Use this method to find $\mathbf{N}$ in the following cases:

(a) $\mathbf{r}(t) = \langle \cos t, t, t^2 \rangle$ at $t = 0$

(b) $\mathbf{r}(t) = \langle t^2, t^{-1}, t \rangle$ at $t = 1$

FIGURE 1 The flight of the space shuttle is analyzed using vector calculus. (*NASA*)

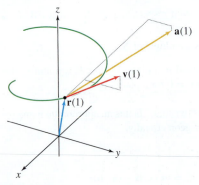

DF **FIGURE 2**

13.5 Motion in 3-Space

In this section, we study the motion of an object traveling along a path $\mathbf{r}(t)$. That object could be a variety of things, including a particle, a baseball, or the space shuttle (Figure 1). Recall that the velocity vector is the derivative

$$\mathbf{v}(t) = \mathbf{r}'(t) = \lim_{h \to 0} \frac{\mathbf{r}(t+h) - \mathbf{r}(t)}{h}$$

As we have seen, $\mathbf{v}(t)$ points in the direction of motion (if it is nonzero), and its magnitude $v(t) = \|\mathbf{v}(t)\|$ is the particle's speed. The **acceleration vector** is the second derivative $\mathbf{r}''(t)$, which we shall denote as $\mathbf{a}(t)$. In summary,

$$\boxed{\mathbf{v}(t) = \mathbf{r}'(t), \qquad v(t) = \|\mathbf{v}(t)\|, \qquad \mathbf{a}(t) = \mathbf{r}''(t)}$$

■ **EXAMPLE 1** Calculate and plot the velocity and acceleration vectors at $t = 1$ of $\mathbf{r}(t) = \langle \sin 2t, -\cos 2t, \sqrt{t+1} \rangle$. Then find the speed at $t = 1$ (Figure 2).

Solution

$$\mathbf{v}(t) = \mathbf{r}'(t) = \left\langle 2\cos 2t, 2\sin 2t, \frac{1}{2}(t+1)^{-1/2} \right\rangle, \qquad \mathbf{v}(1) \approx \langle -0.83, 1.82, 0.35 \rangle$$

$$\mathbf{a}(t) = \mathbf{r}''(t) = \left\langle -4\sin 2t, 4\cos 2t, -\frac{1}{4}(t+1)^{-3/2} \right\rangle, \qquad \mathbf{a}(1) \approx \langle -3.64, -1.66, -0.089 \rangle$$

The speed at $t = 1$ is

$$\|\mathbf{v}(1)\| \approx \sqrt{(-0.83)^2 + (1.82)^2 + (0.35)^2} \approx 2.03 \qquad ■$$

If an object's acceleration is given, we can solve for $\mathbf{v}(t)$ and $\mathbf{r}(t)$ by integrating twice:

$$\mathbf{v}(t) = \int \mathbf{a}(t)\, dt$$

$$\mathbf{r}(t) = \int_0^t \mathbf{v}(t)\, dt$$

The arbitrary constants that result are determined by initial conditions.

■ **EXAMPLE 2** Find $\mathbf{r}(t)$ if

$$\mathbf{a}(t) = 2\mathbf{i} + 12t\mathbf{j}, \qquad \mathbf{v}(0) = 7\mathbf{i}, \qquad \mathbf{r}(0) = 2\mathbf{i} + 9\mathbf{k}$$

Solution We have

$$\mathbf{v}(t) = \int \mathbf{a}(t)\, dt = 2t\mathbf{i} + 6t^2\mathbf{j} + \mathbf{C}_0$$

The initial condition $\mathbf{v}(0) = 0 + \mathbf{C}_0 = 7\mathbf{i}$ gives us $\mathbf{v}(t) = 2t\mathbf{i} + 6t^2\mathbf{j} + 7\mathbf{i} = (2t + 7)\mathbf{i} + 6t^2\mathbf{j}$. Then we have

$$\mathbf{r}(t) = \int \mathbf{v}(t)\, dt = (t^2 + 7t)\mathbf{i} + 2t^3\mathbf{j} + \mathbf{C}_1$$

The initial condition $\mathbf{r}(0) = 0 + \mathbf{C}_1 = 2\mathbf{i} + 9\mathbf{k}$ yields

$$\mathbf{r}(t) = (t^2 + 7t)\mathbf{i} + 2t^3\mathbf{j} + (2\mathbf{i} + 9\mathbf{k}) = (t^2 + 7t + 2)\mathbf{i} + 2t^3\mathbf{j} + 9\mathbf{k} \qquad ■$$

■ **EXAMPLE 3** Find $\mathbf{r}(t)$ if

$$\mathbf{a}(t) = (\cos t)\mathbf{i} + e^t\mathbf{j} + t\mathbf{k}, \qquad \mathbf{v}(0) = \mathbf{i}, \qquad \mathbf{r}(0) = \mathbf{k}$$

CAUTION *Observe that in Example 3, when we determine* $\mathbf{v}(t)$*, the arbitrary constant* $C_0 \neq \mathbf{v}(0)$*. The same holds when determining* $\mathbf{r}(t)$ *The resulting arbitrary constant does not equal the initial value* $\mathbf{r}(0)$.

Solution We obtain

$$\mathbf{v}(t) = \int \mathbf{a}(t)\, dt = (\sin t)\mathbf{i} + e^t\mathbf{j} + \frac{t^2}{2}\mathbf{k} + \mathbf{C}_0$$

Note that here, we must be careful, as it is not true that when $\mathbf{v}$ is evaluated at 0, all that remains is the arbitrary constant vector. The initial condition yields

$$\mathbf{v}(0) = \mathbf{j} + \mathbf{C}_0 = \mathbf{i}$$

Hence, $\mathbf{C}_0 = \mathbf{i} - \mathbf{j}$. Therefore,

$$\mathbf{v}(t) = (\sin t)\mathbf{i} + e^t\mathbf{j} + \frac{t^2}{2}\mathbf{k} + \mathbf{i} - \mathbf{j} = (\sin t + 1)\mathbf{i} + (e^t - 1)\mathbf{j} + \frac{t^2}{2}\mathbf{k}$$

Then we have

$$\mathbf{r}(t) = \int \mathbf{v}(t)\, dt = (-\cos t + t)\mathbf{i} + (e^t - t)\mathbf{j} + \frac{t^3}{6}\mathbf{i} + \mathbf{C}_1$$

The initial condition $\mathbf{r}(0) = -1\mathbf{i} + 1\mathbf{j} + \mathbf{C}_1 = \mathbf{k}$ yields $\mathbf{C}_1 = \mathbf{i} - \mathbf{j} + \mathbf{k}$. Thus,

$$\mathbf{r}(t) = (-\cos t + t)\mathbf{i} + (e^t - t)\mathbf{j} + \frac{t^3}{6}\mathbf{k} + \mathbf{i} - \mathbf{j} + \mathbf{k}$$

$$= (-\cos t + t + 1)\mathbf{i} + (e^t - t - 1)\mathbf{j} + \left(\frac{t^3}{6} + 1\right)\mathbf{k} \qquad\blacksquare$$

Near the surface of the earth, gravity imparts an acceleration of -9.8 m/s^2 in the z-direction. If units are in feet, it is -32 ft/s^2. This means that if we have a projectile moving near the surface of the earth that has no additional means of acquiring acceleration, we know that its acceleration vector is given by $\mathbf{a}(t) = -9.8\mathbf{k}$. Note that in the case where the projectile stays in a vertical plane, we can assume it is above the x-axis, so the entire problem occurs in the xy-plane. We can take the vertical direction to be in the direction of the y-axis, and take the acceleration vector to be $\mathbf{a}(t) = -9.8\mathbf{j}$.

■ **EXAMPLE 4** A bullet is fired from the ground at an angle of $60°$ above the horizontal. What initial speed v_0 must the bullet have in order to hit a point 150 m high on a tower located 250 m away (ignoring air resistance)?

Solution Place the gun at the origin, and let $\mathbf{r}(t)$ be the position vector of the bullet (Figure 3). Then we know $\mathbf{r}_0 = \mathbf{0}$. Furthermore, since the angle of the gun is $60°$, even though we do not know the initial speed v_0, we can say that the initial velocity vector is
$\mathbf{v}_0 = v_0 \cos 60°\mathbf{i} + v_0 \sin 60°\mathbf{j} = \frac{1}{2}v_0\mathbf{i} + v_0\frac{\sqrt{3}}{2}\mathbf{j}$.

Step 1. **Use the acceleration due to gravity.**
Since we can assume that the entire path lies in the xy-plane, we can assume that gravity exerts a downward force generating an acceleration vector $\mathbf{a}(t) = -9.8\mathbf{j}$. We determine $\mathbf{r}(t)$ by integrating twice:

$$\mathbf{v}(t) = \int \mathbf{a}(t)\, dt = -9.8t\mathbf{j} + \mathbf{C}_0$$

Step 2. **Use the initial velocity.**
We know that $\mathbf{v}(0) = \mathbf{v}_0$. Hence, $\mathbf{0} + \mathbf{C}_0 = \frac{1}{2}v_0\mathbf{i} + v_0\frac{\sqrt{3}}{2}\mathbf{j}$. Plugging this expression in for $\mathbf{C}_0$, we obtain

$$\mathbf{v}(t) = -9.8t\mathbf{j} + \frac{1}{2}v_0\mathbf{i} + v_0\frac{\sqrt{3}}{2}\mathbf{j} = \frac{1}{2}v_0\mathbf{i} + (v_0\frac{\sqrt{3}}{2} - 9.8t)\mathbf{j}$$

Step 3. **Integrate again.**

$$\mathbf{r}(t) = \int \mathbf{v}(t)\, dt = \frac{1}{2}v_0 t\mathbf{i} + (v_0\frac{\sqrt{3}}{2}t - 4.9t^2)\mathbf{j} + \mathbf{C}_1$$

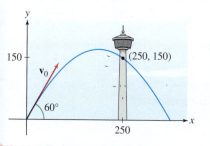

FIGURE 3 Trajectory of the bullet.

Step 4. **Use the initial position.**

We know that $\mathbf{r}(0) = \mathbf{r}_0 = \mathbf{0}$. Hence, $\mathbf{0} + \mathbf{C}_1 = \mathbf{0}$. Plugging this expression in for $\mathbf{C}_1$, we obtain

$$\mathbf{r}(t) = \frac{1}{2}v_0 t \mathbf{i} + \left(v_0 \frac{\sqrt{3}}{2}t - 4.9t^2 \right)\mathbf{j}$$

Step 5. **Solve for v_0.**

The bullet hits the point $250\mathbf{i} + 150\mathbf{j}$ on the tower if there exists a time t such that $\mathbf{r}(t) = 250\mathbf{i} + 150\mathbf{j}$; that is,

$$\frac{1}{2}v_0 t \mathbf{i} + \left(v_0 \frac{\sqrt{3}}{2}t - 4.9t^2 \right)\mathbf{j} = 250\mathbf{i} + 150\mathbf{j}$$

Equating components, we obtain

$$\frac{1}{2}t v_0 = 250, \qquad v_0 \frac{\sqrt{3}}{2}t - 4.9t^2 = 150$$

The first equation yields $t = 500/v_0$. Now substitute in the second equation and solve:

$$\frac{\sqrt{3}}{2}\left(\frac{500}{v_0} \right)v_0 - 4.9\left(\frac{500}{v_0} \right)^2 = 150$$

$$\left(\frac{500}{v_0} \right)^2 = \frac{250\sqrt{3} - 150}{4.9}$$

$$\left(\frac{v_0}{500} \right)^2 = \frac{4.9}{250\sqrt{3} - 150} \approx 0.0173$$

We obtain $v_0 \approx 500\sqrt{0.0173} \approx 66$ m/s. ∎

In linear motion, acceleration is the rate at which an object is speeding up or slowing down. The acceleration is zero if the speed is constant. By contrast, in two or three dimensions, the acceleration can be nonzero even when the object's speed is constant. This happens when $v(t) = \|\mathbf{v}(t)\|$ is constant but the *direction* of $\mathbf{v}(t)$ is changing. The simplest example is **uniform circular motion**, in which an object travels in a circular path at constant speed (Figure 4).

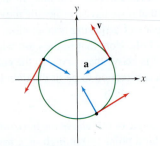

y / x

FIGURE 4 In uniform circular motion, $\mathbf{v}$ has constant length but turns continuously. The acceleration $\mathbf{a}$ is centripetal, pointing toward the center of the circle.

■ **EXAMPLE 5** **Uniform Circular Motion** Find $\mathbf{a}(t)$ and $\|\mathbf{a}(t)\|$ for motion around a circle of radius R with constant speed v.

Solution Assume that the particle follows the circular path $\mathbf{r}(t) = R\langle \cos \omega t, \sin \omega t \rangle$ for some constant ω. Then the velocity and speed of the particle are

$$\mathbf{v}(t) = R\omega\langle -\sin \omega t, \cos \omega t \rangle, \qquad v = \|\mathbf{v}(t)\| = R|\omega|$$

Thus, $|\omega| = v/R$; accordingly,

> The constant ω (lowercase Greek "omega") is called the angular speed *because the particle's angle along the circle changes at a rate of ω radians per unit time.*

$$\mathbf{a}(t) = \mathbf{v}'(t) = -R\omega^2\langle \cos \omega t, \sin \omega t \rangle, \qquad \|\mathbf{a}(t)\| = R\omega^2 = R\left(\frac{v}{R} \right)^2 = \frac{v^2}{R}$$

The vector $\mathbf{a}(t)$ is called the **centripetal acceleration**: It has length v^2/R and points in toward the origin [because $\mathbf{a}(t)$ is a negative multiple of the position vector $\mathbf{r}(t)$], as in Figure 4. ■

Understanding the Acceleration Vector

We have noted that $\mathbf{v}(t)$ can change in two ways: in magnitude and in direction. To understand how the acceleration vector $\mathbf{a}(t)$ "encodes" both types of change, we decompose $\mathbf{a}(t)$ into a sum of tangential and normal components.

Recall the definition of unit tangent and unit normal vectors:

$$\mathbf{T}(t) = \frac{\mathbf{v}(t)}{\|\mathbf{v}(t)\|}, \qquad \mathbf{N}(t) = \frac{\mathbf{T}'(t)}{\|\mathbf{T}'(t)\|}$$

Thus, $\mathbf{v}(t) = v(t)\mathbf{T}(t)$, where $v(t) = \|\mathbf{v}(t)\|$, so by the Product Rule,

$$\mathbf{a}(t) = \frac{d\mathbf{v}}{dt} = \frac{d}{dt}v(t)\mathbf{T}(t) = v'(t)\mathbf{T}(t) + v(t)\mathbf{T}'(t)$$

Furthermore, $\mathbf{T}'(t) = v(t)\kappa(t)\mathbf{N}(t)$ by Eq. (7) of Section 13.4, where $\kappa(t)$ is the curvature. Thus, we can write

$$\boxed{\mathbf{a} = a_{\mathbf{T}}\mathbf{T} + a_{\mathbf{N}}\mathbf{N}, \qquad a_{\mathbf{T}} = v'(t), \quad a_{\mathbf{N}} = \kappa(t)v(t)^2}$$ **1**

The coefficient $a_{\mathbf{T}}(t)$ is called the **tangential component** and $a_{\mathbf{N}}(t)$ the **normal component** of acceleration (Figure 5).

When you make a left turn in an automobile at constant speed, your tangential acceleration is zero [because $v'(t) = 0$] and you will not be pushed back against your seat. But the car seat (via friction) pushes you to the left toward the car door, causing you to accelerate in the normal direction. Due to inertia, you feel as if you are being pushed to the right toward the passenger's seat. This force is proportional to κv^2, so a sharp turn (large κ) or high speed (large v) produces a strong normal force.

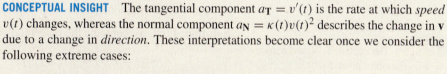

DF **FIGURE 5** Decomposition of **a** into tangential and normal components.

*The normal component $a_{\mathbf{N}}$ is often called the **centripetal acceleration**, especially in the case of circular motion where it is directed toward the center of the circle.*

CONCEPTUAL INSIGHT The tangential component $a_{\mathbf{T}} = v'(t)$ is the rate at which *speed* $v(t)$ changes, whereas the normal component $a_{\mathbf{N}} = \kappa(t)v(t)^2$ describes the change in **v** due to a change in *direction*. These interpretations become clear once we consider the following extreme cases:

• A particle travels in a straight line. Then direction does not change [$\kappa(t) = 0$] and $\mathbf{a}(t) = v'(t)\mathbf{T}$ is parallel to the direction of motion.
• A particle travels with constant speed along a curved path. Then $v'(t) = 0$ and the acceleration vector $\mathbf{a}(t) = \kappa(t)v(t)^2\mathbf{N}$ is normal to the direction of motion.

General motion combines both tangential and normal acceleration.

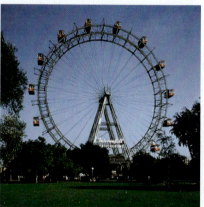

■ **EXAMPLE 6** The Giant Ferris Wheel in Vienna has radius $R = 30$ m (Figure 6). Assume that at time $t = t_0$, the wheel rotates counterclockwise with a speed of 40 m/min and is slowing at a rate of 15 m/min². Find the acceleration vector **a** for a person seated in a car at the lowest point of the wheel.

Solution At the bottom of the wheel, $\mathbf{T} = \langle 1, 0 \rangle$ and $\mathbf{N} = \langle 0, 1 \rangle$. We are told that $a_{\mathbf{T}} = v' = -15$ at time t_0. The curvature of the wheel is $\kappa = 1/R = 1/30$, so the normal component is $a_{\mathbf{N}} = \kappa v^2 = v^2/R = (40)^2/30 \approx 53.3$. Therefore (Figure 7),

$$\mathbf{a} \approx -15\mathbf{T} + 53.3\mathbf{N} = \langle -15, 53.3 \rangle \text{ m/min}^2$$ ■

FIGURE 6 The Giant Ferris Wheel in Vienna, Austria, erected in 1897 to celebrate the 50th anniversary of the coronation of Emperor Franz Joseph I.
(© Peter M. Wilson/Alamy)

The following theorem provides useful formulas for the tangential and normal components.

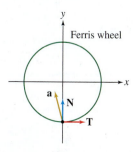

THEOREM 1 Tangential and Normal Components of Acceleration In the decomposition $\mathbf{a} = a_{\mathbf{T}}\mathbf{T} + a_{\mathbf{N}}\mathbf{N}$, we have

$$\boxed{a_{\mathbf{T}} = \mathbf{a} \cdot \mathbf{T} = \frac{\mathbf{a} \cdot \mathbf{v}}{\|\mathbf{v}\|}, \qquad a_{\mathbf{N}} = \mathbf{a} \cdot \mathbf{N} = \sqrt{\|\mathbf{a}\|^2 - |a_{\mathbf{T}}|^2}}$$ **2**

and

$$\boxed{a_{\mathbf{T}}\mathbf{T} = \left(\frac{\mathbf{a} \cdot \mathbf{v}}{\mathbf{v} \cdot \mathbf{v}}\right)\mathbf{v}, \qquad a_{\mathbf{N}}\mathbf{N} = \mathbf{a} - a_{\mathbf{T}}\mathbf{T} = \mathbf{a} - \left(\frac{\mathbf{a} \cdot \mathbf{v}}{\mathbf{v} \cdot \mathbf{v}}\right)\mathbf{v}}$$ **3**

FIGURE 7

Proof We have $\mathbf{T} \cdot \mathbf{T} = 1$ and $\mathbf{N} \cdot \mathbf{T} = 0$. Thus,

$$\mathbf{a} \cdot \mathbf{T} = (a_\mathbf{T}\mathbf{T} + a_\mathbf{N}\mathbf{N}) \cdot \mathbf{T} = a_\mathbf{T}$$

$$\mathbf{a} \cdot \mathbf{N} = (a_\mathbf{T}\mathbf{T} + a_\mathbf{N}\mathbf{N}) \cdot \mathbf{N} = a_\mathbf{N}$$

and since $\mathbf{T} = \dfrac{\mathbf{v}}{\|\mathbf{v}\|}$, we have

$$a_\mathbf{T}\mathbf{T} = (\mathbf{a} \cdot \mathbf{T})\mathbf{T} = \left(\dfrac{\mathbf{a} \cdot \mathbf{v}}{\|\mathbf{v}\|}\right)\dfrac{\mathbf{v}}{\|\mathbf{v}\|} = \left(\dfrac{\mathbf{a} \cdot \mathbf{v}}{\mathbf{v} \cdot \mathbf{v}}\right)\mathbf{v}$$

and

$$a_\mathbf{N}\mathbf{N} = \mathbf{a} - a_\mathbf{T}\mathbf{T} = \mathbf{a} - \left(\dfrac{\mathbf{a} \cdot \mathbf{v}}{\|\mathbf{v}\|}\right)\mathbf{v}$$

Finally, the vectors $a_\mathbf{T}\mathbf{T}$ and $a_\mathbf{N}\mathbf{N}$ are the sides of a right triangle with hypotenuse $\mathbf{a}$ as in Figure 5, so by the Pythagorean Theorem,

$$\|\mathbf{a}\|^2 = |a_\mathbf{T}|^2 + |a_\mathbf{N}|^2 \quad \Rightarrow \quad a_\mathbf{N} = \sqrt{\|\mathbf{a}\|^2 - |a_\mathbf{T}|^2} \qquad \blacksquare$$

Keep in mind that $a_\mathbf{N} \geq 0$ but $a_\mathbf{T}$ is positive or negative, depending on whether the object is speeding up or slowing down.

■ **EXAMPLE 7** Decompose the acceleration vector $\mathbf{a}$ of $\mathbf{r}(t) = \left\langle t^2, 2t, \ln t \right\rangle$ into tangential and normal components at $t = \frac{1}{2}$ (Figure 8).

Solution First, we compute the tangential components $\mathbf{T}$ and $a_\mathbf{T}$. We have

$$\mathbf{v}(t) = \mathbf{r}'(t) = \left\langle 2t, 2, t^{-1}\right\rangle, \qquad \mathbf{a}(t) = \mathbf{r}''(t) = \left\langle 2, 0, -t^{-2}\right\rangle$$

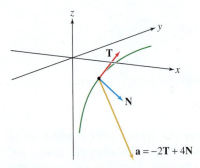

DF **FIGURE 8** The vectors $\mathbf{T}$, $\mathbf{N}$, and $\mathbf{a}$ at $t = \frac{1}{2}$ on the curve given by $\mathbf{r}(t) = \left\langle t^2, 2t, \ln t \right\rangle$.

At $t = \frac{1}{2}$,

$$\mathbf{v} = \mathbf{r}'\left(\dfrac{1}{2}\right) = \left\langle 2\left(\dfrac{1}{2}\right), 2, \left(\dfrac{1}{2}\right)^{-1}\right\rangle = \langle 1, 2, 2\rangle$$

$$\mathbf{a} = \mathbf{r}''\left(\dfrac{1}{2}\right) = \left\langle 2, 0, -\left(\dfrac{1}{2}\right)^{-2}\right\rangle = \langle 2, 0, -4\rangle$$

Thus,

$$\mathbf{T} = \dfrac{\mathbf{v}}{\|\mathbf{v}\|} = \dfrac{\langle 1, 2, 2\rangle}{\sqrt{1^2 + 2^2 + 2^2}} = \left\langle \dfrac{1}{3}, \dfrac{2}{3}, \dfrac{2}{3}\right\rangle$$

and by Eq. (2),

$$a_\mathbf{T} = \mathbf{a} \cdot \mathbf{T} = \langle 2, 0, -4\rangle \cdot \left\langle \dfrac{1}{3}, \dfrac{2}{3}, \dfrac{2}{3}\right\rangle = -2$$

Next, we use Eq. (3):

$$a_\mathbf{N}\mathbf{N} = \mathbf{a} - a_\mathbf{T}\mathbf{T} = \langle 2, 0, -4\rangle - (-2)\left\langle \dfrac{1}{3}, \dfrac{2}{3}, \dfrac{2}{3}\right\rangle = \left\langle \dfrac{8}{3}, \dfrac{4}{3}, -\dfrac{8}{3}\right\rangle$$

This vector has length

$$a_\mathbf{N} = \|a_\mathbf{N}\mathbf{N}\| = \sqrt{\dfrac{64}{9} + \dfrac{16}{9} + \dfrac{64}{9}} = 4$$

and thus,

$$\mathbf{N} = \dfrac{a_\mathbf{N}\mathbf{N}}{a_\mathbf{N}} = \dfrac{\left\langle \frac{8}{3}, \frac{4}{3}, -\frac{8}{3}\right\rangle}{4} = \left\langle \dfrac{2}{3}, \dfrac{1}{3}, -\dfrac{2}{3}\right\rangle$$

Summary of steps in Example 7:

$$\mathbf{T} = \dfrac{\mathbf{v}}{\|\mathbf{v}\|}$$

$$a_\mathbf{T} = \mathbf{a} \cdot \mathbf{T}$$

$$a_\mathbf{N}\mathbf{N} = \mathbf{a} - a_\mathbf{T}\mathbf{T}$$

$$a_\mathbf{N} = \|a_\mathbf{N}\mathbf{N}\|$$

$$\mathbf{N} = \dfrac{a_\mathbf{N}\mathbf{N}}{a_\mathbf{N}}$$

Finally, we obtain the decomposition

$$\mathbf{a} = \langle 2, 0, -4\rangle = a_\mathbf{T}\mathbf{T} + a_\mathbf{N}\mathbf{N} = -2\mathbf{T} + 4\mathbf{N} \qquad \blacksquare$$

◀·· *REMINDER*

• By Eq. (2), $v' = a_{\mathbf{T}} = \mathbf{a} \cdot \mathbf{T}$
• $\mathbf{v} \cdot \mathbf{w} = \|\mathbf{v}\| \|\mathbf{w}\| \cos\theta$

where θ is the angle between $\mathbf{v}$ and $\mathbf{w}$.

■ **EXAMPLE 8** **Nonuniform Circular Motion** Figure 9 shows the acceleration vectors of three particles moving *counterclockwise* around a circle. In each case, state whether the particle's speed v is increasing, decreasing, or momentarily constant.

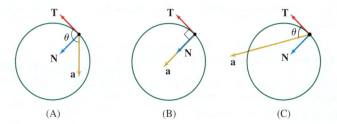

FIGURE 9 Acceleration vectors of particles moving counterclockwise (in the direction of $\mathbf{T}$) around a circle.

(A)　　　(B)　　　(C)

Solution The rate of change of speed depends on the angle θ between $\mathbf{a}$ and $\mathbf{T}$:

$$v' = a_{\mathbf{T}} = \mathbf{a} \cdot \mathbf{T} = \|\mathbf{a}\| \|\mathbf{T}\| \cos\theta = \|\mathbf{a}\| \cos\theta$$

• In (A), θ is obtuse, so $\cos\theta < 0$ and $v' < 0$. The particle's speed is decreasing.
• In (B), $\theta = \frac{\pi}{2}$, so $\cos\theta = 0$ and $v' = 0$. The particle's speed is momentarily constant.
• In (C), θ is acute, so $\cos\theta > 0$ and $v' > 0$. The particle's speed is increasing.　■

■ **EXAMPLE 9** Find the curvature $\kappa\left(\frac{1}{2}\right)$ for the path $\mathbf{r}(t) = \langle t^2, 2t, \ln t \rangle$ in Example 7.

Solution By Eq. (1), the normal component is

$$a_{\mathbf{N}} = \kappa v^2$$

In Example 7, we showed that $a_{\mathbf{N}} = 4$ and $\mathbf{v} = \langle 1, 2, 2 \rangle$ at $t = \frac{1}{2}$. Therefore, $v^2 = \mathbf{v} \cdot \mathbf{v} = 9$ and the curvature is $\kappa\left(\frac{1}{2}\right) = a_{\mathbf{N}}/v^2 = \frac{4}{9}$.　■

13.5 SUMMARY

• For an object whose path is described by a vector-valued function $\mathbf{r}(t)$,

$$\mathbf{v}(t) = \mathbf{r}'(t), \qquad v(t) = \|\mathbf{v}(t)\|, \qquad \mathbf{a}(t) = \mathbf{r}''(t)$$

• The *velocity vector* $\mathbf{v}(t)$ points in the direction of motion. Its length $v(t) = \|\mathbf{v}(t)\|$ is the object's speed.
• The *acceleration vector* $\mathbf{a}$ is the sum of a tangential component (reflecting change in speed) and a normal component (reflecting change in direction):

$$\mathbf{a}(t) = a_{\mathbf{T}}(t)\mathbf{T}(t) + a_{\mathbf{N}}(t)\mathbf{N}(t)$$

Unit tangent vector	$\mathbf{T}(t) = \dfrac{\mathbf{v}(t)}{\|\mathbf{v}(t)\|}$
Unit normal vector	$\mathbf{N}(t) = \dfrac{\mathbf{T}'(t)}{\|\mathbf{T}'(t)\|}$
Tangential component	$a_{\mathbf{T}} = v'(t) = \mathbf{a} \cdot \mathbf{T} = \dfrac{\mathbf{a} \cdot \mathbf{v}}{\|\mathbf{v}\|}$
	$a_{\mathbf{T}}\mathbf{T} = \left(\dfrac{\mathbf{a} \cdot \mathbf{v}}{\mathbf{v} \cdot \mathbf{v}}\right)\mathbf{v}$
Normal component	$a_{\mathbf{N}} = \kappa(t)v(t)^2 = \sqrt{\|\mathbf{a}\|^2 - |a_{\mathbf{T}}|^2}$
	$a_{\mathbf{N}}\mathbf{N} = \mathbf{a} - a_{\mathbf{T}}\mathbf{T} = \mathbf{a} - \left(\dfrac{\mathbf{a} \cdot \mathbf{v}}{\mathbf{v} \cdot \mathbf{v}}\right)\mathbf{v}$

13.5 EXERCISES

Preliminary Questions

1. If a particle travels with constant speed, must its acceleration vector be zero? Explain.

2. For a particle in uniform circular motion around a circle, which of the vectors $\mathbf{v}(t)$ or $\mathbf{a}(t)$ always points toward the center of the circle?

3. Two objects travel to the right along the parabola $y = x^2$ with nonzero speed. Which of the following statements must be true?

(a) Their velocity vectors point in the same direction.

(b) Their velocity vectors have the same length.

(c) Their acceleration vectors point in the same direction.

4. Use the decomposition of acceleration into tangential and normal components to explain the following statement: If the speed is constant, then the acceleration and velocity vectors are orthogonal.

5. If a particle travels along a straight line, then the acceleration and velocity vectors are (choose the correct description):

(a) orthogonal.   **(b)** parallel.

6. What is the length of the acceleration vector of a particle traveling around a circle of radius 2 cm with constant velocity 4 cm/s?

7. Two cars are racing around a circular track. If, at a certain moment, both of their speedometers read 110 mph. then the two cars have the same (choose one):

(a) $a_{\mathbf{T}}$ **(b)** $a_{\mathbf{N}}$

Exercises

1. Use the table below to calculate the difference quotients $\dfrac{\mathbf{r}(1+h) - \mathbf{r}(1)}{h}$ for $h = -0.2, -0.1, 0.1, 0.2$. Then estimate the velocity and speed at $t = 1$.

$\mathbf{r}(0.8)$	$\langle 1.557, 2.459, -1.970 \rangle$
$\mathbf{r}(0.9)$	$\langle 1.559, 2.634, -1.740 \rangle$
$\mathbf{r}(1)$	$\langle 1.540, 2.841, -1.443 \rangle$
$\mathbf{r}(1.1)$	$\langle 1.499, 3.078, -1.035 \rangle$
$\mathbf{r}(1.2)$	$\langle 1.435, 3.342, -0.428 \rangle$

2. Draw the vectors $\mathbf{r}(2+h) - \mathbf{r}(2)$ and $\dfrac{\mathbf{r}(2+h) - \mathbf{r}(2)}{h}$ for $h = 0.5$ for the path in Figure 10. Draw $\mathbf{v}(2)$ (using a rough estimate for its length).

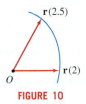

FIGURE 10

In Exercises 3–6, calculate the velocity and acceleration vectors and the speed at the time indicated.

3. $\mathbf{r}(t) = \langle t^3, 1 - t, 4t^2 \rangle, \quad t = 1$

4. $\mathbf{r}(t) = e^t \mathbf{j} - \cos(2t)\mathbf{k}, \quad t = 0$

5. $\mathbf{r}(\theta) = \langle \sin\theta, \cos\theta, \cos 3\theta \rangle, \quad \theta = \frac{\pi}{3}$

6. $\mathbf{r}(s) = \left(\dfrac{1}{1+s^2}, \dfrac{s}{1+s^2} \right), \quad s = 2$

7. Find $\mathbf{a}(t)$ for a particle moving around a circle of radius 8 cm at a constant speed of $v = 4$ cm/s (see Example 5). Draw the path and acceleration vector at $t = \frac{\pi}{4}$.

8. Sketch the path $\mathbf{r}(t) = \langle 1 - t^2, 1 - t \rangle$ for $-2 \le t \le 2$, indicating the direction of motion. Draw the velocity and acceleration vectors at $t = 0$ and $t = 1$.

9. Sketch the path $\mathbf{r}(t) = \langle t^2, t^3 \rangle$ together with the velocity and acceleration vectors at $t = 1$.

10. The paths $\mathbf{r}(t) = \langle t^2, t^3 \rangle$ and $\mathbf{r}_1(t) = \langle t^4, t^6 \rangle$ trace the same curve, and $\mathbf{r}_1(1) = \mathbf{r}(1)$. Do you expect either the velocity vectors or the acceleration vectors of these paths at $t = 1$ to point in the same direction? Compute these vectors and draw them on a single plot of the curve.

In Exercises 11–14, find $\mathbf{v}(t)$ given $\mathbf{a}(t)$ and the initial velocity.

11. $\mathbf{a}(t) = \langle t, 4 \rangle, \quad \mathbf{v}(0) = \langle \frac{1}{3}, -2 \rangle$

12. $\mathbf{a}(t) = \langle e^t, 0, t + 1 \rangle, \quad \mathbf{v}(0) = \langle 1, -3, \sqrt{2} \rangle$

13. $\mathbf{a}(t) = \mathbf{k}, \quad \mathbf{v}(0) = \mathbf{i}$ **14.** $\mathbf{a}(t) = t^2 \mathbf{k}, \quad \mathbf{v}(0) = \mathbf{i} - \mathbf{j}$

In Exercises 15–18, find $\mathbf{r}(t)$ and $\mathbf{v}(t)$ given $\mathbf{a}(t)$ and the initial velocity and position.

15. $\mathbf{a}(t) = \langle t, 4 \rangle, \quad \mathbf{v}(0) = \langle 3, -2 \rangle, \quad \mathbf{r}(0) = \langle 0, 0 \rangle$

16. $\mathbf{a}(t) = \langle e^t, 2t, t + 1 \rangle, \quad \mathbf{v}(0) = \langle 1, 0, 1 \rangle, \quad \mathbf{r}(0) = \langle 2, 1, 1 \rangle$

17. $\mathbf{a}(t) = t\mathbf{k}, \quad \mathbf{v}(0) = \mathbf{i}, \quad \mathbf{r}(0) = \mathbf{j}$

18. $\mathbf{a}(t) = \cos t\mathbf{k}, \quad \mathbf{v}(0) = \mathbf{i} - \mathbf{j}, \quad \mathbf{r}(0) = \mathbf{i}$

In Exercises 19–26, recall that $g = 9.8$ m/s^2 $= 32$ ft/s^2 is the acceleration due to gravity on the earth's surface.

19. A bullet is fired from the ground at an angle of $45°$. What initial speed must the bullet have in order to hit the top of a 120-m tower located 180 m away?

20. Find the initial velocity vector $\mathbf{v}_0$ of a projectile released with initial speed 100 m/s that reaches a maximum height of 300 m.

21. Show that a projectile fired at an angle θ with initial speed v_0 travels a total distance $(v_0^2/g) \sin 2\theta$ before hitting the ground. Conclude that the maximum distance (for a given v_0) is attained for $\theta = 45°$.

22. One player throws a baseball to another player standing 25 m away, with initial speed 18 m/s. Use the result of Exercise 21 to find two angles θ at which the ball can be released. Which angle gets the ball there faster?

23. A bullet is fired at an angle $\theta = \frac{\pi}{4}$ at a tower located $d = 600$ m away, with initial speed $v_0 = 120$ m/s. Find the height H at which the bullet hits the tower.

24. Show that a bullet fired at an angle θ will hit the top of an h-meter tower located d meters away if its initial speed is

$$v_0 = \frac{\sqrt{g/2}\,d\sec\theta}{\sqrt{d\tan\theta - h}}$$

25. In the Superbowl, a quarterback throws a football while standing at the very center of the field on the 50 yard line. The ball leaves his hand at a height of 5 ft and has initial velocity $v_0 = 40\mathbf{i} + 35\mathbf{j} + 32\mathbf{k}$ ft/s. Assume an acceleration of 32 ft/s² due to gravity and that the $\mathbf{i}$ vector points down the field toward the endzone and the $\mathbf{j}$ vector points to the sideline. The field is 150 ft in width and 300 ft in length.

(a) Determine the position function that gives the position of the ball t seconds after it is thrown.

(b) The ball is caught by a player 5 ft above the ground. Is the player in bounds or out of bounds when he receives the ball? Assume the player is standing vertically with both toes on the ground at the time of reception.

26. In the Women's World Cup, Brazilian soccer star Marta has a penalty kick in the quarter-final match. She kicks the soccer ball from ground level with (x, y)-coordinates $(85, 20)$ on the soccer field shown in Figure 11 and with an initial velocity $v_0 = 10\mathbf{i} - 5\mathbf{j} + 25\mathbf{k}$ ft/s. Assume an acceleration of 32 ft/s² due to gravity and that the goal net has a height of 8 ft and a total width of 24 ft.

(a) Determine the position function that gives the position of the ball t seconds after it is hit.

(b) Does the ball go in the goal before hitting the ground? Explain why or why not.

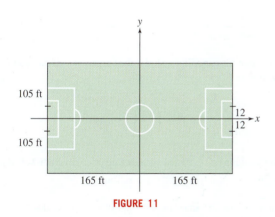

105 ft

105 ft

165 ft   165 ft

FIGURE 11

27. A constant force $\mathbf{F} = \langle 5, 2\rangle$ (in newtons) acts on a 10-kg mass. Find the position of the mass at $t = 10$ s if it is located at the origin at $t = 0$ and has initial velocity $v_0 = \langle 2, -3\rangle$ (in meters per second).

28. A force $\mathbf{F} = \langle 24t, 16 - 8t\rangle$ (in newtons) acts on a 4-kg mass. Find the position of the mass at $t = 3$ s if it is located at $(10, 12)$ at $t = 0$ and has zero initial velocity.

29. A particle follows a path $\mathbf{r}(t)$ for $0 \le t \le T$, beginning at the origin O. The vector $\overline{\mathbf{v}} = \dfrac{1}{T}\displaystyle\int_0^T \mathbf{r}'(t)\,dt$ is called the **average velocity** vector. Suppose that $\overline{\mathbf{v}} = \mathbf{0}$. Answer and explain the following:

(a) Where is the particle located at time T if $\overline{\mathbf{v}} = \mathbf{0}$?

(b) Is the particle's average speed necessarily equal to zero?

30. At a certain moment, a moving particle has velocity $\mathbf{v} = \langle 2, 2, -1\rangle$ and $\mathbf{a} = \langle 0, 4, 3\rangle$. Find $\mathbf{T}, \mathbf{N}$, and the decomposition of $\mathbf{a}$ into tangential and normal components.

31. At a certain moment, a particle moving along a path has velocity $\mathbf{v} = \langle 12, 20, 20\rangle$ and acceleration $\mathbf{a} = \langle 2, 1, -3\rangle$. Is the particle speeding up or slowing down?

In Exercises 32–35, use Eq. (2) to find the coefficients $a_{\mathbf{T}}$ and $a_{\mathbf{N}}$ as a function of t (or at the specified value of t).

32. $\mathbf{r}(t) = \langle t^2, t^3\rangle$ **33.** $\mathbf{r}(t) = \langle t, \cos t, \sin t\rangle$

34. $\mathbf{r}(t) = \langle t^{-1}, \ln t, t^2\rangle$, $t = 1$

35. $\mathbf{r}(t) = \langle e^{2t}, t, e^{-t}\rangle$, $t = 0$

In Exercise 36–43, find the decomposition of $\mathbf{a}(t)$ into tangential and normal components at the point indicated, as in Example 7.

36. $\mathbf{r}(t) = \langle e^t, 1 - t\rangle$, $t = 0$

37. $\mathbf{r}(t) = \left\langle \frac{1}{3}t^3, 1 - 3t\right\rangle$, $t = -1$

38. $\mathbf{r}(t) = \left\langle t, \frac{1}{2}t^2, \frac{1}{6}t^3\right\rangle$, $t = 1$

39. $\mathbf{r}(t) = \left\langle t, \frac{1}{2}t^2, \frac{1}{6}t^3\right\rangle$, $t = 4$

40. $\mathbf{r}(t) = \langle 4 - t, t + 1, t^2\rangle$, $t = 2$

41. $\mathbf{r}(t) = \langle t, e^t, te^t\rangle$, $t = 0$

42. $\mathbf{r}(\theta) = \langle \cos\theta, \sin\theta, \theta\rangle$, $\theta = 0$

43. $\mathbf{r}(t) = \langle t, \cos t, t\sin t\rangle$, $t = \frac{\pi}{2}$

44. Let $\mathbf{r}(t) = \langle t^2, 4t - 3\rangle$. Find $\mathbf{T}(t)$ and $\mathbf{N}(t)$, and show that the decomposition of $\mathbf{a}(t)$ into tangential and normal components is

$$\mathbf{a}(t) = \left(\frac{2t}{\sqrt{t^2 + 4}}\right)\mathbf{T} + \left(\frac{4}{\sqrt{t^2 + 4}}\right)\mathbf{N}$$

45. Find the components $a_{\mathbf{T}}$ and $a_{\mathbf{N}}$ of the acceleration vector of a particle moving along a circular path of radius $R = 100$ cm with constant velocity $v_0 = 5$ cm/s.

46. In the notation of Example 6, find the acceleration vector for a person seated in a car at (a) the highest point of the Ferris wheel and (b) the two points level with the center of the wheel.

47. Suppose that the Ferris wheel in Example 6 is rotating clockwise and that the point P at angle $45°$ has acceleration vector $\mathbf{a} = \langle 0, -50\rangle$ m/min² pointing down, as in Figure 12. Determine the speed and tangential acceleration of the Ferris wheel.

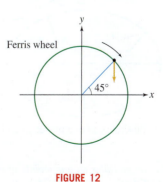

Ferris wheel

$45°$

FIGURE 12

48. At time t_0, a moving particle has velocity vector $\mathbf{v} = 2\mathbf{i}$ and acceleration vector $\mathbf{a} = 3\mathbf{i} + 18\mathbf{k}$. Determine the curvature $\kappa(t_0)$ of the particle's path at time t_0.

49. A space shuttle orbits the earth at an altitude 400 km above the earth's surface, with constant speed $v = 28,000$ km/h. Find the magnitude of the shuttle's acceleration (in kilometers per square hour), assuming that the radius of the earth is 6378 km (Figure 13).

FIGURE 13 Space shuttle orbit.

50. A car proceeds along a circular path of radius $R = 300$ m centered at the origin. Starting at rest, its speed increases at a rate of t m/s^2. Find the acceleration vector **a** at time $t = 3$ s and determine its decomposition into normal and tangential components.

51. A runner runs along the helix $\mathbf{r}(t) = \langle \cos t, \sin t, t \rangle$. When he is at position $\mathbf{r}\left(\frac{\pi}{2}\right)$, his speed is 3 m/s and he is accelerating at a rate of $\frac{1}{2}$ m/s^2. Find his acceleration vector **a** at this moment. *Note:* The runner's acceleration vector does not coincide with the acceleration vector of $\mathbf{r}(t)$.

52. 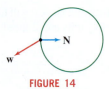   Explain why the vector **w** in Figure 14 cannot be the acceleration vector of a particle moving along the circle. *Hint:* Consider the sign of $\mathbf{w} \cdot \mathbf{N}$.

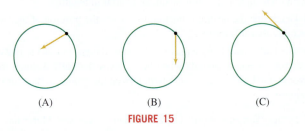

FIGURE 14

53. Figure 15 shows the acceleration vectors of a particle moving clockwise around a circle. In each case, state whether the particle is speeding up, slowing down, or momentarily at constant speed. Explain.

(A)   (B)   (C)

FIGURE 15

54. Prove that $a_N = \dfrac{\|\mathbf{a} \times \mathbf{v}\|}{\|\mathbf{v}\|}$.

55. Suppose that $\mathbf{r} = \mathbf{r}(t)$ lies on a sphere of radius R for all t. Let $\mathbf{J} = \mathbf{r} \times \mathbf{r}'$. Show that $\mathbf{r}' = (\mathbf{J} \times \mathbf{r})/\|\mathbf{r}\|^2$. *Hint:* Observe that $\mathbf{r}$ and $\mathbf{r}'$ are perpendicular.

Further Insights and Challenges

56. The orbit of a planet is an ellipse with the sun at one focus. The sun's gravitational force acts along the radial line from the planet to the sun (the dashed lines in Figure 16), and by Newton's Second Law, the acceleration vector points in the same direction. Assuming that the orbit has positive eccentricity (the orbit is not a circle), explain why the planet must slow down in the upper half of the orbit (as it moves away from the sun) and speed up in the lower half. Kepler's Second Law, discussed in the next section, is a precise version of this qualitative conclusion. *Hint:* Consider the decomposition of **a** into normal and tangential components.

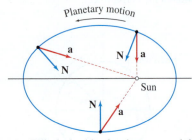

FIGURE 16 Elliptical orbit of a planet around the sun.

In Exercises 57–61, we consider an automobile of mass m traveling along a curved but level road. To avoid skidding, the road must supply a frictional force $\mathbf{F} = m\mathbf{a}$, where $\mathbf{a}$ is the car's acceleration vector. The maximum magnitude of the frictional force is $\mu m g$, where μ is the coefficient of friction and $g = 9.8$ m/s^2. Let v be the car's speed in meters per second.

57. Show that the car will not skid if the curvature κ of the road is such that (with $R = 1/\kappa$)

$$(v')^2 + \left(\frac{v^2}{R}\right)^2 \leq (\mu g)^2 \qquad \boxed{4}$$

Note that braking ($v' < 0$) and speeding up ($v' > 0$) contribute equally to skidding.

58. Suppose that the maximum radius of curvature along a curved highway is $R = 180$ m. How fast can an automobile travel (at constant speed) along the highway without skidding if the coefficient of friction is $\mu = 0.5$?

59. Beginning at rest, an automobile drives around a circular track of radius $R = 300$ m, accelerating at a rate of 0.3 m/s^2. After how many seconds will the car begin to skid if the coefficient of friction is $\mu = 0.6$?

60. You want to reverse your direction in the shortest possible time by driving around a semicircular bend (Figure 17). If you travel at the maximum possible *constant speed* v that will not cause skidding, is it faster to hug the inside curve (radius r) or the outside curb (radius R)? *Hint:* Use Eq. (4) to show that at maximum speed, the time required to drive around the semicircle is proportional to the square root of the radius.

61. What is the smallest radius R about which an automobile can turn without skidding at 100 km/h if $\mu = 0.75$ (a typical value)?

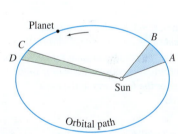

FIGURE 17 Car going around the bend.

13.6 Planetary Motion According to Kepler and Newton

In this section, we derive Kepler's laws of planetary motion, a feat first accomplished by Isaac Newton and published by him in 1687. No event was more emblematic of the scientific revolution. It demonstrated the power of mathematics to make the natural world comprehensible and it led succeeding generations of scientists to seek and discover mathematical laws governing other phenomena, such as electricity and magnetism, thermodynamics, and atomic processes.

According to Kepler, the planetary orbits are ellipses with the sun at one focus. Furthermore, if we imagine a radial vector $\mathbf{r}(t)$ pointing from the sun to the planet, as in Figure 1, then this radial vector sweeps out area at a constant rate or, as Kepler stated in his Second Law, the radial vector sweeps out equal areas in equal times (Figure 2). Kepler's Third Law determines the **period** T of the orbit, defined as the time required to complete one full revolution. These laws are valid not just for planets orbiting the sun, but for any body orbiting about another body according to the inverse-square law of gravitation.

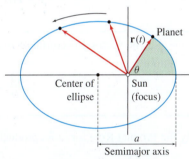

FIGURE 1 The planet travels along an ellipse with the sun at one focus.

DF FIGURE 2 The two shaded regions have equal areas, and by Kepler's Second Law, the planet sweeps them out in equal times. To do so, the planet must travel faster going from A to B than from C to D.

Kepler's version of the Third Law stated only that T^2 is proportional to a^3. Newton discovered that the constant of proportionality is equal to $4\pi^2/(GM)$, and he observed that if you can measure T and a through observation, then you can use the Third Law to solve for the mass M. This method is used by astronomers to find the masses of the planets (by measuring T and a for moons revolving around the planet) as well as the masses of binary stars and galaxies. See Exercises 2–5.

Kepler's Three Laws

(i) **Law of Ellipses:** The orbit of a planet is an ellipse with the sun at one focus.

(ii) **Law of Equal Area in Equal Time:** The position vector pointing from the sun to the planet sweeps out equal areas in equal times.

(iii) **Law of the Period of Motion:** $T^2 = \left(\dfrac{4\pi^2}{GM}\right)a^3$, where

- a is the semimajor axis of the ellipse (Figure 1) in meters.
- G is the universal gravitational constant: 6.673×10^{-11} m^3 kg^{-1} s^{-2}.
- M is the mass of the sun, approximately 1.989×10^{30} kg.
- T is the period of the orbit, in seconds.

Our derivation makes a few simplifying assumptions. We treat the sun and planet as point masses and ignore the gravitational attraction of the planets on each other. And although both the sun and the planet revolve around their mutual center of mass, we ignore the sun's motion and assume that the planet revolves around the center of the sun. This is justified because the sun is much more massive than the planet.

We place the sun at the origin of the coordinate system. Let $\mathbf{r} = \mathbf{r}(t)$ be the position vector of a planet of mass m, as in Figure 1, and let (Figure 3)

$$\mathbf{e}_r = \frac{\mathbf{r}(t)}{\|\mathbf{r}(t)\|}$$

be the unit radial vector at time t ($\mathbf{e}_r$ is the unit vector that points to the planet as it moves around the sun). By Newton's Universal Law of Gravitation (the inverse-square law), the sun attracts the planet with a gravitational force

$$\mathbf{F}(\mathbf{r}(t)) = -\left(\frac{km}{\|\mathbf{r}(t)\|^2}\right)\mathbf{e}_r$$

where $k = GM$ (Figure 3). Combining the Law of Gravitation with Newton's Second Law of Motion $\mathbf{F}(\mathbf{r}(t)) = m\mathbf{r}''(t)$, we obtain

$$\mathbf{r}''(t) = -\frac{k}{\|\mathbf{r}(t)\|^2}\mathbf{e}_r \qquad \boxed{1}$$

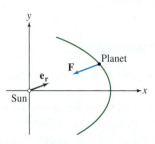

FIGURE 3 The gravitational force $\mathbf{F}$, directed from the planet to the sun, is a negative multiple of $\mathbf{e}_r$.

Kepler's Laws are a consequence of this *differential equation*.

Kepler's Second Law

The key to Kepler's Second Law is the fact that the following cross product is a constant vector [even though both $\mathbf{r}(t)$ and $\mathbf{r}'(t)$ are changing in time]:

$$\mathbf{J} = \mathbf{r}(t) \times \mathbf{r}'(t)$$

In physics, $m\mathbf{J}$ is called the **angular momentum** vector. In situations where $\mathbf{J}$ is constant, we say that angular momentum is conserved. *This conservation law is valid whenever the force acts in the radial direction.*

THEOREM 1 The vector $\mathbf{J}$ is constant—that is,

$$\frac{d\mathbf{J}}{dt} = \mathbf{0} \qquad \boxed{2}$$

Proof By the Product Rule for cross products (Theorem 3 in Section 13.2),

$$\frac{d\mathbf{J}}{dt} = \frac{d}{dt}\big(\mathbf{r}(t) \times \mathbf{r}'(t)\big) = \mathbf{r}'(t) \times \mathbf{r}'(t) + \mathbf{r}(t) \times \mathbf{r}''(t)$$

◄··· **REMINDER**

• $\mathbf{a} \times \mathbf{b}$ is orthogonal to both $\mathbf{a}$ and $\mathbf{b}$.

• $\mathbf{a} \times \mathbf{b} = \mathbf{0}$ if $\mathbf{a}$ and $\mathbf{b}$ are parallel, that is, one is a multiple of the other.

The cross product of parallel vectors is zero, so the first term is certainly zero. The second term is also zero because $\mathbf{r}''(t)$ is a multiple of $\mathbf{e}_r$ by Eq. (1), and hence also of $\mathbf{r}(t)$. ■

How can we use Eq. (2)? First of all, the cross product $\mathbf{J}$ is orthogonal to both $\mathbf{r}(t)$ and $\mathbf{r}'(t)$. Because $\mathbf{J}$ is constant, $\mathbf{r}(t)$ and $\mathbf{r}'(t)$ are confined to the fixed plane orthogonal to $\mathbf{J}$. This proves that the *motion of a planet around the sun takes place in a plane.*

We can choose coordinates so that the sun is at the origin and the planet moves in the counterclockwise direction (Figure 4). Let (r, θ) be the polar coordinates of the planet, where $r = r(t)$ and $\theta = \theta(t)$ are functions of time. Note that $r(t) = \|\mathbf{r}(t)\|$.

Recall from Section 11.4 (Theorem 1) that the area swept out by the planet's radial vector is

$$A = \frac{1}{2}\int_0^\theta r^2\,d\theta$$

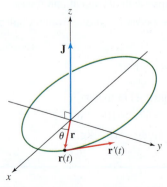

DF **FIGURE 4** The orbit is contained in the plane orthogonal to $\mathbf{J}$. Of course, we have not yet shown that the orbit is an ellipse.

Kepler's Second Law states that this area is swept out at a constant rate. But, this rate is simply dA/dt. By the Fundamental Theorem of Calculus, $\dfrac{dA}{d\theta} = \dfrac{1}{2}r^2$, and by the Chain Rule,

$$\frac{dA}{dt} = \frac{dA}{d\theta}\frac{d\theta}{dt} = \frac{1}{2}\theta'(t)r(t)^2 = \frac{1}{2}r(t)^2\theta'(t)$$

Thus, Kepler's Second Law follows from the next theorem, which tells us that dA/dt has the constant value $\frac{1}{2}\|\mathbf{J}\|$.

> **THEOREM 2** Let $J = \|\mathbf{J}\|$ ($\mathbf{J}$ and hence J are constant by Theorem 1). Then
>
> $$r(t)^2 \theta'(t) = J \qquad \boxed{3}$$

Proof We note that in polar coordinates, $\mathbf{e}_r = \langle \cos\theta, \sin\theta \rangle$. We also define the unit vector $\mathbf{e}_\theta = \langle -\sin\theta, \cos\theta \rangle$ that is orthogonal to $\mathbf{e}_r$ (Figure 5). In summary,

$$\boxed{r(t) = \|\mathbf{r}(t)\|, \quad \mathbf{e}_r = \langle \cos\theta, \sin\theta \rangle, \quad \mathbf{e}_\theta = \langle -\sin\theta, \cos\theta \rangle, \quad \mathbf{e}_r \cdot \mathbf{e}_\theta = 0}$$

We see directly that the derivatives of $\mathbf{e}_r$ and $\mathbf{e}_\theta$ with respect to θ are

$$\boxed{\frac{d}{d\theta}\mathbf{e}_r = \mathbf{e}_\theta, \qquad \frac{d}{d\theta}\mathbf{e}_\theta = -\mathbf{e}_r} \qquad \boxed{4}$$

The time derivative of $\mathbf{e}_r$ is computed using the Chain Rule:

$$\mathbf{e}_r' = \left(\frac{d\theta}{dt}\right)\left(\frac{d}{d\theta}\mathbf{e}_r\right) = \theta'(t)\,\mathbf{e}_\theta \qquad \boxed{5}$$

Now apply the Product Rule to $\mathbf{r} = r\mathbf{e}_r$:

$$\mathbf{r}' = \frac{d}{dt}r\mathbf{e}_r = r'\mathbf{e}_r + r\mathbf{e}_r' = r'\mathbf{e}_r + r\theta'\mathbf{e}_\theta$$

Using $\mathbf{e}_r \times \mathbf{e}_r = \mathbf{0}$, we obtain

$$\mathbf{J} = \mathbf{r} \times \mathbf{r}' = r\mathbf{e}_r \times (r'\mathbf{e}_r + r\theta'\mathbf{e}_\theta) = r^2\theta'(\mathbf{e}_r \times \mathbf{e}_\theta)$$

To compute cross products of vectors in the plane, such as $\mathbf{r}$, $\mathbf{e}_r$, and $\mathbf{e}_\theta$, we treat them as vectors in 3-space with a z-component equal to zero. The cross product is then a multiple of $\mathbf{k}$.

It is straightforward to check that $\mathbf{e}_r \times \mathbf{e}_\theta = \mathbf{k}$, and since $\mathbf{k}$ is a unit vector, $J = \|\mathbf{J}\| = |r^2\theta'|$. However, $\theta' > 0$ because the planet moves in the counterclockwise direction, so $J = r^2\theta'$. This proves Theorem 2. ∎

Proof of the Law of Ellipses

We show that the orbit of a planet is indeed an ellipse with the sun as one of the foci.

Let $\mathbf{v} = \mathbf{r}'(t)$ be the velocity vector. Then $\mathbf{r}'' = \mathbf{v}'$ and Eq. (1) may be written

◀·· **REMINDER** Eq. (1) states

$$\mathbf{r}''(t) = -\frac{k}{r(t)^2}\mathbf{e}_r$$

where $\mathbf{r}(t) = \|\mathbf{r}(t)\|$.

$$\frac{d\mathbf{v}}{dt} = -\frac{k}{r(t)^2}\mathbf{e}_r \qquad \boxed{6}$$

On the other hand, by the Chain Rule and the relation $r(t)^2\theta'(t) = J$ of Eq. (3),

$$\frac{d\mathbf{v}}{dt} = \frac{d\theta}{dt}\frac{d\mathbf{v}}{d\theta} = \theta'(t)\frac{d\mathbf{v}}{d\theta} = \frac{J}{r(t)^2}\frac{d\mathbf{v}}{d\theta}$$

Together with Eq. (6), this yields $J\dfrac{d\mathbf{v}}{d\theta} = -k\mathbf{e}_r$, or

$$\frac{d\mathbf{v}}{d\theta} = -\frac{k}{J}\mathbf{e}_r = -\frac{k}{J}\langle \cos\theta, \sin\theta \rangle$$

This is a first-order differential equation that no longer involves time t. We can solve it by integration:

$$\mathbf{v} = -\frac{k}{J}\int \langle \cos\theta, \sin\theta \rangle \, d\theta = \frac{k}{J}\langle -\sin\theta, \cos\theta \rangle + \mathbf{c} = \frac{k}{J}\mathbf{e}_\theta + \mathbf{c} \qquad \boxed{7}$$

where $\mathbf{c}$ is an arbitrary constant vector.

We are still free to rotate our coordinate system in the plane of motion, so we may assume that $\mathbf{c}$ points along the y-axis. We can then write $\mathbf{c} = \langle 0, (k/J)e \rangle$ for some constant e. We finish the proof by computing $\mathbf{J} = \mathbf{r} \times \mathbf{v}$:

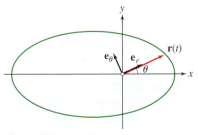

FIGURE 5 The unit vectors $\mathbf{e}_r$ and $\mathbf{e}_\theta$ are orthogonal, and rotate around the origin along with the planet.

$$\mathbf{J} = \mathbf{r} \times \mathbf{v} = r\mathbf{e}_r \times \left(\frac{k}{J}\mathbf{e}_\theta + \mathbf{c} \right) = \frac{k}{J}r\left(\mathbf{e}_r \times \mathbf{e}_\theta + \mathbf{e}_r \times \langle 0, e \rangle \right)$$

Direct calculation yields

$$\mathbf{e}_r \times \mathbf{e}_\theta = \mathbf{k}, \qquad \mathbf{e}_r \times \langle 0, e \rangle = (e \cos \theta)\mathbf{k}$$

so our equation becomes $\mathbf{J} = \frac{k}{J}r(1 + e \cos \theta)\mathbf{k}$. Since $\mathbf{k}$ is a unit vector,

$$J = \|\mathbf{J}\| = \frac{k}{J}r(1 + e \cos \theta)$$

◀·· **REMINDER** The equation of a conic section in polar coordinates is discussed in Section 11.5.

Solving for r, we obtain the polar equation of a conic section of eccentricity e (an ellipse, parabola, or hyperbola):

$$r = \frac{J^2/k}{1 + e \cos \theta}$$

This result shows that if a planet travels around the sun in a bounded orbit, then the orbit must be an ellipse. There are also "open orbits" that are either parabolic or hyperbolic. They describe comets that pass by the sun and then continue into space, never to return. In our derivation, we assumed implicitly that $\mathbf{J} \neq \mathbf{0}$. If $\mathbf{J} = \mathbf{0}$, then $\theta'(t) = 0$. In this case, the orbit is a straight line, and the planet falls directly into the sun.

Kepler's Third Law is verified in Exercises 23 and 24.

CONCEPTUAL INSIGHT We exploited the fact that $\mathbf{J}$ is constant to prove the Law of Ellipses without ever finding a formula for the position vector $\mathbf{r}(t)$ of the planet as a function of time t. In fact, $\mathbf{r}(t)$ cannot be expressed in terms of elementary functions. This illustrates an important principle: Sometimes it is possible to describe solutions of a differential equation even if we cannot write them down explicitly.

The Hubble Space Telescope produced this image of the Antenna galaxies, a pair of spiral galaxies that began to collide hundreds of millions of years ago.

(NASA, ESA, and the Hubble Heritage Team (STScI/AURA)-ESA/Hubble Collaboration)

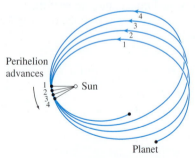

FIGURE 6 The perihelion of an orbit shifts slowly over time. For Mercury, the semimajor axis makes a full revolution approximately once every 24,000 years.

HISTORICAL PERSPECTIVE

The astronomers of the ancient world (Babylon, Egypt, and Greece) mapped out the nighttime sky with impressive accuracy, but their models of planetary motion were based on the erroneous assumption that the planets revolve around the earth. Although the Greek astronomer Aristarchus (310–230 BCE) had suggested that the earth revolves around the sun, this idea was rejected and forgotten for nearly eighteen centuries, until the Polish astronomer Nicolaus Copernicus (1473–1543) introduced a revolutionary set of ideas about the solar system, including the hypothesis that the planets revolve around the sun. Copernicus paved the way for the next generation, most notably Tycho Brahe (1546–1601), Galileo Galilei (1564–1642), and Johannes Kepler (1571–1630).

The German astronomer Johannes Kepler was the son of a mercenary soldier who apparently left his family when Johannes was 5 and may have died at war. Johannes was raised by his mother in his grandfather's inn. Kepler's mathematical brilliance earned him a scholarship at the

University of Tübingen, and at age of 29, he went to work for the Danish astronomer Tycho Brahe, who had compiled the most complete and accurate data on planetary motion then available. When Brahe died in 1601, Kepler succeeded him as "Imperial Mathematician" to the Holy Roman Emperor, and in 1609 he formulated the first two of his laws of planetary motion in a work entitled *Astronomia Nova* (*New Astronomy*).

In the centuries since Kepler's death, as observational data improved, astronomers found that the planetary orbits are not exactly elliptical. Furthermore, the perihelion (the point on the orbit closest to the sun) shifts slowly over time (Figure 6). Most of these deviations can be explained by the mutual pull of the planets, but the perihelion shift of Mercury is larger than can be accounted for by Newton's Laws. On November 18, 1915, Albert Einstein made a discovery about which he later wrote to a friend, "I was beside myself with ecstasy for days." He had been working for a decade on his famous **General Theory of Relativity**, a theory that would replace Newton's Law of Gravitation with a new set of much more complicated equations called the Einstein Field Equations. On that 18th of November, Einstein showed that Mercury's perihelion shift was accurately explained by his new theory. At the time, this was the only substantial piece of evidence that the General Theory of Relativity was correct.

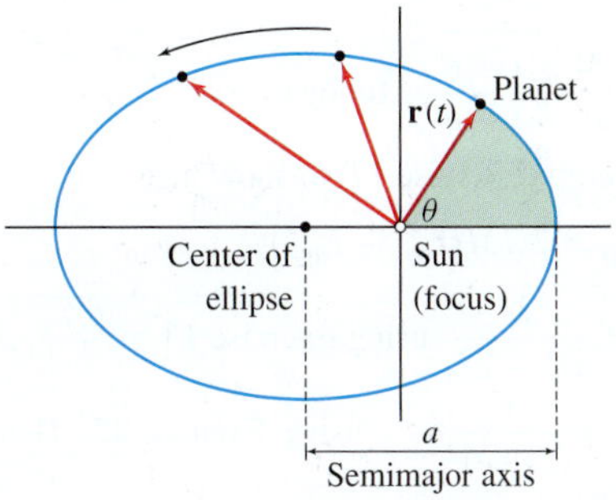

13.6 SUMMARY

- Kepler's three laws of planetary motion:

  - Law of Ellipses
  - Law of Equal Area in Equal Time
  - Law of the Period $T^2 = \left(\dfrac{4\pi^2}{GM}\right) a^3$, where T is the period (time to complete one full revolution) and a is the semimajor axis (Figure 7)

FIGURE 7 Planetary orbit.

Constants:

- *Gravitational constant:*

  $G \approx 6.673 \times 10^{-11} \text{ m}^3 \text{ kg}^{-1} \text{ s}^{-2}$

- *Mass of the sun:*

  $M \approx 1.989 \times 10^{30} \text{ kg}$

- $k = GM \approx 1.327 \times 10^{20}$

- According to Newton's Universal Law of Gravitation and Second Law of Motion, the position vector $\mathbf{r}(t)$ of a planet satisfies the differential equation

$$\mathbf{r}''(t) = -\frac{k}{r(t)^2}\mathbf{e}_r, \qquad \text{where } r(t) = \|\mathbf{r}(t)\|, \quad \mathbf{e}_r = \frac{\mathbf{r}(t)}{\|\mathbf{r}(t)\|}$$

- Properties of $\mathbf{J} = \mathbf{r}(t) \times \mathbf{r}'(t)$:

  - $\mathbf{J}$ is a constant of planetary motion.
  - Let $J = \|\mathbf{J}\|$. Then $J = r(t)^2 \theta'(t)$.
  - The planet sweeps out area at the rate $\dfrac{dA}{dt} = \dfrac{1}{2}J$.

- A planetary orbit has polar equation $r = \dfrac{J^2/k}{1 + e\cos\theta}$, where e is the eccentricity of the orbit.

13.6 EXERCISES

Preliminary Questions

1. Describe the relation between the vector $\mathbf{J} = \mathbf{r} \times \mathbf{r}'$ and the rate at which the radial vector sweeps out area.

2. Equation (1) shows that $\mathbf{r}''$ is proportional to $\mathbf{r}$. Explain how this fact is used to prove Kepler's Second Law.

3. How is the period T affected if the semimajor axis a is increased four-fold?

Exercises

1. Kepler's Third Law states that T^2/a^3 has the same value for each planetary orbit. Do the data in the following table support this conclusion? Estimate the length of Jupiter's period, assuming that $a = 77.8 \times 10^{10}$ m.

Planet	Mercury	Venus	Earth	Mars
a (10^{10} m)	5.79	10.8	15.0	22.8
T (years)	0.241	0.615	1.00	1.88

2. Finding the Mass of a Star Using Kepler's Third Law, show that if a planet revolves around a star with period T and semimajor axis a, then the mass of the star is $M = \left(\dfrac{4\pi^2}{G}\right)\left(\dfrac{a^3}{T^2}\right)$.

3. Ganymede, one of Jupiter's moons discovered by Galileo, has an orbital period of 7.154 days and a semimajor axis of 1.07×10^9 m. Use Exercise 2 to estimate the mass of Jupiter.

4. An astronomer observes a planet orbiting a star with a period of 9.5 years and a semimajor axis of 3×10^8 km. Find the mass of the star using Exercise 2.

5. Mass of the Milky Way The sun revolves around the center of mass of the Milky Way galaxy in an orbit that is approximately circu-

lar, of radius $a \approx 2.8 \times 10^{17}$ km and velocity $v \approx 250$ km/s. Use the result of Exercise 2 to estimate the mass of the portion of the Milky Way inside the sun's orbit (place all of this mass at the center of the orbit).

6. A satellite orbiting above the equator of the earth is **geosynchronous** if the period is $T = 24$ hours (in this case, the satellite stays over a fixed point on the equator). Use Kepler's Third Law to show that in a circular geosynchronous orbit, the distance from the center of the earth is $R \approx 42{,}246$ km. Then compute the altitude h of the orbit above the earth's surface. The earth has mass $M \approx 5.974 \times 10^{24}$ kg and radius $R \approx 6371$ km.

7. Show that a planet in a circular orbit travels at constant speed. *Hint:* Use the facts that $\mathbf{J}$ is constant and that $\mathbf{r}(t)$ is orthogonal to $\mathbf{r}'(t)$ for a circular orbit.

8. Verify that the circular orbit

$$\mathbf{r}(t) = \langle R\cos\omega t, R\sin\omega t\rangle$$

satisfies the differential equation, Eq. (1), provided that $\omega^2 = kR^{-3}$. Then deduce Kepler's Third Law $T^2 = \left(\dfrac{4\pi^2}{k}\right)R^3$ for this orbit.

9. Prove that if a planetary orbit is circular of radius R, then $vT = 2\pi R$, where v is the planet's speed (constant by Exercise 7) and T is the period. Then use Kepler's Third Law to prove that $v = \sqrt{\dfrac{k}{R}}$.

10. Find the velocity of a satellite in geosynchronous orbit about the earth. *Hint:* Use Exercises 6 and 9.

11. A communications satellite orbiting the earth has initial position $\mathbf{r} = \langle 29{,}000, 20{,}000, 0 \rangle$ (in kilometers) and initial velocity $\mathbf{r}' = \langle 1, 1, 1 \rangle$ (in kilometers per second), where the origin is the earth's center. Find the equation of the plane containing the satellite's orbit. *Hint:* This plane is orthogonal to $\mathbf{J}$.

12. Assume that the earth's orbit is circular of radius $R = 150 \times 10^6$ km (it is nearly circular with eccentricity $e = 0.017$). Find the rate at which the earth's radial vector sweeps out area in units of square kilometers per second. What is the magnitude of the vector $\mathbf{J} = \mathbf{r} \times \mathbf{r}'$ for the earth (in units of square kilometers per second)?

In Exercises 13–19, the perihelion and aphelion are the points on the orbit closest to and farthest from the sun, respectively (Figure 8). The distance from the sun at the perihelion is denoted r_{per} and the speed at this point is denoted v_{per}. Similarly, we write r_{ap} and v_{ap} for the distance and speed at the aphelion. The semimajor axis is denoted a.

13. Use the polar equation of an ellipse

$$r = \frac{p}{1 + e\cos\theta}$$

to show that $r_{per} = a(1 - e)$ and $r_{ap} = a(1 + e)$. *Hint:* Use the fact that $r_{per} + r_{ap} = 2a$.

14. Use the result of Exercise 13 to prove the formulas

$$e = \frac{r_{ap} - r_{per}}{r_{ap} + r_{per}}, \qquad p = \frac{2r_{ap}r_{per}}{r_{ap} + r_{per}}$$

15. Use the fact that $\mathbf{J} = \mathbf{r} \times \mathbf{r}'$ is constant to prove

$$v_{per}(1 - e) = v_{ap}(1 + e)$$

Hint: $\mathbf{r}$ is perpendicular to $\mathbf{r}'$ at the perihelion and aphelion.

16. Compute r_{per} and r_{ap} for the orbit of Mercury, which has eccentricity $e = 0.244$ (see the table in Exercise 1 for the semimajor axis).

17. Conservation of Energy The total mechanical energy (kinetic energy plus potential energy) of a planet of mass m orbiting a sun of mass M with position $\mathbf{r}$ and speed $v = \|\mathbf{r}'\|$ is

$$E = \frac{1}{2}mv^2 - \frac{GMm}{\|\mathbf{r}\|} \qquad \boxed{8}$$

(a) Prove the equations

$$\frac{d}{dt}\frac{1}{2}mv^2 = \mathbf{v} \cdot (m\mathbf{a}), \qquad \frac{d}{dt}\frac{GMm}{\|\mathbf{r}\|} = \mathbf{v} \cdot \left(-\frac{GMm}{\|\mathbf{r}\|^3}\mathbf{r}\right)$$

(b) Then use Newton's Law $\mathbf{F} = m\mathbf{a}$ and Eq. (1) to prove that energy is conserved—that is, $\dfrac{dE}{dt} = 0$.

18. Show that the total energy [Eq. (8)] of a planet in a circular orbit of radius R is $E = -\dfrac{GMm}{2R}$. *Hint:* Use Exercise 9.

19. Prove that $v_{per} = \sqrt{\left(\dfrac{GM}{a}\right)\dfrac{1 + e}{1 - e}}$ as follows:

(a) Use Conservation of Energy (Exercise 17) to show that

$$v_{per}^2 - v_{ap}^2 = 2GM\left(r_{per}^{-1} - r_{ap}^{-1}\right)$$

(b) Show that $r_{per}^{-1} - r_{ap}^{-1} = \dfrac{2e}{a(1 - e^2)}$ using Exercise 13.

(c) Show that $v_{per}^2 - v_{ap}^2 = 4\dfrac{e}{(1 + e)^2}v_{per}^2$ using Exercise 15. Then solve for v_{per} using (a) and (b).

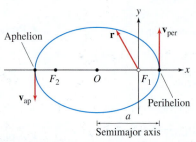

FIGURE 8 $\mathbf{r}$ and $\mathbf{v} = \mathbf{r}'$ are perpendicular at the perihelion and aphelion.

20. Show that a planet in an elliptical orbit has total mechanical energy $E = -\dfrac{GMm}{2a}$, where a is the semimajor axis. *Hint:* Use Exercise 19 to compute the total energy at the perihelion.

21. Prove that $v^2 = GM\left(\dfrac{2}{r} - \dfrac{1}{a}\right)$ at any point on an elliptical orbit, where $r = \|\mathbf{r}\|$, v is the velocity, and a is the semimajor axis of the orbit.

22.   Two space shuttles A and B orbit the earth along the solid trajectory in Figure 9. Hoping to catch up to B, the pilot of A applies a forward thrust to increase her shuttle's kinetic energy. Use Exercise 20 to show that shuttle A will move off into a larger orbit as shown in the figure. Then use Kepler's Third Law to show that A's orbital period T will increase (and she will fall farther and farther behind B)!

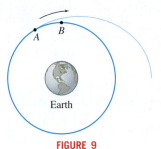

FIGURE 9

Further Insights and Challenges

Exercises 23 and 24 prove Kepler's Third Law. Figure 10 shows an elliptical orbit with polar equation

$$r = \frac{p}{1 + e\cos\theta}$$

where $p = J^2/k$. The origin of the polar coordinates occurs at F_1. Let a and b be the semimajor and semiminor axes, respectively.

23. This exercise shows that $b = \sqrt{pa}$.

(a) Show that $CF_1 = ae$. *Hint:* $r_{per} = a(1 - e)$ by Exercise 13.

(b) Show that $a = \dfrac{p}{1 - e^2}$.

(c) Show that $F_1A + F_2A = 2a$. Conclude that $F_1B + F_2B = 2a$ and hence $F_1B = F_2B = a$.

(d) Use the Pythagorean Theorem to prove that $b = \sqrt{pa}$.

24. The area A of the ellipse is $A = \pi ab$.

(a) Prove, using Kepler's First Law, that $A = \frac{1}{2}JT$, where T is the period of the orbit.

(b) Use Exercise 23 to show that $A = (\pi \sqrt{p})a^{3/2}$.

(c) Deduce Kepler's Third Law: $T^2 = \dfrac{4\pi^2}{GM}a^3$.

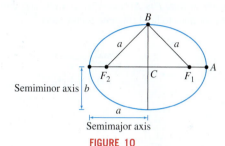

FIGURE 10

25. ![] According to Eq. (7), the velocity vector of a planet as a function of the angle θ is

$$\mathbf{v}(\theta) = \frac{k}{J}\mathbf{e}_\theta + \mathbf{c}$$

Use this to explain the following statement: As a planet revolves around the sun, its velocity vector traces out a circle of radius k/J with its center at the terminal point of $\mathbf{c}$ (Figure 11). This beautiful but hidden property of orbits was discovered by William Rowan Hamilton in 1847.

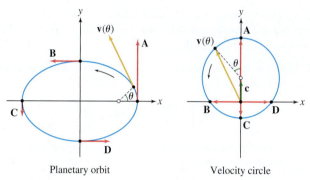

Planetary orbit   Velocity circle

FIGURE 11 The velocity vector traces out a circle as the planet travels along its orbit.

CHAPTER REVIEW EXERCISES

1. Determine the domains of the vector-valued functions.
(a) $\mathbf{r}_1(t) = \langle t^{-1}, (t+1)^{-1}, \sin^{-1} t \rangle$
(b) $\mathbf{r}_2(t) = \langle \sqrt{8-t^3}, \ln t, e^{\sqrt{t}} \rangle$

2. Sketch the paths $\mathbf{r}_1(\theta) = \langle \theta, \cos\theta \rangle$ and $\mathbf{r}_2(\theta) = \langle \cos\theta, \theta \rangle$ in the xy-plane.

3. Find a vector parametrization of the intersection of the surfaces $x^2 + y^4 + 2z^3 = 6$ and $x = y^2$ in $\mathbf{R}^3$.

4. Find a vector parametrization using trigonometric functions of the intersection of the plane $x + y + z = 1$ and the elliptical cylinder $\left(\dfrac{y}{3}\right)^2 + \left(\dfrac{z}{8}\right)^2 = 1$ in $\mathbf{R}^3$.

In Exercises 5–10, calculate the derivative indicated.

5. $\mathbf{r}'(t)$, $\mathbf{r}(t) = \langle 1-t, t^{-2}, \ln t \rangle$

6. $\mathbf{r}'''(t)$, $\mathbf{r}(t) = \langle t^3, 4t^2, 7t \rangle$

7. $\mathbf{r}'(0)$, $\mathbf{r}(t) = \langle e^{2t}, e^{-4t^2}, e^{6t} \rangle$

8. $\mathbf{r}''(-3)$, $\mathbf{r}(t) = \langle t^{-2}, (t+1)^{-1}, t^3 - t \rangle$

9. $\dfrac{d}{dt} e^t \langle 1, t, t^2 \rangle$

10. $\dfrac{d}{d\theta} \mathbf{r}(\cos\theta)$, $\mathbf{r}(s) = \langle s, 2s, s^2 \rangle$

In Exercises 11–14, calculate the derivative at $t = 3$, assuming that

$$\mathbf{r}_1(3) = \langle 1, 1, 0 \rangle, \qquad \mathbf{r}_2(3) = \langle 1, 1, 0 \rangle$$
$$\mathbf{r}_1'(3) = \langle 0, 0, 1 \rangle, \qquad \mathbf{r}_2'(3) = \langle 0, 2, 4 \rangle$$

11. $\dfrac{d}{dt}(6\mathbf{r}_1(t) - 4 \cdot \mathbf{r}_2(t))$

12. $\dfrac{d}{dt}(e^t \mathbf{r}_2(t))$

13. $\dfrac{d}{dt}(\mathbf{r}_1(t) \cdot \mathbf{r}_2(t))$

14. $\dfrac{d}{dt}(\mathbf{r}_1(t) \times \mathbf{r}_2(t))$

15. Calculate $\displaystyle\int_0^3 \langle 4t+3, t^2, -4t^3 \rangle \, dt$.

16. Calculate $\displaystyle\int_0^\pi \langle \sin\theta, \theta, \cos 2\theta \rangle \, d\theta$.

17. A particle located at $(1, 1, 0)$ at time $t = 0$ follows a path whose velocity vector is $\mathbf{v}(t) = \langle 1, t, 2t^2 \rangle$. Find the particle's location at $t = 2$.

18. Find the vector-valued function $\mathbf{r}(t) = \langle x(t), y(t) \rangle$ in $\mathbf{R}^2$ satisfying $\mathbf{r}'(t) = -\mathbf{r}(t)$ with initial conditions $\mathbf{r}(0) = \langle 1, 2 \rangle$.

19. Calculate $\mathbf{r}(t)$, assuming that

$$\mathbf{r}''(t) = \langle 4 - 16t, 12t^2 - t \rangle, \qquad \mathbf{r}'(0) = \langle 1, 0 \rangle, \qquad \mathbf{r}(0) = \langle 0, 1 \rangle$$

20. Solve $\mathbf{r}''(t) = \langle t^2 - 1, t + 1, t^3 \rangle$ subject to the initial conditions $\mathbf{r}(0) = \langle 1, 0, 0 \rangle$ and $\mathbf{r}'(0) = \langle -1, 1, 0 \rangle$

21. Compute the length of the path

$$\mathbf{r}(t) = \langle \sin 2t, \cos 2t, 3t - 1 \rangle \quad \text{for } 1 \le t \le 3$$

22. **CAS** Express the length of the path $\mathbf{r}(t) = \langle \ln t, t, e^t \rangle$ for $1 \le t \le 2$ as a definite integral, and use a computer algebra system to find its value to two decimal places.

23. Find an arc length parametrization of a helix of height 20 cm that makes four full rotations over a circle of radius 5 cm.

24. Find the minimum speed of a particle with trajectory $\mathbf{r}(t) = \langle t, e^{t-3}, e^{4-t} \rangle$.

25. A projectile fired at an angle of $60°$ lands 400 m away. What was its initial speed?

26. A specially trained mouse runs counterclockwise in a circle of radius 0.6 m on the floor of an elevator with speed 0.3 m/s, while the elevator ascends from ground level (along the z-axis) at a speed of 12 m/s. Find the mouse's acceleration vector as a function of time. Assume that the circle is centered at the origin of the xy-plane and the mouse is at $(2, 0, 0)$ at $t = 0$.

27. During a short time interval $[0.5, 1.5]$, the path of an unmanned spy plane is described by

$$\mathbf{r}(t) = \left\langle -\frac{100}{t^2}, 7 - t, 40 - t^2 \right\rangle$$

A laser is fired (in the tangential direction) toward the yz-plane at time $t = 1$. Which point in the yz-plane does the laser beam hit?

28. A force $\mathbf{F} = \langle 12t + 4, 8 - 24t \rangle$ (in newtons) acts on a 2-kg mass. Find the position of the mass at $t = 2$ s if it is located at $(4, 6)$ at $t = 0$ and has initial velocity $\langle 2, 3 \rangle$ in meters per second.

29. Find the unit tangent vector to $\mathbf{r}(t) = \langle \sin t, t, \cos t \rangle$ at $t = \pi$.

30. Find the unit tangent vector to $\mathbf{r}(t) = \langle t^2, \tan^{-1} t, t \rangle$ at $t = 1$.

31. Calculate $\kappa(1)$ for $\mathbf{r}(t) = \langle \ln t, t \rangle$.

32. Calculate $\kappa\left(\frac{\pi}{4}\right)$ for $\mathbf{r}(t) = \langle \tan t, \sec t, \cos t \rangle$.

In Exercises 33 and 34, write the acceleration vector $\mathbf{a}$ at the point indicated as a sum of tangential and normal components.

33. $\mathbf{r}(\theta) = \langle \cos \theta, \sin 2\theta \rangle$, $\theta = \frac{\pi}{4}$

34. $\mathbf{r}(t) = \langle t^2, 2t - t^2, t \rangle$, $t = 2$

35. At a certain time t_0, the path of a moving particle is tangent to the y-axis in the positive direction. The particle's speed at time t_0 is 4 m/s, and its acceleration vector is $\mathbf{a} = \langle 5, 4, 12 \rangle$. Determine the curvature of the path at t_0.

36. Parametrize the osculating circle to $y = x^2 - x^3$ at $x = 1$.

37. Parametrize the osculating circle to $y = \sqrt{x}$ at $x = 4$.

38. Let $\mathbf{r}(t) = \langle \cos t, \sin t, 2t \rangle$.
(a) Find $\mathbf{T}$, $\mathbf{N}$, and $\mathbf{B}$ at the point corresponding to $t = \frac{\pi}{2}$. *Hint:* Evaluate $\mathbf{T}$ at $t = \frac{\pi}{2}$ before finding $\mathbf{N}$ and $\mathbf{B}$.

(b) Find the equation of the osculating plane at the point corresponding to $t = \frac{\pi}{2}$.

39. Let $\mathbf{r}(t) = \left\langle \ln t, t, \frac{t^2}{2} \right\rangle$. Find the equation of the osculating plane corresponding to $t = 1$.

40. If a planet has zero mass $(m = 0)$, then Newton's laws of motion reduce to $\mathbf{r}''(t) = \mathbf{0}$ and the orbit is a straight line $\mathbf{r}(t) = \mathbf{r}_0 + t\mathbf{v}_0$, where $\mathbf{r}_0 = \mathbf{r}(0)$ and $\mathbf{v}_0 = \mathbf{r}'(0)$ (Figure 1). Show that the area swept out by the radial vector at time t is $A(t) = \frac{1}{2}\|\mathbf{r}_0 \times \mathbf{v}_0\|t$, and thus Kepler's Second Law continues to hold (the rate is constant).

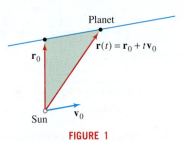

FIGURE 1

41. Suppose the orbit of a planet is an ellipse of eccentricity $e = c/a$ and period T (Figure 2). Use Kepler's Second Law to show that the time required to travel from A' to B' is equal to

$$\left(\frac{1}{4} + \frac{e}{2\pi} \right) T$$

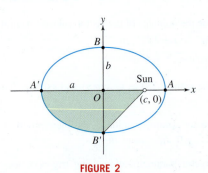

FIGURE 2

42. The period of Mercury is approximately 88 days, and its orbit has eccentricity 0.205. How much longer does it take Mercury to travel from A' to B' than from B' to A (Figure 2)?

The slope experienced by a mountain climber, which is given by the appropriate generalization of the derivative of a function of one variable, depends on the direction of motion. (*Philip and Karen Smith/Getty Images*)

14 DIFFERENTIATION IN SEVERAL VARIABLES

In this chapter, we extend the concepts and techniques of differential calculus to functions of several variables. As we will see, a function f that depends on two or more variables has not just one derivative but rather a set of *partial derivatives*, one for each variable. The partial derivatives are the components of the gradient vector, which provides valuable insight into the function's behavior. In the last two sections, we apply the tools we have developed to optimization in several variables.

14.1 Functions of Two or More Variables

A familiar example of a function of two variables is the area A of a rectangle, equal to the product xy of the base x and height y. We write

$$A(x, y) = xy$$

or $A = f(x, y)$, where $f(x, y) = xy$. An example in three variables is the distance from a point $P = (x, y, z)$ to the origin:

$$g(x, y, z) = \sqrt{x^2 + y^2 + z^2}$$

An important but less familiar example is the density of seawater, denoted ρ, which is a function of salinity S and temperature T (Figure 1). Although there is no simple formula for $\rho(S, T)$, scientists determine values of the function experimentally (Figure 2). According to Table 1, if $S = 32$ (in parts per thousand or ppt) and $T = 10°C$, then

$$\rho(32, 10) = 1.0246 \text{ kg/m}^3$$

FIGURE 1 The global climate is influenced by the ocean "conveyer belt," a system of deep currents driven by variations in seawater density.

FIGURE 2 A Conductivity-Temperature-Depth (CDT) instrument is used to measure seawater variables such as density, temperature, pressure, and salinity. (*NOAA*)

TABLE 1   Seawater Density ρ (kg/m^3) as a Function of Temperature and Salinity.

°C	Salinity (ppt) 32	32.5	33
5	1.0253	1.0257	1.0261
10	1.0246	1.0250	1.0254
15	1.0237	1.0240	1.0244
20	1.0224	1.0229	1.0232

A function of n variables is a function $f(x_1, \ldots, x_n)$ that assigns a real number to each n-tuple $(x_1, \ldots, x_n)$ in a domain in $\mathbf{R}^n$. Sometimes we write $f(P)$ for the value of f at a point $P = (x_1, \ldots, x_n)$. When f is defined by a formula, we usually take as domain the set of all n-tuples for which $f(x_1, \ldots, x_n)$ is defined. The range of f is the set of all values $f(x_1, \ldots, x_n)$ for $(x_1, \ldots, x_n)$ in the domain. Since we focus on functions of two or three variables, we shall often use the variables x, y, and z (rather than x_1, x_2, x_3).

■ **EXAMPLE 1** Sketch the domains of:

(a) $f(x, y) = \sqrt{9 - x^2 - y}$ **(b)** $g(x, y, z) = x\sqrt{y} + \ln(z - 1)$

What are the ranges of these functions?

Solution

(a) $f(x, y) = \sqrt{9 - x^2 - y}$ is defined only when $9 - x^2 - y \geq 0$, or $y \leq 9 - x^2$. Thus, the domain consists of all points (x, y) lying on or below the parabola $y = 9 - x^2$ [Figure 3(A)]:

$$\mathcal{D} = \{(x, y) : y \leq 9 - x^2\}$$

To determine the range, note that f is a nonnegative function and that $f(0, y) = \sqrt{9 - y}$. Since $9 - y$ can be any positive number, $f(0, y)$ takes on all nonnegative values. Therefore, the range of f is the infinite interval $[0, \infty)$.

(b) $g(x, y, z) = x\sqrt{y} + \ln(z - 1)$ is defined only when both $\sqrt{y}$ and $\ln(z - 1)$ are defined. Therefore, both $y \geq 0$ and $z > 1$ are required, so the domain of the function is given by $\{(x, y, z) : y \geq 0, z > 1\}$ [Figure 3(B)]. The range of g is the entire real line $\mathbf{R}$. Indeed, for the particular choices $y = 1$ and $z = 2$, we have $g(x, 1, 2) = x\sqrt{1} + \ln 1 = x$, and since x is arbitrary, we see that g takes on all values. ■

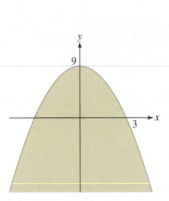

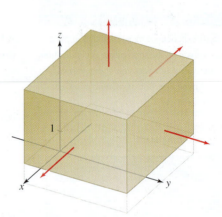

(A) The domain of $f(x, y) = \sqrt{9 - x^2 - y}$ is the set of all points lying on or below the parabola $y = 9 - x^2$.

(B) Domain of $g(x, y, z) = x\sqrt{y} + \ln(z - 1)$ is the set of points with $y \geq 0$ and $z > 1$. The domain continues out to infinity in the directions indicated by the arrows.

FIGURE 3

Graphing Functions of Two Variables

In single-variable calculus, we use graphs to visualize the important features of a function. Graphs play a similar role for functions of two variables. The graph of a function f of two variables consists of all points $(a, b, f(a, b))$ in $\mathbf{R}^3$ for (a, b) in the domain $\mathcal{D}$ of f. Assuming that f is continuous (as defined in the next section), the graph is a surface whose *height* above or below the xy-plane at (a, b) is the value of the function $f(a, b)$ (Figure 4). We often write $z = f(x, y)$ to stress that the z-coordinate of a point on the graph is a function of x and y.

■ **EXAMPLE 2** Sketch the graph of $f(x, y) = 2x^2 + 5y^2$.

Solution The graph is a paraboloid (Figure 5), which we saw in Section 12.6. We sketch the graph using the fact that the horizontal cross section (called the horizontal "trace" below) at height $z = c$ is the ellipse $2x^2 + 5y^2 = c$. ■

Plotting more complicated graphs by hand can be difficult. Fortunately, computer algebra systems eliminate the labor and greatly enhance our ability to explore functions graphically. Graphs can be rotated and viewed from different perspectives (Figure 6).

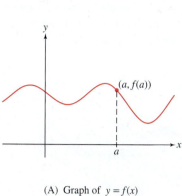

(A) Graph of $y = f(x)$

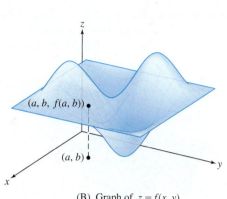

(B) Graph of $z = f(x, y)$

FIGURE 4

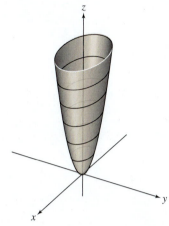

FIGURE 5 Graph of
$f(x, y) = 2x^2 + 5y^2$.

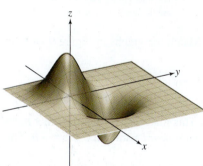

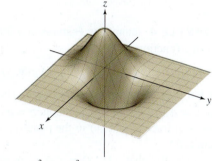

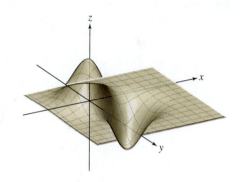

DF **FIGURE 6** Different views of $z = e^{-x^2-y^2} - e^{-(x-1)^2-(y-1)^2}$.

Traces and Level Curves

One way of analyzing the graph of $f(x, y)$ is to freeze the x-coordinate by setting $x = a$ and examine the resulting curve given by $z = f(a, y)$. Similarly, we may set $y = b$ and consider the curve $z = f(x, b)$. Curves of this type are called **vertical traces**. They are obtained by intersecting the graph with planes parallel to a vertical coordinate plane (Figure 7):

- **Vertical trace in the plane $x = a$:** Intersection of the graph with the vertical plane $x = a$, consisting of all points $(a, y, f(a, y))$
- **Vertical trace in the plane $y = b$:** Intersection of the graph with the vertical plane $y = b$, consisting of all points $(x, b, f(x, b))$

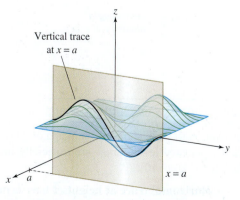

(A) Vertical traces parallel to yz-plane

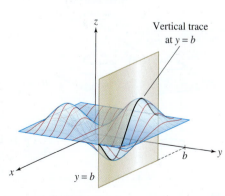

(B) Vertical traces parallel to xz-plane

DF **FIGURE 7**

■ **EXAMPLE 3** Describe the vertical traces of $f(x, y) = x \sin y$.

Solution When we freeze the x-coordinate by setting $x = a$, we obtain the trace curve $z = a \sin y$ (see Figure 8). This is a sine curve located in the plane $x = a$. When we set $y = b$, we obtain a line $z = x(\sin b)$ of slope $\sin b$, located in the plane $y = b$. ■

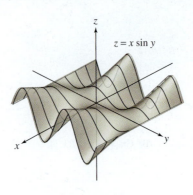

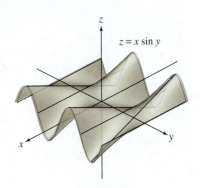

(A) The traces in the planes $x = a$
are the curves $z = a(\sin y)$.

(B) The traces in the planes $y = b$
are the lines $z = x(\sin b)$.

DF FIGURE 8 Vertical traces of $f(x, y) = x \sin y$.

■ **EXAMPLE 4** **Identifying Features of a Graph** Match the graphs in Figure 9 with the following functions:

(i) $f(x, y) = x - y^2$ (ii) $g(x, y) = x^2 - y$

Solution Let's compare vertical traces. The vertical trace of $f(x, y) = x - y^2$ in the plane $x = a$ is a *downward* parabola $z = a - y^2$. This matches (B). On the other hand, the vertical trace of $g(x, y)$ in the plane $y = b$ is an *upward* parabola $z = x^2 - b$. This matches (A).

Notice also that $f(x, y) = x - y^2$ is an increasing function of x [i.e., $f(x, y)$ increases as x increases] as in (B), whereas $g(x, y) = x^2 - y$ is a decreasing function of y as in (A). ■

Upward parabolas
$y = b, z = x^2 - b$

Downward parabolas
$x = a, z = a - y^2$

Decreasing in
positive y-direction

Increasing in
positive x-direction

(A)

(B)

FIGURE 9

Level Curves and Contour Maps

In addition to vertical traces, the graph of $f(x, y)$ has horizontal traces. These traces and their associated level curves are especially important in analyzing the behavior of the function (Figure 10):

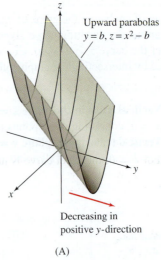

Horizontal trace
at $z = c$

$z = c$

$z = f(x, y)$

Level curve $f(x, y) = c$

DF FIGURE 10 The level curve consists of all points (x, y) where the function takes on the value c.

• **Horizontal trace at height c:** Intersection of the graph with the horizontal plane $z = c$, consisting of the points $(x, y, f(x, y))$ such that $f(x, y) = c$
• **Level curve:** The curve $f(x, y) = c$ in the xy-plane

Thus, the level curve corresponding to c consists of all points (x, y) in the plane where the function takes the value c. Each level curve is the projection onto the xy-plane of the horizontal trace on the graph that lies above it.

A **contour map** is a plot in the xy-plane that shows the level curves $f(x, y) = c$ for equally spaced values of c. The interval m between the values is called the **contour interval**. When you move from one level curve to next, the value of $f(x, y)$ (and hence the height of the graph) changes by $\pm m$.

Figure 11 compares the graph of a function $f(x, y)$ in (A) and its horizontal traces in (B) with the contour map in (C). The contour map in (C) has contour interval $m = 100$.

It is important to understand how the contour map indicates the steepness of the graph. If the level curves are close together, then a small move from one level curve to the next in the xy-plane leads to a large change in height. In other words, *the level curves are close together if the graph is steep* (Figure 11). Similarly, the graph is flatter when the level curves are farther apart.

> On contour maps, level curves are often referred to as **contour lines**.

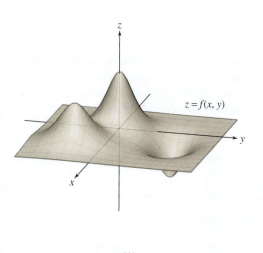

(A)

FIGURE 11

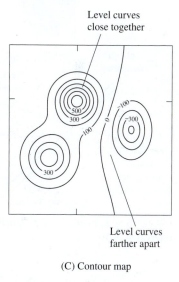

(C) Contour map

(B) Horizontal traces

■ **EXAMPLE 5** **Elliptic Paraboloid** Sketch the contour map of $f(x, y) = x^2 + 3y^2$ and comment on the spacing of the contour curves.

Solution The level curves have equation $f(x, y) = c$, or

$$x^2 + 3y^2 = c$$

- For $c > 0$, the level curve is an ellipse.
- For $c = 0$, the level curve is just the point $(0, 0)$ because $x^2 + 3y^2 = 0$ only for $(x, y) = (0, 0)$.
- The level curve is empty if $c < 0$ because $f(x, y)$ is never negative.

The graph of $f(x, y)$ is an elliptic paraboloid (Figure 12). As we move away from the origin, $f(x, y)$ increases more rapidly. The graph gets steeper, and the level curves become closer together. ■

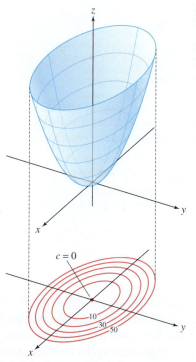

FIGURE 12 $f(x, y) = x^2 + 3y^2$. Contour interval $m = 10$.

■ **EXAMPLE 6** **Hyperbolic Paraboloid** Sketch the contour map of $g(x, y) = x^2 - 3y^2$.

Solution The level curves have equation $g(x, y) = c$, or

$$x^2 - 3y^2 = c$$

- For $c \neq 0$, the level curve is the hyperbola $x^2 - 3y^2 = c$.
- For $c = 0$, the level curve consists of the two lines $x = \pm\sqrt{3}y$ because the equation $g(x, y) = 0$ factors as follows:

$$x^2 - 3y^2 = 0 = (x - \sqrt{3}y)(x + \sqrt{3}y) = 0$$

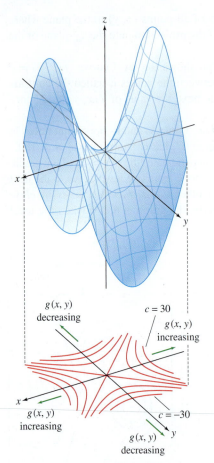

FIGURE 13 $g(x, y) = x^2 - 3y^2$. Contour interval $m = 10$.

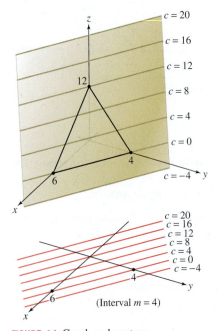

FIGURE 14 Graph and contour map of $f(x, y) = 12 - 2x - 3y$.

The graph of $g(x, y)$ is a hyperbolic paraboloid (Figure 13). When you stand at the origin, $g(x, y)$ increases as you move along the x-axis in either direction and decreases as you move along the y-axis in either direction. Furthermore, the graph gets steeper as you move out from the origin, so the level curves grow closer together. ∎

■ **EXAMPLE 7** Contour Map of a Linear Function Sketch the graph of $f(x, y) = 12 - 2x - 3y$ and the associated contour map with contour interval $m = 4$.

Solution Note that if we set $z = f(x, y)$, we can write the equation as $2x + 3y + z = 12$. As we discussed in Section 12.5, this is the equation of a plane. To plot the graph, we find the intercepts of the plane with the axes (Figure 14). The graph intercepts the z-axis at $z = f(0, 0) = 12$. To find the x-intercept, we set $y = z = 0$ to obtain $12 - 2x - 3(0) = 0$, or $x = 6$. Similarly, solving $12 - 3y = 0$ gives the y-intercept $y = 4$. The graph is the plane determined by the three intercepts.

In general, the level curves of a linear function $f(x, y) = qx + ry + s$ are the lines with equation $qx + ry + s = c$. Therefore, *the contour map of a linear function consists of equally spaced parallel lines*. In our case, the level curves are the lines $12 - 2x - 3y = c$, or $2x + 3y = 12 - c$ (Figure 14). ∎

How can we measure steepness quantitatively? Let's imagine the surface given by $z = f(x, y)$ as a mountain range. In fact, contour maps (also called topographical maps) are used extensively to describe terrain (Figure 15). We place the xy-plane at sea level, so that $f(a, b)$ is the height (also called altitude or elevation) of the mountain above sea level at the point (a, b) in the plane.

FIGURE 15 Mount Whitney Range in California, with contour map.
(*left: © Rachid Dahnoun/Aurora Photos; right: USGS/http://www.topoquest.com*)

Figure 16(A) shows two points P and Q in the xy-plane, together with the points $\widetilde{P}$ and $\widetilde{Q}$ on the graph that lie above them. We define the **average rate of change**:

$$\text{Average rate of change from } P \text{ to } Q = \frac{\Delta \text{ altitude}}{\Delta \text{ horizontal}}$$

where

Δ altitude = change in the height from $\widetilde{P}$ and $\widetilde{Q}$

Δ horizontal = distance from P to Q

CONCEPTUAL INSIGHT We will discuss the idea that rates of change depend on direction when we come to directional derivatives in Section 14.5. In single-variable calculus, we measure the rate of change by the derivative $f'(a)$. In the multivariable case, there is no single rate of change because the change in $f(x, y)$ depends on the direction: The rate is zero along a level curve [because $f(x, y)$ is constant along level curves], and the rate is nonzero in directions pointing from one level curve to the next [Figure 16(B)].

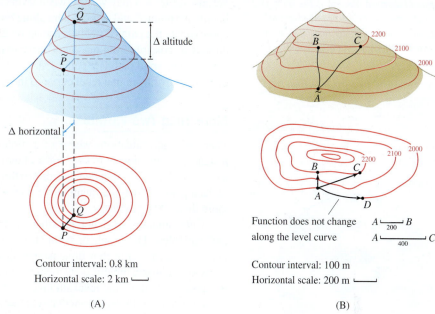

FIGURE 16

Function does not change
along the level curve

$A \; \underset{200}{\rule{1.5cm}{0.5pt}} \; B$

$A \; \underset{400}{\rule{2.5cm}{0.5pt}} \; C$

Contour interval: 0.8 km
Horizontal scale: 2 km ⊢—⊣

(A)

Contour interval: 100 m
Horizontal scale: 200 m ⊢—⊣

(B)

■ **EXAMPLE 8** **Average Rate of Change Depends on Direction** Compute the average rate of change from A to the points B, C, and D in Figure 16(B).

Solution The contour interval in Figure 16(B) is $m = 100$ m. Segments $\overline{AB}$ and $\overline{AC}$ both span two level curves, so the change in altitude is 200 m in both cases. The horizontal scale shows that AB corresponds to a horizontal change of 200 m, and $\overline{AC}$ corresponds to a horizontal change of 400 m. On the other hand, there is no change in altitude from A to D. Therefore,

$$\text{Average rate of change from } A \text{ to } B = \frac{\Delta \text{ altitude}}{\Delta \text{ horizontal}} = \frac{200}{200} = 1.0$$

$$\text{Average rate of change from } A \text{ to } C = \frac{\Delta \text{ altitude}}{\Delta \text{ horizontal}} = \frac{200}{400} = 0.5$$

$$\text{Average rate of change from } A \text{ to } D = \frac{\Delta \text{ altitude}}{\Delta \text{ horizontal}} = 0$$

We see here explicitly that the average rate varies according to the direction.

When we walk up a mountain, the incline at each moment depends on the path we choose. If we walk "around" the mountain, our altitude does not change at all. On the other hand, at each point there is a *steepest* direction in which the altitude increases most rapidly. On a contour map, the steepest direction is approximately the direction that takes us to the closest point on the next highest level curve [Figure 17(A)]. We say "approximately"

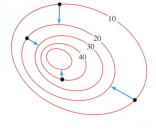

(A) Vectors pointing approximately
in the direction of steepest ascent

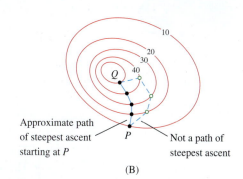

Approximate path
of steepest ascent
starting at P

Not a path of
steepest ascent

(B)

FIGURE 17

A path of steepest descent is the same as a path of steepest ascent traversed in the opposite direction. Water flowing down a mountain follows a path of steepest descent.

because the terrain may vary between level curves. A **path of steepest ascent** is a path that begins at a point P and, everywhere along the way, points in the steepest direction. We can approximate the path of steepest ascent by drawing a sequence of segments that move as directly as possible from one level curve to the next. Figure 17(B) shows two paths from P to Q. The solid path is a path of steepest ascent, but the dashed path is not, because it does not move from one level curve to the next along the shortest possible segment.

More Than Two Variables

There are many situations where a function of more than two variables is necessary. For instance, we might want to keep track of temperature at the various points in a room using a function $T(x, y, z)$ that depends on the three variables corresponding to the coordinates of each point. In making quantitative models of the economy, functions often depend on more than 100 variables.

Unfortunately, it is not possible to draw the graph of a function of more than two variables. The graph of a function $f(x, y, z)$ would consist of the set of points $(x, y, z, f(x, y, z))$ in four-dimensional space $\mathbf{R}^4$. However, just as we can use contour maps to visualize a three-dimensional mountain using curves on a two-dimensional plane, it is possible to draw the **level surfaces** of a function of three variables $f(x, y, z)$. These are the surfaces with equation $f(x, y, z) = c$. For example, the level surfaces of

$$f(x, y, z) = x^2 + y^2 + z^2$$

are the spheres with equation $x^2 + y^2 + z^2 = c$ (Figure 18). In the case of a function $T(x, y, z)$ that represents temperature of points in space, we call the level surfaces corresponding to $T(x, y, z) = k$ the **isotherms**. These are the collections of points, all of which have the same temperature k.

For functions of four or more variables, we can no longer visualize the graph or the level surfaces. We must rely on intuition developed through the study of functions of two and three variables.

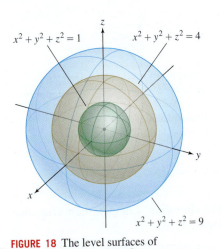

FIGURE 18 The level surfaces of $f(x, y, z) = x^2 + y^2 + z^2$ are spheres.

■ **EXAMPLE 9** Describe the level surfaces of $g(x, y, z) = x^2 + y^2 - z^2$.

Solution The level surface for $c = 0$ is the cone $x^2 + y^2 - z^2 = 0$. For $c \neq 0$, the level surfaces are the hyperboloids $x^2 + y^2 - z^2 = c$. The hyperboloid has one sheet if $c > 0$ and two sheets if $c < 0$ (Figure 19).   ■

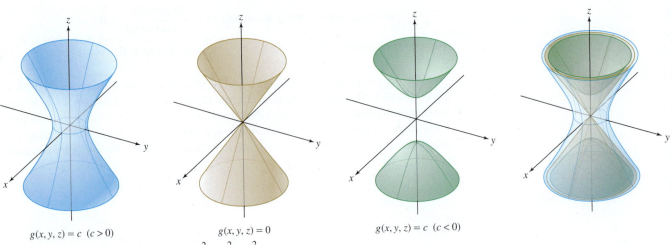

$g(x, y, z) = c \ (c > 0)$ $g(x, y, z) = 0$ $g(x, y, z) = c \ (c < 0)$

FIGURE 19 Level surfaces of $g(x, y, z) = x^2 + y^2 - z^2$.

14.1 SUMMARY

- The domain $\mathcal{D}$ of a function $f(x_1, \dots, x_n)$ of n variables is the set of n-tuples $(a_1, \dots, a_n)$ in $\mathbf{R}^n$ for which $f(a_1, \dots, a_n)$ is defined. The range of f is the set of values taken by f.

- The graph of a continuous real-valued function $f(x, y)$ is the surface in $\mathbf{R}^3$ consisting of the points $(a, b, f(a, b))$ for (a, b) in the domain $\mathcal{D}$ of f.
- A *vertical trace* is a curve obtained by intersecting the graph with a vertical plane $x = a$ or $y = b$.
- A *level curve* is a curve in the xy-plane defined by an equation $f(x, y) = c$. The level curve $f(x, y) = c$ is the projection onto the xy-plane of the horizontal trace curve, obtained by intersecting the graph with the horizontal plane $z = c$.
- A *contour map* shows the level curves $f(x, y) = c$ for equally spaced values of c. The spacing m is called the *contour interval*.
- When reading a contour map, keep in mind:

  - Your altitude does not change when you hike along a level curve.

  - Your altitude increases or decreases by m (the contour interval) when you hike from one level curve to the next.

- The spacing of the level curves indicates steepness: They are closer together where the graph is steeper.
- The *average rate of change* from P to Q is the ratio $\dfrac{\Delta\text{altitude}}{\Delta\text{horizontal}}$.
- A direction of steepest ascent at a point P is a direction along which $f(x, y)$ increases most rapidly. The steepest direction is obtained (approximately) by drawing the segment from P to the nearest point on the next level curve.
- Level surfaces can be used to understand a function $f(x, y, z)$. In the case where the function represents temperature, we call the level surfaces isotherms.

14.1 EXERCISES

Preliminary Questions

1. What is the difference between a horizontal trace and a level curve? How are they related?

2. Describe the trace of $f(x, y) = x^2 - \sin(x^3 y)$ in the xz-plane.

3. Is it possible for two different level curves of a function to intersect? Explain.

4. Describe the contour map of $f(x, y) = x$ with contour interval 1.

5. How will the contour maps of

$$f(x, y) = x \quad \text{and} \quad g(x, y) = 2x$$

with contour interval 1 look different?

Exercises

In Exercises 1–4, evaluate the function at the specified points.

1. $f(x, y) = x + yx^3$, (2, 2), (−1, 4)

2. $g(x, y) = \dfrac{y}{x^2 + y^2}$, (1, 3), (3, −2)

3. $h(x, y, z) = xyz^{-2}$, (3, 8, 2), (3, −2, −6)

4. $Q(y, z) = y^2 + y \sin z$, $(y, z) = \left(2, \frac{\pi}{2}\right), \left(-2, \frac{\pi}{6}\right)$

In Exercises 5–12, sketch the domain of the function.

5. $f(x, y) = 12x - 5y$

6. $f(x, y) = \sqrt{81 - x^2}$

7. $f(x, y) = \ln(4x^2 - y)$

8. $h(x, t) = \dfrac{1}{x + t}$

9. $g(y, z) = \dfrac{1}{z + y^2}$

10. $f(x, y) = \sin \dfrac{y}{x}$

11. $F(I, R) = \sqrt{IR}$

12. $f(x, y) = \cos^{-1}(x + y)$

In Exercises 13–16, describe the domain and range of the function.

13. $f(x, y, z) = xz + e^y$

14. $f(x, y, z) = x\sqrt{y + z}e^{z/x}$

15. $P(r, s, t) = \sqrt{16 - r^2 s^2 t^2}$

16. $g(r, s) = \cos^{-1}(rs)$

17. Match graphs (A) and (B) in Figure 20 with the functions:

(i) $f(x, y) = -x + y^2$ (ii) $g(x, y) = x + y^2$

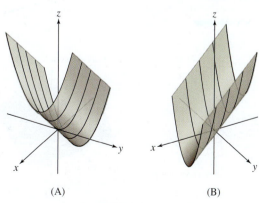

(A)   (B)

FIGURE 20

18. Match each of graphs (A) and (B) in Figure 21 with one of the following functions:

(i) $f(x, y) = (\cos x)(\cos y)$

(ii) $g(x, y) = \cos(x^2 + y^2)$

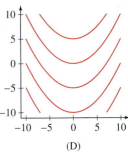

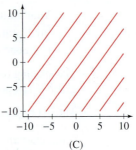

(A)   (B)

FIGURE 21

19. Match the functions (a)–(f) with their graphs (A)–(F) in Figure 22.

(a) $f(x, y) = |x| + |y|$

(b) $f(x, y) = \cos(x - y)$

(c) $f(x, y) = \dfrac{-1}{1 + 9x^2 + y^2}$

(d) $f(x, y) = \cos(y^2)e^{-0.1(x^2+y^2)}$

(e) $f(x, y) = \dfrac{-1}{1 + 9x^2 + 9y^2}$

(f) $f(x, y) = \cos(x^2 + y^2)e^{-0.1(x^2+y^2)}$

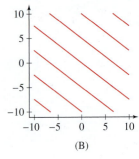

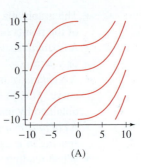

(A)   (B)

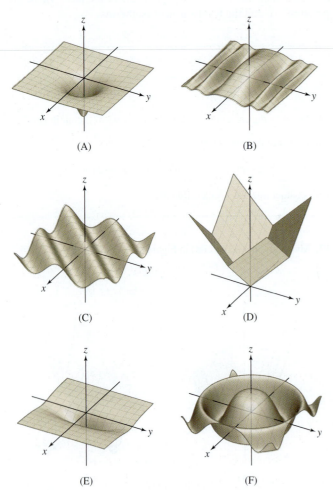

(A)   (B)

(C)   (D)

(E)   (F)

FIGURE 22

20. Match the functions (a)–(d) with their contour maps (A)–(D) in Figure 23.

(a) $f(x, y) = 3x + 4y$

(b) $g(x, y) = x^3 - y$

(c) $h(x, y) = 4x - 3y$

(d) $k(x, y) = x^2 - y$

(A)   (B)

(C)   (D)

FIGURE 23

In Exercises 21–26, sketch the graph and draw several vertical and horizontal traces.

21. $f(x, y) = 12 - 3x - 4y$

22. $f(x, y) = \sqrt{4 - x^2 - y^2}$

23. $f(x, y) = x^2 + 4y^2$

24. $f(x, y) = y^2$

25. $f(x, y) = \sin(x - y)$

26. $f(x, y) = \dfrac{1}{x^2 + y^2 + 1}$

27. Sketch contour maps of $f(x, y) = x + y$ with contour intervals $m = 1$ and 2.

28. Sketch the contour map of $f(x, y) = x^2 + y^2$ with level curves $c = 0, 4, 8, 12, 16$.

In Exercises 29–36, draw a contour map of $f(x, y)$ with an appropriate contour interval, showing at least six level curves.

29. $f(x, y) = x^2 - y$

30. $f(x, y) = \dfrac{y}{x^2}$

31. $f(x, y) = \dfrac{y}{x}$

32. $f(x, y) = xy$

33. $f(x, y) = x^2 + 4y^2$

34. $f(x, y) = x + 2y - 1$

35. $f(x, y) = x^2$

36. $f(x, y) = 3x^2 - y^2$

37. [image: pencil icon] Find the linear function whose contour map (with contour interval $m = 6$) is shown in Figure 24. What is the linear function if $m = 3$ (and the curve labeled $c = 6$ is relabeled $c = 3$)?

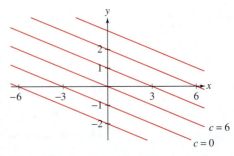

FIGURE 24 Contour map with contour interval $m = 6$.

38. Use the contour map in Figure 25 to calculate the average rate of change:

(a) from A to B.   **(b)** from A to C.

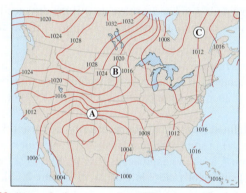

FIGURE 25

39. Referring to Figure 26, answer the following questions:

(a) At which of (A)–(C) is pressure increasing in the northern direction?

(b) At which of (A)–(C) is pressure increasing in the easterly direction?

(c) In which direction at (B) is pressure increasing most rapidly?

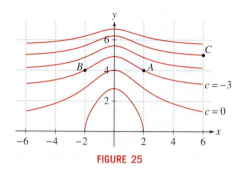

FIGURE 26 Atmospheric pressure (in millibars) over North America on March 26, 2009

In Exercises 40–43, let $T(x, y, z)$ denote temperature at each point in space. Draw level surfaces (also called isotherms) corresponding to the fixed temperatures given.

40. $T(x, y, z) = 2x + 3y - z$. $T = 0, 1, 2$

41. $T(x, y, z) = x - y + 2z$. $T = 0, 1, 2$

42. $T(x, y, z) = x^2 + y^2 - z$. $T = 0, 1, 2$

43. $T(x, y, z) = x^2 - y^2 + z^2$. $T = 0, 1, 2, -1, -2$

In Exercises 44–47, $\rho(S, T)$ is seawater density (kilograms per cubic meter) as a function of salinity S (parts per thousand) and temperature T (degrees Celsius). Refer to the contour map in Figure 27.

44. Calculate the average rate of change of ρ with respect to T from B to A.

45. Calculate the average rate of change of ρ with respect to S from B to C.

46. At a fixed level of salinity, is seawater density an increasing or a decreasing function of temperature?

47. Does water density appear to be more sensitive to a change in temperature at point A or point B?

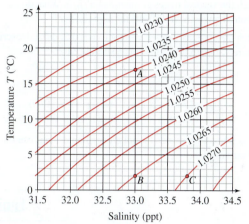

FIGURE 27 Contour map of seawater density $\rho(S, T)$ (kilograms per cubic meter).

In Exercises 48–51, refer to Figure 28.

48. Find the change in elevation from A and B.

49. Estimate the average rate of change from A and B and from A to C.

50. Estimate the average rate of change from A to points i, ii, and iii.

51. Sketch the path of steepest ascent beginning at D.

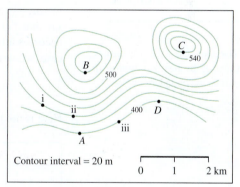

FIGURE 28

52. Let temperature in 3-space be given by $T(x, y, z) = x^2 + y^2 - z$. Draw isotherms corresponding to temperatures $T = -2, -1, 0, 1, 2$.

53. Let temperature in 3-space be given by $T(x, y, z) = \frac{x^2}{4} + \frac{y^2}{9} + z^2$. Draw isotherms corresponding to temperatures $T = 0, 1, 4$.

54. Let temperature in 3-space be given by $T(x, y, z) = x^2 - y^2 - z$. Draw isotherms corresponding to temperatures $T = -1, 0, 1$.

55. Let temperature in 3-space be given by $T(x, y, z) = x^2 - y^2 - z^2$. Draw isotherms corresponding to temperatures $T = -2, -1, 0, 1, 2$.

Further Insights and Challenges

56. The function $f(x, t) = t^{-1/2}e^{-x^2/t}$, whose graph is shown in Figure 29, models the temperature along a metal bar after an intense burst of heat is applied at its center point.

(a) Sketch the vertical traces at times $t = 1, 2, 3$. What do these traces tell us about the way heat diffuses through the bar?

(b) Sketch the vertical traces $x = c$ for $c = \pm 0.2, \pm 0.4$. Describe how temperature varies in time at points near the center.

57. Let

$$f(x, y) = \frac{x}{\sqrt{x^2 + y^2}} \quad \text{for } (x, y) \neq (0, 0)$$

Write f as a function $f(r, \theta)$ in polar coordinates, and use this to find the level curves of f.

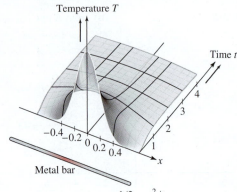

FIGURE 29 Graph of $f(x, t) = t^{-1/2}e^{-x^2/t}$ beginning shortly after $t = 0$.

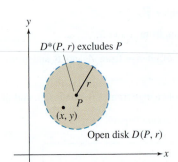

FIGURE 1 The open disk $D(P, r)$ consists of points (x, y) at distance $< r$ from P. It does not include the boundary circle.

14.2 Limits and Continuity in Several Variables

This section develops limits and continuity in the multivariable setting. We focus on functions of two variables, but similar definitions and results apply to functions of three or more variables.

Recall that on the real number line, a number x is close to a if the distance $|x - a|$ is small. In the plane, a point (x, y) is close to another point $P = (a, b)$ if the distance $d((x, y), (a, b)) = \sqrt{(x - a)^2 + (y - b)^2}$ between them is small.

Note that if we take all the points that are a distance of less than r from $P = (a, b)$, as in Figure 1, this is a disk $D(P, r)$ centered at P that does not include its boundary. If we insist also that $d((x, y), (a, b)) \neq 0$, then we get a punctured disk that does not include P and that we denote $D^*(P, r)$.

Now assume that $f(x, y)$ is **defined near** P but not necessarily at P itself. In other words, $f(x, y)$ is defined for all (x, y) in some punctured disk $D^*(P, r)$ with $r > 0$. We say that $f(x, y)$ approaches the limit L as (x, y) approaches $P = (a, b)$ if $|f(x, y) - L|$ becomes arbitrarily small for (x, y) sufficiently close to $P = (a, b)$ [Figure 2(C)]. In this case, we write

$$\lim_{(x,y) \to P} f(x, y) = \lim_{(x,y) \to (a,b)} f(x, y) = L$$

Here is the formal definition.

DEFINITION **Limit** Assume that $f(x, y)$ is defined near $P = (a, b)$. Then

$$\lim_{(x,y) \to P} f(x, y) = L$$

if, for any $\epsilon > 0$, there exists $\delta > 0$ such that if (x, y) satisfies

$$0 < d((x, y), (a, b)) < \delta, \quad \text{then} \quad |f(x, y) - L| < \epsilon$$

This is similar to the definition of the limit in one variable, but there is an important difference. In a one-variable limit, we require that $f(x)$ tend to L as x approaches a from the left or right [Figure 2(A)]. In a multivariable limit, $f(x, y)$ must tend to L no matter how (x, y) approaches P [Figure 2(B)].

■ **EXAMPLE 1** Show that **(a)** $\displaystyle\lim_{(x,y) \to (a,b)} x = a$ and **(b)** $\displaystyle\lim_{(x,y) \to (a,b)} y = b$.

Solution Let $P = (a, b)$. To verify (a), let $f(x, y) = x$ and $L = a$. We must show that for any $\epsilon > 0$, we can find $\delta > 0$ such that

$$\text{If} \quad 0 < d((x, y), (a, b)) < \delta, \quad \text{then} \quad |f(x, y) - L| = |x - a| < \epsilon \qquad \boxed{1}$$

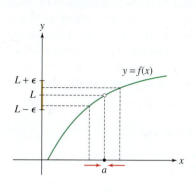

(A) In one variable, we can approach a from only two possible directions.

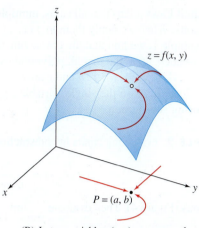

(B) In two variables, (x, y) can approach $P = (a, b)$ along any direction or path.

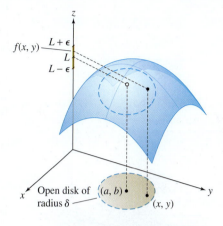

(C) $|f(x, y) - L| < \epsilon$ for all (x, y) inside the punctured disk

FIGURE 2

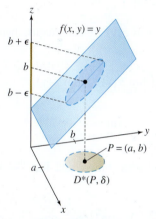

DF FIGURE 3 If $|y - b| < \delta$ for $\delta = \epsilon$, then $|f(x, y) - b| < \epsilon$. Therefore,

$$\lim_{(x,y)\to(a,b)} y = b.$$

In fact, we can choose $\delta = \epsilon$, for if $d((x, y), (a, b)) < \epsilon$, then

$$(x - a)^2 + (y - b)^2 < \epsilon^2 \quad \Rightarrow \quad (x - a)^2 < \epsilon^2 \quad \Rightarrow \quad |x - a| < \epsilon$$

In other words, for any $\epsilon > 0$, if $0 < d((x, y), (a, b)) < \epsilon$, then $|x - a| < \epsilon$. This proves (a). The limit (b) is similar (see Figure 3). ■

The following theorem lists the basic laws for limits. We omit the proofs, which are similar to the proofs of the single-variable Limit Laws.

THEOREM 1 Limit Laws Assume that $\lim_{(x,y)\to P} f(x, y)$ and $\lim_{(x,y)\to P} g(x, y)$ exist. Then:

(i) Sum Law:

$$\lim_{(x,y)\to P} (f(x, y) + g(x, y)) = \lim_{(x,y)\to P} f(x, y) + \lim_{(x,y)\to P} g(x, y)$$

(ii) Constant Multiple Law: For any number k,

$$\lim_{(x,y)\to P} kf(x, y) = k \lim_{(x,y)\to P} f(x, y)$$

(iii) Product Law:

$$\lim_{(x,y)\to P} f(x, y)\, g(x, y) = \left(\lim_{(x,y)\to P} f(x, y)\right)\left(\lim_{(x,y)\to P} g(x, y)\right)$$

(iv) Quotient Law: If $\lim_{(x,y)\to P} g(x, y) \neq 0$, then

$$\lim_{(x,y)\to P} \frac{f(x, y)}{g(x, y)} = \frac{\lim_{(x,y)\to P} f(x, y)}{\lim_{(x,y)\to P} g(x, y)}$$

As in the single-variable case, we say that f is continuous at $P = (a, b)$ if $f(x, y)$ approaches the value of the function $f(a, b)$ as $(x, y) \to (a, b)$.

DEFINITION Continuity A function f of two variables is **continuous** at $P = (a, b)$ if

$$\lim_{(x,y)\to(a,b)} f(x, y) = f(a, b)$$

We say that f is continuous if it is continuous at each point (a, b) in its domain.

The Limit Laws tell us that all sums, multiples, and products of continuous functions are continuous. When we apply them to $f(x, y) = x$ and $g(x, y) = y$, which are continuous by Example 1, we find that the power functions $f(x, y) = x^m y^n$ are continuous for all whole numbers m, n and that all polynomials are continuous. Furthermore, a rational function $h(x, y)/g(x, y)$, where h and g are polynomials, is continuous at all points (a, b) where $g(a, b) \neq 0$. As in the single-variable case, we can evaluate limits of continuous functions using substitution.

■ **EXAMPLE 2**　**Evaluating Limits by Substitution**　Show that

$$f(x, y) = \frac{3x + y}{x^2 + y^2 + 1}$$

is continuous (Figure 4). Then evaluate $\lim\limits_{(x,y)\to(1,2)} f(x, y)$.

Solution The function f is continuous at all points (a, b) because it is a rational function whose denominator $Q(x, y) = x^2 + y^2 + 1$ is never zero. Therefore, we can evaluate the limit by substitution:

$$\lim_{(x,y)\to(1,2)} \frac{3x + y}{x^2 + y^2 + 1} = \frac{3(1) + 2}{1^2 + 2^2 + 1} = \frac{5}{6}$$　■

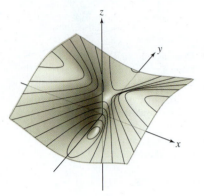

FIGURE 4 Top view of the graph $f(x, y) = \dfrac{3x + y}{x^2 + y^2 + 1}$.

If f is a product $f(x, y) = h(x)g(y)$, where $h(x)$ and $g(y)$ are continuous, then the limit is a product of limits by the Product Law:

$$\lim_{(x,y)\to(a,b)} f(x, y) = \lim_{(x,y)\to(a,b)} h(x)g(y) = \left(\lim_{x\to a} h(x)\right)\left(\lim_{y\to b} g(y)\right)$$

■ **EXAMPLE 3**　**Product Functions**　Evaluate $\lim\limits_{(x,y)\to(3,0)} x^3 \dfrac{\sin y}{y}$.

Solution The limit is equal to a product of limits, and since $\lim\limits_{x\to 3} x^3$ and $\lim\limits_{y\to 0} \frac{\sin y}{y}$ both exist:

$$\lim_{(x,y)\to(3,0)} x^3 \frac{\sin y}{y} = \left(\lim_{x\to 3} x^3\right)\left(\lim_{y\to 0} \frac{\sin y}{y}\right) = (3^3)(1) = 27$$　■

Composition is another important way to build functions. If f is a function of two variables and $G(u)$ a function of one variable, then the composite function $G \circ f$ is the function of two variables given by $G(f(x, y))$. According to the next theorem, a composition of continuous functions is again continuous.

> **THEOREM 2**　**A Composition of Continuous Functions Is Continuous**　If a function of two variables f is continuous at (a, b) and a function of one variable G is continuous at $c = f(a, b)$, then the composite function $G(f(x, y))$ is continuous at (a, b).

■ **EXAMPLE 4**　Write $H(x, y) = e^{-x^2 + 2y}$ as a composite function and evaluate

$$\lim_{(x,y)\to(1,2)} H(x, y)$$

Solution We have $H(x, y) = G \circ f$, where $G(u) = e^u$ and $f(x, y) = -x^2 + 2y$. Both f and G are continuous, so H is also continuous and

$$\lim_{(x,y)\to(1,2)} H(x, y) = \lim_{(x,y)\to(1,2)} e^{-x^2 + 2y} = e^{-(1)^2 + 2(2)} = e^3$$　■

As occurs for limits of functions of one variable, the most difficult limits are those that are of an indeterminate form. If we want to determine $\lim\limits_{(x,y)\to(a,b)} f(x, y)$, and plugging (a, b) into $f(x, y)$ yields the indeterminate form of type $\frac{0}{0}$, then in some cases the limit exists and in other cases, it does not. We know that if a limit $\lim\limits_{(x,y)\to(a,b)} f(x,y)$ exists and equals L, then $f(x,y)$ tends to L as (x,y) approaches (a,b) along any path. In the next example, we prove that a limit *does not exist* by showing that $f(x,y)$ approaches *different limits* when $(0,0)$ is approached along different lines through the origin. We use three different methods on the problem to demonstrate a variety of approaches one can take.

■ **EXAMPLE 5** **Showing a Limit Does Not Exist** Examine $\lim\limits_{(x,y)\to(0,0)} \dfrac{x^2}{x^2+y^2}$ numerically. Then prove that the limit does not exist.

Solution If the limit existed, we would expect the values of $f(x,y)$ in Table 1 to get closer to a limiting value L as (x,y) gets close to $(0,0)$. However, the table suggests that $f(x,y)$ takes on all values between 0 and 1, no matter how close (x,y) gets to $(0,0)$. For example,

$$f(0.1,0)=1, \qquad f(0.1,0.1)=0.5, \qquad f(0,0.1)=0$$

Thus, $f(x,y)$ does not seem to approach any fixed value L as $(x,y)\to(0,0)$.

TABLE 1 Values of $f(x,y)=\dfrac{x^2}{x^2+y^2}$

y \ x	−0.5	−0.4	−0.3	−0.2	−0.1	0	0.1	0.2	0.3	0.4	0.5
0.5	**0.5**	0.39	0.265	0.138	0.038	**0**	0.038	0.138	0.265	0.39	**0.5**
0.4	0.61	**0.5**	0.36	0.2	0.059	**0**	0.059	0.2	0.36	**0.5**	0.61
0.3	0.735	0.64	**0.5**	0.308	0.1	**0**	0.1	0.308	**0.5**	0.64	0.735
0.2	0.862	0.8	0.692	**0.5**	0.2	**0**	0.2	**0.5**	0.692	0.8	0.862
0.1	0.962	0.941	0.9	0.8	**0.5**	**0**	**0.5**	0.8	0.9	0.941	0.962
0	1	1	1	1	1		1	1	1	1	1
−0.1	0.962	0.941	0.9	0.8	**0.5**	**0**	**0.5**	0.8	0.9	0.941	0.962
−0.2	0.862	0.8	0.692	**0.5**	0.2	**0**	0.2	**0.5**	0.692	0.8	0.862
−0.3	0.735	0.640	**0.5**	0.308	0.1	**0**	0.1	0.308	**0.5**	0.640	0.735
−0.4	0.610	**0.5**	0.360	0.2	0.059	**0**	0.059	0.2	0.36	**0.5**	0.61
−0.5	**0.5**	0.39	0.265	0.138	0.038	**0**	0.038	0.138	0.265	0.390	**0.5**

Now let's prove that the limit does not exist.

First Method We show that $f(x,y)$ approaches different limits along the x- and y-axes (Figure 5):

$$\text{Limit along } x\text{-axis:} \qquad \lim_{x\to0} f(x,0)=\lim_{x\to0}\frac{x^2}{x^2+0^2}=\lim_{x\to0}1=1$$

$$\text{Limit along } y\text{-axis:} \qquad \lim_{y\to0} f(0,y)=\lim_{y\to0}\frac{0^2}{0^2+y^2}=\lim_{y\to0}0=0$$

These two limits are different, and hence, $\lim\limits_{(x,y)\to(0,0)} f(x,y)$ does not exist.

Second Method If we set $y=mx$, we have restricted ourselves to the line through the origin with slope m. Then the limit becomes

$$\lim_{x\to0} f(x,mx)=\lim_{x\to0}\frac{x^2}{x^2+(mx)^2}=\frac{1}{1+m^2}$$

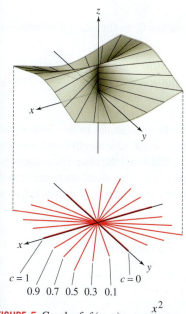

FIGURE 5 Graph of $f(x,y)=\dfrac{x^2}{x^2+y^2}$.

$c=1$ $\qquad$ $c=0$

0.9 0.7 0.5 0.3 0.1

This clearly depends on the slope m, and therefore gives different values when the origin is approached along lines of differing slope. For instance, when $m = 0$, so that we are approaching along the x-axis, we have a limit of 1. But when $m = 1$, so that we are approaching along the diagonal line, the limit is $\frac{1}{2}$. Hence, the limit does not exist. The contour map in Figure 5 shows the variety of limits that occur as we approach the origin along various lines.

Third Method We convert to polar coordinates, setting $x = r \cos \theta$ and $y = r \sin \theta$. Then for any path that approaches $(0, 0)$, it must be the case that r approaches 0. Different paths come from fixing θ at various values as r approaches 0.

Hence, we can consider

$$\lim_{r \to 0} \frac{x^2}{x^2 + y^2} = \lim_{r \to 0} \frac{(r \cos \theta)^2}{(r \cos \theta)^2 + (r \sin \theta)^2} = \lim_{r \to 0} \cos^2 \theta$$

This result depends on θ. For instance, fixing θ at 0 would mean we are approaching $(0, 0)$ along the positive x-axis, and that gives a limit of 1. Fixing θ at $\pi/2$ would mean we are approaching $(0, 0)$ along the positive y-axis, and that gives a limit of 0. Since different values of θ yield different results, the limit does not exist. ∎

■ **EXAMPLE 6**   **Verifying a Limit**   Calculate $\displaystyle\lim_{(x,y) \to (0,0)} f(x, y)$, where $f(x, y)$ is defined for $(x, y) \neq (0, 0)$ by

$$f(x, y) = \frac{xy^2}{x^2 + y^2}$$

as in Figure 6.

Solution Since substitution yields the indeterminate form of type $\frac{0}{0}$, we will try converting to polar coordinates:

$$x = r \cos \theta, \qquad y = r \sin \theta$$

Keep in mind that for any path approaching $(0, 0)$, r will also have to approach 0. Then $x^2 + y^2 = r^2$ and for $r \neq 0$,

$$0 \le \left| \frac{xy^2}{x^2 + y^2} \right| = \left| \frac{(r \cos \theta)(r \sin \theta)^2}{r^2} \right| = r|\cos \theta \sin^2 \theta| \le r$$

As (x, y) approaches $(0, 0)$, the variable r also approaches 0, so the desired conclusion follows from the Squeeze Theorem:

$$0 \le \lim_{(x,y) \to (0,0)} \left| \frac{xy^2}{x^2 + y^2} \right| \le \lim_{r \to 0} r = 0$$ ∎

Note that in the last two examples, we have used polar coordinates to first, prove that a limit does not exist and second, prove that a limit does exist. One might conclude that polar coordinates can always be applied. Unfortunately, that is not the case. They can only work if we are taking a limit as (x, y) approaches $(0, 0)$. Furthermore, there are many situations for which they will not work to prove either the existence or nonexistence of a limit even if (x, y) is approaching $(0, 0)$.

■ **EXAMPLE 7**   Determine whether or not the following limit exists:

$$\lim_{(x,y) \to (0,0)} \frac{x^2 y}{x^4 + y^2}$$

Solution We first consider paths along lines through the origin of the form $y = mx$.

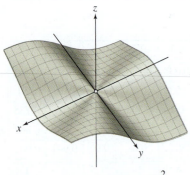

FIGURE 6 Graph of $f(x, y) = \dfrac{xy^2}{x^2 + y^2}$.

Then the limit becomes

$$\lim_{x \to 0} \frac{x^2(mx)}{x^4 + (mx)^2} = \lim_{x \to 0} \frac{xm}{x^2 + m^2} = 0$$

Thus, all paths along lines through the origin yield the same limit. However, this does not mean that all paths through the origin yield the same limit. Consider the path along the parabola $y = x^2$. Then the limit becomes

$$\lim_{x \to 0} \frac{x^2(x^2)}{x^4 + (x^2)^2} = \frac{1}{2}$$

Since this does not equal the limit we obtained when we approached along straight lines, the limit does not exist. ■

To prove a limit does not exist, we need only find two paths that yield different limits. However, as Example 7 demonstrates, to prove that a limit *does* exist at a point, just considering the limit along any finite set of lines through the point or, for that matter, all the lines through the point, is not sufficient.

14.2 SUMMARY

• Suppose that $f(x, y)$ is defined near $P = (a, b)$. Then

$$\lim_{(x,y) \to (a,b)} f(x, y) = L$$

if, for any $\epsilon > 0$, there exists $\delta > 0$ such that if (x, y) satisfies

$$0 < d((x, y), (a, b)) < \delta, \quad \text{then} \quad |f(x, y) - L| < \epsilon$$

• The limit of a product $f(x, y) = h(x)g(y)$ is a product of limits:

$$\lim_{(x,y) \to (a,b)} f(x, y) = \left(\lim_{x \to a} h(x) \right) \left(\lim_{y \to b} g(y) \right)$$

• A function f of two variables is *continuous* at $P = (a, b)$ if

$$\lim_{(x,y) \to (a,b)} f(x, y) = f(a, b)$$

• To prove that a limit does not exist, it is enough to show that the limits obtained along two different paths are not equal.

14.2 EXERCISES

Preliminary Questions

1. What is the difference between $D(P, r)$ and $D^*(P, r)$?

2. Suppose that $f(x, y)$ is continuous at $(2, 3)$ and that $f(2, y) = y^3$ for $y \neq 3$. What is the value $f(2, 3)$?

3. Suppose that $Q(x, y)$ is a function such that $1/Q(x, y)$ is continuous for all (x, y). Which of the following statements are true?

(a) $Q(x, y)$ is continuous for all (x, y).
(b) $Q(x, y)$ is continuous for $(x, y) \neq (0, 0)$.
(c) $Q(x, y) \neq 0$ for all (x, y).

4. Suppose that $f(x, 0) = 3$ for all $x \neq 0$ and $f(0, y) = 5$ for all $y \neq 0$. What can you conclude about $\lim_{(x,y) \to (0,0)} f(x, y)$?

Exercises

In Exercises 1–8, evaluate the limit using continuity.

1. $\lim_{(x,y) \to (1,2)} (x^2 + y)$

2. $\lim_{(x,y) \to (\frac{4}{9}, \frac{2}{9})} \frac{x}{y}$

3. $\lim_{(x,y) \to (2,-1)} (xy - 3x^2 y^3)$

4. $\lim_{(x,y) \to (-2,1)} \frac{2x^2}{4x + y}$

5. $\lim_{(x,y) \to (\frac{\pi}{4}, 0)} \tan x \cos y$

6. $\lim_{(x,y) \to (2,3)} \tan^{-1}(x^2 - y)$

7. $\lim_{(x,y) \to (1,1)} \frac{e^{x^2} - e^{-y^2}}{x + y}$

8. $\lim_{(x,y) \to (1,0)} \ln(x - y)$

In Exercises 9–12, assume that

$$\lim_{(x,y)\to(2,5)} f(x,y) = 3, \qquad \lim_{(x,y)\to(2,5)} g(x,y) = 7$$

to find the limit.

9. $\displaystyle\lim_{(x,y)\to(2,5)} \big(g(x,y) - 2f(x,y)\big)$

10. $\displaystyle\lim_{(x,y)\to(2,5)} f(x,y)^2 g(x,y)$ **11.** $\displaystyle\lim_{(x,y)\to(2,5)} e^{f(x,y)^2 - g(x,y)}$

12. $\displaystyle\lim_{(x,y)\to(2,5)} \frac{f(x,y)}{f(x,y) + g(x,y)}$

13. Does $\displaystyle\lim_{(x,y)\to(0,0)} \frac{y^2}{x^2 + y^2}$ exist? Explain.

14. Let $f(x,y) = xy/(x^2 + y^2)$. Show that $f(x,y)$ approaches zero along the x- and y-axes. Then prove that $\displaystyle\lim_{(x,y)\to(0,0)} f(x,y)$ does not exist by showing that the limit along the line $y = x$ is nonzero.

15. Let $f(x,y) = \dfrac{x^3 + y^3}{xy^2}$. Set $y = mx$ and show that the resulting limit depends on m, and therefore the limit $\displaystyle\lim_{(x,y)\to(0,0)} f(x,y)$ does not exist.

16. Let $f(x,y) = \dfrac{2x^2 + 3y^2}{xy}$. Set $y = mx$ and show that the resulting limit depends on m, and therefore the limit $\displaystyle\lim_{(x,y)\to(0,0)} f(x,y)$ does not exist.

17. Prove that

$$\lim_{(x,y)\to(0,0)} \frac{x}{x^2 + y^2}$$

does not exist by considering the limit along the x-axis.

18. Let $f(x,y) = x^3/(x^2 + y^2)$ and $g(x,y) = x^2/(x^2 + y^2)$. Using polar coordinates, prove that

$$\lim_{(x,y)\to(0,0)} f(x,y) = 0$$

and that $\displaystyle\lim_{(x,y)\to(0,0)} g(x,y)$ does not exist. *Hint:* Show that $g(x,y) = \cos^2\theta$ and observe that $\cos\theta$ can take on any value between -1 and 1 as $(x,y) \to (0,0)$.

In Exercises 19–22, use any method to evaluate the limit or show that it does not exist.

19. $\displaystyle\lim_{(x,y)\to(0,0)} \frac{x^2 - y^2}{\sqrt{x^2 + y^2}}$ **20.** $\displaystyle\lim_{(x,y)\to(0,0)} \frac{x^2 - y^2}{x^2 + y^2}$

21. $\displaystyle\lim_{(x,y)\to(0,0)} \frac{xy}{3x^2 + 2y^2}$

22. $\displaystyle\lim_{(x,y)\to(0,0)} \frac{x^4 - y^4}{x^4 + x^2 y^2 + y^4}$

In Exercises 23–24, show that the limit does not exist by approaching the origin along one or more of the coordinate axes.

23. $\displaystyle\lim_{(x,y,z)\to(0,0,0)} \frac{x + y + z}{x^2 + y^2 + z^2}$ **24.** $\displaystyle\lim_{(x,y)\to(0,0)} \frac{x^2 - y^2 + z^2}{x^2 + y^2 + z^2}$

25. Use the Squeeze Theorem to evaluate

$$\lim_{(x,y)\to(4,0)} (x^2 - 16)\cos\left(\frac{1}{(x-4)^2 + y^2}\right)$$

26. Evaluate $\displaystyle\lim_{(x,y)\to(0,0)} \tan x \sin\left(\frac{1}{|x| + |y|}\right)$.

In Exercises 27–40, evaluate the limit or determine that it does not exist.

27. $\displaystyle\lim_{(z,w)\to(-2,1)} \frac{z^4 \cos(\pi w)}{e^{z+w}}$ **28.** $\displaystyle\lim_{(z,w)\to(-1,2)} (z^2 w - 9z)$

29. $\displaystyle\lim_{(x,y)\to(4,2)} \frac{y - 2}{\sqrt{x^2 - 4}}$ **30.** $\displaystyle\lim_{(x,y)\to(0,0)} \frac{x^2 + y^2}{1 + y^2}$

31. $\displaystyle\lim_{(x,y)\to(3,4)} \frac{1}{\sqrt{x^2 + y^2}}$ **32.** $\displaystyle\lim_{(x,y)\to(0,0)} \frac{xy}{\sqrt{x^2 + y^2}}$

33. $\displaystyle\lim_{(x,y)\to(1,-3)} e^{x-y} \ln(x - y)$ **34.** $\displaystyle\lim_{(x,y)\to(0,0)} \frac{|x|}{|x| + |y|}$

35. $\displaystyle\lim_{(x,y)\to(-3,-2)} (x^2 y^3 + 4xy)$ **36.** $\displaystyle\lim_{(x,y)\to(2,1)} e^{x^2 - y^2}$

37. $\displaystyle\lim_{(x,y)\to(0,0)} \tan(x^2 + y^2)\tan^{-1}\left(\frac{1}{x^2 + y^2}\right)$

38. $\displaystyle\lim_{(x,y)\to(0,0)} (x + y + 2)e^{-1/(x^2 + y^2)}$

39. $\displaystyle\lim_{(x,y)\to(0,0)} \frac{x^2 + y^2}{\sqrt{x^2 + y^2 + 1} - 1}$

40. $\displaystyle\lim_{(x,y)\to(1,1)} \frac{x^2 + y^2 - 2}{|x - 1| + |y - 1|}$

Hint: Rewrite the limit in terms of $u = x - 1$ and $v = y - 1$.

41. Let $f(x,y) = \dfrac{x^3 + y^3}{x^2 + y^2}$.

(a) Show that

$$|x^3| \le |x|(x^2 + y^2), \quad |y^3| \le |y|(x^2 + y^2)$$

(b) Show that $|f(x,y)| \le |x| + |y|$.

(c) Use the Squeeze Theorem to prove that $\displaystyle\lim_{(x,y)\to(0,0)} f(x,y) = 0$.

42. Let $a, b \ge 0$. Show that $\displaystyle\lim_{(x,y)\to(0,0)} \frac{x^a y^b}{x^2 + y^2} = 0$ if $a + b > 2$ and that the limit does not exist if $a + b \le 2$.

43. [figure icon] Figure 7 shows the contour maps of two functions. Explain why the limit $\displaystyle\lim_{(x,y)\to P} f(x,y)$ does not exist. Does $\displaystyle\lim_{(x,y)\to Q} g(x,y)$ appear to exist in (B)? If so, what is its limit?

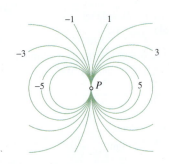

(A) Contour map of $f(x,y)$

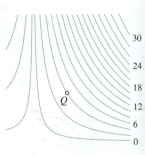

(B) Contour map of $g(x,y)$

FIGURE 7

Further Insights and Challenges

44. Evaluate $\lim\limits_{(x,y)\to(0,2)} (1+x)^{y/x}$.

45. Is the following function continuous?

$$f(x, y) = \begin{cases} x^2 + y^2 & \text{if } x^2 + y^2 < 1 \\ 1 & \text{if } x^2 + y^2 \geq 1 \end{cases}$$

46. *CAS* The function $f(x, y) = \sin(xy)/xy$ is defined for $xy \neq 0$.

(a) Is it possible to extend the domain of f to all of $\mathbf{R}^2$ so that the result is a continuous function?

(b) Use a computer algebra system to plot f. Does the result support your conclusion in (a)?

47. Prove that the function

$$f(x, y) = \begin{cases} \dfrac{(2^x - 1)(\sin y)}{xy} & \text{if } xy \neq 0 \\ \ln 2 & \text{if } xy = 0 \end{cases}$$

is continuous at $(0, 0)$.

48. Prove that if $f(x)$ is continuous at $x = a$ and $g(y)$ is continuous at $y = b$, then $F(x, y) = f(x)g(y)$ is continuous at (a, b).

49. Consider the function $f(x, y) = \dfrac{x^3 y}{x^6 + 2y^2}$.

(a) Show that as $(x, y) \to (0, 0)$ along any line $y = mx$, the limit equals 0.

(b) Show that as $(x, y) \to (0, 0)$ along the curve $y = x^3$, the limit does not equal 0, and hence, $\lim\limits_{(x,y)\to(0,0)} f(x, y)$ does not exist.

14.3 Partial Derivatives

We have stressed that a function f of two or more variables does not have a unique rate of change because each variable may affect f in different ways. For example, the current I in a circuit is a function of both voltage V and resistance R given by Ohm's Law:

$$I(V, R) = \frac{V}{R}$$

The current I is *increasing* as a function of V (when R is fixed) but *decreasing* as a function of R (when V is fixed).

The **partial derivatives** are the rates of change with respect to each variable separately. A function $f(x, y)$ of two variables has two partial derivatives, denoted f_x and f_y, defined by the following limits (if they exist):

$$f_x(a, b) = \lim_{h\to 0} \frac{f(a + h, b) - f(a, b)}{h}, \qquad f_y(a, b) = \lim_{k\to 0} \frac{f(a, b + k) - f(a, b)}{k}$$

Thus, f_x is the derivative of $f(x, b)$ as a function of x alone, and f_y is the derivative at $f(a, y)$ as a function of y alone. The Leibniz notation for partial derivatives is

$$\frac{\partial f}{\partial x} = f_x, \qquad\qquad \frac{\partial f}{\partial y} = f_y$$

$$\left.\frac{\partial f}{\partial x}\right|_{(a,b)} = f_x(a, b), \qquad\qquad \left.\frac{\partial f}{\partial y}\right|_{(a,b)} = f_y(a, b)$$

The partial derivative symbol ∂ is a rounded "d." The symbols $\partial f/\partial x$ and $\partial f/\partial y$ are read as follows: "dee-eff dee-ex" and "dee-eff dee-why."

If $z = f(x, y)$, then we also write $\partial z/\partial x$ and $\partial z/\partial y$.

Partial derivatives are computed just like ordinary derivatives in one variable with this difference: To compute f_x, treat y as a constant, and to compute f_y, treat x as a constant.

■ **EXAMPLE 1** Compute the partial derivatives of $f(x, y) = x^2 y^5$.

Solution

$$\frac{\partial f}{\partial x} = \underbrace{\frac{\partial}{\partial x}(x^2 y^5) = y^5 \frac{\partial}{\partial x}(x^2)}_{\text{Treat } y^5 \text{ as a constant}} = y^5(2x) = 2xy^5$$

$$\frac{\partial f}{\partial y} = \underbrace{\frac{\partial}{\partial y}(x^2 y^5) = x^2 \frac{\partial}{\partial x}(y^5)}_{\text{Treat } x^2 \text{ as a constant}} = x^2(5y^4) = 5x^2 y^4$$

■

The partial derivatives at $P = (a, b)$ are the slopes of the tangent lines to the vertical trace curves through the point $(a, b, f(a, b))$ in Figure 1(A). To compute $f_x(a, b)$, we set $y = b$ and differentiate in the x-direction. This gives us the slope of the tangent line to the trace curve in the plane $y = b$ [Figure 1(B)]. Similarly, $f_y(a, b)$ is the slope of the trace curve in the vertical plane $x = a$ [Figure 1(C)].

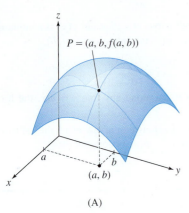

(A)

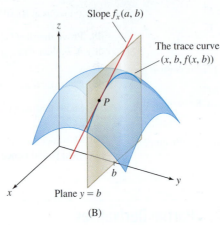

(B)

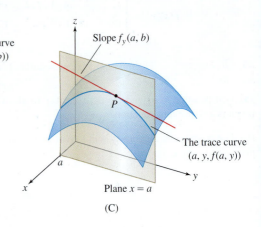

(C)

DF **FIGURE 1** The partial derivatives are the slopes of the vertical trace curves.

The differentiation rules from calculus of one variable (the Product, Quotient, and Chain Rules) are valid for partial derivatives.

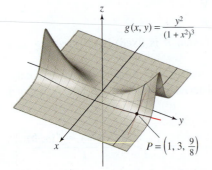

$g(x, y) = \dfrac{y^2}{(1 + x^2)^3}$

FIGURE 2 The slopes of the tangent lines to the trace curves are $g_x(1, 3)$ and $g_y(1, 3)$.

CAUTION *It is not necessary to use the Quotient Rule to compute the partial derivative in Eq. (1). The denominator does not depend on y, so we treat it as a constant when differentiating with respect to y.*

■ **EXAMPLE 2** Calculate $g_x(1, 3)$ and $g_y(1, 3)$, where $g(x, y) = \dfrac{y^2}{(1 + x^2)^3}$.

Solution To calculate g_x, treat y (and therefore y^2) as a constant:

$$g_x(x, y) = \frac{\partial}{\partial x}\left(\frac{y^2}{(1 + x^2)^3}\right) = y^2 \frac{\partial}{\partial x}(1 + x^2)^{-3} = \frac{-6xy^2}{(1 + x^2)^4}$$

$$g_x(1, 3) = \frac{-6(1)3^2}{(1 + 1^2)^4} = -\frac{27}{8}$$

To calculate g_y, treat x (and therefore $1 + x^2$) as a constant:

$$g_y(x, y) = \frac{\partial}{\partial y}\left(\frac{y^2}{(1 + x^2)^3}\right) = \frac{1}{(1 + x^2)^3}\frac{\partial}{\partial y}y^2 = \frac{2y}{(1 + x^2)^3} \qquad \boxed{1}$$

$$g_y(1, 3) = \frac{2(3)}{(1 + 1^2)^3} = \frac{3}{4}$$

These partial derivatives are the slopes of the trace curves through the point $\left(1, 3, \frac{9}{8}\right)$ shown in Figure 2. ■

We use the Chain Rule to compute partial derivatives of a composite function $f(x, y) = F(g(x, y))$, where $F(u)$ is a function of one variable and $u = g(x, y)$:

$$\frac{\partial f}{\partial x} = \frac{dF}{du}\frac{\partial u}{\partial x}, \qquad \frac{\partial f}{\partial y} = \frac{dF}{du}\frac{\partial u}{\partial y}$$

■ **EXAMPLE 3** **Chain Rule for Partial Derivatives** Compute $\dfrac{\partial}{\partial x}\sin(x^2 y^5)$.

Solution Write $\sin(x^2 y^5) = F(u)$, where $F(u) = \sin u$ and $u = x^2 y^5$. Then we have $\dfrac{dF}{du} = \cos u$ and the Chain Rule give us

$$\frac{\partial}{\partial x}\sin(x^2 y^5) = \underbrace{\frac{dF}{du}\frac{\partial u}{\partial x} = \cos(x^2 y^5)\frac{\partial}{\partial x}x^2 y^5 = 2xy^5\cos(x^2 y^5)}_{\text{Chain Rule}}$$

■

Partial derivatives are defined for functions of any number of variables. We compute the partial derivative with respect to any one of the variables by holding the remaining variables constant.

■ **EXAMPLE 4** **More Than Two Variables** Calculate $f_z(0, 0, 1, 1)$, where

$$f(x, y, z, w) = \frac{e^{xz+y}}{z^2 + w}$$

In Example 4, the calculation

$$\frac{\partial}{\partial z}e^{xz+y} = xe^{xz+y}$$

follows from the Chain Rule, just like

$$\frac{d}{dz}e^{az+b} = ae^{az+b}$$

Solution Use the Quotient Rule, treating x, y, and w as constants:

$$f_z(x, y, z, w) = \frac{\partial}{\partial z}\left(\frac{e^{xz+y}}{z^2 + w}\right) = \frac{(z^2 + w)\frac{\partial}{\partial z}e^{xz+y} - e^{xz+y}\frac{\partial}{\partial z}(z^2 + w)}{(z^2 + w)^2}$$

$$= \frac{(z^2 + w)xe^{xz+y} - 2ze^{xz+y}}{(z^2 + w)^2} = \frac{(z^2x + wx - 2z)e^{xz+y}}{(z^2 + w)^2}$$

$$f_z(0, 0, 1, 1) = \frac{-2e^0}{(1^2 + 1)^2} = -\frac{1}{2} \qquad ■$$

Because the partial derivative $f_x(a, b)$ is the derivative $f(x, b)$, viewed as a function of x alone, we can estimate the change Δf when x changes from a to $a + \Delta x$ as in the single-variable case. Similarly, we can estimate the change when y changes by Δy. For small Δx and Δy (just how small depends on f and the accuracy required):

$$f(a + \Delta x, b) - f(a, b) \approx f_x(a, b)\Delta x$$

$$f(a, b + \Delta y) - f(a, b) \approx f_y(a, b)\Delta y$$

This applies to functions f in any number of variables. For example, $\Delta f \approx f_w \Delta w$ if one of the variables w changes by Δw and all other variables remain fixed.

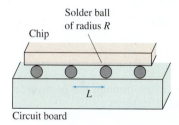

FIGURE 3 A BGA package. Temperature variations strain the BGA and may cause it to fail because the chip and board expand at different rates.

■ **EXAMPLE 5** **Testing Microchips** A **ball grid array** (BGA) is a microchip joined to a circuit board by small solder balls of radius R millimeters separated by a distance L millimeters (Figure 3). Manufacturers test the reliability of BGAs by subjecting them to repeated cycles in which the temperature is varied from 0°C to 100°C over a 40-minute period. According to one model, the average number N of cycles before the chip fails is

$$N = \left(\frac{2200R}{Ld}\right)^{1.9}$$

where d is the difference between the coefficients of expansion of the chip and the board. Estimate the change ΔN when $R = 0.12$, $d = 10$, and L is increased from 0.4 to 0.42.

Solution We use the approximation

$$\Delta N \approx \frac{\partial N}{\partial L}\Delta L$$

with $\Delta L = 0.42 - 0.4 = 0.02$. Since R and d are constant, the partial derivative is

$$\frac{\partial N}{\partial L} = \frac{\partial}{\partial L}\left(\frac{2200R}{Ld}\right)^{1.9} = \left(\frac{2200R}{d}\right)^{1.9}\frac{\partial}{\partial L}L^{-1.9} = -1.9\left(\frac{2200R}{d}\right)^{1.9}L^{-2.9}$$

Now evaluate at $L = 0.4$, $R = 0.12$, and $d = 10$:

$$\left.\frac{\partial N}{\partial L}\right|_{(L,R,d)=(0.4,0.12,10)} = -1.9\left(\frac{2200(0.12)}{10}\right)^{1.9}(0.4)^{-2.9} \approx -13{,}609$$

The decrease in the average number of cycles before a chip fails is

$$\Delta N \approx \frac{\partial N}{\partial L}\Delta L = -13{,}609(0.02) \approx -272 \text{ cycles} \qquad ■$$

In the next example, we estimate a partial derivative numerically. Since f_x and f_y are limits of difference quotients, we have the following approximations when h and k are "small":

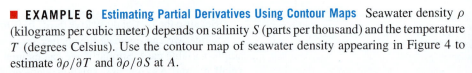

$$f_x(a,b) \approx \frac{\Delta f}{\Delta x} = \frac{f(a+h,b) - f(a,b)}{h}$$

$$f_y(a,b) \approx \frac{\Delta f}{\Delta y} = \frac{f(a,b+k) - f(a,b)}{k}$$

Similar approximations are valid in any number of variables.

■ **EXAMPLE 6** Estimating Partial Derivatives Using Contour Maps Seawater density ρ (kilograms per cubic meter) depends on salinity S (parts per thousand) and the temperature T (degrees Celsius). Use the contour map of seawater density appearing in Figure 4 to estimate $\partial\rho/\partial T$ and $\partial\rho/\partial S$ at A.

Solution Point A has coordinates $(S,T) = (33,15)$ and lies on the level curve $\rho = 1.0245$. We estimate $\partial\rho/\partial T$ at A in two steps.

Step 1. **Move vertically from A.**

Since T varies in the vertical direction, we move up vertically from point A to point B on the next higher-level curve, where $\rho = 1.0240$. Point B has coordinates $(S,T) = (33,17)$. Note that in moving from A to B, we have kept S constant because both points have salinity $S = 33$.

Step 2. **Compute the difference quotient.**

$$\Delta\rho = 1.0240 - 1.0245 = -0.0005 \text{ kg/m}^3$$

$$\Delta T = 17 - 15 = 2°C$$

This gives us the approximation

$$\left.\frac{\partial\rho}{\partial T}\right|_A \approx \frac{\Delta\rho}{\Delta T} = \frac{-0.0005}{2} = -0.00025 \text{ kg-m}^{-3}/°C$$

We estimate $\partial\rho/\partial S$ in a similar way, by moving to the right horizontally to point C with coordinates $(S,T) \approx (33.7,15)$, where $\rho = 1.0250$:

$$\left.\frac{\partial\rho}{\partial S}\right|_A \approx \frac{\Delta\rho}{\Delta S} = \frac{1.0250 - 1.0245}{33.7 - 33} = \frac{0.0005}{0.7} \approx 0.0007 \text{ kg-m}^{-3}/\text{ppt} \qquad ■$$

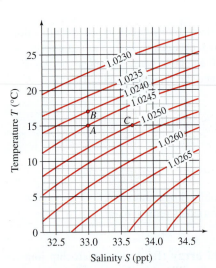

FIGURE 4 Contour map of seawater density as a function of temperature and salinity.

For greater accuracy, we can estimate $f_x(a,b)$ by taking the average of the difference quotients for Δx and $-\Delta x$. A similar remark applies to $f_y(a,b)$.

Higher Order Partial Derivatives

The higher order partial derivatives are the derivatives of derivatives. The *second-order* partial derivatives of f are the partial derivatives of f_x and f_y. We write f_{xx} for the x-derivative of f_x and f_{yy} for the y-derivative of f_y:

$$f_{xx} = \frac{\partial}{\partial x}\left(\frac{\partial f}{\partial x}\right), \qquad f_{yy} = \frac{\partial}{\partial y}\left(\frac{\partial f}{\partial y}\right)$$

We also have the *mixed partials*:

$$f_{xy} = \frac{\partial}{\partial y}\left(\frac{\partial f}{\partial x}\right), \qquad f_{yx} = \frac{\partial}{\partial x}\left(\frac{\partial f}{\partial y}\right)$$

The process can be continued. For example, f_{xyx} is the x-derivative of f_{xy}, and f_{xyy} is the y-derivative of f_{xy} (perform the differentiation in the order of the subscripts from left to right). The Leibniz notation for higher order partial derivatives is

$$f_{xx} = \frac{\partial^2 f}{\partial x^2}, \qquad f_{xy} = \frac{\partial^2 f}{\partial y \partial x}, \qquad f_{yx} = \frac{\partial^2 f}{\partial x \partial y}, \qquad f_{yy} = \frac{\partial^2 f}{\partial y^2}$$

Higher partial derivatives are defined for functions of three or more variables in a similar manner.

■ **EXAMPLE 7** Calculate the second-order partials of $f(x, y) = x^3 + y^2 e^x$.

Solution First, we compute the first-order partial derivatives:

$$f_x(x, y) = \frac{\partial}{\partial x}(x^3 + y^2 e^x) = 3x^2 + y^2 e^x, \qquad f_y(x, y) = \frac{\partial}{\partial y}(x^3 + y^2 e^x) = 2y e^x$$

Then we can compute the second-order derivatives:

$$f_{xx}(x, y) = \frac{\partial}{\partial x} f_x = \frac{\partial}{\partial x}(3x^2 + y^2 e^x) \qquad\qquad f_{yy}(x, y) = \frac{\partial}{\partial y} f_y = \frac{\partial}{\partial y} 2y e^x$$
$$= 6x + y^2 e^x, \qquad\qquad\qquad\qquad = 2e^x$$
$$f_{xy}(x, y) = \frac{\partial f_x}{\partial y} = \frac{\partial}{\partial y}(3x^2 + y^2 e^x) \qquad\qquad f_{yx}(x, y) = \frac{\partial f_y}{\partial x} = \frac{\partial}{\partial x} 2y e^x$$
$$= 2y e^x, \qquad\qquad\qquad\qquad\qquad = 2y e^x \qquad ■$$

■ **EXAMPLE 8** Calculate f_{xyy} for $f(x, y) = x^3 + y^2 e^x$.

Remember how the subscripts are used in partial derivatives. The notation f_{xyy} means "first differentiate with respect to x and then differentiate twice with respect to y."

Solution By the previous example, $f_{xy} = 2y e^x$. Therefore,

$$f_{xyy} = \frac{\partial}{\partial y} f_{xy} = \frac{\partial}{\partial y} 2y e^x = 2e^x \qquad ■$$

Observe in Example 7 that f_{xy} and f_{yx} are both equal to $2y e^x$. It is a pleasant circumstance that the equality $f_{xy} = f_{yx}$ holds in general, provided that the mixed partials are continuous. See Appendix D for a proof of the following theorem named for the French mathematician Alexis Clairaut (Figure 5).

The hypothesis of Clairaut's Theorem, that f_{xy} and f_{yx} are continuous, is almost always satisfied in practice, but see Exercise 84 for an example where the mixed partials are not equal.

THEOREM 1 Clairaut's Theorem: Equality of Mixed Partials If f_{xy} and f_{yx} are both continuous functions on a disk D, then $f_{xy}(a, b) = f_{yx}(a, b)$ for all $(a, b) \in D$. In other words,

$$\frac{\partial^2 f}{\partial x \partial y} = \frac{\partial^2 f}{\partial y \partial x}$$

FIGURE 5 Alexis Clairaut (1713–1765) was a brilliant French mathematician who presented his first paper to the Paris Academy of Sciences at the age of 13. In 1752 Clairaut won a prize for an essay on lunar motion that Euler praised (surely an exaggeration) as "the most important and profound discovery that has ever been made in mathematics." (© *SSPL/The Image Works*)

■ **EXAMPLE 9** Check that $\dfrac{\partial^2 W}{\partial U \partial T} = \dfrac{\partial^2 W}{\partial T \partial U}$ for $W = e^{U/T}$.

Solution We compute both derivatives and observe that they are equal:

$$\frac{\partial W}{\partial T} = e^{U/T} \frac{\partial}{\partial T}\left(\frac{U}{T}\right) = -U T^{-2} e^{U/T}, \qquad \frac{\partial W}{\partial U} = e^{U/T} \frac{\partial}{\partial U}\left(\frac{U}{T}\right) = T^{-1} e^{U/T}$$

$$\frac{\partial}{\partial U} \frac{\partial W}{\partial T} = -T^{-2} e^{U/T} - U T^{-3} e^{U/T}, \qquad \frac{\partial}{\partial T} \frac{\partial W}{\partial U} = -T^{-2} e^{U/T} - U T^{-3} e^{U/T} ■$$

Although Clairaut's Theorem is stated for f_{xy} and f_{yx}, it implies more generally that partial differentiation may be carried out in any order, provided that the derivatives in question are continuous (see Exercise 75). For example, we can compute f_{xyxy} by differentiating f twice with respect to x and twice with respect to y, in any order. Thus,

$$f_{xyxy} = f_{xxyy} = f_{yyxx} = f_{yxyx} = f_{xyyx} = f_{yxxy}$$

■ **EXAMPLE 10** **Choosing the Order Wisely** Calculate the partial derivative g_{zzwx}, where $g(x, y, z, w) = x^3 w^2 z^2 + \sin\left(\dfrac{xy}{z^2}\right)$.

Solution Let's take advantage of the fact that the derivatives may be calculated in any order. If we differentiate with respect to w first, the second term disappears because it does not depend on w:

$$g_w = \frac{\partial}{\partial w}\left(x^3 w^2 z^2 + \sin\left(\frac{xy}{z^2}\right)\right) = 2x^3 w z^2$$

Next, differentiate twice with respect to z and once with respect to x:

$$g_{wz} = \frac{\partial}{\partial z} 2x^3 w z^2 = 4x^3 w z$$

$$g_{wzz} = \frac{\partial}{\partial z} 4x^3 w z = 4x^3 w$$

$$g_{wzzx} = \frac{\partial}{\partial x} 4x^3 w = 12x^2 w$$

We conclude that $g_{zzwx} = g_{wzzx} = 12x^2 w$. ■

A **partial differential equation** (PDE) is a differential equation involving functions of several variables and their partial derivatives. A solution to a PDE is a function that satisfies the equation. The heat equation in the next example is a PDE that models temperature as heat spreads through an object. There are infinitely many solutions, but the particular function in the example describes temperature at times $t > 0$ along a metal rod when the center point is given a burst of heat at $t = 0$ (Figure 6).

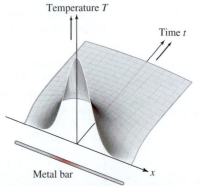

FIGURE 6 The plot of

$$u(x, t) = \frac{1}{2\sqrt{\pi t}} e^{-(x^2/4t)}$$

illustrates the diffusion of a burst of heat over time.

■ **EXAMPLE 11** **The Heat Equation** Show that $u(x, t) = \dfrac{1}{2\sqrt{\pi t}} e^{-(x^2/4t)}$, defined for $t > 0$, satisfies the heat equation

$$\frac{\partial u}{\partial t} = \frac{\partial^2 u}{\partial x^2} \qquad \boxed{2}$$

Solution We write $u(x, t) = \dfrac{1}{2\sqrt{\pi}} t^{-1/2} e^{-(x^2/4t)}$. We first compute $\dfrac{\partial^2 u}{\partial x^2}$:

$$\frac{\partial u}{\partial x} = \frac{\partial}{\partial x}\left(\frac{1}{2\sqrt{\pi}} t^{-1/2} e^{-(x^2/4t)}\right) = -\frac{1}{4\sqrt{\pi}} x t^{-3/2} e^{-(x^2/4t)}$$

$$\frac{\partial^2 u}{\partial x^2} = \frac{\partial}{\partial x}\left(-\frac{1}{4\sqrt{\pi}} x t^{-3/2} e^{-(x^2/4t)}\right) = -\frac{1}{4\sqrt{\pi}} t^{-3/2} e^{-(x^2/4t)} + \frac{1}{8\sqrt{\pi}} x^2 t^{-5/2} e^{-(x^2/4t)}$$

Then compute $\partial u/\partial t$ and observe that it equals $\partial^2 u/\partial x^2$ as required:

$$\frac{\partial u}{\partial t} = \frac{\partial}{\partial t}\left(\frac{1}{2\sqrt{\pi}} t^{-1/2} e^{-(x^2/4t)}\right) = -\frac{1}{4\sqrt{\pi}} t^{-3/2} e^{-(x^2/4t)} + \frac{1}{8\sqrt{\pi}} x^2 t^{-5/2} e^{-(x^2/4t)} \qquad ■$$

HISTORICAL PERSPECTIVE

Joseph Fourier
(1768–1830)
(Hulton Archive/Getty Images)

Adolf Fick
(1829–1901)
(Science Source)

The general heat equation, of which Eq. (2) is a special case, was first introduced in 1807 by French mathematician Jean Baptiste Joseph Fourier. As a young man, Fourier was unsure whether to enter the priesthood or pursue mathematics, but he must have been very ambitious. He wrote in a letter, "Yesterday was my 21st birthday, at that age Newton and Pascal had already acquired many claims to immortality." In his twenties, Fourier got involved in the French Revolution and was imprisoned briefly in 1794 over an incident involving different factions. In 1798 he was summoned, along with more than 150 other scientists, to join Napoleon on his unsuccessful campaign in Egypt.

Fourier's true impact, however, lay in his mathematical contributions. The heat equation is applied throughout the physical sciences and engineering, from the study of heat flow through the earth's oceans and atmosphere to the use of heat probes to destroy tumors and treat heart disease.

Fourier also introduced a striking new technique—known as the **Fourier transform**—for solving his equation, based on the idea that a periodic function can be expressed as a (possibly infinite) sum of sines and cosines. Leading mathematicians of the day, including Lagrange and Laplace, initially raised objections because this technique was not easy to justify rigorously. Nevertheless, the Fourier transform turned out to be one of the most important mathematical discoveries of the nineteenth century. A Web search on the term "Fourier transform" reveals its vast range of modern applications.

In 1855 the German physiologist Adolf Fick showed that the heat equation describes not only heat conduction but also a wide range of diffusion processes, such as osmosis, ion transport at the cellular level, and the motion of pollutants through air or water. The heat equation thus became a basic tool in chemistry, molecular biology, and environmental science, where it is often called **Fick's Second Law**.

14.3 SUMMARY

• The partial derivatives of $f(x, y)$ are defined as the limits

$$f_x(a, b) = \frac{\partial f}{\partial x}\bigg|_{(a,b)} = \lim_{h \to 0} \frac{f(a + h, b) - f(a, b)}{h}$$

$$f_y(a, b) = \frac{\partial f}{\partial y}\bigg|_{(a,b)} = \lim_{k \to 0} \frac{f(a, b + k) - f(a, b)}{k}$$

- Compute f_x by holding y constant, and compute f_y by holding x constant.
- $f_x(a, b)$ is the slope at $x = a$ of the tangent line to the trace curve $z = f(x, b)$. Similarly, $f_y(a, b)$ is the slope at $y = b$ of the tangent line to the trace curve $z = f(a, y)$.
- For small changes Δx and Δy,

$$f(a + \Delta x, b) - f(a, b) \approx f_x(a, b)\Delta x$$

$$f(a, b + \Delta y) - f(a, b) \approx f_y(a, b)\Delta y$$

More generally, if f is a function of n variables and w is one of the variables, then $\Delta f \approx f_w \Delta w$ if w changes by Δw and all other variables remain fixed.

- The second-order partial derivatives are

$$\frac{\partial^2}{\partial x^2} f = f_{xx}, \qquad \frac{\partial^2}{\partial y \, \partial x} f = f_{xy}, \qquad \frac{\partial^2}{\partial x \, \partial y} f = f_{yx}, \qquad \frac{\partial^2}{\partial y^2} f = f_{yy}$$

- Clairaut's Theorem states that mixed partials are equal—that is, $f_{xy} = f_{yx}$ provided that f_{xy} and f_{yx} are continuous.
- More generally, higher order partial derivatives may be computed in any order. For example, $f_{xyyz} = f_{yxzy}$ if f is a function of x, y, z whose fourth-order partial derivatives are continuous.

14.3 EXERCISES

Preliminary Questions

1. Patricia derived the following *incorrect* formula by misapplying the Product Rule:

$$\frac{\partial}{\partial x}(x^2 y^2) = x^2(2y) + y^2(2x)$$

What was her mistake and what is the correct calculation?

2. Explain why it is not necessary to use the Quotient Rule to compute $\dfrac{\partial}{\partial x}\left(\dfrac{x+y}{y+1}\right)$. Should the Quotient Rule be used to compute $\dfrac{\partial}{\partial y}\left(\dfrac{x+y}{y+1}\right)$?

3. Which of the following partial derivatives should be evaluated without using the Quotient Rule?

(a) $\dfrac{\partial}{\partial x}\dfrac{xy}{y^2+1}$ (b) $\dfrac{\partial}{\partial y}\dfrac{xy}{y^2+1}$ (c) $\dfrac{\partial}{\partial x}\dfrac{y^2}{y^2+1}$

4. What is f_x, where $f(x, y, z) = (\sin yz)e^{z^3-z^{-1}\sqrt{y}}$?

5. Assuming the hypotheses of Clairaut's Theorem are satisfied, which of the following partial derivatives are equal to f_{xxy}?

(a) f_{xyx} (b) f_{yyx} (c) f_{xyy} (d) f_{yxx}

Exercises

1. Use the limit definition of the partial derivative to verify the formulas

$$\frac{\partial}{\partial x}xy^2 = y^2, \qquad \frac{\partial}{\partial y}xy^2 = 2xy$$

2. Use the Product Rule to compute $\dfrac{\partial}{\partial y}(x^2 + y)(x + y^4)$.

3. Use the Quotient Rule to compute $\dfrac{\partial}{\partial y}\dfrac{y}{x+y}$.

4. Use the Chain Rule to compute $\dfrac{\partial}{\partial u}\ln(u^2 + uv)$.

5. Calculate $f_z(2, 3, 1)$, where $f(x, y, z) = xyz$.

6. ✏ Explain the relation between the following two formulas (c is a constant):

$$\frac{d}{dx}\sin(cx) = c\cos(cx), \qquad \frac{\partial}{\partial x}\sin(xy) = y\cos(xy)$$

7. The plane $y = 1$ intersects the surface $z = x^4 + 6xy - y^4$ in a certain curve. Find the slope of the tangent line to this curve at the point $P = (1, 1, 6)$.

8. Determine whether the partial derivatives $\partial f/\partial x$ and $\partial f/\partial y$ are positive or negative at the point P on the graph in Figure 7.

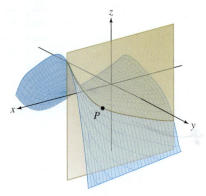

FIGURE 7

In Exercises 9–12, refer to Figure 8.

9. Estimate f_x and f_y at point A.

10. Is f_x positive or negative at B?

11. Starting at point B, in which compass direction (N, NE, SW, etc.) does f increase most rapidly?

12. At which of A, B, or C is f_y smallest?

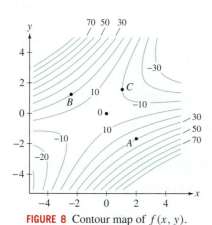

FIGURE 8 Contour map of $f(x, y)$.

In Exercises 13–40, compute the first-order partial derivatives.

13. $z = x^2 + y^2$

14. $z = x^4 y^3$

15. $z = x^4 y + xy^{-2}$

16. $V = \pi r^2 h$

17. $z = \dfrac{x}{y}$

18. $z = \dfrac{x}{x - y}$

19. $z = \sqrt{9 - x^2 - y^2}$

20. $z = \dfrac{x}{\sqrt{x^2 + y^2}}$

21. $z = (\sin x)(\sin y)$

22. $z = \sin(u^2 v)$

23. $z = \tan \dfrac{x}{y}$

24. $S = \tan^{-1}(wz)$

25. $z = \ln(x^2 + y^2)$

26. $A = \sin(4\theta - 9t)$

27. $W = e^{r+s}$

28. $Q = re^\theta$

29. $z = e^{xy}$

30. $R = e^{-v^2/k}$

31. $z = e^{-x^2 - y^2}$

32. $P = e^{\sqrt{y^2 + z^2}}$

33. $U = \dfrac{e^{-rt}}{r}$

34. $z = y^x$

35. $z = \sinh(x^2 y)$

36. $z = \cosh(t - \cos x)$

37. $w = xy^2 z^3$

38. $w = \dfrac{x}{y + z}$

39. $Q = \dfrac{L}{M} e^{-Lt/M}$

40. $w = \dfrac{x}{(x^2 + y^2 + z^2)^{3/2}}$

In Exercises 41–44, compute the given partial derivatives.

41. $f(x, y) = 3x^2 y + 4x^3 y^2 - 7xy^5, \quad f_x(1, 2)$

42. $f(x, y) = \sin(x^2 - y), \quad f_y(0, \pi)$

43. $g(u, v) = u \ln(u + v), \quad g_u(1, 2)$

44. $h(x, z) = e^{xz - x^2 z^3}, \quad h_z(3, 0)$

Exercises 45 and 46 refer to Example 5.

45. Calculate N for $L = 0.4$, $R = 0.12$, and $d = 10$, and use the linear approximation to estimate ΔN if d is increased from 10 to 10.4.

46. Estimate ΔN if $(L, R, d) = (0.5, 0.15, 8)$ and R is increased from 0.15 to 0.17.

47. The **heat index** I is a measure of how hot it feels when the relative humidity is H (as a percentage) and the actual air temperature is T (in degrees Fahrenheit). An approximate formula for the heat index that is valid for (T, H) near $(90, 40)$ is

$$I(T, H) = 45.33 + 0.6845T + 5.758H - 0.00365T^2$$
$$- 0.1565HT + 0.001HT^2$$

(a) Calculate I at $(T, H) = (95, 50)$.

(b) Which partial derivative tells us the increase in I per degree increase in T when $(T, H) = (95, 50)$? Calculate this partial derivative.

48. The **wind-chill temperature** W measures how cold people feel (based on the rate of heat loss from exposed skin) when the outside temperature is $T°C$ (with $T \le 10$) and wind velocity is v m/s (with $v \ge 2$):

$$W = 13.1267 + 0.6215T - 13.947v^{0.16} + 0.486Tv^{0.16}$$

Calculate $\partial W/\partial v$ at $(T, v) = (-10, 15)$ and use this value to estimate ΔW if $\Delta v = 2$.

49. The volume of a right-circular cone of radius r and height h is $V = \frac{\pi}{3}r^2 h$. Suppose that $r = h = 12$ cm. What leads to a greater increase in V, a 1-cm increase in r or a 1-cm increase in h? Argue using partial derivatives.

50. Use the linear approximation to estimate the percentage change in volume of a right-circular cone of radius $r = 40$ cm if the height is increased from 40 to 41 cm.

51. Calculate $\partial W/\partial E$ and $\partial W/\partial T$, where $W = e^{-E/kT}$, where k is a constant.

52. Calculate $\partial P/\partial T$ and $\partial P/\partial V$, where pressure P, volume V, and temperature T are related by the Ideal Gas Law, $PV = nRT$ (R and n are constants).

53. Use the contour map of $f(x, y)$ in Figure 9 to explain the following statements:

(a) f_y is larger at P than at Q, and f_x is smaller (more negative) at P than at Q.

(b) $f_x(x, y)$ is decreasing as a function of y; that is, for any fixed value $x = a$, $f_x(a, y)$ is decreasing in y.

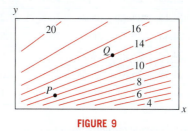

FIGURE 9

54. Estimate the partial derivatives at P of the function whose contour map is shown in Figure 10.

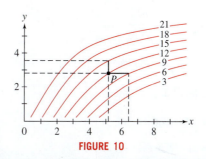

FIGURE 10

55. Over most of the earth, a magnetic compass does not point to true (geographic) north; instead, it points at some angle east or west of true north. The angle D between magnetic north and true north is called the **magnetic declination**. Use Figure 11 to determine which of the following statements is true:

(a) $\dfrac{\partial D}{\partial y}\bigg|_A > \dfrac{\partial D}{\partial y}\bigg|_B$ **(b)** $\dfrac{\partial D}{\partial x}\bigg|_C > 0$ **(c)** $\dfrac{\partial D}{\partial y}\bigg|_C > 0$

Note that the horizontal axis increases from right to left because of the way longitude is measured.

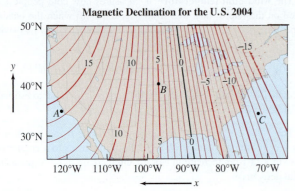

Magnetic Declination for the U.S. 2004

FIGURE 11 Contour interval 1°.

56. Refer to Table 1.

(a) Estimate $\partial\rho/\partial T$ and $\partial\rho/\partial S$ at the points $(S, T) = (34, 2)$ and $(35, 10)$ by computing the average of left-hand and right-hand difference quotients.

(b) For fixed salinity $S = 33$, is ρ concave up or concave down as a function of T? *Hint:* Determine whether the quotients $\Delta\rho/\Delta T$ are increasing or decreasing. What can you conclude about the sign of $\partial^2\rho/\partial T^2$?

TABLE 1   Seawater Density ρ as a Function of Temperature T and Salinity S

T \ S	30	31	32	33	34	35	36
12	22.75	23.51	24.27	25.07	25.82	26.6	27.36
10	23.07	23.85	24.62	25.42	26.17	26.99	27.73
8	23.36	24.15	24.93	25.73	26.5	27.28	29.09
6	23.62	24.44	25.22	26	26.77	27.55	28.35
4	23.85	24.62	25.42	26.23	27	27.8	28.61
2	24	24.78	25.61	26.38	27.18	28.01	28.78
0	24.11	24.92	25.72	26.5	27.34	28.12	28.91

In Exercises 57–62, compute the derivatives indicated.

57. $f(x, y) = 3x^2 y - 6xy^4$, $\dfrac{\partial^2 f}{\partial x^2}$ and $\dfrac{\partial^2 f}{\partial y^2}$

58. $g(x, y) = \dfrac{xy}{x - y}$, $\dfrac{\partial^2 g}{\partial x\, \partial y}$

59. $h(u, v) = \dfrac{u}{u + 4v}$, $h_{vv}(u, v)$

60. $h(x, y) = \ln(x^3 + y^3)$, $h_{xy}(x, y)$

61. $f(x, y) = x \ln(y^2)$, $f_{yy}(2, 3)$

62. $g(x, y) = xe^{-xy}$, $g_{xy}(-3, 2)$

63. Compute f_{xyxzy} for

$$f(x, y, z) = y\sin(xz)\sin(x + z) + (x + z^2)\tan y + x\tan\left(\frac{z + z^{-1}}{y - y^{-1}}\right)$$

Hint: Use a well-chosen order of differentiation on each term.

64. Let

$$f(x, y, u, v) = \frac{x^2 + e^y v}{3y^2 + \ln(2 + u^2)}$$

What is the fastest way to show that $f_{uvxyvu}(x, y, u, v) = 0$ for all (x, y, u, v)?

In Exercises 65–72, compute the derivative indicated.

65. $f(u, v) = \cos(u + v^2)$, f_{uuv}

66. $g(x, y, z) = x^4 y^5 z^6$, g_{xxyz}

67. $F(r, s, t) = r(s^2 + t^2)$, F_{rst}

68. $u(x, t) = t^{-1/2} e^{-(x^2/4t)}$, u_{xx}

69. $F(\theta, u, v) = \sinh(uv + \theta^2)$, $F_{uu\theta}$

70. $R(u, v, w) = \dfrac{u}{v + w}$, R_{uvw}

71. $g(x, y, z) = \sqrt{x^2 + y^2 + z^2}$, g_{xyz}

72. $u(x, t) = \operatorname{sech}^2(x - t)$, u_{xxx}

73. Find a function such that $\dfrac{\partial f}{\partial x} = 2xy$ and $\dfrac{\partial f}{\partial y} = x^2$.

74. Prove that there does not exist any function $f(x, y)$ such that $\dfrac{\partial f}{\partial x} = xy$ and $\dfrac{\partial f}{\partial y} = x^2$. *Hint:* Show that f cannot satisfy Clairaut's Theorem.

75. Assume that f_{xy} and f_{yx} are continuous and that f_{yxx} exists. Show that f_{xyx} also exists and that $f_{yxx} = f_{xyx}$.

76. Show that $u(x, t) = \sin(nx)\, e^{-n^2 t}$ satisfies the heat equation for any constant n:

$$\frac{\partial u}{\partial t} = \frac{\partial^2 u}{\partial x^2} \qquad \boxed{3}$$

77. Find all values of A and B such that $f(x, t) = e^{Ax + Bt}$ satisfies Eq. (3).

78. The function

$$f(x, t) = \frac{1}{2\sqrt{\pi t}} e^{-x^2/4t}$$

describes the temperature profile along a metal rod at time $t > 0$ when a burst of heat is applied at the origin (see Example 11). A small bug sitting on the rod at distance x from the origin feels the temperature rise and fall as heat diffuses through the bar. Show that the bug feels the maximum temperature at time $t = \frac{1}{2}x^2$.

*In Exercises 79–82, the **Laplace operator** Δ is defined by $\Delta f = f_{xx} + f_{yy}$. A function $u(x, y)$ satisfying the Laplace equation $\Delta u = 0$ is called **harmonic**.*

79. Show that the following functions are harmonic:

(a) $u(x, y) = x$

(b) $u(x, y) = e^x \cos y$

(c) $u(x, y) = \tan^{-1} \dfrac{y}{x}$

(d) $u(x, y) = \ln(x^2 + y^2)$

80. Find all harmonic polynomials $u(x, y)$ of degree 3, that is, $u(x, y) = ax^3 + bx^2 y + cxy^2 + dy^3$.

81. Show that if $u(x, y)$ is harmonic, then the partial derivatives $\partial u/\partial x$ and $\partial u/\partial y$ are harmonic.

82. Find all constants a, b such that $u(x, y) = \cos(ax)e^{by}$ is harmonic.

83. Show that $u(x, t) = \text{sech}^2(x - t)$ satisfies the **Korteweg–deVries equation** (which arises in the study of water waves):

$$4u_t + u_{xxx} + 12uu_x = 0$$

Further Insights and Challenges

84. Assumptions Matter This exercise shows that the hypotheses of Clairaut's Theorem are needed. Let

$$f(x, y) = xy \frac{x^2 - y^2}{x^2 + y^2}$$

for $(x, y) \neq (0, 0)$ and $f(0, 0) = 0$.

(a) Verify for $(x, y) \neq (0, 0)$:

$$f_x(x, y) = \frac{y(x^4 + 4x^2 y^2 - y^4)}{(x^2 + y^2)^2}$$

$$f_y(x, y) = \frac{x(x^4 - 4x^2 y^2 - y^4)}{(x^2 + y^2)^2}$$

(b) Use the limit definition of the partial derivative to show that $f_x(0, 0) = f_y(0, 0) = 0$ and that $f_{yx}(0, 0)$ and $f_{xy}(0, 0)$ both exist but are not equal.

(c) Show that for $(x, y) \neq (0, 0)$:

$$f_{xy}(x, y) = f_{yx}(x, y) = \frac{x^6 + 9x^4 y^2 - 9x^2 y^4 - y^6}{(x^2 + y^2)^3}$$

Show that f_{xy} is not continuous at $(0, 0)$. *Hint:* Show that $\lim_{h \to 0} f_{xy}(h, 0) \neq \lim_{h \to 0} f_{xy}(0, h)$.

(d) Explain why the result of part (b) does not contradict Clairaut's Theorem.

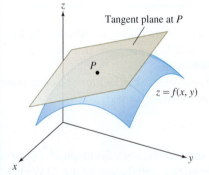

DF **FIGURE 1** Tangent plane to the graph of $z = f(x, y)$.

14.4 Differentiability and Tangent Planes

In this section, we generalize two basic concepts from single-variable calculus: differentiability and the tangent line. The tangent line becomes the *tangent plane* for functions of two variables (Figure 1).

Intuitively, we would like to say that a continuous function $f(x, y)$ is differentiable if it is **locally linear**—that is, if its graph looks flatter and flatter as we zoom in on a point $P = (a, b, f(a, b))$ and eventually becomes indistinguishable from the tangent plane (Figure 2).

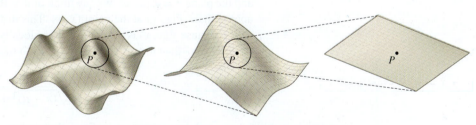

FIGURE 2 The graph looks flatter and flatter as we zoom in on a point P.

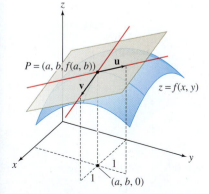

FIGURE 3 Finding the equation of the tangent plane.

Let us momentarily assume that our continuous function $f(x, y)$ is locally linear, and let's find the equation for the tangent plane at a point $P = (a, b, f(a, b))$ on its graph. We already have a point, namely P, on the plane. Hence, all we need is a normal vector to determine the plane. We currently know the slopes of two tangent lines at P, the first with slope $f_x(a, b)$ in the vertical plane given by $x = a$, and the second with slope $f_y(a, b)$ in the vertical plane given by $y = b$. We will convert this information into two vectors $\mathbf{u}$ and $\mathbf{v}$ and then take their cross product to obtain a normal vector to the plane.

Consider first the tangent line in the plane $x = a$ with slope $f_x(a, b)$. To stay on that line, we can move 1 unit in the positive x-direction from $(a, b, f(a, b))$, and then $f_x(a, b)$ units in the positive z-direction from the point $(a, b, f(a, b))$. In other words, the vector defined by $\mathbf{u} = \langle 1, 0, f_x(a, b) \rangle$ lies in that tangent line and therefore lies in the tangent plane, as in Figure 3.

Now consider the tangent line in the plane $y = b$, with slope $f_y(a, b)$. To stay on that line, we can move 1 unit in the positive y-direction from $(a, b, f(a, b))$, and then $f_y(a, b)$ units in the positive z-direction from the point $(a, b, f(a, b))$. That is to say the vector defined by $\mathbf{v} = \langle 0, 1, f_y(a, b) \rangle$ lies in that tangent line and therefore lies in the tangent plane, as in Figure 3.

Thus, we have two vectors that both lie in the tangent plane. They can never be parallel since one lies in the plane given by $x = a$ and the other lies in the plane given by $y = b$, and parallel vectors in these two planes would have to be vertical, which these are not. We can therefore take their cross product as a normal vector to the tangent plane:

$$\mathbf{n} = \mathbf{v} \times \mathbf{u} = \begin{vmatrix} \mathbf{i} & \mathbf{j} & \mathbf{k} \\ 0 & 1 & f_y(a, b) \\ 1 & 0 & f_x(a, b) \end{vmatrix} = \langle f_x(a, b), f_y(a, b), -1 \rangle$$

◄··· REMINDER *A plane through the point*
$P = (x_0, y_0, z_0)$ with normal vector
$\mathbf{n} = \langle A, B, C \rangle$ has equation
$A(x - x_0) + B(y - y_0) + C(z - z_0) =$
0.

Now, we have a normal vector for the tangent plane and a point P on the plane, so we can write down the equation of the plane:

$$f_x(a, b)(x - a) + f_y(a, b)(y - b) - (z - f(a, b)) = 0$$

This gives us the following:

THEOREM 1 Equation of the Tangent Plane   If $f(x, y)$ is locally linear at (a, b), then its tangent plane is given by the equation

$$z = f(a, b) + f_x(a, b)(x - a) + f_y(a, b)(y - b)$$

We define $L(x, y) = f(a, b) + f_x(a, b)(x - a) + f_y(a, b)(y - b)$, which is the linear function that yields the tangent plane. Note that it is linear since it is of the form $L(x, y) = Ax + By + C$, where A, B, and C are constants.

Before we can say that the tangent plane exists, however, we must impose a condition on $f(x, y)$ guaranteeing that the graph looks flat as we zoom in on P. Set

$$e(x, y) = f(x, y) - L(x, y).$$

As we see in Figure 4(B), $|e(x, y)|$ is the vertical distance between the graph of $f(x, y)$ and the plane $z = L(x, y)$. We can think of it as the error we produce if we use $L(x, y)$ to approximate $f(x, y)$ at the point (x, y). This distance tends to zero as (x, y) approaches (a, b) because $f(x, y)$ is continuous. To be locally linear, we require that this distance tend to zero *faster* than the distance from (x, y) to (a, b). We express this by the requirement

$$\lim_{(x,y) \to (a,b)} \frac{e(x, y)}{\sqrt{(x - a)^2 + (y - b)^2}} = 0$$

◄··· REMINDER

$$L(x, y) = f(a, b) + f_x(a, b)(x - a)$$
$$+ f_y(a, b)(y - b)$$

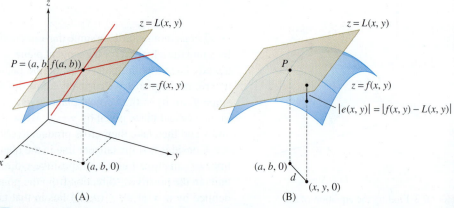

(A)   (B)

DF FIGURE 4

> **DEFINITION Differentiability**   Assume that $f(x, y)$ is defined in a disk D containing (a, b) and that $f_x(a, b)$ and $f_y(a, b)$ exist.
>
> - $f(x, y)$ is **differentiable** at (a, b) if it is **locally linear**—that is, if
>
> $$f(x, y) = L(x, y) + e(x, y) \qquad \boxed{1}$$
>
> where $e(x, y)$ satisfies
>
> $$\lim_{(x,y) \to (a,b)} \frac{e(x, y)}{\sqrt{(x - a)^2 + (y - b)^2}} = 0$$
>
> - In this case, the **tangent plane** to the graph at $(a, b, f(a, b))$ is the plane with equation $z = L(x, y)$. Explicitly,
>
> $$\boxed{z = f(a, b) + f_x(a, b)(x - a) + f_y(a, b)(y - b)} \qquad \boxed{2}$$

If $f(x, y)$ is differentiable at all points in a domain $\mathcal{D}$, we say that $f(x, y)$ is differentiable on $\mathcal{D}$.

It is cumbersome to check the local linearity condition directly (see Exercise 41), but fortunately, this is rarely necessary. The following theorem provides a criterion for differentiability that is easy to apply. It assures us that most functions arising in practice are differentiable on their domains. See Appendix D for a proof.

The definition of differentiability extends to functions of n-variables, and Theorem 2 holds in this setting: If all of the partial derivatives of $f(x_1, \ldots, x_n)$ exist and are continuous on an open domain $\mathcal{D}$, then $f(x_1, \ldots, x_n)$ is differentiable on $\mathcal{D}$.

> **THEOREM 2 Criterion for Differentiability**   If $f_x(x, y)$ and $f_y(x, y)$ exist and are continuous on an open disk D, then $f(x, y)$ is differentiable on D.

■ **EXAMPLE 1**   Show that $f(x, y) = 5x + 4y^2$ is differentiable (Figure 5). Find the equation of the tangent plane at $(a, b) = (2, 1)$.

Solution   The partial derivatives exist and are continuous functions:

$$f(x, y) = 5x + 4y^2, \qquad f_x(x, y) = 5, \qquad f_y(x, y) = 8y$$

Therefore, $f(x, y)$ is differentiable for all (x, y) by Theorem 2. To find the tangent plane, we evaluate the partial derivatives at $(2, 1)$:

$$f(2, 1) = 14, \qquad f_x(2, 1) = 5, \qquad f_y(2, 1) = 8$$

According to Eq. (2), the equation of the tangent plane at $(2, 1)$ is

$$z = \underbrace{14 + 5(x - 2) + 8(y - 1)}_{f(a,b) + f_x(a,b)(x-a) + f_y(a,b)(y-b)} = -4 + 5x + 8y$$

The tangent plane through $P = (2, 1, 14)$ has equation $z = -4 + 5x + 8y$.   ■

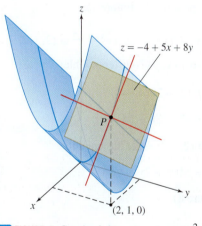

DF FIGURE 5   Graph of $f(x, y) = 5x + 4y^2$ and the tangent plane at $P = (2, 1, 14)$.

Local linearity is used in the next section to prove the Chain Rule for Paths, upon which the fundamental properties of the gradient are based.

Assumptions Matter   Local linearity plays a key role, and although most reasonable functions are locally linear, the mere existence of the partial derivatives does not guarantee local linearity. This is in contrast to the one-variable case, where $f(x)$ is automatically locally linear at $x = a$ if $f'(a)$ exists (Exercise 44).

The function $g(x, y)$ in Figure 6(A) shows what can go wrong. The graph contains the x- and y-axes—in other words, $g(x, 0) = 0$ and $g(0, y) = 0$—and therefore, the partial derivatives $g_x(0, 0)$ and $g_y(0, 0)$ are both zero. The tangent plane at the origin $(0, 0)$, if it existed, would have to be the xy-plane. However, Figure 6(B) shows that the graph also contains lines through the origin that do not lie in the xy-plane (in fact, the graph is composed entirely of lines through the origin). As we zoom in on the origin, these lines remain at an angle to the xy-plane, and the surface does not get any flatter. Thus, $g(x, y)$ cannot be locally linear at $(0, 0)$, and the tangent plane does not exist. In particular, $g(x, y)$ cannot satisfy the assumptions of Theorem 1, so the partial derivatives $g_x(x, y)$ and $g_y(x, y)$ cannot be continuous at the origin (see Exercise 45 for details).

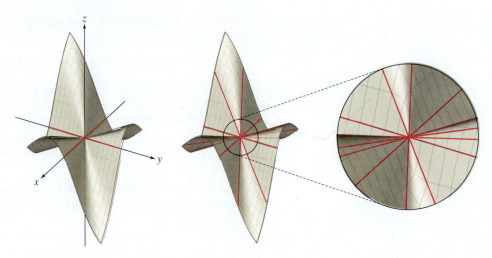

(A) The horizontal trace at $z = 0$ includes the x- and y-axes.

(B) But the graph also contains nonhorizontal lines through the origin.

(C) Thus, the graph does not appear any flatter as we zoom in on the origin.

DF **FIGURE 6** Graphs of $g(x, y) = \dfrac{2xy(x + y)}{x^2 + y^2}$.

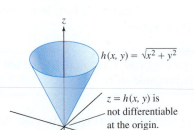

$h(x, y) = \sqrt{x^2 + y^2}$

$z = h(x, y)$ is not differentiable at the origin.

FIGURE 7 The function $h(x, y) = \sqrt{x^2 + y^2}$ is differentiable except at the origin.

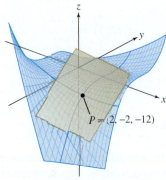

$P = (2, -2, -12)$

FIGURE 8 Tangent plane to the surface $f(x, y) = xy^3 + x^2$ passing through $P = (2, -2, -12)$.

■ EXAMPLE 2 Where is $h(x, y) = \sqrt{x^2 + y^2}$ differentiable?

Solution The partial derivatives exist and are continuous for all $(x, y) \neq (0, 0)$:

$$h_x(x, y) = \frac{x}{\sqrt{x^2 + y^2}}, \qquad h_y(x, y) = \frac{y}{\sqrt{x^2 + y^2}}$$

However, the partial derivatives do not exist at $(0, 0)$. Indeed, $h_x(0, 0)$ does not exist because $h(x, 0) = \sqrt{x^2} = |x|$ is not differentiable at $x = 0$. Similarly, $h_y(0, 0)$ does not exist. By Theorem 2, $h(x, y)$ is differentiable except at $(0, 0)$ (Figure 7). ■

■ EXAMPLE 3 Find a tangent plane of the graph of $f(x, y) = xy^3 + x^2$ at $(2, -2)$.

Solution The partial derivatives are continuous, so $f(x, y)$ is differentiable:

$$f_x(x, y) = y^3 + 2x, \qquad f_x(2, -2) = -4$$

$$f_y(x, y) = 3xy^2, \qquad f_y(2, -2) = 24$$

Since $f(2, -2) = -12$, the tangent plane through $(2, -2, -12)$ has equation

$$z = -12 - 4(x - 2) + 24(y + 2)$$

This can be rewritten as $z = 44 - 4x + 24y$ (Figure 8). ■

Linear Approximation and Differentials

By definition, if $f(x, y)$ is differentiable at (a, b), then it is locally linear and the **linear approximation** is

$$f(x, y) \approx f(a, b) + f_x(a, b)(x - a) + f_y(a, b)(y - b) \qquad \text{for } (x, y) \text{ near } (a, b)$$

Letting $x = a + \Delta x$ and $y = b + \Delta y$, and thinking of Δx and Δy as small changes in x and y, we obtain

$$\boxed{f(a + \Delta x, b + \Delta y) \approx f(a, b) + f_x(a, b)\Delta x + f_y(a, b)\Delta y} \qquad \boxed{3}$$

We can also write the linear approximation in terms of the *change in f*:

$$\Delta f = f(x, y) - f(a, b) \qquad \boxed{4}$$

$$\boxed{\Delta f \approx f_x(a, b)\Delta x + f_y(a, b)\Delta y} \qquad \boxed{5}$$

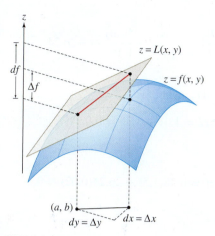

FIGURE 9 The quantity df is the change in height of the tangent plane.

Finally, it is convenient to define the function called the **differential** of f. Letting $dx = \Delta x$ and $dy = \Delta y$, we get

$$df = f_x(x, y)\,dx + f_y(x, y)\,dy = \frac{\partial f}{\partial x}dx + \frac{\partial f}{\partial y}dy$$

As shown in Figure 9, df represents the change in height of the tangent plane for given changes dx and dy in x and y (when we work with differentials, we call them dx and dy instead of Δx and Δy), whereas Δf is the change in the function itself. The linear approximation tells us that the two changes are approximately equal:

$$\Delta f \approx df$$

These approximations apply in any number of variables. In three variables,

$$f(a + \Delta x, b + \Delta y, c + \Delta z) \approx f(a, b, c) + f_x(a, b, c)\Delta x$$
$$+ f_y(a, b, c)\Delta y + f_z(a, b, c)\Delta z$$

■ **EXAMPLE 4** Use the linear approximation to estimate

$$(3.99)^3(1.01)^4(1.98)^{-1}$$

◀·· **REMINDER** The percentage error is equal to

$$\left|\frac{\text{Error}}{\text{Actual value}}\right| \times 100\%$$

Then use a calculator to find the percentage error.

Solution Think of $(3.99)^3(1.01)^4(1.98)^{-1}$ as a value of $f(x, y, z) = x^3 y^4 z^{-1}$:

$$f(3.99, 1.01, 1.98) = (3.99)^3(1.01)^4(1.98)^{-1}$$

Then it makes sense to use the linear approximation at $(a, b, c) = (4, 1, 2)$:

$$f(x, y, z) = x^3 y^4 z^{-1}, \qquad f(4, 1, 2) = (4^3)(1^4)(2^{-1}) = 32$$
$$f_x(x, y, z) = 3x^2 y^4 z^{-1}, \qquad f_x(4, 1, 2) = 24$$
$$f_y(x, y, z) = 4x^3 y^3 z^{-1}, \qquad f_y(4, 1, 2) = 128$$
$$f_z(x, y, z) = -x^3 y^4 z^{-2}, \qquad f_z(4, 1, 2) = -16$$

The linear approximation in three variables stated above, with $a = 4, b = 1, c = 2$, gives us

$$\underbrace{(4 + \Delta x)^3(1 + \Delta y)^4(2 + \Delta z)^{-1}}_{f(4+\Delta x, 1+\Delta y, 2+\Delta z)} \approx 32 + 24\Delta x + 128\Delta y - 16\Delta z$$

For $\Delta x = -0.01$, $\Delta y = 0.01$, and $\Delta z = -0.02$, we obtain the desired estimate

$$(3.99)^3(1.01)^4(1.98)^{-1} \approx 32 + 24(-0.01) + 128(0.01) - 16(-0.02) = 33.36$$

The calculator value is $(3.99)^3(1.01)^4(1.98)^{-1} \approx 33.38$, so the error in our estimate is less than 0.025. The percentage error is

$$\text{Percentage error} \approx \frac{|33.38 - 33.36|}{33.38} \times 100 \approx 0.075\% \qquad ■$$

BMI is one factor used to assess the risk of certain diseases such as diabetes and high blood pressure. The range $18.5 \leq I \leq 24.9$ is considered normal for adults over 20 years of age.

■ **EXAMPLE 5** **Body Mass Index** A person's BMI is $I = W/H^2$, where W is the body weight (in kilograms) and H is the body height (in meters). Estimate the change in a child's BMI if (W, H) changes from $(40, 1.45)$ to $(41.5, 1.47)$.

Solution

Step 1. **Compute the partial derivatives at $(W, H) = (40, 1.45)$.**

$$\frac{\partial I}{\partial W} = \frac{\partial}{\partial W}\left(\frac{W}{H^2}\right) = \frac{1}{H^2}, \qquad \frac{\partial I}{\partial H} = \frac{\partial}{\partial H}\left(\frac{W}{H^2}\right) = -\frac{2W}{H^3}$$

At $(W, H) = (40, 1.45)$, we have

$$\frac{\partial I}{\partial W}\bigg|_{(40,1.45)} = \frac{1}{1.45^2} \approx 0.48, \qquad \frac{\partial I}{\partial H}\bigg|_{(40,1.45)} = -\frac{2(40)}{1.45^3} \approx -26.24$$

Step 2. Estimate the change.

If (W, H) changes from $(40, 1.45)$ to $(41.5, 1.47)$, then

$$\Delta W = 41.5 - 40 = 1.5, \qquad \Delta H = 1.47 - 1.45 = 0.02$$

Therefore, by Eq. (5),

$$\Delta I \approx \frac{\partial I}{\partial W}\bigg|_{(40,1.45)} \Delta W + \frac{\partial I}{\partial H}\bigg|_{(40,1.45)} \Delta H = 0.48(1.5) - 26.24(0.02) \approx 0.2$$

We find that BMI increases by approximately 0.2. ∎

14.4 SUMMARY

- The *linearization* of f in two and three variables:

$$L(x, y) = f(a, b) + f_x(a, b)(x - a) + f_y(a, b)(y - b)$$

$$L(x, y, z) = f(a, b, c) + f_x(a, b, c)(x - a) + f_y(a, b, c)(y - b) + f_z(a, b, c)(z - c)$$

- $f(x, y)$ is *differentiable* at (a, b) if $f_x(a, b)$ and $f_y(a, b)$ exist and

$$f(x, y) = L(x, y) + e(x, y)$$

where $e(x, y)$ is a function such that

$$\lim_{(x,y)\to(a,b)} \frac{e(x, y)}{\sqrt{(x - a)^2 + (y - b)^2}} = 0$$

- Result used in practice: *If $f_x(x, y)$ and $f_y(x, y)$ exist and are continuous in a disk D containing (a, b), then $f(x, y)$ is differentiable at (a, b).*
- Equation of the tangent plane to $z = f(x, y)$ at (a, b):

$$z = f(a, b) + f_x(a, b)(x - a) + f_y(a, b)(y - b)$$

- Linear approximation:

$$f(a + \Delta x, b + \Delta y) \approx f(a, b) + f_x(a, b)\Delta x + f_y(a, b)\Delta y$$

$$\Delta f \approx f_x(a, b)\, \Delta x + f_y(a, b)\, \Delta y$$

- In differential form, $\Delta f \approx df$, where

$$df = f_x(x, y)\, dx + f_y(x, y)\, dy = \frac{\partial f}{\partial x}dx + \frac{\partial f}{\partial y}dy$$

$$df = f_x(x, y, z)\, dx + f_y(x, y, z)\, dy + f_z(x, y, z)\, dz = \frac{\partial f}{\partial x}dx + \frac{\partial f}{\partial y}dy + \frac{\partial f}{\partial z}dz$$

14.4 EXERCISES

Preliminary Questions

1. How is the linearization of $f(x, y)$ at (a, b) defined?

2. Define local linearity for functions of two variables.

In Exercises 3–5, assume that

$$f(2, 3) = 8, \qquad f_x(2, 3) = 5, \qquad f_y(2, 3) = 7$$

3. Which of (a)–(b) is the linearization of f at $(2, 3)$?

(a) $L(x, y) = 8 + 5x + 7y$

(b) $L(x, y) = 8 + 5(x - 2) + 7(y - 3)$

4. Estimate $f(2, 3.1)$.

5. Estimate Δf at $(2, 3)$ if $\Delta x = -0.3$ and $\Delta y = 0.2$.

6. Which theorem allows us to conclude that $f(x, y) = x^3 y^8$ is differentiable?

Exercises

1. Use Eq. (2) to find an equation of the tangent plane to the graph of $f(x, y) = 2x^2 - 4xy^2$ at $(-1, 2)$.

2. Find the equation of the plane in Figure 10, which is tangent to the graph at $(x, y) = (1, 0.8)$.

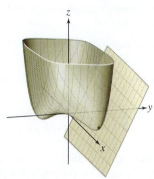

FIGURE 10 Graph of $f(x, y) = 0.2x^4 + y^6 - xy$.

In Exercises 3–10, find an equation of the tangent plane at the given point.

3. $f(x, y) = x^2 y + xy^3$, $(2, 1)$

4. $f(x, y) = \dfrac{x}{\sqrt{y}}$, $(4, 4)$

5. $f(x, y) = x^2 + y^{-2}$, $(4, 1)$

6. $G(u, w) = \sin(uw)$, $\left(\frac{\pi}{6}, 1\right)$

7. $F(r, s) = r^2 s^{-1/2} + s^{-3}$, $(2, 1)$

8. $g(x, y) = e^{x/y}$, $(2, 1)$

9. $f(x, y) = \operatorname{sech}(x - y)$, $(\ln 4, \ln 2)$

10. $f(x, y) = \ln(4x^2 - y^2)$, $(1, 1)$

11. Find the points on the graph of $z = 3x^2 - 4y^2$ at which the vector $\mathbf{n} = \langle 3, 2, 2 \rangle$ is normal to the tangent plane.

12. Find the points on the graph of $z = xy^3 + 8y^{-1}$ where the tangent plane is parallel to $2x + 7y + 2z = 0$.

13. Find the linearization $L(x, y)$ of $f(x, y) = x^2 y^3$ at $(a, b) = (2, 1)$. Use it to estimate $f(2.01, 1.02)$ and $f(1.97, 1.01)$, and compare with values obtained using a calculator.

14. Write the linear approximation to $f(x, y) = x(1 + y)^{-1}$ at $(a, b) = (8, 1)$ in the form

$$f(a + h, b + k) \approx f(a, b) + f_x(a, b)h + f_y(a, b)k$$

Use it to estimate $\dfrac{7.98}{2.02}$ and compare with the value obtained using a calculator.

15. Let $f(x, y) = x^3 y^{-4}$. Use Eq. (5) to estimate the change

$$\Delta f = f(2.03, 0.9) - f(2, 1)$$

16. Use the linear approximation to $f(x, y) = \sqrt{x/y}$ at $(9, 4)$ to estimate $\sqrt{9.1/3.9}$.

17. Use the linear approximation of $f(x, y) = e^{x^2 + y}$ at $(0, 0)$ to estimate $f(0.01, -0.02)$. Compare with the value obtained using a calculator.

18. Let $f(x, y) = x^2/(y^2 + 1)$. Use the linear approximation at an appropriate point (a, b) to estimate $f(4.01, 0.98)$.

19. Find the linearization of $f(x, y, z) = z\sqrt{x + y}$ at $(8, 4, 5)$.

20. Find the linearization to $f(x, y, z) = xy/z$ at the point $(2, 1, 2)$. Use it to estimate $f(2.05, 0.9, 2.01)$ and compare with the value obtained from a calculator.

21. Estimate $f(2.1, 3.8)$ assuming that

$$f(2, 4) = 5, \qquad f_x(2, 4) = 0.3, \qquad f_y(2, 4) = -0.2$$

22. Estimate $f(1.02, 0.01, -0.03)$ assuming that

$$f(1, 0, 0) = -3, \qquad f_x(1, 0, 0) = -2$$

$$f_y(1, 0, 0) = 4, \qquad f_z(1, 0, 0) = 2$$

In Exercises 23–28, use the linear approximation to estimate the value. Compare with the value given by a calculator.

23. $(2.01)^3 (1.02)^2$

24. $\dfrac{4.1}{7.9}$

25. $\sqrt{3.01^2 + 3.99^2}$

26. $\dfrac{0.98^2}{2.01^3 + 1}$

27. $\sqrt{(1.9)(2.02)(4.05)}$

28. $\dfrac{8.01}{\sqrt{(1.99)(2.01)}}$

29. Suppose that the plane tangent to $z = f(x, y)$ at $(-2, 3, 4)$ has equation $4x + 2y + z = 2$. Estimate $f(-2.1, 3.1)$.

30. In the derivation of the equation for the tangent plane that appears in Theorem 1, we chose $\mathbf{n} = \mathbf{v} \times \mathbf{u}$ for the normal vector to the plane. How would the choice of $\mathbf{n} = \mathbf{u} \times \mathbf{v}$ for the normal vector have affected the resultant equation?

In Exercises 31–34, let $I = W/H^2$ denote the BMI described in Example 5.

31. A boy has weight $W = 34$ kg and height $H = 1.3$ m. Use the linear approximation to estimate the change in I if (W, H) changes to $(36, 1.32)$.

32. Suppose that $(W, H) = (34, 1.3)$. Use the linear approximation to estimate the increase in H required to keep I constant if W increases to 35.

33. (a) Show that $\Delta I \approx 0$ if $\Delta H/\Delta W \approx H/2W$.

(b) Suppose that $(W, H) = (25, 1.1)$. What increase in H will leave I (approximately) constant if W is increased by 1 kg?

34. Estimate the change in height that will decrease I by 1 if $(W, H) = (25, 1.1)$, assuming that W remains constant.

35. A cylinder of radius r and height h has volume $V = \pi r^2 h$.

(a) Use the linear approximation to show that

$$\frac{\Delta V}{V} \approx \frac{2\Delta r}{r} + \frac{\Delta h}{h}$$

(b) Estimate the percentage increase in V if r and h are each increased by 2%.

(c) The volume of a certain cylinder V is determined by measuring r and h. Which will lead to a greater error in V: a 1% error in r or a 1% error in h?

36. Use the linear approximation to show that if $I = x^a y^b$, then

$$\frac{\Delta I}{I} \approx a\frac{\Delta x}{x} + b\frac{\Delta y}{y}$$

37. The monthly payment for a home loan is given by a function $f(P, r, N)$, where P is the principal (initial size of the loan), r the interest rate, and N the length of the loan in months. Interest rates are expressed as a decimal: A 6% interest rate is denoted by $r = 0.06$. If $P = \$100,000$, $r = 0.06$, and $N = 240$ (a 20-year loan), then the monthly payment is $f(100,000, 0.06, 240) = 716.43$. Furthermore, at these values, we have

$$\frac{\partial f}{\partial P} = 0.0071, \qquad \frac{\partial f}{\partial r} = 5769, \qquad \frac{\partial f}{\partial N} = -1.5467$$

Estimate:

(a) The change in monthly payment per $1000 increase in loan principal

(b) The change in monthly payment if the interest rate increases to $r = 6.5\%$ and $r = 7\%$

(c) The change in monthly payment if the length of the loan increases to 24 years

38. Automobile traffic passes a point P on a road of width w feet (ft) at an average rate of R vehicles per second. Although the arrival of automobiles is irregular, traffic engineers have found that the average waiting time T until there is a gap in traffic of at least t seconds is approximately $T = te^{Rt}$ seconds (s). A pedestrian walking at a speed of 3.5 ft/s (5.1 mph) requires $t = w/3.5$ s to cross the road. Therefore, the average time the pedestrian will have to wait before crossing is $f(w, R) = (w/3.5)e^{wR/3.5}$ s.

(a) What is the pedestrian's average waiting time if $w = 25$ ft and $R = 0.2$ vehicle per second?

(b) Use the linear approximation to estimate the increase in waiting time if w is increased to 27 ft.

(c) Estimate the waiting time if the width is increased to 27 ft and R decreases to 0.18.

(d) What is the rate of increase in waiting time per 1-ft increase in width when $w = 30$ ft and $R = 0.3$ vehicle per second?

39. The volume V of a right-circular cylinder is computed using the values 3.5 m for diameter and 6.2 m for height. Use the linear approximation to estimate the maximum error in V if each of these values has a possible error of at most 5%. Recall that $V = \pi r^2 h$.

Further Insights and Challenges

40. Show that if $f(x, y)$ is differentiable at (a, b), then the function of one variable $f(x, b)$ is differentiable at $x = a$. Use this to prove that $f(x, y) = \sqrt{x^2 + y^2}$ is *not* differentiable at $(0, 0)$.

41. This exercise shows directly (without using Theorem 2) that the function $f(x, y) = 5x + 4y^2$ from Example 1 is locally linear at $(a, b) = (2, 1)$.

(a) Show that $f(x, y) = L(x, y) + e(x, y)$ with $e(x, y) = 4(y - 1)^2$.

(b) Show that

$$0 \le \frac{e(x, y)}{\sqrt{(x - 2)^2 + (y - 1)^2}} \le 4|y - 1|$$

(c) Verify that $f(x, y)$ is locally linear.

42. Show directly, as in Exercise 41, that $f(x, y) = xy^2$ is differentiable at $(0, 2)$.

43. Differentiability Implies Continuity Use the definition of differentiability to prove that if f is differentiable at (a, b), then f is continuous at (a, b).

44. Let $f(x)$ be a function of one variable defined near $x = a$. Given a number M, set

$$L(x) = f(a) + M(x - a), \qquad e(x) = f(x) - L(x)$$

Thus, $f(x) = L(x) + e(x)$. We say that f is locally linear at $x = a$ if M can be chosen so that $\lim\limits_{x \to a} \dfrac{e(x)}{|x - a|} = 0$.

(a) Show that if $f(x)$ is differentiable at $x = a$, then $f(x)$ is locally linear with $M = f'(a)$.

(b) Show conversely that if f is locally linear at $x = a$, then $f(x)$ is differentiable and $M = f'(a)$.

45. Assumptions Matter Define $g(x, y) = 2xy(x + y)/(x^2 + y^2)$ for $(x, y) \ne 0$ and $g(0, 0) = 0$. In this exercise, we show that $g(x, y)$ is continuous at $(0, 0)$ and that $g_x(0, 0)$ and $g_y(0, 0)$ exist, but $g(x, y)$ is not differentiable at $(0, 0)$.

(a) Show using polar coordinates that $g(x, y)$ is continuous at $(0, 0)$.

(b) Use the limit definitions to show that $g_x(0, 0)$ and $g_y(0, 0)$ exist and that both are equal to zero.

(c) Show that the linearization of $g(x, y)$ at $(0, 0)$ is $L(x, y) = 0$.

(d) Show that if $g(x, y)$ were locally linear at $(0, 0)$, we would have $\lim\limits_{h \to 0} \dfrac{g(h, h)}{h} = 0$. Then observe that this is not the case because $g(h, h) = 2h$. This shows that $g(x, y)$ is not locally linear at $(0, 0)$ and hence not differentiable at $(0, 0)$.

14.5 The Gradient and Directional Derivatives

We have seen that the rate of change of a function f of several variables depends on a choice of direction. Since directions are indicated by vectors, it is natural to use vectors to describe the derivative of f in a specified direction.

To do this, we introduce the **gradient** ∇f_P, which is the vector whose components are the partial derivatives of f at P.

The gradient of a function of n variables is the vector

$$\nabla f = \left\langle \frac{\partial f}{\partial x_1}, \frac{\partial f}{\partial x_2}, \ldots, \frac{\partial f}{\partial x_n} \right\rangle$$

Initially, we can think of it as a convenient method for keeping track of the collection of first partial derivatives. We will soon see, though, it is much more than just that.

The symbol ∇, called "del," is an upside-down Greek delta. It was popularized by the Scottish physicist P. G. Tait (1831–1901), who called the symbol "nabla," because of its resemblance to an ancient Assyrian harp. The great physicist James Clerk Maxwell was reluctant to adopt this term and would refer to the gradient simply as the "slope." He wrote jokingly to his friend Tait in 1871, "Still harping on that nabla?"

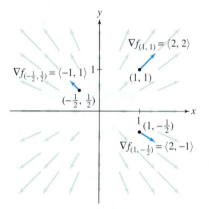

FIGURE 1 Gradient vectors of $f(x, y) = x^2 + y^2$ at several points (vectors not drawn to scale).

DEFINITION **The Gradient**   The gradient of a function $f(x, y)$ at a point $P = (a, b)$ is the vector

$$\nabla f_P = \langle f_x(a, b), f_y(a, b) \rangle$$

In three variables, if $P = (a, b, c)$,

$$\nabla f_P = \langle f_x(a, b, c), f_y(a, b, c), f_z(a, b, c) \rangle$$

We also write $\nabla f_{(a,b)}$ or $\nabla f(a, b)$ for the gradient. Sometimes, we omit reference to the point P and write

$$\nabla f = \left\langle \frac{\partial f}{\partial x}, \frac{\partial f}{\partial y} \right\rangle \qquad \text{or} \qquad \nabla f = \left\langle \frac{\partial f}{\partial x}, \frac{\partial f}{\partial y}, \frac{\partial f}{\partial z} \right\rangle$$

The gradient ∇f "assigns" a vector ∇f_P to each point in the domain of f, as in Figure 1.

■ **EXAMPLE 1**   **Drawing Gradient Vectors**   Let $f(x, y) = x^2 + y^2$. Calculate the gradient ∇f, draw several gradient vectors, and compute ∇f_P at $P = (1, 1)$.

Solution   The partial derivatives are $f_x(x, y) = 2x$ and $f_y(x, y) = 2y$, so

$$\nabla f = \langle 2x, 2y \rangle$$

The gradient attaches the vector $\langle 2x, 2y \rangle$ to the point (x, y). As we see in Figure 1, these vectors point away from the origin. At the particular point $(1, 1)$,

$$\nabla f_P = \nabla f(1, 1) = \langle 2, 2 \rangle \qquad\qquad ■$$

■ **EXAMPLE 2**   **Gradient in Three Variables**   Calculate $\nabla f_{(3,-2,4)}$, where

$$f(x, y, z) = ze^{2x+3y}$$

Solution   The partial derivatives and the gradient are

$$\frac{\partial f}{\partial x} = 2ze^{2x+3y}, \qquad \frac{\partial f}{\partial y} = 3ze^{2x+3y}, \qquad \frac{\partial f}{\partial z} = e^{2x+3y}$$

$$\nabla f = \langle 2ze^{2x+3y}, 3ze^{2x+3y}, e^{2x+3y} \rangle$$

Therefore, $\nabla f_{(3,-2,4)} = \langle 2 \cdot 4e^0, 3 \cdot 4e^0, e^0 \rangle = \langle 8, 12, 1 \rangle$. ■

The following theorem lists some useful properties of the gradient. The proofs are left as exercises (see Exercises 64–66).

THEOREM 1   **Properties of the Gradient**   If $f(x, y, z)$ and $g(x, y, z)$ are differentiable and c is a constant, then

(i) $\nabla(f + g) = \nabla f + \nabla g$

(ii) $\nabla(cf) = c\nabla f$

(iii) **Product Rule for Gradients:** $\nabla(fg) = f\nabla g + g\nabla f$

(iv) **Chain Rule for Gradients:** If $F(t)$ is a differentiable function of one variable, then

$$\nabla(F(f(x, y, z))) = F'(f(x, y, z))\nabla f \qquad\qquad \boxed{1}$$

Note that ∇f is behaving just as we would expect a derivative to behave.

■ **EXAMPLE 3** **Using the Chain Rule for Gradients** Find the gradient of

$$g(x, y, z) = (x^2 + y^2 + z^2)^8$$

Solution The function g is a composite $g(x, y, z) = F(f(x, y, z))$ with $F(t) = t^8$ and $f(x, y, z) = x^2 + y^2 + z^2$. Apply Eq. (1):

$$\nabla g = \nabla\big((x^2 + y^2 + z^2)^8\big) = 8(x^2 + y^2 + z^2)^7 \nabla(x^2 + y^2 + z^2)$$

$$= 8(x^2 + y^2 + z^2)^7 \langle 2x, 2y, 2z \rangle$$

$$= 16(x^2 + y^2 + z^2)^7 \langle x, y, z \rangle \qquad ■$$

The Chain Rule for Paths

Our first application of the gradient is the Chain Rule for Paths. In Chapter 13, we represented a path in $\mathbf{R}^3$ by a vector-valued function $\mathbf{r}(t) = \langle x(t), y(t), z(t) \rangle$. When there is no harm in doing so, we will sometimes think of $\mathbf{r}(t)$ as the path of endpoints of the vectors, rather than as the path of vectors themselves (Figure 2). Thus, we might write $\mathbf{r}(t) = (x(t), y(t), z(t))$ rather than $\mathbf{r}(t) = \langle x(t), y(t), z(t) \rangle$.

By definition, $\mathbf{r}'(t)$ is the vector of derivatives as before:

$$\boxed{\mathbf{r}(t) = (x(t), y(t), z(t)), \qquad \mathbf{r}'(t) = \langle x'(t), y'(t), z'(t) \rangle}$$

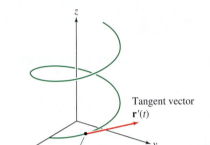

FIGURE 2 Tangent vector $\mathbf{r}'(t)$ to a path $\mathbf{r}(t) = \langle x(t), y(t), z(t) \rangle$.

Recall from Section 13.2 that $\mathbf{r}'(t)$ is the tangent or "velocity" vector that is tangent to the path and points in the direction of motion. We use similar notation for paths in $\mathbf{R}^2$.

The Chain Rule for Paths deals with composite functions of the type $f(\mathbf{r}(t))$. What is the idea behind a composite function of this type? As an example, suppose that $T(x, y)$ is the temperature at location (x, y) (Figure 3). Now imagine a biker—we'll call her Alexa—riding along a path $\mathbf{r}(t)$. We suppose that Alexa carries a thermometer with her and checks it as she rides. Her location at time t is $\mathbf{r}(t)$, so her temperature reading at time t is the composite function

$$T(\mathbf{r}(t)) = \text{Alexa's temperature at time } t$$

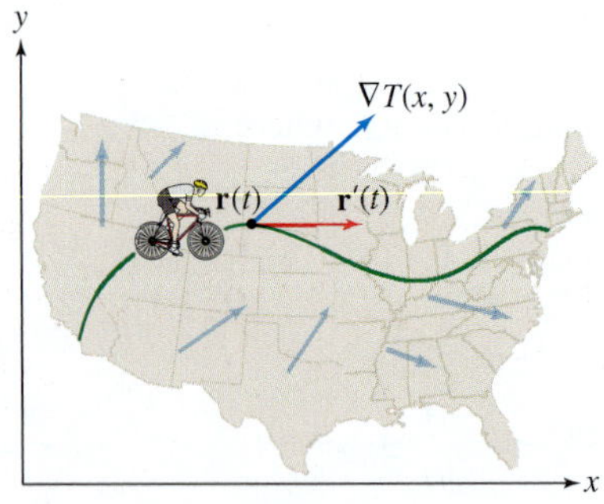

The temperature reading varies as Alexa's location changes, and the rate at which it changes is the derivative

$$\frac{d}{dt} T(\mathbf{r}(t))$$

FIGURE 3 Alexa's temperature changes at the rate $\nabla T_{\mathbf{r}(t)} \cdot \mathbf{r}'(t)$.

The Chain Rule for Paths tells us that this derivative is simply the dot product of the temperature gradient ∇T evaluated at $\mathbf{r}(t)$ and Alexa's velocity vector $\mathbf{r}'(t)$.

CAUTION Do not confuse the Chain Rule for Paths with the more elementary Chain Rule for Gradients stated in Theorem 1 above.

THEOREM 2 **Chain Rule for Paths** If f and $\mathbf{r}(t)$ are differentiable, then

$$\boxed{\frac{d}{dt} f(\mathbf{r}(t)) = \nabla f_{\mathbf{r}(t)} \cdot \mathbf{r}'(t)}$$

Explicitly, in the case of three variables, if $\mathbf{r}(t) = \langle x(t), y(t), z(t) \rangle$, then

$$\frac{d}{dt} f(\mathbf{r}(t)) = \left\langle \frac{\partial f}{\partial x}, \frac{\partial f}{\partial y}, \frac{\partial f}{\partial z} \right\rangle \cdot \langle x'(t), y'(t), z'(t) \rangle = \frac{\partial f}{\partial x}\frac{dx}{dt} + \frac{\partial f}{\partial y}\frac{dy}{dt} + \frac{\partial f}{\partial z}\frac{dz}{dt}$$

Proof By definition,

$$\frac{d}{dt} f(\mathbf{r}(t)) = \lim_{h \to 0} \frac{f(x(t+h), y(t+h)) - f(x(t), y(t))}{h}$$

To calculate this derivative, set

$$\Delta f = f(x(t+h), y(t+h)) - f(x(t), y(t))$$

$$\Delta x = x(t+h) - x(t), \qquad \Delta y = y(t+h) - y(t)$$

The proof is based on the local linearity of f. As in Section 14.4, we write

$$\Delta f = f_x(x(t), y(t))\Delta x + f_y(x(t), y(t))\Delta y + e(x(t+h), y(t+h))$$

Now set $h = \Delta t$ and divide by Δt:

$$\frac{\Delta f}{\Delta t} = f_x(x(t), y(t))\frac{\Delta x}{\Delta t} + f_y(x(t), y(t))\frac{\Delta y}{\Delta t} + \frac{e(x(t+\Delta t), y(t+\Delta t))}{\Delta t}$$

Suppose for a moment that the last term tends to zero as $\Delta t \to 0$. Then we obtain the desired result:

$$\frac{d}{dt}f(\mathbf{r}(t)) = \lim_{\Delta t \to 0}\frac{\Delta f}{\Delta t}$$

$$= f_x(x(t), y(t))\lim_{\Delta t \to 0}\frac{\Delta x}{\Delta t} + f_y(x(t), y(t))\lim_{\Delta t \to 0}\frac{\Delta y}{\Delta t}$$

$$= f_x(x(t), y(t))\frac{dx}{dt} + f_y(x(t), y(t))\frac{dy}{dt}$$

$$= \nabla f_{\mathbf{r}(t)} \cdot \mathbf{r}'(t)$$

We verify that the last term tends to zero as follows:

$$\lim_{\Delta t \to 0}\frac{e(x(t+\Delta t), y(t+\Delta t))}{\Delta t} = \lim_{\Delta t \to 0}\frac{e(x(t+\Delta t), y(t+\Delta t))}{\sqrt{(\Delta x)^2 + (\Delta y)^2}}\left(\frac{\sqrt{(\Delta x)^2 + (\Delta y)^2}}{\Delta t}\right)$$

$$= \underbrace{\left(\lim_{\Delta t \to 0}\frac{e(x(t+\Delta t), y(t+\Delta t))}{\sqrt{(\Delta x)^2 + (\Delta y)^2}}\right)}_{\text{Zero}}\lim_{\Delta t \to 0}\left(\sqrt{\left(\frac{\Delta x}{\Delta t}\right)^2 + \left(\frac{\Delta y}{\Delta t}\right)^2}\right) = 0$$

The first limit is zero because a differentiable function is locally linear (Section 14.4). The second limit is equal to $\sqrt{x'(t)^2 + y'(t)^2}$, so the product is zero. ■

■ **EXAMPLE 4** The temperature at location (x, y) is $T(x, y) = 20 + 10e^{-0.3(x^2+y^2)}$ °C. A bug carries a tiny thermometer along the path

$$\mathbf{r}(t) = \langle \cos(t - 2), \sin 2t \rangle$$

(t in seconds) as in Figure 4. How fast is the temperature changing at $t = 0.6$ s?

Solution At $t = 0.6$ s, the bug is at location

$$\mathbf{r}(0.6) = \langle \cos(-1.4), \sin 0.6 \rangle \approx \langle 0.170, 0.932 \rangle$$

By the Chain Rule for Paths, the rate of change of temperature is the dot product

$$\left.\frac{dT}{dt}\right|_{t=0.6} = \nabla T_{\mathbf{r}(0.6)} \cdot \mathbf{r}'(0.6)$$

We compute the vectors

$$\nabla T = \left\langle -6xe^{-0.3(x^2+y^2)}, -6ye^{-0.3(x^2+y^2)} \right\rangle$$

$$\mathbf{r}'(t) = \langle -\sin(t - 2), 2\cos 2t \rangle$$

and evaluate at $\mathbf{r}(0.6) = \langle 0.170, 0.932 \rangle$ using a calculator:

$$\nabla T_{\mathbf{r}(0.6)} \approx \langle -0.779, -4.272 \rangle$$

$$\mathbf{r}'(0.6) \approx \langle 0.985, 0.725 \rangle$$

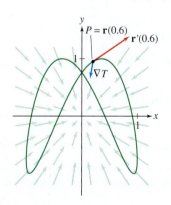

FIGURE 4 Gradient vectors ∇T and the path $\mathbf{r}(t) = \langle \cos(t - 2), \sin 2t \rangle$.

Therefore, the rate of change is

$$\frac{dT}{dt}\bigg|_{t=0.6} \nabla T_{\mathbf{r}(0.6)} \cdot \mathbf{r}'(t) \approx \langle -0.779, -4.272 \rangle \cdot \langle 0.985, 0.725 \rangle \approx -3.87°\text{C/s} \quad \blacksquare$$

In the next example, we apply the Chain Rule for Paths to a function of three variables. In general, if $f(x_1, \ldots, x_n)$ is a differentiable function of n variables and $\mathbf{r}(t) = \langle x_1(t), \ldots, x_n(t) \rangle$ is a differentiable path, then
indexpath!Chain Rule for

$$\frac{d}{dt} f(\mathbf{r}(t)) = \nabla f \cdot \mathbf{r}'(t) = \frac{\partial f}{\partial x_1} \frac{dx_1}{dt} + \frac{\partial f}{\partial x_2} \frac{dx_2}{dt} + \cdots + \frac{\partial f}{\partial x_n} \frac{dx_n}{dt}$$

■ **EXAMPLE 5** Calculate $\dfrac{d}{dt} f(\mathbf{r}(t))\bigg|_{t=\pi/2}$, where

$$f(x, y, z) = xy + z^2 \qquad \text{and} \qquad \mathbf{r}(t) = \langle \cos t, \sin t, t \rangle$$

Solution We have $\mathbf{r}\left(\frac{\pi}{2}\right) = \langle \cos \frac{\pi}{2}, \sin \frac{\pi}{2}, \frac{\pi}{2} \rangle = \left(0, 1, \frac{\pi}{2}\right)$. Compute the gradient:

$$\nabla f = \left(\frac{\partial f}{\partial x}, \frac{\partial f}{\partial y}, \frac{\partial f}{\partial z} \right) = \langle y, x, 2z \rangle, \qquad \nabla f_{\mathbf{r}(\pi/2)} = \nabla f\left(0, 1, \frac{\pi}{2}\right) = \langle 1, 0, \pi \rangle$$

Then compute the tangent vector:

$$\mathbf{r}'(t) = \langle -\sin t, \cos t, 1 \rangle, \qquad \mathbf{r}'\left(\frac{\pi}{2}\right) = \langle -\sin \frac{\pi}{2}, \cos \frac{\pi}{2}, 1 \rangle = \langle -1, 0, 1 \rangle$$

By the Chain Rule,

$$\frac{d}{dt} f(\mathbf{r}(t))\bigg|_{t=\pi/2} = \nabla f_{\mathbf{r}(\pi/2)} \cdot \mathbf{r}'\left(\frac{\pi}{2}\right) = \langle 1, 0, \pi \rangle \cdot \langle -1, 0, 1 \rangle = \pi - 1 \quad \blacksquare$$

Directional Derivatives

We come now to one of the most important applications of the Chain Rule for Paths. Consider a line through a point $P = (a, b)$ in the direction of a unit vector $\mathbf{u} = \langle h, k \rangle$ (see Figure 5):

$$\mathbf{r}(t) = \langle a + th, b + tk \rangle$$

The derivative of $f(\mathbf{r}(t))$ at $t = 0$ is called the **directional derivative of f with respect to u at P**, and is denoted $D_{\mathbf{u}} f(P)$ or $D_{\mathbf{u}} f(a, b)$:

$$D_{\mathbf{u}} f(a, b) = \frac{d}{dt} f(\mathbf{r}(t))\bigg|_{t=0} = \lim_{t \to 0} \frac{f(a + th, b + tk) - f(a, b)}{t}$$

Directional derivatives of functions of three or more variables are defined in a similar way.

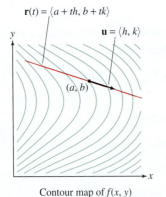

$\mathbf{r}(t) = \langle a + th, b + tk \rangle$

$\mathbf{u} = \langle h, k \rangle$

(a, b)

Contour map of $f(x, y)$

FIGURE 5 The directional derivative $D_{\mathbf{u}} f(a, b)$ is the rate of change of f along the linear path through P with direction vector **u**.

DEFINITION   Directional Derivative   The directional derivative in the direction of a unit vector $\mathbf{u} = \langle h, k \rangle$ is the limit (assuming it exists)

$$D_{\mathbf{u}} f(P) = D_{\mathbf{u}} f(a, b) = \lim_{t \to 0} \frac{f(a + th, b + tk) - f(a, b)}{t}$$

Note that the partial derivatives are the directional derivatives with respect to the standard unit vectors $\mathbf{i} = \langle 1, 0 \rangle$ and $\mathbf{j} = \langle 0, 1 \rangle$. For example,

$$D_{\mathbf{i}} f(a, b) = \lim_{t \to 0} \frac{f(a + t(1), b + t(0)) - f(a, b)}{t} = \lim_{t \to 0} \frac{f(a + t, b) - f(a, b)}{t}$$

$$= f_x(a, b)$$

Thus, we have

$$f_x(a, b) = D_{\mathbf{i}} f(a, b), \qquad f_y(a, b) = D_{\mathbf{j}} f(a, b)$$

CONCEPTUAL INSIGHT The directional derivative $D_{\mathbf{u}} f(P)$ is the rate of change of f per *unit change* in the horizontal direction of $\mathbf{u}$ at P (Figure 6). This is the slope of the tangent line at Q to the trace curve obtained when we intersect the graph with the vertical plane through P in the direction $\mathbf{u}$.

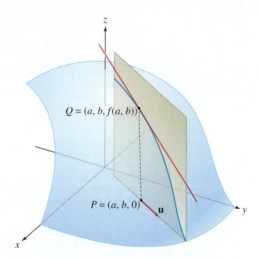

FIGURE 6 $D_{\mathbf{u}} f(a, b)$ is the slope of the tangent line to the trace curve through Q in the vertical plane through P in the direction $\mathbf{u}$.

To compute the directional derivative, we set $\mathbf{r}(t) = (a + th, b + tk)$. Then $D_{\mathbf{u}} f(a, b)$ is the derivative at $t = 0$ of the composite function $f(\mathbf{r}(t))$, and we can evaluate it using the Chain Rule for Paths. We have $\mathbf{r}'(t) = \langle h, k \rangle = \mathbf{u}$, so

$$D_{\mathbf{u}} f(a, b) = \nabla f_{(a,b)} \cdot \mathbf{r}'(0) = \nabla f_{(a,b)} \cdot \mathbf{u}$$

This yields the basic formula

$$D_{\mathbf{u}} f(a, b) = \nabla f_{(a,b)} \cdot \mathbf{u} \qquad \boxed{2}$$

Similarly, in three variables, $D_{\mathbf{u}} f(a, b, c) = \nabla f_{(a,b,c)} \cdot \mathbf{u}$.

THEOREM 3 Computing the Directional Derivative If $\mathbf{u}$ is a unit vector, then the directional derivative in the direction of $\mathbf{u}$ is given by

$$\boxed{D_{\mathbf{u}} f(P) = \nabla f_P \cdot \mathbf{u}} \qquad \boxed{3}$$

■ **EXAMPLE 6** Let $f(x, y) = xe^y$, $P = (2, -1)$, and $\mathbf{v} = \langle 2, 3 \rangle$. Calculate the directional derivative in the direction of $\mathbf{v}$.

Solution First note that $\mathbf{v}$ is NOT a unit vector. So, we first replace it with the vector

$$\mathbf{u} = \frac{\mathbf{v}}{\|\mathbf{v}\|} = \frac{\langle 2, 3 \rangle}{\sqrt{13}} = \left\langle \frac{2}{\sqrt{13}}, \frac{3}{\sqrt{13}} \right\rangle$$

Then compute the gradient at $P = (2, -1)$:

$$\nabla f = \left\langle \frac{\partial f}{\partial x}, \frac{\partial f}{\partial y} \right\rangle = \langle e^y, xe^y \rangle \quad \Rightarrow \quad \nabla f_P = \nabla f_{(2,-1)} = \left\langle e^{-1}, 2e^{-1} \right\rangle$$

Next use Theorem 3:

$$D_{\mathbf{u}} f(P) = \nabla f_P \cdot \mathbf{u} = \left\langle e^{-1}, 2e^{-1} \right\rangle \cdot \left\langle \frac{2}{\sqrt{13}}, \frac{3}{\sqrt{13}} \right\rangle = \frac{8e^{-1}}{\sqrt{13}} \approx 0.82. \qquad ■$$

This means that if we think of this function as representing a mountain, then at coordinate $(x, y) = (2, -1)$, we should expect that if we head 1 unit in the direction of vector $\mathbf{v}$, we would have to step up in the vertical direction by approximately 0.82 units.

■ **EXAMPLE 7** Find the rate of change of pressure at the point $Q = (1, 2, 1)$ in the direction of $\mathbf{v} = \langle 0, 1, 1 \rangle$, assuming that the pressure (in millibars) is given by

$$f(x, y, z) = 1000 + 0.01(yz^2 + x^2z - xy^2) \qquad (x, y, z \text{ in kilometers})$$

Solution First compute the gradient at $Q = (1, 2, 1)$:

$$\nabla f = 0.01 \left\langle 2xz - y^2, z^2 - 2xy, 2yz + x^2 \right\rangle$$

$$\nabla f_Q = \nabla f_{(1,2,1)} = \langle -0.02, -0.03, 0.05 \rangle$$

Then we replace $\mathbf{v}$ with a unit vector $\mathbf{u}$ in the same direction:

$$\mathbf{u} = \frac{\mathbf{v}}{\|\mathbf{v}\|} = \left\langle 0, \frac{1}{\sqrt{2}}, \frac{1}{\sqrt{2}} \right\rangle$$

Next

$$D_{\mathbf{u}}f(Q) = \nabla f_Q \cdot \mathbf{u} = \langle -0.02, -0.03, 0.05 \rangle \cdot \left\langle 0, \frac{1}{\sqrt{2}}, \frac{1}{\sqrt{2}} \right\rangle \approx 0.014 \text{ millibars/km}$$

Thus, we expect that as we move in the direction of $\mathbf{v}$ from Q, the pressure should increase by about 0.014 millibars/km. ■

Properties of the Gradient

We are now in a position to draw some interesting and important conclusions about the gradient. First, suppose that $\nabla f_P \neq \mathbf{0}$ and let $\mathbf{u}$ be a unit vector (Figure 7). By the properties of the dot product,

$$D_{\mathbf{u}}f(P) = \nabla f_P \cdot \mathbf{u} = \|\nabla f_P\| \cos \theta \qquad \boxed{4}$$

where θ is the angle between ∇f_P and $\mathbf{u}$. In other words, *the rate of change in a given direction varies with the cosine of the angle θ between the gradient and the direction.*

Because the cosine takes values between -1 and 1, we have

$$-\|\nabla f_P\| \leq D_{\mathbf{u}}f(P) \leq \|\nabla f_P\|$$

Since $\cos 0 = 1$, the maximum value of $D_{\mathbf{u}}f(P)$ occurs for $\theta = 0$—that is, when $\mathbf{u}$ points in the direction of ∇f_P. In other words the *gradient vector points in the direction of the maximum rate of increase, and this maximum rate is $\|\nabla f_P\|$.* Similarly, f decreases most rapidly in the opposite direction, $-\nabla f_P$, because $\cos \theta = -1$ for $\theta = \pi$. The rate of maximum decrease is $-\|\nabla f_P\|$. The directional derivative is zero in directions orthogonal to the gradient because $\cos \frac{\pi}{2} = 0$.

In the earlier scenario where the biker Alexa rides along a path (Figure 8), the temperature T changes at a rate that depends on the cosine of the angle θ between ∇T and the direction of motion.

*◀··· **REMINDER** For any vectors $\mathbf{u}$ and $\mathbf{v}$,*

$$\mathbf{v} \cdot \mathbf{u} = \|\mathbf{v}\| \|\mathbf{u}\| \cos \theta$$

where θ is the angle between $\mathbf{v}$ and $\mathbf{u}$. If $\mathbf{u}$ is a unit vector, then

$$\mathbf{v} \cdot \mathbf{u} = \|\mathbf{v}\| \cos \theta$$

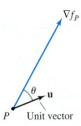

FIGURE 7 $D_{\mathbf{u}}f(P) = \|\nabla f_P\| \cos \theta$.

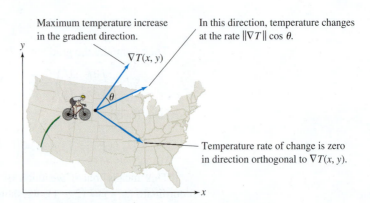

Maximum temperature increase in the gradient direction.

In this direction, temperature changes at the rate $\|\nabla T\| \cos \theta$.

$\nabla T(x, y)$

Temperature rate of change is zero in direction orthogonal to $\nabla T(x, y)$.

FIGURE 8

◄·· REMINDER

- *The words "normal" and "orthogonal" both mean "perpendicular."*
- *We say that a vector is normal to a curve at a point P if it is normal to the tangent line to the curve at P.*

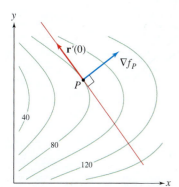

FIGURE 9 Contour map of $f(x, y)$. The gradient at P is orthogonal to the level curve through P and points in the direction of maximum increase of $f(x, y)$.

Another key property is that gradient vectors are normal to level curves (Figure 9). To prove this, suppose that P lies on the level curve $f(x, y) = k$. We parametrize this level curve by a path $\mathbf{r}(t)$ such that $\mathbf{r}(0) = P$ and $\mathbf{r}'(0) \neq \mathbf{0}$ (this is possible whenever $\nabla f_P \neq \mathbf{0}$). Then $f(\mathbf{r}(t)) = k$ for all t, so by the Chain Rule,

$$\nabla f_P \cdot \mathbf{r}'(0) = \frac{d}{dt} f(\mathbf{r}(t)) \Big|_{t=0} = \frac{d}{dt} k = 0$$

This proves that ∇f_P is orthogonal to $\mathbf{r}'(0)$, and since $\mathbf{r}'(0)$ is tangent to the level curve, we conclude that ∇f_P is normal to the level curve (Figure 9). We encapsulate these remarks in the following theorem.

THEOREM 4 Interpretation of the Gradient Assume that $\nabla f_P \neq \mathbf{0}$. Let $\mathbf{u}$ be a unit vector making an angle θ with ∇f_P. Then

$$\boxed{D_{\mathbf{u}} f(P) = \|\nabla f_P\| \cos \theta} \qquad 5$$

- ∇f_P points in the direction of maximum rate of increase of f at P.
- $-\nabla f_P$ points in the direction of maximum rate of decrease at P.
- ∇f_P is normal to the level curve (or surface) of f at P.
- $\|\nabla f_P\|$ yields the maximum slope of a tangent line to the surface given by $z = f(x, y)$ at $(P, f(P))$.

GRAPHICAL INSIGHT At each point P, there is a unique direction in which $f(x, y)$ increases most rapidly (per unit distance). Theorem 4 tells us that this direction of fastest increase is perpendicular to the level curves and that it is specified by the gradient vector (Figure 9). For most functions, however, the direction of maximum rate of increase varies from point to point.

■ **EXAMPLE 8** Let $f(x, y) = x^4 y^{-2}$ and $P = (2, 1)$. Find the unit vector that points in the direction of maximum rate of increase at P and determine that maximum rate.

Solution The gradient points in the direction of maximum rate of increase, so we evaluate the gradient at P:

$$\nabla f = \left\langle 4x^3 y^{-2}, -2x^4 y^{-3} \right\rangle, \qquad \nabla f_{(2,1)} = \langle 32, -32 \rangle$$

The unit vector in this direction is

$$\mathbf{u} = \frac{\langle 32, -32 \rangle}{\|\langle 32, -32 \rangle\|} = \frac{\langle 32, -32 \rangle}{32\sqrt{2}} = \left\langle \frac{\sqrt{2}}{2}, -\frac{\sqrt{2}}{2} \right\rangle$$

The maximum rate, which is the rate in this direction, is given by

$$\|\nabla f_{(2,1)}\| = \sqrt{32^2 + (-32)^2} = 32\sqrt{2} \qquad ■$$

■ **EXAMPLE 9** Draw the contour map and gradient vectors at various points in the plane corresponding to the function $f(x, y) = y^2 - x^2$.

Solution The equation $z = y^2 - x^2$ yields a saddle, or, as we have called it, a hyperbolic paraboloid. We can set z equal to various values, and determine the level curves and contour map as appears in Figure 10. At each point on any given level curve, the gradient vector must point in the horizontal direction that yields the steepest increase on the saddle. This will always be perpendicular to the level curve, since in the direction of the level curve, the altitude does not change. Thus, we get a picture as in Figure 10. If we were actually on the saddle, each gradient vector tells us the horizontal direction we should head in to go most steeply up the saddle surface. Note that at $(0, 0)$ the gradient vector is the zero vector, since the tangent plane to the saddle at that point is horizontal, and there is no direction that would correspond to the steepest ascent on the surface. ■

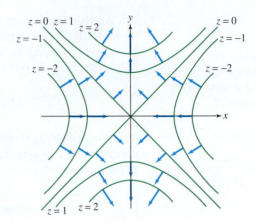

FIGURE 10 Contour map of $f(x, y) = y^2 - x^2$ and the corresponding gradient vectors at each point.

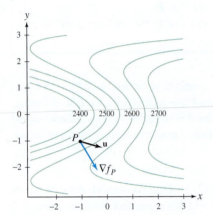

FIGURE 11 Contour map of the function $f(x, y)$ in Example 10.

■ **EXAMPLE 10** The altitude of a mountain at (x, y) is

$$f(x, y) = 2500 + 100(x + y^2)e^{-0.3y^2}$$

where x, y are in units of 100 m.

(a) Find the directional derivative of f at $P = (-1, -1)$ in the direction of unit vector $\mathbf{u}$ making an angle of $\theta = \frac{\pi}{4}$ with the gradient (Figure 11).

(b) What is the interpretation of this derivative?

Solution First compute $\|\nabla f_P\|$:

$$f_x(x, y) = 100e^{-0.3y^2}, \qquad f_y(x, y) = 100y(2 - 0.6x - 0.6y^2)e^{-0.3y^2}$$

$$f_x(-1, -1) = 100e^{-0.3} \approx 74, \qquad f_y(-1, -1) = -200e^{-0.3} \approx -148$$

Hence, $\nabla f_P \approx \langle 74, -148 \rangle$ and

$$\|\nabla f_P\| \approx \sqrt{74^2 + (-148)^2} \approx 165.5$$

Apply Eq. (5) with $\theta = \pi/4$:

$$D_{\mathbf{u}} f(P) = \|\nabla f_P\| \cos\theta \approx 165.5 \left(\frac{\sqrt{2}}{2}\right) \approx 117$$

Recall that x and y are measured in units of 100 m. Therefore, the interpretation is the following: If you stand on the mountain at the point lying above $(-1, -1)$ and begin climbing so that your horizontal displacement is in the direction of $\mathbf{u}$, then your altitude increases at a rate of 117 m per 100 m of horizontal displacement, or 1.17 m per meter of horizontal displacement. ■

The symbol ψ (pronounced "p-sigh" or "p-see") is the lowercase Greek letter psi.

CONCEPTUAL INSIGHT The directional derivative is related to the **angle of inclination** ψ in Figure 12. Think of the graph of $z = f(x, y)$ as a mountain lying over the xy-plane. Let Q be the point on the mountain lying above a point $P = (a, b)$ in the xy-plane. If you start moving up the mountain so that your horizontal displacement is in the direction of $\mathbf{u}$, then you will actually be moving up the mountain at an angle of inclination ψ defined by

$$\tan\psi = D_{\mathbf{u}} f(P) \qquad \boxed{6}$$

The steepest direction up the mountain is the direction for which the horizontal displacement is in the direction of ∇f_P.

■ **EXAMPLE 11** **Angle of Inclination** You are standing on the side of a mountain in the shape $z = f(x, y)$, at a point $Q = (a, b, f(a, b))$, where $\nabla f_{(a,b)} = \langle 0.4, 0.02 \rangle$. Find the angle of inclination in a direction making an angle of $\theta = \frac{\pi}{3}$ with the gradient.

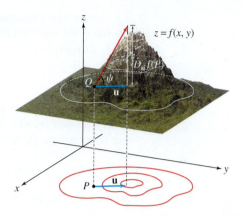

FIGURE 12

Solution The gradient has length $\|\nabla f_{(a,b)}\| = \sqrt{(0.4)^2 + (0.02)^2} \approx 0.4$. If $\mathbf{u}$ is a unit vector making an angle of $\theta = \frac{\pi}{3}$ with $\nabla f_{(a,b)}$, then

$$D_{\mathbf{u}} f(a, b) = \|\nabla f_{(a,b)}\| \cos \frac{\pi}{3} \approx (0.4)(0.5) = 0.2$$

The angle of inclination at Q in the direction of $\mathbf{u}$ satisfies $\tan \psi = 0.2$. It follows that $\psi \approx \tan^{-1} 0.2 \approx 0.197$ radians or approximately $11.3°$. ∎

Another use of the gradient is in finding normal vectors on a surface with equation $F(x, y, z) = k$, where k is a constant.

> **THEOREM 5** **Gradient as a Normal Vector** Let $P = (a, b, c)$ be a point on the surface given by $F(x, y, z) = k$ and assume that $\nabla F_P \neq \mathbf{0}$. Then ∇F_P is a vector normal to the tangent plane to the surface at P. Moreover, the tangent plane to the surface at P has equation
>
> $$F_x(a, b, c)(x - a) + F_y(a, b, c)(y - b) + F_z(a, b, c)(z - c) = 0$$

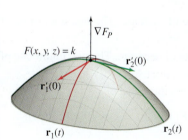

FIGURE 13 ∇F_P is normal to the surface $F(x, y, z) = k$ at P.

Proof Let $\mathbf{r}(t)$ be any path on the surface such that $\mathbf{r}(0) = P$ and $\mathbf{r}'(0) \neq \mathbf{0}$. Then $F(\mathbf{r}(t)) = k$ since all points on the curve must satisfy the equation $F(x, y, z) = k$. Differentiating both sides of this equation and applying the Chain Rule for Paths, we have

$$\nabla F_P \cdot \mathbf{r}'(0) = 0$$

Hence, ∇F_P is perpendicular to $\mathbf{r}'0)$, which we know to be tangent to the curve given by $\mathbf{r}(t)$ at P and thus tangent to the surface at P. However, we can take $\mathbf{r}(t)$ to pass through P from any direction, as in Figure 13, and hence, ∇F_P must be perpendicular to tangent vectors pointing in any direction, and therefore perpendicular to the entire tangent plane at P.

Since $\nabla F_P = \langle F_x(a, b, c), F_y(a, b, c), F_z(a, b, c) \rangle$ is a normal vector to the tangent plane and $P(a, b, c)$ is a point on the plane, an equation of the tangent plane is given by

$$F_x(a, b, c)(x - a) + F_y(a, b, c)(y - b) + F_z(a, b, c)(z - c) = 0 \quad ∎$$

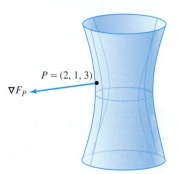

FIGURE 14 The gradient vector ∇F_P is normal to the surface at P.

■ **EXAMPLE 12** **Normal Vector and Tangent Plane** Find an equation of the tangent plane to the surface $4x^2 + 9y^2 - z^2 = 16$ at $P = (2, 1, 3)$.

Solution Let $F(x, y, z) = 4x^2 + 9y^2 - z^2$. Then

$$\nabla F = \langle 8x, 18y, -2z \rangle, \qquad \nabla F_P = \nabla F_{(2,1,3)} = \langle 16, 18, -6 \rangle$$

The vector $\langle 16, 18, -6 \rangle$ is normal to the surface $F(x, y, z) = 16$ (Figure 14), so the tangent plane at P has equation

$$16(x - 2) + 18(y - 1) - 6(z - 3) = 0 \quad \text{or} \quad 16x + 18y - 6z = 32 \quad ■$$

Notice how this equation for the tangent plane relates to Eq. (2) of Section 14.4, where we found the equation for the tangent plane to a surface given by $z = f(x, y)$ at a point $(a, b, f(a, b))$ to be given by

$$z = f(a, b) + f_x(a, b)(x - a) + f_y(a, b)(y - b)$$

If we want to apply our new formula for the tangent plane to this situation, we must take $F(x, y, z) = f(x, y) - z = 0$. Then note that $F_x = f_x$, $F_y = f_y$ and $F_z = -1$. Hence, our new formula yields

$$f_x(a, b)(x - a) + f_y(a, b)(y - b) + (-1)(z - c) = 0$$

where $c = f(a, b)$. This agrees exactly with Eq. (2) from Section 14.4.

14.5 SUMMARY

- The *gradient* of a function f is the vector of partial derivatives:

$$\nabla f = \left\langle \frac{\partial f}{\partial x}, \frac{\partial f}{\partial y} \right\rangle \qquad \text{or} \qquad \nabla f = \left\langle \frac{\partial f}{\partial x}, \frac{\partial f}{\partial y}, \frac{\partial f}{\partial z} \right\rangle$$

- Chain Rule for Paths:

$$\frac{d}{dt} f(\mathbf{r}(t)) = \nabla f_{\mathbf{r}(t)} \cdot \mathbf{r}'(t)$$

- For $\mathbf{u}$ a unit vector, $D_{\mathbf{u}} f$ is called the *directional derivative*.
- *Directional derivative with respect to* $\mathbf{u} = \langle h, k \rangle$:

$$D_{\mathbf{u}} f(a, b) = \lim_{t \to 0} \frac{f(a + th, b + tk) - f(a, b)}{t}$$

This definition extends to three or more variables.
- Formula for the directional derivative with respect to $\mathbf{u}$: $D_{\mathbf{u}} f(a, b) = \nabla f_{(a,b)} \cdot \mathbf{u}$.
- $D_{\mathbf{u}} f(a, b) = \| \nabla f_{(a,b)} \| \cos \theta$, where θ is the angle between $\nabla f_{(a,b)}$ and $\mathbf{u}$.
- Basic geometric properties of the gradient (assume $\nabla f_P \neq \mathbf{0}$):

  - ∇f_P points in the direction of maximum rate of increase. The maximum rate of increase is $\| \nabla f_P \|$.

  - $-\nabla f_P$ points in the direction of maximum rate of decrease. The maximum rate of decrease is $-\| \nabla f_P \|$.

  - ∇f_P is orthogonal to the level curve (or surface) through P.

- Equation of the tangent plane to the level surface $F(x, y, z) = k$ at $P = (a, b, c)$:

$$\nabla F_P \cdot \langle x - a, y - b, z - c \rangle = 0$$

$$F_x(a, b, c)(x - a) + F_y(a, b, c)(y - b) + F_z(a, b, c)(z - c) = 0$$

14.5 EXERCISES

Preliminary Questions

1. Which of the following is a possible value of the gradient ∇f of a function $f(x, y)$ of two variables?

(a) 5   **(b)** $\langle 3, 4 \rangle$ **(c)** $\langle 3, 4, 5 \rangle$

2. True or false? A differentiable function increases at the rate $\| \nabla f_P \|$ in the direction of ∇f_P.

3. Describe the two main geometric properties of the gradient ∇f.

4. You are standing at a point where the temperature gradient vector is pointing in the northeast (NE) direction. In which direction(s) should you walk to avoid a change in temperature?

(a) NE   **(b)** NW   **(c)** SE   **(d)** SW

5. What is the rate of change of $f(x, y)$ at $(0, 0)$ in the direction making an angle of $45°$ with the x-axis if $\nabla f(0, 0) = \langle 2, 4 \rangle$?

Exercises

1. Let $f(x, y) = xy^2$ and $\mathbf{r}(t) = \left\langle \frac{1}{2}t^2, t^3 \right\rangle$.

(a) Calculate ∇f and $\mathbf{r}'(t)$.

(b) Use the Chain Rule for Paths to evaluate $\dfrac{d}{dt} f(\mathbf{r}(t))$ at $t = 1$ and $t = -1$.

2. Let $f(x, y) = e^{xy}$ and $\mathbf{r}(t) = \left\langle t^3, 1 + t \right\rangle$.

(a) Calculate ∇f and $\mathbf{r}'(t)$.

(b) Use the Chain Rule for Paths to calculate $\dfrac{d}{dt} f(\mathbf{r}(t))$.

(c) Write out the composite $f(\mathbf{r}(t))$ as a function of t and differentiate. Check that the result agrees with part (b).

3. Figure 15 shows the level curves of a function $f(x, y)$ and a path $\mathbf{r}(t)$, traversed in the direction indicated. State whether the derivative $\dfrac{d}{dt} f(\mathbf{r}(t))$ is positive, negative, or zero at points A–D.

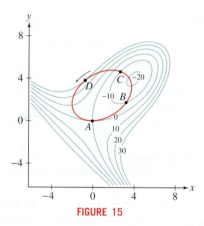

FIGURE 15

4. Let $f(x, y) = x^2 + y^2$ and $\mathbf{r}(t) = \langle \cos t, \sin t \rangle$.

(a) Find $\dfrac{d}{dt} f(\mathbf{r}(t))$ without making any calculations. Explain.

(b) Verify your answer to (a) using the Chain Rule.

In Exercises 5–8, calculate the gradient.

5. $f(x, y) = \cos(x^2 + y)$

6. $g(x, y) = \dfrac{x}{x^2 + y^2}$

7. $h(x, y, z) = xyz^{-3}$

8. $r(x, y, z, w) = xze^{yw}$

In Exercises 9–20, use the Chain Rule to calculate $\dfrac{d}{dt} f(\mathbf{r}(t))$.

9. $f(x, y) = 3x - 7y$, $\mathbf{r}(t) = \langle \cos t, \sin t \rangle$, $t = 0$

10. $f(x, y) = 3x - 7y$, $\mathbf{r}(t) = \left\langle t^2, t^3 \right\rangle$, $t = 2$

11. $f(x, y) = x^2 - 3xy$, $\mathbf{r}(t) = \langle \cos t, \sin t \rangle$, $t = 0$

12. $f(x, y) = x^2 - 3xy$, $\mathbf{r}(t) = \langle \cos t, \sin t \rangle$, $t = \frac{\pi}{2}$

13. $f(x, y) = \sin(xy)$, $\mathbf{r}(t) = \left\langle e^{2t}, e^{3t} \right\rangle$, $t = 0$

14. $f(x, y) = \cos(y - x)$, $\mathbf{r}(t) = \left\langle e^t, e^{2t} \right\rangle$, $t = \ln 3$

15. $f(x, y) = x - xy$, $\mathbf{r}(t) = \left\langle t^2, t^2 - 4t \right\rangle$, $t = 4$

16. $f(x, y) = xe^y$, $\mathbf{r}(t) = \left\langle t^2, t^2 - 4t \right\rangle$, $t = 0$

17. $f(x, y) = \ln x + \ln y$, $\mathbf{r}(t) = \left\langle \cos t, t^2 \right\rangle$, $t = \frac{\pi}{4}$

18. $g(x, y, z) = xye^z$, $\mathbf{r}(t) = \left\langle t^2, t^3, t - 1 \right\rangle$, $t = 1$

19. $g(x, y, z) = xyz^{-1}$, $\mathbf{r}(t) = \left\langle e^t, t, t^2 \right\rangle$, $t = 1$

20. $g(x, y, z, w) = x + 2y + 3z + 5w$, $\mathbf{r}(t) = \left\langle t^2, t^3, t, t-2 \right\rangle$, $t = 1$

In Exercises 21–30, calculate the directional derivative in the direction of $\mathbf{v}$ at the given point. Remember to normalize the direction vector.

21. $f(x, y) = x^2 + y^3$, $\mathbf{v} = \langle 4, 3 \rangle$, $P = (1, 2)$

22. $f(x, y) = x^2 y^3$, $\mathbf{v} = \mathbf{i} + \mathbf{j}$, $P = (-2, 1)$

23. $f(x, y) = x^2 y^3$, $\mathbf{v} = \mathbf{i} + \mathbf{j}$, $P = \left(\frac{1}{6}, 3 \right)$

24. $f(x, y) = \sin(x - y)$, $\mathbf{v} = \langle 1, 1 \rangle$, $P = \left(\frac{\pi}{2}, \frac{\pi}{6} \right)$

25. $f(x, y) = \tan^{-1}(xy)$, $\mathbf{v} = \langle 1, 1 \rangle$, $P = (3, 4)$

26. $f(x, y) = e^{xy - y^2}$, $\mathbf{v} = \langle 12, -5 \rangle$, $P = (2, 2)$

27. $f(x, y) = \ln(x^2 + y^2)$, $\mathbf{v} = 3\mathbf{i} - 2\mathbf{j}$, $P = (1, 0)$

28. $g(x, y, z) = z^2 - xy^2$, $\mathbf{v} = \langle -1, 2, 2 \rangle$, $P = (2, 1, 3)$

29. $g(x, y, z) = xe^{-yz}$, $\mathbf{v} = \langle 1, 1, 1 \rangle$, $P = (1, 2, 0)$

30. $g(x, y, z) = x \ln(y + z)$, $\mathbf{v} = 2\mathbf{i} - \mathbf{j} + \mathbf{k}$, $P = (2, e, e)$

31. Find the directional derivative of $f(x, y) = x^2 + 4y^2$ at the point $P = (3, 2)$ in the direction pointing to the origin.

32. Find the directional derivative of $f(x, y, z) = xy + z^3$ at the point $P = (3, -2, -1)$ in the direction pointing to the origin.

33. A bug located at $(3, 9, 4)$ begins walking in a straight line toward $(5, 7, 3)$. At what rate is the bug's temperature changing if the temperature is $T(x, y, z) = xe^{y-z}$? Units are in meters and degrees Celsius.

In Exercises 34–35, assume that the positive x-axis points East and the positive y-axis points North.

34. Suppose you are hiking on a terrain modeled by $z = x^2 + y^2 - y$. You are at the point $(1, 2, 3)$.

(a) Determine the slope you would encounter if you headed due East from your position. What angle of inclination does that correspond to?

(b) Determine the slope you would encounter if you headed due North from your position. What angle of inclination does that correspond to?

(c) Determine the slope you would encounter if you headed due North-East from your position. What angle of inclination does that correspond to?

(d) Determine the steepest slope you could encounter from your position.

35. Suppose you are hiking on a terrain modeled by $z = xy + y^3 - x^2$. You are at the point $(2, 1, -1)$.

(a) Determine the slope you would encounter if you headed due West from your position. What angle of inclination does that correspond to?

(b) Determine the slope you would encounter if you headed due North-West from your position. What angle of inclination does that correspond to?

(c) Determine the slope you would encounter if you headed due South-East from your position. What angle of inclination does that correspond to?

(d) Determine the steepest slope you could encounter from your position, and the compass direction measured in degrees from East that you would head to realize this steepest slope.

36. The temperature at location (x, y) is $T(x, y) = 20 + 0.1(x^2 - xy)$ (degrees Celsius). Beginning at $(200, 0)$ at time $t = 0$ (seconds), a bug travels counterclockwise along a circle of radius 200 cm centered at the origin, at a speed of 3 cm/s. How fast is the temperature changing when the bug is at a position on the circle corresponding to an angle of $\theta = \pi/3$?

37. Suppose that $\nabla f_P = \langle 2, -4, 4 \rangle$. Is f increasing or decreasing at P in the direction $\mathbf{v} = \langle 2, 1, 3 \rangle$?

38. Let $f(x, y) = xe^{x^2 - y}$ and $P = (1, 1)$.
(a) Calculate $\|\nabla f_P\|$.
(b) Find the rate of change of f in the direction ∇f_P.
(c) Find the rate of change of f in the direction of a vector making an angle of $45°$ with ∇f_P.

39. Let $f(x, y, z) = \sin(xy + z)$ and $P = (0, -1, \pi)$. Calculate $D_{\mathbf{u}} f(P)$, where $\mathbf{u}$ is a unit vector making an angle $\theta = 30°$ with ∇f_P.

40. Let $T(x, y)$ be the temperature at location (x, y) on a thin sheet of metal. Assume that $\nabla T = \langle y - 4, x + 2y \rangle$. Let $\mathbf{r}(t) = \langle t^2, t \rangle$ be a path on the sheet. Find the values of t such that

$$\frac{d}{dt} T(\mathbf{r}(t)) = 0$$

41. Find a vector normal to the surface $x^2 + y^2 - z^2 = 6$ at $P = (3, 1, 2)$.

42. Find a vector normal to the surface $3z^3 + x^2 y - y^2 x = 1$ at $P = (1, -1, 1)$.

43. Find the two points on the ellipsoid

$$\frac{x^2}{4} + \frac{y^2}{9} + z^2 = 1$$

where the tangent plane is normal to $\mathbf{v} = \langle 1, 1, -2 \rangle$.

In Exercises 44–47, find an equation of the tangent plane to the surface at the given point.

44. $x^2 + 3y^2 + 4z^2 = 20$, $P = (2, 2, 1)$

45. $xz + 2x^2 y + y^2 z^3 = 11$, $P = (2, 1, 1)$

46. $x^2 + z^2 e^{y-x} = 13$, $P = \left(2, 3, \dfrac{3}{\sqrt{e}}\right)$

47. $\ln[1 + 4x^2 + 9y^4] - 0.1z^2 = 0$, $P = (3, 1, 6.1876)$

48. Verify what is clear from Figure 16: Every tangent plane to the cone $x^2 + y^2 - z^2 = 0$ passes through the origin.

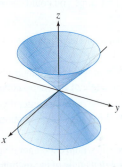

FIGURE 16 Graph of $x^2 + y^2 - z^2 = 0$.

49. *CAS* Use a computer algebra system to produce a contour plot of $f(x, y) = x^2 - 3xy + y - y^2$ together with its gradient vector field on the domain $[-4, 4] \times [-4, 4]$.

50. Find a function $f(x, y, z)$ such that ∇f is the constant vector $\langle 1, 3, 1 \rangle$.

51. Find a function $f(x, y, z)$ such that $\nabla f = \langle 2x, 1, 2 \rangle$.

52. Find a function $f(x, y, z)$ such that $\nabla f = \left\langle x, y^2, z^3 \right\rangle$.

53. Find a function $f(x, y, z)$ such that $\nabla f = \langle z, 2y, x \rangle$.

54. Find a function $f(x, y)$ such that $\nabla f = \langle y, x \rangle$.

55. Show that there does not exist a function $f(x, y)$ such that $\nabla f = \left\langle y^2, x \right\rangle$. *Hint:* Use Clairaut's Theorem $f_{xy} = f_{yx}$.

56. Let $\Delta f = f(a + h, b + k) - f(a, b)$ be the change in f at $P = (a, b)$. Set $\Delta \mathbf{v} = \langle h, k \rangle$. Show that the linear approximation can be written

$$\Delta f \approx \nabla f_P \cdot \Delta \mathbf{v} \qquad \boxed{7}$$

57. Use Eq. (7) to estimate

$$\Delta f = f(3.53, 8.98) - f(3.5, 9)$$

assuming that $\nabla f_{(3.5,9)} = \langle 2, -1 \rangle$.

58. Find a unit vector $\mathbf{n}$ that is normal to the surface $z^2 - 2x^4 - y^4 = 16$ at $P = (2, 2, 8)$ that points in the direction of the xy-plane (in other words, if you travel in the direction of $\mathbf{n}$, you will eventually cross the xy-plane).

59. Suppose, in the previous exercise, that a particle located at the point $P = (2, 2, 8)$ travels toward the xy-plane in the direction normal to the surface.
(a) Through which point Q on the xy-plane will the particle pass?
(b) Suppose the axes are calibrated in centimeters. Determine the path $\mathbf{r}(t)$ of the particle if it travels at a constant speed of 8 cm/s. How long will it take the particle to reach Q?

60. Let $f(x, y) = \tan^{-1} \dfrac{x}{y}$ and $\mathbf{u} = \left\langle \dfrac{\sqrt{2}}{2}, \dfrac{\sqrt{2}}{2} \right\rangle$.
(a) Calculate the gradient of f.
(b) Calculate $D_{\mathbf{u}} f(1, 1)$ and $D_{\mathbf{u}} f(\sqrt{3}, 1)$.
(c) Show that the lines $y = mx$ for $m \neq 0$ are level curves for f.
(d) Verify that ∇f_P is orthogonal to the level curve through P for $P = (x, y) \neq (0, 0)$.

61. Suppose that the intersection of two surfaces $F(x, y, z) = 0$ and $G(x, y, z) = 0$ is a curve $\mathcal{C}$, and let P be a point on $\mathcal{C}$. Explain why the vector $\mathbf{v} = \nabla F_P \times \nabla G_P$ is a direction vector for the tangent line to $\mathcal{C}$ at P.

62. Let $\mathcal{C}$ be the curve of intersection of the spheres $x^2 + y^2 + z^2 = 3$ and $(x - 2)^2 + (y - 2)^2 + z^2 = 3$. Use the result of Exercise 61 to find parametric equations of the tangent line to $\mathcal{C}$ at $P = (1, 1, 1)$.

63. Let $\mathcal{C}$ be the curve obtained by intersecting the two surfaces $x^3 + 2xy + yz = 7$ and $3x^2 - yz = 1$. Find the parametric equations of the tangent line to $\mathcal{C}$ at $P = (1, 2, 1)$.

64. Verify the linearity relations for gradients:
(a) $\nabla(f + g) = \nabla f + \nabla g$
(b) $\nabla(cf) = c\nabla f$

65. Prove the Chain Rule for Gradients (Theorem 1).

66. Prove the Product Rule for Gradients (Theorem 1).

Further Insights and Challenges

67. Let **u** be a unit vector. Show that the directional derivative $D_{\mathbf{u}} f$ is equal to the component of ∇f along **u**.

68. Let $f(x, y) = (xy)^{1/3}$.

(a) Use the limit definition to show that $f_x(0, 0) = f_y(0, 0) = 0$.

(b) Use the limit definition to show that the directional derivative $D_{\mathbf{u}} f(0, 0)$ does not exist for any unit vector **u** other than **i** and **j**.

(c) Is f differentiable at $(0, 0)$?

69. Use the definition of differentiability to show that if $f(x, y)$ is differentiable at $(0, 0)$ and

$$f(0, 0) = f_x(0, 0) = f_y(0, 0) = 0$$

then

$$\lim_{(x,y) \to (0,0)} \frac{f(x, y)}{\sqrt{x^2 + y^2}} = 0 \qquad \boxed{8}$$

70. This exercise shows that there exists a function that is not differentiable at $(0, 0)$ even though all directional derivatives at $(0, 0)$ exist. Define $f(x, y) = x^2 y/(x^2 + y^2)$ for $(x, y) \neq 0$ and $f(0, 0) = 0$.

(a) Use the limit definition to show that $D_{\mathbf{v}} f(0, 0)$ exists for all vectors **v**. Show that $f_x(0, 0) = f_y(0, 0) = 0$.

(b) Prove that f is *not* differentiable at $(0, 0)$ by showing that Eq. (8) does not hold.

71. Prove that if $f(x, y)$ is differentiable and $\nabla f_{(x,y)} = \mathbf{0}$ for all (x, y), then f is constant.

72. Prove the following Quotient Rule, where f, g are differentiable:

$$\nabla \left(\frac{f}{g} \right) = \frac{g \nabla f - f \nabla g}{g^2}$$

In Exercises 73–75, a path $\mathbf{r}(t) = \langle x(t), y(t) \rangle$ follows the gradient of a function $f(x, y)$ if the tangent vector $\mathbf{r}'(t)$ points in the direction of ∇f for all t. In other words, $\mathbf{r}'(t) = k(t) \nabla f_{\mathbf{r}(t)}$ for some positive function $k(t)$. Note that in this case, $\mathbf{r}(t)$ crosses each level curve of $f(x, y)$ at a right angle.

73. Show that if the path $\mathbf{r}(t) = \langle x(t), y(t) \rangle$ follows the gradient of $f(x, y)$, then

$$\frac{y'(t)}{x'(t)} = \frac{f_y}{f_x}$$

74. Find a path of the form $\mathbf{r}(t) = (t, g(t))$ passing through $(1, 2)$ that follows the gradient of $f(x, y) = 2x^2 + 8y^2$ (Figure 17). *Hint:* Use Separation of Variables.

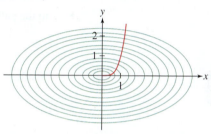

FIGURE 17 The path $\mathbf{r}(t)$ is orthogonal to the level curves of $f(x, y) = 2x^2 + 8y^2$.

75. [CAS] Find the curve $y = g(x)$ passing through $(0, 1)$ that crosses each level curve of $f(x, y) = y \sin x$ at a right angle. If you have a computer algebra system, graph $y = g(x)$ together with the level curves of f.

14.6 The Chain Rule

The Chain Rule for Paths that we derived in the previous section can be extended to general composite functions. Suppose, for example, that x, y, z are differentiable functions of s and t—say, $x = x(s, t)$, $y = y(s, t)$, and $z = z(s, t)$. The composition

$$f(x(s, t), y(s, t), z(s, t)) \qquad \boxed{1}$$

is then a function of s and t. We refer to s and t as the **independent variables**.

■ **EXAMPLE 1** Find the composite function where $f(x, y, z) = xy + z$ and $x = s^2$, $y = st$, $z = t^2$.

Solution We can keep track of which variable depends on which other variable by using a chart as in Figure 1. The composite function is given by

$$f(x(s, t), y(s, t), z(s, t)) = xy + z = (s^2)(st) + t^2 = s^3 t + t^2 \qquad ■$$

The Chain Rule expresses the derivatives of f with respect to the independent variables. For example, the partial derivatives of $f(x(s, t), y(s, t), z(s, t))$ are

$$\frac{\partial f}{\partial s} = \frac{\partial f}{\partial x} \frac{\partial x}{\partial s} + \frac{\partial f}{\partial y} \frac{\partial y}{\partial s} + \frac{\partial f}{\partial z} \frac{\partial z}{\partial s} \qquad \boxed{2}$$

$$\frac{\partial f}{\partial t} = \frac{\partial f}{\partial x} \frac{\partial x}{\partial t} + \frac{\partial f}{\partial y} \frac{\partial y}{\partial t} + \frac{\partial f}{\partial z} \frac{\partial z}{\partial t} \qquad \boxed{3}$$

FIGURE 1 Keeping track of the relationships between the variables.

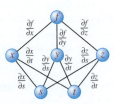

FIGURE 2 Keeping track of the relationships between the variables.

Note that we can recover the formula for $\dfrac{\partial f}{\partial s}$ by labeling each edge in Figure 2 with the partial derivative of the top variable with respect to the bottom variable as in Figure 2. Then to obtain the formula for $\dfrac{\partial f}{\partial s}$, we consider each of the paths along the edges down from f to s: the first through x, the second through y, and the third through z. Each path contributes a term to the formula. The first is the product of the partial derivatives labeling its edges, giving $\dfrac{\partial f}{\partial x}\dfrac{\partial x}{\partial s}$. The second yields $\dfrac{\partial f}{\partial y}\dfrac{\partial y}{\partial s}$. And the third yields $\dfrac{\partial f}{\partial z}\dfrac{\partial z}{\partial s}$. Thus, we obtain the formula

$$\frac{\partial f}{\partial s} = \frac{\partial f}{\partial x}\frac{\partial x}{\partial s} + \frac{\partial f}{\partial y}\frac{\partial y}{\partial s} + \frac{\partial f}{\partial z}\frac{\partial z}{\partial s}$$

We recover the formula for $\dfrac{\partial f}{\partial t}$ in a similar manner. To prove these formulas, we observe that $\dfrac{\partial f}{\partial s}$, when evaluated at a point (s_0, t_0), is equal to the derivative with respect to the path

$$\mathbf{r}(s) = \langle x(s, t_0), y(s, t_0), z(s, t_0) \rangle$$

In other words, we fix $t = t_0$ and take the derivative with respect to s:

$$\frac{\partial f}{\partial s}(s_0, t_0) = \frac{d}{ds} f(\mathbf{r}(s)) \bigg|_{s=s_0}$$

The tangent vector is

$$\mathbf{r}'(s) = \left\langle \frac{\partial x}{\partial s}(s, t_0), \frac{\partial y}{\partial s}(s, t_0), \frac{\partial z}{\partial s}(s, t_0) \right\rangle$$

Therefore, by the Chain Rule for Paths,

$$\frac{\partial f}{\partial s}\bigg|_{(s_0,t_0)} = \frac{d}{ds}f(\mathbf{r}(s))\bigg|_{s=s_0} = \nabla f \cdot \mathbf{r}'(s_0) = \frac{\partial f}{\partial x}\frac{\partial x}{\partial s} + \frac{\partial f}{\partial y}\frac{\partial y}{\partial s} + \frac{\partial f}{\partial z}\frac{\partial z}{\partial s}$$

The derivatives on the right are evaluated at (s_0, t_0). This proves Eq. (2). A similar argument proves Eq. (3), as well as the general case of a function $f(x_1, \ldots, x_n)$, where the variables x_i depend on independent variables $t_1, \ldots, t_m$.

> **THEOREM 1 General Version of the Chain Rule** Let $f(x_1, \ldots, x_n)$ be a differentiable function of n variables. Suppose that each of the variables $x_1, \ldots, x_n$ is a differentiable function of m independent variables $t_1, \ldots, t_m$. Then, for $k = 1, \ldots, m$,
>
> $$\frac{\partial f}{\partial t_k} = \frac{\partial f}{\partial x_1}\frac{\partial x_1}{\partial t_k} + \frac{\partial f}{\partial x_2}\frac{\partial x_2}{\partial t_k} + \cdots + \frac{\partial f}{\partial x_n}\frac{\partial x_n}{\partial t_k} \qquad \boxed{4}$$

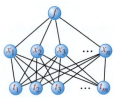

FIGURE 3 Keeping track of the dependencies between the variables.

We keep track of the dependencies between the variables as in Figure 3. As an aid to remembering the Chain Rule, we will refer to

$$\frac{\partial f}{\partial x_1}, \quad \cdots, \quad \frac{\partial f}{\partial x_n}$$

as the **primary derivatives**. They are the components of the gradient ∇f. By Eq. (4), the derivative of f with respect to the independent variable t_k is equal to a sum of n terms:

$$j\text{th term:} \qquad \frac{\partial f}{\partial x_j}\frac{\partial x_j}{\partial t_k} \quad \text{for } j = 1, 2, \ldots, n$$

The term "primary derivative" is not standard. We use it in this section only, to clarify the structure of the Chain Rule.

Note that we can write Eq. (4) as a dot product:

$$\frac{\partial f}{\partial t_k} = \left\langle \frac{\partial f}{\partial x_1}, \frac{\partial f}{\partial x_2}, \ldots, \frac{\partial f}{\partial x_n} \right\rangle \cdot \left\langle \frac{\partial x_1}{\partial t_k}, \frac{\partial x_2}{\partial t_k}, \ldots, \frac{\partial x_n}{\partial t_k} \right\rangle$$

$$\boxed{\frac{\partial f}{\partial t_k} = \nabla f \cdot \left\langle \frac{\partial x_1}{\partial t_k}, \frac{\partial x_2}{\partial t_k}, \ldots, \frac{\partial x_n}{\partial t_k} \right\rangle} \qquad \boxed{5}$$

■ **EXAMPLE 2** **Using the Chain Rule** Let $f(x, y, z) = xy + z$. Calculate $\dfrac{\partial f}{\partial s}$, where

$$x = s^2, \quad y = st, \quad z = t^2$$

Solution

We keep track of the dependencies of the variables as in Figure 2.

Step 1. **Compute the primary derivatives.**

$$\frac{\partial f}{\partial x} = y, \qquad \frac{\partial f}{\partial y} = x, \qquad \frac{\partial f}{\partial z} = 1$$

Step 2. **Apply the Chain Rule.**

$$\frac{\partial f}{\partial s} = \frac{\partial f}{\partial x}\frac{\partial x}{\partial s} + \frac{\partial f}{\partial y}\frac{\partial y}{\partial s} + \frac{\partial f}{\partial z}\frac{\partial z}{\partial s} = y\frac{\partial}{\partial s}(s^2) + x\frac{\partial}{\partial s}(st) + \frac{\partial}{\partial s}(t^2)$$

$$= (y)(2s) + (x)(t) + 0$$

$$= 2sy + xt$$

This expresses the derivative in terms of both sets of variables. If desired, we can substitute $x = s^2$ and $y = st$ to write the derivative in terms of s and t:

$$\frac{\partial f}{\partial s} = 2ys + xt = 2(st)s + (s^2)t = 3s^2 t$$

To check this result, recall that in Example 1, we computed the composite function:

$$f(x(s, t), y(s, t), z(s, t)) = f(s^2, st, t^2) = s^3 t + t^2$$

From this, we see directly that $\dfrac{\partial f}{\partial s} = 3s^2 t$, confirming our result. ■

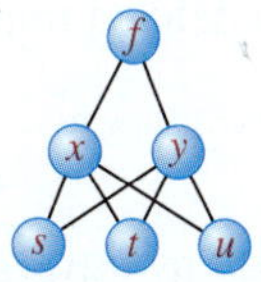

FIGURE 4 Keeping track of the dependencies between the variables.

■ **EXAMPLE 3** **Evaluating the Derivative** Let $f(x, y) = e^{xy}$. Evaluate $\dfrac{\partial f}{\partial t}$ at $(s, t, u) = (2, 3, -1)$, where $x = st$, $y = s - ut^2$.

Solution We keep track of the dependencies of the variables as in Figure 4. We can use either Eq. (4) or Eq. (5). We'll use the dot product form in Eq. (5). We have

$$\nabla f = \left\langle \frac{\partial f}{\partial x}, \frac{\partial f}{\partial y} \right\rangle = \left\langle ye^{xy}, xe^{xy} \right\rangle, \qquad \left\langle \frac{\partial x}{\partial t}, \frac{\partial y}{\partial t} \right\rangle = \langle s, -2ut \rangle$$

and the Chain Rule gives us

$$\frac{\partial f}{\partial t} = \nabla f \cdot \left\langle \frac{\partial x}{\partial t}, \frac{\partial y}{\partial t} \right\rangle = \left\langle ye^{xy}, xe^{xy} \right\rangle \cdot \langle s, -2ut \rangle$$

$$= ye^{xy}(s) + xe^{xy}(-2ut)$$

$$= (ys - 2xut)e^{xy}$$

To finish the problem, we do not have to rewrite $\dfrac{\partial f}{\partial t}$ in terms of s, t, u. For $(s, t, u) = (2, 3, -1)$, we obtain

$$x = st = 2(3) = 6, \qquad y = s - ut^2 = 2 - (-1)(3^2) = 11$$

With $(s, t, u) = (2, 3, -1)$ and $(x, y) = (6, 11)$, we obtain

$$\left. \frac{\partial f}{\partial t} \right|_{(2,3,-1)} = (ys - 2xut)e^{xy} \bigg|_{(2,3,-1)} = \Big((11)(2) - 2(6)(-1)(3) \Big) e^{6(11)} = 58e^{66} \quad ■$$

■ **EXAMPLE 4** **Polar Coordinates** Let $f(x, y)$ be a function of two variables, and let (r, θ) be polar coordinates.

(a) Express $\dfrac{\partial f}{\partial \theta}$ in terms of $\dfrac{\partial f}{\partial x}$ and $\dfrac{\partial f}{\partial y}$.

(b) Evaluate $\dfrac{\partial f}{\partial \theta}$ at $(x, y) = (1, 1)$ for $f(x, y) = x^2 y$.

Solution

(a) Since $x = r \cos \theta$ and $y = r \sin \theta$,

$$\frac{\partial x}{\partial \theta} = -r \sin \theta, \qquad \frac{\partial y}{\partial \theta} = r \cos \theta$$

By the Chain Rule,

$$\frac{\partial f}{\partial \theta} = \frac{\partial f}{\partial x}\frac{\partial x}{\partial \theta} + \frac{\partial f}{\partial y}\frac{\partial y}{\partial \theta} = -r \sin \theta \frac{\partial f}{\partial x} + r \cos \theta \frac{\partial f}{\partial y}$$

Since $x = r \cos \theta$ and $y = r \sin \theta$, we can write $\dfrac{\partial f}{\partial \theta}$ in terms of x and y alone:

$$\frac{\partial f}{\partial \theta} = x \frac{\partial f}{\partial y} - y \frac{\partial f}{\partial x} \qquad \boxed{6}$$

*If you have studied quantum mechanics, you may recognize the right-hand side of Eq. (6) as the **angular momentum** operator (with respect to the z-axis) applied to the function f.*

(b) Apply Eq. (6) to $f(x, y) = x^2 y$:

$$\frac{\partial f}{\partial \theta} = x \frac{\partial}{\partial y}(x^2 y) - y \frac{\partial}{\partial x}(x^2 y) = x^3 - 2xy^2$$

$$\left. \frac{\partial f}{\partial \theta} \right|_{(x,y)=(1,1)} = 1^3 - 2(1)(1^2) = -1$$ ■

Notice that the General Version of the Chain Rule encompasses the Chain Rule for Paths, which appeared as Theorem 2 in Section 14.5. In the Chain Rule for Paths, there is just one independent variable, which is the parameter for the path.

Implicit Differentiation

In single-variable calculus, we used implicit differentiation to compute dy/dx when y is defined implicitly as a function of x through an equation $f(x, y) = 0$. This method also works for functions of several variables. Suppose that z is defined implicitly by an equation

$$F(x, y, z) = 0$$

Thus, $z = z(x, y)$ is a function of x and y. We may not be able to solve explicitly for $z(x, y)$, but we can treat $F(x, y, z)$ as a composite function with x and y as independent variables, and use the Chain Rule to differentiate with respect to x:

$$\frac{\partial F}{\partial x}\frac{\partial x}{\partial x} + \frac{\partial F}{\partial y}\frac{\partial y}{\partial x} + \frac{\partial F}{\partial z}\frac{\partial z}{\partial x} = 0$$

We have $\partial x/\partial x = 1$ and also $\partial y/\partial x = 0$, since y does not depend on x. Thus,

$$\frac{\partial F}{\partial x} + \frac{\partial F}{\partial z}\frac{\partial z}{\partial x} = F_x + F_z \frac{\partial z}{\partial x} = 0$$

If $F_z \neq 0$, we may solve for $\partial z/\partial x$ (we compute $\partial z/\partial y$ similarly):

$$\boxed{\frac{\partial z}{\partial x} = -\frac{F_x}{F_z}, \qquad \frac{\partial z}{\partial y} = -\frac{F_y}{F_z}} \qquad \boxed{7}$$

■ **EXAMPLE 5** Calculate $\partial z/\partial x$ and $\partial z/\partial y$ at $P = (1, 1, 1)$, where

$$F(x, y, z) = x^2 + y^2 - 2z^2 + 12x - 8z - 4 = 0$$

What is the graphical interpretation of these partial derivatives?

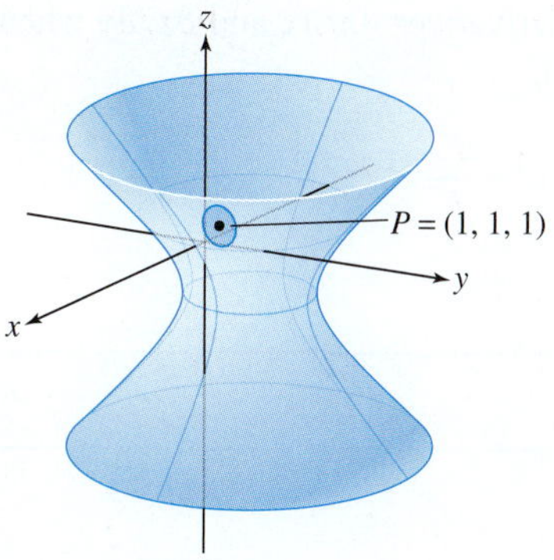

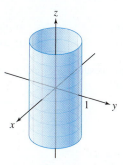

FIGURE 5 The surface
$x^2 + y^2 - 2z^2 + 12x - 8z - 4 = 0$.
A small patch of the surface around P can
be represented as the graph of a function of
x and y.

Solution We have

$$F_x = 2x + 12, \qquad F_y = 2y, \qquad F_z = -4z - 8$$

Hence,

$$\frac{\partial z}{\partial x} = -\frac{F_x}{F_z} = \frac{2x + 12}{4z + 8}, \qquad \frac{\partial z}{\partial y} = -\frac{F_y}{F_z} = \frac{2y}{4z + 8}$$

The derivatives at $P = (1, 1, 1)$ are

$$\frac{\partial z}{\partial x}\bigg|_{(1,1,1)} = \frac{2(1) + 12}{4(1) + 8} = \frac{14}{12} = \frac{7}{6}, \qquad \frac{\partial z}{\partial y}\bigg|_{(1,1,1)} = \frac{2(1)}{4(1) + 8} = \frac{2}{12} = \frac{1}{6}$$

Figure 5 shows the surface $F(x, y, z) = 0$. The surface as a whole is not the graph of a function because it fails the Vertical Line Test. However, a small patch near P may be represented as a graph of a function $z = f(x, y)$, and the partial derivatives $\dfrac{\partial z}{\partial x}$ and $\dfrac{\partial z}{\partial y}$ are equal to f_x and f_y. Implicit differentiation has enabled us to compute these partial derivatives without finding $f(x, y)$ explicitly. ■

Assumptions Matter Implicit differentiation is based on the assumption that we can solve the equation $F(x, y, z) = 0$ for z in the form $z = f(x, y)$. Otherwise, the partial derivatives $\dfrac{\partial z}{\partial x}$ and $\dfrac{\partial z}{\partial y}$ would have no meaning. The Implicit Function Theorem of advanced calculus guarantees that this can be done (at least near a point P) if F has continuous partial derivatives and $F_z(P) \neq 0$. Why is this condition necessary? Recall that the gradient vector $\nabla F_P = \langle F_x(P), F_y(P), F_z(P)\rangle$ is normal to the surface at P, so $F_z(P) = 0$ means that the tangent plane at P is vertical. To see what can go wrong, consider the cylinder (shown in Figure 6):

$$F(x, y, z) = x^2 + y^2 - 1 = 0$$

In this extreme case, F_z is always 0. The z-coordinate on the cylinder does not depend on x or y, so it is impossible to represent the cylinder as a graph $z = f(x, y)$ and the derivatives $\dfrac{\partial z}{\partial x}$ and $\dfrac{\partial z}{\partial y}$ do not exist.

FIGURE 6 Graph of the cylinder
$x^2 + y^2 - 1 = 0$.

14.6 SUMMARY

- If $f(x, y, z)$ is a function of x, y, z, and if x, y, z depend on two other variables, say, s and t, then

$$f(x, y, z) = f(x(s, t), y(s, t), z(s, t))$$

is a composite function of s and t. We refer to s and t as the *independent variables*.
- The *Chain Rule* expresses the partial derivatives with respect to the independent variables s and t in terms of the *primary derivatives*:

$$\frac{\partial f}{\partial x}, \qquad \frac{\partial f}{\partial y}, \qquad \frac{\partial f}{\partial z}$$

Namely,

$$\frac{\partial f}{\partial s} = \frac{\partial f}{\partial x}\frac{\partial x}{\partial s} + \frac{\partial f}{\partial y}\frac{\partial y}{\partial s} + \frac{\partial f}{\partial z}\frac{\partial z}{\partial s}, \qquad \frac{\partial f}{\partial t} = \frac{\partial f}{\partial x}\frac{\partial x}{\partial t} + \frac{\partial f}{\partial y}\frac{\partial y}{\partial t} + \frac{\partial f}{\partial z}\frac{\partial z}{\partial t}$$

- In general, if $f(x_1, \ldots, x_n)$ is a function of n variables and if $x_1, \ldots, x_n$ depend on the independent variables $t_1, \ldots, t_m$, then

$$\frac{\partial f}{\partial t_k} = \frac{\partial f}{\partial x_1}\frac{\partial x_1}{\partial t_k} + \frac{\partial f}{\partial x_2}\frac{\partial x_2}{\partial t_k} + \cdots + \frac{\partial f}{\partial x_n}\frac{\partial x_n}{\partial t_k}$$

- The Chain Rule can be expressed as a dot product:

$$\frac{\partial f}{\partial t_k} = \underbrace{\left\langle \frac{\partial f}{\partial x_1}, \frac{\partial f}{\partial x_2}, \ldots, \frac{\partial f}{\partial x_n}\right\rangle}_{\nabla f} \cdot \left\langle \frac{\partial x_1}{\partial t_k}, \frac{\partial x_2}{\partial t_k}, \ldots, \frac{\partial x_n}{\partial t_k}\right\rangle$$

- Implicit differentiation is used to find the partial derivatives $\partial z/\partial x$ and $\partial z/\partial y$ when z is defined implicitly by an equation $F(x, y, z) = 0$:

$$\frac{\partial z}{\partial x} = -\frac{F_x}{F_z}, \qquad \frac{\partial z}{\partial y} = -\frac{F_y}{F_z}$$

14.6 EXERCISES

Preliminary Questions

1. Let $f(x, y) = xy$, where $x = uv$ and $y = u + v$.

(a) What are the primary derivatives of f?

(b) What are the independent variables?

In Questions 2 and 3, suppose that $f(u, v) = ue^v$, where $u = rs$ and $v = r + s$.

2. The composite function $f(u, v)$ is equal to:

(a) rse^{r+s} **(b)** re^s **(c)** rse^{rs}

3. What is the value of $f(u, v)$ at $(r, s) = (1, 1)$?

4. According to the Chain Rule, $\partial f/\partial r$ is equal to (choose the correct answer):

(a) $\dfrac{\partial f}{\partial x}\dfrac{\partial x}{\partial r} + \dfrac{\partial f}{\partial x}\dfrac{\partial x}{\partial s}$

(b) $\dfrac{\partial f}{\partial x}\dfrac{\partial x}{\partial r} + \dfrac{\partial f}{\partial y}\dfrac{\partial y}{\partial r}$

(c) $\dfrac{\partial f}{\partial r}\dfrac{\partial r}{\partial x} + \dfrac{\partial f}{\partial s}\dfrac{\partial s}{\partial x}$

5. Suppose that x, y, z are functions of the independent variables u, v, w. Which of the following terms appear in the Chain Rule expression for $\partial f/\partial w$?

(a) $\dfrac{\partial f}{\partial v}\dfrac{\partial x}{\partial v}$ **(b)** $\dfrac{\partial f}{\partial w}\dfrac{\partial w}{\partial x}$ **(c)** $\dfrac{\partial f}{\partial z}\dfrac{\partial z}{\partial w}$

6. With notation as in the previous question, does $\partial x/\partial v$ appear in the Chain Rule expression for $\partial f/\partial u$?

Exercises

1. Let $f(x, y, z) = x^2 y^3 + z^4$ and $x = s^2$, $y = st^2$, and $z = s^2 t$.

(a) Calculate the primary derivatives $\dfrac{\partial f}{\partial x}, \dfrac{\partial f}{\partial y}, \dfrac{\partial f}{\partial z}$.

(b) Calculate $\dfrac{\partial x}{\partial s}, \dfrac{\partial y}{\partial s}, \dfrac{\partial z}{\partial s}$.

(c) Compute $\dfrac{\partial f}{\partial s}$ using the Chain Rule:

$$\frac{\partial f}{\partial s} = \frac{\partial f}{\partial x}\frac{\partial x}{\partial s} + \frac{\partial f}{\partial y}\frac{\partial y}{\partial s} + \frac{\partial f}{\partial z}\frac{\partial z}{\partial s}$$

Express the answer in terms of the independent variables s, t.

2. Let $f(x, y) = x\cos(y)$ and $x = u^2 + v^2$ and $y = u - v$.

(a) Calculate the primary derivatives $\dfrac{\partial f}{\partial x}, \dfrac{\partial f}{\partial y}$.

(b) Use the Chain Rule to calculate $\partial f/\partial v$. Leave the answer in terms of both the dependent and the independent variables.

(c) Determine (x, y) for $(u, v) = (2, 1)$ and evaluate $\partial f/\partial v$ at $(u, v) = (2, 1)$.

In Exercises 3–10, use the Chain Rule to calculate the partial derivatives. Express the answer in terms of the independent variables.

3. $\dfrac{\partial f}{\partial s}, \dfrac{\partial f}{\partial r}$; $f(x, y, z) = xy + z^2$, $x = s^2$, $y = 2rs$, $z = r^2$

4. $\dfrac{\partial f}{\partial r}, \dfrac{\partial f}{\partial t}$; $f(x, y, z) = xy + z^2$, $x = r + s - 2t$, $y = 3rt$, $z = s^2$

5. $\dfrac{\partial g}{\partial u}, \dfrac{\partial g}{\partial v}$; $g(x, y) = \cos(x - y)$, $x = 3u - 5v$, $y = -7u + 15v$

6. $\dfrac{\partial R}{\partial u}, \dfrac{\partial R}{\partial v}$; $R(x, y) = (3x + 4y)^5$, $x = u^2$, $y = uv$

7. $\dfrac{\partial F}{\partial y}$; $F(u, v) = e^{u+v}$, $u = x^2$, $v = xy$

8. $\dfrac{\partial f}{\partial u}$; $f(x, y) = x^2 + y^2$, $x = e^{u+v}$, $y = u + v$

9. $\dfrac{\partial h}{\partial t_2}$; $h(x, y) = \dfrac{x}{y}$, $x = t_1 t_2$, $y = t_1^2 t_2$

10. $\dfrac{\partial f}{\partial \theta}$; $f(x, y, z) = xy - z^2$, $x = r\cos\theta$, $y = \cos^2\theta$, $z = r$

In Exercises 11–16, use the Chain Rule to evaluate the partial derivative at the point specified.

11. $\partial f/\partial u$ and $\partial f/\partial v$ at $(u, v) = (-1, -1)$, where $f(x, y, z) = x^3 + yz^2$, $x = u^2 + v$, $y = u + v^2$, $z = uv$

12. $\partial f/\partial s$ at $(r, s) = (1, 0)$, where $f(x, y) = \ln(xy)$, $x = 3r + 2s$, $y = 5r + 3s$

13. $\partial g/\partial \theta$ at $(r, \theta) = (2\sqrt{2}, \frac{\pi}{4})$, where $g(x, y) = 1/(x + y^2)$, $x = r\cos\theta$, $y = r\sin\theta$

14. $\partial g/\partial s$ at $s = 4$, where $g(x, y) = x^2 - y^2$, $x = s^2 + 1$, $y = 1 - 2s$

15. $\partial g/\partial u$ at $(u, v) = (0, 1)$, where $g(x, y) = x^2 - y^2$, $x = e^u \cos v$, $y = e^u \sin v$

16. $\dfrac{\partial h}{\partial q}$ at $(q, r) = (3, 2)$, where $h(u, v) = ue^v$, $u = q^3$, $v = qr^2$

17. A baseball player hits the ball and then runs down the first base line at 20 ft/s. The first baseman fields the ball and then runs toward first base along the second base line at 18 ft/s as in Figure 7.

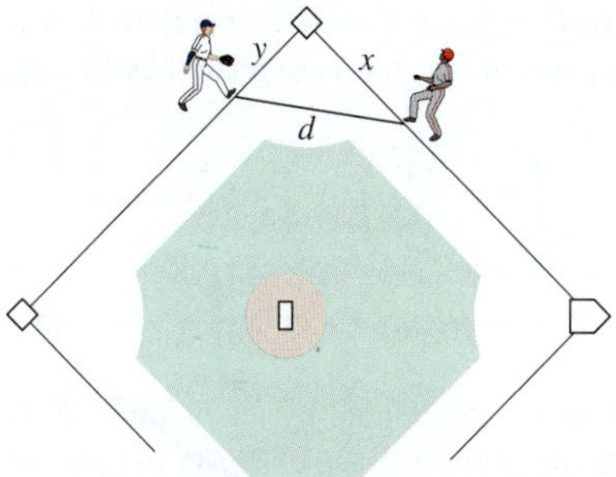

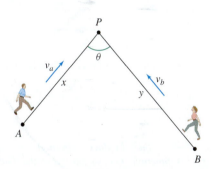

FIGURE 7

Determine how fast the distance between the two players is changing at a moment when the hitter is 8 ft from first base and the first baseman is 6 ft from first base.

18. Jessica and Matthew are running toward the point P along the straight paths that make a fixed angle of θ (Figure 8). Suppose that Matthew runs with velocity v_a meters per second and Jessica with velocity v_b meters per second. Let $f(x, y)$ be the distance from Matthew to Jessica when Matthew is x meters from P and Jessica is y meters from P.

(a) Show that $f(x, y) = \sqrt{x^2 + y^2 - 2xy \cos \theta}$.

(b) Assume that $\theta = \pi/3$. Use the Chain Rule to determine the rate at which the distance between Matthew and Jessica is changing when $x = 30$, $y = 20$, $v_a = 4$ m/s, and $v_b = 3$ m/s.

FIGURE 8

19. Two spacecraft are following paths in space given by $\mathbf{r}_1 = \left\langle \sin t, t, t^2 \right\rangle$ and $\mathbf{r}_2 = \left\langle \cos t, 1 - t, t^3 \right\rangle$ If the temperature for points in space are given by $T(x, y, z) = x^2 y(1 - z)$, use the Chain Rule to determine the rate of change of the difference D in the temperatures the two spacecraft experience at time $t = \pi$.

20. The Law of Cosines states that $c^2 = a^2 + b^2 - 2ab \cos \theta$, where a, b, c are the sides of a triangle and θ is the angle opposite the side of length c.

(a) Compute $\partial\theta/\partial a$, $\partial\theta/\partial b$, and $\partial\theta/\partial c$ using implicit differentiation.

(b) Suppose that $a = 10$, $b = 16$, $c = 22$. Estimate the change in θ if a and b are increased by 1 and c is increased by 2.

21. Let $u = u(x, y)$, and let (r, θ) be polar coordinates. Verify the relation

$$\|\nabla u\|^2 = u_r^2 + \frac{1}{r^2} u_\theta^2 \qquad \boxed{8}$$

Hint: Compute the right-hand side by expressing u_θ and u_r in terms of u_x and u_y.

22. Let $u(r, \theta) = r^2 \cos^2 \theta$. Use Eq. (8) to compute $\|\nabla u\|^2$. Then compute $\|\nabla u\|^2$ directly by observing that $u(x, y) = x^2$, and compare.

23. Let $x = s + t$ and $y = s - t$. Show that for any differentiable function $f(x, y)$,

$$\left(\frac{\partial f}{\partial x}\right)^2 - \left(\frac{\partial f}{\partial y}\right)^2 = \frac{\partial f}{\partial s}\frac{\partial f}{\partial t}$$

24. Express the derivatives

$$\frac{\partial f}{\partial \rho}, \frac{\partial f}{\partial \theta}, \frac{\partial f}{\partial \phi} \quad \text{in terms of} \quad \frac{\partial f}{\partial x}, \frac{\partial f}{\partial y}, \frac{\partial f}{\partial z}$$

where (ρ, θ, ϕ) are spherical coordinates.

25. Suppose that z is defined implicitly as a function of x and y by the equation $F(x, y, z) = xz^2 + y^2 z + xy - 1 = 0$.

(a) Calculate F_x, F_y, F_z.

(b) Use Eq. (7) to calculate $\dfrac{\partial z}{\partial x}$ and $\dfrac{\partial z}{\partial y}$.

26. Calculate $\partial z/\partial x$ and $\partial z/\partial y$ at the points $(3, 2, 1)$ and $(3, 2, -1)$, where z is defined implicitly by the equation $z^4 + z^2 x^2 - y - 8 = 0$.

In Exercises 27–32, calculate the partial derivative using implicit differentiation.

27. $\dfrac{\partial z}{\partial x}$, $x^2 y + y^2 z + xz^2 = 10$

28. $\dfrac{\partial w}{\partial z}$, $x^2 w + w^3 + wz^2 + 3yz = 0$

29. $\dfrac{\partial z}{\partial y}$, $e^{xy} + \sin(xz) + y = 0$

30. $\dfrac{\partial r}{\partial t}$ and $\dfrac{\partial t}{\partial r}$, $r^2 = t e^{s/r}$

31. $\dfrac{\partial w}{\partial y}$, $\dfrac{1}{w^2 + x^2} + \dfrac{1}{w^2 + y^2} = 1$ at $(x, y, w) = (1, 1, 1)$

32. $\partial U/\partial T$ and $\partial T/\partial U$, $(TU - V)^2 \ln(W - UV) = 1$ at $(T, U, V, W) = (1, 1, 2, 4)$

33. Let $\mathbf{r} = \langle x, y, z \rangle$ and $e_{\mathbf{r}} = \mathbf{r}/\|\mathbf{r}\|$. Show that if a function $f(x, y, z) = F(r)$ depends only on the distance from the origin $r = \|\mathbf{r}\| = \sqrt{x^2 + y^2 + z^2}$, then

$$\nabla f = F'(r)e_{\mathbf{r}} \qquad \boxed{9}$$

34. Let $f(x, y, z) = e^{-x^2 - y^2 - z^2} = e^{-r^2}$, with r as in Exercise 33. Compute ∇f directly and using Eq. (9).

35. Use Eq. (9) to compute $\nabla\left(\dfrac{1}{r}\right)$.

36. Use Eq. (9) to compute $\nabla(\ln r)$.

37. Figure 9 shows the graph of the equation

$$F(x, y, z) = x^2 + y^2 - z^2 - 12x - 8z - 4 = 0$$

(a) Use the quadratic formula to solve for z as a function of x and y. This gives two formulas, depending on the choice of sign.

(b) Which formula defines the portion of the surface satisfying $z \geq -4$? Which formula defines the portion satisfying $z \leq -4$?

(c) Calculate $\partial z/\partial x$ using the formula $z = f(x, y)$ (for both choices of sign) and again via implicit differentiation. Verify that the two answers agree.

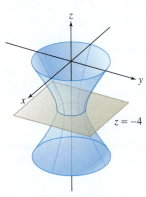

FIGURE 9 Graph of $x^2 + y^2 - z^2 - 12x - 8z - 4 = 0$.

38. For all $x > 0$, there is a unique value $y = r(x)$ that solves the equation $y^3 + 4xy = 16$.

(a) Show that $dy/dx = -4y/(3y^2 + 4x)$.

(b) Let $g(x) = f(x, r(x))$, where $f(x, y)$ is a function satisfying

$$f_x(1, 2) = 8, \quad f_y(1, 2) = 10$$

Use the Chain Rule to calculate $g'(1)$. Note that $r(1) = 2$ because $(x, y) = (1, 2)$ satisfies $y^3 + 4xy = 16$.

39. The pressure P, volume V, and temperature T of a van der Waals gas with n molecules (n constant) are related by the equation

$$\left(P + \frac{an^2}{V^2}\right)(V - nb) = nRT$$

where a, b, and R are constant. Calculate $\partial P/\partial T$ and $\partial V/\partial P$.

40. When x, y, and z are related by an equation $F(x, y, z) = 0$, we sometimes write $(\partial z/\partial x)_y$ in place of $\partial z/\partial x$ to indicate that in the differentiation, z is treated as a function of x with y held constant (and similarly for the other variables).

(a) Use Eq. (7) to prove the **cyclic relation**

$$\left(\frac{\partial z}{\partial x}\right)_y \left(\frac{\partial x}{\partial y}\right)_z \left(\frac{\partial y}{\partial z}\right)_x = -1 \qquad \boxed{10}$$

(b) Verify Eq. (10) for $F(x, y, z) = x + y + z = 0$.

(c) Verify the cyclic relation for the variables P, V, T in the Ideal Gas Law $PV - nRT = 0$ (n and R are constants).

41. Show that if $f(x)$ is differentiable and $c \neq 0$ is a constant, then $u(x, t) = f(x - ct)$ satisfies the so-called **advection equation**

$$\frac{\partial u}{\partial t} + c\frac{\partial u}{\partial x} = 0$$

Further Insights and Challenges

*In Exercises 42–45, a function $f(x, y, z)$ is called **homogeneous of degree n** if $f(\lambda x, \lambda y, \lambda z) = \lambda^n f(x, y, z)$ for all $\lambda \in \mathbf{R}$.*

42. Show that the following functions are homogeneous and determine their degree:

(a) $f(x, y, z) = x^2 y + xyz$

(b) $f(x, y, z) = 3x + 2y - 8z$

(c) $f(x, y, z) = \ln\left(\dfrac{xy}{z^2}\right)$

(d) $f(x, y, z) = z^4$

43. Prove that if $f(x, y, z)$ is homogeneous of degree n, then $f_x(x, y, z)$ is homogeneous of degree $n - 1$. *Hint:* Either use the limit definition or apply the Chain Rule to $f(\lambda x, \lambda y, \lambda z)$.

44. Prove that if $f(x, y, z)$ is homogeneous of degree n, then

$$x\frac{\partial f}{\partial x} + y\frac{\partial f}{\partial y} + z\frac{\partial f}{\partial z} = nf \qquad \boxed{11}$$

Hint: Let $F(t) = f(tx, ty, tz)$ and calculate $F'(1)$ using the Chain Rule.

45. Verify Eq. (11) for the functions in Exercise 42.

46. Suppose that f is a function of x and y, where $x = g(t, s)$, $y = h(t, s)$. Show that f_{tt} is equal to

$$f_{xx}\left(\frac{\partial x}{\partial t}\right)^2 + 2f_{xy}\left(\frac{\partial x}{\partial t}\right)\left(\frac{\partial y}{\partial t}\right) + f_{yy}\left(\frac{\partial y}{\partial t}\right)^2$$

$$+ f_x\frac{\partial^2 x}{\partial t^2} + f_y\frac{\partial^2 y}{\partial t^2} \qquad \boxed{12}$$

47. Let $r = \sqrt{x_1^2 + \cdots + x_n^2}$ and let $g(r)$ be a function of r. Prove the formulas

$$\frac{\partial g}{\partial x_i} = \frac{x_i}{r}g_r, \qquad \frac{\partial^2 g}{\partial x_i^2} = \frac{x_i^2}{r^2}g_{rr} + \frac{r^2 - x_i^2}{r^3}g_r$$

48. Prove that if $g(r)$ is a function of r as in Exercise 47, then

$$\frac{\partial^2 g}{\partial x_1^2} + \cdots + \frac{\partial^2 g}{\partial x_n^2} = g_{rr} + \frac{n-1}{r}g_r$$

*In Exercises 49–53, the **Laplace operator** is defined by $\Delta f = f_{xx} + f_{yy}$. A function $f(x, y)$ satisfying the Laplace equation $\Delta f = 0$ is called **harmonic**. A function $f(x, y)$ is called **radial** if $f(x, y) = g(r)$, where $r = \sqrt{x^2 + y^2}$.*

49. Use Eq. (12) to prove that in polar coordinates (r, θ),

$$\Delta f = f_{rr} + \frac{1}{r^2}f_{\theta\theta} + \frac{1}{r}f_r \qquad \boxed{13}$$

50. Use Eq. (13) to show that $f(x, y) = \ln r$ is harmonic.

51. Verify that $f(x, y) = x$ and $f(x, y) = y$ are harmonic using both the rectangular and polar expressions for Δf.

52. Verify that $f(x, y) = \tan^{-1}\frac{y}{x}$ is harmonic using both the rectangular and polar expressions for Δf.

53. Use the Product Rule to show that

$$f_{rr} + \frac{1}{r}f_r = r^{-1}\frac{\partial}{\partial r}\left(r\frac{\partial f}{\partial r}\right)$$

Use this formula to show that if f is a radial harmonic function, then $rf_r = C$ for some constant C. Conclude that $f(x, y) = C\ln r + b$ for some constant b.

14.7 Optimization in Several Variables

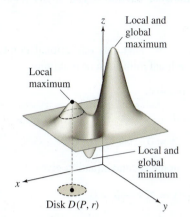

FIGURE 1 $f(x, y)$ has a local maximum at P.

Recall that optimization is the process of finding the extreme values of a function. This amounts to finding the highest and lowest points on the graph over a given domain. As we saw in the one-variable case, it is important to distinguish between *local* and *global* extreme values. A local extreme value is a value $f(a, b)$ that is a maximum or minimum in some small open disk around (a, b) (Figure 1).

> **DEFINITION** **Local Extreme Values** A function $f(x, y)$ has a **local extremum** at $P = (a, b)$ if there exists an open disk $D(P, r)$ such that
>
> - **Local maximum:** $f(x, y) \leq f(a, b)$ for all $(x, y) \in D(P, r)$
> - **Local minimum:** $f(x, y) \geq f(a, b)$ for all $(x, y) \in D(P, r)$

<-- *REMINDER The term "extremum" (the plural is "extrema") means a minimum or maximum value.*

Fermat's Theorem for functions of one variable states that if $f(a)$ is a local extreme value, then a is a critical point and thus the tangent line (if it exists) is horizontal at $x = a$. A similar result holds for functions of two variables, but in this case, it is the *tangent plane* that must be horizontal (Figure 2). The tangent plane to $z = f(x, y)$ at $P = (a, b)$ has equation

$$z = f(a, b) + f_x(a, b)(x - a) + f_y(a, b)(y - b)$$

Thus, the tangent plane is horizontal if $f_x(a, b) = f_y(a, b) = 0$—that is, if the equation reduces to $z = f(a, b)$. This leads to the following definition of a critical point, where we take into account the possibility that one or both partial derivatives do not exist.

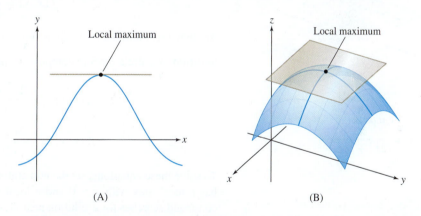

(A) (B)

FIGURE 2 The tangent line or plane is horizontal at a local extremum.

- *More generally, $(a_1, \ldots, a_n)$ is a critical point of $f(x_1, \ldots, x_n)$ if each partial derivative satisfies*

$$f_{x_j}(a_1, \ldots, a_n) = 0$$

or does not exist.
- *Theorem 1 holds in any number of variables: Local extrema occur at critical points.*

> **DEFINITION** **Critical Point** A point $P = (a, b)$ in the domain of $f(x, y)$ is called a **critical point** if:
>
> - $f_x(a, b) = 0$ or $f_x(a, b)$ does not exist, and
> - $f_y(a, b) = 0$ or $f_y(a, b)$ does not exist.

As in the single-variable case, we have the following:

> **THEOREM 1** **Fermat's Theorem** If $f(x, y)$ has a local minimum or maximum at $P = (a, b)$, then (a, b) is a critical point of $f(x, y)$.

Proof If $f(x, y)$ has a local minimum at $P = (a, b)$, then $f(x, y) \geq f(a, b)$ for all (x, y) near (a, b). In particular, there exists $r > 0$ such that $f(x, b) \geq f(a, b)$ if $|x - a| < r$. In other words, $g(x) = f(x, b)$ has a local minimum at $x = a$. By Fermat's Theorem for functions of one variable, either $g'(a) = 0$ or $g'(a)$ does not exist. Since $g'(a) = f_x(a, b)$, we conclude that either $f_x(a, b) = 0$ or $f_x(a, b)$ does not exist. Similarly, $f_y(a, b) = 0$ or $f_y(a, b)$ does not exist. Therefore, $P = (a, b)$ is a critical point. The case of a local maximum is similar. ∎

Usually, we deal with functions whose partial derivatives exist. In this case, finding the critical points amounts to solving the simultaneous equations $f_x(x, y) = 0$ and $f_y(x, y) = 0$.

■ **EXAMPLE 1** Show that $f(x, y) = 11x^2 - 2xy + 2y^2 + 3y$ has one critical point. Use Figure 3 to determine whether it corresponds to a local minimum or maximum.

Solution Set the partial derivatives equal to zero and solve:

$$f_x(x, y) = 22x - 2y = 0$$
$$f_y(x, y) = -2x + 4y + 3 = 0$$

By the first equation, $y = 11x$. Substituting $y = 11x$ in the second equation gives

$$-2x + 4y + 3 = -2x + 4(11x) + 3 = 42x + 3 = 0$$

Thus, $x = -\frac{1}{14}$ and $y = -\frac{11}{14}$. There is just one critical point, $P = \left(-\frac{1}{14}, -\frac{11}{14}\right)$. Figure 3 shows that $f(x, y)$ has a local minimum at P. ■

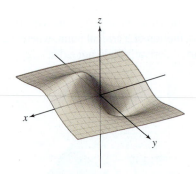

DF **FIGURE 3** Graph of $f(x, y) = 11x^2 - 2xy + 2y^2 + 3y$.

It is not always possible to find the solutions exactly, but we can use a computer to find numerical approximations.

■ **EXAMPLE 2** ⌂⌐⌐ **Numerical Example** Use a computer algebra system to approximate the critical points of

$$f(x, y) = \frac{x - y}{2x^2 + 8y^2 + 3}$$

Are they local minima or maxima? Refer to Figure 4.

Solution We use a CAS to compute the partial derivatives and solve

$$f_x(x, y) = \frac{-2x^2 + 8y^2 + 4xy + 3}{(2x^2 + 8y^2 + 3)^2} = 0$$

$$f_y(x, y) = \frac{-2x^2 + 8y^2 - 16xy - 3}{(2x^2 + 8y^2 + 3)^2} = 0$$

FIGURE 4 Graph of $f(x, y) = \dfrac{x - y}{2x^2 + 8y^2 + 3}$.

To solve these equations, set the numerators equal to zero. Figure 4 suggests that $f(x, y)$ has a local max with $x > 0$ and a local min with $x < 0$. The following Mathematica command searches for a solution near $(1, 0)$:

```
FindRoot[{-2x^2+8y^2+4xy+3 == 0, -2x^2+8y^2-16xy-3 == 0},
  {{x,1},{y,0}}]
```

The result is

```
{x -> 1.095, y -> -0.274}
```

Thus, $(1.095, -0.274)$ is an approximate critical point where, by Figure 4, f takes on a local maximum. A second search near $(-1, 0)$ yields $(-1.095, 0.274)$, which approximates the critical point where $f(x, y)$ takes on a local minimum. ■

We know that in one variable, a function $f(x)$ may have a point of inflection rather than a local extremum at a critical point. A similar phenomenon occurs in several variables. Each of the functions in Figure 5 has a critical point at $(0, 0)$. However, the function in Figure 5(C) has a saddle point, which is neither a local minimum nor a local maximum. If you stand at the saddle point and begin walking, some directions such as the $+\mathbf{j}$ or $-\mathbf{j}$ directions take you uphill and other directions such as the $+\mathbf{i}$ or $-\mathbf{i}$ directions take you downhill.

As in the one-variable case, there is a Second Derivative Test determining the type of a critical point (a, b) of a function $f(x, y)$ in two variables. This test relies on the sign of the **discriminant** $D = D(a, b)$, defined as follows:

The discriminant is also referred to as the "Hessian determinant."

$$D = D(a, b) = f_{xx}(a, b) f_{yy}(a, b) - f_{xy}^2(a, b)$$

We can remember the formula for the discriminant by recognizing it as a determinant:

$$D = \begin{vmatrix} f_{xx}(a, b) & f_{xy}(a, b) \\ f_{yx}(a, b) & f_{yy}(a, b) \end{vmatrix}$$

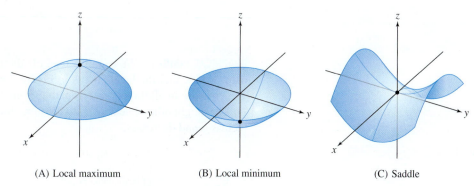

(A) Local maximum   (B) Local minimum   (C) Saddle

FIGURE 5

If $D > 0$, then $f_{xx}(a, b)$ and $f_{yy}(a, b)$ must have the same sign, so the sign of $f_{yy}(a, b)$ also determines whether $f(a, b)$ is a local minimum or a local maximum.

THEOREM 2 Second Derivative Test for $f(x, y)$ Let $P = (a, b)$ be a critical point of $f(x, y)$. Assume that f_{xx}, f_{yy}, f_{xy} are continuous near P. Then:

(i) If $D > 0$ and $f_{xx}(a, b) > 0$, then $f(a, b)$ is a local minimum.

(ii) If $D > 0$ and $f_{xx}(a, b) < 0$, then $f(a, b)$ is a local maximum.

(iii) If $D < 0$, then f has a saddle point at (a, b).

(iv) If $D = 0$, the test is inconclusive.

A proof of this theorem is discussed at the end of this section.

■ **EXAMPLE 3 Applying the Second Derivative Test**   Find the critical points of

$$f(x, y) = (x^2 + y^2)e^{-x}$$

and analyze them using the Second Derivative Test.

Solution

Step 1. **Find the critical points.**

Set the partial derivatives equal to zero and solve:

$$f_x(x, y) = -(x^2 + y^2)e^{-x} + 2xe^{-x} = (2x - x^2 - y^2)e^{-x} = 0$$

$$f_y(x, y) = 2ye^{-x} = 0 \quad \Rightarrow \quad y = 0$$

Substituting $y = 0$ in the first equation then gives

$$(2x - x^2 - y^2)e^{-x} = (2x - x^2)e^{-x} = 0 \quad \Rightarrow \quad x = 0, 2$$

The critical points are $(0, 0)$ and $(2, 0)$ (Figure 6).

Step 2. **Compute the second-order partials.**

$$f_{xx}(x, y) = \frac{\partial}{\partial x}\left((2x - x^2 - y^2)e^{-x}\right) = (2 - 4x + x^2 + y^2)e^{-x}$$

$$f_{yy}(x, y) = \frac{\partial}{\partial y}(2ye^{-x}) = 2e^{-x}$$

$$f_{xy}(x, y) = f_{yx}(x, y) = \frac{\partial}{\partial x}(2ye^{-x}) = -2ye^{-x}$$

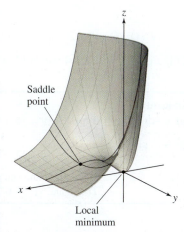

Saddle point

Local minimum

FIGURE 6 Graph of $f(x, y) = (x^2 + y^2)e^{-x}$.

Step 3. **Apply the Second Derivative Test.**

Critical Point	f_{xx}	f_{yy}	f_{xy}	Discriminant $D = f_{xx}f_{yy} - f_{xy}^2$	Type
$(0, 0)$	2	2	0	$2(2) - 0^2 = 4$	Local minimum since $D > 0$ and $f_{xx} > 0$
$(2, 0)$	$-2e^{-2}$	$2e^{-2}$	0	$-2e^{-2}(2e^{-2}) - 0^2 = -4e^{-4}$	Saddle since $D < 0$

■

GRAPHICAL INSIGHT   We can also read off the type of critical point from the contour map. Notice that the level curves in Figure 7 encircle the local minimum at P, with f increasing in all directions emanating from P. By contrast, f has a saddle point at Q: The neighborhood near Q is divided into four regions in which $f(x, y)$ alternately increases and decreases.

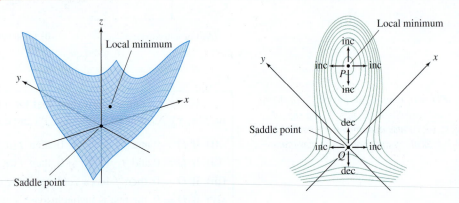

FIGURE 7 $f(x, y) = x^3 + y^3 - 12xy.$

■ **EXAMPLE 4**   Analyze the critical points of $f(x, y) = x^3 + y^3 - 12xy.$

Solution   Again, we set the partial derivatives equal to zero and solve:

$$f_x(x, y) = 3x^2 - 12y = 0 \quad \Rightarrow \quad y = \frac{1}{4}x^2$$

$$f_y(x, y) = 3y^2 - 12x = 0$$

Substituting $y = \frac{1}{4}x^2$ in the second equation yields

$$3y^2 - 12x = 3\left(\frac{1}{4}x^2\right)^2 - 12x = \frac{3}{16}x(x^3 - 64) = 0 \quad \Rightarrow \quad x = 0, 4$$

Since $y = \frac{1}{4}x^2$, the critical points are $(0, 0)$ and $(4, 4)$.
   We have

$$f_{xx}(x, y) = 6x, \qquad f_{yy}(x, y) = 6y, \qquad f_{xy}(x, y) = -12$$

The Second Derivative Test confirms what we see in Figure 7: f has a local min at $(4, 4)$ and a saddle at $(0, 0)$.

Critical Point	f_{xx}	f_{yy}	f_{xy}	Discriminant $D = f_{xx}f_{yy} - f_{xy}^2$	Type
$(0, 0)$	0	0	-12	$0(0) - 12^2 = -144$	Saddle since $D < 0$
$(4, 4)$	24	24	-12	$24(24) - 12^2 = 432$	Local minimum since $D > 0$ and $f_{xx} > 0$

■

■ **EXAMPLE 5**   **When the Second Derivative Test Fails**   Analyze the critical points of $f(x, y) = 3xy^2 - x^3.$

Solution We set the partial derivatives equal to zero and solve:

$$f_x(x, y) = 3y^2 - 3x^2 = 0$$

$$f_y(x, y) = 6xy = 0$$

From the second equation, either $x = 0$ or $y = 0$. From the first equation, we find that the only critical point is $(0, 0)$. We have

$$f_{xx}(x, y) = -6x, \qquad f_{yy}(x, y) = 6x, \qquad f_{xy}(x, y) = 6y$$

Applying the Second Derivative Test, we obtain

Critical Point	f_{xx}	f_{yy}	f_{xy}	Discriminant $D = f_{xx}f_{yy} - f_{xy}^2$	Type
$(0, 0)$	0	0	0	0	No information since $D = 0$

Thus, we need to analyze this critical point by examining the graph more carefully. We will slice the graph with the xz-plane by setting $y = 0$. Then $f(x, 0) = -x^3$. A graph of this function appears as in Figure 8. When it passes through $(0, 0)$, it has neither a local maximum nor local minimum, so the critical point $(0, 0)$ is not an extreme point. It is a saddle point.

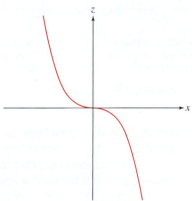

FIGURE 8 Intersection of the graph of $f(x, y) = 3xy^2 - x^3$ with the xz-plane.

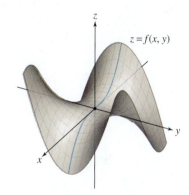

FIGURE 9 Graph of a "monkey saddle" with equation $f(x, y) = 3xy^2 - x^3$.

However, graphs can take on a variety of different shapes at a saddle point. The graph of $f(x, y)$ appears in Figure 9. The curve $f(x, 0) = -x^3$ is shown in blue. This surface is called a "monkey saddle" because a monkey can sit on this saddle with room for its tail in the back. ■

Global Extrema

Often we are interested in finding the minimum or maximum value of a function f on a given domain $\mathcal{D}$. These are called **global** or **absolute extreme values**. However, global extrema do not always exist. The function $f(x, y) = x + y$ has a maximum value on the unit square $\mathcal{D}_1$ in Figure 10 [the max is $f(1, 1) = 2$], but it has no maximum value on the entire plane $\mathbf{R}^2$.

To state conditions that guarantee the existence of global extrema, we need a few definitions. First, we say that a domain $\mathcal{D}$ is **bounded** if there is a number $M > 0$ such that $\mathcal{D}$ is contained in a disk of radius M centered at the origin. In other words, no point of $\mathcal{D}$ is more than a distance M from the origin [Figures 12(A) and 12(B)]. Next, a point P is called:

- An **interior point** of $\mathcal{D}$ if $\mathcal{D}$ contains some open disk $D(P, r)$ centered at P.
- A **boundary point** of $\mathcal{D}$ if every disk centered at P contains points in $\mathcal{D}$ and points not in $\mathcal{D}$.

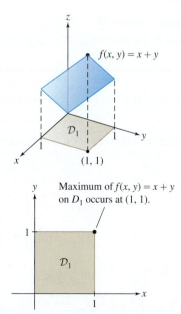

FIGURE 10

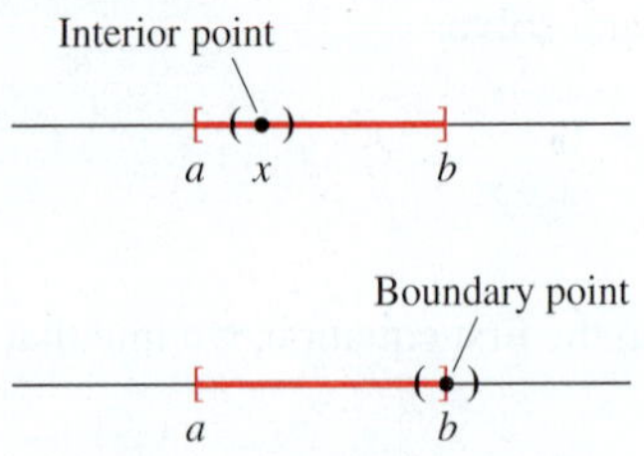

Interior point

FIGURE 11 Interior and boundary points of an interval $[a, b]$.

CONCEPTUAL INSIGHT To understand the concept of interior and boundary points, think of the familiar case of an interval $I = [a, b]$ in the real line **R** (Figure 11). Every point x in the open interval (a, b) is an *interior point* of I (because there exists a small open interval around x entirely contained in I). The two endpoints a and b are *boundary points* (because every open interval containing a or b also contains points not in I).

The **interior** of $\mathcal{D}$ is the set of all interior points, and the **boundary** of $\mathcal{D}$ is the set of all boundary points. In Figure 12(C), the boundary is the curve surrounding the domain. The interior consists of all points in the domain not lying on the boundary curve.

A domain $\mathcal{D}$ is called **closed** if $\mathcal{D}$ contains all its boundary points (like a closed interval in **R**). A domain $\mathcal{D}$ is called **open** if every point of $\mathcal{D}$ is an interior point (like an open interval in **R**). The domain in Figure 12(A) is closed because the domain includes its boundary curve. In Figure 12(C), some boundary points are included and some are excluded, so the domain is neither open nor closed.

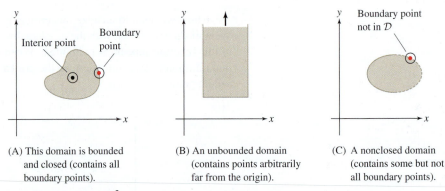

(A) This domain is bounded and closed (contains all boundary points).

(B) An unbounded domain (contains points arbitrarily far from the origin).

(C) A nonclosed domain (contains some but not all boundary points).

FIGURE 12 Domains in **R**^{2}.

In Section 4.2, we stated two basic results. First, a continuous function $f(x)$ on a *closed, bounded interval* $[a, b]$ takes on both a minimum and a maximum value on $[a, b]$. Second, these extreme values occur either at critical points in the interior (a, b) or at the endpoints. Analogous results are valid in several variables.

THEOREM 3 Existence and Location of Global Extrema Let $f(x, y)$ be a continuous function on a closed, bounded domain $\mathcal{D}$ in **R**^{2}. Then:

(i) $f(x, y)$ takes on both a minimum and a maximum value on $\mathcal{D}$.

(ii) The extreme values occur either at critical points in the interior of $\mathcal{D}$ or at points on the boundary of $\mathcal{D}$.

■ **EXAMPLE 6** Find the maximum value of $f(x, y) = 2x + y - 3xy$ on the unit square $\mathcal{D} = \{(x, y) : 0 \le x, y \le 1\}$.

Solution By Theorem 3, the maximum occurs either at a critical point or on the boundary of the square (Figure 13).

Step 1. **Examine the critical points.**
Set the partial derivatives equal to zero and solve:

$$f_x(x, y) = 2 - 3y = 0 \quad \Rightarrow \quad y = \frac{2}{3}, \qquad f_y(x, y) = 1 - 3x = 0 \quad \Rightarrow \quad x = \frac{1}{3}$$

There is a unique critical point $P = \left(\frac{1}{3}, \frac{2}{3}\right)$ and

$$f(P) = f\left(\frac{1}{3}, \frac{2}{3}\right) = 2\left(\frac{1}{3}\right) + \left(\frac{2}{3}\right) - 3\left(\frac{1}{3}\right)\left(\frac{2}{3}\right) = \frac{2}{3}$$

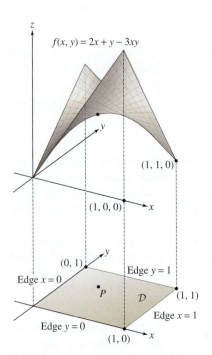

FIGURE 13

Step 2. **Check the boundary.**

We do this by checking each of the four edges of the square separately. The bottom edge is described by $y = 0, 0 \le x \le 1$. On this edge, $f(x, 0) = 2x$, and the maximum value occurs at $x = 1$, where $f(1, 0) = 2$. Proceeding in a similar fashion with the other edges, we obtain

Edge	Restriction of $f(x, y)$ to Edge	Maximum of $f(x, y)$ on Edge
Lower: $y = 0, 0 \le x \le 1$	$f(x, 0) = 2x$	$f(1, 0) = 2$
Upper: $y = 1, 0 \le x \le 1$	$f(x, 1) = 1 - x$	$f(0, 1) = 1$
Left: $x = 0, 0 \le y \le 1$	$f(0, y) = y$	$f(0, 1) = 1$
Right: $x = 1, 0 \le y \le 1$	$f(1, y) = 2 - 2y$	$f(1, 0) = 2$

Step 3. **Compare.**

The maximum of f on the boundary is $f(1, 0) = 2$. This is larger than the value $f(P) = \frac{2}{3}$ at the critical point, so the maximum of f on the unit square is 2. ∎

■ **EXAMPLE 7** Find the maximum and minimum values of the function $f(x, y) = xy$ on the disk $\mathcal{D} = \{(x, y) : x^2 + y^2 \le 1\}$.

Solution

Step 1. **Examine the critical points.**

$$f_x(x, y) = y = 0, \qquad f_y(x, y) = x = 0$$

There is a unique critical point $P = (0, 0)$ in the interior of the disk, and $f(0, 0) = 0$.

Step 2. **Check the boundary.**

As in Figure 14, we subdivide the boundary into two arcs labeled I and II. The first is given by $y = +\sqrt{1 - x^2}, -1 \le x \le 1$. Restricting f to this part of the boundary, we have $f(x, \sqrt{1 - x^2}) = x\sqrt{1 - x^2}$. Thus,

$$f'(x) = \sqrt{1 - x^2} - x \frac{x}{\sqrt{1 - x^2}} = 0$$

$$1 - 2x^2 = 0$$

$$x = \pm \frac{1}{\sqrt{2}}$$

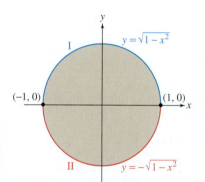

y

I $y = \sqrt{1 - x^2}$

$(-1, 0)$ $(1, 0)$ x

II $y = -\sqrt{1 - x^2}$

FIGURE 14 Dividing the boundary of the domain into arcs.

Since $y = \sqrt{1 - x^2}$, we have the two critical points on Arc I given by $\left(\frac{1}{\sqrt{2}}, \frac{1}{\sqrt{2}}\right)$ and $\left(-\frac{1}{\sqrt{2}}, \frac{1}{\sqrt{2}}\right)$.

Restricting f to Arc II, which is given by $y = -\sqrt{1 - x^2}, -1 \le x \le 1$, our function becomes $f(x, -\sqrt{1 - x^2}) = -x\sqrt{1 - x^2}$. Since this only differs by a sign from the restriction of our function to Arc I, we know that the critical points along this arc are $\left(\frac{1}{\sqrt{2}}, -\frac{1}{\sqrt{2}}\right)$ and $\left(-\frac{1}{\sqrt{2}}, -\frac{1}{\sqrt{2}}\right)$.

Step 3. **Compare.**

Evaluating f at the interior critical point, each of these four critical points and at the endpoints of the arcs, we find

$$f(0, 0) = 0, \; f\left(\frac{1}{\sqrt{2}}, \frac{1}{\sqrt{2}}\right) = \frac{1}{2}, \; f\left(-\frac{1}{\sqrt{2}}, \frac{1}{\sqrt{2}}\right) = -\frac{1}{2}, \; f\left(\frac{1}{\sqrt{2}}, -\frac{1}{\sqrt{2}}\right) = -\frac{1}{2},$$

$$f\left(-\frac{1}{\sqrt{2}}, -\frac{1}{\sqrt{2}}\right) = \frac{1}{2}, \; f(1, 0) = 0, \; f(-1, 0) = 0$$

Comparing these values, we see that the maximum value of $\frac{1}{2}$ over the disk occurs at the two boundary points $\left(\frac{1}{\sqrt{2}}, \frac{1}{\sqrt{2}}\right)$ and $\left(-\frac{1}{\sqrt{2}}, -\frac{1}{\sqrt{2}}\right)$, and the minimum value of $-\frac{1}{2}$ occurs at the two boundary points $\left(-\frac{1}{\sqrt{2}}, \frac{1}{\sqrt{2}}\right)$ and $\left(\frac{1}{\sqrt{2}}, -\frac{1}{\sqrt{2}}\right)$. ∎

■ **EXAMPLE 8　Box of Maximum Volume**　Find the maximum volume of a box inscribed in the tetrahedron bounded by the coordinate planes and the plane $\frac{1}{3}x + y + z = 1$.

Solution

Step 1. Find a function to be maximized.

Let $P = (x, y, z)$ be the corner of the box lying on the front face of the tetrahedron (Figure 15). Then the box has sides of lengths x, y, z and volume $V = xyz$. Using $\frac{1}{3}x + y + z = 1$, or $z = 1 - \frac{1}{3}x - y$, we express V in terms of x and y:

$$V(x, y) = xyz = xy\left(1 - \frac{1}{3}x - y\right) = xy - \frac{1}{3}x^2 y - xy^2$$

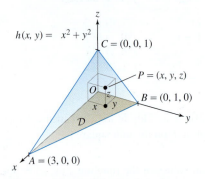

$h(x, y) = x^2 + y^2$

$C = (0, 0, 1)$

$P = (x, y, z)$

O

$B = (0, 1, 0)$

$\mathcal{D}$

$A = (3, 0, 0)$

FIGURE 15 The shaded triangle is the domain of $V(x, y)$.

Our problem is to maximize V, but which domain $\mathcal{D}$ should we choose? We let $\mathcal{D}$ be the shaded triangle $\triangle OAB$ in the xy-plane in Figure 15. Then the corner point $P = (x, y, z)$ of each possible box lies above a point (x, y) in $\mathcal{D}$. Because $\mathcal{D}$ is closed and bounded, the maximum occurs at a critical point inside $\mathcal{D}$ or on the boundary of $\mathcal{D}$.

Step 2. Examine the critical points.

First, set the partial derivatives equal to zero and solve:

$$\frac{\partial V}{\partial x} = y - \frac{2}{3}xy - y^2 = y\left(1 - \frac{2}{3}x - y\right) = 0$$

$$\frac{\partial V}{\partial y} = x - \frac{1}{3}x^2 - 2xy = x\left(1 - \frac{1}{3}x - 2y\right) = 0$$

If $x = 0$ or $y = 0$, then (x, y) lies on the boundary of $\mathcal{D}$, so assume that x and y are both nonzero. Then the first equation gives us

$$1 - \frac{2}{3}x - y = 0 \quad \Rightarrow \quad y = 1 - \frac{2}{3}x$$

The second equation yields

$$1 - \frac{1}{3}x - 2y = 1 - \frac{1}{3}x - 2\left(1 - \frac{2}{3}x\right) = 0 \quad \Rightarrow \quad x - 1 = 0 \quad \Rightarrow \quad x = 1$$

For $x = 1$, we have $y = 1 - \frac{2}{3}x = \frac{1}{3}$. Therefore, $\left(1, \frac{1}{3}\right)$ is a critical point, and

$$V\left(1, \frac{1}{3}\right) = (1)\frac{1}{3} - \frac{1}{3}(1)^2 \frac{1}{3} - (1)\left(\frac{1}{3}\right)^2 = \frac{1}{9}$$

Step 3. Check the boundary.

We have $V(x, y) = 0$ for all points on the boundary of $\mathcal{D}$ (because the three edges of the boundary are defined by $x = 0$, $y = 0$, and $1 - \frac{1}{3}x - y = 0$). Clearly, then, the maximum occurs at the critical point, and the maximum volume is $\frac{1}{9}$. ■

Proof of the Second Derivative Test　The proof is based on "completing the square" for quadratic forms. A **quadratic form** is a function

$$\boxed{Q(h, k) = ah^2 + 2bhk + ck^2}$$

where a, b, c are constants (not all zero). The discriminant of Q is the quantity

$$\boxed{D = ac - b^2}$$

Some quadratic forms take on only positive values for $(h, k) \neq (0, 0)$, and others take on both positive and negative values. According to the next theorem, the sign of the discriminant determines which of these two possibilities occurs.

To illustrate Theorem 4, consider

$$Q(h, k) = h^2 + 2hk + 2k^2$$

It has a positive discriminant

$$D = (1)(2) - 1 = 1$$

We can see directly that $Q(h, k)$ takes on only positive values for $(h, k) \neq (0, 0)$ by writing $Q(h, k)$ as

$$Q(h, k) = (h + k)^2 + k^2$$

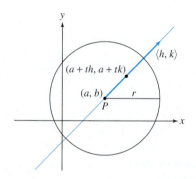

FIGURE 16 Line through P in the direction of $\langle h, k \rangle$.

THEOREM 4   With $Q(h, k)$ and D as above:

  (i) If $D > 0$ and $a > 0$, then $Q(h, k) > 0$ for $(h, k) \neq (0, 0)$.
  (ii) If $D > 0$ and $a < 0$, then $Q(h, k) < 0$ for $(h, k) \neq (0, 0)$.
  (iii) If $D < 0$, then $Q(h, k)$ takes on both positive and negative values.

Proof Assume first that $a \neq 0$ and rewrite $Q(h, k)$ by "completing the square":

$$Q(h, k) = ah^2 + 2bhk + ck^2 = a\left(h + \frac{b}{a}k\right)^2 + \left(c - \frac{b^2}{a}\right)k^2$$

$$= a\left(h + \frac{b}{a}k\right)^2 + \frac{D}{a}k^2 \qquad \boxed{1}$$

If $D > 0$ and $a > 0$, then $D/a > 0$ and both terms in Eq. (1) are nonnegative. Furthermore, if $Q(h, k) = 0$, then each term in Eq. (1) must equal zero. Thus, $k = 0$ and $h + \frac{b}{a}k = 0$, and then, necessarily, $h = 0$. This shows that $Q(h, k) > 0$ if $(h, k) \neq 0$, and (i) is proved. Part (ii) follows similarly. To prove (iii), note that if $a \neq 0$ and $D < 0$, then the coefficients of the squared terms in Eq. (1) have opposite signs and $Q(h, k)$ takes on both positive and negative values. Finally, if $a = 0$ and $D < 0$, then $Q(h, k) = 2bhk + ck^2$ with $b \neq 0$. In this case, $Q(h, k)$ again takes on both positive and negative values.

Now assume that $f(x, y)$ has a critical point at $P = (a, b)$. We shall analyze f by considering the restriction of $f(x, y)$ to the line (Figure 16) through $P = (a, b)$ in the direction of a unit vector $\langle h, k \rangle$:

$$F(t) = f(a + th, b + tk)$$

Then $F(0) = f(a, b)$. By the Chain Rule,

$$F'(t) = f_x(a + th, b + tk)h + f_y(a + th, b + tk)k$$

Because P is a critical point, we have $f_x(a, b) = f_y(a, b) = 0$, and therefore,

$$F'(0) = f_x(a, b)h + f_y(a, b)k = 0$$

Thus, $t = 0$ is a critical point of $F(t)$.

Now apply the Chain Rule again:

$$F''(t) = \frac{d}{dt}\left(f_x(a + th, b + tk)h + f_y(a + th, b + tk)k\right)$$

$$= \left(f_{xx}(a + th, b + tk)h^2 + f_{xy}(a + th, b + tk)hk\right)$$

$$+ \left(f_{yx}(a + th, b + tk)kh + f_{yy}(a + th, b + tk)k^2\right)$$

$$= f_{xx}(a + th, b + tk)h^2 + 2f_{xy}(a + th, b + tk)hk + f_{yy}(a + th, b + tk)k^2 \quad \boxed{2}$$

We see that $F''(t)$ is the value at (h, k) of a quadratic form whose discriminant is equal to $D(a + th, b + tk)$. Here, we set

$$D(r, s) = f_{xx}(r, s)f_{yy}(r, s) - f_{xy}(r, s)^2$$

Note that the discriminant of $f(x, y)$ at the critical point $P = (a, b)$ is $D = D(a, b)$.

Case 1: $D(a, b) > 0$ and $f_{xx}(a, b) > 0$. We must prove that $f(a, b)$ is a local minimum. Consider a small disk of radius r around P (Figure 16). Because the second derivatives are continuous near P, we can choose $r > 0$ so that for every unit vector $\langle h, k \rangle$,

$$D(a + th, b + tk) > 0 \qquad \text{for } |t| < r$$

$$f_{xx}(a + th, b + tk) > 0 \qquad \text{for } |t| < r$$

Then $F''(t)$ is positive for $|t| < r$ by Theorem 4(i). This tells us that $F(t)$ is concave up, and hence, $F(0) < F(t)$ if $0 < |t| < |r|$ (see Exercise 68 in Section 4.4). Because $F(0) = f(a, b)$, we may conclude that $f(a, b)$ is the minimum value of f along each segment of radius r through (a, b). Therefore, $f(a, b)$ is a local minimum value of f as claimed. The case that $D(a, b) > 0$ and $f_{xx}(a, b) < 0$ is similar.

Case 2: $D(a, b) < 0$. For $t = 0$, Eq. (2) yields

$$F''(0) = f_{xx}(a, b)h^2 + 2f_{xy}(a, b)hk + f_{yy}(a, b)k^2$$

Since $D(a, b) < 0$, this quadratic form takes on both positive and negative values by Theorem 4(iii). Choose $\langle h, k \rangle$ for which $F''(0) > 0$. By the Second Derivative Test in one variable, $F(0)$ is a local minimum of $F(t)$, and hence, there is a value $r > 0$ such that $F(0) < F(t)$ for all $0 < |t| < r$. However, we can also choose $\langle h, k \rangle$ so that $F''(0) < 0$, in which case $F(0) > F(t)$ for $0 < |t| < r$ for some $r > 0$. Because $F(0) = f(a, b)$, we conclude that $f(a, b)$ is a local min in some directions and a local max in other directions. Therefore, f has a saddle point at $P = (a, b)$. ■

14.7 SUMMARY

- We say that $P = (a, b)$ is a *critical point* of $f(x, y)$ if

  - $f_x(a, b) = 0$ or $f_x(a, b)$ does not exist, and
  - $f_y(a, b) = 0$ or $f_y(a, b)$ does not exist.

  In n-variables, $P = (a_1, \ldots, a_n)$ is a critical point of $f(x_1, \ldots, x_n)$ if each partial derivative $f_{x_j}(a_1, \ldots, a_n)$ either is zero or does not exist.
- The local minimum or maximum values of f occur at critical points.
- The *discriminant* of $f(x, y)$ at $P = (a, b)$ is the quantity

$$D(a, b) = f_{xx}(a, b)f_{yy}(a, b) - f_{xy}^2(a, b)$$

- Second Derivative Test: If $P = (a, b)$ is a critical point of $f(x, y)$, then

$$D(a, b) > 0, \quad f_{xx}(a, b) > 0 \quad \Rightarrow \quad f(a, b) \text{ is a local minimum}$$

$$D(a, b) > 0, \quad f_{xx}(a, b) < 0 \quad \Rightarrow \quad f(a, b) \text{ is a local maximum}$$

$$D(a, b) < 0 \quad \Rightarrow \quad \text{saddle point}$$

$$D(a, b) = 0 \quad \Rightarrow \quad \text{test inconclusive}$$

- A point P is an *interior* point of a domain $\mathcal{D}$ if $\mathcal{D}$ contains some open disk $D(P, r)$ centered at P. A point P is a *boundary point* of $\mathcal{D}$ if every open disk $D(P, r)$ contains points in $\mathcal{D}$ and points not in $\mathcal{D}$. The *interior* of $\mathcal{D}$ is the set of all interior points, and the *boundary* is the set of all boundary points. A domain is *closed* if it contains all of its boundary points and *open* if it is equal to its interior.
- Existence and location of global extrema: If f is continuous and $\mathcal{D}$ is closed and bounded, then

  - f takes on both a minimum and a maximum value on $\mathcal{D}$.
  - The extreme values occur either at critical points in the interior of $\mathcal{D}$ or at points on the boundary of $\mathcal{D}$.

  To determine the extreme values, first find the critical points in the interior of $\mathcal{D}$. Then compare the values of f at the critical points with the minimum and maximum values of f on the boundary.

14.7 EXERCISES

Preliminary Questions

1. The functions $f(x, y) = x^2 + y^2$ and $g(x, y) = x^2 - y^2$ both have a critical point at $(0, 0)$. How is the behavior of the two functions at the critical point different?

2. Identify the points indicated in the contour maps as local minima, local maxima, saddle points, or neither (Figure 17).

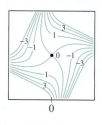
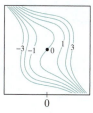
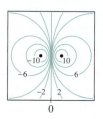

FIGURE 17

3. Let $f(x, y)$ be a continuous function on a domain $\mathcal{D}$ in $\mathbf{R}^2$. Determine which of the following statements are true:

(a) If $\mathcal{D}$ is closed and bounded, then f takes on a maximum value on $\mathcal{D}$.

(b) If $\mathcal{D}$ is neither closed nor bounded, then f does not take on a maximum value of $\mathcal{D}$.

(c) $f(x, y)$ need not have a maximum value on the domain $\mathcal{D}$ defined by $0 \le x \le 1, 0 \le y \le 1$.

(d) A continuous function takes on neither a minimum nor a maximum value on the open quadrant

$$\{(x, y) : x > 0, y > 0\}$$

Exercises

1. Let $P = (a, b)$ be a critical point of $f(x, y) = x^2 + y^4 - 4xy$.

(a) First use $f_x(x, y) = 0$ to show that $a = 2b$. Then use $f_y(x, y) = 0$ to show that $P = (0, 0)$, $(2\sqrt{2}, \sqrt{2})$, or $(-2\sqrt{2}, -\sqrt{2})$.

(b) Referring to Figure 18, determine the local minima and saddle points of $f(x, y)$ and find the absolute minimum value of $f(x, y)$.

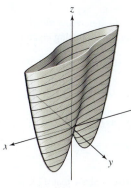

FIGURE 18

2. Find the critical points of the functions

$$f(x, y) = x^2 + 2y^2 - 4y + 6x, \qquad g(x, y) = x^2 - 12xy + y$$

Use the Second Derivative Test to determine the local minimum, local maximum, and saddle points. Match $f(x, y)$ and $g(x, y)$ with their graphs in Figure 19.

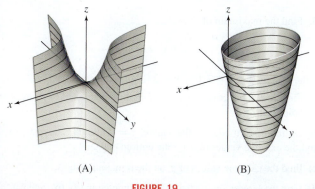

FIGURE 19

3. Find the critical points of

$$f(x, y) = 8y^4 + x^2 + xy - 3y^2 - y^3$$

Use the contour map in Figure 20 to determine their nature (local minimum, local maximum, or saddle point).

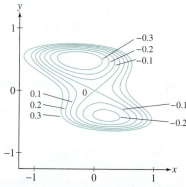

FIGURE 20 Contour map of $f(x, y) = 8y^4 + x^2 + xy - 3y^2 - y^3$.

4. Use the contour map in Figure 21 to determine whether the critical points A, B, C, D are local minima, local maxima, or saddle points.

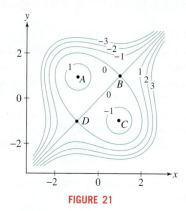

FIGURE 21

5. Let $f(x, y) = y^2 x - yx^2 + xy$.

(a) Show that the critical points (x, y) satisfy the equations

$$y(y - 2x + 1) = 0, \qquad x(2y - x + 1) = 0$$

(b) Show that f has three critical points where $x = 0$ or $y = 0$ (or both) and one critical point where x and y are nonzero.

(c) Use the Second Derivative Test to determine the nature of the critical points.

6. Show that $f(x, y) = \sqrt{x^2 + y^2}$ has one critical point P and that f is nondifferentiable at P. Does f have a minimum, maximum, or saddle point at P?

In Exercises 7–23, find the critical points of the function. Then use the Second Derivative Test to determine whether they are local minima, local maxima, or saddle points (or state that the test fails).

7. $f(x, y) = x^2 + y^2 - xy + x$ **8.** $f(x, y) = x^3 - xy + y^3$

9. $f(x, y) = x^3 + 2xy - 2y^2 - 10x$

10. $f(x, y) = x^3 y + 12x^2 - 8y$

11. $f(x, y) = 4x - 3x^3 - 2xy^2$

12. $f(x, y) = x^3 + y^4 - 6x - 2y^2$

13. $f(x, y) = x^4 + y^4 - 4xy$ **14.** $f(x, y) = e^{x^2 - y^2 + 4y}$

15. $f(x, y) = xye^{-x^2 - y^2}$ **16.** $f(x, y) = e^x - xe^y$

17. $f(x, y) = \sin(x + y) - \cos x$ **18.** $f(x, y) = x \ln(x + y)$

19. $f(x, y) = \ln x + 2 \ln y - x - 4y$

20. $f(x, y) = (x + y) \ln(x^2 + y^2)$

21. $f(x, y) = x - y^2 - \ln(x + y)$ **22.** $f(x, y) = (x - y)e^{x^2 - y^2}$

23. $f(x, y) = (x + 3y)e^{y - x^2}$

24. Show that $f(x, y) = x^2$ has infinitely many critical points (as a function of two variables) and that the Second Derivative Test fails for all of them. What is the minimum value of f? Does $f(x, y)$ have any local maxima?

25. Prove that the function $f(x, y) = \frac{1}{3}x^3 + \frac{2}{3}y^{3/2} - xy$ satisfies $f(x, y) \geq 0$ for $x \geq 0$ and $y \geq 0$.

(a) First, verify that the set of critical points of f is the parabola $y = x^2$ and that the Second Derivative Test fails for these points.

(b) Show that for fixed b, the function $g(x) = f(x, b)$ is concave up for $x > 0$ with a critical point at $x = b^{1/2}$.

(c) Conclude that $f(a, b) \geq f(b^{1/2}, b) = 0$ for all $a, b \geq 0$.

26. Let $f(x, y) = (x^2 + y^2)e^{-x^2 - y^2}$.

(a) Where does f take on its minimum value? Do not use calculus to answer this question.

(b) Verify that the set of critical points of f consists of the origin $(0, 0)$ and the unit circle $x^2 + y^2 = 1$.

(c) The Second Derivative Test fails for points on the unit circle (this can be checked by some lengthy algebra). Prove, however, that f takes on its maximum value on the unit circle by analyzing the function $g(t) = te^{-t}$ for $t > 0$.

27. *CAS* Use a computer algebra system to find a numerical approximation to the critical point of

$$f(x, y) = (1 - x + x^2)e^{y^2} + (1 - y + y^2)e^{x^2}$$

Apply the Second Derivative Test to confirm that it corresponds to a local minimum as in Figure 22.

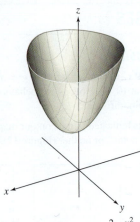

FIGURE 22 Plot of $f(x, y) = (1 - x + x^2)e^{y^2} + (1 - y + y^2)e^{x^2}$.

28. Which of the following domains are closed and which are bounded?

(a) $\{(x, y) \in \mathbf{R}^2 : x^2 + y^2 \leq 1\}$

(b) $\{(x, y) \in \mathbf{R}^2 : x^2 + y^2 < 1\}$

(c) $\{(x, y) \in \mathbf{R}^2 : x \geq 0\}$

(d) $\{(x, y) \in \mathbf{R}^2 : x > 0, y > 0\}$

(e) $\{(x, y) \in \mathbf{R}^2 : 1 \leq x \leq 4, 5 \leq y \leq 10\}$

(f) $\{(x, y) \in \mathbf{R}^2 : x > 0, x^2 + y^2 \leq 10\}$

In Exercises 29–32, determine the global extreme values of the function on the given set without using calculus.

29. $f(x, y) = x + y, \quad 0 \leq x \leq 1, \quad 0 \leq y \leq 1$

30. $f(x, y) = 2x - y, \quad 0 \leq x \leq 1, \quad 0 \leq y \leq 3$

31. $f(x, y) = (x^2 + y^2 + 1)^{-1}, \quad 0 \leq x \leq 3, \quad 0 \leq y \leq 5$

32. $f(x, y) = e^{-x^2 - y^2}, \quad x^2 + y^2 \leq 1$

33. Assumptions Matter Show that $f(x, y) = xy$ does not have a global minimum or a global maximum on the domain

$$\mathcal{D} = \{(x, y) : 0 < x < 1, 0 < y < 1\}$$

Explain why this does not contradict Theorem 3.

34. Find a continuous function that does not have a global maximum on the domain $\mathcal{D} = \{(x, y) : x + y \geq 0, x + y \leq 1\}$. Explain why this does not contradict Theorem 3.

35. Find the maximum of

$$f(x, y) = x + y - x^2 - y^2 - xy$$

on the square, $0 \leq x \leq 2, 0 \leq y \leq 2$ (Figure 23).

(a) First, locate the critical point of f in the square, and evaluate f at this point.

(b) On the bottom edge of the square, $y = 0$ and $f(x, 0) = x - x^2$. Find the extreme values of f on the bottom edge.

(c) Find the extreme values of f on the remaining edges.

(d) Find the largest among the values computed in (a), (b), and (c).

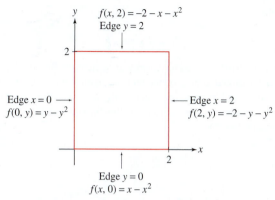

FIGURE 23 The function $f(x, y) = x + y - x^2 - y^2 - xy$ on the boundary segments of the square $0 \le x \le 2, 0 \le y \le 2$.

36. Find the maximum of $f(x, y) = y^2 + xy - x^2$ on the square $0 \le x \le 2, 0 \le y \le 2$.

In Exercises 37–45, determine the global extreme values of the function on the given domain.

37. $f(x, y) = x^3 - 2y$, $0 \le x \le 1$, $0 \le y \le 1$

38. $f(x, y) = 5x - 3y$, $y \ge x - 2$, $y \ge -x - 2$, $y \le 3$

39. $f(x, y) = x^2 + 2y^2$, $0 \le x \le 1$, $0 \le y \le 1$

40. $f(x, y) = x^3 + x^2y + 2y^2$, $x, y \ge 0$, $x + y \le 1$

41. $f(x, y) = x^2 + xy^3 + y^2$, $x, y \ge 0$, $x + y \le 1$

42. $f(x, y) = x^3 + y^3 - 3xy$, $0 \le x \le 1$, $0 \le y \le 1$

43. $f(x, y) = x^2 + y^2 - 2x - 4y$, $x \ge 0$, $0 \le y \le 3$, $y \ge x$

44. $f(x, y) = (4y^2 - x^2)e^{-x^2 - y^2}$, $x^2 + y^2 \le 2$

45. $f(x, y) = x^2 + 2xy^2$, $x^2 + y^2 \le 1$

46. Find the maximum volume of a box inscribed in the tetrahedron bounded by the coordinate planes and the plane

$$x + \frac{1}{2}y + \frac{1}{3}z = 1$$

47. Find the maximum volume of the largest box of the type shown in Figure 24, with one corner at the origin and the opposite corner at a point $P = (x, y, z)$ on the paraboloid

$$z = 1 - \frac{x^2}{4} - \frac{y^2}{9} \quad \text{with } x, y, z \ge 0$$

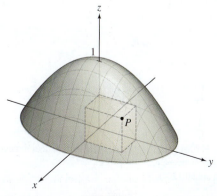

FIGURE 24

48. Find the point on the plane

$$z = x + y + 1$$

closest to the point $P = (1, 0, 0)$. *Hint:* Minimize the square of the distance.

49. Show that the sum of the squares of the distances from a point $P = (c, d)$ to n fixed points $(a_1, b_1), \dots, (a_n, b_n)$ is minimized when c is the average of the x-coordinates a_i and d is the average of the y-coordinates b_i.

50. Show that the rectangular box (including the top and bottom) with fixed volume $V = 27$ m^3 and smallest possible surface area is a cube (Figure 25).

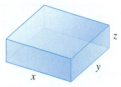

FIGURE 25 Rectangular box with sides x, y, z.

51. Consider a rectangular box B that has a bottom and sides but no top and has minimal surface area among all boxes with fixed volume V.

(a) Do you think B is a cube as in the solution to Exercise 50? If not, how would its shape differ from a cube?

(b) Find the dimensions of B and compare with your response to (a).

52. Find three positive numbers that sum to 150 with the largest possible product of the three.

53. A 120-m long fence is to be cut into pieces to make three enclosures, each of which is square. How should the fence be cut up in order to minimize the total area enclosed by the fence?

54. A box with a volume of 8 m^3 is to be constructed with a gold-plated top, silver-plated bottom, and copper-plated sides. If gold plate costs $120 per square meter, silver plate costs $40 per square meter, and copper plate costs $10 per square meter, find the dimensions that will minimize the cost of the materials for the box.

55. Find the maximum volume of a cylindrical soda can such that the sum of its height and its circumference is 120 cm.

56. Given n data points $(x_1, y_1), \dots, (x_n, y_n)$, the **linear least-squares fit** is the linear function

$$f(x) = mx + b$$

that minimizes the sum of the squares (Figure 26):

$$E(m, b) = \sum_{j=1}^{n} (y_j - f(x_j))^2$$

Show that the minimum value of E occurs for m and b satisfying the two equations

$$m \left(\sum_{j=1}^{n} x_j \right) + bn = \sum_{j=1}^{n} y_j$$

$$m \sum_{j=1}^{n} x_j^2 + b \sum_{j=1}^{n} x_j = \sum_{j=1}^{n} x_j y_j$$

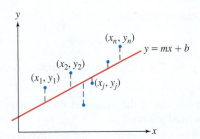

FIGURE 26 The linear least-squares fit minimizes the sum of the squares of the vertical distances from the data points to the line.

57. The power (in microwatts) of a laser is measured as a function of current (in milliamps). Find the linear least-squares fit (Exercise 56) for the data points.

Current (milliamps)	1.0	1.1	1.2	1.3	1.4	1.5
Laser power (microwatts)	0.52	0.56	0.82	0.78	1.23	1.50

58. Let $A = (a, b)$ be a fixed point in the plane, and let $f_A(P)$ be the distance from A to the point $P = (x, y)$. For $P \neq A$, let $\mathbf{e}_{AP}$ be the unit vector pointing from A to P (Figure 27):

$$\mathbf{e}_{AP} = \frac{\overrightarrow{AP}}{\|\overrightarrow{AP}\|}$$

Show that

$$\nabla f_A(P) = \mathbf{e}_{AP}$$

Note that we can derive this result without calculation: Because $\nabla f_A(P)$ points in the direction of maximal increase, it must point directly away from A at P, and because the distance $f_A(x, y)$ increases at a rate of 1 as you move away from A along the line through A and P, $\nabla f_A(P)$ must be a unit vector.

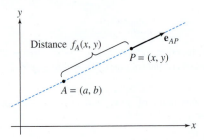

FIGURE 27 The distance from A to P increases most rapidly in the direction $\mathbf{e}_{AP}$.

Further Insights and Challenges

59. In this exercise, we prove that for all $x, y \geq 0$:

$$\frac{1}{\alpha}x^\alpha + \frac{1}{\beta}x^\beta \geq xy$$

where $\alpha \geq 1$ and $\beta \geq 1$ are numbers such that $\alpha^{-1} + \beta^{-1} = 1$. To do this, we prove that the function

$$f(x, y) = \alpha^{-1}x^\alpha + \beta^{-1}y^\beta - xy$$

satisfies $f(x, y) \geq 0$ for all $x, y \geq 0$.

(a) Show that the set of critical points of $f(x, y)$ is the curve $y = x^{\alpha-1}$ (Figure 28). Note that this curve can also be described as $x = y^{\beta-1}$. What is the value of $f(x, y)$ at points on this curve?

(b) Verify that the Second Derivative Test fails. Show, however, that for fixed $b > 0$, the function $g(x) = f(x, b)$ is concave up with a critical point at $x = b^{\beta-1}$.

(c) Conclude that for all $x > 0$, $f(x, b) \geq f(b^{\beta-1}, b) = 0$.

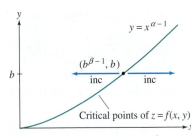

FIGURE 28 The critical points of $f(x, y) = \alpha^{-1}x^\alpha + \beta^{-1}y^\beta - xy$ form a curve $y = x^{\alpha-1}$.

60. The following problem was posed by Pierre de Fermat: Given three points $A = (a_1, a_2)$, $B = (b_1, b_2)$, and $C = (c_1, c_2)$ in the plane, find the point $P = (x, y)$ that minimizes the sum of the distances

$$f(x, y) = AP + BP + CP$$

Let $\mathbf{e}, \mathbf{f}, \mathbf{g}$ be the unit vectors pointing from P to the points A, B, C as in Figure 29.

(a) Use Exercise 58 to show that the condition $\nabla f(P) = 0$ is equivalent to

$$\mathbf{e} + \mathbf{f} + \mathbf{g} = 0 \qquad \boxed{3}$$

(b) Show that $f(x, y)$ is differentiable except at points A, B, C. Conclude that the minimum of $f(x, y)$ occurs either at a point P satisfying Eq. (3) or at one of the points A, B, or C.

(c) Prove that Eq. (3) holds if and only if P is the **Fermat point**, defined as the point P for which the angles between the segments $\overline{AP}$, $\overline{BP}$, $\overline{CP}$ are all $120°$ (Figure 29).

(d) Show that the Fermat point does not exist if one of the angles in $\triangle ABC$ is greater than $120°$. Where does the minimum occur in this case?

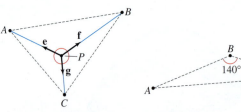

(A) P is the Fermat point (the angles between $\mathbf{e}$, $\mathbf{f}$, and $\mathbf{g}$ are all $120°$).

(B) Fermat point does not exist.

FIGURE 29

14.8 Lagrange Multipliers: Optimizing with a Constraint

Some optimization problems involve finding the extreme values of a function $f(x, y)$ subject to a constraint $g(x, y) = 0$. Suppose that we want to find the point on the line $2x + 3y = 6$ closest to the origin (Figure 1). The distance from (x, y) to the origin is $f(x, y) = \sqrt{x^2 + y^2}$, so our problem is

$$\text{Minimize} \quad f(x, y) = \sqrt{x^2 + y^2} \quad \text{subject to} \quad g(x, y) = 2x + 3y - 6 = 0$$

We are not seeking the minimum value of $f(x, y)$ (which is 0), but rather the minimum among all points (x, y) that lie on the line.

The method of **Lagrange multipliers** is a general procedure for solving optimization problems with a constraint. Here is a description of the main idea.

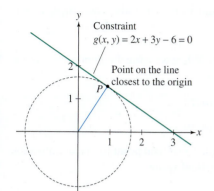

DF **FIGURE 1** Finding the minimum of

$$f(x, y) = \sqrt{x^2 + y^2}$$

on the line $2x + 3y = 6$.

GRAPHICAL INSIGHT   Imagine standing at point Q in Figure 2(A). We want to increase the value of f while remaining on the constraint curve. The gradient vector ∇f_Q points in the direction of *maximum* increase, but we cannot move in the gradient direction because that would take us off the constraint curve. However, the gradient points to the right, so we can still increase f somewhat by moving to the right along the constraint curve.

We keep moving to the right until we arrive at the point P, where ∇f_P is orthogonal to the constraint curve [Figure 2(B)]. Once at P, we cannot increase f further by moving either to the right or to the left along the constraint curve. Thus, $f(P)$ is a local maximum subject to the constraint.

Now, the vector ∇g_P is also orthogonal to the constraint curve, so ∇f_P and ∇g_P must point in the same or opposite directions. In other words, $\nabla f_P = \lambda \nabla g_P$ for some scalar λ (called a **Lagrange multiplier**). Graphically, this means that a local max subject to the constraint occurs at points P where the level curves of f and g are tangent.

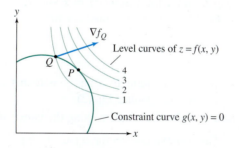

(A) f increases as we move to the right along the constraint curve.

(B) The local maximum of f on the constraint curve occurs where ∇f_P and ∇g_P are parallel.

DF **FIGURE 2**

In Theorem 1, the assumption $\nabla g_P \neq \mathbf{0}$ guarantees (by the Implicit Function Theorem of advanced calculus) that we can parametrize the curve $g(x, y) = 0$ near P by a path $\mathbf{r}(t)$ such that $\mathbf{r}(0) = P$ and $\mathbf{r}'(0) \neq \mathbf{0}$.

THEOREM 1 Lagrange Multipliers   Assume that $f(x, y)$ and $g(x, y)$ are differentiable functions. If $f(x, y)$ has a local minimum or a local maximum on the constraint curve $g(x, y) = 0$ at $P = (a, b)$, and if $\nabla g_P \neq \mathbf{0}$, then there is a scalar λ such that

$$\nabla f_P = \lambda \nabla g_P \qquad \boxed{1}$$

Proof   Let $\mathbf{r}(t)$ be a parametrization of the constraint curve $g(x, y) = 0$ near P, chosen so that $\mathbf{r}(0) = P$ and $\mathbf{r}'(0) \neq \mathbf{0}$. Then $f(\mathbf{r}(0)) = f(P)$, and by assumption, $f(\mathbf{r}(t))$ has a local min or max at $t = 0$. Thus, $t = 0$ is a critical point of $f(\mathbf{r}(t))$ and

$$\underbrace{\frac{d}{dt} f(\mathbf{r}(t)) \Big|_{t=0} = \nabla f_P \cdot \mathbf{r}'(0) = 0}_{\text{Chain Rule}}$$

This shows that ∇f_P is orthogonal to the tangent vector $\mathbf{r}'(0)$ to the curve $g(x, y) = 0$. The gradient ∇g_P is also orthogonal to $\mathbf{r}'(0)$ [because ∇g_P is orthogonal to the level curve $g(x, y) = 0$ at P]. We conclude that ∇f_P and ∇g_P are parallel, and hence, ∇f_P is a multiple of ∇g_P as claimed. ■

⬅·· REMINDER Eq. (1) states that if a local min or max of $f(x, y)$ subject to a constraint $g(x, y) = 0$ occurs at $P = (a, b)$, then

$$\nabla f_P = \lambda \nabla g_P$$

provided that $\nabla g_P \neq \mathbf{0}$.

We refer to Eq. (1) as the **Lagrange condition**. When we write this condition in terms of components, we obtain the **Lagrange equations**:

$$f_x(a, b) = \lambda g_x(a, b)$$
$$f_y(a, b) = \lambda g_y(a, b)$$

A point $P = (a, b)$ satisfying these equations is called a **critical point** for the optimization problem with constraint and $f(a, b)$ is called a **critical value**.

■ **EXAMPLE 1** Find the extreme values of $f(x, y) = 2x + 5y$ on the ellipse

$$\left(\frac{x}{4}\right)^2 + \left(\frac{y}{3}\right)^2 = 1$$

Solution

Step 1. **Write out the Lagrange equations.**

The constraint curve is $g(x, y) = 0$, where $g(x, y) = (x/4)^2 + (y/3)^2 - 1$. We have

$$\nabla f = \langle 2, 5 \rangle, \qquad \nabla g = \left\langle \frac{x}{8}, \frac{2y}{9} \right\rangle$$

The Lagrange equations $\nabla f_P = \lambda \nabla g_P$ are

$$\langle 2, 5 \rangle = \lambda \left\langle \frac{x}{8}, \frac{2y}{9} \right\rangle \quad \Rightarrow \quad 2 = \frac{\lambda x}{8}, \quad 5 = \frac{\lambda(2y)}{9} \qquad \boxed{2}$$

Step 2. **Solve for λ in terms of x and y.**

Eq. (2) gives us two equations for λ:

$$\lambda = \frac{16}{x}, \qquad \lambda = \frac{45}{2y} \qquad \boxed{3}$$

To justify dividing by x and y, note that x and y must be nonzero, because $x = 0$ or $y = 0$ would violate Eq. (2).

Step 3. **Solve for x and y using the constraint.**

The two expressions for λ must be equal, so we obtain $\dfrac{16}{x} = \dfrac{45}{2y}$ or $y = \dfrac{45}{32}x$. Now substitute this in the constraint equation and solve for x:

$$\left(\frac{x}{4}\right)^2 + \left(\frac{\frac{45}{32}x}{3}\right)^2 = 1$$

$$x^2\left(\frac{1}{16} + \frac{225}{1024}\right) = x^2\left(\frac{289}{1024}\right) = 1$$

Thus, $x = \pm\sqrt{\dfrac{1024}{289}} = \pm\dfrac{32}{17}$, and since $y = \dfrac{45x}{32}$, the critical points are $P = \left(\dfrac{32}{17}, \dfrac{45}{17}\right)$ and $Q = \left(-\dfrac{32}{17}, -\dfrac{45}{17}\right)$.

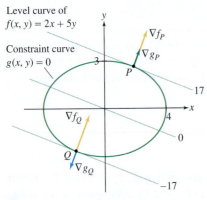

Level curve of
$f(x, y) = 2x + 5y$

Constraint curve
$g(x, y) = 0$

DF FIGURE 3 The min and max occur where a level curve of f is tangent to the constraint curve $g(x, y) = \left(\dfrac{x}{4}\right)^2 + \left(\dfrac{y}{3}\right)^2 - 1 = 0$.

Step 4. **Calculate the critical values.**

$$f(P) = f\left(\frac{32}{17}, \frac{45}{17}\right) = 2\left(\frac{32}{17}\right) + 5\left(\frac{45}{17}\right) = 17$$

and $f(Q) = -17$. We conclude that the maximum of $f(x, y)$ on the ellipse is 17 and the minimum is -17 (Figure 3). ■

Assumptions Matter According to Theorem 3 in Section 14.7, a continuous function on a closed, bounded domain takes on extreme values. This tells us that if the constraint curve is *bounded* (as in the previous example, where the constraint curve is an ellipse), then every continuous function $f(x, y)$ takes on both a minimum and a maximum value subject to the constraint. Be aware, however, that extreme values need not exist if the constraint curve is not bounded. For example, the constraint $x - y = 0$ is an unbounded line. The function $f(x, y) = x$ has neither a minimum nor a maximum subject to $x - y = 0$ because $P = (a, a)$ satisfies the constraint, yet $f(a, a) = a$ can be arbitrarily large or small.

■ **EXAMPLE 2** **Cobb–Douglas Production Function** By investing x units of labor and y units of capital, a watch manufacturer can produce $P(x, y) = 50x^{0.4}y^{0.6}$ watches. (See Figure 4.) Find the maximum number of watches that can be produced on a budget of $20,000 if labor costs $100 per unit and capital costs $200 per unit.

FIGURE 4 Economist Paul Douglas, working with mathematician Charles Cobb, arrived at the production functions $P(x, y) = Cx^a y^b$ by fitting data gathered on the relationships between labor, capital, and output in an industrial economy. Douglas was a professor at the University of Chicago and also served as U.S. senator for Illinois from 1949 to 1967. (© *Bettmann/CORBIS*)

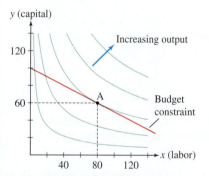

FIGURE 5 Contour plot of the Cobb–Douglas production function $P(x, y) = 50x^{0.4}y^{0.6}$. The level curves of a production function are called *isoquants*.

Solution The total cost of x units of labor and y units of capital is $100x + 200y$. Our task is to maximize the function $P(x, y) = 50x^{0.4}y^{0.6}$ subject to the following budget constraint (Figure 5):

$$g(x, y) = 100x + 200y - 20{,}000 = 0 \qquad \boxed{4}$$

Step 1. Write out the Lagrange equations.

$$P_x(x, y) = \lambda g_x(x, y): \quad 20x^{-0.6}y^{0.6} = 100\lambda$$

$$P_y(x, y) = \lambda g_y(x, y): \quad 30x^{0.4}y^{-0.4} = 200\lambda$$

Step 2. Solve for λ in terms of x and y.

These equations yield two expressions for λ that must be equal:

$$\lambda = \frac{1}{5}\left(\frac{y}{x}\right)^{0.6} = \frac{3}{20}\left(\frac{y}{x}\right)^{-0.4} \qquad \boxed{5}$$

Step 3. Solve for x and y using the constraint.

Multiply Eq. (5) by $5(y/x)^{0.4}$ to obtain $y/x = 15/20$, or $y = \frac{3}{4}x$. Then substitute in Eq. (4):

$$100x + 200y = 100x + 200\left(\frac{3}{4}x\right) = 20{,}000 \quad \Rightarrow \quad 250x = 20{,}000$$

We obtain $x = \dfrac{20{,}000}{250} = 80$ and $y = \frac{3}{4}x = 60$. The critical point is $A = (80, 60)$.

Step 4. Calculate the critical values.

Since $P(x, y)$ is increasing as a function of x and y, ∇P points to the northeast, and it is clear that $P(x, y)$ takes on a maximum value at A (Figure 5). The maximum is $P(80, 60) = 50(80)^{0.4}(60)^{0.6} = 3365.87$, or roughly 3365 watches, with a cost per watch of $\dfrac{20{,}000}{3365}$ or about $5.94. ■

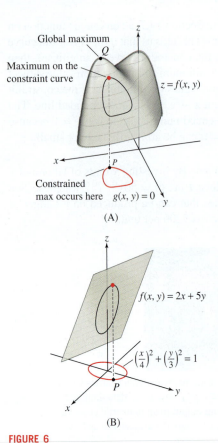

Global maximum Q

Maximum on the constraint curve

$z = f(x, y)$

x

P

Constrained max occurs here $g(x, y) = 0$ y

(A)

z

$f(x, y) = 2x + 5y$

$\left(\frac{x}{4}\right)^2 + \left(\frac{y}{3}\right)^2 = 1$

P

y

x

(B)

FIGURE 6

GRAPHICAL INSIGHT   In an ordinary optimization problem without constraint, the global maximum value is the height of the highest point on the surface $z = f(x, y)$ [point Q in Figure 6(A)]. When a constraint is given, we restrict our attention to the curve on the surface lying above the constraint curve $g(x, y) = 0$. The maximum value subject to the constraint is the height of the highest point on this curve. Figure 6(B) shows the optimization problem solved in Example 1.

The method of Lagrange multipliers is valid in any number of variables. Imagine, for instance, that we are trying to find the maximum temperature $f(x, y, z)$ for points on a surface S in 3-space given by $g(x, y, z) = 0$, as in Figure 7. This surface is a level set for the function g, and therefore, ∇g_P is perpendicular to the normal planes to this surface at every point P on the surface. Consider the level sets for temperature, which we have called the isotherms. They appear as surfaces in 3-space, and their intersections with S yield the level sets of temperature on S. If, as in the figure, the temperature increases as we move to the right on the surface, then it is apparent that the maximum temperature for the surface occurs when the last isotherm intersects the surface in just a single point and hence that isotherm is tangent to the surface. That is to say, the last isotherm and the surface share the same tangent plane at their single point of intersection.

However, as we know, ∇f_P is always perpendicular to the tangent plane to the level sets for f at each point P on a level set. So at the hottest point on the surface, ∇g_P and ∇f_P are both perpendicular to the same tangent plane. Hence, they must be parallel, and one must be a multiple of the other. Thus, at that point, $\nabla f_P = \lambda \nabla g_P$. A similar argument holds for the minimum temperature on the surface.

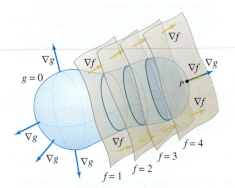

FIGURE 7 As we move to the right, temperature increases, attaining a maximum on the surface of $f = 4$ at P.

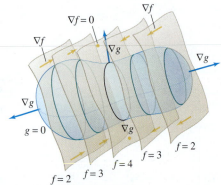

FIGURE 8 As we move to the right, temperature increases and then decreases.

There is one other situation to consider. Imagine that as we move left to right across our surface, temperature first increases to $f = 4$ and then it decreases again, as in Figure 8. There are a whole collection of points with the maximal temperature of $f = 4$. In this case, ∇f must point to the right on isotherms to the left of $f = 4$ since this is the direction of increasing temperature, and they must point to the left on isotherms to the right of $f = 4$ since this is the direction of increasing temperature. Hence, in order for the left-pointing gradient vectors to become right-pointing gradient vectors in a continuous manner, they must be equal to $\mathbf{0}$ on the $f = 4$ isotherm. This makes sense as on that isotherm, there is no direction of increasing temperature. So for all of the points on the surface with maximal temperature, of which there are now many, the equation $\nabla f = \lambda \nabla g$ is still satisfied, but by taking $\lambda = 0$.

In the next example, we consider a problem in three variables.

■ **EXAMPLE 3   Lagrange Multipliers in Three Variables**   Find the point on the plane $\frac{x}{2} + \frac{y}{4} + \frac{z}{4} = 1$ closest to the origin in $\mathbf{R}^3$.

Solution   Our task is to minimize the distance $d = \sqrt{x^2 + y^2 + z^2}$ subject to the constraint $\frac{x}{2} + \frac{y}{4} + \frac{z}{4} = 1$. But, finding the minimum distance d is the same as finding the minimum square of the distance d^2, so our problem can be stated:

Minimize $f(x, y, z) = x^2 + y^2 + z^2$ subject to $g(x, y, z) = \dfrac{x}{2} + \dfrac{y}{4} + \dfrac{z}{4} - 1 = 0$

The Lagrange condition is

$$\underbrace{\langle 2x, 2y, 2z \rangle}_{\nabla f} = \lambda \underbrace{\left\langle \frac{1}{2}, \frac{1}{4}, \frac{1}{4} \right\rangle}_{\nabla g}$$

This yields

$$\lambda = 4x = 8y = 8z \quad \Rightarrow \quad z = y = \frac{x}{2}$$

Substituting in the constraint equation, we obtain

$$\frac{x}{2} + \frac{y}{4} + \frac{z}{4} = \frac{2z}{2} + \frac{z}{4} + \frac{z}{4} = \frac{3z}{2} = 1 \quad \Rightarrow \quad z = \frac{2}{3}$$

Thus, $x = 2z = \frac{4}{3}$ and $y = z = \frac{2}{3}$. This critical point must correspond to the minimum of f. There is no maximum of f on the plane since there are points on the plane that are arbitrarily far from the origin. Hence, the point on the plane closest to the origin is $P = \left(\frac{4}{3}, \frac{2}{3}, \frac{2}{3} \right)$ (Figure 9). ■

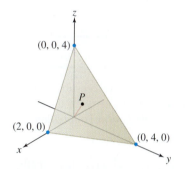

FIGURE 9 Point P closest to the origin on the plane.

The method of Lagrange multipliers can be used when there is more than one constraint equation, but we must add another multiplier for each additional constraint. For example, if the problem is to minimize $f(x, y, z)$ subject to constraints $g(x, y, z) = 0$ and $h(x, y, z) = 0$, then the Lagrange condition is

$$\nabla f = \lambda \nabla g + \mu \nabla h$$

*The intersection of a sphere with a plane through its center is called a **great circle**.*

■ EXAMPLE 4 **Lagrange Multipliers with Multiple Constraints** The intersection of the plane $x + \frac{1}{2}y + \frac{1}{3}z = 0$ with the unit sphere $x^2 + y^2 + z^2 = 1$ is a great circle (Figure 10). Find the point on this great circle with the largest x-coordinate.

Solution Our task is to maximize the function $f(x, y, z) = x$ subject to the two constraint equations

$$g(x, y, z) = x + \frac{1}{2}y + \frac{1}{3}z = 0, \qquad h(x, y, z) = x^2 + y^2 + z^2 - 1 = 0$$

The Lagrange condition is

$$\nabla f = \lambda \nabla g + \mu \nabla h$$

$$\langle 1, 0, 0 \rangle = \lambda \left\langle 1, \frac{1}{2}, \frac{1}{3} \right\rangle + \mu \langle 2x, 2y, 2z \rangle$$

Note that μ cannot be zero, since, if it were, the Lagrange condition would become $\langle 1, 0, 0 \rangle = \lambda \langle 1, \frac{1}{2}, \frac{1}{3} \rangle$, and this equation is not satisfied for any value of λ. Now, the Lagrange condition gives us three equations:

$$\lambda + 2\mu x = 1, \qquad \frac{1}{2}\lambda + 2\mu y = 0, \qquad \frac{1}{3}\lambda + 2\mu z = 0$$

The last two equations yield $\lambda = -4\mu y$ and $\lambda = -6\mu z$. Because $\mu \neq 0$,

$$-4\mu y = -6\mu z \quad \Rightarrow \quad \boxed{y = \frac{3}{2}z}$$

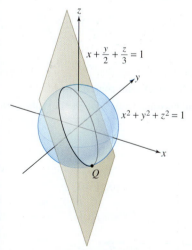

FIGURE 10 The plane intersects the sphere in a great circle. Q is the point on this great circle with the largest x-coordinate.

Now use this relation in the first constraint equation:

$$x + \frac{1}{2}y + \frac{1}{3}z = x + \frac{1}{2}\left(\frac{3}{2}z\right) + \frac{1}{3}z = 0 \quad \Rightarrow \quad \boxed{x = -\frac{13}{12}z}$$

Finally, we can substitute in the second constraint equation:

$$x^2 + y^2 + z^2 - 1 = \left(-\frac{13}{12}z\right)^2 + \left(\frac{3}{2}z\right)^2 + z^2 - 1 = 0$$

to obtain $\dfrac{637}{144}z^2 = 1$ or $z = \pm\dfrac{12}{7\sqrt{13}}$. Since $x = -\frac{13}{12}z$ and $y = \frac{3}{2}z$, the critical points are

$$P = \left(-\frac{\sqrt{13}}{7}, \frac{18}{7\sqrt{13}}, \frac{12}{7\sqrt{13}}\right), \qquad Q = \left(\frac{\sqrt{13}}{7}, -\frac{18}{7\sqrt{13}}, -\frac{12}{7\sqrt{13}}\right)$$

The critical point with the largest x-coordinate [the maximum value of $f(x, y, z)$] is Q with x-coordinate $\dfrac{\sqrt{13}}{7} \approx 0.515$.   ■

14.8 SUMMARY

- Method of Lagrange multipliers: The local extreme values of $f(x, y)$ subject to a constraint $g(x, y) = 0$ occur at points P (called critical points) satisfying the Lagrange condition $\nabla f_P = \lambda \nabla g_P$. This condition is equivalent to the *Lagrange equations*

$$f_x(x, y) = \lambda g_x(x, y), \qquad f_y(x, y) = \lambda g_y(x, y)$$

- If the constraint curve $g(x, y) = 0$ is bounded [e.g., if $g(x, y) = 0$ is a circle or ellipse], then global minimum and maximum values of f subject to the constraint exist.
- Lagrange condition for a function of three variables $f(x, y, z)$ subject to two constraints $g(x, y, z) = 0$ and $h(x, y, z) = 0$:

$$\nabla f = \lambda \nabla g + \mu \nabla h$$

14.8 EXERCISES

Preliminary Questions

1. Suppose that the maximum of $f(x, y)$ subject to the constraint $g(x, y) = 0$ occurs at a point $P = (a, b)$ such that $\nabla f_P \neq 0$. Which of the following statements is true?

(a) ∇f_P is tangent to $g(x, y) = 0$ at P.

(b) ∇f_P is orthogonal to $g(x, y) = 0$ at P.

2. Figure 11 shows a constraint $g(x, y) = 0$ and the level curves of a function f. In each case, determine whether f has a local minimum, a local maximum, or neither at the labeled point.

3. On the contour map in Figure 12:

(a) Identify the points where $\nabla f = \lambda \nabla g$ for some scalar λ.

(b) Identify the minimum and maximum values of $f(x, y)$ subject to $g(x, y) = 0$.

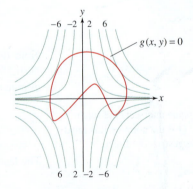

Contour plot of $f(x, y)$
(contour interval 2)

FIGURE 12 Contour map of $f(x, y)$; contour interval 2.

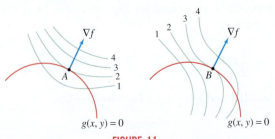

FIGURE 11

Exercises

In this exercise set, use the method of Lagrange multipliers unless otherwise stated.

1. Find the extreme values of the function $f(x, y) = 2x + 4y$ subject to the constraint $g(x, y) = x^2 + y^2 - 5 = 0$.

(a) Show that the Lagrange equation $\nabla f = \lambda \nabla g$ gives $\lambda x = 1$ and $\lambda y = 2$.

(b) Show that these equations imply $\lambda \neq 0$ and $y = 2x$.

(c) Use the constraint equation to determine the possible critical points (x, y).

(d) Evaluate $f(x, y)$ at the critical points and determine the minimum and maximum values.

2. Find the extreme values of $f(x, y) = x^2 + 2y^2$ subject to the constraint $g(x, y) = 4x - 6y = 25$.

(a) Show that the Lagrange equations yield $2x = 4\lambda$, $4y = -6\lambda$.

(b) Show that if $x = 0$ or $y = 0$, then the Lagrange equations give $x = y = 0$. Since $(0, 0)$ does not satisfy the constraint, you may assume that x and y are nonzero.

(c) Use the Lagrange equations to show that $y = -\frac{3}{4}x$.

(d) Substitute in the constraint equation to show that there is a unique critical point P.

(e) Does P correspond to a minimum or maximum value of f? Refer to Figure 13 to justify your answer. *Hint:* Do the values of $f(x, y)$ increase or decrease as (x, y) moves away from P along the line $g(x, y) = 0$?

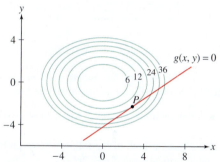

FIGURE 13 Level curves of $f(x, y) = x^2 + 2y^2$ and graph of the constraint $g(x, y) = 4x - 6y - 25 = 0$.

3. Apply the method of Lagrange multipliers to the function $f(x, y) = (x^2 + 1)y$ subject to the constraint $x^2 + y^2 = 5$. *Hint:* First show that $y \neq 0$; then treat the cases $x = 0$ and $x \neq 0$ separately.

In Exercises 4–15, find the minimum and maximum values of the function subject to the given constraint.

4. $f(x, y) = 2x + 3y$, $\quad x^2 + y^2 = 4$

5. $f(x, y) = x^2 + y^2$, $\quad 2x + 3y = 6$

6. $f(x, y) = 4x^2 + 9y^2$, $\quad xy = 4$

7. $f(x, y) = xy$, $\quad 4x^2 + 9y^2 = 32$

8. $f(x, y) = x^2 y + x + y$, $\quad xy = 4$

9. $f(x, y) = x^2 + y^2$, $\quad x^4 + y^4 = 1$

10. $f(x, y) = x^2 y^4$, $\quad x^2 + 2y^2 = 6$

11. $f(x, y, z) = 3x + 2y + 4z$, $\quad x^2 + 2y^2 + 6z^2 = 1$

12. $f(x, y, z) = x^2 - y - z$, $\quad x^2 - y^2 + z = 0$

13. $f(x, y, z) = xy + 2z$, $\quad x^2 + y^2 + z^2 = 36$

14. $f(x, y, z) = x^2 + y^2 + z^2$, $\quad x + 3y + 2z = 36$

15. $f(x, y, z) = xy + xz$, $\quad x^2 + y^2 + z^2 = 4$

16. Let

$$f(x, y) = x^3 + xy + y^3, \qquad g(x, y) = x^3 - xy + y^3$$

(a) Show that there is a unique point $P = (a, b)$ on $g(x, y) = 1$ where $\nabla f_P = \lambda \nabla g_P$ for some scalar λ.

(b) Refer to Figure 14 to determine whether $f(P)$ is a local minimum or a local maximum of f subject to the constraint.

(c) Does Figure 14 suggest that $f(P)$ is a global extremum subject to the constraint?

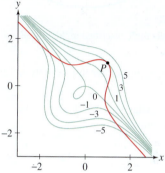

FIGURE 14 Contour map of $f(x, y) = x^3 + xy + y^3$ and graph of the constraint $g(x, y) = x^3 - xy + y^3 = 1$.

17. Find the point (a, b) on the graph of $y = e^x$ where the value ab is as small as possible.

18. Find the rectangular box of maximum volume if the sum of the lengths of the edges is 300 cm.

19. The surface area of a right-circular cone of radius r and height h is $S = \pi r \sqrt{r^2 + h^2}$, and its volume is $V = \frac{1}{3}\pi r^2 h$.

(a) Determine the ratio h/r for the cone with given surface area S and maximum volume V.

(b) What is the ratio h/r for a cone with given volume V and minimum surface area S?

(c) Does a cone with given volume V and maximum surface area exist?

20. In Example 1, we found the maximum of $f(x, y) = 2x + 5y$ on the ellipse $(x/4)^2 + (y/3)^2 = 1$. Solve this problem again without using Lagrange multipliers. First, show that the ellipse is parametrized by $x = 4\cos t$, $y = 3\sin t$. Then find the maximum value of $f(4\cos t, 3\sin t)$ using single-variable calculus. Is one method easier than the other?

21. Find the point on the ellipse

$$x^2 + 6y^2 + 3xy = 40$$

with the largest x-coordinate (Figure 15).

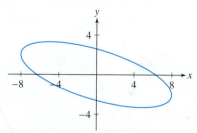

FIGURE 15 Graph of $x^2 + 6y^2 + 3xy = 40$.

22. Use Lagrange multipliers to find the maximum area of a rectangle inscribed in the ellipse (Figure 16):

$$\frac{x^2}{a^2} + \frac{y^2}{b^2} = 1$$

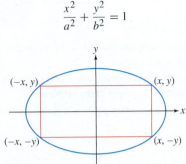

FIGURE 16 Rectangle inscribed in the ellipse $\frac{x^2}{a^2} + \frac{y^2}{b^2} = 1$.

23. Find the point (x_0, y_0) on the line $4x + 9y = 12$ that is closest to the origin.

24. Show that the point (x_0, y_0) closest to the origin on the line $ax + by = c$ has coordinates

$$x_0 = \frac{ac}{a^2 + b^2}, \qquad y_0 = \frac{bc}{a^2 + b^2}$$

25. Find the maximum value of $f(x, y) = x^a y^b$ for $x \geq 0$, $y \geq 0$ on the line $x + y = 1$, where $a, b > 0$ are constants.

26. Show that the maximum value of $f(x, y) = x^2 y^3$ on the unit circle is $\frac{6}{25}\sqrt{\frac{3}{5}}$.

27. Find the maximum value of $f(x, y) = x^a y^b$ for $x \geq 0$, $y \geq 0$ on the unit circle, where $a, b > 0$ are constants.

28. Find the maximum value of $f(x, y, z) = x^a y^b z^c$ for $x, y, z \geq 0$ on the unit sphere, where $a, b, c > 0$ are constants.

29. Show that the minimum distance from the origin to a point on the plane $ax + by + cz = d$ is

$$\frac{|d|}{\sqrt{a^2 + b^2 + c^2}}$$

30. Antonio has $5.00 to spend on a lunch consisting of hamburgers ($1.50 each) and french fries ($1.00 per order). Antonio's satisfaction from eating x_1 hamburgers and x_2 orders of french fries is measured by a function $U(x_1, x_2) = \sqrt{x_1 x_2}$. How much of each type of food should he purchase to maximize his satisfaction? (Assume that fractional amounts of each food can be purchased.)

31. Let Q be the point on an ellipse closest to a given point P outside the ellipse. It was known to the Greek mathematician Apollonius (third century BCE) that $\overline{PQ}$ is perpendicular to the tangent to the ellipse at Q (Figure 17). Explain in words why this conclusion is a consequence of the method of Lagrange multipliers. *Hint:* The circles centered at P are level curves of the function to be minimized.

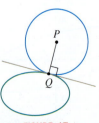

FIGURE 17

32.   In a contest, a runner starting at A must touch a point P along a river and then run to B in the shortest time possible (Figure 18). The runner should choose the point P that minimizes the total length of the path.

(a) Define a function

$$f(x, y) = AP + PB, \quad \text{where } P = (x, y)$$

Rephrase the runner's problem as a constrained optimization problem, assuming that the river is given by an equation $g(x, y) = 0$.

(b) Explain why the level curves of $f(x, y)$ are ellipses.

(c) Use Lagrange multipliers to justify the following statement: The ellipse through the point P minimizing the length of the path is tangent to the river.

(d) Identify the point on the river in Figure 18 for which the length is minimal.

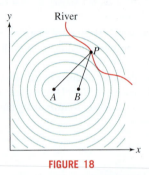

FIGURE 18

In Exercises 33 and 34, let V be the volume of a can of radius r and height h, and let S be its surface area (including the top and bottom).

33. Find r and h that minimize S subject to the constraint $V = 54\pi$.

34.   Show that for both of the following two problems, $P = (r, h)$ is a Lagrange critical point if $h = 2r$:

- Minimize surface area S for fixed volume V.
- Maximize volume V for fixed surface area S.

Then use the contour plots in Figure 19 to explain why S has a minimum for fixed V but no maximum and, similarly, V has a maximum for fixed S but no minimum.

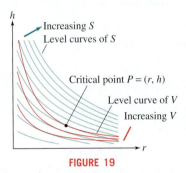

FIGURE 19

35. A plane with equation $\frac{x}{a} + \frac{y}{b} + \frac{z}{c} = 1$ $(a, b, c > 0)$ together with the positive coordinate planes forms a tetrahedron of volume $V = \frac{1}{6}abc$ (Figure 20). Find the minimum value of V among all planes passing through the point $P = (1, 1, 1)$.

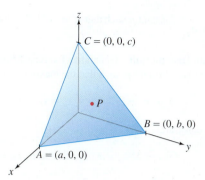

FIGURE 20

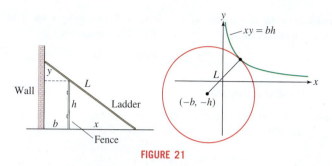

FIGURE 21

36. With the same set-up as in the previous problem, find the plane that minimizes V if the plane is constrained to pass through a point $P = (\alpha, \beta, \gamma)$ with $\alpha, \beta, \gamma > 0$.

37. Show that the Lagrange equations for $f(x, y) = x + y$ subject to the constraint $g(x, y) = x + 2y = 0$ have no solution. What can you conclude about the minimum and maximum values of f subject to $g = 0$? Show this directly.

38. Show that the Lagrange equations for $f(x, y) = 2x + y$ subject to the constraint $g(x, y) = x^2 - y^2 = 1$ have a solution but that f has no min or max on the constraint curve. Does this contradict Theorem 1?

39. Let L be the minimum length of a ladder that can reach over a fence of height h to a wall located a distance b behind the wall.

(a) Use Lagrange multipliers to show that $L = (h^{2/3} + b^{2/3})^{3/2}$ (Figure 21). *Hint:* Show that the problem amounts to minimizing $f(x, y) = (x + b)^2 + (y + h)^2$ subject to $y/b = h/x$ or $xy = bh$.

(b) Show that the value of L is also equal to the radius of the circle with center $(-b, -h)$ that is tangent to the graph of $xy = bh$.

40. Find the maximum value of $f(x, y, z) = xy + xz + yz - xyz$ subject to the constraint $x + y + z = 1$, for $x \geq 0$, $y \geq 0$, $z \geq 0$.

41. Find the minimum of $f(x, y, z) = x^2 + y^2 + z^2$ subject to the two constraints $x + y + z = 1$ and $x + 2y + 3z = 6$.

42. Find the maximum of $f(x, y, z) = z$ subject to the two constraints $x^2 + y^2 = 1$ and $x + y + z = 1$.

43. Find the point lying on the intersection of the plane $x + \frac{1}{2}y + \frac{1}{4}z = 0$ and the sphere $x^2 + y^2 + z^2 = 9$ with the largest z-coordinate.

44. Find the maximum of $f(x, y, z) = x + y + z$ subject to the two constraints $x^2 + y^2 + z^2 = 9$ and $\frac{1}{4}x^2 + \frac{1}{4}y^2 + 4z^2 = 9$.

45. The cylinder $x^2 + y^2 = 1$ intersects the plane $x + z = 1$ in an ellipse. Find the point on such an ellipse that is farthest from the origin.

46. Find the minimum and maximum of $f(x, y, z) = y + 2z$ subject to two constraints, $2x + z = 4$ and $x^2 + y^2 = 1$.

47. Find the minimum value of $f(x, y, z) = x^2 + y^2 + z^2$ subject to two constraints, $x + 2y + z = 3$ and $x - y = 4$.

Further Insights and Challenges

48. Suppose that both $f(x, y)$ and the constraint function $g(x, y)$ are linear. Use contour maps to explain why $f(x, y)$ does not have a maximum subject to $g(x, y) = 0$ unless $g = af + b$ for some constants a, b.

49. Assumptions Matter Consider the problem of minimizing $f(x, y) = x$ subject to $g(x, y) = (x - 1)^3 - y^2 = 0$.

(a) Show, without using calculus, that the minimum occurs at $P = (1, 0)$.

(b) Show that the Lagrange condition $\nabla f_P = \lambda \nabla g_P$ is not satisfied for any value of λ.

(c) Does this contradict Theorem 1?

50. Marginal Utility Goods 1 and 2 are available at dollar prices of p_1 per unit of Good 1 and p_2 per unit of Good 2. A utility function $U(x_1, x_2)$ is a function representing the **utility** or benefit of consuming x_j units of good j. The **marginal utility** of the jth good is $\partial U / \partial x_j$, the rate of increase in utility per unit increase in the jth good. Prove the following law of economics: Given a budget of L dollars, utility is maximized at the consumption level (a, b) where the ratio of marginal utility is equal to the ratio of prices:

$$\frac{\text{Marginal utility of Good 1}}{\text{Marginal utility of Good 2}} = \frac{U_{x_1}(a, b)}{U_{x_2}(a, b)} = \frac{p_1}{p_2}$$

51. Consider the utility function $U(x_1, x_2) = x_1 x_2$ with budget constraint $p_1 x_1 + p_2 x_2 = c$.

(a) Show that the maximum of $U(x_1, x_2)$ subject to the budget constraint is equal to $c^2 / (4 p_1 p_2)$.

(b) Calculate the value of the Lagrange multiplier λ occurring in (a).

(c) Prove the following interpretation: λ is the rate of increase in utility per unit increase in total budget c.

52. This exercise shows that the multiplier λ may be interpreted as a rate of change in general. Assume that the maximum of $f(x, y)$ subject to $g(x, y) = c$ occurs at a point P. Then P depends on the value of c, so we may write $P = (x(c), y(c))$ and we have $g(x(c), y(c)) = c$.

(a) Show that

$$\nabla g(x(c), y(c)) \cdot \left\langle x'(c), y'(c) \right\rangle = 1$$

Hint: Differentiate the equation $g(x(c), y(c)) = c$ with respect to c using the Chain Rule.

(b) Use the Chain Rule and the Lagrange condition $\nabla f_P = \lambda \nabla g_P$ to show that

$$\frac{d}{dc} f(x(c), y(c)) = \lambda$$

(c) Conclude that λ is the rate of increase in f per unit increase in the "budget level" c.

53. Let $B > 0$. Show that the maximum of

$$f(x_1, \ldots, x_n) = x_1 x_2 \cdots x_n$$

subject to the constraints $x_1 + \cdots + x_n = B$ and $x_j \geq 0$ for $j = 1, \ldots, n$ occurs for $x_1 = \cdots = x_n = B/n$. Use this to conclude that

$$(a_1 a_2 \cdots a_n)^{1/n} \leq \frac{a_1 + \cdots + a_n}{n}$$

for all positive numbers $a_1, \ldots, a_n$.

54. Let $B > 0$. Show that the maximum of $f(x_1, \ldots, x_n) = x_1 + \cdots + x_n$ subject to $x_1^2 + \cdots + x_n^2 = B^2$ is $\sqrt{n}B$. Conclude that

$$|a_1| + \cdots + |a_n| \leq \sqrt{n}(a_1^2 + \cdots + a_n^2)^{1/2}$$

for all numbers $a_1, \ldots, a_n$.

55. Given constants E, E_1, E_2, E_3, consider the maximum of

$$S(x_1, x_2, x_3) = x_1 \ln x_1 + x_2 \ln x_2 + x_3 \ln x_3$$

subject to two constraints:

$$x_1 + x_2 + x_3 = N, \qquad E_1 x_1 + E_2 x_2 + E_3 x_3 = E$$

Show that there is a constant μ such that $x_i = A^{-1} e^{\mu E_i}$ for $i = 1, 2, 3$, where $A = N^{-1}(e^{\mu E_1} + e^{\mu E_2} + e^{\mu E_3})$.

56. Boltzmann Distribution Generalize Exercise 55 to n variables: Show that there is a constant μ such that the maximum of

$$S = x_1 \ln x_1 + \cdots + x_n \ln x_n$$

subject to the constraints

$$x_1 + \cdots + x_n = N, \qquad E_1 x_1 + \cdots + E_n x_n = E$$

occurs for $x_i = A^{-1} e^{\mu E_i}$, where

$$A = N^{-1}(e^{\mu E_1} + \cdots + e^{\mu E_n})$$

This result lies at the heart of statistical mechanics. It is used to determine the distribution of velocities of gas molecules at temperature T; x_i is the number of molecules with kinetic energy E_i; $\mu = -(kT)^{-1}$, where k is Boltzmann's constant. The quantity S is called the **entropy**.

CHAPTER REVIEW EXERCISES

1. Given $f(x, y) = \dfrac{\sqrt{x^2 - y^2}}{x + 3}$:

(a) Sketch the domain of f.

(b) Calculate $f(3, 1)$ and $f(-5, -3)$.

(c) Find a point satisfying $f(x, y) = 1$.

2. Find the domain and range of:

(a) $f(x, y, z) = \sqrt{x - y} + \sqrt{y - z}$

(b) $f(x, y) = \ln(4x^2 - y)$

3. Sketch the graph $f(x, y) = x^2 - y + 1$ and describe its vertical and horizontal traces.

4. *CAS* Use a graphing utility to draw the graph of the function $\cos(x^2 + y^2)e^{1-xy}$ in the domains $[-1, 1] \times [-1, 1]$, $[-2, 2] \times [-2, 2]$, and $[-3, 3] \times [-3, 3]$, and explain its behavior.

5. Match the functions (a)–(d) with their graphs in Figure 1.

(a) $f(x, y) = x^2 + y$

(b) $f(x, y) = x^2 + 4y^2$

(c) $f(x, y) = \sin(4xy)e^{-x^2 - y^2}$

(d) $f(x, y) = \sin(4x)e^{-x^2 - y^2}$

6. Referring to the contour map in Figure 2:

(a) Estimate the average rate of change of elevation from A to B and from A to D.

(b) Estimate the directional derivative at A in the direction of $\mathbf{v}$.

(c) What are the signs of f_x and f_y at D?

(d) At which of the labeled points are both f_x and f_y negative?

7. Describe the level curves of:

(a) $f(x, y) = e^{4x - y}$ **(b)** $f(x, y) = \ln(4x - y)$

(c) $f(x, y) = 3x^2 - 4y^2$ **(d)** $f(x, y) = x + y^2$

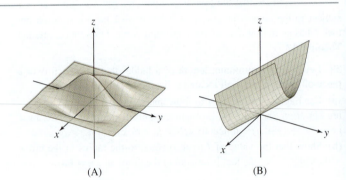

(A)   (B)

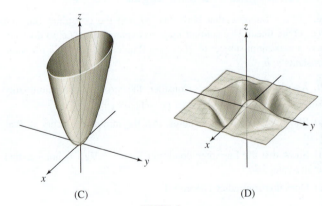

(C)   (D)

FIGURE 1

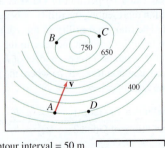

Contour interval = 50 m   0   1   2 km

FIGURE 2

8. Match each function (a)–(c) with its contour graph (i)–(iii) in Figure 3:

(a) $f(x, y) = xy$

(b) $f(x, y) = e^{xy}$

(c) $f(x, y) = \sin(xy)$

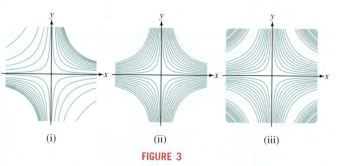

(i)　　　　　　(ii)　　　　　　(iii)

FIGURE 3

In Exercises 9–14, evaluate the limit or state that it does not exist.

9. $\displaystyle\lim_{(x,y)\to(1,-3)} (xy + y^2)$

10. $\displaystyle\lim_{(x,y)\to(1,-3)} \ln(3x + y)$

11. $\displaystyle\lim_{(x,y)\to(0,0)} \frac{xy + xy^2}{x^2 + y^2}$

12. $\displaystyle\lim_{(x,y)\to(0,0)} \frac{x^3y^2 + x^2y^3}{x^4 + y^4}$

13. $\displaystyle\lim_{(x,y)\to(1,-3)} (2x + y)e^{-x+y}$

14. $\displaystyle\lim_{(x,y)\to(0,2)} \frac{(e^x - 1)(e^y - 1)}{x}$

15. Let

$$f(x, y) = \begin{cases} \dfrac{(xy)^p}{x^4 + y^4} & (x, y) \neq (0, 0) \\ 0 & (x, y) = (0, 0) \end{cases}$$

Use polar coordinates to show that $f(x, y)$ is continuous at all (x, y) if $p > 2$ but is discontinuous at $(0, 0)$ if $p \leq 2$.

16. Calculate $f_x(1, 3)$ and $f_y(1, 3)$ for $f(x, y) = \sqrt{7x + y^2}$.

In Exercises 17–20, compute f_x and f_y.

17. $f(x, y) = 2x + y^2$

18. $f(x, y) = 4xy^3$

19. $f(x, y) = \sin(xy)e^{-x-y}$

20. $f(x, y) = \ln(x^2 + xy^2)$

21. Calculate f_{xxyz} for $f(x, y, z) = y \sin(x + z)$.

22. Fix $c > 0$. Show that for any constants α, β, the function $u(t, x) = \sin(\alpha ct + \beta)\sin(\alpha x)$ satisfies the wave equation

$$\frac{\partial^2 u}{\partial t^2} = c^2 \frac{\partial^2 u}{\partial x^2}$$

23. Find an equation of the tangent plane to the graph of $f(x, y) = xy^2 - xy + 3x^3y$ at $P = (1, 3)$.

24. Suppose that $f(4, 4) = 3$ and $f_x(4, 4) = f_y(4, 4) = -1$. Use the linear approximation to estimate $f(4.1, 4)$ and $f(3.88, 4.03)$.

25. Use a linear approximation of $f(x, y, z) = \sqrt{x^2 + y^2 + z}$ to estimate $\sqrt{7.1^2 + 4.9^2 + 69.5}$. Compare with a calculator value.

26. The plane $z = 2x - y - 1$ is tangent to the graph of $z = f(x, y)$ at $P = (5, 3)$.

a) Determine $f(5, 3)$, $f_x(5, 3)$, and $f_y(5, 3)$.

b) Approximate $f(5.2, 2.9)$.

27. Figure 4 shows the contour map of a function $f(x, y)$ together with a path $\mathbf{r}(t)$ in the counterclockwise direction. The points $\mathbf{r}(1)$, $\mathbf{r}(2)$, and $\mathbf{r}(3)$ are indicated on the path. Let $g(t) = f(\mathbf{r}(t))$. Which of statements (i)–(iv) are true? Explain.

(i) $g'(1) > 0$.

(ii) $g(t)$ has a local minimum for some $1 \leq t \leq 2$.

(iii) $g'(2) = 0$.

(iv) $g'(3) = 0$.

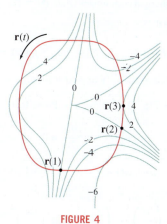

FIGURE 4

28. Jason earns $S(h, c) = 20h\left(1 + \frac{c}{100}\right)^{1.5}$ dollars per month at a used car lot, where h is the number of hours worked and c is the number of cars sold. He has already worked 160 hours and sold 69 cars. Right now Jason wants to go home but wonders how much more he might earn if he stays another 10 minutes with a customer who is considering buying a car. Use the linear approximation to estimate how much extra money Jason will earn if he sells his 70th car during these 10 minutes.

In Exercises 29–32, compute $\dfrac{d}{dt} f(\mathbf{r}(t))$ at the given value of t.

29. $f(x, y) = x + e^y$, $\mathbf{r}(t) = \left\langle 3t - 1, t^2 \right\rangle$ at $t = 2$

30. $f(x, y, z) = xz - y^2$, $\mathbf{r}(t) = \left\langle t, t^3, 1 - t \right\rangle$ at $t = -2$

31. $f(x, y) = xe^{3y} - ye^{3x}$, $\mathbf{r}(t) = \left\langle e^t, \ln t \right\rangle$ at $t = 1$

32. $f(x, y) = \tan^{-1}\frac{y}{x}$, $\mathbf{r}(t) = \langle \cos t, \sin t \rangle$, $t = \frac{\pi}{3}$

In Exercises 33–36, compute the directional derivative at P in the direction of $\mathbf{v}$.

33. $f(x, y) = x^3y^4$, $P = (3, -1)$, $\mathbf{v} = 2\mathbf{i} + \mathbf{j}$

34. $f(x, y, z) = zx - xy^2$, $P = (1, 1, 1)$, $\mathbf{v} = \langle 2, -1, 2 \rangle$

35. $f(x, y) = e^{x^2 + y^2}$, $P = \left(\frac{\sqrt{2}}{2}, \frac{\sqrt{2}}{2}\right)$, $\mathbf{v} = \langle 3, -4 \rangle$

36. $f(x, y, z) = \sin(xy + z)$, $P = (0, 0, 0)$, $\mathbf{v} = \mathbf{j} + \mathbf{k}$

37. Find the unit vector $\mathbf{e}$ at $P = (0, 0, 1)$ pointing in the direction along which $f(x, y, z) = xz + e^{-x^2 + y}$ increases most rapidly.

38. Find an equation of the tangent plane at $P = (0, 3, -1)$ to the surface with equation

$$ze^x + e^{z+1} = xy + y - 3$$

39. Let $n \neq 0$ be an integer and r an arbitrary constant. Show that the tangent plane to the surface $x^n + y^n + z^n = r$ at $P = (a, b, c)$ has equation

$$a^{n-1}x + b^{n-1}y + c^{n-1}z = r$$

40. Let $f(x, y) = (x - y)e^x$. Use the Chain Rule to calculate $\partial f/\partial u$ and $\partial f/\partial v$ (in terms of u and v), where $x = u - v$ and $y = u + v$.

41. Let $f(x, y, z) = x^2 y + y^2 z$. Use the Chain Rule to calculate $\partial f/\partial s$ and $\partial f/\partial t$ (in terms of s and t), where

$$x = s + t, \quad y = st, \quad z = 2s - t$$

42. Let P have spherical coordinates $(\rho, \theta, \phi) = \left(2, \frac{\pi}{4}, \frac{\pi}{4}\right)$. Calculate $\left.\frac{\partial f}{\partial \phi}\right|_P$ assuming that

$$f_x(P) = 4, \quad f_y(P) = -3, \quad f_z(P) = 8$$

Recall that $x = \rho \cos\theta \sin\phi$, $y = \rho \sin\theta \sin\phi$, $z = \rho \cos\phi$.

43. Let $g(u, v) = f(u^3 - v^3, v^3 - u^3)$. Prove that

$$v^2 \frac{\partial g}{\partial u} - u^2 \frac{\partial g}{\partial v} = 0$$

44. Let $f(x, y) = g(u)$, where $u = x^2 + y^2$ and $g(u)$ is differentiable. Prove that

$$\left(\frac{\partial f}{\partial x}\right)^2 + \left(\frac{\partial f}{\partial y}\right)^2 = 4u \left(\frac{dg}{du}\right)^2$$

45. Calculate $\partial z/\partial x$, where $xe^z + ze^y = x + y$.

46. Let $f(x, y) = x^4 - 2x^2 + y^2 - 6y$.

(a) Find the critical points of f and use the Second Derivative Test to determine whether they are a local minima or a local maxima.

(b) Find the minimum value of f without calculus by completing the square.

In Exercises 47–50, find the critical points of the function and analyze them using the Second Derivative Test.

47. $f(x, y) = x^4 - 4xy + 2y^2$

48. $f(x, y) = x^3 + 2y^3 - xy$

49. $f(x, y) = e^{x+y} - xe^{2y}$

50. $f(x, y) = \sin(x + y) - \frac{1}{2}(x + y^2)$

51. Prove that $f(x, y) = (x + 2y)e^{xy}$ has no critical points.

52. Find the global extrema of $f(x, y) = x^3 - xy - y^2 + y$ on the square $[0, 1] \times [0, 1]$.

53. Find the global extrema of $f(x, y) = 2xy - x - y$ on the domain $\{y \leq 4, y \geq x^2\}$.

54. Find the maximum of $f(x, y, z) = xyz$ subject to the constraint $g(x, y, z) = 2x + y + 4z = 1$.

55. Use Lagrange multipliers to find the minimum and maximum values of $f(x, y) = 3x - 2y$ on the circle $x^2 + y^2 = 4$.

56. Find the minimum value of $f(x, y) = xy$ subject to the constraint $5x - y = 4$ in two ways: using Lagrange multipliers and setting $y = 5x - 4$ in $f(x, y)$.

57. Find the minimum and maximum values of $f(x, y) = x^2 y$ on the ellipse $4x^2 + 9y^2 = 36$.

58. Find the point in the first quadrant on the curve $y = x + x^{-1}$ closest to the origin.

59. Find the extreme values of $f(x, y, z) = x + 2y + 3z$ subject to the two constraints $x + y + z = 1$ and $x^2 + y^2 + z^2 = 1$.

60. Find the minimum and maximum values of $f(x, y, z) = x - z$ on the intersection of the cylinders $x^2 + y^2 = 1$ and $x^2 + z^2 = 1$ (Figure 5).

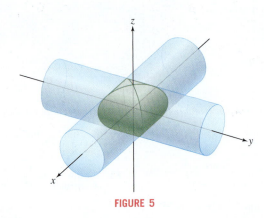

FIGURE 5

61. Use Lagrange multipliers to find the dimensions of a cylindrical can with a bottom but no top, of fixed volume V with minimum surface area.

62. Find the dimensions of the box of maximum volume with its sides parallel to the coordinate planes that can be inscribed in the ellipsoid (Figure 6)

$$\left(\frac{x}{a}\right)^2 + \left(\frac{y}{b}\right)^2 + \left(\frac{z}{c}\right)^2 = 1$$

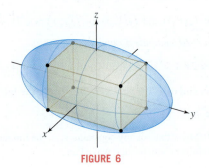

FIGURE 6

63. Given n nonzero numbers $\sigma_1, \ldots, \sigma_n$, show that the minimum value of

$$f(x_1, \ldots, x_n) = x_1^2 \sigma_1^2 + \cdots + x_n^2 \sigma_n^2$$

subject to $x_1 + \cdots + x_n = 1$ is c, where $c = \left(\sum_{j=1}^{n} \sigma_j^{-2}\right)^{-1}$.

These terraced fields illustrate how volume under a graph can be computed using iterated integration. (© bbbar/age fotostock)

15 MULTIPLE INTEGRATION

Integrals of functions of several variables, called **multiple integrals**, are a natural extension of the single-variable integrals studied in the first part of the text. They are used to compute many quantities that appear in applications, such as volumes, surface areas, centers of mass, probabilities, and average values.

15.1 Integration in Two Variables

The integral of a function of two variables $f(x, y)$, called a **double integral**, is denoted

$$\iint_{\mathcal{D}} f(x, y)\, dA$$

It represents the *signed volume* of the solid region between the graph of $f(x, y)$ and a domain $\mathcal{D}$ in the xy-plane (Figure 1), where the volume is positive for regions above the xy-plane and negative for regions below.

There are many similarities between double integrals and the single integrals:

- Double integrals are defined as limits of sums.
- Double integrals are evaluated using the Fundamental Theorem of Calculus (but we have to use it twice—see the discussion of iterated integrals below).

An important difference, however, is that the domain of integration is potentially substantially more complicated in the multivariable case. In one variable, the domain of integration is simply an interval $[a, b]$. In two variables, the domain $\mathcal{D}$ is a plane region whose boundary may be curved (Figure 1).

In this section, we focus on the simplest case where the domain is a rectangle, leaving more general domains for Section 15.2. Let

$$\mathcal{R} = [a, b] \times [c, d]$$

denote the rectangle in the plane (Figure 2) consisting of all points (x, y) such that

$$\mathcal{R}: \quad a \le x \le b, \qquad c \le y \le d$$

Like integrals in one variable, double integrals are defined through a three-step process: subdivision, summation, and passage to the limit. Figure 3 illustrates how the rectangle $\mathcal{R}$ is subdivided:

1. Subdivide $[a, b]$ and $[c, d]$ by choosing partitions:

$$a = x_0 < x_1 < \cdots < x_N = b, \qquad c = y_0 < y_1 < \cdots < y_M = d$$

where N and M are positive integers.

2. Create an $N \times M$ grid of subrectangles $\mathcal{R}_{ij}$.

3. Choose a sample point P_{ij} in each $\mathcal{R}_{ij}$.

Note that $\mathcal{R}_{ij} = [x_{i-1}, x_i] \times [y_{j-1}, y_j]$, so $\mathcal{R}_{ij}$ has area

$$\Delta A_{ij} = \Delta x_i \, \Delta y_j$$

where $\Delta x_i = x_i - x_{i-1}$ and $\Delta y_j = y_j - y_{j-1}$.

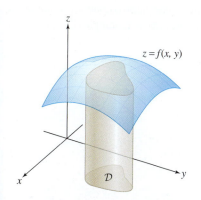

FIGURE 1 The double integral of $f(x, y)$ over the domain $\mathcal{D}$ yields the volume of the solid region between the graph of $f(x, y)$ and the xy-plane over a domain $\mathcal{D}$.

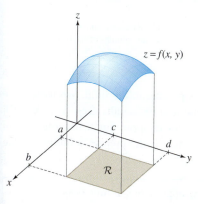

FIGURE 2

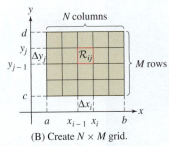
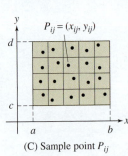

(A) Rectangle $\mathcal{R} = [a, b] \times [c, d]$ (B) Create $N \times M$ grid.   (C) Sample point P_{ij}

FIGURE 3

Keep in mind that a Riemann sum depends on the choice of partition and sample points. It would be more proper to write

$$S_{N,M}(\{P_{ij}\}, \{x_i\}, \{y_j\})$$

but we write $S_{N,M}$ to keep the notation simple.

Next, we form the Riemann sum with the function values $f(P_{ij})$:

$$S_{N,M} = \sum_{i=1}^{N} \sum_{j=1}^{M} f(P_{ij}) \,\Delta A_{ij} = \sum_{i=1}^{N} \sum_{j=1}^{M} f(P_{ij}) \,\Delta x_i \,\Delta y_j$$

The double summation runs over all i and j in the ranges $1 \le i \le N$ and $1 \le j \le M$, a total of NM terms.

The geometric interpretation of $S_{N,M}$ is shown in Figure 4. Each individual term $f(P_{ij}) \,\Delta A_{ij}$ of the sum is equal to the signed volume of the narrow box of height $f(P_{ij})$ above $\mathcal{R}_{ij}$:

$$f(P_{ij}) \,\Delta A_{ij} = f(P_{ij}) \,\Delta x_i \,\Delta y_j = \underbrace{\text{height} \times \text{area}}_{\text{Signed volume of box}}$$

When $f(P_{ij})$ is negative, the box lies below the xy-plane and has negative signed volume. The sum $S_{N,M}$ of the signed volumes of these narrow boxes approximates volume in the same way that Riemann sums in one variable approximate area by rectangles [Figure 4(A)].

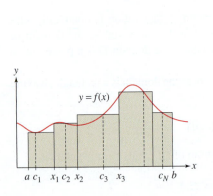

(A) In one variable, a Riemann sum approximates the area under the curve by a sum of areas of rectangles.

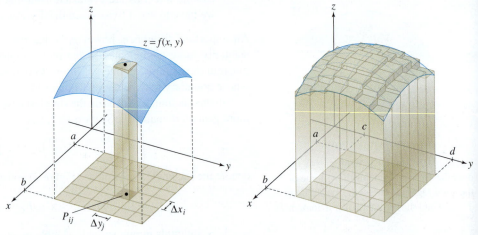

(B) The volume of the box is $f(P_{ij})\Delta A_{ij}$, where $\Delta A_{ij} = \Delta x_i \Delta y_j$.

(C) The Riemann sum $S_{N,M}$ is the sum of the volumes of the boxes.

DF **FIGURE 4**

The final step in defining the double integral is passing to the limit. We write $\mathcal{P} = \{\{x_i\}, \{y_j\}\}$ for the partition and $\|\mathcal{P}\|$ for the maximum of the widths Δx_i, Δy_j. As $\|\mathcal{P}\|$ tends to zero (and both M and N tend to infinity), the boxes approximate the solid region under the graph more and more closely (Figure 5). Here is the precise definition of the limit:

> **Limit of Riemann Sums** The Riemann sum $S_{N,M}$ approaches a limit L as $\|\mathcal{P}\| \to 0$ if, for all $\epsilon > 0$, there exists $\delta > 0$ such that
>
> $$|L - S_{N,M}| < \epsilon$$
>
> for all partitions satisfying $\|\mathcal{P}\| < \delta$ and all choices of sample points.

(A) $N = 4$, $M = 6$ (B) $N = 8$, $M = 12$ (C) $N = 20$, $M = 30$

DF **FIGURE 5** Midpoint approximations to the volume under $z = 24 - 3x^2 - y^2$.

In this case, we write

$$\lim_{\|\mathcal{P}\| \to 0} S_{N,M} = \lim_{\|\mathcal{P}\| \to 0} \sum_{i=1}^{N} \sum_{j=1}^{M} f(P_{ij}) \, \Delta A_{ij} = L$$

This limit L, if it exists, is the double integral $\displaystyle\iint_{\mathcal{R}} f(x, y) \, dA$.

DEFINITION **Double Integral Over a Rectangle** The double integral of $f(x, y)$ over a rectangle $\mathcal{R}$ is defined as the limit

$$\iint_{\mathcal{R}} f(x, y) \, dA = \lim_{\|\mathcal{P}\| \to 0} \sum_{i=1}^{N} \sum_{j=1}^{M} f(P_{ij}) \Delta A_{ij}$$

If this limit exists, we say that $f(x, y)$ is **integrable** over $\mathcal{R}$.

The double integral enables us to define the volume V of the solid region between the graph of a positive function $f(x, y)$ and the rectangle $\mathcal{R}$ by

$$V = \iint_{\mathcal{R}} f(x, y) \, dA$$

If $f(x, y)$ takes on both positive and negative values, the double integral defines the signed volume. So in Figure 6, $\displaystyle\iint_{\mathcal{R}} f(x, y) \, dA = V_1 + V_2 - V_3 - V_4$.

In computations, we often assume that the partition $\mathcal{P}$ is **regular**, meaning that the intervals $[a, b]$ and $[c, d]$ are each divided into subintervals of equal length. In other words, the partition is regular if $\Delta x_i = \Delta x$ and $\Delta y_j = \Delta y$, where

$$\Delta x = \frac{b - a}{N}, \qquad \Delta y = \frac{d - c}{M}$$

For a regular partition, $\|\mathcal{P}\|$ tends to zero as N and M tend to ∞.

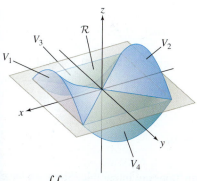

FIGURE 6 $\displaystyle\iint_{\mathcal{R}} f(x, y) \, dA$ is the signed volume of the region between the graph of $z = f(x, y)$ and the rectangle $\mathcal{R}$.

■ **EXAMPLE 1** **Estimating a Double Integral** Let $\mathcal{R} = [1, 2.5] \times [1, 2]$. Calculate $S_{3,2}$ for the integral (Figure 7)

$$\iint_{\mathcal{R}} xy \, dA$$

using the following two choices of sample points:

(a) Lower-left vertex   **(b)** Midpoint of rectangle

Solution Since we use the regular partition to compute $S_{3,2}$, each subrectangle (in this case they are squares) has sides of length

$$\Delta x = \frac{2.5 - 1}{3} = \frac{1}{2}, \qquad \Delta y = \frac{2 - 1}{2} = \frac{1}{2}$$

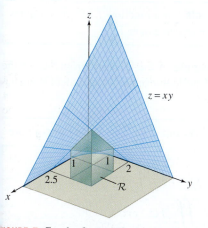

FIGURE 7 Graph of $z = xy$.

and area $\Delta A = \Delta x \, \Delta y = \frac{1}{4}$. The corresponding Riemann sum is

$$S_{3,2} = \sum_{i=1}^{3} \sum_{j=1}^{2} f(P_{ij}) \Delta A = \frac{1}{4} \sum_{i=1}^{3} \sum_{j=1}^{2} f(P_{ij})$$

where $f(x, y) = xy$.

(a) If we use the lower-left vertices shown in Figure 8(A), the Riemann sum is

$$S_{3,2} = \frac{1}{4}\left(f(1, 1) + f(1, \tfrac{3}{2}) + f(\tfrac{3}{2}, 1) + f(\tfrac{3}{2}, \tfrac{3}{2}) + f(2, 1) + f(2, \tfrac{3}{2})\right)$$

$$= \frac{1}{4}\left(1 + \tfrac{3}{2} + \tfrac{3}{2} + \tfrac{9}{4} + 2 + 3\right) = \frac{1}{4}\left(\tfrac{45}{4}\right) = 2.8125$$

(b) Using the midpoints of the rectangles shown in Figure 8(B), we obtain

$$S_{3,2} = \frac{1}{4}\left(f(\tfrac{5}{4}, \tfrac{5}{4}) + f(\tfrac{5}{4}, \tfrac{7}{4}) + f(\tfrac{7}{4}, \tfrac{5}{4}) + f(\tfrac{7}{4}, \tfrac{7}{4}) + f(\tfrac{9}{4}, \tfrac{5}{4}) + f(\tfrac{9}{4}, \tfrac{7}{4})\right)$$

$$= \frac{1}{4}\left(\tfrac{25}{16} + \tfrac{35}{16} + \tfrac{35}{16} + \tfrac{49}{16} + \tfrac{45}{16} + \tfrac{63}{16}\right) = \frac{1}{4}\left(\tfrac{252}{16}\right) = 3.9375 \quad \blacksquare$$

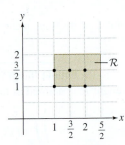 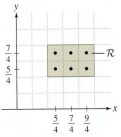

(A) Sample points are the lower-left vertices. (B) Sample points are midpoints.

FIGURE 8

■ **EXAMPLE 2** Use geometry to evaluate $\iint_{\mathcal{R}} (8 - 2y) \, dA$, where $\mathcal{R} = [0, 3] \times [0, 4]$.

Solution Figure 9 shows the graph of $z = 8 - 2y$. The double integral is equal to the volume V of the solid wedge underneath the graph. The triangular face of the wedge has area $A = \frac{1}{2}(8)4 = 16$. The volume of the wedge is equal to the area A times the length $\ell = 3$; that is, $V = \ell A = 3(16) = 48$. Therefore,

$$\iint_{\mathcal{R}} (8 - 2y) \, dA = 48$$

■

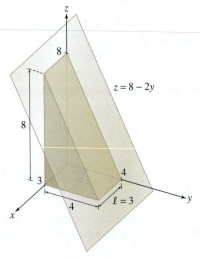

DF **FIGURE 9** Solid wedge under the graph of $z = 8 - 2y$.

The next theorem assures us that continuous functions are integrable. Since we have not yet defined continuity at boundary points of a domain, for the purposes of the next theorem, we define continuity on $\mathcal{R}$ to mean that f is defined and continuous on some open set containing $\mathcal{R}$. We omit the proof, which is similar to the single-variable case.

CAUTION The converse of Theorem 1 need not hold. There are integrable functions that are not continuous.

THEOREM 1 **Continuous Functions Are Integrable** If a function f of two variables is continuous on a rectangle $\mathcal{R}$, then $f(x, y)$ is integrable over $\mathcal{R}$.

As in the single-variable case, we often make use of the linearity properties of the double integral. They follow from the definition of the double integral as a limit of Riemann sums.

THEOREM 2 **Linearity of the Double Integral** Assume that $f(x, y)$ and $g(x, y)$ are integrable over a rectangle $\mathcal{R}$. Then:

(i) $\displaystyle \iint_{\mathcal{R}} \left(f(x, y) + g(x, y) \right) dA = \iint_{\mathcal{R}} f(x, y) \, dA + \iint_{\mathcal{R}} g(x, y) \, dA$

(ii) For any constant C, $\displaystyle \iint_{\mathcal{R}} Cf(x, y) \, dA = C \iint_{\mathcal{R}} f(x, y) \, dA$

If $f(x, y) = C$ is a constant function, then

$$\iint_{\mathcal{R}} C \, dA = C \cdot \text{Area}(\mathcal{R})$$

The double integral is the signed volume of the box of base $\mathcal{R}$ and height C (Figure 10). If $C < 0$, then the rectangle lies below the xy-plane, and the integral is equal to the signed volume, which is negative.

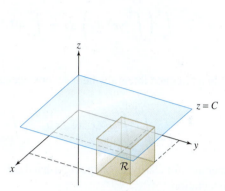

FIGURE 10 The double integral of $f(x, y) = C$ over a rectangle $\mathcal{R}$ is $C \cdot \text{Area}(\mathcal{R})$.

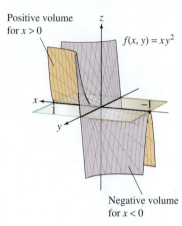

FIGURE 11

■ **EXAMPLE 3** **Arguing by Symmetry** Use symmetry to show that $\iint_{\mathcal{R}} xy^2 \, dA = 0$, where $\mathcal{R} = [-1, 1] \times [-1, 1]$.

Solution The double integral is the signed volume of the region between the graph of $f(x, y) = xy^2$ and the xy-plane (Figure 11). However, $f(x, y)$ takes opposite values at (x, y) and $(-x, y)$:

$$f(-x, y) = -xy^2 = -f(x, y)$$

Because of symmetry, the (negative) signed volume of the region below the xy-plane where $-1 \leq x \leq 0$ cancels with the (positive) signed volume of the region above the xy-plane where $0 \leq x \leq 1$. The net result is $\iint_{\mathcal{R}} xy^2 \, dA = 0$. ■

Iterated Integrals

Our main tool for evaluating double integrals is the Fundamental Theorem of Calculus (FTC), as in the single-variable case. To use the FTC, we express the double integral as an **iterated integral**, which is an expression of the form

$$\int_a^b \left(\int_c^d f(x, y) \, dy \right) dx$$

Iterated integrals are evaluated in two steps.

We often omit the parentheses in the notation for an iterated integral:

$$\int_a^b \int_c^d f(x, y) \, dy \, dx$$

The order of the variables in $dy \, dx$ tells us to integrate first with respect to y between the limits $y = c$ and $y = d$.

Step 1. Hold x constant and evaluate the inner integral with respect to y. This gives us a function of x alone:

$$S(x) = \int_c^d f(x, y) \, dy$$

Step 2. Integrate the resulting function $S(x)$ with respect to x.

■ **EXAMPLE 4** Evaluate $\displaystyle\int_2^4 \left(\int_1^9 y e^x \, dy \right) dx.$

Solution First evaluate the inner integral, treating x as a constant:

$$S(x) = \int_1^9 y e^x \, dy = e^x \int_1^9 y \, dy = e^x \left(\frac{1}{2} y^2 \right) \Big|_{y=1}^9 = e^x \left(\frac{81-1}{2} \right) = 40 e^x$$

Then integrate $S(x)$ with respect to x:

$$\int_2^4 \left(\int_1^9 y e^x \, dy \right) dx = \int_2^4 40 e^x \, dx = 40 e^x \Big|_2^4 = 40 (e^4 - e^2) \qquad ■$$

In an iterated integral where dx precedes dy, integrate first with respect to x:

$$\int_c^d \int_a^b f(x, y) \, dx \, dy = \int_{y=c}^d \left(\int_{x=a}^b f(x, y) \, dx \right) dy$$

Sometimes for clarity, as on the right-hand side here, we include the variables in the limits of integration.

■ **EXAMPLE 5** Evaluate $\displaystyle\int_{y=0}^4 \int_{x=0}^3 \frac{dx \, dy}{\sqrt{3x+4y}}.$

Solution We evaluate the inner integral first, treating y as a constant. Since we are integrating with respect to x, we need an antiderivative of $1/\sqrt{3x+4y}$ as a function of x. We can use $\frac{2}{3}\sqrt{3x+4y}$ because

$$\frac{\partial}{\partial x} \left(\frac{2}{3} \sqrt{3x+4y} \right) = \frac{1}{\sqrt{3x+4y}}$$

Thus, we have

$$\int_{x=0}^3 \frac{dx}{\sqrt{3x+4y}} = \frac{2}{3} \sqrt{3x+4y} \Big|_{x=0}^3 = \frac{2}{3} \left(\sqrt{4y+9} - \sqrt{4y} \right)$$

$$\int_{y=0}^4 \int_{x=0}^3 \frac{dx \, dy}{\sqrt{3x+4y}} = \frac{2}{3} \int_{y=0}^4 \left(\sqrt{4y+9} - \sqrt{4y} \right) dy$$

Therefore, we obtain

$$\int_{y=0}^4 \int_{x=0}^3 \frac{dx \, dy}{\sqrt{3x+4y}} = \frac{2}{3} \left(\frac{1}{6} (4y+9)^{3/2} - \frac{1}{6} (4y)^{3/2} \right) \Big|_{y=0}^4$$

$$= \frac{1}{9} \left(25^{3/2} - 16^{3/2} - 9^{3/2} \right) = \frac{34}{9} \qquad ■$$

■ **EXAMPLE 6** **Reversing the Order of Integration** Verify that

$$\int_{y=0}^4 \int_{x=0}^3 \frac{dx \, dy}{\sqrt{3x+4y}} = \int_{x=0}^3 \int_{y=0}^4 \frac{dy \, dx}{\sqrt{3x+4y}}$$

Solution We evaluated the iterated integral on the left in the previous example. We compute the integral on the right and verify that the result is also $\frac{34}{9}$:

$$\int_{y=0}^{4} \frac{dy}{\sqrt{3x+4y}} = \frac{1}{2}\sqrt{3x+4y}\,\bigg|_{y=0}^{4} = \frac{1}{2}\left(\sqrt{3x+16}-\sqrt{3x}\right)$$

$$\int_{x=0}^{3}\int_{y=0}^{4} \frac{dy\,dx}{\sqrt{3x+4y}} = \frac{1}{2}\int_{0}^{3}\left(\sqrt{3x+16}-\sqrt{3x}\right)dy$$

$$= \frac{1}{2}\left(\frac{2}{9}(3x+16)^{3/2}-\frac{2}{9}(3x)^{3/2}\right)\bigg|_{x=0}^{3}$$

$$= \frac{1}{9}\left(25^{3/2}-9^{3/2}-16^{3/2}\right) = \frac{34}{9} \qquad\blacksquare$$

The previous example illustrates a general fact: The value of an iterated integral does not depend on the order in which the integration is performed. This is part of Fubini's Theorem. Even more important, Fubini's Theorem states that a double integral over a rectangle can be evaluated as an iterated integral.

THEOREM 3 Fubini's Theorem   The double integral of a continuous function $f(x,y)$ over a rectangle $\mathcal{R} = [a,b]\times[c,d]$ is equal to the iterated integral (in either order):

$$\iint_{\mathcal{R}} f(x,y)\,dA = \int_{x=a}^{b}\int_{y=c}^{d} f(x,y)\,dy\,dx = \int_{y=c}^{d}\int_{x=a}^{b} f(x,y)\,dx\,dy$$

CAUTION When you reverse the order of integration in an iterated integral over a rectangle, remember to interchange the limits of integration (the inner limits become the outer limits).

Proof We sketch the proof. We can compute the double integral as a limit of Riemann sums that use a regular partition of $\mathcal{R}$ and sample points $P_{ij} = (x_i, y_j)$, where $\{x_i\}$ are sample points for a regular partition on $[a,b]$, and $\{y_j\}$ are sample points for a regular partition of $[c,d]$:

$$\iint_{\mathcal{R}} f(x,y)\,dA = \lim_{N,M\to\infty}\sum_{i=1}^{N}\sum_{j=1}^{M} f(x_i,y_j)\Delta y\,\Delta x$$

Here, $\Delta x = (b-a)/N$ and $\Delta y = (d-c)/M$. Fubini's Theorem stems from the elementary fact that we can add up the values in the sum in any order. So if we list the values $f(P_{ij})$ in an $N\times M$ array as shown in the margin, we can add up the columns first and then add up the column sums. This yields

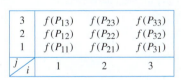

j			
3	$f(P_{13})$	$f(P_{23})$	$f(P_{33})$
2	$f(P_{12})$	$f(P_{22})$	$f(P_{32})$
1	$f(P_{11})$	$f(P_{21})$	$f(P_{31})$
i	1	2	3

$$\iint_{\mathcal{R}} f(x,y)\,dA = \lim_{N,M\to\infty}\sum_{i=1}^{N}\underbrace{\left(\sum_{j=1}^{M} f(x_i,y_j)\Delta y\right)}\,\Delta x$$

First sum the columns;
then add up the column sums.

For fixed i, $f(x_i,y)$ is a continuous function of y and the inner sum on the right is a Riemann sum that approaches the single integral $\int_{c}^{d} f(x_i,y)\,dy$. In other words, setting $S(x) = \int_{c}^{d} f(x,y)\,dy$, we have

$$\lim_{M\to\infty}\sum_{j=1}^{M} f(x_i,y_j) = \int_{c}^{d} f(x_i,y)\,dy = S(x_i)$$

To complete the proof, we take two facts for granted. First, $S(x)$ is a continuous function for $a \leq x \leq b$. Second, the limit as $N, M \to \infty$ may be computed by taking the limit first with respect to M and then with respect to N. Granting this, we get

$$\iint_{\mathcal{R}} f(x, y) \, dA = \lim_{N \to \infty} \sum_{i=1}^{N} \left(\lim_{M \to \infty} \sum_{j=1}^{M} f(x_i, y_j) \Delta y \right) \Delta x = \lim_{N \to \infty} \sum_{i=1}^{N} S(x_i) \Delta x$$

$$= \int_{a}^{b} S(x) \, dx = \int_{a}^{b} \left(\int_{c}^{d} f(x, y) \, dy \right) dx$$

Note that the sums on the right in the first line are Riemann sums for $S(x)$ that converge to the integral of $S(x)$ in the second line. This proves Fubini's Theorem for the order $dy \, dx$. A similar argument applies to the order $dx \, dy$. ∎

GRAPHICAL INSIGHT When we write a double integral as an iterated integral in the order $dy \, dx$, then for each fixed value $x = x_0$, the inner integral is the area of the cross section of S in the vertical plane $x = x_0$ perpendicular to the x-axis [Figure 12(A)]:

$$S(x_0) = \int_{c}^{d} f(x_0, y) \, dy = \frac{\text{area of cross section in vertical plane}}{x = x_0 \text{ perpendicular to the } x\text{-axis}}$$

What Fubini's Theorem says is that the volume V of S can be calculated as the integral of cross-sectional area $S(x)$:

$$V = \int_{a}^{b} \int_{c}^{d} f(x, y) \, dy \, dx = \int_{a}^{b} S(x) \, dx = \text{integral of cross-sectional area}$$

Similarly, the iterated integral in the order $dx \, dy$ calculates V as the integral of cross sections perpendicular to the y-axis [Figure 12(B)].

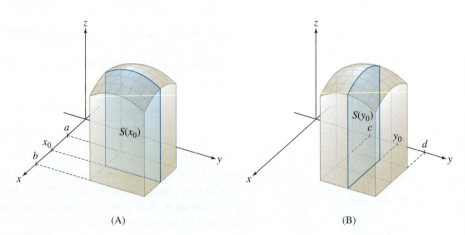

(A)   (B)

FIGURE 12

■ **EXAMPLE 7** Find the volume V between the graph of $f(x, y) = 16 - x^2 - 3y^2$ and the rectangle $\mathcal{R} = [0, 3] \times [0, 1]$ (Figure 13).

Solution The volume V is equal to the double integral of $f(x, y)$, which we write as an iterated integral:

$$V = \iint_{\mathcal{R}} (16 - x^2 - 3y^2) \, dA = \int_{x=0}^{3} \int_{y=0}^{1} (16 - x^2 - 3y^2) \, dy \, dx$$

We evaluate the inner integral first and then compute V:

$$\int_{y=0}^{1} (16 - x^2 - 3y^2) \, dy = (16y - x^2 y - y^3) \Big|_{y=0}^{1} = 15 - x^2$$

$$V = \int_{x=0}^{3} (15 - x^2) \, dx = \left(15x - \frac{1}{3}x^3 \right) \Big|_{0}^{3} = 36 \quad ■$$

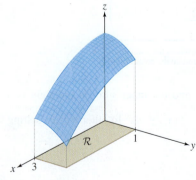

FIGURE 13 Graph of
$f(x, y) = 16 - x^2 - 3y^2$ over
$\mathcal{R} = [0, 3] \times [0, 1]$.

■ EXAMPLE 8 Calculate $\displaystyle\iint_{\mathcal{R}} \frac{dA}{(x+y)^2}$, where $\mathcal{R} = [1, 2] \times [0, 1]$ (Figure 14).

Solution

$$\iint_{\mathcal{R}} \frac{dA}{(x+y)^2} = \int_{x=1}^{2} \left(\int_{y=0}^{1} \frac{dy}{(x+y)^2} \right) dx = \int_{1}^{2} \left(-\frac{1}{x+y} \Big|_{y=0}^{1} \right) dx$$

$$= \int_{1}^{2} \left(-\frac{1}{x+1} + \frac{1}{x} \right) dx = \left(\ln x - \ln(x+1) \right) \Big|_{1}^{2}$$

$$= \left(\ln 2 - \ln 3 \right) - \left(\ln 1 - \ln 2 \right) = 2 \ln 2 - \ln 3 = \ln \frac{4}{3} \qquad ■$$

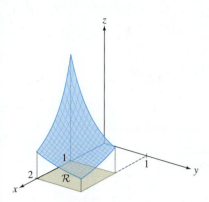

FIGURE 14 Graph of $z = (x + y)^{-2}$ over $\mathcal{R} = [1, 2] \times [0, 1]$.

15.1 SUMMARY

- A *Riemann sum* for $f(x, y)$ on a rectangle $\mathcal{R} = [a, b] \times [c, d]$ is a sum of the form

$$S_{N,M} = \sum_{i=1}^{N} \sum_{j=1}^{M} f(P_{ij}) \, \Delta x_i \, \Delta y_j$$

corresponding to partitions of $[a, b]$ and $[c, d]$, and choice of sample points P_{ij} in the subrectangle $\mathcal{R}_{ij}$.
- The double integral of $f(x, y)$ over $\mathcal{R}$ is defined as the limit (if it exists):

$$\iint_{\mathcal{R}} f(x, y) \, dA = \lim_{M, N \to \infty} \sum_{i=1}^{N} \sum_{j=1}^{M} f(P_{ij}) \, \Delta x_i \, \Delta y_j$$

We say that $f(x, y)$ is *integrable* over $\mathcal{R}$ if this limit exists.
- A continuous function on a rectangle $\mathcal{R}$ is integrable.
- The double integral is equal to the *signed volume* of the region between the graph of $z = f(x, y)$ and the rectangle $\mathcal{R}$. The signed volume of a region is positive if it lies above the xy-plane and negative if it lies below the xy-plane.
- If $f(x, y) = C$ is a constant function, then

$$\iint_{\mathcal{R}} C \, dA = C \cdot \text{Area}(\mathcal{R})$$

- Fubini's Theorem: The double integral of a continuous function $f(x, y)$ over a rectangle $\mathcal{R} = [a, b] \times [c, d]$ can be evaluated as an iterated integral (in either order):

$$\iint_{\mathcal{R}} f(x, y) \, dA = \int_{x=a}^{b} \int_{y=c}^{d} f(x, y) \, dy \, dx = \int_{y=c}^{d} \int_{x=a}^{b} f(x, y) \, dx \, dy$$

15.1 EXERCISES

Preliminary Questions

1. If $S_{8,4}$ is a Riemann sum for a double integral over $\mathcal{R} = [1, 5] \times [2, 10]$ using a regular partition, what is the area of each subrectangle? How many subrectangles are there?

2. Estimate the double integral of a continuous function f over the small rectangle $\mathcal{R} = [0.9, 1.1] \times [1.9, 2.1]$ if $f(1, 2) = 4$.

3. What is the integral of the constant function $f(x, y) = 5$ over the rectangle $[-2, 3] \times [2, 4]$?

4. What is the interpretation of $\displaystyle\iint_{\mathcal{R}} f(x, y) \, dA$ if $f(x, y)$ takes on both positive and negative values on $\mathcal{R}$?

5. Which of (a) or (b) is equal to $\displaystyle\int_{1}^{2} \int_{4}^{5} f(x, y) \, dy \, dx$?

(a) $\displaystyle\int_{1}^{2} \int_{4}^{5} f(x, y) \, dx \, dy$ 　　　 **(b)** $\displaystyle\int_{4}^{5} \int_{1}^{2} f(x, y) \, dx \, dy$

6. For which of the following functions is the double integral over the rectangle in Figure 15 equal to zero? Explain your reasoning.

(a) $f(x, y) = x^2 y$

(b) $f(x, y) = xy^2$

(c) $f(x, y) = \sin x$

(d) $f(x, y) = e^x$

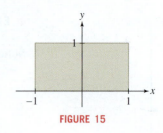

FIGURE 15

Exercises

1. Compute the Riemann sum $S_{4,3}$ to estimate the double integral of $f(x, y) = xy$ over $\mathcal{R} = [1, 3] \times [1, 2.5]$. Use the regular partition and upper-right vertices of the subrectangles as sample points.

2. Compute the Riemann sum with $N = M = 2$ to estimate the integral of $\sqrt{x + y}$ over $\mathcal{R} = [0, 1] \times [0, 1]$. Use the regular partition and midpoints of the subrectangles as sample points.

In Exercises 3–6, compute the Riemann sums for the double integral $\iint_{\mathcal{R}} f(x, y)\, dA$, where $\mathcal{R} = [1, 4] \times [1, 3]$, for the grid and two choices of sample points shown in Figure 16.

3. $f(x, y) = 2x + y$

4. $f(x, y) = 7$

5. $f(x, y) = 4x$

6. $f(x, y) = x - 2y$

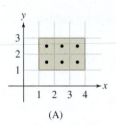 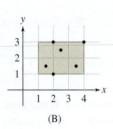

FIGURE 16

7. Let $\mathcal{R} = [0, 1] \times [0, 1]$. Estimate $\iint_{\mathcal{R}} (x + y)\, dA$ by computing two different Riemann sums, each with at least six rectangles.

8. Evaluate $\iint_{\mathcal{R}} 4\, dA$, where $\mathcal{R} = [2, 5] \times [4, 7]$.

9. Evaluate $\iint_{\mathcal{R}} (15 - 3x)\, dA$, where $\mathcal{R} = [0, 5] \times [0, 3]$, and sketch the corresponding solid region (see Example 2).

10. Evaluate $\iint_{\mathcal{R}} (-5)\, dA$, where $\mathcal{R} = [2, 5] \times [4, 7]$.

11. The following table gives the approximate height at quarter-meter intervals of a mound of gravel. Estimate the volume of the mound by computing the average of the two Riemann sums $S_{4,3}$ with lower-left and upper-right vertices of the subrectangles as sample points.

0.75	0.1	0.2	0.2	0.15	0.1
0.5	0.2	0.3	0.5	0.4	0.2
0.25	0.15	0.2	0.4	0.3	0.2
0	0.1	0.15	0.2	0.15	0.1
y / x	0	0.25	0.5	0.75	1

12. Use the following table to compute a Riemann sum $S_{3,3}$ for $f(x, y)$ on the square $\mathcal{R} = [0, 1.5] \times [0.5, 2]$. Use the regular partition and sample points of your choosing.

Values of $f(x, y)$

2	2.6	2.17	1.86	1.62	1.44
1.5	2.2	1.83	1.57	1.37	1.22
1	1.8	1.5	1.29	1.12	1
0.5	1.4	1.17	1	0.87	0.78
0	1	0.83	0.71	0.62	0.56
y / x	0	0.5	1	1.5	2

13. **CAS** Let $S_{N,N}$ be the Riemann sum for $\int_0^1 \int_0^1 e^{x^3 - y^3}\, dy\, dx$ using the regular partition and the lower left-hand vertex of each subrectangle as sample points. Use a computer algebra system to calculate $S_{N,N}$ for $N = 25, 50, 100$.

14. **CAS** Let $S_{N,M}$ be the Riemann sum for

$$\int_0^4 \int_0^2 \ln(1 + x^2 + y^2)\, dy\, dx$$

using the regular partition and the upper right-hand vertex of each subrectangle as sample points. Use a computer algebra system to calculate $S_{2N,N}$ for $N = 25, 50, 100$.

In Exercises 15–18, use symmetry to evaluate the double integral.

15. $\iint_{\mathcal{R}} x^3\, dA, \quad \mathcal{R} = [-4, 4] \times [0, 5]$

16. $\iint_{\mathcal{R}} 1\, dA, \quad \mathcal{R} = [2, 4] \times [-7, 7]$

17. $\iint_{\mathcal{R}} \sin x\, dA, \quad \mathcal{R} = [0, 2\pi] \times [0, 2\pi]$

18. $\iint_{\mathcal{R}} (2 + x^2 y)\, dA, \quad \mathcal{R} = [0, 1] \times [-1, 1]$

In Exercises 19–36, evaluate the iterated integral.

19. $\int_1^3 \int_0^2 x^3 y\, dy\, dx$

20. $\int_0^2 \int_1^3 x^3 y\, dx\, dy$

21. $\int_4^9 \int_{-3}^8 1\, dx\, dy$

22. $\int_{-4}^{-1} \int_4^8 (-5)\, dx\, dy$

23. $\int_{-1}^1 \int_0^\pi x^2 \sin y\, dy\, dx$

24. $\int_{-1}^1 \int_0^\pi x^2 \sin y\, dx\, dy$

25. $\int_2^6 \int_1^4 x^2\, dx\, dy$

26. $\int_2^6 \int_1^4 y^2\, dx\, dy$

27. $\int_0^1 \int_0^2 (x + 4y^3)\, dx\, dy$

28. $\int_0^2 \int_0^2 (x^2 - y^2)\, dy\, dx$

29. $\int_0^4 \int_0^9 \sqrt{x + 4y}\, dx\, dy$

30. $\int_0^{\pi/4} \int_{\pi/4}^{\pi/2} \cos(2x + y)\, dy\, dx$

31. $\int_1^2 \int_0^4 \frac{dy\, dx}{x + y}$

32. $\int_1^2 \int_2^4 e^{3x - y}\, dy\, dx$

33. $\int_0^4 \int_0^5 \frac{dy\, dx}{\sqrt{x + y}}$

34. $\int_0^8 \int_1^2 \frac{x\, dx\, dy}{\sqrt{x^2 + y}}$

35. $\int_1^2 \int_1^3 \frac{\ln(xy)\, dy\, dx}{y}$

36. $\int_0^1 \int_2^3 \frac{1}{(x + 4y)^3}\, dx\, dy$

In Exercises 37–42, evaluate the integral.

37. $\iint_{\mathcal{R}} \frac{x}{y}\, dA, \quad \mathcal{R} = [-2, 4] \times [1, 3]$

38. $\iint_{\mathcal{R}} x^2 y\, dA, \quad \mathcal{R} = [-1, 1] \times [0, 2]$

39. $\iint_{\mathcal{R}} \cos x \sin 2y\, dA, \quad \mathcal{R} = \left[0, \frac{\pi}{2}\right] \times \left[0, \frac{\pi}{2}\right]$

40. $\iint_{\mathcal{R}} \frac{y}{x + 1}\, dA, \quad \mathcal{R} = [0, 2] \times [0, 4]$

41. $\iint_{\mathcal{R}} e^x \sin y\, dA, \quad \mathcal{R} = [0, 2] \times \left[0, \frac{\pi}{4}\right]$

42. $\iint_{\mathcal{R}} e^{3x + 4y}\, dA, \quad \mathcal{R} = [0, 1] \times [1, 2]$

43. Let $f(x, y) = mxy^2$, where m is a constant. Find a value of m such that $\iint_{\mathcal{R}} f(x, y)\, dA = 1$, where $\mathcal{R} = [0, 1] \times [0, 2]$.

44. Evaluate $I = \int_1^3 \int_0^1 ye^{xy}\, dy\, dx$. You will need Integration by Parts and the formula

$$\int e^x(x^{-1} - x^{-2})\, dx = x^{-1}e^x + C$$

Then evaluate I again using Fubini's Theorem to change the order of integration (i.e., integrate first with respect to x). Which method is easier?

45. Evaluate $\int_0^1 \int_0^1 y\sqrt{1 + xy}\, dy\, dx$. *Hint:* Change the order of integration.

46. Evaluate $\int_0^1 \int_0^1 xe^{xy}\, dx\, dy$. *Hint:* Change the order of integration.

47. Evaluate $\int_0^1 \int_0^1 \frac{y}{1 + xy}\, dy\, dx$. *Hint:* Change the order of integration.

48. Calculate a Riemann sum $S_{3,3}$ on the square $\mathcal{R} = [0, 3] \times [0, 3]$ for the function $f(x, y)$ whose contour plot is shown in Figure 17. Choose sample points and use the plot to find the values of $f(x, y)$ at these points.

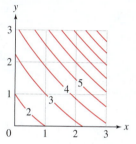

FIGURE 17 Contour plot of $f(x, y)$.

49. Using Fubini's Theorem, argue that the solid in Figure 18 has volume AL, where A is the area of the front face of the solid.

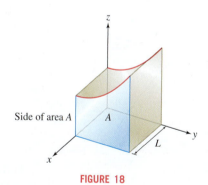

Side of area A

FIGURE 18

Further Insights and Challenges

50. Prove the following extension of the Fundamental Theorem of Calculus to two variables: If $\frac{\partial^2 F}{\partial x\, \partial y} = f(x, y)$, then

$$\iint_{\mathcal{R}} f(x, y)\, dA = F(b, d) - F(a, d) - F(b, c) + F(a, c)$$

where $\mathcal{R} = [a, b] \times [c, d]$.

51. Let $F(x, y) = x^{-1}e^{xy}$. Show that $\frac{\partial^2 F}{\partial x\, \partial y} = ye^{xy}$ and use the result of Exercise 50 to evaluate $\iint_{\mathcal{R}} ye^{xy}\, dA$ for the rectangle $\mathcal{R} = [1, 3] \times [0, 1]$.

52. Find a function $F(x, y)$ satisfying $\frac{\partial^2 F}{\partial x\, \partial y} = 6x^2 y$ and use the result of Exercise 50 to evaluate $\iint_{\mathcal{R}} 6x^2 y\, dA$ for the rectangle $\mathcal{R} = [0, 1] \times [0, 4]$.

53. In this exercise, we use double integration to evaluate the following improper integral for $a > 0$ a positive constant:

$$I(a) = \int_0^\infty \frac{e^{-x} - e^{-ax}}{x}\, dx$$

(a) Use L'Hôpital's Rule to show that $f(x) = \frac{e^{-x} - e^{-ax}}{x}$, though not defined at $x = 0$, can be made continuous by assigning the value $f(0) = a - 1$.

(b) Prove that $|f(x)| \leq e^{-x} + e^{-ax}$ for $x > 1$ (use the triangle inequality), and apply the Comparison Theorem to show that $I(a)$ converges.

(c) Show that $I(a) = \int_0^\infty \int_1^a e^{-xy}\, dy\, dx$.

(d) Prove, by interchanging the order of integration, that

$$I(a) = \ln a - \lim_{T \to \infty} \int_1^a \frac{e^{-Ty}}{y} \, dy \qquad \boxed{1}$$

(e) Use the Comparison Theorem to show that the limit in Eq. (1) is zero. *Hint:* If $a \geq 1$, show that $e^{-Ty}/y \leq e^{-T}$ for $y \geq 1$, and if $a < 1$, show that $e^{-Ty}/y \leq e^{-aT}/a$ for $a \leq y \leq 1$. Conclude that $I(a) = \ln a$ (Figure 19).

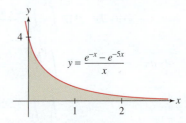

FIGURE 19 The shaded region has area $\ln 5$.

15.2 Double Integrals over More General Regions

In the previous section, we restricted our attention to rectangular domains. Now we shall treat the more general case of domains $\mathcal{D}$ whose boundaries are simple closed curves (a curve is *simple* if it does not intersect itself). We assume that the boundary of $\mathcal{D}$ is smooth as in Figure 1(A) or consists of finitely many smooth curves, joined together with possible corners, as in Figure 1(B). A boundary curve of this type is called **piecewise smooth**. We also assume that $\mathcal{D}$ is a closed domain; that is, $\mathcal{D}$ contains its boundary.

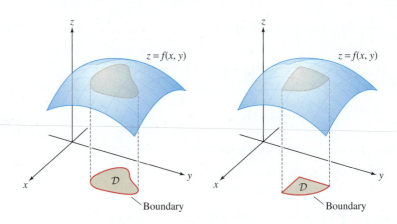

(A) $\mathcal{D}$ has a smooth boundary.

(B) $\mathcal{D}$ has a piecewise smooth boundary, consisting of three smooth curves joined at the corners.

FIGURE 1

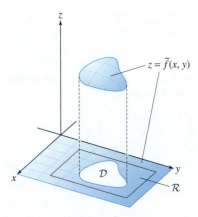

FIGURE 2 The function $\tilde{f}$ is zero outside of $\mathcal{D}$.

Fortunately, we do not need to start from the beginning to define the double integral over a domain $\mathcal{D}$ of this type. Given a function $f(x, y)$ on $\mathcal{D}$, we choose a rectangle $\mathcal{R} = [a, b] \times [c, d]$ containing $\mathcal{D}$ and define a new function $\tilde{f}(x, y)$ that agrees with $f(x, y)$ on $\mathcal{D}$ and is zero outside of $\mathcal{D}$ (Figure 2):

$$\tilde{f}(x, y) = \begin{cases} f(x, y) & \text{if } (x, y) \in \mathcal{D} \\ 0 & \text{if } (x, y) \notin \mathcal{D} \end{cases}$$

The double integral of f over $\mathcal{D}$ is defined as the integral of $\tilde{f}$ over $\mathcal{R}$:

$$\boxed{\iint_{\mathcal{D}} f(x, y) \, dA = \iint_{\mathcal{R}} \tilde{f}(x, y) \, dA} \qquad \boxed{1}$$

We say that f is **integrable** over $\mathcal{D}$ if the integral of $\tilde{f}$ over $\mathcal{R}$ exists. The value of the integral does not depend on the particular choice of $\mathcal{R}$ because $\tilde{f}$ is zero outside of $\mathcal{D}$.

This definition seems reasonable because the integral of $\tilde{f}$ only "picks up" the values of f on $\mathcal{D}$. However, $\tilde{f}$ is likely to be discontinuous because its values jump suddenly to zero beyond the boundary. Despite this possible discontinuity, the next theorem guarantees that the integral of $\tilde{f}$ over $\mathcal{R}$ exists if our original function f is continuous.

> **THEOREM 1** If $f(x, y)$ is continuous on a closed domain $\mathcal{D}$ whose boundary is a closed, simple, piecewise smooth curve, then $\iint_{\mathcal{D}} f(x, y)\, dA$ exists.

In Theorem 1, we define continuity on $\mathcal{D}$ to mean that f is defined and continuous on some open set containing $\mathcal{D}$.

As in the previous section, the double integral defines the signed volume between the graph of $f(x, y)$ and the xy-plane, where regions below the xy-plane are assigned negative volume.

We can approximate the double integral by Riemann sums for the function $\tilde{f}$ on a rectangle $\mathcal{R}$ containing $\mathcal{D}$. Because $\tilde{f}(P) = 0$ for points P in $\mathcal{R}$ that do not belong to $\mathcal{D}$, any such Riemann sum reduces to a sum over those sample points that lie in $\mathcal{D}$:

$$\iint_{\mathcal{D}} f(x, y)\, dA \approx \sum_{i=1}^{N} \sum_{j=1}^{M} \tilde{f}(P_{ij})\, \Delta x_i\, \Delta y_j = \underbrace{\sum f(P_{ij})\, \Delta x_i\, \Delta y_j}_{\substack{\text{Sum only over points} \\ P_{ij} \text{ that lie in } \mathcal{D}}} \qquad \boxed{2}$$

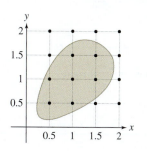

FIGURE 3 Domain $\mathcal{D}$.

■ **EXAMPLE 1** Compute $S_{4,4}$ for the integral $\iint_{\mathcal{D}} (x + y)\, dA$, where $\mathcal{D}$ is the shaded domain in Figure 3. Use the upper right-hand corners of the squares as sample points.

Solution Let $f(x, y) = x + y$. The subrectangles in Figure 3 have sides of length $\Delta x = \Delta y = \frac{1}{2}$ and area $\Delta A = \frac{1}{4}$. Only 7 of the 16 sample points lie in $\mathcal{D}$, so

$$S_{4,4} = \sum_{i=1}^{4} \sum_{j=1}^{4} \tilde{f}(P_{ij})\, \Delta x\, \Delta y = \frac{1}{4}\big(f(0.5, 0.5) + f(1, 0.5) + f(0.5, 1) + f(1, 1)$$

$$+ f(1.5, 1) + f(1, 1.5) + f(1.5, 1.5)\big)$$

$$= \frac{1}{4}\big(1 + 1.5 + 1.5 + 2 + 2.5 + 2.5 + 3\big) = \frac{7}{2} \qquad ■$$

The linearity properties of the double integral carry over to general domains: If $f(x, y)$ and $g(x, y)$ are integrable and C is a constant, then

$$\iint_{\mathcal{D}} (f(x, y) + g(x, y))\, dA = \iint_{\mathcal{D}} f(x, y)\, dA + \iint_{\mathcal{D}} g(x, y)\, dA$$

$$\iint_{\mathcal{D}} Cf(x, y)\, dA = C \iint_{\mathcal{D}} f(x, y)\, dA$$

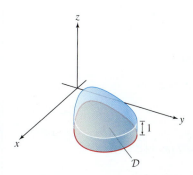

FIGURE 4 The volume of the cylinder of height 1 with $\mathcal{D}$ as its base is equal to the area of $\mathcal{D}$.

Although we usually think of double integrals as representing volumes, it is worth noting that we can express the *area* of a domain $\mathcal{D}$ in the plane as the double integral of the constant function $f(x, y) = 1$:

$$\boxed{\text{Area}(\mathcal{D}) = \iint_{\mathcal{D}} 1\, dA} \qquad \boxed{3}$$

Indeed, as we see in Figure 4, the the area of $\mathcal{D}$ is equal to the volume of the "cylinder" of height 1 with $\mathcal{D}$ as base. More generally, for any constant C,

$$\iint_{\mathcal{D}} C\, dA = C\, \text{Area}(\mathcal{D}) \qquad \boxed{4}$$

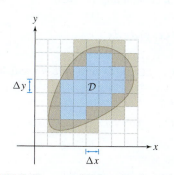

FIGURE 5 The area of $\mathcal{D}$ is approximated by the sum of the areas of the rectangles contained in $\mathcal{D}$.

CONCEPTUAL INSIGHT Eq. (3) tells us that we can approximate the area of a domain $\mathcal{D}$ by a Riemann sum for $\iint_{\mathcal{D}} 1\, dA$. In this case, $f(x, y) = 1$, and we obtain a Riemann sum by adding up the areas $\Delta x_i\, \Delta y_j$ of those rectangles in a grid that are contained in $\mathcal{D}$ or that intersects the boundary of $\mathcal{D}$ (Figure 5). The finer the grid, the better the approximation. The exact area is the limit as the sides of the rectangles tend to zero.

Regions Between Two Graphs

When $\mathcal{D}$ is a region between two graphs in the xy-plane, we can evaluate double integrals over $\mathcal{D}$ as iterated integrals. We say that $\mathcal{D}$ is **vertically simple** if it is the region between the graphs of two continuous functions $y = g_1(x)$ and $y = g_2(x)$ (Figure 6):

$$\mathcal{D} = \{(x, y) : a \le x \le b, \quad g_1(x) \le y \le g_2(x)\}$$

Similarly, $\mathcal{D}$ is **horizontally simple** if

$$\mathcal{D} = \{(x, y) : c \le y \le d, \quad g_1(y) \le x \le g_2(y)\}$$

When you write a double integral over a vertically simple region as an iterated integral, the inner integral is an integral over the dashed segment shown in Figure 6(A). For a horizontally simple region, the inner integral is an integral over the dashed segment shown in Figure 6(B).

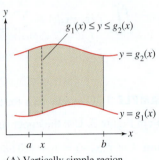
(A) Vertically simple region

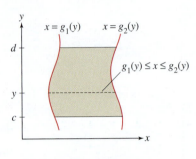
(B) Horizontally simple region

FIGURE 6

THEOREM 2 If $\mathcal{D}$ is vertically simple with description

$$a \le x \le b, \qquad g_1(x) \le y \le g_2(x)$$

then

$$\iint_{\mathcal{D}} f(x, y) \, dA = \int_a^b \int_{g_1(x)}^{g_2(x)} f(x, y) \, dy \, dx$$

If $\mathcal{D}$ is a horizontally simple region with description

$$c \le y \le d, \qquad g_1(y) \le x \le g_2(y)$$

then

$$\iint_{\mathcal{D}} f(x, y) \, dA = \int_c^d \int_{g_1(y)}^{g_2(y)} f(x, y) \, dx \, dy$$

Proof We sketch the proof, assuming that $\mathcal{D}$ is vertically simple (the horizontally simple case is similar). Choose a rectangle $\mathcal{R} = [a, b] \times [c, d]$ containing $\mathcal{D}$. Then

$$\iint_{\mathcal{D}} f(x, y) \, dA = \int_a^b \int_c^d \tilde{f}(x, y) \, dy \, dx \qquad \boxed{5}$$

Although $\tilde{f}$, which was defined to be f inside $\mathcal{D}$ and 0 outside $\mathcal{D}$, need not be continuous, the use of Fubini's Theorem in Eq. (5) can be justified. In particular, the integral $\int_c^d \tilde{f}(x, y) \, dy$ exists and is a continuous function of x.

By definition, $\tilde{f}(x, y)$ is zero outside $\mathcal{D}$, so for fixed x, $\tilde{f}(x, y)$ is zero unless y satisfies $g_1(x) \le y \le g_2(x)$. Therefore,

$$\int_c^d \tilde{f}(x, y) \, dy = \int_{g_1(x)}^{g_2(x)} f(x, y) \, dy$$

Substituting in Eq. (5), we obtain the desired equality:

$$\iint_{\mathcal{D}} f(x, y) \, dA = \int_a^b \int_{g_1(x)}^{g_2(x)} f(x, y) \, dy \, dx \qquad ■$$

Integration over a simple region is similar to integration over a rectangle with one difference: The limits of the inner integral may be functions instead of constants.

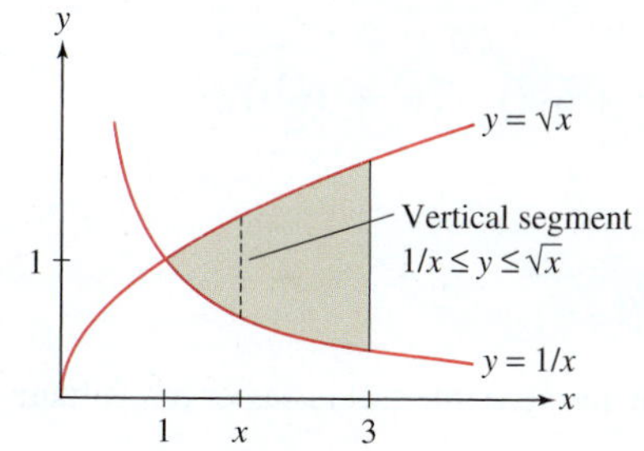

■ EXAMPLE 2 Evaluate $\iint_{\mathcal{D}} x^2 y \, dA$, where $\mathcal{D}$ is the region in Figure 7.

Solution

Step 1. Describe $\mathcal{D}$ as a vertically simple region.

$$\underbrace{1 \le x \le 3}_{\substack{\text{Limits of outer}\\\text{integral}}}, \qquad \underbrace{\frac{1}{x} \le y \le \sqrt{x}}_{\substack{\text{Limits of inner}\\\text{integral}}}$$

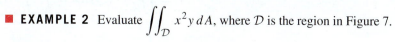

In this case, $g_1(x) = 1/x$ and $g_2(x) = \sqrt{x}$.

Step 2. Set up the iterated integral.

$$\iint_{\mathcal{D}} x^2 y \, dA = \int_1^3 \int_{y=1/x}^{\sqrt{x}} x^2 y \, dy \, dx$$

Notice that the inner integral is an integral over a vertical segment between the graphs of $y = 1/x$ and $y = \sqrt{x}$.

Step 3. Compute the iterated integral.

As usual, we evaluate the inner integral by treating x as a constant, but now the upper and lower limits depend on x:

$$\int_{y=1/x}^{\sqrt{x}} x^2 y \, dy = \frac{1}{2} x^2 y^2 \Big|_{y=1/x}^{\sqrt{x}} = \frac{1}{2} x^2 (\sqrt{x})^2 - \frac{1}{2} x^2 \left(\frac{1}{x} \right)^2 = \frac{1}{2} x^3 - \frac{1}{2}$$

We complete the calculation by integrating with respect to x:

$$\iint_{\mathcal{D}} x^2 y \, dA = \int_1^3 \left(\frac{1}{2} x^3 - \frac{1}{2} \right) dx = \left(\frac{1}{8} x^4 - \frac{1}{2} x \right) \Big|_1^3$$
$$= \frac{69}{8} - \left(-\frac{3}{8} \right) = 9 \qquad \blacksquare$$

■ EXAMPLE 3   Horizontally Simple Description Better Find the volume V of the region between the plane $z = 2x + 3y$ and the triangle $\mathcal{D}$ in Figure 8.

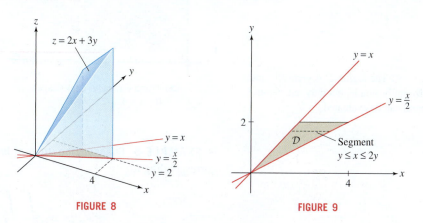

FIGURE 8

FIGURE 9

Solution The triangle $\mathcal{D}$ is bounded by the lines $y = x/2$, $y = x$, and $y = 2$. We see in Figure 9 that $\mathcal{D}$ is vertically simple, but the upper curve is not given by a single formula: The formula switches from $y = x$ to $y = 2$. Therefore, it is more convenient to describe $\mathcal{D}$ as a horizontally simple region (Figure 9):

$$\mathcal{D} : 0 \le y \le 2, \quad y \le x \le 2y$$

On the left margin:

y

$y = \sqrt{x}$

Vertical segment
$1/x \le y \le \sqrt{x}$

1

$y = 1/x$

$1 \quad x \quad 3$

x

FIGURE 7 Domain between $y = \sqrt{x}$ and $y = 1/x$.

The volume is equal to the double integral of $f(x, y) = 2x + 3y$ over $\mathcal{D}$:

$$V = \iint_{\mathcal{D}} f(x, y)\, dA = \int_0^2 \int_{x=y}^{2y} (2x + 3y)\, dx\, dy$$

$$= \int_0^2 \left(x^2 + 3yx \right)\Big|_{x=y}^{2y} dy = \int_0^2 \left((4y^2 + 6y^2) - (y^2 + 3y^2) \right) dy$$

$$= \int_0^2 6y^2\, dy = 2y^3\Big|_0^2 = 16 \qquad \blacksquare$$

The next example shows that in some cases, one iterated integral is easier to evaluate than the other.

■ **EXAMPLE 4** **Choosing the Best Iterated Integral** Evaluate $\displaystyle\iint_{\mathcal{D}} e^{y^2}\, dA$ for $\mathcal{D}$ in Figure 10.

Solution First, let's try describing $\mathcal{D}$ as a vertically simple domain. Referring to Figure 10(A), we have

$$\mathcal{D}: 0 \le x \le 4, \quad \frac{1}{2}x \le y \le 2 \quad \Rightarrow \quad \iint_{\mathcal{D}} e^{y^2}\, dA = \int_{x=0}^4 \int_{y=x/2}^2 e^{y^2}\, dy\, dx$$

The inner integral cannot be evaluated because we have no explicit antiderivative for e^{y^2}. Therefore, we try describing $\mathcal{D}$ as horizontally simple [Figure 10(B)]:

$$\mathcal{D}: 0 \le y \le 2, \quad 0 \le x \le 2y$$

This leads to an iterated integral that can be evaluated:

$$\int_0^2 \int_{x=0}^{2y} e^{y^2}\, dx\, dy = \int_0^2 \left(x e^{y^2}\Big|_{x=0}^{2y} \right) dy = \int_0^2 2y e^{y^2}\, dy$$

$$= e^{y^2}\Big|_0^2 = e^4 - 1 \qquad \blacksquare$$

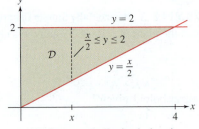

(A) $\mathcal{D}$ as a vertically simple domain:
$0 \le x \le 4,\ x/2 \le y \le 2$

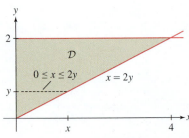

(B) $\mathcal{D}$ as a horizontally simple domain:
$0 \le y \le 2,\ 0 \le x \le 2y$

FIGURE 10 The region $\mathcal{D}$ is vertically and horizontally simple, but only one iterated integral can be evaluated.

■ **EXAMPLE 5** **Changing the Order of Integration** Sketch the domain of integration $\mathcal{D}$ corresponding to

$$\int_1^9 \int_{\sqrt{y}}^3 x e^y\, dx\, dy$$

Then change the order of integration and evaluate.

Solution The limits of integration give us inequalities that describe the domain $\mathcal{D}$ (as a horizontally simple region since dx precedes dy):

$$1 \le y \le 9, \qquad \sqrt{y} \le x \le 3$$

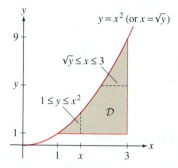

FIGURE 11 Describing $\mathcal{D}$ as a horizontally or vertically simple region.

We sketch the region in Figure 11. Now observe that $\mathcal{D}$ is also vertically simple:

$$1 \le x \le 3, \qquad 1 \le y \le x^2$$

so we can rewrite our integral and evaluate:

$$\int_1^9 \int_{x=\sqrt{y}}^3 xe^y \, dx \, dy = \int_1^3 \int_{y=1}^{x^2} xe^y \, dy \, dx = \int_1^3 \left(\int_{y=1}^{x^2} xe^y \, dy \right) dx$$

$$= \int_1^3 \left(xe^y \Big|_{y=1}^{x^2} \right) dx = \int_1^3 (xe^{x^2} - ex) \, dx = \frac{1}{2}(e^{x^2} - ex^2) \Big|_1^3$$

$$= \frac{1}{2}(e^9 - 9e) - 0 = \frac{1}{2}(e^9 - 9e)$$ ∎

If we want to compute the volume of a solid region in space that is sandwiched between two surfaces and that lies above a domain $\mathcal{D}$ in the xy-plane as in Figure 12, we can use the linearity of the double integral to find a formula.

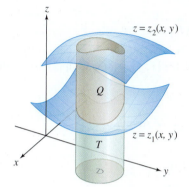

FIGURE 12 Finding the volume of a solid Q sandwiched between two surfaces above a domain $\mathcal{D}$.

> **THEOREM 3**   Suppose $z_1(x, y)$ and $z_2(x, y)$ are integrable functions on $\mathcal{D}$ and $z_1(x, y) \le z_2(x, y)$ for all points in $\mathcal{D}$. Then the volume V of the solid region between the surfaces given by $z = z_1(x, y)$ and $z = z_2(x, y)$ is given by
>
> $$V = \iint_{\mathcal{D}} (z_2(x, y) - z_1(x, y)) \, dA$$

Proof   First assume that $0 \le z_1(x, y) \le z_2(x, y)$ for all (x, y) in $\mathcal{D}$, as in Figure 12. Then, letting Q be the region sandwiched between the two surfaces and T the region between the bottom surface and the xy-plane, we have

$$\text{Vol}(Q) = \text{Vol}(Q \cup T) - \text{Vol}(T)$$

$$= \iint_{\mathcal{D}} z_2(x, y) \, dA - \iint_{\mathcal{D}} z_1(x, y) \, dA = \iint_{\mathcal{D}} (z_2(x, y) - z_1(x, y)) \, dA$$

A similar argument applies even when the two surfaces are not above the xy-plane. ∎

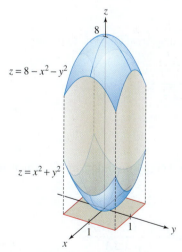

FIGURE 13 Finding the volume of a solid sandwiched between two paraboloids above a square.

■ **EXAMPLE 6**   Find the volume V of the solid bounded above and below by the paraboloids corresponding to $z = 8 - x^2 - y^2$ and $z = x^2 + y^2$ and lying over the domain $\mathcal{D} = \{(x, y) : -1 \le x \le 1, -1 \le y \le 1\}$.

Solution   The solid region appears as in Figure 13. Since it is true that over the domain $\mathcal{D}$, the paraboloid given by $z = 8 - x^2 - y^2$ is always above the paraboloid given by $z = x^2 + y^2$, Theorem 3 applies.

$$V = \iint_{\mathcal{D}} (h_2(x, y) - h_1(x, y)) \, dA = \int_{-1}^1 \int_{-1}^1 ((8 - x^2 - y^2) - (x^2 + y^2)) \, dy \, dx$$

$$= \int_{-1}^1 \int_{-1}^1 (8 - 2x^2 - 2y^2) \, dy \, dx = \int_{-1}^1 \left(8y - 2x^2y - \frac{2y^3}{3} \right) \Big|_{-1}^1 dx$$

$$= \int_{-1}^1 \left(8 - 2x^2 - \frac{2}{3} \right) - \left(-8 + 2x^2 + \frac{2}{3} \right) dx = \int_{-1}^1 \left(16 - 4x^2 - \frac{4}{3} \right) dx$$

$$= 16x - \frac{4x^3}{3} - \frac{4x}{3} \Big|_{-1}^1 = \left(16 - \frac{4}{3} - \frac{4}{3} \right) - \left(-16 + \frac{4}{3} + \frac{4}{3} \right) = 26\frac{2}{3}$$ ■

In the next theorem, part (a) is a formal statement of the fact that larger functions have larger integrals, a fact that we also noted in the single-variable case. Part (b) is useful for estimating integrals.

> **THEOREM 4** Let $f(x, y)$ and $g(x, y)$ be integrable functions on $\mathcal{D}$.
>
> (a) If $f(x, y) \leq g(x, y)$ for all $(x, y) \in \mathcal{D}$, then
>
> $$\iint_{\mathcal{D}} f(x, y)\, dA \leq \iint_{\mathcal{D}} g(x, y)\, dA \qquad \boxed{6}$$
>
> (b) If $m \leq f(x, y) \leq M$ for all $(x, y) \in \mathcal{D}$, then
>
> $$m \cdot \text{Area}(\mathcal{D}) \leq \iint_{\mathcal{D}} f(x, y)\, dA \leq M \cdot \text{Area}(\mathcal{D}) \qquad \boxed{7}$$

Proof If $f(x, y) \leq g(x, y)$, then every Riemann sum for $f(x, y)$ is less than or equal to the corresponding Riemann sum for g:

$$\sum f(P_{ij})\, \Delta x_i\, \Delta y_j \leq \sum g(P_{ij})\, \Delta x_i\, \Delta y_j$$

We obtain (6) by taking the limit. Now suppose that $f(x, y) \leq M$ and apply (6) with $g(x, y) = M$:

$$\iint_{\mathcal{D}} f(x, y)\, dA \leq \iint_{\mathcal{D}} M\, dA = M\, \text{Area}(\mathcal{D})$$

This proves half of (7). The other half follows similarly. ∎

■ **EXAMPLE 7** Estimate $\displaystyle\iint_{\mathcal{D}} \frac{dA}{\sqrt{x^2 + (y-2)^2}}$, where $\mathcal{D}$ is the disk of radius 1 centered at the origin.

Solution The quantity $\sqrt{x^2 + (y-2)^2}$ is the distance d from (x, y) to $(0, 2)$, and we see from Figure 14 that $1 \leq d \leq 3$. Taking reciprocals, we have

$$\frac{1}{3} \leq \frac{1}{\sqrt{x^2 + (y-2)^2}} \leq 1$$

We apply (7) with $m = \frac{1}{3}$ and $M = 1$, using the fact that $\text{Area}(\mathcal{D}) = \pi$, to obtain

$$\frac{\pi}{3} \leq \iint_{\mathcal{D}} \frac{dA}{\sqrt{x^2 + (y-2)^2}} \leq \pi \qquad ■$$

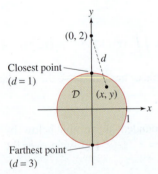

DF FIGURE 14 The distance d from (x, y) to $(0, 2)$ varies from 1 to 3 for (x, y) in the unit disk.

Average Value

The **average value** (or **mean value**) of a function $f(x, y)$ on a domain $\mathcal{D}$, which we denote by $\overline{f}$, is the quantity

←·· *REMINDER Eq. (8) is similar to the definition of an average value in one variable:*

$$\overline{f} = \frac{1}{b-a} \int_a^b f(x)\, dx = \frac{\int_a^b f(x)\, dx}{\int_a^b 1\, dx}$$

$$\boxed{\overline{f} = \frac{1}{\text{Area}(\mathcal{D})} \iint_{\mathcal{D}} f(x, y)\, dA = \frac{\iint_{\mathcal{D}} f(x, y)\, dA}{\iint_{\mathcal{D}} 1\, dA}} \qquad \boxed{8}$$

Equivalently, $\overline{f}$ is the value satisfying the relation

$$\boxed{\iint_{\mathcal{D}} f(x, y)\, dA = \overline{f} \cdot \text{Area}(\mathcal{D})}$$

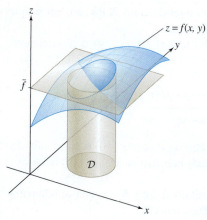

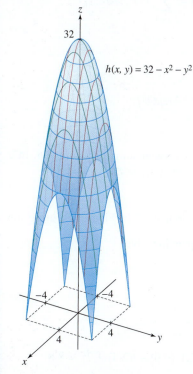

FIGURE 15

FIGURE 16 Pagoda with ceiling
$H(x, y) = 32 - x^2 - y^2$.

GRAPHICAL INSIGHT The solid region under the graph has the same (signed) volume as the cylinder with base $\mathcal{D}$ of height $\overline{f}$ (Figure 15).

■ **EXAMPLE 8** An architect needs to know the average height $\overline{H}$ of the ceiling of a pagoda whose base $\mathcal{D}$ is the square $[-4, 4] \times [-4, 4]$ and roof is the graph of

$$H(x, y) = 32 - x^2 - y^2$$

where distances are in feet (Figure 16). Calculate $\overline{H}$.

Solution First, we compute the integral of $H(x, y)$ over $\mathcal{D}$:

$$\iint_{\mathcal{D}} (32 - x^2 - y^2)\, dA = \int_{-4}^{4} \int_{-4}^{4} (32 - x^2 - y^2)\, dy\, dx$$

$$= \int_{-4}^{4} \left(\left(32y - x^2 y - \frac{1}{3} y^3 \right) \Big|_{-4}^{4} \right) dx = \int_{-4}^{4} \left(\frac{640}{3} - 8x^2 \right) dx$$

$$= \left(\frac{640}{3} x - \frac{8}{3} x^3 \right) \Big|_{-4}^{4} = \frac{4096}{3}$$

The area of $\mathcal{D}$ is $8 \times 8 = 64$, so the average height of the pagoda's ceiling is

$$\overline{H} = \frac{1}{\text{Area}(\mathcal{D})} \iint_{\mathcal{D}} H(x, y)\, dA = \frac{1}{64} \left(\frac{4096}{3} \right) = \frac{64}{3} \approx 21.3 \text{ ft} \quad ■$$

The Mean Value Theorem states that a continuous function on a domain $\mathcal{D}$ must take on its average value at some point P in $\mathcal{D}$, provided that $\mathcal{D}$ is closed, bounded, and also **connected** (see Exercise 67 for a proof). By definition, $\mathcal{D}$ is connected if any two points in $\mathcal{D}$ can be joined by a curve in $\mathcal{D}$ (Figure 17).

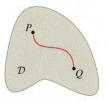

(A) Connected domain: Any two points can be joined by a curve lying entirely in $\mathcal{D}$.

(B) Nonconnected domain.

FIGURE 17

THEOREM 5 Mean Value Theorem for Double Integrals If $f(x, y)$ is continuous and $\mathcal{D}$ is closed, bounded, and connected, then there exists a point $P \in \mathcal{D}$ such that

$$\iint_{\mathcal{D}} f(x, y)\, dA = f(P)\, \text{Area}(\mathcal{D}) \qquad \boxed{9}$$

Equivalently, $f(P) = \overline{f}$, where $\overline{f}$ is the average value of f on $\mathcal{D}$.

Decomposing the Domain into Smaller Domains

Double integrals are additive with respect to the domain: If $\mathcal{D}$ is the union of domains $\mathcal{D}_1, \mathcal{D}_2, \ldots, \mathcal{D}_N$ that do not overlap except possibly on boundary curves (Figure 18), then

$$\boxed{\iint_{\mathcal{D}} f(x, y)\, dA = \iint_{\mathcal{D}_1} f(x, y)\, dA + \cdots + \iint_{\mathcal{D}_N} f(x, y)\, dA}$$

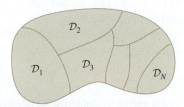

FIGURE 18 The region $\mathcal{D}$ is a union of smaller domains.

Additivity may be used to evaluate double integrals over domains $\mathcal{D}$ that are not simple but can be decomposed into finitely many simple domains.

We close this section with a simple but useful remark. If $f(x, y)$ is a continuous function on a *small* domain $\mathcal{D}$, then

$$\iint_{\mathcal{D}} f(x, y) \, dA \approx \underbrace{f(P) \, \text{Area}(\mathcal{D})}_{\text{Function value} \times \text{area}}$$

10

where P is any sample point in $\mathcal{D}$. In fact, we can choose P so that (10) is an equality by Theorem 5. But if $\mathcal{D}$ is small enough, then f is nearly constant on $\mathcal{D}$, and (10) holds as a good approximation for all $P \in \mathcal{D}$.

If the domain $\mathcal{D}$ is not small, we may partition it into N smaller subdomains $\mathcal{D}_1, \ldots, \mathcal{D}_N$ and choose sample points P_j in $\mathcal{D}_j$. By additivity,

$$\iint_{\mathcal{D}} f(x, y) \, dA = \sum_{j=1}^{N} \iint_{\mathcal{D}_j} f(x, y) \, dA \approx \sum_{j=1}^{N} f(P_j) \, \text{Area}(\mathcal{D}_j)$$

and thus we have the approximation

$$\iint_{\mathcal{D}} f(x, y) \, dA \approx \sum_{j=1}^{N} f(P_j) \, \text{Area}(\mathcal{D}_j)$$

11

We can think of Eq. (11) as a generalization of the Riemann sum approximation. In a Riemann sum, $\mathcal{D}$ is partitioned by rectangles $\mathcal{R}_{ij}$ of area $\Delta A_{ij} = \Delta x_i \, \Delta y_j$.

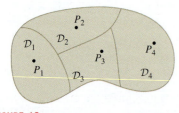

FIGURE 19

j	1	2	3	4
Area($\mathcal{D}_j$)	1	1	0.9	1.2
$f(P_j)$	1.8	2.2	2.1	2.4

■ **EXAMPLE 9** Estimate $\iint_{\mathcal{D}} f(x, y) \, dA$ for the domain $\mathcal{D}$ in Figure 19, using the areas and function values given there and the accompanying table.

Solution

$$\iint_{\mathcal{D}} f(x, y) \, dA \approx \sum_{j=1}^{4} f(P_j) \, \text{Area}(\mathcal{D}_j)$$

$$= (1.8)(1) + (2.2)(1) + (2.1)(0.9) + (2.4)(1.2) \approx 8.8 \qquad ■$$

15.2 SUMMARY

- We assume that $\mathcal{D}$ is a closed, bounded domain whose boundary is a simple closed curve that either is smooth or has a finite number of corners. The double integral is defined by

$$\iint_{\mathcal{D}} f(x, y) \, dA = \iint_{\mathcal{R}} \tilde{f}(x, y) \, dA$$

where $\mathcal{R}$ is a rectangle containing $\mathcal{D}$ and $\tilde{f}(x, y) = f(x, y)$ if $(x, y) \in \mathcal{D}$, and $\tilde{f}(x, y) = 0$ otherwise. The value of the integral does not depend on the choice of $\mathcal{R}$.

- The double integral defines the signed volume between the graph of $f(x, y)$ and the xy-plane, where regions below the xy-plane are assigned negative volume.

- For any constant C, $\iint_{\mathcal{D}} C \, dA = C \cdot \text{Area}(\mathcal{D})$.

- If $\mathcal{D}$ is vertically or horizontally simple, $\iint_{\mathcal{D}} f(x, y) \, dA$ can be evaluated as an iterated integral:

Vertically simple domain $a \le x \le b$, $g_1(x) \le y \le g_2(x)$	$\displaystyle\int_a^b \int_{g_1(x)}^{g_2(x)} f(x, y)\, dy\, dx$
Horizontally simple domain $c \le y \le d$, $g_1(y) \le x \le g_2(y)$	$\displaystyle\int_c^d \int_{g_1(y)}^{g_2(y)} f(x, y)\, dx\, dy$

- If $f(x, y) \le g(x, y)$ on $\mathcal{D}$, then $\displaystyle\iint_{\mathcal{D}} f(x, y)\, dA \le \iint_{\mathcal{D}} g(x, y)\, dA$.
- If m is the minimum value and M the maximum value of f on $\mathcal{D}$, then

$$m \cdot \text{Area}(\mathcal{D}) \le \iint_{\mathcal{D}} f(x, y)\, dA \le M \cdot \text{Area}(\mathcal{D})$$

- If $z_1(x, y) \le z_2(x, y)$ for all points in $\mathcal{D}$, then the volume V of the solid region between the surfaces given by $z = z_1(x, y)$ and $z = z_2(x, y)$ over $\mathcal{D}$ is given by

$$V = \iint_{\mathcal{D}} (z_2(x, y) - z_1(x, y))\, dA$$

- The *average value* of f on $\mathcal{D}$ is

$$\overline{f} = \frac{1}{\text{Area}(\mathcal{D})} \iint_{\mathcal{D}} f(x, y)\, dA = \frac{\iint_{\mathcal{D}} f(x, y)\, dA}{\iint_{\mathcal{D}} 1\, dA}$$

- Mean Value Theorem for Integrals: If $f(x, y)$ is continuous and $\mathcal{D}$ is closed, bounded, and connected, then there exists a point $P \in \mathcal{D}$ such that

$$\iint_{\mathcal{D}} f(x, y)\, dA = f(P) \cdot \text{Area}(\mathcal{D})$$

Equivalently, $f(P) = \overline{f}$, where $\overline{f}$ is the average value of f on $\mathcal{D}$.
- Additivity with respect to the domain: If $\mathcal{D}$ is a union of nonoverlapping (except possibly on their boundaries) domains $\mathcal{D}_1, \ldots, \mathcal{D}_N$, then

$$\iint_{\mathcal{D}} f(x, y)\, dA = \sum_{j=1}^{N} \iint_{\mathcal{D}_j} f(x, y)\, dA$$

- If the domains $\mathcal{D}_1, \ldots, \mathcal{D}_N$ are small and P_j is a sample point in $\mathcal{D}_j$, then

$$\iint_{\mathcal{D}} f(x, y)\, dA \approx \sum_{j=1}^{N} f(P_j)\text{Area}(\mathcal{D}_j)$$

15.2 EXERCISES

Preliminary Questions

1. Which of the following expressions do not make sense?

(a) $\displaystyle\int_0^1 \int_1^x f(x, y)\, dy\, dx$ **(b)** $\displaystyle\int_0^1 \int_1^y f(x, y)\, dy\, dx$

(c) $\displaystyle\int_0^1 \int_x^y f(x, y)\, dy\, dx$ **(d)** $\displaystyle\int_0^1 \int_x^1 f(x, y)\, dy\, dx$

2. Draw a domain in the plane that is neither vertically nor horizontally simple.

3. Which of the four regions in Figure 20 is the domain of integration for $\displaystyle\int_{-\sqrt{2}/2}^{0} \int_{-x}^{\sqrt{1-x^2}} f(x, y)\, dy\, dx$?

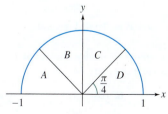

FIGURE 20

4. Let $\mathcal{D}$ be the unit disk. If the maximum value of $f(x, y)$ on $\mathcal{D}$ is 4, then the largest possible value of $\displaystyle\iint_{\mathcal{D}} f(x, y)\, dA$ is (choose the correct answer):

(a) 4   **(b)** 4π **(c)** $\dfrac{4}{\pi}$

Exercises

1. Calculate the Riemann sum for $f(x, y) = x - y$ and the shaded domain $\mathcal{D}$ in Figure 21 with two choices of sample points, • and ∘. Which do you think is a better approximation to the integral of f over $\mathcal{D}$? Why?

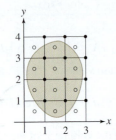

FIGURE 21

2. Approximate values of $f(x, y)$ at sample points on a grid are given in Figure 22. Estimate $\displaystyle\iint_{\mathcal{D}} f(x, y)\,dx\,dy$ for the shaded domain by computing the Riemann sum with the given sample points.

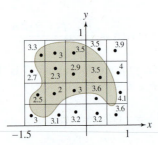

FIGURE 22

3. Express the domain $\mathcal{D}$ in Figure 23 as both a vertically simple region and a horizontally simple region, and evaluate the integral of $f(x, y) = xy$ over $\mathcal{D}$ as an iterated integral in two ways.

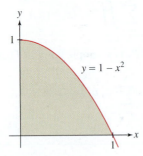

FIGURE 23

4. Sketch the domain

$$\mathcal{D}: 0 \le x \le 1, \quad x^2 \le y \le 4 - x^2$$

and evaluate $\displaystyle\iint_{\mathcal{D}} y\,dA$ as an iterated integral.

In Exercises 5–7, compute the double integral of $f(x, y) = x^2 y$ over the given shaded domain in Figure 24.

5. (A) **6.** (B) **7.** (C)

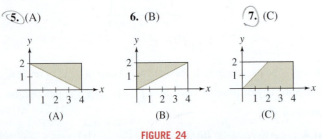

FIGURE 24

8. Sketch the domain $\mathcal{D}$ defined by $x + y \le 12$, $x \ge 4$, $y \ge 4$ and compute $\displaystyle\iint_{\mathcal{D}} e^{x+y}\,dA$.

9. Integrate $f(x, y) = x$ over the region bounded by $y = x^2$ and $y = x + 2$.

10. Sketch the region $\mathcal{D}$ between $y = x^2$ and $y = x(1 - x)$. Express $\mathcal{D}$ as a simple region and calculate the integral of $f(x, y) = 2y$ over $\mathcal{D}$.

11. Evaluate $\displaystyle\iint_{\mathcal{D}} \frac{y}{x}\,dA$, where $\mathcal{D}$ is the shaded part of the semicircle of radius 2 in Figure 25.

12. Calculate the double integral of $f(x, y) = y^2$ over the rhombus $\mathcal{R}$ in Figure 26.

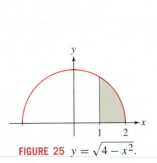

FIGURE 25 $y = \sqrt{4 - x^2}$.

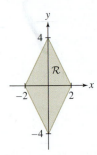

FIGURE 26
$|x| + \frac{1}{2}|y| \le 1$.

13. Calculate the double integral of $f(x, y) = x + y$ over the domain $\mathcal{D} = \{(x, y) : x^2 + y^2 \le 4, y \ge 0\}$.

14. Integrate $f(x, y) = (x + y + 1)^{-2}$ over the triangle with vertices $(0, 0)$, $(4, 0)$, and $(0, 8)$.

15. Calculate the integral of $f(x, y) = x$ over the region $\mathcal{D}$ bounded above by $y = x(2 - x)$ and below by $x = y(2 - y)$. *Hint:* Apply the quadratic formula to the lower boundary curve to solve for y as a function of x.

16. Integrate $f(x, y) = x$ over the region bounded by $y = x$, $y = 4x - x^2$, and $y = 0$ in two ways: as a vertically simple region and as a horizontally simple region.

In Exercises 17–24, compute the double integral of $f(x, y)$ over the domain $\mathcal{D}$ indicated.

17. $f(x, y) = x^2 y$; $\quad 1 \le x \le 3$, $\quad x \le y \le 2x + 1$

18. $f(x, y) = 1$; $\quad 0 \le x \le 1$, $\quad 1 \le y \le e^x$

19. $f(x, y) = x$; $\quad 0 \le x \le 1$, $\quad 1 \le y \le e^{x^2}$

20. $f(x, y) = \cos(2x + y)$; $\quad \frac{1}{2} \le x \le \frac{\pi}{2}$, $\quad 1 \le y \le 2x$

21. $f(x, y) = 2xy$; bounded by $x = y$, $x = y^2$

22. $f(x, y) = \sin x$; bounded by $x = 0$, $x = 1$, $y = \cos x$

23. $f(x, y) = e^{x+y}$; bounded by $y = x - 1$, $y = 12 - x$ for $2 \le y \le 4$

24. $f(x, y) = (x + y)^{-1}$; bounded by $y = x$, $y = 1$, $y = e$, $x = 0$

In Exercises 25–28, sketch the domain of integration and express as an iterated integral in the opposite order.

25. $\int_0^4 \int_x^4 f(x, y)\, dy\, dx$

26. $\int_4^9 \int_{\sqrt{y}}^3 f(x, y)\, dx\, dy$

27. $\int_4^9 \int_2^{\sqrt{y}} f(x, y)\, dx\, dy$

28. $\int_0^1 \int_{e^x}^e f(x, y)\, dy\, dx$

29. Sketch the domain $\mathcal{D}$ corresponding to

$$\int_0^4 \int_{\sqrt{y}}^2 \sqrt{4x^2 + 5y}\, dx\, dy$$

Then change the order of integration and evaluate.

30. Change the order of integration and evaluate

$$\int_0^1 \int_0^{\pi/2} x \cos(xy)\, dx\, dy$$

Explain the simplification achieved by changing the order.

31. Compute the integral of $f(x, y) = (\ln y)^{-1}$ over the domain $\mathcal{D}$ bounded by $y = e^x$ and $y = e^{\sqrt{x}}$. *Hint:* Choose the order of integration that enables you to evaluate the integral.

32. Evaluate by changing the order of integration:

$$\int_0^4 \int_{\sqrt{x}}^2 \sin y^3\, dy\, dx$$

In Exercises 33–36, sketch the domain of integration. Then change the order of integration and evaluate. Explain the simplification achieved by changing the order.

33. $\int_0^1 \int_y^1 \frac{\sin x}{x}\, dx\, dy$

34. $\int_0^4 \int_{\sqrt{y}}^2 \sqrt{x^3 + 1}\, dx\, dy$

35. $\int_0^1 \int_{y=x}^1 x e^{y^3}\, dy\, dx$

36. $\int_0^1 \int_{y=x^{2/3}}^1 x e^{y^4}\, dy\, dx$

37. Sketch the domain $\mathcal{D}$ where $0 \le x \le 2$, $0 \le y \le 2$, and x *or* y is greater than 1. Then compute $\iint_{\mathcal{D}} e^{x+y}\, dA$.

38. Calculate $\iint_{\mathcal{D}} e^x\, dA$, where $\mathcal{D}$ is bounded by the lines $y = x + 1$, $y = x$, $x = 0$, and $x = 1$.

In Exercises 39–42, calculate the double integral of $f(x, y)$ over the triangle indicated in Figure 27.

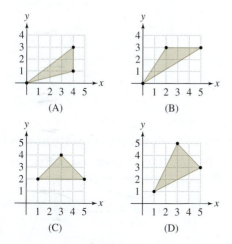

(A) (B) (C) (D)

FIGURE 27

39. $f(x, y) = e^{x^2}$, (A)

40. $f(x, y) = 1 - 2x$, (B)

41. $f(x, y) = \dfrac{x}{y^2}$, (C)

42. $f(x, y) = x + 1$, (D)

43. Calculate the double integral of $f(x, y) = \dfrac{\sin y}{y}$ over the region $\mathcal{D}$ in Figure 28.

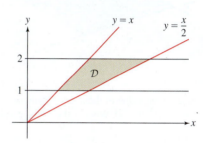

FIGURE 28

44. Evaluate $\iint_{\mathcal{D}} x\, dA$ for $\mathcal{D}$ in Figure 29.

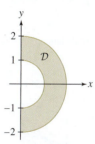

FIGURE 29

45. Find the volume of the region bounded by $z = 40 - 10y$, $z = 0$, $y = 0$, and $y = 4 - x^2$.

46. Find the volume of the region enclosed by $z = 1 - y^2$ and $z = y^2 - 1$ for $0 \le x \le 2$.

47. Find the volume of the region bounded by $z = 16 - y$, $z = y$, $y = x^2$, and $y = 8 - x^2$.

48. Find the volume of the region bounded by $y = 1 - x^2$, $z = 1$, $y = 0$, and $z + y = 2$.

49. Set up a double integral that gives the volume of the region bounded by the two paraboloids $z = x^2 + y^2$ and $z = 8 - x^2 - y^2$. (Do not evaluate the double integral.)

50. Set up a double integral that gives the volume of the region bounded by $z = 1 - y^2$, $z = y$, $x = 0$, $y = 0$, and $x + y = 1$. (Do not evaluate the double integral.)

51. Calculate the average value of $f(x, y) = e^{x+y}$ on the square $[0, 1] \times [0, 1]$.

52. Calculate the average height above the x-axis of a point in the region $0 \le x \le 1$, $0 \le y \le x^2$.

53. Find the average height of the "ceiling" in Figure 30 defined by $z = y^2 \sin x$ for $0 \le x \le \pi, 0 \le y \le 1$.

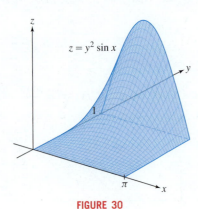

$z = y^2 \sin x$

FIGURE 30

54. Calculate the average value of the x-coordinate of a point on the semicircle $x^2 + y^2 \le R^2$, $x \ge 0$. What is the average value of the y-coordinate?

55. What is the average value of the linear function

$$f(x, y) = mx + ny + p$$

on the ellipse $\left(\dfrac{x}{a}\right)^2 + \left(\dfrac{y}{b}\right)^2 \le 1$? Argue by symmetry rather than calculation.

56. Find the average square distance from the origin to a point in the domain $\mathcal{D}$ in Figure 31.

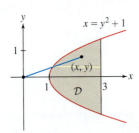

$x = y^2 + 1$

(x, y)

$\mathcal{D}$

FIGURE 31

57. Let $\mathcal{D}$ be the rectangle $0 \le x \le 2, -\frac{1}{8} \le y \le \frac{1}{8}$, and let $f(x, y) = \sqrt{x^3 + 1}$. Prove that

$$\iint_{\mathcal{D}} f(x, y)\, dA \le \frac{3}{2}$$

58. (a) Use the inequality $0 \le \sin x \le x$ for $x \ge 0$ to show that

$$\int_0^1 \int_0^1 \sin(xy)\, dx\, dy \le \frac{1}{4}$$

(b) Use a computer algebra system to evaluate the double integral to three decimal places.

59. Prove the inequality $\iint_{\mathcal{D}} \dfrac{dA}{4 + x^2 + y^2} \le \pi$, where $\mathcal{D}$ is the disk $x^2 + y^2 \le 4$.

60. Let $\mathcal{D}$ be the domain bounded by $y = x^2 + 1$ and $y = 2$. Prove the inequality

$$\frac{4}{3} \le \iint_{\mathcal{D}} (x^2 + y^2)\, dA \le \frac{20}{3}$$

61. Let $\overline{f}$ be the average of $f(x, y) = xy^2$ on $\mathcal{D} = [0, 1] \times [0, 4]$. Find a point $P \in \mathcal{D}$ such that $f(P) = \overline{f}$ (the existence of such a point is guaranteed by the Mean Value Theorem for Double Integrals).

62. Verify the Mean Value Theorem for Double Integrals for $f(x, y) = e^{x-y}$ on the triangle bounded by $y = 0$, $x = 1$, and $y = x$.

In Exercises 63 and 64, use (11) to estimate the double integral.

63. The following table lists the areas of the subdomains $\mathcal{D}_j$ of the domain $\mathcal{D}$ in Figure 32 and the values of a function $f(x, y)$ at sample points $P_j \in \mathcal{D}_j$. Estimate $\iint_{\mathcal{D}} f(x, y)\, dA$.

j	1	2	3	4	5	6
Area($\mathcal{D}_j$)	1.2	1.1	1.4	0.6	1.2	0.8
$f(P_j)$	9	9.1	9.3	9.1	8.9	8.8

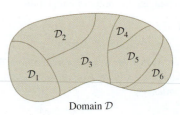

Domain $\mathcal{D}$

FIGURE 32

64. The domain $\mathcal{D}$ between the circles of radii 5 and 5.2 in the first quadrant in Figure 33 is divided into six subdomains of angular width $\Delta\theta = \frac{\pi}{12}$, and the values of a function $f(x, y)$ at sample points are given. Compute the area of the subdomains and estimate

$$\iint_{\mathcal{D}} f(x, y)\, dA.$$

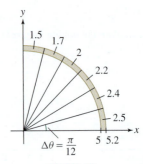

$\Delta\theta = \dfrac{\pi}{12}$

FIGURE 33

65. According to Eq. (3), the area of a domain $\mathcal{D}$ is equal to $\iint_{\mathcal{D}} 1\, dA$. Prove that if $\mathcal{D}$ is the region between two curves $y = g_1(x)$ and $y = g_2(x)$ with $g_2(x) \le g_1(x)$ for $a \le x \le b$, then

$$\iint_{\mathcal{D}} 1\, dA = \int_a^b (g_1(x) - g_2(x))\, dx$$

Further Insights and Challenges

66. Let $\mathcal{D}$ be a closed connected domain and let $P, Q \in \mathcal{D}$. The Intermediate Value Theorem (IVT) states that if f is continuous on $\mathcal{D}$, then $f(x, y)$ takes on every value between $f(P)$ and $f(Q)$ at some point in $\mathcal{D}$.

(a) Show, by constructing a counterexample, that the IVT is false if $\mathcal{D}$ is not connected.

(b) Prove the IVT as follows: Let $\mathbf{c}(t)$ be a path such that $\mathbf{c}(0) = P$ and $\mathbf{c}(1) = Q$ (such a path exists because $\mathcal{D}$ is connected). Apply the IVT in one variable to the composite function $f(\mathbf{c}(t))$.

67. Use the fact that a continuous function on a closed domain $\mathcal{D}$ attains both a minimum value m and a maximum value M, together with Theorem 3, to prove that the average value $\overline{f}$ lies between m and M.

Then use the IVT in Exercise 62 to prove the Mean Value Theorem for Double Integrals.

68. Let $f(y)$ be a function of y alone and set $G(t) =$
$$\int_0^t \int_0^x f(y)\, dy\, dx.$$

(a) Use the Fundamental Theorem of Calculus to prove that $G''(t) = f(t)$.

(b) Show, by changing the order in the double integral, that $G(t) = \int_0^t (t - y) f(y)\, dy$. This illustrates that the "second antiderivative" of $f(y)$ can be expressed as a single integral.

15.3 Triple Integrals

Triple integrals of functions $f(x, y, z)$ of three variables are a fairly straightforward generalization of double integrals. Instead of a rectangle in the plane, our domain is a box (Figure 1)

$$B = [a, b] \times [c, d] \times [p, q]$$

consisting of all points (x, y, z) in $\mathbf{R}^3$ such that

$$a \leq x \leq b, \qquad c \leq y \leq d, \qquad p \leq z \leq q$$

To integrate over this box, we subdivide the box (as usual) into "sub"-boxes

$$\mathcal{B}_{ijk} = [x_{i-1}, x_i] \times [y_{j-1}, y_j] \times [z_{k-1}, z_k]$$

by choosing partitions of the three intervals

$$a = x_0 < x_1 < \cdots < x_N = b$$
$$c = y_0 < y_1 < \cdots < y_M = d$$
$$p = z_0 < z_1 < \cdots < z_L = q$$

Here N, M, and L are positive integers. The volume of $\mathcal{B}_{ijk}$ is $\Delta V_{ijk} = \Delta x_i\, \Delta y_j\, \Delta z_k$, where

$$\Delta x_i = x_i - x_{i-1}, \qquad \Delta y_j = y_j - y_{j-1}, \qquad \Delta z_k = z_k - z_{k-1}$$

Then we choose a sample point P_{ijk} in each box $\mathcal{B}_{ijk}$ and form the Riemann sum:

$$S_{N,M,L} = \sum_{i=1}^{N} \sum_{j=1}^{M} \sum_{k=1}^{L} f(P_{ijk})\, \Delta V_{ijk}$$

As in the previous section, we write $\mathcal{P} = \{\{x_i\}, \{y_j\}, \{z_k\}\}$ for the partition and let $\|\mathcal{P}\|$ be the maximum of the widths $\Delta x_i, \Delta y_j, \Delta z_k$. If the sums $S_{N,M,L}$ approach a limit as $\|\mathcal{P}\| \to 0$ for arbitrary choices of sample points, we say that f is **integrable** over $\mathcal{B}$. The limit value is denoted

$$\iiint_{\mathcal{B}} f(x, y, z)\, dV = \lim_{\|\mathcal{P}\| \to 0} S_{N,M,L}$$

Triple integrals have many of the same properties as double and single integrals. The linear properties are satisfied, and continuous functions are integrable over a box $\mathcal{B}$. Furthermore, triple integrals can be evaluated as iterated integrals.

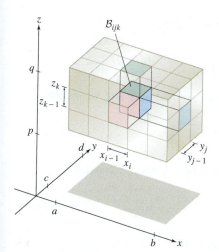

FIGURE 1 The box $\mathcal{B} = [a, b] \times [c, d] \times [p, q]$ decomposed into smaller boxes.

The notation dA, used in the previous section, suggests area and occurs in double integrals over domains in the plane. Similarly, dV suggests volume and is used in the notation for triple integrals.

THEOREM 1 Fubini's Theorem for Triple Integrals The triple integral of a continuous function $f(x, y, z)$ over a box $\mathcal{B} = [a, b] \times [c, d] \times [p, q]$ is equal to the iterated integral:

$$\iiint_{\mathcal{B}} f(x, y, z)\, dV = \int_{x=a}^{b} \int_{y=c}^{d} \int_{z=p}^{q} f(x, y, z)\, dz\, dy\, dx$$

Furthermore, the iterated integral may be evaluated in any order.

As noted in the theorem, we are free to evaluate the iterated integral in any order (there are six different orders). For instance,

$$\int_{x=a}^{b} \int_{y=c}^{d} \int_{z=p}^{q} f(x, y, z)\, dz\, dy\, dx = \int_{z=p}^{q} \int_{y=c}^{d} \int_{x=a}^{b} f(x, y, z)\, dx\, dy\, dz$$

■ **EXAMPLE 1 Integration over a Box** Calculate the integral $\iiint_{\mathcal{B}} x^2 e^{y+3z}\, dV$, where $\mathcal{B} = [1, 4] \times [0, 3] \times [2, 6]$.

Solution We write this triple integral as an iterated integral:

$$\iiint_{\mathcal{B}} x^2 e^{y+3z}\, dV = \int_{1}^{4} \int_{0}^{3} \int_{2}^{6} x^2 e^{y+3z}\, dz\, dy\, dx$$

Step 1. Evaluate the inner integral with respect to z, holding x and y constant.

$$\int_{z=2}^{6} x^2 e^{y+3z}\, dz = \frac{1}{3} x^2 e^{y+3z} \Big|_{2}^{6} = \frac{1}{3} x^2 e^{y+18} - \frac{1}{3} x^2 e^{y+6} = \frac{1}{3}(e^{18} - e^{6}) x^2 e^{y}$$

Step 2. Evaluate the middle integral with respect to y, holding x constant.

$$\int_{y=0}^{3} \frac{1}{3}(e^{18} - e^{6}) x^2 e^{y}\, dy = \frac{1}{3}(e^{18} - e^{6}) x^2 \int_{y=0}^{3} e^{y}\, dy = \frac{1}{3}(e^{18} - e^{6})(e^{3} - 1) x^2$$

Step 3. Evaluate the outer integral with respect to x.

$$\iiint_{\mathcal{B}} (x^2 e^{y+3z})\, dV = \frac{1}{3}(e^{18} - e^{6})(e^{3} - 1) \int_{x=1}^{4} x^2\, dx = 7(e^{18} - e^{6})(e^{3} - 1) \quad ■$$

Note that in the previous example, the integrand factors as a product of three functions $f(x, y, z) = g(x)h(y)k(z)$—namely,

$$f(x, y, z) = x^2 e^{y+3z} = x^2 e^{y} e^{3z}$$

Because of this, the triple integral can be evaluated simply as the product of three single integrals:

$$\iiint_{\mathcal{B}} x^2 e^{y} e^{3z}\, dV = \left(\int_{1}^{4} x^2\, dx \right) \left(\int_{0}^{3} e^{y}\, dy \right) \left(\int_{2}^{6} e^{3z}\, dz \right)$$

$$= (21)(e^{3} - 1) \left(\frac{e^{18} - e^{6}}{3} \right) = 7(e^{18} - e^{6})(e^{3} - 1)$$

Next, instead of a box, we integrate over a solid region $\mathcal{W}$ that is *z-simple* as in Figure 2. In other words, $\mathcal{W}$ is the region between two surfaces $z = z_1(x, y)$ and $z = z_2(x, y)$ over a domain $\mathcal{D}$ in the xy-plane. In this case,

$$\mathcal{W} = \{(x, y, z) : (x, y) \in \mathcal{D} \quad \text{and} \quad z_1(x, y) \leq z \leq z_2(x, y)\} \qquad \boxed{1}$$

The domain $\mathcal{D}$ is the **projection** of $\mathcal{W}$ onto the xy-plane.

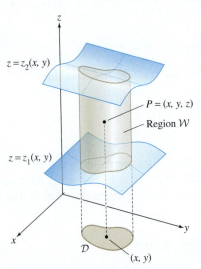

FIGURE 2 The point $P = (x, y, z)$ in the z-simple region $\mathcal{W}$ if $(x, y) \in \mathcal{D}$ and $z_1(x, y) \leq z \leq z_2(x, y)$.

As a formal matter, as in the case of double integrals, we define the triple integral of $f(x, y, z)$ over $\mathcal{W}$ by

$$\iiint_{\mathcal{W}} f(x, y, z)\, dV = \iiint_{\mathcal{B}} \tilde{f}(x, y, z)\, dV$$

where $\mathcal{B}$ is a box containing $\mathcal{W}$, and $\tilde{f}$ is the function that is equal to f on $\mathcal{W}$ and equal to zero outside of $\mathcal{W}$. The triple integral exists, assuming that $z_1(x, y)$, $z_2(x, y)$, and the integrand f are continuous. In practice, however, we evaluate triple integrals as iterated integrals. This is justified by the following theorem, whose proof is similar to that of Theorem 2 in Section 15.2.

THEOREM 2 The triple integral of a continuous function f over the region

$$\mathcal{W} : (x, y) \in \mathcal{D}, \quad z_1(x, y) \le z \le z_2(x, y)$$

is equal to the iterated integral

$$\iiint_{\mathcal{W}} f(x, y, z)\, dV = \iint_{\mathcal{D}} \left(\int_{z=z_1(x,y)}^{z_2(x,y)} f(x, y, z)\, dz \right) dA$$

More generally, integrals of functions of n variables (for any n) arise naturally in many different contexts. For example, the average distance between two points in a ball is expressed as a six-fold integral because we integrate over all possible coordinates of the two points. Each point has three coordinates for a total of six variables.

One thing missing from our discussion so far is a geometric interpretation of triple integrals. A double integral represents the signed volume of the three-dimensional region between a graph $z = f(x, y)$ and the xy-plane. The graph of a function $f(x, y, z)$ of three variables lives in *four-dimensional space*, and thus, a triple integral represents a four-dimensional volume. This volume is hard or impossible to visualize. On the other hand, triple integrals represent many other types of quantities. Some examples are total mass, average value, probabilities, and centers of mass (see Section 15.5).

Furthermore, the volume V of a region $\mathcal{W}$ is defined as the triple integral of the constant function $f(x, y, z) = 1$:

Equation 2 is analogous to the fact we have already seen that the area A of a region $\mathcal{R}$ in the xy-plane is given by taking the double integral of 1 over the region: $A = \iint_{\mathcal{R}} 1\, dA$.

$$V = \iiint_{\mathcal{W}} 1\, dV \qquad \boxed{2}$$

In particular, if $\mathcal{W}$ is a z-simple region between $z = h_1(x, y)$ and $z = h_2(x, y)$, then

$$\iiint_{\mathcal{W}} 1\, dV = \iint_{\mathcal{D}} \left(\int_{z=h_1(x,y)}^{h_2(x,y)} 1\, dz \right) dA = \iint_{\mathcal{D}} (h_2(x, y) - h_1(x, y))\, dA$$

Thus, the triple integral is equal to the double integral defining the volume of the region between the two surfaces.

■ **EXAMPLE 2** **Solid Region with a Rectangular Base** Evaluate $\iiint_{\mathcal{W}} z\, dV$, where $\mathcal{W}$ is the region between the planes $z = x + y$ and $z = 3x + 5y$ lying over the rectangle $\mathcal{D} = [0, 3] \times [0, 2]$ (Figure 3).

Solution Apply Theorem 2 with $z_1(x, y) = x + y$ and $z_2(x, y) = 3x + 5y$:

$$\iiint_{\mathcal{W}} z\, dV = \iint_{\mathcal{D}} \left(\int_{z=x+y}^{3x+5y} z\, dz \right) dA = \int_{x=0}^{3} \int_{y=0}^{2} \int_{z=x+y}^{3x+5y} z\, dz\, dy\, dx$$

Step 1. Evaluate the inner integral with respect to z.

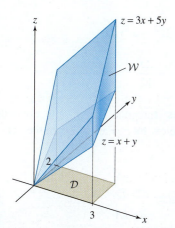

FIGURE 3 Region $\mathcal{W}$ between the planes $z = x + y$ and $z = 3x + 5y$ lying over $\mathcal{D} = [0, 3] \times [0, 2]$.

$$\int_{z=x+y}^{3x+5y} z\, dz = \frac{1}{2} z^2 \Big|_{z=x+y}^{3x+5y} = \frac{1}{2}(3x + 5y)^2 - \frac{1}{2}(x + y)^2 = 4x^2 + 14xy + 12y^2 \qquad \boxed{3}$$

Step 2. Evaluate the result with respect to y.

$$\int_{y=0}^{2} (4x^2 + 14xy + 12y^2)\,dy = (4x^2y + 7xy^2 + 4y^3)\Big|_{y=0}^{2} = 8x^2 + 28x + 32$$

Step 3. Evaluate the result with respect to x.

$$\iiint_{\mathcal{W}} z\,dV = \int_{x=0}^{3}(8x^2 + 28x + 32)\,dx = \left(\frac{8}{3}x^3 + 14x^2 + 32x\right)\Big|_{0}^{3}$$

$$= 72 + 126 + 96 = 294 \qquad \blacksquare$$

■ **EXAMPLE 3** **Solid Region with a Triangular Base** Evaluate $\iiint_{\mathcal{W}} z\,dV$, where $\mathcal{W}$ is the region in Figure 4.

Solution This is similar to the previous example, but now $\mathcal{W}$ lies over the triangle $\mathcal{D}$ in the xy-plane defined by

$$0 \le x \le 1, \qquad 0 \le y \le 1 - x$$

Thus, the triple integral is equal to the iterated integral:

$$\iiint_{\mathcal{W}} z\,dV = \iint_{\mathcal{D}} \left(\int_{z=x+y}^{3x+5y} z\,dz\right) dA = \underbrace{\int_{x=0}^{1}\int_{y=0}^{1-x}}_{\substack{\text{Integral}\\\text{over triangle}}} \int_{z=x+y}^{3x+5y} z\,dz\,dy\,dx$$

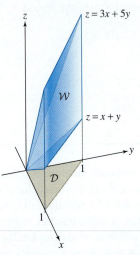

DF FIGURE 4 Region $\mathcal{W}$ between the planes $z = x + y$ and $z = 3x + 5y$ lying over the triangle $\mathcal{D}$.

We computed the inner integral in the previous example [see Eq. (3)]:

$$\int_{z=x+y}^{3x+5y} z\,dz = \frac{1}{2}z^2\Big|_{x+y}^{3x+5y} = 4x^2 + 14xy + 12y^2$$

Next, we integrate with respect to y:

$$\int_{y=0}^{1-x}(4x^2 + 14xy + 12y^2)\,dy = (4x^2y + 7xy^2 + 4y^3)\Big|_{y=0}^{1-x}$$

$$= 4x^2(1 - x) + 7x(1 - x)^2 + 4(1 - x)^3$$

$$= 4 - 5x + 2x^2 - x^3$$

And finally,

$$\iiint_{\mathcal{W}} z\,dV = \int_{x=0}^{1}(4 - 5x + 2x^2 - x^3)\,dx$$

$$= 4 - \frac{5}{2} + \frac{2}{3} - \frac{1}{4} = \frac{23}{12} \qquad \blacksquare$$

■ **EXAMPLE 4** **Region Between Intersecting Surfaces** Integrate $f(x, y, z) = x$ over the region $\mathcal{W}$ bounded above by $z = 4 - x^2 - y^2$ and below by $z = x^2 + 3y^2$ in the octant $x \ge 0, y \ge 0, z \ge 0$.

Solution The region $\mathcal{W}$ is z-simple, so

$$\iiint_{\mathcal{W}} x\,dV = \iint_{\mathcal{D}} \int_{z=x^2+3y^2}^{4-x^2-y^2} x\,dz\,dA$$

where $\mathcal{D}$ is the projection of $\mathcal{W}$ onto the xy-plane. To evaluate the integral over $\mathcal{D}$, we must find the equation of the curved part of the boundary of $\mathcal{D}$.

Step 1. **Find the boundary of** $\mathcal{D}$.

The upper and lower surfaces intersect where they have the same height:

$$4 - x^2 - y^2 = z = x^2 + 3y^2 \qquad \text{or} \qquad x^2 + 2y^2 = 2$$

Therefore, as we see in Figure 5, $\mathcal{W}$ projects onto the domain $\mathcal{D}$ consisting of the quarter of the ellipse $x^2 + 2y^2 = 2$ in the first quadrant. This ellipse hits the axes at $(\sqrt{2}, 0)$ and $(0, 1)$.

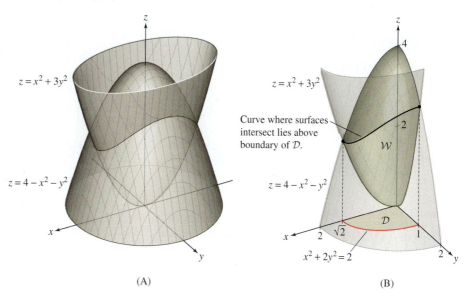

$z = x^2 + 3y^2$

$z = 4 - x^2 - y^2$

Curve where surfaces intersect lies above boundary of $\mathcal{D}$.

$z = x^2 + 3y^2$

$z = 4 - x^2 - y^2$

$x^2 + 2y^2 = 2$

FIGURE 5 Region $x^2 + 3y^2 \le z \le 4 - x^2 - y^2$.

(A)   (B)

Step 2. **Express** $\mathcal{D}$ **as a simple domain.**

Since $\mathcal{D}$ is both vertically and horizontally simple, we can integrate in either the order $dy\,dx$ or $dx\,dy$. If we choose $dx\,dy$, then y varies from 0 to 1 and the domain is described by

$$\mathcal{D}: 0 \le y \le 1, \quad 0 \le x \le \sqrt{2 - 2y^2}$$

Step 3. **Write the triple integral as an iterated integral.**

$$\iiint_{\mathcal{W}} x\,dV = \int_{y=0}^{1} \int_{x=0}^{\sqrt{2-2y^2}} \int_{z=x^2+3y^2}^{4-x^2-y^2} x\,dz\,dx\,dy$$

Step 4. **Evaluate.**

Here are the results of evaluating the integrals in order:

Inner integral: $\displaystyle\int_{z=x^2+y^2}^{4-x^2-y^2} x\,dz = xz\,\Big|_{z=x^2+3y^2}^{4-x^2-y^2} = 4x - 2x^3 - 4y^2 x$

Middle integral:

$$\int_{x=0}^{\sqrt{2-2y^2}} (4x - 2x^3 - 4y^2 x)\,dx = \left(2x^2 - \frac{1}{2}x^4 - 2x^2 y^2\right)\Big|_{x=0}^{\sqrt{2-2y^2}}$$

$$= 2 - 4y^2 + 2y^4$$

Triple integral: $\displaystyle\iiint_{\mathcal{W}} x\,dV = \int_{0}^{1} (2 - 4y^2 + 2y^4)\,dy = 2 - \frac{4}{3} + \frac{2}{5} = \frac{16}{15}$ ■

So far, we have evaluated triple integrals by projecting the region $\mathcal{W}$ onto a domain $\mathcal{D}$ in the xy-plane. We can integrate equally well by projecting onto domains in the xz- or yz-plane. For example, if $\mathcal{W}$ is the x-simple region between the graphs of $x = x_1(y, z)$ and $x = x_2(y, z)$ lying over a domain $\mathcal{D}$ in the yz-plane (Figure 6), then

$$\iiint_{\mathcal{W}} f(x, y, z)\,dV = \iint_{\mathcal{D}} \left(\int_{x=x_1(y,z)}^{x_2(y,z)} f(x, y, z)\,dx\right) dA$$

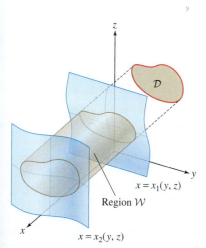

$x = x_1(y, z)$

Region $\mathcal{W}$

$x = x_2(y, z)$

FIGURE 6 $\mathcal{D}$ is the projection of $\mathcal{W}$ onto the yz-plane.

■ **EXAMPLE 5** **Writing a Triple Integral in Three Ways** The region $\mathcal{W}$ in Figure 7 is bounded by

$$z = 4 - y^2, \qquad y = 2x, \qquad z = 0, \qquad x = 0$$

Express $\iiint_{\mathcal{W}} xyz\, dV$ as an iterated integral in three ways, by projecting onto each of the three coordinate planes (but do not evaluate).

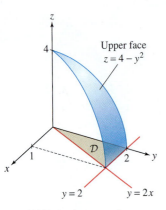

(A) Projection to xy-plane

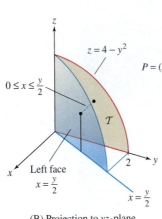

(B) Projection to yz-plane

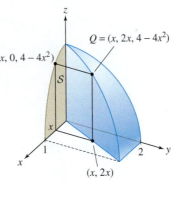

(C) Projection to xz-plane

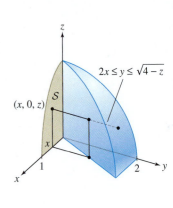

(D) The y-coordinates of points in the solid satisfy $2x \le y \le \sqrt{4-z}$.

DF FIGURE 7

Solution We consider each coordinate plane separately.

Step 1. The xy-plane.
This is a z-simple region. The upper face $z = 4 - y^2$ intersects the first quadrant of the xy-plane ($z = 0$) in the line $y = 2$ [Figure 7(A)]. Therefore, the projection of $\mathcal{W}$ onto the xy-plane is a triangle $\mathcal{D}$ defined by $0 \le x \le 1$, $2x \le y \le 2$, and

$$\mathcal{W} : 0 \le x \le 1, \quad 2x \le y \le 2, \quad 0 \le z \le 4 - y^2$$

You can check that all three ways of writing the triple integral in Example 5 yield the same answer:

$$\iiint_{\mathcal{W}} xyz\, dV = \frac{2}{3}$$

$$\iiint_{\mathcal{W}} xyz\, dV = \int_{x=0}^{1} \int_{y=2x}^{2} \int_{z=0}^{4-y^2} xyz\, dz\, dy\, dx \qquad \boxed{4}$$

Step 2. The yz-plane.
This is also an x-simple region. The projection of $\mathcal{W}$ onto the yz-plane is the domain $\mathcal{T}$ [Figure 7(B)]:

$$\mathcal{T} : 0 \le y \le 2, \quad 0 \le z \le 4 - y^2$$

The region $\mathcal{W}$ consists of all points lying between $\mathcal{T}$ and the "left face" $x = \frac{1}{2}y$. In other words, the x-coordinate must satisfy $0 \le x \le \frac{1}{2}y$. Thus,

$$\mathcal{W} : 0 \le y \le 2, \quad 0 \le z \le 4 - y^2, \quad 0 \le x \le \frac{1}{2}y$$

$$\iiint_{\mathcal{W}} xyz\, dV = \int_{y=0}^{2} \int_{z=0}^{4-y^2} \int_{x=0}^{y/2} xyz\, dx\, dz\, dy$$

Step 3. The xz-plane.
This is a y-simple region. The challenge in this case is to determine the projection of $\mathcal{W}$ onto the xz-plane, that is, the region $\mathcal{S}$ in Figure 7(C). We need to find the equation of the boundary curve of $\mathcal{S}$. A point P on this curve is the projection of a point $Q = (x, y, z)$ on the boundary of the left face. Since Q lies on both the plane $y = 2x$ and the surface $z = 4 - y^2$, $Q = (x, 2x, 4 - 4x^2)$. The projection of Q is $P = (x, 0, 4 - 4x^2)$. We see that the projection of $\mathcal{W}$ onto the xz-plane is the domain

$$\mathcal{S} : 0 \le x \le 1, \quad 0 \le z \le 4 - 4x^2$$

This gives us limits for x and z variables, so the triple integral can be written

$$\iiint_{\mathcal{W}} xyz \, dV = \int_{x=0}^{1} \int_{z=0}^{4-4x^2} \int_{y=??}^{??} xyz \, dy \, dz \, dx$$

What are the limits for y? The equation of the upper face $z = 4 - y^2$ can be written $y = \sqrt{4-z}$. Referring to Figure 7(D), we see that $\mathcal{W}$ is bounded by the left face $y = 2x$ and the upper face $y = \sqrt{4-z}$. In other words, the y-coordinate of a point in $\mathcal{W}$ satisfies

$$2x \leq y \leq \sqrt{4-z}$$

Now we can write the triple integral as the following iterated integral:

$$\iiint_{\mathcal{W}} xyz \, dV = \int_{x=0}^{1} \int_{z=0}^{4-4x^2} \int_{y=2x}^{\sqrt{4-z}} xyz \, dy \, dz \, dx \qquad \blacksquare$$

The **average value** of a function of three variables is defined as in the case of two variables:

$$\overline{f} = \frac{1}{\text{Volume}(\mathcal{W})} \iiint_{\mathcal{W}} f(x, y, z) \, dV \qquad \boxed{5}$$

where $\text{Volume}(\mathcal{W}) = \iiint_{\mathcal{W}} 1 \, dV$. And, as in the case of two variables, $\overline{f}$ lies between the minimum and maximum values of f on $\mathcal{D}$, and the Mean Value Theorem holds: If $\mathcal{W}$ is connected and f is continuous on $\mathcal{W}$, then there exists a point $P \in \mathcal{W}$ such that $f(P) = \overline{f}$.

■ **EXAMPLE 6** A solid piece of crystal $\mathcal{W}$ in the first octant of space is bounded by the five planes given by $z = 0, y = 0, x = 0, x + z = 1$, and $x + y + z = 3$. The temperature at every point in the crystal is given by $T(x, y, z) = x$ in degrees centigrade. Find the average temperature for all points in the crystal.

Solution To find the average temperature, we first need to find the volume. The region $\mathcal{W}$, which appears in Figure 8, is not z-simple since there is no single top surface defining the region. Neither is it x-simple, since there is no single leftmost surface defining the region. However, it is y-simple, since it lies between the planes given by $x + y + z = 3$ and $y = 0$, and projects onto the triangle in the xz-plane given by $0 \leq x \leq 1, 0 \leq z \leq 1 - x$. Then $\mathcal{W}$ is described by

$$0 \leq x \leq 1, \quad 0 \leq z \leq 1 - x, \quad 0 \leq y \leq 3 - x - z$$

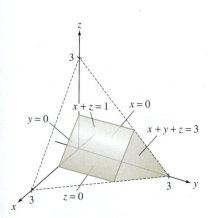

FIGURE 8 Finding the average temperature of a crystal.

$$\text{Volume}(\mathcal{W}) = \iiint_{\mathcal{W}} dV = \int_{x=0}^{1} \int_{z=0}^{1-x} \int_{y=0}^{3-x-z} dy \, dz \, dx$$

$$= \int_{x=0}^{1} \int_{z=0}^{1-x} y \Big|_{y=0}^{3-x-z} dz \, dx$$

$$= \int_{x=0}^{1} \int_{z=0}^{1-x} (3 - x - z) \, dz \, dx$$

$$= \int_{x=0}^{1} \left(3z - xz - \frac{z^2}{2} \right) \Big|_{z=0}^{1-x} dx$$

$$= \int_{x=0}^{1} \left(3(1-x) - x(1-x) - \frac{(1-x)^2}{2} \right) dx$$

$$= 3x - \frac{3x^2}{2} - \frac{x^2}{2} + \frac{x^3}{3} + \frac{(1-x)^3}{6} \Big|_{x=0}^{1} = \frac{7}{6}$$

To obtain the average temperature, we must next compute $\iiint_{\mathcal{W}} T(x, y, z)\,dV$. The only difference from our previous calculation is that we now insert $T(x, y, z)$ into the triple integral:

$$
\begin{aligned}
\iiint_{\mathcal{W}} T(x, y, z)\,dV &= \int_{x=0}^{1} \int_{z=0}^{1-x} \int_{y=0}^{3-x-z} x\,dy\,dz\,dx \\
&= \int_{x=0}^{1} \int_{z=0}^{1-x} xy \Big|_{y=0}^{3-x-z} dz\,dx \\
&= \int_{x=0}^{1} \int_{z=0}^{1-x} \left(3x - x^2 - zx\right) dz\,dx \\
&= \int_{x=0}^{1} \left(3xz - x^2 z - \frac{xz^2}{2}\right)\Big|_{z=0}^{1-x} dx \\
&= \int_{x=0}^{1} \left(3x(1-x) - x^2(1-x) - \frac{x(1-x)^2}{2}\right) dx \\
&= \frac{3x^2}{2} - x^3 - \frac{x^3}{3} + \frac{x^4}{4} - \frac{x^2}{4} + \frac{x^3}{3} - \frac{x^4}{8}\Big|_{x=0}^{1} = \frac{3}{8}
\end{aligned}
$$

Therefore,

$$
\overline{T} = \frac{1}{\text{Volume}(\mathcal{W})} \iiint_{\mathcal{W}} T(x, y, z)\,dV = \left(\frac{6}{7}\right)\left(\frac{3}{8}\right) = \frac{9}{28}\,°\text{C}
$$

Note that the mean value theorem implies that there is some point in the crystal where the temperature is exactly $9/28\,°\text{C}$. ∎

Excursion: Volume of the Sphere in Higher Dimensions

Archimedes (287–212 BCE) proved the beautiful formula $V = \frac{4}{3}\pi r^3$ for the volume of a sphere nearly 2000 years before calculus was invented, by means of a brilliant geometric argument showing that the volume of a sphere is equal to two-thirds the volume of the circumscribed cylinder. According to Plutarch (ca. 45–120 CE), Archimedes valued this achievement so highly that he requested that a sphere with a circumscribed cylinder be engraved on his tomb.

We can use integration to generalize Archimedes's formula to n dimensions. The ball of radius r in $\mathbf{R}^n$, denoted $B_n(r)$, is the set of points $(x_1, \ldots, x_n)$ in $\mathbf{R}^n$ such that

$$
x_1^2 + x_2^2 + \cdots + x_n^2 \leq r^2
$$

$B_1(r)$　Interval of radius r

$B_2(r)$　Disk of radius r

$B_3(r)$　Ball of radius r

FIGURE 9 Balls of radius r in dimensions $n = 1, 2, 3$.

The balls $B_n(r)$ in dimensions 1, 2, and 3 are the interval, disk, and ball shown in Figure 9. In dimensions $n \geq 4$, the ball $B_n(r)$ is difficult, if not impossible, to visualize, but we can compute its volume. Denote this volume by $V_n(r)$. For $n = 1$, the "volume" $V_1(r)$ is the length of the interval $B_1(r)$, and for $n = 2$, $V_2(r)$ is the area of the disk $B_2(r)$. We know that

$$
V_1(r) = 2r, \qquad V_2(r) = \pi r^2, \qquad V_3(r) = \frac{4}{3}\pi r^3
$$

For $n \geq 4$, $V_n(r)$ is sometimes called the **hypervolume**.

The key idea is to determine $V_n(r)$ from the formula for $V_{n-1}(r)$ by integrating cross-sectional volume. Consider the case $n = 3$, where the horizontal slice at height $z = c$ is a two-dimensional ball (a disk) of radius $\sqrt{r^2 - c^2}$ (Figure 10). The volume $V_3(r)$ is equal to the integral of the areas of these horizontal slices:

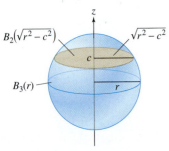

FIGURE 10 The volume $V_3(r)$ is the integral of cross-sectional area $V_2(\sqrt{r^2 - c^2})$.

$$
V_3(r) = \int_{z=-r}^{r} V_2\left(\sqrt{r^2 - z^2}\right) dz = \int_{z=-r}^{r} \pi(r^2 - z^2)\,dz = \frac{4}{3}\pi r^3
$$

Next, we show by induction that for all $n \geq 1$, there is a constant A_n (equal to the volume of the n-dimensional unit ball) such that

$$V_n(r) = A_n r^n \qquad \boxed{6}$$

The slice of $B_n(r)$ at height $x_n = c$ has equation

$$x_1^2 + x_2^2 + \cdots + x_{n-1}^2 + c^2 = r^2$$

This slice is the ball $B_{n-1}\left(\sqrt{r^2 - c^2}\right)$ of radius $\sqrt{r^2 - c^2}$, and $V_n(r)$ is obtained by integrating the volume of these slices:

$$V_n(r) = \int_{x_n=-r}^{r} V_{n-1}\left(\sqrt{r^2 - x_n^2}\right) dx_n = A_{n-1} \int_{x_n=-r}^{r} \left(\sqrt{r^2 - x_n^2}\right)^{n-1} dx_n$$

Using the substitution $x_n = r \sin \theta$ and $dx_n = r \cos \theta \, d\theta$, we have

$$V_n(r) = A_{n-1} r^n \int_{-\pi/2}^{\pi/2} \cos^n \theta \, d\theta = A_{n-1} C_n r^n$$

where $C_n = \displaystyle\int_{\theta=-\pi/2}^{\pi/2} \cos^n \theta \, d\theta$. This proves Eq. (6) with

$$A_n = A_{n-1} C_n \qquad \boxed{7}$$

In Exercise 43, you are asked to use Integration by Parts to verify the relation

$$C_n = \left(\frac{n-1}{n}\right) C_{n-2} \qquad \boxed{8}$$

It is easy to check directly that $C_0 = \pi$ and $C_1 = 2$. By Eq. (8), $C_2 = \frac{1}{2}C_0 = \frac{\pi}{2}$, $C_3 = \frac{2}{3}(2) = \frac{4}{3}$, and so on. Here are the first few values of C_n:

n	0	1	2	3	4	5	6	7
C_n	π	2	$\dfrac{\pi}{2}$	$\dfrac{4}{3}$	$\dfrac{3\pi}{8}$	$\dfrac{16}{15}$	$\dfrac{5\pi}{16}$	$\dfrac{32}{35}$

We also have $A_1 = 2$ and $A_2 = \pi$, so we can use the values of C_n together with Eq. (7) to obtain the values of A_n in Table 1. We see, for example, that the ball of radius r in four dimensions has volume $V_4(r) = \frac{1}{2}\pi^2 r^4$. The general formula depends on whether n is even or odd. Using induction and formulas (7) and (8), we can prove that

$$A_{2m} = \frac{\pi^m}{m!}, \qquad A_{2m+1} = \frac{2^{m+1}\pi^m}{1 \cdot 3 \cdot 5 \cdots (2m+1)}$$

This sequence of numbers A_n has a curious property. Setting $r = 1$ in Eq. (6), we see that A_n is the volume of the unit ball in n dimensions. From Table 1, it appears that the volumes increase up to dimension 5 and then begin to decrease. In Exercise 44, you are asked to verify that the five-dimensional unit ball has the largest volume. Furthermore, the volumes A_n tend to 0 as $n \to \infty$.

TABLE 1 **Volume of an n-dimensional unit ball is $V_n = A_n$**

n	A_n
1	2
2	$\pi \approx 3.14$
3	$\dfrac{4}{3}\pi \approx 4.19$
4	$\dfrac{\pi^2}{2} \approx 4.93$
5	$\dfrac{8\pi^2}{15} \approx 5.26$
6	$\dfrac{\pi^3}{6} \approx 5.17$
7	$\dfrac{16\pi^3}{105} \approx 4.72$

15.3 SUMMARY

- The triple integral over a box $\mathcal{B} = [a, b] \times [c, d] \times [p, q]$ is equal to the iterated integral

$$\iiint_{\mathcal{B}} f(x, y, z)\, dV = \int_{x=a}^{b} \int_{y=c}^{d} \int_{z=p}^{q} f(x, y, z)\, dz\, dy\, dx$$

The iterated integral may be written in any one of six possible orders—for example,

$$\int_{z=p}^{q} \int_{y=c}^{d} \int_{x=a}^{b} f(x, y, z)\, dx\, dy\, dz$$

- A *z-simple region* $\mathcal{W}$ in $\mathbf{R}^3$ is a region consisting of the points (x, y, z) between two surfaces $z = z_1(x, y)$ and $z = z_2(x, y)$, where $z_1(x, y) \le z_2(x, y)$, lying over a domain $\mathcal{D}$ in the xy-plane. In other words, $\mathcal{W}$ is defined by

$$(x, y) \in \mathcal{D}, \qquad z_1(x, y) \le z \le z_2(x, y)$$

Similarly, we have x-simple regions and y-simple regions.
- The triple integral over $\mathcal{W}$ is equal to an iterated integral:

$$\iiint_{\mathcal{W}} f(x, y, z)\, dV = \iint_{\mathcal{D}} \left(\int_{z=z_1(x,y)}^{z_2(x,y)} f(x, y, z)\, dz \right) dA$$

- The *average value* of $f(x, y, z)$ on a region $\mathcal{W}$ of volume V is the quantity

$$\overline{f} = \frac{1}{V} \iiint_{\mathcal{W}} f(x, y, z)\, dV, \qquad V = \iiint_{\mathcal{W}} 1\, dV$$

15.3 EXERCISES

Preliminary Questions

1. Which of (a)–(c) is not equal to $\int_0^1 \int_3^4 \int_6^7 f(x, y, z)\, dz\, dy\, dx$?

(a) $\int_6^7 \int_0^1 \int_3^4 f(x, y, z)\, dy\, dx\, dz$

(b) $\int_3^4 \int_0^1 \int_6^7 f(x, y, z)\, dz\, dx\, dy$

(c) $\int_0^1 \int_3^4 \int_6^7 f(x, y, z)\, dx\, dz\, dy$

2. Which of the following is not a meaningful triple integral?

(a) $\int_0^1 \int_0^x \int_{x+y}^{2x+y} e^{x+y+z}\, dz\, dy\, dx$

(b) $\int_0^1 \int_0^z \int_{x+y}^{2x+y} e^{x+y+z}\, dz\, dy\, dx$

3. Describe the projection of the region of integration $\mathcal{W}$ onto the xy-plane:

(a) $\int_0^1 \int_0^x \int_0^{x^2+y^2} f(x, y, z)\, dz\, dy\, dx$

(b) $\int_0^1 \int_0^{\sqrt{1-x^2}} \int_2^4 f(x, y, z)\, dz\, dy\, dx$

Exercises

In Exercises 1–8, evaluate $\iiint_{\mathcal{B}} f(x, y, z)\, dV$ for the specified function f and box $\mathcal{B}$.

1. $f(x, y, z) = z^4;\quad 2 \le x \le 8,\quad 0 \le y \le 5,\quad 0 \le z \le 1$

2. $f(x, y, z) = xz^2;\quad [-2, 3] \times [1, 3] \times [1, 4]$

3. $f(x, y, z) = xe^{y-2z};\quad 0 \le x \le 2,\quad 0 \le y \le 1,\quad 0 \le z \le 1$

4. $f(x, y, z) = \dfrac{x}{(y+z)^2};\quad [0, 2] \times [2, 4] \times [-1, 1]$

5. $f(x, y, z) = (x - y)(y - z);\quad [0, 1] \times [0, 3] \times [0, 3]$

6. $f(x, y, z) = \dfrac{z}{x};\quad 1 \le x \le 3,\quad 0 \le y \le 2,\quad 0 \le z \le 4$

7. $f(x, y, z) = (x + z)^3;\quad [0, a] \times [0, b] \times [0, c]$

8. $f(x, y, z) = (x + y - z)^2;\quad [0, a] \times [0, b] \times [0, c]$

In Exercises 9–14, evaluate $\iiint_{\mathcal{W}} f(x, y, z)\, dV$ for the function f and region $\mathcal{W}$ specified.

9. $f(x, y, z) = x + y;\quad \mathcal{W}: y \le z \le x,\quad 0 \le y \le x,\quad 0 \le x \le 1$

10. $f(x, y, z) = e^{x+y+z};\quad \mathcal{W}: 0 \le z \le 1,\quad 0 \le y \le x,\quad 0 \le x \le 1$

11. $f(x, y, z) = xyz;\quad \mathcal{W}: 0 \le z \le 1,\quad 0 \le y \le \sqrt{1-x^2},\quad 0 \le x \le 1$

12. $f(x, y, z) = x;\quad \mathcal{W}: x^2 + y^2 \le z \le 4$

13. $f(x, y, z) = e^z;\quad \mathcal{W}: x + y + z \le 1,\quad x \ge 0,\quad y \ge 0,\quad z \ge 0$

14. $f(x, y, z) = z;\quad \mathcal{W}: x^2 \le y \le 2,\quad 0 \le x \le 1,\quad x - y \le z \le x + y$

15. Calculate the integral of $f(x, y, z) = z$ over the region $\mathcal{W}$ in Figure 11 below the hemisphere of radius 3 and lying over the triangle $\mathcal{D}$ in the xy-plane bounded by $x = 1$, $y = 0$, and $x = y$.

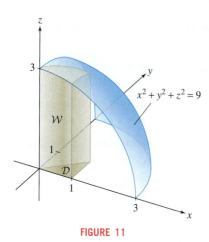

FIGURE 11

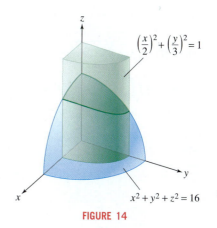

FIGURE 14

16. Calculate the integral of $f(x, y, z) = e^z$ over the tetrahedron $\mathcal{W}$ in Figure 12.

FIGURE 12

17. Integrate $f(x, y, z) = x$ over the region in the first octant ($x \geq 0, y \geq 0, z \geq 0$) above $z = y^2$ and below $z = 8 - 2x^2 - y^2$.

18. Compute the integral of $f(x, y, z) = y^2$ over the region within the cylinder $x^2 + y^2 = 4$, where $0 \leq z \leq y$.

19. Find the triple integral of the function z over the ramp in Figure 13. Here, z is the height above the ground.

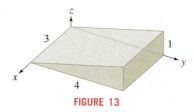

FIGURE 13

20. Find the volume of the solid in $\mathbf{R}^3$ bounded by $y = x^2$, $x = y^2$, $z = x + y + 5$, and $z = 0$.

21. Find the volume of the solid in the octant $x \geq 0, y \geq 0, z \geq 0$ bounded by $x + y + z = 1$ and $x + y + 2z = 1$.

22. Calculate $\iiint_{\mathcal{W}} y \, dV$, where $\mathcal{W}$ is the region above $z = x^2 + y^2$ and below $z = 5$, and bounded by $y = 0$ and $y = 1$.

23. Evaluate $\iiint_{\mathcal{W}} xz \, dV$, where $\mathcal{W}$ is the domain bounded by the elliptic cylinder $\dfrac{x^2}{4} + \dfrac{y^2}{9} = 1$ and the sphere $x^2 + y^2 + z^2 = 16$ in the first octant $x \geq 0, y \geq 0, z \geq 0$ (Figure 14).

24. Describe the domain of integration and evaluate:

$$\int_0^3 \int_0^{\sqrt{9-x^2}} \int_0^{\sqrt{9-x^2-y^2}} xy \, dz \, dy \, dx$$

25. Describe the domain of integration of the following integral:

$$\int_{-2}^2 \int_{-\sqrt{4-z^2}}^{\sqrt{4-z^2}} \int_1^{\sqrt{5-x^2-z^2}} f(x, y, z) \, dy \, dx \, dz$$

26. Let $\mathcal{W}$ be the region below the paraboloid

$$x^2 + y^2 = z - 2$$

that lies above the part of the plane $x + y + z = 1$ in the first octant ($x \geq 0, y \geq 0, z \geq 0$). Express

$$\iiint_{\mathcal{W}} f(x, y, z) \, dV$$

as an iterated integral (for an arbitrary function f).

27. In Example 5, we expressed a triple integral as an iterated integral in the three orders

$$dz \, dy \, dx, \quad dx \, dz \, dy, \quad \text{and} \quad dy \, dz \, dx$$

Write this integral in the three other orders:

$$dz \, dx \, dy, \quad dx \, dy \, dz, \quad \text{and} \quad dy \, dx \, dz$$

28. Let $\mathcal{W}$ be the region bounded by

$$y + z = 2, \quad 2x = y, \quad x = 0, \quad \text{and} \quad z = 0$$

(Figure 15). Express and evaluate the triple integral of $f(x, y, z) = z$ by projecting $\mathcal{W}$ onto the:

(a) xy-plane. **(b)** yz-plane. **(c)** xz-plane.

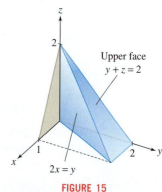

FIGURE 15

29. Let

$$W = \left\{(x, y, z) : \sqrt{x^2 + y^2} \le z \le 1\right\}$$

(see Figure 16). Express $\iiint_W f(x, y, z)\, dV$ as an iterated integral in the order $dz\, dy\, dx$ (for an arbitrary function f).

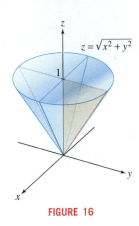

$z = \sqrt{x^2 + y^2}$

FIGURE 16

30. Repeat Exercise 29 for the order $dx\, dy\, dz$.

31. Let W be the region bounded by $z = 1 - y^2$, $y = x^2$, and the plane $z = 0$. Calculate the volume of W as a triple integral in the order $dz\, dy\, dx$.

32. Calculate the volume of the region W in Exercise 31 as a triple integral in the following orders:

(a) $dx\, dz\, dy$ **(b)** $dy\, dz\, dx$

In Exercises 33–36, draw the region W and then set up but do NOT compute a single triple integral that yields the volume of W.

33. The region W is bounded by the surfaces given by $z = 1 - y^2$, $x = 0$ and $z = 0$, $z + x = 3$.

34. The region W is bounded by the surfaces given by $z = x^2$, $z + y = 1$, and $z - y = 1$.

35. The region W is bounded by the surfaces given by $z = y^2$, $y = z^2$ and $x = 0$, $x + y + z = 4$.

36. The region W is bounded by the surfaces given by $z = 1 - x^2$, $z = 0$, and $y = 3 - x^2 - z^2$.

In Exercises 37–40, compute the average value of $f(x, y, z)$ over the region W.

37. $f(x, y, z) = xy \sin(\pi z)$; $W = [0, 1] \times [0, 1] \times [0, 1]$

38. $f(x, y, z) = xyz$; $W : 0 \le z \le y \le x \le 1$

39. $f(x, y, z) = e^y$; $W : 0 \le y \le 1 - x^2$, $0 \le z \le x$

40. $f(x, y, z) = x^2 + y^2 + z^2$; W bounded by the planes $2y + z = 1$, $x = 0$, $x = 1$, $z = 0$, and $y = 0$

In Exercises 41 and 42, let $I = \int_0^1 \int_0^1 \int_0^1 f(x, y, z)\, dV$ and let $S_{N,N,N}$ be the Riemann sum approximation

$$S_{N,N,N} = \frac{1}{N^3} \sum_{i=1}^{N} \sum_{j=1}^{N} \sum_{k=1}^{N} f\left(\frac{i}{N}, \frac{j}{N}, \frac{k}{N}\right)$$

41. **CAS** Calculate $S_{N,N,N}$ for $f(x, y, z) = e^{x^2 - y - z}$ for $N = 10$, 20, 30. Then evaluate I and find an N such that $S_{N,N,N}$ approximates I to two decimal places.

42. **CAS** Calculate $S_{N,N,N}$ for $f(x, y, z) = \sin(xyz)$ for $N = 10$, 20, 30. Then use a computer algebra system to calculate I numerically and estimate the error $|I - S_{N,N,N}|$.

Further Insights and Challenges

43. Use Integration by Parts to verify Eq. (8).

44. Compute the volume A_n of the unit ball in $\mathbf{R}^n$ for $n = 8, 9, 10$.

Show that $C_n \le 1$ for $n \ge 6$ and use this to prove that of all unit balls, the five-dimensional ball has the largest volume. Can you explain why A_n tends to 0 as $n \to \infty$?

15.4 Integration in Polar, Cylindrical, and Spherical Coordinates

In single-variable calculus, a well-chosen substitution (also called a change of variables) often transforms a complicated integral into a simpler one. Change of variables is also useful in multivariable calculus, but the emphasis is different. In the multivariable case, we are usually interested in simplifying not just the integrand, but also the domain of integration.

This section treats three of the most useful changes of variables, in which an integral is expressed in polar, cylindrical, or spherical coordinates. As in Figure 1, certain physical systems are much more easily modeled with the right coordinate system. The general Change of Variables Formula is discussed in Section 15.6.

Double Integrals in Polar Coordinates

Polar coordinates are convenient when the domain of integration is an angular sector or a **polar rectangle** (Figure 2):

$$\mathcal{R} : \theta_1 \le \theta \le \theta_2, \quad r_1 \le r \le r_2 \qquad \boxed{1}$$

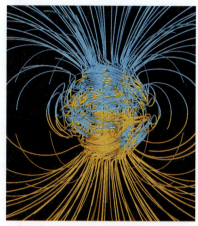

FIGURE 1 Spherical coordinates are used in mathematical models of Earth's magnetic field. This computer simulation, based on the Glatzmaier–Roberts model, shows the magnetic lines of force, representing inward and outward directed field lines in blue and yellow, respectively.

(Gary A. Glatzmaier [University of California, Santa Cruz] and Paul H. Roberts [University of California, Los Angeles])

Eq. (2) expresses the integral of $f(x, y)$ over the polar rectangle in Figure 2 as the integral of a new function $rf(r\cos\theta, r\sin\theta)$ over the ordinary rectangle $[\theta_1, \theta_2] \times [r_1, r_2]$. In this sense, the change of variables "simplifies" the domain of integration.

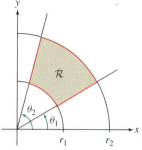

FIGURE 2 Polar rectangle.

We assume throughout that $r_1 \geq 0$ and that all radial coordinates are nonnegative. Recall that rectangular and polar coordinates are related by

$$x = r\cos\theta, \qquad y = r\sin\theta$$

Thus, we write a function $f(x, y)$ in polar coordinates as $f(r\cos\theta, r\sin\theta)$. The Change of Variables Formula for a polar rectangle $\mathcal{R}$ is

$$\iint_{\mathcal{R}} f(x, y)\, dA = \int_{\theta_1}^{\theta_2} \int_{r_1}^{r_2} f(r\cos\theta, r\sin\theta)\, r\, dr\, d\theta \qquad \boxed{2}$$

Notice the extra factor r in the integrand on the right.

To derive Eq. (2), the key step is to estimate the area ΔA of the small polar rectangle shown in Figure 3. If Δr and $\Delta\theta$ are small, then this polar rectangle is very nearly an ordinary rectangle of sides Δr and $r\Delta\theta$, as in the Reminder at left, and therefore, $\Delta A \approx r\, \Delta r\, \Delta\theta$. In fact, ΔA is the difference of areas of two sectors:

$$\Delta A = \frac{1}{2}(r + \Delta r)^2\, \Delta\theta - \frac{1}{2}r^2\, \Delta\theta = r(\Delta r\, \Delta\theta) + \frac{1}{2}(\Delta r)^2 \Delta\theta \approx r\, \Delta r\, \Delta\theta$$

The error in our approximation is the term $\frac{1}{2}(\Delta r)^2 \Delta\theta$, which has a smaller order of magnitude than $\Delta r\, \Delta\theta$ when Δr and $\Delta\theta$ are both small.

←·· REMINDER In Figure 4, the length β of the arc subtended by the angle θ is the fraction of the entire circumference that θ is of the entire angle 2π. Hence, $\beta = \frac{\theta}{2\pi} 2\pi r = r\theta$. Similarly, the area of the sector subtended by θ is $\frac{1}{2}r^2\theta$.

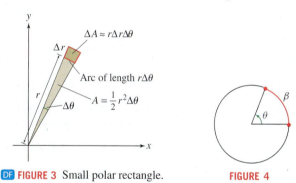

DF **FIGURE 3** Small polar rectangle.   **FIGURE 4**

←·· REMINDER In Eq. (3) we use the approximation (10) in Section 15.2: If f is continuous and $\mathcal{D}$ is a small domain,

$$\iint_{\mathcal{D}} f(x, y)\, dA \approx f(P)\, \text{Area}(\mathcal{D})$$

where P is any sample point in $\mathcal{D}$.

Now, decompose $\mathcal{R}$ into an $N \times M$ grid of small polar subrectangles $\mathcal{R}_{ij}$ as in Figure 5, and choose a sample point P_{ij} in $\mathcal{R}_{ij}$. If $\mathcal{R}_{ij}$ is small and $f(x, y)$ is continuous, then

$$\iint_{\mathcal{R}_{ij}} f(x, y)\, dx\, dy \approx f(P_{ij})\, \text{Area}(\mathcal{R}_{ij}) \approx f(P_{ij})\, r_{ij}\, \Delta r\, \Delta\theta \qquad \boxed{3}$$

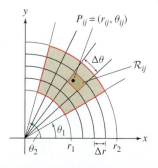

FIGURE 5 Decomposition of a polar rectangle into subrectangles.

Note that each polar rectangle $\mathcal{R}_{ij}$ has angular width $\Delta\theta = (\theta_2 - \theta_1)/N$ and radial width $\Delta r = (r_2 - r_1)/M$. The integral over $\mathcal{R}$ is the sum

$$\iint_{\mathcal{R}} f(x, y) \, dx \, dy = \sum_{i=1}^{N} \sum_{j=1}^{M} \iint_{\mathcal{R}_{ij}} f(x, y) \, dx \, dy$$

$$\approx \sum_{i=1}^{N} \sum_{j=1}^{M} f(P_{ij}) \, \text{Area}(\mathcal{R}_{ij})$$

$$\approx \sum_{i=1}^{N} \sum_{j=1}^{M} f(r_{ij} \cos\theta_{ij}, r_{ij} \sin\theta_{ij}) \, r_{ij} \, \Delta r \, \Delta\theta$$

This is a Riemann sum for the double integral of $rf(r\cos\theta, r\sin\theta)$ over the region $r_1 \le r \le r_2$, $\theta_1 \le \theta \le \theta_2$, and we can prove that it approaches the double integral as $N, M \to \infty$. A similar derivation is valid for domains (Figure 6) that can be described as the region between two polar curves $r = r_1(\theta)$ and $r = r_2(\theta)$ for θ between the values θ_1 and θ_2. This gives us Theorem 1.

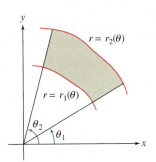

FIGURE 6 General polar region.

Eq. (4) is summarized in the symbolic expression for the "area element" dA in polar coordinates:

$$dA = r \, dr \, d\theta$$

THEOREM 1 **Double Integral in Polar Coordinates**   For a continuous function f on the domain

$$\mathcal{D} : \theta_1 \le \theta \le \theta_2, \quad r_1(\theta) \le r \le r_2(\theta)$$

$$\iint_{\mathcal{D}} f(x, y) \, dA = \int_{\theta_1}^{\theta_2} \int_{r=r_1(\theta)}^{r_2(\theta)} f(r\cos\theta, r\sin\theta) \, r \, dr \, d\theta \qquad \boxed{4}$$

We call such a region $\mathcal{D}$ **radially simple**. It has the property that every ray from the origin intersects the region in a single point or line segment such that the line segment begins on $r = r_1(\theta)$ and ends on $r = r_2(\theta)$.

■ **EXAMPLE 1**   Compute $\displaystyle\iint_{\mathcal{D}} (x + y) \, dA$, where $\mathcal{D}$ is the quarter annulus in Figure 7.

Solution   The quarter annulus is an example of a domain that is radially simple.

Step 1.   Describe $\mathcal{D}$ and f in polar coordinates.

The quarter annulus $\mathcal{D}$ is defined by the inequalities (Figure 7)

$$\mathcal{D} : 0 \le \theta \le \frac{\pi}{2}, \quad 2 \le r \le 4$$

In polar coordinates,

$$f(x, y) = x + y = r\cos\theta + r\sin\theta = r(\cos\theta + \sin\theta)$$

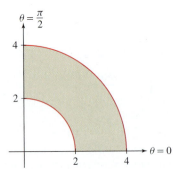

FIGURE 7 Quarter annulus $0 \le \theta \le \frac{\pi}{2}$, $2 \le r \le 4$.

Step 2.   Change variables and evaluate.

To write the integral in polar coordinates, we replace dA by $r \, dr \, d\theta$:

$$\iint_{\mathcal{D}} (x + y) \, dA = \int_{0}^{\pi/2} \int_{r=2}^{4} r(\cos\theta + \sin\theta) \, r \, dr \, d\theta$$

The inner integral is

$$\int_{r=2}^{4} (\cos\theta + \sin\theta) \, r^2 \, dr = (\cos\theta + \sin\theta) \left(\frac{4^3}{3} - \frac{2^3}{3} \right) = \frac{56}{3}(\cos\theta + \sin\theta)$$

and

$$\iint_{\mathcal{D}} (x + y) \, dA = \frac{56}{3} \int_{0}^{\pi/2} (\cos\theta + \sin\theta) \, d\theta = \frac{56}{3}(\sin\theta - \cos\theta) \Big|_{0}^{\pi/2} = \frac{112}{3} \quad ■$$

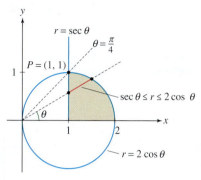

$r = \sec\theta$

$\theta = \frac{\pi}{4}$

$P = (1, 1)$

$\sec\theta \le r \le 2\cos\theta$

$r = 2\cos\theta$

FIGURE 8

■ **EXAMPLE 2** Calculate $\displaystyle\iint_{\mathcal{D}} (x^2 + y^2)^{-2} \, dA$ for the shaded domain $\mathcal{D}$ in Figure 8.

Solution

Step 1. Describe $\mathcal{D}$ and f in polar coordinates.

The quarter circle lies in the angular sector $0 \le \theta \le \frac{\pi}{4}$ because the line through $P = (1, 1)$ makes an angle of $\frac{\pi}{4}$ with the x-axis (Figure 8).

To determine the limits on r, recall from Section 11.3 (Examples 6 and 8) that:

- The vertical line $x = 1$ has polar equation $r\cos\theta = 1$ or $r = \sec\theta$.
- The circle of radius 1 and center $(1, 0)$ has polar equation $r = 2\cos\theta$.

Therefore, a ray of angle θ intersects $\mathcal{D}$ in the segment where r ranges from $\sec\theta$ to $2\cos\theta$. In other words, our domain has polar description

$$\mathcal{D} : 0 \le \theta \le \frac{\pi}{4}, \quad \sec\theta \le r \le 2\cos\theta$$

The function in polar coordinates is

$$f(x, y) = (x^2 + y^2)^{-2} = (r^2)^{-2} = r^{-4}$$

Step 2. Change variables and evaluate.

◄·· *REMINDER*

$$\int \cos^2\theta \, d\theta = \frac{1}{2}\left(\theta + \frac{1}{2}\sin 2\theta\right) + C$$

$$\int \sec^2\theta \, d\theta = \tan\theta + C$$

$$\iint_{\mathcal{D}} (x^2 + y^2)^{-2} \, dA = \int_0^{\pi/4} \int_{r=\sec\theta}^{2\cos\theta} r^{-4} \, r \, dr \, d\theta = \int_0^{\pi/4} \int_{r=\sec\theta}^{2\cos\theta} r^{-3} \, dr \, d\theta$$

The inner integral is

$$\int_{r=\sec\theta}^{2\cos\theta} r^{-3} \, dr = -\frac{1}{2}r^{-2} \Big|_{r=\sec\theta}^{2\cos\theta} = -\frac{1}{8}\sec^2\theta + \frac{1}{2}\cos^2\theta$$

Therefore,

$$\iint_{\mathcal{D}} (x^2 + y^2)^{-2} \, dA = \int_0^{\pi/4} \left(\frac{1}{2}\cos^2\theta - \frac{1}{8}\sec^2\theta\right) d\theta$$

$$= \left(\frac{1}{4}\left(\theta + \frac{1}{2}\sin 2\theta\right) - \frac{1}{8}\tan\theta\right) \Big|_0^{\pi/4}$$

$$= \frac{1}{4}\left(\frac{\pi}{4} + \frac{1}{2}\sin\frac{\pi}{2}\right) - \frac{1}{8}\tan\frac{\pi}{4} = \frac{\pi}{16}$$ ■

Triple Integrals in Cylindrical Coordinates

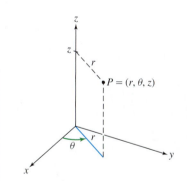

$P = (r, \theta, z)$

FIGURE 9 Cylindrical coordinates.

Cylindrical coordinates, introduced in Section 12.7, are useful when the domain has **axial symmetry**—that is, symmetry with respect to an axis. In cylindrical coordinates (r, θ, z), the axis of symmetry is the z-axis. Recall the relations (Figure 9)

$$x = r\cos\theta, \qquad y = r\sin\theta, \qquad z = z$$

To set up a triple integral in cylindrical coordinates, we assume that the domain of integration $\mathcal{W}$ can be described as the region between two surfaces (Figure 10)

$$z_1(r, \theta) \le z \le z_2(r, \theta)$$

lying over a domain $\mathcal{D}$ in the xy-plane with polar description

$$\mathcal{D} : \theta_1 \le \theta \le \theta_2, \quad r_1(\theta) \le r \le r_2(\theta)$$

A triple integral over $\mathcal{W}$ can be written as an iterated integral (Theorem 2 of Section 15.3):

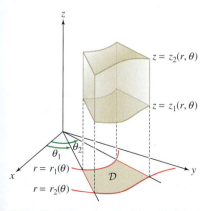

$z = z_2(r, \theta)$

$z = z_1(r, \theta)$

θ_1 θ_2

$r = r_1(\theta)$

$r = r_2(\theta)$

$\mathcal{D}$

FIGURE 10 Region described in cylindrical coordinates.

$$\iiint_{\mathcal{W}} f(x, y, z) \, dV = \iint_{\mathcal{D}} \left(\int_{z=z_1(r,\theta)}^{z_2(r,\theta)} f(x, y, z) \, dz\right) dA$$

By expressing the integral over $\mathcal{D}$ in polar coordinates, we obtain the following Change of Variables Formula.

Eq. (5) is summarized in the symbolic expression for the "volume element" dV in cylindrical coordinates:

$$dV = r\,dz\,dr\,d\theta$$

> **THEOREM 2 Triple Integrals in Cylindrical Coordinates** For a continuous function f on the region
>
> $$\theta_1 \le \theta \le \theta_2, \qquad r_1(\theta) \le r \le r_2(\theta), \qquad z_1(r,\theta) \le z \le z_2(r,\theta),$$
>
> the triple integral $\iiint_{\mathcal{W}} f(x,y,z)\,dV$ is equal to
>
> $$\int_{\theta_1}^{\theta_2} \int_{r=r_1(\theta)}^{r_2(\theta)} \int_{z=z_1(r,\theta)}^{z_2(r,\theta)} f(r\cos\theta, r\sin\theta, z)\,r\,dz\,dr\,d\theta \qquad \boxed{5}$$

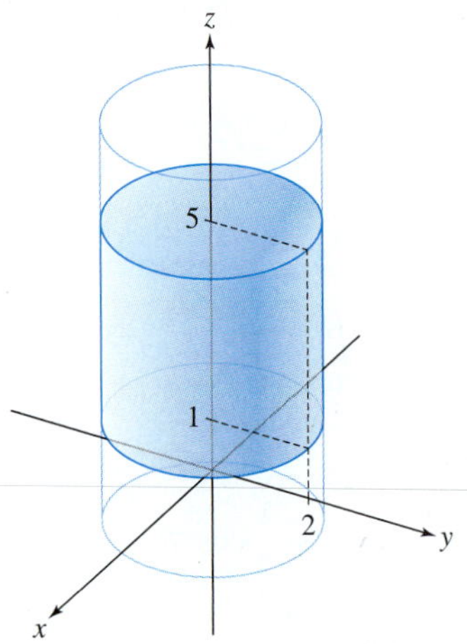

■ **EXAMPLE 3** Integrate $f(x,y,z) = z\sqrt{x^2 + y^2}$ over the cylinder $x^2 + y^2 \le 4$ for $1 \le z \le 5$ (Figure 11).

Solution The domain of integration $\mathcal{W}$ lies above the disk of radius 2, so in cylindrical coordinates,

$$\mathcal{W} : 0 \le \theta \le 2\pi, \quad 0 \le r \le 2, \quad 1 \le z \le 5$$

We write the function in cylindrical coordinates:

$$f(x,y,z) = z\sqrt{x^2 + y^2} = zr$$

and integrate with respect to $dV = r\,dz\,dr\,d\theta$. The function f is a product zr, so the resulting triple integral is a product of single integrals:

$$\iiint_{\mathcal{W}} z\sqrt{x^2 + y^2}\,dV = \int_0^{2\pi} \int_{r=0}^{2} \int_{z=1}^{5} (zr)r\,dz\,dr\,d\theta$$

$$= \left(\int_0^{2\pi} d\theta\right)\left(\int_{r=0}^{2} r^2\,dr\right)\left(\int_{z=1}^{5} z\,dz\right)$$

$$= (2\pi)\left(\frac{2^3}{3}\right)\left(\frac{5^2 - 1^2}{2}\right) = 64\pi \qquad ■$$

FIGURE 11 The cylinder $x^2 + y^2 \le 4$.

■ **EXAMPLE 4** Compute the integral of $f(x,y,z) = z$ over the region $\mathcal{W}$ within the cylinder $x^2 + y^2 \le 4$, where $0 \le z \le y$.

Solution

Step 1. **Express $\mathcal{W}$ in cylindrical coordinates.**
The condition $0 \le z \le y$ tells us that $y \ge 0$, so $\mathcal{W}$ projects onto the semicircle $\mathcal{D}$ in the xy-plane of radius 2, where $y \ge 0$, shown in Figure 12. In polar coordinates,

$$\mathcal{D} : 0 \le \theta \le \pi, \quad 0 \le r \le 2$$

The z-coordinate in $\mathcal{W}$ varies from $z = 0$ to $z = y$, and in polar coordinates $y = r\sin\theta$, so the region has the description

$$\mathcal{W} : 0 \le \theta \le \pi, \quad 0 \le r \le 2, \quad 0 \le z \le r\sin\theta$$

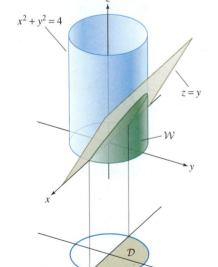

FIGURE 12

Step 2. **Change variables and evaluate.**

$$\iiint_{\mathcal{W}} f(x, y, z)\, dV = \int_0^\pi \int_{r=0}^2 \int_{z=0}^{r\sin\theta} zr\, dz\, dr\, d\theta$$

$$= \int_0^\pi \int_{r=0}^2 \frac{1}{2}(r\sin\theta)^2 r\, dr\, d\theta$$

$$= \frac{1}{2}\left(\int_0^\pi \sin^2\theta\, d\theta\right)\left(\int_0^2 r^3\, dr\right)$$

$$= \frac{1}{2}\left(\frac{\pi}{2}\right)\frac{2^4}{4} = \pi \qquad \blacksquare$$

◀┄ **REMINDER**

$$\int \sin^2\theta\, d\theta = \frac{1}{2}\left(\theta - \frac{1}{2}\sin 2\theta\right) + C$$

$$\int_0^\pi \sin^2\theta\, d\theta = \frac{\pi}{2}$$

Triple Integrals in Spherical Coordinates

We noted that the Change of Variables Formula in cylindrical coordinates is summarized by the symbolic equation $dV = r\, dr\, d\theta\, dz$. In spherical coordinates (ρ, θ, ϕ) (introduced in Section 12.7), the analog is the formula

$$\boxed{dV = \rho^2 \sin\phi\, d\rho\, d\phi\, d\theta}$$

Recall (Figure 13) that

$$x = \rho\sin\phi\cos\theta, \qquad y = \rho\sin\phi\sin\theta, \qquad z = \rho\cos\phi, \qquad r = \rho\sin\phi$$

The key step in deriving this formula is estimating the volume of a small **spherical wedge** $\mathcal{W}$. Suppose it is defined by fixing values for ρ, ϕ, and θ, and varying each coordinate by a small amount given by $\Delta\rho$, $\Delta\phi$, and $\Delta\theta$ as in Figure 14.

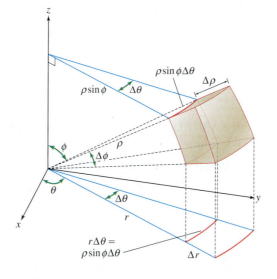

For small increments, the wedge is nearly a rectangular box with dimensions $\rho\sin\phi\Delta\theta \times \rho\Delta\phi \times \Delta\rho$.

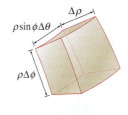

DF FIGURE 13 Spherical coordinates.

$z = \rho\cos\phi$

$r = \rho\sin\phi$

$P = (x, y, z) = (\rho, \theta, \phi)$

DF FIGURE 14 Spherical wedge.

The spherical wedge is nearly a box with sides $\Delta\rho$, $\rho\Delta\phi$, and $r\Delta\theta$ by projecting up from the corresponding length in the xy-plane. Converting $r\Delta\theta$ to spherical coordinates, we obtain $\rho\sin\phi\Delta\theta$ for this third dimension of the box.

Therefore, the volume of the spherical wedge is approximately given by the product of these three dimensions, the accuracy of which improves the smaller we take our changes in the variables:

$$\text{Volume}(\mathcal{W}) \approx \rho^2 \sin\phi\, \Delta\rho\, \Delta\phi\, \Delta\theta \qquad \boxed{6}$$

Following the usual steps, we decompose $\mathcal{W}$ into N^3 spherical subwedges $\mathcal{W}_i$ (Figure 15) with increments

FIGURE 15 Decomposition of a spherical wedge into subwedges.

$$\Delta\theta = \frac{\theta_2 - \theta_1}{N}, \qquad \Delta\phi = \frac{\phi_2 - \phi_1}{N}, \qquad \Delta\rho = \frac{\rho_2 - \rho_1}{N}$$

and choose a sample point $P_i = (\rho_i, \theta_i, \phi_i)$ in each $\mathcal{W}_i$. Assuming f is continuous, the following approximation holds for large N (small $\mathcal{W}_i$):

$$\iiint_{\mathcal{W}_i} f(x, y, z)\, dV \approx f(P_i)\text{Volume}(\mathcal{W}_i)$$

$$\approx f(P_i)\rho_i^2 \sin \phi_i\, \Delta\rho\, \Delta\theta\, \Delta\phi$$

Taking the sum over i, we obtain

$$\iiint_{\mathcal{W}} f(x, y, z)\, dV \approx \sum_i f(P_i)\rho_i^2 \sin \phi_i\, \Delta\rho\, \Delta\theta\, \Delta\phi \qquad \boxed{7}$$

The sum on the right is a Riemann sum for the function

$$f(\rho \cos \theta \sin \phi, \, \rho \sin \theta \sin \phi, \, \rho \cos \phi)\, \rho^2 \sin \phi$$

on the domain $\mathcal{W}$. Eq. (8) below follows by passing to the limit an $N \to \infty$ [and showing that the error in Eq. (7) tends to zero]. This argument applies more generally to regions defined by an inequality $\rho_1(\theta, \phi) \le \rho \le \rho_2(\theta, \phi)$.

THEOREM 3   Triple Integrals in Spherical Coordinates   For a region $\mathcal{W}$ defined by

$$\theta_1 \le \theta \le \theta_2, \qquad \phi_1 \le \phi \le \phi_2, \qquad \rho_1(\theta, \phi) \le \rho \le \rho_2(\theta, \phi)$$

the triple integral $\iiint_{\mathcal{W}} f(x, y, z)\, dV$ is equal to

$$\int_{\theta_1}^{\theta_2} \int_{\phi=\phi_1}^{\phi_2} \int_{\rho=\rho_1(\theta,\phi)}^{\rho_2(\theta,\phi)} f(\rho \sin \phi \cos \theta, \, \rho \sin \phi \sin \theta, \, \rho \cos \phi)\, \rho^2 \sin \phi\, d\rho\, d\phi\, d\theta \qquad \boxed{8}$$

Eq. (8) is summarized in the symbolic expression for the "volume element" dV in spherical coordinates:

$$dV = \rho^2 \sin \phi\, d\rho\, d\phi\, d\theta$$

We call such a region $\mathcal{W}$ **centrally simple** in that every ray from the origin intersects the solid in a single line segment or point such that the first endpoint of the segment lies on the surface $\rho = \rho_1(\theta, \phi)$ and the second endpoint lies on the surface $\rho = \rho_2(\theta, \phi)$.

■ **EXAMPLE 5**   Compute the integral of $f(x, y, z) = x^2 + y^2$ over the sphere $\mathcal{S}$ of radius 4 centered at the origin (Figure 16).

Solution   First, write $f(x, y, z)$ in spherical coordinates:

$$f(x, y, z) = x^2 + y^2 = (\rho \sin \phi \cos \theta)^2 + (\rho \sin \phi \sin \theta)^2$$

$$= \rho^2 \sin^2 \phi (\cos^2 \theta + \sin^2 \theta) = \rho^2 \sin^2 \phi$$

Since we are integrating over the entire sphere $\mathcal{S}$ of radius 4, this is a centrally simple region, with each ray beginning at the origin and ending on the sphere. So, ρ varies from 0 to 4, θ from 0 to 2π, and ϕ from 0 to π. In the following computation, we integrate first with respect to θ:

$$\iiint_{\mathcal{S}} (x^2 + y^2)\, dV = \int_0^{2\pi} \int_{\phi=0}^{\pi} \int_{\rho=0}^{4} (\rho^2 \sin^2 \phi)\, \rho^2 \sin \phi\, d\rho\, d\phi\, d\theta$$

$$= 2\pi \int_{\phi=0}^{\pi} \int_{\rho=0}^{4} \rho^4 \sin^3 \phi\, d\rho\, d\phi = 2\pi \int_0^{\pi} \left(\frac{\rho^5}{5} \Big|_0^4 \right) \sin^3 \phi\, d\phi$$

$$= \frac{2048\pi}{5} \int_0^{\pi} \sin^3 \phi\, d\phi$$

$$= \frac{2048\pi}{5} \left(\frac{1}{3} \cos^3 \phi - \cos \phi \right) \Big|_0^{\pi} = \frac{8192\pi}{15} \qquad ■$$

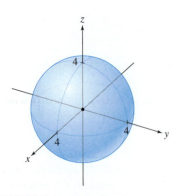

FIGURE 16 Sphere of radius 4.

◀··· *REMINDER*

$$\int \sin^3 \phi\, d\phi = \frac{1}{3} \cos^3 \phi - \cos \phi + C$$

[write $\sin^3 \phi = \sin \phi (1 - \cos^2 \phi)$]

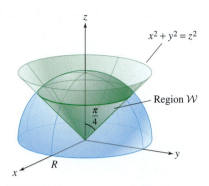

FIGURE 17 Ice cream cone defined by $0 \le \rho \le R, 0 \le \phi \le \pi/4$.

■ **EXAMPLE 6** Integrate $f(x, y, z) = z$ over the ice cream cone–shaped region $\mathcal{W}$ in Figure 17, lying above the cone and below the sphere.

Solution The cone has equation $x^2 + y^2 = z^2$, which in spherical coordinates is

$$(\rho \sin \phi \cos \theta)^2 + (\rho \sin \phi \sin \theta)^2 = (\rho \cos \phi)^2$$
$$\rho^2 \sin^2 \phi (\cos^2 \theta + \sin^2 \theta) = \rho^2 \cos^2 \phi$$
$$\sin^2 \phi = \cos^2 \phi$$
$$\sin \phi = \pm \cos \phi \quad \Rightarrow \quad \phi = \frac{\pi}{4}, \frac{3\pi}{4}$$

The upper branch of the cone has the simple equation $\phi = \frac{\pi}{4}$. On the other hand, the sphere has equation $\rho = R$, so the ice cream cone has the description

$$\mathcal{W} : 0 \le \theta \le 2\pi, \quad 0 \le \phi \le \frac{\pi}{4}, \quad 0 \le \rho \le R$$

We have

$$\iiint_{\mathcal{W}} z \, dV = \int_0^{2\pi} \int_{\phi=0}^{\pi/4} \int_{\rho=0}^{R} (\rho \cos \phi) \rho^2 \sin \phi \, d\rho \, d\phi \, d\theta$$

$$= 2\pi \int_{\phi=0}^{\pi/4} \int_{\rho=0}^{R} \rho^3 \cos \phi \sin \phi \, d\rho \, d\phi = \frac{\pi R^4}{2} \int_0^{\pi/4} \sin \phi \cos \phi \, d\phi = \frac{\pi R^4}{8}$$ ■

15.4 SUMMARY

In symbolic form:

$$\boxed{dA = r \, dr \, d\theta}$$

$$\boxed{dV = r \, dz \, dr \, d\theta}$$

$$\boxed{dV = \rho^2 \sin \phi \, d\rho \, d\phi \, d\theta}$$

• Double integral in *polar coordinates*:

$$\iint_{\mathcal{D}} f(x, y) \, dA = \int_{\theta_1}^{\theta_2} \int_{r=r_1(\theta)}^{r_2(\theta)} f(r \cos \theta, r \sin \theta) r \, dr \, d\theta$$

• Triple integral $\iiint_{\mathcal{R}} f(x, y, z) \, dV$

   – In *cylindrical coordinates*:

$$\int_{\theta_1}^{\theta_2} \int_{r=r_1(\theta)}^{r_2(\theta)} \int_{z=z_1(r,\theta)}^{z_2(r,\theta)} f(r \cos \theta, r \sin \theta, z) r \, dz \, dr \, d\theta$$

   – In *spherical coordinates*:

$$\int_{\theta_1}^{\theta_2} \int_{\phi=\phi_1}^{\phi_2} \int_{\rho=\rho_1(\theta,\phi)}^{\rho_2(\theta,\phi)} f(\rho \sin \phi \cos \theta, \rho \sin \phi \sin \theta, \rho \cos \phi) \rho^2 \sin \phi \, d\rho \, d\phi \, d\theta$$

15.4 EXERCISES

Preliminary Questions

1. Which of the following represent the integral of $f(x, y) = x^2 + y^2$ over the unit circle?

(a) $\int_0^1 \int_0^{2\pi} r^2 \, dr \, d\theta$

(b) $\int_0^{2\pi} \int_0^1 r^2 \, dr \, d\theta$

(c) $\int_0^1 \int_0^{2\pi} r^3 \, dr \, d\theta$

(d) $\int_0^{2\pi} \int_0^1 r^3 \, dr \, d\theta$

2. What are the limits of integration in $\iiint f(r, \theta, z) r \, dr \, d\theta \, dz$ if the integration extends over the following regions?

(a) $x^2 + y^2 \le 4, \quad -1 \le z \le 2$

(b) Lower hemisphere of the sphere of radius 2, center at origin

3. What are the limits of integration in

$$\iiint f(\rho, \phi, \theta) \rho^2 \sin \phi \, d\rho \, d\phi \, d\theta$$

if the integration extends over the following spherical regions centered at the origin?

(a) Sphere of radius 4

(b) Region between the spheres of radii 4 and 5

(c) Lower hemisphere of the sphere of radius 2

4. An ordinary rectangle of sides Δx and Δy has area $\Delta x \, \Delta y$, no matter where it is located in the plane. However, the area of a polar rectangle of sides Δr and $\Delta \theta$ depends on its distance from the origin.

How is this difference reflected in the Change of Variables Formula for polar coordinates?

Exercises

In Exercises 1–6, sketch the region $\mathcal{D}$ indicated and integrate $f(x, y)$ over $\mathcal{D}$ using polar coordinates.

1. $f(x, y) = \sqrt{x^2 + y^2}, \quad x^2 + y^2 \le 2$

2. $f(x, y) = x^2 + y^2; \quad 1 \le x^2 + y^2 \le 4$

3. $f(x, y) = xy; \quad x \ge 0, \quad y \ge 0, \quad x^2 + y^2 \le 4$

4. $f(x, y) = y(x^2 + y^2)^3; \quad y \ge 0, \quad x^2 + y^2 \le 1$

5. $f(x, y) = y(x^2 + y^2)^{-1}; \quad y \ge \frac{1}{2}, \quad x^2 + y^2 \le 1$

6. $f(x, y) = e^{x^2 + y^2}; \quad x^2 + y^2 \le R$

In Exercises 7–14, sketch the region of integration and evaluate by changing to polar coordinates.

7. $\displaystyle\int_{-2}^{2} \int_{0}^{\sqrt{4-x^2}} (x^2 + y^2) \, dy \, dx$

8. $\displaystyle\int_{0}^{3} \int_{0}^{\sqrt{9-y^2}} \sqrt{x^2 + y^2} \, dx \, dy$

9. $\displaystyle\int_{0}^{1/2} \int_{\sqrt{3}x}^{\sqrt{1-x^2}} x \, dy \, dx$

10. $\displaystyle\int_{0}^{4} \int_{0}^{\sqrt{16-x^2}} \tan^{-1} \frac{y}{x} \, dy \, dx$

11. $\displaystyle\int_{0}^{5} \int_{0}^{y} x \, dx \, dy$

12. $\displaystyle\int_{0}^{2} \int_{x}^{\sqrt{3}x} y \, dy \, dx$

13. $\displaystyle\int_{-1}^{2} \int_{0}^{\sqrt{4-x^2}} (x^2 + y^2) \, dy \, dx$

14. $\displaystyle\int_{1}^{2} \int_{0}^{\sqrt{2x-x^2}} \frac{1}{\sqrt{x^2 + y^2}} \, dy \, dx$

In Exercises 15–20, calculate the integral over the given region by changing to polar coordinates.

15. $f(x, y) = (x^2 + y^2)^{-2}; \quad x^2 + y^2 \le 2, \quad x \ge 1$

16. $f(x, y) = x; \quad 2 \le x^2 + y^2 \le 4$

17. $f(x, y) = |xy|; \quad x^2 + y^2 \le 1$

18. $f(x, y) = (x^2 + y^2)^{-3/2}; \quad x^2 + y^2 \le 1, \quad x + y \ge 1$

19. $f(x, y) = x - y; \quad x^2 + y^2 \le 1, \quad x + y \ge 1$

20. $f(x, y) = y; \quad x^2 + y^2 \le 1, \quad (x-1)^2 + y^2 \le 1$

21. Find the volume of the wedge-shaped region (Figure 18) contained in the cylinder $x^2 + y^2 = 9$, bounded above by the plane $z = x$ and below by the xy-plane.

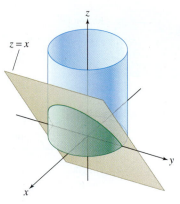

FIGURE 18

22. Let $\mathcal{W}$ be the region above the sphere $x^2 + y^2 + z^2 = 6$ and below the paraboloid $z = 4 - x^2 - y^2$.

(a) Show that the projection of $\mathcal{W}$ on the xy-plane is the disk $x^2 + y^2 \le 2$ (Figure 19).

(b) Compute the volume of $\mathcal{W}$ using polar coordinates.

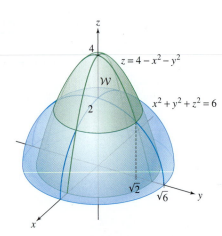

FIGURE 19

23. Evaluate $\displaystyle\iint_{\mathcal{D}} \sqrt{x^2 + y^2} \, dA$, where $\mathcal{D}$ is the domain in Figure 20. *Hint:* Find the equation of the inner circle in polar coordinates and treat the right and left parts of the region separately.

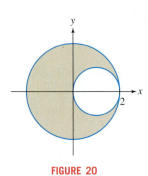

FIGURE 20

24. Evaluate $\displaystyle\iint_{\mathcal{D}} x\sqrt{x^2 + y^2} \, dA$, where $\mathcal{D}$ is the shaded region enclosed by the lemniscate curve $r^2 = \sin 2\theta$ in Figure 21.

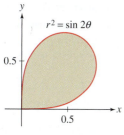

FIGURE 21

25. Let $\mathcal{W}$ be the region above the plane $z = 2$ and below the paraboloid $z = 6 - (x^2 + y^2)$.

(a) Describe $\mathcal{W}$ in cylindrical coordinates.

(b) Use cylindrical coordinates to compute the volume of $\mathcal{W}$.

26. Use cylindrical coordinates to calculate the integral of the function $f(x, y, z) = z$ over the region above the disk $x^2 + y^2 \leq 1$ in the xy-plane and below the surface $z = 4 + x^2 + y^2$.

In Exercises 27–32, use cylindrical coordinates to calculate
$$\iiint_{\mathcal{W}} f(x, y, z)\, dV \text{ for the given function and region.}$$

27. $f(x, y, z) = x^2 + y^2;\quad x^2 + y^2 \leq 9,\quad 0 \leq z \leq 5$

28. $f(x, y, z) = xz;\quad x^2 + y^2 \leq 1,\quad x \geq 0,\quad 0 \leq z \leq 2$

29. $f(x, y, z) = y;\quad x^2 + y^2 \leq 1,\quad x \geq 0,\quad y \geq 0,\quad 0 \leq z \leq 2$

30. $f(x, y, z) = z\sqrt{x^2 + y^2};\quad x^2 + y^2 \leq z \leq 8 - (x^2 + y^2)$

31. $f(x, y, z) = z;\quad x^2 + y^2 \leq z \leq 9$

32. $f(x, y, z) = z;\quad 0 \leq z \leq x^2 + y^2 \leq 9$

In Exercises 33–36, express the triple integral in cylindrical coordinates.

33. $\displaystyle\int_{-1}^{1} \int_{y=-\sqrt{1-x^2}}^{y=\sqrt{1-x^2}} \int_{z=0}^{4} f(x, y, z)\, dz\, dy\, dx$

34. $\displaystyle\int_{0}^{1} \int_{y=-\sqrt{1-x^2}}^{y=\sqrt{1-x^2}} \int_{z=0}^{4} f(x, y, z)\, dz\, dy\, dx$

35. $\displaystyle\int_{-1}^{1} \int_{y=0}^{y=\sqrt{1-x^2}} \int_{z=0}^{x^2+y^2} f(x, y, z)\, dz\, dy\, dx$

36. $\displaystyle\int_{0}^{2} \int_{y=0}^{y=\sqrt{2x-x^2}} \int_{z=0}^{\sqrt{x^2+y^2}} f(x, y, z)\, dz\, dy\, dx$

37. Find the equation of the right-circular cone in Figure 22 in cylindrical coordinates and compute its volume.

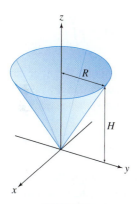

FIGURE 22

38. Use cylindrical coordinates to integrate $f(x, y, z) = z$ over the intersection of the solid hemisphere $x^2 + y^2 + z^2 \leq 4$, $z \geq 0$, and the cylinder $x^2 + y^2 = \leq 1$.

39. Find the volume of the region appearing between the two surfaces in Figure 23.

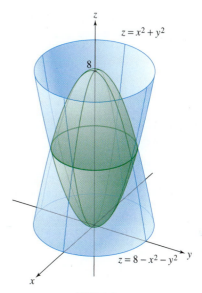

FIGURE 23

40. Use cylindrical coordinates to find the volume of a sphere of radius a from which a central cylinder of radius b has been removed, where $0 < b < a$.

41. Use cylindrical coordinates to show that the volume of a sphere of radius a from which a central cylinder of radius b has been removed, where $0 < b < a$, only depends on the height of the band that results. In particular, this implies that such a band of radius 2 m and height 1 m has the same volume as such a band of radius 6,400 km (the radius of the earth) and height 1 m.

42. Use cylindrical coordinates to find the volume of the region bounded below by the plane $z = 1$ and above by the sphere $x^2 + y^2 + z^2 = 4$.

43. Use spherical coordinates to find the volume of the region bounded below by the plane $z = 1$ and above by the sphere $x^2 + y^2 + z^2 = 4$.

44. Use spherical coordinates to find the volume of a sphere of radius 2 from which a central cylinder of radius 1 has been removed.

In Exercises 45–50, use spherical coordinates to calculate the triple integral of $f(x, y, z)$ over the given region.

45. $f(x, y, z) = y;\quad x^2 + y^2 + z^2 \leq 1,\quad x, y, z \leq 0$

46. $f(x, y, z) = \rho^{-3};\quad 2 \leq x^2 + y^2 + z^2 \leq 4$

47. $f(x, y, z) = x^2 + y^2;\quad \rho \leq 1$

48. $f(x, y, z) = 1;\quad x^2 + y^2 + z^2 \leq 4z,\quad z \geq \sqrt{x^2 + y^2}$

49. $f(x, y, z) = \sqrt{x^2 + y^2 + z^2};\quad x^2 + y^2 + z^2 \leq 2z$

50. $f(x, y, z) = \rho;\quad x^2 + y^2 + z^2 \leq 4,\quad z \leq 1,\quad x \geq 0$

51. Use spherical coordinates to evaluate the triple integral of $f(x, y, z) = z$ over the region

$$0 \le \theta \le \frac{\pi}{3}, \qquad 0 \le \phi \le \frac{\pi}{2}, \qquad 1 \le \rho \le 2$$

52. Find the volume of the region lying above the cone $\phi = \phi_0$ and below the sphere $\rho = R$.

53. Calculate the integral of

$$f(x, y, z) = z(x^2 + y^2 + z^2)^{-3/2}$$

over the part of the ball $x^2 + y^2 + z^2 \le 16$ defined by $z \ge 2$.

54. Calculate the volume of the cone in Figure 22, using spherical coordinates.

55. Calculate the volume of the sphere $x^2 + y^2 + z^2 = a^2$, using both spherical and cylindrical coordinates.

56. Let W be the region within the cylinder $x^2 + y^2 = 2$ between $z = 0$ and the cone $z = \sqrt{x^2 + y^2}$. Calculate the integral of $f(x, y, z) = x^2 + y^2$ over W, using both spherical and cylindrical coordinates.

57. Bell-Shaped Curve   One of the key results in calculus is the computation of the area under the bell-shaped curve (Figure 24):

$$I = \int_{-\infty}^{\infty} e^{-x^2} \, dx$$

This integral appears throughout engineering, physics, and statistics, and although e^{-x^2} does not have an elementary antiderivative, we can compute I using multiple integration.

(a) Show that $I^2 = J$, where J is the improper double integral

$$J = \int_{-\infty}^{\infty} \int_{-\infty}^{\infty} e^{-x^2 - y^2} \, dx \, dy$$

Hint: Use Fubini's Theorem and $e^{-x^2-y^2} = e^{-x^2} e^{-y^2}$.

(b) Evaluate J in polar coordinates.

(c) Prove that $I = \sqrt{\pi}$.

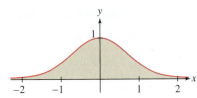

FIGURE 24 The bell-shaped curve $y = e^{-x^2}$.

Further Insights and Challenges

58. An Improper Multiple Integral   Show that a triple integral of $(x^2 + y^2 + z^2 + 1)^{-2}$ over all of $\mathbf{R}^3$ is equal to π^2. This is an improper integral, so integrate first over $\rho \le R$ and let $R \to \infty$.

59. Prove the formula

$$\iint_{\mathcal{D}} \ln r \, dA = -\frac{\pi}{2}$$

where $r = \sqrt{x^2 + y^2}$ and $\mathcal{D}$ is the unit disk $x^2 + y^2 \le 1$. This is an improper integral since $\ln r$ is not defined at $(0, 0)$, so integrate first over the annulus $a \le r \le 1$, where $0 < a < 1$, and let $a \to 0$.

60. Recall that the improper integral $\int_0^1 x^{-a} \, dx$ converges if and only if $a < 1$. For which values of a does $\iint_{\mathcal{D}} r^{-a} \, dA$ converge, where $r = \sqrt{x^2 + y^2}$ and $\mathcal{D}$ is the unit disk $x^2 + y^2 \le 1$?

15.5 Applications of Multiple Integrals

This section discusses some applications of multiple integrals. First, we consider quantities (such as mass, charge, and population) that are distributed with a given density δ in $\mathbf{R}^2$ or $\mathbf{R}^3$. In single-variable calculus, we saw that the "total amount" is defined as the integral of density. Similarly, the total amount of a quantity distributed in $\mathbf{R}^2$ or $\mathbf{R}^3$ is defined as the double or triple integral:

$$\boxed{\text{Total amount} = \iint_{\mathcal{D}} \delta(x, y) \, dA \quad \text{or} \quad \iiint_{\mathcal{W}} \delta(x, y, z) \, dV} \qquad \boxed{1}$$

The density function δ has units of "amount per unit area" (or per unit volume).

The intuition behind Eq. (1) is similar to that of the single-variable case. Suppose, for example, that $\delta(x, y)$ is population density (Figure 1). When density is constant, the total population is simply density times area:

$$\text{Population} = \text{density (people/km}^2) \times \text{area (km}^2)$$

To treat variable density in the case, say, of a rectangle $\mathcal{R}$, we divide $\mathcal{R}$ into smaller rectangles $\mathcal{R}_{ij}$ of area $\Delta x \, \Delta y$ on which δ is nearly constant (assuming that δ is continuous on $\mathcal{R}$). The population in $\mathcal{R}_{ij}$ is approximately $\delta(P_{ij}) \, \Delta x \, \Delta y$ for any sample point P_{ij} in

$\mathcal{R}_{ij}$, and the sum of these approximations is a Riemann sum that converges to the double integral:

$$\int_{\mathcal{R}} \delta(x, y) \, dA \approx \sum_i \sum_j \delta(P_{ij}) \Delta x \, \Delta y$$

■ **EXAMPLE 1** **Population Density** The population in a rural area near a river has density

$$\delta(x, y) = 40xe^{0.1y} \text{ people per km}^2$$

How many people live in the region $\mathcal{R}$: $2 \le x \le 6$, $1 \le y \le 3$ (Figure 1)?

Solution The total population is the integral of population density:

$$\iint_{\mathcal{R}} 40xe^{0.1y} \, dA = \int_2^6 \int_1^3 40xe^{0.1y} \, dx \, dy$$

$$= \left(\int_2^6 40x \, dx \right) \left(\int_1^3 e^{0.1y} \, dy \right)$$

$$= \left(20x^2 \Big|_{x=2}^6 \right) \left(10e^{0.1y} \Big|_{y=1}^3 \right) \approx (640)(2.447) \approx 1566 \text{ people} \quad ■$$

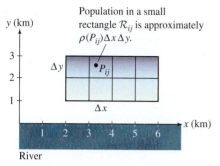

y (km)

Population in a small rectangle $\mathcal{R}_{ij}$ is approximately $\rho(P_{ij})\Delta x \Delta y$.

Δy $\bullet P_{ij}$

Δx

x (km)

River

FIGURE 1 An aerial view of a region near the river.

In the next example, we compute the mass of an object as the integral of mass density. In three dimensions, we justify this computation by dividing $\mathcal{W}$ into boxes $\mathcal{B}_{ijk}$ of volume ΔV that are so small that the mass density is nearly constant on $\mathcal{B}_{ijk}$ (Figure 2). The mass of $\mathcal{B}_{ijk}$ is approximately $\delta(P_{ijk}) \Delta V$, where P_{ijk} is any sample point in $\mathcal{B}_{ijk}$, and the sum of these approximations is a Riemann sum that converges to the triple integral:

$$\iiint_{\mathcal{W}} \delta(x, y, z) \, dV \approx \sum_i \sum_j \sum_k \underbrace{\delta(P_{ijk}) \Delta V}_{\substack{\text{Approximate mass} \\ \text{of } \mathcal{B}_{ijk}}}$$

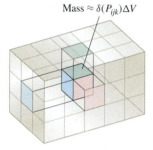

Mass $\approx \delta(P_{ijk})\Delta V$

FIGURE 2 The mass of a small box is approximately $\delta(P_{ijk}) \Delta V$.

When δ is constant, we say that the solid has a **uniform** mass density. In this case, the triple integral has the value δV and the mass is simply $M = \delta V$.

■ **EXAMPLE 2** Let $a > 0$. Find the mass of the "solid bowl" $\mathcal{W}$ consisting of points inside the paraboloid $z = a(x^2 + y^2)$ for $0 \le z \le H$ (Figure 3). Assume a mass density of $\delta(x, y, z) = z$.

Solution Because the bowl is symmetric with respect to the z-axis, we use cylindrical coordinates (r, θ, z). Recall that $r^2 = x^2 + y^2$, so the polar equation of the paraboloid is $z = ar^2$. A point (r, θ, z) lies *above* the paraboloid if $z \ge ar^2$, so it lies *in* the bowl if $ar^2 \le z \le H$. In other words, the bowl is described by

$$0 \le \theta \le 2\pi, \qquad 0 \le r \le \sqrt{\frac{H}{a}}, \qquad ar^2 \le z \le H$$

The mass of the bowl is the integral of mass density:

$$M = \iiint_{\mathcal{W}} \delta(x, y, z) \, dV = \int_{\theta=0}^{2\pi} \int_{r=0}^{\sqrt{H/a}} \int_{z=ar^2}^{H} (z) r \, dz \, dr \, d\theta$$

$$= 2\pi \int_{r=0}^{\sqrt{H/a}} \left(\frac{1}{2}H^2 - \frac{1}{2}a^2r^4 \right) r \, dr$$

$$= 2\pi \left(\frac{H^2 r^2}{4} - \frac{a^2 r^6}{12} \right) \Bigg|_{r=0}^{\sqrt{H/a}}$$

$$= 2\pi \left(\frac{H^3}{4a} - \frac{H^3}{12a} \right) = \frac{\pi H^3}{3a} \qquad ■$$

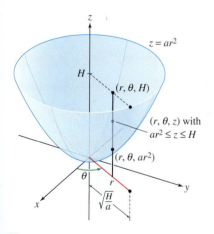

$z = ar^2$

H

(r, θ, H)

(r, θ, z) with $ar^2 \le z \le H$

(r, θ, ar^2)

θ r

$\sqrt{\dfrac{H}{a}}$

x y

DF **FIGURE 3** The paraboloid $z = a(x^2 + y^2)$.

Next, we compute centers of mass. In Section 8.3, we computed centers of mass of laminas (thin plates in the plane), but we had to assume that the mass density is constant. Multiple integration enables us to treat variable mass density. We define the moments of a lamina $\mathcal{D}$ with respect to the coordinate axes:

In $\mathbf{R}^2$, we denote the integral of $x\delta(x, y)$ by M_y and call it the moment with respect to the y-axis because x is the signed distance from (x, y) to the y-axis.

$$M_y = \iint_{\mathcal{D}} x\delta(x, y)\, dA, \qquad M_x = \iint_{\mathcal{D}} y\delta(x, y)\, dA$$

The **center of mass** (COM) is the point $P_{CM} = (x_{CM}, y_{CM})$, where

$$x_{CM} = \frac{M_y}{M}, \qquad y_{CM} = \frac{M_x}{M}$$

$\boxed{2}$

You can think of the coordinates x_{CM} and y_{CM} as **weighted averages**—they are the averages of x and y in which the factor δ assigns a larger coefficient to points with larger mass density.

If $\mathcal{D}$ has uniform mass density (δ constant), then the factors of δ in the numerator and denominator in Eq. (2) cancel, and the center of mass coincides with the **centroid**, defined as the point whose coordinates are the averages of the coordinates over the domain:

$$\overline{x} = \frac{1}{A} \iint_{\mathcal{D}} x\, dA, \qquad \overline{y} = \frac{1}{A} \iint_{\mathcal{D}} y\, dA$$

Here, $A = \iint_{\mathcal{D}} 1\, dA$ is the area of $\mathcal{D}$.

In $\mathbf{R}^3$, we denote the integral of $x\delta(x, y, z)$ by M_{yz} and call it the moment with respect to the yz-plane because x is the signed distance from (x, y, z) to the yz-plane.

In $\mathbf{R}^3$, the moments of a solid region $\mathcal{W}$ are defined not with respect to the axes as in $\mathbf{R}^2$, but with respect to the coordinate planes:

$$M_{yz} = \iiint_{\mathcal{W}} x\delta(x, y, z)\, dV$$

$$M_{xz} = \iiint_{\mathcal{W}} y\delta(x, y, z)\, dV$$

$$M_{xy} = \iiint_{\mathcal{W}} z\delta(x, y, z)\, dV$$

The center of mass is the point $P_{CM} = (x_{CM}, y_{CM}, z_{CM})$ with coordinates

$$x_{CM} = \frac{M_{yz}}{M}, \qquad y_{CM} = \frac{M_{xz}}{M}, \qquad z_{CM} = \frac{M_{xy}}{M}$$

The centroid of $\mathcal{W}$ is the point $P = (\overline{x}, \overline{y}, \overline{z})$, which, as before, coincides with the center of mass when δ is constant:

$$\overline{x} = \frac{1}{V} \iiint_{\mathcal{W}} x\, dV, \qquad \overline{y} = \frac{1}{V} \iiint_{\mathcal{W}} y\, dV, \qquad \overline{z} = \frac{1}{V} \iiint_{\mathcal{W}} z\, dV$$

where $V = \iiint_{\mathcal{W}} 1\, dV$ is the volume of $\mathcal{W}$.

Symmetry can often be used to simplify COM calculations. We say that a region $\mathcal{W}$ in $\mathbf{R}^3$ is symmetric with respect to the xy-plane if $(x, y, -z)$ lies in $\mathcal{W}$ whenever (x, y, z) lies in $\mathcal{W}$. The density δ is symmetric with respect to the xy-plane if

$$\delta(x, y, -z) = \delta(x, y, z)$$

In other words, the mass density is the same at points located symmetrically with respect to the xy-plane. If both $\mathcal{W}$ and δ have this symmetry, then $M_{xy} = 0$ and the COM lies on the xy-plane—that is, $z_{CM} = 0$. Similar remarks apply to the other coordinate axes and to domains in the plane.

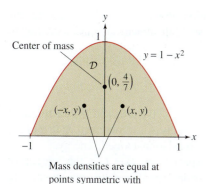

Center of mass

$y = 1 - x^2$

$\left(0, \frac{4}{7}\right)$

$\mathcal{D}$

$(-x, y)$ (x, y)

Mass densities are equal at points symmetric with respect to the y-axis.

FIGURE 4

■ **EXAMPLE 3** **Center of Mass** Find the center of mass of the domain $\mathcal{D}$ bounded by $y = 1 - x^2$ and the x-axis, assuming a mass density of $\delta(x, y) = y$ (Figure 4).

Solution The domain $\mathcal{D}$ is symmetric with respect to the y-axis, and so too is the mass density because $\delta(x, y) = \delta(-x, y) = y$. Therefore, $x_{CM} = 0$. We need only compute y_{CM}:

$$M_x = \iint_{\mathcal{D}} y\delta(x, y)\, dA = \int_{x=-1}^{1} \int_{y=0}^{1-x^2} y^2\, dy\, dx = \int_{x=-1}^{1} \left(\frac{1}{3} y^3 \Big|_{y=0}^{1-x^2}\right) dx$$

$$= \frac{1}{3} \int_{x=-1}^{1} \left(1 - 3x^2 + 3x^4 - x^6\right) dx = \frac{1}{3}\left(2 - 2 + \frac{6}{5} - \frac{2}{7}\right) = \frac{32}{105}$$

$$M = \iint_{\mathcal{D}} \delta(x, y)\, dA = \int_{x=-1}^{1} \int_{y=0}^{1-x^2} y\, dy\, dx = \int_{x=-1}^{1} \left(\frac{1}{2} y^2 \Big|_{y=0}^{1-x^2}\right) dx$$

$$= \frac{1}{2} \int_{x=-1}^{1} \left(1 - 2x^2 + x^4\right) dx = \frac{1}{2}\left(2 - \frac{4}{3} + \frac{2}{5}\right) = \frac{8}{15}$$

Therefore, $y_{CM} = \dfrac{M_x}{M} = \dfrac{32}{105}\left(\dfrac{8}{15}\right)^{-1} = \dfrac{4}{7}$. ■

■ **EXAMPLE 4** Find the center of mass of the solid bowl $\mathcal{W}$ in Example 2 consisting of points inside the paraboloid $z = a(x^2 + y^2)$ for $0 \le z \le H$, assuming a mass density of $\delta(x, y, z) = z$.

Solution The domain is shown in Figure 3 above.

***Step 1.* Use symmetry.**
The bowl $\mathcal{W}$ and the mass density are both symmetric with respect to the z-axis, so we can expect the COM to lie on the z-axis. In fact, the density satisfies both $\delta(-x, y, z) = \delta(x, y, z)$ and $\delta(x, -y, z) = \delta(x, y, z)$, and thus, we have $M_{xz} = M_{yz} = 0$. It remains to compute the moment M_{xy}.

***Step 2.* Compute the moment.**
In Example 2, we described the bowl in cylindrical coordinates as

$$0 \le \theta \le 2\pi, \qquad 0 \le r \le \sqrt{\frac{H}{a}}, \qquad ar^2 \le z \le H$$

and we computed the bowl's mass as $M = \dfrac{\pi H^3}{3a}$. The moment is

$$M_{xy} = \iiint_{\mathcal{W}} z\,\delta(x, y, z)\, dV = \iiint_{\mathcal{W}} z^2\, dV = \int_{\theta=0}^{2\pi} \int_{r=0}^{\sqrt{H/a}} \int_{z=ar^2}^{H} z^2 r\, dz\, dr\, d\theta$$

$$= 2\pi \int_{r=0}^{\sqrt{H/a}} \left(\frac{1}{3} H^3 - \frac{1}{3} a^3 r^6\right) r\, dr$$

$$= 2\pi \left(\frac{1}{6} H^3 r^2 - \frac{1}{24} a^3 r^8\right) \Bigg|_{r=0}^{\sqrt{H/a}}$$

$$= 2\pi \left(\frac{H^4}{6a} - \frac{a^3 H^4}{24a^4}\right) = \frac{\pi H^4}{4a}$$

The z-coordinate of the center of mass is

$$z_{CM} = \frac{M_{xy}}{V} = \frac{\pi H^4/(4a)}{\pi H^3/(3a)} = \frac{3}{4} H$$

and the center of mass itself is $\left(0, 0, \frac{3}{4} H\right)$. ■

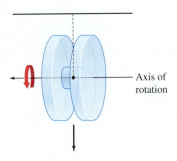

FIGURE 5 A spinning yo-yo has rotational kinetic energy $\frac{1}{2}I\omega^2$, where I is the moment of inertia and ω is the angular velocity. See Exercise 47.

Moments of inertia are used to analyze rotation about an axis. For example, the spinning yo-yo in Figure 5 rotates about its center as it falls downward, and according to physics, it has a rotational kinetic energy equal to

$$\text{Rotational KE} = \frac{1}{2}I\omega^2$$

Here, ω is the angular velocity (in radians per second) about this axis and I is the **moment of inertia** with respect to the axis of rotation. The quantity I is a rotational analog of the mass m, which appears in the expression $\frac{1}{2}mv^2$ for translational kinetic energy.

By definition, the moment of inertia with respect to an axis L is the integral of "distance squared from the axis," weighted by mass density. We confine our attention to the coordinate axes. Thus, for a lamina in the plane $\mathbf{R}^2$, we define the moments of inertia

$$I_x = \iint_{\mathcal{D}} y^2 \delta(x, y)\, dA$$

$$I_y = \iint_{\mathcal{D}} x^2 \delta(x, y)\, dA$$

$$I_0 = \iint_{\mathcal{D}} (x^2 + y^2) \delta(x, y)\, dA$$

3

The quantity I_0 is called the **polar moment of inertia.** It is the moment of inertia relative to the z-axis, because $x^2 + y^2$ is the square of the distance from a point in the xy-plane to the z-axis. Notice that $I_0 = I_x + I_y$.

For a solid object occupying the region $\mathcal{W}$ in $\mathbf{R}^3$,

$$I_x = \iiint_{\mathcal{W}} (y^2 + z^2)\delta(x, y, z)\, dV$$

$$I_y = \iiint_{\mathcal{W}} (x^2 + z^2)\delta(x, y, z)\, dV$$

$$I_z = \iiint_{\mathcal{W}} (x^2 + y^2)\delta(x, y, z)\, dV$$

Moments of inertia have units of mass times length squared.

■ **EXAMPLE 5** A lamina $\mathcal{D}$ of uniform mass density and total mass M kilograms occupies the region between $y = 1 - x^2$ and the x-axis (in meters). Calculate the rotational KE if $\mathcal{D}$ rotates with angular velocity $\omega = 4$ radians per second (rad/s) about:

(a) the x-axis.   **(b)** the z-axis.

Solution The lamina is shown in Figure 6. To find the rotational kinetic energy about the x- and z-axes, we need to compute I_x and I_0, respectively.

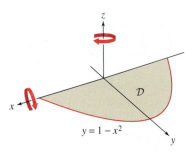

FIGURE 6 Rotating about the z-axis, the plate remains in the xy-plane. About the x-axis, it rotates out of the xy-plane.

Step 1. **Find the mass density.**

The mass density is uniform (i.e., δ is constant), but this does not mean that $\delta = 1$. In fact, the area of $\mathcal{D}$ is $\int_{-1}^{1} (1 - x^2)\, dx = \frac{4}{3}$, so the mass density (mass per unit area) is

$$\delta = \frac{\text{mass}}{\text{area}} = \frac{M}{\frac{4}{3}} = \frac{3M}{4} \text{ kg/m}^2$$

Step 2. **Calculate the moments.**

$$I_x = \int_{-1}^{1} \int_{y=0}^{1-x^2} y^2 \delta\, dy\, dx = \int_{-1}^{1} \frac{1}{3}(1 - x^2)^3 \left(\frac{3M}{4}\right) dx$$

$$= \frac{M}{4} \int_{-1}^{1} (1 - 3x^2 + 3x^4 - x^6)\, dx = \frac{8M}{35} \text{ kg-m}^2$$

To calculate I_0, we use the relation $I_0 = I_x + I_y$. We have

$$I_y = \int_{-1}^{1} \int_{y=0}^{1-x^2} x^2 \delta \, dy \, dx = \left(\frac{3M}{4}\right) \int_{-1}^{1} x^2(1-x^2) \, dx = \frac{M}{5} \text{ kg-m}^2$$

and thus,

$$I_0 = I_x + I_y = \frac{8M}{35} + \frac{M}{5} = \frac{3M}{7} \text{ kg-m}^2 \qquad \boxed{4}$$

CAUTION The relation

$$I_0 = I_x + I_y$$

is valid for a lamina in the xy-plane. However, there is no relation of this type for solid objects in $\mathbf{R}^3$.

Step 3. Calculate kinetic energy.
Assuming an angular velocity of $\omega = 4$ rad/s,

$$\text{Rotational KE about } x\text{-axis} = \frac{1}{2}I_x\omega^2 = \frac{1}{2}\left(\frac{8M}{35}\right)4^2 \approx 1.8M \text{ joules}$$

$$\text{Rotational KE about } z\text{-axis} = \frac{1}{2}I_0\omega^2 = \frac{1}{2}\left(\frac{3M}{7}\right)4^2 \approx 3.4M \text{ joules}$$

The unit of energy is the joule (J), equal to 1 kg-m²/s². ∎

A point mass m located a distance r from an axis has moment of inertia $I = mr^2$ with respect to that axis. Given an extended object of total mass M (not necessarily a point mass) whose moment of inertia with respect to the axis is I, we define the **radius of gyration** by $r_g = (I/M)^{1/2}$. With this definition, the moment of inertia would not change if all of the mass of the object were concentrated at a point located a distance r_g from the axis.

■ **EXAMPLE 6** **Radius of Gyration of a Hemisphere** Find the radius of gyration about the z-axis of the solid hemisphere $\mathcal{W}$ defined by $x^2 + y^2 + z^2 = R^2, 0 \le z \le 1$, assuming a mass density of $\delta(x, y, z) = z$ kg/m³.

Solution To compute the radius of gyration about the z-axis, we must compute I_z and the total mass M. We use spherical coordinates:

$$x^2 + y^2 = (\rho\cos\theta\sin\phi)^2 + (\rho\sin\theta\sin\phi)^2 = \rho^2\sin^2\phi, \qquad z = \rho\cos\phi$$

$$I_z = \iiint_{\mathcal{W}} (x^2+y^2)z \, dV = \int_{\theta=0}^{2\pi}\int_{\phi=0}^{\pi/2}\int_{\rho=0}^{R}(\rho^2\sin^2\phi)(\rho\cos\phi)\rho^2\sin\phi \, d\rho \, d\phi \, d\theta$$

$$= 2\pi\left(\int_0^R \rho^5 \, d\rho\right)\left(\int_{\phi=0}^{\pi/2}\sin^3\phi\cos\phi \, d\phi\right)$$

$$= 2\pi\left(\frac{R^6}{6}\right)\left(\frac{\sin^4\phi}{4}\Big|_0^{\pi/2}\right) = \frac{\pi R^6}{12} \text{ kg-m}^2$$

$$M = \iiint_{\mathcal{W}} z \, dV = \int_{\theta=0}^{2\pi}\int_{\phi=0}^{\pi/2}\int_{\rho=0}^{R}(\rho\cos\phi)\rho^2\sin\phi \, d\rho \, d\phi \, d\theta$$

$$= \left(\int_{\rho=0}^{R}\rho^3 \, d\rho\right)\left(\int_{\phi=0}^{\pi/2}\cos\phi\sin\phi \, d\phi\right)\left(\int_{\theta=0}^{2\pi}d\theta\right) = \frac{\pi R^4}{4} \text{ kg}$$

The radius of gyration is $r_g = (I_z/M)^{1/2} = (R^2/3)^{1/2} = R/\sqrt{3}$ m. ∎

Probability Theory

In Section 7.8, we discussed how probabilities can be represented as areas under curves (Figure 7). Recall that a *random variable* X is defined as the outcome of an experiment or measurement whose value is not known in advance. The probability that the value of X lies between a and b is denoted $P(a \le X \le b)$. Furthermore, X is a *continuous random*

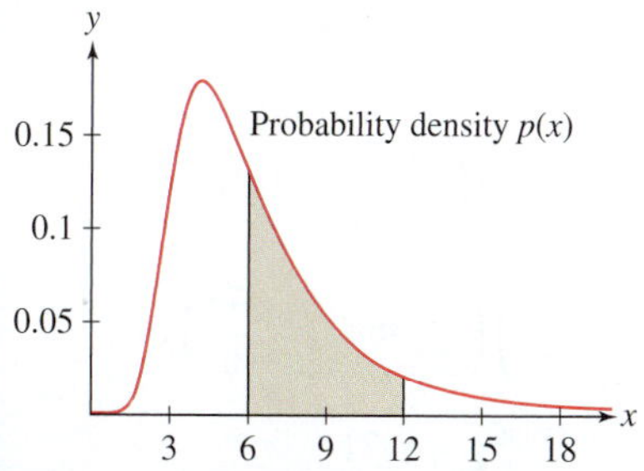

FIGURE 7 The shaded area is the probability that X lies between 6 and 12.

variable if there is a continuous function p of one variable, called the *probability density function*, such that (Figure 7)

$$P(a \leq X \leq b) = \int_a^b p(x)\,dx$$

Double integration enters the picture when we compute "joint probabilities" of two random variables X and Y. We let

$$P(a \leq X \leq b;\ c \leq Y \leq d)$$

denote the probability that X and Y satisfy

$$a \leq X \leq b, \qquad c \leq Y \leq d$$

For example, if X is the height (in centimeters) and Y is the weight (in kilograms) in a certain population, then

$$P(160 \leq X \leq 170;\ 52 \leq Y \leq 63)$$

is the probability that a person chosen at random has height between 160 and 170 cm and weight between 52 and 63 kg.

We say that X and Y are jointly continuous if there is a continuous function $p(x, y)$, called the **joint probability density function** (or simply the joint density), such that for all intervals $[a, b]$ and $[c, d]$ (Figure 8),

$$P(a \leq X \leq b; c \leq Y \leq d) = \int_{x=a}^b \int_{y=c}^d p(x, y)\,dy\,dx$$

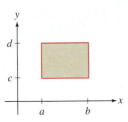

FIGURE 8 The probability $P(a \leq X \leq b; c \leq Y \leq d)$ is equal to the integral of $p(x, y)$ over the rectangle.

← ··· *REMINDER Conditions on a probability density function:*

- $p(x) \geq 0$
- $p(x)$ satisfies $\displaystyle\int_{-\infty}^{\infty} p(x)\,dx = 1$

In the margin, we recall two conditions that a probability density function must satisfy. Joint density functions must satisfy similar conditions: First, $p(x, y) \geq 0$ for all x and y (because probabilities cannot be negative), and second,

$$\int_{-\infty}^{\infty} \int_{-\infty}^{\infty} p(x, y)\,dy\,dx = 1 \qquad \boxed{5}$$

This is often called the **normalization condition**. It holds because it is certain (the probability is 1) that X and Y each take on some value between $-\infty$ and ∞.

■ **EXAMPLE 7** Without proper maintenance, the time to failure (in months) of two sensors in an aircraft are random variables X and Y with joint density

$$p(x, y) = \begin{cases} \dfrac{1}{864} e^{-x/24 - y/36} & \text{for } x \geq 0, y \geq 0 \\[2mm] 0 & \text{otherwise} \end{cases}$$

What is the probability that neither sensor functions after 2 years?

Solution The problem asks for the probability $P(0 \leq X \leq 24;\ 0 \leq Y \leq 24)$:

$$\int_{x=0}^{24} \int_{y=0}^{24} p(x, y)\,dy\,dx = \frac{1}{864} \int_{x=0}^{24} \int_{y=0}^{24} e^{-x/24 - y/36}\,dy\,dx$$

$$= \frac{1}{864} \left(\int_{x=0}^{24} e^{-x/24}\,dx \right) \left(\int_{y=0}^{24} e^{-y/36}\,dy \right)$$

$$= \frac{1}{864} \left(-24 e^{-x/24} \Big|_0^{24} \right) \left(-36 e^{-y/36} \Big|_0^{24} \right)$$

$$= \left(1 - e^{-1} \right)\left(1 - e^{-24/36} \right) \approx 0.31$$

There is a 31% chance that neither sensor will function after 2 years. ■

More generally, we can compute the probability that X and Y satisfy conditions of various types. For example, $P(X + Y \le M)$ denotes the probability that the sum $X + Y$ is at most M. This probability is equal to the integral

$$P(X + Y \le M) = \iint_{\mathcal{D}} p(x, y)\, dy\, dx$$

where $\mathcal{D} = \{(x, y) : x + y \le M\}$.

■ **EXAMPLE 8** Calculate the probability that $X + Y \le 3$, where X and Y have joint probability density

$$p(x, y) = \begin{cases} \dfrac{1}{81}(2xy + 2x + y) & \text{for } 0 \le x \le 3,\ 0 \le y \le 3 \\ 0 & \text{otherwise} \end{cases}$$

Solution The probability density function $p(x, y)$ is nonzero only on the square in Figure 9. Within that square, the inequality $x + y \le 3$ holds only on the shaded triangle, so the probability that $X + Y \le 3$ is equal to the integral of $p(x, y)$ over the triangle:

$$\int_{x=0}^{3} \int_{y=0}^{3-x} p(x, y)\, dy\, dx = \frac{1}{81} \int_{x=0}^{3} \left(xy^2 + \frac{1}{2}y^2 + 2xy \right) \Big|_{y=0}^{3-x} dx$$

$$= \frac{1}{81} \int_{x=0}^{3} \left(x^3 - \frac{15}{2}x^2 + 12x + \frac{9}{2} \right) dx$$

$$= \frac{1}{81} \left(\frac{1}{4}3^4 - \frac{5}{2}3^3 + 6(3^2) + \frac{9}{2}(3) \right) = \frac{1}{4} \qquad ■$$

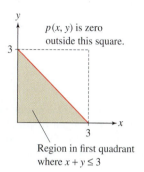

$p(x, y)$ is zero outside this square.

Region in first quadrant where $x + y \le 3$

FIGURE 9

15.5 SUMMARY

• If the mass density is constant, then the center of mass coincides with the *centroid*, whose coordinates $\overline{x}$, $\overline{y}$ (and $\overline{z}$ in three dimensions) are the average values of x, y, and z over the domain. For a domain in $\mathbf{R}^2$,

$$\overline{x} = \frac{1}{A} \iint_{\mathcal{D}} x\, dA, \qquad \overline{y} = \frac{1}{A} \iint_{\mathcal{D}} y\, dA, \qquad A = \iint_{\mathcal{D}} 1\, dA$$

	In $\mathbf{R}^2$	In $\mathbf{R}^3$
Total mass	$M = \iint_{\mathcal{D}} \delta(x, y)\, dA$	$M = \iiint_{\mathcal{W}} \delta(x, y, z)\, dV$
Moments	$M_x = \iint_{\mathcal{D}} y\delta(x, y)\, dA$ $M_y = \iint_{\mathcal{D}} x\delta(x, y)\, dA$	$M_{yz} = \iiint_{\mathcal{W}} x\delta(x, y, z)\, dV$ $M_{xz} = \iiint_{\mathcal{W}} y\delta(x, y, z)\, dV$ $M_{xy} = \iiint_{\mathcal{W}} z\delta(x, y, z)\, dV$
Center of mass	$x_{CM} = \dfrac{M_y}{M}, \quad y_{CM} = \dfrac{M_x}{M}$	$x_{CM} = \dfrac{M_{yz}}{M}, \quad y_{CM} = \dfrac{M_{xz}}{M}, \quad z_{CM} = \dfrac{M_{xy}}{M}$
Moments of inertia	$I_x = \iint_{\mathcal{D}} y^2\delta(x, y)\, dA$ $I_y = \iint_{\mathcal{D}} x^2\delta(x, y)\, dA$ $I_0 = \iint_{\mathcal{D}} (x^2 + y^2)\delta(x, y)\, dA$ $(I_0 = I_x + I_y)$	$I_x = \iiint_{\mathcal{W}} (y^2 + z^2)\delta(x, y, z)\, dV$ $I_y = \iiint_{\mathcal{W}} (x^2 + z^2)\delta(x, y, z)\, dV$ $I_z = \iiint_{\mathcal{W}} (x^2 + y^2)\delta(x, y, z)\, dV$

- Radius of gyration: $r_g = (I/M)^{1/2}$
- Random variables X and Y have joint probability density function $p(x, y)$ if

$$P(a \le X \le b; c \le Y \le d) = \int_{x=a}^{b} \int_{y=c}^{d} p(x, y)\, dy\, dx$$

- A joint probability density function must satisfy $p(x, y) \ge 0$ and

$$\int_{x=-\infty}^{\infty} \int_{y=-\infty}^{\infty} p(x, y)\, dy\, dx = 1$$

15.5 EXERCISES

Preliminary Questions

1. What is the mass density $\delta(x, y, z)$ of a solid of volume 5 m^3 with uniform mass density and total mass 25 kg?

2. A domain $\mathcal{D}$ in $\mathbf{R}^2$ with uniform mass density is symmetric with respect to the y-axis. Which of the following are true?

(a) $x_{CM} = 0$ **(b)** $y_{CM} = 0$ **(c)** $I_x = 0$ **(d)** $I_y = 0$

3. If $p(x, y)$ is the joint probability density function of random variables X and Y, what does the double integral of $p(x, y)$ over $[0, 1] \times [0, 1]$ represent? What does the integral of $p(x, y)$ over the triangle bounded by $x = 0$, $y = 0$, and $x + y = 1$ represent?

Exercises

1. Find the total mass of the square $0 \le x \le 1, 0 \le y \le 1$ assuming a mass density of

$$\delta(x, y) = x^2 + y^2$$

2. Calculate the total mass of a plate bounded by $y = 0$ and $y = x^{-1}$ for $1 \le x \le 4$ (in meters) assuming a mass density of $\delta(x, y) = y/x$ kg/m^2.

3. Find the total charge in the region under the graph of $y = 4e^{-x^2/2}$ for $0 \le x \le 10$ (in centimeters) assuming a charge density of $\delta(x, y) = 10^{-6}xy$ coulombs per square centimeter (C/cm^2).

4. Find the total population within a 4-km radius of the city center (located at the origin) assuming a population density of $\delta(x, y) = 2000(x^2 + y^2)^{-0.2}$ people per square kilometer.

5. Find the total population within the sector $2|x| \le y \le 8$ assuming a population density of $\delta(x, y) = 100e^{-0.1y}$ people per square kilometer.

6. Find the total mass of the solid region $\mathcal{W}$ defined by $x \ge 0, y \ge 0$, $x^2 + y^2 \le 4$, and $x \le z \le 32 - x$ (in centimeters) assuming a mass density of $\delta(x, y, z) = 6y$ g/cm^3.

7. Calculate the total charge of the solid ball $x^2 + y^2 + z^2 \le 5$ (in centimeters) assuming a charge density (in coulombs per cubic centimeter) of

$$\delta(x, y, z) = (3 \cdot 10^{-8})(x^2 + y^2 + z^2)^{1/2}$$

8. Compute the total mass of the plate in Figure 10 assuming a mass density of $f(x, y) = x^2/(x^2 + y^2)$ g/cm^2.

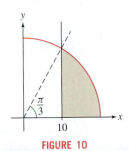

FIGURE 10

9. Assume that the density of the atmosphere as a function of altitude h (in kilometers) above sea level is $\delta(h) = ae^{-bh}$ kg/km^3, where $a = 1.225 \times 10^9$ and $b = 0.13$. Calculate the total mass of the atmosphere contained in the cone-shaped region $\sqrt{x^2 + y^2} \le h \le 3$.

10. Calculate the total charge on a plate $\mathcal{D}$ in the shape of the ellipse with the polar equation

$$r^2 = \left(\frac{1}{6}\sin^2\theta + \frac{1}{9}\cos^2\theta \right)^{-1}$$

with the disk $x^2 + y^2 \le 1$ removed (Figure 11) assuming a charge density of $\rho(r, \theta) = 3r^{-4}$ C/cm^2.

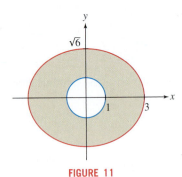

FIGURE 11

In Exercises 11–14, find the centroid of the given region assuming the density $\delta(x, y) = 1$.

11. Region bounded by $y = 1 - x^2$ and $y = 0$

12. Region bounded by $y^2 = x + 4$ and $x = 4$

13. Quarter circle $x^2 + y^2 \le R^2$, $x \ge 0$, $y \ge 0$

14. Infinite lamina bounded by the x- and y-axes and the graph of $y = e^{-x}$

15. *CAS*   Use a computer algebra system to compute numerically the centroid of the shaded region in Figure 12 bounded by $r^2 = \cos 2\theta$ for $x \ge 0$.

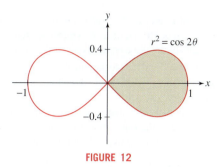

FIGURE 12

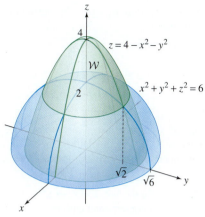

FIGURE 15

16. Show that the centroid of the sector in Figure 13 has y-coordinate

$$\bar{y} = \left(\frac{2R}{3}\right)\left(\frac{\sin \alpha}{\alpha}\right)$$

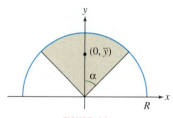

FIGURE 13

In Exercises 17–19, find the centroid of the given solid region assuming a density of $\delta(x, y) = 1$.

17. Hemisphere $x^2 + y^2 + z^2 \le R^2$, $z \ge 0$

18. Region bounded by the xy-plane, the cylinder $x^2 + y^2 = R^2$, and the plane $x/R + z/H = 1$, where $R > 0$ and $H > 0$

19. The "ice cream cone" region $\mathcal{W}$ bounded, in spherical coordinates, by the cone $\phi = \pi/3$ and the sphere $\rho = 2$

20. Show that the z-coordinate of the centroid of the tetrahedron bounded by the coordinate planes and the plane

$$\frac{x}{a} + \frac{y}{b} + \frac{z}{c} = 1$$

in Figure 14 is $\bar{z} = c/4$. Conclude by symmetry that the centroid is $(a/4, b/4, c/4)$.

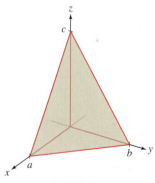

FIGURE 14

21. Find the centroid of the region $\mathcal{W}$ lying above the unit sphere $x^2 + y^2 + z^2 = 6$ and below the paraboloid $z = 4 - x^2 - y^2$ (Figure 15).

22. Let $R > 0$ and $H > 0$, and let $\mathcal{W}$ be the upper half of the ellipsoid $x^2 + y^2 + (Rz/H)^2 = R^2$, where $z \ge 0$ (Figure 16). Find the centroid of $\mathcal{W}$ and show that it depends on the height H but not on the radius R.

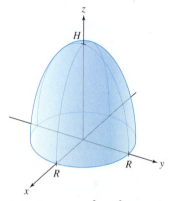

FIGURE 16 Upper half of ellipsoid $x^2 + y^2 + (Rz/H)^2 = R^2$, $z \ge 0$.

In Exercises 23–26, find the center of mass of the region with the given mass density δ.

23. Region bounded by $y = 4 - x$, $x = 0$, $y = 0$; $\quad \delta(x, y) = x$

24. Region bounded by $y^2 = x + 4$ and $x = 0$; $\quad \delta(x, y) = |y|$

25. Region $|x| + |y| \le 1$; $\quad \delta(x, y) = (x + 1)(y + 1)$

26. Semicircle $x^2 + y^2 \le R^2$, $y \ge 0$; $\quad \delta(x, y) = y$

27. Find the z-coordinate of the center of mass of the first octant of the unit sphere with mass density $\delta(x, y, z) = y$ (Figure 17).

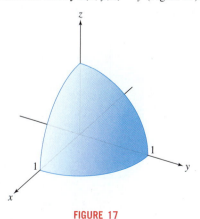

FIGURE 17

28. Find the center of mass of a cylinder of radius 2 and height 4 and mass density e^{-z}, where z is the height above the base.

29. Let $\mathcal{R}$ be the rectangle $[-a, a] \times [b, -b]$ with uniform density and total mass M. Calculate:

(a) The mass density δ of $\mathcal{R}$

(b) I_x and I_0

(c) The radius of gyration about the x-axis

30. Calculate I_x and I_0 for the rectangle in Exercise 29 assuming a mass density of $\delta(x, y) = x$.

31. Calculate I_0 and I_x for the disk $\mathcal{D}$ defined by $x^2 + y^2 \leq 16$ (in meters), with total mass 1000 kg and uniform mass density. *Hint:* Calculate I_0 first and observe that $I_0 = 2I_x$. Express your answer in the correct units.

32. Calculate I_x and I_y for the half-disk $x^2 + y^2 \leq R^2, x \geq 0$ (in meters), with total mass M kilograms and uniform mass density.

In Exercises 33–36, let $\mathcal{D}$ be the triangular domain bounded by the coordinate axes and the line $y = 3 - x$, with mass density $\delta(x, y) = y$. Compute the given quantities.

33. Total mass

34. Center of mass

35. I_x

36. I_0

In Exercises 37–40, let $\mathcal{D}$ be the domain between the line $y = bx/a$ and the parabola $y = bx^2/a^2$, where $a, b > 0$. Assume the mass density is $\delta(x, y) = 1$ for Exercise 37 and $\delta(x, y) = xy$ for Exercises 38–40. Compute the given quantities.

37. Centroid

38. Center of mass

39. I_x

40. I_0

41. Calculate the moment of inertia I_x of the disk $\mathcal{D}$ defined by $x^2 + y^2 \leq R^2$ (in meters), with total mass M kg. How much kinetic energy (in joules) is required to rotate the disk about the x-axis with angular velocity 10 rad/s?

42. Calculate the moment of inertia I_z of the box $\mathcal{W} = [-a, a] \times [-a, a] \times [0, H]$ assuming that $\mathcal{W}$ has total mass M.

43. Show that the moment of inertia of a sphere of radius R of total mass M with uniform mass density about any axis passing through the center of the sphere is $\frac{2}{5}MR^2$. Note that the mass density of the sphere is $\delta = M/\left(\frac{4}{3}\pi R^3\right)$.

44. Use the result of Exercise 43 to calculate the radius of gyration of a uniform sphere of radius R about any axis through the center of the sphere.

In Exercises 45 and 46, prove the formula for the right circular cylinder in Figure 18.

45. $I_z = \frac{1}{2}MR^2$

46. $I_x = \frac{1}{4}MR^2 + \frac{1}{12}MH^2$

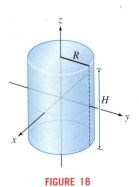

FIGURE 18

47. The yo-yo in Figure 19 is made up of two disks of radius $r = 3$ cm and an axle of radius $b = 1$ cm. Each disk has mass $M_1 = 20$ g, and the axle has mass $M_2 = 5$ g.

(a) Use the result of Exercise 45 to calculate the moment of inertia I of the yo-yo with respect to the axis of symmetry. Note that I is the sum of the moments of the three components of the yo-yo.

(b) The yo-yo is released and falls to the end of a 100-cm string, where it spins with angular velocity ω. The total mass of the yo-yo is $m = 45$ g, so the potential energy lost is PE $= mgh = (45)(980)100$ g-cm^2/s^2. Find ω using the fact that the potential energy is the sum of the rotational kinetic energy and the translational kinetic energy and that the velocity $v = b\omega$ since the string unravels at this rate.

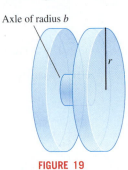

FIGURE 19

48. Calculate I_z for the solid region $\mathcal{W}$ inside the hyperboloid $x^2 + y^2 = z^2 + 1$ between $z = 0$ and $z = 1$.

49. Calculate $P(0 \leq X \leq 2; 1 \leq Y \leq 2)$, where X and Y have joint probability density function

$$p(x, y) = \begin{cases} \frac{1}{72}(2xy + 2x + y) & \text{if } 0 \leq x \leq 4 \text{ and } 0 \leq y \leq 2 \\ 0 & \text{otherwise} \end{cases}$$

50. Calculate the probability that $X + Y \leq 2$ for random variables with joint probability density function as in Exercise 49.

51. The lifetime (in months) of two components in a certain device are random variables X and Y that have joint probability distribution function

$$p(x, y) = \begin{cases} \frac{1}{9216}(48 - 2x - y) & \text{if } x \geq 0, y \geq 0, 2x + y \leq 48 \\ 0 & \text{otherwise} \end{cases}$$

Calculate the probability that both components function for at least 12 months without failing. Note that $p(x, y)$ is nonzero only within the triangle bounded by the coordinate axes and the line $2x + y = 48$ shown in Figure 20.

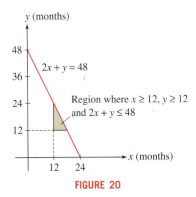

FIGURE 20

52. Find a constant C such that

$$p(x, y) = \begin{cases} Cxy & \text{if } 0 \leq x \text{ and } 0 \leq y \leq 1 - x \\ 0 & \text{otherwise} \end{cases}$$

is a joint probability density function. Then calculate:

(a) $P\left(X \leq \frac{1}{2}; Y \leq \frac{1}{4}\right)$

(b) $P(X \geq Y)$

53. Find a constant C such that

$$p(x, y) = \begin{cases} Cy & \text{if } 0 \le x \le 1 \text{ and } x^2 \le y \le x \\ 0 & \text{otherwise} \end{cases}$$

is a joint probability density function. Then calculate the probability that $Y \ge X^{3/2}$.

54. Numbers X and Y between 0 and 1 are chosen randomly. The joint probability density is $p(x, y) = 1$ if $0 \le x \le 1$ and $0 \le y \le 1$, and $p(x, y) = 0$ otherwise. Calculate the probability P that the product XY is at least $\frac{1}{2}$.

55. According to quantum mechanics, the x- and y-coordinates of a particle confined to the region $\mathcal{R} = [0, 1] \times [0, 1]$ are random variables with joint probability density function

$$p(x, y) = \begin{cases} C \sin^2(2\pi \ell x) \sin^2(2\pi n y) & \text{if } (x, y) \in \mathcal{R} \\ 0 & \text{otherwise} \end{cases}$$

The integers ℓ and n determine the energy of the particle, and C is a constant.

(a) Find the constant C.

(b) Calculate the probability that a particle with $\ell = 2$, $n = 3$ lies in the region $\left[0, \frac{1}{4}\right] \times \left[0, \frac{1}{8}\right]$.

56. The wave function for the 1s state of an electron in the hydrogen atom is

$$\psi_{1s}(\rho) = \frac{1}{\sqrt{\pi a_0^3}} e^{-\rho/a_0}$$

where a_0 is the Bohr radius. The probability of finding the electron in a region $\mathcal{W}$ of $\mathbf{R}^3$ is equal to

$$\iiint_{\mathcal{W}} p(x, y, z) \, dV$$

where, in spherical coordinates,

$$p(\rho) = |\psi_{1s}(\rho)|^2$$

Use integration in spherical coordinates to show that the probability of finding the electron at a distance greater than the Bohr radius is equal to $5/e^2 \approx 0.677$. (The Bohr radius is $a_0 = 5.3 \times 10^{-11}$ m, but this value is not needed.)

57. According to Coulomb's Law, the attractive force between two electric charges of magnitude q_1 and q_2 separated by a distance r is

kq_1q_2/r^2 (k is a constant). Let F be the net force on a charged particle P of charge Q coulombs located d centimeters above the center of a circular disk of radius R, with a uniform charge distribution of density ρ coulombs per square meter (Figure 21). By symmetry, F acts in the vertical direction.

(a) Let $\mathcal{R}$ be a small polar rectangle of size $\Delta r \times \Delta \theta$ located at distance r. Show that $\mathcal{R}$ exerts a force on P whose vertical component is

$$\left(\frac{k\rho Q d}{(r^2 + d^2)^{3/2}} \right) r \, \Delta r \, \Delta \theta$$

(b) Explain why F is equal to the following double integral, and evaluate:

$$F = k\rho Q d \int_0^{2\pi} \int_0^R \frac{r \, dr \, d\theta}{(r^2 + d^2)^{3/2}}$$

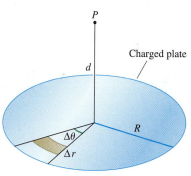

FIGURE 21

58. Let $\mathcal{D}$ be the annular region

$$-\frac{\pi}{2} \le \theta \le \frac{\pi}{2}, \quad a \le r \le b$$

where $b > a > 0$. Assume that $\mathcal{D}$ has a uniform charge distribution of ρ coulombs per square meter. Let F be the net force on a charged particle of charge Q coulombs located at the origin (by symmetry, F acts along the x-axis).

(a) Argue as in Exercise 57 to show that

$$F = k\rho Q \int_{\theta=-\pi/2}^{\pi/2} \int_{r=a}^b \left(\frac{\cos\theta}{r^2} \right) r \, dr \, d\theta$$

(b) Compute F.

Further Insights and Challenges

59. Let $\mathcal{D}$ be the domain in Figure 22. Assume that $\mathcal{D}$ is symmetric with respect to the y-axis; that is, both $g_1(x)$ and $g_2(x)$ are even functions.

(a) Prove that the centroid lies on the y-axis—that is, that $\overline{x} = 0$.

(b) Show that if the mass density satisfies $\delta(-x, y) = \delta(x, y)$, then $M_y = 0$ and $x_{CM} = 0$.

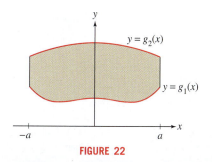

FIGURE 22

60. Pappus's Theorem Let A be the area of the region $\mathcal{D}$ between two graphs $y = g_1(x)$ and $y = g_2(x)$ over the interval $[a, b]$, where $g_2(x) \ge g_1(x) \ge 0$. Prove Pappus's Theorem: The volume of the solid obtained by revolving $\mathcal{D}$ about the x-axis is $V = 2\pi A\overline{y}$, where $\overline{y}$ is the y-coordinate of the centroid of $\mathcal{D}$ (the average of the y-coordinate). *Hint:* Show that

$$A\overline{y} = \int_{x=a}^b \int_{y=g_1(x)}^{g_2(x)} y \, dy \, dx$$

61. Use Pappus's Theorem in Exercise 60 to show that the torus obtained by revolving a circle of radius b centered at $(0, a)$ about the x-axis (where $b < a$) has volume $V = 2\pi^2 ab^2$.

62. Use Pappus's Theorem to compute $\overline{y}$ for the upper half of the disk $x^2 + y^2 \le a^2$, $y \ge 0$. *Hint:* The disk revolved about the x-axis is a sphere.

63. Parallel-Axis Theorem Let $\mathcal{W}$ be a region in $\mathbf{R}^3$ with center of mass at the origin. Let I_z be the moment of inertia of $\mathcal{W}$ about the z-axis, and let I_h be the moment of inertia about the vertical axis through a point $P = (a, b, 0)$, where $h = \sqrt{a^2 + b^2}$. By definition,

$$I_h = \iiint_{\mathcal{W}} ((x - a)^2 + (y - b)^2)\delta(x, y, z)\, dV$$

Prove the Parallel-Axis Theorem: $I_h = I_z + Mh^2$.

64. Let $\mathcal{W}$ be a cylinder of radius 10 cm and height 20 cm, with total mass $M = 500$ g. Use the Parallel-Axis Theorem (Exercise 63) and the result of Exercise 45 to calculate the moment of inertia of $\mathcal{W}$ about an axis that is parallel to and at a distance of 30 cm from the cylinder's axis of symmetry.

15.6 Change of Variables

The formulas for integration in polar, cylindrical, and spherical coordinates are important special cases of the general Change of Variables Formula for multiple integrals. In this section, we discuss the general formula.

Maps from $\mathbf{R}^2$ to $\mathbf{R}^2$

A function $G : X \to Y$ from a set X (the domain) to another set Y is often called a **map** or a **mapping**. For $x \in X$, the element $G(x)$ belongs to Y and is called the **image** of x. The set of all images $G(x)$ is called the image or **range** of G. We denote the image by $G(X)$.

In this section, we consider maps $G : \mathcal{D} \to \mathbf{R}^2$ defined on a domain $\mathcal{D}$ in $\mathbf{R}^2$ (Figure 1). To prevent confusion, we'll often use u, v as our domain variables and x, y for the range. Thus, we will write $G(u, v) = (x(u, v), y(u, v))$, where the components x and y are functions of u and v:

$$x = x(u, v), \qquad y = y(u, v)$$

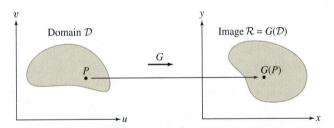

FIGURE 1 G maps $\mathcal{D}$ to $\mathcal{R}$.

One map we are familiar with is the map defining polar coordinates. For this map, we use variables r, θ instead of u, v. The **polar coordinates map** $G : \mathbf{R}^2 \to \mathbf{R}^2$ is defined by

$$G(r, \theta) = (r \cos \theta, r \sin \theta)$$

■ **EXAMPLE 1   Polar Coordinates Map** Describe the image of a polar rectangle $\mathcal{R} = [r_1, r_2] \times [\theta_1, \theta_2]$ under the polar coordinates map.

Solution Referring to Figure 2, we see that:

- A vertical line $r = r_1$ (shown in red) is mapped to the set of points with radial coordinate r_1 and arbitrary angle. This is the circle of radius r_1.

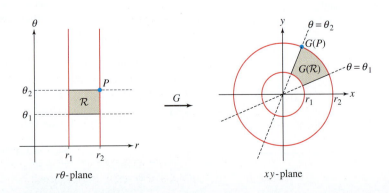

FIGURE 2 The polar coordinates map $G(r, \theta) = (r \cos \theta, r \sin \theta)$.

• A horizontal line $\theta = \theta_1$ (dashed line in the figure) is mapped to the set of points with polar angle θ and arbitrary r-coordinate. This is the line through the origin of angle θ_1.

The image of $\mathcal{R} = [r_1, r_2] \times [\theta_1, \theta_2]$ under the polar coordinates map $G(r, \theta) = (r\cos\theta, r\sin\theta)$ is the polar rectangle in the xy-plane defined by $r_1 \le r \le r_2, \theta_1 \le \theta \le \theta_2$.

■

General mappings can be quite complicated, so it is useful to study the simplest case—linear maps—in detail. A map $G(u, v)$ is **linear** if it has the form

$$G(u, v) = (Au + Cv, \ Bu + Dv) \quad (A, B, C, D \text{ are constants})$$

We can get a clear picture of this linear map by thinking of G as a map from vectors in the uv-plane to vectors in the xy-plane. Then G has the following linearity properties (see Exercise 46):

$$G(u_1 + u_2, v_1 + v_2) = G(u_1, v_1) + G(u_2, v_2) \qquad \boxed{1}$$

$$G(cu, cv) = cG(u, v) \quad (c \text{ any constant}) \qquad \boxed{2}$$

A consequence of these properties is that G maps the parallelogram spanned by any two vectors $\mathbf{a}$ and $\mathbf{b}$ in the uv-plane to the parallelogram spanned by the images $G(\mathbf{a})$ and $G(\mathbf{b})$, as shown in Figure 3.

More generally, *G maps the segment joining any two points P and Q to the segment joining $G(P)$ and $G(Q)$* (see Exercise 47). The grid generated by basis vectors $\mathbf{i} = \langle 1, 0 \rangle$ and $\mathbf{j} = \langle 0, 1 \rangle$ is mapped to the grid generated by the image vectors (Figure 3)

$$\mathbf{r} = G(1, 0) = \langle A, B \rangle$$

$$\mathbf{s} = G(0, 1) = \langle C, D \rangle$$

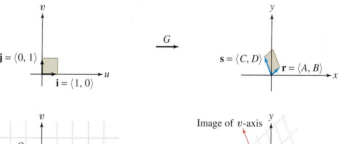

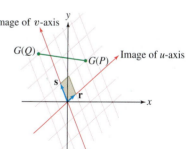

DF **FIGURE 3** A linear mapping G maps a parallelogram to a parallelogram.

■ **EXAMPLE 2** **Image of a Triangle** Find the image of the triangle $\mathcal{T}$ with vertices $(1, 2)$, $(2, 1)$, $(3, 4)$ under the linear map $G(u, v) = (2u - v, u + v)$.

Solution Because G is linear, it maps the segment joining two vertices of $\mathcal{T}$ to the segment joining the images of the two vertices. Therefore, the image of $\mathcal{T}$ is the triangle whose vertices are the images (Figure 4)

$$G(1, 2) = (0, 3), \qquad G(2, 1) = (3, 3), \qquad G(3, 4) = (2, 7) \qquad ■$$

To understand a nonlinear map, it is usually helpful to determine the images of horizontal and vertical lines, as we did for the polar coordinate mapping.

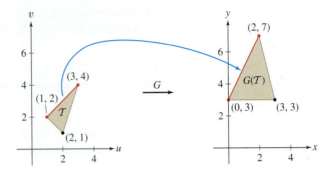

FIGURE 4 The map
$G(u, v) = (2u - v, u + v)$.

■ **EXAMPLE 3** Let $G(u, v) = (uv^{-1}, uv)$ for $u > 0, v > 0$. Determine the images of:

(a) The lines $u = c$ and $v = c$ **(b)** $[1, 2] \times [1, 2]$

Find the inverse map G^{-1}.

Solution In this map, we have $x = uv^{-1}$ and $y = uv$. Thus,

$$xy = u^2, \qquad \frac{y}{x} = v^2 \qquad \boxed{3}$$

(a) By the first part of Eq. (3), G maps a point (c, v) to a point in the xy-plane with $xy = c^2$. In other words, G maps the vertical line $u = c$ to the hyperbola $xy = c^2$. Similarly, by the second part of Eq. (3), the horizontal line $v = c$ is mapped to the set of points where $x/y = c^2$, or $y = c^2 x$, which is the line through the origin of slope c^2. See Figure 5.

The term "curvilinear rectangle" refers to a region bounded on four sides by curves as in Figure 5.

(b) The image of $[1, 2] \times [1, 2]$ is the *curvilinear* rectangle bounded by the four curves that are the images of the lines $u = 1, u = 2$, and $v = 1, v = 2$. By Eq. (3) this region is defined by the inequalities

$$1 \le xy \le 4, \qquad 1 \le \frac{y}{x} \le 4$$

To find G^{-1}, we use Eq. (3) to write $u = \sqrt{xy}$ and $v = \sqrt{y/x}$. Therefore, the inverse map is $G^{-1}(x, y) = \left(\sqrt{xy}, \sqrt{y/x}\right)$. We take positive square roots because $u > 0$ and $v > 0$ on the domain we are considering. ■

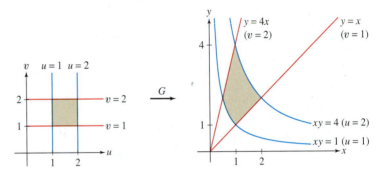

FIGURE 5 The mapping
$G(u, v) = (uv^{-1}, uv)$.

How Area Changes Under a Mapping: The Jacobian Determinant

The **Jacobian determinant** (or simply the Jacobian) of a map

$$G(u, v) = (x(u, v), y(u, v))$$

is the determinant

◀·· *REMINDER The definition of a 2×2 determinant is*

$$\begin{vmatrix} a & b \\ c & d \end{vmatrix} = ad - bc \qquad \boxed{4}$$

$$\mathrm{Jac}(G) = \begin{vmatrix} \dfrac{\partial x}{\partial u} & \dfrac{\partial x}{\partial v} \\[2mm] \dfrac{\partial y}{\partial u} & \dfrac{\partial y}{\partial v} \end{vmatrix} = \frac{\partial x}{\partial u}\frac{\partial y}{\partial v} - \frac{\partial x}{\partial v}\frac{\partial y}{\partial u}$$

The Jacobian $\mathrm{Jac}(G)$ is also denoted $\dfrac{\partial(x, y)}{\partial(u, v)}$. Note that $\mathrm{Jac}(G)$ is a function of u and v.

■ **EXAMPLE 4** Evaluate the Jacobian of $G(u, v) = (u^3 + v, uv)$ at $(u, v) = (2, 1)$.

Solution We have $x = u^3 + v$ and $y = uv$, so

$$\text{Jac}(G) = \frac{\partial(x, y)}{\partial(u, v)} = \begin{vmatrix} \dfrac{\partial x}{\partial u} & \dfrac{\partial x}{\partial v} \\[2mm] \dfrac{\partial y}{\partial u} & \dfrac{\partial y}{\partial v} \end{vmatrix}$$

$$= \begin{vmatrix} 3u^2 & 1 \\ v & u \end{vmatrix} = 3u^3 - v$$

The value of the Jacobian at $(2, 1)$ is $\text{Jac}(G)(2, 1) = 3(2)^3 - 1 = 23$. ■

The Jacobian tells us how area changes under a map G. We can see this most directly in the case of a linear map $G(u, v) = (Au + Cv, Bu + Dv)$.

THEOREM 1 Jacobian of a Linear Map The Jacobian of a linear map

$$G(u, v) = (Au + Cv, Bu + Dv)$$

is *constant* with value

$$\text{Jac}(G) = \begin{vmatrix} A & C \\ B & D \end{vmatrix} = AD - BC \qquad \boxed{5}$$

Under G, the area of a region $\mathcal{D}$ is multiplied by the factor $|\text{Jac}(G)|$; that is,

$$\boxed{\text{Area}(G(\mathcal{D})) = |\text{Jac}(G)|\,\text{Area}(\mathcal{D})} \qquad \boxed{6}$$

Proof Eq. (5) is verified by direct calculation: Because

$$x = Au + Cv \quad \text{and} \quad y = Bu + Dv$$

the partial derivatives in the Jacobian are the constants A, B, C, D.

We sketch a proof of Eq. (6). It certainly holds for the unit rectangle $\mathcal{D} = [1, 0] \times [0, 1]$ because $G(\mathcal{D})$ is the parallelogram spanned by the vectors $\langle A, B \rangle$ and $\langle C, D \rangle$ (Figure 6) and this parallelogram has area

$$|\text{Jac}(G)| = |AD - BC|$$

by Eq. (10) in Section 12.4. Similarly, we can check directly that Eq. (6) holds for arbitrary parallelograms (see Exercise 48). To verify Eq. (6) for a general domain $\mathcal{D}$, we use the fact that $\mathcal{D}$ can be approximated as closely as desired by a union of rectangles in a fine grid of lines parallel to the u- and v-axes. ■

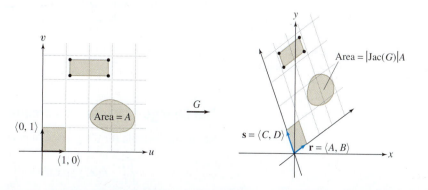

FIGURE 6 A linear map G expands (or shrinks) area by the factor $|\text{Jac}(G)|$.

We cannot expect Eq. (6) to hold for a nonlinear map. In fact, it would not make sense as stated because the value $\text{Jac}(G)(P)$ may vary from point to point. However, it is *approximately true* if the domain $\mathcal{D}$ is small and P is a sample point in $\mathcal{D}$:

$$\text{Area}(G(\mathcal{D})) \approx |\text{Jac}(G)(P)|\text{Area}(\mathcal{D})$$ **7**

This result may be stated more precisely as the limit relation:

$$|\text{Jac}(G)(P)| = \lim_{|\mathcal{D}| \to 0} \frac{\text{Area}(G(\mathcal{D}))}{\text{Area}(\mathcal{D})}$$ **8**

Here, we write $|\mathcal{D}| \to 0$ to indicate the limit as the diameter of $\mathcal{D}$ (the maximum distance between two points in $\mathcal{D}$) tends to zero.

CONCEPTUAL INSIGHT   Although a rigorous proof of Eq. (8) is too technical to include here, we can understand Eq. (7) as an application of linear approximation. Consider a rectangle $\mathcal{R}$ with vertex at $P = (u, v)$ and sides of lengths Δu and Δv, assumed to be small as in Figure 7. The image $G(\mathcal{R})$ is not a parallelogram, but it is approximated well by the parallelogram spanned by the vectors $\mathbf{A}$ and $\mathbf{B}$ in the figure:

$$\mathbf{A} = G(u + \Delta u, v) - G(u, v)$$
$$= (x(u + \Delta u, v) - x(u, v), y(u + \Delta u, v) - y(u, v))$$
$$\mathbf{B} = G(u, v + \Delta v) - G(u, v)$$
$$= (x(u, v + \Delta v) - x(u, v), y(u, v + \Delta v) - y(u, v))$$

The linear approximation applied to the components of G yields

$$\mathbf{A} \approx \left\langle \frac{\partial x}{\partial u}\Delta u, \frac{\partial y}{\partial u}\Delta u \right\rangle$$ **9**

$$\mathbf{B} \approx \left\langle \frac{\partial x}{\partial v}\Delta v, \frac{\partial y}{\partial v}\Delta v \right\rangle$$ **10**

This yields the desired approximation:

$$\text{Area}(G(\mathcal{R})) \approx \left| \det\begin{pmatrix} \mathbf{A} \\ \mathbf{B} \end{pmatrix} \right| = \left| \det\begin{pmatrix} \frac{\partial x}{\partial u}\Delta u & \frac{\partial y}{\partial u}\Delta u \\ \frac{\partial x}{\partial v}\Delta v & \frac{\partial y}{\partial v}\Delta v \end{pmatrix} \right|$$

$$= \left| \frac{\partial x}{\partial u}\frac{\partial y}{\partial v} - \frac{\partial y}{\partial u}\frac{\partial x}{\partial v} \right| \Delta u \, \Delta v$$

$$= |\text{Jac}(G)(P)|\text{Area}(\mathcal{R})$$

since the area of $\mathcal{R}$ is $\Delta u \, \Delta v$.

⟵ REMINDER *Eqs. (9) and (10) use the linear approximations*

$$x(u + \Delta u, v) - x(u, v) \approx \frac{\partial x}{\partial u}\Delta u$$

$$y(u + \Delta u, v) - y(u, v) \approx \frac{\partial y}{\partial u}\Delta u$$

and

$$x(u, v + \Delta v) - x(u, v) \approx \frac{\partial x}{\partial v}\Delta v$$

$$y(u, v + \Delta v) - y(u, v) \approx \frac{\partial y}{\partial v}\Delta v$$

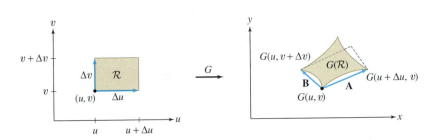

FIGURE 7 The image of a small rectangle under a nonlinear map can be approximated by a parallelogram whose sides are determined by the linear approximation.

The Change of Variables Formula

Recall the formula for integration in polar coordinates:

$$\iint_{\mathcal{D}} f(x, y)\, dx\, dy = \int_{\theta_1}^{\theta_2}\int_{r_1}^{r_2} f(r\cos\theta, r\sin\theta)\, r\, dr\, d\theta$$ **11**

Here, $\mathcal{D}$ is the polar rectangle consisting of points $(x, y) = (r\cos\theta, r\sin\theta)$ in the xy-plane (see Figure 2 above). The domain of integration on the right is the rectangle $\mathcal{R} = [\theta_1, \theta_2] \times [r_1, r_2]$ in the $r\theta$-plane. Thus, $\mathcal{D}$ is the image of the domain on the right under the polar coordinates map.

The general Change of Variables Formula has a similar form. Given a map

$$G : \quad \underset{\text{in } uv\text{-plane}}{\mathcal{D}_0} \quad \to \quad \underset{\text{in } xy\text{-plane}}{\mathcal{D}}$$

◄⋯ **REMINDER** G is called "one-to-one" if $G(P) = G(Q)$ only for $P = Q$.

from a domain in the uv-plane to a domain in the xy-plane (Figure 8), our formula expresses an integral over $\mathcal{D}$ as an integral over $\mathcal{D}_0$. The Jacobian plays the role of the factor r on the right-hand side of Eq. (11).

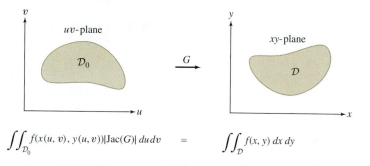

$$\iint_{\mathcal{D}_0} f(x(u, v), y(u, v))|\text{Jac}(G)| \, du \, dv \quad = \quad \iint_{\mathcal{D}} f(x, y) \, dx \, dy$$

FIGURE 8 The Change of Variables Formula expresses a double integral over $\mathcal{D}$ as a double integral over $\mathcal{D}_0$.

A few technical assumptions are necessary. First, we assume that G is one-to-one, at least on the interior of $\mathcal{D}_0$, because we want G to cover the target domain $\mathcal{D}$ just once. We also assume that G is a C^1 **map**, by which we mean that the component functions x and y have continuous partial derivatives. Under these conditions, if $f(x, y)$ is continuous, we have the following result.

> **THEOREM 2** **Change of Variables Formula** Let $G : \mathcal{D}_0 \to \mathcal{D}$ be a C^1 mapping that is one-to-one on the interior of $\mathcal{D}_0$. If $f(x, y)$ is continuous, then
>
> $$\iint_{\mathcal{D}} f(x, y) \, dx \, dy = \iint_{\mathcal{D}_0} f(x(u, v), y(u, v)) \left| \frac{\partial(x, y)}{\partial(u, v)} \right| du \, dv \qquad \boxed{12}$$

Proof We sketch the proof. Observe first that Eq. (12) is *approximately* true if the domains $\mathcal{D}_0$ and $\mathcal{D}$ are small. Let $P = G(P_0)$, where P_0 is any sample point in $\mathcal{D}_0$. Since $f(x, y)$ is continuous, the approximation recalled in the margin together with Eq. (7) yields

$$\iint_{\mathcal{D}} f(x, y) \, dx \, dy \approx f(P)\text{Area}(\mathcal{D})$$
$$\approx f(G(P_0)) \, |\text{Jac}(G)(P_0)| \, \text{Area}(\mathcal{D}_0)$$
$$\approx \iint_{\mathcal{D}_0} f(G(u, v)) \, |\text{Jac}(G)(u, v)| \, du \, dv$$

If $\mathcal{D}$ is not small, divide it into small subdomains $D_j = G(\mathcal{D}_{0j})$ (Figure 9 shows a rectangle divided into smaller rectangles), apply the approximation to each subdomain, and sum:

$$\iint_{\mathcal{D}} f(x, y) \, dx \, dy = \sum_j \iint_{\mathcal{D}_j} f(x, y) \, dx \, dy$$
$$\approx \sum_j \iint_{\mathcal{D}_{0j}} f(G(u, v))) \, |\text{Jac}(G)(u, v)| \, du \, dv$$
$$= \iint_{\mathcal{D}_0} f(G(u, v)) \, |\text{Jac}(G)(u, v)| \, du \, dv$$

Careful estimates show that the error tends to zero as the maximum of the diameters of the subdomains $\mathcal{D}_j$ tends to zero. This yields the Change of Variables Formula. ∎

Eq. (12) is summarized by the symbolic equality

$$dx \, dy = \left| \frac{\partial(x, y)}{\partial(u, v)} \right| du \, dv$$

Recall that $\dfrac{\partial(x, y)}{\partial(u, v)}$ denotes the Jacobian Jac(G).

◄⋯ **REMINDER** If $\mathcal{D}$ is a domain of small diameter, $P \in \mathcal{D}$ is a sample point, and $f(x, y)$ is continuous, then (see Section 15.2)

$$\iint_{\mathcal{D}} f(x, y) \, dx \, dy \approx f(P)\text{Area}(\mathcal{D})$$

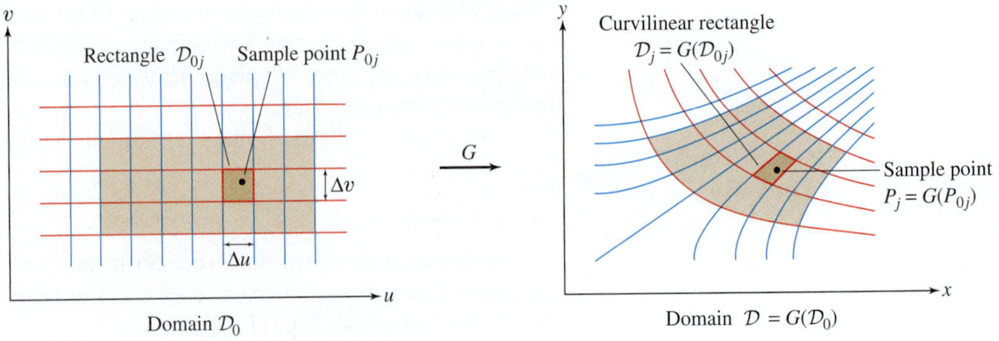

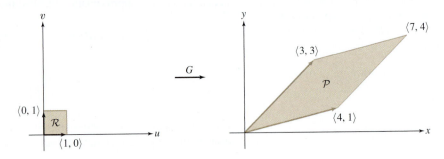

FIGURE 9 G maps a rectangular grid on $\mathcal{D}_0$ to a curvilinear grid on $\mathcal{D}$.

■ **EXAMPLE 5** **Polar Coordinates Revisited** Use the Change of Variables Formula to derive the formula for integration in polar coordinates.

Solution The Jacobian of the polar coordinate map $G(r, \theta) = (r \cos \theta, r \sin \theta)$ is

$$\text{Jac}(G) = \begin{vmatrix} \dfrac{\partial x}{\partial r} & \dfrac{\partial x}{\partial \theta} \\[2mm] \dfrac{\partial y}{\partial r} & \dfrac{\partial y}{\partial \theta} \end{vmatrix} = \begin{vmatrix} \cos \theta & -r \sin \theta \\ \sin \theta & r \cos \theta \end{vmatrix} = r(\cos^2 \theta + \sin^2 \theta) = r$$

Let $\mathcal{D} = G(\mathcal{R})$ be the image under the polar coordinates map G of the rectangle $\mathcal{R}$ defined by $r_0 \leq r \leq r_1, \theta_0 \leq \theta \leq \theta_1$ (see Figure 2). Then Eq. (12) yields the expected formula for polar coordinates:

$$\iint_{\mathcal{D}} f(x, y)\, dx\, dy = \int_{\theta_0}^{\theta_1} \int_{r_0}^{r_1} f(r \cos \theta, r \sin \theta)\, r\, dr\, d\theta \qquad \boxed{\textbf{13}}$$

■

Assumptions Matter In the Change of Variables Formula, we assume that G is one-to-one on the interior but not necessarily on the boundary of the domain. Thus, we can apply Eq. (12) to the polar coordinates map G on the rectangle $\mathcal{D}_0 = [0, 1] \times [0, 2\pi]$. In this case, G is one-to-one on the interior but not on the boundary of $\mathcal{D}_0$ since $G(0, \theta) = (0, 0)$ for all θ and $G(r, 0) = G(r, 2\pi)$ for all r. On the other hand, Eq. (12) cannot be applied to G on the rectangle $[0, 1] \times [0, 4\pi]$ because it is not one-to-one on the interior.

■ **EXAMPLE 6** Use the Change of Variables Formula to calculate $\iint_{\mathcal{P}} e^{4x - y}\, dx\, dy$, where $\mathcal{P}$ is the parallelogram spanned by the vectors $\langle 4, 1 \rangle, \langle 3, 3 \rangle$ in Figure 10.

FIGURE 10 The map
$G(u, v) = (4u + 3v, u + 3v)$.

Recall that the linear map

$$G(u, v) = (Au + Cv, Bu + Dv)$$

satisfies

$$G(1, 0) = (A, B), \quad G(0, 1) = (C, D)$$

Solution

Step 1. **Define a map.**
We can convert our double integral to an integral over the unit square $\mathcal{R} = [0, 1] \times [0, 1]$ if we can find a map that sends $\mathcal{R}$ to $\mathcal{P}$. The following linear map does the job:

$$G(u, v) = (4u + 3v, u + 3v)$$

Indeed, $G(1, 0) = (4, 1)$ and $G(0, 1) = (3, 3)$, so it maps $\mathcal{R}$ to $\mathcal{P}$ because linear maps map parallelograms to parallelograms.

***Step 2.* Compute the Jacobian.**

$$\operatorname{Jac}(G) = \begin{vmatrix} \dfrac{\partial x}{\partial u} & \dfrac{\partial x}{\partial v} \\ \dfrac{\partial y}{\partial u} & \dfrac{\partial y}{\partial v} \end{vmatrix} = \begin{vmatrix} 4 & 3 \\ 1 & 3 \end{vmatrix} = 9$$

***Step 3.* Express $f(x, y)$ in terms of the new variables.**

Since $x = 4u + 3v$ and $y = u + 3v$, we have

$$e^{4x-y} = e^{4(4u+3v)-(u+3v)} = e^{15u+9v}$$

***Step 4.* Apply the Change of Variables Formula.**

The Change of Variables Formula tells us that $dx\,dy = 9\,du\,dv$:

$$\iint_{\mathcal{P}} e^{4x-y}\,dx\,dy = \iint_{\mathcal{R}} e^{15u+9v}\,|\operatorname{Jac}(G)|\,du\,dv = \int_0^1 \int_0^1 e^{15u+9v}\,(9\,du\,dv)$$

$$= 9\left(\int_0^1 e^{15u}\,du\right)\left(\int_0^1 e^{9v}\,dv\right) = \frac{1}{15}(e^{15} - 1)(e^9 - 1) \quad \blacksquare$$

■ **EXAMPLE 7** Use the Change of Variables Formula to compute

$$\iint_{\mathcal{D}} (x^2 + y^2)\,dx\,dy$$

where $\mathcal{D}$ is the domain $1 \le xy \le 4,\ 1 \le y/x \le 4$ (Figure 11).

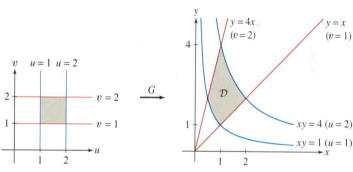

DF **FIGURE 11**

Solution In Example 3, we studied the map $G(u, v) = (uv^{-1}, uv)$, which can be written

$$x = uv^{-1}, \qquad y = uv$$

We showed (Figure 11) that G maps the rectangle $\mathcal{D}_0 = [1, 2] \times [1, 2]$ to our domain $\mathcal{D}$. Indeed, because $xy = u^2$ and $xy^{-1} = v^2$, the two conditions $1 \le xy \le 4$ and $1 \le y/x \le 4$ that define $\mathcal{D}$ become $1 \le u \le 2$ and $1 \le v \le 2$.

The Jacobian is

$$\operatorname{Jac}(G) = \frac{\partial(x, y)}{\partial(u, v)} = \begin{vmatrix} \dfrac{\partial x}{\partial u} & \dfrac{\partial x}{\partial v} \\ \dfrac{\partial y}{\partial u} & \dfrac{\partial y}{\partial v} \end{vmatrix} = \begin{vmatrix} v^{-1} & -uv^{-2} \\ v & u \end{vmatrix} = \frac{2u}{v}$$

To apply the Change of Variables Formula, we write $f(x, y)$ in terms of u and v:

$$f(x, y) = x^2 + y^2 = \left(\frac{u}{v}\right)^2 + (uv)^2 = u^2(v^{-2} + v^2)$$

By the Change of Variables Formula,

$$\iint_{\mathcal{D}} (x^2 + y^2)\, dx\, dy = \iint_{\mathcal{D}_0} u^2(v^{-2} + v^2) \left| \frac{2u}{v} \right| du\, dv$$

$$= 2 \int_{v=1}^{2} \int_{u=1}^{2} u^3(v^{-3} + v)\, du\, dv$$

$$= 2 \left(\int_{u=1}^{2} u^3\, du \right) \left(\int_{v=1}^{2} (v^{-3} + v)\, dv \right)$$

$$= 2 \left(\frac{1}{4} u^4 \Big|_1^2 \right) \left(-\frac{1}{2} v^{-2} + \frac{1}{2} v^2 \Big|_1^2 \right) = \frac{225}{16} \qquad \blacksquare$$

Keep in mind that the Change of Variables Formula turns an xy-integral into a uv-integral, but the map G goes from the uv-domain to the xy-domain. Sometimes, it is easier to find a map F going in the *wrong direction*, from the xy-domain to the uv-domain. The desired map G is then the inverse $G = F^{-1}$. The next example shows that in some cases, we can evaluate the integral without solving for G. The key fact is that the Jacobian of F is the reciprocal of $\mathrm{Jac}(G)$ (see Exercises 49–51):

Eq. (14) can be written in the suggestive form

$$\frac{\partial(x, y)}{\partial(u, v)} = \left(\frac{\partial(u, v)}{\partial(x, y)} \right)^{-1}$$

$$\boxed{\mathrm{Jac}(G) = \mathrm{Jac}(F)^{-1}, \qquad \text{where } F = G^{-1}} \qquad \boxed{14}$$

■ **EXAMPLE 8** **Using the Inverse Map** Integrate $f(x, y) = xy(x^2 + y^2)$ over

$$\mathcal{D} : -3 \le x^2 - y^2 \le 3, \quad 1 \le xy \le 4$$

Solution There is a simple map F that goes in the *wrong* direction. Let $u = x^2 - y^2$ and $v = xy$. Then our domain is defined by the inequalities $-3 \le u \le 3$ and $1 \le v \le 4$, and we can define a map from $\mathcal{D}$ to the rectangle $\mathcal{R} = [-3, 3] \times [1, 4]$ in the uv-plane (Figure 12):

$$F : \mathcal{D} \to \mathcal{R}$$

$$(x, y) \to (x^2 - y^2, xy)$$

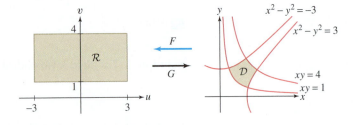

DF **FIGURE 12** The map F goes in the "wrong" direction.

To convert the integral over $\mathcal{D}$ into an integral over the rectangle $\mathcal{R}$, we have to apply the Change of Variables Formula to the inverse mapping:

$$G = F^{-1} : \mathcal{R} \to \mathcal{D}$$

We will see that it is not necessary to find G explicitly. Since $u = x^2 - y^2$ and $v = xy$, the Jacobian of F is

$$\mathrm{Jac}(F) = \begin{vmatrix} \dfrac{\partial u}{\partial x} & \dfrac{\partial u}{\partial y} \\[2mm] \dfrac{\partial v}{\partial x} & \dfrac{\partial v}{\partial y} \end{vmatrix} = \begin{vmatrix} 2x & -2y \\ y & x \end{vmatrix} = 2(x^2 + y^2)$$

By Eq. (14),

$$\text{Jac}(G) = \text{Jac}(F)^{-1} = \frac{1}{2(x^2 + y^2)}$$

Normally, the next step would be to express $f(x, y)$ in terms of u and v. We can avoid doing this in our case by observing that the Jacobian cancels with one factor of $f(x, y)$:

$$\iint_{\mathcal{D}} xy(x^2 + y^2)\, dx\, dy = \iint_{\mathcal{R}} f(x(u, v), y(u, v))\, |\text{Jac}(G)|\, du\, dv$$

$$= \iint_{\mathcal{R}} xy(x^2 + y^2) \frac{1}{2(x^2 + y^2)}\, du\, dv$$

$$= \frac{1}{2} \iint_{\mathcal{R}} xy\, du\, dv$$

$$= \frac{1}{2} \iint_{\mathcal{R}} v\, du\, dv \qquad \text{(because } v = xy\text{)}$$

$$= \frac{1}{2} \int_{-3}^{3} \int_{1}^{4} v\, dv\, du = \frac{1}{2}(6)\left(\frac{1}{2}4^2 - \frac{1}{2}1^2\right) = \frac{45}{2} \qquad \blacksquare$$

Change of Variables in Three Variables

The Change of Variables Formula has the same form in three (or more) variables as in two variables. Let

$$G : \mathcal{W}_0 \to \mathcal{W}$$

be a mapping from a three-dimensional region $\mathcal{W}_0$ in (u, v, w)-space to a region $\mathcal{W}$ in (x, y, z)-space, say,

$$x = x(u, v, w), \qquad y = y(u, v, w), \qquad z = z(u, v, w)$$

←·· *REMINDER* 3×3-*determinants are defined in Eq. (2) of Section 12.4.*

The Jacobian $\text{Jac}(G)$ is the 3×3 determinant:

$$\text{Jac}(G) = \frac{\partial(x, y, z)}{\partial(u, v, w)} = \begin{vmatrix} \dfrac{\partial x}{\partial u} & \dfrac{\partial x}{\partial v} & \dfrac{\partial x}{\partial w} \\[6pt] \dfrac{\partial y}{\partial u} & \dfrac{\partial y}{\partial v} & \dfrac{\partial y}{\partial w} \\[6pt] \dfrac{\partial z}{\partial u} & \dfrac{\partial z}{\partial v} & \dfrac{\partial z}{\partial w} \end{vmatrix} \qquad \boxed{15}$$

The Change of Variables Formula states

$$\boxed{dx\, dy\, dz = \left| \frac{\partial(x, y, z)}{\partial(u, v, w)} \right| du\, dv\, dw}$$

More precisely, if G is C^1 and one-to-one on the interior of $\mathcal{W}_0$, and if f is continuous, then

$$\iiint_{\mathcal{W}} f(x, y, z)\, dx\, dy\, dz$$

$$= \iiint_{\mathcal{W}_0} f(x(u, v, w), y(u, v, w), z(u, v, w)) \left| \frac{\partial(x, y, z)}{\partial(u, v, w)} \right| du\, dv\, dw \qquad \boxed{16}$$

In Exercises 42 and 43, you are asked to use the general Change of Variables Formula to derive the formulas for integration in cylindrical and spherical coordinates developed in Section 15.4.

15.6 SUMMARY

- Let $G(u, v) = (x(u, v), y(u, v))$ be a mapping. We write $x = x(u, v)$ or $x = x(u, v)$ and, similarly, $y = y(u, v)$ or $y = y(u, v)$. The Jacobian of G is the determinant

$$\text{Jac}(G) = \frac{\partial(x, y)}{\partial(u, v)} = \begin{vmatrix} \dfrac{\partial x}{\partial u} & \dfrac{\partial x}{\partial v} \\ \dfrac{\partial y}{\partial u} & \dfrac{\partial y}{\partial v} \end{vmatrix}$$

- $\text{Jac}(G) = \text{Jac}(F)^{-1}$, where $F = G^{-1}$.
- Change of Variables Formula: If $G : \mathcal{D}_0 \to \mathcal{D}$ has component functions with continuous partial derivatives and one-to-one on the interior of $\mathcal{D}_0$, and if f is continuous, then

$$\iint_{\mathcal{D}} f(x, y)\, dx\, dy = \iint_{\mathcal{D}_0} f(x(u, v), y(u, v)) \left| \frac{\partial(x, y)}{\partial(u, v)} \right| du\, dv$$

- The Change of Variables Formula is written symbolically in two and three variables as

$$dx\, dy = \left| \frac{\partial(x, y)}{\partial(u, v)} \right| du\, dv, \qquad dx\, dy\, dz = \left| \frac{\partial(x, y, z)}{\partial(u, v, w)} \right| du\, dv\, dw$$

15.6 EXERCISES

Preliminary Questions

1. Which of the following maps is linear?

(a) (uv, v) (b) $(u + v, u)$ (c) $(3, e^u)$

2. Suppose that G is a linear map such that $G(2, 0) = (4, 0)$ and $G(0, 3) = (-3, 9)$. Find the images of:

(a) $G(1, 0)$ (b) $G(1, 1)$ (c) $G(2, 1)$

3. What is the area of $G(\mathcal{R})$ if $\mathcal{R}$ is a rectangle of area 9 and G is a mapping whose Jacobian has constant value 4?

4. Estimate the area of $G(\mathcal{R})$, where $\mathcal{R} = [1, 1.2] \times [3, 3.1]$ and G is a mapping such that $\text{Jac}(G)(1, 3) = 3$.

Exercises

1. Determine the image under $G(u, v) = (2u, u + v)$ of the following sets:

(a) The u- and v-axes
(b) The rectangle $\mathcal{R} = [0, 5] \times [0, 7]$
(c) The line segment joining $(1, 2)$ and $(5, 3)$
(d) The triangle with vertices $(0, 1)$, $(1, 0)$, and $(1, 1)$

2. Describe [in the form $y = f(x)$] the images of the lines $u = c$ and $v = c$ under the mapping $G(u, v) = (u/v, u^2 - v^2)$.

3. Let $G(u, v) = (u^2, v)$. Is G one-to-one? If not, determine a domain on which G is one-to-one. Find the image under G of:

(a) The u- and v-axes
(b) The rectangle $\mathcal{R} = [-1, 1] \times [-1, 1]$
(c) The line segment joining $(0, 0)$ and $(1, 1)$
(d) The triangle with vertices $(0, 0)$, $(0, 1)$, and $(1, 1)$

4. Let $G(u, v) = (e^u, e^{u+v})$.

(a) Is G one-to-one? What is the image of G?
(b) Describe the images of the vertical lines $u = c$ and the horizontal lines $v = c$.

In Exercises 5–12, let $G(u, v) = (2u + v, 5u + 3v)$ be a map from the uv-plane to the xy-plane.

5. Show that the image of the horizontal line $v = c$ is the line $y = \frac{5}{2}x + \frac{1}{2}c$. What is the image (in slope-intercept form) of the vertical line $u = c$?

6. Describe the image of the line through the points $(u, v) = (1, 1)$ and $(u, v) = (1, -1)$ under G in slope-intercept form.

7. Describe the image of the line $v = 4u$ under G in slope-intercept form.

8. Show that G maps the line $v = mu$ to the line of slope $(5 + 3m)/(2 + m)$ through the origin in the xy-plane.

9. Show that the inverse of G is

$$G^{-1}(x, y) = (3x - y, -5x + 2y)$$

Hint: Show that $G(G^{-1}(x, y)) = (x, y)$ and $G^{-1}(G(u, v)) = (u, v)$.

10. Use the inverse in Exercise 9 to find:

(a) A point in the uv-plane mapping to $(2, 1)$
(b) A segment in the uv-plane mapping to the segment joining $(-2, 1)$ and $(3, 4)$

11. Calculate $\text{Jac}(G) = \dfrac{\partial(x, y)}{\partial(u, v)}$.

12. Calculate $\text{Jac}(G^{-1}) = \dfrac{\partial(u, v)}{\partial(x, y)}$.

In Exercises 13–18, compute the Jacobian (at the point, if indicated).

13. $G(u, v) = (3u + 4v, u - 2v)$

14. $G(r, s) = (rs, r + s)$

15. $G(r, t) = (r \sin t, r - \cos t), \quad (r, t) = (1, \pi)$

16. $G(u, v) = (v \ln u, u^2 v^{-1}), \quad (u, v) = (1, 2)$

17. $G(r, \theta) = (r \cos \theta, r \sin \theta), \quad (r, \theta) = \left(4, \frac{\pi}{6}\right)$

18. $G(u, v) = (ue^v, e^u)$

19. Find a linear mapping G that maps $[0, 1] \times [0, 1]$ to the parallelogram in the xy-plane spanned by the vectors $\langle 2, 3 \rangle$ and $\langle 4, 1 \rangle$.

20. Find a linear mapping G that maps $[0, 1] \times [0, 1]$ to the parallelogram in the xy-plane spanned by the vectors $\langle -2, 5 \rangle$ and $\langle 1, 7 \rangle$.

21. Let $\mathcal{D}$ be the parallelogram in Figure 13. Apply the Change of Variables Formula to the map $G(u, v) = (5u + 3v, u + 4v)$ to evaluate $\iint_{\mathcal{D}} xy \, dx \, dy$ as an integral over $\mathcal{D}_0 = [0, 1] \times [0, 1]$.

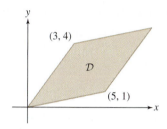

FIGURE 13

22. Let $G(u, v) = (u - uv, uv)$.

(a) Show that the image of the horizontal line $v = c$ is $y = \dfrac{c}{1-c} x$ if $c \neq 1$, and is the y-axis if $c = 1$.

(b) Determine the images of vertical lines in the uv-plane.

(c) Compute the Jacobian of G.

(d) Observe that by the formula for the area of a triangle, the region $\mathcal{D}$ in Figure 14 has area $\frac{1}{2}(b^2 - a^2)$. Compute this area again, using the Change of Variables Formula applied to G.

(e) Calculate $\iint_{\mathcal{D}} xy \, dx \, dy$.

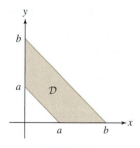

FIGURE 14

23. Let $G(u, v) = (3u + v, u - 2v)$. Use the Jacobian to determine the area of $G(\mathcal{R})$ for:

(a) $\mathcal{R} = [0, 3] \times [0, 5]$ **(b)** $\mathcal{R} = [2, 5] \times [1, 7]$

24. Find a linear map T that maps $[0, 1] \times [0, 1]$ to the parallelogram $\mathcal{P}$ in the xy-plane with vertices $(0, 0), (2, 2), (1, 4), (3, 6)$. Then calculate the double integral of e^{2x-y} over $\mathcal{P}$ via change of variables.

25. With G as in Example 3, use the Change of Variables Formula to compute the area of the image of $[1, 4] \times [1, 4]$.

In Exercises 26–28, let $\mathcal{R}_0 = [0, 1] \times [0, 1]$ be the unit square. The translate of a map $G_0(u, v) = (\phi(u, v), \psi(u, v))$ is a map

$$G(u, v) = (a + \phi(u, v), b + \psi(u, v))$$

where a, b are constants. Observe that the map G_0 in Figure 15 maps $\mathcal{R}_0$ to the parallelogram $\mathcal{P}_0$ and that the translate

$$G_1(u, v) = (2 + 4u + 2v, 1 + u + 3v)$$

maps $\mathcal{R}_0$ to $\mathcal{P}_1$.

26. Find translates G_2 and G_3 of the mapping G_0 in Figure 15 that map the unit square $\mathcal{R}_0$ to the parallelograms $\mathcal{P}_2$ and $\mathcal{P}_3$.

27. Sketch the parallelogram $\mathcal{P}$ with vertices $(1, 1), (2, 4), (3, 6), (4, 9)$ and find the translate of a linear mapping that maps $\mathcal{R}_0$ to $\mathcal{P}$.

28. Find the translate of a linear mapping that maps $\mathcal{R}_0$ to the parallelogram spanned by the vectors $\langle 3, 9 \rangle$ and $\langle -4, 6 \rangle$ based at $(4, 2)$.

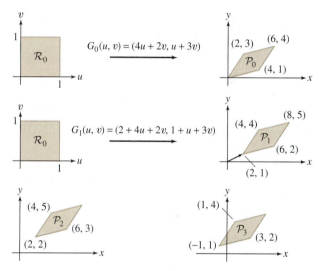

FIGURE 15

29. Let $\mathcal{D} = G(\mathcal{R})$, where $G(u, v) = (u^2, u + v)$ and $\mathcal{R} = [1, 2] \times [0, 6]$. Calculate $\iint_{\mathcal{D}} y \, dx \, dy$. *Note:* It is not necessary to describe $\mathcal{D}$.

30. Let $\mathcal{D}$ be the image of $\mathcal{R} = [1, 4] \times [1, 4]$ under the map $G(u, v) = (u^2/v, v^2/u)$.

(a) Compute $\text{Jac}(G)$.

(b) Sketch $\mathcal{D}$.

(c) Use the Change of Variables Formula to compute $\text{Area}(\mathcal{D})$ and $\iint_{\mathcal{D}} f(x, y) \, dx \, dy$, where $f(x, y) = x + y$.

31. Compute $\iint_{\mathcal{D}} (x + 3y) \, dx \, dy$, where $\mathcal{D}$ is the shaded region in Figure 16. *Hint:* Use the map $G(u, v) = (u - 2v, v)$.

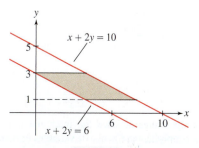

FIGURE 16

32. Use the map $G(u, v) = \left(\dfrac{u}{v+1}, \dfrac{uv}{v+1} \right)$ to compute

$$\iint_{\mathcal{D}} (x + y)\, dx\, dy$$

where $\mathcal{D}$ is the shaded region in Figure 17.

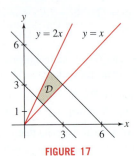

FIGURE 17

33. Show that $T(u, v) = (u^2 - v^2, 2uv)$ maps the triangle $\mathcal{D}_0 = \{(u, v) : 0 \le v \le u \le 1\}$ to the domain $\mathcal{D}$ bounded by $x = 0$, $y = 0$, and $y^2 = 4 - 4x$. Use T to evaluate

$$\iint_{\mathcal{D}} \sqrt{x^2 + y^2}\, dx\, dy$$

34. Find a mapping G that maps the disk $u^2 + v^2 \le 1$ onto the interior of the ellipse $\left(\dfrac{x}{a} \right)^2 + \left(\dfrac{y}{b} \right)^2 \le 1$. Then use the Change of Variables Formula to prove that the area of the ellipse is πab.

35. Calculate $\displaystyle\iint_{\mathcal{D}} e^{9x^2 + 4y^2}\, dx\, dy$, where $\mathcal{D}$ is the interior of the ellipse $\left(\dfrac{x}{2} \right)^2 + \left(\dfrac{y}{3} \right)^2 \le 1$.

36. Compute the area of the region enclosed by the ellipse $x^2 + 2xy + 2y^2 - 4y = 8$ as an integral in the variables $u = x + y$, $v = y - 2$.

37. Sketch the domain $\mathcal{D}$ bounded by $y = x^2$, $y = \frac{1}{2}x^2$, and $y = x$. Use a change of variables with the map $x = uv$, $y = u^2$ to calculate

$$\iint_{\mathcal{D}} y^{-1}\, dx\, dy$$

This is an improper integral since $f(x, y) = y^{-1}$ is undefined at $(0, 0)$, but it becomes proper after changing variables.

38. Find an appropriate change of variables to evaluate

$$\iint_{\mathcal{R}} (x + y)^2 e^{x^2 - y^2}\, dx\, dy$$

where $\mathcal{R}$ is the square with vertices $(1, 0)$, $(0, 1)$, $(-1, 0)$, $(0, -1)$.

39. Let G be the inverse of the map $F(x, y) = (xy, x^2 y)$ from the xy-plane to the uv-plane. Let $\mathcal{D}$ be the domain in Figure 18. Show, by applying the Change of Variables Formula to the inverse $G = F^{-1}$, that

$$\iint_{\mathcal{D}} e^{xy}\, dx\, dy = \int_{10}^{20} \int_{20}^{40} e^u v^{-1}\, dv\, du$$

and evaluate this result. *Hint:* See Example 8.

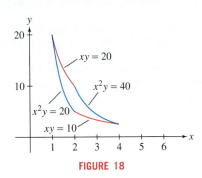

FIGURE 18

40. Sketch the domain

$$\mathcal{D} = \{(x, y) : 1 \le x + y \le 4,\ -4 \le y - 2x \le 1\}$$

(a) Let F be the map $u = x + y$, $v = y - 2x$ from the xy-plane to the uv-plane, and let G be its inverse. Use Eq. (14) to compute $\mathrm{Jac}(G)$.

(b) Compute $\displaystyle\iint_{\mathcal{D}} e^{x+y}\, dx\, dy$ using the Change of Variables Formula with the map G. *Hint:* It is not necessary to solve for G explicitly.

41. Let $I = \displaystyle\iint_{\mathcal{D}} (x^2 - y^2)\, dx\, dy$, where

$$\mathcal{D} = \{(x, y) : 2 \le xy \le 4, 0 \le x - y \le 3, x \ge 0, y \ge 0\}$$

(a) Show that the mapping $u = xy$, $v = x - y$ maps $\mathcal{D}$ to the rectangle $\mathcal{R} = [2, 4] \times [0, 3]$.

(b) Compute $\partial(x, y)/\partial(u, v)$ by first computing $\partial(u, v)/\partial(x, y)$.

(c) Use the Change of Variables Formula to show that I is equal to the integral of $f(u, v) = v$ over $\mathcal{R}$ and evaluate.

42. Derive formula (5) in Section 15.4 for integration in cylindrical coordinates from the general Change of Variables Formula.

43. Derive formula (8) in Section 15.4 for integration in spherical coordinates from the general Change of Variables Formula.

44. Use the Change of Variables Formula in three variables to prove that the volume of the ellipsoid $\left(\dfrac{x}{a} \right)^2 + \left(\dfrac{y}{b} \right)^2 + \left(\dfrac{z}{c} \right)^2 = 1$ is equal to $abc \times$ the volume of the unit sphere.

Further Insights and Challenges

45. Use the map

$$x = \frac{\sin u}{\cos v}, \qquad y = \frac{\sin v}{\cos u}$$

to evaluate the integral

$$\int_0^1 \int_0^1 \frac{dx\, dy}{1 - x^2 y^2}$$

This is an improper integral since the integrand is infinite if $x = \pm 1$, $y = \pm 1$, but applying the Change of Variables Formula shows that the result is finite.

46. Verify properties (1) and (2) for linear functions and show that any map satisfying these two properties is linear.

47. Let P and Q be points in $\mathbf{R}^2$. Show that a linear map $G(u, v) = (Au + Cv, Bu + Dv)$ maps the segment joining P and Q to the segment joining $G(P)$ to $G(Q)$. *Hint:* The segment joining P and Q has parametrization

$$(1 - t)\overrightarrow{OP} + t\overrightarrow{OQ} \quad \text{for} \quad 0 \le t \le 1$$

48. ▭ Let G be a linear map. Prove Eq. (6) in the following steps.

(a) For any set $\mathcal{D}$ in the uv-plane and any vector $\mathbf{u}$, let $\mathcal{D} + \mathbf{u}$ be the set obtained by translating all points in $\mathcal{D}$ by $\mathbf{u}$. By linearity, G maps $\mathcal{D} + \mathbf{u}$

to the translate $G(\mathcal{D}) + G(\mathbf{u})$ [Figure 19(C)]. Therefore, if Eq. (6) holds for $\mathcal{D}$, it also holds for $\mathcal{D} + \mathbf{u}$.

(b) In the text, we verified Eq. (6) for the unit rectangle. Use linearity to show that Eq. (6) also holds for all rectangles with vertex at the origin and sides parallel to the axes. Then argue that it also holds for each triangular half of such a rectangle, as in Figure 19(A).

(c) Figure 19(B) shows that the area of a parallelogram is a difference of the areas of rectangles and triangles covered by steps (a) and (b). Use this to prove Eq. (6) for arbitrary parallelograms.

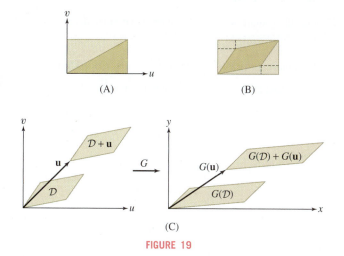

(A)   (B)

(C)

FIGURE 19

49. The product of 2×2 matrices A and B is the matrix AB defined by

$$\underbrace{\begin{pmatrix} a & b \\ c & d \end{pmatrix}}_{A} \underbrace{\begin{pmatrix} a' & b' \\ c' & d' \end{pmatrix}}_{B} = \underbrace{\begin{pmatrix} aa' + bc' & ab' + bd' \\ ca' + dc' & cb' + dd' \end{pmatrix}}_{AB}$$

The (i, j)-entry of A is the **dot product** of the ith row of A and the jth column of B. Prove that $\det(AB) = \det(A)\det(B)$.

50. Let $G_1 : \mathcal{D}_1 \to \mathcal{D}_2$ and $G_2 : \mathcal{D}_2 \to \mathcal{D}_3$ be C^1 maps, and let $G_2 \circ G_1 : \mathcal{D}_1 \to \mathcal{D}_3$ be the composite map. Use the Multivariable Chain Rule and Exercise 49 to show that

$$\mathrm{Jac}(G_2 \circ G_1) = \mathrm{Jac}(G_2)\mathrm{Jac}(G_1)$$

51. Use Exercise 50 to prove that

$$\mathrm{Jac}(G^{-1}) = \mathrm{Jac}(G)^{-1}$$

Hint: Verify that $\mathrm{Jac}(I) = 1$, where I is the identity map $I(u, v) = (u, v)$.

52. Let $(\bar{x}, \bar{y})$ be the centroid of a domain $\mathcal{D}$. For $\lambda > 0$, let $\lambda\mathcal{D}$ be the **dilation** of $\mathcal{D}$, defined by

$$\lambda\mathcal{D} = \{(\lambda x, \lambda y) : (x, y) \in \mathcal{D}\}$$

Use the Change of Variables Formula to prove that the centroid of $\lambda\mathcal{D}$ is $(\lambda\bar{x}, \lambda\bar{y})$.

CHAPTER REVIEW EXERCISES

1. Calculate the Riemann sum $S_{2,3}$ for $\int_1^4 \int_2^6 x^2 y \, dx \, dy$ using two choices of sample points:

(a) Lower-left vertex

(b) Midpoint of rectangle

Then calculate the exact value of the double integral.

2. Let $S_{N,N}$ be the Riemann sum for $\int_0^1 \int_0^1 \cos(xy) \, dx \, dy$ using midpoints as sample points.

(a) Calculate $S_{4,4}$.

(b) *CAS* Use a computer algebra system to calculate $S_{N,N}$ for $N = 10, 50, 100$.

3. Let $\mathcal{D}$ be the shaded domain in Figure 1.

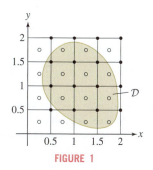

FIGURE 1

Estimate $\iint_{\mathcal{D}} xy \, dA$ by the Riemann sum whose sample points are the midpoints of the squares in the grid.

4. Explain the following:

(a) $\int_{-1}^1 \int_{-1}^1 \sin(xy) \, dx \, dy = 0$

(b) $\int_{-1}^1 \int_{-1}^1 \cos(xy) \, dx \, dy > 0$

In Exercises 5–8, evaluate the iterated integral.

5. $\int_0^2 \int_3^5 y(x - y) \, dx \, dy$

6. $\int_{1/2}^0 \int_0^{\pi/6} e^{2y} \sin(3x) \, dx \, dy$

7. $\int_0^{\pi/3} \int_0^{\pi/6} \sin(x + y) \, dx \, dy$

8. $\int_1^2 \int_1^2 \dfrac{y \, dx \, dy}{x + y^2}$

In Exercises 9–14, sketch the domain $\mathcal{D}$ and calculate $\iint_{\mathcal{D}} f(x, y) \, dA$.

9. $\mathcal{D} = \{0 \le x \le 4, \ 0 \le y \le x\}, \quad f(x, y) = \cos y$

10. $\mathcal{D} = \{0 \le x \le 2, \ 0 \le y \le 2x - x^2\}, \quad f(x, y) = \sqrt{xy}$

11. $\mathcal{D} = \{0 \le x \le 1, \ 1 - x \le y \le 2 - x\}, \quad f(x, y) = e^{x+2y}$

12. $\mathcal{D} = \{1 \le x \le 2, \ 0 \le y \le 1/x\}, \quad f(x, y) = \cos(xy)$

13. $\mathcal{D} = \{0 \le y \le 1, \ 0.5y^2 \le x \le y^2\}, \quad f(x, y) = ye^{1+x}$

14. $\mathcal{D} = \{1 \le y \le e, \ y \le x \le 2y\}, \quad f(x, y) = \ln(x + y)$

15. Express $\displaystyle\int_{-3}^{3}\int_{0}^{9-x^2} f(x,y)\,dy\,dx$ as an iterated integral in the order $dx\,dy$.

16. Let $\mathcal{W}$ be the region bounded by the planes $y = z$, $2y + z = 3$, and $z = 0$ for $0 \le x \le 4$.

(a) Express the triple integral $\displaystyle\iiint_{\mathcal{W}} f(x, y, z)\,dV$ as an iterated integral in the order $dy\,dx\,dz$ (project $\mathcal{W}$ onto the xz-plane).

(b) Evaluate the triple integral for $f(x, y, z) = 1$.

(c) Compute the volume of $\mathcal{W}$ using geometry and check that the result coincides with the answer to (b).

17. Let $\mathcal{D}$ be the domain between $y = x$ and $y = \sqrt{x}$. Calculate $\displaystyle\iint_{\mathcal{D}} xy\,dA$ as an iterated integral in the order $dx\,dy$ and $dy\,dx$.

18. Find the double integral of $f(x, y) = x^3 y$ over the region between the curves $y = x^2$ and $y = x(1 - x)$.

19. Change the order of integration and evaluate $\displaystyle\int_{0}^{9}\int_{0}^{\sqrt{y}} \frac{x\,dx\,dy}{(x^2 + y)^{1/2}}$.

20. Verify directly that

$$\int_{2}^{3}\int_{0}^{2} \frac{dy\,dx}{1 + x - y} = \int_{0}^{2}\int_{2}^{3} \frac{dx\,dy}{1 + x - y}$$

21. Prove the formula

$$\int_{0}^{1}\int_{0}^{y} f(x)\,dx\,dy = \int_{0}^{1} (1 - x)f(x)\,dx$$

Then use it to calculate $\displaystyle\int_{0}^{1}\int_{0}^{y} \frac{\sin x}{1 - x}\,dx\,dy$.

22. Rewrite $\displaystyle\int_{0}^{1}\int_{-\sqrt{1-y^2}}^{\sqrt{1-y^2}} \frac{y\,dx\,dy}{(1 + x^2 + y^2)^2}$ by interchanging the order of integration, and evaluate.

23. Use cylindrical coordinates to compute the volume of the region defined by $4 - x^2 - y^2 \le z \le 10 - 4x^2 - 4y^2$.

24. Evaluate $\displaystyle\iint_{\mathcal{D}} x\,dA$, where $\mathcal{D}$ is the shaded domain in Figure 2.

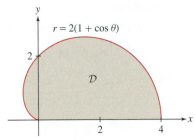

FIGURE 2

25. Find the volume of the region between the graph of the function $f(x, y) = 1 - (x^2 + y^2)$ and the xy-plane.

26. Evaluate $\displaystyle\int_{0}^{3}\int_{1}^{4}\int_{2}^{4} (x^3 + y^2 + z)\,dx\,dy\,dz$.

27. Calculate $\displaystyle\iiint_{\mathcal{B}} (xy + z)\,dV$, where

$$\mathcal{B} = \{0 \le x \le 2,\ 0 \le y \le 1,\ 1 \le z \le 3\}$$

as an iterated integral in two different ways.

28. Calculate $\displaystyle\iiint_{\mathcal{W}} xyz\,dV$, where

$$\mathcal{W} = \{0 \le x \le 1,\ x \le y \le 1,\ x \le z \le x + y\}$$

29. Evaluate $\displaystyle I = \int_{-1}^{1}\int_{0}^{\sqrt{1-x^2}}\int_{0}^{1} (x + y + z)\,dz\,dy\,dx$.

30. Describe a region whose volume is equal to:

(a) $\displaystyle\int_{0}^{2\pi}\int_{0}^{\pi/2}\int_{4}^{9} \rho^2 \sin\phi\,d\rho\,d\phi\,d\theta$

(b) $\displaystyle\int_{-2}^{1}\int_{\pi/3}^{\pi/4}\int_{0}^{2} r\,dr\,d\theta\,dz$

(c) $\displaystyle\int_{0}^{2\pi}\int_{0}^{3}\int_{-\sqrt{9-r^2}}^{0} r\,dz\,dr\,d\theta$

31. Find the volume of the solid contained in the cylinder $x^2 + y^2 = 1$ below the surface $z = (x + y)^2$ and above the surface $z = -(x - y)^2$.

32. Use polar coordinates to evaluate $\displaystyle\iint_{\mathcal{D}} x\,dA$, where $\mathcal{D}$ is the shaded region between the two circles of radius 1 in Figure 3.

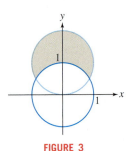

FIGURE 3

33. Use polar coordinates to calculate $\displaystyle\iint_{\mathcal{D}} \sqrt{x^2 + y^2}\,dA$, where $\mathcal{D}$ is the region in the first quadrant bounded by the spiral $r = \theta$, the circle $r = 1$, and the x-axis.

34. Calculate $\displaystyle\iint_{\mathcal{D}} \sin(x^2 + y^2)\,dA$, where

$$\mathcal{D} = \left\{\frac{\pi}{2} \le x^2 + y^2 \le \pi\right\}$$

35. Express in cylindrical coordinates and evaluate:

$$\int_{0}^{1}\int_{0}^{\sqrt{1-x^2}}\int_{0}^{\sqrt{x^2+y^2}} z\,dz\,dy\,dx$$

36. Use spherical coordinates to calculate the triple integral of $f(x, y, z) = x^2 + y^2 + z^2$ over the region

$$1 \le x^2 + y^2 + z^2 \le 4$$

37. Convert to spherical coordinates and evaluate:

$$\int_{-2}^{2}\int_{-\sqrt{4-x^2}}^{\sqrt{4-x^2}}\int_{0}^{\sqrt{4-x^2-y^2}} e^{-(x^2+y^2+z^2)^{3/2}}\,dz\,dy\,dx$$

38. Find the average value of $f(x, y, z) = xy^2z^3$ on the box $[0, 1] \times [0, 2] \times [0, 3]$.

39. Let $\mathcal{W}$ be the ball of radius R in $\mathbf{R}^3$ centered at the origin, and let $P = (0, 0, R)$ be the North Pole. Let $d_P(x, y, z)$ be the distance from P to (x, y, z). Show that the average value of d_P over the sphere $\mathcal{W}$ is equal to $\bar{d} = 6R/5$. *Hint:* Show that

$$\bar{d} = \frac{1}{\frac{4}{3}\pi R^3}\int_{\theta=0}^{2\pi}\int_{\rho=0}^{R}\int_{\phi=0}^{\pi} \rho^2 \sin\phi\sqrt{R^2 + \rho^2 - 2\rho R\cos\phi}\,d\phi\,d\rho\,d\theta$$

and evaluate.

40. $\boxed{CAS}$ Express the average value of $f(x, y) = e^{xy}$ over the ellipse $\dfrac{x^2}{2} + y^2 = 1$ as an iterated integral, and evaluate numerically using a computer algebra system.

41. Use cylindrical coordinates to find the mass of the solid bounded by $z = 8 - x^2 - y^2$ and $z = x^2 + y^2$, assuming a mass density of $f(x, y, z) = (x^2 + y^2)^{1/2}$.

42. Let $\mathcal{W}$ be the portion of the half-cylinder $x^2 + y^2 \le 4$, $x \ge 0$ such that $0 \le z \le 3y$. Use cylindrical coordinates to compute the mass of $\mathcal{W}$ if the mass density is $\rho(x, y, z) = z^2$.

43. Use cylindrical coordinates to find the mass of a cylinder of radius 4 and height 10 if the mass density at a point is equal to the square of the distance from the cylinder's central axis.

44. Find the centroid of the region $\mathcal{W}$ bounded, in spherical coordinates, by $\phi = \phi_0$ and the sphere $\rho = R$.

45. Find the centroid of the solid bounded by the xy-plane, the cylinder $x^2 + y^2 = R^2$, and the plane $x/R + z/H = 1$.

46. Using cylindrical coordinates, prove that the centroid of a right circular cone of height h and radius R is located at height $\frac{h}{4}$ on the central axis.

47. Find the centroid of solid (A) in Figure 4 defined by $x^2 + y^2 \le R^2$, $0 \le z \le H$, and $\frac{\pi}{6} \le \theta \le 2\pi$, where θ is the polar angle of (x, y).

48. Calculate the coordinate y_{CM} of the centroid of solid (B) in Figure 4 defined by $x^2 + y^2 \le 1$ and $0 \le z \le \frac{1}{2}y + \frac{3}{2}$.

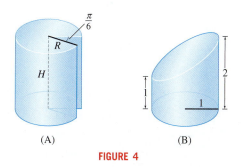

(A)   (B)

FIGURE 4

49. Find the center of mass of the cylinder $x^2 + y^2 = 1$ for $0 \le z \le 1$, assuming a mass density of $\delta(x, y, z) = z$.

50. Find the center of mass of the sector of central angle $2\theta_0$ (symmetric with respect to the y-axis) in Figure 5, assuming that the mass density is $\delta(x, y) = x^2$.

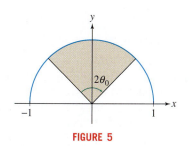

FIGURE 5

51. Find the center of mass of the first octant of the ball $x^2 + y^2 + z^2 = 1$, assuming a mass density of $\rho(x, y, z) = x$.

52. Find a constant C such that

$$p(x, y) = \begin{cases} C(4x - y + 3) & \text{if } 0 \le x \le 2 \text{ and } 0 \le y \le 3 \\ 0 & \text{otherwise} \end{cases}$$

is a probability distribution and calculate $P(X \le 1; Y \le 2)$.

53. Calculate $P(3X + 2Y \ge 6)$ for the probability density in Exercise 52.

54. The lifetimes X and Y (in years) of two machine components have joint probability density

$$p(x, y) = \begin{cases} \frac{6}{125}(5 - x - y) & \text{if } 0 \le x \le 5 - y \text{ and } 0 \le y \le 5 \\ 0 & \text{otherwise} \end{cases}$$

What is the probability that both components are still functioning after 2 years?

55. An insurance company issues two kinds of policies: A and B. Let X be the time until the next claim of type A is filed, and let Y be the time (in days) until the next claim of type B is filed. The random variables have joint probability density

$$p(x, y) = 12e^{-4x - 3y}$$

Find the probability that $X \le Y$.

56. Compute the Jacobian of the map

$$G(r, s) = \left(e^r \cosh(s), e^r \sinh(s) \right)$$

57. Find a linear mapping $G(u, v)$ that maps the unit square to the parallelogram in the xy-plane spanned by the vectors $\langle 3, -1 \rangle$ and $\langle 1, 4 \rangle$. Then use the Jacobian to find the area of the image of the rectangle $\mathcal{R} = [0, 4] \times [0, 3]$ under G.

58. Use the map

$$G(u, v) = \left(\frac{u + v}{2}, \frac{u - v}{2} \right)$$

to compute $\displaystyle\iint_{\mathcal{R}} \left((x - y) \sin(x + y) \right)^2 dx \, dy$, where $\mathcal{R}$ is the square with vertices $(\pi, 0)$, $(2\pi, \pi)$, $(\pi, 2\pi)$, and $(0, \pi)$.

59. Let $\mathcal{D}$ be the shaded region in Figure 6, and let F be the map

$$u = y + x^2, \qquad v = y - x^3$$

(a) Show that F maps $\mathcal{D}$ to a rectangle $\mathcal{R}$ in the uv-plane.

(b) Apply Eq. (7) in Section 15.6 with $P = (1, 7)$ to estimate Area($\mathcal{D}$).

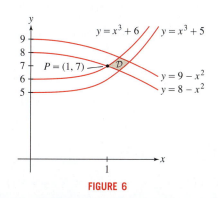

FIGURE 6

60. Calculate the integral of $f(x, y) = e^{3x-2y}$ over the parallelogram in Figure 7.

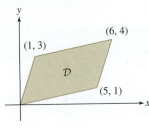

y
(6, 4)
(1, 3)
$\mathcal{D}$
(5, 1)
x

FIGURE 7

61. Sketch the region $\mathcal{D}$ bounded by the curves $y = 2/x$, $y = 1/(2x)$, $y = 2x$, $y = x/2$ in the first quadrant. Let F be the map $u = xy$, $v = y/x$ from the xy-plane to the uv-plane.

(a) Find the image of $\mathcal{D}$ under F.

(b) Let $G = F^{-1}$. Show that $|\text{Jac}(G)| = \dfrac{1}{2|v|}$.

(c) Apply the Change of Variables Formula to prove the formula

$$\iint_{\mathcal{D}} f\left(\frac{y}{x}\right) dx \, dy = \frac{3}{4} \int_{1/2}^{2} \frac{f(v) \, dv}{v}$$

(d) Apply (c) to evaluate $\displaystyle\iint_{\mathcal{D}} \frac{y e^{y/x}}{x} \, dx \, dy$.

Water flow around the hull of a boat can be modeled using a vector field. (© *age fotostock Spain, S.L./Alamy*)

16 LINE AND SURFACE INTEGRALS

In the previous chapter, we generalized integration from one variable to several variables. In this chapter, we generalize still further to include integration over curves and surfaces, and we will integrate not just functions but also vector fields. Integrals of vector fields are used in the study of phenomena such as electromagnetism, fluid dynamics, and heat transfer. To lay the groundwork, the chapter begins with a discussion of vector fields.

16.1 Vector Fields

How can we describe a physical object such as the wind, which consists of a large number of molecules moving in a region of space? What we need is a new type of function called a **vector field**. In this case, a vector field **F** assigns to each point $P = (x, y, z)$ a vector $\mathbf{F}(x, y, z)$ that represents the velocity (speed and direction) of the wind at that point (Figure 1). Another velocity field is shown in Figure 2. However, vector fields describe many other types of quantities, such as forces, and electric and magnetic fields.

> In general, a vector field **F** in $\mathbf{R}^n$ is a function that assigns to each point $(x_1, x_2, \ldots, x_n)$ in $\mathbf{R}^n$ a vector $\mathbf{F}(x_1, x_2, \ldots, x_n)$ in $\mathbf{R}^n$. In this book, we focus on vector fields in $\mathbf{R}^2$ and $\mathbf{R}^3$.

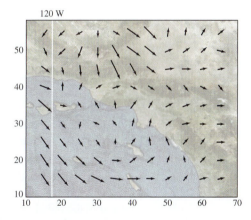

FIGURE 1 Velocity vector field of wind velocity off the coast at Los Angeles.

FIGURE 2 Blood flow in an artery represented by a vector field. (*Michelle Borkin, Harvard University*)

Mathematically, a vector field in $\mathbf{R}^3$ is represented by a vector whose components are functions:

$$\mathbf{F}(x, y, z) = \langle F_1(x, y, z), F_2(x, y, z), F_3(x, y, z) \rangle$$

To each point $P = (a, b, c)$ is associated the vector $\mathbf{F}(a, b, c)$, which we also denote by $\mathbf{F}(P)$. Alternatively,

$$\mathbf{F} = F_1\mathbf{i} + F_2\mathbf{j} + F_3\mathbf{k}$$

When drawing a vector field, we draw $\mathbf{F}(P)$ as a vector based at P (rather than the origin). The **domain** of **F** is the set of points P for which $\mathbf{F}(P)$ is defined. Vector fields in the plane are written in a similar fashion:

$$\mathbf{F}(x, y) = \langle F_1(x, y), F_2(x, y) \rangle = F_1\mathbf{i} + F_2\mathbf{j}$$

Throughout this chapter, we assume that the component functions F_j are smooth—that is, they have partial derivatives of all orders on their domains.

FIGURE 3

■ **EXAMPLE 1** Which vector corresponds to the point $P = (2, 4, 2)$ for the vector field $\mathbf{F}(x, y, z) = \langle y - z, x, z - \sqrt{y} \rangle$?

Solution The vector attached to P is

$$\mathbf{F}(2, 4, 2) = \langle 4 - 2, 2, 2 - \sqrt{4} \rangle = \langle 2, 2, 0 \rangle$$

This is the red vector in Figure 3.

Although it is not practical to sketch complicated vector fields in three dimensions by hand, computer algebra systems can produce useful visual representations (Figure 4). The vector field in Figure 4(B) is an example of a **constant vector field**. It assigns the same vector $\langle 1, -1, 3 \rangle$ to every point in $\mathbf{R}^3$.

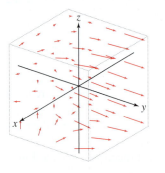

(A) $\mathbf{F} = \langle x \sin z, y^2, x/(z^2 + 1) \rangle$ (B) Constant vector field $\mathbf{F} = \langle 1, -1, 3 \rangle$

FIGURE 4

In the next example, we analyze two vector fields in the plane "qualitatively."

■ **EXAMPLE 2** Describe the following vector fields:

(a) $\mathbf{G} = \mathbf{i} + x\mathbf{j}$ **(b)** $\mathbf{F} = \langle -y, x \rangle$

Solution **(a)** The vector field $\mathbf{G} = \mathbf{i} + x\mathbf{j}$ assigns the vector $\langle 1, a \rangle$ to the point (a, b). In particular, it assigns the same vector to all points with the same x-coordinate [Figure 5(A)]. Notice that $\langle 1, a \rangle$ has slope a and length $\sqrt{1 + a^2}$. We may describe $\mathbf{G}$ as follows: $\mathbf{G}$ assigns a vector of slope a and length $\sqrt{1 + a^2}$ to all points with $x = a$.

(b) To visualize $\mathbf{F}$, observe that $\mathbf{F}(a, b) = \langle -b, a \rangle$ has length $r = \sqrt{a^2 + b^2}$. It is perpendicular to the radial vector $\langle a, b \rangle$ and points counterclockwise. Thus, $\mathbf{F}$ has the following description: The vectors along the circle of radius r all have length r and they are tangent to the circle, pointing counterclockwise [Figure 5(B)]. ■

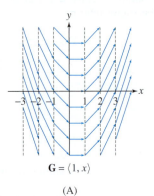

$\mathbf{G} = \langle 1, x \rangle$

(A)

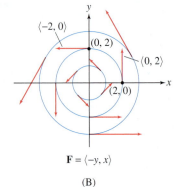

$\mathbf{F} = \langle -y, x \rangle$

(B)

DF **FIGURE 5**

A **unit vector field** is a vector field $\mathbf{F}$ such that $\|\mathbf{F}(P)\| = 1$ for all points P. A vector field $\mathbf{F}$ is called a **radial vector field** if $\mathbf{F}(P)$ depends only on the distance r from P to the origin and is parallel to $\overrightarrow{OP}$. Here, we use the notation $r = (x^2 + y^2)^{1/2}$ for $n = 2$

and $r = (x^2 + y^2 + z^2)^{1/2}$ for $n = 3$. Two important examples are the unit radial vector fields in two and three dimensions [Figures 6(A) and (B)]:

$$\mathbf{e}_r = \left\langle \frac{x}{r}, \frac{y}{r} \right\rangle = \left\langle \frac{x}{\sqrt{x^2 + y^2}}, \frac{y}{\sqrt{x^2 + y^2}} \right\rangle \qquad \boxed{1}$$

$$\mathbf{e}_r = \left\langle \frac{x}{r}, \frac{y}{r}, \frac{z}{r} \right\rangle = \left\langle \frac{x}{\sqrt{x^2 + y^2 + z^2}}, \frac{y}{\sqrt{x^2 + y^2 + z^2}}, \frac{z}{\sqrt{x^2 + y^2 + z^2}} \right\rangle \qquad \boxed{2}$$

Observe that $\mathbf{e}_r(P)$ is a unit vector pointing away from the origin at P. Note, however, that $\mathbf{e}_r$ is not defined at the origin where $r = 0$.

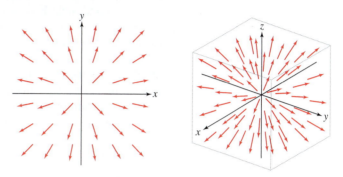

(A) Unit radial vector field in the plane
$\mathbf{e}_r = \langle x/r, y/r \rangle$

(B) Unit radial vector field in 3-space
$\mathbf{e}_r = \langle x/r, y/r, z/r \rangle$

FIGURE 6

Operations on Vector Fields

There are several operations that can be applied to a vector field that are particularly useful in a variety of contexts. The first is the **divergence** of a vector field $\mathbf{F} = \langle F_1, F_2, F_3 \rangle$, defined by

$$\operatorname{div}(\mathbf{F}) = \frac{\partial F_1}{\partial x} + \frac{\partial F_2}{\partial y} + \frac{\partial F_3}{\partial z} \qquad \boxed{3}$$

We often write the divergence as a symbolic dot product of the so-called *del* operator (also called "nabla") given by

$$\nabla = \left\langle \frac{\partial}{\partial x}, \frac{\partial}{\partial y}, \frac{\partial}{\partial z} \right\rangle$$

and the vector field $\mathbf{F}$:

$$\nabla \cdot \mathbf{F} = \left\langle \frac{\partial}{\partial x}, \frac{\partial}{\partial y}, \frac{\partial}{\partial z} \right\rangle \cdot \langle F_1, F_2, F_3 \rangle = \frac{\partial F_1}{\partial x} + \frac{\partial F_2}{\partial y} + \frac{\partial F_3}{\partial z}$$

Note that rather than outputting a vector field, $\operatorname{div}(\mathbf{F})$ is a scalar function. Divergence obeys the **linearity** rules:

$$\operatorname{div}(\mathbf{F} + \mathbf{G}) = \operatorname{div}(\mathbf{F}) + \operatorname{div}(\mathbf{G})$$

$$\operatorname{div}(c\mathbf{F}) = c \operatorname{div}(\mathbf{F}) \qquad (c \text{ any constant})$$

In Section 17.3, we will consider what the divergence tells us about a vector field.

■ **EXAMPLE 3** Evaluate the divergence of $\mathbf{F} = \langle e^{xy}, xy, z^4 \rangle$ at $P = (1, 0, 2)$.

Solution

$$\operatorname{div}(\mathbf{F}) = \frac{\partial}{\partial x} e^{xy} + \frac{\partial}{\partial y} xy + \frac{\partial}{\partial z} z^4 = y e^{xy} + x + 4z^3$$

$$\operatorname{div}(\mathbf{F})(P) = \operatorname{div}(\mathbf{F})(1, 0, 2) = 0 \cdot e^0 + 1 + 4 \cdot 2^3 = 33 \qquad ■$$

Although we will wait until Section 17.3 for the tools to justify this interpretation, we give here a physical significance for divergence. Consider a gas with velocity vector field given by **F**. When $\text{div}(\mathbf{F}) > 0$ at a point P, an outflow of gas occurs near this point. In other words, the gas is expanding around the point, as might occur when the gas is heated. When $\text{div}(\mathbf{F}) < 0$ at a point P, the gas is compressing toward P, as might occur when the gas is cooled. When $\text{div}(\mathbf{F}) = 0$, the gas is neither compressing nor expanding near P.

For example, the vector field $\mathbf{F} = \langle x, y, z \rangle$, appearing as in Figure 7(A), has $\text{div}(\mathbf{F}) = 3$ everywhere. Thinking of this as the velocity vector field for a gas, at every point, the gas is expanding. This is most obvious at the origin, but even at other points, the gas is expanding in the sense that more gas atoms are moving away from any given point than are moving toward it. We say that each of these points is a **source**.

For the vector field $\mathbf{F} = \langle -x, -y, -z \rangle$ appearing in Figure 7(B), the divergence is given by $\text{div}(\mathbf{F}) = -3$ for all points P and the gas is compressing at every point. We say that every point is a **sink**.

For the vector field $\mathbf{F} = \langle 0, 1, 0 \rangle$, appearing in Figure 7(C), the divergence is given by $\text{div}(\mathbf{F}) = 0$. The gas is neither expanding or compressing at any point. Rather, it is simply shifting to the right. In this case, no points are sources or sinks and we say that the vector field is **incompressible**. For other vector fields, there can be points that are sources, points that are sinks, and points that are neither.

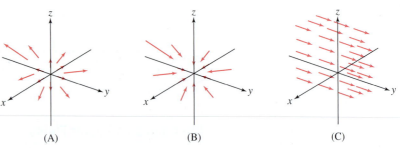

(A) (B) (C)

FIGURE 7

The second operation we consider on a vector field produces another vector field. The **curl** of the vector field $\mathbf{F} = \langle F_1, F_2, F_3 \rangle$ is the vector field defined by the symbolic determinant

$$\text{curl}(\mathbf{F}) = \begin{vmatrix} \mathbf{i} & \mathbf{j} & \mathbf{k} \\ \dfrac{\partial}{\partial x} & \dfrac{\partial}{\partial y} & \dfrac{\partial}{\partial z} \\ F_1 & F_2 & F_3 \end{vmatrix}$$

$$= \left(\frac{\partial F_3}{\partial y} - \frac{\partial F_2}{\partial z} \right) \mathbf{i} - \left(\frac{\partial F_3}{\partial x} - \frac{\partial F_1}{\partial z} \right) \mathbf{j} + \left(\frac{\partial F_2}{\partial x} - \frac{\partial F_1}{\partial y} \right) \mathbf{k}$$

In more compact form, the curl is the symbolic cross product

$$\text{curl}(\mathbf{F}) = \nabla \times \mathbf{F}$$

where ∇ is again the del "operator" $\nabla = \left\langle \dfrac{\partial}{\partial x}, \dfrac{\partial}{\partial y}, \dfrac{\partial}{\partial z} \right\rangle$. In terms of components, $\text{curl}(\mathbf{F})$ is the vector field

$$\text{curl}(\mathbf{F}) = \left\langle \frac{\partial F_3}{\partial y} - \frac{\partial F_2}{\partial z}, \frac{\partial F_1}{\partial z} - \frac{\partial F_3}{\partial x}, \frac{\partial F_2}{\partial x} - \frac{\partial F_1}{\partial y} \right\rangle$$

It is straightforward to check that curl also obeys the **linearity** rules:

$$\text{curl}(\mathbf{F} + \mathbf{G}) = \text{curl}(\mathbf{F}) + \text{curl}(\mathbf{G})$$

$$\text{curl}(c\mathbf{F}) = c\,\text{curl}(\mathbf{F}) \qquad (c \text{ any constant})$$

■ **EXAMPLE 4** **Calculating the Curl** Calculate the curl of $\mathbf{F} = \langle xy, e^x, y + z \rangle$.

Solution We compute the curl as a symbolic determinant:

$$
\text{curl}(\mathbf{F}) = \begin{vmatrix} \mathbf{i} & \mathbf{j} & \mathbf{k} \\ \dfrac{\partial}{\partial x} & \dfrac{\partial}{\partial y} & \dfrac{\partial}{\partial z} \\ xy & e^x & y+z \end{vmatrix}
$$

$$
= \left(\frac{\partial}{\partial y}(y+z) - \frac{\partial}{\partial z} e^x \right) \mathbf{i} - \left(\frac{\partial}{\partial x}(y+z) - \frac{\partial}{\partial z} xy \right) \mathbf{j} + \left(\frac{\partial}{\partial x} e^x - \frac{\partial}{\partial y} xy \right) \mathbf{k}
$$

$$
= \mathbf{i} + (e^x - x)\mathbf{k} \qquad\qquad ■
$$

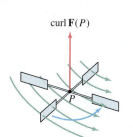

curl $\mathbf{F}(P)$

FIGURE 8 curl $\mathbf{F}(P)$ tells us about the rotation of the fluid.

The magnitude of the vector curl $\mathbf{F}(P)$ is a measure of how fast the vector field $\mathbf{F}$, when considered as the velocity vector field of a fluid flow, would turn a paddle wheel inserted into the fluid as in Figure 8. The paddle wheel is placed perpendicular to the vector curl $\mathbf{F}(P)$ in order to achieve the fastest rotation. We justify this interpretation of curl($\mathbf{F}$) in Sections 17.1 and 17.2, where we investigate the physical significance of the curl.

We have now seen that the "del" operator $\nabla = \left\langle \dfrac{\partial}{\partial x}, \dfrac{\partial}{\partial y}, \dfrac{\partial}{\partial z} \right\rangle$ can be applied in three ways. It can be applied to a scalar function $f(x, y, z)$ to obtain the gradient vector field ∇f, as in Chapter 14. It can be applied to a vector field $\mathbf{F}$ via a symbolic dot product to obtain the divergence div($\mathbf{F}$), a scalar function. And it can be applied to a vector field $\mathbf{F}$ via a symbolic cross product to obtain the vector field curl($\mathbf{F}$). As we will see, these three operations are intimately related. In particular, we now turn to look at those vector fields that result from applying ∇ to a scalar function to obtain what is known as a gradient vector field.

Conservative Vector Fields

- The term "conservative" comes from physics and the Law of Conservation of Energy (see Section 16.3).
- Any letter can be used to denote a potential function. We use f. Some textbooks use V, which suggests "volt," the unit of electric potential. Others use $\phi(x, y, z)$ or $U(x, y, z)$.

Given a differentiable function $f(x, y, z)$, its gradient vector field

$$
\mathbf{F} = \nabla f = \left\langle \frac{\partial f}{\partial x}, \frac{\partial f}{\partial y}, \frac{\partial f}{\partial z} \right\rangle
$$

is called a **conservative vector field**, and the function f is called a **potential function** (or scalar potential function) for $\mathbf{F}$.

The same terms apply in two variables and, more generally, in n variables. Recall that the gradient vectors are orthogonal to the level curves, and thus in a conservative vector field, the vector at every point P is orthogonal to the level curve through P (Figure 9). Conservative vector fields have critically important properties. For instance, in Section 16.3 we will see that the work done by a conservative vector field as a particle travels from point A to point B is independent of the path taken. In physics, conservative vector fields appear naturally as the forces corresponding to physical systems in which energy is conserved.

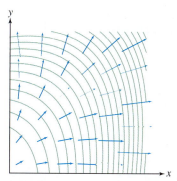

FIGURE 9 A conservative vector field is orthogonal to the level curves of the potential function.

■ **EXAMPLE 5** Verify that $f(x, y, z) = xy + yz^2$ is a potential function for the vector field $\mathbf{F} = \langle y, x + z^2, 2yz \rangle$.

Solution We compute the gradient of f:

$$
\frac{\partial f}{\partial x} = y, \quad \frac{\partial f}{\partial y} = x + z^2, \quad \frac{\partial f}{\partial z} = 2yz
$$

Thus, $\nabla f = \langle y, x + z^2, 2yz \rangle = \mathbf{F}$ as claimed. ■

A conservative vector field always has the property that its curl is the trivial vector field $\mathbf{0}$.

THEOREM 1 Curl of a Conservative Vector Field

1. In dimension 2, if the vector field $\mathbf{F} = \langle F_1, F_2 \rangle$ is conservative, then

$$\boxed{\frac{\partial F_1}{\partial y} = \frac{\partial F_2}{\partial x}}$$

2. In dimension 3, if the vector field $\mathbf{F} = \langle F_1, F_2, F_3 \rangle$ is conservative, then

$$\boxed{\text{curl}(\mathbf{F}) = \mathbf{0}, \quad \text{or equivalently,} \quad \frac{\partial F_1}{\partial y} = \frac{\partial F_2}{\partial x}, \quad \frac{\partial F_2}{\partial z} = \frac{\partial F_3}{\partial y}, \quad \frac{\partial F_3}{\partial x} = \frac{\partial F_1}{\partial z}}$$

Note that we could also write this result as $\text{curl}(\nabla f) = \mathbf{0}$ or $\nabla \times \nabla f = \mathbf{0}$.

Proof We provide the proof for a vector field in dimension 3, but the same idea works for a vector field in dimension 2. If $\mathbf{F} = \nabla f$, then

$$F_1 = \frac{\partial f}{\partial x}, \quad F_2 = \frac{\partial f}{\partial y}, \quad F_3 = \frac{\partial f}{\partial z}$$

Now compare the "cross"-partial derivatives:

$$\frac{\partial F_1}{\partial y} = \frac{\partial}{\partial y}\left(\frac{\partial f}{\partial x}\right) = \frac{\partial^2 f}{\partial y \partial x}$$

$$\frac{\partial F_2}{\partial x} = \frac{\partial}{\partial x}\left(\frac{\partial f}{\partial y}\right) = \frac{\partial^2 f}{\partial x \partial y}$$

Clairaut's Theorem (Section 14.3) tells us that $\dfrac{\partial^2 f}{\partial y \, \partial x} = \dfrac{\partial^2 f}{\partial x \, \partial y}$, and thus,

$$\frac{\partial F_1}{\partial y} = \frac{\partial F_2}{\partial x}$$

Similarly, $\dfrac{\partial F_2}{\partial z} = \dfrac{\partial F_3}{\partial y}$ and $\dfrac{\partial F_3}{\partial x} = \dfrac{\partial F_1}{\partial z}$. It follows that $\text{curl}(\mathbf{F}) = \mathbf{0}$. ∎

From Theorem 1, we can see that most vector fields are *not* conservative. Indeed, an arbitrary triple of functions $\langle F_1, F_2, F_3 \rangle$ does not satisfy the cross-partials condition. Here is an example.

■ **EXAMPLE 6** Show that $\mathbf{F} = \left\langle xy, \dfrac{x^2}{2}, zy \right\rangle$ is not conservative.

Solution We have

$$\frac{\partial F_1}{\partial y} = \frac{\partial}{\partial y}(xy) = x, \qquad \frac{\partial F_2}{\partial x} = \frac{\partial}{\partial x}\left(\frac{x^2}{2}\right) = x$$

Thus, $\dfrac{\partial F_1}{\partial y} = \dfrac{\partial F_2}{\partial x}$. However,

$$\frac{\partial F_2}{\partial z} = \frac{\partial}{\partial z}\left(\frac{x^2}{2}\right) = 0, \qquad \frac{\partial F_3}{\partial y} = \frac{\partial}{\partial y}(zy) = z$$

Thus, $\dfrac{\partial F_2}{\partial z} \neq \dfrac{\partial F_3}{\partial y}$. By Theorem 1, $\mathbf{F}$ is not conservative, even though the last pair of cross partials also agree:

$$\frac{\partial F_3}{\partial x} = \frac{\partial F_1}{\partial z} = 0$$

■

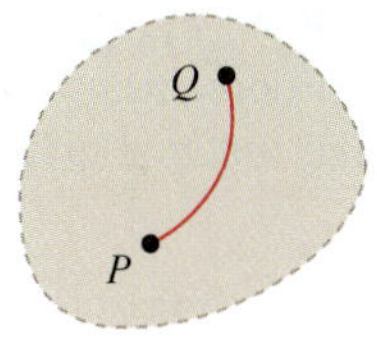

Connected domain

FIGURE 10 In a connected open domain $\mathcal{D}$, any two points in $\mathcal{D}$ can be joined by a path entirely contained in $\mathcal{D}$.

Potential functions, like antiderivatives in one variable, are unique to within an additive constant. To state this precisely, we must assume that the domain $\mathcal{D}$ of the vector field is open and connected (Figure 10). "Connected" means that any two points can be joined by a path within the domain.

> **THEOREM 2 Uniqueness of Potential Functions** If $\mathbf{F}$ is conservative on an open connected domain, then any two potential functions of $\mathbf{F}$ differ by a constant.

Proof If both f_1 and f_2 are potential functions of $\mathbf{F}$, then

$$\nabla(f_1 - f_2) = \nabla f_1 - \nabla f_2 = \mathbf{F} - \mathbf{F} = \mathbf{0}$$

However, a function whose gradient is zero on an open connected domain is a constant function (this generalizes the fact from single-variable calculus that a function on an interval with zero derivative is a constant function—see Exercise 55). Thus, $f_1 - f_2 = C$ for some constant C, and hence, $f_1 = f_2 + C$. ■

The next two examples consider two important radial vector fields.

The result of Example 7 is valid in $\mathbf{R}^2$: The function

$$f(x, y) = \sqrt{x^2 + y^2} = r$$

is a potential function for the unit radial vector field $\mathbf{e}_r = \langle x/r, y/r, z/r \rangle$.

■ **EXAMPLE 7 Unit Radial Vector Fields** Show that

$$f(x, y, z) = r = \sqrt{x^2 + y^2 + z^2}$$

is a potential function for the unit radial vector field $\mathbf{e}_r = \left\langle \frac{x}{r}, \frac{y}{r}, \frac{z}{r} \right\rangle$. That is, $\mathbf{e}_r = \nabla r$.

Solution We have

$$\frac{\partial r}{\partial x} = \frac{\partial}{\partial x}\sqrt{x^2 + y^2 + z^2} = \frac{x}{\sqrt{x^2 + y^2 + z^2}} = \frac{x}{r}$$

Similarly, $\dfrac{\partial r}{\partial y} = \dfrac{y}{r}$ and $\dfrac{\partial r}{\partial z} = \dfrac{z}{r}$. Therefore, $\nabla r = \left\langle \dfrac{x}{r}, \dfrac{y}{r}, \dfrac{z}{r} \right\rangle = \mathbf{e}_r$. ■

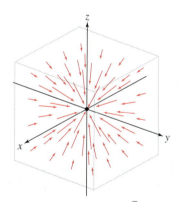

FIGURE 11 The vector field $-\dfrac{Gm\mathbf{e}_r}{r^2}$ represents the force of gravitational attraction due to a point mass located at the origin.

The gravitational force exerted by a point mass m is described by an inverse-square force field (Figure 11). A point mass located at the origin exerts a gravitational force $\mathbf{F}$ on a unit mass located at (x, y, z) equal to

$$\mathbf{F} = -\frac{Gm}{r^2}\mathbf{e}_r = -Gm\left\langle \frac{x}{r^3}, \frac{y}{r^3}, \frac{z}{r^3} \right\rangle$$

where G is the universal gravitation constant. The minus sign indicates that the force is attractive (it pulls in the direction of the origin). The electrostatic force field due to a charged particle is also an inverse-square vector field. The next example shows that these vector fields are conservative.

◄·· **REMINDER**

$$\mathbf{e}_r = \left\langle \frac{x}{r}, \frac{y}{r}, \frac{z}{r} \right\rangle$$

where

$$r = (x^2 + y^2 + z^2)^{1/2}$$

In $\mathbf{R}^2$,

$$\mathbf{e}_r = \left\langle \frac{x}{r}, \frac{y}{r} \right\rangle$$

where $r = (x^2 + y^2)^{1/2}$.

■ **EXAMPLE 8 Inverse-Square Vector Field** Show that

$$\frac{\mathbf{e}_r}{r^2} = \nabla\left(\frac{-1}{r}\right)$$

Solution Use the Chain Rule for gradients (Theorem 1 in Section 14.5) and Example 7:

$$\nabla(-r^{-1}) = r^{-2}\nabla r = r^{-2}\mathbf{e}_r$$ ■

16.1 SUMMARY

- A *vector field* assigns a vector to each point in a domain. A vector field in $\mathbf{R}^3$ is represented by a triple of functions

$$\mathbf{F} = \langle F_1, F_2, F_3 \rangle$$

A vector field in $\mathbf{R}^2$ is represented by a pair of functions $\mathbf{F} = \langle F_1, F_2 \rangle$. We always assume that the components F_j are smooth functions on their domains.

- The *divergence* of a vector field $\mathbf{F} = \langle F_1, F_2, F_3 \rangle$ is the scalar function given by

$$\text{div}(\mathbf{F}) = \nabla \cdot \mathbf{F} = \frac{\partial F_1}{\partial x} + \frac{\partial F_2}{\partial y} + \frac{\partial F_3}{\partial z}$$

- The *curl* of a vector field $\mathbf{F} = \langle F_1, F_2, F_3 \rangle$ is the vector field given by

$$\text{curl}(\mathbf{F}) = \nabla \times \mathbf{F} = \left(\frac{\partial F_3}{\partial y} - \frac{\partial F_2}{\partial z} \right) \mathbf{i} - \left(\frac{\partial F_3}{\partial x} - \frac{\partial F_1}{\partial z} \right) \mathbf{j} + \left(\frac{\partial F_2}{\partial x} - \frac{\partial F_1}{\partial y} \right) \mathbf{k}$$

- If $\mathbf{F} = \nabla f$, then f is called a *potential function* for $\mathbf{F}$.
- $\mathbf{F}$ is called *conservative* if it has a potential function.
- Any two potential functions for a conservative vector field differ by a constant (on an open, connected domain).
- A conservative vector field $\mathbf{F} = \langle F_1, F_2, F_3 \rangle$ satisfies the condition:

$$\text{curl}(\mathbf{F}) = \mathbf{0}, \quad \text{or equivalently,} \quad \frac{\partial F_1}{\partial y} = \frac{\partial F_2}{\partial x}, \quad \frac{\partial F_2}{\partial z} = \frac{\partial F_3}{\partial y}, \quad \frac{\partial F_3}{\partial x} = \frac{\partial F_1}{\partial z}$$

- We define

$$r = \sqrt{x^2 + y^2 + z^2}$$

- The radial unit vector field and the inverse-square vector field are conservative:

$$\mathbf{e}_r = \left\langle \frac{x}{r}, \frac{y}{r}, \frac{z}{r} \right\rangle = \nabla r, \qquad \frac{\mathbf{e}_r}{r^2} = \left\langle \frac{x}{r^3}, \frac{y}{r^3}, \frac{z}{r^3} \right\rangle = \nabla(-r^{-1})$$

16.1 EXERCISES

Preliminary Questions

1. Which of the following is a unit vector field in the plane?

(a) $\mathbf{F} = \langle y, x \rangle$

(b) $\mathbf{F} = \left\langle \dfrac{y}{\sqrt{x^2 + y^2}}, \dfrac{x}{\sqrt{x^2 + y^2}} \right\rangle$

(c) $\mathbf{F} = \left\langle \dfrac{y}{x^2 + y^2}, \dfrac{x}{x^2 + y^2} \right\rangle$

2. Sketch an example of a nonconstant vector field in the plane in which each vector is parallel to $\langle 1, 1 \rangle$.

3. Show that the vector field $\mathbf{F} = \langle -z, 0, x \rangle$ is orthogonal to the position vector $\overrightarrow{OP}$ at each point P. Give an example of another vector field with this property.

4. Give an example of a potential function for $\langle yz, xz, xy \rangle$ other than $f(x, y, z) = xyz$.

Exercises

1. Compute and sketch the vector assigned to the points $P = (1, 2)$ and $Q = (-1, -1)$ by the vector field $\mathbf{F} = \langle x^2, x \rangle$.

2. Compute and sketch the vector assigned to the points $P = (1, 2)$ and $Q = (-1, -1)$ by the vector field $\mathbf{F} = \langle -y, x \rangle$.

3. Compute and sketch the vector assigned to the points $P = (0, 1, 1)$ and $Q = (2, 1, 0)$ by the vector field $\mathbf{F} = \langle xy, z^2, x \rangle$.

4. Compute the vector assigned to the points $P = (1, 1, 0)$ and $Q = (2, 1, 2)$ by the vector fields $\mathbf{e}_r$, $\dfrac{\mathbf{e}_r}{r}$, and $\dfrac{\mathbf{e}_r}{r^2}$.

In Exercises 5–12, sketch the following planar vector fields by drawing the vectors attached to points with integer coordinates in the rectangle $-3 \le x \le 3$, $-3 \le y \le 3$. Instead of drawing the vectors with their true lengths, scale them if necessary to avoid overlap.

5. $\mathbf{F} = \langle 1, 0 \rangle$

6. $\mathbf{F} = \langle 1, 1 \rangle$

7. $\mathbf{F} = x\mathbf{i}$

8. $\mathbf{F} = y\mathbf{i}$

9. $\mathbf{F} = \langle 0, x \rangle$

10. $\mathbf{F} = x^2\mathbf{i} + y\mathbf{j}$

11. $\mathbf{F} = \left\langle \dfrac{x}{x^2 + y^2}, \dfrac{y}{x^2 + y^2} \right\rangle$

12. $\mathbf{F} = \left\langle \dfrac{-y}{\sqrt{x^2 + y^2}}, \dfrac{x}{\sqrt{x^2 + y^2}} \right\rangle$

In Exercises 13–16, match each of the following planar vector fields with the corresponding plot in Figure 12.

13. $\mathbf{F} = \langle 2, x \rangle$

14. $\mathbf{F} = \langle 2x + 2, y \rangle$

15. $\mathbf{F} = \langle y, \cos x \rangle$

16. $\mathbf{F} = \langle x + y, x - y \rangle$

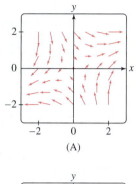
(A)

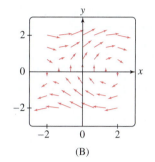
(B)

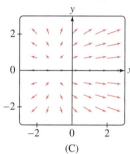
(C)

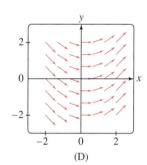
(D)

FIGURE 12

In Exercises 17–20, match each three-dimensional vector field with the corresponding plot in Figure 13.

17. $\mathbf{F} = \langle 1, 1, 1 \rangle$

18. $\mathbf{F} = \langle x, 0, z \rangle$

19. $\mathbf{F} = \langle x, y, z \rangle$

20. $\mathbf{F} = \mathbf{e}_r$

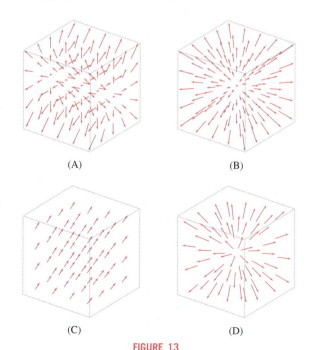
(A)   (B)

(C)   (D)

FIGURE 13

21. A river 200 meters wide is modeled by the region in the xy-plane given by $-100 \le x \le 100$. The velocity vector field on the surface

of the river is given by $\mathbf{F} = \left\langle \dfrac{-x}{20}, 20 - \dfrac{x^2}{1000} \right\rangle$ in meters per second (m/s). Determine the coordinates of those points that have the maximum speed.

22. The velocity vectors in kilometers per hour for the wind speed of a tornado near the ground are given by the vector field $\mathbf{F} = \left\langle \dfrac{-y}{e^{(x^2+y^2-1)^2}}, \dfrac{x}{e^{(x^2+y^2-1)^2}} \right\rangle$. Determine the coordinates of those points where the wind speed is the highest.

In Exercises 23–30, calculate div$(\mathbf{F})$ *and* curl$(\mathbf{F})$.

23. $\mathbf{F} = \langle xy, yz, y^2 - x^3 \rangle$

24. $x\mathbf{i} + y\mathbf{j} + z\mathbf{k}$

25. $\mathbf{F} = \langle x - 2zx^2, z - xy, z^2x^2 \rangle$

26. $\sin(x + z)\mathbf{i} - ye^{xz}\mathbf{k}$

27. $\mathbf{F} = \langle z - y^2, x + z^3, y + x^2 \rangle$

28. $\mathbf{F} = \left\langle \dfrac{y}{x}, \dfrac{y}{z}, \dfrac{z}{x} \right\rangle$

29. $\mathbf{F} = \langle e^y, \sin x, \cos x \rangle$

30. $\mathbf{F} = \left\langle \dfrac{x}{x^2 + y^2}, \dfrac{y}{x^2 + y^2}, 0 \right\rangle$

In Exercises 31–37, prove the identities assuming that the appropriate partial derivatives exist and are continuous.

31. div$(\mathbf{F} + \mathbf{G}) = $ div$(\mathbf{F}) + $ div$(\mathbf{G})$

32. curl$(\mathbf{F} + \mathbf{G}) = $ curl$(\mathbf{F}) + $ curl$(\mathbf{G})$

33. div curl$(\mathbf{F}) = 0$

34. div$(\mathbf{F} \times \mathbf{G}) = \mathbf{G} \cdot $ curl$(\mathbf{F}) - \mathbf{F} \cdot $ curl$(\mathbf{G})$

35. If f is a scalar function, then div$(f\mathbf{F}) = f$ div$(\mathbf{F}) + \mathbf{F} \cdot \nabla f$.

36. curl$(f\mathbf{F}) = f$ curl$(\mathbf{F}) + (\nabla f) \times \mathbf{F}$

37. div$(\nabla f \times \nabla g) = 0$

38. Find (by inspection) a potential function for $\mathbf{F} = \langle x, 0 \rangle$ and prove that $\mathbf{G} = \langle y, 0 \rangle$ is not conservative.

In Exercises 39–45, find a potential function for the vector field $\mathbf{F}$ by inspection or show that one does not exist.

39. $\mathbf{F} = \langle x, y \rangle$

40. $\mathbf{F} = \langle yz, xz, y \rangle$

41. $\mathbf{F} = \langle ye^{xy}, xe^{xy} \rangle$

42. $\mathbf{F} = \langle 2xyz, x^2z, x^2yz \rangle$

43. $\mathbf{F} = \langle yz^2, xz^2, 2xyz \rangle$

44. $\mathbf{F} = \langle 2xze^{x^2}, 0, e^{x^2} \rangle$

45. $\mathbf{F} = \langle yz \cos(xyz), xz \cos(xyz), xy \cos(xyz) \rangle$.

46. Find potential functions for $\mathbf{F} = \dfrac{\mathbf{e}_r}{r^3}$ and $\mathbf{G} = \dfrac{\mathbf{e}_r}{r^4}$ in $\mathbf{R}^3$. *Hint:* See Example 8.

47. Show that $\mathbf{F} = \langle 3, 1, 2 \rangle$ is conservative. Then prove more generally that any constant vector field $\mathbf{F} = \langle a, b, c \rangle$ is conservative.

48. Let $\varphi = \ln r$, where $r = \sqrt{x^2 + y^2}$. Express $\nabla \varphi$ in terms of the unit radial vector $\mathbf{e}_r$ in $\mathbf{R}^2$.

49. For $P = (a, b)$, we define the unit radial vector field based at P:

$$\mathbf{e}_P = \frac{\langle x - a, y - b \rangle}{\sqrt{(x - a)^2 + (y - b)^2}}$$

(a) Verify that $\mathbf{e}_P$ is a unit vector field.

(b) Calculate $\mathbf{e}_P(1, 1)$ for $P = (3, 2)$.

(c) Find a potential function for $\mathbf{e}_P$.

50. Which of (A) or (B) in Figure 14 is the contour plot of a potential function for the vector field **F**? Recall that the gradient vectors are perpendicular to the level curves.

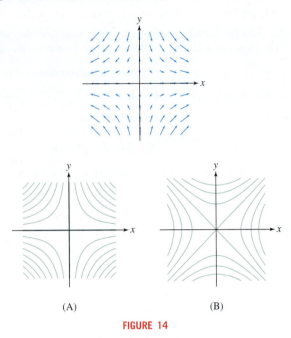

(A)   (B)

FIGURE 14

51. Which of (A) or (B) in Figure 15 is the contour plot of a potential function for the vector field **F**?

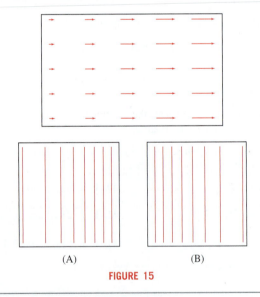

(A)   (B)

FIGURE 15

Further Insights and Challenges

54. Show that any vector field of the form

$$\mathbf{F} = \langle f(x), g(y), h(z) \rangle$$

has a potential function. Assume that f, g, and h are continuous.

55. Let $\mathcal{D}$ be a disk in $\mathbf{R}^2$. This exercise shows that if

$$\nabla f(x, y) = \mathbf{0}$$

for all (x, y) in $\mathcal{D}$, then f is constant. Consider points $P = (a, b)$, $Q = (c, d)$, and $R = (c, b)$ as in Figure 18.

(a) Use single-variable calculus to show that f is constant along the segments $\overline{PR}$ and $\overline{RQ}$.

(b) Conclude that $f(P) = f(Q)$ for any two points $P, Q \in \mathcal{D}$.

52. Match each of these descriptions with a vector field in Figure 16:

(a) The gravitational field created by two planets of equal mass located at P and Q

(b) The electrostatic field created by two equal and opposite charges located at P and Q (representing the force on a negative test charge; opposite charges attract and like charges repel)

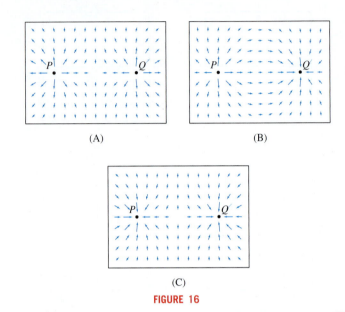

(A)   (B)

(C)

FIGURE 16

53. ✎ In this exercise, we show that the vector field **F** in Figure 17 is not conservative. Explain the following statements:

(a) If a potential function f for **F** exists, then the level curves of f must be vertical lines.

(b) If a potential function f for **F** exists, then the level curves of f must grow farther apart as y increases.

(c) Explain why (a) and (b) are incompatible, and hence f cannot exist.

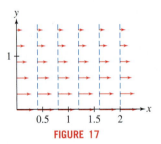

FIGURE 17

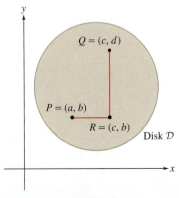

FIGURE 18

16.2 Line Integrals

In this section, we introduce two types of integrals over curves: integrals of functions and integrals of vector fields. These are traditionally called **line integrals**, although it would be more appropriate to call them "curve" or "path" integrals.

Scalar Line Integrals

We begin by defining the **scalar line integral** $\int_{\mathcal{C}} f(x, y, z)\, ds$ of a function f over a curve $\mathcal{C}$. We will see how integrals of this type represent total mass and charge, and how they can be used to find electric potentials.

Like all integrals, this line integral is defined through a process of subdivision, summation, and passage to the limit. We divide $\mathcal{C}$ into N consecutive arcs $\mathcal{C}_1, \ldots, \mathcal{C}_N$, choose a sample point P_i in each arc $\mathcal{C}_i$, and form the Riemann sum (Figure 1)

$$\sum_{i=1}^{N} f(P_i)\, \text{length}(\mathcal{C}_i) = \sum_{i=1}^{N} f(P_i)\, \Delta s_i$$

where Δs_i is the length of $\mathcal{C}_i$.

DF FIGURE 1 The curve $\mathcal{C}$ is divided into N small arcs.

Partition of $\mathcal{C}$ into N small arcs

Choice of sample points P_i in each arc

The line integral of f over $\mathcal{C}$ is the limit (if it exists) of these Riemann sums as the maximum of the lengths Δs_i approaches zero:

In Eq. (1), we write $\{\Delta s_i\} \to 0$ to indicate that the limit is taken over all Riemann sums as the maximum of the lengths Δs_i tends to zero.

$$\int_{\mathcal{C}} f(x, y, z)\, ds = \lim_{\{\Delta s_i\} \to 0} \sum_{i=1}^{N} f(P_i)\, \Delta s_i \qquad \boxed{1}$$

This definition also applies to functions $f(x, y)$ of two variables.

The scalar line integral of the function $f(x, y, z) = 1$ is simply the length of $\mathcal{C}$. In this case, all the Riemann sums have the same value:

$$\sum_{i=1}^{N} 1\, \Delta s_i = \sum_{i=1}^{N} \text{length}(\mathcal{C}_i) = \text{length}(\mathcal{C})$$

and thus,

$$\int_{\mathcal{C}} 1\, ds = \text{length}(\mathcal{C})$$

In practice, line integrals are computed using parametrizations. Suppose that $\mathcal{C}$ has a parametrization $\mathbf{r}(t)$ for $a \le t \le b$ with continuous derivative $\mathbf{r}'(t)$. Recall that the derivative is the tangent vector

$$\mathbf{r}'(t) = \langle x'(t), y'(t), z'(t) \rangle$$

We divide $\mathcal{C}$ into N consecutive arcs $\mathcal{C}_1, \ldots, \mathcal{C}_N$ corresponding to a partition of the interval $[a, b]$:

$$a = t_0 < t_1 < \cdot s < t_{N-1} < t_N = b$$

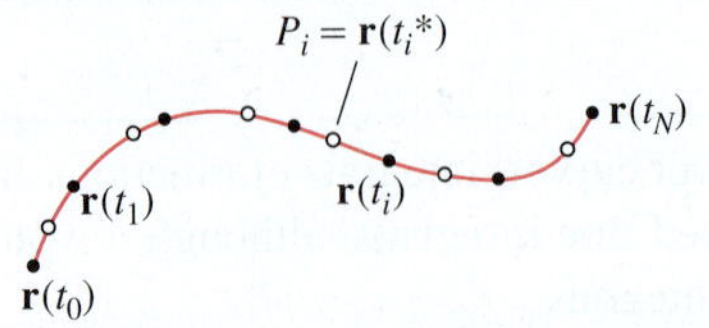

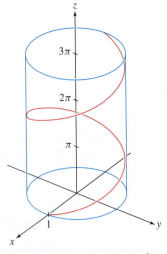

$P_i = \mathbf{r}(t_i^*)$

$\mathbf{r}(t_N)$

$\mathbf{r}(t_1)$ $\mathbf{r}(t_i)$

$\mathbf{r}(t_0)$

FIGURE 2 Partition of parametrized curve $\mathbf{r}(t)$.

◄·· **REMINDER** Arc length formula: The length s of a path $\mathbf{r}(t)$ for $a \leq t \leq b$ is

$$s = \int_a^b \|\mathbf{r}'(t)\| \, dt$$

$$= \int_a^b \sqrt{x'(t)^2 + y'(t)^2 + z'(t)^2} \, dt$$

so that $\mathcal{C}_i$ is parametrized by $\mathbf{r}(t)$ for $t_{i-1} \leq t \leq t_i$ (Figure 2), and choose sample points $P_i = \mathbf{r}(t_i^*)$ with t_i^* in $[t_{i-1}, t_i]$. According to the arc length formula (Section 13.3),

$$\text{Length}(\mathcal{C}_i) = \Delta s_i = \int_{t_{i-1}}^{t_i} \|\mathbf{r}'(t)\| \, dt$$

Because $\mathbf{r}'(t)$ is continuous, the function $\|\mathbf{r}'(t)\|$ is nearly constant on $[t_{i-1}, t_i]$ if the length $\Delta t_i = t_i - t_{i-1}$ is small, and thus, $\int_{t_{i-1}}^{t_i} \|\mathbf{r}'(t)\| \, dt \approx \|\mathbf{r}'(t_i^*)\| \Delta t_i$. This gives us the approximation

$$\sum_{i=1}^{N} f(P_i) \Delta s_i \approx \sum_{i=1}^{N} f(\mathbf{r}(t_i^*)) \|\mathbf{r}'(t_i^*)\| \Delta t_i \qquad \boxed{2}$$

The sum on the right is a Riemann sum that converges to the integral

$$\int_a^b f(\mathbf{r}(t)) \|\mathbf{r}'(t)\| \, dt \qquad \boxed{3}$$

as the maximum of the lengths Δt_i tends to zero. By estimating the errors in this approximation, we can show that the sums on the left-hand side of (2) also approach (3). This gives us the following formula for the scalar line integral.

THEOREM 1 Computing a Scalar Line Integral Let $\mathbf{r}(t)$ be a parametrization of a curve $\mathcal{C}$ for $a \leq t \leq b$. If $f(x, y, z)$ and $\mathbf{r}'(t)$ are continuous, then

$$\int_{\mathcal{C}} f(x, y, z) \, ds = \int_a^b f(\mathbf{r}(t)) \|\mathbf{r}'(t)\| \, dt \qquad \boxed{4}$$

Since arc length along a curve is given by $s(t) = \int_a^t \|\mathbf{r}'(t)\| \, dt$, the Fundamental Theorem of Calculus says that $\dfrac{ds}{dt} = \|\mathbf{r}'(t)\|$. Hence, it makes sense to call $ds = \dfrac{ds}{dt} \, dt = \|\mathbf{r}'(t)\| \, dt$ the arc length differential.

The symbol ds is intended to suggest arc length s and is often referred to as the **line element** or **arc length differential**. In terms of a parametrization, we have the symbolic equation

$$\boxed{ds = \|\mathbf{r}'(t)\| \, dt}$$

where

$$\|\mathbf{r}'(t)\| = \sqrt{x'(t)^2 + y'(t)^2 + z'(t)^2}$$

Eq. (4) tells us that to evaluate a scalar line integral, we replace the integrand $f(x, y, z) \, ds$ with $f(\mathbf{r}(t)) \|\mathbf{r}'(t)\| \, dt$.

■ **EXAMPLE 1 Integrating Along a Helix** Calculate

$$\int_{\mathcal{C}} (x + y + z) \, ds$$

where $\mathcal{C}$ is the helix $\mathbf{r}(t) = (\cos t, \sin t, t)$ for $0 \leq t \leq \pi$ (Figure 3).

Solution

Step 1. Compute ds.

$$\mathbf{r}'(t) = \langle -\sin t, \cos t, 1 \rangle$$

$$\|\mathbf{r}'(t)\| = \sqrt{(-\sin t)^2 + \cos^2 t + 1} = \sqrt{2}$$

$$ds = \|\mathbf{r}'(t)\| \, dt = \sqrt{2} \, dt$$

z

3π

2π

π

1

x

y

FIGURE 3 The helix $\mathbf{r}(t) = (\cos t, \sin t, t)$.

Step 2. Write out the integrand and evaluate.

We have $f(x, y, z) = x + y + z$, and so,

$$f(\mathbf{r}(t)) = f(\cos t, \sin t, t) = \cos t + \sin t + t$$

$$f(x, y, z)\, ds = f(\mathbf{r}(t)) \|\mathbf{r}'(t)\|\, dt = (\cos t + \sin t + t)\sqrt{2}\, dt$$

By Eq. (4),

$$\int_C f(x, y, z)\, ds = \int_0^\pi f(\mathbf{r}(t)) \|\mathbf{r}'(t)\|\, dt = \int_0^\pi (\cos t + \sin t + t)\sqrt{2}\, dt$$

$$= \sqrt{2}\left(\sin t - \cos t + \frac{1}{2}t^2\right)\bigg|_0^\pi$$

$$= \sqrt{2}\left(0 + 1 + \frac{1}{2}\pi^2\right) - \sqrt{2}\,(0 - 1 + 0) = 2\sqrt{2} + \frac{\sqrt{2}}{2}\pi^2 \quad \blacksquare$$

■ **EXAMPLE 2** **Arc Length** Calculate $\int_C 1\, ds$ for the helix in the previous example: $\mathbf{r}(t) = (\cos t, \sin t, t)$ for $0 \leq t \leq \pi$. What does this integral represent?

Solution In the previous example, we showed that $ds = \sqrt{2}\, dt$ and thus,

$$\int_C 1\, ds = \int_0^\pi \sqrt{2}\, dt = \pi\sqrt{2}$$

This is the length of the helix for $0 \leq t \leq \pi$. ■

Applications of the Scalar Line Integral

In Section 15.5, we discussed the general principle that "the integral of a density is the total quantity." This applies to scalar line integrals. For example, we can view the curve $\mathcal{C}$ as a wire with continuous **mass density** $\delta(x, y, z)$, given in units of mass per unit length. The total mass is defined as the integral of mass density:

$$\boxed{\text{Total mass of } \mathcal{C} = \int_C \delta(x, y, z)\, ds} \qquad \boxed{5}$$

A similar formula for total charge is valid if $\delta(x, y, z)$ is the charge density along the curve. As in Section 15.5, we justify this interpretation by dividing $\mathcal{C}$ into N arcs $\mathcal{C}_i$ of length Δs_i with N large. The mass density is nearly constant on $\mathcal{C}_i$, and therefore, the mass of $\mathcal{C}_i$ is approximately $\delta(P_i)\, \Delta s_i$, where P_i is any sample point on $\mathcal{C}_i$ (Figure 4). The total mass is the sum

$$\text{Total mass of } \mathcal{C} = \sum_{i=1}^N \text{mass of } \mathcal{C}_i \approx \sum_{i=1}^N \delta(P_i)\, \Delta s_i$$

As the maximum of the lengths Δs_i tends to zero, the sums on the right approach the line integral in Eq. (5).

■ **EXAMPLE 3** **Scalar Line Integral as Total Mass** Find the total mass of a wire in the shape of the parabola $y = x^2$ for $1 \leq x \leq 4$ (in centimeters) with mass density given by $\delta(x, y) = y/x$ g/cm.

Solution The arc of the parabola is parametrized by $\mathbf{r}(t) = (t, t^2)$ for $1 \leq t \leq 4$.

Step 1. Compute ds.

$$\mathbf{r}'(t) = \langle 1, 2t \rangle$$

$$ds = \|\mathbf{r}'(t)\|\, dt = \sqrt{1 + 4t^2}\, dt$$

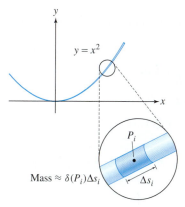

FIGURE 4

Mass $\approx \delta(P_i)\Delta s_i$

Step 2. Write out the integrand and evaluate.

We have $\delta(\mathbf{r}(t)) = \rho(t, t^2) = t^2/t = t$, and thus,

$$\delta(x, y)\, ds = \rho(\mathbf{r}(t))\sqrt{1 + 4t^2}\, dt = t\sqrt{1 + 4t^2}\, dt$$

We evaluate the line integral of mass density using the substitution $u = 1 + 4t^2$, $du = 8t\, dt$:

$$\int_C \delta(x, y)\, ds = \int_1^4 \delta(\mathbf{r}(t)) \|\mathbf{r}'(t)\|\, dt = \int_1^4 t\sqrt{1 + 4t^2}\, dt$$

$$= \frac{1}{8} \int_5^{65} \sqrt{u}\, du = \frac{1}{12} u^{3/2} \Big|_5^{65}$$

$$= \frac{1}{12}(65^{3/2} - 5^{3/2}) \approx 42.74$$

Note that after the substitution, the limits of integration become $u(1) = 5$ and $u(4) = 65$. The total mass of the wire is approximately 42.7 g. ∎

Scalar line integrals are also used to compute electric potentials. When an electric charge is distributed continuously along a curve $\mathcal{C}$, with charge density $\delta(x, y, z)$, the charge distribution sets up an electrostatic field $\mathbf{E}$ that is a conservative vector field. Coulomb's Law tells us that $\mathbf{E} = -\nabla V$, where

> By definition, $\mathbf{E}$ is the vector field with the property that the electrostatic force on a point charge q placed at location $P = (x, y, z)$ is the vector $q\mathbf{E}(x, y, z)$.

$$\boxed{V(P) = k \int_C \frac{\delta(x, y, z)\, ds}{d_P}} \qquad \boxed{6}$$

> The constant k is usually written as $\frac{1}{4\pi\epsilon_0}$, where ϵ_0 is the vacuum permittivity.

In this integral, $d_P = d_P(x, y, z)$ denotes the distance from (x, y, z) to P. The constant k has the value $k = 8.99 \times 10^9$ N-m²/C². In a situation like this, we use V to denote the function and call it the **electric potential**. It is defined for all points P that do not lie on $\mathcal{C}$ and has units of volts [1 volt (or V) is 1 N-m/C].

■ **EXAMPLE 4 Electric Potential** A charged semicircle of radius R centered at the origin in the xy-plane (Figure 5) has charge density

$$\delta(x, y, 0) = 10^{-8}\left(2 - \frac{x}{R}\right) \text{ C/m}$$

Find the electric potential at a point $P = (0, 0, a)$ if $R = 0.1$ m.

Solution Parametrize the semicircle by $\mathbf{r}(t) = (R\cos t, R\sin t, 0)$ for $-\pi/2 \le t \le \pi/2$:

$$\|\mathbf{r}'(t)\| = \|\langle -R\sin t, R\cos t, 0\rangle\| = \sqrt{R^2\sin^2 t + R^2\cos^2 t + 0} = R$$

$$ds = \|\mathbf{r}'(t)\|\, dt = R\, dt$$

$$\delta(\mathbf{r}(t)) = \rho(R\cos t, R\sin t, 0) = 10^{-8}\left(2 - \frac{R\cos t}{R}\right) = 10^{-8}(2 - \cos t)$$

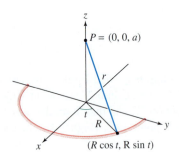

FIGURE 5

In our case, the distance d_P from P to a point $(x, y, 0)$ on the semicircle has the constant value $d_P = \sqrt{R^2 + a^2}$ (Figure 5). Thus,

$$V(P) = k \int_C \frac{\delta(x, y, z)\, ds}{d_P} = k \int_C \frac{10^{-8}(2 - \cos t)\, R\, dt}{\sqrt{R^2 + a^2}}$$

$$= \frac{10^{-8}kR}{\sqrt{R^2 + a^2}} \int_{-\pi/2}^{\pi/2} (2 - \cos t)\, dt = \frac{10^{-8}kR}{\sqrt{R^2 + a^2}}(2\pi - 2)$$

With $R = 0.1$ m and $k = 8.99 \times 10^9$, we then obtain $10^{-8}kR(2\pi - 2) \approx 38.5$ and

$$V(P) \approx \frac{38.5}{\sqrt{0.01 + a^2}} \text{ volts.} \qquad ∎$$

Vector Line Integrals

When you carry a backpack up a mountain, you do work against the earth's gravitational field. The work, or energy expended, is one example of a quantity represented by a vector line integral.

An important difference between vector and scalar line integrals is that vector line integrals depend on the *direction* along the curve. This is reasonable if you think of the vector line integral as work, because the work performed going down the mountain is the negative of the work performed going up.

A specified direction along a curve C is called an **orientation** (Figure 6). We refer to this direction as the **positive** direction along C, the opposite direction is the **negative** direction, and C itself is called an **oriented curve**. In Figure 6(A), if we reversed the orientation, the positive direction would become the direction from Q to P.

*The unit tangent vector **T** varies from point to point along the curve. When it is necessary to stress this dependence, we write **T**(P).*

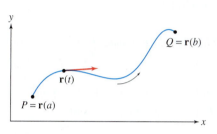

(A) Oriented path from P to Q

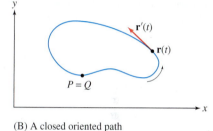

(B) A closed oriented path

FIGURE 6

The line integral of a vector field $\mathbf{F}$ over a curve C is defined as the scalar line integral of the tangential component of $\mathbf{F}$. More precisely, let $\mathbf{T} = \mathbf{T}(P)$ denote the unit tangent vector at a point P on C pointing in the positive direction. The **tangential component** of $\mathbf{F}$ at P is the dot product (Figure 7)

$$\mathbf{F}(P) \cdot \mathbf{T}(P) = \|\mathbf{F}(P)\| \, \|\mathbf{T}(P)\| \cos\theta = \|\mathbf{F}(P)\| \cos\theta$$

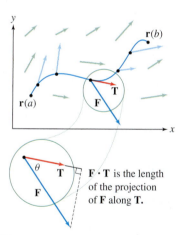

θ $\mathbf{T}$ $\mathbf{F} \cdot \mathbf{T}$ is the length of the projection of $\mathbf{F}$ along $\mathbf{T}$.

DF FIGURE 7 The line integral is the integral of the tangential component of $\mathbf{F}$ along C.

where θ is the angle between $\mathbf{F}(P)$ and $\mathbf{T}(P)$. The vector line integral of $\mathbf{F}$ is the scalar line integral of the scalar function $\mathbf{F} \cdot \mathbf{T}$. We make the standing assumption that C is piecewise smooth (it consists of finitely many smooth curves joined together with possible corners).

DEFINITION Vector Line Integral   The line integral of a vector field $\mathbf{F}$ along an oriented curve C is the integral of the tangential component of $\mathbf{F}$:

$$\int_C \mathbf{F} \cdot d\mathbf{r} = \int_C (\mathbf{F} \cdot \mathbf{T}) \, ds \qquad \boxed{7}$$

We use parametrizations to evaluate vector line integrals, but there is one important difference with the scalar case: The parametrization $\mathbf{r}(t)$ must be *positively oriented*; that is, $\mathbf{r}(t)$ must trace C in the positive direction. We assume also that $\mathbf{r}(t)$ is regular (see Section 13.4); that is, $\mathbf{r}'(t) \neq \mathbf{0}$ for $a \le t \le b$. Then $\mathbf{r}'(t)$ is a nonzero tangent vector pointing in the positive direction, and

$$\mathbf{T} = \frac{\mathbf{r}'(t)}{\|\mathbf{r}'(t)\|}$$

In terms of the arc length differential $ds = \|\mathbf{r}'(t)\| \, dt$, we have

$$(\mathbf{F} \cdot \mathbf{T}) \, ds = \left(\mathbf{F}(\mathbf{r}(t)) \cdot \frac{\mathbf{r}'(t)}{\|\mathbf{r}'(t)\|} \right) \|\mathbf{r}'(t)\| \, dt = \mathbf{F}(\mathbf{r}(t)) \cdot \mathbf{r}'(t) \, dt$$

Therefore, the integral on the right-hand side of Eq. (7) is equal to the right-hand side of Eq. (8) in the next theorem.

> **THEOREM 2 Computing a Vector Line Integral** If $\mathbf{r}(t)$ is a regular parametrization of an oriented curve $\mathcal{C}$ for $a \leq t \leq b$, then
>
> $$\int_{\mathcal{C}} \mathbf{F} \cdot d\mathbf{r} = \int_a^b \mathbf{F}(\mathbf{r}(t)) \cdot \mathbf{r}'(t)\,dt \qquad \boxed{8}$$

It is useful to think of $d\mathbf{r}$ as a "vector line element" or "vector differential" that is related to the parametrization by the symbolic equation

$$d\mathbf{r} = \mathbf{r}'(t)\,dt = \langle x'(t),\, y'(t),\, z'(t) \rangle\,dt$$

Vector line integrals are usually easier to calculate than scalar line integrals, because the length $\|\mathbf{r}'(t)\|$, which involves a square root, does not appear in the integrand.

Eq. (8) tells us that to evaluate a vector line integral, we replace the integrand $\mathbf{F} \cdot d\mathbf{r}$ with $\mathbf{F}(\mathbf{r}(t)) \cdot \mathbf{r}'(t)\,dt$.

■ **EXAMPLE 5** Evaluate $\displaystyle\int_{\mathcal{C}} \mathbf{F} \cdot d\mathbf{r}$, where $\mathbf{F} = \langle z, y^2, x \rangle$ and $\mathcal{C}$ is parametrized (in the positive direction) by $\mathbf{r}(t) = (t + 1, e^t, t^2)$ for $0 \leq t \leq 2$.

Solution There are two steps in evaluating a line integral.

Step 1. **Calculate the integrand.**

$$\mathbf{r}(t) = (t + 1, e^t, t^2)$$

$$\mathbf{F}(\mathbf{r}(t)) = \langle z, y^2, x \rangle = \langle t^2, e^{2t}, t + 1 \rangle$$

$$\mathbf{r}'(t) = \langle 1, e^t, 2t \rangle$$

The integrand (as a differential) is the dot product:

$$\mathbf{F}(\mathbf{r}(t)) \cdot \mathbf{r}'(t)\,dt = \langle t^2, e^{2t}, t + 1 \rangle \cdot \langle 1, e^t, 2t \rangle\,dt = (e^{3t} + 3t^2 + 2t)\,dt$$

Step 2. **Evaluate the line integral.**

$$\int_{\mathcal{C}} \mathbf{F} \cdot d\mathbf{r} = \int_0^2 \mathbf{F}(\mathbf{r}(t)) \cdot \mathbf{r}'(t)\,dt$$

$$= \int_0^2 (e^{3t} + 3t^2 + 2t)\,dt = \left(\frac{1}{3}e^{3t} + t^3 + t^2 \right) \Big|_0^2$$

$$= \left(\frac{1}{3}e^6 + 8 + 4 \right) - \frac{1}{3} = \frac{1}{3}\left(e^6 + 35 \right) \qquad ■$$

Another standard notation for the line integral $\displaystyle\int_{\mathcal{C}} \mathbf{F} \cdot d\mathbf{r}$ is

$$\int_{\mathcal{C}} F_1\,dx + F_2\,dy + F_3\,dz$$

In this notation, we write $d\mathbf{r}$ as a vector differential

$$d\mathbf{r} = \langle dx, dy, dz \rangle$$

so that

$$\mathbf{F} \cdot d\mathbf{r} = \langle F_1, F_2, F_3 \rangle \cdot \langle dx, dy, dz \rangle = F_1\,dx + F_2\,dy + F_3\,dz$$

In terms of a parametrization $\mathbf{r}(t) = (x(t), y(t), z(t))$,

$$d\mathbf{r} = \left\langle \frac{dx}{dt}, \frac{dy}{dt}, \frac{dz}{dt} \right\rangle dt$$

$$\mathbf{F} \cdot d\mathbf{r} = \left(F_1(\mathbf{r}(t))\frac{dx}{dt} + F_2(\mathbf{r}(t))\frac{dy}{dt} + F_3(\mathbf{r}(t))\frac{dz}{dt} \right) dt$$

So, we have the following formula:

$$\int_C F_1\,dx + F_2\,dy + F_3\,dz = \int_a^b \left(F_1(\mathbf{r}(t))\frac{dx}{dt} + F_2(\mathbf{r}(t))\frac{dy}{dt} + F_3(\mathbf{r}(t))\frac{dz}{dt} \right) dt$$

GRAPHICAL INSIGHT The magnitude of a vector line integral (or even whether it is positive or negative) depends on the angles between $\mathbf{F}$ and $\mathbf{T}$ along the path. Consider the line integral of $\mathbf{F} = \langle 2y, -3 \rangle$ around the ellipse in Figure 8.

• In Figure 8(A), the angles θ between $\mathbf{F}$ and $\mathbf{T}$ appear to be mostly obtuse along the top part of the ellipse. Consequently, $\mathbf{F} \cdot \mathbf{T} \leq 0$ and the line integral is negative. We are working against the vector field as we travel along the curve.
• In Figure 8(B), the angles θ appear to be mostly acute along the bottom part of the ellipse. Consequently, $\mathbf{F} \cdot \mathbf{T} \geq 0$ and the line integral is positive. We are working with the vector field as we travel along the curve.
• We can guess that the line integral around the entire ellipse is negative because $\|\mathbf{F}\|$ is larger in the top half, so the negative contribution of $\mathbf{F} \cdot \mathbf{T}$ from the top half appears to dominate the positive contribution of the bottom half. We verify this in Example 6.

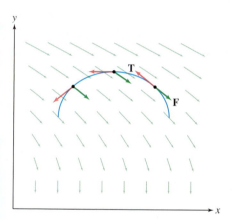

(A) Most of the dot products $\mathbf{F} \cdot \mathbf{T}$ are negative because the angles between the vectors are obtuse.
Therefore, the line integral is negative.

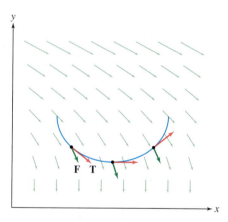

(B) Most of the dot products $\mathbf{F} \cdot \mathbf{T}$ are positive because the angles between the vectors are acute.
Therefore, the line integral is positive.

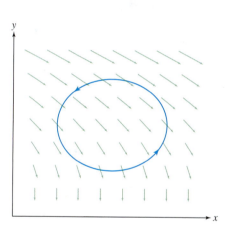

(C) Total line integral is negative.

DF FIGURE 8 The vector field $\mathbf{F} = \langle 2y, -3 \rangle$.

■ EXAMPLE 6 The ellipse C in Figure 8(C) with counterclockwise orientation is parametrized by $\mathbf{r}(\theta) = (5 + 4\cos\theta, 3 + 2\sin\theta)$ for $0 \leq \theta < 2\pi$. Calculate

$$\int_C 2y\,dx - 3\,dy$$

In Example 6, keep in mind that

$$\int_C 2y\,dx - 3\,dy$$

is another notation for the line integral of $\mathbf{F} = \langle 2y, -3 \rangle$ *over* C. *Formally,*

$$\mathbf{F} \cdot d\mathbf{r} = \langle 2y, -3 \rangle \cdot \langle dx, dy \rangle$$
$$= 2y\,dx - 3\,dy$$

Solution We have $x(\theta) = 5 + 4\cos\theta$ and $y(\theta) = 3 + 2\sin\theta$, and

$$\frac{dx}{d\theta} = -4\sin\theta, \qquad \frac{dy}{d\theta} = 2\cos\theta$$

The integrand of the line integral is

$$2y\,dx - 3\,dy = \left(2y\frac{dx}{d\theta} - 3\frac{dy}{d\theta} \right) d\theta$$

$$= \big(2(3 + 2\sin\theta)(-4\sin\theta) - 3(2\cos\theta) \big)\,d\theta$$

$$= -\big(24\sin\theta + 16\sin^2\theta + 6\cos\theta \big)\,d\theta$$

$$\bullet \int \sin^2 \theta \, d\theta = \frac{1}{2}\theta - \frac{1}{4}\sin 2\theta$$

$$\bullet \int_0^{2\pi} \sin^2 \theta \, d\theta = \pi$$

Since the integrals of $\cos \theta$ and $\sin \theta$ over $[0, 2\pi]$ are zero,

$$\int_C 2y \, dx - 3 \, dy = -\int_0^{2\pi} \left(24 \sin \theta + 16 \sin^2 \theta + 6 \cos \theta\right) d\theta$$

$$= -16 \int_0^{2\pi} \sin^2 \theta \, d\theta = -16\pi \qquad \blacksquare$$

We now state some basic properties of vector line integrals. First, given an oriented curve C, we write $-C$ to denote the curve C with the opposite orientation (Figure 9). The unit tangent vector changes sign from **T** to $-$**T** when we change orientation, so the tangential component of **F** and the line integral also change sign:

$$\int_{-C} \mathbf{F} \cdot d\mathbf{r} = -\int_C \mathbf{F} \cdot d\mathbf{r}$$

Next, if we are given n oriented curves $C_1, \ldots, C_n$, we write

$$C = C_1 + \cdot s + C_n$$

to indicate the union of the curves, and we define the line integral over C as the sum

$$\int_C \mathbf{F} \cdot d\mathbf{r} = \int_{C_1} \mathbf{F} \cdot d\mathbf{r} + \cdot s + \int_{C_n} \mathbf{F} \cdot d\mathbf{r}$$

We use this formula to define the line integral when C is **piecewise smooth**, meaning that C is a union of smooth curves $C_1, \ldots, C_n$. For example, the triangle in Figure 10 is piecewise smooth but not smooth. The next theorem summarizes the main properties of vector line integrals.

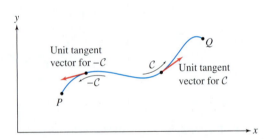

FIGURE 9 The path from P to Q has two possible orientations.

THEOREM 3 Properties of Vector Line Integrals   Let C be a smooth oriented curve, and let **F** and **G** be vector fields.

(i) **Linearity:** $\displaystyle \int_C (\mathbf{F} + \mathbf{G}) \cdot d\mathbf{r} = \int_C \mathbf{F} \cdot d\mathbf{r} + \int_C \mathbf{G} \cdot d\mathbf{r}$

$$\int_C k\mathbf{F} \cdot d\mathbf{r} = k \int_C \mathbf{F} \cdot d\mathbf{r} \quad (k \text{ a constant})$$

(ii) **Reversing orientation:** $\displaystyle \int_{-C} \mathbf{F} \cdot d\mathbf{r} = -\int_C \mathbf{F} \cdot d\mathbf{r}$

(iii) **Additivity:** If C is a union of n smooth curves $C_1 + \cdot s + C_n$, then

$$\int_C \mathbf{F} \cdot d\mathbf{r} = \int_{C_1} \mathbf{F} \cdot d\mathbf{r} + \cdot s + \int_{C_n} \mathbf{F} \cdot d\mathbf{r}$$

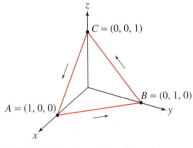

FIGURE 10 The triangle is piecewise smooth: It is the union of its three edges, each of which is smooth.

■ **EXAMPLE 7** Compute $\displaystyle \int_C \mathbf{F} \cdot d\mathbf{r}$, where $\mathbf{F} = \langle e^z, e^y, x + y \rangle$ and C is the triangle joining $(1, 0, 0), (0, 1, 0)$, and $(0, 0, 1)$ oriented in the counterclockwise direction when viewed from above (Figure 10).

Solution The line integral is the sum of the line integrals over the edges of the triangle:

$$\int_C \mathbf{F} \cdot d\mathbf{r} = \int_{\overline{AB}} \mathbf{F} \cdot d\mathbf{r} + \int_{\overline{BC}} \mathbf{F} \cdot d\mathbf{r} + \int_{\overline{CA}} \mathbf{F} \cdot d\mathbf{r}$$

Segment $\overline{AB}$ is parametrized by $\mathbf{r}(t) = (1 - t, t, 0)$ for $0 \le t \le 1$. We have

$$\mathbf{F}(\mathbf{r}(t)) \cdot \mathbf{r}'(t) = \mathbf{F}(1 - t, t, 0) \cdot \langle -1, 1, 0 \rangle = \langle e^0, e^t, 1 \rangle \cdot \langle -1, 1, 0 \rangle = -1 + e^t$$

$$\int_{\overline{AB}} \mathbf{F} \cdot d\mathbf{r} = \int_0^1 (e^t - 1)\, dt = (e^t - t)\Big|_0^1 = (e - 1) - 1 = e - 2$$

Similarly, $\overline{BC}$ is parametrized by $\mathbf{r}(t) = (0, 1 - t, t)$ for $0 \le t \le 1$, and

$$\mathbf{F}(\mathbf{r}(t)) \cdot \mathbf{r}'(t) = \langle e^t, e^{1-t}, 1 - t \rangle \cdot \langle 0, -1, 1 \rangle = -e^{1-t} + 1 - t$$

$$\int_{\overline{BC}} \mathbf{F} \cdot d\mathbf{r} = \int_0^1 (-e^{1-t} + 1 - t)\, dt = \left(e^{1-t} + t - \frac{1}{2}t^2 \right)\Big|_0^1 = \frac{3}{2} - e$$

Finally, $\overline{CA}$ is parametrized by $\mathbf{r}(t) = (t, 0, 1 - t)$ for $0 \le t \le 1$, and

$$\mathbf{F}(\mathbf{r}(t)) \cdot \mathbf{r}'(t) = \langle e^{1-t}, 1, t \rangle \cdot \langle 1, 0, -1 \rangle = e^{1-t} - t$$

$$\int_{\overline{CA}} \mathbf{F} \cdot d\mathbf{r} = \int_0^1 (e^{1-t} - t)\, dt = \left(-e^{1-t} - \frac{1}{2}t^2 \right)\Big|_0^1 = -\frac{3}{2} + e$$

The total line integral is the sum

$$\int_C \mathbf{F} \cdot d\mathbf{r} = (e - 2) + \left(\frac{3}{2} - e \right) + \left(-\frac{3}{2} + e \right) = e - 2 \qquad \blacksquare$$

Applications of the Vector Line Integral

Recall that in physics, "work" refers to the energy expended when a force is applied to an object as it moves along a path. By definition, the work W performed along the straight segment from P to Q by applying a constant force $\mathbf{F}$ at an angle θ [Figure 11(A)] is

$$W = (\text{tangential component of } \mathbf{F}) \times \text{distance} = (\|\mathbf{F}\| \cos\theta) \times PQ$$

When the force acts on the object moving along a curve $\mathcal{C}$, it makes sense to define the work W performed as the line integral [Figure 11(B)]:

$$W = \int_C \mathbf{F} \cdot d\mathbf{r} \qquad \boxed{9}$$

This is the work "performed by the field $\mathbf{F}$." The idea is that we can divide $\mathcal{C}$ into a large number of short consecutive arcs $\mathcal{C}_1, \dots, \mathcal{C}_N$, where $\mathcal{C}_i$ has length Δs_i. The work W_i performed along $\mathcal{C}_i$ is approximately equal to the tangential component $\mathbf{F}(P_i) \cdot \mathbf{T}(P_i)$ times the length Δs_i, where P_i is a sample point in $\mathcal{C}_i$. Thus, we have

$$W = \sum_{i=1}^{N} W_i \approx \sum_{i=1}^{N} (\mathbf{F}(P_i) \cdot \mathbf{T}(P_i)) \Delta s_i$$

The right-hand side approaches $\displaystyle\int_C \mathbf{F} \cdot d\mathbf{r}$ as the lengths Δs_i tend to zero.

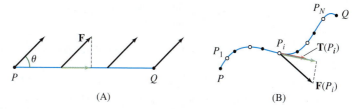

(A) (B)

FIGURE 11

◀··· **REMINDER** *Work has units of energy. The SI unit of force is the newton, and the unit of energy is the joule, defined as 1 newton-meter. The British unit is the foot-pound.*

Often, we are interested in calculating the work required to move an object along a path in the presence of a force field $\mathbf{F}$ (such as an electrical or gravitational field). In this case, $\mathbf{F}$ acts on the object and we must work *against* the force field to move the object. The work required is the negative of the line integral in Eq. (9):

$$\text{Work performed against } \mathbf{F} = -\int_{\mathcal{C}} \mathbf{F} \cdot d\mathbf{r}$$

■ **EXAMPLE 8** **Calculating Work** Calculate the work performed against $\mathbf{F}$ in moving a particle from $P = (0, 0, 0)$ to $Q = (4, 8, 1)$ along the path

$$\mathbf{r}(t) = (t^2, t^3, t) \text{ (in meters)} \qquad \text{for } 1 \leq t \leq 2$$

in the presence of a force field $\mathbf{F} = \langle x^2, -z, -yz^{-1} \rangle$ in newtons.

Solution We have

$$\mathbf{F}(\mathbf{r}(t)) = \mathbf{F}(t^2, t^3, t) = \left\langle t^4, -t, -t^2 \right\rangle$$

$$\mathbf{r}'(t) = \left\langle 2t, 3t^2, 1 \right\rangle$$

$$\mathbf{F} \cdot d\mathbf{r} = \mathbf{F}(\mathbf{r}(t)) \cdot \mathbf{r}'(t)\, dt = \left\langle t^4, -t, -t^2 \right\rangle \cdot \left\langle 2t, 3t^2, 1 \right\rangle dt = (2t^5 - 3t^3 - t^2)\, dt$$

The work performed against the force field in joules is

$$W = -\int_{\mathcal{C}} \mathbf{F} \cdot d\mathbf{r} = -\int_{1}^{2} (2t^5 - 3t^3 - t^2)\, dt = \frac{89}{12}$$ ■

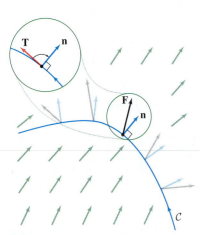

DF FIGURE 12

CAUTION In Sections 13.4 and 13.5, $\mathbf{N}$ was the principal unit normal vector to a curve in space. Here, it represents a normal vector, not necessarily a unit vector, to a curve in the plane, and $\mathbf{n}$ represents the corresponding unit normal vector. This conforms to common usage.

Line integrals are also used to compute the **flux across a plane curve**, defined as the integral of the *normal component* of a vector field, rather than the tangential component (Figure 12). Suppose that a plane curve $\mathcal{C}$ is parametrized by $\mathbf{r}(t)$ for $a \leq t \leq b$, and let

$$\mathbf{N} = \mathbf{N}(t) = \langle y'(t), -x'(t) \rangle, \qquad \mathbf{n}(t) = \frac{\mathbf{N}(t)}{\|\mathbf{N}(t)\|} = \frac{\mathbf{N}(t)}{\|\mathbf{r}'(t)\|}$$

These vectors are normal to $\mathcal{C}$ since the dot product of $\mathbf{N}$ with the tangent vector $\mathbf{r}'(t) = \langle x'(t), y'(t) \rangle$ is 0. Both $\mathbf{N}$ and $\mathbf{n}$ also point to the right as you follow the curve in the direction of $\mathbf{r}$. The flux across $\mathcal{C}$ is the integral of the normal component $\mathbf{F} \cdot \mathbf{n}$, which can be obtained by integrating $\mathbf{F}(\mathbf{r}(t)) \cdot \mathbf{N}(t)$ with respect to t:

$$\text{Flux across } \mathcal{C} = \int_{\mathcal{C}} (\mathbf{F} \cdot \mathbf{n})\, ds = \int_{a}^{b} \mathbf{F}(\mathbf{r}(t)) \cdot \frac{\mathbf{N}(t)}{\|\mathbf{r}'(t)\|} \|\mathbf{r}'(t)\|\, dt = \int_{a}^{b} \mathbf{F}(\mathbf{r}(t)) \cdot \mathbf{N}(t)\, dt$$

10

If $\mathbf{F}$ is the velocity field of a fluid (modeled as a two-dimensional fluid), then the flux is the quantity of fluid flowing across the curve per unit time.

■ **EXAMPLE 9** **Flux Across a Curve** Calculate the flux of the velocity vector field $\mathbf{v} = \langle 3 + 2y - y^2/3, 0 \rangle$ (in centimeters per second) across the quarter ellipse $\mathbf{r}(t) = \langle 3\cos t, 6\sin t \rangle$ for $0 \leq t \leq \frac{\pi}{2}$ (Figure 13).

Solution The vector field along the path is

$$\mathbf{v}(\mathbf{r}(t)) = \left\langle 3 + 2(6\sin t) - (6\sin t)^2/3, 0 \right\rangle = \left\langle 3 + 12\sin t - 12\sin^2 t, 0 \right\rangle$$

The tangent vector is $\mathbf{r}'(t) = \langle -3\sin t, 6\cos t \rangle$, and thus, $\mathbf{N}(t) = \langle 6\cos t, 3\sin t \rangle$. We integrate the dot product

$$\mathbf{v}(\mathbf{r}(t)) \cdot \mathbf{N}(t) = \left\langle 3 + 12\sin t - 12\sin^2 t, 0 \right\rangle \cdot \langle 6\cos t, 3\sin t \rangle$$

$$= (3 + 12\sin t - 12\sin^2 t)(6\cos t)$$

$$= 18\cos t + 72\sin t \cos t - 72\sin^2 t \cos t$$

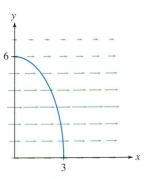

FIGURE 13

to obtain the flux:

$$\int_a^b \mathbf{v}(\mathbf{r}(t)) \cdot \mathbf{N}(t)\, dt = \int_0^{\pi/2} (18\cos t + 72\sin t \cos t - 72\sin^2 t \cos t)\, dt$$

$$= 18 + 36 - 24 = 30 \text{ cm}^2/\text{s} \qquad\blacksquare$$

16.2 SUMMARY

- Line integral over a curve with parametrization $\mathbf{r}(t)$ for $a \le t \le b$:

    Scalar line integral: $\quad \displaystyle\int_C f(x,y,z)\, ds = \int_a^b f(\mathbf{r}(t))\, \|\mathbf{r}'(t)\|\, dt$

    Vector line integral: $\quad \displaystyle\int_C \mathbf{F} \cdot d\mathbf{r} = \int_C (\mathbf{F} \cdot \mathbf{T})\, ds = \int_a^b \mathbf{F}(\mathbf{r}(t)) \cdot \mathbf{r}'(t)\, dt$

    $$= \int_C F_1\, dx + F_2\, dy + F_3\, dz$$

- Arc length differential: $ds = \|\mathbf{r}'(t)\|\, dt$. To evaluate a scalar line integral, replace $f(x,y,z)\, d\mathbf{r}$ with $f(\mathbf{r}(t))\, \|\mathbf{r}'(t)\|\, dt$.
- Vector differential: $d\mathbf{r} = \mathbf{r}'(t)\, dt$. To evaluate a vector line integral, replace $\mathbf{F} \cdot d\mathbf{r}$ with $\mathbf{F}(\mathbf{r}(t)) \cdot \mathbf{r}'(t)\, dt$.
- An *oriented curve* C is a curve in which one of the two possible directions along C (called the *positive direction*) is chosen.
- The vector line integral depends on the orientation of the curve C. The parametrization $\mathbf{r}(t)$ must be regular (that is, $\mathbf{r}'(t) \ne \mathbf{0}$), and it must trace C in the positive direction.
- We write $-C$ for the curve C with the opposite orientation. Then

    $$\int_{-C} \mathbf{F} \cdot d\mathbf{r} = -\int_C \mathbf{F} \cdot d\mathbf{r}$$

- If $\delta(x,y,z)$ is the mass or charge density along C, then the total mass or charge is equal to the scalar line integral $\displaystyle\int_C \delta(x,y,z)\, ds$.
- The vector line integral is used to compute the work W exerted on an object along a curve C:

    $$W = \int_C \mathbf{F} \cdot d\mathbf{r}$$

    The work performed *against* $\mathbf{F}$ is the quantity $-\displaystyle\int_C \mathbf{F} \cdot d\mathbf{r}$.
- Flux across $C = \int_C (\mathbf{F} \cdot \mathbf{n})\, ds = \int_a^b \mathbf{F}(\mathbf{r}(t)) \cdot \mathbf{N}(t)\, dt$, where $\mathbf{N}(t) = \langle y'(t), -x'(t)\rangle$.

16.2 EXERCISES

Preliminary Questions

1. What is the line integral of the constant function $f(x,y,z) = 10$ over a curve C of length 5?

2. Which of the following have a zero line integral over the vertical segment from $(0,0)$ to $(0,1)$?

(a) $f(x,y) = x$ (b) $f(x,y) = y$
(c) $\mathbf{F} = \langle x, 0\rangle$ (d) $\mathbf{F} = \langle y, 0\rangle$
(e) $\mathbf{F} = \langle 0, x\rangle$ (f) $\mathbf{F} = \langle 0, y\rangle$

3. State whether each statement is true or false. If the statement is false, give the correct statement.

(a) The scalar line integral does not depend on how you parametrize the curve.

(b) If you reverse the orientation of the curve, neither the vector line integral nor the scalar line integral changes sign.

4. Suppose that C has length 5. What is the value of $\displaystyle\int_C \mathbf{F} \cdot d\mathbf{r}$ if:

(a) $\mathbf{F}(P)$ is normal to C at all points P on C?

(b) $\mathbf{F}(P)$ is a unit vector pointing in the negative direction along the curve?

Exercises

1. Let $f(x, y, z) = x + yz$, and let C be the line segment from $P = (0, 0, 0)$ to $(6, 2, 2)$.

(a) Calculate $f(\mathbf{r}(t))$ and $ds = \|\mathbf{r}'(t)\| \, dt$ for the parametrization $\mathbf{r}(t) = (6t, 2t, 2t)$ for $0 \le t \le 1$.

(b) Evaluate $\displaystyle\int_C f(x, y, z) \, ds$.

2. Repeat Exercise 1 with the parametrization $\mathbf{r}(t) = (3t^2, t^2, t^2)$ for $0 \le t \le \sqrt{2}$.

3. Let $\mathbf{F} = \langle y^2, x^2 \rangle$, and let C be the curve $y = x^{-1}$ for $1 \le x \le 2$, oriented from left to right.

(a) Calculate $\mathbf{F}(\mathbf{r}(t))$ and $d\mathbf{r} = \mathbf{r}'(t) \, dt$ for the parametrization of C given by $\mathbf{r}(t) = (t, t^{-1})$.

(b) Calculate the dot product $\mathbf{F}(\mathbf{r}(t)) \cdot \mathbf{r}'(t) \, dt$ and evaluate $\displaystyle\int_C \mathbf{F} \cdot d\mathbf{r}$.

4. Let $\mathbf{F}(x, y, z) = \langle z^2, x, y \rangle$, and let C be the curve that is given by $\mathbf{r}(t) = (3 + 5t^2, 3 - t^2, t)$ for $0 \le t \le 2$.

(a) Calculate $\mathbf{F}(\mathbf{r}(t))$ and $d\mathbf{r} = \mathbf{r}'(t) \, dt$.

(b) Calculate the dot product $\mathbf{F}(\mathbf{r}(t)) \cdot \mathbf{r}'(t) \, dt$ and evaluate $\displaystyle\int_C \mathbf{F} \cdot d\mathbf{r}$.

In Exercises 5–8, compute the integral of the scalar function or vector field over $\mathbf{r}(t) = (\cos t, \sin t, t)$ for $0 \le t \le \pi$.

5. $f(x, y, z) = x^2 + y^2 + z^2$

6. $f(x, y, z) = xy + z$

7. $\mathbf{F}(x, y, z) = \langle x, y, z \rangle$

8. $\mathbf{F}(x, y, z) = \langle xy, 2, z^3 \rangle$

In Exercises 9–16, compute $\displaystyle\int_C f \, ds$ for the curve specified.

9. $f(x, y) = \sqrt{1 + 9xy}$, $y = x^3$ for $0 \le x \le 1$

10. $f(x, y) = \dfrac{y^3}{x^7}$, $y = \frac{1}{4}x^4$ for $1 \le x \le 2$

11. $f(x, y, z) = z^2$, $\mathbf{r}(t) = (2t, 3t, 4t)$ for $0 \le t \le 2$

12. $f(x, y, z) = 3x - 2y + z$, $\mathbf{r}(t) = (2 + t, 2 - t, 2t)$ for $-2 \le t \le 1$

13. $f(x, y, z) = xe^{z^2}$, piecewise linear path from $(0, 0, 1)$ to $(0, 2, 0)$ to $(1, 1, 1)$

14. $f(x, y, z) = x^2 z$, $\mathbf{r}(t) = (e^t, \sqrt{2}t, e^{-t})$ for $0 \le t \le 1$

15. $f(x, y, z) = 2x^2 + 8z$, $\mathbf{r}(t) = (e^t, t^2, t)$, $0 \le t \le 1$

16. $f(x, y, z) = 6xz - 2y^2$, $\mathbf{r}(t) = \left(t, \dfrac{t^2}{\sqrt{2}}, \dfrac{t^3}{3}\right)$, $0 \le t \le 2$

17. Calculate $\displaystyle\int_C 1 \, ds$, where the curve C is parametrized by $\mathbf{r}(t) = (4t, -3t, 12t)$ for $2 \le t \le 5$. What does this integral represent?

18. Calculate $\displaystyle\int_C 1 \, ds$, where the curve C is parametrized by $\mathbf{r}(t) = (e^t, \sqrt{2}t, e^{-t})$ for $0 \le t \le 2$.

In Exercises 19–26, compute $\displaystyle\int_C \mathbf{F} \cdot d\mathbf{r}$ for the oriented curve specified.

19. $\mathbf{F}(x, y) = \langle x^2, xy \rangle$, line segment from $(0, 0)$ to $(2, 2)$

20. $\mathbf{F}(x, y) = \langle 4, y \rangle$, quarter circle $x^2 + y^2 = 1$ with $x \le 0$, $y \le 0$, oriented counterclockwise

21. $\mathbf{F}(x, y) = \langle x^2, xy \rangle$, part of circle $x^2 + y^2 = 9$ with $x \le 0$, $y \ge 0$, oriented clockwise

22. $\mathbf{F}(x, y) = \langle e^{y-x}, e^{2x} \rangle$, piecewise linear path from $(1, 1)$ to $(2, 2)$ to $(0, 2)$

23. $\mathbf{F}(x, y) = \langle 3zy^{-1}, 4x, -y \rangle$, $\mathbf{r}(t) = (e^t, e^t, t)$ for $-1 \le t \le 1$

24. $\mathbf{F}(x, y) = \left\langle \dfrac{-y}{(x^2 + y^2)^2}, \dfrac{x}{(x^2 + y^2)^2} \right\rangle$, circle of radius R with center at the origin oriented counterclockwise

25. $\mathbf{F}(x, y, z) = \left\langle \dfrac{1}{y^3 + 1}, \dfrac{1}{z + 1}, 1 \right\rangle$, $\mathbf{r}(t) = (t^3, 2, t^2)$ for $0 \le t \le 1$

26. $\mathbf{F}(x, y, z) = \langle z^3, yz, x \rangle$, quarter of the circle of radius 2 in the yz-plane with center at the origin where $y \ge 0$ and $z \ge 0$, oriented clockwise when viewed from the positive x-axis

In Exercises 27–32, evaluate the line integral.

27. $\displaystyle\int_C y \, dx - x \, dy$, parabola $y = x^2$ for $0 \le x \le 2$

28. $\displaystyle\int_C y \, dx + z \, dy + x \, dz$, $\mathbf{r}(t) = (2 + t^{-1}, t^3, t^2)$ for $0 \le t \le 1$

29. $\displaystyle\int_C (x - y) \, dx + (y - z) \, dy + z \, dz$, line segment from $(0, 0, 0)$ to $(1, 4, 4)$

30. $\displaystyle\int_C z \, dx + x^2 \, dy + y \, dz$, $\mathbf{r}(t) = (\cos t, \tan t, t)$ for $0 \le t \le \frac{\pi}{4}$

31. $\displaystyle\int_C \dfrac{-y \, dx + x \, dy}{x^2 + y^2}$, segment from $(1, 0)$ to $(0, 1)$.

32. $\displaystyle\int_C y^2 \, dx + z^2 \, dy + (1 - x^2) \, dz$, quarter of the circle of radius 1 in the xz-plane with center at the origin in the quadrant $x \ge 0$, $z \le 0$, oriented counterclockwise when viewed from the positive y-axis

33. **CAS** Let $f(x, y, z) = x^{-1}yz$, and let C be the curve parametrized by $\mathbf{r}(t) = (\ln t, t, t^2)$ for $2 \le t \le 4$. Use a computer algebra system to calculate $\displaystyle\int_C f(x, y, z) \, ds$ to four decimal places.

34. **CAS** Use a CAS to calculate $\displaystyle\int_C \langle e^{x-y}, e^{x+y} \rangle \cdot d\mathbf{r}$ to four decimal places, where C is the curve $y = \sin x$ for $0 \le x \le \pi$, oriented from left to right.

In Exercises 35 and 36, calculate the line integral of $\mathbf{F}(x, y, z) = \langle e^z, e^{x-y}, e^y \rangle$ over the given path.

35. The blue path from P to Q in Figure 14

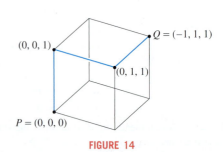

FIGURE 14

36. The closed path *ABCA* in Figure 15

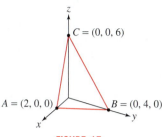

FIGURE 15

In Exercises 37 and 38, C is the path from P to Q in Figure 16 that traces C_1, C_2, and C_3 in the orientation indicated, and $\mathbf{F}$ is a vector field such that

$$\int_C \mathbf{F} \cdot d\mathbf{r} = 5, \qquad \int_{C_1} \mathbf{F} \cdot d\mathbf{r} = 8, \qquad \int_{C_3} \mathbf{F} \cdot d\mathbf{r} = 8$$

37. Determine:

(a) $\displaystyle\int_{-C_3} \mathbf{F} \cdot d\mathbf{r}$ **(b)** $\displaystyle\int_{C_2} \mathbf{F} \cdot d\mathbf{r}$ **(c)** $\displaystyle\int_{-C_1-C_3} \mathbf{F} \cdot d\mathbf{r}$

38. Find the value of $\displaystyle\int_{C'} \mathbf{F} \cdot d\mathbf{r}$, where C' is the path that traverses the loop C_2 four times in the clockwise direction.

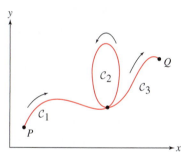

FIGURE 16

39. The values of a function $f(x, y, z)$ and vector field $\mathbf{F}(x, y, z)$ are given at six sample points along the path *ABC* in Figure 17. Estimate the line integrals of f and $\mathbf{F}$ along *ABC*.

Point	$f(x, y, z)$	$\mathbf{F}(x, y, z)$
$\left(1, \frac{1}{6}, 0\right)$	3	$\langle 1, 0, 2 \rangle$
$\left(1, \frac{1}{2}, 0\right)$	3.3	$\langle 1, 1, 3 \rangle$
$\left(1, \frac{5}{6}, 0\right)$	3.6	$\langle 2, 1, 5 \rangle$
$\left(1, 1, \frac{1}{6}\right)$	4.2	$\langle 3, 2, 4 \rangle$
$\left(1, 1, \frac{1}{2}\right)$	4.5	$\langle 3, 3, 3 \rangle$
$\left(1, 1, \frac{5}{6}\right)$	4.2	$\langle 5, 3, 3 \rangle$

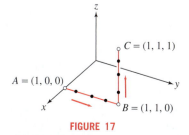

FIGURE 17

40. Estimate the line integrals of $f(x, y)$ and $\mathbf{F}(x, y)$ along the quarter circle (oriented counterclockwise) in Figure 18 using the values at the three sample points along each path.

Point	$f(x, y)$	$\mathbf{F}(x, y)$
A	1	$\langle 1, 2 \rangle$
B	-2	$\langle 1, 3 \rangle$
C	4	$\langle -2, 4 \rangle$

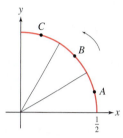

FIGURE 18

41. Determine whether the line integrals of the vector fields around the circle (oriented counterclockwise) in Figure 19 are positive, negative, or zero.

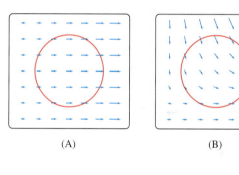

(A)   (B)

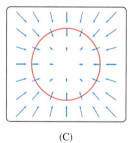

(C)

FIGURE 19

42. Determine whether the line integrals of the vector fields along the oriented curves in Figure 20 are positive or negative.

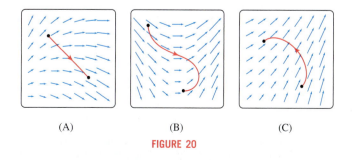

(A)   (B)   (C)

FIGURE 20

43. Calculate the total mass of a circular piece of wire of radius 4 cm centered at the origin whose mass density is $\delta(x, y) = x^2$ g/cm.

44. Calculate the total mass of a metal tube in the helical shape $\mathbf{r}(t) = (\cos t, \sin t, t^2)$ (distance in centimeters) for $0 \le t \le 2\pi$ if the mass density is $\delta(x, y, z) = \sqrt{z}$ g/cm.

45. Find the total charge on the curve $y = x^{4/3}$ for $1 \le x \le 8$ (in centimeters) assuming a charge density of $\delta(x, y) = x/y$ (in units of 10^{-6} C/cm).

46. Find the total charge on the curve $\mathbf{r}(t) = (\sin t, \cos t, \sin^2 t)$ in centimeters for $0 \le t \le \frac{\pi}{8}$ assuming a charge density of $\delta(x, y, z) = xy(y^2 - z)$ (in units of 10^{-6} C/cm).

In Exercises 47–50, use Eq. (6) to compute the electric potential $V(P)$ at the point P for the given charge density (in units of 10^{-6} C).

47. Calculate $V(P)$ at $P = (0, 0, 12)$ if the electric charge is distributed along the quarter circle of radius 4 centered at the origin with charge density $\delta(x, y, z) = xy$.

48. Calculate $V(P)$ at the origin $P = (0, 0)$ if the negative charge is distributed along $y = x^2$ for $1 \le x \le 2$ with charge density $\delta(x, y) = -y\sqrt{x^2 + 1}$.

49. Calculate $V(P)$ at $P = (2, 0, 2)$ if the negative charge is distributed along the y-axis for $1 \le y \le 3$ with charge density $\delta(x, y, z) = -y$.

50. Calculate $V(P)$ at the origin $P = (0, 0)$ if the electric charge is distributed along $y = x^{-1}$ for $\frac{1}{2} \le x \le 2$ with charge density $\delta(x, y) = x^3 y$.

51. Calculate the work done by a field $\mathbf{F} = \langle x + y, x - y \rangle$ when an object moves from $(0, 0)$ to $(1, 1)$ along each of the paths $y = x^2$ and $x = y^2$.

In Exercises 52-54, calculate the work done by the field $\mathbf{F}$ when the object moves along the given path from the initial point to the final point.

52. $\mathbf{F}(x, y, z) = \langle x, y, z \rangle$, $\mathbf{r} = \langle \cos t, \sin t, t \rangle$ for $0 \le t \le 3\pi$.

53. $\mathbf{F}(x, y, z) = \langle xy, yz, xz \rangle$, $\mathbf{r} = \left\langle t, t^2, t^3 \right\rangle$ for $0 \le t \le 1$.

54. $\mathbf{F}(x, y, z) = \left\langle e^x, e^y, xyz \right\rangle$, $\mathbf{r} = \left\langle t^2, t, t/2 \right\rangle$ for $0 \le t \le 1$.

55. Figure 21 shows a force field $\mathbf{F}$.

(a) Over which of the two paths, *ADC* or *ABC*, does $\mathbf{F}$ perform less work?

(b) If you have to work against $\mathbf{F}$ to move an object from C to A, which of the paths, *CBA* or *CDA*, requires less work?

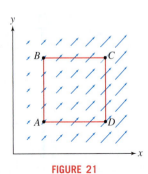

FIGURE 21

56. Verify that the work performed along the segment $\overline{PQ}$ by the constant vector field $\mathbf{F} = \langle 2, -1, 4 \rangle$ is equal to $\mathbf{F} \cdot \overrightarrow{PQ}$ in these cases:

(a) $P = (0, 0, 0)$, $Q = (4, 3, 5)$

(b) $P = (3, 2, 3)$, $Q = (4, 8, 12)$

57. Show that work performed by a constant force field $\mathbf{F}$ over any path C from P to Q is equal to $\mathbf{F} \cdot \overrightarrow{PQ}$.

58. Note that a curve C in polar form $r = f(\theta)$ is parametrized by $\mathbf{r}(\theta) = (f(\theta) \cos \theta, f(\theta) \sin \theta)$ because the x- and y-coordinates are given by $x = r \cos \theta$ and $y = r \sin \theta$.

(a) Show that $\|\mathbf{r}'(\theta)\| = \sqrt{f(\theta)^2 + f'(\theta)^2}$.

(b) Evaluate $\int_C (x - y)^2 \, ds$, where C is the semicircle in Figure 22 with polar equation $r = 2 \cos \theta$, $0 \le \theta \le \frac{\pi}{2}$.

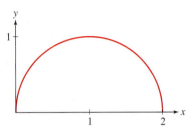

FIGURE 22 Semicircle $r = 2 \cos \theta$.

59. Charge is distributed along the spiral with polar equation $r = \theta$ for $0 \le \theta \le 2\pi$. The charge density is $\delta(r, \theta) = r$ (assume distance is in centimeters and charge in units of 10^{-6} C/cm). Use the result of Exercise 58(a) to compute the total charge.

*In Exercises 60–63, let $\mathbf{F}$ be the **vortex field** (so-called because it swirls around the origin as in Figure 23):*

$$\mathbf{F}(x, y) = \left\langle \frac{-y}{x^2 + y^2}, \frac{x}{x^2 + y^2} \right\rangle$$

60. Calculate $I = \int_C \mathbf{F} \cdot d\mathbf{r}$, where C is the circle of radius 2 centered at the origin. Verify that I changes sign when C is oriented in the clockwise direction.

61. Show that the value of $\int_{C_R} \mathbf{F} \cdot d\mathbf{r}$, where C_R is the circle of radius R centered at the origin and oriented counterclockwise, does not depend on R.

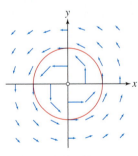

FIGURE 23

62. Let $a > 0$, $b < c$. Show that the integral of $\mathbf{F}$ along the segment [Figure 24(A)] from $P = (a, b)$ to $Q = (a, c)$ is equal to the angle $\angle POQ$.

63. Let C be a curve in polar form $r = f(\theta)$ for $\theta_1 \le \theta \le \theta_2$ [Figure 24(B)], parametrized by $\mathbf{r}(\theta) = (f(\theta) \cos \theta, f(\theta) \sin \theta)$ as in Exercise 58.

(a) Show that the vortex field in polar coordinates is written $\mathbf{F}(r, \theta) = r^{-1} \langle -\sin \theta, \cos \theta \rangle$.

(b) Show that $\mathbf{F} \cdot \mathbf{r}'(\theta) \, d\theta = d\theta$.

(c) Show that $\int_C \mathbf{F} \cdot d\mathbf{r} = \theta_2 - \theta_1$.

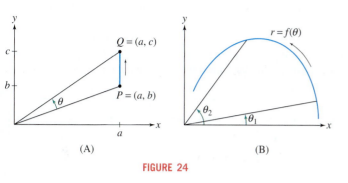

(A)   (B)

FIGURE 24

In Exercises 64–67, use Eq. (10) to calculate the flux of the vector field across the curve specified.

---

64. $\mathbf{F}(x, y) = \langle -y, x \rangle$;   upper half of the unit circle, oriented clockwise

65. $\mathbf{F}(x, y) = \langle x^2, y^2 \rangle$;   segment from $(3, 0)$ to $(0, 3)$, oriented upward

66. $\mathbf{F}(x, y) = \left\langle \dfrac{x + 1}{(x + 1)^2 + y^2}, \dfrac{y}{(x + 1)^2 + y^2} \right\rangle$;   segment $1 \le y \le 4$

along the y-axis, oriented upward

67. $\mathbf{F}(x, y) = \langle e^y, 2x - 1 \rangle$;   parabola $y = x^2$ for $0 \le x \le 1$, oriented left to right

68. ✏️   Let $I = \int_C f(x, y, z) \, ds$. Assume that $f(x, y, z) \ge m$ for some number m and all points (x, y, z) on C. Which of the following conclusions is correct? Explain.

(a) $I \ge m$

(b) $I \ge mL$, where L is the length of C

Further Insights and Challenges

69. Let $\mathbf{F}(x, y) = \langle x, 0 \rangle$. Prove that if C is any path from (a, b) to (c, d), then

$$\int_C \mathbf{F} \cdot d\mathbf{r} = \frac{1}{2}(c^2 - a^2)$$

70. Let $\mathbf{F}(x, y) = \langle y, x \rangle$. Prove that if C is any path from (a, b) to (c, d), then

$$\int_C \mathbf{F} \cdot d\mathbf{r} = cd - ab$$

71. We wish to define the **average value** $\mathrm{Av}(f)$ of a continuous function f along a curve C of length L. Divide C into N consecutive arcs $C_1, \ldots, C_N$, each of length L/N, and let P_i be a sample point in C_i (Figure 25). The sum

$$\frac{1}{N} \sum_{i=1}^{N} f(P_i)$$

may be considered an approximation to $\mathrm{Av}(f)$, so we define

$$\mathrm{Av}(f) = \lim_{N \to \infty} \frac{1}{N} \sum_{i=1}^{N} f(P_i)$$

Prove that

$$\mathrm{Av}(f) = \frac{1}{L} \int_C f(x, y, z) \, ds \qquad \boxed{11}$$

Hint: Show that $\dfrac{L}{N} \sum_{i=1}^{N} f(P_i)$ is a Riemann sum approximation to the line integral of f along C.

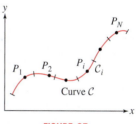

FIGURE 25

72. Use Eq. (11) to calculate the average value of $f(x, y) = x - y$ along the segment from $P = (2, 1)$ to $Q = (5, 5)$.

73. Use Eq. (11) to calculate the average value of $f(x, y) = x$ along the curve $y = x^2$ for $0 \le x \le 1$.

74. The temperature (in degrees centigrade) at a point P on a circular wire of radius 2 cm centered at the origin is equal to the square of the distance from P to $P_0 = (2, 0)$. Compute the average temperature along the wire.

75. The value of a scalar line integral does not depend on the choice of parametrization (because it is defined without reference to a parametrization). Prove this directly. That is, suppose that $\mathbf{r}_1(t)$ and $\mathbf{r}(t)$ are two parametrizations such that $\mathbf{r}_1(t) = \mathbf{r}(\varphi(t))$, where $\varphi(t)$ is an increasing function. Use the Change of Variables Formula to verify that

$$\int_c^d f(\mathbf{r}_1(t)) \|\mathbf{r}_1'(t)\| \, dt = \int_a^b f(\mathbf{r}(t)) \|\mathbf{r}'(t)\| \, dt$$

where $a = \varphi(c)$ and $b = \varphi(d)$.

16.3 Conservative Vector Fields

◄‑‑ REMINDER

• A vector field $\mathbf{F}$ is conservative if
 $\mathbf{F} = \nabla f$ for some function $f(x, y, z)$.
• f is called a potential function.

In this section, we study conservative vector fields in greater depth. When the curve C is *closed*, we often refer to the line integral as the **circulation** of $\mathbf{F}$ around C (Figure 1) and denote it with the symbol $\oint$:

$$\oint_C \mathbf{F} \cdot d\mathbf{r}$$

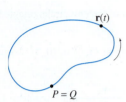

FIGURE 1 The circulation around a closed path is denoted $\oint_C \mathbf{F} \cdot d\mathbf{r}$.

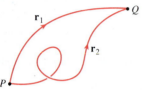

FIGURE 2 Path independence: If $\mathbf{F}$ is conservative, then the line integrals over $\mathbf{r}_1$ and $\mathbf{r}_2$ are equal.

It actually does not matter what point we take as the starting point when we have a closed curve. If we start at point A, the circulation as we travel around the curve back to A is the sum of the line integral as we travel from A to B and the line integral as we travel from B back to A, but switching the order of these two, that is, the circulation starting at B and returning to B.

Our first result establishes the fundamental **path independence** of conservative vector fields, which means that the line integral of $\mathbf{F}$ along a path from P to Q depends only on the endpoints P and Q, not on the particular path followed (Figure 2).

> **THEOREM 1 Fundamental Theorem for Conservative Vector Fields** Assume that $\mathbf{F} = \nabla f$ on a domain $\mathcal{D}$.
>
> **1.** If $\mathbf{r}$ is a path along a curve C from P to Q in $\mathcal{D}$, then
>
> $$\int_C \mathbf{F} \cdot d\mathbf{r} = f(Q) - f(P) \qquad \boxed{1}$$
>
> In particular, $\mathbf{F}$ is path-independent.
>
> **2.** The circulation around a closed curve C (i.e., $P = Q$) is zero:
>
> $$\oint_C \mathbf{F} \cdot d\mathbf{r} = 0$$

Proof Let $\mathbf{r}(t)$ be a path along the curve C in $\mathcal{D}$ for $a \leq t \leq b$ with $\mathbf{r}(a) = P$ and $\mathbf{r}(b) = Q$. Then

$$\int_C \mathbf{F} \cdot d\mathbf{r} = \int_C \nabla f \cdot d\mathbf{r} = \int_a^b \nabla f(\mathbf{r}(t)) \cdot \mathbf{r}'(t) \, dt$$

However, by the Chain Rule for Paths (Theorem 2 in Section 14.5),

$$\frac{d}{dt} f(\mathbf{r}(t)) = \nabla f(\mathbf{r}(t)) \cdot \mathbf{r}'(t)$$

Thus, we can apply the Fundamental Theorem of Calculus:

$$\int_C \mathbf{F} \cdot d\mathbf{r} = \int_a^b \frac{d}{dt} f(\mathbf{r}(t)) \, dt = f(\mathbf{r}(t)) \Big|_a^b = f(\mathbf{r}(b)) - f(\mathbf{r}(a)) = f(Q) - f(P)$$

This proves Eq. (1). It also proves path independence, because the quantity $f(Q) - f(P)$ depends on the endpoints but not on the path $\mathbf{r}$. If $\mathbf{r}$ is a closed path, then $P = Q$ and $f(Q) - f(P) = 0$. ■

■ **EXAMPLE 1** Let $\mathbf{F}(x, y, z) = \langle 2xy + z, x^2, x \rangle$.

(a) Verify that $f(x, y, z) = x^2 y + xz$ is a potential function.

(b) Evaluate $\int_C \mathbf{F} \cdot d\mathbf{r}$, where C is a curve from $P = (1, -1, 2)$ to $Q = (2, 2, 3)$.

Solution **(a)** The partial derivatives of $f(x, y, z) = x^2 y + xz$ are the components of $\mathbf{F}$:

$$\frac{\partial f}{\partial x} = 2xy + z, \qquad \frac{\partial f}{\partial y} = x^2, \qquad \frac{\partial f}{\partial z} = x$$

Therefore, $\nabla f = \langle 2xy + z, x^2, x \rangle = \mathbf{F}$.

(b) By Theorem 1, the line integral over any path $\mathbf{r}(t)$ from $P = (1, -1, 2)$ to $Q = (2, 2, 3)$ (Figure 3) has the value

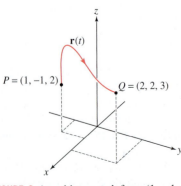

FIGURE 3 An arbitrary path from $(1, -1, 2)$ to $(2, 2, 3)$.

$$\int_{\mathcal{C}} \mathbf{F} \cdot d\mathbf{r} = f(Q) - f(P)$$

$$= f(2, 2, 3) - f(1, -1, 2)$$

$$= \left(2^2(2) + 2(3)\right) - \left(1^2(-1) + 1(2)\right) = 13 \quad \blacksquare$$

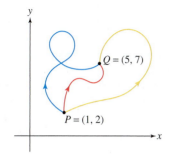

EXAMPLE 2 Find a potential function for $\mathbf{F} = \langle 2x + y, x \rangle$ and use it to evaluate $\int_{\mathcal{C}} \mathbf{F} \cdot d\mathbf{r}$, where $\mathbf{r}$ is any path (Figure 4) from $(1, 2)$ to $(5, 7)$.

Solution We will develop a general method for finding potential functions. At this point, we can see that $f(x, y) = x^2 + xy$ satisfies $\nabla f = \mathbf{F}$:

$$\frac{\partial f}{\partial x} = \frac{\partial}{\partial x}(x^2 + xy) = 2x + y, \qquad \frac{\partial f}{\partial y} = \frac{\partial}{\partial y}(x^2 + xy) = x$$

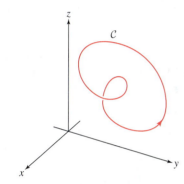

FIGURE 4 Paths from $(1, 2)$ to $(5, 7)$.

Therefore, for any path $\mathbf{r}$ from $(1, 2)$ to $(5, 7)$,

$$\int_{\mathcal{C}} \mathbf{F} \cdot d\mathbf{r} = f(5, 7) - f(1, 2) = (5^2 + 5(7)) - (1^2 + 1(2)) = 57 \quad \blacksquare$$

EXAMPLE 3 **Integral Around a Closed Path** Let $f(x, y, z) = xy \sin(yz)$. Evaluate $\oint_{\mathcal{C}} \nabla f \cdot d\mathbf{r}$, where $\mathcal{C}$ is the closed curve in Figure 5.

Solution By Theorem 1, the integral of a gradient vector around any closed path is zero. In other words, $\oint_{\mathcal{C}} \nabla f \cdot d\mathbf{r} = 0$. $\quad \blacksquare$

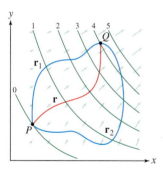

FIGURE 5 The line integral of a conservative vector field around a closed curve is zero.

> **CONCEPTUAL INSIGHT** A good way to think about path independence is in terms of the contour map of the potential function. Consider a vector field $\mathbf{F} = \nabla f$ in the plane (Figure 6). The level curves of f are called **equipotential curves**, and the value $f(P)$ is called the potential at P.
>
> When we integrate $\mathbf{F}$ along a path $\mathbf{r}(t)$ from P to Q, the integrand is
>
> $$\mathbf{F}(\mathbf{r}(t)) \cdot \mathbf{r}'(t) = \nabla f(\mathbf{r}(t)) \cdot \mathbf{r}'(t)$$
>
> Now recall that by the Chain Rule for paths,
>
> $$\nabla f(\mathbf{r}(t)) \cdot \mathbf{r}'(t)) = \frac{d}{dt} f(\mathbf{r}(t))$$
>
> In other words, the integrand is the rate at which the potential changes along the path, and thus, the integral itself is the net change in potential:
>
> $$\int \mathbf{F} \cdot d\mathbf{r} = \underbrace{f(Q) - f(P)}_{\text{Net change in potential}}$$
>
> So, informally speaking, what the line integral does is count the net number of equipotential curves crossed as you move along any path P to Q. By "net number" we mean that crossings in the opposite direction are counted with a minus sign. This net number is independent of the particular path.
>
> We can also interpret the line integral in terms of the graph of the potential function $z = f(x, y)$. The line integral computes the change in height as we move up the surface (Figure 7). Again, this change in height does not depend on the path from P to Q. Of course, these interpretations apply only to conservative vector fields—otherwise, there is no potential function.

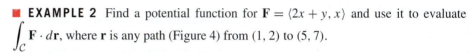

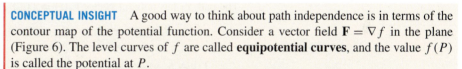

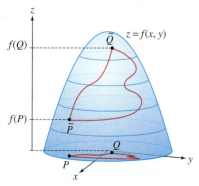

FIGURE 6 Vector field $\mathbf{F} = \nabla f$ with the contour lines of f.

FIGURE 7 The potential surface $z = f(x, y)$.

You might wonder whether there exist any path-independent vector fields other than the conservative ones. The answer is no. By the next theorem, a path-independent vector field is necessarily conservative.

THEOREM 2 A vector field $\mathbf{F}$ on an open connected domain $\mathcal{D}$ is path-independent if and only if it is conservative.

Proof We have already shown that conservative vector fields are path-independent. So, we assume that $\mathbf{F}$ is path-independent and prove that $\mathbf{F}$ has a potential function.

To simplify the notation, we treat the case of a planar vector field $\mathbf{F} = \langle F_1, F_2 \rangle$. The proof for vector fields in $\mathbf{R}^3$ is similar. Choose a point P_0 in $\mathcal{D}$, and for any point $P = (x, y) \in \mathcal{D}$, define

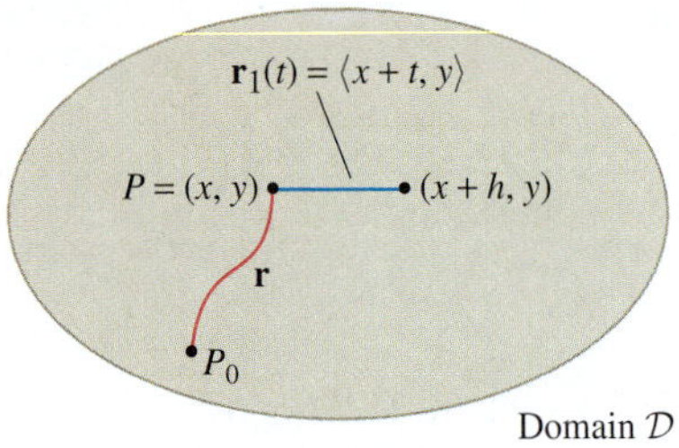

$$f(P) = f(x, y) = \int_C \mathbf{F} \cdot d\mathbf{r}$$

where $\mathbf{r}$ is any path in $\mathcal{D}$ from P_0 to P (Figure 8). Note that this definition of $f(P)$ is meaningful only because we assume that the line integral does not depend on the path $\mathbf{r}$.

We will prove that $\mathbf{F} = \nabla f$, which involves showing that $\dfrac{\partial f}{\partial x} = F_1$ and $\dfrac{\partial f}{\partial y} = F_2$. We will only verify the first equation, as the second can be checked in a similar manner. Let $\mathbf{r}_1$ be the horizontal path $\mathbf{r}_1(t) = (x + t, y)$ for $0 \le t \le h$. For $|h|$ small enough, $\mathbf{r}_1$ lies inside $\mathcal{D}$. Let $\mathbf{r} + \mathbf{r}_1$ denote the path $\mathbf{r}$ followed by $\mathbf{r}_1$. It begins at P_0 and ends at $(x + h, y)$, so

$$f(x + h, y) - f(x, y) = \int_{\mathbf{r}+\mathbf{r}_1} \mathbf{F} \cdot d\mathbf{r} - \int_{\mathbf{r}} \mathbf{F} \cdot d\mathbf{r}$$

$$= \left(\int_{\mathbf{r}} \mathbf{F} \cdot d\mathbf{r} + \int_{\mathbf{r}_1} \mathbf{F} \cdot d\mathbf{r} \right) - \int_{\mathbf{r}} \mathbf{F} \cdot d\mathbf{r} = \int_{\mathbf{r}_1} \mathbf{F} \cdot d\mathbf{r}$$

The path $\mathbf{r}_1$ has tangent vector $\mathbf{r}_1'(t) = \langle 1, 0 \rangle$, so

$$\mathbf{F}(\mathbf{r}_1(t)) \cdot \mathbf{r}_1'(t) = \langle F_1(x + t, y), F_2(x + t, y) \rangle \cdot \langle 1, 0 \rangle = F_1(x + t, y)$$

$$f(x + h, y) - f(x, y) = \int_{\mathbf{r}_1} \mathbf{F} \cdot d\mathbf{r} = \int_0^h F_1(x + t, y)\, dt$$

Using the substitution $u = x + t$, we have

$$\frac{f(x + h, y) - f(x, y)}{h} = \frac{1}{h} \int_0^h F_1(x + t, y)\, dt = \frac{1}{h} \int_x^{x+h} F_1(u, y)\, du$$

The integral on the right is the average value of $F_1(u, y)$ over the interval $[x, x + h]$. It converges to the value $F_1(x, y)$ as $h \to 0$, and this yields the desired result:

$$\frac{\partial f}{\partial x} = \lim_{h \to 0} \frac{f(x + h, y) - f(x, y)}{h} = \lim_{h \to 0} \frac{1}{h} \int_x^{x+h} F_1(u, y)\, du = F_1(x, y) \quad \blacksquare$$

$\mathbf{r}_1(t) = \langle x + t, y \rangle$

$P = (x, y)$ $(x + h, y)$

$\mathbf{r}$

P_0

Domain $\mathcal{D}$

FIGURE 8

Conservative Fields in Physics

The Conservation of Energy principle says that the sum $KE + PE$ of kinetic and potential energy remains constant in an isolated system. For example, a falling object picks up kinetic energy as it falls to Earth, but this gain in kinetic energy is offset by a loss in gravitational potential energy (g times the change in height), such that the sum $KE + PE$ remains unchanged.

We show now that conservation of energy is valid for the motion of a particle of mass m under a force field $\mathbf{F}$ if $\mathbf{F}$ has a potential function. This explains why the term "conservative" is used to describe vector fields that have a potential function.

We follow the convention in physics of writing the potential function with a minus sign and using V rather than f:

$$\mathbf{F} = -\nabla V$$

In a conservative force field, the work W against $\mathbf{F}$ required to move the particle from P to Q is equal to the change in potential energy:

$$W = -\int_C \mathbf{F} \cdot d\mathbf{r} = V(Q) - V(P)$$

Potential functions first appeared in 1774 in the writings of Jean-Louis Lagrange (1736–1813). One of the greatest mathematicians of his time, Lagrange made fundamental contributions to physics, analysis, algebra, and number theory. He was born in Turin, Italy, to a family of French origin but spent most of his career first in Berlin and then in Paris. After the French Revolution, Lagrange was required to teach courses in elementary mathematics, but apparently he spoke above the heads of his audience. A contemporary wrote, "Whatever this great man says deserves the highest degree of consideration, but he is too abstract for youth."

When the particle is located at $P = (x, y, z)$, it is said to have **potential energy** $V(P)$. Suppose that the particle moves along a path $\mathbf{r}(t)$. The particle's velocity is $\mathbf{v} = \mathbf{r}'(t)$, and its kinetic energy is $KE = \frac{1}{2}m\|\mathbf{v}\|^2 = \frac{1}{2}m\mathbf{v} \cdot \mathbf{v}$. By definition, the **total energy** at time t is the sum

$$E = KE + PE = \frac{1}{2}m\mathbf{v} \cdot \mathbf{v} + V(\mathbf{r}(t))$$

THEOREM 3 Conservation of Energy The total energy E of a particle moving under the influence of a conservative force field $\mathbf{F} = -\nabla V$ is constant in time. That is,
$$\frac{dE}{dt} = 0.$$

Proof Let $\mathbf{a} = \mathbf{v}'(t)$ be the particle's acceleration and m its mass. According to Newton's Second Law of Motion, $\mathbf{F}(\mathbf{r}(t)) = m\mathbf{a}(t)$, and thus,

$$\frac{dE}{dt} = \frac{d}{dt}\left(\frac{1}{2}m\mathbf{v} \cdot \mathbf{v} + V(\mathbf{r}(t))\right)$$

$$= \frac{1}{2}m\left(\frac{d\mathbf{v}}{dt} \cdot \mathbf{v} + \mathbf{v} \cdot \frac{d\mathbf{v}}{dt}\right) + \nabla V(\mathbf{r}(t)) \cdot \mathbf{r}'(t) \qquad \text{(Product and Chain Rules)}$$

$$= m\mathbf{v} \cdot \mathbf{a} + \nabla V(\mathbf{r}(t)) \cdot \mathbf{r}'(t)$$

$$= \mathbf{v} \cdot m\mathbf{a} - \mathbf{F} \cdot \mathbf{v} \qquad \text{(since } \mathbf{F} = -\nabla V \text{ and } \mathbf{r}'(t) = \mathbf{v})$$

$$= \mathbf{v} \cdot (m\mathbf{a} - \mathbf{F}) = 0 \qquad \text{(since } \mathbf{F} = m\mathbf{a}) \qquad \blacksquare$$

In Example 8 of Section 16.1, we verified that inverse-square vector fields are conservative:

$$\mathbf{F} = k\frac{\mathbf{e}_r}{r^2} = -\nabla f \quad \text{with} \quad f = \frac{k}{r}$$

Basic examples of inverse-square vector fields are the gravitational and electrostatic forces due to a point mass or charge. By convention, these fields have units of force *per unit mass or unit charge*. Thus, if $\mathbf{F}$ is a gravitational field, the force on a particle of mass m is $m\mathbf{F}$ and its potential energy is mf, where $\mathbf{F} = -\nabla f$.

■ **EXAMPLE 4 Work Against Gravity** Compute the work W against the earth's gravitational field required to move a satellite of mass $m = 600$ kg along any path from an orbit of altitude 2000 km to an orbit of altitude 4000 km.

Solution The earth's gravitational field is the inverse-square field

Example 8 of Section 16.1 showed that

$$\frac{\mathbf{e}_r}{r^2} = -\nabla\left(\frac{1}{r}\right)$$

The constant k is equal to GM_e where $G \approx 6.67 \cdot 10^{-11}$ m^3 kg^{-1} s^{-2} and the mass of the earth is $M_e \approx 5.98 \cdot 10^{24}$ kg:

$$k = GM_e \approx 4 \cdot 10^{14} \text{ m}^3\text{s}^{-2}$$

$$\mathbf{F} = -k\frac{\mathbf{e}_r}{r^2} = -\nabla f, \qquad f = -\frac{k}{r}$$

where r is the distance from the center of the earth and $k = 4 \cdot 10^{14}$ (see marginal note). The radius of the earth is approximately $6.4 \cdot 10^6$ meters, so the satellite must be moved from $r = 8.4 \cdot 10^6$ meters to $r = 10.4 \cdot 10^6$ meters. The force on the satellite is $m\mathbf{F} = 600\mathbf{F}$, and the work W required to move the satellite along a path $\mathbf{r}$ is

$$W = -\int_{\mathbf{r}} m\mathbf{F} \cdot d\mathbf{r} = 600 \int_{\mathbf{r}} \nabla f \cdot d\mathbf{r}$$

$$= -\frac{600k}{r}\bigg|_{8.4 \cdot 10^6}^{10.4 \times 10^6}$$

$$\approx -\frac{2.4 \cdot 10^{17}}{10.4 \cdot 10^6} + \frac{2.4 \cdot 10^{17}}{8.4 \cdot 10^6} \approx 5.5 \cdot 10^9 \text{ joules} \qquad \blacksquare$$

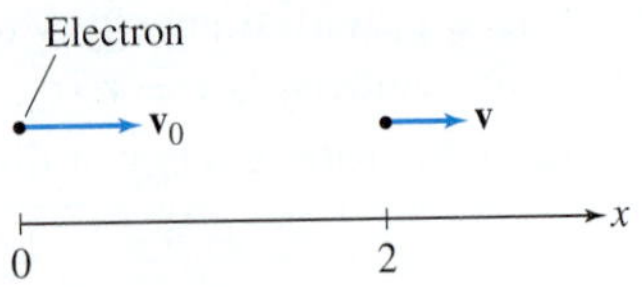

FIGURE 9 An electron moving in an electric field.

■ **EXAMPLE 5** An electron is traveling in the positive x-direction with speed $v_0 = 10^7$ m/s. When it passes $x = 0$, a horizontal electric field $\mathbf{E} = 100x\mathbf{i}$ (in newtons per coulomb) is turned on. Find the electron's velocity after it has traveled 2 meters (see Figure 9). Assume that $q_e/m_e = -1.76 \cdot 10^{11}$ C/kg, where q_e and m_e are the mass and charge of the electron, respectively.

Solution We have $\mathbf{E} = -\nabla V$ where $V(x, y, z) = -50x^2$, so the electric field is conservative. Since V depends only on x, we write $V(x)$ for $V(x, y, z)$. By the Law of Conservation of Energy, the electrons, total energy E is constant and therefore E has the same value when the electron is at $x = 0$ and at $x = 2$:

$$E = \frac{1}{2}m_e v_0^2 + q_e V(0) = \frac{1}{2}m_e v^2 + q_e V(2)$$

Since $V(0) = 0$, we obtain

$$\frac{1}{2}m_e v_0^2 = \frac{1}{2}m_e v^2 + q_e V(2) \quad \Rightarrow \quad v = \sqrt{v_0^2 - 2(q_e/m_e)V(2)}$$

Using the numerical value of q_e/m_e, we have

$$v \approx \sqrt{10^{14} - 2(-1.76 \cdot 10^{11})(-50(2)^2)} \approx \sqrt{2.96 \cdot 10^{13}} \approx 5.4 \cdot 10^6 \text{ m/s}$$

Note that the velocity has decreased. This is because $\mathbf{E}$ exerts a force in the negative x-direction on a negative charge. ■

Finding Potential Functions

We do not yet have an effective way of telling whether a given vector field is conservative. By Theorem 1 in Section 16.1, every conservative vector field satisfies the condition

$$\boxed{\text{curl}(\mathbf{F}) = \mathbf{0}, \quad \text{or equivalently,} \quad \frac{\partial F_1}{\partial y} = \frac{\partial F_2}{\partial x}, \quad \frac{\partial F_2}{\partial z} = \frac{\partial F_3}{\partial y}, \quad \frac{\partial F_3}{\partial x} = \frac{\partial F_1}{\partial z}}$$

$$\boxed{2}$$

But does this condition guarantee that $\mathbf{F}$ is conservative? The answer is a qualified yes; the cross-partials condition does guarantee that $\mathbf{F}$ is conservative, but only on domains $\mathcal{D}$ with a property called simple connectedness.

　　Roughly speaking, a domain $\mathcal{D}$ in the plane is **simply connected** if it is connected and it does not have any "holes" (Figure 10). More precisely, $\mathcal{D}$ is simply connected if every loop in $\mathcal{D}$ can be drawn down, or "contracted," to a point *while staying within* $\mathcal{D}$ as in Figure 11(A). Examples of simply connected regions in $\mathbf{R}^2$ are disks, rectangles, and the entire plane $\mathbf{R}^2$. By contrast, the disk with a point removed in Figure 11(B) is not simply connected: The loop cannot be drawn down to a point without passing through the point that was removed. In $\mathbf{R}^3$, the interiors of balls and boxes are simply connected, as is the entire space $\mathbf{R}^3$.

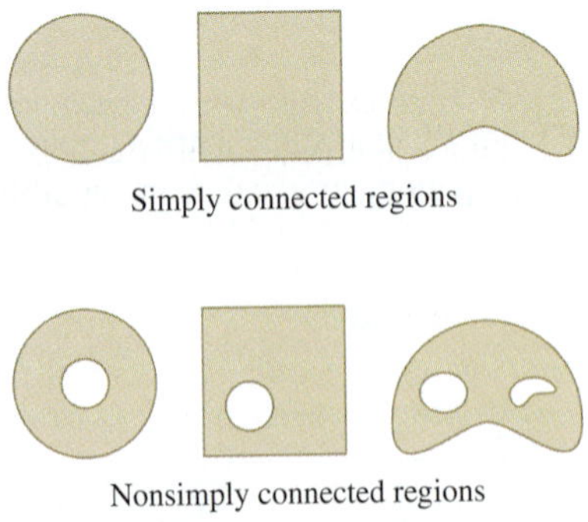

Simply connected regions

Nonsimply connected regions

FIGURE 10 Simple connectedness means "no holes."

(A) Simply connected region:
Any loop can be drawn down to
a point within the region.

(B) Nonsimply connected region:
a loop around the missing hole
cannot be drawn tight without
passing through the hole.

FIGURE 11

> **THEOREM 4** **Existence of a Potential Function** Let $\mathbf{F}$ be a vector field on a simply connected domain $\mathcal{D}$. If $\mathbf{F}$ satisfies the cross-partials condition (2), then $\mathbf{F}$ is conservative.

Rather than prove Theorem 4, we illustrate a practical procedure for finding a potential function when the cross-partials condition is satisfied. The proof itself involves Stokes' Theorem and is somewhat technical because of the role played by the simply-connected property of the domain.

■ **EXAMPLE 6** **Finding a Potential Function** Show that

$$\mathbf{F} = \left\langle 2xy + y^3, x^2 + 3xy^2 + 2y \right\rangle$$

is conservative and find a potential function.

Solution First we observe that the cross-partial derivatives are equal:

$$\frac{\partial F_1}{\partial y} = \frac{\partial}{\partial y}(2xy + y^3) \qquad = 2x + 3y^2$$

$$\frac{\partial F_2}{\partial x} = \frac{\partial}{\partial x}(x^2 + 3xy^2 + 2y) = 2x + 3y^2$$

Furthermore, $\mathbf{F}$ is defined on all of $\mathbf{R}^2$, which is a simply connected domain. Therefore, a potential function exists by Theorem 4.

Now, the potential function f satisfies

$$\frac{\partial f}{\partial x} = F_1(x, y) = 2xy + y^3$$

This tells us that f is an antiderivative of $F_1(x, y)$, regarded as a function of x alone:

$$f(x, y) = \int F_1(x, y)\, dx$$

$$= \int \left(2xy + y^3 \right) dx$$

$$= x^2 y + xy^3 + g(y)$$

Note that to obtain the general antiderivative of $F_1(x, y)$ with respect to x, we must add on an arbitrary function $g(y)$ depending on y alone, rather than the usual constant of integration. Similarly, we have

$$f(x, y) = \int F_2(x, y)\, dy$$

$$= \int \left(x^2 + 3xy^2 + 2y \right) dy$$

$$= x^2 y + xy^3 + y^2 + h(x)$$

The two expressions for $f(x, y)$ must be equal:

$$x^2 y + xy^3 + g(y) = x^2 y + xy^3 + y^2 + h(x)$$

This tells us that $g(y) = y^2$ and $h(x) = 0$, up to the addition of an arbitrary numerical constant C. Thus, we obtain the general potential function

$$f(x, y) = x^2 y + xy^3 + y^2 + C \qquad\qquad ■$$

The same method works for vector fields in 3-space.

■ **EXAMPLE 7** Find a potential function for

$$\mathbf{F} = \left\langle 2xyz^{-1}, z + x^2z^{-1}, y - x^2yz^{-2} \right\rangle$$

Solution If a potential function f exists, then it satisfies

$$f(x, y, z) = \int 2xyz^{-1}\, dx \qquad = x^2yz^{-1} + f(y, z)$$

$$f(x, y, z) = \int \left(z + x^2z^{-1}\right) dy \quad = zy + x^2z^{-1}y + g(x, z)$$

$$f(x, y, z) = \int \left(y - x^2yz^{-2}\right) dz = yz + x^2yz^{-1} + h(x, y)$$

These three ways of writing $f(x, y, z)$ must be equal:

$$x^2yz^{-1} + f(y, z) = zy + x^2z^{-1}y + g(x, z) = yz + x^2yz^{-1} + h(x, y)$$

In Example 7, **F** is only defined for $z \neq 0$, so the domain has two halves: $z > 0$ and $z < 0$. We are free to choose different constants C on the two halves, if desired.

These equalities hold if $f(y, z) = yz$, $g(x, z) = 0$, and $h(x, y) = 0$. Thus, **F** is conservative and, for any constant C, a potential function is

$$f(x, y, z) = x^2yz^{-1} + yz + C \qquad \blacksquare$$

Assumptions Matter We cannot expect the method for finding a potential function to work if **F** does not satisfy the cross-partials condition (because in this case, no potential function exists). What goes wrong? Consider $\mathbf{F} = \langle y, 0 \rangle$. If we attempted to find a potential function, we would calculate

$$f(x, y) = \int y\, dx = xy + g(y)$$

$$f(x, y) = \int 0\, dy = 0 + h(x)$$

However, there is no choice of $g(y)$ and $h(x)$ for which $xy + g(y) = h(x)$. If there were, we could differentiate this equation twice, once with respect to x and once with respect to y. This would yield $1 = 0$, which is a contradiction. The method fails in this case because **F** does not satisfy the cross-partials condition and thus is not conservative.

The Vortex Field

Why does Theorem 4 require that the domain is simply connected? This is an interesting question that we can answer by examining the vortex field (Figure 12):

$$\mathbf{F} = \left\langle \frac{-y}{x^2 + y^2}, \frac{x}{x^2 + y^2} \right\rangle$$

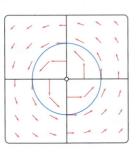

FIGURE 12 The vortex field.

■ **EXAMPLE 8** Show that the vortex field satisfies the cross-partials condition but is not conservative. Does this contradict Theorem 4?

Solution We check the cross-partials condition directly:

$$\frac{\partial}{\partial x}\left(\frac{x}{x^2 + y^2}\right) = \frac{(x^2 + y^2) - x(\partial/\partial x)(x^2 + y^2)}{(x^2 + y^2)^2} = \frac{y^2 - x^2}{(x^2 + y^2)^2}$$

$$\frac{\partial}{\partial y}\left(\frac{-y}{(x^2 + y^2)}\right) = \frac{-(x^2 + y^2) + y(\partial/\partial y)(x^2 + y^2)}{(x^2 + y^2)^2} = \frac{y^2 - x^2}{(x^2 + y^2)^2}$$

FIGURE 13 The domain $\mathcal{D}$ of the vortex $\mathbf{F}$ is the plane with the origin removed. This domain is not simply connected.

Now consider the line integral of $\mathbf{F}$ around the unit circle $\mathcal{C}$ parametrized by $\mathbf{r}(t) = \langle \cos t, \sin t \rangle$:

$$F(\mathbf{r}(t)) \cdot \mathbf{r}'(t) = \langle -\sin t, \cos t \rangle \cdot \langle -\sin t, \cos t \rangle = \sin^2 t + \cos^2 t = 1$$

$$\oint_{\mathcal{C}} \mathbf{F} \cdot d\mathbf{r} = \int_0^{2\pi} \mathbf{F}(\mathbf{r}(t)) \cdot \mathbf{r}'(t)\, dt = \int_0^{2\pi} dt = 2\pi \neq 0 \qquad \boxed{3}$$

If $\mathbf{F}$ were conservative, its circulation around every closed curve would be zero by Theorem 1. Thus, $\mathbf{F}$ cannot be conservative, even though it satisfies the cross-partials condition.

This result does not contradict Theorem 4 because the domain of $\mathbf{F}$ does not satisfy the simply connected condition of the theorem. Because $\mathbf{F}$ is not defined at $(x, y) = (0, 0)$, its domain is $\mathcal{D} = \{(x, y) \neq (0, 0)\}$, and this domain is not simply connected (Figure 13). ■

CONCEPTUAL INSIGHT Although the vortex field $\mathbf{F}$ is not conservative on its domain, it is conservative on any smaller, simply connected domain, such as the upper half-plane $\{(x, y) : y > 0\}$. In fact, we can show (see the marginal note) that $\mathbf{F} = \nabla f$, where

$$f(x, y) = \theta = \tan^{-1} \frac{y}{x} \qquad \text{when} \qquad x \neq 0$$

Since $\tan^{-1} \frac{y}{x}$ yields the angle θ of the point (x, y) in polar coordinates, we can extend the definition of $f(x, y)$ to be $f(x, y) = \theta$, which even holds for points with $x = 0$, with the origin being the only exception (Figure 14). (But keep in mind that θ is only defined up to 2π.)

The line integral of $\mathbf{F}$ along a path $\mathbf{r}$ is equal to the change in potential θ along the path [Figures 15(A) and (B)]:

$$\int_{\mathcal{C}} \mathbf{F} \cdot d\mathbf{r} = \theta_2 - \theta_1 = \text{the change in angle } \theta \text{ along } \mathbf{r}$$

Now we can see what is preventing $\mathbf{F}$ from being conservative on all of its domain. As we just mentioned, the angle θ is defined only up to integer multiples of 2π. The angle along a path that goes all the way around the origin does not return to its original value but rather increases by 2π. This explains why the line integral around the unit circle (Eq. 3) is 2π rather than 0. And it shows that $V(x, y) = \theta$ cannot be defined as a continuous function on the entire domain $\mathcal{D}$. However, θ is continuous on any domain that does not enclose the origin, and on such domains we have $\mathbf{F} = \nabla \theta$.

In general, if a closed path $\mathbf{r}$ winds around the origin n times (where n is negative if the curve winds in the clockwise direction), then [Figures 15(C) and (D)]

$$\oint_{\mathcal{C}} \mathbf{F} \cdot d\mathbf{r} = 2\pi n$$

The number n is called the *winding number* of the path. It plays an important role in the mathematical field of topology.

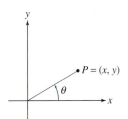

FIGURE 14 The potential function $f(x, y)$ takes the value θ at (x, y).

Using the Chain Rule and the formula

$$\frac{d}{dt} \tan^{-1} t = \frac{1}{1 + t^2}$$

we can check that when $x \neq 0$, $\mathbf{F} = \nabla f$:

$$\frac{\partial \theta}{\partial x} = \frac{\partial}{\partial x} \tan^{-1} \frac{y}{x} = \frac{-y/x^2}{1 + (y/x)^2} = \frac{-y}{x^2 + y^2}$$

$$\frac{\partial \theta}{\partial y} = \frac{\partial}{\partial y} \tan^{-1} \frac{y}{x} = \frac{1/x}{1 + (y/x)^2} = \frac{x}{x^2 + y^2}$$

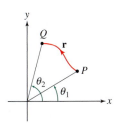

(A) $\int_{\mathcal{C}} \mathbf{F} \cdot d\mathbf{r} = \theta_2 - \theta_1$

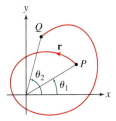

(B) $\int_{\mathcal{C}} \mathbf{F} \cdot d\mathbf{r} = \theta_2 - \theta_1 + 2\pi$

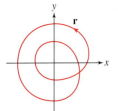

(C) $\mathbf{r}$ goes around the origin twice, so $\int_{\mathcal{C}} \mathbf{F} \cdot d\mathbf{r} = 4\pi$.

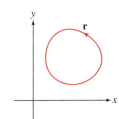

(D) $\mathbf{r}$ does not go around the origin, so $\int_{\mathcal{C}} \mathbf{F} \cdot d\mathbf{r} = 0$.

FIGURE 15 The line integral of the vortex field $\mathbf{F} = \nabla \theta$ is equal to the change in θ along the path.

16.3 SUMMARY

- A vector field $\mathbf{F}$ on a domain $\mathcal{D}$ is conservative if there exists a function f such that $\nabla f = \mathbf{F}$ on $\mathcal{D}$. The function f is called a *potential function* of $\mathbf{F}$.
- A vector field $\mathbf{F}$ on a domain $\mathcal{D}$ is called *path-independent* if for any two points $P, Q \in \mathcal{D}$, we have

$$\int_{\mathcal{C}_1} \mathbf{F} \cdot d\mathbf{r} = \int_{\mathcal{C}_2} \mathbf{F} \cdot d\mathbf{r}$$

for any two paths $\mathcal{C}_1$ and $\mathcal{C}_2$ in $\mathcal{D}$ from P to Q.
- The Fundamental Theorem for Conservative Vector Fields: If $\mathbf{F} = \nabla f$, then

$$\int_{\mathcal{C}} \mathbf{F} \cdot d\mathbf{r} = f(Q) - f(P)$$

for any path $\mathbf{r}$ from P to Q in the domain of $\mathbf{F}$. This shows that conservative vector fields are path-independent. In particular, if $\mathbf{r}$ is a *closed path* ($P = Q$), then

$$\oint_{\mathcal{C}} \mathbf{F} \cdot d\mathbf{r} = 0$$

- The converse is also true: On an open, connected domain, a path-independent vector field is conservative.
- Conservative vector fields satisfy the cross-partial condition

$$\frac{\partial F_1}{\partial y} = \frac{\partial F_2}{\partial x}, \qquad \frac{\partial F_2}{\partial z} = \frac{\partial F_3}{\partial y}, \qquad \frac{\partial F_3}{\partial x} = \frac{\partial F_1}{\partial z} \qquad \boxed{4}$$

- Equality of the cross partials guarantees that $\mathbf{F}$ is conservative if the domain $\mathcal{D}$ is simply connected—that is, if any loop in $\mathcal{D}$ can be drawn down to a point within $\mathcal{D}$.

16.3 EXERCISES

Preliminary Questions

1. The following statement is false. *If $\mathbf{F}$ is a gradient vector field, then the line integral of $\mathbf{F}$ along every curve is zero.* Which single word must be added to make it true?

2. Which of the following statements are true for all vector fields, and which are true only for conservative vector fields?

(a) The line integral along a path from P to Q does not depend on which path is chosen.

(b) The line integral over an oriented curve $\mathcal{C}$ does not depend on how $\mathcal{C}$ is parametrized.

(c) The line integral around a closed curve is zero.

(d) The line integral changes sign if the orientation is reversed.

(e) The line integral is equal to the difference of a potential function at the two endpoints.

(f) The line integral is equal to the integral of the tangential component along the curve.

(g) The cross partials of the components are equal.

3. Let $\mathbf{F}$ be a vector field on an open, connected domain $\mathcal{D}$ with continuous second partial derivatives. Which of the following statements are always true, and which are true under additional hypotheses on $\mathcal{D}$?

(a) If $\mathbf{F}$ has a potential function, then $\mathbf{F}$ is conservative.

(b) If $\mathbf{F}$ is conservative, then the cross partials of $\mathbf{F}$ are equal.

(c) If the cross partials of $\mathbf{F}$ are equal, then $\mathbf{F}$ is conservative.

4. Let $\mathcal{C}$, $\mathcal{D}$, and $\mathcal{E}$ be the oriented curves in Figure 16, and let $\mathbf{F} = \nabla f$ be a gradient vector field such that $\int_{\mathcal{C}} \mathbf{F} \cdot d\mathbf{r} = 4$. What are the values of the following integrals?

(a) $\displaystyle\int_{\mathcal{D}} \mathbf{F} \cdot d\mathbf{r}$ **(b)** $\displaystyle\int_{\mathcal{E}} \mathbf{F} \cdot d\mathbf{r}$

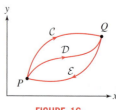

FIGURE 16

Exercises

1. Let $f(x, y, z) = xy \sin(yz)$ and $\mathbf{F} = \nabla f$. Evaluate $\int_{\mathcal{C}} \mathbf{F} \cdot d\mathbf{r}$, where $\mathcal{C}$ is any path from $(0, 0, 0)$ to $(1, 1, \pi)$.

2. Let $\mathbf{F}(x, y, z) = \langle x^{-1}z, y^{-1}z, \ln(xy) \rangle$.

(a) Verify that $\mathbf{F} = \nabla f$, where $f(x, y, z) = z \ln(xy)$.

(b) Evaluate $\int_{\mathcal{C}} \mathbf{F} \cdot d\mathbf{r}$, where $\mathbf{r}(t) = \langle e^t, e^{2t}, t^2 \rangle$ for $1 \le t \le 3$.

(c) Evaluate $\int_{\mathcal{C}} \mathbf{F} \cdot d\mathbf{r}$ for any path $\mathcal{C}$ from $P = (\frac{1}{2}, 4, 2)$ to $Q = (2, 2, 3)$ contained in the region $x > 0$, $y > 0$.

(d) Why is it necessary to specify that the path lies in the region where x and y are positive?

In Exercises 3–6, verify that $\mathbf{F} = \nabla f$ and evaluate the line integral of $\mathbf{F}$ over the given path.

3. $\mathbf{F}(x, y) = \langle 3, 6y \rangle$, $f(x, y) = 3x + 3y^2$; $\mathbf{r}(t) = \langle t, 2t^{-1} \rangle$ for $1 \le t \le 4$

4. $\mathbf{F}(x, y) = \langle \cos y, -x \sin y \rangle$, $f(x, y) = x \cos y$; upper half of the unit circle centered at the origin, oriented counterclockwise

5. $\mathbf{F}(x, y, z) = ye^z \mathbf{i} + xe^z \mathbf{j} + xye^z \mathbf{k}$, $f(x, y, z) = xye^z$; $\mathbf{r}(t) = (t^2, t^3, t - 1)$ for $1 \le t \le 2$

6. $\mathbf{F}(x, y, z) = \dfrac{z}{x}\mathbf{i} + \mathbf{j} + \ln x \mathbf{k}$, $f(x, y, z) = y + z \ln x$; circle $(x - 4)^2 + y^2 = 1$ in the clockwise direction

In Exercises 7–16, find a potential function for $\mathbf{F}$ or determine that $\mathbf{F}$ is not conservative.

7. $\mathbf{F} = \langle z, 1, x \rangle$ **8.** $\mathbf{F} = x\mathbf{j} + y\mathbf{k}$

9. $\mathbf{F} = y^2\mathbf{i} + (2xy + e^z)\mathbf{j} + ye^z\mathbf{k}$ **10.** $\mathbf{F} = \langle y, x, z^3 \rangle$

11. $\mathbf{F} = \langle \cos(xz), \sin(yz), xy \sin z \rangle$ **12.** $\mathbf{F} = \langle \cos z, 2y, -x \sin z \rangle$

13. $\mathbf{F} = \langle z \sec^2 x, z, y + \tan x \rangle$

14. $\mathbf{F} = \langle e^x(z + 1), -\cos y, e^x \rangle$

15. $\mathbf{F} = \langle 2xy + 5, x^2 - 4z, -4y \rangle$

16. $\mathbf{F} = \langle yze^{xy}, xze^{xy} - z, e^{xy} - y \rangle$

17. Evaluate

$$\int_{\mathcal{C}} 2xyz \, dx + x^2z \, dy + x^2y \, dz$$

over the path $\mathbf{r}(t) = (t^2, \sin(\pi t/4), e^{t^2 - 2t})$ for $0 \le t \le 2$.

18. Evaluate

$$\oint_{\mathcal{C}} \sin x \, dx + z \cos y \, dy + \sin y \, dz$$

where $\mathcal{C}$ is the ellipse $4x^2 + 9y^2 = 36$, oriented clockwise.

In Exercises 19–20, let $\mathbf{F} = \nabla f$, and determine directly $\int_{\mathcal{C}} \mathbf{F} \cdot d\mathbf{r}$ for each of the two paths given, showing that they both give the same answer, which is $f(Q) - f(P)$.

19. $f = x^2y - z$, $\mathbf{r}_1 = \langle t, t, 0 \rangle$ for $0 \le t \le 1$, and $\mathbf{r}_2 = (t, t^2, 0)$ for $0 \le t \le 1$

20. $f = zy + xy + xz$, $\mathbf{r}_1 = \langle t, t, t \rangle$ for $0 \le t \le 1$, $\mathbf{r}_2 = (t, t^2, t^3)$ for $0 \le t \le 1$

21. A vector field $\mathbf{F}$ and contour lines of a potential function for $\mathbf{F}$ are shown in Figure 17. Calculate the common value of $\int_{\mathcal{C}} \mathbf{F} \cdot d\mathbf{r}$ for the curves shown in Figure 17 oriented in the direction from P to Q.

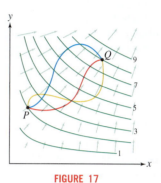

FIGURE 17

22. Give a reason why the vector field $\mathbf{F}$ in Figure 18 is not conservative.

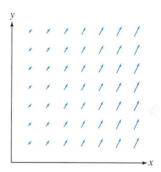

FIGURE 18

23. Calculate the work expended when a particle is moved from O to Q along segments $\overline{OP}$ and $\overline{PQ}$ in Figure 19 in the presence of the force field $\mathbf{F} = \langle x^2, y^2 \rangle$. How much work is expended moving in a complete circuit around the square?

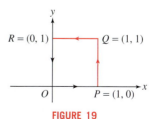

FIGURE 19

24. Let $\mathbf{F}(x, y) = \langle \dfrac{1}{x}, \dfrac{-1}{y} \rangle$. Calculate the work against F required to move an object from $(1, 1)$ to $(3, 4)$ along any path in the first quadrant.

25. Compute the work W against the earth's gravitational field required to move a satellite of mass $m = 1000$ kg along any path from an orbit of altitude 4000 km to an orbit of altitude 6000 km.

26. An electric dipole with dipole moment $p = 4 \times 10^{-5}$ C-m sets up an electric field (in newtons per coulomb)

$$\mathbf{F}(x, y, z) = \frac{kp}{r^5}\left\langle 3xz, 3yz, 2z^2 - x^2 - y^2 \right\rangle$$

where $r = (x^2 + y^2 + z^2)^{1/2}$ with distance in meters and $k = 8.99 \times 10^9$ N-m^2/C^2. Calculate the work against $\mathbf{F}$ required to move a particle of charge $q = 0.01$ C from $(1, -5, 0)$ to $(3, 4, 4)$. *Note:* The force on q is $q\mathbf{F}$ newtons.

27. On the surface of the earth, the gravitational field (with z as vertical coordinate measured in meters) is $\mathbf{F} = \langle 0, 0, -g \rangle$.

(a) Find a potential function for $\mathbf{F}$.

(b) Beginning at rest, a ball of mass $m = 2$ kg moves under the influence of gravity (without friction) along a path from $P = (3, 2, 400)$ to $Q = (-21, 40, 50)$. Find the ball's velocity when it reaches Q.

28. An electron at rest at $P = (5, 3, 7)$ moves along a path ending at $Q = (1, 1, 1)$ under the influence of the electric field (in newtons per coulomb)

$$\mathbf{F}(x, y, z) = 400(x^2 + z^2)^{-1} \langle x, 0, z \rangle$$

(a) Find a potential function for $\mathbf{F}$.

(b) What is the electron's speed at point Q? Use Conservation of Energy and the value $q_e/m_e = -1.76 \times 10^{11}$ C/kg, where q_e and m_e are the charge and mass on the electron, respectively.

29. Let $\mathbf{F} = \left\langle \dfrac{-y}{x^2 + y^2}, \dfrac{x}{x^2 + y^2} \right\rangle$ be the vortex field. Determine $\displaystyle\int_C \mathbf{F} \cdot d\mathbf{r}$ for each of the paths in Figure 20.

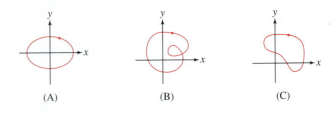

(A)   (B)   (C)

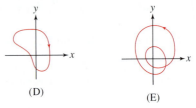

(D)   (E)

FIGURE 20

30. The vector field $\mathbf{F}(x, y) = \left\langle \dfrac{x}{x^2 + y^2}, \dfrac{y}{x^2 + y^2} \right\rangle$ is defined on the domain $\mathcal{D} = \{(x, y) \neq (0, 0)\}$.

(a) Is $\mathcal{D}$ simply connected?

(b) Show that $\mathbf{F}$ satisfies the cross-partial condition. Does this guarantee that $\mathbf{F}$ is conservative?

(c) Show that $\mathbf{F}$ is conservative on $\mathcal{D}$ by finding a potential function.

(d) Do these results contradict Theorem 4?

Further Insights and Challenges

31. Suppose that $\mathbf{F}$ is defined on $\mathbf{R}^3$ and that $\displaystyle\oint_C \mathbf{F} \cdot d\mathbf{r} = 0$ for all closed paths C in $\mathbf{R}^3$. Prove:

(a) $\mathbf{F}$ is path-independent; that is, for any two paths C_1 and C_2 in $\mathcal{D}$ with the same initial and terminal points,

$$\int_{C_1} \mathbf{F} \cdot d\mathbf{r} = \int_{C_2} \mathbf{F} \cdot d\mathbf{r}$$

(b) $\mathbf{F}$ is conservative.

16.4 Parametrized Surfaces and Surface Integrals

The basic idea of an integral appears in several guises. So far, we have defined single, double, and triple integrals and, in the previous section, line integrals over curves. Now we consider one last type on integral: integrals over surfaces. We treat scalar surface integrals in this section and vector surface integrals in the following section.

Just as parametrized curves are a key ingredient in the discussion of line integrals, surface integrals require the notion of a **parametrized surface**—that is, a surface $\mathcal{S}$ whose points are described in the form

$$G(u, v) = (x(u, v), y(u, v), z(u, v))$$

The variables u, v (called parameters) vary in a region $\mathcal{D}$ called the **parameter domain**. Two parameters u and v are needed to parametrize a surface because the surface is two-dimensional.

Figure 1 shows a plot of the surface $\mathcal{S}$ with the parametrization

$$G(u, v) = (u + v, u^3 - v, v^3 - u)$$

This surface consists of all points (x, y, z) in $\mathbf{R}^3$ such that

$$x = u + v, \qquad y = u^3 - v, \qquad z = v^3 - u$$

for (u, v) in $\mathcal{D} = \mathbf{R}^2$.

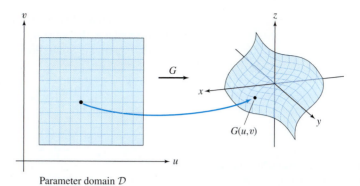

FIGURE 1 The parametric surface
$G(u, v) = (u + v, u^3 - v, v^3 - u)$.

Parameter domain $\mathcal{D}$

■ **EXAMPLE 1** Find a parametrization for the cylinder $x^2 + y^2 = 1$.

Solution The cylinder of radius 1 with equation $x^2 + y^2 = 1$ is conveniently parametrized in cylindrical coordinates (Figure 2). Points on the cylinder have cylindrical coordinates $(1, \theta, z)$, so we use θ and z as parameters.

We obtain

$$G(\theta, z) = (\cos \theta, \sin \theta, z), \qquad 0 \le \theta < 2\pi, \quad -\infty < z < \infty \qquad ■$$

If necessary, review cylindrical and spherical coordinates in Section 12.7.

Similarly, we obtain the parametrization for any vertical cylinder of radius R, given by $x^2 + y^2 = R^2$:

Parametrization of a Cylinder:

$$G(\theta, z) = (R \cos \theta, R \sin \theta, z), \qquad 0 \le \theta < 2\pi, \quad -\infty < z < \infty$$

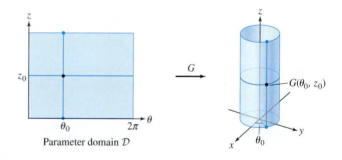

FIGURE 2 The parametrization of a cylinder by cylindrical coordinates amounts to wrapping the rectangle around the cylinder.

Parameter domain $\mathcal{D}$

■ **EXAMPLE 2** Find a parametrization for the sphere of radius 2.

Solution The sphere of radius 2 with its center at the origin is parametrized conveniently using spherical coordinates (ρ, θ, ϕ) with $\rho = 2$ and each of the x, y, and z coordinates given in terms of the equations for them in terms of spherical coordinates (Figure 3).

$$G(\theta, \phi) = (2 \cos \theta \sin \phi, 2 \sin \theta \sin \phi, 2 \cos \phi), \quad 0 \le \theta < 2\pi, \quad 0 \le \phi \le \pi \qquad ■$$

More generally, we can parametrize a sphere of radius R as follows:

Parametrization of a Sphere:

$$G(\theta, \phi) = (R \cos \theta \sin \phi, R \sin \theta \sin \phi, R \cos \phi), \quad 0 \le \theta < 2\pi, \quad 0 \le \phi \le \pi$$

FIGURE 3 Spherical coordinates on a sphere of radius R.

The North and South Poles correspond to $\phi = 0$ and $\phi = \pi$ with any value of θ (the function G fails to be one-to-one at the poles):

North Pole: $G(\theta, 0) = (0, 0, R)$, \qquad South Pole: $G(\theta, \pi) = (0, 0, -R)$

As shown in Figure 4, G sends each horizontal segment $\phi = c$ ($0 < c < \pi$) to a latitude (a circle parallel to the equator) and each vertical segment $\theta = c$ to a longitudinal arc from the North Pole to the South Pole.

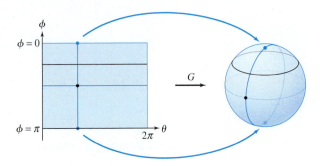

FIGURE 4 The parametrization by spherical coordinates amounts to wrapping the rectangle around the sphere. The top and bottom edges of the rectangle are collapsed to the North and South Poles.

The simplest situation for generating a parametrization of a surface occurs when the surface is the **graph of a function** $z = f(x, y)$, as in Figure 5.

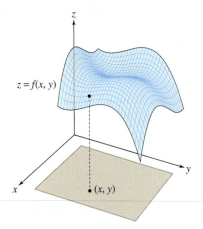

FIGURE 5 Parametrizing the graph of a surface.

Parametrization of a Graph:

$$G(x, y) = (x, y, f(x, y))$$

In this case, the parameters are x and y.

■ **EXAMPLE 3** Find a parametrization of the paraboloid given by $f(x, y) = x^2 + y^2$.

Solution We can immediately define $G(x, y) = (x, y, x^2 + y^2)$. Then G sends the xy-plane onto the paraboloid. ■

Most surfaces in which we are interested do not appear as graphs of functions. In this case, we need to find some other parametrization.

■ **EXAMPLE 4** **Parametrization of a Cone** Find a parametrization of the portion S of the cone with equation $x^2 + y^2 = z^2$ lying above and below the disk $x^2 + y^2 \le 4$. Specify the domain $\mathcal{D}$ of the parametrization.

Solution Notice as in Figure 6 that this portion of the cone is not the graph of a function since it includes that part of the cone that is above the xy-plane with that part that lies below the xy-plane. However, each point on the cone is uniquely determined by its cylindrical coordinates. Because the cone is given by $z^2 = r^2$, the r-coordinate at any point is determined by the z-coordinate, so it is enough to keep track of its z-coordinate and its θ-coordinate. Letting the parameter u correspond to the z-coordinate, and v correspond to the θ-coordinate, a point on the cone at height u has coordinates $(u \cos v, u \sin v, u)$ for some angle v. Thus, the cone has the parametrization

$$G(u, v) = (u \cos v, u \sin v, u)$$

Since we are interested in the portion of the cone where $x^2 + y^2 = u^2 \le 4$, the height variable u satisfies $-2 \le u \le 2$. The angular variable v varies in the interval $[0, 2\pi)$, and therefore, the parameter domain is $\mathcal{D} = [-2, 2] \times [0, 2\pi)$. ■

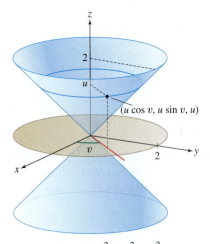

FIGURE 6 The cone $x^2 + y^2 = z^2$.

Grid Curves, Normal Vectors, and the Tangent Plane

Suppose that a surface S has a parametrization

$$G(u, v) = (x(u, v), y(u, v), z(u, v))$$

that is one-to-one on a domain $\mathcal{D}$. We shall always assume that G is **continuously differentiable**, meaning that the functions $x(u, v)$, $y(u, v)$, and $z(u, v)$ have continuous partial derivatives.

*In essence, a parametrization labels each point P on S by a unique pair (u_0, v_0) in the parameter domain. We can think of (u_0, v_0) as the "coordinates" of P determined by the parametrization. They are sometimes called **curvilinear coordinates**.*

In the uv-plane, we can form a grid of lines parallel to the coordinates axes. These grid lines correspond under G to a system of **grid curves** on the surface (Figure 7). More precisely, the horizontal and vertical lines through (u_0, v_0) in the domain correspond to the grid curves $G(u, v_0)$ and $G(u_0, v)$ that intersect at the point $P = G(u_0, v_0)$.

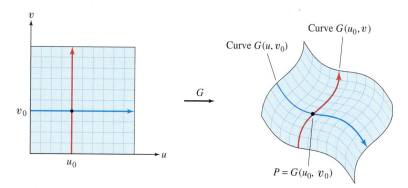

FIGURE 7 Grid curves.

Now consider the tangent vectors to these grid curves (Figure 8):

For $G(u, v_0)$: $\mathbf{T}_u(P) = \dfrac{\partial G}{\partial u}(u_0, v_0) = \left\langle \dfrac{\partial x}{\partial u}(u_0, v_0), \dfrac{\partial y}{\partial u}(u_0, v_0), \dfrac{\partial z}{\partial u}(u_0, v_0) \right\rangle$

For $G(u_0, v)$: $\mathbf{T}_v(P) = \dfrac{\partial G}{\partial v}(u_0, v_0) = \left\langle \dfrac{\partial x}{\partial v}(u_0, v_0), \dfrac{\partial y}{\partial v}(u_0, v_0), \dfrac{\partial z}{\partial v}(u_0, v_0) \right\rangle$

The parametrization G is called **regular** at P if the following cross product is nonzero:

$$\mathbf{N}(P) = \mathbf{N}(u_0, v_0) = \mathbf{T}_u(P) \times \mathbf{T}_v(P)$$

In this case, $\mathbf{T}_u$ and $\mathbf{T}_v$ span the tangent plane to $\mathcal{S}$ at P and $\mathbf{N}(P)$ is a **normal vector** to the tangent plane. We call $\mathbf{N}(P)$ a normal to the surface $\mathcal{S}$.

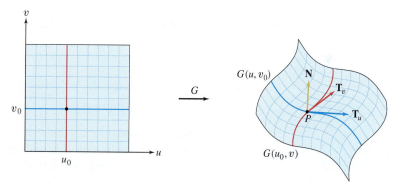

FIGURE 8 The vectors $\mathbf{T}_u$ and $\mathbf{T}_v$ are tangent to the grid curves through $P = G(u_0, v_0)$.

At each point on a surface, the normal vector points in one of two opposite directions. If we change the parametrization, the length of $\mathbf{N}$ may change and its direction may be reversed.

We often write $\mathbf{N}$ instead of $\mathbf{N}(P)$ or $\mathbf{N}(u, v)$, but it is understood that the vector $\mathbf{N}$ varies from point to point on the surface. Similarly, we often denote the tangent vectors by $\mathbf{T}_u$ and $\mathbf{T}_v$. Note that $\mathbf{T}_u$, $\mathbf{T}_v$, and $\mathbf{N}$ need not be unit vectors (thus, the notation here differs from that in Sections 13.4 and 13.5, where $\mathbf{N}$ denotes a unit vector).

■ **EXAMPLE 5** Consider the parametrization $G(\theta, z) = (2\cos\theta, 2\sin\theta, z)$ of the cylinder $x^2 + y^2 = 4$:

(a) Describe the grid curves.

(b) Compute $\mathbf{T}_\theta$, $\mathbf{T}_z$, and $\mathbf{N}(\theta, z)$.

(c) Find an equation of the tangent plane at $P = G(\frac{\pi}{4}, 5)$.

Solution

(a) The grid curves on the cylinder through $P = (\theta_0, z_0)$ are (Figure 9)

θ-grid curve: $G(\theta, z_0) = (2\cos\theta, 2\sin\theta, z_0)$ (circle of radius 2 at height $z = z_0$)

z-grid curve: $G(\theta_0, z) = (2\cos\theta_0, 2\sin\theta_0, z)$ (vertical line through P with $\theta = \theta_0$)

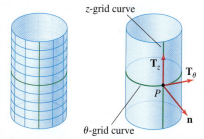

FIGURE 9 Grid curves on the cylinder.

(b) The partial derivatives of $G(\theta, z) = (2\cos\theta, 2\sin\theta, z)$ give us the tangent vectors at P:

$$\theta\text{-grid curve:}\quad \mathbf{T}_\theta = \frac{\partial G}{\partial \theta} = \frac{\partial}{\partial \theta}(2\cos\theta, 2\sin\theta, z) = \langle -2\sin\theta, 2\cos\theta, 0 \rangle$$

$$z\text{-grid curve:}\quad \mathbf{T}_z = \frac{\partial G}{\partial z} = \frac{\partial}{\partial z}(2\cos\theta, 2\sin\theta, z) = \langle 0, 0, 1 \rangle$$

Observe in Figure 9 that $\mathbf{T}_\theta$ is tangent to the θ-grid curve and $\mathbf{T}_z$ is tangent to the z-grid curve. The normal vector is

$$\mathbf{N}(\theta, z) = \mathbf{T}_\theta \times \mathbf{T}_z = \begin{vmatrix} \mathbf{i} & \mathbf{j} & \mathbf{k} \\ -2\sin\theta & 2\cos\theta & 0 \\ 0 & 0 & 1 \end{vmatrix} = 2\cos\theta\,\mathbf{i} + 2\sin\theta\,\mathbf{j}$$

The coefficient of $\mathbf{k}$ is zero, so $\mathbf{N}$ points directly out of the cylinder.

(c) For $\theta = \frac{\pi}{4}$, $z = 5$,

$$P = G\left(\frac{\pi}{4}, 5\right) = \langle \sqrt{2}, \sqrt{2}, 5 \rangle, \qquad \mathbf{N} = \mathbf{N}\left(\frac{\pi}{4}, 5\right) = \langle \sqrt{2}, \sqrt{2}, 0 \rangle$$

> **← ∙∙ REMINDER** An equation of the plane through $P = (x_0, y_0, z_0)$ with normal vector $\mathbf{N}$ is
>
> $$\langle x - x_0, y - y_0, z - z_0 \rangle \cdot \mathbf{N} = 0$$

The tangent plane through P has normal vector $\mathbf{N}$ and thus has equation

$$\langle x - \sqrt{2}, y - \sqrt{2}, z - 5 \rangle \cdot \langle \sqrt{2}, \sqrt{2}, 0 \rangle = 0$$

This can be written

$$\sqrt{2}(x - \sqrt{2}) + \sqrt{2}(y - \sqrt{2}) = 0 \qquad \text{or} \qquad x + y = 2\sqrt{2}$$

The tangent plane is vertical (because z does not appear in the equation).   ∎

■ **EXAMPLE 6** ⊏⋿⟀ **Helicoid Surface** Describe the surface $\mathcal{S}$ with parametrization

$$G(u, v) = (u\cos v, u\sin v, v), \qquad -1 \le u \le 1, \quad 0 \le v < 2\pi$$

(a) Use a computer algebra system to plot $\mathcal{S}$.

(b) Compute $\mathbf{N}(u, v)$ at $u = \frac{1}{2}$, $v = \frac{\pi}{2}$.

Solution For each fixed value $u = a$, the curve $G(a, v) = (a\cos v, a\sin v, v)$ is a helix of radius a. Therefore, as u varies from -1 to 1, $G(u, v)$ describes a family of helices of radius u. The resulting surface is a "helical ramp."

(a) Here is a typical command for a computer algebra system that generates the helicoid surface shown on the right-hand side of Figure 10:

```
ParametricPlot3D[{u*Cos[v],u*Sin[v],v},{u,-1,1},{v,0,2Pi}]
```

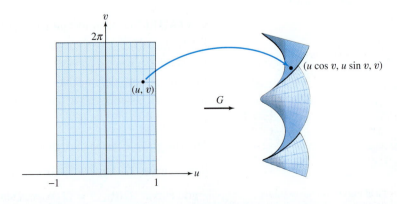

DF **FIGURE 10** Helicoid.

(b) The tangent and normal vectors are

$$\mathbf{T}_u = \frac{\partial G}{\partial u} = \langle \cos v, \sin v, 0 \rangle$$

$$\mathbf{T}_v = \frac{\partial G}{\partial v} = \langle -u \sin v, u \cos v, 1 \rangle$$

$$\mathbf{N}(u, v) = \mathbf{T}_u \times \mathbf{T}_v = \begin{vmatrix} \mathbf{i} & \mathbf{j} & \mathbf{k} \\ \cos v & \sin v & 0 \\ -u \sin v & u \cos v & 1 \end{vmatrix} = (\sin v)\mathbf{i} - (\cos v)\mathbf{j} + u\mathbf{k}$$

At $u = \frac{1}{2}$, $v = \frac{\pi}{2}$, we have $\mathbf{N} = \mathbf{i} + \frac{1}{2}\mathbf{k}$. ■

For future reference, we compute the outward-pointing normal vector in the standard parametrization of the sphere of radius R centered at the origin (Figure 11):

$$G(\theta, \phi) = (R \cos \theta \sin \phi, R \sin \theta \sin \phi, R \cos \phi)$$

Note first that since the distance from $G(\theta, \phi)$ to the origin is R, the *unit* radial vector at $G(\theta, \phi)$ is obtained by dividing by R:

$$\mathbf{e}_r = \langle \cos \theta \sin \phi, \sin \theta \sin \phi, \cos \phi \rangle$$

Furthermore,

$$\mathbf{T}_\theta = \langle -R \sin \theta \sin \phi, R \cos \theta \sin \phi, 0 \rangle$$

$$\mathbf{T}_\phi = \langle R \cos \theta \cos \phi, R \sin \theta \cos \phi, -R \sin \phi \rangle$$

$$\mathbf{N} = \mathbf{T}_\theta \times \mathbf{T}_\phi = \begin{vmatrix} \mathbf{i} & \mathbf{j} & \mathbf{k} \\ -R \sin \theta \sin \phi & R \cos \theta \sin \phi & 0 \\ R \cos \theta \cos \phi & R \sin \theta \cos \phi & -R \sin \phi \end{vmatrix}$$

$$= -R^2 \cos \theta \sin^2 \phi \, \mathbf{i} - R^2 \sin \theta \sin^2 \phi \, \mathbf{j} - R^2 \cos \phi \sin \phi \, \mathbf{k}$$

$$= -R^2 \sin \phi \langle \cos \theta \sin \phi, \sin \theta \sin \phi, \cos \phi \rangle \qquad \boxed{1}$$

$$= -(R^2 \sin \phi) \, \mathbf{e}_r$$

This is an inward-pointing normal vector. However, in most computations it is standard to use the outward-pointing normal vector:

$$\boxed{\mathbf{N} = \mathbf{T}_\phi \times \mathbf{T}_\theta = (R^2 \sin \phi) \, \mathbf{e}_r, \qquad \|\mathbf{N}\| = R^2 \sin \phi} \qquad \boxed{2}$$

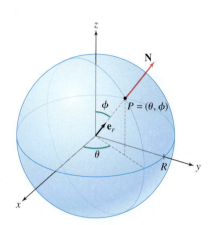

FIGURE 11 The normal vector **n** points in the radial direction $\mathbf{e}_r$.

Surface Area

The length $\|\mathbf{N}\|$ of the normal vector in a parametrization has an important interpretation in terms of area. Assume, for simplicity, that $\mathcal{D}$ is a rectangle (the argument also applies to more general domains). Divide $\mathcal{D}$ into a grid of small rectangles $\mathcal{R}_{ij}$ of size $\Delta u \times \Delta v$, as in Figure 12, and compare the area of $\mathcal{R}_{ij}$ with the area of its image under G. This image is a "curved" parallelogram $\mathcal{S}_{ij} = G(\mathcal{R}_{ij})$.

First, we note that if Δu and Δv in Figure 12 are small, then the curved parallelogram $\mathcal{S}_{ij}$ has approximately the same area as the "genuine" parallelogram with sides $\overrightarrow{PQ}$ and $\overrightarrow{PS}$. Recall that the area of the parallelogram spanned by two vectors is the length of their cross product, so

$$\text{Area}(\mathcal{S}_{ij}) \approx \|\overrightarrow{PQ} \times \overrightarrow{PS}\|$$

◄ ∙∙ REMINDER *By Theorem 3 in Section 12.4, the area of the parallelogram spanned by vectors* **v** *and* **w** *in* $\mathbf{R}^3$ *is equal to* $\|\mathbf{v} \times \mathbf{w}\|$.

Next, we use the linear approximation to estimate the vectors $\overrightarrow{PQ}$ and $\overrightarrow{PS}$:

$$\overrightarrow{PQ} = G(u_{ij} + \Delta u, v_{ij}) - G(u_{ij}, v_{ij}) \approx \frac{\partial G}{\partial u}(u_{ij}, v_{ij})\Delta u = \mathbf{T}_u \Delta u$$

$$\overrightarrow{PS} = G(u_{ij}, v_{ij} + \Delta v) - G(u_{ij}, v_{ij}) \approx \frac{\partial G}{\partial v}(u_{ij}, v_{ij})\Delta v = \mathbf{T}_v \Delta v$$

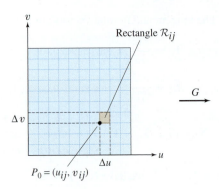

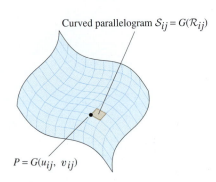

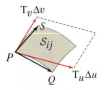

FIGURE 12

The approximation (3) is valid for any small region $\mathcal{R}$ in the uv-plane:

$$\text{Area}(\mathcal{S}) \approx \|\mathbf{N}(u_0, v_0))\|\text{Area}(\mathcal{R})$$

where $\mathcal{S} = G(\mathcal{R})$ and (u_0, v_0) is any sample point in $\mathcal{R}$. Here, "small" means contained in a small disk. We do not allow $\mathcal{R}$ to be very thin and wide.

Note: We require only that G be one-to-one on the interior of $\mathcal{D}$. Many common parametrizations (such as the parametrizations by cylindrical and spherical coordinates) fail to be one-to-one on the boundary of their domains.

Thus, we have

$$\text{Area}(\mathcal{S}_{ij})) \approx \|\mathbf{T}_u \Delta u \times \mathbf{T}_v \Delta v\| = \|\mathbf{T}_u \times \mathbf{T}_v\|\, \Delta u\, \Delta v$$

Since $\mathbf{N}(u_{ij}, v_{ij}) = \mathbf{T}_u \times \mathbf{T}_v$ and $\text{Area}(\mathcal{R}_{ij}) = \Delta u \Delta v$, we obtain

$$\boxed{\text{Area}(\mathcal{S}_{ij}) \approx \|\mathbf{N}(u_{ij}, v_{ij})\|\text{Area}(\mathcal{R}_{ij})} \qquad \boxed{3}$$

Our conclusion: $\|\mathbf{N}\|$ *is a distortion factor that measures how the area of a small rectangle* $\mathcal{R}_{ij}$ *is altered under the map* G.

To compute the surface area of $\mathcal{S}$, we assume that G is one-to-one, except possibly on the boundary of $\mathcal{D}$. We also assume that G is regular, except possibly on the boundary of $\mathcal{D}$. Recall that "regular" means $\mathbf{N}(u, v)$ is nonzero.

The entire surface $\mathcal{S}$ is the union of the small patches $\mathcal{S}_{ij}$, so we can apply the approximation on each patch to obtain

$$\text{Area}(\mathcal{S}) = \sum_{i,j}\text{Area}(\mathcal{S}_{ij}) \approx \sum_{i,j}\|\mathbf{N}(u_{ij}, v_{ij})\|\Delta u\, \Delta v \qquad \boxed{4}$$

The sum on the right is a Riemann sum for the double integral of $\|\mathbf{N}(u, v)\|$ over the parameter domain $\mathcal{D}$. As Δu and Δv tend to zero, these Riemann sums converge to a double integral, which we take as the definition of surface area:

$$\boxed{\text{Area}(\mathcal{S}) = \iint_{\mathcal{D}} \|\mathbf{N}(u, v)\|\, du\, dv}$$

Surface Integral

Now we can define the surface integral of a function $f(x, y, z)$:

$$\iint_{\mathcal{S}} f(x, y, z)\, dS$$

This is similar to the definition of the line integral of a function over a curve. Choose a sample point $P_{ij} = G(u_{ij}, v_{ij})$ in each small patch $\mathcal{S}_{ij}$ and form the sum:

$$\sum_{i,j} f(P_{ij})\text{Area}(\mathcal{S}_{ij}) \qquad \boxed{5}$$

The limit of these sums as Δu and Δv tend to zero (if it exists) is the **surface integral**:

$$\iint_{\mathcal{S}} f(x, y, z)\, dS = \lim_{\Delta u, \Delta v \to 0} \sum_{i,j} f(P_{ij}) \text{Area}(\mathcal{S}_{ij})$$

To evaluate the surface integral, we use Eq. (3) to write

$$\sum_{i,j} f(P_{ij}) \text{Area}(\mathcal{S}_{ij}) \approx \sum_{i,j} f(G(u_{ij}, v_{ij})) \|\mathbf{N}(u_{ij}, v_{ij})\| \, \Delta u \, \Delta v \qquad \boxed{6}$$

On the right, we have a Riemann sum for the double integral of

$$f(G(u, v)) \|\mathbf{N}(u, v)\|$$

over the parameter domain $\mathcal{D}$. Under the assumption that G is continuously differentiable, we can show these sums in Eq. (6) approach the same limit. This yields the next theorem.

It is interesting to note that Eq. (7) includes the Change of Variables Formula for double integrals (Theorem 2 in Section 15.6) as a special case. If the surface $\mathcal{S}$ is a domain in the xy-plane [in other words, $z(u, v) = 0$], then the integral over $\mathcal{S}$ reduces to the double integral of the function $f(x, y, 0)$. We may view $G(u, v)$ as a mapping from the uv-plane to the xy-plane, and we find that $\|\mathbf{N}(u, v)\|$ is the Jacobian of this mapping.

THEOREM 1 Surface Integrals and Surface Area Let $G(u, v)$ be a parametrization of a surface $\mathcal{S}$ with parameter domain $\mathcal{D}$. Assume that G is continuously differentiable, one-to-one, and regular (except possibly at the boundary of $\mathcal{D}$). Then

$$\iint_{\mathcal{S}} f(x, y, z)\, dS = \iint_{\mathcal{D}} f(G(u, v)) \|\mathbf{N}(u, v)\| \, du\, dv \qquad \boxed{7}$$

For $f(x, y, z) = 1$, we obtain the surface area of $\mathcal{S}$:

$$\text{Area}(\mathcal{S}) = \iint_{\mathcal{D}} \|\mathbf{N}(u, v)\| \, du\, dv$$

Eq. (7) is summarized by the symbolic expression for the "surface element":

$$dS = \|\mathbf{N}(u, v)\| \, du\, dv$$

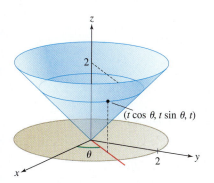

FIGURE 13 Portion $\mathcal{S}$ of the cone $x^2 + y^2 = z^2$ lying over the disk $x^2 + y^2 \le 4$.

◄··· **REMINDER** In this example,

$$G(\theta, t) = (t \cos\theta, t \sin\theta, t)$$

■ **EXAMPLE 7** Calculate the surface area of the portion $\mathcal{S}$ of the cone $x^2 + y^2 = z^2$ lying above the disk $x^2 + y^2 \le 4$ (Figure 13). Then calculate $\displaystyle\iint_{\mathcal{S}} x^2 z\, dS$.

Solution A parametrization of the cone was found in Example 4. Using the variables θ and t, this parametrization is

$$G(\theta, t) = (t \cos\theta, t \sin\theta, t), \qquad 0 \le t \le 2, \quad 0 \le \theta < 2\pi$$

***Step 1.* Compute the tangent and normal vectors.**

$$\mathbf{T}_\theta = \frac{\partial G}{\partial \theta} = \langle -t \sin\theta, t \cos\theta, 0 \rangle, \qquad \mathbf{T}_t = \frac{\partial G}{\partial t} = \langle \cos\theta, \sin\theta, 1 \rangle$$

$$\mathbf{N} = \mathbf{T}_\theta \times \mathbf{T}_t = \begin{vmatrix} \mathbf{i} & \mathbf{j} & \mathbf{k} \\ -t \sin\theta & t \cos\theta & 0 \\ \cos\theta & \sin\theta & 1 \end{vmatrix} = t \cos\theta\, \mathbf{i} + t \sin\theta\, \mathbf{j} - t \mathbf{k}$$

The normal vector has length

$$\|\mathbf{N}\| = \sqrt{t^2 \cos^2\theta + t^2 \sin^2\theta + (-t)^2} = \sqrt{2t^2} = \sqrt{2}\, |t|$$

Thus, $dS = \sqrt{2}\, |t|\, d\theta\, dt$. Since $t \ge 0$ on our domain, we drop the absolute value.

***Step 2.* Calculate the surface area.**

$$\text{Area}(\mathcal{S}) = \iint_{\mathcal{D}} \|\mathbf{N}\| \, du\, dv = \int_0^2 \int_0^{2\pi} \sqrt{2}\, t\, d\theta\, dt = \sqrt{2}\pi t^2 \Big|_0^2 = 4\sqrt{2}\pi$$

Step 3. **Calculate the surface integral.**

We express $f(x, y, z) = x^2 z$ in terms of the parameters t and θ and evaluate:

$$f(G(\theta, t)) = f(t \cos \theta, t \sin \theta, t) = (t \cos \theta)^2 t = t^3 \cos^2 \theta$$

$$\iint_S f(x, y, z) \, dS = \int_{t=0}^{2} \int_{\theta=0}^{2\pi} f(G(\theta, t)) \, \|N(\theta, t)\| \, d\theta \, dt$$

$$= \int_{t=0}^{2} \int_{\theta=0}^{2\pi} (t^3 \cos^2 \theta)(\sqrt{2}\,t) \, d\theta \, dt$$

$$= \sqrt{2} \left(\int_0^2 t^4 \, dt \right) \left(\int_0^{2\pi} \cos^2 \theta \, d\theta \right)$$

$$= \sqrt{2} \left(\frac{32}{5} \right) (\pi) = \frac{32\sqrt{2}\pi}{5} \qquad \blacksquare$$

◀┅ REMINDER

$$\int_0^{2\pi} \cos^2 \theta \, d\theta = \int_0^{2\pi} \frac{1 + \cos 2\theta}{2} \, d\theta = \pi$$

In previous discussions of multiple and line integrals, we applied the principle that the integral of a density is the total quantity. This applies to surface integrals as well. For example, a surface with mass density $\delta(x, y, z)$ (in units of mass per area) is the surface integral of the mass density:

$$\text{Mass of } \mathcal{S} = \iint_S \delta(x, y, z) \, dS$$

Similarly, if an electric charge is distributed over $\mathcal{S}$ with charge density $\delta(x, y, z)$, then the surface integral of $\delta(x, y, z)$ is the total charge on $\mathcal{S}$.

■ **EXAMPLE 8** **Total Charge on a Surface** Find the total charge (in coulombs) on a sphere S of radius 5 cm whose charge density in spherical coordinates is $\delta(\theta, \phi) = 0.003 \cos^2 \phi$ C/cm^2.

Solution We parametrize S in spherical coordinates:

$$G(\theta, \phi) = (5 \cos \theta \sin \phi, 5 \sin \theta \sin \phi, 5 \cos \phi)$$

By Eq. (2), $\|N\| = 5^2 \sin \phi$ and

$$\text{Total charge} = \iint_S \delta(\theta, \phi) \, dS = \int_{\theta=0}^{2\pi} \int_{\phi=0}^{\pi} \delta(\theta, \phi) \|N\| \, d\phi \, d\theta$$

$$= \int_{\theta=0}^{2\pi} \int_{\phi=0}^{\pi} (0.003 \cos^2 \phi)(25 \sin \phi) \, d\phi \, d\theta$$

$$= (0.075)(2\pi) \int_{\phi=0}^{\pi} \cos^2 \phi \sin \phi \, d\phi$$

$$= 0.15\pi \left(-\frac{\cos^3 \phi}{3} \right) \Big|_0^{\pi} = 0.15\pi \left(\frac{2}{3} \right) \approx 0.1\pi \text{ coulombs} \qquad \blacksquare$$

When a graph $z = g(x, y)$ is parametrized by $G(x, y) = (x, y, g(x, y))$, the tangent and normal vectors are

$$T_x = (1, 0, g_x), \qquad T_y = (0, 1, g_y)$$

$$N = T_x \times T_y = \begin{vmatrix} i & j & k \\ 1 & 0 & g_x \\ 0 & 1 & g_y \end{vmatrix} = -g_x i - g_y j + k, \quad \|N\| = \sqrt{1 + g_x^2 + g_y^2} \qquad \boxed{8}$$

The surface integral over the portion of a graph lying over a domain $\mathcal{D}$ in the xy-plane is

$$\text{Surface integral over a graph} = \iint_{\mathcal{D}} f(x, y, g(x, y)) \sqrt{1 + g_x^2 + g_y^2} \, dx \, dy \qquad \boxed{9}$$

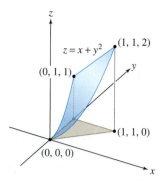

FIGURE 14 The surface $z = x + y^2$ over $0 \leq x \leq y \leq 1$.

■ **EXAMPLE 9** Calculate $\iint_{S} (z - x) \, dS$, where S is the portion of the graph of $z = x + y^2$ where $0 \leq x \leq y$, $0 \leq y \leq 1$ (Figure 14).

Solution Let $z = g(x, y) = x + y^2$. Then $g_x = 1$ and $g_y = 2y$, and

$$dS = \sqrt{1 + g_x^2 + g_y^2} \, dx \, dy = \sqrt{1 + 1 + 4y^2} \, dx \, dy = \sqrt{2 + 4y^2} \, dx \, dy$$

On the surface S, we have $z = x + y^2$, and thus,

$$f(x, y, z) = z - x = (x + y^2) - x = y^2$$

By Eq. (9),

$$\iint_{S} f(x, y, z) \, dS = \int_{y=0}^{1} \int_{x=0}^{y} y^2 \sqrt{2 + 4y^2} \, dx \, dy$$

$$= \int_{y=0}^{1} \left(y^2 \sqrt{2 + 4y^2} \right) x \Big|_{x=0}^{y} dy = \int_{0}^{1} y^3 \sqrt{2 + 4y^2} \, dy$$

Now use the substitution $u = 2 + 4y^2$, $du = 8y \, dy$. Then $y^2 = \frac{1}{4}(u - 2)$, and

$$\int_{0}^{1} y^3 \sqrt{2 + 4y^2} \, dy = \frac{1}{8} \int_{2}^{6} \frac{1}{4} (u - 2) \sqrt{u} \, du = \frac{1}{32} \int_{2}^{6} (u^{3/2} - 2u^{1/2}) \, du$$

$$= \frac{1}{32} \left(\frac{2}{5} u^{5/2} - \frac{4}{3} u^{3/2} \right) \Big|_{2}^{6} = \frac{1}{30} (6\sqrt{6} + \sqrt{2}) \approx 0.54 \quad ■$$

Excursion

The French mathematician Pierre Simon, Marquis de Laplace (1749–1827) showed that the gravitational potential satisfies the Laplace equation $\Delta V = 0$, where Δ is the Laplace operator

$$\Delta V = \frac{\partial^2 V}{\partial x^2} + \frac{\partial^2 V}{\partial y^2} + \frac{\partial^2 V}{\partial z^2}$$

This equation plays an important role in more advanced branches of math and physics.

In physics, it is an important fact that the gravitational field **F** corresponding to any arrangement of masses is conservative; that is, $\mathbf{F} = -\nabla V$ (recall that the minus sign is a convention of physics). The field at a point P due to a mass m located at point Q is $\mathbf{F} = -\dfrac{Gm}{r^2} \mathbf{e}_r$, where $\mathbf{e}_r$ is the unit vector pointing from Q to P, and r is the distance from P to Q, which we denote by $|P - Q|$. As we saw in Example 4 of Section 16.3,

$$V(P) = -\frac{Gm}{r} = -\frac{Gm}{|P - Q|}$$

If, instead of a single mass, we have K point masses $m_1, \ldots, m_K$ located at $Q_1, \ldots, Q_K$, then the gravitational potential is the sum

$$V(P) = -G \sum_{i=1}^{K} \frac{m_i}{|P - Q_i|} \qquad \boxed{10}$$

If mass is distributed continuously over a thin surface S with mass density function $\delta(x, y, z)$, we replace the sum by the surface integral

$$V(P) = -G \iint_{S} \frac{\delta(x, y, z) \, dS}{|P - Q|} = -G \iint_{S} \frac{\delta(x, y, z) \, dS}{\sqrt{(x - a)^2 + (y - b)^2 + (z - c)^2}} \qquad \boxed{11}$$

where $P = (a, b, c)$. However, this surface integral cannot usually be evaluated explicitly unless the surface and mass distribution are sufficiently symmetric, as in the case of a hollow sphere of uniform mass density (Figure 15).

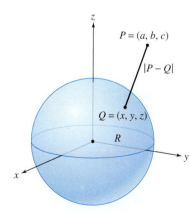

FIGURE 15

THEOREM 2 Gravitational Potential of a Uniform Hollow Sphere The gravitational potential V due to a hollow sphere of radius R with uniform mass distribution of total mass m at a point P located at a distance r from the center of the sphere is

$$V(P) = \begin{cases} -\dfrac{Gm}{r} & \text{if } r > R \quad (P \text{ outside the sphere}) \\[2mm] -\dfrac{Gm}{R} & \text{if } r < R \quad (P \text{ inside the sphere}) \end{cases} \qquad \boxed{12}$$

We leave this calculation as an exercise (Exercise 48), because we will derive it again with much less effort using Gauss's Law in Section 17.3.

In his magnum opus, *Principia Mathematica*, Isaac Newton proved that a sphere of uniform mass density (whether hollow or solid) attracts a particle outside the sphere as if the entire mass were concentrated at the center. In other words, a uniform sphere behaves like a point mass as far as gravity is concerned. Furthermore, if the sphere is hollow, then the sphere exerts no gravitational force on a particle inside it. Newton's result follows from Eq. (12). Outside the sphere, V has the same formula as the potential due to a point mass. Inside the sphere, the potential is *constant* with value $-Gm/R$. But constant potential means zero force because the force is the (negative) gradient of the potential. This discussion applies equally well to the electrostatic force. In particular, a uniformly charged sphere behaves like a point charge (when viewed from outside the sphere).

16.4 SUMMARY

- A *parametrized surface* is a surface S whose points are described in the form

$$G(u, v) = (x(u, v), y(u, v), z(u, v))$$

where the *parameters* u and v vary in a domain $\mathcal{D}$ in the uv-plane.
- Tangent and normal vectors:

$$\mathbf{T}_u = \frac{\partial G}{\partial u} = \left\langle \frac{\partial x}{\partial u}, \frac{\partial y}{\partial u}, \frac{\partial z}{\partial u} \right\rangle, \qquad \mathbf{T}_v = \frac{\partial G}{\partial v} = \left\langle \frac{\partial x}{\partial v}, \frac{\partial y}{\partial v}, \frac{\partial z}{\partial v} \right\rangle$$

$$\mathbf{N} = \mathbf{N}(u, v) = \mathbf{T}_u \times \mathbf{T}_v$$

The parametrization is *regular* at (u, v) if $\mathbf{N}(u, v) \neq \mathbf{0}$.
- The quantity $\|\mathbf{N}\|$ is an "area distortion factor." If $\mathcal{D}$ is a small region in the uv-plane and $S = G(\mathcal{D})$, then

$$\text{Area}(S) \approx \|\mathbf{N}(u_0, v_0)\| \text{Area}(\mathcal{D})$$

where (u_0, v_0) is any sample point in $\mathcal{D}$.
- Surface integrals and surface area:

$$\iint_S f(x, y, z) \, dS = \iint_{\mathcal{D}} f(G(u, v)) \, \|\mathbf{N}(u, v)\| \, du \, dv$$

$$\text{Area}(S) = \iint_{\mathcal{D}} \|\mathbf{N}(u, v)\| \, du \, dv$$

- Some standard parametrizations:

  - Cylinder of radius R (z-axis as central axis):

    $$G(\theta, z) = (R \cos \theta, R \sin \theta, z)$$

    Outward normal: $\mathbf{N} = \mathbf{T}_\theta \times \mathbf{T}_z = R \langle \cos \theta, \sin \theta, 0 \rangle$

    $dS = \|\mathbf{N}\| \, d\theta \, dz = R \, d\theta \, dz$

  - Sphere of radius R, centered at the origin:

    $$G(\theta, \phi) = (R \cos \theta \sin \phi, R \sin \theta \sin \phi, R \cos \phi)$$

    Unit radial vector: $\mathbf{e}_r = \langle \cos \theta \sin \phi, \sin \theta \sin \phi, \cos \phi \rangle$

    Outward normal: $\mathbf{N} = \mathbf{T}_\phi \times \mathbf{T}_\theta = (R^2 \sin \phi) \mathbf{e}_r$

    $dS = \|\mathbf{N}\| \, d\phi \, d\theta = R^2 \sin \phi \, d\phi \, d\theta$

– Graph of $z = g(x, y)$:

$$G(x, y) = (x, y, g(x, y))$$

$$\mathbf{N} = \mathbf{T}_x \times \mathbf{T}_y = \langle -g_x, -g_y, 1 \rangle$$

$$dS = \|\mathbf{N}\| \, dx \, dy = \sqrt{1 + g_x^2 + g_y^2} \, dx \, dy$$

16.4 EXERCISES

Preliminary Questions

1. What is the surface integral of the function $f(x, y, z) = 10$ over a surface of total area 5?

2. What interpretation can we give to the length $\|\mathbf{N}\|$ of the normal vector for a parametrization $G(u, v)$?

3. A parametrization maps a rectangle of size 0.01×0.02 in the uv-plane onto a small patch S of a surface. Estimate Area(S) if $\mathbf{T}_u \times \mathbf{T}_v = \langle 1, 2, 2 \rangle$ at a sample point in the rectangle.

4. A small surface S is divided into three small pieces, each of area 0.2. Estimate $\iint_S f(x, y, z) \, dS$ if $f(x, y, z)$ takes the values 0.9, 1, and 1.1 at sample points in these three pieces.

5. A surface S has a parametrization whose domain is the square $0 \le u, v \le 2$ such that $\|\mathbf{N}(u, v)\| = 5$ for all (u, v). What is Area(S)?

6. What is the outward-pointing unit normal to the sphere of radius 3 centered at the origin at $P = (2, 2, 1)$?

Exercises

1. Match each parametrization with the corresponding surface in Figure 16.

(a) $(u, \cos v, \sin v)$

(b) $(u, u + v, v)$

(c) (u, v^3, v)

(d) $(\cos u \sin v, 3 \cos u \sin v, \cos v)$

(e) $(u, u(2 + \cos v), u(2 + \sin v))$

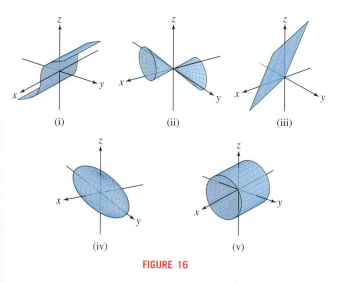

(i)   (ii)   (iii)

(iv)   (v)

FIGURE 16

2. Show that $G(r, \theta) = (r \cos \theta, r \sin \theta, 1 - r^2)$ parametrizes the paraboloid $z = 1 - x^2 - y^2$. Describe the grid curves of this parametrization.

3. Show that $G(u, v) = (2u + 1, u - v, 3u + v)$ parametrizes the plane $2x - y - z = 2$. Then:

(a) Calculate $\mathbf{T}_u, \mathbf{T}_v,$ and $\mathbf{N}(u, v)$.

(b) Find the area of $S = G(\mathcal{D})$, where $\mathcal{D} = \{(u, v) : 0 \le u \le 2, 0 \le v \le 1\}$.

(c) Express $f(x, y, z) = yz$ in terms of u and v, and evaluate $\iint_S f(x, y, z) \, dS$.

4. Let $S = G(\mathcal{D})$, where $\mathcal{D} = \{(u, v) : u^2 + v^2 \le 1, u \ge 0, v \ge 0\}$ and G is as defined in Exercise 3.

(a) Calculate the surface area of S.

(b) Evaluate $\iint_S (x - y) \, dS$. *Hint:* Use polar coordinates.

5. Let $G(x, y) = (x, y, xy)$.

(a) Calculate $\mathbf{T}_x, \mathbf{T}_y,$ and $\mathbf{N}(x, y)$.

(b) Let S be the part of the surface with parameter domain $\mathcal{D} = \{(x, y) : x^2 + y^2 \le 1, x \ge 0, y \ge 0\}$. Verify the following formula and evaluate using polar coordinates:

$$\iint_S 1 \, dS = \iint_{\mathcal{D}} \sqrt{1 + x^2 + y^2} \, dx \, dy$$

(c) Verify the following formula and evaluate:

$$\iint_S z \, dS = \int_0^{\pi/2} \int_0^1 (\sin \theta \cos \theta) r^3 \sqrt{1 + r^2} \, dr \, d\theta$$

6. A surface S has a parametrization $G(u, v)$ whose domain $\mathcal{D}$ is the square in Figure 17. Suppose that G has the following normal vectors:

$$\mathbf{N}(A) = \langle 2, 1, 0 \rangle, \quad \mathbf{N}(B) = \langle 1, 3, 0 \rangle$$

$$\mathbf{N}(C) = \langle 3, 0, 1 \rangle, \quad \mathbf{N}(D) = \langle 2, 0, 1 \rangle$$

Estimate $\iint_S f(x, y, z) \, dS$, where f is a function such that $f(G(u, v)) = u + v$.

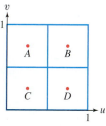

FIGURE 17

In Exercises 7–10, calculate $\mathbf{T}_u$, $\mathbf{T}_v$, *and* $\mathbf{N}(u, v)$ *for the parametrized surface at the given point. Then find the equation of the tangent plane to the surface at that point.*

7. $G(u, v) = (2u + v, u - 4v, 3u);$ $u = 1,$ $v = 4$

8. $G(u, v) = (u^2 - v^2, u + v, u - v);$ $u = 2,$ $v = 3$

9. $G(\theta, \phi) = (\cos \theta \sin \phi, \sin \theta \sin \phi, \cos \phi);$ $\theta = \frac{\pi}{2},$ $\phi = \frac{\pi}{4}$

10. $G(r, \theta) = (r \cos \theta, r \sin \theta, 1 - r^2);$ $r = \frac{1}{2},$ $\theta = \frac{\pi}{4}$

11. Use the normal vector computed in Exercise 8 to estimate the area of the small patch of the surface $G(u, v) = (u^2 - v^2, u + v, u - v)$ defined by

$$2 \le u \le 2.1, \qquad 3 \le v \le 3.2$$

12. Sketch the small patch of the sphere whose spherical coordinates satisfy

$$\frac{\pi}{2} - 0.15 \le \theta \le \frac{\pi}{2} + 0.15, \qquad \frac{\pi}{4} - 0.1 \le \phi \le \frac{\pi}{4} + 0.1$$

Use the normal vector computed in Exercise 9 to estimate its area.

In Exercises 13–26, calculate $\displaystyle\iint_S f(x, y, z) \, dS$ *for the given surface and function.*

13. $G(u, v) = (u \cos v, u \sin v, u),$ $0 \le u \le 1,$ $0 \le v \le 1;$
$f(x, y, z) = z(x^2 + y^2)$

14. $G(r, \theta) = (r \cos \theta, r \sin \theta, \theta),$ $0 \le r \le 1,$ $0 \le \theta \le 2\pi;$
$f(x, y, z) = \sqrt{x^2 + y^2}$

15. $y = 9 - z^2,$ $0 \le x \le 3, 0 \le z \le 3;$ $f(x, y, z) = z$

16. $y = 9 - z^2,$ $0 \le x \le z \le 3;$ $f(x, y, z) = 1$

17. $x^2 + y^2 + z^2 = 1,$ $x, y, z \ge 0;$ $f(x, y, z) = x^2.$

18. $z = 4 - x^2 - y^2,$ $0 \le z \le 3;$ $f(x, y, z) = x^2/(4 - z)$

19. $x^2 + y^2 = 4,$ $0 \le z \le 4;$ $f(x, y, z) = e^{-z}$

20. $G(u, v) = (u, v^3, u + v),$ $0 \le u \le 1, 0 \le v \le 1;$
$f(x, y, z) = y$

21. Part of the plane $x + y + z = 1$, where $x, y, z \ge 0$;
$f(x, y, z) = z$

22. Part of the plane $x + y + z = 0$ contained in the cylinder $x^2 + y^2 = 1$; $f(x, y, z) = z^2$

23. $x^2 + y^2 + z^2 = 4, 1 \le z \le 2;$
$f(x, y, z) = z^2(x^2 + y^2 + z^2)^{-1}$

24. $x^2 + y^2 + z^2 = 4, 0 \le y \le 1;$ $f(x, y, z) = y$

25. Part of the surface $z = x^3$, where $0 \le x \le 1,$ $0 \le y \le 1;$
$f(x, y, z) = z$

26. Part of the unit sphere centered at the origin, where $x \ge 0$ and $|y| \le x;$ $f(x, y, z) = x$

27. A surface S has a parametrization $G(u, v)$ with domain $0 \le u \le 2, 0 \le v \le 4$ such that the following partial derivatives are constant:

$$\frac{\partial G}{\partial u} = \langle 2, 0, 1 \rangle, \qquad \frac{\partial G}{\partial v} = \langle 4, 0, 3 \rangle$$

What is the surface area of S?

28. Let S be the sphere of radius R centered at the origin. Explain using symmetry:

$$\iint_S x^2 \, dS = \iint_S y^2 \, dS = \iint_S z^2 \, dS$$

Then show that $\displaystyle\iint_S x^2 \, dS = \frac{4}{3}\pi R^4$ by adding the integrals.

29. Calculate $\displaystyle\iint_S (xy + e^z) \, dS$, where S is the triangle in Figure 18 with vertices $(0, 0, 3)$, $(1, 0, 2)$, and $(0, 4, 1)$.

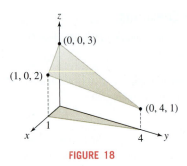

FIGURE 18

30. Use spherical coordinates to compute the surface area of a sphere of radius R.

31. Use cylindrical coordinates to compute the surface area of a sphere of radius R.

32. [CAS] Let S be the surface with parametrization

$$G(u, v) = \big((3 + \sin v) \cos u, (3 + \sin v) \sin u, v\big)$$

for $0 \le u \le 2\pi, 0 \le v \le 2\pi$. Using a computer algebra system:
(a) plot S from several different viewpoints. Is S best described as a "vase that holds water" or a "bottomless vase"?
(b) calculate the normal vector $\mathbf{N}(u, v)$.
(c) calculate the surface area of S to four decimal places.

33. [CAS] Let S be the surface $z = \ln(5 - x^2 - y^2)$ for $0 \le x \le 1,$ $0 \le y \le 1$. Using a computer algebra system:
(a) calculate the surface area of S to four decimal places.
(b) calculate $\displaystyle\iint_S x^2 y^3 \, dS$ to four decimal places.

34. Find the area of the portion of the plane $2x + 3y + 4z = 28$ lying above the rectangle $1 \le x \le 3, 2 \le y \le 5$ in the xy-plane.

35. What is the area of the portion of the plane $2x + 3y + 4z = 28$ lying above the domain $\mathcal{D}$ in the xy-plane in Figure 19 if Area$(\mathcal{D}) = 5$?

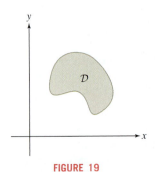

FIGURE 19

36. Find the surface area of the part of the cone $x^2 + y^2 = z^2$ between the planes $z = 2$ and $z = 5$.

37. Find the surface area of the portion S of the cone $z^2 = x^2 + y^2$, where $z \geq 0$, contained within the cylinder $y^2 + z^2 \leq 1$.

38. Calculate the integral of ze^{2x+y} over the surface of the box in Figure 20.

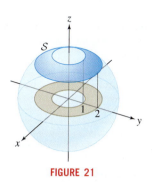

FIGURE 21

39. Calculate $\displaystyle\iint_G x^2 z \, dS$, where G is the cylinder (including the top and bottom) $x^2 + y^2 = 4$, $0 \leq z \leq 3$.

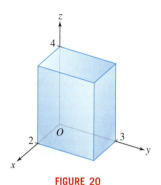

FIGURE 20

40. Let S be the portion of the sphere $x^2 + y^2 + z^2 = 9$ where $1 \leq x^2 + y^2 \leq 4$ and $z \geq 0$ (Figure 21). Find a parametrization of S in polar coordinates and use it to compute:

(a) The area of S **(b)** $\displaystyle\iint_S z^{-1} \, dS$

41. Prove a famous result of Archimedes: The surface area of the portion of the sphere of radius R between two horizontal planes $z = a$ and $z = b$ is equal to the surface area of the corresponding portion of the circumscribed cylinder (Figure 22).

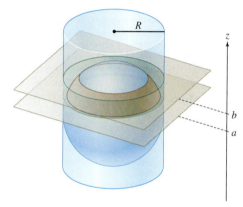

FIGURE 22

Further Insights and Challenges

42. Surfaces of Revolution Let S be the surface formed by rotating the region under the graph $z = g(y)$ in the yz-plane for $c \leq y \leq d$ about the z-axis, where $c \geq 0$ (Figure 23).

(a) Show that the circle generated by rotating a point $(0, a, b)$ about the z-axis is parametrized by

$$(a \cos \theta, a \sin \theta, b), \quad 0 \leq \theta \leq 2\pi$$

(b) Show that S is parametrized by

$$G(y, \theta) = (y \cos \theta, y \sin \theta, g(y)) \qquad \boxed{13}$$

for $c \leq y \leq d, 0 \leq \theta \leq 2\pi$.

(c) Use Eq. (13) to prove the formula

$$\text{Area}(S) = 2\pi \int_c^d y\sqrt{1 + g'(y)^2} \, dy \qquad \boxed{14}$$

43. Use Eq. (14) to compute the surface area of $z = 4 - y^2$ for $0 \leq y \leq 2$ rotated about the z-axis.

44. Describe the upper half of the cone $x^2 + y^2 = z^2$ for $0 \leq z \leq d$ as a surface of revolution (Figure 6) and use Eq. (14) to compute its surface area.

45. Area of a Torus Let $\mathcal{T}$ be the torus obtained by rotating the circle in the yz-plane of radius a centered at $(0, b, 0)$ about the z-axis (Figure 24). We assume that $b > a > 0$.

(a) Use Eq. (14) to show that

$$\text{Area}(\mathcal{T}) = 4\pi \int_{b-a}^{b+a} \frac{ay}{\sqrt{a^2 - (b-y)^2}} \, dy$$

(b) Show that $\text{Area}(\mathcal{T}) = 4\pi^2 ab$.

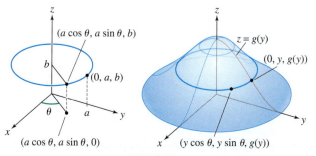

FIGURE 23

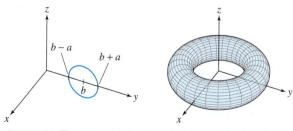

FIGURE 24 The torus obtained by rotating a circle of radius a.

46. Pappus's Theorem (also called **Guldin's Rule**), which we introduced in Section 8.3, states that the area of a surface of revolution S is equal to the length L of the generating curve times the distance traversed by the center of mass. Use Eq. (14) to prove Pappus's Theorem. If C is the graph $z = g(y)$ for $c \le y \le d$, then the center of mass is defined as the point $(\bar{y}, \bar{z})$ with

$$\bar{y} = \frac{1}{L} \int_C y \, ds, \qquad \bar{z} = \frac{1}{L} \int_C z \, ds$$

47. Compute the surface area of the torus in Exercise 45 using Pappus's Theorem.

48. 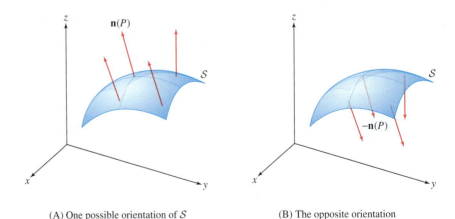 **Potential Due to a Uniform Sphere** Let S be a hollow sphere of radius R with its center at the origin with a uniform mass distribution of total mass m [since S has surface area $4\pi R^2$, the mass density is $\delta = m/(4\pi R^2)$]. The gravitational potential $V(P)$ due to S at a point $P = (a, b, c)$ is equal to

$$-G \iint_S \frac{\delta \, dS}{\sqrt{(x-a)^2 + (y-b)^2 + (z-c)^2}}$$

(a) Use symmetry to conclude that the potential depends only on the distance r from P to the center of the sphere. Therefore, it suffices to compute $V(P)$ for a point $P = (0, 0, r)$ on the z-axis (with $r \ne R$).

(b) Use spherical coordinates to show that $V(0, 0, r)$ is equal to

$$\frac{-Gm}{4\pi} \int_0^\pi \int_0^{2\pi} \frac{\sin\phi \, d\theta \, d\phi}{\sqrt{R^2 + r^2 - 2Rr\cos\phi}}$$

(c) Use the substitution $u = R^2 + r^2 - 2Rr\cos\phi$ to show that

$$V(0, 0, r) = \frac{-mG}{2Rr}\left(|R+r| - |R-r|\right)$$

(d) Verify Eq. (12) for V.

49. Calculate the gravitational potential V for a hemisphere of radius R with uniform mass distribution.

50. The surface of a cylinder of radius R and length L has a uniform mass distribution δ (the top and bottom of the cylinder are excluded). Use Eq. (11) to find the gravitational potential at a point P located along the axis of the cylinder.

51. Let S be the part of the graph $z = g(x, y)$ lying over a domain $\mathcal{D}$ in the xy-plane. Let $\phi = \phi(x, y)$ be the angle between the normal to S and the vertical. Prove the formula

$$\text{Area}(S) = \iint_{\mathcal{D}} \frac{dA}{|\cos\phi|}$$

16.5 Surface Integrals of Vector Fields

*The word **flux** is derived from the Latin word **fluere**, which means "to flow."*

The last integrals we will consider are surface integrals of vector fields. These integrals represent flux or rates of flow through a surface. One example is the flux of molecules across a cell membrane (number of molecules per unit time).

Because flux through a surface goes from one side of the surface to the other, we need to specify a *positive direction* of flow. This is done by means of an **orientation**, which is a choice of unit normal vector $\mathbf{n}(P)$ at each point P of S, chosen in a continuously varying manner (Figure 1). There are two normal directions at each point, so the orientation serves to specify one of the two "sides" of the surface in a consistent manner. The unit vectors $-\mathbf{n}(P)$ define the *opposite orientation*. For example, if the $\mathbf{n}$ vectors are outward-pointing unit normal vectors on a sphere, then a flow from the inside of the sphere to the outside is a positive flux.

FIGURE 1 The surface S has two possible orientations.

(A) One possible orientation of S

(B) The opposite orientation

The **normal component** of a vector field $\mathbf{F}$ at a point P on an oriented surface S is the dot product

$$\text{Normal component at } P = \mathbf{F}(P) \cdot \mathbf{n}(P) = \|\mathbf{F}(P)\| \cos\theta$$

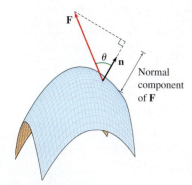

FIGURE 2 The normal component of a vector to a surface.

⬅·· *REMINDER Formula for a scalar surface integral in terms of an oriented parametrization:*

$$\iint_{\mathcal{S}} f(x, y, z)\, dS$$

$$= \iint f(G(u, v)) \|\mathbf{N}(u, v)\|\, du\, dv \quad \boxed{1}$$

where θ is the angle between $\mathbf{F}(P)$ and $\mathbf{n}(P)$ (Figure 2). Often, we write $\mathbf{n}$ instead of $\mathbf{n}(P)$, but it is understood that $\mathbf{n}$ varies from point to point on the surface. The vector surface integral, denoted $\iint_{\mathcal{S}} \mathbf{F} \cdot d\mathbf{S}$, is defined as the integral of the normal component:

> Vector surface integral: $\quad \iint_{\mathcal{S}} \mathbf{F} \cdot d\mathbf{S} = \iint_{\mathcal{S}} (\mathbf{F} \cdot \mathbf{n})\, dS$

This quantity is also called the **flux** of $\mathbf{F}$ across or through $\mathcal{S}$.

An oriented parametrization $G(u, v)$ is a regular parametrization [meaning that $\mathbf{N}(u, v)$ is nonzero for all u, v] whose unit normal vector defines the orientation:

$$\mathbf{n} = \mathbf{n}(u, v) = \frac{\mathbf{N}(u, v)}{\|\mathbf{N}(u, v)\|}$$

Applying Eq. (1) in the margin to $\mathbf{F} \cdot \mathbf{n}$, we obtain

$$\iint_{\mathcal{S}} \mathbf{F} \cdot d\mathbf{S} = \iint_{\mathcal{D}} (\mathbf{F} \cdot \mathbf{n}) \|\mathbf{N}(u, v)\|\, du\, dv$$

$$= \iint_{\mathcal{D}} \mathbf{F}(G(u, v)) \cdot \left(\frac{\mathbf{N}(u, v)}{\|\mathbf{N}(u, v)\|} \right) \|\mathbf{N}(u, v)\|\, du\, dv$$

$$= \iint_{\mathcal{D}} \mathbf{F}(G(u, v)) \cdot \mathbf{N}(u, v)\, du\, dv \quad \boxed{2}$$

This formula remains valid even if $\mathbf{N}(u, v)$ is zero at points on the boundary of the parameter domain $\mathcal{D}$. If we reverse the orientation of $\mathcal{S}$ in a vector surface integral, $\mathbf{N}(u, v)$ is replaced by $-\mathbf{N}(u, v)$ and the integral changes sign.

We can think of $d\mathbf{S}$ as a "vector surface element" that is related to a parametrization by the symbolic equation

> $d\mathbf{S} = \mathbf{N}(u, v)\, du\, dv$

Thus, we obtain the following theorem.

> **THEOREM 1 Vector Surface Integral** Let $G(u, v)$ be an oriented parametrization of an oriented surface $\mathcal{S}$ with parameter domain $\mathcal{D}$. Assume that G is one-to-one and regular, except possibly at points on the boundary of $\mathcal{D}$. Then
>
> $$\iint_{\mathcal{S}} \mathbf{F} \cdot d\mathbf{S} = \iint_{\mathcal{D}} \mathbf{F}(G(u, v)) \cdot \mathbf{N}(u, v)\, du\, dv \quad \boxed{3}$$
>
> If the orientation of $\mathcal{S}$ is reversed, the surface integral changes sign.

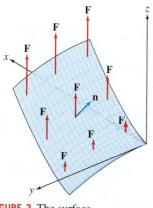

FIGURE 3 The surface $G(u, v) = (u^2, v, u^3 - v^2)$ with an upward-pointing normal. The vector field $\mathbf{F} = \langle 0, 0, x \rangle$ points in the vertical direction.

■ **EXAMPLE 1** Calculate $\displaystyle\iint_{\mathcal{S}} \mathbf{F} \cdot d\mathbf{S}$, where $\mathbf{F} = \langle 0, 0, x \rangle$ and $\mathcal{S}$ is the surface with parametrization $G(u, v) = (u^2, v, u^3 - v^2)$ for $0 \le u \le 1$, $0 \le v \le 1$ and oriented by upward-pointing normal vectors.

Solution

***Step 1.* Compute the tangent and normal vectors.**

$$\mathbf{T}_u = \langle 2u, 0, 3u^2 \rangle, \qquad \mathbf{T}_v = \langle 0, 1, -2v \rangle$$

$$\mathbf{N}(u, v) = \mathbf{T}_u \times \mathbf{T}_v = \begin{vmatrix} \mathbf{i} & \mathbf{j} & \mathbf{k} \\ 2u & 0 & 3u^2 \\ 0 & 1 & -2v \end{vmatrix}$$

$$= -3u^2 \mathbf{i} + 4uv \mathbf{j} + 2u \mathbf{k} = \langle -3u^2, 4uv, 2u \rangle$$

The z-component of $\mathbf{N}$ is positive on the domain $0 \le u \le 1$, so $\mathbf{N}$ is the upward-pointing normal (Figure 3).

Step 2. **Evaluate F · N.**

Write **F** in terms of the parameters u and v. Since $x = u^2$,

$$\mathbf{F}(G(u, v)) = \langle 0, 0, x \rangle = \langle 0, 0, u^2 \rangle$$

and

$$\mathbf{F}(G(u, v)) \cdot \mathbf{N}(u, v) = \langle 0, 0, u^2 \rangle \cdot \langle -3u^2, 4uv, 2u \rangle = 2u^3$$

Step 3. **Evaluate the surface integral.**

The parameter domain is $0 \le u \le 1, 0 \le v \le 1$, so

$$\iint_{\mathcal{S}} \mathbf{F} \cdot d\mathbf{S} = \int_{u=0}^{1} \int_{v=0}^{1} \mathbf{F}(G(u, v)) \cdot \mathbf{N}(u, v) \, dv \, du$$

$$= \int_{u=0}^{1} \int_{v=0}^{1} 2u^3 \, dv \, du = \int_{u=0}^{1} 2u^3 \, du = \frac{1}{2} \quad \blacksquare$$

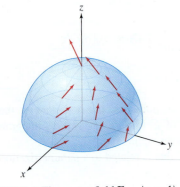

FIGURE 4 The vector field $\mathbf{F} = \langle z, x, 1 \rangle$.

■ **EXAMPLE 2** **Integral over a Hemisphere** Calculate the flux of $\mathbf{F} = \langle z, x, 1 \rangle$ across the upper hemisphere $\mathcal{S}$ of the sphere $x^2 + y^2 + z^2 = 1$, oriented with outward-pointing normal vectors (Figure 4).

Solution Parametrize the hemisphere by spherical coordinates:

$$G(\theta, \phi) = (\cos\theta\sin\phi, \sin\theta\sin\phi, \cos\phi), \qquad 0 \le \phi \le \frac{\pi}{2}, \quad 0 \le \theta < 2\pi$$

Step 1. **Compute the normal vector.**

According to Eq. (2) in Section 16.4, the outward-pointing normal vector is

$$\mathbf{N} = \mathbf{T}_\phi \times \mathbf{T}_\theta = \sin\phi \langle \cos\theta\sin\phi, \sin\theta\sin\phi, \cos\phi \rangle$$

Step 2. **Evaluate F · N.**

$$\mathbf{F}(G(\theta, \phi)) = \langle z, x, 1 \rangle = \langle \cos\phi, \cos\theta\sin\phi, 1 \rangle$$

$$\mathbf{F}(G(\theta, \phi)) \cdot \mathbf{N}(\theta, \phi) = \langle \cos\phi, \cos\theta\sin\phi, 1 \rangle \cdot \langle \cos\theta\sin^2\phi, \sin\theta\sin^2\phi, \cos\phi\sin\phi \rangle$$

$$= \cos\theta\sin^2\phi\cos\phi + \cos\theta\sin\theta\sin^3\phi + \cos\phi\sin\phi$$

Step 3. **Evaluate the surface integral.**

$$\iint_{\mathcal{S}} \mathbf{F} \cdot d\mathbf{S} = \int_{\phi=0}^{\pi/2} \int_{\theta=0}^{2\pi} \mathbf{F}(G(\theta, \phi)) \cdot \mathbf{N}(\theta, \phi) \, d\theta \, d\phi$$

$$= \int_{\phi=0}^{\pi/2} \int_{\theta=0}^{2\pi} \underbrace{(\cos\theta\sin^2\phi\cos\phi + \cos\theta\sin\theta\sin^3\phi}_{\text{Integral over } \theta \text{ is zero}} + \cos\phi\sin\phi) \, d\theta \, d\phi$$

The integrals of $\cos\theta$ and $\cos\theta\sin\theta$ over $[0, 2\pi]$ are both zero, so we are left with

$$\int_{\phi=0}^{\pi/2} \int_{\theta=0}^{2\pi} \cos\phi\sin\phi \, d\theta \, d\phi = 2\pi \int_{\phi=0}^{\pi/2} \cos\phi\sin\phi \, d\phi = -2\pi \frac{\cos^2\phi}{2}\bigg|_{0}^{\pi/2} = \pi \quad \blacksquare$$

■ **EXAMPLE 3** **Surface Integral over a Graph** Calculate the flux of $\mathbf{F} = x^2\mathbf{j}$ through the surface $\mathcal{S}$ defined by $y = 1 + x^2 + z^2$ for $1 \le y \le 5$. Orient $\mathcal{S}$ with normal pointing in the negative y-direction.

Solution This surface is the graph of the function $y = 1 + x^2 + z^2$, where x and z are the independent variables (Figure 5).

Step 1. **Find a parametrization.**

It is convenient to use x and z because y is given explicitly as a function of x and z. Thus, we define

$$G(x, z) = (x, 1 + x^2 + z^2, z)$$

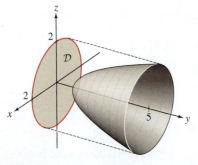

FIGURE 5

What is the parameter domain? The condition $1 \leq y \leq 5$ is equivalent to $1 \leq 1 + x^2 + z^2 \leq 5$ or $0 \leq x^2 + z^2 \leq 4$. Therefore, the parameter domain is the disk of radius 2 in the xz-plane—that is, $\mathcal{D} = \{(x, z) : x^2 + z^2 \leq 4\}$.

Because the parameter domain is a disk, it makes sense to use the polar variables r and θ in the xz-plane. In other words, we write $x = r \cos \theta$, $z = r \sin \theta$. Then

$$y = 1 + x^2 + z^2 = 1 + r^2$$

$$G(r, \theta) = (r \cos \theta, 1 + r^2, r \sin \theta), \quad 0 \leq \theta \leq 2\pi, \quad 0 \leq r \leq 2$$

Step 2. Compute the tangent and normal vectors.

$$\mathbf{T}_r = \langle \cos \theta, 2r, \sin \theta \rangle, \qquad \mathbf{T}_\theta = \langle -r \sin \theta, 0, r \cos \theta \rangle$$

$$\mathbf{N} = \mathbf{T}_r \times \mathbf{T}_\theta = \begin{vmatrix} \mathbf{i} & \mathbf{j} & \mathbf{k} \\ \cos \theta & 2r & \sin \theta \\ -r \sin \theta & 0 & r \cos \theta \end{vmatrix} = 2r^2 \cos \theta \, \mathbf{i} - r \mathbf{j} + 2r^2 \sin \theta \, \mathbf{k}$$

The coefficient of $\mathbf{j}$ is $-r$. Because it is negative, $\mathbf{N}$ points in the negative y-direction, as required.

Step 3. Evaluate $\mathbf{F} \cdot \mathbf{N}$.

CAUTION In Step 3, we integrate $\mathbf{F} \cdot \mathbf{N}$ with respect to $dr\, d\theta$, and <u>not</u> $r \, dr \, d\theta$. The factor of r in $r \, dr \, d\theta$ is a Jacobian factor that we add only when changing variables in a double integral. In surface integrals, the Jacobian factor is incorporated into the magnitude of $\mathbf{N}$ (recall that $\|\mathbf{N}\|$ is the area "distortion factor").

$$\mathbf{F}(G(r, \theta)) = x^2 \mathbf{j} = r^2 \cos^2 \theta \, \mathbf{j} = \langle 0, r^2 \cos^2 \theta, 0 \rangle$$

$$\mathbf{F}(G(r, \theta)) \cdot \mathbf{N} = \langle 0, r^2 \cos^2 \theta, 0 \rangle \cdot \langle 2r^2 \cos \theta, -r, 2r^2 \sin \theta \rangle = -r^3 \cos^2 \theta$$

$$\iint_{\mathcal{S}} \mathbf{F} \cdot d\mathbf{S} = \iint_{\mathcal{D}} \mathbf{F}(G(r, \theta)) \cdot \mathbf{N} \, dr \, d\theta = \int_0^{2\pi} \int_0^2 (-r^3 \cos^2 \theta) \, dr \, d\theta$$

$$= -\left(\int_0^{2\pi} \cos^2 \theta \, d\theta \right) \left(\int_0^2 r^3 \, dr \right)$$

$$= -(\pi)\left(\frac{2^4}{4} \right) = -4\pi$$

It is not surprising that the flux is negative since the outward-pointing normal points in the negative y-direction but the force field $\mathbf{F}$ points in the positive y-direction. ∎

CONCEPTUAL INSIGHT Since a vector surface integral depends on the orientation of the surface, this integral is defined only for surfaces that have two sides. However, some surfaces, such as the Möbius strip (discovered in 1858 independently by August Möbius and Johann Listing), cannot be oriented because they are one-sided. You can construct a Möbius strip M with a rectangular strip of paper: Join the two ends of the strip together with a 180° twist. Unlike an ordinary two-sided strip, the Möbius strip M has only one side, and it is impossible to specify an outward direction in a consistent manner (Figure 6). If you choose a unit normal vector at a point P and carry that unit vector continuously around M, when you return to P, the vector will point in the opposite direction. Therefore, we cannot integrate a vector field over a Möbius strip, and it is not meaningful to speak of the "flux" across M. On the other hand, it is possible to integrate a scalar function. For example, the integral of mass density would equal the total mass of the Möbius strip.

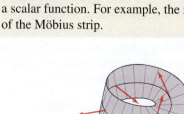

Möbius strip

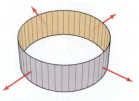

Ordinary (untwisted) band

FIGURE 7 Velocity field of a fluid flow.

Fluid Flux

Imagine dipping a net into a stream of flowing water (Figure 7). The **flow rate** is the volume of water that flows through the net per unit time.

To compute the flow rate, let $\mathbf{v}$ be the velocity vector field. At each point P, $\mathbf{v}(P)$ is the velocity vector of the fluid particle located at the point P. We claim that *the flow rate through a surface S is equal to the surface integral of $\mathbf{v}$ over S.*

To explain why, suppose first that S is a rectangle of area A and that $\mathbf{v}$ is a constant vector field with value $\mathbf{v}_0$ perpendicular to the rectangle. The particles travel at speed $\|\mathbf{v}_0\|$, say, in meters per second, so a given particle flows through S within a 1-second time interval if its distance to S is at most $\|\mathbf{v}_0\|$ meters—in other words, if its velocity vector passes through S (see Figure 8). Thus, the block of fluid passing through S in a 1-second interval is a box of volume $\|\mathbf{v}_0\| A$ (Figure 9), and

$$\text{Flow rate} = (\text{velocity})(\text{area}) = \|\mathbf{v}_0\| A$$

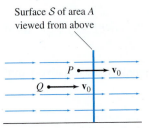

FIGURE 8 The particle P flows through S within a 1-second interval, but Q does not.

Surface S of area A

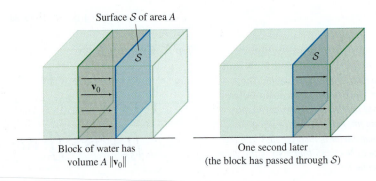

Block of water has volume $A \|\mathbf{v}_0\|$ One second later (the block has passed through S)

FIGURE 9

If the fluid flows at an angle θ relative to S, then the block of water is a parallelepiped (rather than a box) of volume $A\|\mathbf{v}_0\| \cos\theta$ (Figure 10). If $\mathbf{N}$ is a vector normal to S of length equal to the area A, then we can write the flow rate as a dot product:

$$\text{Flow rate} = A\|\mathbf{v}_0\| \cos\theta = \mathbf{v}_0 \cdot \mathbf{N}$$

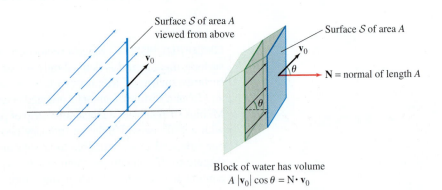

Block of water has volume $A |\mathbf{v}_0| \cos\theta = \mathbf{N} \cdot \mathbf{v}_0$

FIGURE 10 Water flowing at constant velocity $\mathbf{v}_0$, making an angle θ with a rectangular surface.

In the general case, the velocity field $\mathbf{v}$ is not constant, and the surface S may be curved. To compute the flow rate, we choose a parametrization $G(u, v)$ and we consider a small rectangle of size $\Delta u \times \Delta v$ that is mapped by G to a small patch S_0 of S (Figure 11). For any sample point $G(u_0, v_0)$ in S_0, the vector $\mathbf{N}(u_0, v_0) \Delta u \, \Delta v$ is a normal vector of length approximately equal to the area of S_0 [Eq. (3) in Section 16.4]. This patch is nearly rectangular, so we have the approximation

$$\text{Flow rate through } S_0 \approx \mathbf{v}(u_0, v_0) \cdot \mathbf{N}(u_0, v_0) \, \Delta u \, \Delta v$$

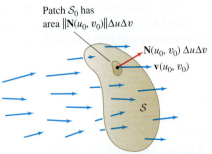

Patch S_0 has area $\|\mathbf{N}(u_0, v_0)\| \Delta u \Delta v$

$\mathbf{N}(u_0, v_0) \Delta u \Delta v$
$\mathbf{v}(u_0, v_0)$

FIGURE 11 The flow rate across the small patch S_0 is approximately $\mathbf{v}(u_0, v_0) \cdot \mathbf{N}(u_0, v_0) \Delta u \Delta v$.

The total flow per second is the sum of the flows through the small patches. As usual, the limit of the sums as Δu and Δv tend to zero is the integral of $\mathbf{v}(u, v) \cdot \mathbf{N}(u, v)$, which is the surface integral of $\mathbf{v}$ over S.

Flow Rate Through a Surface   For a fluid with velocity vector field **v**,

$$\text{Flow rate across the } S \text{ (volume per unit time)} = \iint_{S} \mathbf{v} \cdot d\mathbf{S} \qquad \boxed{4}$$

■ **EXAMPLE 4**   Let $\mathbf{v} = \langle x^2 + y^2, 0, z^2 \rangle$ be the velocity field (in centimeters per second) of a fluid in $\mathbf{R}^3$. Compute the flow rate through the upper hemisphere S of the unit sphere centered at the origin.

Solution   We use spherical coordinates:

$$x = \cos\theta \sin\phi, \qquad y = \sin\theta \sin\phi, \qquad z = \cos\phi$$

The upper hemisphere corresponds to the ranges $0 \le \phi \le \frac{\pi}{2}$ and $0 \le \theta \le 2\pi$. By Eq. (2) in Section 16.4, the upward-pointing normal is

$$\mathbf{N} = \mathbf{T}_\phi \times \mathbf{T}_\theta = \sin\phi \langle \cos\theta \sin\phi, \sin\theta \sin\phi, \cos\phi \rangle$$

We have $x^2 + y^2 = \sin^2\phi$, so

$$\mathbf{v} = \langle x^2 + y^2, 0, z^2 \rangle = \langle \sin^2\phi, 0, \cos^2\phi \rangle$$

$$\mathbf{v} \cdot \mathbf{N} = \sin\phi \langle \sin^2\phi, 0, \cos^2\phi \rangle \cdot \langle \cos\theta \sin\phi, \sin\theta \sin\phi, \cos\phi \rangle$$

$$= \sin^4\phi \cos\theta + \sin\phi \cos^3\phi$$

$$\iint_{S} \mathbf{v} \cdot d\mathbf{S} = \int_{\phi=0}^{\pi/2} \int_{\theta=0}^{2\pi} (\sin^4\phi \cos\theta + \sin\phi \cos^3\phi) \, d\theta \, d\phi$$

The integral of $\sin^4\phi \cos\theta$ with respect to θ is zero, so we are left with

$$\int_{\phi=0}^{\pi/2} \int_{\theta=0}^{2\pi} \sin\phi \cos^3\phi \, d\theta \, d\phi = 2\pi \int_{\phi=0}^{\pi/2} \cos^3\phi \sin\phi \, d\phi$$

$$= 2\pi \left(-\frac{\cos^4\phi}{4} \right) \Bigg|_{\phi=0}^{\pi/2} = \frac{\pi}{2} \text{ cm}^3/\text{s}$$

Since $\mathbf{N}$ is an upward-pointing normal, this is the rate at which fluid flows across the hemisphere from below to above.   ■

Electric and Magnetic Fields

The laws of electricity and magnetism are expressed in terms of two vector fields, the electric field **E** and the magnetic field **B**, whose properties are summarized in Maxwell's four equations. One of these equations is **Faraday's Law of Induction**, which can be formulated either as a partial differential equation or in the following "integral form":

$$\int_{C} \mathbf{E} \cdot d\mathbf{r} = -\frac{d}{dt} \iint_{S} \mathbf{B} \cdot d\mathbf{S} \qquad \boxed{5}$$

In this equation, S is an oriented surface with boundary curve C, oriented as indicated in Figure 12. The line integral of **E** is equal to the voltage drop around the boundary curve (the work performed by **E** moving a positive unit charge around C).

To illustrate Faraday's Law, consider an electric current of i amperes flowing through a straight wire. According to the Biot-Savart Law, this current produces a magnetic field **B** of magnitude $B(r) = \dfrac{\mu_0 |i|}{2\pi r}$ T, where r is the distance (in meters) from the wire and $\mu_0 = 4\pi \cdot 10^{-7}$ T-m/A. At each point P, **B** is tangent to the circle through P perpendicular to the wire as in Figure 13(A), with the direction determined by the right-hand rule: If the thumb of your right hand points in the direction of the current, then your fingers curl in the direction of **B**.

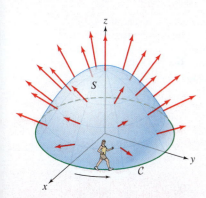

FIGURE 12   The positive direction along the boundary curve C is defined so that if a pedestrian walks in the positive direction with the surface to her left, then her head points in the outward (normal) direction.

*The **tesla** (T) is the SI unit of magnetic field strength. A 1-coulomb point charge passing through a magnetic field of 1 tesla at 1 m/s experiences a force of 1 newton.*

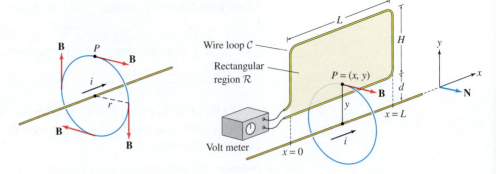

(A) Magnetic field **B** due to
a current in a wire

(B) The magnetic field **B** points in
the direction **N** normal to $\mathcal{R}$

FIGURE 13

*The electric field **E** is conservative when the charges are stationary or, more generally, when the magnetic field **B** is constant. When **B** varies in time, the integral on the right in Eq. (5) is nonzero for some surface, and hence, the circulation of **E** around the boundary curve $\mathcal{C}$ is also nonzero. This shows that **E** is not conservative when **B** varies in time.*

■ **EXAMPLE 5** A varying current of magnitude (t in seconds)

$$i = 28\cos(400t) \text{ amperes}$$

flows through a straight wire [Figure 13(B)]. A rectangular wire loop $\mathcal{C}$ of length $L = 1.2$ m and width $H = 0.7$ m is located a distance $d = 0.1$ m from the wire as in the figure. The loop encloses a rectangular surface $\mathcal{R}$, which is oriented by normal vectors pointing out of the page.

(a) Calculate the flux $\Phi(t)$ of **B** through $\mathcal{R}$.

(b) Use Faraday's Law to determine the voltage drop (in volts) around the loop $\mathcal{C}$.

Magnetic flux as a function of time is often denoted by the Greek letter Φ:

$$\Phi(t) = \iint_S \mathbf{B} \cdot d\mathbf{S}$$

Solution We choose coordinates (x, y) on rectangle $\mathcal{R}$ as in Figure 13, so that y is the distance from the wire and $\mathcal{R}$ is the region

$$0 \le x \le L, \qquad d \le y \le H + d$$

Our parametrization of $\mathcal{R}$ is simply $G(x, y) = (x, y)$, for which the normal vector **N** is the unit vector perpendicular to $\mathcal{R}$, pointing out of the page. The magnetic field **B** at $P = (x, y)$ has magnitude $\dfrac{\mu_0 |i|}{2\pi y}$. It points out of the page in the direction of **N** when i is positive and into the page when i is negative. Thus,

$$\mathbf{B} = \frac{\mu_0 i}{2\pi y}\mathbf{N} \qquad \text{and} \qquad \mathbf{B} \cdot \mathbf{N} = \frac{\mu_0 i}{2\pi y}$$

(a) The flux $\Phi(t)$ of **B** through $\mathcal{R}$ at time t is

$$\Phi(t) = \iint_{\mathcal{R}} \mathbf{B} \cdot d\mathbf{S} = \int_{x=0}^{L} \int_{y=d}^{H+d} \mathbf{B} \cdot \mathbf{N}\, dy\, dx$$

$$= \int_{x=0}^{L} \int_{y=d}^{H+d} \frac{\mu_0 i}{2\pi y}\, dy\, dx = \frac{\mu_0 L i}{2\pi} \int_{y=d}^{H+d} \frac{dy}{y}$$

$$= \frac{\mu_0 L}{2\pi}\left(\ln \frac{H+d}{d}\right) i$$

$$= \frac{\mu_0 (1.2)}{2\pi}\left(\ln \frac{0.8}{0.1}\right) 28 \cos(400t)$$

With $\mu_0 = 4\pi \cdot 10^{-7}$, we obtain

$$\Phi(t) \approx 1.4 \times 10^{-5} \cos(400t) \text{ T-m}^2$$

(b) By Faraday's Law [Eq. (5)], the voltage drop around the rectangular loop $\mathcal{C}$, oriented in the counterclockwise direction, is

$$\int_{\mathcal{C}} \mathbf{E} \cdot d\mathbf{s} = -\frac{d\Phi}{dt} \approx -(1.4 \times 10^{-5})(400)\sin(400t) = -0.0056\sin(400t) \text{ volts} \quad ■$$

Types of Integrals

We end with a list of the types of integrals we have introduced in this chapter.

1. Scalar line integral along a curve $\mathcal{C}$ given by $\mathbf{r}(t)$ for $a \leq t \leq b$ (can be used to compute arc length, mass, electric potential):

$$\int_{\mathcal{C}} f(x, y, z)\, d\mathbf{r} = \int_a^b f(\mathbf{r}(t)) \|\mathbf{r}'(t)\|\, dt$$

2. Vector line integral to calculate work along a curve $\mathcal{C}$ given by $\mathbf{r}(t)$ for $a \leq t \leq b$:

$$\int_{\mathcal{C}} \mathbf{F} \cdot d\mathbf{r} = \int_a^b \mathbf{F}(\mathbf{r}(t)) \cdot \mathbf{r}'(t)\, dt = \int_{\mathcal{C}} F_1\, dx + F_2\, dy + F_3\, dz$$

3. Vector line integral to calculate flux across a curve $\mathcal{C}$ given by $\mathbf{r}(t)$ for $a \leq t \leq b$:

$$\int_{\mathcal{C}} \mathbf{F} \cdot \mathbf{n}\, ds = \int_a^b \mathbf{F}(\mathbf{r}(t)) \cdot \mathbf{N}(t)\, dt$$

4. Surface integral over a surface with parametrization $G(u, v)$ and parameter domain $\mathcal{D}$ (can be used to calculate surface area, total charge, gravitational potential):

$$\iint_{\mathcal{S}} f(x, y, z)\, dS = \iint_{\mathcal{D}} f(G(u, v)) \|\mathbf{N}(u, v)\|\, du\, dv$$

5. Vector surface integral to calculate flux of a vector field $\mathbf{F}$ across a surface S with parametrization $G(u, v)$ and parameter domain $\mathcal{D}$:

$$\iint_{\mathcal{S}} \mathbf{F} \cdot d\mathbf{S} = \iint_{\mathcal{D}} \mathbf{F}(G(u, v)) \cdot \mathbf{N}(u, v)\, du\, dv$$

16.5 SUMMARY

- A surface $\mathcal{S}$ is *oriented* if a continuously varying unit normal vector $\mathbf{n}(P)$ is specified at each point on $\mathcal{S}$. This distinguishes an "outward" direction on the surface.
- The integral of a vector field $\mathbf{F}$ over an oriented surface $\mathcal{S}$ is defined as the integral of the normal component $\mathbf{F} \cdot \mathbf{n}$ over $\mathcal{S}$.
- Vector surface integrals are computed using the formula

$$\iint_{\mathcal{S}} \mathbf{F} \cdot d\mathbf{S} = \iint_{\mathcal{D}} \mathbf{F}(G(u, v)) \cdot \mathbf{N}(u, v)\, du\, dv$$

Here, $G(u, v)$ is a parametrization of $\mathcal{S}$ such that $\mathbf{N}(u, v) = \mathbf{T}_u \times \mathbf{T}_v$ points in the direction of the unit normal vector specified by the orientation.
- The surface integral of a vector field $\mathbf{F}$ over $\mathcal{S}$ is also called the *flux* of $\mathbf{F}$ through G. If $\mathbf{F}$ is the velocity field of a fluid, then $\iint_{\mathcal{S}} \mathbf{F} \cdot d\mathbf{S}$ is the rate at which fluid flows through $\mathcal{S}$ per unit time.

16.5 EXERCISES

Preliminary Questions

1. Let $\mathbf{F}$ be a vector field and $G(u, v)$ a parametrization of a surface $\mathcal{S}$, and set $\mathbf{N} = \mathbf{T}_u \times \mathbf{T}_v$. Which of the following is the normal component of $\mathbf{F}$?

(a) $\mathbf{F} \cdot \mathbf{N}$ **(b)** $\mathbf{F} \cdot \mathbf{n}$

2. The vector surface integral $\iint_{\mathcal{S}} \mathbf{F} \cdot d\mathbf{S}$ is equal to the scalar surface integral of the function (choose the correct answer):

(a) $\|\mathbf{F}\|$.

(b) $\mathbf{F} \cdot \mathbf{N}$, where $\mathbf{N}$ is a normal vector.

(c) $\mathbf{F} \cdot \mathbf{n}$, where $\mathbf{n}$ is the unit normal vector.

3. $\iint_{\mathcal{S}} \mathbf{F} \cdot d\mathbf{S}$ is zero if (choose the correct answer):

(a) $\mathbf{F}$ is tangent to $\mathcal{S}$ at every point.

(b) $\mathbf{F}$ is perpendicular to $\mathcal{S}$ at every point.

4. If $\mathbf{F}(P) = \mathbf{n}(P)$ at each point on $\mathcal{S}$, then $\iint_{\mathcal{S}} \mathbf{F} \cdot d\mathbf{S}$ is equal to which of the following?

(a) Zero   **(b)** Area($\mathcal{S}$)   **(c)** Neither

5. Let $\mathcal{S}$ be the disk $x^2 + y^2 \leq 1$ in the xy-plane oriented with normal in the positive z-direction. Determine $\iint_{\mathcal{S}} \mathbf{F} \cdot d\mathbf{S}$ for each of the following vector constant fields:

(a) $\mathbf{F} = \langle 1, 0, 0 \rangle$ **(b)** $\mathbf{F} = \langle 0, 0, 1 \rangle$ **(c)** $\mathbf{F} = \langle 1, 1, 1 \rangle$

6. Estimate $\iint_{\mathcal{S}} \mathbf{F} \cdot d\mathbf{S}$, where $\mathcal{S}$ is a tiny oriented surface of area 0.05 and the value of $\mathbf{F}$ at a sample point in $\mathcal{S}$ is a vector of length 2 making an angle $\frac{\pi}{4}$ with the normal to the surface.

7. A small surface $\mathcal{S}$ is divided into three pieces of area 0.2. Estimate $\iint_{\mathcal{S}} \mathbf{F} \cdot d\mathbf{S}$ if $\mathbf{F}$ is a unit vector field making angles of 85°, 90°, and 95° with the normal at sample points in these three pieces.

Exercises

1. Let $\mathbf{F} = \langle z, 0, y \rangle$, and let $\mathcal{S}$ be the oriented surface parametrized by $G(u, v) = (u^2 - v, u, v^2)$ for $0 \leq u \leq 2, -1 \leq v \leq 4$. Calculate:

(a) $\mathbf{N}$ and $\mathbf{F} \cdot \mathbf{N}$ as functions of u and v

(b) The normal component of $\mathbf{F}$ to the surface at $P = (3, 2, 1) = G(2, 1)$

(c) $\iint_{\mathcal{S}} \mathbf{F} \cdot d\mathbf{S}$

2. Let $\mathbf{F} = \langle y, -x, x^2 + y^2 \rangle$, and let $\mathcal{S}$ be the portion of the paraboloid $z = x^2 + y^2$ where $x^2 + y^2 \leq 3$.

(a) Show that if $\mathcal{S}$ is parametrized in polar variables $x = r\cos\theta$, $y = r\sin\theta$, then $\mathbf{F} \cdot \mathbf{N} = r^3$.

(b) Show that $\iint_{\mathcal{S}} \mathbf{F} \cdot d\mathbf{S} = \int_0^{2\pi} \int_0^3 r^3 \, dr \, d\theta$ and evaluate.

3. Let $\mathcal{S}$ be the unit square in the xy-plane shown in Figure 14, oriented with the normal pointing in the positive z-direction. Estimate

$$\iint_{\mathcal{S}} \mathbf{F} \cdot d\mathbf{S}$$

where $\mathbf{F}$ is a vector field whose values at the labeled points are

$$\mathbf{F}(A) = \langle 2, 6, 4 \rangle, \qquad \mathbf{F}(B) = \langle 1, 1, 7 \rangle$$
$$\mathbf{F}(C) = \langle 3, 3, -3 \rangle, \qquad \mathbf{F}(D) = \langle 0, 1, 8 \rangle$$

4. Suppose that $\mathcal{S}$ is a surface in $\mathbf{R}^3$ with a parametrization G whose domain $\mathcal{D}$ is the square in Figure 14. The values of a function f, a vector field $\mathbf{F}$, and the normal vector $\mathbf{N} = \mathbf{T}_u \times \mathbf{T}_v$ at $G(P)$ are given for the four sample points in $\mathcal{D}$ in the following table. Estimate the surface integrals of f and $\mathbf{F}$ over $\mathcal{S}$.

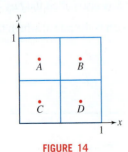

FIGURE 14

Point P in $\mathcal{D}$	f	$\mathbf{F}$	$\mathbf{N}$
A	3	$\langle 2, 6, 4 \rangle$	$\langle 1, 1, 1 \rangle$
B	1	$\langle 1, 1, 7 \rangle$	$\langle 1, 1, 0 \rangle$
C	2	$\langle 3, 3, -3 \rangle$	$\langle 1, 0, -1 \rangle$
D	5	$\langle 0, 1, 8 \rangle$	$\langle 2, 1, 0 \rangle$

In Exercises 5–17, compute $\iint_{\mathcal{S}} \mathbf{F} \cdot d\mathbf{S}$ *for the given oriented surface.*

5. $\mathbf{F} = \langle y, z, x \rangle$,   plane $3x - 4y + z = 1$,
$0 \leq x \leq 1, 0 \leq y \leq 1$,   upward-pointing normal

6. $\mathbf{F} = \langle e^z, z, x \rangle$, $G(r, s) = (rs, r + s, r)$,
$0 \leq r \leq 1, 0 \leq s \leq 1$,   oriented by $\mathbf{T}_r \times \mathbf{T}_s$

7. $\mathbf{F} = \langle 0, 3, x \rangle$,   part of sphere $x^2 + y^2 + z^2 = 9$,
where $x \geq 0, y \geq 0, z \geq 0$,   outward-pointing normal

8. $\mathbf{F} = \langle x, y, z \rangle$,   part of sphere $x^2 + y^2 + z^2 = 1$,
where $\dfrac{1}{2} \leq z \leq \dfrac{\sqrt{3}}{2}$,   inward-pointing normal

9. $\mathbf{F} = \langle z, z, x \rangle$, $z = 9 - x^2 - y^2, x \geq 0, y \geq 0, z \geq 0$,   upward-pointing normal

10. $\mathbf{F} = \langle \sin y, \sin z, yz \rangle$,   rectangle $0 \leq y \leq 2, 0 \leq z \leq 3$ in the (y, z)-plane,   normal pointing in negative x-direction

11. $\mathbf{F} = y^2\mathbf{i} + 2\mathbf{j} - x\mathbf{k}$,   portion of the plane $x + y + z = 1$ in the octant $x, y, z \geq 0$,   upward-pointing normal

12. $\mathbf{F} = \langle x, y, e^z \rangle$,   cylinder $x^2 + y^2 = 4$, $1 \leq z \leq 5$,   outward-pointing normal

13. $\mathbf{F} = \langle xz, yz, z^{-1} \rangle$,   disk of radius 3 at height 4 parallel to the xy-plane,   upward-pointing normal

14. $\mathbf{F} = \langle xy, y, 0 \rangle$,   cone $z^2 = x^2 + y^2$, $x^2 + y^2 \leq 4$, $z \geq 0$,   downward-pointing normal

15. $\mathbf{F} = \langle 0, 0, e^{y+z} \rangle$,   boundary of unit cube $0 \leq x \leq 1, 0 \leq y \leq 1$, $0 \leq z \leq 1$,   outward-pointing normal

16. $\mathbf{F} = \langle 0, 0, z^2 \rangle$, $G(u, v) = (u\cos v, u\sin v, v)$, $0 \leq u \leq 1$, $0 \leq v \leq 2\pi$,   upward-pointing normal

17. $\mathbf{F} = \langle y, z, 0 \rangle$, $G(u, v) = (u^3 - v, u + v, v^2)$, $0 \leq u \leq 2$, $0 \leq v \leq 3$,   downward-pointing normal

18. ✍ Let $\mathcal{S}$ be the oriented half-cylinder in Figure 15. In (a)–(f), determine whether $\iint_{\mathcal{S}} \mathbf{F} \cdot d\mathbf{S}$ is positive, negative, or zero. Explain your reasoning.

(a) $\mathbf{F} = \mathbf{i}$ **(b)** $\mathbf{F} = \mathbf{j}$ **(c)** $\mathbf{F} = \mathbf{k}$

(d) $\mathbf{F} = y\mathbf{i}$ **(e)** $\mathbf{F} = -y\mathbf{j}$ **(f)** $\mathbf{F} = x\mathbf{j}$

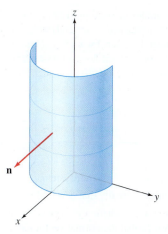

FIGURE 15

19. Let $\mathbf{e_r} = \langle x/r, y/r, z/r \rangle$ be the unit radial vector, where $r = \sqrt{x^2 + y^2 + z^2}$. Calculate the integral of $\mathbf{F} = e^{-r} \mathbf{e_r}$ over:

(a) the upper hemisphere of $x^2 + y^2 + z^2 = 9$, outward-pointing normal.

(b) the octant $x \geq 0$, $y \geq 0$, $z \geq 0$ of the unit sphere centered at the origin.

20. Show that the flux of $\mathbf{F} = \dfrac{\mathbf{e_r}}{r^2}$ through a sphere centered at the origin does not depend on the radius of the sphere.

21. The electric field due to a point charge located at the origin in $\mathbf{R}^3$ is $\mathbf{E} = k\dfrac{\mathbf{e_r}}{r^2}$, where $r = \sqrt{x^2 + y^2 + z^2}$ and k is a constant. Calculate the flux of $\mathbf{E}$ through the disk D of radius 2 parallel to the xy-plane with center $(0, 0, 3)$.

22. Let S be the ellipsoid $\left(\dfrac{x}{4}\right)^2 + \left(\dfrac{y}{3}\right)^2 + \left(\dfrac{z}{2}\right)^2 = 1$. Calculate the flux of $\mathbf{F} = z\mathbf{i}$ over the portion of S where $x, y, z \leq 0$ with upward-pointing normal. *Hint:* Parametrize S using a modified form of spherical coordinates (θ, ϕ).

23. Let $\mathbf{v} = z\mathbf{k}$ be the velocity field (in meters per second) of a fluid in $\mathbf{R}^3$. Calculate the flow rate (in cubic meters per second) through the upper hemisphere ($z \geq 0$) of the sphere $x^2 + y^2 + z^2 = 1$.

24. Calculate the flow rate of a fluid with velocity field $\mathbf{v} = \langle x, y, x^2 y \rangle$ (in meters per second) through the portion of the ellipse $\left(\dfrac{x}{2}\right)^2 + \left(\dfrac{y}{3}\right)^2 = 1$ in the xy-plane, where $x, y \geq 0$, oriented with the normal in the positive z-direction.

In Exercises 25–28, a net is dipped in a river. Determine the flow rate of water across the net if the velocity vector field for the river is given by $\mathbf{v}$ and the net is described by the given equations.

25. $\mathbf{v} = \left(x - y, z + y + 4, z^2\right)$, net given by $x^2 + z^2 \leq 1$, $y = 0$, oriented in the positive y-direction

26. $\mathbf{v} = \left(x - y, z + y + 4, z^2\right)$, net given by $y = 1 - x^2 - z^2$, $y \geq 0$, oriented in the positive y-direction

27. $\mathbf{v} = \left(x - y, z + y + 4, z^2\right)$, net given by $y = \sqrt{1 - x^2 - z^2}$, $y \geq 0$, oriented in the positive y-direction

28. $\mathbf{v} = \langle zy, xz, xy \rangle$, net given by $y = 1 - x - z$, for $x, y, z \geq 0$ oriented in the positive y-direction

In Exercises 29–30, let $\mathcal{T}$ be the triangular region with vertices $(1, 0, 0)$, $(0, 1, 0)$, and $(0, 0, 1)$ oriented with upward-pointing normal vector (Figure 16). Assume distances are in meters.

29. A fluid flows with constant velocity field $\mathbf{v} = 2\mathbf{k}$ (meters per second or m/s). Calculate:

(a) the flow rate through $\mathcal{T}$.

(b) the flow rate through the projection of $\mathcal{T}$ onto the xy-plane [the triangle with vertices $(0, 0, 0)$, $(1, 0, 0)$, and $(0, 1, 0)$].

30. Calculate the flow rate through $\mathcal{T}$ if $\mathbf{v} = -\mathbf{j}$ m/s.

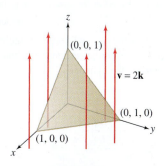

FIGURE 16

31. Prove that if S is the part of a graph $z = g(x, y)$ lying over a domain $\mathcal{D}$ in the xy-plane, then

$$\iint_S \mathbf{F} \cdot d\mathbf{S} = \iint_{\mathcal{D}} \left(-F_1 \frac{\partial g}{\partial x} - F_2 \frac{\partial g}{\partial y} + F_3\right) dx\, dy$$

In Exercises 32–33, a varying current $i(t)$ flows through a long straight wire in the xy-plane as in Example 5. The current produces a magnetic field $\mathbf{B}$ whose magnitude at a distance r from the wire is $B = \dfrac{\mu_0 i}{2\pi r}$ T, where $\mu_0 = 4\pi \cdot 10^{-7}$ T-m/A. Furthermore, $\mathbf{B}$ points into the page at points P in the xy-plane.

32. Assume that $i(t) = t(12 - t)$ A (t in seconds). Calculate the flux $\Phi(t)$, at time t, of $\mathbf{B}$ through a rectangle of dimensions $L \times H = 3 \times 2$ m whose top and bottom edges are parallel to the wire and whose bottom edge is located $d = 0.5$ m above the wire, similar to Figure 13(B). Then use Faraday's Law to determine the voltage drop around the rectangular loop (the boundary of the rectangle) at time t.

33. Assume that $i = 10e^{-0.1t}$ A (t in seconds). Calculate the flux $\Phi(t)$, at time t, of $\mathbf{B}$ through the isosceles triangle of base 12 cm and height 6 cm whose bottom edge is 3 cm from the wire, as in Figure 17. Assume the triangle is oriented with normal vector pointing out of the page. Use Faraday's Law to determine the voltage drop around the triangular loop (the boundary of the triangle) at time t.

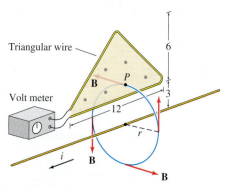

FIGURE 17

In Exercises 34–35, a solid material that has thermal conductivity K in kilowatts per meter-kelvin and temperature given at each point by $w(x, y, z)$ has heat flow given by the vector field $\mathbf{F} = -K\nabla w$ and rate of heat flow across a surface $\mathcal{S}$ within the solid given by $-K \int \int_{\mathcal{S}} \nabla w \, dS$.

34. Find the rate of heat flow out of a sphere of radius 1 meter inside a large cube of copper ($K = 400$ kilowatts/m-k) with temperature function given by $w(x, y, z) = 20 - 5(x^2 + y^2 + z^2)°$C.

35. A insulated cylinder of solid gold ($K = 310$ kW/m-k) of radius $\sqrt{2}$ m and height 5 m is heated at one end until the temperature at each point in the cylinder is given by $w(x, y, z) = (30 - z^2)(2 - (x^2 + y^2))$. Determine the rate of heat flow across each horizontal disk given by $z = 1$, $z = 2$, and $z = 3$, identifying which has the greatest rate of heat flow across it.

Further Insights and Challenges

36. A point mass m is located at the origin. Let Q be the flux of the gravitational field $\mathbf{F} = -Gm\dfrac{\mathbf{e}_r}{r^2}$ through the cylinder $x^2 + y^2 = R^2$ for $a \le z \le b$, including the top and bottom (Figure 18). Show that $Q = -4\pi Gm$ if $a < 0 < b$ (m lies inside the cylinder) and $Q = 0$ if $0 < a < b$ (m lies outside the cylinder).

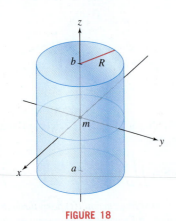

FIGURE 18

In Exercises 37 and 38, let $\mathcal{S}$ be the surface with parametrization

$$G(u, v) = \left(\left(1 + v\cos\frac{u}{2}\right)\cos u, \left(1 + v\cos\frac{u}{2}\right)\sin u, v\sin\frac{u}{2}\right)$$

for $0 \le u \le 2\pi$, $-\frac{1}{2} \le v \le \frac{1}{2}$.

37. *CAS* Use a computer algebra system.

(a) Plot $\mathcal{S}$ and confirm visually that $\mathcal{S}$ is a Möbius strip.

(b) The intersection of $\mathcal{S}$ with the xy-plane is the unit circle $G(u, 0) = (\cos u, \sin u, 0)$. Verify that the normal vector along this circle is

$$\mathbf{N}(u, 0) = \left\langle\cos u \sin\frac{u}{2}, \sin u \sin\frac{u}{2}, -\cos\frac{u}{2}\right\rangle$$

(c) As u varies from 0 to 2π, the point $G(u, 0)$ moves once around the unit circle, beginning and ending at $G(0, 0) = G(2\pi, 0) = (1, 0, 0)$. Verify that $\mathbf{N}(u, 0)$ is a unit vector that varies continuously but that $\mathbf{N}(2\pi, 0) = -\mathbf{N}(0, 0)$. This shows that $\mathcal{S}$ is not orientable—that is, it is not possible to choose a nonzero normal vector at each point on $\mathcal{S}$ in a continuously varying manner (if it were possible, the unit normal vector would return to itself rather than to its negative when carried around the circle).

38. *CAS* We cannot integrate vector fields over $\mathcal{S}$ because $\mathcal{S}$ is not orientable, but it is possible to integrate functions over $\mathcal{S}$. Using a computer algebra system:

(a) Verify that

$$\|\mathbf{N}(u, v)\|^2 = 1 + \frac{3}{4}v^2 + 2v\cos\frac{u}{2} + \frac{1}{2}v^2\cos u$$

(b) Compute the surface area of $\mathcal{S}$ to four decimal places.

(c) Compute $\displaystyle\iint_{\mathcal{S}}(x^2 + y^2 + z^2)\,dS$ to four decimal places.

CHAPTER REVIEW EXERCISES

1. Compute the vector assigned to the point $P = (-3, 5)$ by the vector field:

(a) $\mathbf{F}(x, y) = \langle xy, y - x\rangle$

(b) $\mathbf{F}(x, y) = \langle 4, 8\rangle$

(c) $\mathbf{F}(x, y) = \langle 3^{x+y}, \log_2(x + y)\rangle$

2. Find a vector field $\mathbf{F}$ in the plane such that $\|\mathbf{F}(x, y)\| = 1$ and $\mathbf{F}(x, y)$ is orthogonal to $\mathbf{G}(x, y) = \langle x, y\rangle$ for all x, y.

In Exercises 3–6, sketch the vector field.

3. $\mathbf{F}(x, y) = \langle y, 1\rangle$

4. $\mathbf{F}(x, y) = \langle 4, 1\rangle$

5. ∇f, where $f(x, y) = x^2 - y$

6. $\mathbf{F}(x, y) = \left\langle\dfrac{4y}{\sqrt{x^2 + 4y^2}}, \dfrac{-x}{\sqrt{x^2 + 16y^2}}\right\rangle$

Hint: Show that $\mathbf{F}$ is a unit vector field tangent to the family of ellipses $x^2 + 4y^2 = c^2$.

In Exercises 7–14, calculate div($\mathbf{F}$) and curl($\mathbf{F}$).

7. $\mathbf{F} = \langle x^2, y^2, z^2\rangle$

8. $\mathbf{F} = \langle yz, xz, xy\rangle$

9. $\mathbf{F} = \langle x^3y, xz^2, y^2z\rangle$

10. $\mathbf{F} = \langle \sin xy, \cos yz, \sin xz\rangle$

11. $\mathbf{F} = y\mathbf{i} - z\mathbf{k}$

12. $\mathbf{F} = \langle e^{x+y}, e^{y+z}, xyz\rangle$

13. $\mathbf{F} = \nabla(e^{-x^2-y^2-z^2})$

14. $\mathbf{e}_r = r^{-1}\langle x, y, z\rangle$ $\left(r = \sqrt{x^2 + y^2 + z^2}\right)$

15. Show that if F_1, F_2, and F_3 are differentiable functions of one variable, then

$$\text{curl}\left(\langle F_1(x), F_2(y), F_3(z)\rangle\right) = \mathbf{0}$$

Use this to calculate the curl of

$$\mathbf{F}(x, y, z) = \langle x^2 + y^2, \ln y + z^2, z^3\sin(z^2)e^{z^3}\rangle$$

16. Give an example of a nonzero vector field $\mathbf{F}$ such that curl($\mathbf{F}$) = $\mathbf{0}$ and div($\mathbf{F}$) = 0.

17. Verify the identity div(curl$\mathbf{F}$) = 0 for the vector fields $\mathbf{F} = \langle xz, ye^x, yz\rangle$ and $\mathbf{G} = \langle z^2, xy^3, x^2y\rangle$.

In Exercises 18–26, determine whether the vector field is conservative, and if so, find a potential function.

18. $\mathbf{F}(x, y) = \langle x^2 y, y^2 x \rangle$

19. $\mathbf{F}(x, y) = \langle 4x^3 y^5, 5x^4 y^4 \rangle$

20. $\mathbf{F}(x, y, z) = \langle \sin x, e^y, z \rangle$

21. $\mathbf{F}(x, y, z) = \langle 2, 4, e^z \rangle$

22. $\mathbf{F}(x, y, z) = \langle xyz, \frac{1}{2}x^2 z, 2z^2 y \rangle$

23. $\mathbf{F}(x, y) = \langle y^4 x^3, x^4 y^3 \rangle$

24. $\mathbf{F}(x, y, z) = \left\langle \dfrac{y}{1 + x^2}, \tan^{-1} x, 2z \right\rangle$

25. $\mathbf{F}(x, y, z) = \left\langle \dfrac{2xy}{x^2 + z}, \ln(x^2 + z), \dfrac{y}{x^2 + z} \right\rangle$

26. $\mathbf{F}(x, y, z) = \langle xe^{2x}, ye^{2z}, ze^{2y} \rangle$

27. Find a conservative vector field of the form $\mathbf{F} = \langle g(y), h(x) \rangle$ such that $\mathbf{F}(0, 0) = \langle 1, 1 \rangle$, where $g(y)$ and $h(x)$ are differentiable functions. Determine all such vector fields.

In Exercises 28–31, compute the line integral $\displaystyle\int_{\mathcal{C}} f(x, y)\, ds$ for the given function and path or curve.

28. $f(x, y) = xy$, the path $\mathbf{r}(t) = \langle t, 2t - 1 \rangle$ for $0 \leq t \leq 1$

29. $f(x, y) = x - y$, the unit semicircle $x^2 + y^2 = 1$, $y \geq 0$

30. $f(x, y, z) = e^x - \dfrac{y}{2\sqrt{2z}}$, the path $\mathbf{r}(t) = \left\langle \ln t, \sqrt{2}t, \frac{1}{2}t^2 \right\rangle$ for $1 \leq t \leq 2$

31. $f(x, y, z) = x + 2y + z$, the helix $\mathbf{r}(t) = \langle \cos t, \sin t, t \rangle$ for $0 \leq t \leq \pi/2$

32. Find the total mass of an L-shaped rod consisting of the segments $(2t, 2)$ and $(2, 2 - 2t)$ for $0 \leq t \leq 1$ (length in centimeters) with mass density $\rho(x, y) = x^2 y$ g/cm.

33. Calculate $\mathbf{F} = \nabla f$, where $f(x, y, z) = xye^z$, and compute $\displaystyle\int_{\mathcal{C}} \mathbf{F} \cdot d\mathbf{r}$, where:

(a) $\mathcal{C}$ is any curve from $(1, 1, 0)$ to $(3, e, -1)$.

(b) $\mathcal{C}$ is the boundary of the square $0 \leq x \leq 1$, $0 \leq y \leq 1$ oriented counterclockwise.

34. Calculate $\displaystyle\int_{\mathcal{C}_1} y\, dx + x^2 y\, dy$, where $\mathcal{C}_1$ is the oriented curve in Figure 1(A).

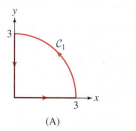
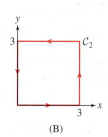

FIGURE 1

35. Let $\mathbf{F}(x, y) = \langle 9y - y^3, e^{\sqrt{y}}(x^2 - 3x) \rangle$, and let $\mathcal{C}_2$ be the oriented curve in Figure 1(B).

(a) Show that $\mathbf{F}$ is not conservative.

(b) Show that $\displaystyle\int_{\mathcal{C}_2} \mathbf{F} \cdot d\mathbf{r} = 0$ without explicitly computing the integral. *Hint:* Show that $\mathbf{F}$ is orthogonal to the edges along the square.

In Exercises 36–39, compute the line integral $\displaystyle\int_{\mathbf{c}} \mathbf{F} \cdot d\mathbf{r}$ for the given vector field and path.

36. $\mathbf{F}(x, y) = \left\langle \dfrac{2y}{x^2 + 4y^2}, \dfrac{x}{x^2 + 4y^2} \right\rangle$, the path $\mathbf{r}(t) = \left\langle \cos t, \frac{1}{2}\sin t \right\rangle$ for $0 \leq t \leq 2\pi$

37. $\mathbf{F}(x, y) = \langle 2xy, x^2 + y^2 \rangle$, the part of the unit circle in the first quadrant oriented counterclockwise

38. $\mathbf{F}(x, y) = \langle x^2 y, y^2 z, z^2 x \rangle$, the path $\mathbf{r}(t) = \langle e^{-t}, e^{-2t}, e^{-3t} \rangle$ for $0 \leq t < \infty$

39. $\mathbf{F} = \nabla f$, where $f(x, y, z) = 4x^2 \ln(1 + y^4 + z^2)$, the path $\mathbf{r}(t) = \langle t^3, \ln(1 + t^2), e^t \rangle$ for $0 \leq t \leq 1$

40. Consider the line integrals $\displaystyle\int_{\mathcal{C}} \mathbf{F} \cdot d\mathbf{r}$ for the vector fields $\mathbf{F}$ and paths $\mathbf{r}$ in Figure 2. Which two of the line integrals appear to have a value of zero? Which of the other two appears to have a negative value?

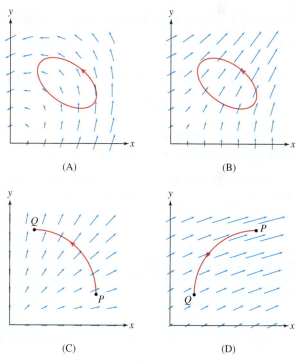

FIGURE 2

41. Calculate the work required to move an object from $P = (1, 1, 1)$ to $Q = (3, -4, -2)$ against the force field $\mathbf{F}(x, y, z) = -12r^{-4} \langle x, y, z \rangle$ (distance in meters, force in newtons), where $r = \sqrt{x^2 + y^2 + z^2}$. *Hint:* Find a potential function for $\mathbf{F}$.

42. Find constants a, b, c such that

$$G(u, v) = (u + av, bu + v, 2u - c)$$

parametrizes the plane $3x - 4y + z = 5$. Calculate $\mathbf{T}_u$, $\mathbf{T}_v$, and $\mathbf{N}(u, v)$.

43. Calculate the integral of $f(x, y, z) = e^z$ over the portion of the plane $x + 2y + 2z = 3$, where $x, y, z \geq 0$.

44. Let S be the surface parametrized by

$$G(u, v) = \left(2u \sin \frac{v}{2}, 2u \cos \frac{v}{2}, 3v\right)$$

for $0 \leq u \leq 1$ and $0 \leq v \leq 2\pi$.

(a) Calculate the tangent vectors $\mathbf{T}_u$ and $\mathbf{T}_v$ and the normal vector $\mathbf{N}(u, v)$ at $P = G(1, \frac{\pi}{3})$.

(b) Find the equation of the tangent plane at P.

(c) Compute the surface area of S.

45. `CAS` Plot the surface with parametrization

$$G(u, v) = (u + 4v, 2u - v, 5uv)$$

for $-1 \leq v \leq 1$, $-1 \leq u \leq 1$. Express the surface area as a double integral and use a computer algebra system to compute the area numerically.

46. `CAS` Express the surface area of the surface $z = 10 - x^2 - y^2$ for $-1 \leq x \leq 1$, $-3 \leq y \leq 3$ as a double integral. Evaluate the integral numerically using a CAS.

47. Evaluate $\iint_S x^2 y \, dS$, where S is the surface $z = \sqrt{3}x + y^2$, $-1 \leq x \leq 1$, $0 \leq y \leq 1$.

48. Calculate $\iint_S (x^2 + y^2) e^{-z} \, dS$, where S is the cylinder with equation $x^2 + y^2 = 9$ for $0 \leq z \leq 10$.

49. Let S be the upper hemisphere $x^2 + y^2 + z^2 = 1$, $z \geq 0$. For each of the functions (a)–(d), determine whether $\iint_S f \, dS$ is positive, zero, or negative (without evaluating the integral). Explain your reasoning.

(a) $f(x, y, z) = y^3$　　　　　**(b)** $f(x, y, z) = z^3$

(c) $f(x, y, z) = xyz$　　　　**(d)** $f(x, y, z) = z^2 - 2$

50. Let S be a small patch of surface with a parametrization $G(u, v)$ for $0 \leq u \leq 0.1$, $0 \leq v \leq 0.1$ such that the normal vector $\mathbf{N}(u, v)$ for $(u, v) = (0, 0)$ is $\mathbf{N} = \langle 2, -2, 4 \rangle$. Use Eq. (3) in Section 16.4 to estimate the surface area of S.

51. The upper half of the sphere $x^2 + y^2 + z^2 = 9$ has parametrization $G(r, \theta) = (r \cos \theta, r \sin \theta, \sqrt{9 - r^2})$ in cylindrical coordinates (Figure 3).

(a) Calculate the normal vector $\mathbf{N} = \mathbf{T}_r \times \mathbf{T}_\theta$ at the point $G(2, \frac{\pi}{3})$.

(b) Use Eq. (3) in Section 16.4 to estimate the surface area of $G(\mathcal{R})$, where $\mathcal{R}$ is the small domain defined by

$$2 \leq r \leq 2.1, \qquad \frac{\pi}{3} \leq \theta \leq \frac{\pi}{3} + 0.05$$

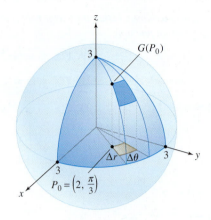

FIGURE 3

In Exercises 52–57, compute $\iint_S \mathbf{F} \cdot d\mathbf{S}$ for the given oriented surface or parametrized surface.

52. $\mathbf{F}(x, y, z) = \langle y, x, e^{xz} \rangle$, $\quad x^2 + y^2 = 9$, $x \geq 0$, $y \geq 0$, $-3 \leq z \leq 3$, outward-pointing normal

53. $\mathbf{F}(x, y, z) = \langle -y, z, -x \rangle$, $\quad G(u, v) = (u + 3v, v - 2u, 2v + 5)$, $0 \leq u \leq 1$, $0 \leq v \leq 1$, upward-pointing normal

54. $\mathbf{F}(x, y, z) = \langle 0, 0, x^2 + y^2 \rangle$, $\quad x^2 + y^2 + z^2 = 4$, $z \geq 0$, outward-pointing normal

55. $\mathbf{F}(x, y, z) = \langle z, 0, z^2 \rangle$, $\quad G(u, v) = (v \cosh u, v \sinh u, v)$, $0 \leq u \leq 1$, $0 \leq v \leq 1$, upward-pointing normal

56. $\mathbf{F}(x, y, z) = \langle 0, 0, xze^{xy} \rangle$, $\quad z = xy$, $0 \leq x \leq 1$, $0 \leq y \leq 1$, upward-pointing normal

57. $\mathbf{F}(x, y, z) = \langle 0, 0, z \rangle$, $\quad 3x^2 + 2y^2 + z^2 = 1$, $z \geq 0$, upward-pointing normal

58. Calculate the total charge on the cylinder

$$x^2 + y^2 = R^2, \qquad 0 \leq z \leq H$$

if the charge density in cylindrical coordinates is $\rho(\theta, z) = Kz^2 \cos^2 \theta$, where K is a constant.

59. Find the flow rate of a fluid with velocity field $\mathbf{v} = \langle 2x, y, xy \rangle$ m/s across the part of the cylinder $x^2 + y^2 = 9$ where $x \geq 0$, $y \geq 0$, and $0 \leq z \leq 4$ (distance in meters).

60. With $\mathbf{v}$ as in Exercise 59, calculate the flow rate across the part of the elliptic cylinder $\dfrac{x^2}{4} + y^2 = 1$ where $x \geq 0$, $y \geq 0$, and $0 \leq z \leq 4$.

61. Calculate the flux of the vector field $\mathbf{E}(x, y, z) = \langle 0, 0, x \rangle$ through the part of the ellipsoid

$$4x^2 + 9y^2 + z^2 = 36$$

where $z \geq 3$, $x \geq 0$, $y \geq 0$. *Hint:* Use the parametrization

$$G(r, \theta) = \left(3r \cos \theta, 2r \sin \theta, 6\sqrt{1 - r^2}\right)$$

17 FUNDAMENTAL THEOREMS OF VECTOR ANALYSIS

Fluid flows, such as this water vortex, are analyzed using the fundamental theorems of vector analysis. (© *Adafir/Alamy*)

I n this final chapter, we study three generalizations of the Fundamental Theorem of Calculus, which we have seen is given by $\int_a^b F'(x)\,dx = F(b) - F(a)$. If we think of the boundary of the interval $[a, b]$ as being given by the two points $\{a, b\}$, then the Fundamental Theorem says that we can find the integral of the derivative of a function over an interval just by evaluating that function on the boundary of the interval. The first of these new theorems, Green's Theorem, says that we can find a double integral of a certain type of derivative over a region in the xy-plane by finding a line integral around the boundary of the region. The second theorem, Stokes' Theorem, allows us to find a surface integral of a certain derivative over a surface with boundary curves in space by evaluating a line integral on the boundary curves. The third theorem, the Divergence Theorem, allows us to find the triple integral of another certain kind of derivative over a solid in space by evaluating a surface integral over the boundary surface of the solid.

This is a culmination of our efforts to extend the ideas of single-variable calculus to the multivariable setting. However, vector analysis is not so much an endpoint as a gateway to a host of applications, not only in the traditional domains of physics and engineering, but also in biological, earth, and environmental sciences, where an understanding of fluid and aerodynamics, and continuous matter is required.

17.1 Green's Theorem

In Section 16.3, we showed that the circulation of a conservative vector field $\mathbf{F}$ around every closed path is zero. For vector fields in the plane, Green's Theorem tells us what happens when $\mathbf{F}$ is not conservative.

To formulate Green's Theorem, we need some notation. Consider a domain $\mathcal{D}$ whose boundary $\mathcal{C}$ is a **simple closed curve**—that is, a closed curve that does not intersect itself (Figure 1). We follow standard usage and denote the boundary curve $\mathcal{C}$ by $\partial\mathcal{D}$. The **boundary orientation** of $\partial\mathcal{D}$ is the direction to traverse the boundary such that the region is always to your left, as in Figure 1. When there is a single boundary curve, the boundary orientation is the counterclockwise orientation.

Recall that we have two notations for the line integral of $\mathbf{F} = \langle F_1, F_2 \rangle$:

$$\int_{\mathcal{C}} \mathbf{F} \cdot d\mathbf{r} \qquad \text{and} \qquad \int_{\mathcal{C}} F_1\,dx + F_2\,dy$$

FIGURE 1 The boundary of $\mathcal{D}$ is a simple closed curve $\mathcal{C}$ that is denoted $\partial\mathcal{D}$. The boundary is oriented in the counterclockwise direction.

If $\mathcal{C}$ is parametrized by $\mathbf{r}(t) = \langle x(t), y(t) \rangle$ for $a \le t \le b$, then

$$dx = x'(t)\,dt, \qquad dy = y'(t)\,dt$$

$$\int_{\mathcal{C}} F_1\,dx + F_2\,dy = \int_a^b \Big(F_1(x(t), y(t))x'(t) + F_2(x(t), y(t))y'(t) \Big)\,dt \qquad \boxed{1}$$

Throughout this chapter, we assume that the components of all vector fields have continuous second-order partial derivatives, and also that $\mathcal{C}$ is smooth ($\mathcal{C}$ has a parametrization with derivatives of all orders) or piecewise smooth (a finite union of smooth curves joined together at endpoints).

←·· REMINDER The line integral of a vector
field over a closed curve is called the
"circulation" and is often denoted with the
symbol $\oint$.

Green's theorem can also be written

$$\oint_{\partial D} \mathbf{F} \cdot d\mathbf{r} = \iint_D \left(\frac{\partial F_2}{\partial x} - \frac{\partial F_1}{\partial y} \right) dA$$

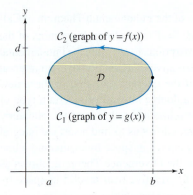

FIGURE 2 The boundary curve ∂D is the
union of the graphs of $y = g(x)$ and
$y = f(x)$ oriented counterclockwise.

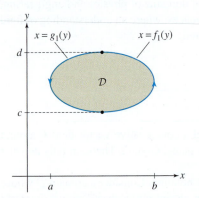

FIGURE 3 The boundary curve ∂D is also
the union of the graphs of $x = g_1(x)$ and
$y = f(x)$ oriented counterclockwise.

THEOREM 1 Green's Theorem Let $\mathcal{D}$ be a domain whose boundary $\partial \mathcal{D}$ is a simple closed curve, oriented counterclockwise. Then

$$\oint_{\partial D} F_1 \, dx + F_2 \, dy = \iint_D \left(\frac{\partial F_2}{\partial x} - \frac{\partial F_1}{\partial y} \right) dA \qquad \boxed{2}$$

Proof Because a complete proof is quite technical, we shall make the simplifying assumption that the boundary of $\mathcal{D}$ can be described as the union of two graphs $y = g(x)$ and $y = f(x)$ with $g(x) \le f(x)$ as in Figure 2 and also as the union of two graphs $x = g_1(y)$ and $x = f_1(y)$, with $g_1(y) \le f_1(y)$ as in Figure 3.

From the terms in Eq. (2), we construct two equations, one for F_1 and one for F_2:

$$\oint_{\partial D} F_1 \, dx = -\iint_D \frac{\partial F_1}{\partial y} \, dA \qquad \boxed{3}$$

$$\oint_{\partial D} F_2 \, dy = \iint_D \frac{\partial F_2}{\partial x} \, dA \qquad \boxed{4}$$

If we can show that both these equations hold, then we obtain a proof of Green's Theorem by adding them together. To prove Eq. (3), we write

$$\oint_{\partial D} F_1 \, dx = \int_{C_1} F_1 \, dx + \int_{C_2} F_1 \, dx$$

where C_1 is the graph of $y = g(x)$ and C_2 is the graph of $y = f(x)$, oriented as in Figure 2. To compute these line integrals, we parameterize the graphs from left to right using $t = x$ as the parameter:

$$\text{Graph of } y = g(x): \qquad \mathbf{r}_1(t) = \langle t, g(t) \rangle, \qquad a \le t \le b$$

$$\text{Graph of } y = f(x): \qquad \mathbf{r}_2(t) = \langle t, f(t) \rangle, \qquad a \le t \le b$$

Since C_2 is oriented from right to left, the line integral over $\partial \mathcal{D}$ is the difference

$$\oint_{\partial D} F_1 \, dx = \int_{C_1} F_1 \, dx - \int_{C_2} F_1 \, dx$$

In both parametrizations, $x = t$, so $dx = dt$, and by Eq. (1)

$$\oint_{\partial D} F_1 \, dx = \int_{t=a}^{b} F_1(t, g(t)) \, dt - \int_{t=a}^{b} F_1(t, f(t)) \, dt \qquad \boxed{5}$$

Now, the key step is to apply the Fundamental Theorem of Calculus to $\dfrac{\partial F_1}{\partial y}(t, y)$ as a function of y with t held constant:

$$F_1(t, f(t)) - F_1(t, g(t)) = \int_{y=g(t)}^{f(t)} \frac{\partial F_1}{\partial y}(t, y) \, dy$$

Substituting the integral on the right in Eq. (5), we obtain Eq. (3):

$$\oint_{\partial D} F_1 \, dx = -\int_{t=a}^{b} \int_{y=g(t)}^{f(t)} \frac{\partial F_1}{\partial y}(t, y) \, dy \, dt = -\iint_D \frac{\partial F_1}{\partial y} \, dA$$

Eq. (4) is proved in a similar fashion, by expressing $\partial \mathcal{D}$ as the union of the graphs of $x = f_1(y)$ and $x = g_1(y)$ (Figure 3). ∎

Recall from Section 16.1 that if **F** is a conservative vector field, i.e., $\mathbf{F} = \nabla f$, then the cross-partial condition is satisfied:

$$\frac{\partial F_2}{\partial x} - \frac{\partial F_1}{\partial y} = 0$$

In this case, Green's Theorem merely confirms what we already know: The line integral of a conservative vector field around any closed curve is zero.

■ **EXAMPLE 1** **Verifying Green's Theorem** Verify Green's Theorem for the line integral along the unit circle C, oriented counterclockwise (Figure 4):

$$\oint_C xy^2\, dx + x\, dy$$

Solution

Step 1. **Evaluate the line integral directly.**
We use the standard parametrization of the unit circle:

$$x = \cos\theta, \qquad\qquad y = \sin\theta$$
$$dx = -\sin\theta\, d\theta, \qquad dy = \cos\theta\, d\theta$$

The integrand in the line integral is

$$xy^2\, dx + x\, dy = \cos\theta \sin^2\theta(-\sin\theta\, d\theta) + \cos\theta(\cos\theta\, d\theta)$$
$$= \left(-\cos\theta \sin^3\theta + \cos^2\theta\right) d\theta$$

and

$$\oint_C xy^2\, dx + x\, dy = \int_0^{2\pi} \left(-\cos\theta \sin^3\theta + \cos^2\theta\right) d\theta$$
$$= -\frac{\sin^4\theta}{4}\bigg|_0^{2\pi} + \frac{1}{2}\left(\theta + \frac{1}{2}\sin 2\theta\right)\bigg|_0^{2\pi}$$
$$= 0 + \frac{1}{2}(2\pi + 0) = \boxed{\pi}$$

Step 2. **Evaluate the line integral using Green's Theorem.**
In this example, $F_1 = xy^2$ and $F_2 = x$, so

$$\frac{\partial F_2}{\partial x} - \frac{\partial F_1}{\partial y} = \frac{\partial}{\partial x}x - \frac{\partial}{\partial y}xy^2 = 1 - 2xy$$

According to Green's Theorem,

$$\oint_C xy^2\, dx + x\, dy = \iint_{\mathcal{D}} \left(\frac{\partial F_2}{\partial x} - \frac{\partial F_1}{\partial y}\right) dA = \iint_{\mathcal{D}} (1 - 2xy)\, dA$$

where $\mathcal{D}$ is the disk $x^2 + y^2 \le 1$ enclosed by C. The integral of $2xy$ over $\mathcal{D}$ is zero by symmetry—the contributions for positive and negative x cancel. We can check this directly:

$$\iint_{\mathcal{D}} (-2xy)\, dA = -2\int_{x=-1}^1 \int_{y=-\sqrt{1-x^2}}^{\sqrt{1-x^2}} xy\, dy\, dx = -\int_{x=-1}^1 xy^2\bigg|_{y=-\sqrt{1-x^2}}^{\sqrt{1-x^2}} dx = 0$$

Therefore,

$$\iint_{\mathcal{D}} \left(\frac{\partial F_2}{\partial x} - \frac{\partial F_1}{\partial y}\right) dA = \iint_{\mathcal{D}} 1\, dA = \text{Area}(\mathcal{D}) = \boxed{\pi}$$

This agrees with the value in Step 1. So Green's Theorem is verified in this case. ■

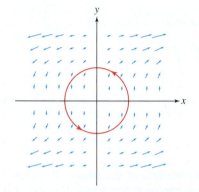

FIGURE 4 The vector field
$\mathbf{F}(x, y) = \langle xy^2, x\rangle$.

Green's Theorem states

$$\oint_{\partial \mathcal{D}} F_1\, dx + F_2\, dy$$
$$= \iint_{\mathcal{D}} \left(\frac{\partial F_2}{\partial x} - \frac{\partial F_1}{\partial y}\right) dA$$

◀··· **REMINDER** To integrate $\cos^2\theta$, use the identity $\cos^2\theta = \frac{1}{2}(1 + \cos 2\theta)$.

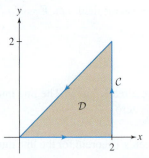

FIGURE 5 The region $\mathcal{D}$ is described by $0 \le x \le 2, 0 \le y \le x$.

■ EXAMPLE 2 Computing a Line Integral Using Green's Theorem Compute the circulation of $\mathbf{F}(x, y) = \left\langle \sin x, x^2 y^3 \right\rangle$ around the triangular path C with the counterclockwise orientation shown in Figure 5.

Solution To compute the line integral directly, we would have to parametrize all three sides of the triangle. Instead, we apply Green's Theorem to the domain $\mathcal{D}$ enclosed by the triangle. This domain is described by $0 \le x \le 2, 0 \le y \le x$.

Applying Green's Theorem, we obtain

$$\frac{\partial F_2}{\partial x} - \frac{\partial F_1}{\partial y} = \frac{\partial}{\partial x} x^2 y^3 - \frac{\partial}{\partial y} \sin x = 2xy^3$$

$$\oint_C \sin x \, dx + x^2 y^3 \, dy = \iint_{\mathcal{D}} 2xy^3 \, dA = \int_0^2 \int_{y=0}^x 2xy^3 \, dy \, dx$$

$$= \int_0^2 \left(\frac{1}{2} x y^4 \Big|_0^x \right) dx = \frac{1}{2} \int_0^2 x^5 \, dx = \frac{1}{12} x^6 \Big|_0^2 = \frac{16}{3} \qquad ■$$

Area via Green's Theorem

We can use Green's Theorem to obtain formulas for the area of the domain $\mathcal{D}$ enclosed by a simple closed curve C (Figure 6). The trick is to choose a vector field $\mathbf{F} = \langle F_1, F_2 \rangle$ such that $\frac{\partial F_2}{\partial x} - \frac{\partial F_1}{\partial y} = 1$.

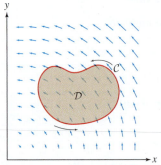

FIGURE 6 The line integrals

$$\oint_C x \, dy = \oint_C -y \, dx = \frac{1}{2} \oint_C x \, dy - y \, dx$$

are equal to the area enclosed by C.

If we choose $\mathbf{F}(x, y) = \langle 0, x \rangle$, then $\dfrac{\partial F_2}{\partial x} - \dfrac{\partial F_1}{\partial y} = \dfrac{\partial}{\partial x} x - \dfrac{\partial}{\partial y} 0 = 1$

If we choose $\mathbf{F}(x, y) = \langle -y, 0 \rangle$, then $\dfrac{\partial F_2}{\partial x} - \dfrac{\partial F_1}{\partial y} = \dfrac{\partial}{\partial x} 0 - \dfrac{\partial}{\partial y} (-y) = 1$

If we choose $\mathbf{F}(x, y) = \langle -y/2, x/2 \rangle$, then $\dfrac{\partial F_2}{\partial x} - \dfrac{\partial F_1}{\partial y} = \dfrac{\partial}{\partial x} \left(\dfrac{x}{2} \right) - \dfrac{\partial}{\partial y} \left(\dfrac{-y}{2} \right)$

$$= \frac{1}{2} + \frac{1}{2} = 1$$

In all three cases, the expression on the right side of the equality in Green's Theorem becomes

$$\iint_{\mathcal{D}} \left(\frac{\partial F_2}{\partial x} - \frac{\partial F_1}{\partial y} \right) dA = \iint_{\mathcal{D}} 1 \, dA = \text{area}(\mathcal{D})$$

By Green's Theorem, this equals $\oint_C F_1 \, dx + F_2 \, dy$. Plugging in F_1 and F_2 for each of these three cases, we obtain the following three formulas for the area of the domain $\mathcal{D}$.

$$\boxed{\text{Area enclosed by } C = \oint_C x \, dy = \oint_C -y \, dx = \frac{1}{2} \oint_C x \, dy - y \, dx} \qquad \boxed{6}$$

These remarkable formulas tell us how to compute an enclosed area by making measurements only along the boundary. It is the mathematical basis of the **planimeter**, a device that computes the area of an irregular shape when you trace the boundary with a pointer at the end of a movable arm (Figure 7).

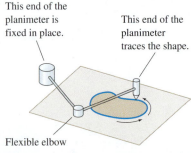

This end of the planimeter is fixed in place.

This end of the planimeter traces the shape.

Flexible elbow

FIGURE 7 A planimeter is a mechanical device used for measuring the areas of irregular shapes. *(Courtesy of John D. Eggers, UCSD; photo by Adriene Hughes, UC San Diego Media Lab)*

■ EXAMPLE 3 Computing Area via Green's Theorem Compute the area of the ellipse $\left(\dfrac{x}{a}\right)^2 + \left(\dfrac{y}{b}\right)^2 = 1$ using a line integral.

Solution We parametrize the boundary of the ellipse by

$$x = a\cos\theta, \qquad y = b\sin\theta, \qquad 0 \le \theta < 2\pi$$

We calculate the area in each of the three possible ways. Using the first formula in Eq. (6):

$$\text{Enclosed area} = \oint_C x\,dy = \int_0^{2\pi} (a\cos\theta)(b\cos\theta)\,d\theta$$

$$= ab\int_0^{2\pi} \cos^2\theta\,d\theta$$

$$= \pi ab$$

Using the second formula in Eq. (6), we get

$$\text{Enclosed area} = \oint_C -y\,dx = \int_0^{2\pi} (-b\sin\theta)(-a\sin\theta)\,d\theta$$

$$= ab\int_0^{2\pi} \sin^2\theta\,d\theta$$

$$= \pi ab$$

And using the third formula in Eq. (6) yields

$$\text{Enclosed area} = \frac{1}{2}\oint_C x\,dy - y\,dx$$

$$= \frac{1}{2}\int_0^{2\pi} [(a\cos\theta)(b\cos\theta) - (b\sin\theta)(-a\sin\theta)]\,d\theta$$

$$= \frac{ab}{2}\int_0^{2\pi} (\cos^2\theta + \sin^2\theta)\,d\theta$$

$$= \frac{ab}{2}\int_0^{2\pi} d\theta = \pi ab$$

All three methods yield the standard formula for the area of an ellipse. ■

←·· **REMINDER** We use the fact that

$$\int_0^{2\pi} \sin^2\theta\,d\theta = \int_0^{2\pi} \cos^2\theta\,d\theta = \pi$$

which follows immediately from the two identities

$$\sin^2\theta = \frac{1 - \cos 2\theta}{2}$$

and

$$\cos^2\theta = \frac{1 + \cos 2\theta}{2}$$

as in Section 7.2.

For convenience, we can think of a two-dimensional vector field $\mathbf{F} = \langle F_1, F_2 \rangle$ as a three-dimensional vector field with third component 0. So, $\mathbf{F} = \langle F_1, F_2, 0 \rangle$. Then when we take the curl, keeping in mind that F_1 and F_2 only depend on x and y, we find

$$\text{curl}(\mathbf{F}) = \begin{vmatrix} \mathbf{i} & \mathbf{j} & \mathbf{k} \\ \dfrac{\partial}{\partial x} & \dfrac{\partial}{\partial y} & \dfrac{\partial}{\partial z} \\ F_1 & F_2 & 0 \end{vmatrix}$$

$$= 0\mathbf{i} + 0\mathbf{j} + \left(\frac{\partial F_2}{\partial x} - \frac{\partial F_1}{\partial y} \right)\mathbf{k}$$

The z-component of the result is $\dfrac{\partial F_2}{\partial x} - \dfrac{\partial F_1}{\partial y}$, which is the integrand that appears in Green's Theorem. Thus, we define

$$\text{curl}_z(\mathbf{F}) = \text{curl}(\mathbf{F}) \cdot \mathbf{k} = \frac{\partial F_2}{\partial x} - \frac{\partial F_1}{\partial y}$$

Then Green's Theorem becomes

$$\oint_{\mathcal{C}} \mathbf{F} \cdot d\mathbf{r} = \iint_{\mathcal{D}} \text{curl}_z(\mathbf{F}) \, dA$$

7

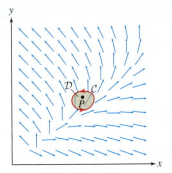

FIGURE 8 The circulation of **F** around $\mathcal{C}$ is approximately $\text{curl}_z(\mathbf{F})(P) \cdot \text{Area}(\mathcal{D})$.

CONCEPTUAL INSIGHT   What is the meaning of the integrand $\text{curl}_z(\mathbf{F})$ in Green's Theorem? Apply Green's Theorem to a small domain $\mathcal{D}$ with a simple closed boundary curve and let P be a point in $\mathcal{D}$. Because $\text{curl}_z(\mathbf{F})$ is a continuous function, its value does not change much on D if $\mathcal{C}$ is sufficiently small, so to a first approximation, we can replace $\text{curl}_z(\mathbf{F})$ by the constant value $\text{curl}_z(\mathbf{F})(P)$ (Figure 8). Green's Theorem yields the following approximation for the circulation:

$$\oint_{\mathcal{C}} \mathbf{F} \cdot d\mathbf{r} = \iint_{\mathcal{D}} \text{curl}_z(\mathbf{F}) \, dA$$

$$\approx \text{curl}_z(\mathbf{F})(P) \iint_{\mathcal{D}} dA$$ **8**

$$\approx \text{curl}_z(\mathbf{F})(P) \cdot \text{Area}(\mathcal{D})$$

In other words, *the circulation around a small, simple closed curve $\mathcal{C}$ is, to a first-order approximation, equal to the curl times the enclosed area.* Thus, we can think of $\text{curl}_z(\mathbf{F})$ as the **circulation per unit of enclosed area at each point P**.

GRAPHICAL INSIGHT   If we think of **F** as the velocity field of a fluid, then we can measure the curl by placing a small paddle wheel in the stream at a point P and observing how fast it rotates (Figure 9). Because the fluid pushes each paddle to move with a velocity equal to the tangential component of **F**, we can assume that the wheel itself rotates with a velocity v_a equal to the *average tangential component* of **F**. If the paddle is a circle $\mathcal{C}_r$ of radius r (and hence length $2\pi r$), then the average tangential component of velocity is

$$v_a = \frac{1}{2\pi r} \oint_{\mathcal{C}_r} \mathbf{F} \cdot d\mathbf{r}$$

On the other hand, the paddle encloses an area of πr^2, and for small r, we can apply the approximation (8):

$$v_a \approx \frac{1}{2\pi r}(\pi r^2)\text{curl}_z(\mathbf{F})(P) = \left(\frac{1}{2}r\right)\text{curl}_z(\mathbf{F})(P)$$

Now if an object moves along a circle of radius r with speed v_a, then its angular velocity (in radians per unit time) is $v_a/r \approx \frac{1}{2}\text{curl}_z(\mathbf{F})(P)$. Therefore, *the angular velocity of the paddle wheel is approximately one-half the curl.*

Angular Velocity   An arc of ℓ meters on a circle of radius r meters has radian measure ℓ/r. Therefore, an object moving along the circle with a speed of v meters per second travels v/r radians per second. In other words, the object has angular velocity v/r.

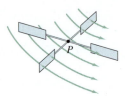
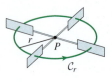

FIGURE 9 The curl is approximately equal to one-half the angular velocity of a small paddle wheel placed at P.

Figure 10 shows vector fields such that $\text{curl}_z(\mathbf{F})$ is constant. Field (A) describes a fluid rotating counterclockwise around the origin, and field (B) describes a fluid that spirals into the origin. In both these cases, a small paddle wheel placed anywhere in the fluid rotates counterclockwise. However, a nonzero curl does not mean that the fluid itself is necessarily rotating. It means only that a small paddle wheel rotates if placed in the fluid. Field (C) is an example of **shear flow** (also known as a Couette flow). It has nonzero curl, but unlike cases (A) and (B), the fluid does not rotate about any point. However, the paddle wheel does rotate clockwise wherever it is placed. Compare with the fields in Figures (D) and (E), which have zero curl. In either case, a paddle wheel placed anywhere in the plane does not rotate. Unlike these examples, for most vector fields $\mathbf{F}$, $\mathrm{curl}_z(\mathbf{F})$ varies over the plane, having points where the paddle wheel turns counterclockwise, clockwise, or not at all.

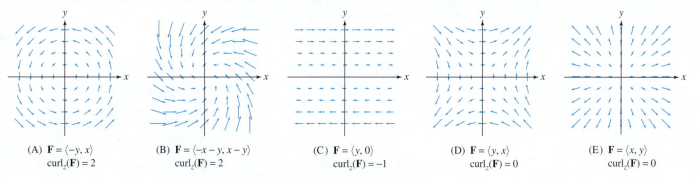

(A) $\mathbf{F} = \langle -y, x \rangle$
$\mathrm{curl}_z(\mathbf{F}) = 2$

(B) $\mathbf{F} = \langle -x - y, x - y \rangle$
$\mathrm{curl}_z(\mathbf{F}) = 2$

(C) $\mathbf{F} = \langle y, 0 \rangle$
$\mathrm{curl}_z(\mathbf{F}) = -1$

(D) $\mathbf{F} = \langle y, x \rangle$
$\mathrm{curl}_z(\mathbf{F}) = 0$

(E) $\mathbf{F} = \langle x, y \rangle$
$\mathrm{curl}_z(\mathbf{F}) = 0$

FIGURE 10 Examples of vector fields and the corresponding function $\mathrm{curl}_z(\mathbf{F})$.

Additivity of Circulation

Circulation around a closed curve has an important additivity property: If we decompose a domain $\mathcal{D}$ into two (or more) nonoverlapping domains $\mathcal{D}_1$ and $\mathcal{D}_2$ that intersect only on part of their boundaries as in Figure 11, then

$$\oint_{\partial\mathcal{D}} \mathbf{F} \cdot d\mathbf{r} = \oint_{\partial\mathcal{D}_1} \mathbf{F} \cdot d\mathbf{r} + \oint_{\partial\mathcal{D}_2} \mathbf{F} \cdot d\mathbf{r}$$

9

To verify this equation, note first that

$$\oint_{\partial\mathcal{D}} \mathbf{F} \cdot d\mathbf{r} = \int_{\mathcal{C}_{\mathrm{top}}} \mathbf{F} \cdot d\mathbf{r} + \int_{\mathcal{C}_{\mathrm{bot}}} \mathbf{F} \cdot d\mathbf{r}$$

with $\mathcal{C}_{\mathrm{top}}$ and $\mathcal{C}_{\mathrm{bot}}$ as in Figure 11, with the orientations shown. Then observe that the dashed segment $\mathcal{C}_{\mathrm{middle}}$ occurs in both $\partial\mathcal{D}_1$ and $\partial\mathcal{D}_2$ but with opposite orientations. If $\mathcal{C}_{\mathrm{middle}}$ is oriented right to left, then

$$\oint_{\partial\mathcal{D}_1} \mathbf{F} \cdot d\mathbf{r} = \int_{\mathcal{C}_{\mathrm{top}}} \mathbf{F} \cdot d\mathbf{r} - \int_{\mathcal{C}_{\mathrm{middle}}} \mathbf{F} \cdot d\mathbf{r}$$

$$\oint_{\partial\mathcal{D}_2} \mathbf{F} \cdot d\mathbf{r} = \int_{\mathcal{C}_{\mathrm{bot}}} \mathbf{F} \cdot d\mathbf{r} + \int_{\mathcal{C}_{\mathrm{middle}}} \mathbf{F} \cdot d\mathbf{r}$$

We obtain Eq. (9) by adding these two equations:

$$\oint_{\partial\mathcal{D}_1} \mathbf{F} \cdot d\mathbf{r} + \oint_{\partial\mathcal{D}_2} \mathbf{F} \cdot d\mathbf{r} = \int_{\mathcal{C}_{\mathrm{top}}} \mathbf{F} \cdot d\mathbf{r} + \int_{\mathcal{C}_{\mathrm{bot}}} \mathbf{F} \cdot d\mathbf{r} = \oint_{\partial\mathcal{D}} \mathbf{F} \cdot d\mathbf{r}$$

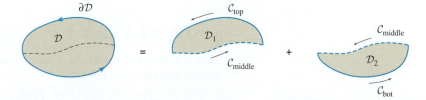

FIGURE 11 The domain $\mathcal{D}$ is the union of $\mathcal{D}_1$ and $\mathcal{D}_2$.

More General Form of Green's Theorem

Consider a domain $\mathcal{D}$ whose boundary consists of more than one simple closed curve as in Figure 12. As before, $\partial\mathcal{D}$ denotes the boundary of $\mathcal{D}$ with its boundary orientation. In other words, *the region lies to the left as the curve is traversed in the direction specified by the orientation.* For the domains in Figure 12,

$$\partial\mathcal{D}_1 = \mathcal{C}_1 + \mathcal{C}_2, \qquad \partial\mathcal{D}_2 = \mathcal{C}_3 + \mathcal{C}_4 - \mathcal{C}_5$$

The curve $\mathcal{C}_5$ occurs with a minus sign because it is oriented counterclockwise, but the boundary orientation requires a clockwise orientation.

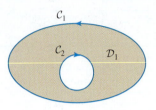 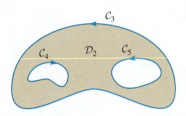

(A) Oriented boundary of $\mathcal{D}_1$ is $\mathcal{C}_1 + \mathcal{C}_2$.   (B) Oriented boundary of $\mathcal{D}_2$ is $\mathcal{C}_3 + \mathcal{C}_4 - \mathcal{C}_5$.

FIGURE 12

Green's Theorem remains valid for more general domains of this type:

$$\oint_{\partial\mathcal{D}} \mathbf{F} \cdot d\mathbf{r} = \iint_{\mathcal{D}} \left(\frac{\partial F_2}{\partial x} - \frac{\partial F_1}{\partial y} \right) dA \qquad \boxed{10}$$

This equality is proved by decomposing $\mathcal{D}$ into smaller domains, each of which is bounded by a simple closed curve. To illustrate, consider the region $\mathcal{D}$ in Figure 13. We decompose $\mathcal{D}$ into domains $\mathcal{D}_1$ and $\mathcal{D}_2$. Then

$$\partial\mathcal{D} = \partial\mathcal{D}_1 + \partial\mathcal{D}_2$$

because the edges common to $\partial\mathcal{D}_1$ and $\partial\mathcal{D}_2$ occur with opposite orientation and therefore cancel. The previous version of Green's Theorem applies to both $\mathcal{D}_1$ and $\mathcal{D}_2$, and thus,

$$\oint_{\partial\mathcal{D}} \mathbf{F} \cdot d\mathbf{r} = \int_{\partial\mathcal{D}_1} \mathbf{F} \cdot d\mathbf{r} + \int_{\partial\mathcal{D}_2} \mathbf{F} \cdot d\mathbf{r}$$

$$= \iint_{\mathcal{D}_1} \left(\frac{\partial F_2}{\partial x} - \frac{\partial F_1}{\partial y} \right) dA + \iint_{\mathcal{D}_2} \left(\frac{\partial F_2}{\partial x} - \frac{\partial F_1}{\partial y} \right) dA$$

$$= \iint_{\mathcal{D}} \left(\frac{\partial F_2}{\partial x} - \frac{\partial F_1}{\partial y} \right) dA$$

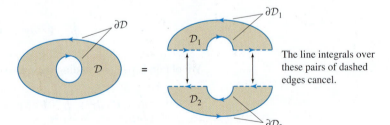

The line integrals over these pairs of dashed edges cancel.

FIGURE 13 The boundary of $\partial\mathcal{D}$ is the sum $\partial\mathcal{D}_1 + \partial\mathcal{D}_2$ because the straight edges cancel.

■ **EXAMPLE 4** Calculate $\oint_{\mathcal{C}_1} \mathbf{F} \cdot d\mathbf{r}$, where $\mathbf{F}(x, y) = \langle x - y, x + y^3 \rangle$ and $\mathcal{C}_1$ is the outer boundary curve oriented counterclockwise of the domain $\mathcal{D}$ appearing in Figure 14. Assume that the area of $\mathcal{D}$ is 8.

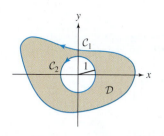

FIGURE 14 $\mathcal{D}$ has area 8, and C_2 is a circle of radius 1.

Solution We cannot compute the line integral over C_1 directly because the curve C_1 is not specified. However, $\partial \mathcal{D} = C_1 - C_2$, so Green's Theorem yields

$$\oint_{C_1} \mathbf{F} \cdot d\mathbf{r} - \oint_{C_2} \mathbf{F} \cdot d\mathbf{r} = \iint_{\mathcal{D}} \left(\frac{\partial F_2}{\partial x} - \frac{\partial F_1}{\partial y} \right) dA \qquad \boxed{11}$$

We have

$$\frac{\partial F_2}{\partial x} - \frac{\partial F_1}{\partial y} = \frac{\partial}{\partial x}(x + y^3) - \frac{\partial}{\partial y}(x - y) = 1 - (-1) = 2$$

$$\iint_{\mathcal{D}} \left(\frac{\partial F_2}{\partial x} - \frac{\partial F_1}{\partial y} \right) dA = \iint_{\mathcal{D}} 2\, dA = 2\,\text{Area}(\mathcal{D}) = 2(8) = 16$$

Thus, Eq. (11) gives us

$$\oint_{C_1} \mathbf{F} \cdot d\mathbf{r} - \oint_{C_2} \mathbf{F} \cdot d\mathbf{r} = 16 \qquad \boxed{12}$$

To compute the second integral, parametrize the unit circle C_2 by $\mathbf{r}(\theta) = \langle \cos\theta, \sin\theta \rangle$. Then

$$\mathbf{F} \cdot \mathbf{r}'(\theta) = \langle \cos\theta - \sin\theta, \cos\theta + \sin^3\theta \rangle \cdot \langle -\sin\theta, \cos\theta \rangle$$

$$= -\sin\theta \cos\theta + \sin^2\theta + \cos^2\theta + \sin^3\theta \cos\theta$$

$$= 1 - \sin\theta \cos\theta + \sin^3\theta \cos\theta$$

The integrals of $\sin\theta \cos\theta$ and $\sin^3\theta \cos\theta$ over $[0, 2\pi]$ are both zero, so

$$\oint_{C_2} \mathbf{F} \cdot d\mathbf{r} = \int_0^{2\pi} (1 - \sin\theta \cos\theta + \sin^3\theta \cos\theta)\, d\theta = \int_0^{2\pi} d\theta = 2\pi$$

Eq. (12) yields $\displaystyle\oint_{C_1} \mathbf{F} \cdot d\mathbf{r} = 16 + 2\pi.$ ∎

Vector Form of Green's Theorem

As we have seen, we can write Green's Theorem in the form

$$\oint_{\partial \mathcal{D}} \mathbf{F} \cdot d\mathbf{r} = \iint_{\mathcal{D}} \text{curl}_z(\mathbf{F})\, dA$$

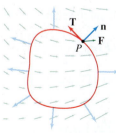

FIGURE 15 The flux of $\mathbf{F}$ is the integral of the normal component $\mathbf{F} \cdot \mathbf{n}$ around the curve.

This allows us to calculate the integral of the tangential component of $\mathbf{F}$ around the curve C, which is the circulation, using a double integral rather than doing it directly. But what if instead, we would like to compute the flux of the vector field $\mathbf{F}$ across the curve $\mathbf{F}$, as in Figure 15? That is, we want to integrate the normal component of $\mathbf{F}$ around the curve C, as we discussed at the end of Section 16.2.

There, we said that if the curve is parametrized by $\mathbf{r}(t) = \langle x(t), y(t) \rangle$ for $a \le t \le b$, such that $\mathbf{r}'(t) \ne \mathbf{0}$, the unit tangent vector is given by $\mathbf{T} = \dfrac{\mathbf{r}'(t)}{\|\mathbf{r}'(t)\|} = \left\langle \dfrac{x'(t)}{\|\mathbf{r}'(t)\|}, \dfrac{y'(t)}{\|\mathbf{r}'(t)\|} \right\rangle$

and the outward unit normal vector is given by $\mathbf{n}(t) = \left\langle \dfrac{y'(t)}{\|\mathbf{r}'(t)\|}, \dfrac{-x'(t)}{\|\mathbf{r}'(t)\|} \right\rangle$ since its dot product with $\mathbf{T}$ is 0 and $\mathbf{n}$ points to the right as we travel around the curve.

Thus, the flux of $\mathbf{F}$ out of C is given by

$$
\oint_C \mathbf{F} \cdot \mathbf{n}\, ds = \int_a^b (\mathbf{F} \cdot \mathbf{n})(t)\|\mathbf{r}'(t)\|\, dt
$$

$$
= \int_a^b \left[\frac{F_1\, y'(t)}{\|\mathbf{r}(t)\|} - \frac{F_2\, x'(t)}{\|\mathbf{r}(t)\|} \right] \|\mathbf{r}(t)\|\, dt
$$

$$
= \int_a^b F_1 y'(t)\, dt - F_2 x'(t)\, dt
$$

$$
= \int_a^b F_1\, dy - F_2\, dx
$$

This is in a form to which we can apply Green's Theorem, but we have switched the roles of F_1 and F_2 and added a negative sign to the second term. So, Green's Theorem gives us

$$
\int_{\partial D} F_1\, dy - F_2\, dx = \iint_D \left(\frac{\partial F_1}{\partial x} + \frac{\partial F_2}{\partial y} \right) dA = \iint_D \operatorname{div}(\mathbf{F})\, dA
$$

Thus, we obtain the Vector Form of Green's Theorem for flux.

$$
\boxed{\;\oint_{\partial D} \mathbf{F} \cdot \mathbf{n}\, ds = \iint_D \operatorname{div}(\mathbf{F})\, dA\;}
\qquad \boxed{13}
$$

Note how both curl and divergence make appearances in the two different forms of Green's Theorem.

■ **EXAMPLE 5** Calculate the flux of $\mathbf{F}(x, y) = \langle x^3, y^3 + y \rangle$ out of the unit circle.

Solution We find $\operatorname{div}\mathbf{F} = \dfrac{\partial F_1}{\partial x} + \dfrac{\partial F_2}{\partial y} = 3x^2 + 3y^2 + 1$. Therefore, the flux of $\mathbf{F}$ out of the unit circle is given by

$$
\text{Flux} = \iint_D \operatorname{div}(\mathbf{F})\, dA = \iint_D (3x^2 + 3y^2 + 1)\, dA
$$

Converting to polar coordinates, we have

$$
\text{Flux} = \int_0^{2\pi} \int_0^1 (3r^2 + 1) r\, dr\, d\theta = \int_0^{2\pi} \int_0^1 (3r^3 + r)\, dr\, d\theta
$$

$$
= 2\pi \left(\frac{3r^4}{4} + \frac{r^2}{2} \right) \Big|_0^1 = \frac{5\pi}{2} \qquad ■
$$

17.1 SUMMARY

- We have two notations for the line integral of a vector field on the plane

$$
\int_C \mathbf{F} \cdot d\mathbf{r} \qquad \text{and} \qquad \int_C F_1\, dx + F_2\, dy
$$

- ∂D denotes the boundary of D with its boundary orientation (Figure 16).
- Green's Theorem:

$$
\oint_{\partial D} F_1\, dx + F_2\, dy = \iint_D \left(\frac{\partial F_2}{\partial x} - \frac{\partial F_1}{\partial y} \right) dA
$$

or

$$
\oint_{\partial D} \mathbf{F} \cdot d\mathbf{r} = \iint_D \operatorname{curl}_z(\mathbf{F})\, dA
$$

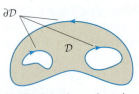

FIGURE 16 The boundary orientation is chosen so that the region lies to your left as you walk along the curve.

• Formulas for the area of the region $\mathcal{D}$ enclosed by $\mathcal{C}$:

$$\text{Area}(\mathcal{D}) = \oint_{\mathcal{C}} x \, dy = \oint_{\mathcal{C}} -y \, dx = \frac{1}{2} \oint_{\mathcal{C}} x \, dy - y \, dx$$

• The quantity

$$\text{curl}_z(\mathbf{F}) = \frac{\partial F_2}{\partial x} - \frac{\partial F_1}{\partial y}$$

is interpreted as *circulation per unit area*. If $\mathcal{D}$ is a small domain with boundary $\mathcal{C}$, then for any $P \in \mathcal{D}$,

$$\oint_{\mathcal{C}} F_1 \, dx + F_2 \, dy \approx \text{curl}_z(\mathbf{F})(P) \cdot \text{Area}(\mathcal{D})$$

• Vector Form of Green's Theorem:

$$\oint_{\partial \mathcal{D}} \mathbf{F} \cdot \mathbf{n} \, ds = \iint_{\mathcal{D}} \text{div}(\mathbf{F}) \, dA$$

17.1 EXERCISES

Preliminary Questions

1. Which vector field $\mathbf{F}$ is being integrated in the line integral $\oint x^2 \, dy - e^y \, dx$?

2. Draw a domain in the shape of an ellipse and indicate with an arrow the boundary orientation of the boundary curve. Do the same for the annulus (the region between two concentric circles).

3. The circulation of a conservative vector field around a closed curve is zero. Is this fact consistent with Green's Theorem? Explain.

4. Indicate which of the following vector fields possess this property: For every simple closed curve $\mathcal{C}$, $\oint_{\mathcal{C}} \mathbf{F} \cdot d\mathbf{r}$ is equal to the area enclosed by $\mathcal{C}$.

(a) $\mathbf{F}(x, y) = \langle -y, 0 \rangle$

(b) $\mathbf{F}(x, y) = \langle x, y \rangle$

(c) $\mathbf{F}(x, y) = \langle \sin(x^2), x + e^{y^2} \rangle$

Exercises

1. Verify Green's Theorem for the line integral $\oint_{\mathcal{C}} xy \, dx + y \, dy$, where $\mathcal{C}$ is the unit circle, oriented counterclockwise.

2. Let $I = \oint_{\mathcal{C}} \mathbf{F} \cdot d\mathbf{r}$, where $\mathbf{F}(x, y) = \langle y + \sin x^2, x^2 + e^{y^2} \rangle$ and $\mathcal{C}$ is the circle of radius 4 centered at the origin.

(a) Which is easier, evaluating I directly or using Green's Theorem?

(b) Evaluate I using the easier method.

In Exercises 3–10, use Green's Theorem to evaluate the line integral. Orient the curve counterclockwise unless otherwise indicated.

3. $\oint_{\mathcal{C}} y^2 \, dx + x^2 \, dy$, where $\mathcal{C}$ is the boundary of the unit square $0 \le x \le 1, 0 \le y \le 1$

4. $\oint_{\mathcal{C}} e^{2x+y} \, dx + e^{-y} \, dy$, where $\mathcal{C}$ is the triangle with vertices $(0, 0)$, $(1, 0)$, and $(1, 1)$

5. $\oint_{\mathcal{C}} x^2 y \, dx$, where $\mathcal{C}$ is the unit circle centered at the origin

6. $\oint_{\mathcal{C}} \mathbf{F} \cdot d\mathbf{r}$, where $\mathbf{F}(x, y) = \langle x + y, x^2 - y \rangle$ and $\mathcal{C}$ is the boundary of the region enclosed by $y = x^2$ and $y = \sqrt{x}$ for $0 \le x \le 1$

7. $\oint_{\mathcal{C}} \mathbf{F} \cdot d\mathbf{r}$, where $\mathbf{F}(x, y) = \langle x^2, x^2 \rangle$ and $\mathcal{C}$ consists of the arcs $y = x^2$ and $y = x$ for $0 \le x \le 1$

8. $\oint_{\mathcal{C}} (\ln x + y) \, dx - x^2 \, dy$, where $\mathcal{C}$ is the rectangle with vertices $(1, 1)$, $(3, 1)$, $(1, 4)$, and $(3, 4)$

9. The line integral of $\mathbf{F}(x, y) = \langle e^{x+y}, e^{x-y} \rangle$ along the curve (oriented clockwise) consisting of the line segments by joining the points $(0, 0)$, $(2, 2)$, $(4, 2)$, $(2, 0)$, and back to $(0, 0)$ (Note the orientation.)

10. $\int_{\mathcal{C}} xy \, dx + (x^2 + x) \, dy$, where $\mathcal{C}$ is the path in Figure 17

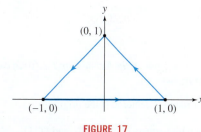

FIGURE 17

11. Let $\mathbf{F}(x, y) = \langle 2xe^y, x + x^2e^y \rangle$ and let C be the quarter-circle path from A to B in Figure 18. Evaluate $I = \oint_C \mathbf{F} \cdot d\mathbf{r}$ as follows:

(a) Find a function $f(x, y)$ such that $\mathbf{F} = \mathbf{G} + \nabla f$, where $\mathbf{G} = \langle 0, x \rangle$.

(b) Show that the line integrals of $\mathbf{G}$ along the segments $\overline{OA}$ and $\overline{OB}$ are zero.

(c) Evaluate I. *Hint:* Use Green's Theorem to show that

$$I = f(B) - f(A) + 4\pi$$

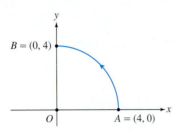

FIGURE 18

12. Compute the line integral of $\mathbf{F}(x, y) = \langle x^3, 4x \rangle$ along the path from A to B in Figure 19. To save work, use Green's Theorem to relate this line integral to the line integral along the vertical path from B to A.

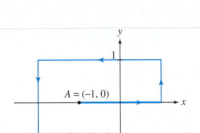

FIGURE 19

13. Evaluate $I = \int_C (\sin x + y)\, dx + (3x + y)\, dy$ for the nonclosed path $ABCD$ in Figure 20. Use the method of Exercise 12.

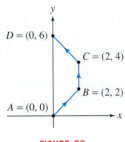

FIGURE 20

In Exercises 14–17, use one of the formulas in Eq. (6) to calculate the area of the given region.

14. The circle of radius 3 centered at the origin

15. The triangle with vertices $(0, 0)$, $(1, 0)$, and $(1, 1)$

16. The region between the x-axis and the cycloid parametrized by $\mathbf{r}(t) = \langle t - \sin t, 1 - \cos t \rangle$ for $0 \le t \le 2\pi$ (Figure 21)

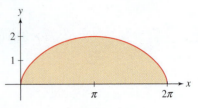

FIGURE 21 Cycloid.

17. The region between the graph of $y = x^2$ and the x-axis for $0 \le x \le 2$

18. A square with vertices $(1, 1)$, $(-1, 1)$, $(-1, -1)$, and $(1, -1)$ has area 4. Calculate this area three times using the formulas in Eq. (6).

19. Let $x^3 + y^3 = 3xy$ be the **folium of Descartes** (Figure 22).

(a) Show that the folium has a parametrization in terms of $t = y/x$ given by

$$x = \frac{3t}{1 + t^3}, \qquad y = \frac{3t^2}{1 + t^3} \qquad (-\infty < t < \infty) \quad (t \ne -1)$$

(b) Show that

$$x\, dy - y\, dx = \frac{9t^2}{(1 + t^3)^2}\, dt$$

Hint: By the Quotient Rule,

$$x^2 d\left(\frac{y}{x}\right) = x\, dy - y\, dx$$

(c) Find the area of the loop of the folium. *Hint:* The limits of integration are 0 and ∞.

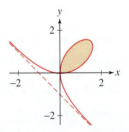

FIGURE 22 Folium of Descartes.

20. Find a parametrization of the lemniscate $(x^2 + y^2)^2 = xy$ (see Figure 23) by using $t = y/x$ as a parameter (see Exercise 19). Then use Eq. (6) to find the area of one loop of the lemniscate.

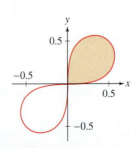

FIGURE 23 Lemniscate.

21. The Centroid via Boundary Measurements The centroid (see Section 15.5) of a domain $\mathcal{D}$ enclosed by a simple closed curve $\mathcal{C}$ is the point with coordinates $(\overline{x}, \overline{y}) = (M_y/M, M_x/M)$, where M is the area of $\mathcal{D}$ and the moments are defined by

$$M_x = \iint_{\mathcal{D}} y \, dA, \qquad M_y = \iint_{\mathcal{D}} x \, dA$$

Show that $M_x = \oint_{\mathcal{C}} xy \, dy$. Find a similar expression for M_y.

22. Use the result of Exercise 21 to compute the moments of the semicircle $x^2 + y^2 = R^2$, $y \geq 0$ as line integrals. Verify that the centroid is $(0, 4R/(3\pi))$.

23. Let $\mathcal{C}_R$ be the circle of radius R centered at the origin. Use the general form of Green's Theorem to determine $\oint_{\mathcal{C}_2} \mathbf{F} \cdot d\mathbf{r}$, where $\mathbf{F}$ is a vector field such that $\oint_{\mathcal{C}_1} \mathbf{F} \cdot d\mathbf{r} = 9$ and $\dfrac{\partial F_2}{\partial x} - \dfrac{\partial F_1}{\partial y} = x^2 + y^2$ for (x, y) in the annulus $1 \leq x^2 + y^2 \leq 4$.

24. Referring to Figure 24, suppose that $\oint_{\mathcal{C}_2} \mathbf{F} \cdot d\mathbf{r} = 12$. Use Green's Theorem to determine $\oint_{\mathcal{C}_1} \mathbf{F} \cdot d\mathbf{r}$, assuming that $\dfrac{\partial F_2}{\partial y} - \dfrac{\partial F_1}{\partial y} = -3$ in $\mathcal{D}$.

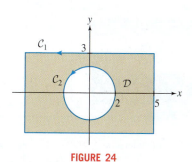

FIGURE 24

25. Referring to Figure 25, suppose that

$$\oint_{\mathcal{C}_2} \mathbf{F} \cdot d\mathbf{r} = 3\pi, \qquad \oint_{\mathcal{C}_3} \mathbf{F} \cdot d\mathbf{r} = 4\pi$$

Use Green's Theorem to determine the circulation of $\mathbf{F}$ around $\mathcal{C}_1$, assuming that $\dfrac{\partial F_2}{\partial x} - \dfrac{\partial F_1}{\partial y} = 9$ on the shaded region.

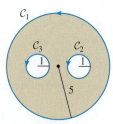

FIGURE 25

26. Let $\mathbf{F}$ be the vortex vector field

$$\mathbf{F}(x, y) = \left\langle \frac{-y}{x^2 + y^2}, \frac{x}{x^2 + y^2} \right\rangle$$

In Section 16.3, we verified that $\displaystyle\int_{\mathcal{C}_R} \mathbf{F} \cdot d\mathbf{r} = 2\pi$, where $\mathcal{C}_R$ is the circle of radius R centered at the origin. Prove that $\displaystyle\oint_{\mathcal{C}} \mathbf{F} \cdot d\mathbf{r} = 2\pi$ for any simple closed curve $\mathcal{C}$ whose interior contains the origin (Figure 26). *Hint:* Apply the general form of Green's Theorem to the domain between $\mathcal{C}$ and $\mathcal{C}_R$, where R is so small that $\mathcal{C}_R$ is contained in $\mathcal{C}$.

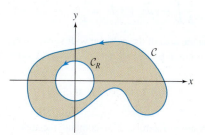

FIGURE 26

In Exercises 27–30, we refer to the integrand that occurs in Green's Theorem and that appears as

$$\operatorname{curl}_z(\mathbf{F}) = \frac{\partial F_2}{\partial x} - \frac{\partial F_1}{\partial y}$$

27. For the vector fields (A)–(D) in Figure 27, state whether the curl_z at the origin appears to be positive, negative, or zero.

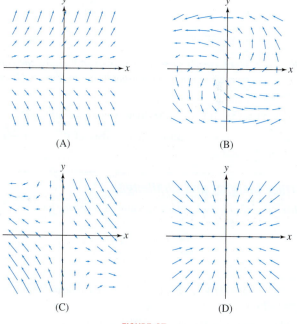

(A)   (B)

(C)   (D)

FIGURE 27

28. Estimate the circulation of a vector field $\mathbf{F}$ around a circle of radius $R = 0.1$, assuming that $\operatorname{curl}_z(\mathbf{F})$ takes the value 4 at the center of the circle.

29. Estimate $\oint_{\mathcal{C}} \mathbf{F} \cdot d\mathbf{r}$, where $\mathbf{F}(x, y) = \left\langle x + 0.1y^2, y - 0.1x^2 \right\rangle$ and $\mathcal{C}$ encloses a small region of area 0.25 containing the point $P = (1, 1)$.

30. Let $\mathbf{F}$ be the velocity field. Estimate the circulation of $\mathbf{F}$ around a circle of radius $R = 0.05$ with center P, assuming that $\operatorname{curl}_z(\mathbf{F})(P) = -3$. In which direction would a small paddle placed at P rotate? How fast would it rotate (in radians per second) if $\mathbf{F}$ is expressed in meters per second?

31. Let C_R be the circle of radius R centered at the origin. Use Green's Theorem to find the value of R that maximizes $\oint_{C_R} y^3 \, dx + x \, dy$.

32. Area of a Polygon Green's Theorem leads to a convenient formula for the area of a polygon.

(a) Let C be the line segment joining (x_1, y_1) to (x_2, y_2). Show that

$$\frac{1}{2} \int_C -y \, dx + x \, dy = \frac{1}{2}(x_1 y_2 - x_2 y_1)$$

(b) Prove that the area of the polygon with vertices (x_1, y_1), (x_2, y_2), $\ldots, (x_n, y_n)$ is equal [where we set $(x_{n+1}, y_{n+1}) = (x_1, y_1)$] to

$$\frac{1}{2} \sum_{i=1}^{n} (x_i y_{i+1} - x_{i+1} y_i)$$

33. Use the result of Exercise 32 to compute the areas of the polygons in Figure 28. Check your result for the area of the triangle in (A) using geometry.

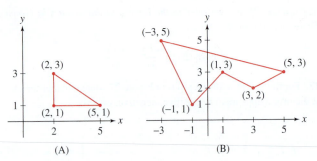

(A)   (B)

FIGURE 28

In Exercises 34–39, compute the flux $\oint \mathbf{F} \cdot \mathbf{n} \, ds$ of $\mathbf{F}$ across the curve C for the given vector field and curve using the vector form of Green's Theorem.

34. $\mathbf{F}(x, y) = \langle 3x, 2y \rangle$ across the circle given by $x^2 + y^2 = 9$

35. $\mathbf{F}(x, y) = \langle xy, x - y \rangle$ across the boundary of the square $-1 \le x \le 1, -1 \le y \le 1$

36. $\mathbf{F}(x, y) = \langle x^2, y^2 \rangle$ across the boundary of the triangle with vertices $(0, 0)$, $(1, 0)$, and $(0, 1)$

37. $\mathbf{F}(x, y) = \langle 2x + y^3, 3y - x^4 \rangle$ across the unit circle

38. $\mathbf{F}(x, y) = \langle \cos y, \sin y \rangle$ across the boundary of the square $0 \le x \le 2, 0 \le y \le \frac{\pi}{2}$

39. $\mathbf{F}(x, y) = \langle xy^2 + 2x, x^2 y - 2y \rangle$ across the simple closed curve that is the boundary of the half-disk given by $x^2 + y^2 \le 3, y \ge 0$

40. If $\mathbf{v}$ is the velocity field of a fluid, the flux of $\mathbf{v}$ across C is equal to the flow rate (amount of fluid flowing across C in square meters per second). Find the flow rate across the circle of radius 2 centered at the origin if $\text{div}(\mathbf{v}) = x^2$.

41. A buffalo (Figure 29) stampede is described by a velocity vector field $\mathbf{F} = \langle xy - y^3, x^2 + y \rangle$ km/h in the region $\mathcal{D}$ defined by $2 \le x \le 3$, $2 \le y \le 3$ in units of kilometers (Figure 30). Assuming a density is $\rho = 500$ buffalo per square kilometer, use Eq. (13) to determine the net number of buffalo leaving or entering $\mathcal{D}$ per minute (equal to ρ times the flux of $\mathbf{F}$ across the boundary of $\mathcal{D}$).

FIGURE 29 Buffalo stampede. (C. K. Lorenz/Science Source)

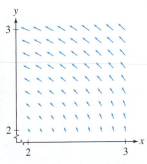

FIGURE 30 The vector field $\mathbf{F} = \langle xy - y^3, x^2 + y \rangle$.

Further Insights and Challenges

*In Exercises 42–45, the **Laplace operator** Δ is defined by*

$$\Delta \varphi = \frac{\partial^2 \varphi}{\partial x^2} + \frac{\partial^2 \varphi}{\partial y^2} \qquad \boxed{14}$$

For any vector field $\mathbf{F} = \langle F_1, F_2 \rangle$, define the conjugate vector field $\mathbf{F}^* = \langle -F_2, F_1 \rangle$.

42. Show that if $\mathbf{F} = \nabla \varphi$, then $\text{curl}_z(\mathbf{F}^*) = \Delta \varphi$.

43. Let $\mathbf{n}$ be the outward-pointing unit normal vector to a simple closed curve C. The **normal derivative** of a function φ, denoted $\dfrac{\partial \varphi}{\partial \mathbf{n}}$, is the directional derivative $D_{\mathbf{n}}(\varphi) = \nabla \varphi \cdot \mathbf{n}$. Prove that

$$\oint_C \frac{\partial \varphi}{\partial \mathbf{n}} \, ds = \iint_{\mathcal{D}} \Delta \varphi \, dA$$

where $\mathcal{D}$ is the domain enclosed by a simple closed curve C. *Hint:* Let $\mathbf{F} = \nabla \varphi$. Show that $\dfrac{\partial \varphi}{\partial \mathbf{n}} = \mathbf{F}^* \cdot \mathbf{T}$, where $\mathbf{T}$ is the unit tangent vector, and apply Green's Theorem.

44. Let $P = (a, b)$ and let C_r be the circle of radius r centered at P. The average value of a continuous function φ on C_r is defined as the integral

$$I_\varphi(r) = \frac{1}{2\pi} \int_0^{2\pi} \varphi(a + r \cos \theta, b + r \sin \theta) \, d\theta$$

(a) Show that

$$\frac{\partial \varphi}{\partial \mathbf{n}}(a + r \cos \theta, b + r \sin \theta)$$

$$= \frac{\partial \varphi}{\partial r}(a + r \cos \theta, b + r \sin \theta)$$

(b) Use differentiation under the integral sign to prove that

$$\frac{d}{dr} I_\varphi(r) = \frac{1}{2\pi r} \int_{C_r} \frac{\partial \varphi}{\partial \mathbf{n}} \, ds$$

(c) Use Exercise 43 to conclude that

$$\frac{d}{dr} I_\varphi(r) = \frac{1}{2\pi r} \iint_{\mathcal{D}(r)} \Delta \varphi \, dA$$

where $\mathcal{D}(r)$ is the interior of C_r.

45. Prove that $m(r) \leq I_\varphi(r) \leq M(r)$, where $m(r)$ and $M(r)$ are the minimum and maximum values of φ on C_r. Then use the continuity of φ to prove that $\lim\limits_{r \to 0} I_\varphi(r) = \varphi(P)$.

In Exercises 46 and 47, let $\mathcal{D}$ be the region bounded by a simple closed curve C. A function $\varphi(x, y)$ on $\mathcal{D}$ (whose second-order partial derivatives exist and are continuous) is called **harmonic** *if $\Delta\varphi = 0$, where $\Delta\varphi$ is the Laplace operator defined in Eq. (14).*

46. Use the results of Exercises 44 and 45 to prove the **mean-value property** of harmonic functions: If φ is harmonic, then $I_\varphi(r) = \varphi(P)$ for all r.

47. Show that $f(x, y) = x^2 - y^2$ is harmonic. Verify the mean-value property for $f(x, y)$ directly [expand $f(a + r\cos\theta, b + r\sin\theta)$ as a function of θ and compute $I_\varphi(r)$]. Show that $x^2 + y^2$ is not harmonic and does not satisfy the mean-value property.

17.2 Stokes' Theorem

Stokes' Theorem is an extension of Green's Theorem to three dimensions in which circulation is related to a surface integral over a surface in $\mathbf{R}^3$ (rather than to a double integral over a region in the plane). In order to state it, we introduce some definitions and terminology.

Figure 1 shows three surfaces with different types of boundaries. The boundary of a surface S is denoted ∂S. Observe that the boundary in (A) is a single, simple closed curve and the boundary in (B) consists of three simple closed curves. The surface in (C) is called a **closed surface** because its boundary is empty. In this case, we write $\partial S = \varnothing$.

(A) Boundary consists of a single closed curve.	(B) Boundary consists of three closed curves.	(C) Closed surface (the boundary is empty)

FIGURE 1 Surfaces and their boundaries.

Recall from Section 16.5 that an orientation of a surface is a continuously varying choice of unit normal vector at each point of a surface S. When S is oriented, we can specify an orientation of ∂S, called the **boundary orientation**. Imagine that you are a unit normal vector walking along the boundary curve with your head at the head end of the vector and your feet at the tail end. The boundary orientation is the direction for which the surface is on your left as you walk. For example, the boundary of the surface in Figure 2 consists of two curves, C_1 and C_2. In (A), the normal vector points to the outside. The woman (representing the normal vector) is walking along C_1 and has the surface to her left, so she is walking in the positive direction. The curve C_2 is oriented in the opposite direction because she would have to walk along C_2 in that direction to keep the surface on her left. The boundary orientations in (B) are reversed because the opposite normal has been selected to orient the surface.

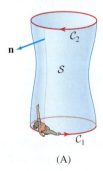

FIGURE 2 The orientation of the boundary ∂S for each of the two possible orientations of the surface S.

(A)　　　　　(B)

In the next theorem, we assume that S is an oriented surface with parametrization $G : \mathcal{D} \to S$, where $\mathcal{D}$ is a domain in the plane bounded by smooth, simple closed curves, and G is one-to-one and regular, except possibly on the boundary of $\mathcal{D}$. More generally, S may be a finite union of surfaces of this type. The surfaces in applications we consider, such as spheres, cubes, and graphs of functions, satisfy these conditions.

THEOREM 1 Stokes' Theorem   Let S be a surface as described above, and let $\mathbf{F}$ be a vector field whose components have continuous partial derivatives on an open region containing S.

$$\oint_{\partial S} \mathbf{F} \cdot d\mathbf{r} = \iint_S \operatorname{curl}(\mathbf{F}) \cdot d\mathbf{S}$$ **1**

The integral on the left is defined relative to the boundary orientation of ∂S. If S is closed (i.e., ∂S is empty), then the surface integral on the right is zero.

The curl measures the extent to which $\mathbf{F}$ fails to be conservative. If $\mathbf{F}$ is conservative, then $\operatorname{curl}(\mathbf{F}) = \mathbf{0}$ and Stokes' Theorem merely confirms what we already know: The circulation of a conservative vector field around a closed path is zero.

Stokes' Theorem is often written in the form

$$\oint_{\partial S} \mathbf{F} \cdot d\mathbf{r} = \iint_S (\nabla \times \mathbf{F}) \cdot d\mathbf{S}$$

Again, we see the analogy with the Fundamental Theorem of Calculus. A double integral over a surface of a derivative, in this case the curl, yields a single integral over the boundary of the surface.

Proof   The left side of Eq. (1) is equal to a sum over the components of $\mathbf{F}$:

$$\oint_C \mathbf{F} \cdot d\mathbf{r} = \oint_C F_1 \, dx + F_2 \, dy + F_3 \, dz$$ **2**

Considering $\mathbf{F} = F_1 \mathbf{i} + F_2 \mathbf{j} + F_3 \mathbf{k}$, and using the additivity property of the curl operator, we have $\operatorname{curl}(\mathbf{F}) = \operatorname{curl}(F_1 \mathbf{i}) + \operatorname{curl}(F_2 \mathbf{j}) + \operatorname{curl}(F_3 \mathbf{k})$ and therefore:

$$\iint_S \operatorname{curl}(\mathbf{F}) \cdot d\mathbf{S} = \iint_S \operatorname{curl}(F_1 \mathbf{i}) \cdot d\mathbf{S} + \iint_S \operatorname{curl}(F_2 \mathbf{j}) \cdot d\mathbf{S} + \iint_S \operatorname{curl}(F_3 \mathbf{k}) \cdot d\mathbf{S}$$ **3**

The proof consists of showing that the F_1, F_2, and F_3 terms in Eqs. (2) and (3) are separately equal.

Because a complete proof is quite technical, we will prove it under the simplifying assumption that S is the graph of a function $z = f(x, y)$ lying over a domain $\mathcal{D}$ in the xy-plane. Furthermore, we will carry the details only for the F_1 terms. The calculation for F_2-components is similar, and we leave as an exercise the equality of the F_3 terms (Exercise 35). Thus, we shall prove that

$$\oint_C F_1 \, dx = \iint_S \operatorname{curl}(F_1(x, y, z)\mathbf{i}) \cdot d\mathbf{S}$$ **4**

Orient S with an upward-pointing normal as in Figure 3, and let $C = \partial S$ be the boundary curve with orientation determined by the orientation of S. Let C_0 be the boundary of $\mathcal{D}$ in the xy-plane, and let $\mathbf{r}_0(t) = \langle x(t), y(t) \rangle$ (for $a \le t \le b$) be a counterclockwise parametrization of C_0 as in Figure 3. The boundary curve C projects onto C_0, so C has parametrization

$$\mathbf{r}(t) = \langle x(t), y(t), f(x(t), y(t)) \rangle$$

and thus,

$$\oint_C F_1(x, y, z) \, dx = \int_a^b F_1\big(x(t), y(t), f(x(t), y(t))\big) \frac{dx}{dt} \, dt$$

The integral on the right is precisely the integral we obtain by integrating $F_1\big(x, y, f(x, y)\big) \, dx$ over the curve C_0 in the plane $\mathbf{R}^2$. In other words,

$$\oint_C F_1(x, y, z) \, dx = \oint_{C_0} F_1\big(x, y, f(x, y)\big) \, dx$$

By Green's Theorem applied to the integral on the right, we get

$$\oint_C F_1(x, y, z) \, dx = -\iint_{\mathcal{D}} \frac{\partial}{\partial y} F_1(x, y, f(x, y)) \, dA$$

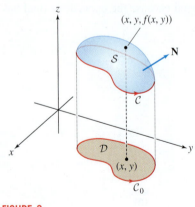

FIGURE 3

By the Chain Rule,

$$\frac{\partial}{\partial y} F_1\big(x, y, f(x, y)\big) = F_{1y}\big(x, y, f(x, y)\big) + F_{1z}\big(x, y, f(x, y)\big) f_y(x, y)$$

so finally we obtain

$$\oint_C F_1 \, dx = -\iint_{\mathcal{D}} \Big(F_{1y}\big(x, y, f(x, y)\big) + F_{1z}\big(x, y, f(x, y)\big) f_y(x, y) \Big) \, dA \qquad \boxed{5}$$

> **← REMINDER** *Calculating a surface integral:*
>
> $$\iint_{\mathcal{S}} \mathbf{F} \cdot d\mathbf{S} = \iint_{\mathcal{D}} \mathbf{F}(u, v) \cdot \mathbf{N}(u, v) \, du \, dv$$
>
> *If $\mathcal{S}$ is a graph $z = f(x, y)$, parametrized by $G(x, y) = (x, y, f(x, y))$, then*
>
> $$\mathbf{N}(x, y) = \langle -f_x(x, y), -f_y(x, y), 1 \rangle$$

To finish the proof, we compute the surface integral of $\text{curl}(F_1\mathbf{i})$ using the parametrization $G(x, y) = (x, y, f(x, y))$ of $\mathcal{S}$:

$$\mathbf{N} = \langle -f_x(x, y), -f_y(x, y), 1 \rangle \qquad \text{(upward-pointing normal)}$$

$$\text{curl}(F_1\mathbf{i}) \cdot \mathbf{N} = \langle 0, F_{1z}, -F_{1y} \rangle \cdot \langle -f_x(x, y), -f_y(x, y), 1 \rangle$$

$$= -F_{1z}\big(x, y, f(x, y)\big) f_y(x, y) - F_{1y}\big(x, y, f(x, y)\big)$$

$$\iint_{\mathcal{S}} \text{curl}(F_1\mathbf{i}) \cdot d\mathbf{S} = -\iint_{\mathcal{D}} \Big(F_{1z}(x, y, z) f_y(x, y) + F_{1y}\big(x, y, f(x, y)\big) \Big) \, dA \qquad \boxed{6}$$

The right-hand sides of Eq. (5) and Eq. (6) are equal. This proves Eq. (4). ∎

■ **EXAMPLE 1** **Verifying Stokes' Theorem** Verify Stokes' Theorem for

$$\mathbf{F}(x, y, z) = \langle -y, 2x, x + z \rangle$$

and the upper hemisphere with outward-pointing normal vectors (Figure 4):

$$\mathcal{S} = \{ (x, y, z) : x^2 + y^2 + z^2 = 1, z \geq 0 \}$$

Solution We will show that both the line integral and the surface integral in Stokes' Theorem are equal to 3π.

Step 1. Compute the line integral around the boundary curve.
The boundary of $\mathcal{S}$ is the unit circle oriented in the counterclockwise direction with parametrization $\mathbf{r}(t) = \langle \cos t, \sin t, 0 \rangle$. Thus,

$$\mathbf{r}'(t) = \langle -\sin t, \cos t, 0 \rangle$$

$$\mathbf{F}(\mathbf{r}(t)) = \langle -\sin t, 2\cos t, \cos t \rangle$$

$$\mathbf{F}(\mathbf{r}(t)) \cdot \mathbf{r}'(t) = \langle -\sin t, 2\cos t, \cos t \rangle \cdot \langle -\sin t, \cos t, 0 \rangle$$

$$= \sin^2 t + 2\cos^2 t = 1 + \cos^2 t$$

> **← REMINDER** *In Eq. (7), we use*
>
> $$\int_0^{2\pi} \cos^2 t \, dt = \int_0^{2\pi} \frac{1 + \cos 2t}{2} \, dt = \pi$$

$$\oint_{\partial \mathcal{S}} \mathbf{F} \cdot d\mathbf{r} = \int_0^{2\pi} (1 + \cos^2 t) \, dt = 2\pi + \pi = \boxed{3\pi} \qquad \boxed{7}$$

Step 2. Compute the curl.

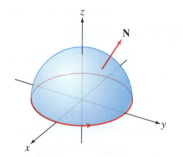

FIGURE 4 Upper hemisphere with oriented boundary.

$$\text{curl}(\mathbf{F}) = \begin{vmatrix} \mathbf{i} & \mathbf{j} & \mathbf{k} \\ \dfrac{\partial}{\partial x} & \dfrac{\partial}{\partial y} & \dfrac{\partial}{\partial z} \\ -y & 2x & x + z \end{vmatrix}$$

$$= \left(\frac{\partial}{\partial y}(x + z) - \frac{\partial}{\partial z}2x \right)\mathbf{i} - \left(\frac{\partial}{\partial x}(x + z) - \frac{\partial}{\partial z}(-y) \right)\mathbf{j}$$

$$+ \left(\frac{\partial}{\partial x}2x - \frac{\partial}{\partial y}(-y) \right)\mathbf{k}$$

$$= \langle 0, -1, 3 \rangle$$

◄┄ **REMINDER** *Stokes' Theorem states*

$$\oint_{\partial S} \mathbf{F} \cdot d\mathbf{r} = \iint_S \text{curl}(\mathbf{F}) \cdot d\mathbf{S}$$

Step 3. Compute the surface integral of the curl.

We parametrize the hemisphere using spherical coordinates:

$$G(\theta, \phi) = (\cos\theta \sin\phi, \sin\theta \sin\phi, \cos\phi)$$

By Eq. (1) of Section 16.4, the outward-pointing normal vector is

$$\mathbf{N} = \sin\phi \, \langle \cos\theta \sin\phi, \sin\theta \sin\phi, \cos\phi \rangle$$

Therefore,

$$\text{curl}(\mathbf{F}) \cdot \mathbf{N} = \sin\phi \, \langle 0, -1, 3 \rangle \cdot \langle \cos\theta \sin\phi, \sin\theta \sin\phi, \cos\phi \rangle$$

$$= -\sin\theta \sin^2\phi + 3\cos\phi \sin\phi$$

The upper hemisphere $\mathcal{S}$ corresponds to $0 \le \phi \le \frac{\pi}{2}$, so

$$\iint_S \text{curl}(\mathbf{F}) \cdot d\mathbf{S} = \int_{\phi=0}^{\pi/2} \int_{\theta=0}^{2\pi} (-\sin\theta \sin^2\phi + 3\cos\phi \sin\phi) \, d\theta \, d\phi$$

$$= 0 + 2\pi \int_{\phi=0}^{\pi/2} 3\cos\phi \sin\phi \, d\phi = 2\pi \left(\frac{3}{2}\sin^2\phi \right) \Big|_{\phi=0}^{\pi/2}$$

$$= \boxed{3\pi} \quad\blacksquare$$

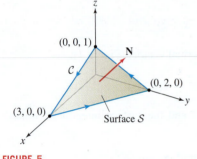

(0, 0, 1)

$\mathbf{N}$

$\mathcal{C}$

(0, 2, 0)

(3, 0, 0)

Surface $\mathcal{S}$

FIGURE 5

■ **EXAMPLE 2** Use Stokes' Theorem to show that $\oint_{\mathcal{C}} \mathbf{F} \cdot d\mathbf{r} = 0$, where

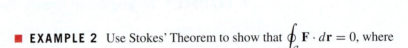

$$\mathbf{F}(x, y, z) = \langle \sin(x^2), e^{y^2} + x^2, z^4 + 2x^2 \rangle$$

and $\mathcal{C}$ is the boundary of the triangle in Figure 5 with the indicated orientation.

Solution Note that if we wanted to evaluate $\oint \mathbf{F} \cdot d\mathbf{r}$ directly, we would need to parametrize each of the three edges of $\mathcal{C}$, and do three integrals. Instead, we apply Stokes' Theorem:

$$\oint_{\mathcal{C}} \mathbf{F} \cdot d\mathbf{r} = \iint_S \text{curl}(\mathbf{F}) \cdot d\mathbf{S}$$

and show that the integral on the right is zero. We first compute the curl:

$$\text{curl}\left(\langle \sin x^2, e^{y^2} + x^2, z^4 + 2x^2 \rangle \right) = \begin{vmatrix} \mathbf{i} & \mathbf{j} & \mathbf{k} \\ \dfrac{\partial}{\partial x} & \dfrac{\partial}{\partial y} & \dfrac{\partial}{\partial z} \\ \sin x^2 & e^{y^2} + x^2 & z^4 + 2x^2 \end{vmatrix} = \langle 0, -4x, 2x \rangle$$

Now, it turns out (by the author's design) that we can show the surface integral is zero without actually computing it. Referring to Figure 5, we see that $\mathcal{C}$ is the boundary of the triangular surface $\mathcal{S}$ contained in the plane

$$\frac{x}{3} + \frac{y}{2} + z = 1$$

Therefore, $\mathbf{N} = \left\langle \frac{1}{3}, \frac{1}{2}, 1 \right\rangle$ is a normal vector to this plane. But $\mathbf{N}$ and $\text{curl}(\mathbf{F})$ are orthogonal:

$$\text{curl}(\mathbf{F}) \cdot \mathbf{N} = \langle 0, -4x, 2x \rangle \cdot \left\langle \frac{1}{3}, \frac{1}{2}, 1 \right\rangle = -2x + 2x = 0$$

In other words, the normal component of $\text{curl}(\mathbf{F})$ along $\mathcal{S}$ is zero. Since the surface integral of a vector field is equal to the surface integral of the normal component, we conclude that $\iint_S \text{curl}(\mathbf{F}) \cdot d\mathbf{S} = 0$. ■

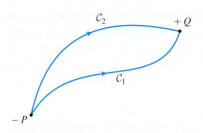

FIGURE 6 Two paths with the same boundary $Q - P$.

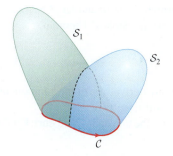

FIGURE 7 Surfaces $\mathcal{S}_1$ and $\mathcal{S}_2$ have the same oriented boundary.

Vector potentials are not unique: If $\mathbf{F} = \mathrm{curl}(\mathbf{A})$, *then* $\mathbf{F} = \mathrm{curl}(\mathbf{A} + \mathbf{B})$ *for any vector field* $\mathbf{B}$ *such that* $\mathrm{curl}(\mathbf{B}) = \mathbf{0}$.

←·· *REMINDER By the **flux** of a vector field through a surface, we mean the surface integral of the vector field.*

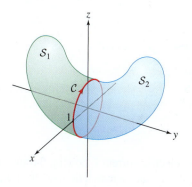

FIGURE 8

CONCEPTUAL INSIGHT Recall that if $\mathbf{F}$ is conservative—that is, $\mathbf{F} = \nabla f$—then for any two paths $\mathcal{C}_1$ and $\mathcal{C}_2$ from P to Q (Figure 6),

$$\int_{\mathcal{C}_1} \mathbf{F} \cdot d\mathbf{r} = \int_{\mathcal{C}_2} \mathbf{F} \cdot d\mathbf{r} = f(Q) - f(P)$$

In other words, the line integral is path-independent. In particular, $\oint_{\mathcal{C}} \mathbf{F} \cdot d\mathbf{r}$ is zero if $\mathcal{C}$ is closed ($P = Q$).

Analogous facts are true for surface integrals when $\mathbf{F} = \mathrm{curl}(\mathbf{A})$. The vector field $\mathbf{A}$ is called a **vector potential** for $\mathbf{F}$. Stokes' Theorem tells us that for any two surfaces $\mathcal{S}_1$ and $\mathcal{S}_2$ with the same oriented boundary $\mathcal{C}$ (Figure 7),

$$\iint_{\mathcal{S}_1} \mathbf{F} \cdot d\mathbf{S} = \iint_{\mathcal{S}_2} \mathbf{F} \cdot d\mathbf{S} = \oint_{\mathcal{C}} \mathbf{A} \cdot d\mathbf{r}$$

In other words, *the surface integral of a vector field with vector potential* $\mathbf{A}$ *is surface-independent*, just as a vector field with a potential function f is path-independent.

If the surface is closed, then the boundary curve is empty and the surface integral is zero:

$$\iint_{\mathcal{S}} \mathbf{F} \cdot d\mathbf{S} = 0 \qquad \text{if} \quad \mathbf{F} = \mathrm{curl}(\mathbf{A}) \text{ and } \mathcal{S} \text{ is closed}$$

THEOREM 2 Surface Independence for Curl Vector Fields If $\mathbf{F} = \mathrm{curl}(\mathbf{A})$, then the flux of $\mathbf{F}$ through a surface $\mathcal{S}$ depends only on the oriented boundary $\partial \mathcal{S}$ and not on the surface itself:

$$\iint_{\mathcal{S}} \mathbf{F} \cdot d\mathbf{S} = \oint_{\partial \mathcal{S}} \mathbf{A} \cdot d\mathbf{r} \qquad \boxed{8}$$

In particular, if $\mathcal{S}$ is *closed* (i.e., $\partial \mathcal{S}$ is empty), then $\iint_{\mathcal{S}} \mathbf{F} \cdot d\mathbf{S} = 0$.

We can write this theorem in our other notation as

$$\iint_{\mathcal{S}} (\nabla \times \mathbf{A}) \cdot d\mathbf{S} = \oint \mathbf{A} \cdot d\mathbf{r}$$

■ **EXAMPLE 3** Let $\mathbf{F} = \mathrm{curl}(\mathbf{A})$, where $\mathbf{A}(x, y, z) = \langle y + z, \sin(xy), e^{xyz} \rangle$. Find the flux of $\mathbf{F}$ outward through the surfaces $\mathcal{S}_1$ and $\mathcal{S}_2$ in Figure 8 whose common boundary $\mathcal{C}$ is the unit circle in the xz-plane.

Solution With $\mathcal{C}$ oriented in the direction of the arrow, $\mathcal{S}_1$ lies to the left, and by Eq. (8),

$$\iint_{\mathcal{S}_1} \mathbf{F} \cdot d\mathbf{S} = \oint_{\mathcal{C}} \mathbf{A} \cdot d\mathbf{r}$$

We shall compute the line integral on the right. The parametrization $\mathbf{r}(t) = \langle \cos t, 0, \sin t \rangle$ traces $\mathcal{C}$ in the direction of the arrow because it begins at $\mathbf{r}(0) = \langle 1, 0, 0 \rangle$ and moves in the direction of $\mathbf{r}(\frac{\pi}{2}) = \langle 0, 0, 1 \rangle$. We have

$$\mathbf{A}(\mathbf{r}(t)) = \langle 0 + \sin t, \sin(0), e^0 \rangle = \langle \sin t, 0, 1 \rangle$$

$$\mathbf{A}(\mathbf{r}(t)) \cdot \mathbf{r}'(t) = \langle \sin t, 0, 1 \rangle \cdot \langle -\sin t, 0, \cos t \rangle = -\sin^2 t + \cos t$$

$$\oint_{\mathcal{C}} \mathbf{A} \cdot d\mathbf{r} = \int_0^{2\pi} (-\sin^2 t + \cos t) \, dt = -\pi$$

We conclude that $\iint_{\mathcal{S}_1} \mathbf{F} \cdot d\mathbf{S} = -\pi$. On the other hand, $\mathcal{S}_2$ lies on the right as you traverse $\mathcal{C}$. Therefore, $\mathcal{S}_2$ has oriented boundary $-\mathcal{C}$, and

$$\iint_{\mathcal{S}_2} \mathbf{F} \cdot d\mathbf{S} = \oint_{-\mathcal{C}} \mathbf{A} \cdot d\mathbf{r} = -\oint_{\mathcal{C}} \mathbf{A} \cdot d\mathbf{r} = \pi$$

Note that as Theorem 2 tells us it must, the total flux of $\mathbf{F} = \mathrm{curl}(\mathbf{A})$ through the closed surface that is the union of $\mathcal{S}_1$, and $\mathcal{S}_2$ is $\pi - \pi = 0$. If instead, we had oriented $\mathcal{S}_2$ with the inward pointing normal, then both $\mathcal{S}_1$ and $\mathcal{S}_2$ would have a boundary given by the curve $\mathcal{C}$ with the orientation shown. Therefore, by the Surface Independence of Curl Vector Fields, the two surfaces would generate the same flux, which in this case is $-\pi$. ■

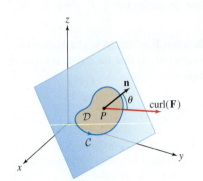

FIGURE 9 The curve $\mathcal{C}$ around P lies in the plane through P with normal vector $\mathbf{n}$.

CONCEPTUAL INSIGHT **Interpretation of the Curl** In Section 17.1, we showed that the quantity $\dfrac{\partial F_2}{\partial x} - \dfrac{\partial F_1}{\partial y}$ in Green's Theorem is the "circulation per unit of enclosed area." A similar interpretation is valid in $\mathbf{R}^3$.

Consider a plane through a point P with unit normal vector $\mathbf{n}$, and let $\mathcal{D}$ be a small domain containing P with boundary curve $\mathcal{C}$ (Figure 9). By Stokes' Theorem,

$$\oint_{\mathcal{C}} \mathbf{F} \cdot d\mathbf{r} \approx \iint_{\mathcal{D}} (\mathrm{curl}(\mathbf{F}) \cdot \mathbf{n}) \, dS \qquad \boxed{9}$$

The vector field $\mathrm{curl}(\mathbf{F})$ is continuous (its components are derivatives of the components of $\mathbf{F}$), so its value does not change much on $\mathcal{D}$ if $\mathcal{D}$ is sufficiently small.

To a first approximation, we can replace $\mathrm{curl}(\mathbf{F})$ by the constant value $\mathrm{curl}(\mathbf{F})(P)$, giving us the approximation

$$\iint_{\mathcal{D}} (\mathrm{curl}(\mathbf{F}) \cdot \mathbf{n}) \, dS \approx \iint_{\mathcal{D}} (\mathrm{curl}(\mathbf{F})(P) \cdot \mathbf{n}) \, dS$$

$$\approx (\mathrm{curl}(\mathbf{F})(P) \cdot \mathbf{n}) \, \mathrm{Area}(\mathcal{D}) \qquad \boxed{10}$$

Furthermore, $\mathrm{curl}(\mathbf{F})(P) \cdot \mathbf{n} = \|\mathrm{curl}(\mathbf{F})(P)\| \cos\theta$, where θ is the angle between $\mathrm{curl}(\mathbf{F})$ and $\mathbf{n}$. Together, Eqs. (9) and (10) give us

$$\oint_{\mathcal{C}} \mathbf{F} \cdot d\mathbf{r} \approx \|\mathrm{curl}(\mathbf{F})(P)\| (\cos\theta) \, \mathrm{Area}(\mathcal{D}) \qquad \boxed{11}$$

This is a remarkable result. It tells us that $\mathrm{curl}(\mathbf{F})$ encodes the the circulation per unit of enclosed area in every plane through P in a simple way—namely, as the dot product $\mathrm{curl}(\mathbf{F})(P) \cdot \mathbf{n}$. In particular, the circulation rate varies (to a first-order approximation) as the cosine of the angle θ between $\mathrm{curl}(\mathbf{F})(P)$ and $\mathbf{n}$.

We can also argue (as in Section 17.1 for vector fields in the plane) that if $\mathbf{F}$ is the velocity field of a fluid, then a small paddle wheel with normal $\mathbf{n}$ will rotate with an angular velocity of approximately $\frac{1}{2}\mathrm{curl}(\mathbf{F})(P) \cdot \mathbf{n}$ (see Figure 10). At any given point, the angular velocity is maximized when the normal vector to the paddle wheel $\mathbf{n}$ points in the direction of $\mathrm{curl}(\mathbf{F})$.

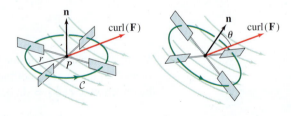

FIGURE 10 The paddle wheel can be oriented in different ways, as specified by the normal vector $\mathbf{n}$.

■ **EXAMPLE 4** **Vector Potential for a Solenoid** An electric current I flowing through a solenoid (a tightly wound spiral of wire; see Figure 11) creates a magnetic field $\mathbf{B}$. If we assume that the solenoid is infinitely long, with radius R and the z-axis as the central axis, then

$$\mathbf{B}(r) = \begin{cases} \mathbf{0} & \text{if } r > R \\ B\mathbf{k} & \text{if } r < R \end{cases}$$

where $r = (x^2 + y^2)^{1/2}$ is the distance to the z-axis, and B is a constant that depends on the current strength I and the spacing of the turns of wire.

(a) Show that a vector potential for **B** is

$$\mathbf{A}(r) = \begin{cases} \dfrac{1}{2}R^2 B\left\langle -\dfrac{y}{r^2}, \dfrac{x}{r^2}, 0 \right\rangle & \text{if } r > R \\[2ex] \dfrac{1}{2}B\left\langle -y, x, 0 \right\rangle & \text{if } r < R \end{cases}$$

(b) Calculate the flux of **B** through the surface $\mathcal{S}$ (with an upward-pointing normal) in Figure 11 whose boundary is a circle of radius r, where $r > R$.

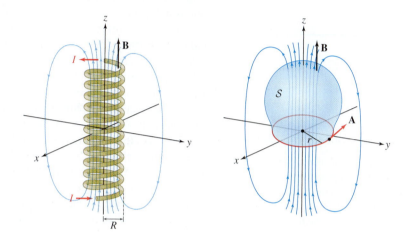

FIGURE 11 The magnetic field of a long solenoid is nearly uniform inside and weak outside. In practice, we treat the solenoid as "infinitely long" if it is very long in comparison with its radius.

Solution

(a) For any functions f and g,

$$\text{curl}(\langle f, g, 0 \rangle) = \langle -g_z, f_z, g_x - f_y \rangle$$

*The vector potential **A** is continuous but not differentiable on the cylinder $r = R$, that is, on the solenoid itself (Figure 12). The magnetic field **B** = curl(**A**) has a jump discontinuity where $r = R$. We take for granted the fact that Stokes' Theorem remains valid in this setting.*

Applying this to **A** for $r < R$, we obtain

$$\text{curl}(\mathbf{A}) = \frac{1}{2}B\left\langle 0, 0, \frac{\partial}{\partial x}x - \frac{\partial}{\partial y}(-y) \right\rangle = \langle 0, 0, B \rangle = B\mathbf{k} = \mathbf{B}$$

We leave it as an exercise (Exercise 33) to show that $\text{curl}(\mathbf{A}) = \mathbf{B} = \mathbf{0}$ for $r > R$.

(b) The boundary of S is a circle with counterclockwise parametrization $\mathbf{r}(t) = \langle r\cos t, r\sin t, 0 \rangle$, so

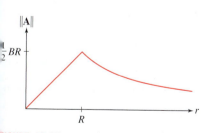

FIGURE 12 The magnitude $\|\mathbf{A}\|$ of the vector potential as a function of distance r to the z-axis.

$$\mathbf{r}'(t) = \langle -r\sin t, r\cos t, 0 \rangle$$

$$\mathbf{A}(\mathbf{r}(t)) = \frac{1}{2}R^2 B r^{-1}\langle -\sin t, \cos t, 0 \rangle$$

$$\mathbf{A}(\mathbf{r}(t)) \cdot \mathbf{r}'(t) = \frac{1}{2}R^2 B\left((-\sin t)^2 + \cos^2 t\right) = \frac{1}{2}R^2 B$$

By Stokes' Theorem, the flux of **B** through $\mathcal{S}$ is equal to

$$\iint_{\mathcal{S}} \mathbf{B} \cdot d\mathbf{S} = \oint_{\partial \mathcal{S}} \mathbf{A} \cdot d\mathbf{r} = \int_0^{2\pi} \mathbf{A}(\mathbf{r}(t)) \cdot \mathbf{r}'(t)\, dt = \frac{1}{2}R^2 B \int_0^{2\pi} dt = \pi R^2 B \quad \blacksquare$$

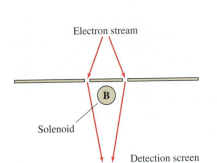

Electron stream

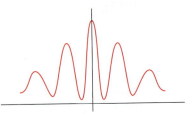

B

Solenoid

Detection screen

FIGURE 13 A stream of electrons passing through a double slit produces an interference pattern on the detection screen. The pattern shifts slightly when an electric current flows through the solenoid.

CONCEPTUAL INSIGHT There is an interesting difference between scalar and vector potentials. If $\mathbf{F} = \nabla f$, then the scalar potential f is constant in regions where the field $\mathbf{F}$ is zero (since a function with zero gradient is constant). This is not true for vector potentials. As we saw in Example 4, the magnetic field $\mathbf{B}$ produced by a solenoid is zero everywhere outside the solenoid, but the vector potential $\mathbf{A}$ is not constant outside the solenoid. In fact, $\mathbf{A}$ is proportional to $\left\langle -\dfrac{y}{r^2}, \dfrac{x}{r^2}, 0 \right\rangle$. This is related to an intriguing phenomenon in physics called the *Aharonov–Bohm (AB) effect*, first proposed on theoretical grounds in the 1940s.

According to electromagnetic theory, a magnetic field $\mathbf{B}$ exerts a force on a moving electron, causing a deflection in the electron's path. We do not expect any deflection when an electron moves past a solenoid because $\mathbf{B}$ is zero outside the solenoid (in practice, the field is not actually zero for a solenoid of finite length, but it is very small—we ignore this difficulty). However, according to quantum mechanics, electrons have both particle and wave properties. In a double-slit experiment, a stream of electrons passing through two small slits creates a wavelike interference pattern on a detection screen (Figure 13).

The AB effect predicts that if we place a small solenoid between the slits as in the figure (the solenoid is so small that the electrons never pass through it), then the interference pattern will shift slightly. It is as if the electrons are "aware" of the magnetic field inside the solenoid, even though they never encounter the field directly.

The AB effect was hotly debated until it was confirmed definitively in 1985, in experiments carried out by a team of Japanese physicists led by Akira Tonomura. The AB effect appeared to contradict "classical" electromagnetic theory, according to which the trajectory of an electron is determined by $\mathbf{B}$ alone. There is no such contradiction in quantum mechanics, because the behavior of the electrons is governed not by $\mathbf{B}$ but by a "wave function" derived from the nonconstant vector potential $\mathbf{A}$.

17.2 SUMMARY

- The *boundary* of a surface $\mathcal{S}$ is denoted $\partial \mathcal{S}$. We say that $\mathcal{S}$ is *closed* if $\partial \mathcal{S}$ is empty.
- Suppose that $\mathcal{S}$ is oriented (a continuously varying unit normal is specified at each point of $\mathcal{S}$). The *boundary orientation* of $\partial \mathcal{S}$ is defined as follows: If you walk along the boundary in the positive direction with your head pointing in the normal direction, then the surface is on your left.
- Stokes' Theorem relates the circulation around the boundary to the surface integral of the curl:

$$\oint_{\partial \mathcal{S}} \mathbf{F} \cdot d\mathbf{r} = \iint_{\mathcal{S}} \operatorname{curl}(\mathbf{F}) \cdot d\mathbf{S}$$

- Surface independence: If $\mathbf{F} = \operatorname{curl}(\mathbf{A})$, then the flux of $\mathbf{F}$ through a surface $\mathcal{S}$ depends only on the oriented boundary $\partial \mathcal{S}$ and not on the surface itself:

$$\iint_{\mathcal{S}} \mathbf{F} \cdot d\mathbf{S} = \oint_{\partial \mathcal{S}} \mathbf{A} \cdot d\mathbf{r}$$

In particular, if $\mathcal{S}$ is *closed* (i.e., $\partial \mathcal{S}$ is empty) and $\mathbf{F} = \operatorname{curl}(\mathbf{A})$, then $\iint_{\mathcal{S}} \mathbf{F} \cdot d\mathbf{S} = 0$. If $\mathcal{S}_1$ and $\mathcal{S}_2$ are oriented surfaces that share an oriented boundary and $\mathbf{F} = \operatorname{curl}(\mathbf{A})$, then

$$\iint_{\mathcal{S}_1} \mathbf{F} \cdot d\mathbf{S} = \iint_{\mathcal{S}_2} \mathbf{F} \cdot d\mathbf{S}$$

- The curl is interpreted as a vector that encodes circulation per unit area: If P is any point and $\mathbf{n}$ is a unit normal vector, then

$$\int_{\mathcal{C}} \mathbf{F} \cdot d\mathbf{r} \approx (\operatorname{curl}(\mathbf{F})(P) \cdot \mathbf{n}) \operatorname{Area}(\mathcal{D})$$

where $\mathcal{C}$ is a small, simple closed curve around P in the plane through P with normal vector $\mathbf{n}$, and $\mathcal{D}$ is the region enclosed by $\mathcal{C}$.

17.2 EXERCISES

Preliminary Questions

1. Indicate with an arrow the boundary orientation of the boundary curves of the surfaces in Figure 14, oriented by the outward-pointing normal vectors.

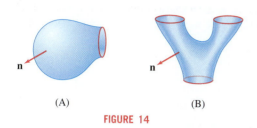

(A)   (B)

FIGURE 14

2. Let $\mathbf{F} = \mathrm{curl}(\mathbf{A})$. Which of the following are related by Stokes' Theorem?

(a) The circulation of $\mathbf{A}$ and flux of $\mathbf{F}$

(b) The circulation of $\mathbf{F}$ and flux of $\mathbf{A}$

3. What is the definition of a vector potential?

4. Which of the following statements is correct?

(a) The flux of $\mathrm{curl}(\mathbf{A})$ through every oriented surface is zero.

(b) The flux of $\mathrm{curl}(\mathbf{A})$ through every closed, oriented surface is zero.

5. Which condition on $\mathbf{F}$ guarantees that the flux through $\mathcal{S}_1$ is equal to the flux through $\mathcal{S}_2$ for any two oriented surfaces $\mathcal{S}_1$ and $\mathcal{S}_2$ with the same oriented boundary?

Exercises

In Exercises 1–4, verify Stokes' Theorem for the given vector field and surface, oriented with an upward-pointing normal.

1. $\mathbf{F} = \langle 2xy, x, y + z \rangle$,   the surface $z = 1 - x^2 - y^2$ for $x^2 + y^2 \le 1$

2. $\mathbf{F} = \langle yz, 0, x \rangle$,   the portion of the plane $\frac{x}{2} + \frac{y}{3} + z = 1$ where $x, y, z \ge 0$

3. $\mathbf{F} = \langle e^{y-z}, 0, 0 \rangle$,   the square with vertices $(1, 0, 1)$, $(1, 1, 1)$, $(0, 1, 1)$, and $(0, 0, 1)$

4. $\mathbf{F} = \langle y, x, x^2 + y^2 \rangle$,   the upper hemisphere $x^2 + y^2 + z^2 = 1, z \ge 0$

In Exercises 5–10, calculate $\mathrm{curl}(\mathbf{F})$ and then apply Stokes' Theorem to compute the flux of $\mathrm{curl}(\mathbf{F})$ through the given surface using a line integral.

5. $\mathbf{F} = \langle e^{z^2} - y, e^{z^3} + x, \cos(xz) \rangle$,   the upper hemisphere $x^2 + y^2 + z^2 = 1, z \ge 0$ with outward-pointing normal

6. $\mathbf{F} = \langle x + y, z^2 - 4, x\sqrt{y^2 + 1} \rangle$,   surface of the wedge-shaped box in Figure 15 (bottom included, top excluded) with outward-pointing normal

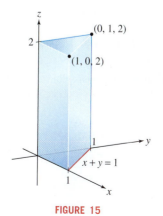

FIGURE 15

7. $\mathbf{F} = \langle 3z, 5x, -2y \rangle$,   that part of the paraboloid $z = x^2 + y^2$ that lies below the plane $z = 4$ with upward-pointing unit normal vector

8. $\mathbf{F} = \langle yz, -xz, z^3 \rangle$,   that part of the cone $z = \sqrt{x^2 + y^2}$ that lies between the two planes $z = 1$ and $z = 3$ with upward-pointing unit normal vector

9. $\mathbf{F} = \langle yz, xz, xy \rangle$,   that part of the cylinder $x^2 + y^2 = 1$ that lies between the two planes $z = 1$ and $z = 4$ with outward-pointing unit normal vector

10. $\mathbf{F} = \langle 2y, e^z, -\arctan x \rangle$,   that part of the paraboloid $z = 4 - x^2 - y^2$ cut off by the xy-plane with upward-pointing unit normal vector

In Exercises 11–14, apply Stokes' Theorem to evaluate $\oint_C \mathbf{F} \cdot d\mathbf{r}$ by finding the flux of $\mathrm{curl}(\mathbf{F})$ across an appropriate surface.

11. $\mathbf{F} = \langle 3y, -2x, 3y \rangle$, C is the circle $x^2 + y^2 = 9, z = 2$, oriented counterclockwise as viewed from above.

12. $\mathbf{F} = \langle yz, xy, xz \rangle$, C is the square with vertices $(0, 0, 2), (1, 0, 2),$ $(1, 1, 2),$ and $(0, 1, 2)$, oriented counterclockwise as viewed from above.

13. $\mathbf{F} = \langle y, z, x \rangle$, C is the triangle with vertices $(0, 0, 0), (3, 0, 0),$ and $(0, 3, 3)$, oriented counterclockwise as viewed from above.

14. $\mathbf{F} = \langle y, -2z, 4x \rangle$, C is the boundary of that portion of the plane $x + 2y + 3z = 1$ that is in the first octant of space, oriented counterclockwise as viewed from above.

15. Let $\mathcal{S}$ be the surface of the cylinder (not including the top and bottom) of radius 2 for $1 \le z \le 6$, oriented with outward-pointing normal (Figure 16).

(a) Indicate with an arrow the orientation of $\partial\mathcal{S}$ (the top and bottom circles).

(b) Verify Stokes' Theorem for $\mathcal{S}$ and $\mathbf{F} = \langle yz^2, 0, 0 \rangle$.

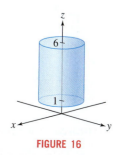

FIGURE 16

16. Let S be the portion of the plane $z = x$ contained in the half-cylinder of radius R depicted in Figure 17. Use Stokes' Theorem to calculate the circulation of $\mathbf{F} = \langle z, x, y + 2z \rangle$ around the boundary of S (a half-ellipse) in the counterclockwise direction when viewed from above. *Hint:* Show that curl($\mathbf{F}$) is orthogonal to the normal vector to the plane.

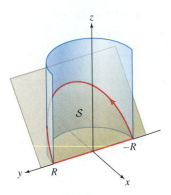

FIGURE 17

17. Let I be the flux of $\mathbf{F} = \langle e^y, 2xe^{x^2}, z^2 \rangle$ through the upper hemisphere S of the unit sphere.

(a) Let $\mathbf{G} = \langle e^y, 2xe^{x^2}, 0 \rangle$. Find a vector field $\mathbf{A}$ such that curl($\mathbf{A}$) = $\mathbf{G}$.

(b) Use Stokes' Theorem to show that the flux of $\mathbf{G}$ through S is zero. *Hint:* Calculate the circulation of $\mathbf{A}$ around ∂S.

(c) Calculate I. *Hint:* Use (b) to show that I is equal to the flux of $\langle 0, 0, z^2 \rangle$ through S.

18. Let $\mathbf{F} = \langle 0, -z, 1 \rangle$. Let S be the spherical cap $x^2 + y^2 + z^2 \le 1$, where $z \ge \frac{1}{2}$. Evaluate $\iint_S \mathbf{F} \cdot d\mathbf{S}$ directly as a surface integral. Then verify that $\mathbf{F} = $ curl($\mathbf{A}$), where $\mathbf{A} = (0, x, xz)$ and evaluate the surface integral again using Stokes' Theorem.

19. Let $\mathbf{A}$ be the vector potential and $\mathbf{B}$ the magnetic field of the infinite solenoid of radius R in Example 4. Use Stokes' Theorem to compute:

(a) The flux of $\mathbf{B}$ through a circle in the xy-plane of radius $r < R$

(b) The circulation of $\mathbf{A}$ around the boundary C of a surface lying outside the solenoid

20. The magnetic field $\mathbf{B}$ due to a small current loop (which we place at the origin) is called a **magnetic dipole** (Figure 18). Let $\rho = (x^2 + y^2 + z^2)^{1/2}$. For ρ large, $\mathbf{B} = $ curl($\mathbf{A}$), where

$$\mathbf{A} = \left\langle -\frac{y}{\rho^3}, \frac{x}{\rho^3}, 0 \right\rangle$$

(a) Let C be a horizontal circle of radius R with center $(0, 0, c)$, where c is large. Show that $\mathbf{A}$ is tangent to C.

(b) Use Stokes' Theorem to calculate the flux of $\mathbf{B}$ through C.

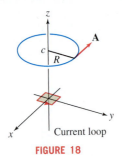

FIGURE 18

21. A uniform magnetic field $\mathbf{B}$ has constant strength b in the z-direction [i.e., $\mathbf{B} = \langle 0, 0, b \rangle$].

(a) Verify that $\mathbf{A} = \frac{1}{2}\mathbf{B} \times \mathbf{r}$ is a vector potential for $\mathbf{B}$, where $\mathbf{r} = \langle x, y, 0 \rangle$.

(b) Calculate the flux of $\mathbf{B}$ through the rectangle with vertices A, B, C, and D in Figure 19.

22. Let $\mathbf{F} = \langle -x^2 y, x, 0 \rangle$. Referring to Figure 19, let C be the closed path $ABCD$. Use Stokes' Theorem to evaluate $\int_C \mathbf{F} \cdot d\mathbf{r}$ in two ways. First, regard C as the boundary of the rectangle with vertices A, B, C, and D. Then treat C as the boundary of the wedge-shaped box with an open top.

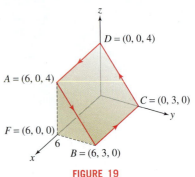

FIGURE 19

23. Let $\mathbf{F} = \langle y^2, 2z + x, 2y^2 \rangle$. Use Stokes' Theorem to find a plane with equation $ax + by + cz = 0$ (where a, b, c are not all zero) such that $\oint_C \mathbf{F} \cdot d\mathbf{r} = 0$ for every closed C lying in the plane. *Hint:* Choose a, b, c so that curl($\mathbf{F}$) lies in the plane.

24. Let $\mathbf{F} = \langle -z^2, 2zx, 4y - x^2 \rangle$, and let C be a simple closed curve in the plane $x + y + z = 4$ that encloses a region of area 16 (Figure 20). Calculate $\oint_C \mathbf{F} \cdot d\mathbf{r}$, where C is oriented in the counterclockwise direction (when viewed from above the plane).

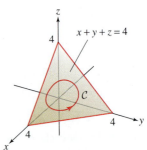

FIGURE 20

25. Let $\mathbf{F} = \langle y^2, x^2, z^2 \rangle$. Show that

$$\int_{C_1} \mathbf{F} \cdot d\mathbf{r} = \int_{C_2} \mathbf{F} \cdot d\mathbf{r}$$

for any two closed curves lying on a cylinder whose central axis is the z-axis (Figure 21).

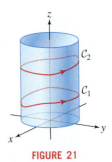

FIGURE 21

26. The curl of a vector field $\mathbf{F}$ at the origin is $\mathbf{v}_0 = \langle 3, 1, 4 \rangle$. Estimate the circulation around the small parallelogram spanned by the vectors $\mathbf{A} = \langle 0, \frac{1}{2}, \frac{1}{2} \rangle$ and $\mathbf{B} = \langle 0, 0, \frac{1}{3} \rangle$.

27. You know two things about a vector field $\mathbf{F}$:

 (i) $\mathbf{F}$ has a vector potential $\mathbf{A}$ (but $\mathbf{A}$ is unknown).

 (ii) The circulation of $\mathbf{A}$ around the unit circle (oriented counterclockwise) is 25.

Determine the flux of $\mathbf{F}$ through the surface $\mathcal{S}$ in Figure 22, oriented with an upward-pointing normal.

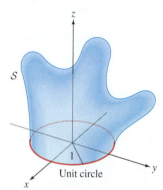

FIGURE 22 Surface $\mathcal{S}$ whose boundary is the unit circle.

28. Suppose that $\mathbf{F}$ has a vector potential and that $\mathbf{F}(x, y, 0) = \mathbf{k}$. Find the flux of $\mathbf{F}$ through the surface $\mathcal{S}$ in Figure 22, oriented with an upward-pointing normal.

29. Prove that $\mathrm{curl}(f\mathbf{a}) = \nabla f \times \mathbf{a}$, where f is a differentiable function and $\mathbf{a}$ is a constant vector.

30. Show that $\mathrm{curl}(\mathbf{F}) = \mathbf{0}$ if $\mathbf{F}$ is **radial**, meaning that $\mathbf{F} = f(\rho) \langle x, y, z \rangle$ for some function $f(\rho)$, where $\rho = \sqrt{x^2 + y^2 + z^2}$. *Hint:* It is enough to show that one component of $\mathrm{curl}(\mathbf{F})$ is zero, because it will then follow for the other two components by symmetry.

31. Prove the following Product Rule:

$$\mathrm{curl}(f\mathbf{F}) = f\,\mathrm{curl}(\mathbf{F}) + \nabla f \times \mathbf{F}$$

32. Assume that f and g have continuous partial derivatives of order 2. Prove that

$$\oint_{\partial \mathcal{S}} f \nabla(g) \cdot d\mathbf{r} = \iint_{\mathcal{S}} \nabla(f) \times \nabla(g) \cdot d\mathbf{S}$$

33. Verify that $\mathbf{B} = \mathrm{curl}(\mathbf{A})$ for $r > R$ in the setting of Example 4.

34. [✎] Explain carefully why Green's Theorem is a special case of Stokes' Theorem.

Further Insights and Challenges

35. In this exercise, we use the notation of the proof of Theorem 1 and prove

$$\oint_{\mathcal{C}} F_3(x, y, z)\mathbf{k} \cdot d\mathbf{r} = \iint_{\mathcal{S}} \mathrm{curl}(F_3(x, y, z)\mathbf{k}) \cdot d\mathbf{S} \qquad \boxed{12}$$

In particular, $\mathcal{S}$ is the graph of $z = f(x, y)$ over a domain $\mathcal{D}$, and $\mathcal{C}$ is the boundary of $\mathcal{S}$ with parametrization $(x(t), y(t), f(x(t), y(t)))$.

(a) Use the Chain Rule to show that

$$F_3(x, y, z)\mathbf{k} \cdot d\mathbf{r} = F_3(x(t), y(t), f(x(t), y(t)))$$

$$\left(f_x(x(t), y(t))x'(t) + f_y(x(t), y(t))y'(t) \right) dt$$

and verify that

$$\oint_{\mathcal{C}} F_3(x, y, z)\mathbf{k} \cdot d\mathbf{r} =$$

$$\oint_{\mathcal{C}_0} \left\langle F_3(x, y, z) f_x(x, y),\ F_3(x, y, z) f_y(x, y) \right\rangle \cdot d\mathbf{r}$$

where $\mathcal{C}_0$ has parametrization $(x(t), y(t))$.

(b) Apply Green's Theorem to the line integral over $\mathcal{C}_0$ and show that the result is equal to the right-hand side of Eq. (12).

36. Let $\mathbf{F}$ be a continuously differentiable vector field in $\mathbf{R}^3$, Q a point, and $\mathcal{S}$ a plane containing Q with unit normal vector $\mathbf{e}$. Let $\mathcal{C}_r$ be a circle of radius r centered at Q in $\mathcal{S}$, and let $\mathcal{S}_r$ be the disk enclosed by $\mathcal{C}_r$. Assume $\mathcal{S}_r$ is oriented with unit normal vector $\mathbf{e}$.

(a) Let $m(r)$ and $M(r)$ be the minimum and maximum values of $\mathrm{curl}(\mathbf{F}(P)) \cdot \mathbf{e}$ for $P \in \mathcal{S}_r$. Prove that

$$m(r) \le \frac{1}{\pi r^2} \iint_{\mathcal{S}_r} \mathrm{curl}(\mathbf{F}) \cdot d\mathbf{S} \le M(r)$$

(b) Prove that

$$\mathrm{curl}(\mathbf{F}(Q)) \cdot \mathbf{e} = \lim_{r \to 0} \frac{1}{\pi r^2} \int_{\mathcal{C}_r} \mathbf{F} \cdot d\mathbf{r}$$

This proves that $\mathrm{curl}(\mathbf{F}(Q)) \cdot \mathbf{e}$ is the circulation per unit area in the plane $\mathcal{S}$.

17.3 Divergence Theorem

We have studied several "Fundamental Theorems." Each of these is a relation of the type:

$$\begin{matrix} \text{Integral of a derivative} \\ \text{on an oriented domain} \end{matrix} = \begin{matrix} \text{Integral over the } \textit{oriented} \\ \textit{boundary} \text{ of the domain} \end{matrix}$$

Here are the examples we have seen so far:

- In single-variable calculus, the Fundamental Theorem of Calculus (FTC) relates the integral of $f'(x)$ over an interval $[a, b]$ to the "integral" of $f(x)$ over the boundary of $[a, b]$ consisting of two points a and b:

$$\underbrace{\int_a^b f'(x)\,dx}_{\text{Integral of derivative over } [a, b]} = \underbrace{f(b) - f(a)}_{\text{"Integral" over the boundary of } [a, b]}$$

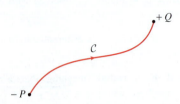

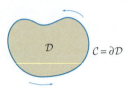

FIGURE 1 The oriented boundary of C is $\partial C = Q - P$.

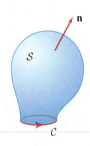

FIGURE 2 Domain $\mathcal{D}$ in $\mathbf{R}^2$ with boundary curve $C = \partial \mathcal{D}$.

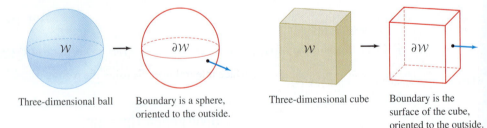

FIGURE 3 The oriented boundary of S is $C = \partial S$.

The boundary of $[a, b]$ is oriented by assigning a plus sign to b and a minus sign to a.

- The Fundamental Theorem for Line Integrals generalizes the FTC: Instead of taking an integral over an interval $[a, b]$ (a path from a to b along the x-axis), we take an integral along any path from points P to Q in $\mathbf{R}^3$ (Figure 1), and instead of $f'(x)$, we use the gradient:

$$\underbrace{\int_C \nabla f \cdot d\mathbf{r}}_{\text{Integral of derivative over a curve}} = \underbrace{f(Q) - f(P)}_{\substack{\text{"Integral" over the} \\ \text{boundary } \partial C = Q - P}}$$

- Green's Theorem is a two-dimensional version of the FTC that relates the integral of a certain derivative over a domain $\mathcal{D}$ in the plane to an integral over its boundary curve $C = \partial \mathcal{D}$ (Figure 2):

$$\underbrace{\iint_{\mathcal{D}} \left(\frac{\partial F_2}{\partial x} - \frac{\partial F_1}{\partial y} \right) dA}_{\text{Integral of derivative over domain}} = \underbrace{\int_C \mathbf{F} \cdot d\mathbf{r}}_{\text{Integral over boundary curve}}$$

- Stokes' Theorem extends Green's Theorem: Instead of a domain in the plane (a flat surface), we allow any surface in $\mathbf{R}^3$ (Figure 3). The appropriate derivative is the curl:

$$\underbrace{\iint_{S} \text{curl}(\mathbf{F}) \cdot d\mathbf{S}}_{\text{Integral of derivative over surface}} = \underbrace{\int_C \mathbf{F} \cdot d\mathbf{r}}_{\text{Integral over boundary curve}}$$

Our last theorem—the Divergence Theorem—also follows this pattern:

$$\underbrace{\iiint_{\mathcal{W}} \text{div}(\mathbf{F}) \, dV}_{\text{Integral of derivative over 3D region}} = \underbrace{\iint_{S} \mathbf{F} \cdot d\mathbf{S}}_{\text{Integral over boundary surface}}$$

Here, S is a *closed* surface that encloses a three-dimensional (or 3D) region $\mathcal{W}$. In other words, S is the boundary of $\mathcal{W}$: $S = \partial \mathcal{W}$. Recall that a closed surface is a surface that "holds air." Figure 4 shows two examples of regions and boundary surfaces that we will consider.

Three-dimensional ball   Boundary is a sphere, oriented to the outside.

Three-dimensional cube   Boundary is the surface of the cube, oriented to the outside.

FIGURE 4

We consider a piecewise smooth closed surface S, which means S consists of one smooth surface or at most finitely many smooth surfaces that have been glued together along their boundaries, as in the example of the cube.

THEOREM 1 Divergence Theorem Let S be a closed surface that encloses a region $\mathcal{W}$ in $\mathbf{R}^3$. Assume that S is piecewise smooth and is oriented by normal vectors pointing to the outside of $\mathcal{W}$. Let $\mathbf{F}$ be a vector field whose domain contains $\mathcal{W}$. Then

$$\iint_{S} \mathbf{F} \cdot d\mathbf{S} = \iiint_{\mathcal{W}} \text{div}(\mathbf{F}) \, dV$$

$\boxed{1}$

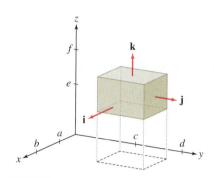

FIGURE 5 A box
$\mathcal{W} = [a, b] \times [c, d] \times [e, f]$.

◄·· **REMINDER** *The Divergence Theorem states*

$$\iint_{S} \mathbf{F} \cdot d\mathbf{S} = \iiint_{\mathcal{W}} \operatorname{div}(\mathbf{F}) \, dV$$

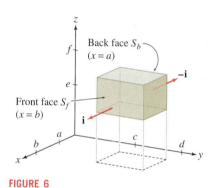

FIGURE 6

This theorem often appears in the form

$$\iint_{S} \mathbf{F} \cdot d\mathbf{S} = \iiint_{\mathcal{W}} \nabla \cdot \mathbf{F} \, dV$$

Proof We prove the Divergence Theorem in the special case that $\mathcal{W}$ is a box $[a, b] \times [c, d] \times [e, f]$ as in Figure 5. The proof can be modified to treat more general regions such as the interiors of spheres and cylinders.

We write each side of Eq. (1) as a sum over components:

$$\iint_{\partial\mathcal{W}} (F_1\mathbf{i} + F_2\mathbf{j} + F_3\mathbf{k}) \cdot d\mathbf{S} = \iint_{\partial\mathcal{W}} F_1\mathbf{i} \cdot d\mathbf{S} + \iint_{\partial\mathcal{W}} F_2\mathbf{j} \cdot d\mathbf{S} + \iint_{\partial\mathcal{W}} F_3\mathbf{k} \cdot d\mathbf{S}$$

$$\iiint_{\mathcal{W}} \operatorname{div}(F_1\mathbf{i} + F_2\mathbf{j} + F_3\mathbf{k}) \, dV = \iiint_{\mathcal{W}} \operatorname{div}(F_1\mathbf{i}) \, dV + \iiint_{\mathcal{W}} \operatorname{div}(F_2\mathbf{j}) \, dV$$
$$+ \iiint_{\mathcal{W}} \operatorname{div}(F_3\mathbf{k}) \, dV$$

As in the proofs of Green's and Stokes' Theorems, we show that corresponding terms are equal. It will suffice to carry out the argument for the **i**-component (the other two components are similar). Thus, we assume that $\mathbf{F} = F_1\mathbf{i}$.

The surface integral over boundary S of the box is the sum of the integrals over the six faces. However, $\mathbf{F} = F_1\mathbf{i}$ is orthogonal to the normal vectors to the top and bottom as well as the two side faces because $\mathbf{F} \cdot \mathbf{j} = \mathbf{F} \cdot \mathbf{k} = 0$. Therefore, the surface integrals over these faces are zero. Nonzero contributions come only from the front and back faces, which we denote S_f and S_b (Figure 6):

$$\iint_{S} \mathbf{F} \cdot d\mathbf{S} = \iint_{S_f} \mathbf{F} \cdot d\mathbf{S} + \iint_{S_b} \mathbf{F} \cdot d\mathbf{S}$$

To evaluate these integrals, we parametrize S_f and S_b by

$$G_f(y, z) = (b, y, z), \qquad c \le y \le d, \ e \le z \le f$$
$$G_b(y, z) = (a, y, z), \qquad c \le y \le d, \ e \le z \le f$$

The normal vectors for these parametrizations are

$$\frac{\partial G_f}{\partial y} \times \frac{\partial G_f}{\partial z} = \mathbf{j} \times \mathbf{k} = \mathbf{i}$$

$$\frac{\partial G_b}{\partial y} \times \frac{\partial G_b}{\partial z} = \mathbf{j} \times \mathbf{k} = \mathbf{i}$$

However, the outward-pointing normal for S_b is $-\mathbf{i}$, so a minus sign is needed in the surface integral over S_b using the parametrization G_b:

$$\iint_{S_f} \mathbf{F} \cdot d\mathbf{S} + \iint_{S_b} \mathbf{F} \cdot d\mathbf{S} = \int_e^f \int_c^d F_1(b, y, z) \, dy \, dz - \int_e^f \int_c^d F_1(a, y, z) \, dy \, dz$$

$$= \int_e^f \int_c^d \Big(F_1(b, y, z) - F_1(a, y, z) \Big) \, dy \, dz$$

By the FTC in one variable,

$$F_1(b, y, z) - F_1(a, y, z) = \int_a^b \frac{\partial F_1}{\partial x}(x, y, z) \, dx$$

Since $\operatorname{div}(\mathbf{F}) = \operatorname{div}(F_1\mathbf{i}) = \dfrac{\partial F_1}{\partial x}$, we obtain the desired result:

$$\iint_{S} \mathbf{F} \cdot d\mathbf{S} = \int_e^f \int_c^d \int_a^b \frac{\partial F_1}{\partial x}(x, y, z) \, dx \, dy \, dz = \iiint_{\mathcal{W}} \operatorname{div}(\mathbf{F}) \, dV \qquad ■$$

The names attached to mathematical theorems often conceal a more complex historical development. What we call Green's Theorem was stated by Augustin Cauchy in 1846, but it was never stated by George Green himself (he published a result that implies Green's Theorem in 1828). Stokes' Theorem first appeared as a problem on a competitive exam written by George Stokes at Cambridge University, but William Thomson (Lord Kelvin) had previously stated the theorem in a letter to Stokes. Gauss published special cases of the Divergence Theorem in 1813 and later in 1833 and 1839, while the general theorem was stated and proved by the Russian mathematician Michael Ostrogradsky in 1826. For this reason, the Divergence Theorem is also referred to as "Gauss's Theorem" or the "Gauss–Ostrogradsky Theorem."

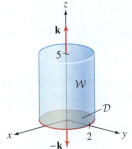

DF **FIGURE 7** Cylinder of radius 2 and
height 5.

■ **EXAMPLE 1** **Verifying the Divergence Theorem** Verify Theorem 1 for $\mathbf{F}(x, y, z) = \langle y, yz, z^2 \rangle$ and the cylinder in Figure 7.

Solution We must verify that the flux $\iint_{\mathcal{S}} \mathbf{F} \cdot d\mathbf{S}$, where $\mathcal{S}$ is the surface of the cylinder, is equal to the integral of div($\mathbf{F}$) over the cylinder. We compute the flux through $\mathcal{S}$ first: It is the sum of three surface integrals over the side, the top, and the bottom.

Step 1. Integrate over the side of the cylinder.
We use the standard parametrization of the cylinder:

$$G(\theta, z) = (2\cos\theta, 2\sin\theta, z), \qquad 0 \le \theta < 2\pi, \quad 0 \le z \le 5$$

The normal vector is

$$\mathbf{N} = \mathbf{T}_\theta \times \mathbf{T}_z = \langle -2\sin\theta, 2\cos\theta, 0 \rangle \times \langle 0, 0, 1 \rangle = \langle 2\cos\theta, 2\sin\theta, 0 \rangle$$

and $\mathbf{F}(G(\theta, z)) = \langle y, yz, z^2 \rangle = \langle 2\sin\theta, 2z\sin\theta, z^2 \rangle$. Thus,

$$\mathbf{F} \cdot d\mathbf{S} = \langle 2\sin\theta, 2z\sin\theta, z^2 \rangle \cdot \langle 2\cos\theta, 2\sin\theta, 0 \rangle \, d\theta \, dz$$

$$= 4\cos\theta\sin\theta + 4z\sin^2\theta \, d\theta \, dz$$

$$\iint_{\text{side}} \mathbf{F} \cdot d\mathbf{S} = \int_0^5 \int_0^{2\pi} (4\cos\theta\sin\theta + 4z\sin^2\theta) \, d\theta \, dz$$

◄·· **REMINDER** In Eq. (2), we use

$$\int_0^{2\pi} \cos\theta\sin\theta \, d\theta = 0$$

$$\int_0^{2\pi} \sin^2\theta \, d\theta = \pi$$

$$= 0 + 4\pi \int_0^5 z \, dz = 4\pi \left(\frac{25}{2} \right) = 50\pi \qquad \boxed{2}$$

Step 2. Integrate over the top and bottom of the cylinder.
The top of the cylinder is at height $z = 5$, so we can parametrize the top by $G(x, y) = (x, y, 5)$ for (x, y) in the disk $\mathcal{D}$ of radius 2:

$$\mathcal{D} = \{(x, y) : x^2 + y^2 \le 4\}$$

Then

$$\mathbf{N} = \mathbf{T}_x \times \mathbf{T}_y = \langle 1, 0, 0 \rangle \times \langle 0, 1, 0 \rangle = \langle 0, 0, 1 \rangle$$

and since $\mathbf{F}(G(x, y)) = \mathbf{F}(x, y, 5) = \langle y, 5y, 5^2 \rangle$, we have

$$\mathbf{F}(G(x, y)) \cdot \mathbf{N} = \langle y, 5y, 5^2 \rangle \cdot \langle 0, 0, 1 \rangle = 25$$

$$\iint_{\text{top}} \mathbf{F} \cdot d\mathbf{S} = \iint_{\mathcal{D}} 25 \, dA = 25 \, \text{Area}(\mathcal{D}) = 25(4\pi) = 100\pi$$

Along the bottom disk of the cylinder, we have $z = 0$ and $\mathbf{F}(x, y, 0) = \langle y, 0, 0 \rangle$. Thus, $\mathbf{F}$ is orthogonal to the vector $-\mathbf{k}$ normal to the bottom disk, and the integral along the bottom is zero.

Step 3. Find the total flux.

$$\iint_{\mathcal{S}} \mathbf{F} \cdot d\mathbf{S} = \text{sides} + \text{top} + \text{bottom} = 50\pi + 100\pi + 0 = \boxed{150\pi}$$

Step 4. Compare with the integral of divergence.

$$\text{div}(\mathbf{F}) = \text{div}(\langle y, yz, z^2 \rangle) = \frac{\partial}{\partial x} y + \frac{\partial}{\partial y}(yz) + \frac{\partial}{\partial z} z^2 = 0 + z + 2z = 3z$$

The cylinder $\mathcal{W}$ consists of all points (x, y, z) for $0 \le z \le 5$ and (x, y) in the disk $\mathcal{D}$. We see that the integral of the divergence is equal to the total flux as required:

$$\iiint_{\mathcal{W}} \text{div}(\mathbf{F}) \, dV = \iint_{\mathcal{D}} \int_{z=0}^5 3z \, dV = \iint_{\mathcal{D}} \frac{75}{2} \, dA$$

$$= \left(\frac{75}{2} \right) (\text{Area}(\mathcal{D})) = \left(\frac{75}{2} \right) (4\pi) = \boxed{150\pi} \qquad ■$$

In many applications, the Divergence Theorem is used to compute flux. In the next example, we reduce a flux computation (that would involve integrating over six sides of a box) to a more simple triple integral.

■ **EXAMPLE 2** **Using the Divergence Theorem** Use the Divergence Theorem to evaluate $\iint_{\mathcal{S}} \langle x^2, z^4, e^z \rangle \cdot d\mathbf{S}$, where $\mathcal{S}$ is the boundary of the box $\mathcal{W}$ in Figure 8.

Solution First, compute the divergence:

$$\operatorname{div}\left(\langle x^2, z^4, e^z \rangle\right) = \frac{\partial}{\partial x} x^2 + \frac{\partial}{\partial y} z^4 + \frac{\partial}{\partial z} e^z = 2x + e^z$$

Then apply the Divergence Theorem and use Fubini's Theorem (Section 15.1):

$$\iint_{\mathcal{S}} \langle x^2, z^4, e^z \rangle \cdot d\mathbf{S} = \iiint_{\mathcal{W}} (2x + e^z)\, dV = \int_0^2 \int_0^3 \int_0^1 (2x + e^z)\, dz\, dy\, dx$$

$$= 3 \int_0^2 2x\, dx + 6 \int_0^1 e^z\, dz = 12 + 6(e - 1) = 6e + 6 \qquad ■$$

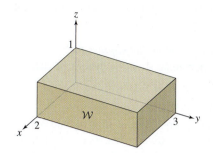

FIGURE 8

■ **EXAMPLE 3** **A Vector Field with Zero Divergence** Compute the flux of

$$\mathbf{F} = \left\langle z^2 + xy^2, \cos(x + z), e^{-y} - zy^2 \right\rangle$$

through the boundary of the surface $\mathcal{S}$ in Figure 9.

Solution Although $\mathbf{F}$ is rather complicated, its divergence is zero:

$$\operatorname{div}(\mathbf{F}) = \frac{\partial}{\partial x}(z^2 + xy^2) + \frac{\partial}{\partial y}\cos(x + z) + \frac{\partial}{\partial z}(e^{-y} - zy^2) = y^2 - y^2 = 0$$

The Divergence Theorem shows that the flux is zero. Letting $\mathcal{W}$ be the region enclosed by $\mathcal{S}$, we have

$$\iint_{\mathcal{S}} \mathbf{F} \cdot d\mathbf{S} = \iiint_{\mathcal{W}} \operatorname{div}(\mathbf{F})\, dV = \iiint_{\mathcal{W}} 0\, dV = 0 \qquad ■$$

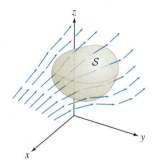

FIGURE 9

FIGURE 10 For a velocity field, the flux through a surface is the flow rate (in volume per unit time) of fluid across the surface.

GRAPHICAL INSIGHT **Interpretation of Divergence** Let's assume again that $\mathbf{F}$ is the velocity field of a fluid (Figure 10). Then the flux of $\mathbf{F}$ through a surface $\mathcal{S}$ is the flow rate (volume of fluid passing through $\mathcal{S}$ per unit time). If $\mathcal{S}$ encloses the region $\mathcal{W}$, then by the Divergence Theorem,

$$\text{Flow rate across } \mathcal{S} = \iiint_{\mathcal{W}} \operatorname{div}(\mathbf{F})\, dV \qquad \boxed{3}$$

Now assume that $\mathcal{S}$ is a small surface containing a point P. Because $\operatorname{div}(\mathbf{F})$ is continuous (it is a sum of derivatives of the components of $\mathbf{F}$), its value does not change much on $\mathcal{W}$ if $\mathcal{S}$ is sufficiently small, and to a first approximation, we can replace $\operatorname{div}(\mathbf{F})$ by the constant value $\operatorname{div}(\mathbf{F})(P)$. This gives us the approximation

$$\text{Flow rate across } \mathcal{S} = \iiint_{\mathcal{W}} \operatorname{div}(\mathbf{F})\, dV \approx \operatorname{div}(\mathbf{F})(P) \cdot \operatorname{Vol}(\mathcal{W}) \qquad \boxed{4}$$

In other words, the *flow rate through a small closed surface containing P is approximately equal to the divergence at P times the enclosed volume*, and thus, $\operatorname{div}(\mathbf{F})(P)$ has an interpretation as "flow rate (or flux) per unit volume":

• If $\operatorname{div}(\mathbf{F})(P) > 0$, there is a net outflow of fluid across any small closed surface enclosing P, or, in other words, a net "creation" of fluid near P. In this case, we call P a *source*.

Because of this, $\operatorname{div}(\mathbf{F})$ is sometimes called the *source density* of the field.

Do the units match up in Eq. (4)? The flow rate has units of volume per unit time. On the other hand, the divergence is a sum of derivatives of velocity with respect to distance. Therefore, the divergence has units of "distance per unit time per distance," or unit time^{-1}, and the right-hand side of Eq. (4) also has units of volume per unit time.

- If $\text{div}(\mathbf{F})(P) < 0$, there is a net inflow of fluid across any small closed surface enclosing P, or, in other words, a net "destruction" of fluid near P. In this case, we call P a *sink*.
- If $\text{div}(\mathbf{F})(P) = 0$, then to a first-order approximation, the net flow across any small closed surface enclosing P is equal to zero.

A vector field such that $\text{div}(\mathbf{F}) = 0$ everywhere is called *incompressible*.

To visualize these cases, consider the two-dimensional situation, where we define

$$\text{div}(\langle F_1, F_2 \rangle) = \frac{\partial F_1}{\partial x} + \frac{\partial F_2}{\partial y}$$

In Figure 11, field (A) has positive divergence. There is a positive net flow of fluid across every circle per unit time. Similarly, field (B) has negative divergence. By contrast, field (C) is *incompressible*. The fluid flowing into every circle is balanced by the fluid flowing out.

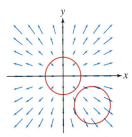

(A) The field $\mathbf{F} = \langle x, y \rangle$ with $\text{div}(\mathbf{F}) = 2$. There is a net outflow through every circle.

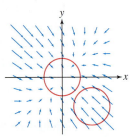

(B) The field $\mathbf{F} = \langle y - 2x, x - 2y \rangle$ with $\text{div}(\mathbf{F}) = -4$. There is a net inflow into every circle.

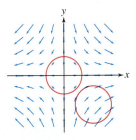

(C) The field $\mathbf{F} = \langle x, -y \rangle$ with $\text{div}(\mathbf{F}) = 0$. The flux through every circle is zero.

FIGURE 11

Applications to Electrostatics

The Divergence Theorem is a powerful tool for computing the flux of electrostatic fields. This is due to the fact that the electrostatic field of a point charge is the inverse-square vector field, which has special properties. In this section, we denote the inverse-square vector field by $\mathbf{F}_{\text{IS}}$:

$$\mathbf{F}_{\text{IS}} = \frac{\mathbf{e}_r}{r^2} = \frac{\mathbf{r}}{r^3}$$

The unit radial vector field $\mathbf{e}_r$ appears in Figure 12. Note that $\mathbf{F}_{\text{IS}}$ is only defined for $r \neq 0$. The next example verifies the key property that $\text{div}(\mathbf{F}_{\text{IS}}) = 0$.

■ **EXAMPLE 4　The Inverse-Square Vector Field**　Verify that $\mathbf{F}_{\text{IS}} = \dfrac{\mathbf{e}_r}{r^2}$ has zero divergence:

$$\text{div}\left(\frac{\mathbf{e}_r}{r^2}\right) = 0$$

Solution　Write the field as

$$\mathbf{F}_{\text{IS}} = \langle F_1, F_2, F_3 \rangle = \frac{1}{r^2}\left\langle \frac{x}{r}, \frac{y}{r}, \frac{z}{r} \right\rangle = \left\langle xr^{-3}, yr^{-3}, zr^{-3} \right\rangle$$

We have

$$\frac{\partial r}{\partial x} = \frac{\partial}{\partial x}(x^2 + y^2 + z^2)^{1/2} = \frac{1}{2}(x^2 + y^2 + z^2)^{-1/2}(2x) = \frac{x}{r}$$

$$\frac{\partial F_1}{\partial x} = \frac{\partial}{\partial x}xr^{-3} = r^{-3} - 3xr^{-4}\frac{\partial r}{\partial x} = r^{-3} - (3xr^{-4})\frac{x}{r} = \frac{r^2 - 3x^2}{r^5}$$

◄·· *REMINDER*

$$r = \sqrt{x^2 + y^2 + z^2}$$

For $r \neq 0$,

$$\mathbf{e}_r = \frac{\langle x, y, z \rangle}{r} = \frac{\langle x, y, z \rangle}{\sqrt{x^2 + y^2 + z^2}}$$

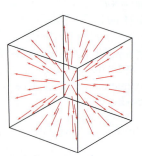

FIGURE 12 Unit radial vector field $\mathbf{e}_r$.

The derivatives $\dfrac{\partial F_2}{\partial y}$ and $\dfrac{\partial F_3}{\partial z}$ are similar, so

$$\text{div}(\mathbf{F}_{\text{IS}}) = \frac{r^2 - 3x^2}{r^5} + \frac{r^2 - 3y^2}{r^5} + \frac{r^2 - 3z^2}{r^5} = \frac{3r^2 - 3(x^2 + y^2 + z^2)}{r^5} = 0 \quad \blacksquare$$

The next theorem shows that the flux of $\mathbf{F}_{\text{IS}}$ through a closed surface $\mathcal{S}$ depends only on whether $\mathcal{S}$ contains the origin.

> **THEOREM 2** **Flux of the Inverse-Square Field** The flux of $\mathbf{F}_{\text{IS}} = \dfrac{\mathbf{e}_r}{r^2}$ through closed surfaces has the following remarkable description:
>
> $$\iint_{\mathcal{S}} \left(\frac{\mathbf{e}_r}{r^2}\right) \cdot d\mathbf{S} = \begin{cases} 4\pi & \text{if } \mathcal{S} \text{ encloses the origin} \\ 0 & \text{if } \mathcal{S} \text{ does not enclose the origin} \end{cases}$$

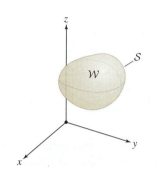

FIGURE 13 $\mathcal{S}$ is contained in the domain of $\mathbf{F}_{\text{IS}}$ (away from the origin).

Proof First, assume that $\mathcal{S}$ does not contain the origin (Figure 13). Then the region $\mathcal{W}$ enclosed by $\mathcal{S}$ is contained in the domain of $\mathbf{F}_{\text{IS}}$ and we can apply the Divergence Theorem. By Example 4, $\text{div}(\mathbf{F}_{\text{IS}}) = 0$, and therefore,

$$\iint_{\mathcal{S}} \left(\frac{\mathbf{e}_r}{r^2}\right) \cdot d\mathbf{S} = \iiint_{\mathcal{W}} \text{div}(\mathbf{F}_{\text{IS}}) \, dV = \iiint_{\mathcal{W}} 0 \, dV = 0$$

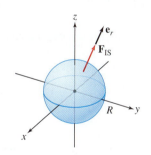

FIGURE 14

Next, let $\mathcal{S}_R$ be the sphere of radius R centered at the origin (Figure 14). We cannot use the Divergence Theorem because $\mathcal{S}_R$ contains a point (the origin) where $\mathbf{F}_{\text{IS}}$ is not defined. However, we can compute the flux of $\mathbf{F}_{\text{IS}}$ through $\mathcal{S}_R$ using spherical coordinates. Recall from Section 16.4 [Eq. (5)] that the outward-pointing normal vector to the sphere in spherical coordinates is

$$\mathbf{N} = \mathbf{T}_\phi \times \mathbf{T}_\theta = (R^2 \sin \phi) \mathbf{e}_r$$

The inverse-square field on $\mathcal{S}_R$ is simply $\mathbf{F}_{\text{IS}} = R^{-2} \mathbf{e}_r$, and thus,

$$\mathbf{F}_{\text{IS}} \cdot \mathbf{N} = (R^{-2} \mathbf{e}_r) \cdot (R^2 \sin \phi \, \mathbf{e}_r) = \sin \phi (\mathbf{e}_r \cdot \mathbf{e}_r) = \sin \phi$$

$$\iint_{\mathcal{S}_R} \mathbf{F}_{\text{IS}} \cdot d\mathbf{S} = \int_0^{2\pi} \int_0^{\pi} \mathbf{F}_{\text{IS}} \cdot \mathbf{N} \, d\phi \, d\theta$$

$$= \int_0^{2\pi} \int_0^{\pi} \sin \phi \, d\phi \, d\theta$$

$$= 2\pi \int_0^{\pi} \sin \phi \, d\phi = 4\pi$$

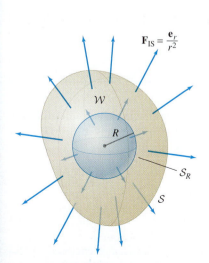

FIGURE 15 $\mathcal{W}$ is the region between $\mathcal{S}$ and the sphere $\mathcal{S}_R$.

To extend this result to *any* surface $\mathcal{S}$ containing the origin, choose a sphere $\mathcal{S}_R$ whose radius $R > 0$ is so small that $\mathcal{S}_R$ is contained inside $\mathcal{S}$. Let $\mathcal{W}$ be the region *between* $\mathcal{S}_R$ and $\mathcal{S}$ (Figure 15). The oriented boundary of $\mathcal{W}$ is the difference

$$\partial \mathcal{W} = \mathcal{S} - \mathcal{S}_R$$

This means that $\mathcal{S}$ is oriented by outward-pointing normals and $\mathcal{S}_R$ by inward-pointing normals. By the Divergence Theorem,

$$\iint_{\partial \mathcal{W}} \mathbf{F}_{\text{IS}} \cdot d\mathbf{S} = \iint_{\mathcal{S}} \mathbf{F}_{\text{IS}} \cdot d\mathbf{S} - \iint_{\mathcal{S}_R} \mathbf{F}_{\text{IS}} \cdot d\mathbf{S}$$

$$= \iiint_{\mathcal{W}} \text{div}(\mathbf{F}_{\text{IS}}) \, dV \qquad \text{(Divergence Theorem)}$$

$$= \iiint_{\mathcal{W}} 0 \, dV = 0 \qquad [\text{Because div}(\mathbf{F}_{\text{IS}}) = 0]$$

This proves that the fluxes through $\mathcal{S}$ and $\mathcal{S}_R$ are equal, and hence, both equal 4π.

To verify that the Divergence Theorem remains valid for regions between two surfaces, such as the region $\mathcal{W}$ in Figure 15, we cut $\mathcal{W}$ down the middle. Each half is a region enclosed by a surface, so the the Divergence Theorem as we have stated it applies. By adding the results for the two halves, we obtain the Divergence Theorem for $\mathcal{W}$. This uses the fact that the fluxes through the common face of the two halves cancel since the common faces have opposite orientations.

Notice that we just applied the Divergence Theorem to a region $\mathcal{W}$ that lies *between two surfaces, one contained in the other*. This is a more general form of the theorem than the one we stated formally in Theorem 1 above. The marginal comment explains why this is justified. ∎

This result applies directly to the electric field $\mathbf{E}$ of a point charge, which is a multiple of the inverse-square vector field. For a charge of q coulombs at the origin,

$$\mathbf{E} = \left(\frac{q}{4\pi\epsilon_0} \right) \frac{\mathbf{e}_r}{r^2}$$

where $\epsilon_0 = 8.85 \times 10^{-12}$ C^2/N-m^2 is the permittivity constant. Therefore,

$$\text{Flux of } \mathbf{E} \text{ through } \mathcal{S} = \begin{cases} \dfrac{q}{\epsilon_0} & \text{if } q \text{ is inside } \mathcal{S} \\ 0 & \text{if } q \text{ is outside } \mathcal{S} \end{cases}$$

Now, instead of placing just one point charge at the origin, we may distribute a finite number M of point charges q_i at different points in space. The resulting electric field $\mathbf{E}$ is the sum of the fields $\mathbf{E}_i$ due to the individual charges, and

$$\iint_{\mathcal{S}} \mathbf{E} \cdot d\mathbf{S} = \iint_{\mathcal{S}} \mathbf{E}_1 \cdot d\mathbf{S} + \cdots + \iint_{\mathcal{S}} \mathbf{E}_M \cdot d\mathbf{S}$$

Each integral on the right is either 0 or q_i/ϵ_0, according to whether or not $\mathcal{S}$ contains q_i, so we conclude that

$$\iint_{\mathcal{S}} \mathbf{E} \cdot d\mathbf{S} = \frac{\text{total charge enclosed by } \mathcal{S}}{\epsilon_0} \qquad \boxed{5}$$

This fundamental relation is called **Gauss's Law**. A limiting argument shows that Eq. (5) remains valid for the electric field due to a *continuous* distribution of charge.

The next theorem, describing the electric field due to a uniformly charged sphere, is a classic application of Gauss's Law.

We proved Theorem 3 in the analogous case of a gravitational field (also a radial inverse-square field) by a laborious calculation in Exercise 48 of Section 16.4. Here, we have derived it from Gauss's Law and a simple appeal to symmetry.

> **THEOREM 3 Uniformly Charged Sphere**   The electric field due to a uniformly charged hollow sphere $\mathcal{S}_R$ of radius R, centered at the origin and of total charge Q, is
>
> $$\mathbf{E} = \begin{cases} \left(\dfrac{Q}{4\pi\epsilon_0} \right) \dfrac{\mathbf{e}_r}{r^2} & \text{if } r > R \\ \mathbf{0} & \text{if } r < R \end{cases} \qquad \boxed{6}$$
>
> where $\epsilon_0 = 8.85 \times 10^{-12}$ C^2/N-m^2.

Proof By symmetry (Figure 16), the electric field $\mathbf{E}$ must be directed in the radial direction $\mathbf{e}_r$ with magnitude depending only on the distance r to the origin. Thus, $\mathbf{E} = E(r)\mathbf{e}_r$ for some function $E(r)$. The flux of $\mathbf{E}$ through the sphere $\mathcal{S}_r$ of radius r is

$$\iint_{\mathcal{S}_r} \mathbf{E} \cdot d\mathbf{S} = E(r) \underbrace{\iint_{\mathcal{S}_r} \mathbf{e}_r \cdot d\mathbf{S}}_{\text{Surface area of sphere}} = 4\pi r^2 E(r)$$

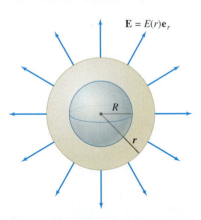

$\mathbf{E} = E(r)\mathbf{e}_r$

R

r

FIGURE 16 The electric field due to a uniformly charged sphere.

By Gauss's Law, this flux is equal to C/ϵ_0, where C is the charge enclosed by $\mathcal{S}_r$. If $r < R$, then $C = 0$ and $\mathbf{E} = \mathbf{0}$. If $r > R$, then $C = Q$ and $4\pi r^2 E(r) = Q/\epsilon_0$, or $E(r) = Q/(\epsilon_0 4\pi r^2)$. This proves Eq. (6). ∎

CONCEPTUAL INSIGHT Here is a summary of the basic operations on functions and vector fields:

$$
f \xrightarrow{\;\nabla\;} \mathbf{F} \xrightarrow{\;\text{curl}\;} \mathbf{G} \xrightarrow{\;\text{div}\;} g
$$
function   vector field   vector field   function

One basic fact is that the result of two consecutive operations in this diagram is zero:

$$
\text{curl}(\nabla(f)) = \mathbf{0}, \qquad \text{div}(\text{curl}(\mathbf{F})) = 0
$$

We saw that the first identity holds in Theorem 1 of Section 16.1. The second identity appeared as Exercise 33 in Section 16.1. An interesting question is whether every vector field satisfying $\text{curl}(\mathbf{F}) = \mathbf{0}$ is necessarily conservative—that is, $\mathbf{F} = \nabla f$ for some function f. The answer is yes, but only if the domain $\mathcal{D}$ is simply connected (every path can be shrunk to a point in $\mathcal{D}$). We saw in Section 16.3 that the vortex vector satisfies $\text{curl}(\mathbf{F}) = \mathbf{0}$ and yet cannot be conservative because its circulation around the unit circle is nonzero (the circulation of a conservative vector field is always zero). However, the domain of the vortex vector field is $\mathbf{R}^2$ with the origin removed, and this domain is not simply connected.

The situation for vector potentials is similar. Can every vector field $\mathbf{G}$ satisfying $\text{div}(\mathbf{G}) = 0$ be written in the form $\mathbf{G} = \text{curl}(\mathbf{A})$ for some vector potential $\mathbf{A}$? Again, the answer is yes—provided that the domain is a region $\mathcal{W}$ in $\mathbf{R}^3$ that has "no holes," like a ball, cube, or all of $\mathbf{R}^3$. The inverse-square field $\mathbf{F}_{\text{IS}} = \mathbf{e}_r/r^2$ plays the role of vortex field in this setting: Although $\text{div}(\mathbf{F}_{\text{IS}}) = 0$, $\mathbf{F}_{\text{IS}}$ cannot have a vector potential because its flux through the unit sphere is nonzero as shown in Theorem 2 (the flux over a closed surface of a vector field with a vector potential is always zero by Theorem 2 of Section 17.2). In this case, the domain of $\mathbf{F}_{\text{IS}} = \mathbf{e}_r/r^2$ is $\mathbf{R}^3$ with the origin removed, which "has a hole."

These properties of the vortex and inverse-square vector fields are significant because they relate line and surface integrals to "topological" properties of the domain, such as whether the domain is simply connected or has holes. They are a first hint of the important and fascinating connections between vector analysis and the area of mathematics called topology.

17.3 SUMMARY

- The Divergence Theorem: If $\mathcal{W}$ is a region in $\mathbf{R}^3$ whose boundary $\partial\mathcal{W}$ is a surface, oriented by normal vectors pointing outside $\mathcal{W}$, then

$$
\iint_{\partial\mathcal{W}} \mathbf{F} \cdot d\mathbf{S} = \iiint_{\mathcal{W}} \text{div}(\mathbf{F})\, dV
$$

- Corollary: If $\text{div}(\mathbf{F}) = 0$, then $\mathbf{F}$ has zero flux through the boundary $\partial\mathcal{W}$ of any $\mathcal{W}$ contained in the domain of $\mathbf{F}$.
- The divergence $\text{div}(\mathbf{F})$ is interpreted as "flux per unit volume," which means that the flux through a small closed surface containing a point P is approximately equal to $\text{div}(\mathbf{F})(P)$ times the enclosed volume.
- Basic operations on functions and vector fields:

$$
f \xrightarrow{\;\nabla\;} \mathbf{F} \xrightarrow{\;\text{curl}\;} \mathbf{G} \xrightarrow{\;\text{div}\;} g
$$
function   vector field   vector field   function

- The result of two consecutive operations is zero:

$$
\text{curl}(\nabla(f)) = \mathbf{0}, \qquad \text{div}(\text{curl}(\mathbf{F})) = 0
$$

- The inverse-square field $\mathbf{F}_{\text{IS}} = \mathbf{e}_r/r^2$, defined for $r \neq 0$, satisfies $\text{div}(\mathbf{F}_{\text{IS}}) = 0$. The flux of $\mathbf{F}_{\text{IS}}$ through a closed surface $\mathcal{S}$ is 4π if $\mathcal{S}$ contains the origin and is zero otherwise.

(© SSPL/The Image Works)

HISTORICAL PERSPECTIVE

James Clerk Maxwell (1831–1879)

Vector analysis was developed in the nineteenth century, in large part, to express the laws of electricity and magnetism. Electromagnetism was studied intensively in the period 1750–1890, culminating in the famous Maxwell Equations, which provide a unified understanding in terms of two vector fields: the electric field **E** and the magnetic field **B**. In a region of empty space (where there are no charged particles), the Maxwell Equations are

> *This is not just mathematical elegance... but beauty.*
> *It is so simple and yet it describes something so complex.*
>
> Francis Collins (1950–), leading geneticist and former director of the Human Genome Project, speaking of the Maxwell Equations.

$$\text{div}(\mathbf{E}) = 0, \qquad \text{div}(\mathbf{B}) = 0$$
$$\text{curl}(\mathbf{E}) = -\frac{\partial \mathbf{B}}{\partial t}, \qquad \text{curl}(\mathbf{B}) = \mu_0 \epsilon_0 \frac{\partial \mathbf{E}}{\partial t}$$

where μ_0 and ϵ_0 are experimentally determined constants. In SI units,

$$\mu_0 = 4\pi \times 10^{-7} \text{ henries/m}$$
$$\epsilon_0 \approx 8.85 \times 10^{-12} \text{ farads/m}$$

These equations led Maxwell to make two predictions of fundamental importance: (1) that electromagnetic waves exist (this was confirmed by H. Hertz in 1887), and (2) that light is an electromagnetic wave.

How do the Maxwell Equations suggest that electromagnetic waves exist? And why did Maxwell conclude that light is an electromagnetic wave? It was known to mathematicians in the eighteenth century that waves traveling with velocity c may be described by functions $\varphi(x, y, z, t)$ that satisfy the *wave equation*

$$\Delta \varphi = \frac{1}{c^2} \frac{\partial^2 \varphi}{\partial t^2} \qquad \boxed{7}$$

where Δ is the Laplace operator (or "Laplacian")

$$\Delta \varphi = \frac{\partial^2 \varphi}{\partial x^2} + \frac{\partial^2 \varphi}{\partial y^2} + \frac{\partial^2 \varphi}{\partial z^2}$$

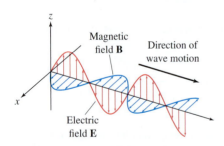

FIGURE 17 The **E** and **B** fields of an electromagnetic wave along an axis of motion.

We will show that the components of **E** satisfy this wave equation. Take the curl of both sides of Maxwell's third equation:

$$\text{curl}(\text{curl}(\mathbf{E})) = \text{curl}\left(-\frac{\partial \mathbf{B}}{\partial t}\right) = -\frac{\partial}{\partial t} \text{curl}(\mathbf{B})$$

Then apply Maxwell's fourth equation to obtain

$$\text{curl}(\text{curl}(\mathbf{E})) = -\frac{\partial}{\partial t}\left(\mu_0 \epsilon_0 \frac{\partial \mathbf{E}}{\partial t}\right)$$
$$= -\mu_0 \epsilon_0 \frac{\partial^2 \mathbf{E}}{\partial t^2} \qquad \boxed{8}$$

Finally, let us define the Laplacian of a vector field

$$\mathbf{F} = \langle F_1, F_2, F_3 \rangle$$

by applying the Laplacian Δ to each component, $\Delta \mathbf{F} = \langle \Delta F_1, \Delta F_2, \Delta F_3 \rangle$. Then the following identity holds (see Exercise 34):

$$\text{curl}(\text{curl}(\mathbf{F})) = \nabla(\text{div}(\mathbf{F})) - \Delta \mathbf{F}$$

Applying this identity to **E**, we obtain $\text{curl}(\text{curl}(\mathbf{E})) = -\Delta \mathbf{E}$ because $\text{div}(\mathbf{E}) = 0$ by Maxwell's first equation. Thus, Eq. (8) yields

$$\boxed{\Delta \mathbf{E} = \mu_0 \epsilon_0 \frac{\partial^2 \mathbf{E}}{\partial t^2}}$$

In other words, each component of the electric field satisfies the wave equation (7), with $c = (\mu_0 \epsilon_0)^{-1/2}$. This tells us that the **E**-field (and similarly the **B**-field) can propagate through space like a wave, giving rise to electromagnetic radiation (Figure 17).

Maxwell computed the velocity c of an electromagnetic wave:

$$c = (\mu_0 \epsilon_0)^{-1/2} \approx 3 \times 10^8 \text{ m/s}$$

and observed that the value is suspiciously close to the velocity of light (first measured by Olaf Römer in 1676). This had to be more than a coincidence, as Maxwell wrote in 1862: "We can scarcely avoid the conclusion that light consists in the transverse undulations of the same medium which is the cause of electric and magnetic phenomena." Needless to say, the wireless technologies that drive our modern society rely on the unseen electromagnetic radiation whose existence Maxwell first predicted on mathematical grounds.

17.3 EXERCISES

Preliminary Questions

1. What is the flux of $\mathbf{F} = \langle 1, 0, 0 \rangle$ through a closed surface?

2. Justify the following statement: The flux of $\mathbf{F} = \langle x^3, y^3, z^3 \rangle$ through every closed surface is positive.

3. Which of the following expressions are meaningful (where $\mathbf{F}$ is a vector field and f is a function)? Of those that are meaningful, which are automatically zero?

(a) $\operatorname{div}(\nabla f)$ **(b)** $\operatorname{curl}(\nabla f)$ **(c)** $\nabla \operatorname{curl}(f)$
(d) $\operatorname{div}(\operatorname{curl}(\mathbf{F}))$ **(e)** $\operatorname{curl}(\operatorname{div}(\mathbf{F}))$ **(f)** $\nabla(\operatorname{div}(\mathbf{F}))$

4. Which of the following statements is correct (where $\mathbf{F}$ is a continuously differentiable vector field defined everywhere)?

(a) The flux of $\operatorname{curl}(\mathbf{F})$ through all surfaces is zero.
(b) If $\mathbf{F} = \nabla\varphi$, then the flux of $\mathbf{F}$ through all surfaces is zero.
(c) The flux of $\operatorname{curl}(\mathbf{F})$ through all closed surfaces is zero.

5. How does the Divergence Theorem imply that the flux of $\mathbf{F} = \langle x^2, y - e^z, y - 2zx \rangle$ through a closed surface is equal to the enclosed volume?

Exercises

In Exercises 1–4, verify the Divergence Theorem for the vector field and region.

1. $\mathbf{F}(x, y, z) = \langle z, x, y \rangle$, the box $[0, 4] \times [0, 2] \times [0, 3]$

2. $\mathbf{F}(x, y, z) = \langle y, x, z \rangle$, the region $x^2 + y^2 + z^2 \le 4$

3. $\mathbf{F}(x, y, z) = \langle 2x, 3z, 3y \rangle$, the region $x^2 + y^2 \le 1, 0 \le z \le 2$

4. $\mathbf{F}(x, y, z) = \langle x, 0, 0 \rangle$, the region $x^2 + y^2 \le z \le 4$

In Exercises 5–16, use the Divergence Theorem to evaluate the flux $\iint_{\mathcal{S}} \mathbf{F} \cdot d\mathbf{S}$.

5. $\mathbf{F}(x, y, z) = \left\langle 0, 0, z^3/3 \right\rangle$, $\mathcal{S}$ is the sphere $x^2 + y^2 + z^2 = 1$.

6. $\mathbf{F}(x, y, z) = \langle y, z, x \rangle$, $\mathcal{S}$ is the sphere $x^2 + y^2 + z^2 = 1$.

7. $\mathbf{F}(x, y, z) = \left\langle xy^2, yz^2, zx^2 \right\rangle$, $\mathcal{S}$ is the boundary of the cylinder given by $x^2 + y^2 \le 4, 0 \le z \le 3$.

8. $\mathbf{F}(x, y, z) = \left\langle x^2z, yx, xyz \right\rangle$, $\mathcal{S}$ is the boundary of the tetrahedron given by $x + y + z \le 1, 0 \le x, 0 \le y, 0 \le z$.

9. $\mathbf{F}(x, y, z) = \left\langle x + z^2, xz + y^2, zx - y \right\rangle$, $\mathcal{S}$ is the surface that bounds the solid region with boundary given by the parabolic cylinder $z = 1 - x^2$, and the planes $z = 0$, $y = 0$ and $z + y = 5$.

10. $\mathbf{F}(x, y, z) = \left\langle zx, yx^3, x^2z \right\rangle$, $\mathcal{S}$ is the surface that bounds the solid region with boundary given by $y = 4 - x^2 + z^2$ and $y = 0$.

11. $\mathbf{F}(x, y, z) = \left\langle x^3, 0, z^3 \right\rangle$, $\mathcal{S}$ is boundary of the region in the first octant of space given by $x^2 + y^2 + z^2 \le 4$, and $x \ge 0, y \ge 0, z \ge 0$.

12. $\mathbf{F}(x, y, z) = \left\langle e^{x+y}, e^{x+z}, e^{x+y} \right\rangle$, $\mathcal{S}$ is the boundary of the unit cube $0 \le x \le 1, 0 \le y \le 1, 0 \le z \le 1$.

13. $\mathbf{F}(x, y, z) = \left\langle x, y^2, z + y \right\rangle$, $\mathcal{S}$ is the boundary of the region contained in the cylinder $x^2 + y^2 = 4$ between the planes $z = x$ and $z = 8$.

14. $\mathbf{F}(x, y, z) = \left\langle x^2 - z^2, e^{z^2} - \cos x, y^3 \right\rangle$, $\mathcal{S}$ is the boundary of the region bounded by $x + 2y + 4z = 12$ and the coordinate planes in the first octant.

15. $\mathbf{F}(x, y, z) = \langle x + y, z, z - x \rangle$, $\mathcal{S}$ is the boundary of the region between the paraboloid $z = 9 - x^2 - y^2$ and the xy-plane.

16. $\mathbf{F}(x, y, z) = \left\langle e^{z^2}, 2y + \sin(x^2z), 4z + \sqrt{x^2 + 9y^2} \right\rangle$, $\mathcal{S}$ is the region $x^2 + y^2 \le z \le 8 - x^2 - y^2$.

17. Calculate the flux of the vector field $\mathbf{F} = 2xy\mathbf{i} - y^2\mathbf{j} + \mathbf{k}$ through the surface $\mathcal{S}$ in Figure 18. *Hint:* Apply the Divergence Theorem to the closed surface consisting of $\mathcal{S}$ and the unit disk.

18. Let $\mathcal{S}_1$ be the closed surface consisting of $\mathcal{S}$ in Figure 18 together with the unit disk. Find the volume enclosed by $\mathcal{S}_1$, assuming that

$$\iint_{\mathcal{S}_1} \langle x, 2y, 3z \rangle \cdot d\mathbf{S} = 72$$

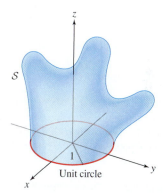

FIGURE 18 Surface $\mathcal{S}$ whose boundary is the unit circle.

19. Let $\mathcal{S}$ be the half-cylinder $x^2 + y^2 = 1, x \ge 0, 0 \le z \le 1$. Assume that $\mathbf{F}$ is a horizontal vector field (the z-component is zero) such that $\mathbf{F}(0, y, z) = zy^2\mathbf{i}$. Let $\mathcal{W}$ be the solid region enclosed by $\mathcal{S}$, and assume that

$$\iiint_{\mathcal{W}} \operatorname{div}(\mathbf{F}) \, dV = 4$$

Find the flux of $\mathbf{F}$ through the curved side of $\mathcal{S}$.

20. Volume as a Surface Integral Let $\mathbf{F}(x, y, z) = \langle x, y, z \rangle$. Prove that if $\mathcal{W}$ is a region $\mathbf{R}^3$ with a smooth boundary $\mathcal{S}$, then

$$\operatorname{Volume}(\mathcal{W}) = \frac{1}{3} \iint_{\mathcal{S}} \mathbf{F} \cdot d\mathbf{S} \qquad \boxed{9}$$

21. Use Eq. (9) to calculate the volume of the unit ball as a surface integral over the unit sphere.

22. Verify that Eq. (9) applied to the box $[0, a] \times [0, b] \times [0, c]$ yields the volume $V = abc$.

23. Let $\mathcal{W}$ be the region in Figure 19 bounded by the cylinder $x^2 + y^2 = 4$, the plane $z = x + 1$, and the xy-plane. Use the Divergence Theorem to compute the flux of $\mathbf{F}(x, y, z) = \left\langle z, x, y + z^2 \right\rangle$ through the boundary of $\mathcal{W}$.

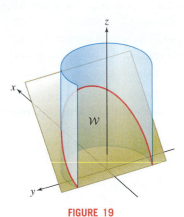

FIGURE 19

24. Let $I = \displaystyle\iint_{\mathcal{S}} \mathbf{F} \cdot d\mathbf{S}$, where

$$\mathbf{F}(x, y, z) = \left\langle \frac{2yz}{r^2}, -\frac{xz}{r^2}, -\frac{xy}{r^2} \right\rangle$$

($r = \sqrt{x^2 + y^2 + z^2}$) and $\mathcal{S}$ is the boundary of a region $\mathcal{W}$.

(a) Check that $\mathbf{F}$ is divergence-free.

(b) Show that $I = 0$ if $\mathcal{S}$ is a sphere centered at the origin. Explain, however, why the Divergence Theorem cannot be used to prove this.

25. The velocity field of a fluid $\mathbf{v}$ (in meters per second) has divergence $\operatorname{div}(\mathbf{v})(P) = 3$ at the point $P = (2, 2, 2)$. Estimate the flow rate out of the sphere of radius 0.5 centered at P.

26. A hose feeds into a small screen box of volume 10 cm^3 that is suspended in a swimming pool. Water flows across the surface of the box at a rate of 12 cm^3/s. Estimate $\operatorname{div}(\mathbf{v})(P)$, where $\mathbf{v}$ is the velocity field of the water in the pool and P is the center of the box. What are the units of $\operatorname{div}(\mathbf{v})(P)$?

27. The electric field due to a unit electric dipole oriented in the $\mathbf{k}$-direction is $\mathbf{E} = \nabla(z/r^3)$, where $r = (x^2 + y^2 + z^2)^{1/2}$ (Figure 20). Let $\mathbf{e}_r = r^{-1} \langle x, y, z \rangle$.

(a) Show that $\mathbf{E} = r^{-3}\mathbf{k} - 3zr^{-4}\mathbf{e}_r$.

(b) Calculate the flux of $\mathbf{E}$ through a sphere centered at the origin.

(c) Calculate $\operatorname{div}(\mathbf{E})$.

(d) Can we use the Divergence Theorem to compute the flux of $\mathbf{E}$ through a sphere centered at the origin?

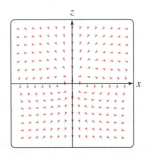

FIGURE 20 The dipole vector field restricted to the xz-plane.

28. Let $\mathbf{E}$ be the electric field due to a long, uniformly charged rod of radius R with charge density δ per unit length (Figure 21). By symmetry, we may assume that $\mathbf{E}$ is everywhere perpendicular to the rod and its magnitude $E(d)$ depends only on the distance d to the rod (strictly speaking, this would hold only if the rod were infinite, but it is nearly true if the rod is long enough). Show that $E(d) = \delta/2\pi\epsilon_0 d$ for $d > R$. *Hint:* Apply Gauss's Law to a cylinder of radius R and of unit length with its axis along the rod.

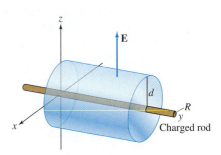

FIGURE 21

29. Let $\mathcal{W}$ be the region between the sphere of radius 4 and the cube of side 1, both centered at the origin. What is the flux through the boundary $\mathcal{S} = \partial\mathcal{W}$ of a vector field $\mathbf{F}$ whose divergence has the constant value $\operatorname{div}(\mathbf{F}) = -4$?

30. Let $\mathcal{W}$ be the region between the sphere of radius 3 and the sphere of radius 2, both centered at the origin. Use the Divergence Theorem to calculate the flux of $\mathbf{F} = x\mathbf{i}$ through the boundary $\mathcal{S} = \partial\mathcal{W}$.

31. Find and prove a Product Rule expressing $\operatorname{div}(f\mathbf{F})$ in terms of $\operatorname{div}(\mathbf{F})$ and ∇f.

32. Prove the identity

$$\operatorname{div}(\mathbf{F} \times \mathbf{G}) = \operatorname{curl}(\mathbf{F}) \cdot \mathbf{G} - \mathbf{F} \cdot \operatorname{curl}(\mathbf{G})$$

Then prove that the cross product of two irrotational vector fields is incompressible [$\mathbf{F}$ is called **irrotational** if $\operatorname{curl}(\mathbf{F}) = 0$ and **incompressible** if $\operatorname{div}(\mathbf{F}) = 0$].

33. Prove that $\operatorname{div}(\nabla f \times \nabla g) = 0$.

In Exercises 34–36, Δ denotes the Laplace operator defined by

$$\Delta\varphi = \frac{\partial^2\varphi}{\partial x^2} + \frac{\partial^2\varphi}{\partial y^2} + \frac{\partial^2\varphi}{\partial z^2}$$

34. Prove the identity

$$\operatorname{curl}(\operatorname{curl}(\mathbf{F})) = \nabla(\operatorname{div}(\mathbf{F})) - \Delta\mathbf{F}$$

where $\Delta\mathbf{F}$ denotes $\langle \Delta F_1, \Delta F_2, \Delta F_3 \rangle$.

35. A function φ satisfying $\Delta\varphi = 0$ is called **harmonic**.

(a) Show that $\Delta\varphi = \operatorname{div}(\nabla\varphi)$ for any function φ.

(b) Show that φ is harmonic if and only if $\operatorname{div}(\nabla\varphi) = 0$.

(c) Show that if $\mathbf{F}$ is the gradient of a harmonic function, then $\operatorname{curl}(F) = 0$ and $\operatorname{div}(F) = 0$.

(d) Show that $\mathbf{F}(x, y, z) = \left\langle xz, -yz, \frac{1}{2}(x^2 - y^2) \right\rangle$ is the gradient of a harmonic function. What is the flux of $\mathbf{F}$ through a closed surface?

36. Let $\mathbf{F} = r^n\mathbf{e}_r$, where n is any number, $r = (x^2 + y^2 + z^2)^{1/2}$, and $\mathbf{e}_r = r^{-1}\langle x, y, z \rangle$ is the unit radial vector.

(a) Calculate $\operatorname{div}(\mathbf{F})$.

(b) Calculate the flux of $\mathbf{F}$ through the surface of a sphere of radius R centered at the origin. For which values of n is this flux independent of R?

(c) Prove that $\nabla(r^n) = n\,r^{n-1}\mathbf{e}_r$.

(d) Use (c) to show that $\mathbf{F}$ is conservative for $n \neq -1$. Then show that $\mathbf{F} = r^{-1}\mathbf{e}_r$ is also conservative by computing the gradient of $\ln r$.

(e) What is the value of $\displaystyle\int_C \mathbf{F} \cdot d\mathbf{s}$, where C is a closed curve that does not pass through the origin?

(f) Find the values of n for which the function $\varphi = r^n$ is harmonic.

Further Insights and Challenges

37. Let $\mathcal{S}$ be the boundary surface of a region $\mathcal{W}$ in $\mathbf{R}^3$, and let $D_{\mathbf{n}}\varphi$ denote the directional derivative of φ, where $\mathbf{n}$ is the outward unit normal vector. Let Δ be the Laplace operator defined earlier.

(a) Use the Divergence Theorem to prove that

$$\iint_{\mathcal{S}} D_{\mathbf{n}}\varphi \, dS = \iiint_{\mathcal{W}} \Delta\varphi \, dV$$

(b) Show that if φ is a harmonic function (defined in Exercise 35), then

$$\iint_{\mathcal{S}} D_{\mathbf{n}}\varphi \, dS = 0$$

38. Assume that φ is harmonic. Show that $\operatorname{div}(\varphi\nabla\varphi) = \|\nabla\varphi\|^2$ and conclude that

$$\iint_{\mathcal{S}} \varphi D_{\mathbf{n}}\varphi \, dS = \iiint_{\mathcal{W}} \|\nabla\varphi\|^2 \, dV$$

39. Let $\mathbf{F} = \langle P, Q, R \rangle$ be a vector field defined on $\mathbf{R}^3$ such that $\operatorname{div}(\mathbf{F}) = 0$. Use the following steps to show that $\mathbf{F}$ has a vector potential.

(a) Let $\mathbf{A} = \langle f, 0, g \rangle$. Show that

$$\operatorname{curl}(\mathbf{A}) = \left\langle \frac{\partial g}{\partial y}, \frac{\partial f}{\partial z} - \frac{\partial g}{\partial x}, -\frac{\partial f}{\partial y} \right\rangle$$

(b) Fix any value y_0 and show that if we define

$$f(x, y, z) = -\int_{y_0}^{y} R(x, t, z)\, dt + \alpha(x, z)$$

$$g(x, y, z) = \int_{y_0}^{y} P(x, t, z)\, dt + \beta(x, z)$$

where α and β are any functions of x and z, then $\partial g/\partial y = P$ and $-\partial f/\partial y = R$.

(c) It remains for us to show that α and β can be chosen so $Q = \partial f/\partial z - \partial g/\partial x$. Verify that the following choice works (for any choice of z_0):

$$\alpha(x, z) = \int_{z_0}^{z} Q(x, y_0, t)\, dt, \qquad \beta(x, z) = 0$$

Hint: You will need to use the relation $\operatorname{div}(\mathbf{F}) = 0$.

40. Show that

$$\mathbf{F}(x, y, z) = \langle 2y - 1, 3z^2, 2xy \rangle$$

has a vector potential and find one.

41. Show that

$$\mathbf{F}(x, y, z) = \langle 2ye^z - xy, y, yz - z \rangle$$

has a vector potential and find one.

42. In the text, we observed that although the inverse-square radial vector field $\mathbf{F} = \dfrac{\mathbf{e}_r}{r^2}$ satisfies $\operatorname{div}(\mathbf{F}) = 0$, $\mathbf{F}$ cannot have a vector potential on its domain $\{(x, y, z) \neq (0, 0, 0)\}$ because the flux of $\mathbf{F}$ through a sphere containing the origin is nonzero.

(a) Show that the method of Exercise 39 produces a vector potential $\mathbf{A}$ such that $\mathbf{F} = \operatorname{curl}(\mathbf{A})$ on the restricted domain $\mathcal{D}$ consisting of $\mathbf{R}^3$ with the y-axis removed.

(b) Show that $\mathbf{F}$ also has a vector potential on the domains obtained by removing either the x-axis or the z-axis from $\mathbf{R}^3$.

(c) Does the existence of a vector potential on these restricted domains contradict the fact that the flux of $\mathbf{F}$ through a sphere containing the origin is nonzero?

CHAPTER REVIEW EXERCISES

1. Let $\mathbf{F}(x, y) = \langle x + y^2, x^2 - y \rangle$, and let C be the unit circle, oriented counterclockwise. Evaluate $\displaystyle\oint_C \mathbf{F} \cdot d\mathbf{r}$ directly as a line integral and using Green's Theorem.

2. Let $\partial\mathcal{R}$ be the boundary of the rectangle in Figure 1, and let $\partial\mathcal{R}_1$ and $\partial\mathcal{R}_2$ be the boundaries of the two triangles, all oriented counterclockwise.

(a) Determine $\displaystyle\oint_{\partial\mathcal{R}_1} \mathbf{F} \cdot d\mathbf{r}$ if $\displaystyle\oint_{\partial\mathcal{R}} \mathbf{F} \cdot d\mathbf{r} = 4$ and $\displaystyle\oint_{\partial\mathcal{R}_2} \mathbf{F} \cdot d\mathbf{r} = -2$.

(b) What is the value of $\displaystyle\oint_{\partial\mathcal{R}} \mathbf{F}\, d\mathbf{r}$ if $\partial\mathcal{R}$ is oriented clockwise?

In Exercises 3–6, use Green's Theorem to evaluate the line integral around the given closed curve.

3. $\displaystyle\oint_C xy^3\, dx + x^3 y\, dy$, where C is the rectangle $-1 \le x \le 2$, $-2 \le y \le 3$, oriented counterclockwise

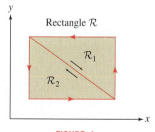

FIGURE 1

4. $\displaystyle\oint_C (3x + 5y - \cos y)\, dx + x \sin y\, dy$, where C is any closed curve enclosing a region with area 4, oriented counterclockwise

5. $\displaystyle\oint_C y^2\, dx - x^2\, dy$, where C consists of the arcs $y = x^2$ and $y = \sqrt{x}$, $0 \le x \le 1$, oriented clockwise

6. $\displaystyle\oint_C ye^x\, dx + xe^y\, dy$, where C is the triangle with vertices $(-1, 0)$, $(0, 4)$, and $(0, 1)$, oriented counterclockwise

7. Let $\mathbf{r}(t) = \left\langle t^2(1-t), t(t-1)^2 \right\rangle$.

(a) [GU] Plot the path $\mathbf{r}(t)$ for $0 \le t \le 1$.

(b) Calculate the area A of the region enclosed by $\mathbf{r}(t)$ for $0 \le t \le 1$ using the formula $A = \dfrac{1}{2} \oint_C (x\,dy - y\,dx)$.

8. Calculate the area of the region bounded by the two curves $y = x^2$ and $y = 4$ using the formula $A = \oint_C x\,dy$.

9. Calculate the area of the region bounded by the two curves $y = x^2$ and $y = \sqrt{x}$ for $x \ge 0$ using the formula $A = \oint_C x\,dy$.

10. Calculate the area of the region bounded by the two curves $y = x^2$ and $y = 4$ using the formula $A = \oint_C -y\,dx$.

11. Calculate the area of the region bounded by the two curves $y = x^2$ and $y = \sqrt{x}$ for $x \ge 0$ using the formula $A = \oint_C -y\,dx$.

12. In (a)–(d), state whether the equation is an identity (valid for all $\mathbf{F}$ or f). If it is not, provide an example in which the equation does not hold.

(a) $\operatorname{curl}(\nabla f) = \mathbf{0}$ **(b)** $\operatorname{div}(\nabla f) = 0$

(c) $\operatorname{div}(\operatorname{curl}(\mathbf{F})) = 0$ **(d)** $\nabla(\operatorname{div}(\mathbf{F})) = \mathbf{0}$

13. Let $\mathbf{F}(x, y) = \left\langle x^2 y, xy^2 \right\rangle$ be the velocity vector field for a fluid in the plane. Find all points where the angular velocity of a small paddle wheel inserted into the fluid would be 0.

14. Compute the flux $\oint_{\partial \mathcal{D}} \mathbf{F} \cdot \mathbf{n}\,ds$ of $\mathbf{F}(x, y) = \left\langle x^3, yx^2 \right\rangle$ across the unit square $\mathcal{D}$ using the vector form of Green's Theorem.

15. Compute the flux $\oint_{\partial \mathcal{D}} \mathbf{F} \cdot \mathbf{n}\,ds$ of $\mathbf{F}(x, y) = \left\langle x^3 + 2x, y^3 + y \right\rangle$ across the circle $\mathcal{D}$ given by $x^2 + y^2 = 4$ using the vector form of Green's Theorem.

16. Suppose that $\mathcal{S}_1$ and $\mathcal{S}_2$ are surfaces with the same oriented boundary curve $\mathcal{C}$. Which of the following conditions guarantees that the flux of $\mathbf{F}$ through $\mathcal{S}_1$ is equal to the flux of $\mathbf{F}$ through $\mathcal{S}_2$?

(i) $\mathbf{F} = \nabla f$ for some function f

(ii) $\mathbf{F} = \operatorname{curl}(\mathbf{G})$ for some vector field $\mathbf{G}$

17. Prove that if $\mathbf{F}$ is a gradient vector field, then the flux of $\operatorname{curl}(\mathbf{F})$ through a smooth surface $\mathcal{S}$ (whether closed or not) is equal to zero.

18. Verify Stokes' Theorem for $\mathbf{F}(x, y, z) = \langle y, z - x, 0 \rangle$ and the surface $z = 4 - x^2 - y^2$, $z \ge 0$, oriented by outward-pointing normals.

19. Let $\mathbf{F}(x, y, z) = \left\langle z^2, x + z, y^2 \right\rangle$, and let $\mathcal{S}$ be the upper half of the ellipsoid

$$\frac{x^2}{4} + y^2 + z^2 = 1$$

oriented by outward-pointing normals. Use Stokes' Theorem to compute $\iint_{\mathcal{S}} \operatorname{curl}(\mathbf{F}) \cdot d\mathbf{S}$.

20. Use Stokes' Theorem to evaluate $\oint_C \langle y, z, x \rangle \cdot d\mathbf{r}$, where $\mathcal{C}$ is the curve in Figure 2.

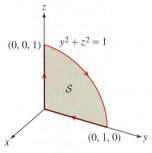

FIGURE 2

21. Let $\mathcal{S}$ be the side of the cylinder $x^2 + y^2 = 4$, $0 \le z \le 2$ (not including the top and bottom of the cylinder). Use Stokes' Theorem to compute the flux of $\mathbf{F}(x, y, z) = \langle 0, y, -z \rangle$ through $\mathcal{S}$ (with outward-pointing normal) by finding a vector potential $\mathbf{A}$ such that $\operatorname{curl}(\mathbf{A}) = \mathbf{F}$.

22. Verify the Divergence Theorem for $\mathbf{F}(x, y, z) = \langle 0, 0, z \rangle$ and the region $x^2 + y^2 + z^2 = 1$.

In Exercises 23–26, use the Divergence Theorem to calculate $\iint_{\mathcal{S}} \mathbf{F} \cdot d\mathbf{S}$ *for the given vector field and surface.*

23. $\mathbf{F}(x, y, z) = \left\langle xy, yz, x^2 z + z^2 \right\rangle$, $\mathcal{S}$ is the boundary of the box $[0, 1] \times [2, 4] \times [1, 5]$.

24. $\mathbf{F}(x, y, z) = \left\langle xy, yz, x^2 z + z^2 \right\rangle$, $\mathcal{S}$ is the boundary of the unit sphere.

25. $\mathbf{F}(x, y, z) = \left\langle xyz + xy, \frac{1}{2} y^2 (1 - z) + e^x, e^{x^2 + y^2} \right\rangle$, $\mathcal{S}$ is the boundary of the solid bounded by the cylinder $x^2 + y^2 = 16$ and the planes $z = 0$ and $z = y - 4$.

26. $\mathbf{F}(x, y, z) = \left\langle \sin(yz), \sqrt{x^2 + z^4}, x \cos(x - y) \right\rangle$, $\mathcal{S}$ is any smooth closed surface that is the boundary of a region in $\mathbf{R}^3$.

27. Find the volume of a region $\mathcal{W}$ if

$$\iint_{\partial \mathcal{W}} \left\langle x + xy + z, x + 3y - \frac{1}{2} y^2, 4z \right\rangle \cdot d\mathbf{S} = 16$$

28. Show that the circulation of $\mathbf{F}(x, y, z) = \left\langle x^2, y^2, z(x^2 + y^2) \right\rangle$ around any curve $\mathcal{C}$ on the surface of the cone $z^2 = x^2 + y^2$ is equal to zero (Figure 3).

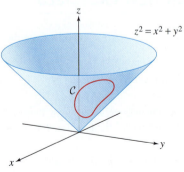

FIGURE 3

In Exercises 29–32, let $\mathbf{F}$ be a vector field whose curl and divergence at the origin are

$$\operatorname{curl}(\mathbf{F})(0, 0, 0) = \langle 2, -1, 4 \rangle, \qquad \operatorname{div}(\mathbf{F})(0, 0, 0) = -2$$

29. Estimate $\oint_C \mathbf{F} \cdot d\mathbf{r}$, where $\mathcal{C}$ is the circle of radius 0.03 in the xy-plane centered at the origin.

30. Estimate $\oint_{\mathcal{C}} \mathbf{F} \cdot d\mathbf{r}$, where $\mathcal{C}$ is the boundary of the square of side 0.03 in the yz-plane centered at the origin. Does the estimate depend on how the square is oriented within the yz-plane? Might the actual circulation depend on how it is oriented?

31. Suppose that $\mathbf{v}$ is the velocity field of a fluid and imagine placing a small paddle wheel at the origin. Find the equation of the plane in which the paddle wheel should be placed to make it rotate as quickly as possible.

32. Estimate the flux of $\mathbf{F}$ through the box of side 0.5 in Figure 4. Does the result depend on how the box is oriented relative to the coordinate axes?

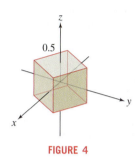

FIGURE 4

33. The velocity vector field of a fluid (in meters per second) is

$$\mathbf{F}(x, y, z) = \langle x^2 + y^2, 0, z^2 \rangle$$

Let $\mathcal{W}$ be the region between the hemisphere

$$S = \{(x, y, z) : x^2 + y^2 + z^2 = 1, \quad x, y, z \geq 0\}$$

and the disk $\mathcal{D} = \{(x, y, 0) : x^2 + y^2 \leq 1\}$ in the xy-plane. Recall that the flow rate of a fluid across a surface is equal to the flux of $\mathbf{F}$ through the surface.

(a) Show that the flow rate across $\mathcal{D}$ is zero.

(b) Use the Divergence Theorem to show that the flow rate across S, oriented with an outward-pointing normal, is equal to

$$\iiint_{\mathcal{W}} \operatorname{div}(\mathbf{F}) \, dV.$$ Then compute this triple integral.

34. The velocity field of a fluid (in meters per second) is

$$\mathbf{F} = (3y - 4)\mathbf{i} + e^{-y(z+1)}\mathbf{j} + (x^2 + y^2)\mathbf{k}$$

(a) Estimate the flow rate (in cubic meters per second) through a small surface S around the origin if S encloses a region of volume 0.01 m³.

(b) Estimate the circulation of $\mathbf{F}$ about a circle in the xy-plane of radius $r = 0.1$ m centered at the origin (oriented counterclockwise when viewed from above).

(c) Estimate the circulation of $\mathbf{F}$ about a circle in the yz-plane of radius $r = 0.1$ m centered at the origin (oriented counterclockwise when viewed from the positive x-axis).

35. Let $f(x, y) = x + \dfrac{x}{x^2 + y^2}$. The vector field $\mathbf{F} = \nabla f$ (Figure 5) provides a model in the plane of the velocity field of an incompressible, irrotational fluid flowing past a cylindrical obstacle (in this case, the obstacle is the unit circle $x^2 + y^2 = 1$).

(a) Verify that $\mathbf{F}$ is irrotational [by definition, $\mathbf{F}$ is irrotational if curl($\mathbf{F}$) = $\mathbf{0}$].

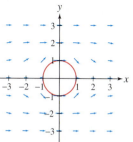

FIGURE 5 The vector field ∇f for $f(x, y) = x + \dfrac{x}{x^2 + y^2}$.

(b) Verify that $\mathbf{F}$ is tangent to the unit circle at each point along the unit circle except $(1, 0)$ and $(-1, 0)$ (where $\mathbf{F} = \mathbf{0}$).

(c) What is the circulation of $\mathbf{F}$ around the unit circle?

(d) Calculate the line integral of $\mathbf{F}$ along the upper and lower halves of the unit circle separately.

36. Figure 6 shows the vector field $\mathbf{F} = \nabla f$, where

$$f(x, y) = \ln\left(x^2 + (y - 1)^2\right) + \ln\left(x^2 + (y + 1)^2\right)$$

which is the velocity field for the flow of a fluid with sources of equal strength at $(0, \pm 1)$ (note that f is undefined at these two points). Show that $\mathbf{F}$ is both irrotational and incompressible—that is, $\operatorname{curl}_z(\mathbf{F}) = 0$ and $\operatorname{div}(\mathbf{F}) = 0$ [in computing $\operatorname{div}(\mathbf{F})$, treat $\mathbf{F}$ as a vector field in $\mathbf{R}^3$ with a zero z-component]. Is it necessary to compute $\operatorname{curl}_z(\mathbf{F})$ to conclude that it is zero?

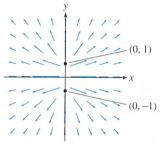

FIGURE 6 The vector field ∇f for $f(x, y) = \ln(x^2 + (y - 1)^2) + \ln(x^2 + (y + 1)^2)$.

37. In Section 17.1, we showed that if $\mathcal{C}$ is a simple closed curve, oriented counterclockwise, then the area enclosed by $\mathcal{C}$ is given by

$$\text{Area enclosed by } \mathcal{C} = \frac{1}{2} \oint_{\mathcal{C}} x \, dy - y \, dx \qquad \boxed{1}$$

Suppose that $\mathcal{C}$ is a path from P to Q that is not closed but has the property that every line through the origin intersects $\mathcal{C}$ in at most one point, as in Figure 7. Let $\mathcal{R}$ be the region enclosed by $\mathcal{C}$ and the two radial segments joining P and Q to the origin. Show that the line integral in Eq. (1) is equal to the area of $\mathcal{R}$. *Hint:* Show that the line integral of $\mathbf{F} = \langle -y, x \rangle$ along the two radial segments is zero and apply Green's Theorem.

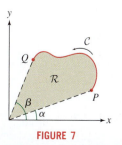

FIGURE 7

38. Suppose that the curve C in Figure 7 has the polar equation $r = f(\theta)$.

(a) Show that $\mathbf{r}(\theta) = \langle f(\theta)\cos\theta, f(\theta)\sin\theta\rangle$ is a counterclockwise parametrization of C.

(b) In Section 11.4, we showed that the area of the region R is given by the formula

$$\text{Area of } R = \frac{1}{2}\int_\alpha^\beta f(\theta)^2\, d\theta$$

Use the result of Exercise 37 to give a new proof of this formula. *Hint:* Evaluate the line integral in Eq. (1) using $\mathbf{r}(\theta)$.

39. Prove the following generalization of Eq. (1). Let C be a simple closed curve in the plane (Figure 8)

$$\mathcal{S}: \quad ax + by + cz + d = 0$$

Then the area of the region R enclosed by C is equal to

$$\frac{1}{2\|\mathbf{N}\|}\oint_C (bz - cy)\, dx + (cx - az)\, dy + (ay - bx)\, dz$$

where $\mathbf{N} = \langle a, b, c\rangle$ is the normal to $\mathcal{S}$, and C is oriented as the boundary of R (relative to the normal vector $\mathbf{N}$). *Hint:* Apply Stokes' Theorem to $\mathbf{F} = \langle bz - cy, cx - az, ay - bx\rangle$.

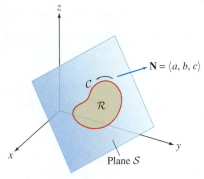

FIGURE 8

40. Use the result of Exercise 39 to calculate the area of the triangle with vertices $(1, 0, 0)$, $(0, 1, 0)$, and $(0, 0, 1)$ as a line integral. Verify your result using geometry.

41. Show that $G(\theta, \phi) = (a\cos\theta\sin\phi, b\sin\theta\sin\phi, c\cos\phi)$ is a parametrization of the ellipsoid

$$\left(\frac{x}{a}\right)^2 + \left(\frac{y}{b}\right)^2 + \left(\frac{z}{c}\right)^2 = 1$$

Then calculate the volume of the ellipsoid as the surface integral of $\mathbf{F} = \frac{1}{3}\langle x, y, z\rangle$ (this surface integral is equal to the volume by the Divergence Theorem).

A THE LANGUAGE OF MATHEMATICS

One of the challenges in learning calculus is growing accustomed to its precise language and terminology, especially in the statements of theorems. In this section, we analyze a few details of logic that are helpful, and indeed essential, in understanding and applying theorems properly.

Many theorems in mathematics involve an **implication**. If A and B are statements, then the implication $A \implies B$ is the assertion that A implies B:

$$A \implies B: \qquad \text{If } A \text{ is true, then } B \text{ is true.}$$

Statement A is called the **hypothesis** (or premise) and statement B the **conclusion** of the implication. Here is an example: *If m and n are even integers, then m + n is an even integer*. This statement may be divided into a hypothesis and conclusion:

$$\underbrace{m \text{ and } n \text{ are even integers}}_{A} \implies \underbrace{m + n \text{ is an even integer}}_{B}$$

In everyday speech, implications are often used in a less precise way. An example is: *If you work hard, then you will succeed*. Furthermore, some statements that do not initially have the form $A \implies B$ may be restated as implications. For example, the statement, "Cats are mammals," can be rephrased as follows:

$$\text{Let } X \text{ be an animal.} \quad \underbrace{X \text{ is a cat}}_{A} \implies \underbrace{X \text{ is a mammal}}_{B}$$

When we say that an implication $A \implies B$ is true, we do not claim that A or B is necessarily true. Rather, we are making the conditional statement that *if* A happens to be true, *then* B is also true. In the above, if X does not happen to be a cat, the implication tells us nothing.

The **negation** of a statement A is the assertion that A is false and is denoted $\neg A$.

Statement A	Negation $\neg A$
X lives in California.	X does not live in California.
$\triangle ABC$ is a right triangle.	$\triangle ABC$ is not a right triangle.

The negation of the negation is the original statement: $\neg(\neg A) = A$. To say that X does *not not live in California* is the same as saying that X lives in California.

■ **EXAMPLE 1** State the negation of each statement.

(a) The door is open and the dog is barking.

(b) The door is open or the dog is barking (or both).

Solution

(a) The first statement is true if two conditions are satisfied (door open and dog barking), and it is false if at least one of these conditions is not satisfied. So the negation is

Either the door is not open *OR* the dog is not barking *(or both)*.

(b) The second statement is true if at least one of the conditions (door open or dog barking) is satisfied, and it is false if neither condition is satisfied. So the negation is

The door is not open *AND* the dog is not barking. ■

Contrapositive and Converse

Two important operations are the formation of the contrapositive and the formation of the converse of a statement. The **contrapositive** of $A \implies B$ is the statement "If B is false, then A is false":

Keep in mind that when we form the contrapositive, we reverse the order of A and B. The contrapositive of $A \implies B$ is NOT $\neg A \implies \neg B$.

$$\text{The contrapositive of} \quad A \implies B \quad \text{is} \quad \neg B \implies \neg A.$$

Here are some examples:

Statement	Contrapositive
If X is a cat, then X is a mammal.	If X is not a mammal, then X is not a cat.
If you work hard, then you will succeed.	If you did not succeed, then you did not work hard.
If m and n are both even, then $m + n$ is even.	If $m + n$ is not even, then m and n are not both even.

A key observation is this:

The contrapositive and the original implication are equivalent.

The fact that $A \implies B$ is equivalent to its contrapositive $\neg B \implies \neg A$ is a general rule of logic that does not depend on what A and B happen to mean. This rule belongs to the subject of "formal logic," which deals with logical relations between statements without concern for the actual content of these statements.

In other words, if an implication is true, then its contrapositive is automatically true, and vice versa. In essence, an implication and its contrapositive are two ways of saying the same thing. For example, the contrapositive, "If X is not a mammal, then X is not a cat," is a roundabout way of saying that cats are mammals.

The **converse** of $A \implies B$ is the *reverse* implication $B \implies A$:

Implication: $A \implies B$	Converse $B \implies A$
If A is true, then B is true.	If B is true, then A is true.

The converse plays a very different role than the contrapositive because *the converse is NOT equivalent to the original implication*. The converse may be true or false, even if the original implication is true. Here are some examples:

True Statement	Converse	Converse True or False?
If X is a cat, then X is a mammal.	If X is a mammal, then X is a cat.	False
If m is even, then m^2 is even.	If m^2 is even, then m is even.	True

A counterexample is an example that satisfies the hypothesis but not the conclusion of a statement. If a single counterexample exists, then the statement is false. However, we cannot prove that a statement is true merely by giving an example.

■ **EXAMPLE 2   An Example Where the Converse Is False**   Show that the converse of, "If m and n are even, then $m + n$ is even," is false.

Solution   The converse is, "If $m + n$ is even, then m and n are even." To show that the converse is false, we display a counterexample. Take $m = 1$ and $n = 3$ (or any other pair of odd numbers). The sum is even (since $1 + 3 = 4$) but neither 1 nor 3 is even. Therefore, the converse is false.   ■

■ **EXAMPLE 3   An Example Where the Converse Is True**   State the contrapositive and converse of the Pythagorean Theorem. Are either or both of these true?

Solution   Consider a triangle with sides a, b, and c, and let θ be the angle opposite the side of length c, as in Figure 1. The Pythagorean Theorem states that if $\theta = 90°$, then $a^2 + b^2 = c^2$. Here are the contrapositive and converse:

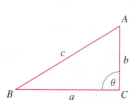

FIGURE 1

Pythagorean Theorem	$\theta = 90° \implies a^2 + b^2 = c^2$	True
Contrapositive	$a^2 + b^2 \neq c^2 \implies \theta \neq 90°$	Automatically true
Converse	$a^2 + b^2 = c^2 \implies \theta = 90°$	True (but not automatic)

The contrapositive is automatically true because it is just another way of stating the original theorem. The converse is not automatically true since there could conceivably exist a nonright triangle that satisfies $a^2 + b^2 = c^2$. However, the converse of the Pythagorean Theorem is, in fact, true. This follows from the Law of Cosines (see Exercise 38). ∎

When both a statement $A \implies B$ and its converse $B \implies A$ are true, we write $A \Longleftrightarrow B$. In this case, A and B are **equivalent**. We often express this with the phrase

$$A \Longleftrightarrow B \qquad A \text{ is true } \textit{if and only if} \text{ } B \text{ is true.}$$

For example,

$$a^2 + b^2 = c^2 \qquad \text{if and only if} \qquad \theta = 90°$$

$$\text{It is morning} \qquad \text{if and only if} \qquad \text{the sun is rising.}$$

We mention the following variations of terminology involving implications that you may come across:

Statement	Is Another Way of Saying
A is true <u>if</u> B is true.	$B \implies A$
A is true <u>only if</u> B is true.	$A \implies B$ (A cannot be true unless B is also true.)
For A to be true, <u>it is necessary</u> that B be true.	$A \implies B$ (A cannot be true unless B is also true.)
For A to be true, <u>it is sufficient</u> that B be true.	$B \implies A$
For A to be true, it is <u>necessary and sufficient</u> that B be true.	$B \Longleftrightarrow A$

Analyzing a Theorem

To see how these rules of logic arise in calculus, consider the following result from Section 4.2:

> **THEOREM 1   Existence of Extrema on a Closed Interval**   A continuous function f on a closed (bounded) interval $I = [a, b]$ takes on both a minimum and a maximum value on I (Figure 2).

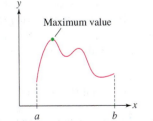

FIGURE 2 A continuous function on a closed interval $I = [a, b]$ has a maximum value.

To analyze this theorem, let's write out the hypotheses and conclusion separately:

Hypotheses A: f is continuous and I is closed.

Conclusion B: f takes on a minimum and a maximum value on I.

A first question to ask is: "Are the hypotheses necessary?" Is the conclusion still true if we drop one or both assumptions? To show that both hypotheses are necessary, we provide counterexamples:

- **The continuity of f is a necessary hypothesis.** Figure 3(A) shows the graph of a function on a closed interval $[a, b]$ that is not continuous. This function has no maximum value on $[a, b]$, which shows that the conclusion may fail if the continuity hypothesis is not satisfied.
- **The hypothesis that I is closed is necessary.** Figure 3(B) shows the graph of a continuous function on an *open interval* (a, b). This function has no maximum value, which shows that the conclusion may fail if the interval is not closed.

We see that both hypotheses in Theorem 1 are necessary. In stating this, we do not claim that the conclusion *always* fails when one or both of the hypotheses are not satisfied. We claim only that the conclusion *may* fail when the hypotheses are not satisfied. Next, let's analyze the contrapositive and converse:

- **Contrapositive $\neg B \Longrightarrow \neg A$ (automatically true):** If f does not have a minimum and a maximum value on I, then either f is not continuous or I is not closed (or both).
- **Converse $B \Longrightarrow A$ (in this case, false):** If f has a minimum and a maximum value on I, then f is continuous and I is closed. We prove this statement false with a counterexample [Figure 3(C)].

As we know, the contrapositive is merely a way of restating the theorem, so it is automatically true. The converse is not automatically true, and in fact, in this case it is false. The function in Figure 3(C) provides a counterexample to the converse: f has a maximum value on $I = (a, b)$, but f is not continuous and I is not closed.

The technique of proof by contradiction is also known by its Latin name reductio ad absurdum or "reduction to the absurd." The ancient Greek mathematicians used proof by contradiction as early as the fifth century BCE, and Euclid (325–265 BCE) employed it in his classic treatise on geometry entitled The Elements. A famous example is the proof that $\sqrt{2}$ is irrational in Example 4. The philosopher Plato (427–347 BCE) wrote: "He is unworthy of the name of man who is ignorant of the fact that the diagonal of a square is incommensurable with its side."

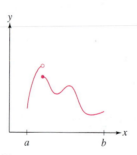

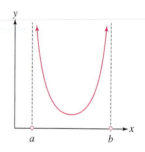

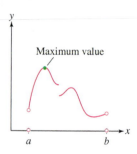

(A) The interval is closed but the function is not continuous. The function has no maximum value.

(B) The function is continuous but the interval is open. The function has no maximum value.

(C) This function is not continuous and the interval is not closed, but the function does have a maximum value.

FIGURE 3

Mathematicians have devised various general strategies and methods for proving theorems. The method of proof by induction is discussed in Appendix C. Another important method is **proof by contradiction**, also called **indirect proof**. Suppose our goal is to prove statement A. In a proof by contradiction, we start by assuming that A is false, and then show that this leads to a contradiction. Therefore, A must be true (to avoid the contradiction).

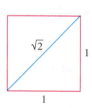

FIGURE 4 The diagonal of the unit square has length $\sqrt{2}$.

■ **EXAMPLE 4** **Proof by Contradiction** The number $\sqrt{2}$ is irrational (Figure 4).

Solution Assume that the theorem is false, namely that $\sqrt{2} = p/q$, where p and q are whole numbers. We may assume that p/q is in lowest terms, and therefore, at most one of p and q is even. Note that if the square m^2 of a whole number is even, then m itself must be even.

The relation $\sqrt{2} = p/q$ implies that $2 = p^2/q^2$ or $p^2 = 2q^2$. This shows that p must be even. But if p is even, then $p = 2m$ for some whole number m, and $p^2 = 4m^2$. Because $p^2 = 2q^2$, we obtain $4m^2 = 2q^2$, or $q^2 = 2m^2$. This shows that q is also even. But we chose p and q so that at most one of them is even. This contradiction shows that our original assumption, that $\sqrt{2} = p/q$, must be false. Therefore, $\sqrt{2}$ is irrational. ■

One of the most famous problems in mathematics is known as "Fermat's Last Theorem." It states that the equation

$$x^n + y^n = z^n$$

has no solutions in positive integers if $n \geq 3$. In a marginal note written around 1630, Fermat claimed to have a proof, and over the centuries, that assertion was verified for many values of the exponent n. However, only in 1994 did the British-American mathematician Andrew Wiles, working at Princeton University, find a complete proof.

CONCEPTUAL INSIGHT The hallmark of mathematics is precision and rigor. A theorem is established, not through observation or experimentation, but by a proof that consists of a chain of reasoning with no gaps.

This approach to mathematics comes down to us from the ancient Greek mathematicians, especially Euclid, and it remains the standard in contemporary research. In recent decades, the computer has become a powerful tool for mathematical experimentation and data analysis. Researchers may use experimental data to discover potential new mathematical facts, but the title "theorem" is not bestowed until someone writes down a proof.

This insistence on theorems and proofs distinguishes mathematics from the other sciences. In the natural sciences, facts are established through experiment and are subject to change or modification as more knowledge is acquired. In mathematics, theories are also developed and expanded, but previous results are not invalidated. The Pythagorean Theorem was discovered in antiquity and is a cornerstone of plane geometry. In the nineteenth century, mathematicians began to study more general types of geometry (of the type that eventually led to Einstein's four-dimensional space-time geometry in the Theory of Relativity). The Pythagorean Theorem does not hold in these more general geometries, but its status in plane geometry is unchanged.

A. SUMMARY

- The implication $A \Longrightarrow B$ is the assertion, "If A is true, then B is true."
- The *contrapositive* of $A \Longrightarrow B$ is the implication $\neg B \Longrightarrow \neg A$, which says, "If B is false, then A is false." An implication and its contrapositive are equivalent (one is true if and only if the other is true).
- The *converse* of $A \Longrightarrow B$ is $B \Longrightarrow A$. An implication and its converse are not necessarily equivalent. One may be true and the other false.
- A and B are *equivalent* if $A \Longrightarrow B$ and $B \Longrightarrow A$ are both true.
- In a proof by contradiction (in which the goal is to prove statement A), we start by assuming that A is false and show that this assumption leads to a contradiction.

A. EXERCISES

Preliminary Questions

1. Which is the contrapositive of $A \Longrightarrow B$?

(a) $B \Longrightarrow A$ **(b)** $\neg B \Longrightarrow A$

(c) $\neg B \Longrightarrow \neg A$ **(d)** $\neg A \Longrightarrow \neg B$

2. Which of the choices in Question 1 is the converse of $A \Longrightarrow B$?

3. Suppose that $A \Longrightarrow B$ is true. Which is then automatically true, the converse or the contrapositive?

4. Restate as an implication: "A triangle is a polygon."

Exercises

1. Which is the negation of the statement, "The car and the shirt are both blue"?

(a) Neither the car nor the shirt is blue.

(b) The car is not blue and/or the shirt is not blue.

2. Which is the contrapositive of the implication, "If the car has gas, then it will run"?

(a) If the car has no gas, then it will not run.

(b) If the car will not run, then it has no gas.

In Exercises 3–8, state the negation.

3. The time is 4 o'clock.

4. $\triangle ABC$ is an isosceles triangle.

5. m and n are odd integers.

6. Either m is odd or n is odd.

7. x is a real number and y is an integer.

8. f is a linear function.

In Exercises 9–14, state the contrapositive and converse.

9. If m and n are odd integers, then mn is odd.

10. If today is Tuesday, then we are in Belgium.

11. If today is Tuesday, then we are not in Belgium.

12. If $x > 4$, then $x^2 > 16$.

13. If m^2 is divisible by 3, then m is divisible by 3.

14. If $x^2 = 2$, then x is irrational.

In Exercise 15–18, give a counterexample to show that the converse of the statement is false.

15. If m is odd, then $2m + 1$ is also odd.

16. If $\triangle ABC$ is equilateral, then it is an isosceles triangle.

17. If m is divisible by 9 and 4, then m is divisible by 12.

18. If m is odd, then $m^3 - m$ is divisible by 3.

In Exercise 19–22, determine whether the converse of the statement is false.

19. If $x > 4$ and $y > 4$, then $x + y > 8$.

20. If $x > 4$, then $x^2 > 16$.

21. If $|x| > 4$, then $x^2 > 16$.

22. If m and n are even, then mn is even.

In Exercises 23 and 24, state the contrapositive and converse (it is not necessary to know what these statements mean).

23. If f and g are differentiable, then fg is differentiable.

24. If the force field is radial and decreases as the inverse square of the distance, then all closed orbits are ellipses.

*In Exercises 25–28, the **inverse** of $A \implies B$ is the implication $\neg A \implies \neg B$.*

25. Which of the following is the inverse of the implication, "If she jumped in the lake, then she got wet"?

(a) If she did not get wet, then she did not jump in the lake.

(b) If she did not jump in the lake, then she did not get wet.

Is the inverse true?

26. State the inverses of these implications:

(a) If X is a mouse, then X is a rodent.

(b) If you sleep late, you will miss class.

(c) If a star revolves around the sun, then it's a planet.

27. [✏] Explain why the inverse is equivalent to the converse.

28. [✏] State the inverse of the Pythagorean Theorem. Is it true?

29. Theorem 1 in Section 2.4 states the following: "If f and g are continuous functions, then $f + g$ is continuous." Does it follow logically that if f and g are not continuous, then $f + g$ is not continuous?

30. Write out a proof by contradiction for this fact: There is no smallest positive rational number. Base your proof on the fact that if $r > 0$, then $0 < r/2 < r$.

31. Use proof by contradiction to prove that if $x + y > 2$, then $x > 1$ or $y > 1$ (or both).

In Exercises 32–35, use proof by contradiction to show that the number is irrational.

32. $\sqrt{\frac{1}{2}}$ **33.** $\sqrt{3}$ **34.** $\sqrt[3]{2}$ **35.** $\sqrt[4]{11}$

36. An isosceles triangle is a triangle with two equal sides. The following theorem holds: If $\triangle$ is a triangle with two equal angles, then $\triangle$ is an isosceles triangle.

(a) What is the hypothesis?

(b) Show by providing a counterexample that the hypothesis is necessary.

(c) What is the contrapositive?

(d) What is the converse? Is it true?

37. Consider the following theorem: Let f be a quadratic polynomial with a positive leading coefficient. Then f has a minimum value.

(a) What are the hypotheses?

(b) What is the contrapositive?

(c) What is the converse? Is it true?

Further Insights and Challenges

38. Let a, b, and c be the sides of a triangle and let θ be the angle opposite c. Use the Law of Cosines (Theorem 1 in Section 1.4) to prove the converse of the Pythagorean Theorem.

39. Carry out the details of the following proof by contradiction that $\sqrt{2}$ is irrational (this proof is due to R. Palais). If $\sqrt{2}$ is rational, then $n\sqrt{2}$ is a whole number for some whole number n. Let n be the smallest such whole number and let $m = n\sqrt{2} - n$.

(a) Prove that $m < n$.

(b) Prove that $m\sqrt{2}$ is a whole number.

Explain why (a) and (b) imply that $\sqrt{2}$ is irrational.

40. Generalize the argument of Exercise 39 to prove that $\sqrt{A}$ is irrational if A is a whole number but not a perfect square. *Hint:* Choose n as before and let $m = n\sqrt{A} - n\lfloor\sqrt{A}\rfloor$, where $\lfloor x \rfloor$ is the greatest integer function.

41. Generalize further and show that for any whole number r, the rth root $\sqrt[r]{A}$ is irrational unless A is an rth power. *Hint:* Let $x = \sqrt[r]{A}$. Show that if x is rational, then we may choose a smallest whole number n such that nx^j is a whole number for $j = 1, \ldots, r - 1$. Then consider $m = nx - n[x]$ as before.

42. [✏] Given a finite list of prime numbers $p_1, \ldots, p_N$, let $M = p_1 \cdot p_2 \cdots p_N + 1$. Show that M is not divisible by any of the primes $p_1, \ldots, p_N$. Use this and the fact that every number has a prime factorization to prove that there exist infinitely many prime numbers. This argument was advanced by Euclid in *The Elements*.

B PROPERTIES OF REAL NUMBERS

The ingenious method of expressing every possible number using a set of ten symbols (each symbol having a place value and an absolute value) emerged in India. The idea seems so simple nowadays that its significance and profound importance is no longer appreciated. Its simplicity lies in the way it facilitated calculation and placed arithmetic foremost amongst useful inventions. The importance of this invention is more readily appreciated when one considers that it was beyond the two greatest men of Antiquity, Archimedes and Apollonius.

—Pierre-Simon Laplace, one of the great French mathematicians of the eighteenth century

In this appendix, we discuss the basic properties of real numbers. First, let us recall that a real number is a number that may be represented by a finite or infinite decimal (also called a decimal expansion). The set of all real numbers is denoted **R** and is often visualized as the "number line" (Figure 1).

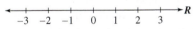

FIGURE 1 The real number line.

Thus, a real number a is represented as

$$a = \pm n.a_1a_2a_3a_4\ldots,$$

where n is any whole number and each digit a_j is a whole number between 0 and 9. For example, $10\pi = 31.41592\ldots$. Recall that a is rational if its expansion is finite or repeating, and is irrational if its expansion is nonrepeating. Furthermore, the decimal expansion is unique apart from the following exception: Every finite expansion is equal to an expansion in which the digit 9 repeats. For example, $0.5 = 0.4999\cdots = 0.4\bar{9}$.

We shall take for granted that the operations of addition and multiplication are defined on **R**—that is, on the set of all decimals. Roughly speaking, addition and multiplication of infinite decimals are defined in terms of finite decimals. For $d \geq 1$, define the dth truncation of $a = n.a_1a_2a_3a_4\ldots$ to be the finite decimal $a(d) = a.a_1a_2\ldots a_d$ obtained by truncating at the dth place. To form the sum $a+b$, assume that both a and b are infinite (possibly ending with repeated nines). This eliminates any possible ambiguity in the expansion. Then the nth digit of $a+b$ is equal to the nth digit of $a(d)+b(d)$ for d sufficiently large [from a certain point onward, the nth digit of $a(d)+b(d)$ no longer changes, and this value is the nth digit of $a+b$]. Multiplication is defined similarly. Furthermore, the Commutative, Associative, and Distributive Laws hold (Table 1).

TABLE 1 Algebraic Laws

Commutative Laws:	$a+b=b+a, \quad ab=ba$
Associative Laws:	$(a+b)+c=a+(b+c), \quad (ab)c=a(bc)$
Distributive Law:	$a(b+c)=ab+ac$

Every real number x has an additive inverse $-x$ such that $x+(-x)=0$, and every nonzero real number x has a multiplicative inverse x^{-1} such that $x(x^{-1})=1$. We do not regard subtraction and division as separate algebraic operations because they are defined in terms of inverses. By definition, the difference $x-y$ is equal to $x+(-y)$, and the quotient x/y is equal to $x(y^{-1})$ for $y \neq 0$.

In addition to the algebraic operations, there is an **order relation** on **R**: For any two real numbers a and b, precisely one of the following is true:

$$\text{Either} \quad a=b, \quad \text{or} \quad a<b, \quad \text{or} \quad a>b$$

To distinguish between the conditions $a \leq b$ and $a<b$, we often refer to $a<b$ as a **strict inequality**. Similar conventions hold for $>$ and $\geq$. The rules given in Table 2 allow us to manipulate inequalities. The last order property says that an inequality reverses direction when multiplied by a negative number c. For example,

$$-2 < 5 \quad \text{but} \quad (-3)(-2) > (-3)5$$

TABLE 2 Order Properties

If $a<b$ and $b<c$,	then $a<c$.
If $a<b$ and $c<d$,	then $a+c<b+d$.
If $a<b$ and $c>0$,	then $ac<bc$.
If $a<b$ and $c<0$,	then $ac>bc$.

The algebraic and order properties of real numbers are certainly familiar. We now discuss the less familiar **Least Upper Bound (LUB) Property** of the real numbers. This property is one way of expressing the so-called **completeness** of the real numbers. There are other ways of formulating completeness (such as the so-called nested interval property discussed in any book on analysis) that are equivalent to the LUB Property and serve the same purpose. Completeness is used in calculus to construct rigorous proofs of basic theorems about continuous functions, such as the Intermediate Value Theorem, (IVT) or the existence of extreme values on a closed interval. The underlying idea is that the real number line "has no holes." We elaborate on this idea below. First, we introduce the necessary definitions.

Suppose that S is a nonempty set of real numbers. A number M is called an **upper bound** for S if

$$x \leq M \qquad \text{for all } x \in S$$

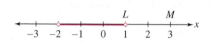

FIGURE 2 $M = 3$ is an upper bound for the set $S = (-2, 1)$. The LUB is $L = 1$.

If S has an upper bound, we say that S is **bounded above**. A **least upper bound** L is an upper bound for S such that every other upper bound M satisfies $M \geq L$. For example (Figure 2),

- $M = 3$ is an upper bound for the open interval $S = (-2, 1)$.
- $L = 1$ is the LUB for $S = (-2, 1)$.

We now state the LUB Property of the real numbers.

THEOREM 1 Existence of a Least Upper Bound   Let S be a nonempty set of real numbers that is bounded above. Then S has an LUB.

In a similar fashion, we say that a number B is a **lower bound** for S if $x \geq B$ for all $x \in S$. We say that S is **bounded below** if S has a lower bound. A **greatest lower bound** (GLB) is a lower bound M such that every other lower bound B satisfies $B \leq M$. The set of real numbers also has the GLB Property: If S is a nonempty set of real numbers that is bounded below, then S has a GLB. This may be deduced immediately from Theorem 1. For any nonempty set of real numbers S, let $-S$ be the set of numbers of the form $-x$ for $x \in S$. Then $-S$ has an upper bound if S has a lower bound. Consequently, $-S$ has an LUB L by Theorem 1, and $-L$ is a GLB for S.

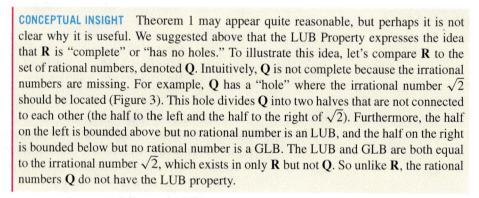

FIGURE 3 The rational numbers have a "hole" at the location $\sqrt{2}$.

CONCEPTUAL INSIGHT   Theorem 1 may appear quite reasonable, but perhaps it is not clear why it is useful. We suggested above that the LUB Property expresses the idea that **R** is "complete" or "has no holes." To illustrate this idea, let's compare **R** to the set of rational numbers, denoted **Q**. Intuitively, **Q** is not complete because the irrational numbers are missing. For example, **Q** has a "hole" where the irrational number $\sqrt{2}$ should be located (Figure 3). This hole divides **Q** into two halves that are not connected to each other (the half to the left and the half to the right of $\sqrt{2}$). Furthermore, the half on the left is bounded above but no rational number is an LUB, and the half on the right is bounded below but no rational number is a GLB. The LUB and GLB are both equal to the irrational number $\sqrt{2}$, which exists in only **R** but not **Q**. So unlike **R**, the rational numbers **Q** do not have the LUB property.

■ **EXAMPLE 1**   Show that 2 has a square root by applying the LUB Property to the set

$$S = \{x : x^2 < 2\}$$

Solution   First, we note that S is bounded with the upper bound $M = 2$. Indeed, if $x > 2$, then x satisfies $x^2 > 4$, and hence x does not belong to S. By the LUB Property, S has a least upper bound. Call it L. We claim that $L = \sqrt{2}$, or, equivalently, that $L^2 = 2$. We prove this by showing that $L^2 \geq 2$ and $L^2 \leq 2$.

If $L^2 < 2$, let $b = L + h$, where $h > 0$. Then

$$b^2 = L^2 + 2Lh + h^2 = L^2 + h(2L + h) \qquad \boxed{1}$$

We can make the quantity $h(2L + h)$ as small as desired by choosing $h > 0$ small enough. In particular, we may choose a positive h so that $h(2L + h) < 2 - L^2$. For this choice, $b^2 < L^2 + (2 - L^2) = 2$ by Eq. (1). Therefore, $b \in S$. But $b > L$ since $h > 0$, and thus

L is not an upper bound for S, in contradiction to our hypothesis on L. We conclude that $L^2 \geq 2$.

If $L^2 > 2$, let $b = L - h$, where $h > 0$. Then

$$b^2 = L^2 - 2Lh + h^2 = L^2 - h(2L - h)$$

Now choose h positive but small enough so that $0 < h(2L - h) < L^2 - 2$. Then $b^2 > L^2 - (L^2 - 2) = 2$. But $b < L$, so b is a smaller lower bound for S. Indeed, if $x \geq b$, then $x^2 \geq b^2 > 2$, and x does not belong to S. This contradicts our hypothesis that L is the LUB. We conclude that $L^2 \leq 2$, and since we have already shown that $L^2 \geq 2$, we have $L^2 = 2$ as claimed. ∎

We now prove three important theorems, the third of which is used in the proof of the LUB Property below.

> **THEOREM 2 Bolzano–Weierstrass Theorem**   Let S be a bounded, infinite set of real numbers. Then there exists a sequence of distinct elements $\{a_n\}$ in S such that the limit $L = \lim\limits_{n \to \infty} a_n$ exists.

Proof   For simplicity of notation, we assume that S is contained in the unit interval $[0, 1]$ (a similar proof works in general). If $k_1, k_2, \ldots, k_n$ is a sequence of n digits (i.e., each k_j is a whole number and $0 \leq k_j \leq 9$), let

$$S(k_1, k_2, \ldots, k_n)$$

be the set of $x \in S$ whose decimal expansion begins $0.k_1 k_2 \ldots k_n$. The set S is the union of the subsets $S(0), S(1), \ldots, S(9)$, and since S is infinite, at least one of these subsets must be infinite. Therefore, we may choose k_1 so that $S(k_1)$ is infinite. In a similar fashion, at least one of the set $S(k_1, 0), S(k_2, 1), \ldots, S(k_1, 9)$ must be infinite, so we may choose k_2 so that $S(k_1, k_2)$ is infinite. Continuing in this way, we obtain an infinite sequence $\{k_n\}$ such that $S(k_1, k_2, \ldots, k_n)$ is infinite for all n. We may choose a sequence of elements $a_n \in S(k_1, k_2, \ldots, k_n)$ with the property that a_n differs from $a_1, \ldots, a_{n-1}$ for all n. Let L be the infinite decimal $0.k_1 k_2 k_3 \ldots$. Then $\lim\limits_{n \to \infty} a_n = L$ since $|L - a_n| < 10^{-n}$ for all n. ∎

We use the Bolzano–Weierstrass Theorem to prove two important results about sequences $\{a_n\}$. Recall that an upper bound for $\{a_n\}$ is a number M such that $a_j \leq M$ for all j. If an upper bound exists, $\{a_n\}$ is said to be bounded from above. Lower bounds are defined similarly and $\{a_n\}$ is said to be bounded from below if a lower bound exists. A sequence is bounded if it is bounded from above and below. A **subsequence** of $\{a_n\}$ is a sequence of elements $a_{n_1}, a_{n_2}, a_{n_3}, \ldots$, where $n_1 < n_2 < n_3 < \cdots$.

Now consider a bounded sequence $\{a_n\}$. If infinitely many of the a_n are distinct, the Bolzano–Weierstrass Theorem implies that there exists a subsequence $\{a_{n_1}, a_{n_2}, \ldots\}$ such that $\lim\limits_{n \to \infty} a_{n_k}$ exists. Otherwise, infinitely many of the a_n must coincide, and these terms form a convergent subsequence. This proves the next result.

Section 10.1

> **THEOREM 3**   Every bounded sequence has a convergent subsequence.

> **THEOREM 4 Bounded Monotonic Sequences Converge**
>
> - If $\{a_n\}$ is increasing and $a_n \leq M$ for all n, then $\{a_n\}$ converges and $\lim\limits_{n \to \infty} a_n \leq M$.
> - If $\{a_n\}$ is decreasing and $a_n \geq M$ for all n, then $\{a_n\}$ converges and $\lim\limits_{n \to \infty} a_n \geq M$.

Proof   Suppose that $\{a_n\}$ is increasing and bounded above by M. Then $\{a_n\}$ is automatically bounded below by $m = a_1$ since $a_1 \leq a_2 \leq a_3 \cdots$. Hence, $\{a_n\}$ is bounded, and by

Theorem 3, we may choose a convergent subsequence $a_{n_1}, a_{n_2}, \ldots$. Let

$$L = \lim_{k \to \infty} a_{n_k}$$

Observe that $a_n \leq L$ for all n. For if not, then $a_n > L$ for some n and then $a_{n_k} \geq a_n > L$ for all k such that $n_k \geq n$. But this contradicts that $a_{n_k} \to L$. Now, by definition, for any $\epsilon > 0$, there exists $N_\epsilon > 0$ such that

$$|a_{n_k} - L| < \epsilon \qquad \text{if } n_k > N_\epsilon$$

Choose m such that $n_m > N_\epsilon$. If $n \geq n_m$, then $a_{n_m} \leq a_n \leq L$, and therefore,

$$|a_n - L| \leq |a_{n_m} - L| < \epsilon \qquad \text{for all } n \geq n_m$$

This proves that $\lim_{n \to \infty} a_n = L$, as desired. It remains to prove that $L \leq M$. If $L > M$, let $\epsilon = (L - M)/2$ and choose N so that

$$|a_n - L| < \epsilon \qquad \text{if } k > N$$

Then $a_n > L - \epsilon = M + \epsilon$. This contradicts our assumption that M is an upper bound for $\{a_n\}$. Therefore, $L \leq M$ as claimed. ∎

Proof of Theorem 1 We now use Theorem 4 to prove the LUB Property (Theorem 1). As above, if x is a real number, let $x(d)$ be the truncation of x of length d. For example,

$$\text{If } x = 1.41569, \text{ then } x(3) = 1.415.$$

We say that x is a *decimal of length d* if $x = x(d)$. Any two distinct decimals of length d differ by at least 10^{-d}. It follows that for any two real numbers $A < B$, there are at most finitely many decimals of length d between A and B.

Now let S be a nonempty set of real numbers with an upper bound M. We shall prove that S has an LUB. Let $S(d)$ be the set of truncations of length d:

$$S(d) = \{x(d) : x \in S\}$$

We claim that $S(d)$ has a maximum element. To verify this, choose any $a \in S$. If $x \in S$ and $x(d) > a(d)$, then

$$a(d) \leq x(d) \leq M$$

Thus, by the remark of the previous paragraph, there are at most finitely many values of $x(d)$ in $S(d)$ larger than $a(d)$. The largest of these is the maximum element in $S(d)$.

For $d = 1, 2, \ldots$, choose an element x_d such that $x_d(d)$ is the maximum element in $S(d)$. By construction, $\{x_d(d)\}$ is an increasing sequence (since the largest dth truncation cannot get smaller as d increases). Furthermore, $x_d(d) \leq M$ for all d. We now apply Theorem 4 to conclude that $\{x_d(d)\}$ converges to a limit L. We claim that L is the LUB of S. Observe first that L is an upper bound for S. Indeed, if $x \in S$, then $x(d) \leq L$ for all d and thus $x \leq L$. To show that L is the LUB, suppose that M is an upper bound such that $M < L$. Then $x_d \leq M$ for all d and hence $x_d(d) \leq M$ for all d. But then

$$L = \lim_{d \to \infty} x_d(d) \leq M$$

This is a contradiction since $M < L$. Therefore, L is the LUB of S. ∎

As mentioned above, the LUB Property is used in calculus to establish certain basic theorems about continuous functions. As an example, we prove the IVT. Another example is the theorem on the existence of extrema on a closed interval (see Appendix D).

THEOREM 5 Intermediate Value Theorem If f is continuous on a closed interval $[a, b]$ and $f(a) \neq f(b)$, then for every value M between $f(a)$ and $f(b)$, there exists at least one value $c \in (a, b)$ such that $f(c) = M$.

Proof Assume first that $M = 0$. Replacing $f(x)$ by $-f(x)$ if necessary, we may assume that $f(a) < 0$ and $f(b) > 0$. Now let

$$S = \{x \in [a, b] : f(x) < 0\}$$

Then $a \in S$ since $f(a) < 0$ and thus S is nonempty. Clearly, b is an upper bound for S. Therefore, by the LUB Property, S has an LUB L. We claim that $f(L) = 0$. If not, set $r = f(L)$. Assume first that $r > 0$.

Since f is continuous, there exists a number $\delta > 0$ such that

$$\text{if } |x - L| < \delta, \text{ then } |f(x) - f(L)| = |f(x) - r| < \frac{1}{2}r.$$

Equivalently,

$$\text{if } |x - L| < \delta, \text{ then } \frac{1}{2}r < f(x) < \frac{3}{2}r.$$

The number $\frac{1}{2}r$ is positive, so we conclude that

$$\text{if } L - \delta < x < L + \delta, \text{ then } f(x) > 0.$$

By definition of L, $f(x) \geq 0$ for all $x \in [a, b]$ such that $x > L$, and thus $f(x) \geq 0$ for all $x \in [a, b]$ such that $x > L - \delta$. Thus, $L - \delta$ is an upper bound for S. This is a contradiction since L is the LUB of S, and it follows that $r = f(L)$ cannot satisfy $r > 0$. Similarly, r cannot satisfy $r < 0$. We conclude that $f(L) = 0$ as claimed.

Now, if M is nonzero, let $g(x) = f(x) - M$. Then 0 lies between $g(a)$ and $g(b)$, and by what we have proved, there exists $c \in (a, b)$ such that $g(c) = 0$. But then $f(c) = g(c) + M = M$, as desired. ◼

C INDUCTION AND THE BINOMIAL THEOREM

The Principle of Induction is a method of proof that is widely used to prove that a given statement $P(n)$ is valid for all natural numbers $n = 1, 2, 3, \ldots$. Here are two statements of this kind:

- $P(n)$: The sum of the first n odd numbers is equal to n^2.
- $P(n)$: $\dfrac{d}{dx} x^n = n x^{n-1}$.

The first statement claims that for all natural numbers n,

$$\underbrace{1 + 3 + \cdots + (2n - 1)}_{\text{Sum of first } n \text{ odd numbers}} = n^2 \qquad \boxed{1}$$

We can check directly that $P(n)$ is true for the first few values of n:

$$P(1) \text{ is the equality:} \qquad 1 = 1^2 \quad \text{(true)}$$
$$P(2) \text{ is the equality:} \qquad 1 + 3 = 2^2 \quad \text{(true)}$$
$$P(3) \text{ is the equality:} \qquad 1 + 3 + 5 = 3^2 \quad \text{(true)}$$

The Principle of Induction may be used to establish $P(n)$ for all n.

The Principle of Induction applies if $P(n)$ is an assertion defined for $n \geq n_0$, where n_0 is a fixed integer. Assume that

*(i) **Initial step:** $P(n_0)$ is true.*

*(ii) **Induction step:** If $P(n)$ is true for $n = k$, then $P(n)$ is also true for $n = k + 1$.*

Then $P(n)$ is true for all $n \geq n_0$.

THEOREM 1 Principle of Induction Let $P(n)$ be an assertion that depends on a natural number n. Assume that

(i) Initial step: $P(1)$ is true.

(ii) Induction step: If $P(n)$ is true for $n = k$, then $P(n)$ is also true for $n = k + 1$.

Then $P(n)$ is true for all natural numbers $n = 1, 2, 3, \ldots$.

■ **EXAMPLE 1** Prove that $1 + 3 + \cdots + (2n - 1) = n^2$ for all natural numbers n.

Solution As above, we let $P(n)$ denote the equality

$$P(n): \qquad 1 + 3 + \cdots + (2n - 1) = n^2$$

Step 1. Initial step: Show that $P(1)$ is true.
We checked this above. $P(1)$ is the equality $1 = 1^2$.

Step 2. Induction step: Show that if $P(n)$ is true for $n = k$, then $P(n)$ is also true for $n = k + 1$.
Assume that $P(k)$ is true. Then

$$1 + 3 + \cdots + (2k - 1) = k^2$$

Add $2k + 1$ to both sides:

$$\left[1 + 3 + \cdots + (2k - 1) \right] + (2k + 1) = k^2 + 2k + 1 = (k + 1)^2$$

$$1 + 3 + \cdots + (2k + 1) = (k + 1)^2$$

This is precisely the statement $P(k + 1)$. Thus, $P(k + 1)$ is true whenever $P(k)$ is true. By the Principle of Induction, $P(k)$ is true for all k. ■

The intuition behind the Principle of Induction is the following. If $P(n)$ were not true for all n, then there would exist a smallest natural number k such that $P(k)$ is false. Furthermore, $k > 1$ since $P(1)$ is true. Thus, $P(k - 1)$ is true [otherwise, $P(k)$ would not be the smallest "counterexample"]. On the other hand, if $P(k - 1)$ is true, then $P(k)$ is also true by the induction step. This is a contradiction. So $P(k)$ must be true for all k.

■ **EXAMPLE 2** Use Induction and the Product Rule to prove that for all whole numbers n,

$$\frac{d}{dx}x^n = nx^{n-1}$$

Solution Let $P(n)$ be the formula $\dfrac{d}{dx}x^n = nx^{n-1}$.

Step 1. Initial step: Show that $P(1)$ is true.
We use the limit definition to verify $P(1)$:

$$\frac{d}{dx}x = \lim_{h \to 0} \frac{(x+h) - x}{h} = \lim_{h \to 0} \frac{h}{h} = \lim_{h \to 0} 1 = 1$$

Step 2. Induction step: Show that if $P(n)$ is true for $n = k$, then $P(n)$ is also true for $n = k + 1$.
To carry out the induction step, assume that $\dfrac{d}{dx}x^k = kx^{k-1}$, where $k \geq 1$. Then, by the Product Rule,

$$\frac{d}{dx}x^{k+1} = \frac{d}{dx}(x \cdot x^k) = x\frac{d}{dx}x^k + x^k\frac{d}{dx}x = x(kx^{k-1}) + x^k$$

$$= kx^k + x^k = (k+1)x^k$$

This shows that $P(k + 1)$ is true.

By the Principle of Induction, $P(n)$ is true for all $n \geq 1$. ■

As another application of induction, we prove the Binomial Theorem, which describes the expansion of the binomial $(a + b)^n$. The first few expansions are familiar:

$$(a + b)^1 = a + b$$

$$(a + b)^2 = a^2 + 2ab + b^2$$

$$(a + b)^3 = a^3 + 3a^2b + 3ab^2 + b^3$$

In general, we have an expansion

$$(a + b)^n = a^n + \binom{n}{1}a^{n-1}b + \binom{n}{2}a^{n-2}b^2 + \binom{n}{3}a^{n-3}b^3$$

$$+ \cdots + \binom{n}{n-1}ab^{n-1} + b^n$$

2

where the coefficient of $x^{n-k}x^k$, denoted $\dbinom{n}{k}$, is called the **binomial coefficient**. Note that the first term in Eq. (2) corresponds to $k = 0$ and the last term to $k = n$; thus, $\dbinom{n}{0} = \dbinom{n}{n} = 1$. In summation notation,

$$(a + b)^n = \sum_{k=0}^{n} \binom{n}{k}a^k b^{n-k}$$

In Pascal's Triangle, the nth row displays the coefficients in the expansion of $(a + b)^n$:

n								
0				1				
1			1		1			
2			1	2	1			
3		1	3		3	1		
4		1	4	6	4	1		
5	1	5	10	10	5	1		
6	1	6	15	20	15	6	1	

The triangle is constructed as follows: Each entry is the sum of the two entries above it in the previous line. For example, the entry 15 in line $n = 6$ is the sum $10 + 5$ of the entries above it in line $n = 5$. The recursion relation guarantees that the entries in the triangle are the binomial coefficients.

Pascal's Triangle (described in the marginal note on page A14) can be used to compute binomial coefficients if n and k are not too large. The Binomial Theorem provides the following general formula:

$$\binom{n}{k} = \frac{n!}{k!\,(n-k)!} = \frac{n(n-1)(n-2)\cdots(n-k+1)}{k(k-1)(k-2)\cdots 2 \cdot 1}$$

3

Before proving this formula, we prove a recursion relation for binomial coefficients. Note, however, that Eq. (3) is certainly correct for $k = 0$ and $k = n$ (recall that by convention, $0! = 1$):

$$\binom{n}{0} = \frac{n!}{(n-0)!\,0!} = \frac{n!}{n!} = 1, \qquad \binom{n}{n} = \frac{n!}{(n-n)!\,n!} = \frac{n!}{n!} = 1$$

THEOREM 2 Recursion Relation for Binomial Coefficients

$$\binom{n}{k} = \binom{n-1}{k} + \binom{n-1}{k-1} \qquad \text{for } 1 \le k \le n-1$$

Proof We write $(a+b)^n$ as $(a+b)(a+b)^{n-1}$ and expand in terms of binomial coefficients:

$$(a+b)^n = (a+b)(a+b)^{n-1}$$

$$\sum_{k=0}^{n} \binom{n}{k} a^{n-k} b^k = (a+b) \sum_{k=0}^{n-1} \binom{n-1}{k} a^{n-1-k} b^k$$

$$= a \sum_{k=0}^{n-1} \binom{n-1}{k} a^{n-1-k} b^k + b \sum_{k=0}^{n-1} \binom{n-1}{k} a^{n-1-k} b^k$$

$$= \sum_{k=0}^{n-1} \binom{n-1}{k} a^{n-k} b^k + \sum_{k=0}^{n-1} \binom{n-1}{k} a^{n-(k+1)} b^{k+1}$$

Replacing k by $k-1$ in the second sum, we obtain

$$\sum_{k=0}^{n} \binom{n}{k} a^{n-k} b^k = \sum_{k=0}^{n-1} \binom{n-1}{k} a^{n-k} b^k + \sum_{k=1}^{n} \binom{n-1}{k-1} a^{n-k} b^k$$

On the right-hand side, the first term in the first sum is a^n and the last term in the second sum is b^n. Thus, we have

$$\sum_{k=0}^{n} \binom{n}{k} a^{n-k} b^k = a^n + \left(\sum_{k=1}^{n-1} \left(\binom{n-1}{k} + \binom{n-1}{k-1} \right) a^{n-k} b^k \right) + b^n$$

The recursion relation follows because the coefficients of $a^{n-k} b^k$ on the two sides of the equation must be equal. ∎

We now use induction to prove Eq. (3). Let $P(n)$ be the claim

$$\binom{n}{k} = \frac{n!}{k!\,(n-k)!} \qquad \text{for } 0 \le k \le n$$

We have $\binom{1}{0} = \binom{1}{1} = 1$ since $(a+b)^1 = a + b$, so $P(1)$ is true. Furthermore, $\binom{n}{n} = \binom{n}{0} = 1$ as observed above, since a^n and b^n have coefficient 1 in the expansion of $(a + b)^n$. For the inductive step, assume that $P(n)$ is true. By the recursion relation, for $1 \le k \le n$, we have

$$\binom{n+1}{k} = \binom{n}{k} + \binom{n}{k-1} = \frac{n!}{k!\,(n-k)!} + \frac{n!}{(k-1)!\,(n-k+1)!}$$

$$= n!\left(\frac{n+1-k}{k!\,(n+1-k)!} + \frac{k}{k!\,(n+1-k)!}\right) = n!\left(\frac{n+1}{k!\,(n+1-k)!}\right)$$

$$= \frac{(n+1)!}{k!\,(n+1-k)!}$$

Thus, $P(n+1)$ is also true and the Binomial Theorem follows by induction.

■ **EXAMPLE 3** Use the Binomial Theorem to expand $(x + y)^5$ and $(x + 2)^3$.

Solution The fifth row in Pascal's Triangle yields

$$(x+y)^5 = x^5 + 5x^4 y + 10x^3 y^2 + 10x^2 y^3 + 5xy^4 + y^5$$

The third row in Pascal's Triangle yields

$$(x+2)^3 = x^3 + 3x^2(2) + 3x(2)^2 + 2^3 = x^3 + 6x^2 + 12x + 8 \qquad ■$$

C. EXERCISES

In Exercises 1–4, use the Principle of Induction to prove the formula for all natural numbers n.

1. $1 + 2 + 3 + \cdots + n = \dfrac{n(n+1)}{2}$

2. $1^3 + 2^3 + 3^3 + \cdots + n^3 = \dfrac{n^2(n+1)^2}{4}$

3. $\dfrac{1}{1 \cdot 2} + \dfrac{1}{2 \cdot 3} + \cdots + \dfrac{1}{n(n+1)} = \dfrac{n}{n+1}$

4. $1 + x + x^2 + \cdots + x^n = \dfrac{1 - x^{n+1}}{1 - x}$ for any $x \ne 1$

5. Let $P(n)$ be the statement $2^n > n$.

(a) Show that $P(1)$ is true.

(b) Observe that if $2^n > n$, then $2^n + 2^n > 2n$. Use this to show that if $P(n)$ is true for $n = k$, then $P(n)$ is true for $n = k + 1$. Conclude that $P(n)$ is true for all n.

6. Use induction to prove that $n! > 2^n$ for $n \ge 4$.

Let $\{F_n\}$ be the Fibonacci sequence, defined by the recursion formula

$$F_n = F_{n-1} + F_{n-2}, \qquad F_1 = F_2 = 1$$

The first few terms are $1, 1, 2, 3, 5, 8, 13, \ldots$. In Exercises 7–10, use induction to prove the identity.

7. $F_1 + F_2 + \cdots + F_n = F_{n+2} - 1$

8. $F_1^2 + F_2^2 + \cdots + F_n^2 = F_{n+1} F_n$

9. $F_n = \dfrac{R_+^n - R_-^n}{\sqrt{5}}$, where $R_\pm = \dfrac{1 \pm \sqrt{5}}{2}$

10. $F_{n+1} F_{n-1} = F_n^2 + (-1)^n$. *Hint:* For the induction step, show that

$$F_{n+2} F_n = F_{n+1} F_n + F_n^2$$

$$F_{n+1}^2 = F_{n+1} F_n + F_{n+1} F_{n-1}$$

11. Use induction to prove that $f(n) = 8^n - 1$ is divisible by 7 for all natural numbers n. *Hint:* For the induction step, show that

$$8^{k+1} - 1 = 7 \cdot 8^k + (8^k - 1)$$

12. Use induction to prove that $n^3 - n$ is divisible by 3 for all natural numbers n.

13. Use induction to prove that $5^{2n} - 4^n$ is divisible by 7 for all natural numbers n.

14. Use Pascal's Triangle to write out the expansions of $(a + b)^6$ and $(a - b)^4$.

15. Expand $(x + x^{-1})^4$.

16. What is the coefficient of x^9 in $(x^3 + x)^5$?

17. Let $S(n) = \displaystyle\sum_{k=0}^{n} \binom{n}{k}$.

(a) Use Pascal's Triangle to compute $S(n)$ for $n = 1, 2, 3, 4$.

(b) Prove that $S(n) = 2^n$ for all $n \ge 1$. *Hint:* Expand $(a + b)^n$ and evaluate at $a = b = 1$.

18. Let $T(n) = \displaystyle\sum_{k=0}^{n} (-1)^k \binom{n}{k}$.

(a) Use Pascal's Triangle to compute $T(n)$ for $n = 1, 2, 3, 4$.

(b) Prove that $T(n) = 0$ for all $n \ge 1$. *Hint:* Expand $(a + b)^n$ and evaluate at $a = 1, b = -1$.

D ADDITIONAL PROOFS

In this appendix, we provide proofs of several theorems that were stated or used in the text.

| Section 2.3

> **THEOREM 1 Basic Limit Laws** Assume that $\lim\limits_{x \to c} f(x)$ and $\lim\limits_{x \to c} g(x)$ exist. Then:
>
> **(i)** $\lim\limits_{x \to c} \big(f(x) + g(x) \big) = \lim\limits_{x \to c} f(x) + \lim\limits_{x \to c} g(x)$
>
> **(ii)** For any number k, $\lim\limits_{x \to c} kf(x) = k \lim\limits_{x \to c} f(x)$
>
> **(iii)** $\lim\limits_{x \to c} f(x)g(x) = \left(\lim\limits_{x \to c} f(x) \right) \left(\lim\limits_{x \to c} g(x) \right)$
>
> **(iv)** If $\lim\limits_{x \to c} g(x) \neq 0$, then
>
> $$\lim_{x \to c} \frac{f(x)}{g(x)} = \frac{\lim\limits_{x \to c} f(x)}{\lim\limits_{x \to c} g(x)}$$

Proof Let $L = \lim\limits_{x \to c} f(x)$ and $M = \lim\limits_{x \to c} g(x)$. The Sum Law (i) was proved in Section 2.9. Observe that (ii) is a special case of (iii), where $g(x) = k$ is a constant function. Thus, it will suffice to prove the Product Law (iii). We write

$$f(x)g(x) - LM = f(x)(g(x) - M) + M(f(x) - L)$$

and apply the Triangle Inequality to obtain

$$|f(x)g(x) - LM| \leq |f(x)(g(x) - M)| + |M(f(x) - L)| \qquad \boxed{1}$$

By the limit definition, we may choose $\delta > 0$ so that

$$\text{if } 0 < |x - c| < \delta, \text{ then } |f(x) - L| < 1.$$

If follows that $|f(x)| < |L| + 1$ for $0 < |x - c| < \delta$. Now choose any number $\epsilon > 0$. Applying the limit definition again, we see that by choosing a smaller δ if necessary, we may also ensure that if $0 < |x - c| < \delta$, then

$$|f(x) - L| \leq \frac{\epsilon}{2(|M| + 1)} \qquad \text{and} \qquad |g(x) - M| \leq \frac{\epsilon}{2(|L| + 1)}$$

Using Eq. (1), we see that if $0 < |x - c| < \delta$, then

$$|f(x)g(x) - LM| \leq |f(x)| \, |g(x) - M| + |M| \, |f(x) - L|$$

$$\leq (|L| + 1)\frac{\epsilon}{2(|L| + 1)} + |M|\frac{\epsilon}{2(|M| + 1)}$$

$$\leq \frac{\epsilon}{2} + \frac{\epsilon}{2} = \epsilon$$

Since ϵ is arbitrary, this proves that $\lim\limits_{x \to c} f(x)g(x) = LM$. To prove the Quotient Law (iv), it suffices to verify that if $M \neq 0$, then

$$\lim_{x \to c} \frac{1}{g(x)} = \frac{1}{M} \qquad \boxed{2}$$

For if Eq. (2) holds, then we may apply the Product Law to $f(x)$ and $g(x)^{-1}$ to obtain the Quotient Law:

$$\lim_{x \to c} \frac{f(x)}{g(x)} = \lim_{x \to c} f(x)\frac{1}{g(x)} = \left(\lim_{x \to c} f(x) \right) \left(\lim_{x \to c} \frac{1}{g(x)} \right)$$

$$= L\left(\frac{1}{M} \right) = \frac{L}{M}$$

We now verify Eq. (2). Since $g(x)$ approaches M and $M \neq 0$, we may choose $\delta > 0$ so that if $0 < |x - c| < \delta$, then $|g(x)| \geq |M|/2$. Now choose any number $\epsilon > 0$. By choosing a smaller δ if necessary, we may also ensure that

$$\text{for } 0 < |x - c| < \delta, \text{ then } |M - g(x)| < \epsilon |M| \left(\frac{|M|}{2} \right).$$

Then

$$\left| \frac{1}{g(x)} - \frac{1}{M} \right| = \left| \frac{M - g(x)}{Mg(x)} \right| \leq \left| \frac{M - g(x)}{M(M/2)} \right| \leq \frac{\epsilon |M|(|M|/2)}{|M|(|M|/2)} = \epsilon$$

Since ϵ is arbitrary, the limit in Eq. (2) is proved.   ■

The following result was used in the text.

THEOREM 2 Limits Preserve Inequalities   Let (a, b) be an open interval and let $c \in (a, b)$. Suppose that $f(x)$ and $g(x)$ are defined on (a, b), except possibly at c. Assume that

$$f(x) \leq g(x) \qquad \text{for } x \in (a, b), \quad x \neq c$$

and that the limits $\lim_{x \to c} f(x)$ and $\lim_{x \to c} g(x)$ exist. Then

$$\lim_{x \to c} f(x) \leq \lim_{x \to c} g(x)$$

Proof   Let $L = \lim_{x \to c} f(x)$ and $M = \lim_{x \to c} g(x)$. To show that $L \leq M$, we use proof by contradiction. If $L > M$, let $\epsilon = \frac{1}{2}(L - M)$. By the formal definition of limits, we may choose $\delta > 0$ so that the following two conditions are satisfied:

$$\text{If } |x - c| < \delta, \text{ then } |M - g(x)| < \epsilon.$$

$$\text{If } |x - c| < \delta, \text{ then } |L - f(x)| < \epsilon.$$

But then

$$f(x) > L - \epsilon = M + \epsilon > g(x)$$

This is a contradiction since $f(x) \leq g(x)$. We conclude that $L \leq M$.   ■

THEOREM 3 Limit of a Composite Function   Assume that the following limits exist:

$$L = \lim_{x \to c} g(x) \qquad \text{and} \qquad M = \lim_{x \to L} f(x)$$

Then $\lim_{x \to c} f(g(x)) = M$.

Proof   Let $\epsilon > 0$ be given. By the limit definition, there exists $\delta_1 > 0$ such that

$$\text{if } 0 < |x - L| < \delta_1, \text{ then } |f(x) - M| < \epsilon. \qquad \boxed{3}$$

Similarly, there exists $\delta > 0$ such that

$$\text{if } 0 < |x - c| < \delta, \text{ then } |g(x) - L| < \delta_1. \qquad \boxed{4}$$

We replace x by $g(x)$ in Eq. (3) and apply Eq. (4) to obtain:

$$\text{If } 0 < |x - c| < \delta, \text{ then } |f(g(x)) - M| < \epsilon.$$

Since ϵ is arbitrary, this proves that $\lim_{x \to c} f(g(x)) = M$.   ■

| Section 2.4

> **THEOREM 4 Continuity of Composite Functions**   Let $F(x) = f(g(x))$ be a composite function. If g is continuous at $x = c$ and f is continuous at $x = g(c)$, then F is continuous at $x = c$.

Proof By definition of continuity,

$$\lim_{x \to c} g(x) = g(c) \qquad \text{and} \qquad \lim_{x \to g(c)} f(x) = f(g(c))$$

Therefore, we may apply Theorem 3 to obtain

$$\lim_{x \to c} f(g(x)) = f(g(c))$$

This proves that $F(x) = f(g(x))$ is continuous at $x = c$. ■

| Section 2.6

> **THEOREM 5 Squeeze Theorem**   Assume that for $x \neq c$ (in some open interval containing c),
>
> $$l(x) \leq f(x) \leq u(x) \qquad \text{and} \qquad \lim_{x \to c} l(x) = \lim_{x \to c} u(x) = L$$
>
> Then $\lim_{x \to c} f(x)$ exists and
>
> $$\lim_{x \to c} f(x) = L$$

Proof Let $\epsilon > 0$ be given. We may choose $\delta > 0$ such that

$$\text{if } 0 < |x - c| < \delta, \text{ then } |l(x) - L| < \epsilon \text{ and } |u(x) - L| < \epsilon.$$

In principle, a different δ may be required to obtain the two inequalities for $l(x)$ and $u(x)$, but we may choose the smaller of the two deltas. Thus, if $0 < |x - c| < \delta$, we have

$$L - \epsilon < l(x) < L + \epsilon$$

and

$$L - \epsilon < u(x) < L + \epsilon$$

Since $f(x)$ lies between $l(x)$ and $u(x)$, it follows that

$$L - \epsilon < l(x) \leq f(x) \leq u(x) < L + \epsilon$$

and therefore $|f(x) - L| < \epsilon$ if $0 < |x - c| < \delta$. Since ϵ is arbitrary, this proves that $\lim_{x \to c} f(x) = L$, as desired. ■

| Section 4.2

> **THEOREM 6 Existence of Extrema on a Closed Interval**   A continuous function f on a closed (bounded) interval $I = [a, b]$ takes on both a minimum and a maximum value on I.

Proof We prove that f takes on a maximum value in two steps (the case of a minimum is similar).

***Step 1.* Prove that f is bounded from above.**
   We use proof by contradiction. If f is not bounded from above, then there exist points $a_n \in [a, b]$ such that $f(a_n) \geq n$ for $n = 1, 2, \ldots$. By Theorem 3 in Appendix B, we may choose a subsequence of elements $a_{n_1}, a_{n_2}, \ldots$ that converges to a limit in $[a, b]$—say, $\lim_{k \to \infty} a_{n_k} = L$. Since f is continuous, there exists $\delta > 0$ such that

$$\text{if } x \in [a, b] \text{ and } |x - L| < \delta, \text{ then } |f(x) - f(L)| < 1.$$

Therefore,

$$\text{if } x \in [a, b] \text{ and } x \in (L - \delta, L + \delta), \text{ then } f(x) < f(L) + 1. \qquad \boxed{5}$$

For k sufficiently large, a_{n_k} lies in $(L - \delta, L + \delta)$ because $\lim\limits_{k \to \infty} a_{n_k} = L$. By Eq. (5), $f(a_{n_k})$ is bounded by $f(L) + 1$. However, $f(a_{n_k}) = n_k$ tends to infinity as $k \to \infty$. This is a contradiction. Hence, our assumption that f is not bounded from above is false.

Step 2. **Prove that f takes on a maximum value.**

The range of f on $I = [a, b]$ is the set

$$S = \{ f(x) : x \in [a, b] \}$$

By the previous step, S is bounded from above and therefore has a least upper bound M by the LUB Property. Thus, $f(x) \le M$ for all $x \in [a, b]$. To complete the proof, we show that $f(c) = M$ for some $c \in [a, b]$. This will show that f attains the maximum value M on $[a, b]$.

By definition, $M - 1/n$ is not an upper bound for $n \ge 1$, and therefore, we may choose a point b_n in $[a, b]$ such that

$$M - \frac{1}{n} \le f(b_n) \le M$$

Again by Theorem 3 in Appendix B, there exists a subsequence of elements $\{ b_{n_1}, b_{n_2}, \dots \}$ in $\{ b_1, b_2, \dots \}$ that converges to a limit—say,

$$\lim_{k \to \infty} b_{n_k} = c$$

Furthermore, this limit c belongs to $[a, b]$ because $[a, b]$ is closed. Let $\epsilon > 0$. Since f is continuous, we may choose k so large that the following two conditions are satisfied: $|f(c) - f(b_{n_k})| < \epsilon/2$ and $n_k > 2/\epsilon$. Then

$$|f(c) - M| \le |f(c) - f(b_{n_k})| + |f(b_{n_k}) - M| \le \frac{\epsilon}{2} + \frac{1}{n_k} \le \frac{\epsilon}{2} + \frac{\epsilon}{2} = \epsilon$$

Thus, $|f(c) - M|$ is smaller than ϵ for all positive numbers ϵ. But this is not possible unless $|f(c) - M| = 0$. Thus, $f(c) = M$, as desired.   ∎

Section 5.2

> **THEOREM 7 Continuous Functions Are Integrable**   If f is continuous on $[a, b]$, then f is integrable over $[a, b]$.

Proof   We shall make the simplifying assumption that f is differentiable and that its derivative f' is bounded. In other words, we assume that $|f'(x)| \le K$ for some constant K. This assumption is used to show that f cannot vary too much in a small interval. More precisely, let us prove that if $[a_0, b_0]$ is any closed interval contained in $[a, b]$ and if m and M are the minimum and maximum values of f on $[a_0, b_0]$, then

$$|M - m| \le K |b_0 - a_0| \qquad \boxed{6}$$

Figure 1 illustrates the idea behind this inequality. Suppose that $f(x_1) = m$ and $f(x_2) = M$, where x_1 and x_2 lie in $[a_0, b_0]$. If $x_1 \ne x_2$, then by the Mean Value Theorem (MVT), there is a point c between x_1 and x_2 such that

$$\frac{M - m}{x_2 - x_1} = \frac{f(x_2) - f(x_1)}{x_2 - x_1} = f'(c)$$

Since x_1, x_2 lie in $[a_0, b_0]$, we have $|x_2 - x_1| \le |b_0 - a_0|$, and thus,

$$|M - m| = |f'(c)| \, |x_2 - x_1| \le K |b_0 - a_0|$$

This proves Eq. (6).

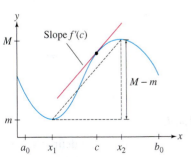

FIGURE 1 Since $M - m = f'(c)(x_2 - x_1)$, we conclude that $M - m \le K(b_0 - a_0)$.

We divide the rest of the proof into two steps. Consider a partition P:

$$P: \quad x_0 = a < x_1 < \quad \cdots \quad < x_{N-1} < x_N = b$$

Let m_i be the minimum value of f on $[x_{i-1}, x_i]$ and M_i the maximum on $[x_{i-1}, x_i]$. We define the *lower* and *upper* Riemann sums

$$L(f, P) = \sum_{i=1}^{N} m_i \, \Delta x_i, \qquad U(f, P) = \sum_{i=1}^{N} M_i \, \Delta x_i$$

These are the particular Riemann sums in which the intermediate point in $[x_{i-1}, x_i]$ is the point where f takes on its minimum or maximum on $[x_{i-1}, x_i]$. Figure 2 illustrates the case $N = 4$.

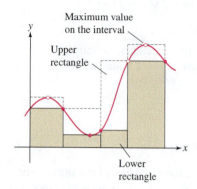

Maximum value on the interval

Upper rectangle

Lower rectangle

FIGURE 2 Lower and upper rectangles for a partition of length $N = 4$.

Step 1. **Prove that the lower and upper sums approach a limit.**
We observe that

$$L(f, P_1) \le U(f, P_2) \quad \text{for any two partitions } P_1 \text{ and } P_2 \qquad \boxed{7}$$

Indeed, if a subinterval I_1 of P_1 overlaps with a subinterval I_2 of P_2, then the minimum of f on I_1 is less than or equal to the maximum of f on I_2 (Figure 3). In particular, the lower sums are bounded above by $U(f, P)$ for all partitions P. Let L be the least upper bound of the lower sums. Then for all partitions P,

$$L(f, P) \le L \le U(f, P) \qquad \boxed{8}$$

According to Eq. (6), $|M_i - m_i| \le K \Delta x_i$ for all i. Since $\|P\|$ is the largest of the widths Δx_i, we see that $|M_i - m_i| \le K\|P\|$ and

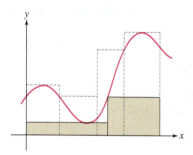

FIGURE 3 The lower rectangles always lie below the upper rectangles, even when the partitions are different.

$$|U(f, P) - L(f, P)| \le \sum_{i=1}^{N} |M_i - m_i| \, \Delta x_i$$

$$\le K \|P\| \sum_{i=1}^{N} \Delta x_i = K \|P\| \, |b - a| \qquad \boxed{9}$$

Let $c = K \, |b - a|$. Using Eq. (8) and Eq. (9), we obtain

$$|L - U(f, P)| \le |U(f, P) - L(f, P)| \le c\|P\|$$

We conclude that $\lim\limits_{\|P\| \to 0} |L - U(f, P)| = 0$. Similarly,

$$|L - L(f, P)| \le c\|P\|$$

and

$$\lim_{\|P\| \to 0} |L - L(f, P)| = 0$$

Thus, we have

$$\lim_{\|P\| \to 0} U(f, P) = \lim_{\|P\| \to 0} L(f, P) = L$$

Step 2. **Prove that $\displaystyle\int_a^b f(x) \, dx$ exists and has value L.**
Recall that for any choice C of intermediate points $c_i \in [x_{i-1}, x_i]$, we define the Riemann sum

$$R(f, P, C) = \sum_{i=1}^{N} f(c_i) \Delta x_i$$

We have

$$L(f, P) \le R(f, P, C) \le U(f, P)$$

Indeed, since $c_i \in [x_{i-1}, x_i]$, we have $m_i \le f(c_i) \le M_i$ for all i and

$$\sum_{i=1}^{N} m_i \, \Delta x_i \le \sum_{i=1}^{N} f(c_i) \, \Delta x_i \le \sum_{i=1}^{N} M_i \, \Delta x_i$$

It follows that

$$|L - R(f, P, C)| \le |U(f, P) - L(f, P)| \le c\|P\|$$

This shows that $R(f, P, C)$ converges to L as $\|P\| \to 0$. ∎

| Section 10.1

> **THEOREM 8**   If f is continuous and $\{a_n\}$ is a sequence such that the limit $\lim_{n \to \infty} a_n = L$ exists, then
>
> $$\lim_{n \to \infty} f(a_n) = f(L)$$

Proof   Choose any $\epsilon > 0$. Since f is continuous, there exists $\delta > 0$ such that

$$\text{if } 0 < |x - L| < \delta, \text{ then } |f(x) - f(L)| < \epsilon.$$

Since $\lim_{n \to \infty} a_n = L$, there exists $N > 0$ such that $|a_n - L| < \delta$ for $n > N$. Thus,

$$|f(a_n) - f(L)| < \epsilon \qquad \text{for } n > N$$

It follows that $\lim_{n \to \infty} f(a_n) = f(L)$. ∎

| Section 14.3

> **THEOREM 9 Clairaut's Theorem**   If f_{xy} and f_{yx} are both continuous functions on a disk D, then $f_{xy}(a, b) = f_{yx}(a, b)$ for all $(a, b) \in D$.

Proof   We prove that both $f_{xy}(a, b)$ and $f_{yx}(a, b)$ are equal to the limit

$$L = \lim_{h \to 0} \frac{f(a + h, b + h) - f(a + h, b) - f(a, b + h) + f(a, b)}{h^2}$$

Let $F(x) = f(x, b + h) - f(x, b)$. The numerator in the limit is equal to

$$F(a + h) - F(a)$$

and $F'(x) = f_x(x, b + h) - f_x(x, b)$. By the MVT, there exists a_1 between a and $a + h$ such that

$$F(a + h) - F(a) = h F'(a_1) = h(f_x(a_1, b + h) - f_x(a_1, b))$$

By the MVT applied to f_x, there exists b_1 between b and $b + h$ such that

$$f_x(a_1, b + h) - f_x(a_1, b) = h f_{xy}(a_1, b_1)$$

Thus,

$$F(a + h) - F(a) = h^2 f_{xy}(a_1, b_1)$$

and

$$L = \lim_{h \to 0} \frac{h^2 f_{xy}(a_1, b_1)}{h^2} = \lim_{h \to 0} f_{xy}(a_1, b_1) = f_{xy}(a, b)$$

The last equality follows from the continuity of f_{xy} since (a_1, b_1) approaches (a, b) as $h \to 0$. To prove that $L = f_{yx}(a, b)$, repeat the argument using the function $F(y) = f(a + h, y) - f(a, y)$, with the roles of x and y reversed. ∎

> **THEOREM 10 Criterion for Differentiability**   If $f_x(x, y)$ and $f_y(x, y)$ exist and are continuous on an open disk D, then $f(x, y)$ is differentiable on D.

Proof   Let $(a, b) \in D$ and set

$$L(x, y) = f(a, b) + f_x(a, b)(x - a) + f_y(a, b)(y - b)$$

It is convenient to switch to the variables h and k, where $x = a + h$ and $y = b + k$. Set

$$\Delta f = f(a + h, b + k) - f(a, b)$$

Then

$$L(x, y) = f(a, b) + f_x(a, b)h + f_y(a, b)k$$

and we may define the function

$$e(h, k) = f(x, y) - L(x, y) = \Delta f - (f_x(a, b)h + f_y(a, b)k)$$

To prove that $f(x, y)$ is differentiable, we must show that

$$\lim_{(h,k) \to (0,0)} \frac{e(h, k)}{\sqrt{h^2 + k^2}} = 0$$

To do this, we write Δf as a sum of two terms:

$$\Delta f = (f(a + h, b + k) - f(a, b + k)) + (f(a, b + k) - f(a, b))$$

and apply the MVT to each term separately. We find that there exist a_1 between a and $a + h$, and b_1 between b and $b + k$, such that

$$f(a + h, b + k) - f(a, b + k) = h f_x(a_1, b + k)$$

$$f(a, b + k) - f(a, b) = k f_y(a, b_1)$$

Therefore,

$$e(h, k) = h(f_x(a_1, b + k) - f_x(a, b)) + k(f_y(a, b_1) - f_y(a, b))$$

and for $(h, k) \neq (0, 0)$,

$$\left| \frac{e(h, k)}{\sqrt{h^2 + k^2}} \right| = \left| \frac{h(f_x(a_1, b + k) - f_x(a, b)) + k(f_y(a, b_1) - f_y(a, b))}{\sqrt{h^2 + k^2}} \right|$$

$$\leq \left| \frac{h(f_x(a_1, b + k) - f_x(a, b))}{\sqrt{h^2 + k^2}} \right| + \left| \frac{k(f_y(a, b_1) - f_y(a, b))}{\sqrt{h^2 + k^2}} \right|$$

$$= |f_x(a_1, b + k) - f_x(a, b)| + |f_y(a, b_1) - f_y(a, b)|$$

In the second line, we use the Triangle Inequality [see Eq. (1) in Section 1.1], and we may pass to the third line because $\left| h/\sqrt{h^2 + k^2} \right|$ and $\left| k/\sqrt{h^2 + k^2} \right|$ are both less than 1. Both terms in the last line tend to zero as $(h, k) \to (0, 0)$ because f_x and f_y are assumed to be continuous. This completes the proof that $f(x, y)$ is differentiable.   ∎

E TAYLOR POLYNOMIALS

English mathematician Brook Taylor (1685–1731) made important contributions to calculus and physics, as well as to the theory of linear perspective used in drawing. (*The Granger Collection, NYC. All rights reserved.*)

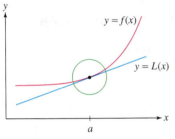

FIGURE 1 The linear approximation $L(x)$ is a first-order approximation to f.

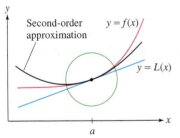

DF FIGURE 2 A second-order approximation is more accurate over a larger interval.

◄·· REMINDER k-factorial *is the number* $k! = k(k-1)(k-2) \cdots (2)(1)$. *Thus,*

$$1! = 1, \quad 2! = (2)1 = 2$$

$$3! = (3)(2)1 = 6$$

By convention, we define $0! = 1$.

Many functions are difficult to work with. For instance, $\sin(x^2)$ cannot be integrated using elementary functions. Nor can e^{-x^2}. In fact, even simple functions like $\sin x$, $\cos x$, e^x and $\ln x$ can only be evaluated exactly at relatively few values of x and otherwise they must be numerically approximated. On the other hand, polynomials such as $f(x) = 3x^4 - 7x^3 + 2x - 4$ can be easily differentiated and integrated. They can be evaluated at any value of x using just multiplication and addition. In this section, we introduce a method for approximating a function f by a polynomial, that behaves similarly to the way f behaves.

We have done this before. In Section 4.1, we used the linearization $L(x) = f(a) + f'(a)(x - a)$ to approximate $f(x)$ near a point $x = a$:

$$f(x) \approx f(a) + f'(a)(x - a)$$

We refer to $L(x)$ as a "first-order" approximation to $f(x)$ at $x = a$ because $f(x)$ and $L(x)$ have the same value and the same first derivative at $x = a$ (Figure 1):

$$L(a) = f(a), \qquad L'(a) = f'(a)$$

A first-order approximation is useful only in a small interval around $x = a$. In this section, we learn how to achieve greater accuracy over larger intervals using higher order approximations (Figure 2). These higher order approximations will simply be polynomials with higher powers. Ultimately, we will see that the errors in utilizing these higher-order approximations are determined by integrals.

In what follows, assume that f is defined on an open interval I and that all higher derivatives $f^{(k)}$ exist on I. Let $a \in I$. We say that two functions f and g **agree to order n** at $x = a$ if their derivatives up to order n at $x = a$ are equal:

$$f(a) = g(a), \quad f'(a) = g'(a), \quad f''(a) = g''(a), \quad \ldots, \quad f^{(n)}(a) = g^{(n)}(a)$$

We also say that g "approximates f to order n" at $x = a$.

Define the nth **Taylor polynomial T_n of f centered at $x = a$** as follows:

$$T_n(x) = f(a) + \frac{f'(a)}{1!}(x - a) + \frac{f''(a)}{2!}(x - a)^2 + \cdots + \frac{f^{(n)}(a)}{n!}(x - a)^n$$

The first few Taylor polynomials are

$$T_0(x) = f(a)$$

$$T_1(x) = f(a) + f'(a)(x - a)$$

$$T_2(x) = f(a) + f'(a)(x - a) + \frac{1}{2}f''(a)(x - a)^2$$

$$T_3(x) = f(a) + f'(a)(x - a) + \frac{1}{2}f''(a)(x - a)^2 + \frac{1}{6}f'''(a)(x - a)^3$$

Note that T_1 is the linearization of f at a. Note also that T_n is obtained from T_{n-1} by adding on a term of degree n:

$$T_n(x) = T_{n-1}(x) + \frac{f^{(n)}(a)}{n!}(x - a)^n$$

The next theorem justifies our definition of T_n.

A23

> **THEOREM 1**   The polynomial T_n centered at a agrees with f to order n at $x = a$, and it is the only polynomial of degree at most n with this property.

The verification of Theorem 1 is left to the exercises (Exercises 70–71), but we'll illustrate the idea by checking that T_2 agrees with f to order $n = 2$:

$$T_2(x) = f(a) + f'(a)(x - a) + \frac{1}{2}f''(a)(x - a)^2, \quad T_2(a) = f(a)$$

$$T_2'(x) = f'(a) + f''(a)(x - a), \qquad\qquad T_2'(a) = f'(a)$$

$$T_2''(x) = f''(a), \qquad\qquad\qquad\qquad T_2''(a) = f''(a)$$

This shows that the value and the derivatives of order up to $n = 2$ at $x = a$ are equal.

Before proceeding to the examples, we write T_n in summation notation:

$$T_n(x) = \sum_{j=0}^{n} \frac{f^{(j)}(a)}{j!} (x - a)^j$$

By convention, we regard f as the *zeroth* derivative, and thus $f^{(0)}$ is f itself. When $a = 0$, T_n is also called the nth **Maclaurin polynomial**.

Scottish mathematician Colin Maclaurin (1698–1746) was a professor in Edinburgh. Newton was so impressed by his work that he once offered to pay part of Maclaurin's salary. *(The Granger Collection, NYC. All rights reserved.)*

■ **EXAMPLE 1**   Maclaurin Polynomials for $f(x) = e^x$ Plot the third and fourth Maclaurin polynomials for $f(x) = e^x$. Compare with the linear approximation.

Solution   All higher derivatives coincide with f itself: $f^{(k)}(x) = e^x$. Therefore,

$$f(0) = f'(0) = f''(0) = f'''(0) = f^{(4)}(0) = e^0 = 1$$

The third Maclaurin polynomial (the case $a = 0$) is

$$T_3(x) = f(0) + f'(0)x + \frac{1}{2}f''(0)x^2 + \frac{1}{3!}f'''(0)x^3 = 1 + x + \frac{1}{2}x^2 + \frac{1}{6}x^3$$

We obtain $T_4(x)$ by adding the term of degree 4 to $T_3(x)$:

$$T_4(x) = T_3(x) + \frac{1}{4!}f^{(4)}(0)x^4 = 1 + x + \frac{1}{2}x^2 + \frac{1}{6}x^3 + \frac{1}{24}x^4$$

Figure 3 shows that T_3 and T_4 approximate $f(x) = e^x$ much more closely than the linear approximation T_1 on an interval around $a = 0$. Higher-degree Maclaurin polynomials would provide even better approximations on larger intervals.   ■

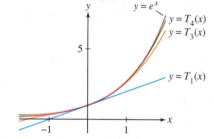

FIGURE 3 Maclaurin polynomials for $f(x) = e^x$.

■ **EXAMPLE 2**   Computing Taylor Polynomials   Compute the Taylor polynomial T_4 centered at $a = 3$ for $f(x) = \sqrt{x + 1}$.

Solution   First evaluate the derivatives up to degree 4 at $a = 3$:

$$f(x) = (x + 1)^{1/2}, \qquad\qquad f(3) = 2$$

$$f'(x) = \frac{1}{2}(x + 1)^{-1/2}, \qquad\qquad f'(3) = \frac{1}{4}$$

$$f''(x) = -\frac{1}{4}(x + 1)^{-3/2}, \qquad\qquad f''(3) = -\frac{1}{32}$$

$$f'''(x) = \frac{3}{8}(x + 1)^{-5/2}, \qquad\qquad f'''(3) = \frac{3}{256}$$

$$f^{(4)}(x) = -\frac{15}{16}(x + 1)^{-7/2}, \qquad\qquad f^{(4)}(3) = -\frac{15}{2048}$$

Then compute the coefficients $\dfrac{f^{(j)}(3)}{j!}$:

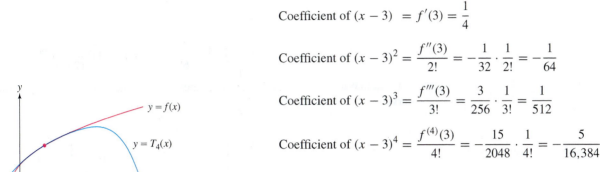

The first term $f(a)$ in the Taylor polynomial T_n is called the constant term.

$$\text{Constant term} \quad = f(3) = 2$$

$$\text{Coefficient of } (x-3) \; = f'(3) = \frac{1}{4}$$

$$\text{Coefficient of } (x-3)^2 = \frac{f''(3)}{2!} = -\frac{1}{32} \cdot \frac{1}{2!} = -\frac{1}{64}$$

$$\text{Coefficient of } (x-3)^3 = \frac{f'''(3)}{3!} = \frac{3}{256} \cdot \frac{1}{3!} = \frac{1}{512}$$

$$\text{Coefficient of } (x-3)^4 = \frac{f^{(4)}(3)}{4!} = -\frac{15}{2048} \cdot \frac{1}{4!} = -\frac{5}{16{,}384}$$

The Taylor polynomial T_4 centered at $a = 3$ is (see Figure 4):

$$T_4(x) = 2 + \frac{1}{4}(x-3) - \frac{1}{64}(x-3)^2 + \frac{1}{512}(x-3)^3 - \frac{5}{16{,}384}(x-3)^4 \quad\blacksquare$$

FIGURE 4 Graph of $f(x) = \sqrt{x+1}$ and T_4 centered at $x = 3$.

■ **EXAMPLE 3** Finding a General Formula for T_n Find the Taylor polynomials T_n of $f(x) = \ln x$ centered at $a = 1$.

Solution For $f(x) = \ln x$, the constant term of T_n at $a = 1$ is zero because $f(1) = \ln 1 = 0$. Next, we compute the derivatives:

After computing several derivatives of $f(x) = \ln x$, we begin to discern the pattern. For many functions of interest, however, the derivatives follow no simple pattern and there is no convenient formula for the general Taylor polynomial.

$$f'(x) = x^{-1}, \qquad f''(x) = -x^{-2}, \qquad f'''(x) = 2x^{-3}, \qquad f^{(4)}(x) = -3 \cdot 2x^{-4}$$

Similarly, $f^{(5)}(x) = 4 \cdot 3 \cdot 2x^{-5}$. The general pattern is that $f^{(k)}(x)$ is a multiple of x^{-k}, with a coefficient $\pm(k-1)!$ that alternates in sign:

$$f^{(k)}(x) = (-1)^{k-1}(k-1)!\,x^{-k} \qquad \boxed{1}$$

The coefficient of $(x-1)^k$ in T_n is

Taylor polynomials for $\ln x$ at $a = 1$:

$$T_1(x) = (x-1)$$

$$T_2(x) = (x-1) - \frac{1}{2}(x-1)^2$$

$$T_3(x) = (x-1) - \frac{1}{2}(x-1)^2 + \frac{1}{3}(x-1)^3$$

$$\frac{f^{(k)}(1)}{k!} = \frac{(-1)^{k-1}(k-1)!}{k!} = \frac{(-1)^{k-1}}{k} \qquad \text{(for } k \geq 1)$$

Thus, the coefficients for $k \geq 1$ form a sequence $1, -\frac{1}{2}, \frac{1}{3}, -\frac{1}{4}, \ldots$, and

$$T_n(x) = (x-1) - \frac{1}{2}(x-1)^2 + \frac{1}{3}(x-1)^3 - \cdots + (-1)^{n-1}\frac{1}{n}(x-1)^n \quad\blacksquare$$

■ **EXAMPLE 4** Cosine Find the Maclaurin polynomials of $f(x) = \cos x$.

Solution The derivatives form a repeating pattern of period 4:

$$f(x) = \cos x, \qquad f'(x) = -\sin x, \qquad f''(x) = -\cos x, \qquad f'''(x) = \sin x,$$

$$f^{(4)}(x) = \cos x, \qquad f^{(5)}(x) = -\sin x, \qquad \cdots$$

In general, $f^{(j+4)}(x) = f^{(j)}(x)$. The derivatives at $x = 0$ also form a pattern:

$f(0)$	$f'(0)$	$f''(0)$	$f'''(0)$	$f^{(4)}(0)$	$f^{(5)}(0)$	$f^{(6)}(0)$	$f^{(7)}(0)$	$\cdots$
1	0	−1	0	1	0	−1	0	$\cdots$

Therefore, the coefficients of the odd powers x^{2k+1} are zero, and the coefficients of the even powers x^{2k} alternate in sign with value $(-1)^k/(2k)!$:

$$T_0(x) = T_1(x) = 1, \qquad T_2(x) = T_3(x) = 1 - \frac{1}{2!}x^2$$

$$T_4(x) = T_5(x) = 1 - \frac{x^2}{2} + \frac{x^4}{4!}$$

$$T_{2n}(x) = T_{2n+1}(x) = 1 - \frac{1}{2}x^2 + \frac{1}{4!}x^4 - \frac{1}{6!}x^6 + \cdots + (-1)^n \frac{1}{(2n)!}x^{2n}$$

Figure 5 shows that as n increases, T_n approximates $f(x) = \cos x$ well over larger and larger intervals, but outside this interval, the approximation fails.   ∎

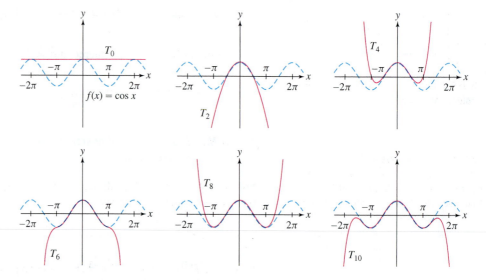

DF FIGURE 5 Maclaurin polynomials for $f(x) = \cos x$. The graph of f is shown as a dashed curve.

∎ **EXAMPLE 5   How Far Is the Horizon?**   Valerie is at the beach, looking out over the ocean (Figure 6). How far can she see? Use Maclaurin polynomials to estimate the distance d, assuming that Valerie's eye level is $h = 1.7$ meters (m) above ground. What if she looks out from a window where her eye level is 20 m?

FIGURE 6 View from the beach. *(Alexandr Ozerov/iStockphoto)*

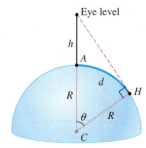

FIGURE 7 Valerie can see a distance $d = R\theta$, the length of arc AH.

Solution   Let R be the radius of the earth. Figure 7 shows that Valerie can see a distance $d = R\theta$, the length of the circular arc AH in Figure 7. We have

$$\cos\theta = \frac{R}{R+h}$$

This calculation ignores the bending of light (called refraction) as it passes through the atmosphere. Refraction typically increases d by around 10%, although the actual effect is complex and varies with atmospheric temperature.

Our key observation is that θ is close to zero (both θ and h are much smaller than shown in the figure), so we lose very little accuracy if we replace $\cos\theta$ by its second Maclaurin polynomial $T_2(\theta) = 1 - \frac{1}{2}\theta^2$, as computed in Example 4:

$$1 - \frac{1}{2}\theta^2 \approx \frac{R}{R+h} \quad \Rightarrow \quad \theta^2 \approx 2 - \frac{2R}{R+h} \quad \Rightarrow \quad \theta \approx \sqrt{\frac{2h}{R+h}}$$

Furthermore, h is very small relative to R, so we may replace $R + h$ by R to obtain

$$d = R\theta \approx R\sqrt{\frac{2h}{R}} = \sqrt{2Rh}$$

The earth's radius is approximately $R \approx 6.37 \times 10^6$ m, so

$$d = \sqrt{2Rh} \approx \sqrt{2(6.37 \times 10^6)h} \approx 3569\sqrt{h}\ \text{m}$$

In particular, we see that d is proportional to $\sqrt{h}$.

If Valerie's eye level is $h = 1.7$ m, then $d \approx 3569\sqrt{1.7} \approx 4653$ m, or roughly 4.7 km. If $h = 20$ m, then $d \approx 3569\sqrt{20} \approx 15.96$ m, or nearly 16 km. ∎

The Error Bound

To use Taylor polynomials effectively, we need a way to estimate the size of the error. This is provided by the next theorem, which shows that the size of this error depends on the size of the $(n + 1)$st derivative.

A proof of Theorem 2 is presented at the end of this section.

THEOREM 2 Error Bound Assume that $f^{(n+1)}$ exists and is continuous. Let K be a number such that $|f^{(n+1)}(u)| \le K$ for all u between a and x. Then

$$|f(x) - T_n(x)| \le K\frac{|x - a|^{n+1}}{(n+1)!}$$

where T_n is the nth Taylor polynomial centered at $x = a$.

∎ **EXAMPLE 6 Using the Error Bound** Apply the error bound to

$$|\ln 1.2 - T_3(1.2)|$$

where T_3 is the third Taylor polynomial for $f(x) = \ln x$ at $a = 1$. Check your result with a calculator.

Solution

Step 1. **Find a value of K.**

To use the error bound with $n = 3$, we must find a value of K such that $|f^{(4)}(u)| \le K$ for all u between $a = 1$ and $x = 1.2$. As we computed in Example 3, $f^{(4)}(x) = -6x^{-4}$. The absolute value $|f^{(4)}(x)|$ is decreasing for $x > 0$, so its maximum value on $[1, 1.2]$ is $|f^{(4)}(1)| = 6$. Therefore, we may take $K = 6$.

Step 2. **Apply the error bound.**

$$|\ln 1.2 - T_3(1.2)| \le K\frac{|x - a|^{n+1}}{(n+1)!} = 6\frac{|1.2 - 1|^4}{4!} \approx 0.0004$$

Step 3. **Check the result.**

Recall from Example 3 that

$$T_3(x) = (x - 1) - \frac{1}{2}(x - 1)^2 + \frac{1}{3}(x - 1)^3$$

The following values from a calculator confirm that the error is at most 0.0004:

$$|\ln 1.2 - T_3(1.2)| \approx |0.182667 - 0.182322| \approx 0.00035 < 0.0004$$

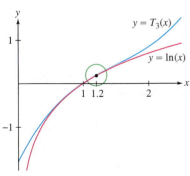

DF FIGURE 8 $y = \ln x$ and $y = T_3(x)$ are indistinguishable near $x = 1.2$.

Observe in Figure 8 that $y = \ln x$ and $y = T_3(x)$ are indistinguishable near $x = 1.2$. ∎

■ **EXAMPLE 7** Approximating with a Given Accuracy Let T_n be the nth Maclaurin polynomial for $f(x) = \cos x$. Find a value of n such that

$$|\cos 0.2 - T_n(0.2)| < 10^{-5}$$

Solution

Step 1. **Find a value of K.**

Since $|f^{(n)}(x)|$ is $|\cos x|$ or $|\sin x|$, depending on whether n is even or odd, we have $|f^{(n)}(u)| \le 1$ for all u. Thus, we may apply the error bound with $K = 1$.

Step 2. **Find a value of n.**

The error bound gives us

> To use the error bound, it is not necessary to find the smallest possible value of K. In this example, we take $K = 1$. This works for all n, but for odd n we could have used the smaller value $K = \sin 0.2 \approx 0.2$.

$$|\cos 0.2 - T_n(0.2)| \le K \frac{|0.2 - 0|^{n+1}}{(n+1)!} = \frac{|0.2|^{n+1}}{(n+1)!}$$

To make the error less than 10^{-5}, we must choose n so that

$$\frac{|0.2|^{n+1}}{(n+1)!} < 10^{-5}$$

It's not possible to solve this inequality for n, but we can find a suitable n by checking several values:

n	2	3	4
$\dfrac{\|0.2\|^{n+1}}{(n+1)!}$	$\dfrac{0.2^3}{3!} \approx 0.0013$	$\dfrac{0.2^4}{4!} \approx 6.67 \times 10^{-5}$	$\dfrac{0.2^5}{5!} \approx 2.67 \times 10^{-6} < 10^{-5}$

We see that the error is less than 10^{-5} for $n = 4$.   ■

The rest of this section is devoted to a proof of the error bound (Theorem 2). Define the *nth remainder*:

$$\boxed{R_n(x) = f(x) - T_n(x)}$$

The error in $T_n(x)$ is the absolute value $|R_n(x)|$. As a first step in proving the error bound, we show that $R_n(x)$ can be represented as an integral.

Taylor's Theorem   Assume that $f^{(n+1)}$ exists and is continuous. Then

$$\boxed{R_n(x) = \frac{1}{n!} \int_a^x (x-u)^n f^{(n+1)}(u)\, du} \qquad \boxed{2}$$

Proof   Set

$$I_n(x) = \frac{1}{n!} \int_a^x (x-u)^n f^{(n+1)}(u)\, du$$

Our goal is to show that $R_n(x) = I_n(x)$. For $n = 0$, $R_0(x) = f(x) - f(a)$ and the desired result is just a restatement of the Fundamental Theorem of Calculus:

$$I_0(x) = \int_a^x f'(u)\, du = f(x) - f(a) = R_0(x)$$

> Exercise 64 reviews this proof for the special case $n = 2$.

To prove the formula for $n > 0$, we apply Integration by Parts to $I_n(x)$ with

$$h(u) = \frac{1}{n!}(x-u)^n, \qquad g(u) = f^{(n)}(u)$$

Then $g'(u) = f^{(n+1)}(u)$, and so

$$I_n(x) = \int_a^x h(u)\, g'(u)\, du = h(u)g(u)\Big|_a^x - \int_a^x h'(u)g(u)\, du$$

$$= \frac{1}{n!}(x-u)^n f^{(n)}(u)\Big|_a^x - \frac{1}{n!}\int_a^x (-n)(x-u)^{n-1} f^{(n)}(u)\, du$$

$$= -\frac{1}{n!}(x-a)^n f^{(n)}(a) + I_{n-1}(x)$$

This can be rewritten as

$$I_{n-1}(x) = \frac{f^{(n)}(a)}{n!}(x-a)^n + I_n(x)$$

Now apply this relation n times, noting that $I_0(x) = f(x) - f(a)$:

$$f(x) = f(a) + I_0(x)$$

$$= f(a) + \frac{f'(a)}{1!}(x-a) + I_1(x)$$

$$= f(a) + \frac{f'(a)}{1!}(x-a) + \frac{f''(a)}{2!}(x-a)^2 + I_2(x)$$

$$\vdots$$

$$= f(a) + \frac{f'(a)}{1!}(x-a) + \cdots + \frac{f^{(n)}(a)}{n!}(x-a)^n + I_n(x)$$

This shows that $f(x) = T_n(x) + I_n(x)$ and hence $I_n(x) = R_n(x)$, as desired.   ■

Proof Now we can prove Theorem 2. Assume first that $x \geq a$. Then

In Eq. (3), we use the inequality

$$\left| \int_a^b f(x)\, dx \right| \leq \int_a^b |f(x)|\, dx$$

which is valid for all integrable functions.

$$|f(x) - T_n(x)| = |R_n(x)| = \left| \frac{1}{n!} \int_a^x (x-u)^n f^{(n+1)}(u)\, du \right|$$

$$\leq \frac{1}{n!} \int_a^x \left| (x-u)^n f^{(n+1)}(u) \right|\, du \qquad \boxed{3}$$

$$\leq \frac{K}{n!} \int_a^x |x-u|^n\, du \qquad \boxed{4}$$

$$= \frac{K}{n!} \frac{-(x-u)^{n+1}}{n+1}\Big|_{u=a}^x = K\frac{|x-a|^{n+1}}{(n+1)!}$$

Note that the absolute value is not needed in Eq. (4) because $x - u \geq 0$ for $a \leq u \leq x$. If $x \leq a$, we must interchange the upper and lower limits of the integral in Eqs. (3) and (4).   ■

E. SUMMARY

- The nth *Taylor polynomial* centered at $x = a$ for the function f is

$$T_n(x) = f(a) + \frac{f'(a)}{1!}(x-a)^1 + \frac{f''(a)}{2!}(x-a)^2 + \cdots + \frac{f^{(n)}(a)}{n!}(x-a)^n$$

When $a = 0$, T_n is also called the nth *Maclaurin polynomial*.
- If $f^{(n+1)}$ exists and is continuous, then we have the *error bound*

$$|T_n(x) - f(x)| \leq K\frac{|x-a|^{n+1}}{(n+1)!}$$

where K is a number such that $|f^{(n+1)}(u)| \leq K$ for all u between a and x.

• For reference, we include a table of standard Maclaurin and Taylor polynomials.

$f(x)$	a	Maclaurin or Taylor Polynomial
e^x	0	$T_n(x) = 1 + x + \dfrac{x^2}{2!} + \dfrac{x^3}{3!} + \cdots + \dfrac{x^n}{n!}$
$\sin x$	0	$T_{2n+1}(x) = T_{2n+2}(x) = x - \dfrac{x^3}{3!} + \cdots + (-1)^n \dfrac{x^{2n+1}}{(2n+1)!}$
$\cos x$	0	$T_{2n}(x) = T_{2n+1}(x) = 1 - \dfrac{x^2}{2!} + \dfrac{x^4}{4!} - \cdots + (-1)^n \dfrac{x^{2n}}{(2n)!}$
$\ln x$	1	$T_n(x) = (x-1) - \dfrac{1}{2}(x-1)^2 + \cdots + \dfrac{(-1)^{n-1}}{n}(x-1)^n$
$\dfrac{1}{1-x}$	0	$T_n(x) = 1 + x + x^2 + \cdots + x^n$

E. EXERCISES

Preliminary Questions

1. What is T_3 centered at $a = 3$ for a function f such that $f(3) = 9$, $f'(3) = 8$, $f''(3) = 4$, and $f'''(3) = 12$?

2. The dashed graphs in Figure 9 are Taylor polynomials for a function f. Which of the two is a Maclaurin polynomial?

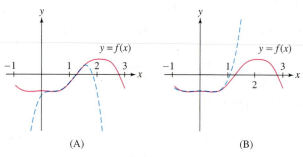

(A)   (B)

FIGURE 9

3. For which value of x does the Maclaurin polynomial T_n satisfy $T_n(x) = f(x)$, no matter what f is?

4. Let T_n be the Maclaurin polynomial of a function f satisfying $|f^{(4)}(x)| \le 1$ for all x. Which of the following statements follow from the error bound?

(a) $|T_4(2) - f(2)| \le \frac{2}{3}$

(b) $|T_3(2) - f(2)| \le \frac{2}{3}$

(c) $|T_3(2) - f(2)| \le \frac{1}{3}$

Exercises

In Exercises 1–14, calculate the Taylor polynomials T_2 and T_3 centered at $x = a$ for the given function and value of a.

1. $f(x) = \sin x,\quad a = 0$

2. $f(x) = \sin x,\quad a = \dfrac{\pi}{2}$

3. $f(x) = \dfrac{1}{1+x},\quad a = 2$

4. $f(x) = \dfrac{1}{1+x^2},\quad a = -1$

5. $f(x) = x^4 - 2x,\quad a = 3$

6. $f(x) = \dfrac{x^2+1}{x+1},\quad a = -2$

7. $f(x) = \tan x,\quad a = 0$

8. $f(x) = \tan x,\quad a = \dfrac{\pi}{4}$

9. $f(x) = e^{-x} + e^{-2x},\quad a = 0$

10. $f(x) = e^{2x},\quad a = \ln 2$

11. $f(x) = x^2 e^{-x},\quad a = 1$

12. $f(x) = \cosh 2x,\quad a = 0$

13. $f(x) = \dfrac{\ln x}{x},\quad a = 1$

14. $f(x) = \ln(x+1),\quad a = 0$

15. Show that the nth Maclaurin polynomial for $f(x) = e^x$ is

$$T_n(x) = 1 + \dfrac{x}{1!} + \dfrac{x^2}{2!} + \cdots + \dfrac{x^n}{n!}$$

16. Show that the nth Taylor polynomial for $f(x) = \dfrac{1}{x+1}$ at $a = 1$ is

$$T_n(x) = \dfrac{1}{2} - \dfrac{(x-1)}{4} + \dfrac{(x-1)^2}{8} + \cdots + (-1)^n \dfrac{(x-1)^n}{2^{n+1}}$$

17. Show that the Maclaurin polynomials for $f(x) = \sin x$ are

$$T_{2n+1}(x) = T_{2n+2}(x) = x - \dfrac{x^3}{3!} + \dfrac{x^5}{5!} - \cdots + (-1)^n \dfrac{x^{2n+1}}{(2n+1)!}$$

18. Show that the Maclaurin polynomials for $f(x) = \ln(1+x)$ are

$$T_n(x) = x - \dfrac{x^2}{2} + \dfrac{x^3}{3} + \cdots + (-1)^{n-1}\dfrac{x^n}{n}$$

In Exercises 19–24, find T_n centered at $x = a$ for all n.

19. $f(x) = \dfrac{1}{1+x},\quad a = 0$

20. $f(x) = \dfrac{1}{x-1},\quad a = 4$

21. $f(x) = e^x,\quad a = 1$

22. $f(x) = x^{-2},\quad a = 2$

23. $f(x) = \cos x,\quad a = \dfrac{\pi}{4}$

24. $f(\theta) = \sin 3\theta,\quad a = 0$

In Exercises 25–28, find T_2 and use a calculator to compute the error $|f(x) - T_2(x)|$ for the given values of a and x.

25. $y = e^x$, $a = 0$, $x = -0.5$

26. $y = \cos x$, $a = 0$, $x = \dfrac{\pi}{12}$

27. $y = x^{-2/3}$, $a = 1$, $x = 1.2$

28. $y = e^{\sin x}$, $a = \dfrac{\pi}{2}$, $x = 1.5$

29. [GU] Compute T_3 for $f(x) = \sqrt{x}$ centered at $a = 1$. Then use a plot of the error $|f(x) - T_3(x)|$ to find a value $c > 1$ such that the error on the interval $[1, c]$ is at most 0.25.

30. [CAS] Plot $f(x) = 1/(1 + x)$ together with the Taylor polynomials T_n at $a = 1$ for $1 \le n \le 4$ on the interval $[-2, 8]$ (be sure to limit the upper plot range).
(a) Over which interval does T_4 appear to approximate f closely?
(b) What happens for $x < -1$?
(c) Use your computer algebra system to produce and plot T_{30} together with f on $[-2, 8]$. Over which interval does T_{30} appear to give a close approximation?

31. Let T_3 be the Maclaurin polynomial of $f(x) = e^x$. Use the error bound to find the maximum possible value of $|f(1.1) - T_3(1.1)|$. Show that we can take $K = e^{1.1}$.

32. Let T_2 be the Taylor polynomial of $f(x) = \sqrt{x}$ at $a = 4$. Apply the error bound to find the maximum possible value of the error $|f(3.9) - T_2(3.9)|$.

In Exercises 33–36, compute the Taylor polynomial indicated and use the error bound to find the maximum possible size of the error. Verify your result with a calculator.

33. $f(x) = \cos x$, $a = 0$; $|\cos 0.25 - T_5(0.25)|$

34. $f(x) = x^{11/2}$, $a = 1$; $|f(1.2) - T_4(1.2)|$

35. $f(x) = x^{-1/2}$, $a = 4$; $|f(4.3) - T_3(4.3)|$

36. $f(x) = \sqrt{1 + x}$, $a = 8$; $|\sqrt{9.02} - T_3(8.02)|$

37. Calculate the Maclaurin polynomial T_3 for $f(x) = \tan^{-1} x$. Compute $T_3\left(\frac{1}{2}\right)$ and use the error bound to find a bound for the error $\left|\tan^{-1}\frac{1}{2} - T_3\left(\frac{1}{2}\right)\right|$. Refer to the graph in Figure 10 to find an acceptable value of K. Verify your result by computing $\left|\tan^{-1}\frac{1}{2} - T_3\left(\frac{1}{2}\right)\right|$ using a calculator.

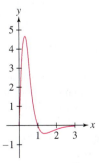

FIGURE 10 Graph of $f^{(4)}(x) = \dfrac{-24x(x^2 - 1)}{(x^2 + 1)^4}$, where $f(x) = \tan^{-1} x$.

38. Let $f(x) = \ln(x^3 - x + 1)$. The third Taylor polynomial at $a = 1$ is
$$T_3(x) = 2(x - 1) + (x - 1)^2 - \frac{7}{3}(x - 1)^3$$

Find the maximum possible value of $|f(1.1) - T_3(1.1)|$, using the graph in Figure 11 to find an acceptable value of K. Verify your result by computing $|f(1.1) - T_3(1.1)|$ using a calculator.

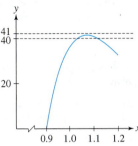

FIGURE 11 Graph of $f^{(4)}$, where $f(x) = \ln(x^3 - x + 1)$.

39. [GU] Let T_2 be the Taylor polynomial at $a = 0.5$ for $f(x) = \cos(x^2)$. Use the error bound to find the maximum possible value of $|f(0.6) - T_2(0.6)|$. Plot $f^{(3)}$ to find an acceptable value of K.

40. [GU] Calculate the Maclaurin polynomial T_2 for $f(x) = \operatorname{sech} x$ and use the error bound to find the maximum possible value of $\left|f\left(\frac{1}{2}\right) - T_2\left(\frac{1}{2}\right)\right|$. Plot f''' to find an acceptable value of K.

In Exercises 41–44, use the error bound to find a value of n for which the given inequality is satisfied. Then verify your result using a calculator.

41. $|\cos 0.1 - T_n(0.1)| \le 10^{-7}$, $a = 0$

42. $|\ln 1.3 - T_n(1.3)| \le 10^{-4}$, $a = 1$

43. $|\sqrt{1.3} - T_n(1.3)| \le 10^{-6}$, $a = 1$

44. $|e^{-0.1} - T_n(-0.1)| \le 10^{-6}$, $a = 0$

45. Let $f(x) = e^{-x}$ and $T_3(x) = 1 - x + \dfrac{x^2}{2} - \dfrac{x^3}{6}$. Use the error bound to show that for all $x \ge 0$,
$$|f(x) - T_3(x)| \le \frac{x^4}{24}$$

If you have a GU, illustrate this inequality by plotting $y = f(x) - T_3(x)$ and $y = x^4/24$ together over $[0, 1]$.

46. Use the error bound with $n = 4$ to show that
$$\left| \sin x - \left(x - \frac{x^3}{6} \right) \right| \le \frac{|x|^5}{120} \quad \text{(for all } x\text{)}$$

47. Let T_n be the Taylor polynomial for $f(x) = \ln x$ at $a = 1$, and let $c > 1$. Show that
$$|\ln c - T_n(c)| \le \frac{|c - 1|^{n+1}}{n + 1}$$

Then find a value of n such that $|\ln 1.5 - T_n(1.5)| \le 10^{-2}$.

48. Let $n \ge 1$. Show that if $|x|$ is small, then
$$(x + 1)^{1/n} \approx 1 + \frac{x}{n} + \frac{1 - n}{2n^2}x^2$$

Use this approximation with $n = 6$ to estimate $1.5^{1/6}$.

49. Verify that the third Maclaurin polynomial for $f(x) = e^x \sin x$ is equal to the product of the third Maclaurin polynomials of $f(x) = e^x$ and $f(x) = \sin x$ (after discarding terms of degree greater than 3 in the product).

50. Find the fourth Maclaurin polynomial for $f(x) = \sin x \cos x$ by multiplying the fourth Maclaurin polynomials for $f(x) = \sin x$ and $f(x) = \cos x$.

51. Find the Maclaurin polynomials T_n for $f(x) = \cos(x^2)$. You may use the fact that $T_n(x)$ is equal to the sum of the terms up to degree n obtained by substituting x^2 for x in the nth Maclaurin polynomial of $\cos x$.

52. Find the Maclaurin polynomials of $1/(1 + x^2)$ by substituting $-x^2$ for x in the Maclaurin polynomials of $1/(1 - x)$.

53. Let $f(x) = 3x^3 + 2x^2 - x - 4$. Calculate T_j for $j = 1, 2, 3, 4, 5$ at both $a = 0$ and $a = 1$. Show that $T_3(x) = f(x)$ in both cases.

54. Let T_n be the nth Taylor polynomial at $x = a$ for a polynomial f of degree n. Based on the result of Exercise 53, guess the value of $|f(x) - T_n(x)|$. Prove that your guess is correct using the error bound.

55. Let $s(t)$ be the distance of a truck to an intersection. At time $t = 0$, the truck is 60 m from the intersection, travels away from it with a velocity of 24 m/s, and begins to slow down with an acceleration of $a = -3$ m/s^2. Determine the second Maclaurin polynomial of s, and use it to estimate the truck's distance from the intersection after 4 s.

56. A bank owns a portfolio of bonds whose value $P(r)$ depends on the interest rate r (measured in percent; e.g., $r = 5$ means a 5% interest rate). The bank's quantitative analyst determines that

$$P(5) = 100{,}000, \quad \left.\frac{dP}{dr}\right|_{r=5} = -40{,}000, \quad \left.\frac{d^2P}{dr^2}\right|_{r=5} = 50{,}000$$

In finance, this second derivative is called **bond convexity**. Find the second Taylor polynomial of $P(r)$ centered at $r = 5$ and use it to estimate the value of the portfolio if the interest rate moves to $r = 5.5\%$.

57. A narrow, negatively charged ring of radius R exerts a force on a positively charged particle P located at distance x above the center of the ring of magnitude

$$F(x) = -\frac{kx}{(x^2 + R^2)^{3/2}}$$

where $k > 0$ is a constant (Figure 12).

(a) Compute the third-degree Maclaurin polynomial for F.

(b) Show that $F \approx -(k/R^3)x$ to second order. This shows that when x is small, $F(x)$ behaves like a restoring force similar to the force exerted by a spring.

(c) Show that $F(x) \approx -k/x^2$ when x is large by showing that

$$\lim_{x \to \infty} \frac{F(x)}{-k/x^2} = 1$$

Thus, $F(x)$ behaves like an inverse square law, and the charged ring looks like a point charge from far away.

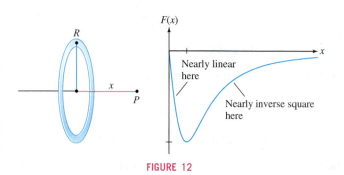

FIGURE 12

58. A light wave of wavelength λ travels from A to B by passing through an aperture (circular region) located in a plane that is perpen-

dicular to $\overline{AB}$ (see Figure 13 for the notation). Let $f(r) = d' + h'$; that is, $f(r)$ is the distance $AC + CB$ as a function of r.

(a) Show that $f(r) = \sqrt{d^2 + r^2} + \sqrt{h^2 + r^2}$, and use the Maclaurin polynomial of order 2 to show that

$$f(r) \approx d + h + \frac{1}{2}\left(\frac{1}{d} + \frac{1}{h}\right)r^2$$

(b) The **Fresnel zones**, used to determine the optical disturbance at B, are the concentric bands bounded by the circles of radius R_n such that $f(R_n) = d + h + n\lambda/2$. Show that $R_n \approx \sqrt{n\lambda L}$, where $L = (d^{-1} + h^{-1})^{-1}$.

(c) Estimate the radii R_1 and R_{100} for blue light ($\lambda = 475 \times 10^{-7}$ cm) if $d = h = 100$ cm.

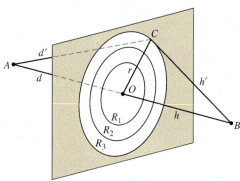

FIGURE 13 The Fresnel zones are the regions between the circles of radius R_n.

59. Referring to Figure 14, let a be the length of the chord $\overline{AC}$ of angle θ of the unit circle. Derive the following approximation for the excess of the arc over the chord:

$$\theta - a \approx \frac{\theta^3}{24}$$

Hint: Show that $\theta - a = \theta - 2\sin(\theta/2)$ and use the third Maclaurin polynomial as an approximation.

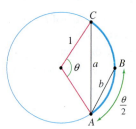

FIGURE 14 Unit circle.

60. To estimate the length θ of a circular arc of the unit circle, the seventeenth-century Dutch scientist Christian Huygens used the approximation $\theta \approx (8b - a)/3$, where a is the length of the chord $\overline{AC}$ of angle θ and b is the length of the chord $\overline{AB}$ of angle $\theta/2$ (Figure 14).

(a) Prove that $a = 2\sin(\theta/2)$ and $b = 2\sin(\theta/4)$, and show that the Huygens approximation amounts to the approximation

$$\theta \approx \frac{16}{3}\sin\frac{\theta}{4} - \frac{2}{3}\sin\frac{\theta}{2}$$

(b) Compute the fifth Maclaurin polynomial of the function on the right.

(c) Use the error bound to show that the error in the Huygens approximation is less than $0.00022|\theta|^5$.

Further Insights and Challenges

61. Show that the nth Maclaurin polynomial of $f(x) = \arcsin x$ for n odd is

$$T_n(x) = x + \frac{1}{2}\frac{x^3}{3} + \frac{1 \cdot 3}{2 \cdot 4}\frac{x^5}{5} + \cdots + \frac{1 \cdot 3 \cdot 5 \cdots (n-2)}{2 \cdot 4 \cdot 6 \cdots (n-1)}\frac{x^n}{n}$$

62. Let $x \geq 0$ and assume that $f^{(n+1)}(t) \geq 0$ for $0 \leq t \leq x$. Use Taylor's Theorem to show that the nth Maclaurin polynomial T_n satisfies

$$T_n(x) \leq f(x), \quad \text{for all } x \geq 0$$

63. Use Exercise 62 to show that for $x \geq 0$ and all n,

$$e^x \geq 1 + x + \frac{x^2}{2!} + \cdots + \frac{x^n}{n!}$$

Sketch the graphs of $y = e^x$, $y = T_1(x)$, and $y = T_2(x)$ on the same coordinate axes. Does this inequality remain true for $x < 0$?

64. This exercise is intended to reinforce the proof of Taylor's Theorem.

(a) Show that $f(x) = T_0(x) + \int_a^x f'(u)\,du$.

(b) Use Integration by Parts to prove the formula

$$\int_a^x (x-u)f^{(2)}(u)\,du = -f'(a)(x-a) + \int_a^x f'(u)\,du$$

(c) Prove the case $n = 2$ of Taylor's Theorem:

$$f(x) = T_1(x) + \int_a^x (x-u)f^{(2)}(u)\,du$$

In Exercises 65–69, we estimate integrals using Taylor polynomials. Exercise 66 is used to estimate the error.

65. Find the fourth Maclaurin polynomial T_4 for $f(x) = e^{-x^2}$, and calculate $I = \int_0^{1/2} T_4(x)\,dx$ as an estimate for $\int_0^{1/2} e^{-x^2}\,dx$. A CAS yields the value $I \approx 0.461281$. How large is the error in your approximation? *Hint:* T_4 is obtained by substituting $-x^2$ in the second Maclaurin polynomial for e^x.

66. Approximating Integrals Let $L > 0$. Show that if two functions f and g satisfy $|f(x) - g(x)| < L$ for all $x \in [a, b]$, then

$$\left| \int_a^b f(x)\,dx - \int_a^b g(x)\,dx \right| dx < L(b - a)$$

67. Let T_4 be the fourth Maclaurin polynomial for $f(x) = \cos x$.

(a) Show that

$$|\cos x - T_4(x)| \leq \frac{\left(\frac{1}{2}\right)^6}{6!} \quad \text{for all} \quad x \in \left[0, \frac{1}{2}\right]$$

Hint: $T_4(x) = T_5(x)$.

(b) Evaluate $\int_0^{1/2} T_4(x)\,dx$ as an approximation to $\int_0^{1/2} \cos x\,dx$. Use Exercise 66 to find a bound for the size of the error.

68. Let $Q(x) = 1 - x^2/6$. Use the error bound for $f(x) = \sin x$ to show that

$$\left| \frac{\sin x}{x} - Q(x) \right| \leq \frac{|x|^4}{5!}$$

Then calculate $\int_0^1 Q(x)\,dx$ as an approximation to $\int_0^1 (\sin x/x)\,dx$ and find a bound for the error.

69. (a) Compute the sixth Maclaurin polynomial T_6 for $f(x) = \sin(x^2)$ by substituting x^2 in $P(x) = x - x^3/6$, the third Maclaurin polynomial for $f(x) = \sin x$.

(b) Show that $|\sin(x^2) - T_6(x)| \leq \frac{|x|^{10}}{5!}$.

Hint: Substitute x^2 for x in the error bound for $|\sin x - P(x)|$, noting that P is also the fourth Maclaurin polynomial for $f(x) = \sin x$.

(c) Use T_6 to approximate $\int_0^{1/2} \sin(x^2)\,dx$ and find a bound for the error.

70. Prove by induction that for all k,

$$\frac{d^j}{dx^j}\left(\frac{(x-a)^k}{k!} \right) = \frac{k(k-1)\cdots(k-j+1)(x-a)^{k-j}}{k!}$$

$$\frac{d^j}{dx^j}\left(\frac{(x-a)^k}{k!} \right)\bigg|_{x=a} = \begin{cases} 1 & \text{for } k = j \\ 0 & \text{for } k \neq j \end{cases}$$

Use this to prove that T_n agrees with f at $x = a$ to order n.

71. Let a be any number and let

$$P(x) = a_n x^n + a_{n-1}x^{n-1} + \cdots + a_1 x + a_0$$

be a polynomial of degree n or less.

(a) Show that if $P^{(j)}(a) = 0$ for $j = 0, 1, \ldots, n$, then $P(x) = 0$, that is, $a_j = 0$ for all j. *Hint:* Use induction, noting that if the statement is true for degree $n - 1$, then $P'(x) = 0$.

(b) Prove that T_n is the only polynomial of degree n or less that agrees with f at $x = a$ to order n. *Hint:* If Q is another such polynomial, apply (a) to $P(x) = T_n(x) - Q(x)$.

ANSWERS TO ODD-NUMBERED EXERCISES

Chapter 10

Section 10.1 Preliminary Questions

1. $a_4 = 12$ **2.** (c) **3.** $\lim\limits_{n\to\infty} a_n = \sqrt{2}$ **4.** (b)

5. (a) False. Counterexample: $a_n = \cos \pi n$
(b) True (c) False. Counterexample: $a_n = (-1)^n$

Section 10.1 Exercises

1. (a) (iv) (b) (i) (c) (iii) (d) (ii)

3. $c_1 = 3, c_2 = \frac{9}{2}, c_3 = \frac{9}{2}, c_4 = \frac{27}{8}$

5. $a_1 = 2, a_2 = 5, a_3 = 47, a_4 = 4415$

7. $b_1 = 4, b_2 = 6, b_3 = 4, b_4 = 6$

9. $c_1 = 1, c_2 = \frac{3}{2}, c_3 = \frac{11}{6}, c_4 = \frac{25}{12}$

11. $b_1 = 2, b_2 = 3, b_3 = 8, b_4 = 19$

13. (a) $a_n = \frac{(-1)^{n+1}}{n^3}$ (b) $a_n = \frac{n+1}{n+5}$

15. $\lim\limits_{n\to\infty} 12 = 12$ **17.** $\lim\limits_{n\to\infty} \frac{5n-1}{12n+9} = \frac{5}{12}$

19. $\lim\limits_{n\to\infty} (-2^{-n}) = 0$ **21.** The sequence diverges.

23. $\lim\limits_{n\to\infty} \frac{n}{\sqrt{n^2+1}} = 1$ **25.** $\lim\limits_{n\to\infty} \ln\left(\frac{12n+2}{-9+4n}\right) = \ln 3$

27. $\lim\limits_{n\to\infty} \sqrt{4+\frac{1}{n}} = 2$ **29.** $\lim\limits_{n\to\infty} \cos^{-1}\left(\frac{n^3}{2n^3+1}\right) = \frac{\pi}{3}$

31. (a) $M = 999$ (b) $M = 99,999$

35. $\lim\limits_{n\to\infty}\left(10 + \left(-\frac{1}{9}\right)^n\right) = 10$ **37.** The sequence diverges.

39. $\lim\limits_{n\to\infty} 2^{1/n} = 1$ **41.** $\lim\limits_{n\to\infty} \frac{9^n}{n!} = 0$

43. $\lim\limits_{n\to\infty} \frac{3n^2+n+2}{2n^2-3} = \frac{3}{2}$ **45.** $\lim\limits_{n\to\infty} \frac{\cos n}{n} = 0$

47. The sequence diverges. **49.** $\lim\limits_{n\to\infty}\left(2 + \frac{4}{n^2}\right)^{1/3} = 2^{1/3}$

51. $\lim\limits_{n\to\infty} \ln\left(\frac{2n+1}{3n+4}\right) = \ln\frac{2}{3}$ **53.** The sequence diverges.

55. $\lim\limits_{n\to\infty} \frac{e^n+(-3)^n}{5^n} = 0$ **57.** $\lim\limits_{n\to\infty} n \sin\frac{\pi}{n} = \pi$

59. $\lim\limits_{n\to\infty} \frac{3-4^n}{2+7\cdot4^n} = -\frac{1}{7}$ **61.** $\lim\limits_{n\to\infty}\left(1+\frac{1}{n}\right)^n = e$

63. $\lim\limits_{n\to\infty} \frac{(\ln n)^2}{n} = 0$ **65.** $\lim\limits_{n\to\infty} n\left(\sqrt{n^2+1}-n\right) = \frac{1}{2}$

67. $\lim\limits_{n\to\infty} \frac{1}{\sqrt{n^4+n^8}} = 0$ **69.** $\lim\limits_{n\to\infty} (2^n+3^n)^{1/n} = 3$ **71.** (b)

73. Any number greater than or equal to 3 is an upper bound.

75. Example: $a_n = (-1)^n$ **79.** Example: $f(x) = \sin \pi x$

87. (e) $AGM\left(1, \sqrt{2}\right) \approx 1.198$

Section 10.2 Preliminary Questions

1. The sum of an infinite series is defined as the limit of the sequence of partial sums. If the limit of this sequence does not exist, the series is said to diverge.

2. $S = \frac{1}{2}$

3. The result is negative, so the result is not valid: a series with all positive terms cannot have a negative sum. The formula is not valid because a geometric series with $|r| \geq 1$ diverges.

4. No **5.** No **6.** $N = 13$

7. No, S_N is increasing and converges to 1, so $S_N \leq 1$ for all N.

8. Example: $\sum\limits_{n=1}^{\infty} \frac{1}{n^{9/10}}$

Section 10.2 Exercises

1. (a) $a_n = \frac{1}{3^n}$ (b) $a_n = \left(\frac{5}{2}\right)^{n-1}$

(c) $a_n = (-1)^{n+1} \frac{n^n}{n!}$ (d) $a_n = \frac{1+\frac{(-1)^{n+1}+1}{2}}{n^2+1}$

3. $S_2 = \frac{5}{4}, S_4 = \frac{205}{144}, S_6 = \frac{5369}{3600}$

5. $S_2 = \frac{2}{3}, S_4 = \frac{4}{5}, S_6 = \frac{6}{7}$ **7.** $S_6 = 1.24992$

9. $S_{10} = 0.03535167962, S_{100} = 0.03539810274,$
$S_{500} = 0.03539816290, S_{1000} = 0.03539816334.$ Yes.

11. $S_3 = \frac{3}{10}, S_4 = \frac{1}{3}, S_5 = \frac{5}{14}, \sum\limits_{n=1}^{\infty}\left(\frac{1}{n+1} - \frac{1}{n+2}\right) = \frac{1}{2}$

13. $S_3 = \frac{3}{7}, S_4 = \frac{4}{9}, S_5 = \frac{5}{11}, \sum\limits_{n=1}^{\infty} \frac{1}{4n^2-1} = \frac{1}{2}$

15. $S = \frac{1}{2}$ **17.** $\lim\limits_{n\to\infty} \frac{n}{10n+12} = \frac{1}{10} \neq 0$

19. $\lim\limits_{n\to\infty} (-1)^n\left(\frac{n-1}{n}\right)$ does not exist.

21. $\lim\limits_{n\to\infty} a_n = \lim\limits_{n\to\infty} \cos\frac{1}{n+1} = 1 \neq 0$

23. $S = \frac{8}{7}$ **25.** The series diverges. **27.** $S = \frac{59,049}{3328}$

29. $S = \frac{1}{e-1}$ **31.** $S = \frac{35}{3}$ **33.** $S = 4$ **35.** $S = \frac{7}{15}$ **37.** $\frac{2}{9}$

39. $\frac{31}{99}$

41. The fractions $S_a = \frac{a}{9}$, where $a = 1, 2, \ldots, 8$ have repeating decimals of the form $0.aaa\ldots$

43. (b) and (c)

47. (a) Counterexample: $\sum\limits_{n=1}^{\infty} \left(\frac{1}{2}\right)^n = 1.$

(b) Counterexample: If $a_n = 1$, then $S_N = N$.

(c) Counterexample: $\sum\limits_{n=1}^{\infty} \frac{1}{n}$ diverges.

(d) Counterexample: $\sum\limits_{n=1}^{\infty} \cos 2\pi n \neq 1.$

49. The total area is $\frac{1}{4}$.

51. (a) $De^{-k} + De^{-2k} + De^{-3k} + \cdots = \frac{De^{-k}}{1-e^{-k}}$

(b) $De^{-kt} + De^{-2kt} + De^{-3kt} + \cdots = \frac{De^{-kt}}{1-e^{-kt}}$

(c) $t \geq -\frac{1}{k}\ln\left(1 - \frac{D}{S}\right)$

53. The total length of the path is $2 + \sqrt{2}$.   **57.** 42 ft

Section 10.3 Preliminary Questions

1. (b)

2. A function f such that $a_n = f(n)$ must be positive, decreasing, and continuous for $x \geq 1$.

3. Convergence of p-series or integral test

4. Comparison Test

5. No; $\sum\limits_{n=1}^{\infty} \frac{1}{n}$ diverges, but since $\frac{e^{-n}}{n} < \frac{1}{n}$ for $n \geq 1$, the

Comparison Test tells us nothing about the convergence of $\sum\limits_{n=1}^{\infty} \frac{e^{-n}}{n}$.

Section 10.3 Exercises

1. $\int_1^{\infty} \frac{dx}{x^4}\,dx$ converges, so the series converges.

3. $\int_1^{\infty} x^{-1/3}\,dx = \infty$, so the series diverges.

5. $\int_{25}^{\infty} \frac{x^2}{(x^3+9)^{5/2}}\,dx$ converges, so the series converges.

7. $\int_1^{\infty} \frac{dx}{x^2+1}$ converges, so the series converges.

9. $\int_1^{\infty} \frac{dx}{x(x+1)}$ converges, so the series converges.

11. $\int_2^{\infty} \frac{1}{x(\ln x)^2}\,dx$ converges, so the series converges.

13. $\int_1^{\infty} \frac{dx}{2^{\ln x}} = \infty$, so the series diverges.

15. $\frac{1}{n^3+8n} \leq \frac{1}{n^3}$, so the series converges.

19. $\frac{1}{n2^n} \leq \left(\frac{1}{2}\right)^n$, so the series converges.

21. $\frac{1}{n^{1/3}+2^n} \leq \left(\frac{1}{2}\right)^n$, so the series converges.

23. $\frac{4}{m!+4^m} \leq 4\left(\frac{1}{4}\right)^m$, so the series converges.

25. $0 \leq \frac{\sin^2 k}{k^2} \leq \frac{1}{k^2}$, so the series converges.

27. $\frac{2}{3^n+3^{-n}} \leq 2\left(\frac{1}{3}\right)^n$, so the series converges.

29. $\frac{1}{(n+1)!} \leq \frac{1}{n^2}$, so the series converges.

31. $\frac{\ln n}{n^3} \leq \frac{1}{n^2}$ for $n \geq 1$, so the series converges.

33. $\frac{(\ln n)^{100}}{n^{1.1}} \leq \frac{1}{n^{1.09}}$ for n sufficiently large, so the series converges.

35. $\frac{n}{3^n} \leq \left(\frac{2}{3}\right)^n$ for $n \geq 1$, so the series converges.

39. The series converges.   **41.** The series diverges.

43. The series converges.   **45.** The series diverges.

47. The series converges.   **49.** The series converges.

51. The series diverges.   **53.** The series converges.

55. The series diverges.   **57.** The series converges.

59. The series diverges.   **61.** The series diverges.

63. The series diverges.   **65.** The series converges.

67. The series diverges.   **69.** The series diverges.

71. The series converges.   **73.** The series converges.

75. The series diverges.   **77.** The series converges.

79. The series converges for $a > 1$ and diverges for $a \leq 1$.

81. The series converges for $p > 1$ and diverges for $p \leq 1$.

89. $\sum\limits_{n=1}^{\infty} n^{-5} \approx 1.0369540120$.

93. $\sum\limits_{n=1}^{1000} \frac{1}{n^2} = 1.6439345667$ and $1 + \sum\limits_{n=1}^{100} \frac{1}{n^2(n+1)} = 1.6448848903$.

The second sum is a better approximation to $\frac{\pi^2}{6} \approx 1.6449340668$.

Section 10.4 Preliminary Questions

1. Example: $\sum \frac{(-1)^n}{\sqrt[3]{n}}$ **2.** (b)   **3.** No

4. $|S - S_{100}| \leq 10^{-3}$, and S is larger than S_{100}.

Section 10.4 Exercises

3. Converges conditionally.   **5.** Converges absolutely.

7. Converges conditionally.   **9.** Converges conditionally.

11. (a)

n	S_n	n	S_n
1	1	6	0.899782407
2	0.875	7	0.902697859
3	0.912037037	8	0.900744734
4	0.896412037	9	0.902116476
5	0.904412037	10	0.901116476

13. $S_5 = 0.947$ **15.** $S_{44} = 0.06567457397$

17. Converges (by geometric series).

19. Converges (by Comparison Test).

21. Converges (by Limit Comparison Test).

23. Diverges (by Limit Comparison Test).

25. Converges (by geometric series and linearity).

27. Converges absolutely (by Integral Test).

29. Converges (by Alternating Series Test).

31. Converges (by Integral Test).   **33.** Converges conditionally.

Section 10.5 Preliminary Questions

1. $\rho = \lim\limits_{n\to\infty} \left|\frac{a_{n+1}}{a_n}\right|$

2. The Ratio Test is conclusive for $\sum\limits_{n=1}^{\infty} \frac{1}{2^n}$ and inconclusive for $\sum\limits_{n=1}^{\infty} \frac{1}{n}$.

3. No

Section 10.5 Exercises

1. Converges absolutely.   **3.** Converges absolutely.

5. The Ratio Test is inconclusive.   **7.** Diverges.

9. Converges absolutely.   **11.** Converges absolutely.

13. Diverges.   **15.** The Ratio Test is inconclusive.

17. Converges absolutely.   **19.** Converges absolutely.

21. $\rho = \frac{1}{3} < 1$ **23.** $\rho = 2|x|$

25. $\rho = |r|$ **29.** Converges absolutely.

31. The Ratio Test is inconclusive, so the series may converge or diverge.

33. Converges absolutely.   **35.** The Ratio Test is inconclusive.

37. Converges absolutely.   **39.** Converges absolutely.

41. Converges absolutely.

43. Converges (by geometric series and linearity).

45. Diverges (by the Divergence Test).

47. Converges (by the Direct Comparison Test).

49. Diverges (by the Direct Comparison Test).

51. Converges (by the Ratio Test).

53. Converges (by the Limit Comparison Test).

55. Diverges (by p-series).   **57.** Converges (by geometric series).

59. Converges (by Limit Comparison Test).

61. Diverges (by Divergence Test).

65. (b) $\sqrt{2\pi} \approx 2.50663$

n	$\dfrac{e^n n!}{n^{n+1/2}}$
1000	2.506837
1500	2.506768
2000	2.506733
2500	2.506712
3000	2.506698

Section 10.6 Preliminary Questions

1. Yes. The series must converge for both $x = 4$ and $x = -3$.

2. (a), (c)   **3.** $R = 4$

4. $F'(x) = \sum\limits_{n=1}^{\infty} n^2 x^{n-1}$; $R = 1$

Section 10.6 Exercises

1. $R = 2$. It does not converge at the endpoints.

3. $R = 3$ for all three series.

9. $(-1, 1)$ **11.** $[-\sqrt{2}, \sqrt{2}]$ **13.** $[-1, 1]$ **15.** $(-\infty, \infty)$

17. $(-\infty, \infty)$ **19.** $(-1, 1]$ **21.** $(-1, 1)$ **23.** $[-1, 1)$

25. $(2, 4)$ **27.** $(6, 8)$ **29.** $\left[-\frac{7}{2}, -\frac{5}{2}\right)$ **31.** $(-\infty, \infty)$

33. $\left(2 - \frac{1}{e}, 2 + \frac{1}{e}\right)$

35. $\sum\limits_{n=0}^{\infty} 3^n x^n$ on the interval $\left(-\frac{1}{3}, \frac{1}{3}\right)$

37. $\sum\limits_{n=0}^{\infty} \frac{x^n}{3^{n+1}}$ on the interval $(-3, 3)$

39. $\sum\limits_{n=0}^{\infty} (-1)^n x^{2n}$ on the interval $(-1, 1)$

43. $\sum\limits_{n=0}^{\infty} (-1)^{n+1} (x - 5)^n$ on the interval $(4, 6)$

47. (c) $S_4 = \frac{69}{640}$ and $|S - S_4| \approx 0.000386 < a_5 = \frac{1}{1920}$

49. $R = 1$ **51.** $\sum\limits_{n=1}^{\infty} \frac{n}{2^n} = 2$ **53.** $F(x) = \frac{1-x-x^2}{1-x^3}$

55. $-1 \le x \le 1$ **57.** $P(x) = \sum\limits_{n=0}^{\infty} (-1)^n \frac{x^n}{n!}$

59. N must be at least 5; $S_5 = 0.3680555556$

61. $P(x) = 1 - \frac{1}{2}x^2 - \sum\limits_{n=2}^{\infty} \frac{1 \cdot 3 \cdot 5 \cdots (2n-3)}{(2n)!} x^{2n}$; $R = \infty$

Section 10.7 Preliminary Questions

1. $f(0) = 3$ and $f'''(0) = 30$

2. $f(-2) = 0$ and $f^{(4)}(-2) = 48$

3. Substitute x^2 for x in the Maclaurin series for $\sin x$.

4. $f(x) = 4 + \sum\limits_{n=1}^{\infty} \frac{(x-3)^{n+1}}{n(n+1)}$ **5.** (c)

Section 10.7 Exercises

1. $f(x) = 2 + 3x + 2x^2 + 2x^3 + \cdots$

3. $\frac{1}{1-2x} = \sum\limits_{n=0}^{\infty} 2^n x^n$ on the interval $\left(-\frac{1}{2}, \frac{1}{2}\right)$

5. $\cos 3x = \sum\limits_{n=0}^{\infty} (-1)^n \frac{9^n x^{2n}}{(2n)!}$ on the interval $(-\infty, \infty)$

7. $\sin(x^2) = \sum\limits_{n=0}^{\infty} (-1)^n \frac{x^{4n+2}}{(2n+1)!}$ on the interval $(-\infty, \infty)$

9. $\ln(1 - x^2) = -\sum\limits_{n=1}^{\infty} \frac{x^{2n}}{n}$ on the interval $(-1, 1)$

11. $\tan^{-1}(x^2) = \sum\limits_{n=0}^{\infty} (-1)^n \frac{x^{4n+2}}{2n+1}$ on the interval $[-1, 1]$

13. $e^{x-2} = \sum\limits_{n=0}^{\infty} \frac{x^n}{e^2 n!}$ on the interval $(-\infty, \infty)$

15. $\ln(1 - 5x) = -\sum\limits_{n=1}^{\infty} \frac{5^n x^n}{n}$ on the interval $\left[-\frac{1}{5}, \frac{1}{5}\right)$

17. $\sinh x = \sum\limits_{k=0}^{\infty} \frac{x^{2k+1}}{(2k+1)!}$ on the interval $(-\infty, \infty)$

19. $e^x \sin x = x + x^2 + \frac{x^3}{3} - \frac{x^5}{30} + \cdots$

21. $\frac{\sin x}{1-x} = x + x^2 + \frac{5x^3}{6} + \frac{5x^4}{6} + \cdots$

23. $(1 + x)^{1/4} = 1 + \frac{1}{4}x - \frac{3}{32}x^2 + \frac{7}{128}x^3 + \cdots$

25. $e^x \tan^{-1} x = x + x^2 + \frac{1}{6}x^3 - \frac{1}{6}x^4 + \cdots$

27. $e^{\sin x} = 1 + x + \frac{1}{2}x^2 - \frac{1}{8}x^4 + \cdots$

29. $\frac{1}{x} = \sum\limits_{n=0}^{\infty} (-1)^n (x - 1)^n$ on the interval $(0, 2)$

31. $\frac{1}{1-x} = \sum\limits_{n=0}^{\infty} (-1)^{n+1} \frac{(x-5)^n}{4^{n+1}}$ on the interval $(1, 9)$

33. $21 + 35(x - 2) + 24(x - 2)^2 + 8(x - 2)^3 + (x - 2)^4$ on the interval $(-\infty, \infty)$

35. $\frac{1}{x^2} = \sum\limits_{n=0}^{\infty} (-1)^n (n + 1) \frac{(x - 4)^n}{4^{n+2}}$ on the interval $(0, 8)$

37. $\frac{1}{1-x^2} = \sum\limits_{n=0}^{\infty} \frac{(-1)^{n+1}(2^{n+1}-1)}{2^{2n+3}} (x - 3)^n$ on the interval $(1, 5)$

39. $\cos^2 x = \frac{1}{2} + \frac{1}{2} \sum\limits_{n=0}^{\infty} (-1)^n \frac{(4)^n x^{2n}}{(2n)!}$ **45.** $S_4 = 0.1822666667$

47. (a) 5   (b) $S_4 = 0.7474867725$

49. $\int_0^1 \cos(x^2)\, dx = \sum\limits_{n=0}^{\infty} \frac{(-1)^n}{(2n)!(4n+1)}$; $S_3 = 0.9045227920$

51. $\int_0^1 e^{-x^3}\, dx = \sum\limits_{n=0}^{\infty} \frac{(-1)^n}{n!(3n+1)}$; $S_5 = 0.8074461996$

53. $\int_0^x \frac{1-\cos(t)}{t}\, dt = \sum\limits_{n=1}^{\infty} (-1)^{n+1} \frac{x^{2n}}{(2n)!2n}$

55. $\int_0^x \ln(1+t^2)\,dt = \sum_{n=1}^{\infty} (-1)^{n-1} \frac{x^{2n+1}}{n(2n+1)}$

57. $\frac{1}{1+2x}$ **59.** $\cos \pi = -1$ **65.** e^{x^3} **67.** $1 - 5x + \sin 5x$

69. $\frac{1}{(1-2x)(1-x)} = \sum_{n=0}^{\infty} \left(2^{n+1} - 1\right) x^n$

71. $I(t) = \frac{V}{R} \sum_{n=1}^{\infty} \frac{(-1)^{n+1}}{n!} \left(\frac{Rt}{L}\right)^n$

73. $f(x) = \sum_{n=0}^{\infty} \frac{(-1)^n x^{6n}}{(2n)!}$ and $f^{(6)}(0) = -360$.

75. $e^{x^{20}} = 1 + x^{20} + \frac{x^{40}}{2} + \cdots$

77. No

n	Series value when $x = 2$
5	2.54297
10	-0.239933
15	41.9276
20	-764.272
25	16,595.8

83. $\lim\limits_{x \to 0} \frac{\sin x - x + \frac{x^3}{6}}{x^5} = \frac{1}{120}$

85. $\lim\limits_{x \to 0} \left(\frac{\sin(x^2)}{x^4} - \frac{\cos x}{x^2} \right) = \frac{1}{2}$

87. $S = \frac{\pi}{4} - \frac{1}{2} \ln 2$ **89.** $L \approx 28.369$

Chapter 10 Review

1. **(a)** $a_1^2 = 4, a_2^2 = \frac{1}{4}, a_3^2 = 0$

(b) $b_1 = \frac{1}{24}, b_2 = \frac{1}{60}, b_3 = \frac{1}{240}$

(c) $a_1 b_1 = -\frac{1}{12}, a_2 b_2 = -\frac{1}{120}, a_3 b_3 = 0$

(d) $2a_2 - 3a_1 = 5, 2a_3 - 3a_2 = \frac{3}{2}, 2a_4 - 3a_3 = \frac{1}{12}$

3. $\lim\limits_{n \to \infty} (5a_n - 2a_n^2) = 2$ **5.** $\lim\limits_{n \to \infty} e^{a_n} = e^2$

7. $\lim\limits_{n \to \infty} (-1)^n a_n$ does not exist.

9. $\lim\limits_{n \to \infty} \left(\sqrt{n+5} - \sqrt{n+2} \right) = 0$ **11.** $\lim\limits_{n \to \infty} 2^{1/n^2} = 1$

13. The sequence diverges.   **15.** $\lim\limits_{n \to \infty} \tan^{-1} \left(\frac{n+2}{n+5} \right) = \frac{\pi}{4}$

17. $\lim\limits_{n \to \infty} \left(\sqrt{n^2 + n} - \sqrt{n^2 + 1} \right) = \frac{1}{2}$

19. $\lim\limits_{m \to \infty} \left(1 + \frac{1}{m} \right)^{3m} = e^3$

21. $\lim\limits_{n \to \infty} \left(n \left(\ln(n+1) - \ln n \right) \right) = 1$

25. $\lim\limits_{n \to \infty} \frac{a_{n+1}}{a_n} = 3$ **27.** $S_4 = -\frac{11}{60}, S_7 = \frac{41}{630}$

29. $\sum_{n=2}^{\infty} \left(\frac{2}{3} \right)^n = \frac{4}{3}$ **31.** $S = \frac{4}{37}$ **33.** $\sum_{n=-1}^{\infty} \frac{2^{n+3}}{3^n} = 36$

35. $a_n = \left(\frac{1}{2} \right)^n + 1 - 2^n, b_n = 2^n - 1$

37. $S = \frac{47}{180}$ **39.** The series diverges.

41. $\int_1^{\infty} \frac{1}{(x+2)(\ln(x+2))^3} \, dx = \frac{1}{2(\ln(3))^2}$, so the series converges.

43. $\frac{1}{(n+1)^2} < \frac{1}{n^2}$, so the series converges.

45. $\sum_{n=0}^{\infty} \frac{1}{n^{1.5}}$ converges, so the series converges.

47. $\frac{n}{\sqrt{n^5+2}} < \frac{1}{n^{3/2}}$, so the series converges.

49. $\sum_{n=0}^{\infty} \left(\frac{10}{11} \right)^n$ converges, so the series converges.

51. Converges.

55. **(b)** $0.3971162690 \le S \le 0.3971172688$, so the maximum size of the error is 10^{-6}.

57. Converges absolutely.   **59.** Diverges.

61. **(a)** 500   **(b)** $K \approx \sum_{n=0}^{499} \frac{(-1)^k}{(2k+1)^2} = 0.9159650942$

63. **(a)** Converges.   **(b)** Converges.   **(c)** Diverges.
(d) Converges.

65. Converges.   **67.** Converges.   **69.** Diverges.

71. Diverges.   **73.** Converges.   **75.** Converges.

77. Converges (by geometric series).

79. Converges (by geometric series).

81. Converges (by the Leibniz Test).

83. Converges (by the Leibniz Test).

85. Diverges (by the Divergence Test).

87. Converges (absolutely, by a Direct Comparison with the p-series $\sum_{n=1}^{\infty} \frac{1}{n^{3/2}}$

89. Converges (by the Root Test).

91. Converges (by the Limit Comparison Test)

93. Converges using partial sums (the series is telescoping).

95. Diverges (by the Direct Comparison Test).

97. Converges (by the Direct Comparison Test).

99. Converges (by the Limit Comparison Test).

101. Converges on the interval $(-\infty, \infty)$.

103. Converges on the interval $[2, 4]$.   **105.** Converges at $x = 0$.

107. $\frac{2}{4-3x} = \frac{1}{2} \sum_{n=0}^{\infty} \left(\frac{3}{4} \right)^n x^n$. The series converges on the interval $\left(\frac{-4}{3}, \frac{4}{3} \right)$.

109. **(c)**

111. $\lim\limits_{x \to 0} \frac{x^2 e^x}{\cos x - 1} = -2$

113. $e^{4x} = \sum_{n=0}^{\infty} \frac{4^n}{n!} x^n$

115. $x^4 = 16 + 32(x - 2) + 24(x - 2)^2 + 8(x - 2)^3 + (x - 2)^4$

117. $\sin x = \sum_{n=0}^{\infty} \frac{(-1)^{n+1}(x-\pi)^{2n+1}}{(2n+1)!}$

119. $\frac{1}{1-2x} = \sum_{n=0}^{\infty} \frac{2^n}{5^{n+1}} (x+2)^n$ **121.** $\ln \frac{x}{2} = \sum_{n=1}^{\infty} \frac{(-1)^{n+1}(x-2)^n}{n 2^n}$

123. $(x^2 - x)e^{x^2} = \sum_{n=0}^{\infty} \left(\frac{x^{2n+2} - x^{2n+1}}{n!} \right)$ so $f^{(3)}(0) = -6$

125. $\frac{1}{1+\tan x} = 1 - x + x^2 - \frac{4}{3}x^3 + \cdots$, so $f^{(3)}(0) = -8$

127. $\frac{\pi}{2} - \frac{\pi^3}{2^3 3!} + \frac{\pi^5}{2^5 5!} - \frac{\pi^7}{2^7 7!} + \cdots = \sin \frac{\pi}{2} = 1$

Chapter 11

Section 11.1 Preliminary Questions

1. A circle of radius 3 centered at the origin

2. The center is at $(4, 5)$.   **3.** Maximum height: 4

4. Yes; No

5. **(a)** The line $y = x$ traversed left to right.

(b) Same path traversed left to right twice as fast.

6. $c_1(t) = (t, t^3)$, $c_2(t) = (\sqrt[3]{t}, t)$, $c_3(t) = (t^3, t^9)$

Section 11.1 Exercises

1. $(t = 0)(1, 9)$; $(t = 2)(9, -3)$; $(t = 4)(65, -39)$

3. $y = 2.5x - .000766x^2$

5. **(a)**
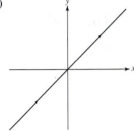
(b)
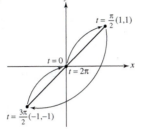

(c)
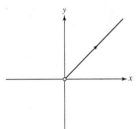
(d)
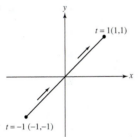

7. $y = 4x - 12$ **9.** $y = \tan^{-1}(x^3 + e^x)$

11. $y = \dfrac{6}{x^2}$ (where $x > 0$)   **13.** $y = 2 - e^x$

15.
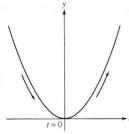
17.
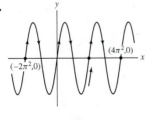

19. (a) ↔ (iv), (b) ↔ (ii), (c) ↔ (iii), (d) ↔ (i)

21. **(a)** $y_{\max} = 100$ cm; **(b)** lands at $x = 2040$ cm from the origin, when $t = 20$ s

23. $c(t) = (t, 9 - 4t)$ **25.** $c(t) = \left(\dfrac{5 + t^2}{4}, t\right)$

27. $c(t) = (-9 + 7\cos t, 4 + 7\sin t)$ **29.** $c(t) = (-4 + t, 9 + 8t)$

31. $c(t) = (3 - 8t, 1 + 3t)$ **33.** $c(t) = (1 + t, 1 + 2t)$ $(0 \le t \le 1)$

35. $c(t) = (3 + 4\cos t, 9 + 4\sin t)$ **37.** $c(t) = \left(-4 + t, -8 + t^2\right)$

39. $c(t) = (2 + t, 2 + 3t)$ **41.** $c(t) = \left(3 + t, (3 + t)^2\right)$

43. $y = \sqrt{x^2 - 1}$ $(1 \le x < \infty)$ **45.** Plot III

47.
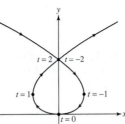

49. $\left.\dfrac{dy}{dx}\right|_{t=-4} = -\dfrac{1}{6}$ **51.** $\left.\dfrac{dy}{dx}\right|_{s=-1} = -\dfrac{3}{4}$ **53.** $-\dfrac{2}{3}$

55. $y = -\dfrac{9}{2}x + \dfrac{11}{2}$; $\dfrac{dy}{dx} = -\dfrac{9}{2}$

57. $y = x^2 + x^{-1}$; $\dfrac{dy}{dx} = 2x - \dfrac{1}{x^2}$ **59.** $(0, 0), (96, 180)$

61.
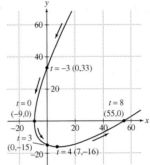

The graph is in: quadrant i for $t < -3$ or $t > 8$, quadrant ii for $-3 < t < 0$, quadrant iii for $0 < t < 3$, quadrant iv for $3 < t < 8$.

63. $(55, 0)$

65. The coordinates of P, $(R\cos\theta, r\sin\theta)$, describe an ellipse for $0 \le \theta \le 2\pi$.

69. $c(t) = (3 - 9t + 24t^2 - 16t^3, 2 + 6t^2 - 4t^3)$, $0 \le t \le 1$

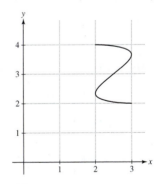

73. $y = -\sqrt{3}x + \dfrac{\sqrt{3}}{2}$

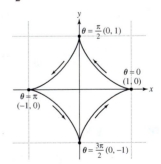

75. $((2k - 1)\pi, 2)$, $k = 0, \pm 1, \pm 2, \ldots$ **85.** $\left.\dfrac{d^2y}{dx^2}\right|_{t=2} = -\dfrac{21}{512}$

87. $\left.\dfrac{d^2y}{dx^2}\right|_{t=-3} = 0$ **89.** Concave up: $t > 0$ **91.** $\dfrac{1}{3}$ **93.** $\dfrac{2}{5}$

95. $\dfrac{2}{3}$

Section 11.2 Preliminary Questions

1. $S = \int_a^b \sqrt{x'(t)^2 + y'(t)^2}\, dt$

2. No. They are equal when the curve traced by $c(t)$ is a line segment from the initial to the final point, and $c(t)$ is a one-to-one function.

3. The speed at time t **4.** Displacement: 5; No

5. $L = 180$ cm   **6.** 4π

Section 11.2 Exercises

1. $S = 10$ **3.** $S = 800$ **5.** $S = \frac{1}{2}(65^{3/2} - 5^{3/2}) \approx 256.43$

7. $S = 3\pi$ **9.** $S = -8\left(\frac{\sqrt{2}}{2} - 1\right) \approx 2.34$

13. $S = \ln(\cosh(A))$ **15.** $\left.\frac{ds}{dt}\right|_{t=2} = 4\sqrt{10} \approx 12.65$ m/s

17. $\left.\frac{ds}{dt}\right|_{t=9} = \sqrt{41} \approx 6.4$ m/s   **19.** $\left.\frac{ds}{dt}\right|_{t=0} = 1$

21. $\left(\frac{ds}{dt}\right)_{\min} \approx \sqrt{4.89} \approx 2.21$ **23.** $\frac{ds}{dt} = 8$

25.

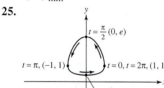

$M_{10} = 6.903734$, $M_{20} = 6.915035$, $M_{30} = 6.914949$,
$M_{50} = 6.914951$

27.

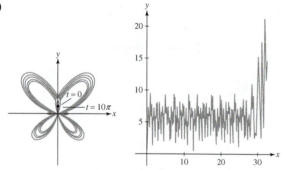

$M_{10} = 25.528309$, $M_{20} = 25.526999$, $M_{30} = 25.526999$,
$M_{50} = 25.526999$

29. $S = 2\pi^2 R$

39. (a)

$$ $$

(b) $L \approx 212.096$

Section 11.3 Preliminary Questions

1. (a)   **2.** Positive: $(r, \theta) = \left(1, \frac{\pi}{2}\right)$; Negative: $(r, \theta) = \left(-1, \frac{3\pi}{2}\right)$

3. (a) Equation of the circle of radius 2 centered at the origin

(b) Equation of the circle of radius $\sqrt{2}$ centered at the origin

(c) Equation of the vertical line through the point $(2, 0)$

4. (a)

Section 11.3 Exercises

1. (A): $\left(3\sqrt{2}, \frac{3\pi}{4}\right)$; (B): $(3, \pi)$; (C): $\left(\sqrt{5}, \pi + 0.46\right) \approx \left(\sqrt{5}, 3.60\right)$; (D): $\left(\sqrt{2}, \frac{5\pi}{4}\right)$; (E): $\left(\sqrt{2}, \frac{\pi}{4}\right)$; (F): $\left(4, \frac{\pi}{6}\right)$; (G): $\left(4, \frac{11\pi}{6}\right)$

3. (a) $(1, 0)$ **(b)** $\left(\sqrt{12}, \frac{\pi}{6}\right)$ **(c)** $\left(\sqrt{8}, \frac{3\pi}{4}\right)$ **(d)** $\left(2, \frac{2\pi}{3}\right)$

5. (a) $\left(\frac{3\sqrt{3}}{2}, \frac{3}{2}\right)$ **(b)** $\left(-\frac{6}{\sqrt{2}}, \frac{6}{\sqrt{2}}\right)$ **(c)** $(0, 0)$ **(d)** $(0, -5)$

7. (A): $0 \le r \le 3, \pi \le \theta \le 2\pi$, (B): $0 \le r \le 3, \frac{\pi}{4} \le \theta \le \frac{\pi}{2}$, (C): $3 \le r \le 5, \frac{3\pi}{4} \le \theta \le \pi$

9. $m = \tan \frac{3\pi}{5} \approx -3.1$ **11.** $x^2 + y^2 = 7^2$

13. $x^2 + (y - 1)^2 = 1$ **15.** $y = x - 1$ **17.** $r = \sqrt{5}$

19. $r = \tan\theta \sec\theta$ **21.** $e^r = 1$

23. (a)↔(iii), (b)↔(iv), (c)↔(i), (d)↔(ii)

25. (a) $(r, 2\pi - \theta)$ **(b)** $(r, \theta + \pi)$ **(c)** $(r, \pi - \theta)$
(d) $\left(r, \frac{\pi}{2} - \theta\right)$

27. $r\cos\left(\theta - \frac{\pi}{3}\right) = d$

29.   **31.**

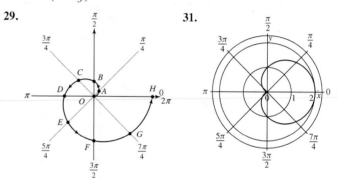

33. (a) A, $\theta = 0, r = 0$; B, $\theta = \frac{\pi}{4}, r = \sin\frac{2\pi}{4} = 1$; C, $\theta = \frac{\pi}{2}$, $r = 0$; D, $\theta = \frac{3\pi}{4}, r = \sin\frac{2\cdot 3\pi}{4} = -1$; E, $\theta = \pi, r = 0$; F, $\theta = \frac{5\pi}{4}$, $r = 1$; G, $\theta = \frac{3\pi}{2}, r = 0$; H, $\theta = \frac{7\pi}{4}, r = -1$; I, $\theta = 2\pi, r = 0$

(b) $0 \le \theta \le \frac{\pi}{2}$ is in the first quadrant. $\frac{\pi}{2} \le \theta \le \pi$ is in the fourth quadrant. $\pi \le \theta \le \frac{3\pi}{2}$ is in the third quadrant. $\frac{3\pi}{2} \le \theta \le 2\pi$ is in the second quadrant.

35.

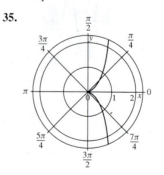

37. $\left(x - \frac{a}{2}\right)^2 + \left(y - \frac{b}{2}\right)^2 = \frac{a^2 + b^2}{4}, r = \frac{\sqrt{x^2 + y^2}}{2}$, centered at the point $\left(\frac{a}{2}, \frac{b}{2}\right)$

39. $r^2 = \sec 2\theta$ **41.** $\left(x^2 + y^2\right)^2 = x^3 - 3y^2 x$

43. $r = 2\sec\left(\theta - \frac{\pi}{9}\right)$ **45.** $r = 2\sqrt{10}\sec(\theta - 4.39)$

49. $r^2 = 2a^2 \cos 2\theta$

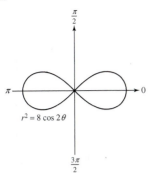

$r^2 = 8 \cos 2\theta$

53. $\theta = \frac{\pi}{2}, m = -\frac{2}{\pi}; \theta = \pi, m = \pi$

55. $\left(\frac{\sqrt{2}}{2}, \frac{\pi}{6}\right), \left(\frac{\sqrt{2}}{2}, \frac{5\pi}{6}\right), \left(\frac{\sqrt{2}}{2}, \frac{7\pi}{6}\right), \left(\frac{\sqrt{2}}{2}, \frac{11\pi}{6}\right)$

57. A: $m = 1$, B: $m = -1$, C: $m = 1$

Section 11.4 Preliminary Questions

1. (b)   **2.** Yes   **3.** (c)

Section 11.4 Exercises

1. $A = \frac{1}{2} \int_{\pi/2}^{\pi} r^2 \, d\theta = \frac{25\pi}{4}$

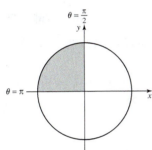

3. $A = \frac{1}{2} \int_0^{\pi} r^2 \, d\theta = 4\pi$ **5.** $A = \frac{\pi^5}{320} + \frac{\pi^4}{16} + \frac{\pi^3}{3}$

7. $A = \frac{3\pi}{2}$ **9.** $A = \frac{\pi}{8} \approx 0.39$

11.

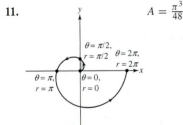

$A = \frac{\pi^3}{48}$

13. $A = \frac{\sqrt{15}}{2} + 7\cos^{-1}\left(\frac{1}{4}\right) \approx 11.163$

15. $A = \pi - \frac{3\sqrt{3}}{2} \approx 0.54$ **17.** $A = \frac{\pi}{8} - \frac{1}{4} \approx 0.14$ **19.** $A = 4\pi$

21. $A = \frac{9\pi}{2} - 4\sqrt{2}$ **23.** $S = 4\pi$

25. $L = \frac{1}{3}\left(\left(\pi^2 + 4\right)^{3/2} - 8\right) \approx 14.55$ **27.** π

29. $L = \sqrt{2}\pi/4 \approx 1.11$

31. $L = \int_0^{2\pi} \sqrt{\cos^4\theta + 4\cos^2\theta \sin^2\theta} \, d\theta \approx 5.52$

33. $\int_0^{\pi/2} \sqrt{2e^{2\theta} + 2e^{\theta} + 1} \, d\theta$ **35.** $\int_0^{\pi/2} \sin^2\theta \sqrt{1 + 8\cos^2\theta} \, d\theta$

37. $L \approx 6.682$ **39.** $L \approx 79.56$

Section 11.5 Preliminary Questions

1. (a) Hyperbola   (b) Parabola   (c) Ellipse
(d) Not a conic section

2. Hyperbolas   **3.** The points $(0, c)$ and $(0, -c)$

4. $\pm\frac{b}{a}$ are the slopes of the two asymptotes of the hyperbola.

Section 11.5 Exercises

1. $F_1 = \left(-\sqrt{65}, 0\right), F_2 = \left(\sqrt{65}, 0\right)$. The vertices are $(9, 0)$, $(-9, 0), (0, 4)$, and $(0, -4)$.

3. $F_1 = \left(\sqrt{97}, 0\right), F_2 = \left(\sqrt{97}, 0\right)$. The vertices are $(4, 0)$ and $(-4, 0)$.

5. $F_1 = \left(\sqrt{65} + 3, -1\right), F_2 = \left(-\sqrt{65} + 3, -1\right)$. The vertices are $(10, -1)$ and $(-4, -1)$.

7. $\frac{x^2}{6^2} + \frac{y^2}{3^2} = 1$ **9.** $\frac{(x-14)^2}{6^2} + \frac{(y+4)^2}{3^2} = 1$

11. $\frac{x^2}{5^2} + \frac{y^2}{7^2} = 1$ **13.** $\frac{x^2}{(40/3)^2} + \frac{y^2}{(50/3)^2} = 1$

15. $\left(\frac{x}{3}\right)^2 - \left(\frac{y}{4}\right)^2 = 1$ **17.** $\frac{x^2}{2^2} - \frac{y^2}{\left(2\sqrt{3}\right)^2} = 1$

19. $\left(\frac{x-2}{5}\right)^2 - \left(\frac{y}{10\sqrt{2}}\right)^2 = 1$ **21.** $y = 3x^2$

23. $y = \frac{1}{20}x^2$ **25.** $y = \frac{1}{16}x^2$ **27.** $x = \frac{1}{8}y^2$

29. Vertices: $(\pm 4, 0), (0, \pm 2)$. Foci: $\left(\pm\sqrt{12}, 0\right)$. Centered at the origin.

31. Vertices: $(7, -5), (-1, -5)$. Foci: $\left(\sqrt{65} + 3, -5\right)$, $\left(-\sqrt{65} + 3, -5\right)$. Center: $(3, -5)$. Asymptotes: $y = \frac{7}{4}x - \frac{41}{4}$ and $y = -\frac{7}{4}x + \frac{1}{4}$.

33. Vertices: $(5, 5), (-7, 5)$. Foci: $\left(\sqrt{84} - 1, 5\right), \left(-\sqrt{84} - 1, 5\right)$. Center: $(-1, 5)$. Asymptotes: $y = \frac{\sqrt{48}}{6}(x + 1) + 5 \approx 1.15x + 6.15$ and $y = -\frac{\sqrt{48}}{6}(x + 1) + 5 \approx -1.15x + 3.85$.

35. Vertex: $(0, 0)$. Focus: $\left(0, \frac{1}{16}\right)$.

37. Vertices: $\left(1 \pm \frac{5}{2}, \frac{1}{5}\right), \left(1, \frac{1}{5} \pm 1\right)$. Foci: $\left(-\frac{\sqrt{21}}{2} + 1, \frac{1}{5}\right)$, $\left(\frac{\sqrt{21}}{2} + 1, \frac{1}{5}\right)$. Centered at $\left(1, \frac{1}{5}\right)$.

39. $D = -87$; ellipse   **41.** $D = 40$; hyperbola

47. Focus: $(0, c)$. Directrix: $y = -c$.   **49.** $A = \frac{8}{3}c^2$

51. $r = \frac{3}{2 + \cos\theta}$ **53.** $r = \frac{4}{1 + \cos\theta}$

55. Hyperbola, $e = 4$, directrix, $x = 2$

57. Ellipse, $e = \frac{3}{4}$, directrix, $x = \frac{8}{3}$ **59.** $r = \frac{-12}{5 + 6\cos\theta}$

61. $\left(\frac{x+3}{5}\right)^2 + \left(\frac{y}{4}\right)^2 = 1$ **63.** 4.5 billion miles

Chapter 11 Review

1. (a), (c)

3. $c(t) = (1 + 2\cos t, 1 + 2\sin t)$. The intersection points with the y-axis are $\left(0, 1 \pm \sqrt{3}\right)$. The intersection points with the x-axis are $\left(1 \pm \sqrt{3}, 0\right)$.

5. $c(\theta) = (\cos(\theta + \pi), \sin(\theta + \pi))$ **7.** $c(t) = (1 + 2t, 3 + 4t)$

9. $y = -\frac{x}{4} + \frac{37}{4}$ **11.** $y = -\frac{8}{(x-3)^3} + \frac{3-x}{2}$

13. $\left.\frac{dy}{dx}\right|_{t=3} = \frac{3}{14}$ **15.** $\left.\frac{dy}{dx}\right|_{t=0} = \frac{\cos 20}{e^{20}}$ **17.** $(1.41, 1.60)$

19. $c(t) = \left(-1 + 6t^2 - 4t^3, -1 + 6t - 6t^2\right)$

21. $\frac{ds}{dt} = \sqrt{3 + 2(\cos t - \sin t)}$; maximal speed: $\sqrt{3 + 2\sqrt{2}}$

23. $s = \sqrt{2}$

25.

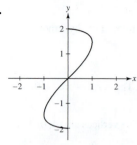

$s = 2 \int_0^\pi \sqrt{\cos^2 2t + \sin^2 t} \, dt \approx 6.0972$

27. $\left(1, \frac{\pi}{6}\right)$ and $\left(3, \frac{5\pi}{4}\right)$ have rectangular coordinates $\left(\frac{\sqrt{3}}{2}, \frac{1}{2}\right)$ and $\left(-\frac{3\sqrt{2}}{2}, -\frac{3\sqrt{2}}{2}\right)$.

29. $\sqrt{x^2 + y^2} = \frac{2x}{x-y}$ **31.** $r = 3 + 2\sin\theta$

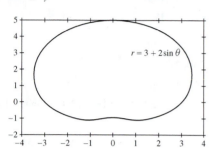

33. $A = \frac{\pi}{16}$ **35.** $e - \frac{1}{e}$
    Note: One needs to double the integral from $-\frac{\pi}{2}$ to $\frac{\pi}{2}$ in order to account for both sides of the graph.

37. $A = \frac{3\pi a^2}{2}$

39. Outer: $L \approx 36.121$, inner: $L \approx 7.5087$, difference: 28.6123

41. Ellipse. Vertices: $(\pm 3, 0)$, $(0, \pm 2)$. Foci: $(\pm\sqrt{5}, 0)$.

43. Ellipse. Vertices: $\left(\pm\frac{2}{\sqrt{5}}, 0\right)$, $\left(0, \pm\frac{4}{\sqrt{5}}\right)$. Foci: $\left(0, \pm\sqrt{\frac{12}{5}}\right)$.

45. $\left(\frac{x}{8}\right)^2 + \left(\frac{y}{\sqrt{61}}\right)^2 = 1$ **47.** $\left(\frac{x}{8}\right)^2 - \left(\frac{y}{6}\right)^2 = 1$ **49.** $x = \frac{1}{32}y^2$

51. $y = \sqrt{3}x + \left(\sqrt{3} - 5\right)$ and $y = -\sqrt{3}x + \left(-\sqrt{3} - 5\right)$

Chapter 12

Section 12.1 Preliminary Questions

1. (a) True (b) False (c) True (d) True
2. $\|-3\mathbf{a}\| = 15$ **3.** The components are not changed.   **4.** $\langle 0, 0 \rangle$
5. (a) True   (b) False

Section 12.1 Exercises

1. $\mathbf{v}_1 = \langle 2, 0 \rangle$, $\|\mathbf{v}_1\| = 2$ $\mathbf{v}_2 = \langle 2, 0 \rangle$, $\|\mathbf{v}_2\| = 2$

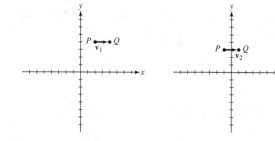

$\mathbf{v}_3 = \langle 3, 1 \rangle$, $\|\mathbf{v}_3\| = \sqrt{10}$ $\mathbf{v}_4 = \langle 2, 2 \rangle$, $\|\mathbf{v}_4\| = 2\sqrt{2}$

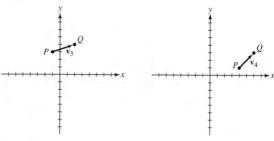

Vectors $\mathbf{v}_1$ and $\mathbf{v}_2$ are equivalent.

3. $\langle 3, 5 \rangle$

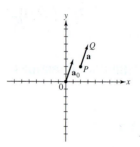

5. $\left\langle \frac{\sqrt{2}}{2} \|\mathbf{u}\|, \frac{\sqrt{2}}{2} \|\mathbf{u}\| \right\rangle$ or $\langle 0.707 \|\mathbf{u}\|, 0.707 \|\mathbf{u}\| \rangle$

7. $\langle \cos(-20°) \|\mathbf{w}\|, \sin(-20°) \|\mathbf{w}\| \rangle$ or
$\langle 0.94 \|\mathbf{w}\|, -0.342 \|\mathbf{w}\| \rangle$

9. $\overrightarrow{PQ} = \langle -1, 5 \rangle$ **11.** $\overrightarrow{PQ} = \langle -2, -9 \rangle$ **13.** $\langle 5, 5 \rangle$
15. $\langle 30, 10 \rangle$ **17.** $\left\langle \frac{5}{2}, 5 \right\rangle$

19. Vector (B)

21. $2\mathbf{v} = \langle 4, 6 \rangle$ $-\mathbf{w} = \langle -4, -1 \rangle$

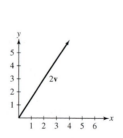

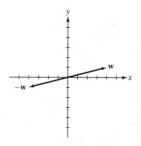

$2\mathbf{v} - \mathbf{w} = \langle 0, 5 \rangle$ $\mathbf{v} + \mathbf{w} = \langle 6, 4 \rangle$

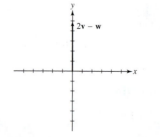

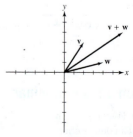

23. $3\mathbf{v} + \mathbf{w} = \langle -2, 10 \rangle, 2\mathbf{v} - 2\mathbf{w} = \langle 4, -4 \rangle$

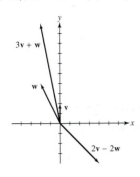

25.

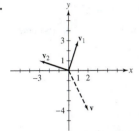

27. (b) and (c)

29. $\overrightarrow{AB} = \langle 2, 6 \rangle$ and $\overrightarrow{PQ} = \langle 2, 6 \rangle$; equivalent

31. $\overrightarrow{AB} = \langle 3, -2 \rangle$ and $\overrightarrow{PQ} = \langle 3, -2 \rangle$; equivalent

33. $\overrightarrow{AB} = \langle 2, 3 \rangle$ and $\overrightarrow{PQ} = \langle 6, 9 \rangle$; parallel and point in the same direction

35. $\overrightarrow{AB} = \langle -8, 1 \rangle$ and $\overrightarrow{PQ} = \langle 8, -1 \rangle$; parallel and point in opposite directions

37. $\left\| \overrightarrow{OR} \right\| = \sqrt{53}$ **39.** $P = (0, 0)$ **41.** $\left(\frac{3}{5}, \frac{4}{5} \right)$

43. $4\mathbf{e_u} = \left\langle -2\sqrt{2}, -2\sqrt{2} \right\rangle$ **45.** $2\mathbf{e_{-v}} = -\sqrt{2}\mathbf{i} + \sqrt{2}\mathbf{j}$

47. $\mathbf{e} = \left\langle \cos \frac{4\pi}{7}, \sin \frac{4\pi}{7} \right\rangle = \langle -0.22, 0.97 \rangle$

49. $\lambda = \pm \frac{1}{\sqrt{13}}$ **51.** $P = (4, 6)$

53. (a) $\rightarrow$ (ii), (b) $\rightarrow$ (iv), (c) $\rightarrow$ (iii), (d) $\rightarrow$ (i)   **55.** $9\mathbf{i} + 7\mathbf{j}$

57. $-5\mathbf{i} - 3\mathbf{j}$

59.

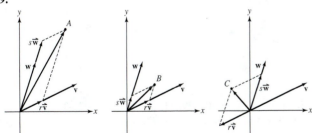

61. $\mathbf{u} = 2\mathbf{v} - \mathbf{w}$

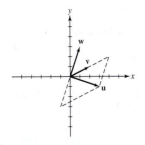

63. The force on cable 1 is $\approx$ 45 lb, and force on cable 2 is $\approx$ 21 lb.

65. 230 km/h   **67.** $\mathbf{r} = \langle 6.45, 0.38 \rangle$

Section 12.2 Preliminary Questions

1. $(4, 3, 2)$ **2.** $\langle 3, 2, 1 \rangle$ **3.** (a)   **4.** (c)

5. Infinitely many direction vectors   **6.** True

Section 12.2 Exercises

1. $\|\mathbf{v}\| = \sqrt{14}$

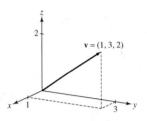

3. The head of $\mathbf{v} = \overrightarrow{PQ}$ is $Q = (1, 2, 1)$.

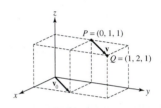

$\mathbf{v}_0 = \overrightarrow{OS}$, where $S = (1, 1, 0)$

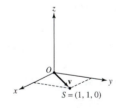

5. $\overrightarrow{PQ} = \langle 1, 1, -1 \rangle$ **7.** $\overrightarrow{PQ} = \left\langle -\frac{9}{2}, -\frac{3}{2}, 1 \right\rangle$

9. $\left\| \overrightarrow{OR} \right\| = \sqrt{26} \approx 5.1$ **11.** $P = (-2, 6, 0)$

13. (a) Parallel and same direction   (b) Not parallel

(c) Parallel and opposite directions   (d) Not parallel

15. Not equivalent   **17.** Not equivalent   **19.** $\langle -8, -18, -2 \rangle$

21. $\langle -2, -2, 3 \rangle$ **23.** $\langle 16, -1, 9 \rangle$

25. Not parallel   **27.** Not parallel

29. $\mathbf{e_w} = \left\langle \frac{4}{\sqrt{21}}, \frac{-2}{\sqrt{21}}, \frac{-1}{\sqrt{21}} \right\rangle$ **31.** $-\mathbf{e_v} = \left\langle \frac{2}{3}, -\frac{2}{3}, -\frac{1}{3} \right\rangle$

33. $\mathbf{r}(t) = \langle 1 + 2t, 2 + t, -8 + 3t \rangle$

35. $\mathbf{r}(t) = \langle 4 + 7t, 0, 8 + 4t \rangle$ **37.** $\mathbf{r}(t) = \langle 1 + 2t, 1 - 6t, 1 + t \rangle$

39. $\mathbf{r}(t) = \langle 4t, t, t \rangle$ **41.** $\mathbf{r}(t) = \langle 0, 0, t \rangle$

43. $\mathbf{r}(t) = \langle -t, -2t, 4 - 2t \rangle$ **45.** (c)

49. $\mathbf{r}_1(t) = \langle 5, 5, 2 \rangle + t \langle 0, -2, 1 \rangle$;

$\mathbf{r}_2(t) = \langle 5, 5, 2 \rangle + t \langle 0, -20, 10 \rangle$

55. 4 min   **57.** $\left(0, \frac{1}{2}, -\frac{1}{2} \right)$ **59.** 2450 newtons

61. $\frac{x+2}{2} = \frac{y-3}{4} = \frac{z-3}{3}$ **63.** $\frac{x-3}{2} = \frac{y-4}{-9} = \frac{z}{12}$

65. $\mathbf{r}(t) = \langle 2t, 7t, 8t \rangle$

Section 12.3 Preliminary Questions

1. Scalar   **2.** Obtuse   **3.** Distributive Law

4. (a)   **5.** (b); (c)   **6.** (c)

Section 12.3 Exercises

1. 15   **3.** 41   **5.** 5   **7.** 0   **9.** 1   **11.** 0   **13.** Obtuse
15. Orthogonal   **17.** Acute   **19.** 0   **21.** $\frac{1}{\sqrt{10}}$ **23.** $\pi/4$
25. ≈ 0.615 **27.** $2\pi/3$
29. (a) $b = -\frac{1}{2}$ (b) $b = 0$ or $b = \frac{1}{2}$
31. $\mathbf{v}_1 = \langle 0, 1, 0 \rangle$, $\mathbf{v}_2 = \langle 3, 2, 2 \rangle$ **33.** $-\frac{3}{2}$ **35.** $\|\mathbf{v}\|^2$
37. $\|\mathbf{v}\|^2 - \|\mathbf{w}\|^2$ **39.** 8   **41.** 2   **43.** π **45.** (b) 7   **49.** 51.91°
51. (a)

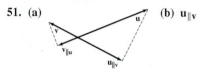

(b) $\mathbf{u}_{\|\mathbf{v}}$

53. $\left(\frac{7}{2}, \frac{7}{2} \right)$ **55.** $\left\langle -\frac{4}{5}, 0, -\frac{2}{5} \right\rangle$ **57.** $-4\mathbf{k}$ **59.** $a\mathbf{i}$ **61.** $2\sqrt{2}$
63. $\sqrt{17}$ **65.** $\mathbf{a} = \left\langle \frac{1}{2}, \frac{1}{2} \right\rangle + \left\langle \frac{1}{2}, -\frac{1}{2} \right\rangle$
67. $\mathbf{a} = \left\langle 0, -\frac{1}{2}, -\frac{1}{2} \right\rangle + \left\langle 4, -\frac{1}{2}, \frac{1}{2} \right\rangle$
69. $\left\langle \frac{x-y}{2}, \frac{y-x}{2} \right\rangle + \left\langle \frac{x+y}{2}, \frac{y+x}{2} \right\rangle$
73. $\approx 35°$ **75.** $\overrightarrow{AD}$ **77.** $\approx 109.5°$ **81.** ≈ 68.07 N
99. $2x + 2y - 2z = 1$

Section 12.4 Preliminary Questions

1. $\begin{vmatrix} -5 & -1 \\ 4 & 0 \end{vmatrix}$ **2.** $\|\mathbf{e} \times \mathbf{f}\| = \frac{1}{2}$ **3.** $\mathbf{u} \times \mathbf{v} = \langle -2, -2, -1 \rangle$

4. (vector notation)   (a) 0   (b) 0

5. $\mathbf{i} \times \mathbf{j} = \mathbf{k}$ and $\mathbf{i} \times \mathbf{k} = -\mathbf{j}$

6. $\mathbf{v} \times \mathbf{w} = \mathbf{0}$ if either $\mathbf{v}$ or $\mathbf{w}$ (or both) is the zero vector or $\mathbf{v}$ and $\mathbf{w}$ are parallel vectors.

7. (a) not meaningful because you cannot find the cross product of a scalar and a vector;

(b) meaningful because it represents the dot product of two vectors;

(c) meaningful because it is the product of two scalars;

(d) meaningful because it is a scalar multiple of a vector.

8. (b)

Section 12.4 Exercises

1. -5 **3.** -15 **5.** -8 **7.** 0   **9.** $\langle 1, 2, -5 \rangle$ **11.** $\langle 6, 0, -8 \rangle$
13. $-\mathbf{j} + \mathbf{i}$ **15.** $\mathbf{i} + \mathbf{j} + \mathbf{k}$ **17.** $\langle -1, -1, 0 \rangle$
19. $\langle -2, -2, -2 \rangle$ **21.** $\langle 4, 4, 0 \rangle$
23. $\mathbf{v} \times \mathbf{i} = c\mathbf{j} - b\mathbf{k}$; $\mathbf{v} \times \mathbf{j} = -c\mathbf{i} + a\mathbf{k}$; $\mathbf{v} \times \mathbf{k} = b\mathbf{i} - a\mathbf{j}$
25. $-\mathbf{u}$ **27.** $\langle 0, 3, 3 \rangle$ **31.** $\mathbf{e}'$ **33.** $\mathbf{F}_1$ **37.** $2\sqrt{138}$
39. The volume is 4.

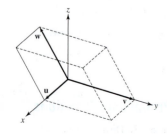

41. $\sqrt{35} \approx 5.92$

43.

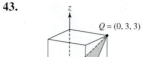

The area of the triangle is $\frac{9\sqrt{3}}{2} \approx 7.8$.

45. 3   **47.** $10\sqrt{3}$ **59.** $\mathbf{X} = \langle a, a, a+1 \rangle$
63. $\tau = 250 \sin 125° \, \mathbf{k} \approx 204.79 \, \mathbf{k}$

Section 12.5 Preliminary Questions

1. $3x + 4y - z = 0$ **2.** (c): $z = 1$ **3.** Plane (c)   **4.** xz-plane
5. (c): $x + y = 0$ **6.** Statement (a)

Section 12.5 Exercises

1. $x + 3y + 2z = 3$ **3.** $-x + 2y + z = 3$ **5.** $x = 3$
7. $z = 2$ **9.** $x = 0$ **11.** Statements (b) and (d)
13. $\langle 9, -4, -11 \rangle$ **15.** $\langle 3, -8, 11 \rangle$ **17.** $6x + 9y + 4z = 19$
19. $x + 2y - z = 1$ **21.** $4x - 9y + z = 0$ **23.** $x = 4$
25. $x + z = 3$ **27.** $13x + y - 5z = 27$ **29.** Yes, the planes are parallel.

31.

33.

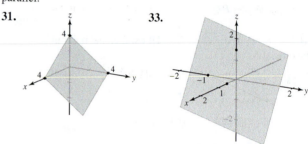

35.

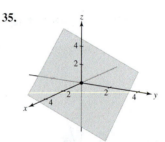

37. $10x + 15y + 6z = 30$ **39.** $(1, 5, 8)$ **41.** $(-2, 3, 12)$
43. $-9y + 4z = 5$ **45.** $x = -\frac{2}{3}$ **47.** $x = -4$
49. The two planes have no common points.
51. $y - 4z = 0$
         $x + y - 4z = 0$
53. $(3\lambda)x + by + (2\lambda)z = 5\lambda$, $\lambda \neq 0$ **55.** $\theta = \pi/2$
57. $\theta = 1.143$ rad or $\theta = 65.49°$ **59.** $\theta \approx 55.0°$
61. $x + y + z = 1$ **63.** $x - y - z = d/a$
65. $x = \frac{9}{5} + 2t$, $y = -\frac{6}{5} - 3t$, $z = 2 + 5t$ **67.** $\pm 24 \langle 1, 2, -2 \rangle$
73. $\left(\frac{2}{3}, -\frac{1}{3}, \frac{2}{3} \right)$ **75.** $\frac{6}{\sqrt{30}} \approx 1.095$ **77.** $|a|$

Section 12.6 Preliminary Questions

1. True, mostly, except at $x = \pm a$, $y = \pm b$, or $z = \pm c$
2. False   **3.** Hyperbolic paraboloid

4. No **5.** Ellipsoid

6. All vertical lines passing through a parabola c in the xy-plane

Section 12.6 Exercises

1. Ellipsoid **3.** Ellipsoid

5. Hyperboloid of one sheet **7.** Elliptic paraboloid

9. Hyperbolic paraboloid **11.** Hyperbolic paraboloid

13. Ellipsoid, the trace is a circle on the xz-plane

15. Ellipsoid, the trace is an ellipse parallel to the xy-plane

17. Hyperboloid of one sheet, the trace is a hyperbola

19. Parabolic cylinder, the trace is the parabola $y = 3x^2$

21. (a) $\leftrightarrow$ Figure b; (b) $\leftrightarrow$ Figure c; (c) $\leftrightarrow$ Figure a

23. $y = \left(\frac{x}{2}\right)^2 + \left(\frac{z}{4}\right)^2$

25.

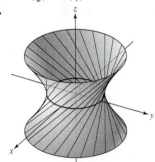

Graph of $x^2 + y^2 + z^2 = 1$

27.

29.

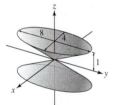

31.

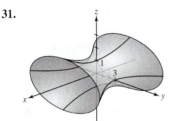

33.

35.

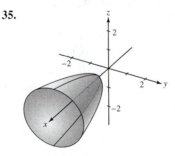

39. $\left(\frac{x}{2}\right)^2 + \left(\frac{y}{4}\right)^2 + \left(\frac{z}{6}\right)^2 = 1$ **41.** $\left(\frac{x}{4}\right)^2 + \left(\frac{y}{6}\right)^2 - \left(\frac{z}{3\sqrt{3}}\right)^2 = 1$

43. One or two vertical lines, or an empty set **45.** An elliptic cone

Section 12.7 Preliminary Questions

1. Cylinder of radius R whose axis is the z-axis, sphere of radius R centered at the origin

2. (b) **3.** (a) **4.** $\phi = 0, \pi$ **5.** $\phi = \frac{\pi}{2}$, the xy-plane

Section 12.7 Exercises

1. $(-4, 0, 4)$ **3.** $\left(0, 0, \frac{1}{2}\right)$ **5.** $\left(\sqrt{2}, \frac{7\pi}{4}, 1\right)$ **7.** $(2, \frac{\pi}{3}, 7)$

9. $\left(5, \frac{\pi}{4}, 2\right)$ **11.** $r^2 \leq 1$ **13.** $r^2 + z^2 \leq 4$, $\theta = \frac{\pi}{2}$ or $\theta = \frac{3\pi}{2}$

15. $r^2 \leq 9$, $\frac{5\pi}{4} \leq \theta \leq 2\pi$ and $0 \leq \theta \leq \frac{\pi}{4}$

17.

19.

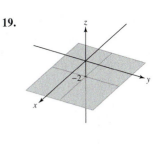

21.

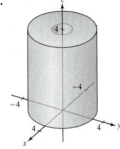

23.

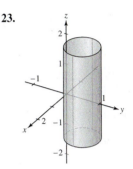

25.

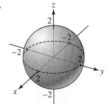

27. $r = \frac{z}{\cos\theta + \sin\theta}$ **29.** $r = \frac{z\tan\theta}{\cos\theta}$ **31.** $r = 2$ **33.** $(3, 0, 0)$

35. $(0, 0, 3)$ **37.** $\left(\frac{3\sqrt{3}}{2}, \frac{3}{2}, -3\sqrt{3}\right)$ **39.** $\left(2, 0, \frac{\pi}{3}\right)$

41. $\left(\sqrt{3}, \frac{\pi}{4}, 0.955\right)$ **43.** $\left(2, \frac{\pi}{3}, \frac{\pi}{6}\right)$ **45.** $\left(2\sqrt{2}, 0, \frac{\pi}{4}\right)$

47. $\left(2\sqrt{2}, 0, 2\sqrt{2}\right)$ **49.** $0 \leq \rho \leq 1$

51. $\rho = 1$, $0 \leq \theta \leq \frac{\pi}{2}$, $0 \leq \phi \leq \frac{\pi}{2}$

53. $\left\{(\rho, \theta, \phi) : 0 \leq \rho \leq 2, \ \theta = \frac{\pi}{2} \text{ or } \theta = \frac{3\pi}{2}\right\}$

55.

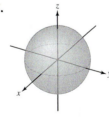

57.

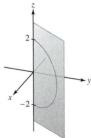

59.

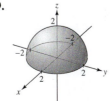

61.

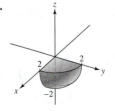

63.

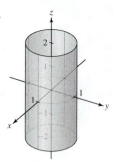

65. $\rho = \frac{2}{\cos\phi}$ **67.** $\rho = \frac{\cos\theta\tan\phi}{\cos\phi}$ **69.** $\rho = \frac{2}{\sin\phi\sqrt{\cos 2\theta}}$ **71.** (b)

73. Helsinki: $(25.0°, \ 29.9°)$, São Paulo: $(313.48°, \ 113.52°)$

75. Sydney: $(-4618.8, \ 2560.3, \ -3562.1)$,
Bogotá: $(1723.7, \ -6111.7, \ 503.1)$

77. $z = \pm r\sqrt{\cos 2\theta}$

79. $\left\{ (r, \theta, z) : -\sqrt{4 - r^2} \le z \le \sqrt{4 - r^2}, 1 \le r \le 2, 0 \le \theta \le 2\pi \right\}$

83. $r = \sqrt{z^2 + 1}$ and $\rho = \sqrt{-\frac{1}{\cos 2\phi}}$; no points; $\frac{\pi}{4} < \phi < \frac{3\pi}{4}$

Chapter 12 Review

1. $\langle 21, \ -25 \rangle$ and $\langle -19, \ 31 \rangle$ **3.** $\left\langle \frac{-2}{\sqrt{29}}, \ \frac{5}{\sqrt{29}} \right\rangle$

5. $\mathbf{i} = \frac{2}{11}\mathbf{v} + \frac{5}{11}\mathbf{w}$ **7.** $\overrightarrow{PQ} = \langle -4, \ 1 \rangle$; $\left\| \overrightarrow{PQ} \right\| = \sqrt{17}$

9. $\left\langle \frac{3}{\sqrt{2}}, \ -\frac{3}{\sqrt{2}} \right\rangle$ **11.** $\beta = \frac{3}{2}$ **13.** $\mathbf{u} = \left\langle \frac{1}{3}, \ -\frac{11}{6}, \ \frac{7}{6} \right\rangle$

15. $\mathbf{r}_1(t) = \langle 1 + 3t, \ 4 + t, \ 5 + 6t \rangle$; $\mathbf{r}_2(t) = \langle 1 + 3t, \ t, \ 6t \rangle$

17. $a = -2$, $b = 2$

19.

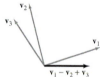

21. $\mathbf{v} \cdot \mathbf{w} = -9$ **23.** $\mathbf{v} \times \mathbf{w} = \langle 10, \ -8, \ -7 \rangle$ **25.** $V = 48$
29. $\frac{5}{3}$ **31.** $\|\mathbf{F}_1\| = \frac{2\|\mathbf{F}_2\|}{\sqrt{3}}$; $\|\mathbf{F}_1\| = 980$ N

33. $\mathbf{v} \times \mathbf{w} = \langle -6, \ 7, \ -2 \rangle$

35. -47 **37.** $5\sqrt{2}$ **41.** $\|\mathbf{e} - 4\mathbf{f}\| = \sqrt{13}$

47. $(x - 0) + 4(y - 1) - 3(z + 1) = 0$

49. $17x - 21y - 13z = -28$ **51.** $3x - 2y = 4$ **53.** Ellipsoid
55. Elliptic paraboloid   **57.** Elliptic cone

59. (a) Empty set   (b) Hyperboloid of one sheet

(c) Hyperboloid of two sheets

61. $(r, \ \theta, \ z) = \left(5, \ \tan^{-1}\frac{4}{3}, \ -1 \right)$, $(\rho, \ \theta, \ \phi) = \left(\sqrt{26}, \ \tan^{-1}\frac{4}{3}, \ \cos^{-1}\left(\frac{-1}{\sqrt{26}}\right) \right)$

63. $(r, \ \theta, \ z) = \left(\frac{3\sqrt{3}}{2}, \ \frac{\pi}{6}, \ \frac{3}{2} \right)$

65. $z = 2x$

69. $A < -1$: Hyperboloid of one sheet
$A = -1$: Cylinder with the z-axis as its central axis
$A > -1$: Ellipsoid
$A = 0$: Sphere

Chapter 13

Section 13.1 Preliminary Questions

1. (c)   **2.** The curve $z = e^x$

3. The projection onto the xz-plane   **4.** The point $(-2, 2, 3)$

5. As t increases from 0 to 2π, a point on $\sin t\,\mathbf{i} + \cos t\,\mathbf{j}$ moves clockwise and a point on $\cos t\,\mathbf{i} + \sin t\,\mathbf{j}$ moves counterclockwise.

6. (a), (c), and (d)

Section 13.1 Exercises

1. $D = \{t \in \mathbf{R}, \ t \ne 0, \ t \ne -1\}$

3. $\mathbf{r}(2) = \left\langle 0, \ 4, \ \frac{1}{5} \right\rangle$; $\mathbf{r}(-1) = \left\langle -1, \ 1, \ \frac{1}{2} \right\rangle$

5. $\mathbf{r}(t) = (3 + 3t)\mathbf{i} - 5\mathbf{j} + (7 + t)\mathbf{k}$

7. Yes, when $t = (2n - 1)\pi$, where n is an integer, at the points $(0, 0, (2n - 1)\pi)$

9. No intersection   **11.** A ↔ ii, B ↔ i, C ↔ iii

13. (a) = (v), (b) = (i), (c) = (ii), (d) = (vi), (e) = (iv), (f) = (iii)

15. C ↔ i, A ↔ ii, B ↔ iii

17. This is a circle of radius 9 centered at the origin lying in the xy-plane.

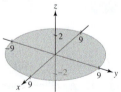

19. Radius 1, center $(0, \ 0, \ 4)$, xz-plane

21.

23. $(0, 1, 0), (0, -1, 0), \left(\frac{1}{\sqrt{2}}, \frac{1}{\sqrt{2}}, 0\right), \left(\frac{1}{\sqrt{2}}, -\frac{1}{\sqrt{2}}, 0\right),$
$\left(-\frac{1}{\sqrt{2}}, -\frac{1}{\sqrt{2}}, 0\right), \left(-\frac{1}{\sqrt{2}}, \frac{1}{\sqrt{2}}, 0\right)$

25. $\mathbf{r}(t) = \left(2t^2 - 7, t, \pm\sqrt{9 - t^2}\right)$, for $-3 \leq t \leq 3$

27. (a) $\mathbf{r}(t) = \left(\pm t\sqrt{1 - t^2}, t^2, t\right)$ for $-1 \leq t \leq 1$

(b) The projection is a circle in the xy-plane with radius $\frac{1}{2}$ and centered at the xy-point $\left(0, \frac{1}{2}\right)$.

29. $\mathbf{r}(t) = \langle\cos t, \pm\sin t, \sin t\rangle$; the projection of the curve onto the xy-plane is traced by $\langle\cos t, \pm\sin t, 0\rangle$, which is the unit circle in this plane; the projection of the curve onto the xz-plane is traced by $\langle\cos t, 0, \sin t\rangle$, which is the unit circle in this plane; the projection of the curve onto the yz-plane is traced by $\langle 0, \pm\sin t, \sin t\rangle$, which is the two segments $z = y$ and $z = -y$ for $-1 \leq y \leq 1$.

31. $\mathbf{r}(t) = \left(\cos t, \sin t, 4\cos t^2\right), 0 \leq t \leq 2\pi$

33. Collide at the point $(12, 4, 2)$ and intersect at the points $(4, 0, -6)$ and $(12, 4, 2)$

35. $\mathbf{r}(t) = \langle 3, 2, t\rangle, \ -\infty < t < \infty$

37. $\mathbf{r}(t) = \langle t, 3t, 15t\rangle, \ -\infty < t < \infty$

39. $\mathbf{r}(t) = \langle 1, 2 + 2\cos t, 5 + 2\sin t\rangle, \ 0 \leq t \leq 2\pi$

41. $\mathbf{r}(t) = \left\langle\frac{\sqrt{3}}{2}\cos t, \frac{1}{2}, \frac{\sqrt{3}}{2}\sin t\right\rangle, \ 0 \leq t \leq 2\pi$

43. $\mathbf{r}(t) = \langle 3 + 2\cos t, 1, 5 + 3\sin t\rangle, \ 0 \leq t \leq 2\pi$

45.

$\mathbf{r}(t) = \langle|t| + t, |t| - t\rangle$

Section 13.2 Preliminary Questions

1. $\frac{d}{dt}(f(t)\mathbf{r}(t)) = f(t)\mathbf{r}'(t) + f'(t)\mathbf{r}(t)$

$\frac{d}{dt}(\mathbf{r}_1(t) \cdot \mathbf{r}_2(t)) = \mathbf{r}_1(t) \cdot \mathbf{r}'_2(t) + \mathbf{r}'_1(t) \cdot \mathbf{r}_2(t)$

$\frac{d}{dt}(\mathbf{r}_1(t) \times \mathbf{r}_2(t)) = \mathbf{r}_1(t) \times \mathbf{r}'_2(t) + \mathbf{r}'_1(t) \times \mathbf{r}_2(t)$

2. True   3. False   4. True   5. False   6. False

7. (a) Vector   (b) Scalar   (c) Vector

Section 13.2 Exercises

1. $\lim_{t\to 3}\left(t^2, 4t, \frac{1}{t}\right) = \left(9, 12, \frac{1}{3}\right)$

3. $\lim_{t\to 0}(e^{2t}\mathbf{i} + \ln(t + 1)\mathbf{j} + 4\mathbf{k}) = \mathbf{i} + 4\mathbf{k}$

5. $\lim_{h\to 0}\frac{\mathbf{r}(t+h) - \mathbf{r}(t)}{h} = \left\langle-\frac{1}{t^2}, \cos t, 0\right\rangle$ 7. $\frac{d\mathbf{r}}{dt} = \left\langle 1, 2t, 3t^2\right\rangle$

9. $\frac{d\mathbf{r}}{ds} = \left\langle 3e^{3s}, -e^{-s}, 4s^3\right\rangle$ 11. $\mathbf{c}'(t) = -t^{-2}\mathbf{i} - 2e^{2t}\mathbf{k}$

13. $\mathbf{r}'(t) = \left\langle 1, 2t, 3t^2\right\rangle, \mathbf{r}''(t) = \langle 0, 2, 6t\rangle$

15.

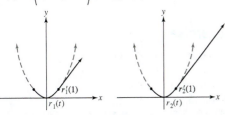

17. $\frac{d}{dt}(\mathbf{r}_1(t) \cdot \mathbf{r}_2(t)) =$
$\quad 2t^3 e^{2t} + 3t^2 e^{3t} + 2te^{3t} + 3t^2 e^{2t} + te^t + e^t$

19. $\frac{d}{dt}(\mathbf{r}_1(t) \times \mathbf{r}_2(t)) =$
$\left\langle\begin{array}{l} 3t^2 e^t - 2te^{2t} - e^{2t} + t^3 e^t, \ e^{3t} + 3te^{3t} - t^2 e^t - 2te^t, \\ 2te^{2t} + 2t^2 e^{2t} - 3t^2 e^{3t} - 3t^3 e^{3t} \end{array}\right\rangle$

21. $2 + 4e$ 23. $\frac{d}{dt}\mathbf{r}(g(t)) = \left\langle 2e^{2t}, -e^t\right\rangle$

25. $\frac{d}{dt}\mathbf{r}(g(t)) = \left\langle 4e^{4t+9}, 8e^{8t+18}, 0\right\rangle$

27. $\frac{d}{dt}(\mathbf{r}(t) \cdot \mathbf{a}(t))|_{t=2} = 13$ 29. $\ell(t) = \langle 4 - 4t, 16 - 32t\rangle$

31. $\ell(t) = \langle -3 - 4t, 10 + 5t, 16 + 24t\rangle$

33. $\ell(t) = \left\langle 2 - t, 0, -\frac{1}{3} + \frac{1}{2}t\right\rangle$

35. $\frac{d}{dt}(\mathbf{r} \times \mathbf{r}') = \left\langle(t^2 - 2)e^t, -te^t, 2t\right\rangle$ 39. $\left\langle\frac{212}{3}, 124\right\rangle$

41. $\langle 0, 0\rangle$ 43. $\left\langle 1, 2, -\frac{\sin 3}{3}\right\rangle$ 45. $(\ln 4)\mathbf{i} + \frac{56}{3}\mathbf{j} - \frac{496}{5}\mathbf{k}$

47. $\mathbf{r}(t) = \left\langle -t^2 + t + c_1, 2t^2 + c_2\right\rangle$; with initial conditions

$\mathbf{r}(t) = \left\langle -t^2 + t + 3, 2t^2 + 1\right\rangle$

49. $\mathbf{r}(t) = \left(\frac{1}{3}t^3\right)\mathbf{i} + \left(\frac{5t^2}{2}\right)\mathbf{j} + t\mathbf{k} + \mathbf{c}$; with initial conditions,

$\mathbf{r}(t) = \left(\frac{1}{3}t^3 - \frac{1}{3}\right)\mathbf{i} + \left(\frac{5}{2}t^2 - \frac{3}{2}\right)\mathbf{j} + (t + 1)\mathbf{k}$

51. $\mathbf{r}(t) = (8t^2)\mathbf{k} + \mathbf{c}_1 t + \mathbf{c}_2$; with initial conditions,
$\mathbf{r}(t) = \mathbf{i} + t\mathbf{j} + (8t^2)\mathbf{k}$

53. $\mathbf{r}(t) = \left\langle 0, t^2, 0\right\rangle + \mathbf{c}_1 t + \mathbf{c}_2$; with initial conditions,

$\mathbf{r}(t) = \left\langle 1, t^2 - 6t + 10, t - 3\right\rangle$

55. $\mathbf{r}(3) = \left\langle\frac{45}{4}, 5\right\rangle$

57. Only at time $t = 3$ can the pilot hit a target located at the origin.

59. $\mathbf{r}(t) = (t - 1)\mathbf{v} + \mathbf{w}$ 61. $\mathbf{r}(t) = e^{2t}\mathbf{c}$

Section 13.3 Preliminary Questions

1. $2\mathbf{r}' = \langle 50, -70, 20\rangle, -\mathbf{r}' = \langle -25, 35, -10\rangle$

2. Statement (b) is true.

3. (a) $L'(2) = 4$

(b) $L(t)$ is the distance along the path traveled, which is usually different from the distance from the origin.

4. 6

Section 13.3 Exercises

1. $L = 3\sqrt{61}$ 3. $L = 15 + \ln 4$ 5. $L = \frac{544\sqrt{34} - 2}{135} \approx 23.48$

7. $L = \pi\sqrt{4\pi^2 + 10} + 5\ln\frac{2\pi + \sqrt{4\pi^2 + 10}}{\sqrt{10}} \approx 29.3$

9. $s(t) = \frac{1}{27}\left((20 + 9t^2)^{3/2} - 20^{3/2}\right)$ 11. $v(4) = \sqrt{21}$

13. $v(1) = \sqrt{2}$ 15. $v\left(\frac{\pi}{2}\right) = 5$ 17. $\mathbf{r}' = \langle 100\sqrt{5}, 200\sqrt{5}\rangle$

19. The bee is at the origin. $\int_0^T \|\mathbf{r}'(u)\| \, du$ represents the total distance the bee traveled on the time interval $[0, T]$.

21. (c) $L_1 \approx 132.0, L_2 \approx 125.7$; the first spring uses more wire.

23. (a) $t = \pi$

25. (a) $s(t) = \sqrt{29}t$ (b) $t = g(s) = \frac{s}{\sqrt{29}}$

27. $\left\langle 1 + \frac{3s}{\sqrt{50}}, 2 + \frac{4s}{\sqrt{50}}, 3 + \frac{5s}{\sqrt{50}}\right\rangle$

29. $\mathbf{r}(s) = \langle 2 + 4\cos(2s), 10, -3 + 4\sin(2s)\rangle$

31. $\mathbf{r}(s) =$
$$\left\langle \cos\left[\left(\tfrac{3}{2}s+1\right)^{2/3}-1\right], \sin\left[\left(\tfrac{3}{2}s+1\right)^{2/3}-1\right], \tfrac{2}{3}\left[\left(\tfrac{3}{2}s+1\right)^{2/3}-1\right]^{3/2}\right\rangle,$$
$s \geq 0$

33. $\mathbf{r}(s) = \left\langle \tfrac{1}{9}(27s+8)^{2/3} - \tfrac{4}{9}, \ \pm\tfrac{1}{27}\left((27s+8)^{2/3}-4\right)^{3/2}\right\rangle$

35. $\left\langle \dfrac{s}{\sqrt{1+m^2}}, \ \dfrac{sm}{\sqrt{1+m^2}}\right\rangle$

37. (a) $\sqrt{17}e^t$ **(b)** $\dfrac{s}{\sqrt{17}}\left\langle \cos\left(4\ln\dfrac{s}{\sqrt{17}}\right), \ \sin\left(4\ln\dfrac{s}{\sqrt{17}}\right)\right\rangle$

39. $L = \int_{-\infty}^{\infty} \|\mathbf{r}'(t)\| = 2\int_{-\infty}^{\infty}\dfrac{dt}{1+t^2} = 2\pi$

Section 13.4 Preliminary Questions

1. $\left\langle \tfrac{2}{3}, \tfrac{1}{3}, -\tfrac{2}{3}\right\rangle$ **2.** $\tfrac{1}{4}$

3. The curvature of a circle of radius 2   **4.** Zero curvature

5. $\kappa = \sqrt{14}$ **6.** 4   **7.** $\tfrac{1}{9}$

Section 13.4 Exercises

1. $\mathbf{r}'(t) = \langle 8t+9\rangle$, $\mathbf{T}(t) = \dfrac{1}{\sqrt{64t^2+81}}\langle 8t, \ 9\rangle$,
$\mathbf{T}(1) = \left\langle \dfrac{8}{\sqrt{145}}, \ \dfrac{9}{\sqrt{145}}\right\rangle$

3. $\mathbf{r}'(t) = \langle 4, -5, 9\rangle$, $\mathbf{T}(t) = \left\langle \dfrac{4}{\sqrt{122}}, \ -\dfrac{5}{\sqrt{122}}, \ \dfrac{9}{\sqrt{122}}\right\rangle$,
$\mathbf{T}(1) = \mathbf{T}(t)$

5. $\mathbf{r}'(t) = \langle -\pi \sin \pi t, \pi \cos \pi t, 1\rangle$,
$\mathbf{T}(t) = \dfrac{1}{\sqrt{\pi^2+1}}\langle -\pi \sin \pi t, \ \pi \cos \pi t, \ 1\rangle$,
$\mathbf{T}(1) = \left\langle 0, \ -\dfrac{\pi}{\sqrt{\pi^2+1}}, \ \dfrac{1}{\sqrt{\pi^2+1}}\right\rangle$

7. $\kappa(t) = \dfrac{e^t}{(1+e^{2t})^{3/2}}$ **9.** $\kappa(t) = 0$ **11.** $\kappa = \dfrac{2\sqrt{74}}{27}$

13. $\kappa = \dfrac{\sqrt{\pi^2+5}}{(\pi^2+1)^{3/2}} \approx 0.108$ **15.** $\kappa(3) = \dfrac{e^3}{(3^6+1)^{3/2}} \approx 0.0025$

17. $\kappa(2) = \dfrac{48\sqrt{41}}{210,125} \approx 0.0015$

19. $\kappa\left(\tfrac{\pi}{3}\right) = \dfrac{\sqrt{330}}{4} \approx 4.54$, $\kappa\left(\tfrac{\pi}{2}\right) = \tfrac{1}{5} = 0.2$ **23.** $\alpha = \pm\sqrt{2}$

29. $\kappa(2) = \dfrac{3\sqrt{10}}{800} \approx 0.012$ **31.** $\kappa(\pi) = \dfrac{\pi\sqrt{2}}{4} \approx 1.11$

35. $\kappa(t) = t^2$

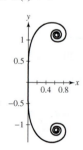

37. $\mathbf{N}(t) = \langle 0, \ -\sin 2t, \ -\cos 2t\rangle$

39. $\mathbf{N}\left(\tfrac{\pi}{4}\right) = \left\langle -\dfrac{\sqrt{2}}{3\sqrt{3}}, \ -\dfrac{2}{3\sqrt{3}}\right\rangle$, $\mathbf{N}\left(\tfrac{3\pi}{4}\right) = \left\langle \dfrac{\sqrt{2}}{3\sqrt{3}}, \ \dfrac{2}{3\sqrt{3}}\right\rangle$

41.
$\mathbf{T}(1) = \left\langle 0, \ \dfrac{\sqrt{5}}{5}, \ \dfrac{2\sqrt{5}}{5}\right\rangle$, $\mathbf{N}(1) = \left\langle 0, \ -\dfrac{2\sqrt{5}}{5}, \ \dfrac{\sqrt{5}}{5}\right\rangle$, $\mathbf{B}(1) = \langle 1, 0, 0\rangle$

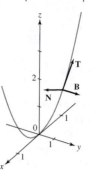

43. $\mathbf{T}(1) = \left\langle \tfrac{1}{3}, \tfrac{2}{3}, \tfrac{2}{3}\right\rangle$, $\mathbf{N}(1) = \left\langle -\tfrac{2}{3}, -\tfrac{1}{3}, \tfrac{2}{3}\right\rangle$, $\mathbf{B}(1) = \left\langle \tfrac{2}{3}, -\tfrac{2}{3}, \tfrac{1}{3}\right\rangle$

45. $\mathbf{N}\left(\pi^{1/3}\right) = \left\langle \tfrac{1}{2}, \ -\tfrac{\sqrt{3}}{2}\right\rangle$ **47.** $\mathbf{N}(1) = \dfrac{1}{\sqrt{13}}\langle -3, \ 2\rangle$

49. $\mathbf{N}(1) = \dfrac{1}{\sqrt{2}}\langle 0, \ 1, \ -1\rangle$

51. $\mathbf{N}(0) = \tfrac{1}{6}\left\langle -\sqrt{6}, \ 2\sqrt{6}, \ -\sqrt{6}\right\rangle$

53. (a)
$\mathbf{T}(1) = \left\langle \tfrac{1}{3}, \tfrac{2}{3}, \tfrac{2}{3}\right\rangle$, $\mathbf{N}(1) = \left\langle -\tfrac{2}{3}, -\tfrac{1}{3}, \tfrac{2}{3}\right\rangle$, $\mathbf{B}(1) = \left\langle \tfrac{2}{3}, -\tfrac{2}{3}, \tfrac{1}{3}\right\rangle$

(b) $6x - 6y + 3z = 1$

55. (a) $\mathbf{T}(t) = \left\langle \dfrac{1}{\sqrt{2+4t^2}}, \ \dfrac{-1}{\sqrt{2+4t^2}}, \ \dfrac{2t}{\sqrt{2+4t^2}}\right\rangle$,
$\mathbf{N}(t) = \left\langle \dfrac{-t\sqrt{2}}{\sqrt{2+4t^2}}, \ \dfrac{t\sqrt{2}}{\sqrt{2+4t^2}}, \ \dfrac{\sqrt{2}}{\sqrt{2+4t^2}}\right\rangle$

(b) $\mathbf{B}(t) = \left\langle -\dfrac{1}{\sqrt{2}}, -\dfrac{1}{\sqrt{2}}, 0\right\rangle$

(c) The osculating planes are all parallel to each other, with equation $x + y = c$ for some c.

59. $\langle \cos t, \ \sin t\rangle$, that is, the unit circle itself.

61. $\mathbf{c}(t) = \left\langle -4, \ \tfrac{7}{2}\right\rangle + \dfrac{5^{3/2}}{2}\langle \cos t, \ \sin t\rangle$

63. $\mathbf{c}(t) = \langle \pi, \ -2\rangle + 4\langle \cos t, \ \sin t\rangle$

65. $\mathbf{c}(t) = \left\langle -1 - 2\cos t, \ \dfrac{2\sin t}{\sqrt{2}}, \ \dfrac{2\sin t}{\sqrt{2}}\right\rangle$

73. $\kappa(\theta) = 1$ **75.** $\kappa(\theta) = \dfrac{1}{\sqrt{2}}e^{-\theta}$

91. (a) $\mathbf{N}(0) = \left\langle -\dfrac{1}{\sqrt{5}}, 0, \dfrac{2}{\sqrt{5}}\right\rangle$ **(b)** $\mathbf{N}(1) = \dfrac{1}{\sqrt{66}}\langle 4, 7, -2\rangle$

Section 13.5 Preliminary Questions

1. No, since the particle may change its direction.   **2.** $\mathbf{a}(t)$

3. Statement (a), their velocity vectors point in the same direction.

4. The velocity vector always points in the direction of motion. Since the vector $\mathbf{N}(t)$ is orthogonal to the direction of motion, the vectors $\mathbf{a}(t)$ and $\mathbf{v}(t)$ are orthogonal.

5. Description (b), parallel   **6.** $\|\mathbf{a}(t)\| = 8$ cm/s²   **7.** $a_{\mathbf{N}}$

Section 13.5 Exercises

1. $h = -0.2$: $\langle -0.085, \ 1.91, \ 2.635\rangle$
$h = -0.1$: $\langle -0.19, \ 2.07, \ 2.97\rangle$
$h = 0.1$: $\langle -0.41, \ 2.37, \ 4.08\rangle$
$h = 0.2$: $\langle -0.525, \ 2.505, \ 5.075\rangle$
$\mathbf{v}(1) \approx \langle -0.3, \ 2.2, \ 3.5\rangle$, $v(1) \approx 4.1$

3. $\mathbf{v}(1) = \langle 3, \ -1, \ 8 \rangle$, $\ \mathbf{a}(1) = \langle 6, \ 0, \ 8 \rangle$, $\ v(1) = \sqrt{74}$

5. $\mathbf{v}(\frac{\pi}{3}) = \langle \frac{1}{2}, \ -\frac{\sqrt{3}}{2}, \ 0 \rangle, \mathbf{a}(\frac{\pi}{3}) = \langle -\frac{\sqrt{3}}{2}, \ -\frac{1}{2}, \ 9 \rangle, v(\frac{\pi}{3}) = 1$

7. $\mathbf{a}(t) = -2 \langle \cos \frac{t}{2}, \ \sin \frac{t}{2} \rangle$; $\ \mathbf{a}(\frac{\pi}{4}) \approx \langle -1.85, \ -0.077 \rangle$

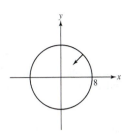

$R(t) = 8 \langle \cos \frac{t}{2}, \sin \frac{t}{2} \rangle$

9.

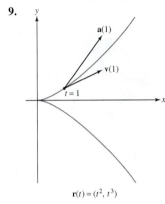

$\mathbf{r}(t) = (t^2, \ t^3)$

11. $\mathbf{v}(t) = \langle \frac{3t^2+2}{6}, \ 4t - 2 \rangle$ **13.** $\mathbf{v}(t) = \mathbf{i} + t\mathbf{k}$

15. $\mathbf{v}(t) = \langle \frac{t^2}{2} + 3, \ 4t - 2 \rangle$, $\ \mathbf{r}(t) = \langle \frac{t^3}{6} + 3t, \ 2t^2 - 2t \rangle$

17. $\mathbf{v}(t) = \mathbf{i} + \frac{t^2}{2}\mathbf{k}$, $\ \mathbf{r}(t) = t\mathbf{i} + \mathbf{j} + \frac{t^3}{6}\mathbf{k}$

19. $v_0 = \sqrt{5292} \approx 72.746$ m/s   **23.** $H = 355$ m

25. **(a)** Assume that $r(0) = \langle 150, 75, 5 \rangle$

$$a(t) = \langle 0, 0, -32 \rangle, v(t) = \langle 40, 35, -32t + 32 \rangle$$
$$r(t) = \langle 40t + 150, 35t + 75, -16t^2 + 32t + 5 \rangle$$

(b) $z = 5$ when $t = 0$ or 2. At $t = 2$, $r(2) = \langle 230, 145, 5 \rangle$, so the player is in bounds, since $(300, 150, z)$ is the maximum possible point to be in bounds.

27. $\mathbf{r}(10) = \langle 45, \ -20 \rangle$

29. **(a)** At its original position   **(b)** No

31. The speed is decreasing.

33. $a_{\mathbf{T}} = 0$, $\ a_{\mathbf{N}} = 1$ **35.** $a_{\mathbf{T}} = \frac{7}{\sqrt{6}}$, $\ a_{\mathbf{N}} = \sqrt{\frac{53}{6}}$

37. $\mathbf{a}(-1) = -\frac{2}{\sqrt{10}}\mathbf{T} + \frac{6}{\sqrt{10}}\mathbf{N}$ with $\mathbf{T} = \frac{1}{\sqrt{10}} \langle 1, \ -3 \rangle$ and $\mathbf{N} = \frac{1}{\sqrt{10}} \langle -3, \ -1 \rangle$

39. $a_{\mathbf{T}}(4) = 4$, $\ a_{\mathbf{N}}(4) = 1$, so $\mathbf{a} = 4\mathbf{T} + \mathbf{N}$, with $\mathbf{T} = \langle \frac{1}{9}, \ \frac{4}{9}, \ \frac{8}{9} \rangle$ and $\mathbf{N} = \langle -\frac{4}{9}, \ -\frac{7}{9}, \ \frac{4}{9} \rangle$

41. $\mathbf{a}(0) = \sqrt{3}\mathbf{T} + \sqrt{2}\mathbf{N}$, with $\mathbf{T} = \frac{1}{\sqrt{3}} \langle 1, \ 1, \ 1 \rangle$ and $\mathbf{N} = \frac{1}{\sqrt{2}} \langle -1, \ 0, \ 1 \rangle$

43. $\mathbf{a}(\frac{\pi}{2}) = -\frac{\pi}{2\sqrt{3}}\mathbf{T} + \frac{\pi}{\sqrt{6}}\mathbf{N}$, with $\mathbf{T} = \frac{1}{\sqrt{3}} \langle 1, \ -1, \ 1 \rangle$ and $\mathbf{N} = \frac{1}{\sqrt{6}} \langle 1, \ -1, \ -2 \rangle$

45. $a_{\mathbf{T}} = 0$, $\ a_{\mathbf{N}} = 0.25$ cm/s^2

47. The tangential acceleration is $\frac{50}{\sqrt{2}} \approx 35.36$ m/min^2, $v = \sqrt{35.36(30)} \approx 32.56$ m/min

49. $\|\mathbf{a}\| = 1.157 \times 10^5$ km/h^2 **51.** $\mathbf{a} = \langle -\frac{1}{6}, \ -1, \ \frac{1}{6} \rangle$

53. (A) slowing down, (B) speeding up, (C) slowing down

57. After 139.91 s, the car will begin to skid.   **61.** $R \approx 105$ m

Section 13.6 Preliminary Questions

1. $\frac{dA}{dt} = \frac{1}{2} \|\mathbf{J}\|$ **3.** The period is increased eight-fold.

Section 13.6 Exercises

1. The data support Kepler's prediction; $T \approx \sqrt{a^3 \cdot 3 \cdot 10^{-4}} \approx 11.9$ years

3. $M \approx 1.897 \times 10^{27}$ kg   **5.** $M \approx 2.6225 \times 10^{41}$ kg

11. The satellite's orbit is in the plane $20x - 29y = 9z = 0$.

Chapter 13 Review

1. **(a)** $-1 < t < 0$ or $0 < t \leq 1$ **(b)** $0 < t \leq 2$

3. $\mathbf{r}(t) = \langle t^2, \ t, \ \sqrt[3]{3 - t^4} \rangle$, $\ -\infty < t < \infty$

5. $\mathbf{r}'(t) = \langle -1, \ -2t^{-3}, \ \frac{1}{t} \rangle$ **7.** $\mathbf{r}'(0) = \langle 2, \ 0, \ 6 \rangle$

9. $\frac{d}{dt} e^t \langle 1, \ t, \ t^2 \rangle = e^t \langle 1, \ 1+t, \ 2t + t^2 \rangle$

11. $\frac{d}{dt} (6\mathbf{r}_1(t) - 4\mathbf{r}_2(t))|_{t=3} = \langle 0, \ -8, \ -10 \rangle$

13. $\frac{d}{dt} (\mathbf{r}_1(t) \cdot \mathbf{r}_2(t))|_{t=3} = 2$

15. $\int_0^3 \langle 4t + 3, \ t^2, \ -4t^3 \rangle \ dt = \langle 27, \ 9, \ -81 \rangle$

17. $\langle 3, \ 3, \ \frac{16}{3} \rangle$ **19.** $\mathbf{r}(t) = \langle 2t^2 - \frac{8}{3}t^3 + t, \ t^4 - \frac{1}{6}t^3 + 1 \rangle$

21. $L = 2\sqrt{13}$ **23.** $\langle 5 \cos \frac{2\pi s}{5\sqrt{1+4\pi^2}}, \ 5 \sin \frac{2\pi s}{5\sqrt{1+4\pi^2}}, \ \frac{s}{\sqrt{1+4\pi^2}} \rangle$

25. $v_0 \approx 67.279$ m/s   **27.** $\langle 0, \ \frac{11}{2}, 38 \rangle$

29. $\mathbf{T}(\pi) = \langle \frac{-1}{\sqrt{2}}, \ \frac{1}{\sqrt{2}}, \ 0 \rangle$ **31.** $\kappa(1) = \frac{1}{2^{3/2}}$

33. $\mathbf{a} = \frac{1}{\sqrt{2}}\mathbf{T} + 4\mathbf{N}$, where $\mathbf{T} = \langle -1, \ 0 \rangle$ and $\mathbf{N} = \langle 0, \ -1 \rangle$

35. $\kappa = \frac{13}{16}$ **37.** $\mathbf{c}(t) = \langle \frac{25}{2} + \frac{17^{3/2}}{2} \cos t, \ -32 + \frac{17^{3/2}}{2} \sin t \rangle$

39. $2x - 4y + 2z = -3$

Chapter 14

Section 14.1 Preliminary Questions

1. Same shape, but located in parallel planes

2. The parabola $z = x^2$ in the xz-plane   **3.** Not possible

4. The vertical lines $x = c$ with distance of 1 unit between adjacent lines

5. In the contour map of $g(x, \ y) = 2x$, the distance between two adjacent vertical lines is $\frac{1}{2}$.

Section 14.1 Exercises

1. $f(2, 2) = 18$, $f(-1, 4) = -5$

3. $h(3, 8, 2) = 6$; $h(3, -2, -6) = -\frac{1}{6}$

5. The domain is the entire xy-plane.

7.

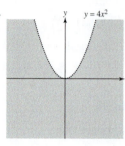

9. $\mathcal{D} = \left\{(y, z) : z \neq -y^2\right\}$

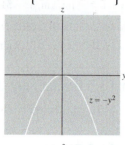

11.

13. Domain: entire (x, y, z)–space; range: entire real line

15. Domain: $\{(r, s, t) : |rst| \leq 4\}$; range: $\{w : 0 \leq w \leq 4\}$

17. $f \leftrightarrow$ (B), $g \leftrightarrow$ (A)

19. (a) D   (b) C   (c) E   (d) B   (e) A   (f) F

21.

Horizontal trace: $3x + 4y = 12 - c$ in the plane $z = c$
Vertical trace: $z = (12 - 3a) - 4y$ and $z = -3x + (12 - 4a)$ in the planes $x = a$, and $y = a$, respectively

23.

The horizontal traces are ellipses for $c > 0$.
The vertical trace in the plane $x = a$ is the parabola $z = a^2 + 4y^2$.
The vertical trace in the plane $y = a$ is the parabola $z = x^2 + 4a^2$.

25.

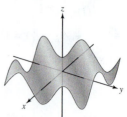

The horizontal traces in the plane $z = c$, $|c| \leq 1$, are the lines $x - y = \sin^{-1} c + 2k\pi$ and $x - y = \pi - \sin^{-1} c + 2k\pi$, for integer k.
The vertical trace in the plane $x = a$ is $z = \sin(a - y)$. .
The vertical trace in the plane $y = a$ is $z = \sin(x - a)$.

27. $m = 1 : m = 2 :$

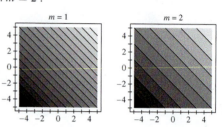

29.

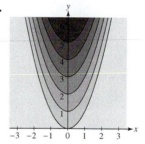

31.

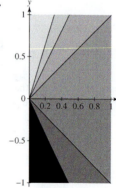

33.

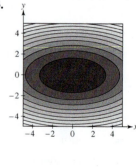

35.

37. $m = 6:\ f(x,y) = 2x + 6y + 6$
$m = 3:\ f(x,y) = x + 3y + 3$

39. (a) Only at (A) (b) Only at (C) (c) West

41.
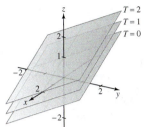

43.
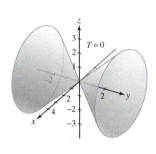

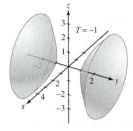

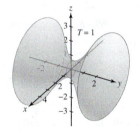

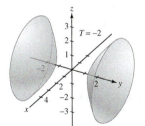

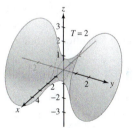

45. Average ROC from B to $C = 0.000625$ kg/m^3 · ppt

47. At point A

49. Average ROC from A to $B \approx 0.0737$, average ROC from A to $C \approx 0.0457$

51.
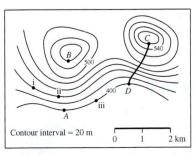

53.
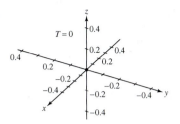

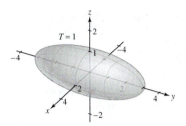

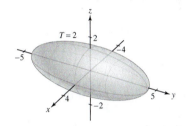

55.
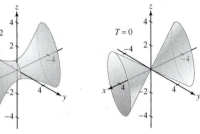

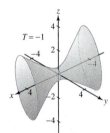
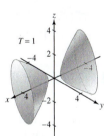

57. $f(r, \theta) = \cos\theta$; the level curves are
$\theta = \pm\cos^{-1}(c)$ for $|c| < 1,\ \ c \neq 0$;
the y-axis for $c = 0$;
the positive x-axis for $c = 1$;
the negative x-axis for $c = -1$.

Section 14.2 Preliminary Questions

1. $D^*(p, r)$ consists of all points in $D(p, r)$ other than p itself.

2. $f(2, 3) = 27$ **3.** All three statements are true.

4. $\lim\limits_{(x, y) \to (0, 0)} f(x, y)$ does not exist.

Section 14.2 Exercises

1. $\lim\limits_{(x, y) \to (1, 2)} (x^2 + y) = 3$ **3.** $\lim\limits_{(x, y) \to (2, -1)} (xy - 3x^2 y^3) = 10$

5. $\lim\limits_{(x, y) \to (\frac{\pi}{4}, 0)} \tan x \cos y = 1$

7. $\lim\limits_{(x,\,y)\to(1,\,1)} \dfrac{e^{x^2}-e^{-y^2}}{x+y} = \frac{1}{2}(e-e^{-1})$

9. $\lim\limits_{(x,\,y)\to(2,\,5)} (g(x,\,y)-2f(x,\,y)) = 1$

11. $\lim\limits_{(x,\,y)\to(2,\,5)} e^{f(x,\,y)^2-g(x,\,y)} = e^2$

13. No; the limit along the x-axis and the limit along the y-axis are different.

15. The limit is $\frac{1+m^3}{m^2}$ for all $m \neq 0$.

17. The limit along the x-axis is $\lim\limits_{(x,y)\to(0,0)} \dfrac{x}{x^2+y^2} = \lim\limits_{x\to 0} \frac{1}{x}$, which does not exist.

19. $\lim\limits_{(x,y)\to(0,0)} \dfrac{x^2-y^2}{\sqrt{x^2+y^2}} = 0$

21. The limit does not exist, because the function values as (x, y) approach $(0, 0)$ along the line $y = mx$ depend on the value of m.

$$\lim\limits_{(x,y)\to(0,0)y=mx} \dfrac{xy}{3x^2+2y^2} = \lim\limits_{x\to 0} \dfrac{mx^2}{3x^2+2m^2x^2} = \dfrac{m}{2m^2+3}$$

23. Along the x-coordinate axis ($y = z = 0$),

$$\lim\limits_{(x,y,z)\to(0,0,0)} \dfrac{x+y+z}{x^2+y^2+z^2} = \lim\limits_{(x,y,z)\to(0,0,0)} \dfrac{1}{x} = \infty$$

25. $\lim\limits_{(x,\,y)\to(4,\,0)} (x^2-16)\cos\left(\dfrac{1}{(x-4)^2+y^2}\right) = 0$

27. $\lim\limits_{(z,\,w)\to(-2,\,1)} \dfrac{z^4\cos(\pi w)}{e^{z+w}} = -16e$

29. $\lim\limits_{(x,\,y)\to(4,\,2)} \dfrac{y-2}{\sqrt{x^2-4}} = 0$ **31.** $\lim\limits_{(x,\,y)\to(3,\,4)} \dfrac{1}{\sqrt{x^2+y^2}} = \frac{1}{5}$

33. $\lim\limits_{(x,\,y)\to(1,\,-3)} e^{x-y}\ln(x-y) = e^4\ln(4)$

35. $\lim\limits_{(x,\,y)\to(-3,\,-2)} (x^2y^3+4xy) = -48$

37. $\lim\limits_{(x,\,y)\to(0,\,0)} \tan(x^2+y^2)\tan^{-1}\left(\dfrac{1}{x^2+y^2}\right) = 0$

39. $\lim\limits_{(x,\,y)\to(0,\,0)} \dfrac{x^2+y^2}{\sqrt{x^2+y^2+1}-1} = 2$

43. $\lim\limits_{(x,\,y)\to Q} g(x,\,y) = 4$ **45.** Yes

Section 14.3 Preliminary Questions

1. $\frac{\partial}{\partial x}(x^2y^2) = 2xy^2$

2. In this case, the Constant Multiple Rule can be used. In the second part, since y appears in both the numerator and the denominator, the Quotient Rule is preferred.

3. (a), (c) **4.** $f_x = 0$ **5.** (a), (d)

Section 14.3 Exercises

3. $\frac{\partial}{\partial y}\frac{y}{x+y} = \dfrac{x}{(x+y)^2}$ **5.** $f_z(2,\,3,\,1) = 6$

7. $m = 10$ **9.** $f_x(A) \approx 10,\ f_y(A) \approx -20$ **11.** NW

13. $\frac{\partial}{\partial x}(x^2+y^2) = 2x,\ \frac{\partial}{\partial y}(x^2+y^2) = 2y$

15. $\frac{\partial}{\partial x}(x^4y+xy^{-2}) = 4x^3y+y^{-2}$, $\frac{\partial}{\partial y}(x^4y+xy^{-2}) = x^4-2xy^{-3}$

17. $\frac{\partial}{\partial x}\left(\frac{x}{y}\right) = \frac{1}{y},\ \frac{\partial}{\partial y}\left(\frac{x}{y}\right) = \frac{-x}{y^2}$

19. $\frac{\partial}{\partial x}\left(\sqrt{9-x^2-y^2}\right) = \dfrac{-x}{\sqrt{9-x^2-y^2}},\ \frac{\partial}{\partial y}\left(\sqrt{9-x^2-y^2}\right) = \dfrac{-y}{\sqrt{9-x^2-y^2}}$

21. $\frac{\partial}{\partial x}(\sin x \sin y) = \sin y \cos x,\ \frac{\partial}{\partial y}(\sin x \sin y) = \sin x \cos y$

23. $\frac{\partial}{\partial x}\left(\tan\frac{x}{y}\right) = \frac{1}{y}\sec^2\frac{x}{y},\ \frac{\partial}{\partial y}\left(\tan\frac{x}{y}\right) = -\frac{x}{y^2}\sec^2\frac{x}{y}$

25. $\frac{\partial}{\partial x}\ln(x^2+y^2) = \dfrac{2x}{x^2+y^2},\ \frac{\partial}{\partial y}\ln(x^2+y^2) = \dfrac{2y}{x^2+y^2}$

27. $\frac{\partial}{\partial r}e^{r+s} = e^{r+s},\ \frac{\partial}{\partial s}e^{r+s} = e^{r+s}$

29. $\frac{\partial}{\partial x}e^{xy} = ye^{xy},\ \frac{\partial}{\partial y}e^{xy} = xe^{xy}$

31. $\frac{\partial z}{\partial y} = -2xe^{-x^2-y^2},\ \frac{\partial z}{\partial y} = -2ye^{-x^2-y^2}$

33. $\frac{\partial U}{\partial t} = -e^{-rt},\ \frac{\partial U}{\partial r} = \dfrac{-e^{-rt}(rt+1)}{r^2}$

35. $\frac{\partial}{\partial x}\sinh(x^2y) = 2xy\cosh(x^2y),\ \frac{\partial}{\partial y}\sinh(x^2y) = x^2\cosh(x^2y)$

37. $\frac{\partial w}{\partial x} = y^2z^3,\ \frac{\partial w}{\partial y} = 2xz^3y,\ \frac{\partial w}{\partial z} = 3xy^2z^2$

39. $\frac{\partial Q}{\partial L} = \dfrac{M-Lt}{M^2}e^{-Lt/M},\ \frac{\partial Q}{\partial M} = \dfrac{L(Lt-M)}{M^3}e^{-Lt/M},$
$\frac{\partial Q}{\partial t} = -\dfrac{L^2}{M^2}e^{-Lt/M}$

41. $f_x(1,\,2) = -164$ **43.** $g_u(1,\,2) = \ln 3 + \frac{1}{3}$

45. $N = 2865.058,\ \Delta N \approx -217.74$

47. (a) $I(95,\,50) \approx 73.1913$ (b) $\frac{\partial I}{\partial T};\ 1.66$

49. A 1-cm increase in r

51. $\frac{\partial W}{\partial E} = -\dfrac{1}{kT}e^{-E/kT},\ \frac{\partial W}{\partial T} = \dfrac{E}{kT^2}e^{-E/kT}$

55. (a), (b) **57.** $\frac{\partial^2 f}{\partial x^2} = 6y,\ \frac{\partial^2 f}{\partial y^2} = -72xy^2$

59. $h_{vv} = \dfrac{32u}{(u+4v)^3}$ **61.** $f_{yy}(2,\,3) = -\frac{4}{9}$

63. $f_{xyxzy} = 0$ **65.** $f_{uuv} = 2v\sin(u+v^2)$

67. $F_{rst} = 0$ **69.** $F_{uu\theta} = \cosh(uv+\theta^2)\cdot 2\theta v^2$

71. $g_{xyz} = \dfrac{3xyz}{(x^2+y^2+z^2)^{5/2}}$ **73.** $f(x,\,y) = x^2y$ **77.** $B = A^2$

Section 14.4 Preliminary Questions

1. $L(x,\,y) = f(a,\,b) + f_x(a,\,b)(x-a) + f_y(a,\,b)(y-b)$

2. f is locally linear at (a, b) if $f(x,\,y) = L(x,\,y) + e(x,\,y)$, where $e(x, y)$ satisfies $\lim\limits_{(x,y)\to(a,b)} \dfrac{e(x,y)}{\sqrt{(x-a)^2+(y-b)^2}} = 0$

3. (b) **4.** $f(2,\,3.1) \approx 8.7$ **5.** $\Delta f \approx -0.1$

6. Criterion for Differentiability

Section 14.4 Exercises

1. $z = -34 - 20x + 16y$ **3.** $z = 5x + 10y - 14$

5. $z = 8x - 2y - 13$ **7.** $z = 4r - 5s + 2$

9. $z = \left(\frac{4}{5} + \frac{12}{25}\ln 2\right) - \frac{12}{25}x + \frac{12}{25}y$ **11.** $\left(-\frac{1}{4},\,\frac{1}{8},\,\frac{1}{8}\right)$

13. (a) $f(x,\,y) = -16 + 4x + 12y$

(b) $f(2.01,\,1.02) \approx 4.28;\ f(1.97,\,1.01) \approx 4$

15. $\Delta f \approx 3.56$ **17.** $f(0.01,\,-0.02) \approx 0.98$

19. $L(x,y,z) = \frac{5}{12}\sqrt{3}x + \frac{5}{12}\sqrt{3}y + 2\sqrt{3}z - 5\sqrt{3}$

21. 5.07 **23.** 8.44 **25.** 4.998 **27.** 3.945

29. $f(-2.1, 3.1) \approx 4.2$ **31.** $\Delta I \approx 0.5644$

33. (b) $\Delta H \approx 0.022$ m

35. (b) 6% (c) 1% error in r

37. (a) $7.10 (b) $28.85, $57.69 (c) -$74.24

39. Maximum error in V is about 8.948 m.

Section 14.5 Preliminary Questions

1. (b) $\langle 3,\,4\rangle$ **2.** False

3. ∇f points in the direction of maximum rate of increase of f and is normal to the level curve of f.

4. (b) NW and (c) SE **5.** $3\sqrt{2}$

Section 14.5 Exercises

1. (a) $\nabla f = \left\langle y^2,\ 2xy \right\rangle, \mathbf{r}'(t) = \left\langle t,\ 3t^2 \right\rangle$

(b) $\frac{d}{dt}\left(f(\mathbf{r}(t))\right)\Big|_{t=1} = 4;\ \frac{d}{dt}\left(f(\mathbf{r}(t))\right)\Big|_{t=-1} = -4$

3. A: zero, B: negative, C: positive, D: zero

5. $\nabla f = -\sin(x^2 + y)\,\langle 2x,\ 1\rangle$

7. $\nabla h = \left\langle yz^{-3},\ xz^{-3},\ -3xyz^{-4}\right\rangle$

9. $\frac{d}{dt}\left(f(\mathbf{r}(t))\right)\Big|_{t=0} = -7$ **11.** $\frac{d}{dt}\left(f(\mathbf{r}(t))\right)\Big|_{t=0} = -3$

13. $\frac{d}{dt}\left(f(\mathbf{r}(t))\right)\Big|_{t=0} = 5\cos 1 \approx 2.702$

15. $\frac{d}{dt}\left(f(\mathbf{r}(t))\right)\Big|_{t=4} = -56$

17. $\frac{d}{dt}\left(f(\mathbf{r}(t))\right)\Big|_{t=\pi/4} = -1 + \frac{\pi}{8} \approx 1.546$

19. $\frac{d}{dt}\left(g(\mathbf{r}(t))\right)\Big|_{t=1} = 0$

21. $D_{\mathbf{u}}f(1,\ 2) = 8.8$ **23.** $D_{\mathbf{u}}f\left(\frac{1}{6},\ 3\right) = \frac{39}{4\sqrt{2}}$

25. $D_{\mathbf{u}}f(3,\ 4) = \frac{7\sqrt{2}}{290}$ **27.** $D_{\mathbf{u}}f(1,\ 0) = \frac{6}{\sqrt{13}}$

29. $D_{\mathbf{u}}f(1,\ 2,\ 0) = -\frac{1}{\sqrt{3}}$ **31.** $D_{\mathbf{u}}f(3,\ 2) = \frac{-50}{\sqrt{13}}$

33. $D_{\mathbf{u}}f(P) = -\frac{e^5}{3} \approx -49.47$

35. (a) $m = 3$, angle of inclination is approximately $1.249 \approx 71.6°$.

(b) $m = 4\sqrt{2} \approx 5.66$, angle of inclination is approximately $1.396 \approx 80.0°$.

(c) $m = -4\sqrt{2} \approx -5.66$, angle of inclination is approximately $-1.396 \approx -80.0°$.

(d) $m = \sqrt{34} \approx 5.83$, angle from due East is approximately $121°$.

37. f is increasing at P in the direction of v.

39. $D_{\mathbf{u}}f(P) = \frac{\sqrt{6}}{2}$ **41.** $\langle 6,\ 2,\ -4\rangle$

43. $\left(\frac{4}{\sqrt{17}},\ \frac{9}{\sqrt{17}},\ -\frac{2}{\sqrt{17}}\right)$ and $\left(-\frac{4}{\sqrt{17}},\ -\frac{9}{\sqrt{17}},\ \frac{2}{\sqrt{17}}\right)$

45. $9x + 10y + 5z = 33$

47. $0.5217x + 0.7826y - 1.2375z = -5.309$

49.

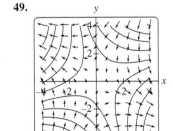

51. $f(x,\ y,\ z) = x^2 + y + 2z$

53. $f(x,\ y,\ z) = xz + y^2$ **57.** $\Delta f \approx 0.08$

59. (a) $\langle 34,\ 18,\ 0\rangle$

(b) $\left\langle 2 + \frac{32}{\sqrt{21}}t,\ 2 + \frac{16}{\sqrt{21}}t,\ 8 - \frac{8}{\sqrt{21}}t\right\rangle;\ \approx 4.58$ s

63. $x = 1 - 4t,\ y = 2 + 26t,\ z = 1 - 25t$

75. $y = \sqrt{1 - \ln(\cos^2 x)}$

Section 14.6 Preliminary Questions

1. (a) $\frac{\partial f}{\partial x}$ and $\frac{\partial f}{\partial y}$ (b) u and v

2. (a) **3.** $f(u,\ v)|_{(r,s)=(1,1)} = e^2$ **4.** (b) **5.** (c) **6.** No

Section 14.6 Exercises

1. (a) $\frac{\partial f}{\partial x} = 2xy^3,\ \frac{\partial f}{\partial y} = 3x^2 y^2,\ \frac{\partial f}{\partial z} = 4z^3$

(b) $\frac{\partial x}{\partial s} = 2s,\ \frac{\partial y}{\partial s} = t^2,\ \frac{\partial z}{\partial s} = 2st$

(c) $\frac{\partial f}{\partial s} = 7s^6 t^6 + 8s^7 t^4$

3. $\frac{\partial f}{\partial s} = 6rs^2,\ \frac{\partial f}{\partial r} = 2s^3 + 4r^3$

5. $\frac{\partial g}{\partial u} = -10\sin(10u - 20v),\ \frac{\partial g}{\partial v} = 20\sin(10u - 20v)$

7. $\frac{\partial F}{\partial y} = xe^{x^2 + xy}$ **9.** $\frac{\partial h}{\partial t_2} = 0$

11. $\frac{\partial f}{\partial u}\Big|_{(u,v)=(-1,-1)} = 1,\ \frac{\partial f}{\partial v}\Big|_{(u,v)=(-1,-1)} = -2$

13. $\frac{\partial g}{\partial \theta}\Big|_{(r,\theta)=\left(2\sqrt{2},\ \pi/4\right)} = -\frac{1}{6}$ **15.** $\frac{\partial g}{\partial u}\Big|_{(u,v)=(0,1)} = 2\cos 2$

17. -26.8 ft/s **19.** $4\pi^3 - 3\pi^2 - 1$

23. (a) $F_x = z^2 + y,\ F_y = 2yz + x,\ F_z = 2xz + y^2$

(b) $\frac{\partial z}{\partial x} = -\frac{z^2 + y}{2xz + y^2},\ \frac{\partial z}{\partial y} = -\frac{2yz + x}{2xz + y^2}$

27. $\frac{\partial z}{\partial x} = -\frac{2xy + z^2}{2xz + y^2}$ **29.** $\frac{\partial z}{\partial y} = -\frac{xe^{xy} + 1}{x\cos(xz)}$

31. $\frac{\partial w}{\partial y} = \frac{-y(w^2 + x^2)^2}{w\left((w^2 + y^2)^2 + (w^2 + x^2)^2\right)};$ at $(1,\ 1,\ 1),\ \frac{\partial w}{\partial y} = -\frac{1}{2}$

35. $\nabla\left(\frac{1}{r}\right) = -\frac{1}{r^3}\mathbf{r}$ **37.** (c) $\frac{\partial z}{\partial x} = \frac{x - 6}{z + 4}$

39. $\frac{\partial P}{\partial T} = -\frac{F_T}{F_P} = -\frac{-nR}{V - nb} = \frac{nR}{V - nb}$

Section 14.7 Preliminary Questions

1. f has a local (and global) min at $(0,\ 0)$; g has a saddle point at $(0,\ 0)$.

2.

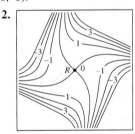

Point R is a saddle point.

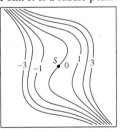

Point S is neither a local extremum nor a saddle point.

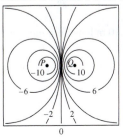

Point P is a local minimum and point Q is a local maximum.

3. Statement (a)

Section 14.7 Exercises

1. (b) $P_1 = (0,0)$ is a saddle point, $P_2 = \left(2\sqrt{2}, \sqrt{2}\right)$ and $P_3 = \left(-2\sqrt{2}, -\sqrt{2}\right)$ are local minima; absolute minimum value of f is -4.

3. $(0, 0)$ saddle point, $\left(\frac{13}{64}, -\frac{13}{32}\right)$ and $\left(-\frac{1}{4}, \frac{1}{2}\right)$ local minima

5. (c) $(0, 0)$, $(1, 0)$, and $(0, -1)$ saddle points, $\left(\frac{1}{3}, -\frac{1}{3}\right)$ local minimum.

7. $\left(-\frac{2}{3}, -\frac{1}{3}\right)$ local minimum

9. $(-2, -1)$ local maximum, $\left(\frac{5}{3}, \frac{5}{6}\right)$ saddle point

11. $\left(0, \pm\sqrt{2}\right)$ saddle points, $\left(\frac{2}{3}, 0\right)$ local maximum, $\left(-\frac{2}{3}, 0\right)$ local minimum

13. $(0,0)$ saddle point, $(1, 1)$ and $(-1, -1)$ local minima

15. $(0, 0)$ saddle point, $\left(\frac{1}{\sqrt{2}}, \frac{1}{\sqrt{2}}\right)$ and $\left(-\frac{1}{\sqrt{2}}, -\frac{1}{\sqrt{2}}\right)$ local maximum, $\left(\frac{1}{\sqrt{2}}, -\frac{1}{\sqrt{2}}\right)$ and $\left(-\frac{1}{\sqrt{2}}, \frac{1}{\sqrt{2}}\right)$ local minimum

17. Critical points are $\left(j\pi, \ k\pi + \frac{\pi}{2}\right)$, for
j,k even: saddle points
j,k odd: local maxima
j even, k odd: local minima
j odd, k even: saddle points

19. $\left(1, \frac{1}{2}\right)$ local maximum **21.** $\left(\frac{3}{2}, -\frac{1}{2}\right)$ saddle point

23. $\left(-\frac{1}{6}, -\frac{17}{18}\right)$ local minimum

27. $x = y = 0.27788$ local minimum

29. Global maximum 2, global minimum 0

31. Global maximum 1, global minimum $\frac{1}{35}$

35. Maximum value $\frac{1}{3}$

37. Global minimum $f(0, 1) = -2$, global maximum $f(1, 0) = 1$

39. Global minimum, $f(0, 0) = 0$, global maximum $f(1, 1) = 3$

41. Global minimum, $f(0, 0) = 0$, global maximum $f(1, 0) = f(0, 1) = 1$

43. Global minimum, $f(1, 2) = -5$, global maximum $f(0, 0) = f(3, 3) = 0$

45. Global minimum, $f(-0.4343, 0.9) = f(-0.4343, -0.9) \approx -0.5161$, global maximum $f(0.7676, 0.6409) = f(0.7676, -0.6409) \approx 1.2199$

47. Maximum volume $\frac{3}{4}$

49. (a) No. In the box B with minimal surface area, z is smaller than $\sqrt[3]{V}$, which is the side of a cube with volume V.

(b) Width: $x = (2V)^{1/3}$; length: $y = (2V)^{1/3}$;
height: $z = \left(\frac{V}{4}\right)^{1/3}$

53. The fence should be cut into 12 pieces 10 m long, forming three 10×10 squares.

55. $V = \frac{64,000}{\pi} \approx 20,372$ cm^3 **57.** $f(x) = 1.9629x - 1.5519$

Section 14.8 Preliminary Questions

1. Statement (b)

2. f had a local maximum 2, under the constraint, at A; $f(B)$ is neither a local minimum nor a local maximum of f.

3. (a)

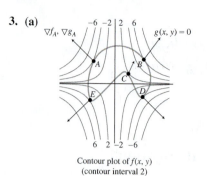

Contour plot of $f(x, y)$
(contour interval 2)

(b) Global minimum -4, global maximum 6

Section 14.8 Exercises

1. (c) Critical points $(-1, -2)$ and $(1, 2)$

(d) Maximum 10, minimum -10

3. Maximum $4\sqrt{2}$, minimum $-4\sqrt{2}$

5. Minimum $\frac{36}{13}$, no maximum value

7. Maximum $\frac{8}{3}$, minimum $-\frac{8}{3}$

9. Maximum $\sqrt{2}$, minimum 1

11. Maximum 3.7, minimum -3.7

13. Maxima at $f(\pm 4, \pm 4, 2) = 20$, and minima at $f(\pm 4, \mp 4, -2) = -20$

15. Maximum $2\sqrt{2}$, minimum $-2\sqrt{2}$ **17.** $(-1, \ e^{-1})$

19. (a) $\frac{h}{r} = \sqrt{2}$ **(b)** $\frac{h}{r} = \sqrt{2}$

(c) There is no cone of fixed V with maximal S.

21. $(8, \ -2)$ **23.** $\left(\frac{48}{97}, \frac{108}{97}\right)$ **25.** $\frac{a^a b^b}{(a+b)^{a+b}}$ **27.** $\sqrt{\frac{a^a b^b}{(a+b)^{a+b}}}$

33. $r = 3$, $h = 6$ **35.** $x + y + z = 3$

41. $\frac{25}{3}$ **43.** $\left(\frac{-6}{\sqrt{105}}, \frac{-3}{\sqrt{105}}, \frac{30}{\sqrt{105}}\right)$ **45.** $(-1, 0, 2)$

47. Minimum $\frac{138}{11} \approx 12.545$, no maximum value

51. (b) $\lambda = \frac{c}{2p_1 p_2}$

Chapter 14 Review

1. (a)

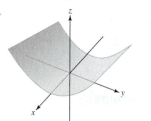

(b) $f(3, 1) = \frac{\sqrt{2}}{3}$, $f(-5, -3) = -2$ **(c)** $\left(-\frac{5}{3}, 1\right)$

3.

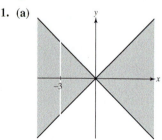

Vertical and horizontal traces: the line $z = (c^2 + 1) - y$ in the plane $x = c$, the parabola $z = x^2 - c + 1$ in the plane $y = c$.

5. (a) Graph (B) (b) Graph (C) (c) Graph (D) (d) Graph (A)

7. (a) Parallel lines $4x - y = \ln c$, $c > 0$, in the xy-plane

(b) Parallel lines $4x - y = e^c$ in the xy-plane

(c) Hyperbolas $3x^2 - 4y^2 = c$ in the xy-plane

(d) Parabolas $x = c - y^2$ in the xy-plane

9. $\displaystyle\lim_{(x,y)\to(1,-3)} (xy + y^2) = 6$

11. The limit does not exist.

13. $\displaystyle\lim_{(x,y)\to(1,-3)} (2x + y)e^{-x+y} = -e^{-4}$

17. $f_x = 2$, $f_y = 2y$

19. $f_x = e^{-x-y}(y\cos(xy) - \sin(xy))$
$f_y = e^{-x-y}(x\cos(yx) - \sin(yx))$

21. $f_{xxyz} = -\cos(x+z)$ **23.** $z = 33x + 8y - 42$

25. Estimate, 12.146; calculator value to three places, 11.996.

27. Statements (ii) and (iv) are true.

29. $\dfrac{d}{dt}\left(f(\mathbf{c}(t))\right)\Big|_{t=2} = 3 + 4e^4 \approx 221.4$

31. $\dfrac{d}{dt}\left(f(\mathbf{c}(t))\right)\Big|_{t=1} = 4e - e^{3e} \approx -3469.3$

33. $D_{\mathbf{u}}f(3, -1) = -\dfrac{54}{\sqrt{5}}$

35. $D_{\mathbf{u}}f(P) = -\dfrac{\sqrt{2}e}{5}$ **37.** $\left\langle \dfrac{1}{\sqrt{2}}, \dfrac{1}{\sqrt{2}}, 0 \right\rangle$

41. $\dfrac{\partial f}{\partial s} = 3s^2t + 4st^2 + t^3 - 2st^3 + 6s^2t^2$
$\dfrac{\partial f}{\partial t} = 4s^2t + 3st^2 + s^3 + 4s^3t - 3s^2t^2$

45. $\dfrac{\partial z}{\partial x} = -\dfrac{e^z - 1}{xe^z + e^y}$

47. $(0, 0)$ saddle point, $(1, 1)$ and $(-1, -1)$ local minima

49. $\left(\dfrac{1}{2}, \dfrac{1}{2}\right)$ saddle point

53. Global maximum $f(2, 4) = 10$, global minimum
$f(-2, 4) = -18$

55. Maximum $\dfrac{26}{\sqrt{13}}$, minimum $-\dfrac{26}{\sqrt{13}}$

57. Maximum $\dfrac{12}{\sqrt{3}}$, minimum $-\dfrac{12}{\sqrt{3}}$

59. Minimum $= f\left(\dfrac{1+\sqrt{3}}{3}, \dfrac{1}{3}, \dfrac{1-\sqrt{3}}{3}\right) = \dfrac{6-2\sqrt{3}}{3}$
Maximum $= f\left(\dfrac{1-\sqrt{3}}{3}, \dfrac{1}{3}, \dfrac{1+\sqrt{3}}{3}\right) = \dfrac{6+2\sqrt{3}}{3}$

61. $r = h = \sqrt[3]{\dfrac{V}{\pi}}$, $S = 3\pi \left(\dfrac{V}{\pi}\right)^{2/3}$

Chapter 15
Section 15.1 Preliminary Questions

1. $\Delta A = 1$, the number of subrectangles is 32.

2. $\iint_R f\, dA \approx S_{1,1} = 0.16$ **3.** $\iint_R 5\, dA = 50$

4. The signed volume between the graph $z = f(x, y)$ and the
xy-plane. The region below the xy-plane is treated as negative volume.

5. (b) **6.** (b), (c)

Section 15.1 Exercises

1. $S_{4,3} = 13.5$ **3.** (A) $S_{3,2} = 42$, (B) $S_{3,2} = 43.5$

5. (A) $S_{3,2} = 60$, (B) $S_{3,2} = 62$

7. Two possible solutions are $S_{3,2} = \dfrac{77}{72}$ and $S_{3,2} = \dfrac{79}{72}$.

9. $\dfrac{225}{2}$

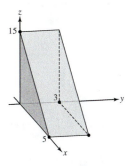

11. 0.19375 **13.** 1.0731, 1.0783, 1.0809 **15.** 0 **17.** 0 **19.** 40

21. 55 **23.** $\dfrac{4}{3}$ **25.** 84 **27.** 4 **29.** $\dfrac{1858}{15}$

31. $6\ln 6 - 2\ln 2 - 5\ln 5 \approx 1.317$ **33.** $\dfrac{4}{3}\left(19 - 5\sqrt{5}\right) \approx 10.426$

35. $\dfrac{1}{2}(\ln 3)(-2 + \ln 48) \approx 1.028$ **37.** $6\ln 3 \approx 6.592$

39. 1 **41.** $\left(e^2 - 1\right)\left(1 - \dfrac{\sqrt{2}}{2}\right) \approx 1.871$ **43.** $m = \dfrac{3}{4}$

45. $\dfrac{2}{15}\left(8\sqrt{2} - 7\right) \approx 0.575$ **47.** $2\ln 2 - 1 \approx 0.386$

51. $\dfrac{e^3}{3} - \dfrac{1}{3} - e + 1 \approx 4.644$

Section 15.2 Preliminary Questions

1. (b), (c)

2. **3.**

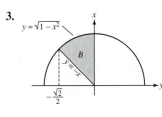

4. (b)

Section 15.2 Exercises

1. (a) Sample points •, $S_{3,4} = -3$

(b) Sample points ○, $S_{3,4} = -4$

3. As a vertically simple region: $0 \le x \le 1$, $0 \le y \le 1 - x^2$; as a
horizontally simple region: $0 \le y \le 1$, $0 \le x \le \sqrt{1 - y}$,
$\displaystyle\int_0^1 \left(\int_0^{1-x^2} (xy)\, dy\right) dx = \dfrac{1}{12}$

5. $\dfrac{192}{5} = 38.4$ **7.** $\dfrac{608}{15} \approx 40.53$ **9.** $2\dfrac{1}{4}$ **11.** $-\dfrac{3}{4} + \ln 4$

13. $\dfrac{16}{3} \approx 5.33$ **15.** $\dfrac{11}{60}$ **17.** $\dfrac{1754}{15} \approx 116.93$ **19.** $\dfrac{e-2}{2} \approx 0.359$

21. $\dfrac{1}{12}$ **23.** $2e^{12} - \dfrac{1}{2}e^9 + \dfrac{1}{2}e^5 \approx 321{,}532.2$

25.

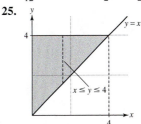

$$\int_0^4 \int_x^4 f(x, y)\, dy\, dx = \int_0^4 \int_0^y f(x, y)\, dx\, dy$$

27.

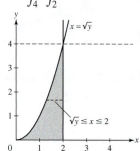

$$\int_4^9 \int_2^{\sqrt{y}} f(x,\,y)\,dx\,dy = \int_2^3 \int_{x^2}^9 f(x,\,y)\,dy\,dx$$

29.

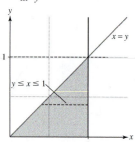

$$\int_0^2 \int_0^{x^2} \sqrt{4x^2 + 5y}\,dy\,dx = \frac{152}{15}$$

31. $\int_1^e \int_{\ln^2 y}^{\ln y} (\ln y)^{-1}\,dx\,dy = e - 2 \approx 0.718$

33.

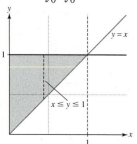

$$\int_0^1 \int_0^x \frac{\sin x}{x}\,dy\,dx = 1 - \cos 1 \approx 0.460$$

35.

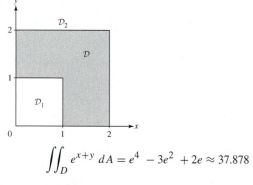

$$\int_0^1 \int_0^y xe^{y^3}\,dx\,dy = \frac{e-1}{6} \approx 0.286$$

37.

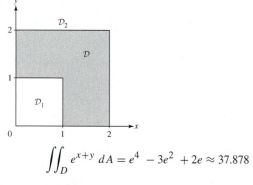

$$\iint_D e^{x+y}\,dA = e^4 - 3e^2 + 2e \approx 37.878$$

39. $\int_0^4 \int_{x/4}^{3x/4} e^{x^2}\,dy\,dx = \frac{1}{4}\left(e^{16} - 1\right)$

41. $\int_2^4 \int_{y-1}^{7-y} \frac{x}{y^2}\,dx\,dy = 6 - 6\ln 2 \approx 1.841$

43. $\iint_D \frac{\sin y}{y}\,dA = \cos 1 - \cos 2 \approx 0.956$

45. $\int_{-2}^2 \int_0^{4-x^2} (40 - 10y)\,dy\,dx = 256$ **47.** $\frac{512}{3}$

49. $\int_{-2}^2 \left(\int_{-\sqrt{4-x^2}}^{\sqrt{4-x^2}} \left[\left(8 - x^2 - y^2\right) - \left(x^2 + y^2\right) \right] dy \right) dx$

51. $\int_0^1 \int_0^1 e^{x+y}\,dx\,dy = e^2 - 2e + 1 \approx 2.952$

53. $\frac{1}{\pi} \int_0^1 \int_0^\pi y^2 \sin x\,dx\,dy = \frac{2}{3\pi}$ **55.** $\bar{f} = p$

61. One possible solution is $P = \left(\frac{2}{3},\, 2\right)$.

63. $\iint_D f(x,\,y)\,dA \approx 57.01$

Section 15.3 Preliminary Questions

1. (c)   **2.** (b)

3. (a) $D = \{(x,\,y) : 0 \le x \le 1,\, 0 \le y \le x\}$

(b) $D = \left\{(x,\,y) : 0 \le x \le 1,\, 0 \le y \le \sqrt{1 - x^2}\right\}$

Section 15.3 Exercises

1. 6   **3.** $(e - 1)(1 - e^{-2})$ **5.** $-\frac{27}{4} = -6.75$

7. $\frac{b}{20}\left[(a+c)^5 - a^5 - c^5\right]$ **9.** $\frac{1}{6}$ **11.** $\frac{1}{16}$ **13.** $e - \frac{5}{2}$

15. $2\frac{1}{12}$ **17.** $\frac{128}{15}$ **19.** $\int_0^3 \int_0^4 \int_0^{y/4} dz\,dy\,dx = 6$

21. $\frac{1}{12}$ **23.** $\frac{126}{5}$

25. Region enclosed by sphere $x^2 + y^2 + z^2 = 5$ to right of plane $y = 1$.

27. $\int_0^2 \int_0^{y/2} \int_0^{4-y^2} xyz\,dz\,dx\,dy$, $\int_0^4 \int_0^{\sqrt{4-z}} \int_0^{y/2} xyz\,dx\,dy\,dz$, and $\int_0^4 \int_0^{\sqrt{1-(z/4)}} \int_{2x}^{\sqrt{4-z}} xyz\,dy\,dx\,dz$

29. $\int_{-1}^1 \int_{-\sqrt{1-x^2}}^{\sqrt{1-x^2}} \int_{\sqrt{x^2+y^2}}^1 f(x,\,y,\,z)\,dz\,dy\,dx$ **31.** $\frac{16}{21}$

33. $\int_{-1}^1 \left(\int_0^{1-y^2} \left(\int_0^{3-z} dx \right) dz \right) dy$

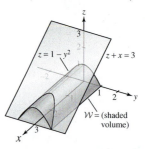

35. $\int_0^1 \left(\int_{\sqrt{y}}^{y^2} \left(\int_0^{4-y-z} dx \right) dz \right) dy$

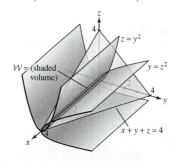

37. $\frac{1}{2\pi}$ **39.** $2e - 4 \approx 1.437$
41. $S_{N,N,N} \approx 0.561, \ 0.572, \ 0.576; \ I \approx 0.584; \ N = 100$

Section 15.4 Preliminary Questions

1. (d)

2. (a) $\int_{-1}^{2} \int_{0}^{2\pi} \int_{0}^{2} f(P) \, r \, dr \, d\theta \, dz$

(b) $\int_{-2}^{0} \int_{0}^{2\pi} \int_{0}^{\sqrt{4-z^2}} r \, dr \, d\theta \, dz$

3. (a) $\int_{0}^{2\pi} \int_{0}^{\pi} \int_{0}^{4} f(P) \, \rho^2 \sin\phi \, d\rho \, d\phi \, d\theta$

(b) $\int_{0}^{2\pi} \int_{0}^{\pi} \int_{4}^{5} f(P) \, \rho^2 \sin\phi \, d\rho \, d\phi \, d\theta$

(c) $\int_{0}^{2\pi} \int_{\pi/2}^{\pi} \int_{0}^{2} f(P) \, \rho^2 \sin\phi \, d\rho \, d\phi \, d\theta$

4. $\Delta A \approx r(\Delta r \, \Delta\theta)$, and the factor r appears in $dA = r \, dr \, d\theta$ in the Change of Variables formula.

Section 15.4 Exercises

1.

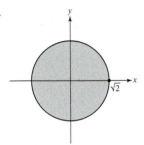

$$\iint_{D} \sqrt{x^2 + y^2} \, dA = \frac{4\sqrt{2}\pi}{3}$$

3.

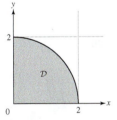

$$\iint_{D} xy \, dA = 2$$

5.

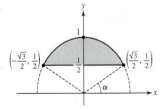

$$\iint_{D} y \left(x^2 + y^2\right)^{-1} dA = \sqrt{3} - \frac{\pi}{3} \approx 0.685$$

7.

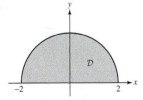

$$\int_{-2}^{2} \int_{0}^{\sqrt{4-x^2}} (x^2 + y^2) \, dy \, dx = 4\pi$$

9.

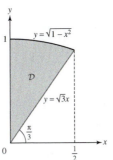

$$\int_{0}^{1/2} \int_{\sqrt{3}x}^{\sqrt{1-x^2}} x \, dy \, dx = \frac{1}{3}\left(1 - \frac{\sqrt{3}}{2}\right) \approx 0.045$$

11.

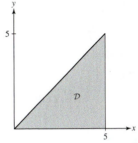

$$\int_{0}^{\pi/4} \left(\int_{0}^{5/\cos(\theta)} \left(r^2 \cos(\theta)\right) dr \right) d\theta = \frac{125}{3}$$

13.

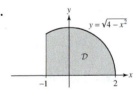

$$\int_{-1}^{2} \int_{0}^{\sqrt{4-x^2}} (x^2 + y^2) \, dy \, dx = \frac{\sqrt{3}}{2} + \frac{8\pi}{3} \approx 9.244$$

15. $\frac{1}{4}$ **17.** $\frac{1}{2}$ **19.** 0   **21.** 18   **23.** $\frac{48\pi - 32}{9} \approx 13.2$

25. (a) $W: 0 \le \theta \le 2\pi, \ 0 \le r \le 2, \ 2 \le z \le 6 - r^2$ (b) 8π

27. $\frac{405\pi}{2} \approx 636.17$ **29.** $\frac{2}{3}$ **31.** 243π

33. $\int_{0}^{2\pi} \int_{0}^{1} \int_{0}^{4} f(r\cos\theta, \ r\sin\theta, \ z) \, r \, dz \, dr \, d\theta$

35. $\int_{0}^{\pi} \int_{0}^{1} \int_{0}^{r^2} f(r\cos\theta, \ r\sin\theta, \ z) \, r \, dz \, dr \, d\theta$

37. $z = \frac{H}{R} r; \ V = \frac{\pi R^2 H}{3}$ **39.** 16π

41. $V =$
$2\int_{0}^{2\pi} \left(\int_{b}^{a} \left(\int_{0}^{\sqrt{a^2 - r^2}} (r) \, dz \right) dr \right) d\theta = \frac{4}{3}\pi \left(a^2 - b^2\right)^{3/2} = \frac{4}{3}\pi r h^3$

43. $V = \int_{0}^{2\pi} \left(\int_{0}^{\pi/3} \left(\int_{\sec(\phi)}^{2} \left(\rho^2 \sin(\phi)\right) d\rho \right) d\phi \right) d\theta = \frac{5\pi}{3}$

45. $-\frac{\pi}{16}$ **47.** $\frac{8\pi}{15}$ **49.** $\frac{8\pi}{5}$ **51.** $\frac{5\pi}{8}$ **53.** π **55.** $\frac{4\pi a^3}{3}$

57. (b) $J = \int_{0}^{2\pi} \left(\int_{0}^{\infty} \left(e^{-r^2} r\right) dr \right) d\theta = \pi$

Section 15.5 Preliminary Questions

1. 5 kg/m³   **2.** (a)

3. The probability that $0 \le X \le 1$ and $0 \le Y \le 1$; the probability that $0 \le X + Y \le 1$

Section 15.5 Exercises

1. $\frac{2}{3}$ **3.** $4\left(1 - e^{-100}\right) \times 10^{-6}$ C $\approx 4 \times 10^{-6}$ C

5. $10{,}000 - 18{,}000e^{-4/5} \approx 1{,}912$

7. $25\pi \left(3 \times 10^{-8} \text{ C}\right) \approx 2.356 \times 10^{-6}$ C

9. $\approx 2.593 \times 10^{10}$ kg   **11.** $\left(0, \frac{2}{5}\right)$ **13.** $\left(\frac{4R}{3\pi}, \frac{4R}{3\pi}\right)$

15. $(0.555, 0)$ **17.** $\left(0, 0, \frac{3R}{8}\right)$ **19.** $\left(0, 0, \frac{9}{8}\right)$

21. $\left(0, 0, \frac{13}{2\left(17 - 6\sqrt{6}\right)}\right)$ **23.** $(2, 1)$ **25.** $\left(\frac{1}{6}, \frac{1}{6}\right)$ **27.** $\frac{16}{15\pi}$

29. (a) $\frac{M}{4ab}$ (b) $I_x = \frac{Mb^2}{3}$; $I_0 = \frac{M(a^2 + b^2)}{3}$ (c) $\frac{b}{\sqrt{3}}$

31. $I_0 = 8{,}000$ kg $\cdot$ m^2; $I_x = 4{,}000$ kg $\cdot$ m^2

33. $\frac{9}{2}$ **35.** $\frac{243}{20}$ **37.** $\left(\frac{a}{2}, \frac{2b}{5}\right)$ **39.** $\frac{a^2b^4}{60}$

41. $I_x = \frac{MR^2}{4}$; kinetic energy required is $\frac{25MR^2}{2}$ J

47. (a) $I = 182.5$ g $\cdot$ cm^2 (b) $\omega \approx 126.92$ rad/s

49. $\frac{13}{72}$ **51.** $\frac{1}{64}$ **53.** $C = 15$; probability is $\frac{5}{8}$.

55. (a) $C = 4$ (b) $\frac{1}{48\pi} + \frac{1}{32} \approx 0.038$

Section 15.6 Preliminary Questions

1. (b)

2. (a) $G(1, 0) = (2, 0)$ (b) $G(1, 1) = (1, 3)$

(c) $G(2, 1) = (3, 3)$

3. Area $(G(R)) = 36$ **4.** Area $(G(R)) = 0.06$

Section 15.6 Exercises

1. (a) Image of the u-axis is the line $y = \frac{1}{2}x$; image of the v-axis is the y-axis.

(b) The parallelogram with vertices $(0, 0), (10, 5)$ $(10, 2), (0, 7)$

(c) The segment joining the points $(2, 3)$ and $(10, 8)$

(d) The triangle with vertices $(0, 1), (2, 1),$ and $(2, 2)$

3. G is not one-to-one; G is one-to-one on the domain $\{(u, v) : u \geq 0\}$, and G is one-to-one on the domain $\{(u, v) : u \leq 0\}$.

(a) The positive x-axis including the origin and the y-axis, respectively

(b) The rectangle $[0, 1] \times [-1, 1]$

(c) The curve $y = \sqrt{x}$ for $0 \leq x \leq 1$

(d)

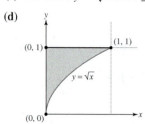

5. $y = 3x - c$ **7.** $y = \frac{17}{6}x$ **11.** $\text{Jac}(G) = 1$

13. $\text{Jac}(G) = -10$ **15.** $\text{Jac}(G) = 1$ **17.** $\text{Jac}(G) = 4$

19. $G(u, v) = (4u + 2v, u + 3v)$ **21.** $\frac{2329}{12} \approx 194.08$

23. (a) Area $(G(R)) = 105$ (b) Area $(G(R)) = 126$

25. $\text{Jac}(G) = \frac{2u}{v}$; for $R = [1, 4] \times [1, 4]$, area $(G(R)) = 15 \ln 4$

27.

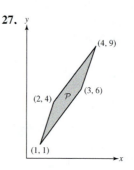

$G(u, v) = (1 + 2u + v, \ 1 + 5u + 3v)$

29. 82   **31.** 80   **33.** $\frac{56}{45}$ **35.** $\frac{\pi(e^{36} - 1)}{6}$

37.

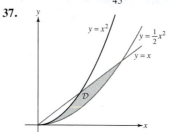

$$\iint_D y^{-1} \, dx \, dy = 1$$

39. $\iint_D e^{xy} \, dA = (e^{20} - e^{10}) \ln 2$

41. (b) $-\frac{1}{x + y}$ (c) $I = 9$ **45.** $\frac{\pi^2}{8}$

Chapter 15 Review

1. (a) $S_{2,3} = 240$ (b) $S_{2,3} = 510$ (c) 520

3. $S_{4,4} = 2.9375$ **5.** $\frac{32}{3}$ **7.** $\frac{\sqrt{3} - 1}{2}$

9.

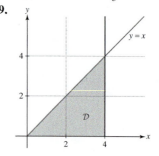

$$\iint_D \cos y \, dA = 1 - \cos 4$$

11.

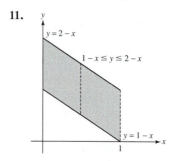

$$\iint_D e^{x+2y} \, dA = \frac{1}{2}e(e + 1)(e - 1)^2$$

37. $\frac{1}{2\pi}$ **39.** $2e - 4 \approx 1.437$

41. $S_{N,N,N} \approx 0.561,\ 0.572,\ 0.576;\ I \approx 0.584;\ N = 100$

Section 15.4 Preliminary Questions

1. (d)

2. (a) $\int_{-1}^{2} \int_{0}^{2\pi} \int_{0}^{2} f(P)\, r\, dr\, d\theta\, dz$

(b) $\int_{-2}^{0} \int_{0}^{2\pi} \int_{0}^{\sqrt{4-z^2}} r\, dr\, d\theta\, dz$

3. (a) $\int_{0}^{2\pi} \int_{0}^{\pi} \int_{0}^{4} f(P)\, \rho^2 \sin\phi\, d\rho\, d\phi\, d\theta$

(b) $\int_{0}^{2\pi} \int_{0}^{\pi} \int_{4}^{5} f(P)\, \rho^2 \sin\phi\, d\rho\, d\phi\, d\theta$

(c) $\int_{0}^{2\pi} \int_{\pi/2}^{\pi} \int_{0}^{2} f(P)\, \rho^2 \sin\phi\, d\rho\, d\phi\, d\theta$

4. $\Delta A \approx r(\Delta r\, \Delta\theta)$, and the factor r appears in $dA = r\, dr\, d\theta$ in the Change of Variables formula.

Section 15.4 Exercises

1.

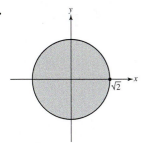

$$\iint_{D} \sqrt{x^2 + y^2}\, dA = \frac{4\sqrt{2}\pi}{3}$$

3.

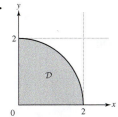

$$\iint_{D} xy\, dA = 2$$

5.

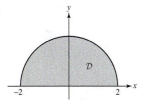

$$\iint_{D} y\left(x^2 + y^2\right)^{-1} dA = \sqrt{3} - \frac{\pi}{3} \approx 0.685$$

7.

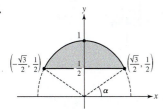

$$\int_{-2}^{2} \int_{0}^{\sqrt{4-x^2}} (x^2 + y^2)\, dy\, dx = 4\pi$$

9.

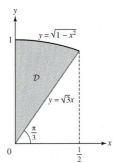

$$\int_{0}^{1/2} \int_{\sqrt{3}x}^{\sqrt{1-x^2}} x\, dy\, dx = \frac{1}{3}\left(1 - \frac{\sqrt{3}}{2}\right) \approx 0.045$$

11.

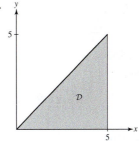

$$\int_{0}^{\pi/4} \left(\int_{0}^{5/\cos(\theta)} \left(r^2 \cos(\theta) \right) dr \right) d\theta = \frac{125}{3}$$

13.

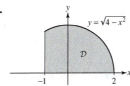

$$\int_{-1}^{2} \int_{0}^{\sqrt{4-x^2}} (x^2 + y^2)\, dy\, dx = \frac{\sqrt{3}}{2} + \frac{8\pi}{3} \approx 9.244$$

15. $\frac{1}{4}$ **17.** $\frac{1}{2}$ **19.** 0   **21.** 18   **23.** $\frac{48\pi - 32}{9} \approx 13.2$

25. (a) $W: 0 \le \theta \le 2\pi,\ 0 \le r \le 2,\ 2 \le z \le 6 - r^2$ **(b)** 8π

27. $\frac{405\pi}{2} \approx 636.17$ **29.** $\frac{2}{3}$ **31.** 243π

33. $\int_{0}^{2\pi} \int_{0}^{1} \int_{0}^{4} f(r\cos\theta,\ r\sin\theta,\ z)\, r\, dz\, dr\, d\theta$

35. $\int_{0}^{\pi} \int_{0}^{1} \int_{0}^{r^2} f(r\cos\theta,\ r\sin\theta,\ z)\, r\, dz\, dr\, d\theta$

37. $z = \frac{H}{R}r;\ V = \frac{\pi R^2 H}{3}$ **39.** 16π

41. $V =$
$2\int_{0}^{2\pi} \left(\int_{b}^{a} \left(\int_{0}^{\sqrt{a^2-r^2}} (r)\, dz \right) dr \right) d\theta = \frac{4}{3}\pi \left(a^2 - b^2\right)^{3/2} = \frac{4}{3}\pi r h^3$

43. $V = \int_{0}^{2\pi} \left(\int_{0}^{\pi/3} \left(\int_{\sec(\phi)}^{2} \left(\rho^2 \sin(\phi) \right) d\rho \right) d\phi \right) d\theta = \frac{5\pi}{3}$

45. $-\frac{\pi}{16}$ **47.** $\frac{8\pi}{15}$ **49.** $\frac{8\pi}{5}$ **51.** $\frac{5\pi}{8}$ **53.** π **55.** $\frac{4\pi a^3}{3}$

57. (b) $J = \int_{0}^{2\pi} \left(\int_{0}^{\infty} \left(e^{-r^2} r \right) dr \right) d\theta = \pi$

Section 15.5 Preliminary Questions

1. 5 kg/m^3 **2. (a)**

3. The probability that $0 \le X \le 1$ and $0 \le Y \le 1$; the probability that $0 \le X + Y \le 1$

Section 15.5 Exercises

1. $\frac{2}{3}$ **3.** $4\left(1 - e^{-100}\right) \times 10^{-6}$ C $\approx 4 \times 10^{-6}$ C

5. $10{,}000 - 18{,}000e^{-4/5} \approx 1{,}912$

7. $25\pi\left(3 \times 10^{-8}\text{ C}\right) \approx 2.356 \times 10^{-6}$ C

9. $\approx 2.593 \times 10^{10}$ kg   **11.** $\left(0, \frac{2}{5}\right)$ **13.** $\left(\frac{4R}{3\pi}, \frac{4R}{3\pi}\right)$

15. $(0.555, 0)$ **17.** $\left(0, 0, \frac{3R}{8}\right)$ **19.** $\left(0, 0, \frac{9}{8}\right)$

21. $\left(0, 0, \frac{13}{2\left(17-6\sqrt{6}\right)}\right)$ **23.** $(2, 1)$ **25.** $\left(\frac{1}{6}, \frac{1}{6}\right)$ **27.** $\frac{16}{15\pi}$

29. (a) $\frac{M}{4ab}$ **(b)** $I_x = \frac{Mb^2}{3}$; $I_0 = \frac{M(a^2 + b^2)}{3}$ **(c)** $\frac{b}{\sqrt{3}}$

31. $I_0 = 8{,}000$ kg $\cdot$ m^2; $I_x = 4{,}000$ kg $\cdot$ m^2

33. $\frac{9}{2}$ **35.** $\frac{243}{20}$ **37.** $\left(\frac{a}{2}, \frac{2b}{5}\right)$ **39.** $\frac{a^2b^4}{60}$

41. $I_x = \frac{MR^2}{4}$; kinetic energy required is $\frac{25MR^2}{2}$ J

47. (a) $I = 182.5$ g $\cdot$ cm^2 **(b)** $\omega \approx 126.92$ rad/s

49. $\frac{13}{72}$ **51.** $\frac{1}{64}$ **53.** $C = 15$; probability is $\frac{5}{8}$.

55. (a) $C = 4$ **(b)** $\frac{1}{48\pi} + \frac{1}{32} \approx 0.038$

Section 15.6 Preliminary Questions

1. (b)

2. (a) $G(1, 0) = (2, 0)$ **(b)** $G(1, 1) = (1, 3)$

(c) $G(2, 1) = (3, 3)$

3. Area $(G(R)) = 36$ **4.** Area $(G(R)) = 0.06$

Section 15.6 Exercises

1. (a) Image of the u-axis is the line $y = \frac{1}{2}x$; image of the v-axis is the y-axis.

(b) The parallelogram with vertices $(0, 0)$, $(10, 5)$ $(10, 2)$, $(0, 7)$

(c) The segment joining the points $(2, 3)$ and $(10, 8)$

(d) The triangle with vertices $(0, 1)$, $(2, 1)$, and $(2, 2)$

3. G is not one-to-one; G is one-to-one on the domain $\{(u, v) : u \geq 0\}$, and G is one-to-one on the domain $\{(u, v) : u \leq 0\}$.

(a) The positive x-axis including the origin and the y-axis, respectively

(b) The rectangle $[0, 1] \times [-1, 1]$

(c) The curve $y = \sqrt{x}$ for $0 \leq x \leq 1$

(d)

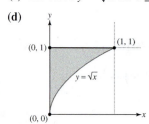

5. $y = 3x - c$ **7.** $y = \frac{17}{6}x$ **11.** $\text{Jac}(G) = 1$

13. $\text{Jac}(G) = -10$ **15.** $\text{Jac}(G) = 1$ **17.** $\text{Jac}(G) = 4$

19. $G(u, v) = (4u + 2v, u + 3v)$ **21.** $\frac{2329}{12} \approx 194.08$

23. (a) Area $(G(R)) = 105$ **(b)** Area $(G(R)) = 126$

25. $\text{Jac}(G) = \frac{2u}{v}$; for $R = [1, 4] \times [1, 4]$, area $(G(R)) = 15\ln 4$

27.

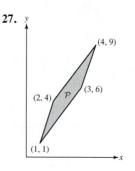

$$G(u, v) = (1 + 2u + v, 1 + 5u + 3v)$$

29. 82   **31.** 80   **33.** $\frac{56}{45}$ **35.** $\frac{\pi(e^{36} - 1)}{6}$

37.

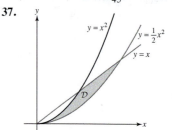

$$\iint_D y^{-1}\, dx\, dy = 1$$

39. $\iint_D e^{xy}\, dA = (e^{20} - e^{10})\ln 2$

41. (b) $-\frac{1}{x + y}$ **(c)** $I = 9$ **45.** $\frac{\pi^2}{8}$

Chapter 15 Review

1. (a) $S_{2,3} = 240$ **(b)** $S_{2,3} = 510$ **(c)** 520

3. $S_{4,4} = 2.9375$ **5.** $\frac{32}{3}$ **7.** $\frac{\sqrt{3} - 1}{2}$

9.

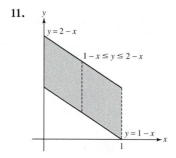

$$\iint_D \cos y\, dA = 1 - \cos 4$$

11.

$$\iint_D e^{x+2y}\, dA = \frac{1}{2}e(e + 1)(e - 1)^2$$

13.

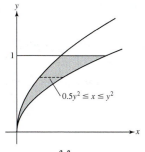

$$\iint_D ye^{1+x}\, dA = 0.5(e^2 - 2e^{1.5} + e)$$

15. $\int_0^9 \int_{-\sqrt{9-y}}^{\sqrt{9-y}} f(x,\ y)\, dx\, dy$ **17.** $\frac{1}{24}$ **19.** $18(\sqrt{2} - 1)$

21. $1 - \cos 1$ **23.** 6π **25.** $\pi/2$ **27.** 10   **29.** $\frac{\pi}{4} + \frac{2}{3}$ **31.** π

33. $\frac{1}{4}$ **35.** $\int_0^{\pi/2} \int_0^1 \int_0^r zr\, dz\, dr\, d\theta = \pi/16$ **37.** $\frac{2\pi(-1 + e^8)}{3e^8}$

41. $\frac{256\pi}{15} \approx 53.62$ **43.** 1280π **45.** $\left(-\frac{1}{4}R,\ 0,\ \frac{5}{8}H\right)$

47. $\left(-\frac{2}{11\pi}R,\ -\frac{2}{11\pi}R(2 - \sqrt{3}),\ \frac{1}{2}H\right)$.   **49.** $\left(0,\ 0,\ \frac{2}{3}\right)$

51. $\left(\frac{8}{15},\ \frac{16}{15\pi},\ \frac{16}{15\pi}\right)$ **53.** $\frac{19}{33}$ **55.** $\frac{4}{7}$

57. $G(u,\ v) = (3u + v,\ -u + 4v)$; Area $(G(R)) = 156$

59. Area$(D) \approx \frac{1}{5}$

61. (a)

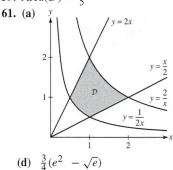

(d) $\frac{3}{4}(e^2 - \sqrt{e})$

Chapter 16

Section 16.1 Preliminary Questions

1. (b)   **2.**

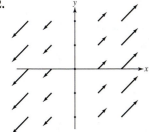

3. $\mathbf{F} = \langle 0,\ -z,\ y \rangle$ **4.** $f_1(x,\ y,\ z) = xyz + 1$

Section 16.1 Exercises

1. $\mathbf{F}(1,\ 2) = \langle 1,\ 1 \rangle$, $\mathbf{F}(-1,\ -1) = \langle 1,\ -1 \rangle$

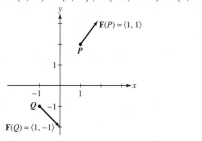

3. $\mathbf{F}(P) = \langle 0,\ 1,\ 0 \rangle$, $\mathbf{F}(Q) = \langle 2,\ 0,\ 2 \rangle$

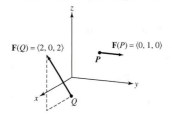

5. $\mathbf{F} = \langle 1,\ 0 \rangle$

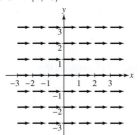

7. $\mathbf{F} = x\mathbf{i}$

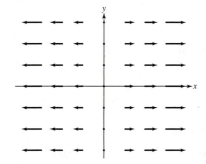

9. $\mathbf{F}(x,\ y) = \langle 0,\ x \rangle$

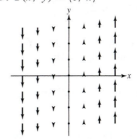

11. $\mathbf{F} = \left\langle \dfrac{x}{x^2 + y^2},\ \dfrac{y}{x^2 + y^2} \right\rangle$

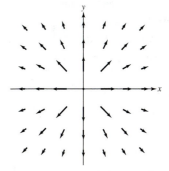

13. Plot (D)   **15.** Plot (B)   **17.** Plot (C)   **19.** Plot (B)

21. $(0,\ y)$ **23.** div$(\mathbf{F}) = y + z$, curl$(\mathbf{F}) = \left\langle y,\ 3x^2,\ -x \right\rangle$

25. div$(\mathbf{F}) = 1 - 4xz - x + 2x^2z$, curl$(\mathbf{F}) = \left\langle -1,\ 2x^2 - 2xz^2,\ -y \right\rangle$

27. div$(\mathbf{F}) = 0$, curl$(\mathbf{F}) = \left\langle 1 - 3z^2,\ 1 - 2x,\ 1 + 2y \right\rangle$

29. div$(\mathbf{F}) = 0$, curl$(\mathbf{F}) = \left\langle 0,\ \sin x,\ \cos x - e^y \right\rangle$

39. $f(x, y) = \frac{1}{2}x^2 + \frac{1}{2}y^2 + K$

41. $f(x, y) = e^{xy} + K$

43. $f(x, y, z) = xyz^2 + K$ **45.** $f(x, y, z) = \sin(xyz) + K$

47. $f(x, y, z) = ax + by + cz + K$

49. (b) $e_P(1, 1) = \langle -2, -1 \rangle / \sqrt{5}$;

(c) $f(x, y, z) = \sqrt{(x-a)^2 + (y-b)^2}$

51. (A)

Section 16.2 Preliminary Questions

1. 50   **2.** (a), (c), (d), (e)

3. (a) True

(b) False. Reversing the orientation of the curve changes the sign of the vector line integral.

4. (a) 0   **(b)** -5

Section 16.2 Exercises

1. (a) $f(\mathbf{r}(t)) = 6t + 4t^2$; $ds = 2\sqrt{11}\, dt$

(b) $\int_0^1 \left(6t + 4t^2\right) 2\sqrt{11}\, dt = \frac{26\sqrt{11}}{3}$

3. (a) $\mathbf{F}(\mathbf{r}(t)) = \left\langle t^{-2}, t^2 \right\rangle$; $d\mathbf{r} = \left\langle 1, -t^{-2} \right\rangle dt$

(b) $\int_1^2 \left(t^{-1} - 1\right) dt = -\frac{1}{2}$

5. $\sqrt{2}\left(\pi + \frac{\pi^3}{3}\right)$ **7.** $\pi^2/2$ **9.** 2.8   **11.** $\frac{128\sqrt{29}}{3} \approx 229.8$

13. $\frac{\sqrt{3}}{2}(e - 1) \approx 1.488$ **15.** $\frac{2}{3}\left((e^2 + 5)^{3/2} - 2^{3/2}\right)$

17. 39; the distance between $(8, -6, 24)$ and $(20, -15, 60)$

19. $\frac{16}{3}$ **21.** 0   **23.** $2(e^2 - e^{-2}) - (e - e^{-1}) \approx 12.157$ **25.** $\frac{10}{9}$

27. $-\frac{8}{3}$ **29.** $\frac{13}{2}$ **31.** $\frac{\pi}{2}$ **33.** 339.5587   **35.** $2 - e - \frac{1}{e}$

37. (a) -8 **(b)** -11 **(c)** -16

39. ≈ 7.6; ≈ 4 **41.** (A) Zero, (B) Negative, (C) Zero   **43.** 64π g

45. $\approx 10.4 \times 10^{-6}$ C   **47.** $\approx 22{,}743.10$ volts   **49.** $\approx -10{,}097$ volts

51. 1   **53.** $\frac{27}{28}$ **55. (a)** ABC   **(b)** CBA

59. $\frac{1}{3}\left((4\pi^2 + 1)^{3/2} - 1\right) \approx 85.5 \times 10^{-6}$ C   **65.** 18   **67.** $e - 1$

Section 16.3 Preliminary Questions

1. Closed

2. (a) Conservative vector fields   **(b)** All vector fields

(c) Conservative vector fields   **(d)** All vector fields

(e) Conservative vector fields   **(f)** All vector fields

(g) Conservative vector fields and *some* other vector fields

3. (a) Always true   **(b)** Always true

(c) True under additional hypotheses on D

4. (a) 4   **(b)** -4

Section 16.3 Exercises

1. 0   **3.** $-\frac{9}{4}$ **5.** $32e - 1$ **7.** $f(x, y, z) = zx + y$

9. $f(x, y, z) = y^2x + e^z y$

11. The vector field is not conservative.

13. $f(x, y, z) = z\tan x + zy$ **15.** $f(x, y, z) = x^2y + 5x - 4zy$

17. 16   **19.** 1   **21.** 6   **23.** $\frac{2}{3}$; 0   **25.** 6.2×10^9 J

27. (a) $f(x, y, z) = -gz$ **(b)** ≈ 82.8 m/s

29. (A) 2π, (B) 2π, (C) 0, (D) -2π, (E) 4π

Section 16.4 Preliminary Questions

1. 50

2. A distortion factor that indicates how much the area of R_{ij} is altered under the map G.

3. Area$(\mathcal{S}) \approx 0.0006$ **4.** $\iint_{\mathcal{S}} f(x, y, z)\, dS \approx 0.6$

5. Area$(\mathcal{S}) = 20$ **6.** $\left\langle \frac{2}{3}, \frac{2}{3}, \frac{1}{3} \right\rangle$

Section 16.4 Exercises

1. (a) v   **(b)** iii   **(c)** i   **(d)** iv   **(e)** ii

3. (a) $\mathbf{T}_u = \langle 2, 1, 3 \rangle$, $\mathbf{T}_v = \langle 0, -1, 1 \rangle$,
$\mathbf{n}(u, v) = \langle 4, -2, -2 \rangle$

(b) Area$(\mathcal{S}) = 4\sqrt{6}$ **(c)** $\iint_{\mathcal{S}} f(x, y, z)\, dS = \frac{32\sqrt{6}}{3}$

5. (a) $\mathbf{T}_x = \langle 1, 0, y \rangle$,
$\mathbf{T}_y = \langle 0, 1, x \rangle$, $\mathbf{N}(x, y) = \langle -y, -x, 1 \rangle$

(b) $\frac{(2\sqrt{2} - 1)\pi}{6}$ **(c)** $\frac{\sqrt{2}+1}{15}$

7. $\mathbf{T}_u = \langle 2, 1, 3 \rangle$, $\mathbf{T}_v = \langle 1, -4, 0 \rangle$, $\mathbf{N}(u, v) = 3\langle 4, 1, -3 \rangle$, $4x + y - 3z = 0$

9. $\mathbf{T}_\theta = \langle -\sin\theta \sin\phi, \cos\theta \sin\phi, 0 \rangle$,
$\mathbf{T}_\phi = \langle \cos\theta \cos\phi, \sin\theta \cos\phi, -\sin\phi \rangle$,
$\mathbf{N}(u, v) = -\cos\theta \sin^2\phi \mathbf{i} - \sin\theta \sin^2\phi \mathbf{j} - \sin\phi \cos\phi \mathbf{k}$,
$y + z = \sqrt{2}$

11. Area$(\mathcal{S}) \approx 0.2078$ **13.** $\frac{\sqrt{2}}{5}$ **15.** $\frac{37\sqrt{37} - 1}{4} \approx 56.02$ **17.** $\frac{\pi}{6}$

19. $4\pi(1 - e^{-4})$ **21.** $\frac{\sqrt{3}}{6}$ **23.** $\frac{7\pi}{3}$ **25.** $\frac{5\sqrt{10}}{27} - \frac{1}{54}$

27. Area$(\mathcal{S}) = 16$ **29.** $3e^3 - 6e^2 + 3e + 1 \approx 25.08$

31. Area$(\mathcal{S}) = 4\pi R^2$

33. (a) Area$(\mathcal{S}) \approx 1.0780$ **(b)** ≈ 0.09814

35. Area$(\mathcal{S}) = \frac{5\sqrt{29}}{4} \approx 6.73$ **37.** Area$(\mathcal{S}) = \pi$ **39.** 48π

43. Area$(\mathcal{S}) = \frac{\pi}{6}\left(17\sqrt{17} - 1\right) \approx 36.18$ **47.** $4\pi^2 ab$

49. $f(r) = -\frac{Gm}{2Rr}\left(\sqrt{R^2 + r^2} - |R - r|\right)$

Section 16.5 Preliminary Questions

1. (b)   **2.** (c)   **3.** (a)   **4.** (b)

5. (a) 0   **(b)** π **(c)** π

6. $\approx 0.05\sqrt{2} \approx 0.0707$ **7.** 0

Section 16.5 Exercises

1. (a) $\mathbf{N} = \langle 2v, -4uv, 1 \rangle$, $\mathbf{F} \cdot \mathbf{N} = 2v^3 + u$

(b) $\frac{4}{\sqrt{69}}$ **(c)** 265

3. 4   **5.** -4 **7.** $\frac{27}{12}(3\pi + 4)$ **9.** $\frac{693}{5}$ **11.** $\frac{11}{12}$ **13.** $\frac{9\pi}{4}$

15. $(e - 1)^2$ **17.** 270

19. (a) $18\pi e^{-3}$ **(b)** $\frac{\pi}{2}e^{-1}$

21. $\left(2 - \frac{6}{\sqrt{13}}\right)\pi k$ **23.** $\frac{2\pi}{3}$ m^3/s   **25.** 4π **27.** $\frac{16\pi}{3}$

29. (a) 1   **(b)** 1

33. $\Phi(t) = -1.56 \times 10^{-5}e^{-0.1t}$ T-m^2;
voltage drop $= -1.56 \times 10^{-6}e^{-0.1t}$ V

35. The flow is greatest at $z = 3$.

13.

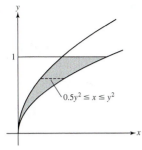

$$\iint_D ye^{1+x}\,dA = 0.5(e^2 - 2e^{1.5} + e)$$

15. $\int_0^9 \int_{-\sqrt{9-y}}^{\sqrt{9-y}} f(x,\ y)\,dx\,dy$ **17.** $\frac{1}{24}$ **19.** $18(\sqrt{2}-1)$

21. $1 - \cos 1$ **23.** 6π **25.** $\pi/2$ **27.** 10   **29.** $\frac{\pi}{4} + \frac{2}{3}$ **31.** π

33. $\frac{1}{4}$ **35.** $\int_0^{\pi/2} \int_0^1 \int_0^r zr\,dz\,dr\,d\theta = \pi/16$ **37.** $\frac{2\pi(-1+e^8)}{3e^8}$

41. $\frac{256\pi}{15} \approx 53.62$ **43.** 1280π **45.** $\left(-\frac{1}{4}R,\ 0,\ \frac{5}{8}H\right)$

47. $\left(-\frac{2}{11\pi}R,\ -\frac{2}{11\pi}R(2-\sqrt{3}),\ \frac{1}{2}H\right)$.   **49.** $\left(0,\ 0,\ \frac{2}{3}\right)$

51. $\left(\frac{8}{15},\ \frac{16}{15\pi},\ \frac{16}{15\pi}\right)$ **53.** $\frac{19}{33}$ **55.** $\frac{4}{7}$

57. $G(u,\ v) = (3u+v,\ -u+4v)$; Area $(G(R)) = 156$

59. Area$(D) \approx \frac{1}{5}$

61. (a)

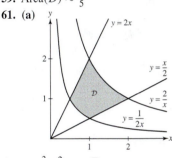

(d) $\frac{3}{4}(e^2 - \sqrt{e})$

Chapter 16

Section 16.1 Preliminary Questions

1. (b)   **2.**

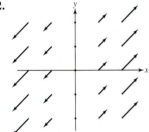

3. $\mathbf{F} = \langle 0,\ -z,\ y \rangle$ **4.** $f_1(x,\ y,\ z) = xyz + 1$

Section 16.1 Exercises

1. $\mathbf{F}(1,\ 2) = \langle 1,\ 1 \rangle$, $\mathbf{F}(-1,\ -1) = \langle 1,\ -1 \rangle$

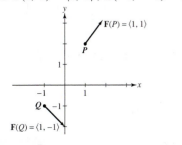

3. $\mathbf{F}(P) = \langle 0,\ 1,\ 0 \rangle$, $\mathbf{F}(Q) = \langle 2,\ 0,\ 2 \rangle$

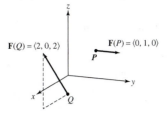

5. $\mathbf{F} = \langle 1,\ 0 \rangle$

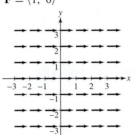

7. $\mathbf{F} = x\mathbf{i}$

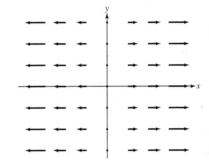

9. $\mathbf{F}(x,\ y) = \langle 0,\ x \rangle$

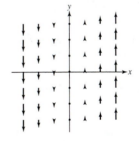

11. $\mathbf{F} = \left\langle \frac{x}{x^2+y^2},\ \frac{y}{x^2+y^2} \right\rangle$

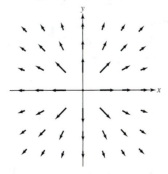

13. Plot (D)   **15.** Plot (B)   **17.** Plot (C)   **19.** Plot (B)

21. $(0, y)$ **23.** $\text{div}(\mathbf{F}) = y + z$, $\text{curl}(\mathbf{F}) = \left\langle y, 3x^2, -x \right\rangle$

25. $\text{div}(\mathbf{F}) = 1 - 4xz - x + 2x^2z$, $\text{curl}(\mathbf{F}) = \left\langle -1, 2x^2 - 2xz^2, -y \right\rangle$

27. $\text{div}(\mathbf{F}) = 0$, $\text{curl}(\mathbf{F}) = \left\langle 1 - 3z^2, 1 - 2x, 1 + 2y \right\rangle$

29. $\text{div}(\mathbf{F}) = 0$, $\text{curl}(\mathbf{F}) = \left\langle 0, \sin x, \cos x - e^y \right\rangle$

39. $f(x, y) = \frac{1}{2}x^2 + \frac{1}{2}y^2 + K$

41. $f(x, y) = e^{xy} + K$

43. $f(x, y, z) = xyz^2 + K$ **45.** $f(x, y, z) = \sin(xyz) + K$

47. $f(x, y, z) = ax + by + cz + K$

49. (b) $e_P(1, 1) = \langle -2, -1 \rangle / \sqrt{5}$;

  (c) $f(x, y, z) = \sqrt{(x-a)^2 + (y-b)^2}$

51. (A)

Section 16.2 Preliminary Questions

1. 50   **2.** (a), (c), (d), (e)

3. (a) True

(b) False. Reversing the orientation of the curve changes the sign of the vector line integral.

4. (a) 0   **(b)** -5

Section 16.2 Exercises

1. (a) $f(\mathbf{r}(t)) = 6t + 4t^2$; $ds = 2\sqrt{11}\,dt$

(b) $\int_0^1 \left(6t + 4t^2 \right) 2\sqrt{11}\,dt = \frac{26\sqrt{11}}{3}$

3. (a) $\mathbf{F}(\mathbf{r}(t)) = \left\langle t^{-2}, t^2 \right\rangle$; $d\mathbf{r} = \left\langle 1, -t^{-2} \right\rangle dt$

(b) $\int_1^2 \left(t^{-1} - 1 \right) dt = -\frac{1}{2}$

5. $\sqrt{2}\left(\pi + \frac{\pi^3}{3}\right)$ **7.** $\pi^2/2$ **9.** 2.8   **11.** $\frac{128\sqrt{29}}{3} \approx 229.8$

13. $\frac{\sqrt{3}}{2}(e - 1) \approx 1.488$ **15.** $\frac{2}{3}\left((e^2 + 5)^{3/2} - 2^{3/2}\right)$

17. 39; the distance between $(8, -6, 24)$ and $(20, -15, 60)$

19. $\frac{16}{3}$ **21.** 0   **23.** $2(e^2 - e^{-2}) - (e - e^{-1}) \approx 12.157$ **25.** $\frac{10}{9}$

27. $-\frac{8}{3}$ **29.** $\frac{13}{2}$ **31.** $\frac{\pi}{2}$ **33.** 339.5587   **35.** $2 - e - \frac{1}{e}$

37. (a) -8 **(b)** -11 **(c)** -16

39. ≈ 7.6; ≈ 4 **41.** (A) Zero, (B) Negative, (C) Zero   **43.** 64π g

45. $\approx 10.4 \times 10^{-6}$ C   **47.** $\approx 22{,}743.10$ volts   **49.** $\approx -10{,}097$ volts

51. 1   **53.** $\frac{27}{28}$ **55. (a)** ABC   **(b)** CBA

59. $\frac{1}{3}\left((4\pi^2 + 1)^{3/2} - 1\right) \approx 85.5 \times 10^{-6}$ C   **65.** 18   **67.** $e - 1$

Section 16.3 Preliminary Questions

1. Closed

2. (a) Conservative vector fields   **(b)** All vector fields

(c) Conservative vector fields   **(d)** All vector fields

(e) Conservative vector fields   **(f)** All vector fields

(g) Conservative vector fields and *some* other vector fields

3. (a) Always true   **(b)** Always true

(c) True under additional hypotheses on D

4. (a) 4   **(b)** -4

Section 16.3 Exercises

1. 0   **3.** $-\frac{9}{4}$ **5.** $32e - 1$ **7.** $f(x, y, z) = zx + y$

9. $f(x, y, z) = y^2 x + e^z y$

11. The vector field is not conservative.

13. $f(x, y, z) = z \tan x + zy$ **15.** $f(x, y, z) = x^2 y + 5x - 4zy$

17. 16   **19.** 1   **21.** 6   **23.** $\frac{2}{3}$; 0   **25.** 6.2×10^9 J

27. (a) $f(x, y, z) = -gz$ **(b)** ≈ 82.8 m/s

29. (A) 2π, (B) 2π, (C) 0, (D) -2π, (E) 4π

Section 16.4 Preliminary Questions

1. 50

2. A distortion factor that indicates how much the area of R_{ij} is altered under the map G.

3. Area$(S) \approx 0.0006$ **4.** $\iint_S f(x, y, z)\,dS \approx 0.6$

5. Area$(S) = 20$ **6.** $\left(\frac{2}{3}, \frac{2}{3}, \frac{1}{3}\right)$

Section 16.4 Exercises

1. (a) v   **(b)** iii   **(c)** i   **(d)** iv   **(e)** ii

3. (a) $\mathbf{T}_u = \langle 2, 1, 3 \rangle$, $\mathbf{T}_v = \langle 0, -1, 1 \rangle$, $\mathbf{n}(u, v) = \langle 4, -2, -2 \rangle$

(b) Area$(S) = 4\sqrt{6}$ **(c)** $\iint_S f(x, y, z)\,dS = \frac{32\sqrt{6}}{3}$

5. (a) $\mathbf{T}_x = \langle 1, 0, y \rangle$, $\mathbf{T}_y = \langle 0, 1, x \rangle$, $\mathbf{N}(x, y) = \langle -y, -x, 1 \rangle$

(b) $\frac{(2\sqrt{2} - 1)\pi}{6}$ **(c)** $\frac{\sqrt{2}+1}{15}$

7. $\mathbf{T}_u = \langle 2, 1, 3 \rangle$, $\mathbf{T}_v = \langle 1, -4, 0 \rangle$, $\mathbf{N}(u, v) = 3\langle 4, 1, -3 \rangle$, $4x + y - 3z = 0$

9. $\mathbf{T}_\theta = \langle -\sin\theta\sin\phi, \cos\theta\sin\phi, 0 \rangle$, $\mathbf{T}_\phi = \langle \cos\theta\cos\phi, \sin\theta\cos\phi, -\sin\phi \rangle$, $\mathbf{N}(u, v) = -\cos\theta\sin^2\phi\,\mathbf{i} - \sin\theta\sin^2\phi\,\mathbf{j} - \sin\phi\cos\phi\,\mathbf{k}$, $y + z = \sqrt{2}$

11. Area$(S) \approx 0.2078$ **13.** $\frac{\sqrt{2}}{5}$ **15.** $\frac{37\sqrt{37} - 1}{4} \approx 56.02$ **17.** $\frac{\pi}{6}$

19. $4\pi(1 - e^{-4})$ **21.** $\frac{\sqrt{3}}{6}$ **23.** $\frac{7\pi}{3}$ **25.** $\frac{5\sqrt{10}}{27} - \frac{1}{54}$

27. Area$(S) = 16$ **29.** $3e^3 - 6e^2 + 3e + 1 \approx 25.08$

31. Area$(S) = 4\pi R^2$

33. (a) Area$(S) \approx 1.0780$ **(b)** ≈ 0.09814

35. Area$(S) = \frac{5\sqrt{29}}{4} \approx 6.73$ **37.** Area$(S) = \pi$ **39.** 48π

43. Area$(S) = \frac{\pi}{6}\left(17\sqrt{17} - 1\right) \approx 36.18$ **47.** $4\pi^2 ab$

49. $f(r) = -\frac{Gm}{2Rr}\left(\sqrt{R^2 + r^2} - |R - r|\right)$

Section 16.5 Preliminary Questions

1. (b)   **2.** (c)   **3.** (a)   **4.** (b)

5. (a) 0   **(b)** π **(c)** π

6. $\approx 0.05\sqrt{2} \approx 0.0707$ **7.** 0

Section 16.5 Exercises

1. (a) $\mathbf{N} = \langle 2v, -4uv, 1 \rangle$, $\mathbf{F} \cdot \mathbf{N} = 2v^3 + u$

(b) $\frac{4}{\sqrt{69}}$ **(c)** 265

3. 4   **5.** -4 **7.** $\frac{27}{12}(3\pi + 4)$ **9.** $\frac{693}{5}$ **11.** $\frac{11}{12}$ **13.** $\frac{9\pi}{4}$

15. $(e - 1)^2$ **17.** 270

19. (a) $18\pi e^{-3}$ **(b)** $\frac{\pi}{2}e^{-1}$

21. $\left(2 - \frac{6}{\sqrt{13}}\right)\pi k$ **23.** $\frac{2\pi}{3}$ m^3/s   **25.** 4π **27.** $\frac{16\pi}{3}$

29. (a) 1   **(b)** 1

33. $\Phi(t) = -1.56 \times 10^{-5} e^{-0.1t}$ T-m^2; voltage drop $= -1.56 \times 10^{-6} e^{-0.1t}$ V

35. The flow is greatest at $z = 3$.

Chapter 16 Review

1. (a) $\langle -15, 8 \rangle$ (b) $\langle 4, 8 \rangle$ (c) $\langle 9, 1 \rangle$

3.

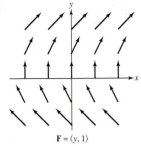

$\mathbf{F} = \langle y, 1 \rangle$

5. $\mathbf{F}(x, y) = \langle 2x, -1 \rangle$

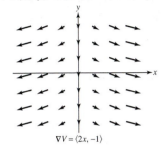

$\nabla V = \langle 2x, -1 \rangle$

7. $\operatorname{div}(\mathbf{F}) = 2x + 2y + 2z$, $\operatorname{curl}(\mathbf{F}) = \langle 0, 0, 0 \rangle$

9. $\operatorname{div}(\mathbf{F}) = 3x^2 y + y^2$, $\operatorname{curl}(\mathbf{F}) = \left\langle 2yz - 2xz, 0, z^2 - x^3 \right\rangle$

11. $\operatorname{div}(\mathbf{F}) = -1$, $\operatorname{curl}(\mathbf{F}) = \langle 0, 0, -1 \rangle$

13. $\operatorname{div}(\mathbf{F}) = 4x^2 e^{-x^2-y^2-z^2} + 4y^2 e^{-x^2-y^2-z^2} + 4z^2 e^{-x^2-y^2-z^2} - 6e^{-x^2-y^2-z^2}$; $\operatorname{curl}(\mathbf{F}) = \langle 0, 0, 0 \rangle$

15. $\operatorname{curl}(\mathbf{F}) = \langle -2z, 0, 2y \rangle$ **19.** $f(x, y) = x^4 y^5$

21. $\mathbf{F}$ is conservative, $f(x, y, z) = 2x + 4y + e^z$

23. $f(x, y) = \frac{x^4 y^4}{4}$

25. $\mathbf{F}$ is conservative, $f(x, y, z) = y \ln \left(x^2 + z \right)$

27. $\mathbf{F} = \langle 1 + by, 1 + bx \rangle$ **29.** -2

31. $\sqrt{2}\,(\sin 3 - 2 \cos 3 + \sin 1 + 2 \cos 1 + 4) \approx 11.375$

33. (a) 2 (b) 0 **37.** $\frac{1}{3}$ **39.** $4 \ln \left(1 + (\ln 2)^4 + e^2 \right) \approx 8.616$

41. $\frac{52}{29} \approx 1.79$ **43.** $3e^{3/2} - \frac{15}{2} \approx 5.945$

45. Area $(S) =$
$$\int_{-1}^{1} \int_{-1}^{1} \sqrt{125u^2 - 100uv + 425v^2 + 81}\, du\, dv \approx 62.911$$

47. $\frac{4}{9} \left(2\sqrt{2} - 1 \right)$

49. (a) Zero since $f(x, y, z) = y^3$ is odd and symmetric about the xz-plane

(b) Positive since $f(x, y, z) = z^3$ is non-negative

(c) Zero since $f(-x, y, z) = -xyz = -f(x, y, z)$ is symmetric about the yz-plane

(d) Negative since $f(x, y, z) = z^2 - 2$ is negative

51. (a) $\mathbf{N} = \left\langle \frac{2}{\sqrt{5}}, \frac{2\sqrt{3}}{\sqrt{5}}, 2 \right\rangle$

(b) Area $(G(R)) = \frac{6}{\sqrt{5}} \cdot 0.1 \cdot 0.05 \approx 0.0134$

53. $\int \int_{S} \mathbf{F} \cdot d\mathbf{S} =$
$$\int_{0}^{1} \left(\int_{0}^{1} (-4(2u - v) - 2(2v + 5) + 7(-u - 3v))\, dv \right) du = -28$$

55. $\frac{1}{4} - \frac{\sinh 1}{3}$ **57.** $\frac{\sqrt{6}\pi}{9}$ **59.** 27π **61.** $\frac{9\sqrt{3}}{4}$

Chapter 17

Section 17.1 Preliminary Questions

1. $\mathbf{F} = \left\langle -e^y, x^2 \right\rangle$

2.

$\mathcal{C}$

3. Yes **4.** (a), (c)

Section 17.1 Exercises

1. $\oint_{\mathcal{C}} xy\, dx + y\, dy =$
$$\int_{0}^{2\pi} (\cos\theta \sin\theta\,(-\sin\theta) + \sin\theta \cos\theta)\, d\theta = 0 =$$
$$\int_{-1}^{1} \int_{-\sqrt{1-x^2}}^{\sqrt{1-x^2}} (0 - x)\, dy\, dx = \int \int_{\mathcal{D}} \left(\frac{\partial}{\partial x} y - \frac{\partial}{\partial y}(xy) \right) dA$$

3. 0 **5.** $-\frac{\pi}{4}$ **7.** $\frac{1}{6}$ **9.** $\frac{(e^2 - 1)(e^4 - 5)}{2}$

11. (a) $V(x, y) = x^2 e^y$ **13.** $I = 34$ **15.** $\frac{1}{2}$ **17.** $\frac{8}{3}$

19. (c) $A = \frac{3}{2}$ **23.** $9 + \frac{15\pi}{2}$ **25.** 214π

27. (A) Zero (B) Positive (C) Negative (D) Zero

29. -0.10 **31.** $R = \sqrt{\frac{2}{3}}$ **33.** Triangle (A), 3; Polygon (B), 12

35. -4 **37.** 5π **39.** $\frac{9\pi}{4}$ **41.** 29.2 buffalo per minute

Section 17.2 Preliminary Questions

1.

(A)   (B)

2. (a)

3. A vector field $\mathbf{A}$ such that $\mathbf{F} = \operatorname{curl}(\mathbf{A})$ is a vector potential for $\mathbf{F}$

4. (b)

5. $\mathbf{F}$ must be the curl of some other vector field $\mathbf{A}$ (see Theorem 2).

Section 17.2 Exercises

1. $\int \int_{\mathcal{C}} \mathbf{F} \cdot d\mathbf{s} = \int \int_{S} \operatorname{curl}(\mathbf{F}) \cdot d\mathbf{S} = \pi$

3. $\int \int_{\mathcal{C}} \mathbf{F} \cdot d\mathbf{s} = \int \int_{S} \operatorname{curl}(\mathbf{F}) \cdot d\mathbf{S} = e^{-1} - 1$

5. $\left\langle -3z^2 e^{z^3}, 2ze^{z^2} + z\sin(xz), 2 \right\rangle$; 2π

7. $\langle -2, 3, 5 \rangle$, 20π **9.** 0 **11.** -45π **13.** 0

15. (a)   (b) 140π

17. (a) $\mathbf{A} = \left\langle 0, 0, e^y - e^{x^2} \right\rangle$ (c) $\int \int_{S} \mathbf{F} \cdot d\mathbf{S} = \frac{\pi}{2}$

19. (a) $\int \int_{S} \mathbf{B} \cdot d\mathbf{S} = r^2 B\pi$ (b) $\int_{\partial S} \mathbf{A} \cdot d\mathbf{s} = 0$

21. $\int \int_{S} B\, dS = 18b$ **23.** $c = 2a$ and b is arbitrary.

27. $\int \int_{S} \mathbf{F} \cdot d\mathbf{S} = 25$

Section 17.3 Preliminary Questions

1. $\iint_S \mathbf{F} \cdot d\mathbf{S} = 0$

2. Since the integrand is positive for all $(x,\ y,\ z) \neq (0,\ 0,\ 0)$, the triple integral, hence also the flux, is positive.

3. (a), (b), (d), (f) are meaningful; (b) and (d) are automatically zero.

4. (c)   **5.** $\operatorname{div}(\mathbf{F}) = 1$ and flux $= \int \operatorname{div}(\mathbf{F})\, dV =$ volume

Section 17.3 Exercises

1. $\iint_S \mathbf{F} \cdot d\mathbf{S} = \iiint_{\mathcal{R}} \operatorname{div}(\mathbf{F})\, dV = \iiint_{\mathcal{R}} 0\, dV = 0$

3. $\iint_S \mathbf{F} \cdot d\mathbf{S} = \iiint_{\mathcal{R}} \operatorname{div}(\mathbf{F})\, dV = 4\pi$ **5.** $\frac{4\pi}{15}$ **7.** 60π

9. $\frac{3{,}616}{105}$ **11.** $\frac{32\pi}{5}$ **13.** 64π **15.** 81π **17.** π **19.** $\frac{13}{3}$

21. $\frac{4\pi}{3}$ **23.** $\frac{16\pi}{3} + \frac{9\sqrt{3}}{2} \approx 24.549$ **25.** $\approx 1.57\ \text{m}^3/\text{s}$

27. (b) 0   (c) 0

(d) Since $\mathbf{E}$ is not defined at the origin, which is inside the ball W, we cannot use the Divergence Theorem.

29. $(-4) \cdot \left[\frac{256\pi}{3} - 1 \right] \approx -1068.33$

31. $\operatorname{Div}(f\mathbf{F}) = f\operatorname{Div}(\mathbf{F}) + \mathbf{F} \cdot \nabla f$

35. (d) Flux through a closed surface is 0.

Chapter 17 Review

1. 0   **3.** -30 **5.** $\frac{3}{5}$

7. (a)   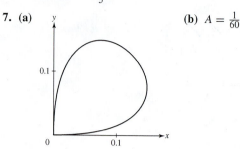   (b) $A = \frac{1}{60}$

9. $\frac{1}{3}$ **11.** $\frac{1}{3}$ **13.** $\{(x, y) \mid y = x \text{ or } y = -x\}$ **15.** 36π **19.** 2π

21. $\mathbf{A} = \langle yz,\ 0,\ 0 \rangle$ and the flux is 8π.   **23.** $\frac{296}{3}$ **25.** -128π

27. Volume$(W) = 2$ **29.** $4 \cdot 0.0009\pi \approx 0.0113$

31. $2x - y + 4z = 0$ **33.** (b) $\frac{\pi}{2}$

35. (c) 0   (d) $\int_{C_1} \mathbf{F} \cdot d\mathbf{s} = -4,\quad \int_{C_2} \mathbf{F} \cdot d\mathbf{s} = 4$

41. $V = \frac{4\pi}{3} abc$

75. For $x \neq 0$, $f'(x) = e^{-1/x^2}\left(\frac{2}{x^3}\right)$. Here, $P(x) = 2$ and $r = 3$.

Assume $f^{(k)}(x) = \frac{P(x)e^{-1/x^2}}{x^r}$. Then

$$f^{(k+1)}(x) = e^{-1/x^2}\left(\frac{x^3 P'(x) + (2 - rx^2)P(x)}{x^{r+3}}\right)$$

which is of the form desired.

Moreover, from Exercise 74, $f'(0) = 0$. Suppose $f^{(k)}(0) = 0$. Then

$$f^{(k+1)}(0) = \lim_{x \to 0}\frac{f^{(k)}(x) - f^{(k)}(0)}{x - 0} = \lim_{x \to 0}\frac{P(x)e^{-1/x^2}}{x^{r+1}}$$

$$= P(0)\lim_{x \to 0}\frac{f(x)}{x^{r+1}} = 0$$

79. $\lim\limits_{x \to 0}\frac{\sin x}{x} = \lim\limits_{x \to 0}\frac{\cos x}{1} = 1$. To use L'Hôpital's Rule to evaluate $\lim_{x \to 0}\frac{\sin x}{x}$, we must know that the derivative of $\sin x$ is $\cos x$, but to determine the derivative of $\sin x$, we must be able to evaluate $\lim_{x \to 0}\frac{\sin x}{x}$.

81. (a) $e^{-1/6} \approx 0.846481724$

x	1	0.1	0.01
$\left(\dfrac{\sin x}{x}\right)^{1/x^2}$	0.841471	0.846435	0.846481

(b) $1/3$

x	± 1	± 0.1	± 0.01
$\dfrac{1}{\sin^2 x} - \dfrac{1}{x^2}$	0.412283	0.334001	0.333340

Section 4.6 Preliminary Questions

1. An arc with the sign combination $++$ (increasing, concave up) is shown below at the left. An arc with the sign combination $-+$ (decreasing, concave up) is shown below at the right.

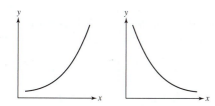

2. (c)

3. $x = 4$ is not in the domain of f.

Section 4.6 Exercises

1.
- In A, f is decreasing and concave up, so $f' < 0$ and $f'' > 0$.
- In B, f is increasing and concave up, so $f' > 0$ and $f'' > 0$.
- In C, f is increasing and concave down, so $f' > 0$ and $f'' < 0$.
- In D, f is decreasing and concave down, so $f' < 0$ and $f'' < 0$.
- In E, f is decreasing and concave up, so $f' < 0$ and $f'' > 0$.
- In F, f is increasing and concave up, so $f' > 0$ and $f'' > 0$.
- In G, f is increasing and concave down, so $f' > 0$ and $f'' < 0$.

3. This function changes from concave up to concave down at $x = -1$ and from increasing to decreasing at $x = 0$.

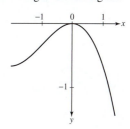

5. The function is decreasing everywhere and changes from concave up to concave down at $x = -1$ and from concave down to concave up at $x = -\frac{1}{2}$.

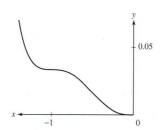

7.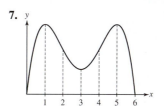

9.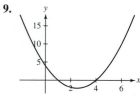

11.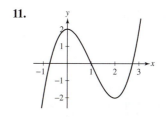

13.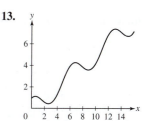

15. Local maximum at $x = -16$, a local minimum at $x = 0$, and an inflection point at $x = -8$

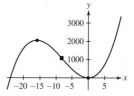

17. $f(0)$ is a local minimum, $f(\frac{1}{6})$ is a local maximum, and there is a point of inflection at $x = \frac{1}{12}$.

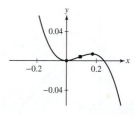

19. f has local minima at $x = \pm\sqrt{6}$, a local maximum at $x = 0$, and inflection points at $x = \pm\sqrt{2}$.

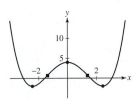

21. Graph has no critical points and is always increasing, inflection point at $(0, 0)$.

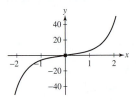

23. $f\left(\frac{1-\sqrt{33}}{8}\right)$ and $f(2)$ are local minima, and $f\left(\frac{1+\sqrt{33}}{8}\right)$ is a local maximum; points of inflection both at $x = 0$ and $x = \frac{3}{2}$.

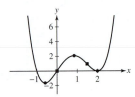

25. $f(0)$ is a local maximum, $f(12)$ is a local minimum, and there is a point of inflection at $x = 10$.

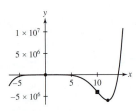

27. $f(4)$ is a local minimum, and the graph is always concave up.

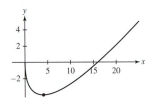

29. f has a local maximum at $x = 6$ and inflection points at $x = 8$ and $x = 12$.

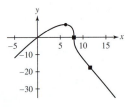

31. f has a local minimum at $x = -\frac{\sqrt{2}}{2}$, a local maximum at $x = \frac{\sqrt{2}}{2}$, inflection points at $x = 0$ and at $x = \pm\frac{\sqrt{3}}{2}$, and a horizontal asymptote at $y = 0$.

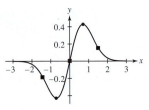

33. $f(2)$ is a local minimum and the graph is always concave up.

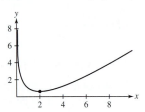

35. f has a local minimum at $x = 1$ and no inflection points. It is concave up everywhere. It has a vertical asymptote at $x = 0$.

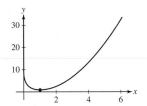

37. Graph has an inflection point at $x = \frac{3}{5}$, a local maximum at $x = 1$ (at which the graph has a cusp), and a local minimum at $x = \frac{9}{5}$.

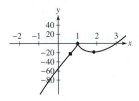

39. f has a local maximum at $x = 0$, local minima at $x = \pm 3$, and points of inflection at $x = \pm\sqrt{-6 + 3\sqrt{5}}$.

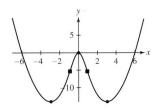

41. f has local minima at $x = -1.473$ and $x = 1.347$, a local maximum at $x = 0.126$, and points of inflection at $x = \pm\sqrt{\frac{2}{3}}$.

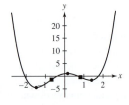

43. Graph has an inflection point at $x = \pi$, and no local maxima or minima.

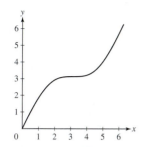

45. Local maximum at $x = \frac{\pi}{2}$, a local minimum at $x = \frac{3\pi}{2}$, and inflection points at $x = \frac{\pi}{6}$ and $x = \frac{5\pi}{6}$.

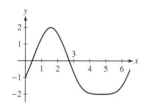

47. Local maximum at $x = \frac{\pi}{6}$ and a point of inflection at $x = \frac{2\pi}{3}$.

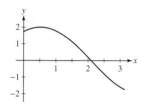

49. In both cases, there is a point where f is not differentiable at the transition from increasing to decreasing or decreasing to increasing.

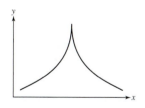

51. Graph (B) cannot be the graph of a polynomial.

53. (B) is the graph of $f(x) = \dfrac{3x^2}{x^2 - 1}$; (A) is the graph of $f(x) = \dfrac{3x}{x^2 - 1}$.

55. f is decreasing for all $x \neq \frac{1}{3}$, concave up for $x > \frac{1}{3}$, concave down for $x < \frac{1}{3}$, has a horizontal asymptote at $y = 0$ and a vertical asymptote at $x = \frac{1}{3}$.

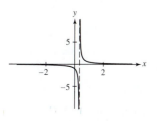

57. f is decreasing for all $x \neq 2$, concave up for $x > 2$, concave down for $x < 2$, has a horizontal asymptote at $y = 1$ and a vertical asymptote at $x = 2$.

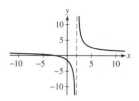

59. f is decreasing for all $x \neq 0, 1$, concave up for $0 < x < \frac{1}{2}$ and $x > 1$, concave down for $x < 0$ and $\frac{1}{2} < x < 1$, has a horizontal asymptote at $y = 0$ and vertical asymptotes at $x = 0$ and $x = 1$.

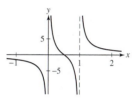

61. f is increasing for $x < 0$ and $0 < x < 1$ and decreasing for $1 < x < 2$ and $x > 2$; f is concave up for $x < 0$ and $x > 2$ and concave down for $0 < x < 2$; f has a horizontal asymptote at $y = 0$ and vertical asymptotes at $x = 0$ and $x = 2$.

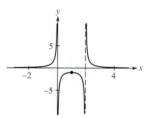

63. f is increasing for $x < 2$ and for $2 < x < 3$, is decreasing for $3 < x < 4$ and for $x > 4$, and has a local maximum at $x = 3$; f is concave up for $x < 2$ and for $x > 4$ and is concave down for $2 < x < 4$; f has a horizontal asymptote at $y = 0$ and vertical asymptotes at $x = 2$ and $x = 4$.

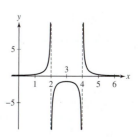

65. f is increasing for $|x| > 2$ and decreasing for $-2 < x < 0$ and for $0 < x < 2$; f is concave down for $-2\sqrt{2} < x < 0$ and for $x > 2\sqrt{2}$ and concave up for $x < -2\sqrt{2}$ and for $0 < x < 2\sqrt{2}$; f has a horizontal asymptote at $y = 1$ and a vertical asymptote at $x = 0$.

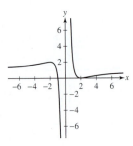

67. f is increasing for $x < 0$ and for $x > 2$ and decreasing for $0 < x < 2$; f is concave up for $x < 0$ and for $0 < x < 1$, is concave down for $1 < x < 2$ and for $x > 2$, and has a point of inflection at $x = 1$; f has a horizontal asymptote at $y = 0$ and vertical asymptotes at $x = 0$ and $x = 2$.

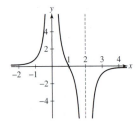

69. f is increasing for $x < 0$, decreasing for $x > 0$, and has a local maximum at $x = 0$; f is concave up for $|x| > 1/\sqrt{5}$, is concave down for $|x| < 1/\sqrt{5}$, and has points of inflection at $x = \pm 1/\sqrt{5}$; f has a horizontal asymptote at $y = 0$ and no vertical asymptotes.

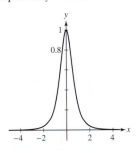

71. f is increasing for $x < 0$ and decreasing for $x > 0$; f is concave down for $|x| < \frac{\sqrt{2}}{2}$ and concave up for $|x| > \frac{\sqrt{2}}{2}$; f has a horizontal asymptote at $y = 0$ and no vertical asymptotes.

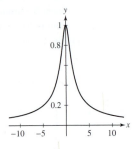

75. f is increasing for $x < -2$ and for $x > 0$, is decreasing for $-2 < x < -1$ and for $-1 < x < 0$, has a local minimum at $x = 0$, has a local maximum at $x = -2$, is concave down on $(-\infty, -1)$ and concave up on $(-1, \infty)$; f has a vertical asymptote at $x = -1$; by polynomial division, $f(x) = x - 1 + \frac{1}{x+1}$ and

$$\lim_{x \to \pm\infty} \left(x - 1 + \frac{1}{x+1} - (x-1) \right) = 0$$

which implies that the slant asymptote is $y = x - 1$.

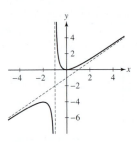

77. $y = x + 2$ is the slant asymptote of f; local minimum at $x = 2 + \sqrt{3}$, a local maximum at $x = 2 - \sqrt{3}$, and f is concave down on $(-\infty, 2)$ and concave up on $(2, \infty)$; vertical asymptote at $x = 2$.

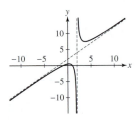

Section 4.7 Preliminary Questions

1. $b + h + \sqrt{b^2 + h^2} = 10$

2. If the function tends to infinity at the endpoints of the interval, then the function must take on a minimum value at a critical point.

3. No

Section 4.7 Exercises

1. (a) $y = \frac{3}{2} - x$ (b) $A = x(\frac{3}{2} - x) = \frac{3}{2}x - x^2$

(c) Closed interval $\left[0, \frac{3}{2}\right]$

(d) The maximum area 0.5625 m^2 is achieved with $x = y = \frac{3}{4}$ m.

3. One side of length 6 and two of length 3   **5.** 4 and 32

7. Allot approximately 5.28 m of the wire to the circle.

9. 20 and 20

11. Not possible. y can be arbitrarily large; then x is positive but close to zero and $x + 2y \approx 2y$ is arbitrarily large.

13. (a) The box should be a cube with side length $12^{1/3}$

(b) The box should be a cube with side length $\dfrac{\sqrt{30}}{3}$.

15. The corral of maximum area has dimensions

$$x = \frac{300}{1 + \pi/4} \text{ m} \qquad \text{and} \qquad y = \frac{150}{1 + \pi/4} \text{ m}$$

where x is the width of the corral and therefore the diameter of the semicircle and y is the height of the rectangular section.

17. Square of side length $4\sqrt{2}$ **19.** About 1.43 m   **21.** $\left(\frac{1}{2}, \frac{1}{2}\right)$

23. $(0.632784, -1.090410)$ **25.** $\theta = \dfrac{\pi}{2}$ **27.** $\dfrac{3\sqrt{3}}{4}r^2$

29. 60 cm wide by 100 cm high for the full poster (48 cm by 80 cm for the printed matter)

31. Radius: $\sqrt{\frac{2}{3}}R$; half-height: $\dfrac{R}{\sqrt{3}}$

33. $x = 10\sqrt{5} \approx 22.36$ m and $y = 20\sqrt{5} \approx 44.72$ m, where x is the length of the brick wall and y is the length of an adjacent side

35. 1.0718 **37.** $LH + \frac{1}{2}(L^2 + H^2)$ **39.** $y = -3x + 24$

43. $s = 3\sqrt[3]{4}$ m and $h = 2\sqrt[3]{4}$ m, where s is the length of the side of the square bottom of the box and h is the height of the box

45. (a) Each compartment has length of 600 m and width of 400 m.

(b) 240,000 m^2

47. $N \approx 58.14$ lb and $P \approx 77.33$ lb **49.** \$990

51. 1.2 million euros in equipment and 600,000 euros in labor

53. Brandon swims diagonally to a point located 20.2 m downstream and then runs the rest of the way.

57. $A = B = 30$ cm

59. $x = \dfrac{x_1 + x_2 + \cdots + x_n}{n}$

63. (a) 900 m^2 when $x = -10$ **(b)** [0, 20]; 800 m^2 when $x = 0$

65. $\tan \theta = \frac{5}{2}$

67. (b) $\left(\dfrac{17}{0.003}\right)^{1/4} \approx 8.676247$

(d) $v_d \approx 11.418583$; $D(v_d) \approx 191.741$ km

69. $s = \left(\dfrac{b^{2/3}}{2^{2/3}} + h^{2/3}\right)^{3/2}$ **71.** $\left(a^{2/3} + b^{2/3}\right)^{3/2}$

73. (a) $\alpha = 0$ corresponds to shooting the ball directly at the basket, while $\alpha = \pi/2$ corresponds to shooting the ball directly upward. In neither case is it possible for the ball to go into the basket. If the angle α is extremely close to 0, the ball is shot almost directly at the basket; on the other hand, if the angle α is extremely close to $\pi/2$, the ball is launched almost vertically. In either one of these cases, the ball has to travel at an enormous speed.

(b) The minimum clearly occurs where $\theta = \pi/3$.

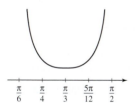

(c) $v^2 = \dfrac{16d}{F(\theta)}$; hence, v^2 is smallest whenever $F(\theta)$ is greatest.

(d) A critical point of F occurs where $\cos(\alpha - 2\theta) = 0$ so that $\alpha - 2\theta = -\frac{\pi}{2}$ (negative because $2\theta > \theta > \alpha$), and this gives us $\theta = \alpha/2 + \pi/4$. The minimum value $F(\theta_0)$ takes place at $\theta_0 = \alpha/2 + \pi/4$.

(e) Plug in $\theta_0 = \alpha/2 + \pi/4$. From Figure 34, we see that

$$\cos \alpha = \frac{d}{\sqrt{d^2 + h^2}} \quad \text{and} \quad \sin \alpha = \frac{h}{\sqrt{d^2 + h^2}}$$

(f) This shows that the minimum velocity required to launch the ball to the basket drops as shooter height increases. This shows one of the ways height is an advantage in free throws; a taller shooter need not shoot the ball as hard to reach the basket.

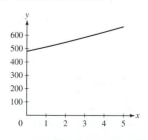

75. (a) From the figure, we see that

$$\theta(x) = \tan^{-1}\frac{c - f(x)}{x} - \tan^{-1}\frac{b - f(x)}{x}$$

Then

$\theta'(x)$

$$= \frac{b - (f(x) - xf'(x))}{x^2 + (b - f(x))^2} - \frac{c - (f(x) - xf'(x))}{x^2 + (c - f(x))^2}$$

$$= (b - c)\frac{x^2 - bc + (b + c)(f(x) - xf'(x)) - (f(x))^2 + 2xf(x)f'(x)}{(x^2 + (b - f(x))^2)(x^2 + (c - f(x))^2)}$$

$$= (b - c)\frac{(x^2 + (xf'(x))^2) - (bc - (b + c)(f(x) - xf'(x)) + (f(x) - xf'(x))^2)}{(x^2 + (b - f(x))^2)(x^2 + (c - f(x))^2)}$$

$$= (b - c)\frac{(x^2 + (xf'(x))^2) - (b - (f(x) - xf'(x)))(c - (f(x) - xf'(x)))}{(x^2 + (b - f(x))^2)(x^2 + (c - f(x))^2)}$$

(b) The point Q is the y-intercept of the line tangent to the graph of f at point P. The equation of this tangent line is

$$Y - f(x) = f'(x)(X - x)$$

The y-coordinate of Q is then $f(x) - xf'(x)$.

(c) From the figure, we see that

$$BQ = b - (f(x) - xf'(x)),$$

$$CQ = c - (f(x) - xf'(x))$$

and

$$PQ = \sqrt{x^2 + (f(x) - (f(x) - xf'(x)))^2} = \sqrt{x^2 + (xf'(x))^2}$$

Comparing these expressions with the numerator of $d\theta/dx$, it follows that $\dfrac{d\theta}{dx} = 0$ is equivalent to

$$PQ^2 = BQ \cdot CQ$$

(d) The equation $PQ^2 = BQ \cdot CQ$ is equivalent to

$$\frac{PQ}{BQ} = \frac{CQ}{PQ}$$

In other words, the sides CQ and PQ from the triangle $\triangle QCP$ are proportional in length to the sides PQ and BQ from the triangle $\triangle QPB$. As $\angle PQB = \angle CQP$, it follows that triangles $\triangle QCP$ and $\triangle QPB$ are similar.

Section 4.8 Preliminary Questions

1. One

2. Every term in the Newton's Method sequence will remain x_0.

3. Newton's Method will fail.

4. Yes, that is a reasonable description. The iteration formula for Newton's Method was derived by solving the equation of the tangent line to $y = f(x)$ at x_0 for its x-intercept.

Section 4.8 Exercises

1.

n	1	2	3
x_n	2.5	2.45	2.44948980

3.

n	1	2	3
x_n	2.16666667	2.15450362	2.15443469

5.

n	1	2	3
x_n	0.28540361	0.24288009	0.24267469

7. We take $x_0 = -1.4$, based on the figure, and then calculate

n	1	2	3
x_n	-1.330964467	-1.328272820	-1.328268856

9. $r_1 \approx 0.25917$ and $r_2 \approx 2.54264$

11. $\sqrt{11} \approx 3.317$; a calculator yields 3.31662479.

13. $2^{7/3} \approx 5.040$; a calculator yields 5.0396842.

15. 2.093064358   **17.** -2.225 **19.** 1.749

21. $x = 4.49341$, which is approximately 1.4303π

23. $(2.7984, -0.941684)$

25. (a) $P \approx \$156.69$

(b) $b \approx 1.02121$; the interest rate is around 25.45%.

27. (a) The sector SAB is the slice OAB with the triangle OBS removed. OAB is a central sector with arc θ and radius $\overline{OA} = a$, and therefore has area $\frac{a^2\theta}{2}$. OBS is a triangle with height $a \sin \theta$ and base length $\overline{OS} = ea$. Hence, the area of the sector is

$$\frac{a^2}{2}\theta - \frac{1}{2}ea^2 \sin \theta = \frac{a^2}{2}(\theta - e \sin \theta)$$

(b) Since Kepler's Second Law indicates that the area of the sector is proportional to the time t since the planet passed point A, we get

$$\pi a^2 (t/T) = a^2/2 (\theta - e \sin \theta)$$

$$2\pi \frac{t}{T} = \theta - e \sin \theta$$

(c) From the point of view of the Sun, Mercury has traversed an angle of approximately 1.76696 radians $= 101.24°$. Mercury has therefore traveled more than one fourth of the way around (from the point of view of central angle) during this time.

29. The sequence of iterates diverges spectacularly, since $x_n = (-2)^n x_0$.

31. (a) Let $f(x) = \frac{1}{x} - c$. Then

$$x - \frac{f(x)}{f'(x)} = x - \frac{\frac{1}{x} - c}{-x^{-2}} = 2x - cx^2$$

(b) For $c = 10.3$, we have $f(x) = \frac{1}{x} - 10.3$ and thus $x_{n+1} = 2x_n - 10.3x_n^2$

• Take $x_0 = 0.1$.

n	1	2	3
x_n	0.097	0.0970873	0.09708738

• Take $x_0 = 0.5$.

n	1	2	3
x_n	-1.575	-28.7004375	-8541.66654

(c) The graph is disconnected. If $x_0 = .5$, $(x_1, f(x_1))$ is on the other portion of the graph, which will never converge to any point under Newton's Method.

33. $\theta \approx 1.2757$; hence, $h = L\dfrac{1 - \cos \theta}{2 \sin \theta} \approx 1.11181$

35. (a) $a = 46.95$ (b) $s = 29.24$

37. (a) $a \approx 28.46$

(b) $\Delta L = 1$ ft yields $\Delta s \approx 0.61$; $\Delta L = 5$ yields $\Delta s \approx 3.05$.

(c) $s(161) - s(160) = 0.62$, very close to the approximation obtained from the Linear Approximation; $s(165) - s(160) = 3.02$, again very close to the approximation obtained from the Linear Approximation.

Chapter 4 Review

1. $8.1^{1/3} - 2 \approx 0.00833333$; error is 3.445×10^{-5}.

3. $625^{1/4} - 624^{1/4} \approx 0.002$; error is 1.201×10^{-6}.

5. $\frac{1}{1.02} \approx 0.98$; error is 3.922×10^{-4}.

7. $L(x) = 5 + \dfrac{1}{10}(x - 25)$ **9.** $L(r) = 36\pi (r - 2)$

11. $L(x) = \dfrac{1}{\sqrt{e}}(2 - x)$ **13.** $\Delta s \approx 0.632$

15. (a) An increase of \$1500 in revenue.

(b) A small increase in price would result in a decrease in revenue.

17. 9%   **21.** $c = \dfrac{3}{\ln 4} \approx 2.164 \in (1, 4)$

23. Let $x > 0$. Because f is continuous on $[0, x]$ and differentiable on $(0, x)$, the Mean Value Theorem guarantees there exists a $c \in (0, x)$ such that

$$f'(c) = \frac{f(x) - f(0)}{x - 0} \qquad \text{or} \qquad f(x) = f(0) + xf'(c)$$

Now, we are given $f(0) = 4$ and $f'(x) \le 2$ for $x > 0$. Therefore, for all $x \ge 0$,

$$f(x) \le 4 + x(2) = 2x + 4$$

25. $x = \frac{2}{3}$ and $x = 2$ are critical points; $f\left(\frac{2}{3}\right)$ is a local maximum, while $f(2)$ is a local minimum.

27. $x = 0$, $x = -2$ and $x = -\frac{4}{5}$ are critical points; $f(-2)$ is neither a local maximum nor a local minimum, $f\left(-\frac{4}{5}\right)$ is a local maximum and $f(0)$ is a local minimum.

29. $\theta = \dfrac{3\pi}{4} + n\pi$ is a critical point for all integers n; $g\left(\dfrac{3\pi}{4} + n\pi\right)$ is neither a local maximum nor a local minimum for any integer n.

31. Maximum value is 21; minimum value is -11.

33. Minimum value is -1; maximum value is $\dfrac{5}{4}$.

35. Minimum value is -1; maximum value is 3.

37. Minimum value is $12 - 12 \ln 12 \approx -17.818880$; maximum value is $40 - 12 \ln 40 \approx -4.266553$.

39. Critical points are $x = 1$ and $x = 3$. The minimum value is 2, occurring at $x = 3$, and the maximum value is 17, occurring at the right endpoint $x = 8$.

41. $x = \dfrac{4}{3}$ **43.** $x = \pm\dfrac{2}{\sqrt{3}}$ **45.** $x = 1$ and $x = 4$

47. No horizontal asymptotes; no vertical asymptotes

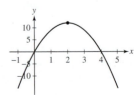

49. No horizontal asymptotes; no vertical asymptotes

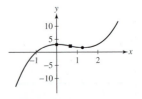

51. $y = 0$ is a horizontal asymptote; $x = -1$ is a vertical asymptote.

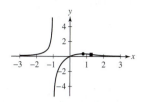

53. Horizontal asymptote of $y = 0$; no vertical asymptotes

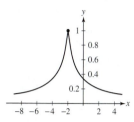

55.

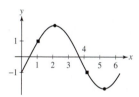

57.

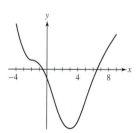

59. $b = \sqrt[3]{12}$ m and $h = \frac{1}{3}\sqrt[3]{12}$ m   **63.** $\frac{16}{9}\pi$ **69.** $\sqrt[3]{25} = 2.9240$

71. $(0, \frac{2}{e})$ is a local minimum.

73. Local minimum at $x = e^{-1}$; no points of inflection; $\lim_{x \to 0+} x \ln x = 0$; $\lim_{x \to \infty} x \ln x = \infty$

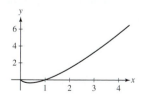

75. Local maximum at $x = e^{-2}$ and a local minimum at $x = 1$; point of inflection at $x = e^{-1}$; $\lim_{x \to 0+} x(\ln x)^2 = 0$; $\lim_{x \to \infty} x(\ln x)^2 = \infty$

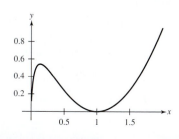

77. As $x \to \infty$, both $2x - \sin x$ and $3x + \cos 2x$ tend toward infinity, so L'Hôpital's Rule applies to $\lim_{x \to \infty} \dfrac{2x - \sin x}{3x + \cos 2x}$; however, the resulting limit, $\lim_{x \to \infty} \dfrac{2 - \cos x}{3 - 2\sin 2x}$, does not exist due to the oscillation of $\sin x$ and $\cos x$. To evaluate the limit, we note

$$\lim_{x \to \infty} \frac{2x - \sin x}{3x + \cos 2x} = \lim_{x \to \infty} \frac{2 - \frac{\sin x}{x}}{3 + \frac{\cos 2x}{x}} = \frac{2}{3}$$

79. 4   **81.** 0   **83.** 3   **85.** $\ln 2$ **87.** $\dfrac{1}{6}$ **89.** 2

Chapter 5

Section 5.1 Preliminary Questions

1. The right endpoints of the subintervals are then $\frac{5}{2}, 3, \frac{7}{2}, 4, \frac{9}{2}, 5$, while the left endpoints are $2, \frac{5}{2}, 3, \frac{7}{2}, 4, \frac{9}{2}$.

2. **(a)** $\dfrac{9}{2}$ **(b)** $\dfrac{3}{2}$ and 2

3. **(a)** *Are* the same   **(b)** *Not* the same

(c) *Are* the same   **(d)** *Are* the same

4. The first term in the sum $\sum_{j=0}^{100} j$ is equal to zero, so it may be dropped; on the other hand, the first term in $\sum_{j=0}^{100} 1$ is not zero.

5. On $[3, 7]$, the function $f(x) = x^{-2}$ is a decreasing function.

Section 5.1 Exercises

1. Over the interval $[0, 3]$: 0.96 km; over the interval $[1, 2.5]$: 0.5 km

3. 28.5 cm; The figure below is a graph of the rainfall as a function of time. The area of the shaded region represents the total rainfall.

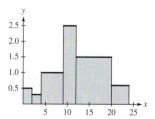

5. $L_5 = 46$; $R_5 = 44$

7. **(a)** $L_6 = 16.5$; $R_6 = 19.5$

(b) Via geometry (see figure below), the exact area is $A = 18$. Thus, L_6 underestimates the true area ($L_6 - A = -1.5$), while R_6 overestimates the true area ($R_6 - A = +1.5$).

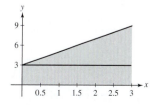

9. $R_3 = 32$; $L_3 = 20$; the area under the graph is larger than L_3 but smaller than R_3.

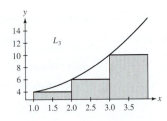

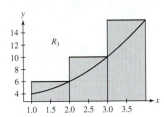

11. $R_3 = 2.5$; $M_3 = 2.875$; $L_6 = 3.4375$

13. (a) $L_4 = 1.75$ **(b)** $R_4 = 3.75$ **(c)** The actual area A under the curve $f(x) = x^2$ over the interval $[0, 2]$ satisfies $L_4 < A < R_4$.

15. $R_3 = \dfrac{16}{3}$ **17.** $M_6 = 87$ **19.** $M_5 \approx 1.30$

21. $L_4 \approx 0.410236$ **23.** $\displaystyle\sum_{k=4}^{8} k^7$ **25.** $\displaystyle\sum_{k=2}^{5}(2^k + 2)$

27. $\displaystyle\sum_{i=1}^{n} \dfrac{i}{(i+1)(i+2)}$

29. (a) 45   **(b)** 24   **(c)** 99

31. (a) -1 **(b)** 13   **(c)** 12

33. 15,050   **35.** 352,800   **37.** 1,093,350   **39.** 41,650

41. $-123,165$ **43.** $\dfrac{1}{2}$ **45.** $\dfrac{1}{3}$

47. 18; the region under the graph is a triangle with base 2 and height 18.

49. 12; the region under the curve is a trapezoid with base width 4 and heights 2 and 4.

51. 2; the region under the curve over $[0, 2]$ is a triangle with base and height 2.

53. $\lim_{N\to\infty} R_N = 16$ **55.** $R_N = \dfrac{1}{3} + \dfrac{1}{2N} + \dfrac{1}{6N^2}; \dfrac{1}{3}$

57. $R_N = 222 + \dfrac{189}{N} + \dfrac{27}{N^2}; 222$ **59.** $R_N = 2 + \dfrac{6}{N} + \dfrac{8}{N^2}; 2$

61. $R_N = (b-a)(2a+1) + (b-a)^2 + \dfrac{(b-a)^2}{N};$

$(b^2 + b) - (a^2 + a)$

63. The area between the graph of $f(x) = x^4$ and the x-axis over the interval $[0, 1]$

65. The area between the graph of $y = e^x$ and the x-axis over the interval $[-2, 3]$

67. $\lim_{N\to\infty} R_N = \lim_{N\to\infty} \dfrac{\pi}{N} \displaystyle\sum_{k=1}^{N} \sin\left(\dfrac{k\pi}{N}\right)$

69. $\lim_{N\to\infty} L_N = \lim_{N\to\infty} \dfrac{4}{N} \displaystyle\sum_{j=0}^{N-1} \sqrt{15 + \dfrac{8j}{N}}$

71. $\lim_{N\to\infty} M_N = \lim_{N\to\infty} \dfrac{1}{2N} \displaystyle\sum_{j=1}^{N} \tan\left(\dfrac{1}{2} + \dfrac{1}{2N}\left(j - \dfrac{1}{2}\right)\right)$

73. Represents the area between the graph of $y = f(x) = \sqrt{1 - x^2}$ and the x-axis over the interval $[0, 1]$. This is the portion of the circular disk $x^2 + y^2 \leq 1$ that lies in the first quadrant. Accordingly, its area is $\frac{\pi}{4}$.

75. Of the three approximations, R_N is the least accurate, and then L_N and finally M_N are the most accurate.

77. The area A under the curve is somewhere between $L_4 \approx 0.518$ and $R_4 \approx 0.768$.

79. f is increasing over the interval $[0, \pi/2]$, so $0.79 \approx L_4 \leq A \leq R_4 \approx 1.18$.

81. $L_{100} = 0.793988$; $R_{100} = 0.80399$; $L_{200} = 0.797074$; $R_{200} = 0.802075$; thus, $A = 0.80$ to two decimal places.

83. (a) Let $f(x) = e^x$ on $[0, 1]$. With $n = N$, $\Delta x = (1 - 0)/N = 1/N$ and

$$x_j = a + j\Delta x = \dfrac{j}{N}$$

for $j = 0, 1, 2, \ldots, N$. Therefore,

$$L_N = \Delta x \sum_{j=0}^{N-1} f(x_j) = \dfrac{1}{N} \sum_{j=0}^{N-1} e^{j/N}$$

(b) Applying Eq. (8) with $r = e^{1/N}$, we have

$$L_N = \dfrac{1}{N} \dfrac{(e^{1/N})^N - 1}{e^{1/N} - 1} = \dfrac{e - 1}{N(e^{1/N} - 1)}$$

(c) $A = e - 1$

85.

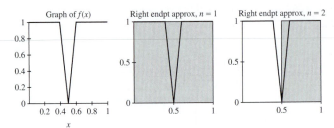

Graph of $f(x)$ Right endpt approx, $n = 1$ Right endpt approx, $n = 2$

87. When f' is large, the graph of f is steeper and hence there is more gap between f and L_N or R_N.

91. $N > 30,000$

Section 5.2 Preliminary Questions

1. 2

2. (a) False. $\int_a^b f(x)\,dx$ is the *signed* area between the graph and the x-axis.

(b) True   **(c)** True

3. Because $\cos(\pi - x) = -\cos x$, the "negative" area between the graph of $y = \cos x$ and the x-axis over $[\frac{\pi}{2}, \pi]$ exactly cancels the "positive" area between the graph and the x-axis over $[0, \frac{\pi}{2}]$.

4. $\displaystyle\int_{-1}^{-5} 8\,dx$

Section 5.2 Exercises

1. The region bounded by the graph of $y = 2x$ and the x-axis over the interval $[-3, 3]$ consists of two right triangles. One has area $\frac{1}{2}(3)(6) = 9$ below the axis, and the other has area $\frac{1}{2}(3)(6) = 9$ above the axis. Hence,

$$\int_{-3}^{3} 2x\,dx = 9 - 9 = 0$$

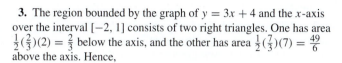

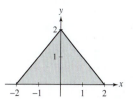

3. The region bounded by the graph of $y = 3x + 4$ and the x-axis over the interval $[-2, 1]$ consists of two right triangles. One has area $\frac{1}{2}(\frac{2}{3})(2) = \frac{2}{3}$ below the axis, and the other has area $\frac{1}{2}(\frac{7}{3})(7) = \frac{49}{6}$ above the axis. Hence,

$$\int_{-2}^{1} (3x + 4)\, dx = \frac{49}{6} - \frac{2}{3} = \frac{15}{2}$$

11. (a) $\lim\limits_{N\to\infty} R_N = \lim\limits_{N\to\infty} \left(30 - \dfrac{50}{N}\right) = 30$

(b) The region bounded by the graph of $y = 8 - x$ and the x-axis over the interval $[0, 10]$ consists of two right triangles. One triangle has area $\frac{1}{2}(8)(8) = 32$ above the axis, and the other has area $\frac{1}{2}(2)(2) = 2$ below the axis. Hence,

$$\int_{0}^{10} (8 - x)\, dx = 32 - 2 = 30$$

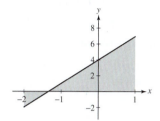

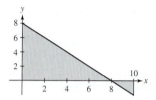

5. The region bounded by the graph of $y = 7 - x$ and the x-axis over the interval $[6, 8]$ consists of two right triangles. One triangle has area $\frac{1}{2}(1)(1) = \frac{1}{2}$ above the axis, and the other has area $\frac{1}{2}(1)(1) = \frac{1}{2}$ below the axis. Hence,

$$\int_{6}^{8} (7 - x)\, dx = \frac{1}{2} - \frac{1}{2} = 0$$

13. (a) $-\dfrac{\pi}{2}$ **(b)** $\dfrac{3\pi}{2}$

15. $\displaystyle\int_{0}^{3} g(t)\, dt = \frac{3}{2};\ \int_{3}^{5} g(t)\, dt = 0$

17. The partition P is defined by

$$x_0 = 0\ <\ x_1 = 1\ <\ x_2 = 2.5\ <\ x_3 = 3.2\ <\ x_4 = 5$$

The set of sample points is given by
$C = \{c_1 = 0.5, c_2 = 2, c_3 = 3, c_4 = 4.5\}$. Finally, the value of the Riemann sum is

$$34.25(1 - 0) + 20(2.5 - 1) + 8(3.2 - 2.5) + 15(5 - 3.2) = 96.85$$

19. $R(f, P, C) = 1.59$; here is a sketch of the graph of f and the rectangles.

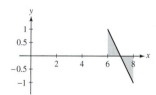

7. The region bounded by the graph of $y = \sqrt{25 - x^2}$ and the x-axis over the interval $[0, 5]$ is one-quarter of a circle of radius 5. Hence,

$$\int_{0}^{5} \sqrt{25 - x^2}\, dx = \frac{1}{4}\pi(5)^2 = \frac{25\pi}{4}$$

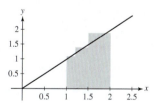

21. $R(f, P, C) = 44.625$; here is a sketch of the graph of f and the rectangles.

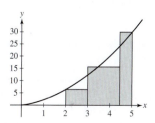

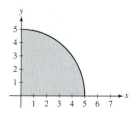

9. The region bounded by the graph of $y = 2 - |x|$ and the x-axis over the interval $[-2, 2]$ is a triangle above the axis with base 4 and height 2. Consequently,

$$\int_{-2}^{2} (2 - |x|)\, dx = \frac{1}{2}(2)(4) = 4$$

23.

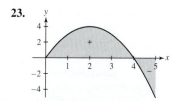

25.

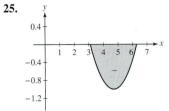

27.

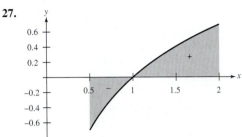

29. The integrand is always positive. The integral must therefore be positive, since the signed area has only a positive part.

31. Since $y = x$ is an increasing function on $[0, 2\pi]$, and $\sin x \geq 0$ on $[0, \pi]$ but $\sin x \leq 0$ on $[\pi, 2\pi]$, it follows that the area below the x-axis is enclosed by $y = x \sin x$ on the interval $[0, \pi]$. Hence, the total area on $[0, 2\pi]$ will be negative, so the definite integral will be negative as well.

33. 36 **35.** 243 **37.** $-\dfrac{2}{3}$ **39.** $\dfrac{196}{3}$ **41.** $\dfrac{1}{3}a^3 - \dfrac{1}{2}a^2 + \dfrac{5}{6}$

43. 17 **45.** -12 **47.** No. **49.** $\dfrac{81}{4}$ **51.** $-\dfrac{63}{4}$ **53.** 7 **55.** 8

57. -7 **59.** $\displaystyle\int_0^7 f(x)\,dx$ **61.** $\displaystyle\int_5^9 f(x)\,dx$ **63.** $\dfrac{4}{5}$ **65.** $-\dfrac{35}{2}$

67. When $f(x)$ takes on both positive and negative values on $[a, b]$, $\int_a^b f(x)\,dx$ represents the signed area between f and the x-axis, whereas $\int_a^b |f(x)|\,dx$ represents the total (unsigned) area between f and the x-axis. Any negatively signed areas that were part of $\int_a^b f(x)\,dx$ are regarded as positive areas in $\int_a^b |f(x)|\,dx$.

69. $[-1, \sqrt{2}]$ or $[-\sqrt{2}, 1]$ **71.** 9 **73.** $\dfrac{1}{2}$

75. On the interval $[0, 1]$, $x^5 \leq x^4$; on the other hand, $x^4 \leq x^5$ for $x \in [1, 2]$.

77. $y = \sin x$ is increasing on $[0.2, 0.3]$. Accordingly, for $0.2 \leq x \leq 0.3$, we have

$$m = 0.198 \leq 0.19867 \approx \sin 0.2 \leq \sin x \leq \sin 0.3$$

$$\approx 0.29552 \leq 0.296 = M$$

Therefore, by the Comparison Theorem, we have

$$0.0198 = m(0.3 - 0.2) = \int_{0.2}^{0.3} m\,dx \leq \int_{0.2}^{0.3} \sin x\,dx \leq \int_{0.2}^{0.3} M\,dx$$

$$= M(0.3 - 0.2) = 0.0296$$

79. f is decreasing and nonnegative on the interval $[\pi/4, \pi/2]$. Therefore, $0 \leq f(x) \leq f(\pi/4) = \dfrac{2\sqrt{2}}{\pi}$ for all x in $[\pi/4, \pi/2]$.

81. The assertion $f'(x) \leq g'(x)$ is false. Consider $a = 0$, $b = 1$, $f(x) = x$, $g(x) = 2$. $f(x) \leq g(x)$ for all x in the interval $[0, 1]$, but $f'(x) = 1$, while $g'(x) = 0$ for all x.

83. If f is an odd function, then $f(-x) = -f(x)$ for all x. Accordingly, for every positively signed area in the right half-plane where f is above the x-axis, there is a corresponding negatively signed area in the left half-plane where f is below the x-axis. Similarly, for every negatively signed area in the right half-plane where f is below the x-axis, there is a corresponding positively signed area in the left half-plane where f is above the x-axis.

Section 5.3 Preliminary Questions

1. Any constant function is an antiderivative for the function $f(x) = 0$.

2. No difference **3.** No

4. (a) False. Even if $f(x) = g(x)$, the antiderivatives F and G may differ by an additive constant.

(b) True. This follows from the fact that the derivative of any constant is 0.

(c) False. If the functions f and g are different, then the antiderivatives F and G differ by a linear function: $F(x) - G(x) = ax + b$ for some constants a and b.

5. No

Section 5.3 Exercises

1. $6x^3 + C$ **3.** $\dfrac{2}{5}x^5 - 8x^3 + 12 \ln|x| + C$

5. $2\sin x + 9\cos x + C$ **7.** $12e^x + 5x^{-1} + C$

9. (a) (ii) **(b)** (iii) **(c)** (i) **(d)** (iv)

11. $4x - 9x^2 + C$ **13.** $\dfrac{11}{5}t^{5/11} + C$ **15.** $3t^6 - 2t^5 - 14t^2 + C$

17. $5z^{1/5} - \dfrac{3}{5}z^{5/3} + \dfrac{4}{9}z^{9/4} + C$ **19.** $\dfrac{3}{2}x^{2/3} + C$ **21.** $-\dfrac{18}{t^2} + C$

23. $\dfrac{2}{5}t^{5/2} + \dfrac{1}{2}t^2 + \dfrac{2}{3}t^{3/2} + t + C$

25. $\dfrac{1}{2}x^2 + 3\ln|x| + 4x^{-1} + C$ **27.** $12\sec x + C$

29. $-\csc t + C$ **31.** $\dfrac{1}{3}\tan(3\theta) + C$ **33.** $\dfrac{25}{3}\tan(3z) + C$

35. $\dfrac{1}{3}\sin(3\theta) - 2\tan\left(\dfrac{\theta}{4}\right) + C$ **37.** $\dfrac{3}{5}e^{5x} + C$

39. $4x^2 + 2e^{5-2x} + C$

41. Graph (B) does not have the same local extrema as indicated by $y = f(x)$ and therefore is *not* an antiderivative of $y = f(x)$.

43. $\dfrac{d}{dx}\left(\dfrac{1}{7}(x+13)^7 + C\right) = (x+13)^6$

45. $\dfrac{d}{dx}\left(\dfrac{1}{12}(4x+13)^3 + C\right) = \dfrac{1}{4}(4x+13)^2(4) = (4x+13)^2$

47. $y = \dfrac{1}{4}x^4 + 4$ **49.** $y = t^2 + 3t^3 - 2$ **51.** $y = \dfrac{2}{3}t^{3/2} + \dfrac{1}{3}$

53. $y = \dfrac{1}{12}(3x+2)^4 - \dfrac{1}{3}$ **55.** $y = 1 - \cos x$

57. $y = 3 + \dfrac{1}{5}\sin 5x$ **59.** $y = e^x - e^2$ **61.** $y = -3e^{12-3t} + 10$

63. $f'(x) = 6x^2 + 1$; $f(x) = 2x^3 + x + 2$

65. $f'(x) = \frac{1}{4}x^4 - x^2 + x + 1$; $f(x) = \dfrac{1}{20}x^5 - \dfrac{1}{3}x^3 + \dfrac{1}{2}x^2 + x$

67. $f'(t) = -2t^{-1/2} + 2$; $f(t) = -4t^{1/2} + 2t + 4$

69. $f'(t) = \dfrac{1}{2}t^2 - \sin t + 2$; $f(t) = \dfrac{1}{6}t^3 + \cos t + 2t - 3$

71. The differential equation satisfied by $s(t)$ is

$$\frac{ds}{dt} = v(t) = 6t^2 - t$$

and the associated initial condition is $s(1) = 0$;

$$s(t) = 2t^3 - \frac{1}{2}t^2 - \frac{3}{2}.$$

73. $v_y = -49$ m/s

75. The differential equation satisfied by $s(t)$ is

$$\frac{ds}{dt} = v(t) = \sin(\pi t/2)$$

and the associated initial condition is $s(0) = 0$;

$$s(t) = \frac{2}{\pi}(1 - \cos(\pi t/2)).$$

77. 6.25 s; 78.125 m   **79.** 300 m/s   **83.** $c_1 = 1$ and $c_2 = -1$

85. (a) By the Chain Rule, we have

$$\frac{d}{dx}\left(\frac{1}{2}F(2x)\right) = \frac{1}{2}F'(2x) \cdot 2 = F'(2x) = f(2x)$$

Thus, $y = \frac{1}{2}F(2x)$ is an antiderivative of $y = f(2x)$.

(b) $\dfrac{1}{k}F(kx) + C$

Section 5.4 Preliminary Questions

1. (a) 4   **(b)** The signed area between $y = f(x)$ and the x-axis

2. 3

3. (a) False. The FTC I is valid for continuous functions.

(b) False. The FTC I works for any antiderivative of the integrand.

(c) False. If you cannot find an antiderivative of the integrand, you cannot use the FTC I to evaluate the definite integral, but the definite integral may still exist.

4. 0

Section 5.4 Exercises

1. $A = \dfrac{1}{3}$

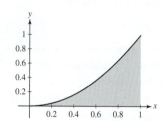

3. $A = \dfrac{1}{2}$

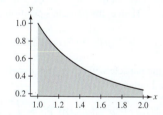

5. $\dfrac{27}{2}$ **7.** -1 **9.** 128   **11.** $\dfrac{27}{2}$ **13.** $\dfrac{16}{3}$ **15.** $\dfrac{31}{40}$ **17.** $\dfrac{2}{3}$

19. 12   **21.** $\dfrac{11}{6}$ **23.** $60\sqrt{3} - \dfrac{8}{3}$ **25.** $\sqrt{2}$ **27.** $\dfrac{3}{2}$ **29.** $\dfrac{4}{3\sqrt{3}}$

31. $\dfrac{1}{5}(\sqrt{2} - 1)$ **33.** $e - 1$ **35.** $\dfrac{1}{6}(e - e^{-17})$ **37.** $\ln 5$ **39.** $\ln 2$

41. $3e^{-6} - 9$ **43.** $\dfrac{5}{2}$ **45.** $\dfrac{97}{4}$ **47.** 2   **49.** $\dfrac{1}{4}\left(b^4 - 1\right)$

51. $\dfrac{1}{6}(b^6 - 1)$ **53.** $\ln 5$ **55.** $\dfrac{707}{12}$

57. Graphically speaking, for an odd function, the positively signed area from $x = 0$ to $x = 1$ cancels the negatively signed area from $x = -1$ to $x = 0$.

59. 24

61. $\int_0^1 x^n \, dx$ represents the area between the positive curve $f(x) = x^n$ and the x-axis over the interval [0, 1]. This area gets smaller as n gets larger, as is readily evident in the following graph, which shows curves for several values of n.

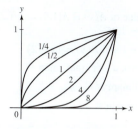

67. Let $a > b$ be real numbers, and let $f(x)$ be such that $|f'(x)| \leq K$ for $x \in [a, b]$. By FTC,

$$\int_a^x f'(t) \, dt = f(x) - f(a)$$

Since $f'(x) \geq -K$ for all $x \in [a, b]$, we get

$$f(x) - f(a) = \int_a^x f'(t) \, dt \geq -K(x - a)$$

Since $f'(x) \leq K$ for all $x \in [a, b]$, we get

$$f(x) - f(a) = \int_a^x f'(t) \, dt \leq K(x - a)$$

Combining these two inequalities yields

$$-K(x - a) \leq f(x) - f(a) \leq K(x - a)$$

so that, by definition,

$$|f(x) - f(a)| \leq K|x - a|$$

Section 5.5 Preliminary Questions

1. (a) No   **(b)** Yes

2. (c)

3. Yes. All continuous functions have an antiderivative, namely, $\displaystyle\int_a^x f(t) \, dt$.

4. (b), (e), and **(f)**

Section 5.5 Exercises

1. $A(x) = \displaystyle\int_{-2}^x (2t + 4) \, dt = (x + 2)^2$

3. $G(1) = 0$; $G'(1) = -1$ and $G'(2) = 2$; $G(x) = \dfrac{1}{3}x^3 - 2x + \dfrac{5}{3}$

5. $G(1) = 0$; $G'(0) = 0$ and $G'(\frac{\pi}{4}) = 1$

7. $\dfrac{1}{5}x^5 - \dfrac{32}{5}$ **9.** $1 - \cos x$ **11.** $\dfrac{1}{3}e^{3x} - \dfrac{1}{3}e^{12}$ **13.** $\dfrac{1}{2}x^4 - \dfrac{1}{2}$

15. $-e^{-9x-2} + e^{-3x}$ **17.** $F(x) = \displaystyle\int_5^x \sqrt{t^3 + 1} \, dt$

19. $F(x) = \displaystyle\int_0^x \sec t \, dt$ **21.** $x^5 - 9x^3$ **23.** $\sec(5t - 9)$

25. (a) $A(2) = 4$; $A(3) = 6.5$; $A'(2) = 2$ and $A'(3) = 3$
(b)

$$A(x) = \begin{cases} 2x, & 0 \leq x < 2 \\ \frac{1}{2}x^2 + 2, & 2 \leq x \leq 4 \end{cases}$$

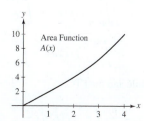

29. $\dfrac{2x^3}{x^2 + 1}$ **31.** $-\cos^4 s \sin s$ **33.** $2x \tan(x^2) - \dfrac{\tan(\sqrt{x})}{2\sqrt{x}}$

35. The minimum value of $A(x)$ is $A(1.5) = -1.25$; the maximum value of A is $A(4.5) = 1.25$.

37. $A(x) = (x - 2) - 1$ and $B(x) = (x - 2)$

39. (a) A does not have a local maximum at P.

(b) A has a local minimum at R.

(c) A has a local maximum at S.

(d) True

41. $g(x) = 2x + 1$; $c = 2$ or $c = -3$

43. (a) If $x = c$ is an inflection point of A, then $A''(c) = f'(c) = 0$.

(b) If A is concave up, then $A''(x) > 0$. Since A is the area function associated with f, $A'(x) = f(x)$ by FTC II, so $A''(x) = f'(x)$. Therefore, $f'(x) > 0$, so f is increasing.

(c) If A is concave down, then $A''(x) < 0$. Since A is the area function associated with $f(x)$, $A'(x) = f(x)$ by FTC II, so $A''(x) = f'(x)$. Therefore, $f'(x) < 0$, so f is decreasing.

45. (a) A is increasing on the intervals $(0, 4)$ and $(8, 12)$ and is decreasing on the intervals $(4, 8)$ and $(12, \infty)$.

(b) Local minimum: $x = 8$; local maximum: $x = 4$ and $x = 12$

(c) A has inflection points at $x = 2$, $x = 6$, and $x = 10$.

(d) A is concave up on the intervals $(0, 2)$ and $(6, 10)$ and is concave down on the intervals $(2, 6)$ and $(10, \infty)$.

47. The graph of one such function is

49. Smallest positive critical point: $x = (\pi/2)^{2/3}$ corresponds to a local maximum; smallest positive inflection point: $x = \pi^{2/3}$, $y = F(x)$ changes from concave down to concave up.

51. (a) Then by the FTC, Part II, $A'(x) = f(x)$ and thus $y = A(x)$ and $y = F(x)$ are both antiderivatives of $y = f(x)$. Hence, $F(x) = A(x) + C$ for some constant C.

(b)

$$F(b) - F(a) = (A(b) + C) - (A(a) + C) = A(b) - A(a)$$

$$= \int_a^b f(t)\, dt - \int_a^a f(t)\, dt$$

$$= \int_a^b f(t)\, dt - 0 = \int_a^b f(t)\, dt$$

which proves the FTC, Part I.

53. Write

$$\int_{u(x)}^{v(x)} f(x)\, dx = \int_{u(x)}^0 f(x)\, dx + \int_0^{v(x)} f(x)\, dx$$

$$= \int_0^{v(x)} f(x)\, dx - \int_0^{u(x)} f(x)\, dx$$

Then, by the Chain Rule and the FTC,

$$\frac{d}{dx} \int_{u(x)}^{v(x)} f(x)\, dx = \frac{d}{dx} \int_0^{v(x)} f(x)\, dx - \frac{d}{dx} \int_0^{u(x)} f(x)\, dx$$

$$= f(v(x))v'(x) - f(u(x))u'(x)$$

Section 5.6 Preliminary Questions

1. The total drop in temperature of the metal object in the first T minutes after being submerged in the cold water

2. 560 km

3. Quantities **(a)** and **(c)** would naturally be represented as derivatives; quantities **(b)** and **(d)** would naturally be represented as integrals.

Section 5.6 Exercises

1. 15,250 gal   **3.** 3,660,000   **5.** 33 m   **7.** 3.675 m

9. Displacement: 10 m; distance: 26 m

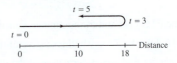

11. Displacement: 0 m; distance: 1 m

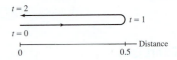

13. 39 m/s   **15.** 9200 cars

17. Total cost: $650; average cost of first 10: $37.50; average cost of last 10: $27.50

19. 112.5 ft

21. The area under the graph in Figure 5 represents the total power consumption over one day in California; 3.627×10^{11} joules

23. (a) 2.916×10^{10}

(b) Approximately 240,526 asteroids of diameter 50 km

25. $\int_0^{365} R(t)\, dt \approx 605.05$ billion ft^3

27. $100 \le t \le 150$: 404.968 families; $350 \le t \le 400$: 245.812 families

29. The particle's velocity is $v(t) = s'(t) = t^{-2}$, an antiderivative for which is $F(t) = -t^{-1}$. Hence, the particle's position at time t is

$$s(t) = \int_1^t s'(u)\, du = F(u)\Big|_1^t = F(t) - F(1) = 1 - \frac{1}{t} < 1$$

for all $t \ge 1$. Thus, the particle will never pass $x = 1$, which implies it will never pass $x = 2$ either.

31. (a) $CS = \int_0^{q^*} [D(q) - p^*]\, dq$ **(b)** $PS = \int_0^{q^*} [p^* - S(q)]\, dq$

Section 5.7 Preliminary Questions

1. (a) and **(b)**

2. (a) $u(x) = x^2 + 9$ **(b)** $u(x) = x^3$ **(c)** $u(x) = \cos x$

3. (c)

Section 5.7 Exercises

1. $du = (3x^2 - 2x)\, dx$ **3.** $du = -2x \sin(x^2)\, dx$

5. $du = 4e^{4x+1}\, dx$

7. $\displaystyle\int (x - 7)^3\, dx = \int u^3\, du = \frac{1}{4}u^4 + C = \frac{1}{4}(x - 7)^4 + C$

9. $\displaystyle\int (3t - 4)^5\, dt = \int \frac{1}{3}u^5\, du = \frac{1}{18}u^6 + C = \frac{1}{18}(3t - 5)^6 + C$

11.

$$\int t\sqrt{t^2 + 1}\, dt = \frac{1}{2} \int u^{1/2}\, du = \frac{1}{3}u^{3/2} + C = \frac{1}{3}(t^2 + 1)^{3/2} + C$$

13. $\displaystyle\int \frac{t^3}{(4-2t^4)^{11}}\,dt = -\frac{1}{8}\int u^{-11}\,du = \frac{1}{80}u^{-10} + C =$
$\displaystyle\frac{1}{80}(4-2t^4)^{-10} + C$

15.

$$\int x(x+1)^9\,dx = \int (u-1)u^9\,du = \int (u^{10}-u^9)\,du$$

$$= \frac{1}{11}u^{11} - \frac{1}{10}u^{10} + C = \frac{1}{11}(x+1)^{11} - \frac{1}{10}(x+1)^{10} + C$$

17.

$$\int x^2\sqrt{x+1}\,dx = \int (u-1)^2 u^{1/2}\,du = \int (u^{5/2} - 2u^{3/2} + u^{1/2})\,du$$

$$= \frac{2}{7}u^{7/2} - \frac{4}{5}u^{5/2} + \frac{2}{3}u^{3/2} + C$$

$$= \frac{2}{7}(x+1)^{7/2} - \frac{4}{5}(x+1)^{5/2} + \frac{2}{3}(x+1)^{3/2} + C$$

19. $\displaystyle\int \sin^2\theta\cos\theta\,d\theta = \int u^2\,du = \frac{1}{3}u^3 + C = \frac{1}{3}\sin^3\theta + C$

21. $\displaystyle\int xe^{-x^2}\,dx = -\frac{1}{2}\int e^u\,du = -\frac{1}{2}e^u + C = -\frac{1}{2}e^{-x^2} + C$

23. $\displaystyle\int \frac{(\ln x)^2}{x}\,dx = \int u^2\,du = \frac{1}{3}u^3 + C = \frac{1}{3}(\ln x)^3 + C$

25. $u = x^4;\ \dfrac{1}{4}\sin(x^4) + C$ **27.** $u = x^{3/2};\ \dfrac{2}{3}\sin(x^{3/2}) + C$

29. $\dfrac{1}{40}(4x+5)^{10} + C$ **31.** $2\sqrt{t+12} + C$

33. $-\dfrac{1}{4(x^2+2x)^2} + C$ **35.** $\sqrt{x^2+9} + C$ **37.** $\dfrac{1}{3}(x^3+x)^3 + C$

39. $\dfrac{1}{36}(3x+8)^{12} + C$ **41.** $\dfrac{2}{9}(x^3+1)^{3/2} + C$

43. $-\dfrac{1}{2}(x+5)^{-2} + C$ **45.** $\dfrac{1}{39}(z^3+1)^{13} + C$

47. $\dfrac{4}{9}(x+1)^{9/4} + \dfrac{4}{5}(x+1)^{5/4} + C$ **49.** $\dfrac{1}{3}\cos(8-3\theta) + C$

51. $2\sin\sqrt{t} + C$ **53.** $\dfrac{1}{4}\ln|\sec(4\theta+9)| + C$ **55.** $\ln|\sin x| + C$

57. $\dfrac{1}{4}\tan(4x+9) + C$ **59.** $2\tan(\sqrt{x}) + C$

61. $-\dfrac{1}{6}(\cos 4x + 1)^{3/2} + C$ **63.** $\dfrac{1}{2}(\sec\theta - 1)^2 + C$

65. $\dfrac{1}{14}e^{14x-7} + C$ **67.** $-\dfrac{1}{3(e^x+1)^3} + C$ **69.** $-\dfrac{1}{e^t+1} + C$

71. $\dfrac{1}{5}(\ln x)^5 + C$ **73.** $-\ln|\cos(\ln x)| + C$

75. $-\dfrac{2}{1+\sqrt{x}} + \dfrac{1}{(1+\sqrt{x})^2} + C$

77. With $u = \sin x$, $\dfrac{1}{2}\sin^2 x + C_1$; with $u = \cos x$, $-\dfrac{1}{2}\cos^2 x + C_2$; the two results differ by a constant.

79. $u = \pi$ and $u = 4\pi$ **81.** 136   **83.** $3 - \sqrt{5}$ **85.** $\dfrac{3}{16}$ **87.** $\dfrac{98}{3}$

89. $\dfrac{243}{4}$ **91.** $\dfrac{1}{2}$ **93.** $\dfrac{1}{2}\ln(\sec 1)$ **95.** $\dfrac{1}{4}$ **97.** $\dfrac{20}{3}\sqrt{5} - \dfrac{32}{5}\sqrt{3}$

99. **(a)** The probability that $v \in [0, b]$ is

$$\int_0^b \frac{1}{32}ve^{-v^2/64}\,dv$$

Let $u = -v^2/64$. Then $du = -v/32\,dv$ and

$$\int_0^b \frac{1}{32}ve^{-v^2/64}\,dv = -\int_0^{-b^2/64} e^u\,du$$

$$= -e^u\Big|_0^{-b^2/64} = -e^{-b^2/64} + 1$$

(b) $e^{-1/16} - e^{-25/64}$

101. $\dfrac{1}{4}f(x)^4 + C$ **103.** $\ln|f(x)| + C$

105. Let $u = \sin\theta$. Then $u(\pi/6) = 1/2$ and $u(0) = 0$, as required. Furthermore, $du = \cos\theta\,d\theta$, so

$$d\theta = \frac{du}{\cos\theta}$$

If $\sin\theta = u$, then $u^2 + \cos^2\theta = 1$, so $\cos\theta = \sqrt{1-u^2}$. Therefore, $d\theta = du/\sqrt{1-u^2}$. This gives

$$\int_0^{\pi/6} f(\sin\theta)\,d\theta = \int_0^{1/2} f(u)\frac{1}{\sqrt{1-u^2}}\,du$$

107. $I = \pi/4$

Section 5.8 Preliminary Questions

1. **(a)** $b = 3$ **(b)** $b = e^3$

2. $b = \sqrt{3}$ **3.** **(b)**   **4.** $x = 4u$

Section 5.8 Exercises

1. $\ln 9$ **3.** 3   **5.** $\dfrac{1}{3}\ln 4$ **7.** $\dfrac{\pi}{12}$ **9.** $\dfrac{\pi}{6}$

11. Let $u = x/3$. Then $x = 3u$, $dx = 3\,du$, $9 + x^2 = 9(1+u^2)$, and

$$\int \frac{dx}{9+x^2} = \int \frac{3\,du}{9(1+u^2)} = \frac{1}{3}\int \frac{du}{1+u^2}$$

$$= \frac{1}{3}\tan^{-1}u + C = \frac{1}{3}\tan^{-1}\frac{x}{3} + C$$

13. $\dfrac{\pi}{3\sqrt{3}}$ **15.** $\dfrac{1}{4}\sin^{-1}(4t) + C$ **17.** $\dfrac{1}{\sqrt{3}}\sin^{-1}\sqrt{\dfrac{3}{5}}t + C$

19. $\dfrac{1}{\sqrt{3}}\sec^{-1}(2x) + C$ **21.** $\dfrac{1}{2}\sec^{-1}x^2 + C$

23. $\dfrac{\pi}{4} - \tan^{-1}(1/2)$ **25.** $\dfrac{(\tan^{-1}x)^2}{2} + C$ **27.** $\dfrac{2}{\ln 3}$ **29.** $\dfrac{1}{\ln 2}$

31. $-\dfrac{1}{\ln 9}\cos(9^x) + C$ **33.** $\dfrac{1}{2}e^{y^2} + C$ **35.** $\dfrac{1}{4}\sqrt{4x^2+9} + C$

37. $-\dfrac{7^{-x}}{\ln 7} + C$ **39.** $\dfrac{1}{8}\tan^8\theta + C$ **41.** $-\sqrt{7-t^2} + C$

43. $\dfrac{3}{2}\ln(x^2+4) + \tan^{-1}(x/2) + C$ **45.** $\dfrac{1}{4}\sin^{-1}(4x) + C$

47. $-e^{-x} - 2x^2 + C$ **49.** $e^x - \dfrac{e^{3x}}{3} + C$

51. $-\sqrt{4-x^2} + 5\sin^{-1}(x/2) + C$ **53.** $\sin(e^x) + C$

55. $\dfrac{1}{4}\sin^{-1}\left(\dfrac{4x}{3}\right) + C$ **57.** $\dfrac{e^{7x}}{7} + \dfrac{3e^{5x}}{5} + e^{3x} + e^x + C$

59. $\dfrac{1}{3}\ln|x^3 + 2| + C$ **61.** $\ln|\sin x| + C$ **63.** $\dfrac{1}{8}(4\ln x + 5)^2 + C$

65. $\dfrac{3^{x^2}}{2\ln 3} + C$ **67.** $\dfrac{(\ln(\sin x))^2}{2} + C$

69. $\dfrac{2}{7}(t-3)^{7/2} + \dfrac{12}{5}(t-3)^{5/2} + 6(t-3)^{3/2} + C$

71. The definite integral $\displaystyle\int_0^x \sqrt{1-t^2}\,dt$ represents the area of the region under the upper half of the unit circle from 0 to x. The region consists of a sector of the circle and a right triangle. The sector has a central angle of $\frac{\pi}{2} - \theta$, where $\cos\theta = x$, and the right triangle has a base of length x and a height of $\sqrt{1-x^2}$.

73. Show that $\dfrac{d}{dt}\left(\sqrt{1-t^2} + t\sin^{-1}t\right) = \sin^{-1}t$.

75. Integrating both sides of the inequality $e^t \geq 1$ yields

$$\int_0^x e^t \, dt = e^x - 1 \geq x \quad \text{or} \quad e^x \geq 1 + x$$

Integrating both sides of this new inequality then gives

$$\int_0^x e^t \, dt = e^x - 1 \geq x + x^2/2 \quad \text{or} \quad e^x \geq 1 + x + x^2/2$$

Finally, integrating both sides again gives

$$\int_0^x e^t \, dt = e^x - 1 \geq x + x^2/2 + x^3/6$$

or

$$e^x \geq 1 + x + x^2/2 + x^3/6$$

as requested.

77. By Exercise 76, $e^x \geq 1 + x + \frac{x^2}{2} + \frac{x^3}{6}$. Thus,

$$\frac{e^x}{x^2} \geq \frac{1}{x^2} + \frac{1}{x} + \frac{1}{2} + \frac{x}{6} \geq \frac{x}{6}$$

Since $\lim\limits_{x \to \infty} x/6 = \infty$, $\lim\limits_{x \to \infty} e^x/x^2 = \infty$. More generally, by Exercise 75,

$$e^x \geq 1 + \frac{x^2}{2} + \cdots + \frac{x^{n+1}}{(n+1)!}$$

Thus,

$$\frac{e^x}{x^n} \geq \frac{1}{x^n} + \cdots + \frac{x}{(n+1)!} \geq \frac{x}{(n+1)!}$$

Since $\lim\limits_{x \to \infty} \frac{x}{(n+1)!} = \infty$, $\lim\limits_{x \to \infty} \frac{e^x}{x^n} = \infty$.

79. (a) The domain of G is $x > 0$ and, by part (i) of the previous exercise, the range of G is $\mathbf{R}$. Now,

$$G'(x) = \frac{1}{x} > 0$$

for all $x > 0$. Thus, G is increasing on its domain, which implies that G has an inverse. The domain of the inverse is $\mathbf{R}$ and the range is $\{x : x > 0\}$. Let F denote the inverse of G.

(b) Let x and y be real numbers and suppose that $x = G(w)$ and $y = G(z)$ for some positive real numbers w and z. Then, using part (b) of the previous exercise,

$$F(x + y) = F(G(w) + G(z)) = F(G(wz)) = wz = F(x) + F(y)$$

(c) Let r be any real number. By part (k) of the previous exercise, $G(E^r) = r$. By definition of an inverse function, it then follows that $F(r) = E^r$.

(d) By the formula for the derivative of an inverse function,

$$F'(x) = \frac{1}{G'(F(x))} = \frac{1}{1/F(x)} = F(x)$$

81.

$$\lim_{n \to -1} \int_1^x t^n \, dt = \lim_{n \to -1} \frac{t^{n+1}}{n+1} \bigg|_1^x = \lim_{n \to -1} \left(\frac{x^{n+1}}{n+1} - \frac{1^{n+1}}{n+1} \right)$$

$$= \lim_{n \to -1} \frac{x^{n+1} - 1}{n+1} = \lim_{n \to -1} (x^{n+1}) \ln x$$

$$= \ln x = \int_1^x t^{-1} \, dt$$

83. (a) Interpreting the graph with y as the independent variable, we see that the function is $x = e^y$. Integrating in y then gives the area of the shaded region as $\int_0^{\ln a} e^y \, dy$.

(b) We can obtain the area under the graph of $y = \ln x$ from $x = 1$ to $x = a$ by computing the area of the rectangle extending from $x = 0$ to $x = a$ horizontally and from $y = 0$ to $y = \ln a$ vertically and then subtracting the area of the shaded region. This yields

$$\int_1^a \ln x \, dx = a \ln a - \int_0^{\ln a} e^y \, dy$$

(c) By direct calculation,

$$\int_0^{\ln a} e^y \, dy = e^y \bigg|_0^{\ln a} = a - 1$$

Thus,

$$\int_1^a \ln x \, dx = a \ln a - (a - 1) = a \ln a - a + 1$$

(d) Based on these results, it appears that

$$\int \ln x \, dx = x \ln x - x + C$$

Section 5.9 Preliminary Questions

1. Doubling time is inversely proportional to the growth constant. Consequently, the quantity with $k = 3.4$ doubles more rapidly.

2. It takes longer for the population to increase from one cell to two cells.

3. $\dfrac{dS}{dn} = -\ln 2 S(n)$ **4. (b)**

5. If the interest rate goes up, the present value of $1 a year from now will decrease.

Section 5.9 Exercises

1. (a) 2000 bacteria initially   **(b)** $t = \dfrac{1}{1.3} \ln 5 \approx 1.24$ h

3. $f(t) = 5e^{t \ln 7}$

5. $N'(t) = \dfrac{\ln 2}{3} N(t)$; 1,048,576 molecules after 1 h

7. $y(t) = Ce^{-5t}$ for some constant C; $y(t) = 3.4e^{-5t}$

9. $y(t) = 1000e^{3(t-2)}$ **11.** 5.33 years

13. $k \approx 0.023$ h^{-1}; $P_0 \approx 332$

15. Double: 11.55 years; triple: 18.31 years; seven-fold: 32.43 years

17. One-half: 1.98 days; one-third: 3.14 days; one-tenth: 6.58 days

19. Set I   **21. (a)** 26.39 years   **(b)** 1969

23. 7600 years   **25.** 2.34×10^{-13} to 2.98×10^{-13} **27.** 2.55 h

29. (a) Yes, the graph looks like an exponential graph especially toward the latter years; $k \approx 0.369$ year^{-1}.

(b)

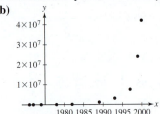

(c) $N(t) = 2250e^{0.369t}$

(d) The doubling time is $\ln 2/0.369 \approx 1.88$ years.

(e) $\approx 2.53 \times 10^{10}$ transistors

(f) No, you can't make a microchip smaller than an atom.

31. With $t_0 = 10$, the doubling time is then 24; with $t_0 = 20$, the doubling time is 44.

33. (a) $P(10) = \$4870.38$ **(b)** $P(10) = \$4902.71$

(c) $P(10) = \$4919.21$

35. (a) 1.0508   **(b)** 1.0513

37. $12,752.56

39. In 3 years:

(a) $PV = \$4176.35$

(b) $PV = \$3594.62$

In 5 years:

(a) $PV = \$3704.09$

(b) $PV = \$2884.75$

41. 9.16%

43. **(a)** The present value of the reduced labor costs is

$$7000(e^{-0.08} + e^{-0.16} + e^{-0.24} + e^{-0.32} + e^{-0.4}) = \$27,708.50$$

This is more than the $25,000 cost of the computer system, so the computer system should be purchased.

(b) The present value of the savings is

$$\$27,708.50 - \$25,000 = \$2708.50$$

45. $39,346.93   **47.** $41,906.75   **51.** $R = \$1200$

53. $71,460.53   **55.** $T = -\dfrac{1}{k}\ln\left(1 - \dfrac{d}{L}\right)$

57. $P(t) = 204e^{ae^{0.15t}}$ with $a \approx -0.02$; 136 rats after 20 months

59. For m-fold growth, $P(t) = mP_0$ for some t. Solving $mP_0 = P_0 e^{kt}$ for t, we find $t = \dfrac{\ln m}{k}$.

61. Start by expressing

$$\ln\left(1 + \frac{x}{n}\right) = \int_1^{1+x/n} \frac{dt}{t}$$

Following the proof in the text, we note that

$$\frac{x}{n+x} \le \ln\left(1 + \frac{x}{n}\right) \le \frac{x}{n}$$

provided $x > 0$, while

$$\frac{x}{n} \le \ln\left(1 + \frac{x}{n}\right) \le \frac{x}{n+x}$$

when $x < 0$. Multiplying both sets of inequalities by n and passing to the limit as $n \to \infty$, the squeeze theorem guarantees that

$$\lim_{n\to\infty}\left(\ln\left(1 + \frac{x}{n}\right)\right)^n = x$$

Finally,

$$\lim_{n\to\infty}\left(1 + \frac{x}{n}\right)^n = e^x$$

63. **(a)** 9.38%

(b) In general,

$$P_0(1 + r/M)^{Mt} = P_0(1 + r_e)^t$$

so $(1 + r/M)^{Mt} = (1 + r_e)^t$ or $r_e = (1 + r/M)^M - 1$. If interest is compounded continuously, then $P_0 e^{rt} = P_0(1 + r_e)^t$, so $e^{rt} = (1 + r_e)^t$ or $r_e = e^r - 1$.

(c) 11.63%

(d) 18.26%

Chapter 5 Review

1. $L_4 = \dfrac{23}{4}$; $M_4 = 7$

3. In general, R_N is larger than $\int_a^b f(x)\,dx$ on any interval $[a, b]$ over which $f(x)$ is increasing. Given the graph of f, we may take $[a, b] = [0, 2]$. In order for L_4 to be larger than $\int_a^b f(x)\,dx$, f must be decreasing over the interval $[a, b]$. We may therefore take $[a, b] = [2, 3]$.

5. $R_6 = \dfrac{625}{8}$

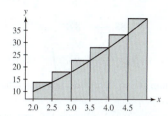

$M_6 = \dfrac{1127}{16}$

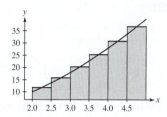

$L_6 = \dfrac{505}{8}$; The rectangles corresponding to this approximation are shown below.

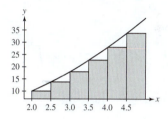

7. $R_N = \dfrac{141}{2} + \dfrac{45}{N} + \dfrac{9}{2N^2}$; $\dfrac{141}{2}$

9. $R_5 \approx 0.733732$; $M_5 \approx 0.786231$; $L_5 \approx 0.833732$

11. The area represented by the shaded rectangles is R_5; $R_5 = 90$; $L_5 = 90$.

13. $\displaystyle\lim_{N\to\infty} \frac{\pi}{6N} \sum_{j=1}^N \sin\left(\frac{\pi}{3} + \frac{\pi j}{6N}\right) = \int_{\pi/3}^{\pi/2} \sin x\,dx = \frac{1}{2}$

15. $\displaystyle\lim_{N\to\infty} \frac{5}{N} \sum_{j=1}^N \sqrt{4 + 5j/N} = \int_4^9 \sqrt{x}\,dx = \frac{38}{3}$

17. $x^4 - \dfrac{2}{3}x^3 + C$ **19.** $-\cos(\theta - 8) + C$

21. $-2t^{-2} + 4t^{-3} + C$ **23.** $\tan x + C$ **25.** $\dfrac{1}{5}(y + 2)^5 + C$

27. $e^x - \dfrac{x^2}{2} + C$ **29.** $4\ln|x| + C$ **31.** $y(x) = x^4 + 3$

33. $y(x) = 2\sqrt{x} - 1$ **35.** $y(x) = -e^{-x} + 4$

37. $f(t) = \dfrac{t^2}{2} - \dfrac{t^3}{3} - t + 2$

39. $\dfrac{1}{4}\ln\dfrac{5}{3}$ **41.** $\dfrac{1}{5}\left(1 - \dfrac{9\sqrt{3}}{32}\right)$

43. $4x^5 - \dfrac{9}{4}x^4 - x^2 + C$ **45.** $\dfrac{4}{5}x^5 - 3x^4 + 3x^3 + C$

47. $\dfrac{1}{4}x^4 + x^3 + C$ **49.** $\dfrac{46}{3}$ **51.** 3

53. $\dfrac{1}{150}(10t - 7)^{15} + C$ **55.** $-\dfrac{1}{24}(3x^4 + 9x^2)^{-4} + C$ **57.** 506

59. $-\dfrac{3\sqrt{3}}{2\pi}$ **61.** $\dfrac{1}{27}\tan(9t^3 + 1) + C$ **63.** $\dfrac{1}{2}\cot(9 - 2\theta) + C$

65. $3 - \dfrac{3\sqrt[3]{4}}{2}$ **67.** $-\dfrac{1}{2}e^{9-2x} + C$ **69.** $\dfrac{1}{3}e^{x^3} + C$

71. $\dfrac{10^x e^x}{\ln 10 + 1} + C$ **73.** $\dfrac{1}{2(e^{-x} + 2)^2} + C$ **75.** $\dfrac{1}{2}\ln 2$

77. $\tan^{-1}(\ln t) + C$ **79.** $\frac{1}{2}$ **81.** $\frac{1}{6}\tan^{-1}\left(\frac{2x}{3}\right) + C$

83. $\sec^{-1}12 - \sec^{-1}4$ **85.** $\frac{\pi}{12}$ **87.** $\frac{1}{2}\sin^{-1}(x^2) + C$

89. $\frac{1}{\sqrt{2}}\tan^{-1}(4\sqrt{2})$ **91.** $\frac{\pi^4}{1024}$ **93.** $\int_{-2}^{6} f(x)\,dx$

95. Local minimum at $x = 0$, no local maxima, inflection points at $x = \pm 1$

97. Daily consumption: 9.312 million gal; from 6 PM to midnight: 1.68 million gal

99. \$208,245 **101.** 0

105. The function $f(x) = 2^x$ is increasing, so $1 \le x \le 2$ implies that $2 = 2^1 \le 2^x \le 2^2 = 4$. Consequently,

$$2 = \int_1^2 2\,dx \le \int_1^2 2^x\,dx \le \int_1^2 4\,dx = 4$$

On the other hand, the function $f(x) = 3^{-x}$ is decreasing, so $1 \le x \le 2$ implies that

$$\frac{1}{9} = 3^{-2} \le 3^{-x} \le 3^{-1} = \frac{1}{3}$$

It then follows that

$$\frac{1}{9} = \int_1^2 \frac{1}{9}\,dx \le \int_1^2 3^{-x}\,dx \le \int_1^2 \frac{1}{3}\,dx = \frac{1}{3}$$

107. $\frac{4}{3} \le \int_0^1 f(x)\,dx \le \frac{5}{3}$ **109.** $-\frac{1}{1+\pi}$

111. $\sin^3 x \cos x$ **113.** -2

115. Consider the figure below, which displays a portion of the graph of a linear function.

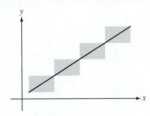

The shaded rectangles represent the differences between the right-endpoint approximation R_N and the left-endpoint approximation L_N. Because the graph of $y = f(x)$ is a line, the lower portion of each shaded rectangle is exactly the same size as the upper portion. Therefore, if we average L_N and R_N, the error in the two approximations will exactly cancel, leaving

$$\frac{1}{2}(R_N + L_N) = \int_a^b f(x)\,dx$$

117. Let

$$F(x) = x\sqrt{x^2 - 1} - 2\int_1^x \sqrt{t^2 - 1}\,dt$$

Then

$$\frac{dF}{dx} = \sqrt{x^2 - 1} + \frac{x^2}{\sqrt{x^2 - 1}} - 2\sqrt{x^2 - 1}$$

$$= \frac{x^2}{\sqrt{x^2 - 1}} - \sqrt{x^2 - 1} = \frac{1}{\sqrt{x^2 - 1}}$$

Also, $\frac{d}{dx}(\cosh^{-1}x) = \frac{1}{\sqrt{x^2-1}}$; therefore, $y = F(x)$ and $y = \cosh^{-1}x$ have the same derivative. We conclude that $y = F(x)$ and $y = \cosh^{-1}x$ differ by a constant:

$$F(x) = \cosh^{-1}x + C$$

Now, let $x = 1$. Because $F(1) = 0$ and $\cosh^{-1}1 = 0$, it follows that $C = 0$. Therefore,

$$F(x) = \cosh^{-1}x$$

121. Approximately 6065.9 years **123.** 5.03% **125.** \$17,979.10

Chapter 6

Section 6.1 Preliminary Questions

1. Area of the region between the graphs of $y = f(x)$ and $y = g(x)$, bounded on the left by the vertical line $x = a$ and on the right by the vertical line $x = b$

2. Yes **3.** $\int_0^3 (f(x) - g(x))\,dx - \int_3^5 (g(x) - f(x))\,dx$

4. Negative

5. The area of the region bounded by the graphs of the functions f and g, and the vertical lines $x = a$ and $x = b$

6.

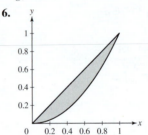

Section 6.1 Exercises

1. 102 **3.** $\frac{32}{3}$

5. $\sqrt{2} - 1$

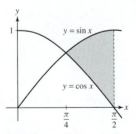

7. $\frac{343}{3}$

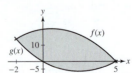

9. $\frac{1}{2}e^2 - e + \frac{1}{2}$

11. $\pi - 2$

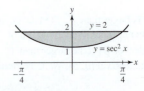

13. Horizontally simple   **15.** Neither   **17.** $\frac{160}{3}$

19. $\frac{12\sqrt{3}-12+\left(\sqrt{3}-2\right)\pi}{24}$ **21.** $2-\frac{\pi}{2}$ **23.** $\frac{1331}{6}$ **25.** 256

27. $\frac{32}{3}$ **29.** $\frac{64}{3}$

31. $\frac{64}{3}$

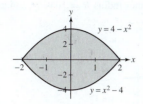

33. 2

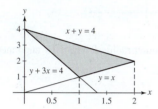

35. $\frac{128}{3}$

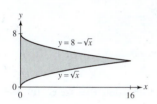

37. $\frac{1}{2}$

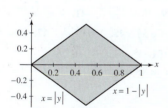

39. $\frac{1225}{8}$

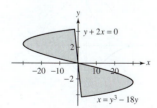

41. $\frac{32}{3}$

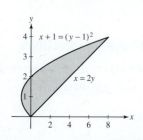

43. $\frac{3\sqrt{3}}{4}$

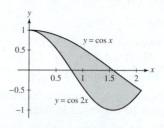

45. $\frac{2-\sqrt{2}}{2}$

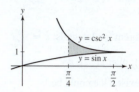

47. $4\ln 2 - 2 \approx 0.77259$

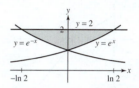

49. ≈ 0.7567130951

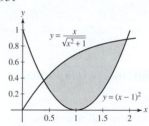

51. (a) (ii)   (b) No   (c) At 10 s, athlete 1; at 25 s, athlete 2

53. $\frac{8}{3}c^{3/2}$; $c = \frac{9^{1/3}}{4} \approx 0.520021$

55. $\int_{-\sqrt{(-1+\sqrt{5})/2}}^{\sqrt{(-1+\sqrt{5})/2}} \left[(1+x^2)^{-1} - x^2\right] dx$

57. 0.8009772242   **59.** 214.75 in.^{2}

61. (b) $\frac{1}{3}$ (c) 0   (d) 1

63. $m = 1 - \left(\frac{1}{2}\right)^{1/3} \approx 0.206299$

Section 6.2 Preliminary Questions

1. 3   **2.** 15

3. Flow rate is the volume of fluid that passes through a cross-sectional area at a given point per unit time.

4. The fluid velocity depended only on the radial distance from the center of the tube.

5. 15

Section 6.2 Exercises

1. (a) $\frac{4}{25}(20-y)^2$ (b) $\frac{1280}{3}$

3. $\frac{\pi R^2 h}{3}$ **5.** $\pi\left(Rh^2 - \frac{h^3}{3}\right)$ **7.** $\frac{1}{6}abc$ **9.** $\frac{8}{3}$ **11.** 36   **13.** 18

15. $\frac{\pi}{3}$ **17.** 96π

21. (a) $2\sqrt{r^2 - y^2}$ **(b)** $4(r^2 - y^2)$ **(c)** $\frac{16}{3}r^3$

23. 160π **25.** $5\,\mathrm{kg}$ **27.** $0.36\,\mathrm{g}$

29. $P \approx 4423.59$ thousand **31.** $L_{10} = 233.86$, $R_6 = 290.56$

33. $P \approx 61$ deer **35.** $Q = 128\pi\,\mathrm{cm^3/s}$ **37.** $Q = \frac{8\pi}{3}\,\mathrm{cm^3/s}$

39. 16 **41.** $\frac{3}{\pi}$ **43.** $\frac{1}{10}$ **45.** -4 **47.** $\frac{1}{n+1}$

49. Over $[0,24]$, the average temperature is 20; over $[2,6]$, the average temperature is $20 + \frac{15}{2\pi} \approx 22.387325$.

51. $\approx 79.56°\mathrm{F}$ **53.** $\frac{100}{\pi}$ **55.** $\frac{17}{2}\,\mathrm{m/s}$ **57.** $159.033\,\mathrm{m/s}$

59. $\frac{3}{5^{1/4}} \approx 2.006221$

61. Mean Value Theorem for Integrals; $c = \frac{A}{\sqrt[3]{4}}$

63. Over $[0, 1]$, $f(x)$; over $[1, 2]$, g

65. Many solutions exist. One could be

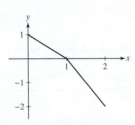

67. $v_0/2$

Section 6.3 Preliminary Questions

1. (a), (c) **2.** True

3. False, the cross sections will be washers.

4. (b)

Section 6.3 Exercises

1. (a)

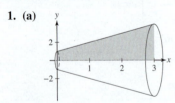

(b) Disk with radius $x + 1$

(c) $V = 21\pi$

3. (a)

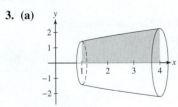

(b) Disk with radius $\sqrt{x + 1}$

(c) $V = \frac{21\pi}{2}$

5. $V = \frac{81\pi}{10}$ **7.** $V = \frac{24{,}573\pi}{13}$ **9.** $V = \pi$

11. $V = \frac{\pi}{2}\left(e^2 - 1\right)$ **13.** (ii)

15. (a)

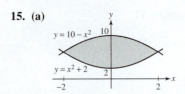

(b) A washer with outer radius $R = 10 - x^2$ and inner radius $r = x^2 + 2$

(c) $V = 256\pi$

17. (a)

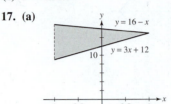

(b) A washer with outer radius $R = 16 - x$ and inner radius $r = 3x + 12$

(c) $V = \frac{656\pi}{3}$

19. (a)

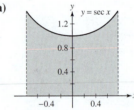

(b) A circular disk with radius $R = \sec x$ **(c)** $V = 2\pi$

21. $V = \frac{15\pi}{2}$ **23.** $V = \frac{3\pi}{10}$ **25.** $V = 32\pi$ **27.** $V = \frac{704\pi}{15}$

29. $V = \frac{128\pi}{5}$ **31.** $V = 40\pi$ **33.** $V = \frac{376\pi}{15}$ **35.** $V = \frac{824\pi}{15}$

37. $V = \frac{32\pi}{3}$ **39.** $V = \frac{1872\pi}{5}$ **41.** $V = \frac{1400\pi}{3}$

43. $V = \pi\left(\frac{7\pi}{9} - \sqrt{3}\right)$ **45.** $V = \frac{96\pi}{5}$ **47.** $V = \frac{16\pi}{35}$

49. $V = \frac{1184\pi}{15}$ **51.** $V = 7\pi\left(1 - \ln 2\right)$

53. $V \approx 12{,}120\pi \approx 38{,}076.1\,\mathrm{cm^3}$ **55.** $V = \frac{1}{3}\pi r^2 h$

57. $V = \frac{32\pi}{105}$

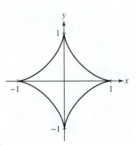

59. $V = 4\pi\sqrt{3}$ **61.** $V = \frac{4}{3}\pi a^2 b$

Section 6.4 Preliminary Questions

1. (a) Radius h and height r **(b)** Radius r and height h

2. (a) With respect to x **(b)** With respect to y

3. $V = 2\pi \int_0^8 y \cdot 1\, dy = 64\pi$

Section 6.4 Exercises

1. $V = \frac{2}{5}\pi$

3. $V = 4\pi$

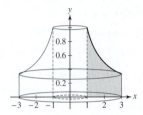

5. $V = 18\pi\left(2\sqrt{2} - 1\right)$

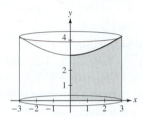

7. $V = \frac{32\pi}{3}$ **9.** $V = 16\pi$ **11.** $V = \frac{32\pi}{5}$

13. The point of intersection is $x = 1.376769504$; $V = 1.321975576$.

15. $V = \frac{3\pi}{5}$

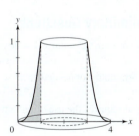

17. $V = \frac{280\pi}{81}$

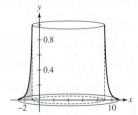

19. $V = \frac{1}{3}\pi a^3 + \pi a^2$

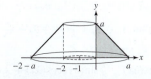

21. $V = \frac{\pi}{3}$

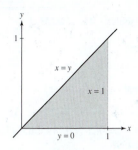

23. $V = \frac{128\pi}{3}$

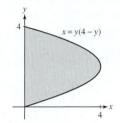

25. $V = \frac{256\pi}{15}$

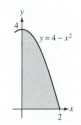

27. (b)

29. (a) $V = \frac{576\pi}{7}$ **(b)** $V = \frac{96\pi}{5}$

31. (a) $\overline{AB}$ generates a disk with radius $R = h(y)$; $\overline{CB}$ generates a shell with radius x and height $f(x)$.

(b) Shell, $V = 2\pi \int_0^2 xf(x)\,dx$; Disk, $V = \pi \int_0^{1.3} (h(y))^2\,dy$

33. $V = \frac{602\pi}{5}$ **35.** $V = 8\pi$ **37.** $V = \frac{40\pi}{3}$ **39.** $V = \frac{1024\pi}{15}$

41. $V = 16\pi$ **43.** $V = \frac{32\pi}{3}$ **45.** $V = \frac{776\pi}{15}$ **47.** $V = \frac{625\pi}{6}$

49. $\frac{3\pi}{10}$ **51.** $V = \frac{121\pi}{525}$ **53.** $V = \frac{563\pi}{30}$

55. $V = \frac{4}{3}\pi r^3$ **57.** $V = 2\pi^2 ab^2$

59. $x^2 + y^2 = 1$ over $[0, 1]$; $y = 1 - x$ over $[0, 1]$

Section 6.5 Preliminary Questions

1. Because the required force is not constant through the stretching process

2. The force involved in lifting the tank is the weight of the tank, which is constant.

3. $\frac{1}{2}kx^2$

4. When the force applied is in the opposite direction to the motion

Section 6.5 Exercises

1. $W = 627.2$ joules   **3.** $W = 5.76$ joules   **5.** $W = 8$ joules

7. $W = 11.25$ joules   **9.** $W = 3.800$ joules

11. $W = 105,840$ joules

13. $W = \frac{56,448\pi}{5}$ joules $\approx 3.547 \times 10^4$ joules

15. $W \approx 1.842 \times 10^{12}$ joules   **17.** $W = 3.92 \times 10^6$ joules

19. $W \approx 1.18 \times 10^8$ joules   **21.** $W = 9800\pi \ell r^3$ joules

23. $W = 2.94 \times 10^6$ joules   **25.** $W \approx 1.222 \times 10^6$ joules

27. $W = 3920$ joules   **29.** $W = 529.2$ joules

31. $W = 1470$ joules **33.** $W = 374.85$ joules

37. $W \approx 5.16 \times 10^9$ joules **41.** $\sqrt{2GM_e \left(\frac{1}{R_e} - \frac{1}{r+R_e} \right)}$ m/s

43. $v_{\text{esc}} = \sqrt{\frac{2GM_e}{R_e}}$ m/s

Chapter 6 Review

1. $\frac{32}{3}$ **3.** $\frac{1}{2}$ **5.** 24 **7.** $\frac{1}{2}$ **9.** $3\sqrt{2} - 1$ **11.** $e - \frac{3}{2}$

13. Intersection points $x = 0$, $x = 0.7145563847$;
Area $= 0.08235024596$

15. $V = 4\pi$ **17.** 2.7552 kg **19.** $\frac{9}{4}$ **21.** $\frac{1}{2} \sinh 1$ **23.** $\frac{3\pi}{4}$

25. 27 **27.** $\frac{2\pi m^5}{15}$ **29.** $V = \frac{162\pi}{5}$ **31.** $V = 64\pi$ **33.** $V = 8\pi$

35. $V = \frac{56\pi}{15}$ **37.** $V = \frac{128\pi}{15}$ **39.** $V = 4\pi \left(1 - \frac{1}{\sqrt{e}} \right)$

41. $V = 2\pi \left(c + \frac{c^3}{3} \right)$ **43.** $V = c\pi$

45. (a) $\int_0^1 \left(\sqrt{1 - (x-1)^2} - \left(1 - \sqrt{1 - x^2} \right) \right) dx$

(b) $\pi \int_0^1 \left[(1 - (x-1)^2) - (1 - \sqrt{1 - x^2})^2 \right] dx$

47. $V = \pi \int_0^{1/a} \left(a\sqrt{x} - ax^2 \right)^2 dx = \frac{\pi}{6}$ **49.** $W = 1.6$ joules

51. $W = 1.93 \times 10^{10}$ joules **53.** $9800\pi \left(h^3 + 2h^2 - \frac{1}{4}h^4 \right)$ joules

Chapter 7

Section 7.1 Preliminary Questions

1. The Integration by Parts formula is derived from the Product Rule.

3. Transforming $v' = x$ into $v = \frac{1}{2}x^2$ increases the power of x and makes the new integral harder than the original.

Section 7.1 Exercises

1. $-x \cos x + \sin x + C$ **3.** $e^x(2x + 7) + C$

5. $\frac{x^4}{16}(4 \ln x - 1) + C$ **7.** $-e^{-x}(4x + 1) + C$

9. $\frac{1}{25}(5x - 1)e^{5x+2} + C$ **11.** $\frac{1}{2}x \sin 2x + \frac{1}{4} \cos 2x + C$

13. $-x^2 \cos x + 2x \sin x + 2 \cos x + C$

15. $-\frac{1}{2}e^{-x}(\sin x + \cos x) + C$

17. $-\frac{1}{26}e^{-5x}(\cos(x) + 5\sin(x)) + C$

19. $\frac{1}{4}x^2(2 \ln x - 1) + C$ **21.** $\frac{x^3}{3} \left(\ln x - \frac{1}{3} \right) + C$

23. $x\left[(\ln x)^2 - 2 \ln x + 2 \right] + C$ **25.** $x \cos^{-1} x - \sqrt{1 - x^2} + C$

27. $x \sec^{-1} x - \ln |x + \sqrt{x^2 - 1}| + C$

29. $\frac{3^x (\sin x + \ln 3 \cos x)}{1 + (\ln 3)^2} + C$

31. $(x^2 + 2) \sinh x - 2x \cosh x + C$

33. $x \tanh^{-1} 4x + \frac{1}{8} \ln |1 - 16x^2| + C$ **35.** $2e^{\sqrt{x}}(\sqrt{x} - 1) + C$

37. $\frac{1}{4}x \sin 4x + \frac{1}{16} \cos 4x + C$

39. $\frac{2}{3}(x + 1)^{3/2} - 2(x + 1)^{1/2} + C$

41. $\sin x \ln(\sin x) - \sin x + C$

43. $2xe^{\sqrt{x}} - 4\sqrt{x}e^{\sqrt{x}} + 4e^{\sqrt{x}} + C$

45. $\frac{1}{4}(\ln x)^2[2 \ln(\ln x) - 1] + C$

47. $\frac{1}{16}(11e^{12} + 1)$

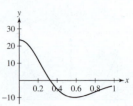

49. $2 \ln 2 - \frac{3}{4}$ **51.** $\frac{e^\pi + 1}{2}$ **51.** $1 - \frac{2}{e}$ **53.** $\frac{3 \ln 3 - 2}{\ln^2 3}$ **55.** $\frac{e^\pi + 1}{2}$

57. $-\frac{1}{5}\sin^4 x \cos x - \frac{4}{15}\sin^2 x \cos x - \frac{8}{15} \cos x + C$

59. $\frac{1}{3}\cos^2 x \sin x + \frac{2}{3} \sin x + C$

61. $e^x(x^3 - 3x^2 + 6x - 6) + C$

63. $\int x^n e^{-x} \, dx = -x^n e^{-x} + n \int x^{n-1} e^{-x} \, dx$ **65.** $\pi \left(\pi^2 - 4 \right)$

67. Use Integration by Parts, with $u = \ln x$ and $dv = \sqrt{x} \, dx$.

69. Use substitution, followed by algebraic manipulation, with $u = 4 - x^2$ and $du = -2x \, dx$.

71. Use substitution with $u = x^2 + 4x + 3$, $\frac{du}{2} = (x + 2) \, dx$.

73. Use Integration by Parts, with $u = x$ and $dv = \sin(3x + 4) \, dx$.

75. $x(\sin^{-1} x)^2 + 2\sqrt{1 - x^2} \sin^{-1} x - 2x + C$

77. $\frac{1}{4}x^4 \sin(x^4) + \frac{1}{4} \cos(x^4) + C$

79. $2\pi(e^2 + 1)$ **81.** \$42,995.10

83. For $k = 2$: $x(\ln x)^2 - 2x \ln x + 2x + C$; for $k = 3$: $x(\ln x)^3 - 3x(\ln x)^2 + 6x \ln x - 6x + C$

85. Use Integration by Parts with $u = x$ and $v' = b^x$.

87. (b) $V(x) = \frac{1}{2}x^2 + \frac{1}{2}$ is simpler and yields $\frac{1}{2}(x^2 \tan^{-1} x - x + \tan^{-1} x) + C$

89. An example of a function satisfying these properties for some λ is $f(x) = \sin \pi x$.

91. (a) $I_n = \frac{1}{2}x^{n-1} \sin(x^2) - \frac{n-1}{2} J_{n-2}$;
(c) $\frac{1}{2}x^2 \sin(x^2) + \frac{1}{2} \cos(x^2) + C$

Section 7.2 Preliminary Questions

1. Rewrite $\sin^5 x = \sin x \sin^4 x = \sin x(1 - \cos^2 x)^2$ and then substitute $u = \cos x$.

3. No, a reduction formula is not needed because the sine function is raised to an odd power.

5. The second integral requires the use of reduction formulas and therefore more work.

Section 7.2 Exercises

1. $\sin x - \frac{1}{3} \sin^3 x + C$ **3.** $-\frac{1}{3} \cos^3 \theta + \frac{1}{5} \cos^5 \theta + C$

5. $-\frac{1}{4} \cos^4 t + \frac{1}{6} \cos^6 t + C$ **7.** 2

9. $\frac{3y}{8} + \frac{1}{4} \sin(2y) + \frac{1}{32} \sin(4y) + C$

11. $\frac{1}{6} \sin^5 x \cos x - \frac{1}{24} \sin^3 x \cos x - \frac{1}{16} \sin x \cos x + \frac{1}{16}x + C$

13. $\frac{1}{5} \sin^4 x \cos x - \frac{4}{15} \sin^2 x \cos x - \frac{2}{15} \cos x + C$

15. $\frac{1}{3} \sec^3 x - \sec x + C$

17. $\frac{1}{5} \tan x \sec^4 x - \frac{1}{15} \tan(x) \sec^2 x - \frac{2}{15} \tan x + C$

19. $-\frac{1}{2} \cot^2 x + \ln |\csc x| + C$ **21.** $-\frac{1}{6} \cot^6 x + C$

23. $-\frac{1}{6} \cos^6 x + C$

25. $\frac{1}{12}\cos^3(3x + 2) \sin(3x + 2) + \frac{1}{8}(3x + 2) + \frac{1}{16} \sin(6x + 4) + C$

27. $\frac{1}{5\pi} \sin^5(\pi\theta) - \frac{1}{7\pi} \sin^7(\pi\theta) + C$

29. $-\frac{1}{12}\sin^3(3x) \cos(3x) - \frac{1}{8} \sin(3x) \cos(3x) + \frac{3}{8}x + C$

31. $\frac{1}{2}\sin^2 x - \frac{1}{2\sin^2 x} - 2\ln|\sin x| + C$

33. $\frac{1}{2}\cot(3 - 2x) + C$ **35.** $\frac{1}{2}\tan^2 x + C$

37. $\frac{1}{8}\sec^8 x - \frac{1}{3}\sec^6 x + \frac{1}{4}\sec^4 x + C$

39. $\frac{1}{9}\tan^9 x + \frac{1}{7}\tan^7 x + C$

41. $-\frac{1}{9}\csc^9 x + \frac{2}{7}\csc^7 x - \frac{1}{5}\csc^5 x + C$ **43.** $\frac{1}{4}\sin^2 2x + C$

45. $\frac{1}{6}\cos^2(t^2)\sin(t^2) + \frac{1}{3}\sin(t^2) + C$

47. $\frac{1}{2}\cos(\sin t)\sin(\sin t) + \frac{1}{2}\sin t + C$ **49.** π **51.** $\frac{8}{15}$

53. $\ln\left(\sqrt{2}+1\right)$ **55.** $\ln 2$ **57.** $\frac{8}{3}$

59. $-\frac{6}{7}$ **61.** $\frac{1}{24}$

63. First, observe $\sin 4x = 2\sin 2x\cos 2x = 2\sin 2x(1 - 2\sin^2 x) = 2\sin 2x - 4\sin 2x\sin^2 x = 2\sin 2x - 8\sin^3 x\cos x$. Then, $\frac{1}{32}(12x - 8\sin 2x + \sin 4x) + C = \frac{3}{8}x - \frac{3}{16}\sin 2x - \frac{1}{4}\sin^3 x\cos x + C = \frac{3}{8}x - \frac{3}{8}\sin x\cos x - \frac{1}{4}\sin^3 x\cos x + C$.

65. $\frac{\pi^2}{2}$ **67.** $\frac{1}{8}x - \frac{1}{16}\sin 2x\cos 2x + C$

69. $\frac{1}{16}x - \frac{1}{48}\sin 2x - \frac{1}{32}\sin 2x\cos 2x + \frac{1}{48}\cos^2 2x\sin 2x + C$

71. Use the identity $\tan^2 x = \sec^2 x - 1$ and the substitution $u = \tan x$, $du = \sec^2 x\,dx$.

73. (a) $I_0 = \int_0^{\pi/2}\sin^0 x\,dx = \frac{\pi}{2}$; $I_1 = \int_0^{\pi/2}\sin x\,dx = 1$

(b) $\frac{m-1}{m}\int_0^{\pi/2}\sin^{m-2} x\,dx$

(c) $I_2 = \frac{\pi}{4}$; $I_3 = \frac{2}{3}$; $I_4 = \frac{3\pi}{16}$; $I_5 = \frac{8}{15}$

75. $\cos(x) - \cos(x)\ln(\sin(x)) + \ln|\csc(x) - \cot(x)| + C$

79. Use Integration by Parts with $u = \sec^{m-2} x$ and $v' = \sec^2 x$.

Section 7.3 Preliminary Questions

1. (a) $x = 3\sin\theta$ (b) $x = 4\sec\theta$ (c) $x = 4\tan\theta$
(d) $x = \sqrt{5}\sec\theta$

3. $2x\sqrt{1-x^2}$

Section 7.3 Exercises

1. (a) $\theta + C$ (b) $\sin^{-1}\left(\frac{x}{3}\right) + C$

3. (a) $\int\frac{dx}{\sqrt{4x^2+9}} = \frac{1}{2}\int\sec\theta\,d\theta$

(b) $\frac{1}{2}\ln|\sec\theta + \tan\theta| + C$

(c) $\frac{1}{2}\ln\left|\sqrt{4x^2+9}+2x\right| + C$

5. $\frac{8}{\sqrt{5}}\arcsin(\frac{x\sqrt{5}}{4}) + \frac{1}{2}x\sqrt{16-5x^2} + C$ **7.** $\frac{1}{3}\sec^{-1}\left(\frac{x}{3}\right) + C$

9. $\frac{-x}{4\sqrt{x^2-4}} + C$ **11.** $\sqrt{x^2-4} + C$

13. (a) $-\sqrt{1-x^2}$ (b) $\frac{1}{8}(\arcsin x - x\sqrt{1-x^2}(1-2x^2))$

(c) $-\frac{1}{3}(1-x^2)^{\frac{3}{2}} + \frac{1}{5}(1-x^2)^{\frac{5}{2}}$

(d) $\sqrt{1-x^2}(-\frac{x^3}{4} - \frac{3x}{8}) + \frac{3}{8}\arcsin(x)$

15. $\frac{9}{2}\sin^{-1}\left(\frac{x}{3}\right) - \frac{1}{2}x\sqrt{9-x^2} + C$

17. $\frac{1}{4}\ln\left|\frac{\sqrt{x^2+16}-4}{x}\right| + C$ **19.** $\ln\left|x + \sqrt{x^2-9}\right| + C$

21. $-\frac{\sqrt{5-y^2}}{5y} + C$ **23.** $\frac{1}{5}\ln\left|5x + \sqrt{25x^2+2}\right| + C$

25. $\frac{1}{16}\sec^{-1}\left(\frac{z}{2}\right) + \frac{\sqrt{z^2-4}}{8z^2} + C$

27. $\frac{1}{12}x\sqrt{6x^2-49} + \frac{49\ln\left(x\sqrt{6}+\sqrt{6x^2-49}\right)}{12\sqrt{6}} + C$

29. $\frac{1}{54}\tan^{-1}\left(\frac{1}{3}\right) + \frac{1}{180}$ **31.** $-\frac{x}{\sqrt{x^2-1}} + \ln\left|x + \sqrt{x^2+1}\right| + C$

33. Use the substitution $x = \sqrt{a}\,u$.

35. (a) $x^2 - 4x + 8 = x^2 - 4x + 4 + 4 = (x-2)^2 + 4$

(b) $\ln\left|\sqrt{u^2+4}+u\right| + C$

(c) $\ln\left|\sqrt{(x-2)^2+4}+x-2\right| + C$

37. $\ln\left|\sqrt{x^2+4x+13}+x+2\right| + C$

39. $\frac{1}{\sqrt{6}}\ln\left|12x + 1 + 2\sqrt{6}\sqrt{x+6x^2}\right| + C$

41. $\frac{1}{2}(x-2)\sqrt{x^2-4x+3} - \frac{1}{2}\ln\left|x-2+\sqrt{x^2-4x+3}\right| + C$

43. $x\sec^{-1} x - \ln\left|x + \sqrt{x^2-1}\right| + C$

45. $x(\ln(x^2+1)-2) + 2\tan^{-1} x + C$

47. $\frac{\pi}{4}$ **49.** $4\pi\left[\sqrt{3} - \ln\left|2+\sqrt{3}\right|\right]$ **51.** $\frac{1}{2}\ln\left(\frac{x-1}{x+1}\right) + C$

53. (a) $1.789 \times 10^6\,\frac{V}{m}$ (b) $3.526 \times 10^6\,\frac{V}{m}$

Section 7.4 Preliminary Questions

1. (a) $x = \sinh t$ (b) $x = 3\sinh t$ (c) $3x = \sinh t$

3. $\frac{1}{2}\ln\left|\frac{1+x}{1-x}\right|$

Section 7.4 Exercises

1. $\frac{1}{3}\sinh(3x) + C$ **3.** $\frac{1}{2}\cosh(x^2+1) + C$

5. $-\frac{1}{2}\tanh(1-2x) + C$ **7.** $\frac{\tanh^2 x}{2} + C$ **9.** $\ln\cosh x + C$

11. $\ln|\sinh x| + C$ **13.** $\frac{1}{16}\sinh(8x-18) - \frac{1}{2}x + C$

15. $\frac{1}{32}\sinh 4x - \frac{1}{8}x + C$ **17.** $\cosh^{-1} x + C$

19. $\frac{1}{5}\sinh^{-1}\left(\frac{5x}{4}\right) + C$ **21.** $\frac{1}{2}x\sqrt{x^2-1} - \frac{1}{2}\cosh^{-1} x + C$

23. $2\tanh^{-1}\left(\frac{1}{2}\right)$ **25.** $\sinh^{-1} 1$

27. $\frac{1}{4}\left(\operatorname{csch}^{-1}\left(-\frac{1}{4}\right) - \operatorname{csch}^{-1}\left(-\frac{3}{4}\right)\right)$

29. $\cosh^{-1} x - \frac{\sqrt{x^2-1}}{x} + C$

31. Let $x = \sinh t$ for the first formula and $x = \cosh t$ for the second.

33. $\frac{1}{2}x\sqrt{x^2+16} + 8\ln\left|\frac{x}{4} + \sqrt{\left(\frac{x}{4}\right)^2+1}\right| + C$

35. Using Integration by Parts with $u = \cosh^{n-1} x$ and $v' = \cosh x$ to begin proof

37. $-\frac{1}{2}\left(\tanh^{-1} x\right)^2 + C$ **39.** $x\tanh^{-1} x + \frac{1}{2}\ln|1-x^2| + C$

41. $u = \sqrt{\frac{\cosh x-1}{\cosh x+1}}$. From this it follows that $\cosh x = \frac{1+u^2}{1-u^2}$, $\sinh x = \frac{2u}{1-u^2}$, and $dx = \frac{2du}{1-u^2}$.

43. $\int du = u + C = \tanh\frac{x}{2} + C$

45. Let $gd(y) = \tan^{-1}(\sinh y)$. Then

$$\frac{d}{dy}gd(y) = \frac{1}{1+\sinh^2 y}\cosh y = \frac{1}{\cosh y} = \operatorname{sech} y$$

where we have used the identity $1 + \sinh^2 y = \cosh^2 y$.

47. Let $x = gd(y) = \tan^{-1}(\sinh y)$. Solving for y yields $y = \sinh^{-1}(\tan x)$. Therefore, $gd^{-1}(y) = \sinh^{-1}(\tan y)$.

49. Let $x = it$. Then $\cosh^2 x = (\cosh(it))^2 = \cos^2 t$ and $\sinh^2 x = (\sinh(it))^2 = i^2\sin^2 t = -\sin^2 t$. Thus, $1 = \cosh^2(it) - \sinh^2(it) = \cos^2 t - (-\sin^2 t) = \cos^2 t + \sin^2 t$, as desired.

Section 7.5 Preliminary Questions

1. No, f cannot be a rational function because the integral of a rational function cannot contain a term with a noninteger exponent such as $\sqrt{x+1}$.

3. (a) Square is already completed; irreducible.

(b) Square is already completed; factors as $(x - \sqrt{5})(x + \sqrt{5})$.

(c) $x^2 + 4x + 6 = (x + 2)^2 + 2$; irreducible.

(d) $x^2 + 4x + 2 = (x + 2)^2 - 2$; factors as $(x + 2 - \sqrt{2})(x + 2 + \sqrt{2})$.

Section 7.5 Exercises

1. (a) $\dfrac{x^2 + 4x + 12}{(x + 2)(x^2 + 4)} = \dfrac{1}{x + 2} + \dfrac{4}{x^2 + 4}$.

(b) $\dfrac{2x^2 + 8x + 24}{(x + 2)^2(x^2 + 4)} = \dfrac{1}{x + 2} + \dfrac{2}{(x + 2)^2} + \dfrac{-x + 2}{x^2 + 4}$.

(c) $\dfrac{x^2 - 4x + 8}{(x - 1)^2(x - 2)^2} = \dfrac{-8}{x - 2} + \dfrac{4}{(x - 2)^2} + \dfrac{8}{x - 1} + \dfrac{5}{(x - 1)^2}$.

(d) $\dfrac{x^4 - 4x + 8}{(x + 2)(x^2 + 4)} = x - 2 + \dfrac{4}{x + 2} - \dfrac{4x - 4}{x^2 + 4}$.

3. -2 **5.** $\frac{1}{9}(3x + 4\ln(3x - 4)) + C$

7. $\frac{x^3}{3} + \ln|x + 2| + C$ **9.** $-\frac{1}{3}\ln|x - 2| + \frac{1}{2}\ln|x - 4| + C$

11. $\ln|x| - \ln|2x + 1| + C$ **13.** $x - 3\arctan\left(\frac{x}{3}\right) + C$

15. $2\ln|x + 3| - \ln|x + 5| - \frac{2}{3}\ln|3x - 2| + C$

17. $3\ln|x - 1| - 2\ln|x + 1| - \frac{5}{x+1} + C$

19. $2\ln|x - 1| - \frac{1}{x-1} - 2\ln|x - 2| - \frac{1}{x-2} + C$

21. $\ln|x| - \ln|x + 2| + \frac{2}{x+2} + \frac{2}{(x+2)^2} + C$

23. $\frac{1}{2\sqrt{6}}\ln\left|\sqrt{2}x - \sqrt{3}\right| - \frac{1}{2\sqrt{6}}\ln\left|\sqrt{2}x + \sqrt{3}\right| + C$

25. $\frac{1}{2(x+1)} + \frac{1}{4}\ln|x - 1| - \frac{1}{4}\ln|x + 1| + C$

27. $\frac{5}{2x+5} - \frac{5}{4(2x+5)^2} + \frac{1}{2}\ln|2x + 5| + C$

29. $-\ln|x| + \ln|x - 1| + \frac{1}{x-1} - \frac{1}{2(x-1)^2} + C$

31. $x + \ln|x| - 3\ln|x + 1| + C$

33. $2\ln|x - 1| + \frac{1}{2}\ln|x^2 + 1| - 3\tan^{-1}x + C$

35. $\frac{1}{25}\ln|x| - \frac{1}{50}\ln|x^2 + 25| + C$

37. $6x - 14\ln|x + 3| + 2\ln|x - 1| + C$

39. $-\frac{1}{5}\ln|x - 1| - \frac{1}{x-1} + \frac{1}{10}\ln|x^2 + 9| - \frac{4}{15}\tan^{-1}\left(\frac{x}{3}\right) + C$

41. $\frac{1}{64}\ln|x| - \frac{1}{128}\ln|x^2 + 8| + \frac{1}{16(x^2+8)} + C$

43. $\frac{1}{6}\ln|x + 2| - \frac{1}{12}\ln|x^2 + 4x + 10| + C$

45. $\ln|x| - \frac{1}{2}\left|x^2 + 2x + 5\right| + \frac{15-5x}{8(x^2+2x+5)} - \frac{13}{16}\tan^{-1}\left(\frac{x+1}{2}\right) + C$

47. $\frac{1}{2}\arctan(x^2) + C$ **49.** $\ln|e^x - 1| - x + C$

51. $2\sqrt{x} + \ln|\sqrt{x} - 1| - \ln|\sqrt{x} + 1| + C$

53. $\ln\left|x^{1/4} - 2\right| - \ln\left|x^{1/4} + 2\right| + C$

55. $\ln\left|\dfrac{x}{\sqrt{x^2-1}} - \dfrac{1}{\sqrt{x^2-1}}\right| + C = \ln\left|\dfrac{x-1}{\sqrt{x^2-1}}\right| + C$

57. If $\theta = 2\tan^{-1}t$, then $d\theta = 2\,dt/(1 + t^2)$. We also have $\cos(\frac{\theta}{2}) = 1/\sqrt{1 + t^2}$ and $\sin(\frac{\theta}{2}) = t/\sqrt{1 + t^2}$. To find $\cos\theta$, we use the double angle identity $\cos\theta = 1 - 2\sin^2(\frac{\theta}{2})$. This gives us $\cos\theta = \frac{1-t^2}{1+t^2}$. To find $\sin\theta$, we use the double angle identity $\sin\theta = 2\sin(\frac{\theta}{2})\cos(\frac{\theta}{2})$. This gives us $\sin\theta = \frac{2t}{1+t^2}$. It follows then

that $\displaystyle\int \dfrac{d\theta}{\cos\theta + \frac{3}{4}\sin\theta} =$
$-\dfrac{4}{5}\ln\left|2 - \tan\left(\dfrac{\theta}{2}\right)\right| + \dfrac{4}{5}\ln\left|1 + 2\tan\left(\dfrac{\theta}{2}\right)\right| + C.$

59. Partial fraction decomposition shows $\dfrac{1}{(x-a)(x-b)} = \dfrac{\frac{1}{a-b}}{x-a} + \dfrac{\frac{1}{b-a}}{x-b}.$
This can be used to show $\displaystyle\int \dfrac{dx}{(x-a)(x-b)} = \dfrac{1}{a-b}\ln\left|\dfrac{x-a}{x-b}\right| + C.$

61. $\frac{2}{x-6} + \frac{1}{x+2}$

Section 7.6 Preliminary Questions

1. Integration by parts with $u = x$, $dv = \sin x\,dx$

2. Trigonometric substitution $x = \tan\theta$, $dx = \sec^2\theta\,d\theta$

3. Partial fractions

4. Substitute $u = \cos x$, $du = -\sin x\,dx$.

5. Integration by parts with $u = \ln x$, $dv = x\,dx$

6. Trigonometric substitution $x = \sin\theta$, $dx = \cos\theta\,d\theta$

7. Trig Method. Rewrite as $\int\left(1 - \sin^2 x\right)\cos^2 x \sin x\,dx$. Then use u-substitution with $u = \cos x$ and $du = -\sin x\,dx$.

8.
$\displaystyle\int \dfrac{u^2\,du}{a + bu} = \dfrac{1}{2b^3}\left[(a + bu)^2 - 4a\,(a + bu) + 2a^2\ln|a + bu|\right] + C$

9. Trig substitution with $x = \frac{5}{4}\tan\theta$ and $dx = \frac{5}{4}\sec^2\theta\,d\theta$

10. $\int \sec^3 u\,du = \frac{1}{2}\sec u\tan u + \frac{1}{2}\ln|\sec u + \tan u| + C$

11. Complete the square, followed by u-substitution with $u = x + 1$ and $du = dx$. Then use trig substitution with $u = 2\tan\theta$ and $du = 2\sec^2\theta\,d\theta$.

Section 7.6 Exercises

1. Complete the square to get $\int \dfrac{x\,dx}{\sqrt{21-(x+3)^2}}$ and then use the substitution $x + 3 = \sqrt{21}\sin u$ so that $dx = \sqrt{21}\cos u\,du$.

3. Trig Method. Rewrite as $\int \sin^3 x\left(1 - \sin^2 x\right)\cos x\,dx$ followed by u-substitution with $u = \sin x$ and $du = \cos x\,dx$.

5. Trig substitution $x = 3\sin\theta$ and $dx = 3\cos\theta\,d\theta$

7. None   **9.** Partial fractions   **11.** $-\dfrac{\sqrt{4-x^2}}{4x} + C$

13. $\dfrac{x}{2} + \dfrac{1}{8}\sin 4x\cos 4x + C$ **15.** $\dfrac{x}{18(9+x^2)} + \dfrac{1}{54}\tan^{-1}\left(\frac{x}{3}\right) + C$

17. $\sec x - \frac{2}{3}\sec^3 x + \frac{1}{5}\sec^5 x + C$

19. $x\ln\left|x^4 + 1\right| - 4x + 2\tan^{-1}x - \ln|x - 1| + \ln|x + 1| + C$

21. $\ln\left|\sqrt{x^2 - 1} + x\right| - \dfrac{x}{\sqrt{x^2-1}} + C$

23. $-\dfrac{x+6}{8(x^2+4x+8)} - \dfrac{1}{16}\tan^{-1}\left(\dfrac{x+2}{2}\right) + C$

25. $6\tan^{-1}\left(x^{1/6}\right) - 6x^{1/6} + 2\sqrt{x} - \dfrac{6}{5}x^{5/6} + \dfrac{6}{7}x^{7/6} + C$

27. $\frac{1}{3}\ln\left|3e^t + 1\right| + C$ **29.** $\frac{1}{3}x^3\ln x - \dfrac{x^3}{9} + C$

31. $\frac{1}{2}\tan^{-1}\left(\dfrac{x+1}{2}\right) + C$ **33.** $\frac{2}{9}\left(1 + x^3\right)^{3/2} + C$

35. $x - \ln\left(1 + e^x\right) + C$

37. $2\sqrt{x} + x + \frac{2}{3}x^{3/2} + 2\ln\left|\sqrt{x} - 1\right| + C$

39. $\frac{2}{5}(x + 1)^{5/2} - \frac{4}{3}(x + 1)^{3/2} + 2\sqrt{x + 1} + C$

41. $x + \cos x - \frac{1}{4}\sin 2x - \frac{1}{3}\cos 3x + \frac{1}{8}\sin 4x + C$

43. $\frac{1}{2}\cos^2 x - \ln|\cos x| + C$

45. $x\ln\left(x^2 + 9\right) + 6\tan^{-1}\left(\frac{x}{3}\right) - 2x + C$

47. $-\frac{1}{3}\cos^3 x + \frac{2}{5}\cos^5 x - \frac{1}{7}\cos^7 x + C$

49. $\sin x - \sin^3 x + \frac{3}{5}\sin^5 x - \frac{1}{7}\sin^7 x + C$

51. $\frac{1}{2}x^2 - \frac{1}{4}\ln\left(x^2+1\right) + \frac{1}{4}\ln|x-1| + \frac{1}{4}\ln|x+1| + C$

53. $\frac{1}{4}\sin 2x + 9\tan x - \frac{11}{2}x + C$

55. $\frac{1}{32}x^4 + \frac{1}{4}x^4(\ln x)^2 - \frac{1}{8}x^4\ln x + C$ **57.** $\cosh^{-1}\left(\frac{x}{6}\right) + C$

Section 7.7 Preliminary Questions

1. (a) The integral converges.

(b) The integral diverges.

(c) The integral diverges.

(d) The integral converges.

3. Any value of b satisfying $|b| \geq 2$ will make this an improper integral.

5. Knowing that an integral is smaller than a divergent integral does not allow us to draw any conclusions using the comparison test.

Section 7.7 Exercises

1. (a) Improper. The function $y = x^{-1/3}$ is infinite at 0.

(b) Improper. Infinite interval of integration.

(c) Improper. Infinite interval of integration.

(d) Proper. The function $y = e^{-x}$ is continuous on the finite interval $[0, 1]$.

(e) Improper. The function $y = \sec x$ is infinite at $\frac{\pi}{2}$.

(f) Improper. Infinite interval of integration.

(g) Proper. The function $y = \sin x$ is continuous on the finite interval $[0, 1]$.

(h) Proper. The function $y = 1/\sqrt{3-x^2}$ is continuous on the finite interval $[0, 1]$.

(i) Improper. Infinite interval of integration.

(j) Improper. The function $y = \ln x$ is infinite at 0.

3. $\int_1^\infty x^{-2/3}\,dx = \lim_{R\to\infty}\int_1^R x^{-2/3}\,dx = $
$\lim_{R\to\infty} 3\left(R^{1/3} - 1\right) = \infty$

5. The integral does not converge.

7. The integral converges; $I = 10{,}000e^{0.0004}$.

9. The integral does not converge.

11. The integral converges; $I = 4$.

13. The integral converges; $I = \frac{1}{8}$.

15. The integral converges; $I = 2$.

17. The integral converges; $I = 0$.

19. The integral converges; $I = \frac{1}{3e^{12}}$.

21. The integral converges; $I = \frac{1}{3}$.

23. The integral converges; $I = 2\sqrt{2}$.

25. The integral does not converge.

27. The integral converges; $I = \frac{1}{2}$.

29. The integral converges; $I = \frac{1}{2}$.

31. The integral converges; $I = \frac{\pi}{2}$.

33. The integral does not converge.

35. The integral does not converge.

37. The integral converges; $I = -1$.

39. The integral does not converge.

41. (a) Partial fractions yield $\frac{dx}{(x-2)(x-3)} = \frac{dx}{x-3} - \frac{dx}{x-2}$. This yields $\int_4^R \frac{dx}{(x-2)(x-3)} = \ln\left|\frac{R-3}{R-2}\right| - \ln\frac{1}{2}$.

(b) $I = \lim_{R\to\infty}\left(\ln\left|\frac{R-3}{R-2}\right| - \ln\frac{1}{2}\right) = \ln 1 - \ln\frac{1}{2} = \ln 2$

43. The integral does not converge.

45. The integral does not converge.

47. The integral converges; $I = 0$.

49. $\int_{-1}^1 \frac{dx}{x^{1/3}} = \int_{-1}^0 \frac{dx}{x^{1/3}} + \int_0^1 \frac{dx}{x^{1/3}} = 0$

51. The integral converges for $a < 0$.

53. $\int_{-\infty}^\infty \frac{dx}{1+x^2} = \pi$.

55. $\frac{1}{x^3+4} \leq \frac{1}{x^3}$. Therefore, by the Comparison Test, the integral converges.

57. For $x \geq 1$, $x^2 \geq x$, so $-x^2 \leq -x$ and $e^{-x^2} \leq e^{-x}$. Now $\int_1^\infty e^{-x}\,dx$ converges,   so $\int_1^\infty e^{-x^2}\,dx$ converges by the Comparison Test. We conclude that our integral converges by writing it as a sum: $\int_0^\infty e^{-x^2}\,dx = \int_0^1 e^{-x^2}\,dx + \int_1^\infty e^{-x^2}\,dx$.

59. Let $f(x) = \dfrac{1 - \sin x}{x^2}$. Since $f(x) \leq \dfrac{2}{x^2}$ and $\int_1^\infty 2x^{-2}\,dx = 2$, it follows that $\displaystyle\int_1^\infty \frac{1 - \sin x}{x^2}\,dx$ converges by the Comparison Test.

61. The integral converges.

63. The integral does not converge.

65. The integral converges.

67. The integral does not converge.

69. The integral converges.

71. The integral converges.

73. The integral converges.

75. The integral does not converge.

77. $\displaystyle\int_0^1 \frac{dx}{x^{1/2}(x+1)}$ and $\displaystyle\int_1^\infty \frac{dx}{x^{1/2}(x+1)}$ both converge; therefore, J converges.

79. $\$\frac{250}{0.07}$ **81.** $\$2{,}000{,}000$

83. $W = \lim_{T\to\infty} CV^2\left(\frac{1}{2} - e^{-T/RC} + \frac{1}{2}e^{-2T/RC}\right) = CV^2\left(\frac{1}{2} - 0 + 0\right) = \frac{1}{2}CV^2$

85. 2π

87. The integrand is infinite at the upper limit of integration, $x = \sqrt{2E/k}$, so the integral is improper.
$T = \lim_{R\to\sqrt{2E/k}} T(R) = 4\sqrt{\frac{m}{k}}\sin^{-1}(1) = 2\pi\sqrt{\frac{m}{k}}$

89. $Lf(s) = \dfrac{-1}{s^2+\alpha^2}\lim_{t\to\infty} e^{-st}(s\sin(\alpha t) + \alpha\cos(\alpha t)) - \alpha$.

91. $\frac{s}{s^2+\alpha^2}$ **93.** $J_n = \frac{n}{\alpha}J_{n-1} = \frac{n}{\alpha}\cdot\frac{(n-1)!}{\alpha^n} = \frac{n!}{\alpha^{n+1}}$

95. $E = \frac{8\pi h}{c^3}\int_0^\infty \frac{v^3}{e^{av}-1}\,dv$. Because $\alpha > 0$ and $8\pi h/c^3$ is a constant, we know E is finite by Exercise 92.

97. Because $t > \ln t$ for $t > 2$, $F(x) = \displaystyle\int_2^x \frac{dt}{\ln t} > \int_2^x \frac{dt}{t} > \ln x$.
Thus, $F(x) \to \infty$ as $x \to \infty$. Moreover,
$\lim_{x\to\infty} G(x) = \lim_{x\to\infty}\frac{1}{1/x} = \lim_{x\to\infty} x = \infty$. Thus, $\lim_{x\to\infty}\dfrac{F(x)}{G(x)}$ is of the form ∞/∞, and L'Hôpital's Rule applies. Finally,
$L = \lim_{x\to\infty}\frac{F(x)}{G(x)} = \lim_{x\to\infty}\frac{\frac{1}{\ln x}}{\frac{\ln x - 1}{(\ln x)^2}} = \lim_{x\to\infty}\frac{\ln x}{\ln x - 1} = 1$.

99. The integral is absolutely convergent. Use the Comparison Test with $\frac{1}{x^2}$.

Section 7.8 Preliminary Questions

1. No, $p(x) \geq 0$ fails.   **3.** $p(x) = 4e^{-4x}$

Section 7.8 Exercises

1. $C = 2$; $P(0 \leq X \leq 1) = \frac{3}{4}$

3. $C = \frac{1}{\pi}$; $P\left(-\frac{1}{2} \leq X \leq \frac{1}{2}\right) = \frac{1}{3}$

5. $C = \frac{2}{\pi}$; $P\left(-\frac{1}{2} \leq X \leq 1\right) = \frac{2}{3} + \frac{\sqrt{3}}{4\pi}$

7. $\int_1^\infty 3x^{-4} = 1$; $\mu = \frac{3}{2}$

9. Integration confirms $\int_0^\infty \frac{1}{50}e^{-t/50} = 1$.

11. $e^{-\frac{3}{2}} \approx 0.2231$ **13.** $\frac{1}{2}\left(2 - 10e^{-2}\right) \approx 0.32$

15. $F(-\frac{2}{3}) - F(-\frac{13}{6}) \approx 0.2374$

17. **(a)** ≈ 0.8849 **(b)** ≈ 0.6554

19. $1 - F(z)$ and $F(-z)$ are the same area on opposite tails of the distribution function. Simple algebra with the standard normal cumulative distribution function shows $P(\mu - r\sigma \leq X \leq \mu + r\sigma) = 2F(r) - 1$.

21. ≈ 0.0062 **25.** $\mu = \frac{5}{3}$; $\sigma = \frac{2\sqrt{5}}{3}$

27. $\mu = 3$; $\sigma = 3$

29. **(a)** $f(t)$ is the fraction of initial atoms present at time t. Therefore, the fraction of atoms that decay is going to be the rate of change of the total number of atoms. Over a small interval, this is simply $-f'(t)\Delta t$.

(b) The fraction of atoms that decay over an arbitrarily small interval is equivalent to the probability that an individual atom will decay over that same interval. Thus, the probability density function becomes $-f'(t)$.

(c) $\int_0^\infty -tf'(t)\,dt = \frac{1}{k}$

Section 7.9 Preliminary Questions

1. $T_1 = 6$; $T_2 = 7$

3. The Trapezoidal Rule integrates linear functions exactly, so the error will be zero.

5. The two graphical interpretations of the Midpoint Rule are the sum of the areas of the midpoint rectangles and the sum of the areas of the tangential trapezoids.

Section 7.9 Exercises

1. $T_4 = 2.75$; $M_4 = 2.625$ **3.** $T_6 = 64.6875$; $M_6 \approx 63.2813$

5. $T_6 \approx 1.4054$; $M_6 \approx 1.3769$ **7.** $T_6 \approx 1.1703$; $M_6 \approx 1.2063$

9. $T_5 \approx 0.3846$; $M_5 \approx 0.3871$ **11.** $T_5 \approx 0.7444$; $M_5 \approx 0.7481$

13. $S_4 \approx 5.2522$ **15.** $S_6 \approx 1.1090$ **17.** $S_4 \approx 0.7469$

19. $S_8 \approx 2.5450$ **21.** $S_{10} \approx 0.3466$ **23.** ≈ 2.4674

25. ≈ 1.8769 **27.** ≈ 608.611

29. **(a)** Assuming the speed of the tsunami is a continuous function, at x miles from the shore, the speed is $\sqrt{15f(x)}$. Covering an infinitesimally small distance, dx, the time T required for the tsunami to cover that distance becomes $\dfrac{dx}{\sqrt{15f(x)}}$. It follows from this that $T = \int_0^M \frac{dx}{\sqrt{15f(x)}}$.

(b) ≈ 3.347 h.

31. **(a)** Since x^3 is concave up on $[0, 2]$, T_6 is too large.

(b) We have $f'(x) = 3x^2$ and $f''(x) = 6x$. Since $|f''(x)| = |6x|$ is *increasing* on $[0, 2]$, its maximum value occurs at $x = 2$ and we may take $K_2 = |f''(2)| = 12$. Thus, Error$(T_6) \leq \frac{2}{9}$.

(c) Error$(T_6) \approx 0.1111 < \frac{2}{9}$

33. T_{10} will overestimate the integral. Error$(T_{10}) \leq 0.045$.

35. M_{10} will overestimate the integral. Error$(M_{10}) \leq 0.0113$

37. $N \geq 10^3$; Error $\approx 3.333 \times 10^{-7}$

39. $N \geq 750$; Error $\approx 2.805 \times 10^{-7}$

41. Error$(T_{10}) \leq 0.0225$; Error$(M_{10}) \leq 0.01125$

43. $S_8 \approx 4.0467$; Error $(S_8) \leq 0.00833$; $N \geq 78$

45. Error$(S_{40}) \leq 1.017 \times 10^{-4}$.   **47.** $N = 306$ **49.** $N = 186$

51. **(a)** The maximum value of $|f^{(4)}(x)|$ on the interval $[0, 1]$ is 24.

(b) $N = 20$; $S_{20} \approx 0.785398$; $\left|0.785398 - \frac{\pi}{4}\right| \approx 1.55 \times 10^{-10}$

53. **(a)** Notice $|f''(x)| = |2\cos(x^2) - 4x^2\sin(x^2)|$; the proof follows.

(b) When $K_2 = 6$, Error$(M_N) \leq \frac{1}{4N^2}$.

(c) $N \geq 16$

55. Error$(T_4) \approx 0.1039$; Error$(T_8) \approx 0.0258$; Error$(T_{16}) \approx 0.0064$; Error$(T_{32}) \approx 0.0016$; Error$(T_{64}) \approx 0.0004$. These are about twice as large as the error in M_N.

57. $S_2 = \frac{1}{4}$. This is the exact value of the integral.

59. $T_N = \dfrac{r(b^2 - a^2)}{2} + s(b - a) = \displaystyle\int_a^b f(x)\,dx$

61. **(a)** This result follows because the even-numbered interior endpoints overlap:

$$\sum_{i=0}^{(N-2)/2} S_2^{2j} = \frac{b-a}{6}\left[(y_0 + 4y_1 + y_2) + (y_2 + 4y_3 + y_4) + \cdots\right]$$

$$= \frac{b-a}{6}\left[y_0 + 4y_1 + 2y_2 + 4y_3 + 2y_4 + \cdots + 4y_{N-1} + y_N\right] = S_N$$

(b) If $f(x)$ is a quadratic polynomial, then by part (a), we have

$$S_N = S_2^0 + S_2^2 + \cdots + S_2^{N-2} = \int_a^b f(x)\,dx$$

63. Let $f(x) = ax^3 + bx^2 + cx + d$, with $a \neq 0$, be any cubic polynomial. Then $f^{(4)}(x) = 0$, so we can take $K_4 = 0$. This yields Error$(S_N) \leq \frac{0}{180N^4} = 0$. In other words, S_N is exact for all cubic polynomials for all N.

Chapter 7 Review

1. **(a)** (v)   **(b)** (iv)   **(c)** (iii)   **(d)** (i)   **(e)** (ii)

3. $\dfrac{\sin^9\theta}{9} - \dfrac{\sin^{11}\theta}{11} + C$

5. $\dfrac{\tan\theta\sec^5\theta}{6} - \dfrac{7\tan\theta\sec^3\theta}{24} + \dfrac{\tan\theta\sec\theta}{16} + \frac{1}{16}\ln|\sec\theta + \tan\theta| + C$

7. $-\dfrac{1}{\sqrt{x^2-1}} - \sec^{-1}x + C$ **9.** $2\tan^{-1}\sqrt{x} + C$

11. $-\dfrac{\tan^{-1}x}{x} + \ln|x| - \frac{1}{2}\ln\left(1 + x^2\right) + C$

13. $\frac{5}{32}e^4 - \frac{1}{32} \approx 8.50$ **15.** $\dfrac{\cos^{12}6\theta}{72} - \dfrac{\cos^{10}6\theta}{60} + C$

17. $5\ln|x - 1| + \ln|x + 1| + C$

19. $\dfrac{\tan^3\theta}{3} + \tan\theta + C$ **21.** $\ln 8 - 1$

23. $-\dfrac{\cos^5\theta}{5} + \dfrac{2\cos^3\theta}{3} - \cos\theta + C$ **25.** $-\frac{1}{4}$

27. $\frac{2}{3}(\tan x)^{3/2} + C$

29. $\dfrac{\sin^6\theta}{6} - \dfrac{\sin^8\theta}{8} + C$ **31.** $-\frac{1}{3}u^3 + C = -\frac{1}{3}\cot^3 x + C$

33.
$$\int_{\pi/4}^{\pi/3} \cot^2(x)\csc^3(x)\,dx = \frac{1}{16}\ln 3 + \frac{1}{8}\ln\left(\sqrt{2} - 1\right) + \frac{3}{8}\sqrt{2} - \frac{5}{56}$$

35. $\frac{1}{49}\ln\left|\frac{t+4}{t-3}\right| - \frac{1}{7}\cdot\frac{1}{t-3} + C$ **37.** $\frac{1}{2}\sec^{-1}\frac{x}{2} + C$

39. $\int \dfrac{dx}{x^{3/2} + ax^{1/2}} = \begin{cases} \dfrac{2}{\sqrt{a}} \tan^{-1} \sqrt{\dfrac{x}{a}} + C & a > 0 \\[2mm] \dfrac{1}{\sqrt{-a}} \ln\left|\dfrac{\sqrt{x}-\sqrt{-a}}{\sqrt{x}+\sqrt{-a}}\right| + C & a < 0 \\[2mm] -\dfrac{2}{\sqrt{x}} + C & a = 0 \end{cases}$

41. $\ln|x+2| + 5/(x+2) - 3/(x+2)^2 + C$

43. $-\ln|x-2| - 2\frac{1}{x-2} + \frac{1}{2}\ln\left(x^2+4\right) + C$

45. $\frac{1}{3}\tan^{-1}\left(\frac{x+4}{3}\right) + C$ **47.** $\ln|x+2| + \frac{5}{x+2} - \frac{3}{(x+2)^2} + C$

49. $-\dfrac{(x^2+4)^{3/2}}{48x^3} + \dfrac{\sqrt{x^2+4}}{16x} + C$ **51.** $-\frac{1}{9}e^{4-3x}(3x+4) + C$

53. $\frac{1}{2}x^2 \sin x^2 + \frac{1}{2}\cos x^2 + C$

55. $\frac{x^2}{2}\tanh^{-1} x + \frac{x}{2} - \frac{1}{4}\ln\left|\frac{1+x}{1-x}\right| + C$

57. $x\ln\left(x^2+9\right) - 2x + 6\tan^{-1}\left(\frac{x}{3}\right) + C$

59. $\frac{1}{2}\sinh 2$ **61.** $t + \frac{1}{4}\coth(1-4t) + C$ **63.** $\frac{\pi}{3}$

65. $\tan^{-1}(\tanh x) + C$

67. (a) $I_n = \int \dfrac{x^n}{x^2+1}\,dx = \int \dfrac{x^{n-2}(x^2+1-1)}{x^2+1}\,dx =$

$\int x^{n-2}\,dx - \int \dfrac{x^{n-2}}{x^2+1}\,dx = \dfrac{x^{n-1}}{n-1} - I_{n-2}$

(b) $I_0 = \tan^{-1}x + C;\ I_1 = \frac{1}{2}\ln\left(x^2+1\right) + C;$

$I_2 = x - \tan^{-1}x + C;\ I_3 = \frac{x^2}{2} - \frac{1}{2}\ln\left(x^2+1\right) + C;$

$I_4 = \frac{x^3}{3} - x + \tan^{-1}x + C;\ I_5 = \frac{x^4}{4} - \frac{x^2}{2} + \frac{1}{2}\ln\left(x^2+1\right) + C$

(c) Prove by induction; show it works for $n = 1$, then assume it works for $n = k$ and use that to show it works for $n = k+1$.

69. $\frac{3}{4}$ **71.** $C = 2;\ p(0 \le X \le 1) = 1 - \frac{2}{e}$

73. (a) 0.1587   **(b)** 0.49997

75. Integral converges; $I = \frac{1}{2}$.   **77.** Integral converges; $I = 3\sqrt[3]{4}$.

79. Integral converges; $I = \frac{\pi}{2}$.

81. The integral does not converge.

83. The integral does not converge.   **85.** The integral converges.

87. The integral converges.   **89.** The integral converges.

91. π **95.** $\dfrac{2}{(s-\alpha)^3}$

97. (a) T_N is smaller and M_N is larger than the integral.

(b) M_N is smaller and T_N is larger than the integral.

(c) M_N is smaller and T_N is larger than the integral.

(d) T_N is smaller and M_N is larger than the integral.

99. $M_5 \approx 0.7481$ **101.** $M_4 \approx 0.7450$ **103.** $S_4 \approx 0.7469$

105. $V \approx T_9 \approx 20$ acre-ft $= 871,200$ ft^3 **107.** Error $\le \frac{3}{128}$.

109. $N \ge 29$

Chapter 8

Section 8.1 Preliminary Questions

1. $\int_0^\pi \sqrt{1+\sin^2 x}\,dx$ **2.** $\int_0^h 2\pi r\,dx = 2\pi rh$ **3.** Yes

4. The graph of $y = f(x) + C$ is a vertical translation of the graph of $y = f(x)$; hence, the two graphs should have the same arc length. We can explicitly establish this as follows:

$$\text{Length of } y = f(x) + C = \int_a^b \sqrt{1 + \left[\frac{d}{dx}(f(x) + C)\right]^2}\,dx$$

$$= \int_a^b \sqrt{1 + [f'(x)]^2}\,dx$$

$$= \text{length of } y = f(x)$$

5. Since $\sqrt{1 + f'(x)^2} \ge 1$ for any function f, we have

$$\text{Length of graph of } y = f(x) \text{ over } [1, 4] = \int_1^4 \sqrt{1 + f'(x)^2}\,dx$$

$$\ge \int_1^4 1\,dx = 3$$

Section 8.1 Exercises

1. $L = \int_2^6 \sqrt{1 + 16x^6}\,dx$ **3.** $\frac{13}{12}$ **5.** $3\sqrt{10}$

7. $\frac{1}{27}(22\sqrt{22} - 13\sqrt{13})$ **9.** $e^2 + \frac{\ln 2}{2} + \frac{1}{4}$

11. $\int_1^2 \sqrt{1+x^6}\,dx \approx 3.957736$ **13.** $\int_1^2 \sqrt{1+\frac{1}{x^4}}\,dx \approx 1.132123$

15. $\int_1^3 \sqrt{1+\frac{1}{x^2}}\,dx = 2.29896$ **17.** 6

21. $a = \sinh^{-1}(5) = \ln(5 + \sqrt{26})$

25. Let s denote the arc length. Then $s = \frac{a}{2}\sqrt{1+4a^2} + \frac{1}{4}\ln|\sqrt{1+4a^2} + 2a|$. Thus, when $a = 1$, $s = \frac{1}{2}\sqrt{5} + \frac{1}{4}\ln(\sqrt{5}+2) \approx 1.478943$.

27. $\sqrt{1+e^{2a}} + \frac{1}{2}\ln\dfrac{\sqrt{1+e^{2a}}-1}{\sqrt{1+e^{2a}}+1} - \sqrt{2} + \frac{1}{2}\ln\dfrac{1+\sqrt{2}}{\sqrt{2}-1}$

29. $\ln(1+\sqrt{2})$ **33.** 1.552248   **35.** $16\pi\sqrt{2}$

37. $\frac{\pi}{27}(145^{3/2} - 1)$ **39.** $\frac{384\pi}{5}$ **41.** $\frac{\pi}{16}(e^4 - 9)$

43. $2\pi \int_1^3 x^{-1}\sqrt{1+x^{-4}}\,dx \approx 7.60306$

45. $2\pi \int_0^2 e^{-x^2/2}\sqrt{1+x^2 e^{-x^2}}\,dx \approx 8.222696$

47. $2\pi \ln 2 + \frac{15\pi}{8}$ **49.** $4\pi^2 br$

51. $2\pi b^2 + \dfrac{2\pi ba^2}{\sqrt{a^2-b^2}}\sin^{-1}\left(\dfrac{\sqrt{a^2-b^2}}{a}\right)$

Section 8.2 Preliminary Questions

1. Pressure is defined as force per unit area.

2. The factor of proportionality is the weight density of the fluid, $w = \rho g$.

3. Fluid force acts in the direction perpendicular to the side of the submerged object.

4. Pressure depends only on depth and does not change horizontally at a given depth.

5. When a plate is submerged vertically, the pressure is not constant along the plate, so the fluid force is not equal to the pressure times the area.

Section 8.2 Exercises

1. (a) Top: $F = 176,400$ newtons; bottom: $F = 705,600$ newtons

(b) $F \approx \sum\limits_{j=1}^N \rho g 3 y_j\,\Delta y$ **(c)** $F = \int_2^8 \rho g 3 y\,dy$

(d) $F = 882,000$ newtons

3. Difference $=$ force on lower plate $-$ force on upper plate $= 19,600$ newtons

5. (a) The width of the triangle varies linearly from 0 at a depth of $y = 3$ m to 1 at a depth of $y = 5$ m. Thus, $f(y) = \frac{1}{2}(y - 3)$.

(b) The area of the strip at depth y is $\frac{1}{2}(y - 3)\Delta y$, and the pressure at depth y is $\rho g y$, where $\rho = 10^3$ kg/m^3 and $g = 9.8$. Thus, the fluid force acting on the strip at depth y is approximately equal to $\rho g \frac{1}{2} y(y - 3)\Delta y$.

(c) $F \approx \sum_{j=1}^{N} \rho g \frac{1}{2} y_j (y_j - 3) \, \Delta y \to \int_3^5 \rho g \frac{1}{2} y(y - 3) \, dy$

(d) $F = \frac{127,400}{3}$ newtons

7. (b) $F = \frac{19,600}{3} r^3$ newtons

9. $F = \frac{19,600}{3} r^3 + 4900\pi m r^2$ newtons   11. $F \approx 328,224,000$ lb

13. $F = \frac{815,360}{3}$ newtons   15. $F \approx 6153.18$ newtons

17. $F \approx 5652.4$ newtons   19. $F = 940,800$ newtons

21. 5.4604×10^{11} newtons   23. $F = (15b + 30a)h^2$ lb

25. Front and back: $F = \frac{62.5\sqrt{3}}{9} H^3$; slanted sides: $F = \frac{62.5\sqrt{3}}{3} \ell H^2$

Section 8.3 Preliminary Questions

1. $M_x = M_y = 0$ 2. $M_x = 21$ 3. $M_x = 5$; $M_y = 10$

4. Because a rectangle is symmetric with respect to both the vertical line and the horizontal line through the center of the rectangle, the Symmetry Principle guarantees that the centroid of the rectangle must lie along both these lines. The only point in common to both lines of symmetry is the center of the rectangle, so the centroid of the rectangle must be the center of the rectangle.

5. If the plate looks like a ring, then the center of mass doesn't occur at any point on the plate.

Section 8.3 Exercises

1. (a) $M_x = 4m$; $M_y = 9m$; center of mass: $\left(\frac{9}{4}, 1\right)$

(b) $\left(\frac{46}{17}, \frac{14}{17}\right)$

5. A sketch of the lamina is shown here.

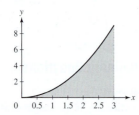

(a) $M_x = \frac{729}{10}$; $M_y = \frac{243}{4}$

(b) Area $= 9$ cm^2; center of mass: $\left(\frac{9}{4}, \frac{27}{10}\right)$

7. $M_x = \frac{64\rho}{7}$; $M_y = \frac{32\rho}{5}$; center of mass: $\left(\frac{8}{5}, \frac{16}{7}\right)$

9. (a) $M_x = 24$

(b) $M = 12$, so $y_{cm} = 2$; center of mass: $(0, 2)$

11. $\left(\frac{93}{35}, \frac{45}{56}\right)$ 13. $\left(\frac{9}{8}, \frac{18}{5}\right)$

15. $\left(\frac{1-5e^{-4}}{1-e^{-4}}, \frac{1-e^{-8}}{4(1-e^{-4})}\right)$ 17. $\left(\frac{\pi}{2}, \frac{\pi}{8}\right)$

19. A sketch of the region is shown here.

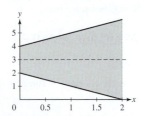

The region is clearly symmetric about the line $y = 3$, so we expect the centroid of the region to lie along this line. We find $M_x = 24$, $M_y = \frac{28}{3}$, centroid: $\left(\frac{7}{6}, 3\right)$.

21. $\left(\frac{9}{20}, \frac{9}{20}\right)$ 23. $\left(\frac{1}{2(e-2)}, \frac{e^2-3}{4(e-2)}\right)$ 25. $\left(\frac{\pi\sqrt{2}-4}{4(\sqrt{2}-1)}, \frac{1}{4(\sqrt{2}-1)}\right)$

27. A sketch of the region is shown here. Centroid: $\left(0, \frac{2}{7}\right)$

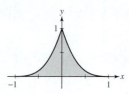

29. $\left(0, \frac{4b}{3\pi}\right)$ 31. $\left(\frac{4}{3\pi}, \frac{4}{3\pi}\right)$

33. $\left(0, \frac{\frac{2}{3}(r^2-h^2)^{3/2}}{r^2 \sin^{-1}\sqrt{1-h^2/r^2} - h\sqrt{r^2-h^2}}\right)$; with $r = 1$ and $h = \frac{1}{2}$:

$\left(0, \frac{3\sqrt{3}}{4\pi - 3\sqrt{3}}\right) \approx (0, 0.71)$

35. $V = \frac{\pi r^2 H}{3}$ 37. $\left(0, \frac{49}{24}\right)$ 39. $\left(-\frac{4}{9\pi}, \frac{4}{9\pi}\right)$

41. For the square on the left: $(4, 4)$; for the square on the right: $\left(4, \frac{25}{7}\right)$

Section 8.4 Preliminary Questions

1. $T_3(x) = 9 + 8(x - 3) + 2(x - 3)^2 + 2(x - 3)^3$

2. The polynomial graphed on the right is a Maclaurin polynomial.

3. A Maclaurin polynomial gives the value of $f(0)$ exactly.

4. The correct statement is (b): $|T_3(2) - f(2)| \leq \frac{2}{3}$.

Section 8.4 Exercises

1. $T_2(x) = x$; $T_3(x) = x - \frac{x^3}{6}$

3. $T_2(x) = \frac{1}{3} - \frac{1}{9}(x - 2) + \frac{1}{27}(x - 2)^2$;
$T_3(x) = \frac{1}{3} - \frac{1}{9}(x - 2) + \frac{1}{27}(x - 2)^2 - \frac{1}{81}(x - 2)^3$

5. $T_2(x) = 75 + 106(x - 3) + 54(x - 3)^2$;
$T_3(x) = 75 + 106(x - 3) + 54(x - 3)^2 + 12(x - 3)^3$

7. $T_2(x) = x$; $T_3(x) = x + \frac{x^3}{3}$

9. $T_2(x) = 2 - 3x + \frac{5x^2}{2}$; $T_3(x) = 2 - 3x + \frac{5x^2}{2} - \frac{3x^3}{2}$

11. $T_2(x) = \frac{1}{e} + \frac{1}{e}(x - 1) - \frac{1}{2e}(x - 1)^2$;
$T_3(x) = \frac{1}{e} + \frac{1}{e}(x - 1) - \frac{1}{2e}(x - 1)^2 - \frac{1}{6e}(x - 1)^3$

13. $T_2(x) = (x - 1) - \frac{3(x-1)^2}{2}$;
$T_3(x) = (x - 1) - \frac{3(x-1)^2}{2} + \frac{11(x-1)^3}{6}$

15. Let $f(x) = e^x$. Then, for all n,

$$f^{(n)}(x) = e^x \quad \text{and} \quad f^{(n)}(0) = 1$$

It follows that

$$T_n(x) = 1 + \frac{x}{1!} + \frac{x^2}{2!} + \cdots + \frac{x^n}{n!}$$

19. $T_n(x) = 1 + x + x^2 + x^3 + \cdots + x^n$

21. $T_n(x) = e + e(x - 1) + \frac{e(x-1)^2}{2!} + \cdots + \frac{e(x-1)^n}{n!}$

23.
$$T_n(x) = \frac{1}{\sqrt{2}} - \frac{1}{\sqrt{2}}\left(x - \frac{\pi}{4}\right) - \frac{1}{2\sqrt{2}}\left(x - \frac{\pi}{4}\right)^2 + \frac{1}{6\sqrt{2}}\left(x - \frac{\pi}{4}\right)^3 \cdots$$
In general, the coefficient of $(x - \pi/4)^n$ is

$$\pm \frac{1}{(\sqrt{2})n!}$$

with the pattern of signs $+, -, -, +, +, -, -, \ldots$.

25. $T_2(x) = 1 + x + \frac{x^2}{2}$; $|T_2(-0.5) - f(-0.5)| \approx 0.018469$

27. $T_2(x) = 1 - \frac{2}{3}(x-1) + \frac{5}{9}(x-1)^2$;
$|f(1.2) - T_2(1.2)| \approx 0.00334008$

29. $T_3(x) = 1 + \frac{1}{2}(x-1) - \frac{1}{8}(x-1)^2 + \frac{1}{16}(x-1)^3$

31. $\frac{e^{1.1}|1.1|^4}{4!}$

33. $T_5(x) = 1 - \frac{x^2}{2} + \frac{x^4}{24}$; maximum error $= \frac{(0.25)^6}{6!}$

35. $T_3(x) = \frac{1}{2} - \frac{1}{16}(x-4) + \frac{3}{256}(x-4)^2 - \frac{5}{2048}(x-4)^3$;
maximum error $= \frac{35(0.3)^4}{65,536}$

37. $T_3(x) = x - \frac{x^3}{3}$; $T_3\left(\frac{1}{2}\right) = \frac{11}{24}$. With $K = 5$,

$$\left| T_3\left(\frac{1}{2}\right) - \tan^{-1}\frac{1}{2} \right| \le \frac{5\left(\frac{1}{2}\right)^4}{4!} = \frac{5}{384}$$

39. $K = 6.25$ is acceptable.   **41.** $n = 4$ **43.** $n = 6$ **47.** $n = 4$

51. $T_{4n}(x) = 1 - \frac{x^4}{2} + \frac{x^8}{4!} + \cdots + (-1)^n \frac{x^{4n}}{(2n)!}$

53. At $a = 0$,

$$T_1(x) = -4 - x$$
$$T_2(x) = -4 - x + 2x^2$$
$$T_3(x) = -4 - x + 2x^2 + 3x^3 = f(x)$$
$$T_4(x) = T_3(x)$$
$$T_5(x) = T_3(x)$$

At $a = 1$,

$$T_1(x) = 12(x-1)$$
$$T_2(x) = 12(x-1) + 11(x-1)^2$$
$$T_3(x) = 12(x-1) + 11(x-1)^2 + 3(x-1)^3$$
$$= -4 - x + 2x^2 + 3x^3 = f(x)$$
$$T_4(x) = T_3(x)$$
$$T_5(x) = T_3(x)$$

55. $T_2(t) = 60 + 24t - \frac{3}{2}t^2$; truck's distance from the intersection after 4 s is ≈ 132 m.

57. (a) $T_3(x) = -\frac{k}{R^3}x + \frac{3k}{2R^5}x^3$

65. $T_4(x) = 1 - x^2 + \frac{1}{2}x^4$; the error is approximately $|0.461458 - 0.461281| = 0.000177$.

67. (b) $\int_0^{1/2} T_4(x)\, dx = \frac{1841}{3840}$; error bound:

$$\left| \int_0^{1/2} \cos x \, dx - \int_0^{1/2} T_4(x)\, dx \right| < \frac{\left(\frac{1}{2}\right)^7}{6!}$$

69. (a) $T_6(x) = x^2 - \frac{1}{6}x^6$

Chapter 8 Review

1. $\frac{779}{240}$ **3.** $4\sqrt{17}$ **7.** $24\pi\sqrt{2}$ **9.** $\frac{67\pi}{36}$

11. $12\pi + 4\pi^2$ **13.** $176,400$ newtons

15. Fluid force on a triangular face: $183,750\sqrt{3} + 306,250$ newtons; fluid force on a slanted rectangular edge: $122,500\sqrt{3} + 294,000$ newtons

17. $M_x = 20,480$; $M_y = 25,600$; center of mass: $\left(2, \frac{8}{5}\right)$

19. $\left(0, \frac{2}{\pi}\right)$ **21.** $\frac{3\pi}{10}$

23. $T_3(x) = 1 + 3(x-1) + 3(x-2)^2 + (x-1)^3$

25. $T_4(x) = (x-1) + \frac{1}{2}(x-1)^2 - \frac{1}{6}(x-1)^3 + \frac{1}{12}(x-1)^4$

27. $T_4(x) = x - x^3$

29. $T_n(x) = 1 + 3x + \frac{1}{2!}(3x)^2 + \frac{1}{3!}(3x)^3 + \cdots + \frac{1}{n!}(3x)^n$

31. $T_3(1.1) = 0.832981496$; $\left| T_3(1.1) - \tan^{-1}1.1 \right| = 2.301 \times 10^{-7}$

33. $n = 11$ is sufficient.

35. The nth Maclaurin polynomial for $g(x) = \frac{1}{1+x}$ is
$T_n(x) = 1 - x + x^2 - x^3 + \cdots + (-x)^n$.

Chapter 9

Section 9.1 Preliminary Questions

1. (a) First order   **(b)** First order   **(c)** Order 3   **(d)** Order 2

2. Yes   **3.** Example: $y' = y^2$ **4.** Example: $y' = y^2$

5. Example: $y' + y = x$

Section 9.1 Exercises

1. (a) First order   **(b)** Not first order   **(c)** First order
(d) First order   **(e)** Not first order   **(f)** First order

3. Let $y = 4x^2$. Then $y' = 8x$ and $y' - 8x = 8x - 8x = 0$.

5. Let $y = 25e^{-2x^2}$. Then $y' = -100xe^{-2x^2}$ and
$y' + 4xy = -100xe^{-2x^2} + 4x(25e^{-2x^2}) = 0$.

7. Let $y = 4x^4 - 12x^2 + 3$. Then
$y'' - 2xy' + 8y = (48x^2 - 24) - 2x(16x^3 - 24x) + 8(4x^4 - 12x^2 + 3)$
$$= 48x^2 - 24 - 32x^4 + 48x^2 + 32x^4 - 96x^2 + 24 = 0$$

9. (a) Separable: $y' = \frac{9}{x}y^2$ **(b)** Separable: $y' = \frac{\sin x}{\sqrt{4 - x^2}}e^{3y}$

(c) Not separable   **(d)** Separable: $y' = (1)(9 - y^2)$

11. $y = \frac{1}{3}x^3 + 1$; $y = \frac{1}{1-x}$ **13. (d)** $y = \ln\left|\frac{1}{x} - \frac{1}{2} + e^4\right|$

15. $y = Ce^{-x^3/3}$, where C is an arbitrary constant.

17. $y = \ln\left(4t^5 + C\right)$, where C is an arbitrary constant.

19. $y = Ce^{-5x/2} + \frac{4}{5}$, where C is an arbitrary constant.

21. $y = Ce^{-\sqrt{1-x^2}}$, where C is an arbitrary constant.

23. $y = \pm\sqrt{x^2 + C}$, where C is an arbitrary constant.

25. $x = \tan(\frac{1}{2}t^2 + t + C)$, where C is an arbitrary constant.

27. $y = \sin^{-1}\left(\frac{1}{2}x^2 + C\right)$, where C is an arbitrary constant.

29. $y = C\sec t$, where C is an arbitrary constant.

31. $y = 75e^{-2x}$ **33.** $y = -\sqrt{\ln\left(x^2 + e^4\right)}$

35. $y = 2 + 2e^{x(x-2)/2}$ **37.** $y = \tan\left(x^2/2\right)$ **39.** $y = e^{1-e^{-t}}$

41. $y = \frac{et}{e^{1/t}} - 1$ **43.** $y = \sin^{-1}\left(\frac{1}{2}e^x\right)$ **45.** $a = -3, 4$

47. $t = \pm\sqrt{\pi + 4}$

49. (a) ≈ 1145 s or 19.1 min   **(b)** ≈ 3910 s or 65.2 min

51. $y = 8 - (8 + 0.0002215t)^{2/3}$; $t_e \approx 66000$ s or 18 hr, 20 min

55. (a) $q(t) = CV\left(1 - e^{-t/RC}\right)$

(b)

$$\lim_{t\to\infty} q(t) = \lim_{t\to\infty} CV\left(1 - e^{-t/RC}\right) = \lim_{t\to\infty} CV(1 - 0) = CV$$

(c) $q(RC) = CV\left(1 - e^{-1}\right) \approx (0.63)\,CV$

57. $V = (kt/3 + C)^3$, V increases roughly with the cube of time.

59. $g(x) = Ce^{(3/2)x}$, where C is an arbitrary constant; $g(x) = \frac{C}{x-1}$, where C is an arbitrary constant.

61. $y = Cx^3$ and $y = \pm\sqrt{A - \frac{x^2}{3}}$

63. (b) $v(t) = -9.8t + 100(\ln(50) - \ln(50 - 4.75t))$;
$v(10) = -98 + 100(\ln(50) - \ln(2.5)) \approx 201.573$ m/s

69. (c) $C = \frac{14\pi}{15B\sqrt{2g}} \cdot R^{5/2}$

Section 9.2 Preliminary Questions

1. $y(t) = 5 - ce^{4t}$ for any positive constant c

2. No   **3.** True

4. The difference in temperature between a cooling object and the ambient temperature is decreasing. Hence, the rate of cooling, which is proportional to this difference, is also decreasing in magnitude.

Section 9.2 Exercises

1. General solution: $y(t) = 10 + ce^{2t}$; solution satisfying $y(0) = 25$: $y(t) = 10 + 15e^{2t}$; solution satisfying $y(0) = 5$: $y(t) = 10 - 5e^{2t}$

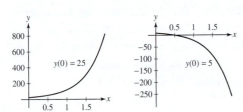

3. $y = -6 + 11e^{4x}$

5. (a) $y' = -0.02(y - 10)$ **(b)** $y = 10 + 90e^{-\frac{1}{50}t}$

(c) $100 \ln 3$ s ≈ 109.8 s

7. $\approx 5{:}50$ AM   **9.** ≈ 0.77 min $= 46.6$ s

11. $500 \ln \frac{3}{2}$ s ≈ 203 s $= 3$ min 23 s

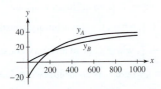

13. -58.8 m/s   **15.** -11.8 m/s

17. (b) $t = \frac{1}{0.09} \ln \left(\frac{13,333.33}{3,333.33} \right) \approx 15.4$ years   **(c)** No, since $C > 0$.

19. (a) $N'(t) = k(1 - N(t)) = -k(N(t) - 1)$

(b) $N(t) = 1 - e^{-kt}$ **(c)** $\approx 64.63\%$

23. (a) $v(t) = \frac{-g}{k} + \left(v_0 + \frac{g}{k}\right)e^{-kt}$

Section 9.3 Preliminary Questions

1. 7   **2.** $y = \pm\sqrt{1+t}$ **3. (b)**   **4.** 20

Section 9.3 Exercises

1. 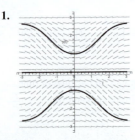   **3.**

5. (a) 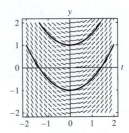   **7.**

9. For $y' = t$, y' only depends on t. The isoclines of any slope c will be the vertical lines $t = c$.

11. (i) C **(ii)** B **(iii)** F **(iv)** D **(v)** A **(vi)** E

13. (a)

15. (a) $y_1 = 3.1$ **(b)** $y_2 = 3.231$
(c) $y_3 = 3.3919$, $y_4 = 3.58171$, $y_5 = 3.799539$, $y_6 = 4.0445851$
(d) $y(2.2) \approx 3.231$, $y(2.5) \approx 3.799539$

17. $y(0.5) \approx 1.7210$ **19.** $y(3.3) \approx 3.3364$

21. $y(2) \approx 2.8838$ **25.** $y(0.5) \approx 1.794894$

27. $y(0.25) \approx 1.094871$

Section 9.4 Preliminary Questions

1. (a) No   **(b)** Yes   **(c)** No   **(d)** Yes

2. No   **3.** Yes

Section 9.4 Exercises

1. $y = \dfrac{5}{1 - e^{-3t}/C}$ and $y = \dfrac{5}{1 + (3/2)e^{-3t}}$

3. $\displaystyle \lim_{t \to \infty} y(t) = 2$

5. (a) $P(t) = \dfrac{2000}{1 + 3e^{-0.6t}}$ **(b)** $t = \frac{1}{0.6} \ln 3 \approx 1.83$ years

7. $k = \ln \frac{81}{31} \approx 0.96$ year^{-1}; $t = \dfrac{\ln 9}{2\ln 9 - \ln 31} \approx 2.29$ years

9. After $t = 7.6$ h, or at 3:36 PM

11. (a) $y_1(t) = \dfrac{10}{10 - 9e^{-t}}$ and $y_2(t) = \dfrac{1}{1 - 2e^{-t}}$

(b) $t = \ln \frac{9}{8}$ **(c)** $t = \ln 2$

13. (a) $A(t) = 16(1 - \frac{5}{3}e^{t/40})^2/(1 + \frac{5}{3}e^{t/40})^2$

(b) $A(10) \approx 2.1$

(c)

15. ≈ 943 million

17. **(d)** $t = -\frac{1}{k}\left(\ln y_0 - \ln\left(A - y_0\right)\right)$

Section 9.5 Preliminary Questions

1. **(a)** Yes   **(b)** No   **(c)** Yes   **(d)** No

2. **(b)**

3. $P(x) = x^{-1}$ **4.** $P(x) = 1$

Section 9.5 Exercises

1. **(c)** $y = \frac{x^4}{5} + \frac{C}{x}$ **(d)** $y = \frac{x^4}{5} - \frac{1}{5x}$

5. $y = \frac{1}{2}x + \frac{C}{x}$ **7.** $y = -\frac{1}{4}x^{-1} + Cx^{1/3}$

9. $y = \frac{1}{5}x^2 + \frac{1}{3} + Cx^{-3}$ **11.** $y = -x\ln x + Cx$

13. $y = \frac{1}{2}e^x + Ce^{-x}$ **15.** $y = x\cos x + C\cos x$

17. $y = x^x + Cx^x e^{-x}$ **19.** $y = \frac{1}{5}e^{2x} - \frac{6}{5}e^{-3x}$

21. $y = \frac{\ln|x|}{x+1} - \frac{1}{x(x+1)} + \frac{5}{x+1}$ **23.** $y = -\cos x + \sin x$

25. $y = \tanh x + 3\operatorname{sech} x$

27. For $m \neq -n$: $y = \frac{1}{m+n}e^{mx} + Ce^{-nx}$; for $m = -n$:
$y = (x + C)e^{-nx}$

29. **(a)** $y' = 4000 - \frac{40y}{500+40t}$; $y = 1000\frac{4t^2+100t+125}{2t+25}$

(b) 40 g/L

31. 50 g/L

33. **(a)** $\frac{dV}{dt} = \frac{20}{1+t} - 5$ and $V(t) = 20\ln(1 + t) - 5t + 100$

(b) The maximum value is $V(3) = 20\ln 4 - 15 + 100 \approx 112.726$.

(c) Estimate empty at ≈ 34 min.

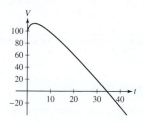

35. $I(t) = \frac{1}{10}\left(1 - e^{-20t}\right)$

37. **(a)** $I(t) = \frac{V}{R} - \frac{V}{R}e^{-(R/L)t}$ **(c)** Approximately 0.0184 s

39. **(b)** $c_1(t) = 10e^{-t/6}$

Chapter 9 Review

1. **(a)** No, first order   **(b)** Yes, first order   **(c)** No, order 3
(d) Yes, second order

3. $y = \pm\left(\frac{4}{3}t^3 + C\right)^{1/4}$, where C is an arbitrary constant.

5. $y = Cx - 1$, where C is an arbitrary constant.

7. $y = \frac{1}{2}\left(x + \frac{1}{2}\sin 2x\right) + \frac{\pi}{4}$ **9.** $y = \frac{2}{2-x^2}$

11.   **13.**

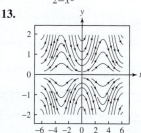

15. $y(t) = \tan t$

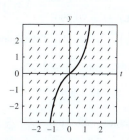

17. $y(0.1) \approx 1.1$; $y(0.2) \approx 1.209890$; $y(0.3) \approx 1.329919$

19. $y = x^2 + 2x$ **21.** $y = \frac{1}{2} + e^{-x} - \frac{11}{2}e^{-2x}$

23. $y = \frac{1}{2}\sin 2x - 2\cos x$ **25.** $y = 1 - \sqrt{t^2 + 15}$

27. $w = \tan\left(k\ln x + \frac{\pi}{4}\right)$

29. $y = -\cos x + \frac{\sin x}{x} + \frac{C}{x}$, where C is an arbitrary constant.

31. Solution satisfying $y(0) = 3$: $y(t) = 4 - e^{-2t}$; solution
satisfying $y(0) = 4$: $y(t) = 4$

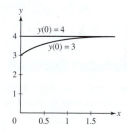

33. **(a)** 12

(b) ∞, if $y(0) > 12$; 12, if $y(0) = 12$; $-\infty$, if $y(0) < 12$

(c) -3

35. $400{,}000 - 200{,}000e^{0.25} \approx \$143{,}194.91$ **37.** $\$400{,}000$

39. Solutions are of the form $y = \frac{B}{A} + Ce^{-At}$ and $\lim\limits_{t\to\infty} y = \frac{B}{A}$.

41. $\frac{-7\sqrt{10}\sqrt{y}}{240y+64{,}800}$; $t = 3225.88$ s or 51 min 56 s

43. 2   **45.** $t = 5\ln 441 \approx 30.45$ days

49. **(a)** $\frac{dc_1}{dt} = -\frac{2}{5}c_1$ **(b)** $c_1(t) = 8e^{(-2/5)t}$ g/L

Chapter 10

Section 10.1 Preliminary Questions

1. $a_4 = 12$ **2.** **(c)**   **3.** $\lim\limits_{n\to\infty} a_n = \sqrt{2}$ **4.** **(b)**

5. **(a)** False. Counterexample: $a_n = \cos\pi n$

(b) True   **(c)** False. Counterexample: $a_n = (-1)^n$

Section 10.1 Exercises

1. **(a)** (iv)   **(b)** (i)   **(c)** (iii)   **(d)** (ii)

3. $c_1 = 3, c_2 = \frac{9}{2}, c_3 = \frac{9}{2}, c_4 = \frac{27}{8}$

5. $a_1 = 2, a_2 = 5, a_3 = 47, a_4 = 4415$

7. $b_1 = 4, b_2 = 6, b_3 = 4, b_4 = 6$

9. $c_1 = 1, c_2 = \frac{3}{2}, c_3 = \frac{11}{6}, c_4 = \frac{25}{12}$

11. $b_1 = 2, b_2 = 3, b_3 = 8, b_4 = 19$

13. **(a)** $a_n = \frac{(-1)^{n+1}}{n^3}$ **(b)** $a_n = \frac{n+1}{n+5}$

15. $\lim\limits_{n\to\infty} 12 = 12$ **17.** $\lim\limits_{n\to\infty} \frac{5n-1}{12n+9} = \frac{5}{12}$

19. $\lim\limits_{n\to\infty}\left(-2^{-n}\right) = 0$ **21.** The sequence diverges.

23. $\lim\limits_{n\to\infty} \frac{n}{\sqrt{n^2+1}} = 1$ **25.** $\lim\limits_{n\to\infty} \ln\left(\frac{12n+2}{-9+4n}\right) = \ln 3$

27. $\lim\limits_{n\to\infty}\sqrt{4+\frac{1}{n}}=2$ **29.** $\lim\limits_{n\to\infty}\cos^{-1}\left(\frac{n^3}{2n^3+1}\right)=\frac{\pi}{3}$

31. (a) $M=999$ (b) $M=99,999$

35. $\lim\limits_{n\to\infty}\left(10+\left(-\frac{1}{9}\right)^n\right)=10$ **37.** The sequence diverges.

39. $\lim\limits_{n\to\infty}2^{1/n}=1$ **41.** $\lim\limits_{n\to\infty}\frac{9^n}{n!}=0$

43. $\lim\limits_{n\to\infty}\frac{3n^2+n+2}{2n^2-3}=\frac{3}{2}$ **45.** $\lim\limits_{n\to\infty}\frac{\cos n}{n}=0$

47. The sequence diverges. **49.** $\lim\limits_{n\to\infty}\left(2+\frac{4}{n^2}\right)^{1/3}=2^{1/3}$

51. $\lim\limits_{n\to\infty}\ln\left(\frac{2n+1}{3n+4}\right)=\ln\frac{2}{3}$ **53.** The sequence diverges.

55. $\lim\limits_{n\to\infty}\frac{e^n+(-3)^n}{5^n}=0$ **57.** $\lim\limits_{n\to\infty}n\sin\frac{\pi}{n}=\pi$

59. $\lim\limits_{n\to\infty}\frac{3-4^n}{2+7\cdot4^n}=-\frac{1}{7}$ **61.** $\lim\limits_{n\to\infty}\left(1+\frac{1}{n}\right)^n=e$

63. $\lim\limits_{n\to\infty}\frac{(\ln n)^2}{n}=0$ **65.** $\lim\limits_{n\to\infty}n\left(\sqrt{n^2+1}-n\right)=\frac{1}{2}$

67. $\lim\limits_{n\to\infty}\frac{1}{\sqrt{n^4+n^8}}=0$ **69.** $\lim\limits_{n\to\infty}(2^n+3^n)^{1/n}=3$ **71.** (b)

73. Any number greater than or equal to 3 is an upper bound.

75. Example: $a_n=(-1)^n$ **79.** Example: $f(x)=\sin\pi x$

87. (e) $AGM\left(1,\sqrt{2}\right)\approx1.198$

Section 10.2 Preliminary Questions

1. The sum of an infinite series is defined as the limit of the sequence of partial sums. If the limit of this sequence does not exist, the series is said to diverge.

2. $S=\frac{1}{2}$

3. The result is negative, so the result is not valid: a series with all positive terms cannot have a negative sum. The formula is not valid because a geometric series with $|r|\geq1$ diverges.

4. No **5.** No **6.** $N=13$

7. No, S_N is increasing and converges to 1, so $S_N\leq1$ for all N.

8. Example: $\sum\limits_{n=1}^{\infty}\frac{1}{n^{9/10}}$

Section 10.2 Exercises

1. (a) $a_n=\frac{1}{3^n}$ (b) $a_n=\left(\frac{5}{2}\right)^{n-1}$

(c) $a_n=(-1)^{n+1}\frac{n^n}{n!}$ (d) $a_n=\frac{1+\frac{(-1)^{n+1}+1}{2}}{n^2+1}$

3. $S_2=\frac{5}{4}$, $S_4=\frac{205}{144}$, $S_6=\frac{5369}{3600}$

5. $S_2=\frac{2}{3}$, $S_4=\frac{4}{5}$, $S_6=\frac{6}{7}$ **7.** $S_6=1.24992$

9. $S_{10}=0.03535167962$, $S_{100}=0.03539810274$, $S_{500}=0.03539816290$, $S_{1000}=0.03539816334$. Yes.

11. $S_3=\frac{3}{10}$, $S_4=\frac{1}{3}$, $S_5=\frac{5}{14}$, $\sum\limits_{n=1}^{\infty}\left(\frac{1}{n+1}-\frac{1}{n+2}\right)=\frac{1}{2}$

13. $S_3=\frac{3}{7}$, $S_4=\frac{4}{9}$, $S_5=\frac{5}{11}$, $\sum\limits_{n=1}^{\infty}\frac{1}{4n^2-1}=\frac{1}{2}$

15. $S=\frac{1}{2}$ **17.** $\lim\limits_{n\to\infty}\frac{n}{10n+12}=\frac{1}{10}\neq0$

19. $\lim\limits_{n\to\infty}(-1)^n\left(\frac{n-1}{n}\right)$ does not exist.

21. $\lim\limits_{n\to\infty}a_n=\lim\limits_{n\to\infty}\cos\frac{1}{n+1}=1\neq0$

23. $S=\frac{8}{7}$ **25.** The series diverges. **27.** $S=\frac{59,049}{3328}$

29. $S=\frac{1}{e-1}$ **31.** $S=\frac{35}{3}$ **33.** $S=4$ **35.** $S=\frac{7}{15}$ **37.** $\frac{2}{9}$

39. $\frac{31}{99}$

41. The fractions $S_a=\frac{a}{9}$, where $a=1,2,\ldots,8$ have repeating decimals of the form $0.aaa\ldots$

43. (b) and (c)

47. (a) Counterexample: $\sum\limits_{n=1}^{\infty}\left(\frac{1}{2}\right)^n=1$.

(b) Counterexample: If $a_n=1$, then $S_N=N$.

(c) Counterexample: $\sum\limits_{n=1}^{\infty}\frac{1}{n}$ diverges.

(d) Counterexample: $\sum\limits_{n=1}^{\infty}\cos2\pi n\neq1$.

49. The total area is $\frac{1}{4}$.

51. (a) $De^{-k}+De^{-2k}+De^{-3k}+\cdots=\frac{De^{-k}}{1-e^{-k}}$

(b) $De^{-kt}+De^{-2kt}+De^{-3kt}+\cdots=\frac{De^{-kt}}{1-e^{-kt}}$

(c) $t\geq-\frac{1}{k}\ln\left(1-\frac{D}{S}\right)$

53. The total length of the path is $2+\sqrt{2}$. **57.** 42 ft

Section 10.3 Preliminary Questions

1. (b)

2. A function f such that $a_n=f(n)$ must be positive, decreasing, and continuous for $x\geq1$.

3. Convergence of p-series or integral test

4. Comparison Test

5. No; $\sum\limits_{n=1}^{\infty}\frac{1}{n}$ diverges, but since $\frac{e^{-n}}{n}<\frac{1}{n}$ for $n\geq1$, the

Comparison Test tells us nothing about the convergence of $\sum\limits_{n=1}^{\infty}\frac{e^{-n}}{n}$.

Section 10.3 Exercises

1. $\int_1^{\infty}\frac{dx}{x^4}dx$ converges, so the series converges.

3. $\int_1^{\infty}x^{-1/3}dx=\infty$, so the series diverges.

5. $\int_{25}^{\infty}\frac{x^2}{(x^3+9)^{5/2}}dx$ converges, so the series converges.

7. $\int_1^{\infty}\frac{dx}{x^2+1}$ converges, so the series converges.

9. $\int_1^{\infty}\frac{dx}{x(x+1)}$ converges, so the series converges.

11. $\int_2^{\infty}\frac{1}{x(\ln x)^2}dx$ converges, so the series converges.

13. $\int_1^{\infty}\frac{dx}{2^{\ln x}}=\infty$, so the series diverges.

15. $\frac{1}{n^3+8n}\leq\frac{1}{n^3}$, so the series converges.

19. $\frac{1}{n2^n}\leq\left(\frac{1}{2}\right)^n$, so the series converges.

21. $\frac{1}{n^{1/3}+2^n}\leq\left(\frac{1}{2}\right)^n$, so the series converges.

23. $\frac{4}{m!+4^m}\leq4\left(\frac{1}{4}\right)^m$, so the series converges.

25. $0\leq\frac{\sin^2 k}{k^2}\leq\frac{1}{k^2}$, so the series converges.

27. $\frac{2}{3^n+3^{-n}}\leq2\left(\frac{1}{3}\right)^n$, so the series converges.

29. $\frac{1}{(n+1)!}\leq\frac{1}{n^2}$, so the series converges.

31. $\frac{\ln n}{n^3}\leq\frac{1}{n^2}$ for $n\geq1$, so the series converges.

33. $\frac{(\ln n)^{100}}{n^{1.1}}\leq\frac{1}{n^{1.09}}$ for n sufficiently large, so the series converges.

35. $\frac{n}{3^n} \le \left(\frac{2}{3}\right)^n$ for $n \ge 1$, so the series converges.

39. The series converges.   **41.** The series diverges.

43. The series converges.   **45.** The series diverges.

47. The series converges.   **49.** The series converges.

51. The series diverges.   **53.** The series converges.

55. The series diverges.   **57.** The series converges.

59. The series diverges.   **61.** The series diverges.

63. The series diverges.   **65.** The series converges.

67. The series diverges.   **69.** The series diverges.

71. The series converges.   **73.** The series converges.

75. The series diverges.   **77.** The series converges.

79. The series converges for $a > 1$ and diverges for $a \le 1$.

81. The series converges for $p > 1$ and diverges for $p \le 1$.

89. $\displaystyle\sum_{n=1}^{\infty} n^{-5} \approx 1.0369540120$.

93. $\displaystyle\sum_{n=1}^{1000} \frac{1}{n^2} = 1.6439345667$ and $1 + \displaystyle\sum_{n=1}^{100} \frac{1}{n^2(n+1)} = 1.6448848903$.

The second sum is a better approximation to $\dfrac{\pi^2}{6} \approx 1.6449340668$.

Section 10.4 Preliminary Questions

1. Example: $\sum \frac{(-1)^n}{\sqrt[3]{n}}$ **2.** (b)   **3.** No

4. $|S - S_{100}| \le 10^{-3}$, and S is larger than S_{100}.

Section 10.4 Exercises

3. Converges conditionally.   **5.** Converges absolutely.

7. Converges conditionally.   **9.** Converges conditionally.

11. (a)

n	S_n	n	S_n
1	1	6	0.899782407
2	0.875	7	0.902697859
3	0.912037037	8	0.900744734
4	0.896412037	9	0.902116476
5	0.904412037	10	0.901116476

13. $S_5 = 0.947$ **15.** $S_{44} = 0.06567457397$

17. Converges (by geometric series).

19. Converges (by Comparison Test).

21. Converges (by Limit Comparison Test).

23. Diverges (by Limit Comparison Test).

25. Converges (by geometric series and linearity).

27. Converges absolutely (by Integral Test).

29. Converges (by Alternating Series Test).

31. Converges (by Integral Test).   **33.** Converges conditionally.

Section 10.5 Preliminary Questions

1. $\rho = \displaystyle\lim_{n \to \infty} \left| \frac{a_{n+1}}{a_n} \right|$

2. The Ratio Test is conclusive for $\displaystyle\sum_{n=1}^{\infty} \frac{1}{2^n}$ and inconclusive

for $\displaystyle\sum_{n=1}^{\infty} \frac{1}{n}$.

3. No

Section 10.5 Exercises

1. Converges absolutely.   **3.** Converges absolutely.

5. The Ratio Test is inconclusive.   **7.** Diverges.

9. Converges absolutely.   **11.** Converges absolutely.

13. Diverges.   **15.** The Ratio Test is inconclusive.

17. Converges absolutely.   **19.** Converges absolutely.

21. $\rho = \frac{1}{3} < 1$ **23.** $\rho = 2|x|$

25. $\rho = |r|$ **29.** Converges absolutely.

31. The Ratio Test is inconclusive, so the series may converge or diverge.

33. Converges absolutely.   **35.** The Ratio Test is inconclusive.

37. Converges absolutely.   **39.** Converges absolutely.

41. Converges absolutely.

43. Converges (by geometric series and linearity).

45. Diverges (by the Divergence Test).

47. Converges (by the Direct Comparison Test).

49. Diverges (by the Direct Comparison Test).

51. Converges (by the Ratio Test).

53. Converges (by the Limit Comparison Test).

55. Diverges (by p-series).   **57.** Converges (by geometric series).

59. Converges (by Limit Comparison Test).

61. Diverges (by Divergence Test).

65. (b) $\sqrt{2\pi} \approx 2.50663$

n	$\frac{e^n n!}{n^{n+1/2}}$
1000	2.506837
1500	2.506768
2000	2.506733
2500	2.506712
3000	2.506698

Section 10.6 Preliminary Questions

1. Yes. The series must converge for both $x = 4$ and $x = -3$.

2. (a), (c)   **3.** $R = 4$

4. $F'(x) = \displaystyle\sum_{n=1}^{\infty} n^2 x^{n-1}$; $R = 1$

Section 10.6 Exercises

1. $R = 2$. It does not converge at the endpoints.

3. $R = 3$ for all three series.

9. $(-1, 1)$ **11.** $[-\sqrt{2}, \sqrt{2}]$ **13.** $[-1, 1]$ **15.** $(-\infty, \infty)$

17. $(-\infty, \infty)$ **19.** $(-1, 1]$ **21.** $(-1, 1)$ **23.** $[-1, 1)$

25. $(2, 4)$ **27.** $(6, 8)$ **29.** $\left[-\frac{7}{2}, -\frac{5}{2} \right)$ **31.** $(-\infty, \infty)$

33. $\left(2 - \frac{1}{e}, 2 + \frac{1}{e} \right)$

35. $\displaystyle\sum_{n=0}^{\infty} 3^n x^n$ on the interval $\left(-\frac{1}{3}, \frac{1}{3} \right)$

37. $\displaystyle\sum_{n=0}^{\infty} \frac{x^n}{3^{n+1}}$ on the interval $(-3, 3)$

39. $\displaystyle\sum_{n=0}^{\infty} (-1)^n x^{2n}$ on the interval $(-1, 1)$

43. $\displaystyle\sum_{n=0}^{\infty} (-1)^{n+1} (x - 5)^n$ on the interval $(4, 6)$

47. (c) $S_4 = \frac{69}{640}$ and $|S - S_4| \approx 0.000386 < a_5 = \frac{1}{1920}$

49. $R = 1$ **51.** $\displaystyle\sum_{n=1}^{\infty} \frac{n}{2^n} = 2$ **53.** $F(x) = \frac{1-x-x^2}{1-x^3}$

55. $-1 \le x \le 1$ **57.** $P(x) = \displaystyle\sum_{n=0}^{\infty} (-1)^n \frac{x^n}{n!}$

59. N must be at least 5; $S_5 = 0.3680555556$

61. $P(x) = 1 - \frac{1}{2}x^2 - \displaystyle\sum_{n=2}^{\infty} \frac{1 \cdot 3 \cdot 5 \cdots (2n-3)}{(2n)!} x^{2n}$; $R = \infty$

Section 10.7 Preliminary Questions

1. $f(0) = 3$ and $f'''(0) = 30$

2. $f(-2) = 0$ and $f^{(4)}(-2) = 48$

3. Substitute x^2 for x in the Maclaurin series for $\sin x$.

4. $f(x) = 4 + \displaystyle\sum_{n=1}^{\infty} \frac{(x-3)^{n+1}}{n(n+1)}$ **5.** (c)

Section 10.7 Exercises

1. $f(x) = 2 + 3x + 2x^2 + 2x^3 + \cdots$

3. $\frac{1}{1-2x} = \displaystyle\sum_{n=0}^{\infty} 2^n x^n$ on the interval $\left(-\frac{1}{2}, \frac{1}{2}\right)$

5. $\cos 3x = \displaystyle\sum_{n=0}^{\infty} (-1)^n \frac{9^n x^{2n}}{(2n)!}$ on the interval $(-\infty, \infty)$

7. $\sin(x^2) = \displaystyle\sum_{n=0}^{\infty} (-1)^n \frac{x^{4n+2}}{(2n+1)!}$ on the interval $(-\infty, \infty)$

9. $\ln(1-x^2) = -\displaystyle\sum_{n=1}^{\infty} \frac{x^{2n}}{n}$ on the interval $(-1, 1)$

11. $\tan^{-1}(x^2) = \displaystyle\sum_{n=0}^{\infty} (-1)^n \frac{x^{4n+2}}{2n+1}$ on the interval $[-1, 1]$

13. $e^{x-2} = \displaystyle\sum_{n=0}^{\infty} \frac{x^n}{e^2 n!}$ on the interval $(-\infty, \infty)$

15. $\ln(1-5x) = -\displaystyle\sum_{n=1}^{\infty} \frac{5^n x^n}{n}$ on the interval $\left[-\frac{1}{5}, \frac{1}{5}\right)$

17. $\sinh x = \displaystyle\sum_{k=0}^{\infty} \frac{x^{2k+1}}{(2k+1)!}$ on the interval $(-\infty, \infty)$

19. $e^x \sin x = x + x^2 + \frac{x^3}{3} - \frac{x^5}{30} + \cdots$

21. $\frac{\sin x}{1-x} = x + x^2 + \frac{5x^3}{6} + \frac{5x^4}{6} + \cdots$

23. $(1+x)^{1/4} = 1 + \frac{1}{4}x - \frac{3}{32}x^2 + \frac{7}{128}x^3 + \cdots$

25. $e^x \tan^{-1} x = x + x^2 + \frac{1}{6}x^3 - \frac{1}{6}x^4 + \cdots$

27. $e^{\sin x} = 1 + x + \frac{1}{2}x^2 - \frac{1}{8}x^4 + \cdots$

29. $\frac{1}{x} = \displaystyle\sum_{n=0}^{\infty} (-1)^n (x-1)^n$ on the interval $(0, 2)$

31. $\frac{1}{1-x} = \displaystyle\sum_{n=0}^{\infty} (-1)^{n+1} \frac{(x-5)^n}{4^{n+1}}$ on the interval $(1, 9)$

33. $21 + 35(x-2) + 24(x-2)^2 + 8(x-2)^3 + (x-2)^4$ on the interval $(-\infty, \infty)$

35. $\frac{1}{x^2} = \displaystyle\sum_{n=0}^{\infty} (-1)^n (n+1) \frac{(x-4)^n}{4^{n+2}}$ on the interval $(0, 8)$

37. $\frac{1}{1-x^2} = \displaystyle\sum_{n=0}^{\infty} \frac{(-1)^{n+1}(2^{n+1}-1)}{2^{2n+3}}(x-3)^n$ on the interval $(1, 5)$

39. $\cos^2 x = \frac{1}{2} + \frac{1}{2}\displaystyle\sum_{n=0}^{\infty} (-1)^n \frac{(4)^n x^{2n}}{(2n)!}$ **45.** $S_4 = 0.1822666667$

47. (a) 5 (b) $S_4 = 0.7474867725$

49. $\int_0^1 \cos(x^2)\,dx = \displaystyle\sum_{n=0}^{\infty} \frac{(-1)^n}{(2n)!(4n+1)}$; $S_3 = 0.9045227920$

51. $\int_0^1 e^{-x^3}\,dx = \displaystyle\sum_{n=0}^{\infty} \frac{(-1)^n}{n!(3n+1)}$; $S_5 = 0.8074461996$

53. $\int_0^x \frac{1-\cos(t)}{t}\,dt = \displaystyle\sum_{n=1}^{\infty} (-1)^{n+1} \frac{x^{2n}}{(2n)!2n}$

55. $\int_0^x \ln(1+t^2)\,dt = \displaystyle\sum_{n=1}^{\infty} (-1)^{n-1} \frac{x^{2n+1}}{n(2n+1)}$

57. $\frac{1}{1+2x}$ **59.** $\cos \pi = -1$ **65.** e^{x^3} **67.** $1 - 5x + \sin 5x$

69. $\frac{1}{(1-2x)(1-x)} = \displaystyle\sum_{n=0}^{\infty} \left(2^{n+1} - 1\right) x^n$

71. $I(t) = \frac{V}{R} \displaystyle\sum_{n=1}^{\infty} \frac{(-1)^{n+1}}{n!} \left(\frac{Rt}{L}\right)^n$

73. $f(x) = \displaystyle\sum_{n=0}^{\infty} \frac{(-1)^n x^{6n}}{(2n)!}$ and $f^{(6)}(0) = -360$.

75. $e^{x^{20}} = 1 + x^{20} + \frac{x^{40}}{2} + \cdots$

77. No

n	Series value when $x = 2$
5	2.54297
10	-0.239933
15	41.9276
20	-764.272
25	16,595.8

83. $\displaystyle\lim_{x \to 0} \frac{\sin x - x + \frac{x^3}{6}}{x^5} = \frac{1}{120}$

85. $\displaystyle\lim_{x \to 0} \left(\frac{\sin(x^2)}{x^4} - \frac{\cos x}{x^2}\right) = \frac{1}{2}$

87. $S = \frac{\pi}{4} - \frac{1}{2}\ln 2$ **89.** $L \approx 28.369$

Chapter 10 Review

1. (a) $a_1^2 = 4$, $a_2^2 = \frac{1}{4}$, $a_3^2 = 0$

(b) $b_1 = \frac{1}{24}$, $b_2 = \frac{1}{60}$, $b_3 = \frac{1}{240}$

(c) $a_1 b_1 = -\frac{1}{12}$, $a_2 b_2 = -\frac{1}{120}$, $a_3 b_3 = 0$

(d) $2a_2 - 3a_1 = 5$, $2a_3 - 3a_2 = \frac{3}{2}$, $2a_4 - 3a_3 = \frac{1}{12}$

3. $\displaystyle\lim_{n \to \infty} (5a_n - 2a_n^2) = 2$ **5.** $\displaystyle\lim_{n \to \infty} e^{a_n} = e^2$

7. $\displaystyle\lim_{n \to \infty} (-1)^n a_n$ does not exist.

9. $\displaystyle\lim_{n \to \infty} \left(\sqrt{n+5} - \sqrt{n+2}\right) = 0$ **11.** $\displaystyle\lim_{n \to \infty} 2^{1/n^2} = 1$

13. The sequence diverges. **15.** $\displaystyle\lim_{n \to \infty} \tan^{-1}\left(\frac{n+2}{n+5}\right) = \frac{\pi}{4}$

17. $\displaystyle\lim_{n \to \infty} \left(\sqrt{n^2 + n} - \sqrt{n^2 + 1}\right) = \frac{1}{2}$

19. $\displaystyle\lim_{m \to \infty} \left(1 + \frac{1}{m}\right)^{3m} = e^3$

21. $\displaystyle\lim_{n \to \infty} \left(n\left(\ln(n+1) - \ln n\right)\right) = 1$

25. $\lim\limits_{n\to\infty}\dfrac{a_{n+1}}{a_n}=3$ **27.** $S_4=-\dfrac{11}{60},\,S_7=\dfrac{41}{630}$

29. $\sum\limits_{n=2}^{\infty}\left(\dfrac{2}{3}\right)^n=\dfrac{4}{3}$ **31.** $S=\dfrac{4}{37}$ **33.** $\sum\limits_{n=-1}^{\infty}\dfrac{2^{n+3}}{3^n}=36$

35. $a_n=\left(\dfrac{1}{2}\right)^n+1-2^n,\,b_n=2^n-1$

37. $S=\dfrac{47}{180}$ **39.** The series diverges.

41. $\int_1^{\infty}\dfrac{1}{(x+2)(\ln(x+2))^3}\,dx=\dfrac{1}{2(\ln(3))^2}$, so the series converges.

43. $\dfrac{1}{(n+1)^2}<\dfrac{1}{n^2}$, so the series converges.

45. $\sum\limits_{n=0}^{\infty}\dfrac{1}{n^{1.5}}$ converges, so the series converges.

47. $\dfrac{n}{\sqrt{n^5+2}}<\dfrac{1}{n^{3/2}}$, so the series converges.

49. $\sum\limits_{n=0}^{\infty}\left(\dfrac{10}{11}\right)^n$ converges, so the series converges.

51. Converges.

55. **(b)** $0.3971162690\le S\le 0.3971172688$, so the maximum size of the error is 10^{-6}.

57. Converges absolutely.   **59.** Diverges.

61. **(a)** 500   **(b)** $K\approx\sum\limits_{n=0}^{499}\dfrac{(-1)^k}{(2k+1)^2}=0.9159650942$

63. **(a)** Converges.   **(b)** Converges.   **(c)** Diverges.
(d) Converges.

65. Converges.   **67.** Converges.   **69.** Diverges.

71. Diverges.   **73.** Converges.   **75.** Converges.

77. Converges (by geometric series).

79. Converges (by geometric series).

81. Converges (by the Leibniz Test).

83. Converges (by the Leibniz Test).

85. Diverges (by the Divergence Test).

87. Converges (absolutely, by a Direct Comparison with the p-series $\sum\limits_{n=1}^{\infty}\dfrac{1}{n^{3/2}}$

89. Converges (by the Root Test).

91. Converges (by the Limit Comparison Test)

93. Converges using partial sums (the series is telescoping).

95. Diverges (by the Direct Comparison Test).

97. Converges (by the Direct Comparison Test).

99. Converges (by the Limit Comparison Test).

101. Converges on the interval $(-\infty,\infty)$.

103. Converges on the interval $[2,4]$.   **105.** Converges at $x=0$.

107. $\dfrac{2}{4-3x}=\dfrac{1}{2}\sum\limits_{n=0}^{\infty}\left(\dfrac{3}{4}\right)^n x^n$. The series converges on the interval $\left(\dfrac{-4}{3},\dfrac{4}{3}\right)$.

109. **(c)**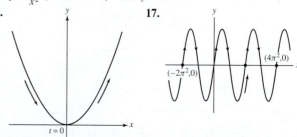

111. $\lim\limits_{x\to 0}\dfrac{x^2 e^x}{\cos x-1}=-2$

113. $e^{4x}=\sum\limits_{n=0}^{\infty}\dfrac{4^n}{n!}x^n$

115. $x^4=16+32(x-2)+24(x-2)^2+8(x-2)^3+(x-2)^4$

117. $\sin x=\sum\limits_{n=0}^{\infty}\dfrac{(-1)^{n+1}(x-\pi)^{2n+1}}{(2n+1)!}$

119. $\dfrac{1}{1-2x}=\sum\limits_{n=0}^{\infty}\dfrac{2^n}{5^{n+1}}(x+2)^n$ **121.** $\ln\dfrac{x}{2}=\sum\limits_{n=1}^{\infty}\dfrac{(-1)^{n+1}(x-2)^n}{n2^n}$

123. $(x^2-x)e^{x^2}=\sum\limits_{n=0}^{\infty}\left(\dfrac{x^{2n+2}-x^{2n+1}}{n!}\right)$ so $f^{(3)}(0)=-6$

125. $\dfrac{1}{1+\tan x}=1-x+x^2-\dfrac{4}{3}x^3+\cdots$, so $f^{(3)}(0)=-8$

127. $\dfrac{\pi}{2}-\dfrac{\pi^3}{2^3 3!}+\dfrac{\pi^5}{2^5 5!}-\dfrac{\pi^7}{2^7 7!}+\cdots=\sin\dfrac{\pi}{2}=1$

Chapter 11

Section 11.1 Preliminary Questions

1. A circle of radius 3 centered at the origin

2. The center is at $(4,5)$.   **3.** Maximum height: 4

4. Yes; No

5. **(a)** The line $y=x$ traversed left to right.

(b) Same path traversed left to right twice as fast.

6. $c_1(t)=(t,t^3),\,c_2(t)=(\sqrt[3]{t},t),\,c_3(t)=(t^3,t^9)$

Section 11.1 Exercises

1. $(t=0)(1,9);\,(t=2)(9,-3);\,(t=4)(65,-39)$

3. $y=2.5x-.000766x^2$

5. **(a)**   **(b)**

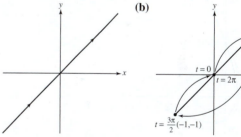

(c)   **(d)**

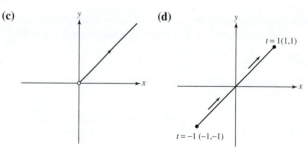

7. $y=4x-12$ **9.** $y=\tan^{-1}\left(x^3+e^x\right)$

11. $y=\dfrac{6}{x^2}$ (where $x>0$)   **13.** $y=2-e^x$

15.   **17.**

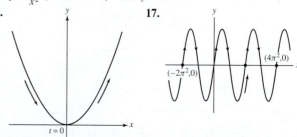

19. (a) ↔ (iv), (b) ↔ (ii), (c) ↔ (iii), (d) ↔ (i)

21. (a) $y_{max} = 100$ cm; (b) lands at $x = 2040$ cm from the origin, when $t = 20$ s

23. $c(t) = (t, 9 - 4t)$ **25.** $c(t) = \left(\frac{5 + t^2}{4}, t\right)$

27. $c(t) = (-9 + 7\cos t, 4 + 7\sin t)$ **29.** $c(t) = (-4 + t, 9 + 8t)$

31. $c(t) = (3 - 8t, 1 + 3t)$ **33.** $c(t) = (1 + t, 1 + 2t) \ (0 \le t \le 1)$

35. $c(t) = (3 + 4\cos t, 9 + 4\sin t)$ **37.** $c(t) = \left(-4 + t, -8 + t^2\right)$

39. $c(t) = (2 + t, 2 + 3t)$ **41.** $c(t) = \left(3 + t, (3 + t)^2\right)$

43. $y = \sqrt{x^2 - 1} \ (1 \le x < \infty)$ **45.** Plot III

47.

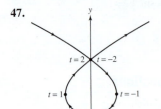

49. $\left.\dfrac{dy}{dx}\right|_{t=-4} = -\dfrac{1}{6}$ **51.** $\left.\dfrac{dy}{dx}\right|_{s=-1} = -\dfrac{3}{4}$ **53.** $-\dfrac{2}{3}$

55. $y = -\dfrac{9}{2}x + \dfrac{11}{2}; \dfrac{dy}{dx} = -\dfrac{9}{2}$

57. $y = x^2 + x^{-1}; \dfrac{dy}{dx} = 2x - \dfrac{1}{x^2}$ **59.** $(0, 0), (96, 180)$

61.

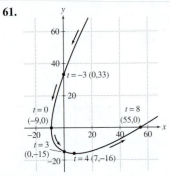

The graph is in: quadrant i for $t < -3$ or $t > 8$, quadrant ii for $-3 < t < 0$, quadrant iii for $0 < t < 3$, quadrant iv for $3 < t < 8$.

63. $(55, 0)$

65. The coordinates of P, $(R\cos\theta, r\sin\theta)$, describe an ellipse for $0 \le \theta \le 2\pi$.

69. $c(t) = (3 - 9t + 24t^2 - 16t^3, 2 + 6t^2 - 4t^3), 0 \le t \le 1$

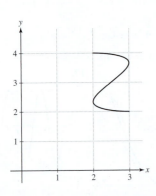

73. $y = -\sqrt{3}x + \dfrac{\sqrt{3}}{2}$

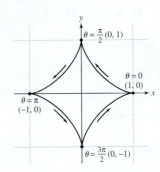

75. $((2k - 1)\pi, 2)$, $k = 0, \pm 1, \pm 2, \dots$ **85.** $\left.\dfrac{d^2 y}{dx^2}\right|_{t=2} = -\dfrac{21}{512}$

87. $\left.\dfrac{d^2 y}{dx^2}\right|_{t=-3} = 0$ **89.** Concave up: $t > 0$ **91.** $\frac{1}{3}$ **93.** $\frac{2}{5}$

95. $\frac{2}{3}$

Section 11.2 Preliminary Questions

1. $S = \int_a^b \sqrt{x'(t)^2 + y'(t)^2}\, dt$

2. No. They are equal when the curve traced by $c(t)$ is a line segment from the initial to the final point, and $c(t)$ is a one-to-one function.

3. The speed at time t **4.** Displacement: 5; No   **5.** $L = 180$ cm

6. 4π

Section 11.2 Exercises

1. $S = 10$ **3.** $S = 800$ **5.** $S = \frac{1}{2}(65^{3/2} - 5^{3/2}) \approx 256.43$

7. $S = 3\pi$ **9.** $S = -8\left(\dfrac{\sqrt{2}}{2} - 1\right) \approx 2.34$

13. $S = \ln(\cosh(A))$ **15.** $\left.\dfrac{ds}{dt}\right|_{t=2} = 4\sqrt{10} \approx 12.65$ m/s

17. $\left.\dfrac{ds}{dt}\right|_{t=9} = \sqrt{41} \approx 6.4$ m/s   **19.** $\left.\dfrac{ds}{dt}\right|_{t=0} = 1$

21. $\left(\dfrac{ds}{dt}\right)_{min} \approx \sqrt{4.89} \approx 2.21$ **23.** $\dfrac{ds}{dt} = 8$

25.

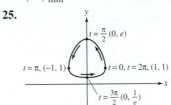

$M_{10} = 6.903734$, $M_{20} = 6.915035$, $M_{30} = 6.914949$, $M_{50} = 6.914951$

27.

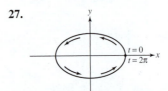

$M_{10} = 25.528309$, $M_{20} = 25.526999$, $M_{30} = 25.526999$, $M_{50} = 25.526999$

29. $S = 2\pi^2 R$

39. (a)

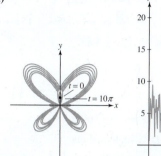

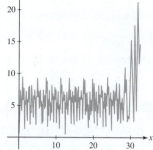

(b) $L \approx 212.096$

Section 11.3 Preliminary Questions

1. (a)   **2.** Positive: $(r, \theta) = \left(1, \frac{\pi}{2}\right)$; Negative: $(r, \theta) = \left(-1, \frac{3\pi}{2}\right)$

3. **(a)** Equation of the circle of radius 2 centered at the origin

(b) Equation of the circle of radius $\sqrt{2}$ centered at the origin

(c) Equation of the vertical line through the point $(2, 0)$

4. (a)

Section 11.3 Exercises

1. (A): $\left(3\sqrt{2}, \frac{3\pi}{4}\right)$; (B): $(3, \pi)$; (C):
$\left(\sqrt{5}, \pi + 0.46\right) \approx \left(\sqrt{5}, 3.60\right)$; (D): $\left(\sqrt{2}, \frac{5\pi}{4}\right)$; (E): $\left(\sqrt{2}, \frac{\pi}{4}\right)$;
(F): $\left(4, \frac{\pi}{6}\right)$; (G): $\left(4, \frac{11\pi}{6}\right)$

3. **(a)** $(1, 0)$ **(b)** $\left(\sqrt{12}, \frac{\pi}{6}\right)$ **(c)** $\left(\sqrt{8}, \frac{3\pi}{4}\right)$ **(d)** $\left(2, \frac{2\pi}{3}\right)$

5. **(a)** $\left(\frac{3\sqrt{3}}{2}, \frac{3}{2}\right)$ **(b)** $\left(-\frac{6}{\sqrt{2}}, \frac{6}{\sqrt{2}}\right)$ **(c)** $(0, 0)$ **(d)** $(0, -5)$

7. (A): $0 \le r \le 3, \pi \le \theta \le 2\pi$, (B): $0 \le r \le 3, \frac{\pi}{4} \le \theta \le \frac{\pi}{2}$, (C):
$3 \le r \le 5, \frac{3\pi}{4} \le \theta \le \pi$

9. $m = \tan \frac{3\pi}{5} \approx -3.1$ **11.** $x^2 + y^2 = 7^2$
13. $x^2 + (y - 1)^2 = 1$ **15.** $y = x - 1$ **17.** $r = \sqrt{5}$
19. $r = \tan \theta \sec \theta$ **21.** $e^r = 1$

23. (a)↔(iii), (b)↔(iv), (c)↔(i), (d)↔(ii)

25. **(a)** $(r, 2\pi - \theta)$ **(b)** $(r, \theta + \pi)$ **(c)** $(r, \pi - \theta)$
(d) $\left(r, \frac{\pi}{2} - \theta\right)$

27. $r \cos \left(\theta - \frac{\pi}{3}\right) = d$

29.

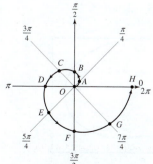

31.

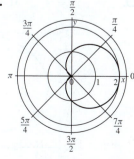

33. (a) A, $\theta = 0, r = 0$; B, $\theta = \frac{\pi}{4}, r = \sin \frac{2\pi}{4} = 1$; C, $\theta = \frac{\pi}{2}$,
$r = 0$; D, $\theta = \frac{3\pi}{4}, r = \sin \frac{2 \cdot 3\pi}{4} = -1$; E, $\theta = \pi, r = 0$; F, $\theta = \frac{5\pi}{4}$,
$r = 1$; G, $\theta = \frac{3\pi}{2}, r = 0$; H, $\theta = \frac{7\pi}{4}, r = -1$; I, $\theta = 2\pi, r = 0$

(b) $0 \le \theta \le \frac{\pi}{2}$ is in the first quadrant. $\frac{\pi}{2} \le \theta \le \pi$ is in the fourth
quadrant. $\pi \le \theta \le \frac{3\pi}{2}$ is in the third quadrant. $\frac{3\pi}{2} \le \theta \le 2\pi$ is in the
second quadrant.

35.

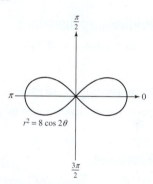

37. $\left(x - \frac{a}{2}\right)^2 + \left(y - \frac{b}{2}\right)^2 = \frac{a^2 + b^2}{4}, r = \frac{\sqrt{x^2 + y^2}}{2}$, centered at the
point $\left(\frac{a}{2}, \frac{b}{2}\right)$

39. $r^2 = \sec 2\theta$ **41.** $\left(x^2 + y^2\right)^2 = x^3 - 3y^2x$
43. $r = 2 \sec \left(\theta - \frac{\pi}{9}\right)$ **45.** $r = 2\sqrt{10} \sec (\theta - 4.39)$
49. $r^2 = 2a^2 \cos 2\theta$

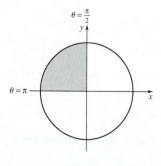

$r^2 = 8 \cos 2\theta$

53. $\theta = \frac{\pi}{2}, m = -\frac{2}{\pi}; \theta = \pi, m = \pi$
55. $\left(\frac{\sqrt{2}}{2}, \frac{\pi}{6}\right), \left(\frac{\sqrt{2}}{2}, \frac{5\pi}{6}\right), \left(\frac{\sqrt{2}}{2}, \frac{7\pi}{6}\right), \left(\frac{\sqrt{2}}{2}, \frac{11\pi}{6}\right)$
57. A: $m = 1$, B: $m = -1$, C: $m = 1$

Section 11.4 Preliminary Questions

1. (b)   **2.** Yes   **3.** (c)

Section 11.4 Exercises

1. $A = \frac{1}{2} \int_{\pi/2}^{\pi} r^2 \, d\theta = \frac{25\pi}{4}$

3. $A = \frac{1}{2}\int_0^\pi r^2 \, d\theta = 4\pi$ **5.** $A = \frac{\pi^5}{320} + \frac{\pi^4}{16} + \frac{\pi^3}{3}$

7. $A = \frac{3\pi}{2}$ **9.** $A = \frac{\pi}{8} \approx 0.39$

11.

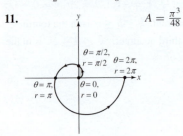

$A = \frac{\pi^3}{48}$

13. $A = \frac{\sqrt{15}}{2} + 7\cos^{-1}\left(\frac{1}{4}\right) \approx 11.163$

15. $A = \pi - \frac{3\sqrt{3}}{2} \approx 0.54$ **17.** $A = \frac{\pi}{8} - \frac{1}{4} \approx 0.14$ **19.** $A = 4\pi$

21. $A = \frac{9\pi}{2} - 4\sqrt{2}$ **23.** $S = 4\pi$

25. $L = \frac{1}{3}\left(\left(\pi^2 + 4\right)^{3/2} - 8\right) \approx 14.55$ **27.** π

29. $L = \sqrt{2}\pi/4 \approx 1.11$

31. $L = \int_0^{2\pi} \sqrt{\cos^4\theta + 4\cos^2\theta\sin^2\theta} \, d\theta \approx 5.52$

33. $\int_0^{\pi/2} \sqrt{2e^{2\theta} + 2e^\theta + 1} \, d\theta$ **35.** $\int_0^{\pi/2} \sin^2\theta\sqrt{1 + 8\cos^2\theta} \, d\theta$
37. $L \approx 6.682$ **39.** $L \approx 79.56$

Section 11.5 Preliminary Questions

1. (a) Hyperbola **(b)** Parabola **(c)** Ellipse

(d) Not a conic section

2. Hyperbolas **3.** The points $(0, c)$ and $(0, -c)$

4. $\pm\frac{b}{a}$ are the slopes of the two asymptotes of the hyperbola.

Section 11.5 Exercises

1. $F_1 = \left(-\sqrt{65}, 0\right)$, $F_2 = \left(\sqrt{65}, 0\right)$. The vertices are $(9, 0)$, $(-9, 0)$, $(0, 4)$, and $(0, -4)$.

3. $F_1 = \left(\sqrt{97}, 0\right)$, $F_2 = \left(\sqrt{97}, 0\right)$. The vertices are $(4, 0)$ and $(-4, 0)$.

5. $F_1 = \left(\sqrt{65} + 3, -1\right)$, $F_2 = \left(-\sqrt{65} + 3, -1\right)$. The vertices are $(10, -1)$ and $(-4, -1)$.

7. $\frac{x^2}{6^2} + \frac{y^2}{3^2} = 1$ **9.** $\frac{(x-14)^2}{6^2} + \frac{(y+4)^2}{3^2} = 1$

11. $\frac{x^2}{5^2} + \frac{y^2}{7^2} = 1$ **13.** $\frac{x^2}{(40/3)^2} + \frac{y^2}{(50/3)^2} = 1$

15. $\left(\frac{x}{3}\right)^2 - \left(\frac{y}{4}\right)^2 = 1$ **17.** $\frac{x^2}{2^2} - \frac{y^2}{\left(2\sqrt{3}\right)^2} = 1$

19. $\left(\frac{x-2}{5}\right)^2 - \left(\frac{y}{10\sqrt{2}}\right)^2 = 1$ **21.** $y = 3x^2$

23. $y = \frac{1}{20}x^2$ **25.** $y = \frac{1}{16}x^2$ **27.** $x = \frac{1}{8}y^2$

29. Vertices: $(\pm 4, 0)$, $(0, \pm 2)$. Foci: $\left(\pm\sqrt{12}, 0\right)$. Centered at the origin.

31. Vertices: $(7, -5)$, $(-1, -5)$. Foci: $\left(\sqrt{65} + 3, -5\right)$, $\left(-\sqrt{65} + 3, -5\right)$. Center: $(3, -5)$. Asymptotes: $y = \frac{7}{4}x - \frac{41}{4}$ and $y = -\frac{7}{4}x + \frac{1}{4}$.

33. Vertices: $(5, 5)$, $(-7, 5)$. Foci: $\left(\sqrt{84} - 1, 5\right)$, $\left(-\sqrt{84} - 1, 5\right)$.

Center: $(-1, 5)$. Asymptotes: $y = \frac{\sqrt{48}}{6}(x + 1) + 5 \approx 1.15x + 6.15$
and $y = -\frac{\sqrt{48}}{6}(x + 1) + 5 \approx -1.15x + 3.85$.

35. Vertex: $(0, 0)$. Focus: $\left(0, \frac{1}{16}\right)$.

37. Vertices: $\left(1 \pm \frac{5}{2}, \frac{1}{5}\right)$, $\left(1, \frac{1}{5} \pm 1\right)$. Foci: $\left(-\frac{\sqrt{21}}{2} + 1, \frac{1}{5}\right)$, $\left(\frac{\sqrt{21}}{2} + 1, \frac{1}{5}\right)$. Centered at $\left(1, \frac{1}{5}\right)$.

39. $D = -87$; ellipse **41.** $D = 40$; hyperbola

47. Focus: $(0, c)$. Directrix: $y = -c$. **49.** $A = \frac{8}{3}c^2$

51. $r = \frac{3}{2+\cos\theta}$ **53.** $r = \frac{4}{1+\cos\theta}$

55. Hyperbola, $e = 4$, directrix, $x = 2$

57. Ellipse, $e = \frac{3}{4}$, directrix, $x = \frac{8}{3}$ **59.** $r = \frac{-12}{5+6\cos\theta}$

61. $\left(\frac{x+3}{5}\right)^2 + \left(\frac{y}{4}\right)^2 = 1$ **63.** 4.5 billion miles

Chapter 11 Review

1. (a), (c)

3. $c(t) = (1 + 2\cos t, 1 + 2\sin t)$. The intersection points with the y-axis are $\left(0, 1 \pm \sqrt{3}\right)$. The intersection points with the x-axis are $\left(1 \pm \sqrt{3}, 0\right)$.

5. $c(\theta) = (\cos(\theta + \pi), \sin(\theta + \pi))$ **7.** $c(t) = (1 + 2t, 3 + 4t)$

9. $y = -\frac{x}{4} + \frac{37}{4}$ **11.** $y = -\frac{8}{(x-3)^3} + \frac{3-x}{2}$

13. $\left.\frac{dy}{dx}\right|_{t=3} = \frac{3}{14}$ **15.** $\left.\frac{dy}{dx}\right|_{t=0} = \frac{\cos 20}{e^{20}}$ **17.** $(1.41, 1.60)$

19. $c(t) = \left(-1 + 6t^2 - 4t^3, -1 + 6t - 6t^2\right)$

21. $\frac{ds}{dt} = \sqrt{3 + 2(\cos t - \sin t)}$; maximal speed: $\sqrt{3 + 2\sqrt{2}}$
23. $s = \sqrt{2}$

25.

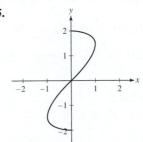

$s = 2\int_0^\pi \sqrt{\cos^2 2t + \sin^2 t} \, dt \approx 6.0972$

27. $\left(1, \frac{\pi}{6}\right)$ and $\left(3, \frac{5\pi}{4}\right)$ have rectangular coordinates $\left(\frac{\sqrt{3}}{2}, \frac{1}{2}\right)$ and $\left(-\frac{3\sqrt{2}}{2}, -\frac{3\sqrt{2}}{2}\right)$.

29. $\sqrt{x^2 + y^2} = \frac{2x}{x-y}$ **31.** $r = 3 + 2\sin\theta$

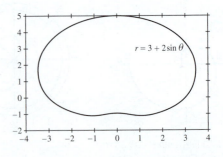

33. $A = \frac{\pi}{16}$ **35.** $e - \frac{1}{e}$

Note: One needs to double the integral from $-\frac{\pi}{2}$ to $\frac{\pi}{2}$ in order to account for both sides of the graph.

37. $A = \frac{3\pi a^2}{2}$

39. Outer: $L \approx 36.121$, inner: $L \approx 7.5087$, difference: 28.6123

41. Ellipse. Vertices: $(\pm 3, 0)$, $(0, \pm 2)$. Foci: $(\pm\sqrt{5}, 0)$.

43. Ellipse. Vertices: $\left(\pm\frac{2}{\sqrt{5}}, 0\right)$, $\left(0, \pm\frac{4}{\sqrt{5}}\right)$. Foci: $\left(0, \pm\sqrt{\frac{12}{5}}\right)$.

45. $\left(\frac{x}{8}\right)^2 + \left(\frac{y}{\sqrt{61}}\right)^2 = 1$ **47.** $\left(\frac{x}{8}\right)^2 - \left(\frac{y}{6}\right)^2 = 1$ **49.** $x = \frac{1}{32}y^2$

51. $y = \sqrt{3}x + \left(\sqrt{3} - 5\right)$ and $y = -\sqrt{3}x + \left(-\sqrt{3} - 5\right)$

Chapter 12

Section 12.1 Preliminary Questions

1. (a) True   **(b)** False   **(c)** True   **(d)** True

2. $\|-3\mathbf{a}\| = 15$ **3.** The components are not changed.   **4.** $\langle 0, 0 \rangle$

5. (a) True   **(b)** False

Section 12.1 Exercises

1. $\mathbf{v}_1 = \langle 2, 0 \rangle$, $\|\mathbf{v}_1\| = 2$ $\mathbf{v}_2 = \langle 2, 0 \rangle$, $\|\mathbf{v}_2\| = 2$

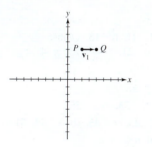

$\mathbf{v}_3 = \langle 3, 1 \rangle$, $\|\mathbf{v}_3\| = \sqrt{10}$ $\mathbf{v}_4 = \langle 2, 2 \rangle$, $\|\mathbf{v}_4\| = 2\sqrt{2}$

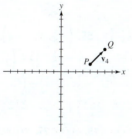

Vectors $\mathbf{v}_1$ and $\mathbf{v}_2$ are equivalent.

3. $(3, 5)$

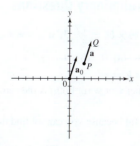

5. $\left\langle \frac{\sqrt{2}}{2}\|\mathbf{u}\|, \frac{\sqrt{2}}{2}\|\mathbf{u}\| \right\rangle$ or $\langle 0.707 \|\mathbf{u}\|, 0.707 \|\mathbf{u}\| \rangle$

7. $\left\langle \cos(-20°)\|\mathbf{w}\|, \sin(-20°)\|\mathbf{w}\| \right\rangle$ or $\langle 0.94 \|\mathbf{w}\|, -0.342 \|\mathbf{w}\| \rangle$

9. $\overrightarrow{PQ} = \langle -1, 5 \rangle$ **11.** $\overrightarrow{PQ} = \langle -2, -9 \rangle$ **13.** $\langle 5, 5 \rangle$

15. $\langle 30, 10 \rangle$ **17.** $\left\langle \frac{5}{2}, 5 \right\rangle$

19. Vector (B)

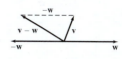

21. $2\mathbf{v} = \langle 4, 6 \rangle$ $-\mathbf{w} = \langle -4, -1 \rangle$

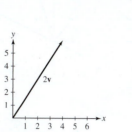

$2\mathbf{v} - \mathbf{w} = \langle 0, 5 \rangle$ $\mathbf{v} + \mathbf{w} = \langle 6, 4 \rangle$

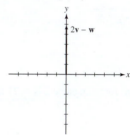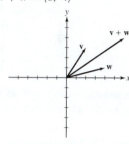

23. $3\mathbf{v} + \mathbf{w} = \langle -2, 10 \rangle$, $2\mathbf{v} - 2\mathbf{w} = \langle 4, -4 \rangle$

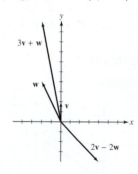

25.

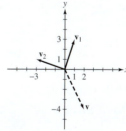

27. (b) and **(c)**

29. $\overrightarrow{AB} = \langle 2, 6 \rangle$ and $\overrightarrow{PQ} = \langle 2, 6 \rangle$; equivalent

31. $\overrightarrow{AB} = \langle 3, -2 \rangle$ and $\overrightarrow{PQ} = \langle 3, -2 \rangle$; equivalent

33. $\overrightarrow{AB} = \langle 2, 3 \rangle$ and $\overrightarrow{PQ} = \langle 6, 9 \rangle$; parallel and point in the same direction

35. $\overrightarrow{AB} = \langle -8, 1 \rangle$ and $\overrightarrow{PQ} = \langle 8, -1 \rangle$; parallel and point in opposite directions

37. $\left\| \overrightarrow{OR} \right\| = \sqrt{53}$ **39.** $P = (0, 0)$ **41.** $\left\langle \frac{3}{5}, \frac{4}{5} \right\rangle$

43. $4\mathbf{e_u} = \left\langle -2\sqrt{2}, -2\sqrt{2} \right\rangle$ **45.** $2\mathbf{e_{-v}} = -\sqrt{2}\mathbf{i} + \sqrt{2}\mathbf{j}$

47. $\mathbf{e} = \left\langle \cos \frac{4\pi}{7}, \; \sin \frac{4\pi}{7} \right\rangle = \langle -0.22, \; 0.97 \rangle$

49. $\lambda = \pm \frac{1}{\sqrt{13}}$ **51.** $P = (4, \; 6)$

53. (a) → (ii), (b) → (iv), (c) → (iii), (d) → (i)   **55.** $9\mathbf{i} + 7\mathbf{j}$

57. $-5\mathbf{i} - 3\mathbf{j}$

59.

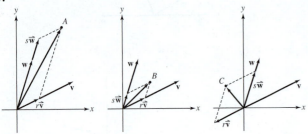

61. $\mathbf{u} = 2\mathbf{v} - \mathbf{w}$

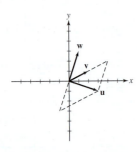

63. The force on cable 1 is ≈ 45 lb, and force on cable 2 is ≈ 21 lb.

65. 230 km/h   **67.** $\mathbf{r} = \langle 6.45, \; 0.38 \rangle$

Section 12.2 Preliminary Questions

1. (4, 3, 2)   **2.** ⟨3, 2, 1⟩   **3.** (a)   **4.** (c)

5. Infinitely many direction vectors   **6.** True

Section 12.2 Exercises

1. $\|\mathbf{v}\| = \sqrt{14}$

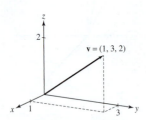

3. The head of $\mathbf{v} = \overrightarrow{PQ}$ is $Q = (1, \; 2, \; 1)$.

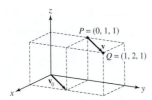

$\mathbf{v}_0 = \overrightarrow{OS}$, where $S = (1, \; 1, \; 0)$

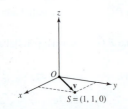

5. $\overrightarrow{PQ} = \langle 1, \; 1, \; -1 \rangle$ **7.** $\overrightarrow{PQ} = \left\langle -\frac{9}{2}, \; -\frac{3}{2}, \; 1 \right\rangle$

9. $\left\| \overrightarrow{OR} \right\| = \sqrt{26} \approx 5.1$ **11.** $P = (-2, \; 6, \; 0)$

13. (a) Parallel and same direction   (b) Not parallel
(c) Parallel and opposite directions   (d) Not parallel

15. Not equivalent   **17.** Not equivalent   **19.** ⟨−8, −18, −2⟩

21. ⟨−2, −2, 3⟩   **23.** ⟨16, −1, 9⟩

25. Not parallel   **27.** Not parallel

29. $\mathbf{e}_w = \left\langle \frac{4}{\sqrt{21}}, \; \frac{-2}{\sqrt{21}}, \; \frac{-1}{\sqrt{21}} \right\rangle$ **31.** $-\mathbf{e}_v = \left\langle \frac{2}{3}, \; -\frac{2}{3}, \; -\frac{1}{3} \right\rangle$

33. $\mathbf{r}(t) = \langle 1 + 2t, \; 2 + t, \; -8 + 3t \rangle$

35. $\mathbf{r}(t) = \langle 4 + 7t, \; 0, \; 8 + 4t \rangle$ **37.** $\mathbf{r}(t) = \langle 1 + 2t, \; 1 - 6t, \; 1 + t \rangle$

39. $\mathbf{r}(t) = \langle 4t, \; t, \; t \rangle$ **41.** $\mathbf{r}(t) = \langle 0, \; 0, \; t \rangle$

43. $\mathbf{r}(t) = \langle -t, \; -2t, \; 4 - 2t \rangle$ **45.** (c)

49. $\mathbf{r}_1(t) = \langle 5, \; 5, \; 2 \rangle + t \langle 0, \; -2, \; 1 \rangle$;
$\mathbf{r}_2(t) = \langle 5, \; 5, \; 2 \rangle + t \langle 0, \; -20, \; 10 \rangle$

55. 4 min   **57.** $\left\langle 0, \; \frac{1}{2}, \; -\frac{1}{2} \right\rangle$ **59.** 2450 newtons

61. $\frac{x+2}{2} = \frac{y-3}{4} = \frac{z-3}{3}$ **63.** $\frac{x-3}{2} = \frac{y-4}{-9} = \frac{z}{12}$

65. $\mathbf{r}(t) = \langle 2t, \; 7t, \; 8t \rangle$

Section 12.3 Preliminary Questions

1. Scalar   **2.** Obtuse   **3.** Distributive Law

4. (a)   **5.** (b); (c)   **6.** (c)

Section 12.3 Exercises

1. 15   **3.** 41   **5.** 5   **7.** 0   **9.** 1   **11.** 0   **13.** Obtuse

15. Orthogonal   **17.** Acute   **19.** 0   **21.** $\frac{1}{\sqrt{10}}$ **23.** $\pi/4$

25. ≈0.615   **27.** $2\pi/3$

29. (a) $b = -\frac{1}{2}$ (b) $b = 0$ or $b = \frac{1}{2}$

31. $\mathbf{v}_1 = \langle 0, \; 1, \; 0 \rangle$, $\mathbf{v}_2 = \langle 3, \; 2, \; 2 \rangle$ **33.** $-\frac{3}{2}$ **35.** $\|\mathbf{v}\|^2$

37. $\|\mathbf{v}\|^2 - \|\mathbf{w}\|^2$ **39.** 8   **41.** 2   **43.** π **45.** (b) 7   **49.** 51.91°

51. (a)   (b) $\mathbf{u}_{\|\mathbf{v}}$

53. $\left\langle \frac{7}{2}, \; \frac{7}{2} \right\rangle$ **55.** $\left\langle -\frac{4}{5}, \; 0, \; -\frac{2}{5} \right\rangle$ **57.** $-4\mathbf{k}$ **59.** $a\mathbf{i}$ **61.** $2\sqrt{2}$

63. $\sqrt{17}$ **65.** $\mathbf{a} = \left\langle \frac{1}{2}, \; \frac{1}{2} \right\rangle + \left\langle \frac{1}{2}, \; -\frac{1}{2} \right\rangle$

67. $\mathbf{a} = \left\langle 0, \; -\frac{1}{2}, \; -\frac{1}{2} \right\rangle + \left\langle 4, \; -\frac{1}{2}, \; \frac{1}{2} \right\rangle$

69. $\left\langle \frac{x-y}{2}, \; \frac{y-x}{2} \right\rangle + \left\langle \frac{x+y}{2}, \; \frac{y+x}{2} \right\rangle$

73. ≈35°   **75.** $\overrightarrow{AD}$ **77.** ≈109.5°   **81.** ≈68.07 N

99. $2x + 2y - 2z = 1$

Section 12.4 Preliminary Questions

1. $\begin{vmatrix} -5 & -1 \\ 4 & 0 \end{vmatrix}$ **2.** $\|\mathbf{e} \times \mathbf{f}\| = \frac{1}{2}$ **3.** $\mathbf{u} \times \mathbf{v} = \langle -2, \; -2, \; -1 \rangle$

4. (vector notation)   (a) 0   (b) 0

5. $\mathbf{i} \times \mathbf{j} = \mathbf{k}$ and $\mathbf{i} \times \mathbf{k} = -\mathbf{j}$

6. $\mathbf{v} \times \mathbf{w} = \mathbf{0}$ if either $\mathbf{v}$ or $\mathbf{w}$ (or both) is the zero vector or $\mathbf{v}$ and $\mathbf{w}$ are parallel vectors.

7. (a) not meaningful because you cannot find the cross product of a scalar and a vector;

(b) meaningful because it represents the dot product of two vectors;

(c) meaningful because it is the product of two scalars;

(d) meaningful because it is a scalar multiple of a vector.

8. (b)

Section 12.4 Exercises

1. -5 **3.** -15 **5.** -8 **7.** 0 **9.** $\langle 1, 2, -5 \rangle$ **11.** $\langle 6, 0, -8 \rangle$
13. $-\mathbf{j} + \mathbf{i}$ **15.** $\mathbf{i} + \mathbf{j} + \mathbf{k}$ **17.** $\langle -1, -1, 0 \rangle$
19. $\langle -2, -2, -2 \rangle$ **21.** $\langle 4, 4, 0 \rangle$
23. $\mathbf{v} \times \mathbf{i} = c\mathbf{j} - b\mathbf{k}$; $\mathbf{v} \times \mathbf{j} = -c\mathbf{i} + a\mathbf{k}$; $\mathbf{v} \times \mathbf{k} = b\mathbf{i} - a\mathbf{j}$
25. $-\mathbf{u}$ **27.** $\langle 0, 3, 3 \rangle$ **31.** e' **33.** $\mathbf{F}_1$ **37.** $2\sqrt{138}$
39. The volume is 4.

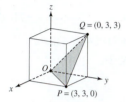

41. $\sqrt{35} \approx 5.92$

43.

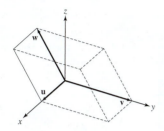

The area of the triangle is $\frac{9\sqrt{3}}{2} \approx 7.8$.
45. 3 **47.** $10\sqrt{3}$ **59.** $\mathbf{X} = \langle a, a, a+1 \rangle$
63. $\tau = 250 \sin 125° \, \mathbf{k} \approx 204.79 \, \mathbf{k}$

Section 12.5 Preliminary Questions

1. $3x + 4y - z = 0$ **2.** (c): $z = 1$ **3.** Plane (c) **4.** xz-plane
5. (c): $x + y = 0$ **6.** Statement (a)

Section 12.5 Exercises

1. $x + 3y + 2z = 3$ **3.** $-x + 2y + z = 3$ **5.** $x = 3$
7. $z = 2$ **9.** $x = 0$ **11.** Statements (b) and (d)
13. $\langle 9, -4, -11 \rangle$ **15.** $\langle 3, -8, 11 \rangle$ **17.** $6x + 9y + 4z = 19$
19. $x + 2y - z = 1$ **21.** $4x - 9y + z = 0$ **23.** $x = 4$
25. $x + z = 3$ **27.** $13x + y - 5z = 27$ **29.** Yes, the planes are parallel.

31.

33.

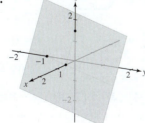

35.

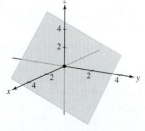

37. $10x + 15y + 6z = 30$ **39.** $(1, 5, 8)$ **41.** $(-2, 3, 12)$
43. $-9y + 4z = 5$ **45.** $x = -\frac{2}{3}$ **47.** $x = -4$
49. The two planes have no common points.
51. $\quad y - 4z = 0$
$\quad x + y - 4z = 0$
53. $(3\lambda)x + by + (2\lambda)z = 5\lambda$, $\lambda \neq 0$ **55.** $\theta = \pi/2$
57. $\theta = 1.143$ rad or $\theta = 65.49°$ **59.** $\theta \approx 55.0°$
61. $x + y + z = 1$ **63.** $x - y - z = d/a$
65. $x = \frac{9}{5} + 2t$, $y = -\frac{6}{5} - 3t$, $z = 2 + 5t$ **67.** $\pm 24 \langle 1, 2, -2 \rangle$
73. $\left(\frac{2}{3}, -\frac{1}{3}, \frac{2}{3} \right)$ **75.** $\frac{6}{\sqrt{30}} \approx 1.095$ **77.** $|a|$

Section 12.6 Preliminary Questions

1. True, mostly, except at $x = \pm a$, $y = \pm b$, or $z = \pm c$
2. False **3.** Hyperbolic paraboloid
4. No **5.** Ellipsoid
6. All vertical lines passing through a parabola c in the xy-plane

Section 12.6 Exercises

1. Ellipsoid **3.** Ellipsoid
5. Hyperboloid of one sheet **7.** Elliptic paraboloid
9. Hyperbolic paraboloid **11.** Hyperbolic paraboloid
13. Ellipsoid, the trace is a circle on the xz-plane
15. Ellipsoid, the trace is an ellipse parallel to the xy-plane
17. Hyperboloid of one sheet, the trace is a hyperbola
19. Parabolic cylinder, the trace is the parabola $y = 3x^2$
21. (a) ↔ Figure b; (b) ↔Figure c; (c) ↔ Figure a
23. $y = \left(\frac{x}{2} \right)^2 + \left(\frac{z}{4} \right)^2$
25.

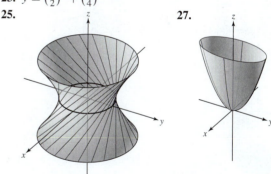

Graph of $x^2 + y^2 + z^2 = 1$

27.

29.

31.

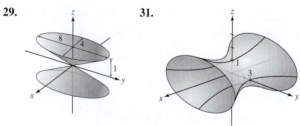

33.

35.

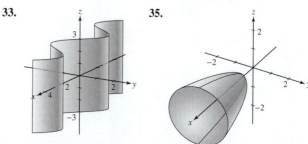

39. $\left(\frac{x}{2}\right)^2 + \left(\frac{y}{4}\right)^2 + \left(\frac{z}{6}\right)^2 = 1$ **41.** $\left(\frac{x}{4}\right)^2 + \left(\frac{y}{6}\right)^2 - \left(\frac{z}{3\sqrt{3}}\right)^2 = 1$

43. One or two vertical lines, or an empty set   **45.** An elliptic cone

Section 12.7 Preliminary Questions

1. Cylinder of radius R whose axis is the z-axis, sphere of radius R centered at the origin

2. (b)   **3.** (a)   **4.** $\phi = 0, \pi$ **5.** $\phi = \frac{\pi}{2}$, the xy-plane

Section 12.7 Exercises

1. $(-4, 0, 4)$ **3.** $\left(0, 0, \frac{1}{2}\right)$ **5.** $\left(\sqrt{2}, \frac{7\pi}{4}, 1\right)$ **7.** $\left(2, \frac{\pi}{3}, 7\right)$

9. $\left(5, \frac{\pi}{4}, 2\right)$ **11.** $r^2 \leq 1$ **13.** $r^2 + z^2 \leq 4, \theta = \frac{\pi}{2}$ or $\theta = \frac{3\pi}{2}$

15. $r^2 \leq 9, \frac{5\pi}{4} \leq \theta \leq 2\pi$ and $0 \leq \theta \leq \frac{\pi}{4}$

17. 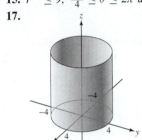   **19.**

21. 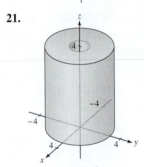   **23.**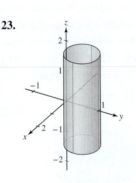

25.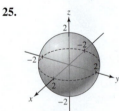

27. $r = \frac{z}{\cos\theta + \sin\theta}$ **29.** $r = \frac{z\tan\theta}{\cos\theta}$ **31.** $r = 2$ **33.** $(3, 0, 0)$

35. $(0, 0, 3)$ **37.** $\left(\frac{3\sqrt{3}}{2}, \frac{3}{2}, -3\sqrt{3}\right)$ **39.** $\left(2, 0, \frac{\pi}{3}\right)$

41. $\left(\sqrt{3}, \frac{\pi}{4}, 0.955\right)$ **43.** $\left(2, \frac{\pi}{3}, \frac{\pi}{6}\right)$ **45.** $\left(2\sqrt{2}, 0, \frac{\pi}{4}\right)$

47. $\left(2\sqrt{2}, 0, 2\sqrt{2}\right)$ **49.** $0 \leq \rho \leq 1$

51. $\rho = 1, 0 \leq \theta \leq \frac{\pi}{2}, 0 \leq \phi \leq \frac{\pi}{2}$

53. $\left\{(\rho, \theta, \phi) : 0 \leq \rho \leq 2, \theta = \frac{\pi}{2} \text{ or } \theta = \frac{3\pi}{2}\right\}$

55.    **57.**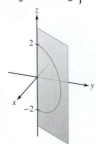

59. 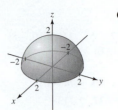   **61.**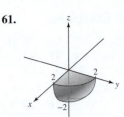

63.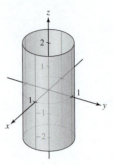

65. $\rho = \frac{2}{\cos\phi}$ **67.** $\rho = \frac{\cos\theta\tan\phi}{\cos\phi}$ **69.** $\rho = \frac{2}{\sin\phi\sqrt{\cos 2\theta}}$ **71.** (b)

73. Helsinki: $(25.0°, 29.9°)$, São Paulo: $(313.48°, 113.52°)$

75. Sydney: $(-4618.8, 2560.3, -3562.1)$,
  Bogotá: $(1723.7, -6111.7, 503.1)$

77. $z = \pm r\sqrt{\cos 2\theta}$

79. $\left\{(r, \theta, z) : -\sqrt{4 - r^2} \leq z \leq \sqrt{4 - r^2}, 1 \leq r \leq 2, 0 \leq \theta \leq 2\pi\right\}$

83. $r = \sqrt{z^2 + 1}$ and $\rho = \sqrt{-\frac{1}{\cos 2\phi}}$; no points; $\frac{\pi}{4} < \phi < \frac{3\pi}{4}$

Chapter 12 Review

1. $\langle 21, -25 \rangle$ and $\langle -19, 31 \rangle$ **3.** $\left\langle \frac{-2}{\sqrt{29}}, \frac{5}{\sqrt{29}} \right\rangle$

5. $\mathbf{i} = \frac{2}{11}\mathbf{v} + \frac{5}{11}\mathbf{w}$ **7.** $\overrightarrow{PQ} = \langle -4, 1 \rangle$; $\left\| \overrightarrow{PQ} \right\| = \sqrt{17}$

9. $\left\langle \frac{3}{\sqrt{2}}, -\frac{3}{\sqrt{2}} \right\rangle$ **11.** $\beta = \frac{3}{2}$ **13.** $\mathbf{u} = \left\langle \frac{1}{3}, -\frac{11}{6}, \frac{7}{6} \right\rangle$

15. $\mathbf{r}_1(t) = \langle 1 + 3t, 4 + t, 5 + 6t \rangle$; $\mathbf{r}_2(t) = \langle 1 + 3t, t, 6t \rangle$

17. $a = -2, b = 2$

19.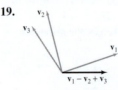

21. $\mathbf{v} \cdot \mathbf{w} = -9$ **23.** $\mathbf{v} \times \mathbf{w} = \langle 10, -8, -7 \rangle$ **25.** $V = 48$

29. $\frac{5}{3}$ **31.** $\|\mathbf{F}_1\| = \frac{2\|\mathbf{F}_2\|}{\sqrt{3}}$; $\|\mathbf{F}_1\| = 980$ N

33. $\mathbf{v} \times \mathbf{w} = \langle -6, 7, -2 \rangle$

35. -47 **37.** $5\sqrt{2}$ **41.** $\|\mathbf{e} - 4\mathbf{f}\| = \sqrt{13}$

47. $(x - 0) + 4(y - 1) - 3(z + 1) = 0$

49. $17x - 21y - 13z = -28$ **51.** $3x - 2y = 4$ **53.** Ellipsoid

55. Elliptic paraboloid   **57.** Elliptic cone

59. (a) Empty set   (b) Hyperboloid of one sheet

(c) Hyperboloid of two sheets

61. $(r, \theta, z) = \left(5, \tan^{-1}\frac{4}{3}, -1\right)$, $(\rho, \theta, \phi) = \left(\sqrt{26}, \tan^{-1}\frac{4}{3}, \cos^{-1}\left(\frac{-1}{\sqrt{26}}\right)\right)$

63. $(r, \theta, z) = \left(\frac{3\sqrt{3}}{2}, \frac{\pi}{6}, \frac{3}{2}\right)$

65. $z = 2x$

69. $A < -1$: Hyperboloid of one sheet
$A = -1$: Cylinder with the z-axis as its central axis
$A > -1$: Ellipsoid
$A = 0$: Sphere

Chapter 13

Section 13.1 Preliminary Questions

1. (c) **2.** The curve $z = e^x$

3. The projection onto the xz-plane **4.** The point $(-2, 2, 3)$

5. As t increases from 0 to 2π, a point on $\sin t\mathbf{i} + \cos t\mathbf{j}$ moves clockwise and a point on $\cos t\mathbf{i} + \sin t\mathbf{j}$ moves counterclockwise.

6. (a), (c), and (d)

Section 13.1 Exercises

1. $D = \{t \in \mathbf{R}, \ t \neq 0, \ t \neq -1\}$

3. $\mathbf{r}(2) = \left\langle 0, \ 4, \ \frac{1}{5} \right\rangle; \mathbf{r}(-1) = \left\langle -1, \ 1, \ \frac{1}{2} \right\rangle$

5. $\mathbf{r}(t) = (3 + 3t)\mathbf{i} - 5\mathbf{j} + (7 + t)\mathbf{k}$

7. Yes, when $t = (2n - 1)\pi$, where n is an integer, at the points $\left(0, 0, (2n - 1)\pi\right)$

9. No intersection **11.** A $\leftrightarrow$ ii, B $\leftrightarrow$ i, C $\leftrightarrow$ iii

13. (a) = (v), (b) = (i), (c) = (ii), (d) = (vi), (e) = (iv), (f) = (iii)

15. C $\leftrightarrow$ i, A $\leftrightarrow$ ii, B $\leftrightarrow$ iii

17. This is a circle of radius 9 centered at the origin lying in the xy-plane.

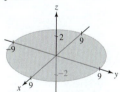

19. Radius 1, center $(0, \ 0, \ 4)$, xz-plane

21.

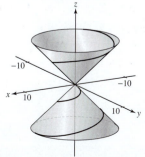

23. $(0, \ 1, \ 0), (0, \ -1, \ 0), \left(\frac{1}{\sqrt{2}}, \ \frac{1}{\sqrt{2}}, \ 0 \right), \left(\frac{1}{\sqrt{2}}, \ -\frac{1}{\sqrt{2}}, \ 0 \right),$
$\left(-\frac{1}{\sqrt{2}}, \ -\frac{1}{\sqrt{2}}, \ 0 \right), \left(-\frac{1}{\sqrt{2}}, \ \frac{1}{\sqrt{2}}, \ 0 \right)$

25. $\mathbf{r}(t) = \left\langle 2t^2 - 7, \ t, \ \pm\sqrt{9 - t^2} \right\rangle$, for $-3 \leq t \leq 3$

27. **(a)** $\mathbf{r}(t) = \left\langle \pm t\sqrt{1 - t^2}, \ t^2, \ t \right\rangle$ for $-1 \leq t \leq 1$

(b) The projection is a circle in the xy-plane with radius $\frac{1}{2}$ and centered at the xy-point $\left(0, \ \frac{1}{2} \right)$.

29. $\mathbf{r}(t) = \langle \cos t, \ \pm \sin t, \ \sin t \rangle$; the projection of the curve onto the xy-plane is traced by $\langle \cos t, \ \pm \sin t, \ 0 \rangle$, which is the unit circle in this plane; the projection of the curve onto the xz-plane is traced by $\langle \cos t, \ 0, \ \sin t \rangle$, which is the unit circle in this plane; the projection of the curve onto the yz-plane is traced by $\langle 0, \ \pm \sin t, \ \sin t \rangle$, which is the two segments $z = y$ and $z = -y$ for $-1 \leq y \leq 1$.

31. $\mathbf{r}(t) = \left\langle \cos t, \ \sin t, \ 4\cos^2 t \right\rangle, 0 \leq t \leq 2\pi$

33. Collide at the point $(12, \ 4, \ 2)$ and intersect at the points $(4, \ 0, \ -6)$ and $(12, \ 4, \ 2)$

35. $\mathbf{r}(t) = \langle 3, \ 2, \ t \rangle, \ -\infty < t < \infty$

37. $\mathbf{r}(t) = \langle t, \ 3t, \ 15t \rangle, \ -\infty < t < \infty$

39. $\mathbf{r}(t) = \langle 1, \ 2 + 2\cos t, \ 5 + 2\sin t \rangle, \ 0 \leq t \leq 2\pi$

41. $\mathbf{r}(t) = \left\langle \frac{\sqrt{3}}{2} \cos t, \ \frac{1}{2}, \ \frac{\sqrt{3}}{2} \sin t \right\rangle, \ 0 \leq t \leq 2\pi$

43. $\mathbf{r}(t) = \langle 3 + 2\cos t, \ 1, \ 5 + 3\sin t \rangle, \ 0 \leq t \leq 2\pi$

45.

$\mathbf{r}(t) = \langle |t| + t, \ |t| - t \rangle$

Section 13.2 Preliminary Questions

1. $\frac{d}{dt} (f(t)\mathbf{r}(t)) = f(t)\mathbf{r}'(t) + f'(t)\mathbf{r}(t)$

$\frac{d}{dt} (\mathbf{r}_1(t) \cdot \mathbf{r}_2(t)) = \mathbf{r}_1(t) \cdot \mathbf{r}'_2(t) + \mathbf{r}'_1(t) \cdot \mathbf{r}_2(t)$

$\frac{d}{dt} (\mathbf{r}_1(t) \times \mathbf{r}_2(t)) = \mathbf{r}_1(t) \times \mathbf{r}'_2(t) + \mathbf{r}'_1(t) \times \mathbf{r}_2(t)$

2. True **3.** False **4.** True **5.** False **6.** False

7. **(a)** Vector **(b)** Scalar **(c)** Vector

Section 13.2 Exercises

1. $\lim\limits_{t \to 3} \left\langle t^2, \ 4t, \ \frac{1}{t} \right\rangle = \left\langle 9, \ 12, \ \frac{1}{3} \right\rangle$

3. $\lim\limits_{t \to 0} (e^{2t}\mathbf{i} + \ln(t + 1)\mathbf{j} + 4\mathbf{k}) = \mathbf{i} + 4\mathbf{k}$

5. $\lim\limits_{h \to 0} \frac{\mathbf{r}(t+h) - \mathbf{r}(t)}{h} = \left\langle -\frac{1}{t^2}, \ \cos t, \ 0 \right\rangle$ **7.** $\frac{d\mathbf{r}}{dt} = \left\langle 1, \ 2t, \ 3t^2 \right\rangle$

9. $\frac{d\mathbf{r}}{ds} = \left\langle 3e^{3s}, \ -e^{-s}, \ 4s^3 \right\rangle$ **11.** $\mathbf{c}'(t) = -t^{-2}\mathbf{i} - 2e^{2t}\mathbf{k}$

13. $\mathbf{r}'(t) = \left\langle 1, \ 2t, \ 3t^2 \right\rangle, \mathbf{r}''(t) = \langle 0, \ 2, \ 6t \rangle$

15.

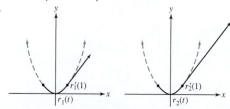

17. $\frac{d}{dt}\left(\mathbf{r}_1(t)\cdot\mathbf{r}_2(t)\right)=$
$$2t^3e^{2t}+3t^2e^{3t}+2te^{3t}+3t^2e^{2t}+te^t+e^t$$

19. $\frac{d}{dt}\left(\mathbf{r}_1(t)\times\mathbf{r}_2(t)\right)=$
$$\left\langle 3t^2e^t-2te^{2t}-e^{2t}+t^3e^t,\ e^{3t}+3te^{3t}-t^2e^t-2te^t,\ 2te^{2t}+2t^2e^{2t}-3t^2e^{3t}-3t^3e^{3t}\right\rangle$$

21. $2+4e$ **23.** $\frac{d}{dt}\mathbf{r}(g(t))=\left\langle 2e^{2t},\ -e^t\right\rangle$

25. $\frac{d}{dt}\mathbf{r}(g(t))=\left\langle 4e^{4t+9},\ 8e^{8t+18},\ 0\right\rangle$

27. $\frac{d}{dt}\left(\mathbf{r}(t)\cdot\mathbf{a}(t)\right)\big|_{t=2}=13$ **29.** $\boldsymbol{\ell}(t)=\langle 4-4t,\ 16-32t\rangle$

31. $\boldsymbol{\ell}(t)=\langle -3-4t,\ 10+5t,\ 16+24t\rangle$

33. $\boldsymbol{\ell}(t)=\left\langle 2-t,\ 0,\ -\frac{1}{3}+\frac{1}{2}t\right\rangle$

35. $\frac{d}{dt}\left(\mathbf{r}\times\mathbf{r}'\right)=\left\langle(t^2-2)e^t,\ -te^t,\ 2t\right\rangle$ **39.** $\left\langle\frac{212}{3},\ 124\right\rangle$

41. $\langle 0,\ 0\rangle$ **43.** $\left\langle 1,\ 2,\ -\frac{\sin 3}{3}\right\rangle$ **45.** $(\ln 4)\mathbf{i}+\frac{56}{3}\mathbf{j}-\frac{496}{5}\mathbf{k}$

47. $\mathbf{r}(t)=\left\langle -t^2+t+c_1,2t^2+c_2\right\rangle$; with initial conditions
$$\mathbf{r}(t)=\left\langle -t^2+t+3,2t^2+1\right\rangle$$

49. $\mathbf{r}(t)=\left(\frac{1}{3}t^3\right)\mathbf{i}+\left(\frac{5t^2}{2}\right)\mathbf{j}+t\mathbf{k}+\mathbf{c}$; with initial conditions,
$$\mathbf{r}(t)=\left(\frac{1}{3}t^3-\frac{1}{3}\right)\mathbf{i}+\left(\frac{5}{2}t^2-\frac{3}{2}\right)\mathbf{j}+(t+1)\mathbf{k}$$

51. $\mathbf{r}(t)=(8t^2)\mathbf{k}+\mathbf{c}_1t+\mathbf{c}_2$; with initial conditions,
$\mathbf{r}(t)=\mathbf{i}+t\mathbf{j}+(8t^2)\mathbf{k}$

53. $\mathbf{r}(t)=\left\langle 0,\ t^2,\ 0\right\rangle+\mathbf{c}_1t+\mathbf{c}_2$; with initial conditions,
$$\mathbf{r}(t)=\left\langle 1,\ t^2-6t+10,\ t-3\right\rangle$$

55. $\mathbf{r}(3)=\left\langle\frac{45}{4},\ 5\right\rangle$

57. Only at time $t=3$ can the pilot hit a target located at the origin.

59. $\mathbf{r}(t)=(t-1)\mathbf{v}+\mathbf{w}$ **61.** $\mathbf{r}(t)=e^{2t}\mathbf{c}$

Section 13.3 Preliminary Questions

1. $2\mathbf{r}'=\langle 50,\ -70,\ 20\rangle$, $-\mathbf{r}'=\langle -25,\ 35,\ -10\rangle$

2. Statement (b) is true.

3. (a) $L'(2)=4$

(b) $L(t)$ is the distance along the path traveled, which is usually different from the distance from the origin.

4. 6

Section 13.3 Exercises

1. $L=3\sqrt{61}$ **3.** $L=15+\ln 4$ **5.** $L=\frac{544\sqrt{34}-2}{135}\approx 23.48$

7. $L=\pi\sqrt{4\pi^2+10}+5\ln\frac{2\pi+\sqrt{4\pi^2+10}}{\sqrt{10}}\approx 29.3$

9. $s(t)=\frac{1}{27}\left((20+9t^2)^{3/2}-20^{3/2}\right)$ **11.** $v(4)=\sqrt{21}$

13. $v(1)=\sqrt{2}$ **15.** $v\left(\frac{\pi}{2}\right)=5$ **17.** $\mathbf{r}'=\langle 100\sqrt{5},200\sqrt{5}\rangle$

19. The bee is at the origin. $\int_0^T\|\mathbf{r}'(u)\|\,du$ represents the total distance the bee traveled on the time interval $[0,T]$.

21. (c) $L_1\approx 132.0,L_2\approx 125.7$; the first spring uses more wire.

23. (a) $t=\pi$

25. (a) $s(t)=\sqrt{29}t$ **(b)** $t=g(s)=\frac{s}{\sqrt{29}}$

27. $\left\langle 1+\frac{3s}{\sqrt{50}},\ 2+\frac{4s}{\sqrt{50}},\ 3+\frac{5s}{\sqrt{50}}\right\rangle$

29. $\mathbf{r}(s)=\langle 2+4\cos(2s),\ 10,\ -3+4\sin(2s)\rangle$

31. $\mathbf{r}(s)=$
$$\left\langle\cos\left[\left(\tfrac{3}{2}s+1\right)^{2/3}-1\right],\ \sin\left[\left(\tfrac{3}{2}s+1\right)^{2/3}-1\right],\ \tfrac{2}{3}\left[\left(\tfrac{3}{2}s+1\right)^{2/3}-1\right]^{3/2}\right\rangle,$$
$s\geq 0$

33. $\mathbf{r}(s)=\left\langle\frac{1}{9}(27s+8)^{2/3}-\frac{4}{9},\ \pm\frac{1}{27}\left((27s+8)^{2/3}-4\right)^{3/2}\right\rangle$

35. $\left\langle\frac{s}{\sqrt{1+m^2}},\ \frac{sm}{\sqrt{1+m^2}}\right\rangle$

37. (a) $\sqrt{17}e^t$ **(b)** $\frac{s}{\sqrt{17}}\left(\cos\left(4\ln\frac{s}{\sqrt{17}}\right),\ \sin\left(4\ln\frac{s}{\sqrt{17}}\right)\right)$

39. $L=\int_{-\infty}^{\infty}\|\mathbf{r}'(t)\|=2\int_{-\infty}^{\infty}\frac{dt}{1+t^2}=2\pi$

Section 13.4 Preliminary Questions

1. $\left\langle\frac{2}{3},\ \frac{1}{3},\ -\frac{2}{3}\right\rangle$ **2.** $\frac{1}{4}$

3. The curvature of a circle of radius 2 **4.** Zero curvature

5. $\kappa=\sqrt{14}$ **6.** 4 **7.** $\frac{1}{9}$

Section 13.4 Exercises

1. $\mathbf{r}'(t)=\langle 8t+9\rangle$, $\mathbf{T}(t)=\frac{1}{\sqrt{64t^2+81}}\langle 8t,\ 9\rangle$,
$\mathbf{T}(1)=\left\langle\frac{8}{\sqrt{145}},\ \frac{9}{\sqrt{145}}\right\rangle$

3. $\mathbf{r}'(t)=\langle 4,-5,9\rangle$, $\mathbf{T}(t)=\left\langle\frac{4}{\sqrt{122}},\ -\frac{5}{\sqrt{122}},\ \frac{9}{\sqrt{122}}\right\rangle$,
$\mathbf{T}(1)=\mathbf{T}(t)$

5. $\mathbf{r}'(t)=\langle-\pi\sin\pi t,\pi\cos\pi t,1\rangle$,
$\mathbf{T}(t)=\frac{1}{\sqrt{\pi^2+1}}\langle-\pi\sin\pi t,\ \pi\cos\pi t,\ 1\rangle$,
$\mathbf{T}(1)=\left\langle 0,\ -\frac{\pi}{\sqrt{\pi^2+1}},\ \frac{1}{\sqrt{\pi^2+1}}\right\rangle$

7. $\kappa(t)=\frac{e^t}{(1+e^{2t})^{3/2}}$ **9.** $\kappa(t)=0$ **11.** $\kappa=\frac{2\sqrt{74}}{27}$

13. $\kappa=\frac{\sqrt{\pi^2+5}}{(\pi^2+1)^{3/2}}\approx 0.108$ **15.** $\kappa(3)=\frac{e^3}{(3^6+1)^{3/2}}\approx 0.0025$

17. $\kappa(2)=\frac{48\sqrt{41}}{210,125}\approx 0.0015$

19. $\kappa\left(\frac{\pi}{3}\right)=\frac{\sqrt{330}}{4}\approx 4.54,\kappa\left(\frac{\pi}{2}\right)=\frac{1}{5}=0.2$ **23.** $\alpha=\pm\sqrt{2}$

29. $\kappa(2)=\frac{3\sqrt{10}}{800}\approx 0.012$ **31.** $\kappa(\pi)=\frac{\pi\sqrt{2}}{4}\approx 1.11$

35. $\kappa(t)=t^2$

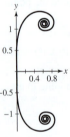

37. $\mathbf{N}(t)=\langle 0,\ -\sin 2t,\ -\cos 2t\rangle$

39. $\mathbf{N}\left(\frac{\pi}{4}\right)=\left\langle-\frac{\sqrt{2}}{3\sqrt{3}},\ -\frac{2}{3\sqrt{3}}\right\rangle$, $\mathbf{N}\left(\frac{3\pi}{4}\right)=\left\langle\frac{\sqrt{2}}{3\sqrt{3}},\ \frac{2}{3\sqrt{3}}\right\rangle$

41.
$\mathbf{T}(1)=\left\langle 0,\ \frac{\sqrt{5}}{5},\ \frac{2\sqrt{5}}{5}\right\rangle$, $\mathbf{N}(1)=\left\langle 0,\ -\frac{2\sqrt{5}}{5},\ \frac{\sqrt{5}}{5}\right\rangle$, $\mathbf{B}(1)=\langle 1,0,0\rangle$

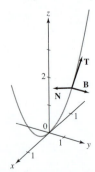

43. $\mathbf{T}(1) = \left\langle \frac{1}{3}, \frac{2}{3}, \frac{2}{3} \right\rangle, \mathbf{N}(1) = \left\langle -\frac{2}{3}, -\frac{1}{3}, \frac{2}{3} \right\rangle, \mathbf{B}(1) = \left\langle \frac{2}{3}, -\frac{2}{3}, \frac{1}{3} \right\rangle$

45. $\mathbf{N}\left(\pi^{1/3}\right) = \left\langle \frac{1}{2}, -\frac{\sqrt{3}}{2} \right\rangle$ **47.** $\mathbf{N}(1) = \frac{1}{\sqrt{13}} \langle -3, 2 \rangle$

49. $\mathbf{N}(1) = \frac{1}{\sqrt{2}} \langle 0, 1, -1 \rangle$

51. $\mathbf{N}(0) = \frac{1}{6} \left\langle -\sqrt{6}, 2\sqrt{6}, -\sqrt{6} \right\rangle$

53. (a)
$\mathbf{T}(1) = \left\langle \frac{1}{3}, \frac{2}{3}, \frac{2}{3} \right\rangle, \mathbf{N}(1) = \left\langle -\frac{2}{3}, -\frac{1}{3}, \frac{2}{3} \right\rangle, \mathbf{B}(1) = \left\langle \frac{2}{3}, -\frac{2}{3}, \frac{1}{3} \right\rangle$

(b) $6x - 6y + 3z = 1$

55. (a) $\mathbf{T}(t) = \left\langle \frac{1}{\sqrt{2+4t^2}}, \frac{-1}{\sqrt{2+4t^2}}, \frac{2t}{\sqrt{2+4t^2}} \right\rangle,$

$\mathbf{N}(t) = \left\langle \frac{-t\sqrt{2}}{\sqrt{2+4t^2}}, \frac{t\sqrt{2}}{\sqrt{2+4t^2}}, \frac{\sqrt{2}}{\sqrt{2+4t^2}} \right\rangle$

(b) $\mathbf{B}(t) = \left\langle -\frac{1}{\sqrt{2}}, -\frac{1}{\sqrt{2}}, 0 \right\rangle$

(c) The osculating planes are all parallel to each other, with equation $x + y = c$ for some c.

59. $\langle \cos t, \sin t \rangle$, that is, the unit circle itself.

61. $\mathbf{c}(t) = \left\langle -4, \frac{7}{2} \right\rangle + \frac{5^{3/2}}{2} \langle \cos t, \sin t \rangle$

63. $\mathbf{c}(t) = \langle \pi, -2 \rangle + 4 \langle \cos t, \sin t \rangle$

65. $\mathbf{c}(t) = \left\langle -1 - 2\cos t, \frac{2\sin t}{\sqrt{2}}, \frac{2\sin t}{\sqrt{2}} \right\rangle$

73. $\kappa(\theta) = 1$ **75.** $\kappa(\theta) = \frac{1}{\sqrt{2}} e^{-\theta}$

91. (a) $\mathbf{N}(0) = \left\langle -\frac{1}{\sqrt{5}}, 0, \frac{2}{\sqrt{5}} \right\rangle$ **(b)** $\mathbf{N}(1) = \frac{1}{\sqrt{66}} \langle 4, 7, -2 \rangle$

Section 13.5 Preliminary Questions

1. No, since the particle may change its direction.   **2.** $\mathbf{a}(t)$

3. Statement (a), their velocity vectors point in the same direction.

4. The velocity vector always points in the direction of motion. Since the vector $\mathbf{N}(t)$ is orthogonal to the direction of motion, the vectors $\mathbf{a}(t)$ and $\mathbf{v}(t)$ are orthogonal.

5. Description (b), parallel   **6.** $\|\mathbf{a}(t)\| = 8$ cm/s^2 **7.** $a_{\mathbf{N}}$

Section 13.5 Exercises

1. $h = -0.2 : \langle -0.085, 1.91, 2.635 \rangle$
$h = -0.1 : \langle -0.19, 2.07, 2.97 \rangle$
$h = 0.1 : \langle -0.41, 2.37, 4.08 \rangle$
$h = 0.2 : \langle -0.525, 2.505, 5.075 \rangle$
$\mathbf{v}(1) \approx \langle -0.3, 2.2, 3.5 \rangle, \quad v(1) \approx 4.1$

3. $\mathbf{v}(1) = \langle 3, -1, 8 \rangle, \quad \mathbf{a}(1) = \langle 6, 0, 8 \rangle, \quad v(1) = \sqrt{74}$

5. $\mathbf{v}\left(\frac{\pi}{3}\right) = \left\langle \frac{1}{2}, -\frac{\sqrt{3}}{2}, 0 \right\rangle, \mathbf{a}\left(\frac{\pi}{3}\right) = \left\langle -\frac{\sqrt{3}}{2}, -\frac{1}{2}, 9 \right\rangle, v\left(\frac{\pi}{3}\right) = 1$

7. $\mathbf{a}(t) = -2 \left\langle \cos \frac{t}{2}, \sin \frac{t}{2} \right\rangle; \quad \mathbf{a}\left(\frac{\pi}{4}\right) \approx \langle -1.85, -0.077 \rangle$

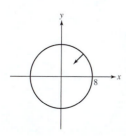

$R(t) = 8 \left\langle \cos \frac{t}{2}, \sin \frac{t}{2} \right\rangle$

9.

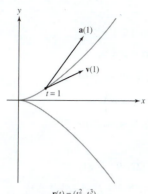

$r(t) = (t^2, t^3)$

11. $\mathbf{v}(t) = \left\langle \frac{3t^2+2}{6}, 4t - 2 \right\rangle$ **13.** $\mathbf{v}(t) = \mathbf{i} + t\mathbf{k}$

15. $\mathbf{v}(t) = \left\langle \frac{t^2}{2} + 3, 4t - 2 \right\rangle, \quad \mathbf{r}(t) = \left\langle \frac{t^3}{6} + 3t, 2t^2 - 2t \right\rangle$

17. $\mathbf{v}(t) = \mathbf{i} + \frac{t^2}{2}\mathbf{k}, \quad \mathbf{r}(t) = t\mathbf{i} + \mathbf{j} + \frac{t^3}{6}\mathbf{k}$

19. $v_0 = \sqrt{5292} \approx 72.746$ m/s   **23.** $H = 355$ m

25. (a) Assume that $r(0) = \langle 150, 75, 5 \rangle$

$a(t) = \langle 0, 0, -32 \rangle, v(t) = \langle 40, 35, -32t + 32 \rangle$

$r(t) = \left\langle 40t + 150, 35t + 75, -16t^2 + 32t + 5 \right\rangle$

(b) $z = 5$ when $t = 0$ or 2. At $t = 2, r(2) = \langle 230, 145, 5 \rangle$, so the player is in bounds, since $(300, 150, z)$ is the maximum possible point to be in bounds.

27. $\mathbf{r}(10) = \langle 45, -20 \rangle$

29. (a) At its original position   **(b)** No

31. The speed is decreasing.

33. $a_{\mathbf{T}} = 0, \quad a_{\mathbf{N}} = 1$ **35.** $a_{\mathbf{T}} = \frac{7}{\sqrt{6}}, \quad a_{\mathbf{N}} = \sqrt{\frac{53}{6}}$

37. $\mathbf{a}(-1) = -\frac{2}{\sqrt{10}}\mathbf{T} + \frac{6}{\sqrt{10}}\mathbf{N}$ with $\mathbf{T} = \frac{1}{\sqrt{10}} \langle 1, -3 \rangle$ and $\mathbf{N} = \frac{1}{\sqrt{10}} \langle -3, -1 \rangle$

39. $a_{\mathbf{T}}(4) = 4, \quad a_{\mathbf{N}}(4) = 1,$ so $\mathbf{a} = 4\mathbf{T} + \mathbf{N}$, with $\mathbf{T} = \left\langle \frac{1}{9}, \frac{4}{9}, \frac{8}{9} \right\rangle$ and $\mathbf{N} = \left\langle -\frac{4}{9}, -\frac{7}{9}, \frac{4}{9} \right\rangle$

41. $\mathbf{a}(0) = \sqrt{3}\mathbf{T} + \sqrt{2}\mathbf{N}$, with $\mathbf{T} = \frac{1}{\sqrt{3}} \langle 1, 1, 1 \rangle$ and $\mathbf{N} = \frac{1}{\sqrt{2}} \langle -1, 0, 1 \rangle$

43. $\mathbf{a}\left(\frac{\pi}{2}\right) = -\frac{\pi}{2\sqrt{3}}\mathbf{T} + \frac{\pi}{\sqrt{6}}\mathbf{N}$, with $\mathbf{T} = \frac{1}{\sqrt{3}} \langle 1, -1, 1 \rangle$ and $\mathbf{N} = \frac{1}{\sqrt{6}} \langle 1, -1, -2 \rangle$

45. $a_{\mathbf{T}} = 0, \quad a_{\mathbf{N}} = 0.25$ cm/s^2

47. The tangential acceleration is $\frac{50}{\sqrt{2}} \approx 35.36$ m/min^2,

$v = \sqrt{35.36(30)} \approx 32.56$ m/min

49. $\|\mathbf{a}\| = 1.157 \times 10^5$ km/h^2 **51.** $\mathbf{a} = \left\langle -\frac{1}{6}, -1, \frac{1}{6} \right\rangle$

53. (A) slowing down, (B) speeding up, (C) slowing down

57. After 139.91 s, the car will begin to skid.   **61.** $R \approx 105$ m

Section 13.6 Preliminary Questions

1. $\frac{dA}{dt} = \frac{1}{2} \|\mathbf{J}\|$ **3.** The period is increased eight-fold.

Section 13.6 Exercises

1. The data support Kepler's prediction;
$T \approx \sqrt{a^3 \cdot 3 \cdot 10^{-4}} \approx 11.9$ years

3. $M \approx 1.897 \times 10^{27}$ kg   **5.** $M \approx 2.6225 \times 10^{41}$ kg

11. The satellite's orbit is in the plane $20x - 29y = 9z = 0$.

Chapter 13 Review

1. (a) $-1 < t < 0$ or $0 < t \leq 1$ (b) $0 < t \leq 2$

3. $\mathbf{r}(t) = \left\langle t^2, \; t, \; \sqrt[3]{3 - t^4} \right\rangle, \; -\infty < t < \infty$

5. $\mathbf{r}'(t) = \left\langle -1, \; -2t^{-3}, \; \frac{1}{t} \right\rangle$ **7.** $\mathbf{r}'(0) = \langle 2, \; 0, \; 6 \rangle$

9. $\frac{d}{dt} e^t \left\langle 1, \; t, \; t^2 \right\rangle = e^t \left\langle 1, \; 1 + t, \; 2t + t^2 \right\rangle$

11. $\frac{d}{dt} (6\mathbf{r}_1(t) - 4\mathbf{r}_2(t))|_{t=3} = \langle 0, \; -8, \; -10 \rangle$

13. $\frac{d}{dt} (\mathbf{r}_1(t) \cdot \mathbf{r}_2(t))|_{t=3} = 2$

15. $\int_0^3 \left\langle 4t + 3, \; t^2, \; -4t^3 \right\rangle \, dt = \langle 27, \; 9, \; -81 \rangle$

17. $\left(3, \; 3, \; \frac{16}{3} \right)$ **19.** $\mathbf{r}(t) = \left\langle 2t^2 - \frac{8}{3}t^3 + t, \; t^4 - \frac{1}{6}t^3 + 1 \right\rangle$

21. $L = 2\sqrt{13}$ **23.** $\left\langle 5 \cos \frac{2\pi s}{5\sqrt{1+4\pi^2}}, \; 5 \sin \frac{2\pi s}{5\sqrt{1+4\pi^2}}, \; \frac{s}{\sqrt{1+4\pi^2}} \right\rangle$

25. $v_0 \approx 67.279$ m/s   **27.** $\left(0, \; \frac{11}{2}, 38 \right)$

29. $\mathbf{T}(\pi) = \left\langle \frac{-1}{\sqrt{2}}, \; \frac{1}{\sqrt{2}}, \; 0 \right\rangle$ **31.** $\kappa(1) = \frac{1}{2^{3/2}}$

33. $\mathbf{a} = \frac{1}{\sqrt{2}}\mathbf{T} + 4\mathbf{N}$, where $\mathbf{T} = \langle -1, \; 0 \rangle$ and $\mathbf{N} = \langle 0, \; -1 \rangle$

35. $\kappa = \frac{13}{16}$ **37.** $\mathbf{c}(t) = \left\langle \frac{25}{2} + \frac{17^{3/2}}{2} \cos t, \; -32 + \frac{17^{3/2}}{2} \sin t \right\rangle$

39. $2x - 4y + 2z = -3$

Chapter 14

Section 14.1 Preliminary Questions

1. Same shape, but located in parallel planes

2. The parabola $z = x^2$ in the xz-plane   **3.** Not possible

4. The vertical lines $x = c$ with distance of 1 unit between adjacent lines

5. In the contour map of $g(x, \; y) = 2x$, the distance between two adjacent vertical lines is $\frac{1}{2}$.

Section 14.1 Exercises

1. $f(2, \; 2) = 18, f(-1, \; 4) = -5$

3. $h(3, \; 8, \; 2) = 6; h(3, \; -2, \; -6) = -\frac{1}{6}$

5. The domain is the entire xy-plane.

7.

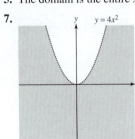

$y = 4x^2$

9. $\mathcal{D} = \left\{ (y, z) : z \neq -y^2 \right\}$

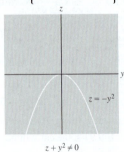

$z = -y^2$

$z + y^2 \neq 0$

11.

$I R \geq 0$

13. Domain: entire $(x, \; y, \; z)$–space; range: entire real line

15. Domain: $\{(r, \; s, \; t) : |rst| \leq 4\}$; range: $\{w : 0 \leq w \leq 4\}$

17. $f \leftrightarrow$ (B), $g \leftrightarrow$ (A)

19. (a) D   (b) C   (c) E   (d) B   (e) A   (f) F

21.

Horizontal trace: $3x + 4y = 12 - c$ in the plane $z = c$
Vertical trace: $z = (12 - 3a) - 4y$ and $z = -3x + (12 - 4a)$ in the planes $x = a$, and $y = a$, respectively

23.

The horizontal traces are ellipses for $c > 0$.
The vertical trace in the plane $x = a$ is the parabola $z = a^2 + 4y^2$.
The vertical trace in the plane $y = a$ is the parabola $z = x^2 + 4a^2$.

25.

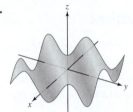

The horizontal traces in the plane $z = c$, $|c| \leq 1$, are the lines $x - y = \sin^{-1} c + 2k\pi$ and $x - y = \pi - \sin^{-1} c + 2k\pi$, for integer k.
The vertical trace in the plane $x = a$ is $z = \sin(a - y)$.
The vertical trace in the plane $y = a$ is $z = \sin(x - a)$.

27. $m = 1 : m = 2 :$

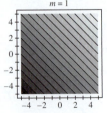
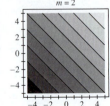

29.

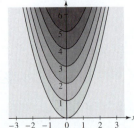

31.

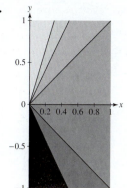

33.

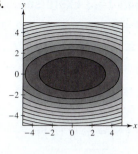

35.

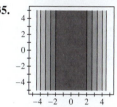

37. $m = 6: \ f(x, y) = 2x + 6y + 6$
  $m = 3: \ f(x, y) = x + 3y + 3$

39. (a) Only at (A) (b) Only at (C) (c) West

41.

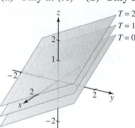

43.

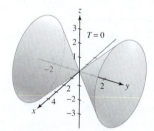
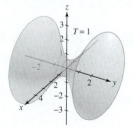
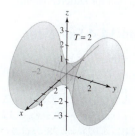

45. Average ROC from B to $C = 0.000625$ kg/m³ · ppt

47. At point A

49. Average ROC from A to $B \approx 0.0737$, average ROC from A to $C \approx 0.0457$

51.

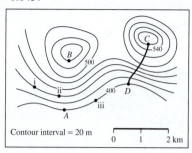

53.

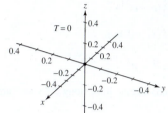
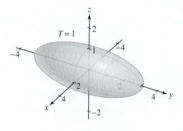
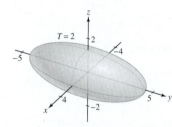

55.

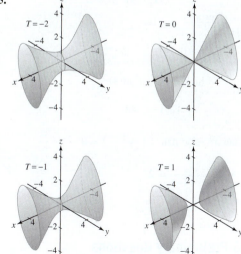

57. $f(r, \theta) = \cos \theta$; the level curves are
$\theta = \pm \cos^{-1}(c)$ for $|c| < 1, \ c \neq 0$;
the y-axis for $c = 0$;
the positive x-axis for $c = 1$;
the negative x-axis for $c = -1$.

Section 14.2 Preliminary Questions

1. $D^*(p, r)$ consists of all points in $D(p, r)$ other than p itself.

2. $f(2, 3) = 27$ **3.** All three statements are true.

4. $\lim\limits_{(x, y) \to (0, 0)} f(x, y)$ does not exist.

Section 14.2 Exercises

1. $\lim\limits_{(x, y) \to (1, 2)} (x^2 + y) = 3$ **3.** $\lim\limits_{(x, y) \to (2, -1)} (xy - 3x^2 y^3) = 10$

5. $\lim\limits_{(x, y) \to (\frac{\pi}{4}, 0)} \tan x \cos y = 1$

7. $\lim\limits_{(x, y) \to (1, 1)} \frac{e^{x^2} - e^{-y^2}}{x + y} = \frac{1}{2}(e - e^{-1})$

9. $\lim\limits_{(x, y) \to (2, 5)} (g(x, y) - 2f(x, y)) = 1$

11. $\lim\limits_{(x, y) \to (2, 5)} e^{f(x, y)^2 - g(x, y)} = e^2$

13. No; the limit along the x-axis and the limit along the y-axis are different.

15. The limit is $\frac{1 + m^3}{m^2}$ for all $m \neq 0$.

17. The limit along the x-axis is $\lim\limits_{(x, y) \to (0,0)} \frac{x}{x^2 + y^2} = \lim\limits_{x \to 0} \frac{1}{x}$, which does not exist.

19. $\lim\limits_{(x, y) \to (0, 0)} \frac{x^2 - y^2}{\sqrt{x^2 + y^2}} = 0$

21. The limit does not exist, because the function values as (x, y) approach $(0, 0)$ along the line $y = mx$ depend on the value of m.

$$\lim\limits_{(x, y) \to (0, 0) y = mx} \frac{xy}{3x^2 + 2y^2} = \lim\limits_{x \to 0} \frac{mx^2}{3x^2 + 2m^2 x^2} = \frac{m}{2m^2 + 3}$$

23. Along the x-coordinate axis ($y = z = 0$),

$$\lim\limits_{(x, y, z) \to (0, 0, 0)} \frac{x + y + z}{x^2 + y^2 + z^2} = \lim\limits_{(x, y, z) \to (0, 0, 0)} \frac{1}{x} = \infty$$

25. $\lim\limits_{(x, y) \to (4, 0)} (x^2 - 16) \cos\left(\frac{1}{(x - 4)^2 + y^2}\right) = 0$

27. $\lim\limits_{(z, w) \to (-2, 1)} \frac{z^4 \cos(\pi w)}{e^{z + w}} = -16e$

29. $\lim\limits_{(x, y) \to (4, 2)} \frac{y - 2}{\sqrt{x^2 - 4}} = 0$ **31.** $\lim\limits_{(x, y) \to (3, 4)} \frac{1}{\sqrt{x^2 + y^2}} = \frac{1}{5}$

33. $\lim\limits_{(x, y) \to (1, -3)} e^{x - y} \ln(x - y) = e^4 \ln(4)$

35. $\lim\limits_{(x, y) \to (-3, -2)} (x^2 y^3 + 4xy) = -48$

37. $\lim\limits_{(x, y) \to (0, 0)} \tan(x^2 + y^2) \tan^{-1}\left(\frac{1}{x^2 + y^2}\right) = 0$

39. $\lim\limits_{(x, y) \to (0, 0)} \frac{x^2 + y^2}{\sqrt{x^2 + y^2 + 1} - 1} = 2$

43. $\lim\limits_{(x, y) \to Q} g(x, y) = 4$ **45.** Yes

Section 14.3 Preliminary Questions

1. $\frac{\partial}{\partial x}(x^2 y^2) = 2xy^2$

2. In this case, the Constant Multiple Rule can be used. In the second part, since y appears in both the numerator and the denominator, the Quotient Rule is preferred.

3. (a), (c) **4.** $f_x = 0$ **5.** (a), (d)

Section 14.3 Exercises

3. $\frac{\partial}{\partial y} \frac{y}{x + y} = \frac{x}{(x + y)^2}$ **5.** $f_z(2, 3, 1) = 6$

7. $m = 10$ **9.** $f_x(A) \approx 10$, $f_y(A) \approx -20$ **11.** NW

13. $\frac{\partial}{\partial x}(x^2 + y^2) = 2x$, $\frac{\partial}{\partial y}(x^2 + y^2) = 2y$

15. $\frac{\partial}{\partial x}(x^4 y + xy^{-2}) = 4x^3 y + y^{-2}$,
$\frac{\partial}{\partial y}(x^4 y + xy^{-2}) = x^4 - 2xy^{-3}$

17. $\frac{\partial}{\partial x}\left(\frac{x}{y}\right) = \frac{1}{y}$, $\frac{\partial}{\partial y}\left(\frac{x}{y}\right) = \frac{-x}{y^2}$

19. $\frac{\partial}{\partial x}\left(\sqrt{9 - x^2 - y^2}\right) = \frac{-x}{\sqrt{9 - x^2 - y^2}}$, $\frac{\partial}{\partial y}\left(\sqrt{9 - x^2 - y^2}\right) = \frac{-y}{\sqrt{9 - x^2 - y^2}}$

21. $\frac{\partial}{\partial x}(\sin x \sin y) = \sin y \cos x$, $\frac{\partial}{\partial y}(\sin x \sin y) = \sin x \cos y$

23. $\frac{\partial}{\partial x}\left(\tan \frac{x}{y}\right) = \frac{1}{y} \sec^2 \frac{x}{y}$, $\frac{\partial}{\partial y}\left(\tan \frac{x}{y}\right) = -\frac{x}{y^2} \sec^2 \frac{x}{y}$

25. $\frac{\partial}{\partial x} \ln(x^2 + y^2) = \frac{2x}{x^2 + y^2}$, $\frac{\partial}{\partial y} \ln(x^2 + y^2) = \frac{2y}{x^2 + y^2}$

27. $\frac{\partial}{\partial r} e^{r + s} = e^{r + s}$, $\frac{\partial}{\partial s} e^{r + s} = e^{r + s}$

29. $\frac{\partial}{\partial x} e^{xy} = ye^{xy}$, $\frac{\partial}{\partial y} e^{xy} = xe^{xy}$

31. $\frac{\partial z}{\partial y} = -2xe^{-x^2 - y^2}$, $\frac{\partial z}{\partial y} = -2ye^{-x^2 - y^2}$

33. $\frac{\partial U}{\partial t} = -e^{-rt}$, $\frac{\partial U}{\partial r} = \frac{-e^{-rt}(rt + 1)}{r^2}$

35. $\frac{\partial}{\partial x} \sinh(x^2 y) = 2xy \cosh(x^2 y)$, $\frac{\partial}{\partial y} \sinh(x^2 y) = x^2 \cosh(x^2 y)$

37. $\frac{\partial w}{\partial x} = y^2 z^3$, $\frac{\partial w}{\partial y} = 2xz^3 y$, $\frac{\partial w}{\partial z} = 3xy^2 z^2$

39. $\frac{\partial Q}{\partial L} = \frac{M - Lt}{M^2} e^{-Lt/M}$, $\frac{\partial Q}{\partial M} = \frac{L(Lt - M)}{M^3} e^{-Lt/M}$,
$\frac{\partial Q}{\partial t} = -\frac{L^2}{M^2} e^{-Lt/M}$

41. $f_x(1, 2) = -164$ **43.** $g_u(1, 2) = \ln 3 + \frac{1}{3}$

45. $N = 2865.058$, $\Delta N \approx -217.74$

47. (a) $I(95, 50) \approx 73.1913$ (b) $\frac{\partial I}{\partial T}$; 1.66

49. A 1-cm increase in r

51. $\frac{\partial W}{\partial E} = -\frac{1}{kT} e^{-E/kT}$, $\frac{\partial W}{\partial T} = \frac{E}{kT^2} e^{-E/kT}$

55. (a), (b) **57.** $\frac{\partial^2 f}{\partial x^2} = 6y$, $\frac{\partial^2 f}{\partial y^2} = -72xy^2$

59. $h_{vv} = \frac{32u}{(u + 4v)^3}$ **61.** $f_{yy}(2, 3) = -\frac{4}{9}$

63. $f_{xyxzy} = 0$ **65.** $f_{uuv} = 2v \sin(u + v^2)$

67. $F_{rst} = 0$ **69.** $F_{uu\theta} = \cosh(uv + \theta^2) \cdot 2\theta v^2$

71. $g_{xyz} = \frac{3xyz}{(x^2 + y^2 + z^2)^{5/2}}$ **73.** $f(x, y) = x^2 y$ **77.** $B = A^2$

Section 14.4 Preliminary Questions

1. $L(x, y) = f(a, b) + f_x(a, b)(x - a) + f_y(a, b)(y - b)$

2. f is locally linear at (a, b) if $f(x, y) = L(x, y) + e(x, y)$, where $e(x, y)$ satisfies $\lim\limits_{(x, y) \to (a, b)} \frac{e(x, y)}{\sqrt{(x - a)^2 + (y - b)^2}} = 0$

3. (b) **4.** $f(2, 3.1) \approx 8.7$ **5.** $\Delta f \approx -0.1$

6. Criterion for Differentiability

Section 14.4 Exercises

1. $z = -34 - 20x + 16y$ **3.** $z = 5x + 10y - 14$

5. $z = 8x - 2y - 13$ **7.** $z = 4r - 5s + 2$

9. $z = \left(\frac{4}{5} + \frac{12}{25} \ln 2\right) - \frac{12}{25} x + \frac{12}{25} y$ **11.** $\left(-\frac{1}{4}, \frac{1}{8}, \frac{1}{8}\right)$

13. (a) $f(x, y) = -16 + 4x + 12y$

(b) $f(2.01, 1.02) \approx 4.28$; $f(1.97, 1.01) \approx 4$

15. $\Delta f \approx 3.56$ **17.** $f(0.01, -0.02) \approx 0.98$

19. $L(x, y, z) = \frac{5}{12}\sqrt{3}x + \frac{5}{12}\sqrt{3}y + 2\sqrt{3}z - 5\sqrt{3}$

21. 5.07 **23.** 8.44 **25.** 4.998 **27.** 3.945

29. $f(-2.1, 3.1) \approx 4.2$ **31.** $\Delta I \approx 0.5644$

33. (b) $\Delta H \approx 0.022$ m

35. (b) 6% (c) 1% error in r

37. (a) $7.10 (b) $28.85, $57.69 (c) $-$74.24

39. Maximum error in V is about 8.948 m.

Section 14.5 Preliminary Questions

1. (b) $\langle 3, 4 \rangle$ **2.** False

3. ∇f points in the direction of maximum rate of increase of f and is normal to the level curve of f.

4. (b) NW and (c) SE **5.** $3\sqrt{2}$

Section 14.5 Exercises

1. (a) $\nabla f = \left\langle y^2, 2xy \right\rangle$, $\mathbf{r}'(t) = \left\langle t, 3t^2 \right\rangle$

(b) $\frac{d}{dt}(f(\mathbf{r}(t)))\Big|_{t=1} = 4$; $\frac{d}{dt}(f(\mathbf{r}(t)))\Big|_{t=-1} = -4$

3. A: zero, B: negative, C: positive, D: zero

5. $\nabla f = -\sin(x^2 + y)\langle 2x, 1 \rangle$

7. $\nabla h = \left\langle yz^{-3}, xz^{-3}, -3xyz^{-4} \right\rangle$

9. $\frac{d}{dt}(f(\mathbf{r}(t)))\Big|_{t=0} = -7$ **11.** $\frac{d}{dt}(f(\mathbf{r}(t)))\Big|_{t=0} = -3$

13. $\frac{d}{dt}(f(\mathbf{r}(t)))\Big|_{t=0} = 5\cos 1 \approx 2.702$

15. $\frac{d}{dt}(f(\mathbf{r}(t)))\Big|_{t=4} = -56$

17. $\frac{d}{dt}(f(\mathbf{r}(t)))\Big|_{t=\pi/4} = -1 + \frac{\pi}{8} \approx 1.546$

19. $\frac{d}{dt}(g(\mathbf{r}(t)))\Big|_{t=1} = 0$

21. $D_{\mathbf{u}}f(1, 2) = 8.8$ **23.** $D_{\mathbf{u}}f\left(\frac{1}{6}, 3\right) = \frac{39}{4\sqrt{2}}$

25. $D_{\mathbf{u}}f(3, 4) = \frac{7\sqrt{2}}{290}$ **27.** $D_{\mathbf{u}}f(1, 0) = \frac{6}{\sqrt{13}}$

29. $D_{\mathbf{u}}f(1, 2, 0) = -\frac{1}{\sqrt{3}}$ **31.** $D_{\mathbf{u}}f(3, 2) = \frac{-50}{\sqrt{13}}$

33. $D_{\mathbf{u}}f(P) = -\frac{e^5}{3} \approx -49.47$

35. (a) $m = 3$, angle of inclination is approximately $1.249 \approx 71.6°$.

(b) $m = 4\sqrt{2} \approx 5.66$, angle of inclination is approximately $1.396 \approx 80.0°$.

(c) $m = -4\sqrt{2} \approx -5.66$, angle of inclination is approximately $-1.396 \approx -80.0°$.

(d) $m = \sqrt{34} \approx 5.83$, angle from due East is approximately $121°$.

37. f is increasing at P in the direction of v.

39. $D_{\mathbf{u}}f(P) = \frac{\sqrt{6}}{2}$ **41.** $\langle 6, 2, -4 \rangle$

43. $\left(\frac{4}{\sqrt{17}}, \frac{9}{\sqrt{17}}, -\frac{2}{\sqrt{17}}\right)$ and $\left(-\frac{4}{\sqrt{17}}, -\frac{9}{\sqrt{17}}, \frac{2}{\sqrt{17}}\right)$

45. $9x + 10y + 5z = 33$

47. $0.5217x + 0.7826y - 1.2375z = -5.309$

49.

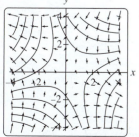

51. $f(x, y, z) = x^2 + y + 2z$

53. $f(x, y, z) = xz + y^2$ **57.** $\Delta f \approx 0.08$

59. (a) $\langle 34, 18, 0 \rangle$

(b) $\left\langle 2 + \frac{32}{\sqrt{21}}t, 2 + \frac{16}{\sqrt{21}}t, 8 - \frac{8}{\sqrt{21}}t \right\rangle$; ≈ 4.58 s

63. $x = 1 - 4t, \quad y = 2 + 26t, \quad z = 1 - 25t$

75. $y = \sqrt{1 - \ln(\cos^2 x)}$

Section 14.6 Preliminary Questions

1. (a) $\frac{\partial f}{\partial x}$ and $\frac{\partial f}{\partial y}$ (b) u and v

2. (a) **3.** $f(u, v)|_{(r,s)=(1,1)} = e^2$ **4.** (b) **5.** (c) **6.** No

Section 14.6 Exercises

1. (a) $\frac{\partial f}{\partial x} = 2xy^3$, $\frac{\partial f}{\partial y} = 3x^2y^2$, $\frac{\partial f}{\partial z} = 4z^3$

(b) $\frac{\partial x}{\partial s} = 2s$, $\frac{\partial y}{\partial s} = t^2$, $\frac{\partial z}{\partial s} = 2st$

(c) $\frac{\partial f}{\partial s} = 7s^6t^6 + 8s^7t^4$

3. $\frac{\partial f}{\partial s} = 6rs^2$, $\frac{\partial f}{\partial r} = 2s^3 + 4r^3$

5. $\frac{\partial g}{\partial u} = -10\sin(10u - 20v)$, $\frac{\partial g}{\partial v} = 20\sin(10u - 20v)$

7. $\frac{\partial F}{\partial y} = xe^{x^2+xy}$ **9.** $\frac{\partial h}{\partial t_2} = 0$

11. $\frac{\partial f}{\partial u}\Big|_{(u,v)=(-1,-1)} = 1$, $\frac{\partial f}{\partial v}\Big|_{(u,v)=(-1,-1)} = -2$

13. $\frac{\partial g}{\partial \theta}\Big|_{(r,\theta)=(2\sqrt{2}, \pi/4)} = -\frac{1}{6}$ **15.** $\frac{\partial g}{\partial u}\Big|_{(u,v)=(0,1)} = 2\cos 2$

17. -26.8 ft/s **19.** $4\pi^3 - 3\pi^2 - 1$

23. (a) $F_x = z^2 + y$, $F_y = 2yz + x$, $F_z = 2xz + y^2$

(b) $\frac{\partial z}{\partial x} = -\frac{z^2 + y}{2xz + y^2}$, $\frac{\partial z}{\partial y} = -\frac{2yz + x}{2xz + y^2}$

27. $\frac{\partial z}{\partial x} = -\frac{2xy + z^2}{2xz + y^2}$ **29.** $\frac{\partial z}{\partial y} = -\frac{xe^{xy} + 1}{x\cos(xz)}$

31. $\frac{\partial w}{\partial y} = \frac{-y(w^2 + x^2)^2}{w\left((w^2 + y^2)^2 + (w^2 + x^2)^2\right)}$; at $(1, 1, 1)$, $\frac{\partial w}{\partial y} = -\frac{1}{2}$

35. $\nabla\left(\frac{1}{r}\right) = -\frac{1}{r^3}\mathbf{r}$ **37.** (c) $\frac{\partial z}{\partial x} = \frac{x - 6}{z + 4}$

39. $\frac{\partial P}{\partial T} = -\frac{F_T}{F_P} = -\frac{-nR}{V - nb} = \frac{nR}{V - nb}$

Section 14.7 Preliminary Questions

1. f has a local (and global) min at $(0, 0)$; g has a saddle point at $(0, 0)$.

2.

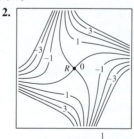

Point R is a saddle point.

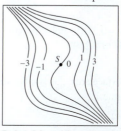

Point S is neither a local extremum nor a saddle point.

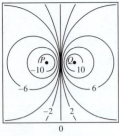

Point P is a local minimum and point Q is a local maximum.

3. Statement (a)

Section 14.7 Exercises

1. (b) $P_1 = (0, 0)$ is a saddle point, $P_2 = \left(2\sqrt{2}, \sqrt{2}\right)$ and
$P_3 = \left(-2\sqrt{2}, -\sqrt{2}\right)$ are local minima; absolute minimum value of
f is -4.

3. $(0, 0)$ saddle point, $\left(\frac{13}{64}, -\frac{13}{32}\right)$ and $\left(-\frac{1}{4}, \frac{1}{2}\right)$ local minima

5. (c) $(0, 0)$, $(1, 0)$, and $(0, -1)$ saddle points, $\left(\frac{1}{3}, -\frac{1}{3}\right)$ local
minimum.

7. $\left(-\frac{2}{3}, -\frac{1}{3}\right)$ local minimum

9. $(-2, -1)$ local maximum, $\left(\frac{5}{3}, \frac{5}{6}\right)$ saddle point

11. $\left(0, \pm\sqrt{2}\right)$ saddle points, $\left(\frac{2}{3}, 0\right)$ local maximum,
$\left(-\frac{2}{3}, 0\right)$ local minimum

13. $(0, 0)$ saddle point, $(1, 1)$ and $(-1, -1)$ local minima

15. $(0, 0)$ saddle point, $\left(\frac{1}{\sqrt{2}}, \frac{1}{\sqrt{2}}\right)$ and $\left(-\frac{1}{\sqrt{2}}, -\frac{1}{\sqrt{2}}\right)$ local
maximum, $\left(\frac{1}{\sqrt{2}}, -\frac{1}{\sqrt{2}}\right)$ and $\left(-\frac{1}{\sqrt{2}}, \frac{1}{\sqrt{2}}\right)$ local minimum

17. Critical points are $\left(j\pi, k\pi + \frac{\pi}{2}\right)$, for
    j, k even: saddle points
    j, k odd: local maxima
    j even, k odd: local minima
    j odd, k even: saddle points

19. $\left(1, \frac{1}{2}\right)$ local maximum   **21.** $\left(\frac{3}{2}, -\frac{1}{2}\right)$ saddle point

23. $\left(-\frac{1}{6}, -\frac{17}{18}\right)$ local minimum

27. $x = y = 0.27788$ local minimum

29. Global maximum 2, global minimum 0

31. Global maximum 1, global minimum $\frac{1}{35}$

35. Maximum value $\frac{1}{3}$

37. Global minimum $f(0, 1) = -2$, global maximum $f(1, 0) = 1$

39. Global minimum, $f(0, 0) = 0$, global maximum $f(1, 1) = 3$

41. Global minimum, $f(0, 0) = 0$, global maximum
$f(1, 0) = f(0, 1) = 1$

43. Global minimum, $f(1, 2) = -5$, global maximum
$f(0, 0) = f(3, 3) = 0$

45. Global minimum,
$f(-0.4343, 0.9) = f(-0.4343, -0.9) \approx -0.5161$, global maximum
$f(0.7676, 0.6409) = f(0.7676, -0.6409) \approx 1.2199$

47. Maximum volume $\frac{3}{4}$

49. (a) No. In the box B with minimal surface area, z is smaller than
$\sqrt[3]{V}$, which is the side of a cube with volume V.

(b) Width: $x = (2V)^{1/3}$; length: $y = (2V)^{1/3}$;

    height: $z = \left(\frac{V}{4}\right)^{1/3}$

53. The fence should be cut into 12 pieces 10 m long, forming three
10×10 squares.

55. $V = \frac{64{,}000}{\pi} \approx 20{,}372 \text{ cm}^3$ **57.** $f(x) = 1.9629x - 1.5519$

Section 14.8 Preliminary Questions

1. Statement (b)

2. f had a local maximum 2, under the constraint, at A; $f(B)$ is
neither a local minimum nor a local maximum of f.

3. (a)

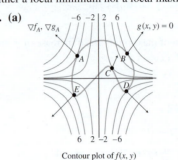

Contour plot of $f(x, y)$
(contour interval 2)

(b) Global minimum -4, global maximum 6

Section 14.8 Exercises

1. (c) Critical points $(-1, -2)$ and $(1, 2)$

(d) Maximum 10, minimum -10

3. Maximum $4\sqrt{2}$, minimum $-4\sqrt{2}$

5. Minimum $\frac{36}{13}$, no maximum value

7. Maximum $\frac{8}{3}$, minimum $-\frac{8}{3}$

9. Maximum $\sqrt{2}$, minimum 1

11. Maximum 3.7, minimum -3.7

13. Maxima at $f(\pm 4, \pm 4, 2) = 20$, and minima
at $f(\pm 4, \mp 4, -2) = -20$

15. Maximum $2\sqrt{2}$, minimum $-2\sqrt{2}$ **17.** $(-1, e^{-1})$

19. (a) $\frac{h}{r} = \sqrt{2}$ **(b)** $\frac{h}{r} = \sqrt{2}$

(c) There is no cone of fixed V with maximal S.

21. $(8, -2)$ **23.** $\left(\frac{48}{97}, \frac{108}{97}\right)$ **25.** $\frac{a^a b^b}{(a+b)^{a+b}}$ **27.** $\sqrt{\frac{a^a b^b}{(a+b)^{a+b}}}$

33. $r = 3$, $h = 6$ **35.** $x + y + z = 3$

41. $\frac{25}{3}$ **43.** $\left(\frac{-6}{\sqrt{105}}, \frac{-3}{\sqrt{105}}, \frac{30}{\sqrt{105}}\right)$ **45.** $(-1, 0, 2)$

47. Minimum $\frac{138}{11} \approx 12.545$, no maximum value

51. (b) $\lambda = \frac{c}{2p_1 p_2}$

Chapter 14 Review

1. (a)

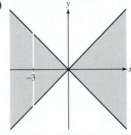

(b) $f(3, 1) = \frac{\sqrt{2}}{3}$, $f(-5, -3) = -2$ **(c)** $\left(-\frac{5}{3}, 1\right)$

3.

Vertical and horizontal traces: the line $z = (c^2 + 1) - y$ in the plane $x = c$, the parabola $z = x^2 - c + 1$ in the plane $y = c$.

5. (a) Graph (B)   (b) Graph (C)   (c) Graph (D)   (d) Graph (A)

7. (a) Parallel lines $4x - y = \ln c$, $c > 0$, in the xy-plane

(b) Parallel lines $4x - y = e^c$ in the xy-plane

(c) Hyperbolas $3x^2 - 4y^2 = c$ in the xy-plane

(d) Parabolas $x = c - y^2$ in the xy-plane

9. $\displaystyle\lim_{(x,y)\to(1,-3)} (xy + y^2) = 6$

11. The limit does not exist.

13. $\displaystyle\lim_{(x,y)\to(1,-3)} (2x + y)e^{-x+y} = -e^{-4}$

17. $f_x = 2$, $f_y = 2y$

19. $f_x = e^{-x-y}(y\cos(xy) - \sin(xy))$
$f_y = e^{-x-y}(x\cos(yx) - \sin(yx))$

21. $f_{xxyz} = -\cos(x + z)$ **23.** $z = 33x + 8y - 42$

25. Estimate, 12.146; calculator value to three places, 11.996.

27. Statements (ii) and (iv) are true.

29. $\dfrac{d}{dt}\big(f(\mathbf{c}(t))\big)\Big|_{t=2} = 3 + 4e^4 \approx 221.4$

31. $\dfrac{d}{dt}\big(f(\mathbf{c}(t))\big)\Big|_{t=1} = 4e - e^{3e} \approx -3469.3$

33. $D_{\mathbf{u}}f(3, -1) = -\dfrac{54}{\sqrt{5}}$

35. $D_{\mathbf{u}}f(P) = -\dfrac{\sqrt{2}e}{5}$ **37.** $\left\langle \dfrac{1}{\sqrt{2}}, \dfrac{1}{\sqrt{2}}, 0 \right\rangle$

41. $\dfrac{\partial f}{\partial s} = 3s^2t + 4st^2 + t^3 - 2st^3 + 6s^2t^2$
$\dfrac{\partial f}{\partial t} = 4s^2t + 3st^2 + s^3 + 4s^3t - 3s^2t^2$

45. $\dfrac{\partial z}{\partial x} = -\dfrac{e^z - 1}{xe^z + e^y}$

47. $(0, 0)$ saddle point, $(1, 1)$ and $(-1, -1)$ local minima

49. $\left(\dfrac{1}{2}, \dfrac{1}{2}\right)$ saddle point

53. Global maximum $f(2, 4) = 10$, global minimum $f(-2, 4) = -18$

55. Maximum $\dfrac{26}{\sqrt{13}}$, minimum $-\dfrac{26}{\sqrt{13}}$

57. Maximum $\dfrac{12}{\sqrt{3}}$, minimum $-\dfrac{12}{\sqrt{3}}$

59. Minimum $= f\left(\dfrac{1+\sqrt{3}}{3}, \dfrac{1}{3}, \dfrac{1-\sqrt{3}}{3}\right) = \dfrac{6-2\sqrt{3}}{3}$
Maximum $= f\left(\dfrac{1-\sqrt{3}}{3}, \dfrac{1}{3}, \dfrac{1+\sqrt{3}}{3}\right) = \dfrac{6+2\sqrt{3}}{3}$

61. $r = h = \sqrt[3]{\dfrac{V}{\pi}}$, $S = 3\pi\left(\dfrac{V}{\pi}\right)^{2/3}$

Chapter 15

Section 15.1 Preliminary Questions

1. $\Delta A = 1$, the number of subrectangles is 32.

2. $\iint_R f\, dA \approx S_{1,1} = 0.16$ **3.** $\iint_R 5\, dA = 50$

4. The signed volume between the graph $z = f(x, y)$ and the xy-plane. The region below the xy-plane is treated as negative volume.

5. (b)   **6.** (b), (c)

Section 15.1 Exercises

1. $S_{4,3} = 13.5$ **3.** (A) $S_{3,2} = 42$, (B) $S_{3,2} = 43.5$

5. (A) $S_{3,2} = 60$, (B) $S_{3,2} = 62$

7. Two possible solutions are $S_{3,2} = \dfrac{77}{72}$ and $S_{3,2} = \dfrac{79}{72}$.

9. $\dfrac{225}{2}$

11. 0.19375   **13.** 1.0731, 1.0783, 1.0809   **15.** 0   **17.** 0   **19.** 40

21. 55   **23.** $\dfrac{4}{3}$ **25.** 84   **27.** 4   **29.** $\dfrac{1858}{15}$

31. $6\ln 6 - 2\ln 2 - 5\ln 5 \approx 1.317$ **33.** $\dfrac{4}{3}\left(19 - 5\sqrt{5}\right) \approx 10.426$

35. $\dfrac{1}{2}(\ln 3)(-2 + \ln 48) \approx 1.028$ **37.** $6\ln 3 \approx 6.592$

39. 1   **41.** $\left(e^2 - 1\right)\left(1 - \dfrac{\sqrt{2}}{2}\right) \approx 1.871$ **43.** $m = \dfrac{3}{4}$

45. $\dfrac{2}{15}\left(8\sqrt{2} - 7\right) \approx 0.575$ **47.** $2\ln 2 - 1 \approx 0.386$

51. $\dfrac{e^3}{3} - \dfrac{1}{3} - e + 1 \approx 4.644$

Section 15.2 Preliminary Questions

1. (b), (c)

2.

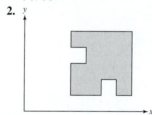

3.

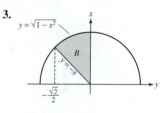

4. (b)

Section 15.2 Exercises

1. (a) Sample points •, $S_{3,4} = -3$

(b) Sample points ○, $S_{3,4} = -4$

3. As a vertically simple region: $0 \le x \le 1$, $0 \le y \le 1 - x^2$; as a horizontally simple region: $0 \le y \le 1$, $0 \le x \le \sqrt{1 - y}$,
$\int_0^1 \left(\int_0^{1-x^2} (xy)\, dy\right) dx = \dfrac{1}{12}$

5. $\dfrac{192}{5} = 38.4$ **7.** $\dfrac{608}{15} \approx 40.53$ **9.** $2\dfrac{1}{4}$ **11.** $-\dfrac{3}{4} + \ln 4$

13. $\dfrac{16}{3} \approx 5.33$ **15.** $\dfrac{11}{60}$ **17.** $\dfrac{1754}{15} \approx 116.93$ **19.** $\dfrac{e-2}{2} \approx 0.359$

21. $\dfrac{1}{12}$ **23.** $2e^{12} - \dfrac{1}{2}e^9 + \dfrac{1}{2}e^5 \approx 321{,}532.2$

25.

$$\int_0^4 \int_x^4 f(x, y)\, dy\, dx = \int_0^4 \int_0^y f(x, y)\, dx\, dy$$

27.

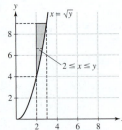

$$\int_4^9 \int_2^{\sqrt{y}} f(x,\ y)\ dx\ dy = \int_2^3 \int_{x^2}^9 f(x,\ y)\ dy\ dx$$

29.

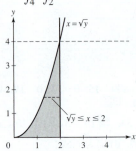

$$\int_0^2 \int_0^{x^2} \sqrt{4x^2 + 5y}\ dy\ dx = \frac{152}{15}$$

31. $\int_1^e \int_{\ln^2 y}^{\ln y} (\ln y)^{-1}\ dx\ dy = e - 2 \approx 0.718$

33.

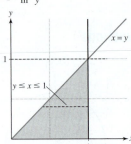

$$\int_0^1 \int_0^x \frac{\sin x}{x}\ dy\ dx = 1 - \cos 1 \approx 0.460$$

35.

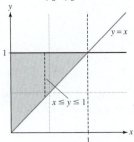

$$\int_0^1 \int_0^y xe^{y^3}\ dx\ dy = \frac{e-1}{6} \approx 0.286$$

37.

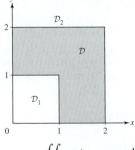

$$\iint_{\mathcal{D}} e^{x+y}\ dA = e^4 - 3e^2 + 2e \approx 37.878$$

39. $\int_0^4 \int_{x/4}^{3x/4} e^{x^2}\ dy\ dx = \frac{1}{4}\left(e^{16} - 1\right)$

41. $\int_2^4 \int_{y-1}^{7-y} \frac{x}{y^2}\ dx\ dy = 6 - 6\ln 2 \approx 1.841$

43. $\iint_{\mathcal{D}} \frac{\sin y}{y}\ dA = \cos 1 - \cos 2 \approx 0.956$

45. $\int_{-2}^2 \int_0^{4-x^2} (40 - 10y)\ dy\ dx = 256$ **47.** $\frac{512}{3}$

49. $\int_{-2}^2 \left(\int_{-\sqrt{4-x^2}}^{\sqrt{4-x^2}} \left[\left(8 - x^2 - y^2\right) - \left(x^2 + y^2\right)\right] dy\right) dx$

51. $\int_0^1 \int_0^1 e^{x+y}\ dx\ dy = e^2 - 2e + 1 \approx 2.952$

53. $\frac{1}{\pi} \int_0^1 \int_0^{\pi} y^2 \sin x\ dx\ dy = \frac{2}{3\pi}$ **55.** $\bar{f} = p$

61. One possible solution is $P = \left(\frac{2}{3},\ 2\right)$.

63. $\iint_{\mathcal{D}} f(x,\ y)\ dA \approx 57.01$

Section 15.3 Preliminary Questions

1. (c)   **2.** (b)

3. (a) $D = \{(x,\ y) : 0 \le x \le 1,\ 0 \le y \le x\}$

(b) $D = \left\{(x,\ y) : 0 \le x \le 1,\ 0 \le y \le \sqrt{1 - x^2}\right\}$

Section 15.3 Exercises

1. 6   **3.** $(e - 1)(1 - e^{-2})$ **5.** $-\frac{27}{4} = -6.75$

7. $\frac{b}{20}\left[(a + c)^5 - a^5 - c^5\right]$ **9.** $\frac{1}{6}$ **11.** $\frac{1}{16}$ **13.** $e - \frac{5}{2}$

15. $2\frac{1}{12}$ **17.** $\frac{128}{15}$ **19.** $\int_0^3 \int_0^4 \int_0^{y/4} dz\ dy\ dx = 6$

21. $\frac{1}{12}$ **23.** $\frac{126}{5}$

25. Region enclosed by sphere $x^2 + y^2 + z^2 = 5$ to right of plane $y = 1$.

27. $\int_0^2 \int_0^{y/2} \int_0^{4-y^2} xyz\ dz\ dx\ dy$, $\int_0^4 \int_0^{\sqrt{4-z}} \int_0^{y/2} xyz\ dx\ dy\ dz$, and $\int_0^4 \int_0^{\sqrt{1-(z/4)}} \int_{2x}^{\sqrt{4-z}} xyz\ dy\ dx\ dz$

29. $\int_{-1}^1 \int_{-\sqrt{1-x^2}}^{\sqrt{1-x^2}} \int_{\sqrt{x^2+y^2}}^1 f(x,\ y,\ z)\ dz\ dy\ dx$ **31.** $\frac{16}{21}$

33. $\int_{-1}^1 \left(\int_0^{1-y^2} \left(\int_0^{3-z} dx\right) dz\right) dy$

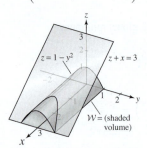

35. $\int_0^1 \left(\int_{\sqrt{y}}^{y^2} \left(\int_0^{4-y-z} dx\right) dz\right) dy$

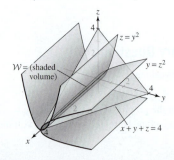

37. $\frac{1}{2\pi}$ **39.** $2e - 4 \approx 1.437$

41. $S_{N,N,N} \approx 0.561,\ 0.572,\ 0.576;\ I \approx 0.584;\ N = 100$

Section 15.4 Preliminary Questions

1. (d)

2. (a) $\int_{-1}^{2} \int_{0}^{2\pi} \int_{0}^{2} f(P)\, r\, dr\, d\theta\, dz$

(b) $\int_{-2}^{0} \int_{0}^{2\pi} \int_{0}^{\sqrt{4-z^2}} r\, dr\, d\theta\, dz$

3. (a) $\int_{0}^{2\pi} \int_{0}^{\pi} \int_{0}^{4} f(P)\, \rho^2 \sin\phi\, d\rho\, d\phi\, d\theta$

(b) $\int_{0}^{2\pi} \int_{0}^{\pi} \int_{4}^{5} f(P)\, \rho^2 \sin\phi\, d\rho\, d\phi\, d\theta$

(c) $\int_{0}^{2\pi} \int_{\pi/2}^{\pi} \int_{0}^{2} f(P)\, \rho^2 \sin\phi\, d\rho\, d\phi\, d\theta$

4. $\Delta A \approx r(\Delta r\, \Delta\theta)$, and the factor r appears in $dA = r\, dr\, d\theta$ in the Change of Variables formula.

Section 15.4 Exercises

1.

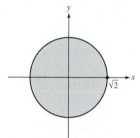

$$\iint_{D} \sqrt{x^2 + y^2}\, dA = \frac{4\sqrt{2}\pi}{3}$$

3.

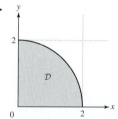

$$\iint_{D} xy\, dA = 2$$

5.

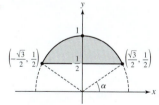

$$\iint_{D} y\left(x^2 + y^2\right)^{-1} dA = \sqrt{3} - \frac{\pi}{3} \approx 0.685$$

7.

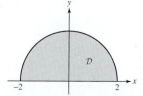

$$\int_{-2}^{2} \int_{0}^{\sqrt{4-x^2}} (x^2 + y^2)\, dy\, dx = 4\pi$$

9.

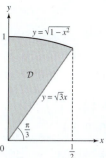

$$\int_{0}^{1/2} \int_{\sqrt{3}x}^{\sqrt{1-x^2}} x\, dy\, dx = \frac{1}{3}\left(1 - \frac{\sqrt{3}}{2}\right) \approx 0.045$$

11.

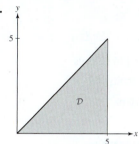

$$\int_{0}^{\pi/4} \left(\int_{0}^{5/\cos(\theta)} \left(r^2 \cos(\theta)\right) dr\right) d\theta = \frac{125}{3}$$

13.

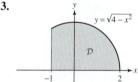

$$\int_{-1}^{2} \int_{0}^{\sqrt{4-x^2}} (x^2 + y^2)\, dy\, dx = \frac{\sqrt{3}}{2} + \frac{8\pi}{3} \approx 9.244$$

15. $\frac{1}{4}$ **17.** $\frac{1}{2}$ **19.** 0   **21.** 18   **23.** $\frac{48\pi - 32}{9} \approx 13.2$

25. (a) $W: 0 \le \theta \le 2\pi,\ 0 \le r \le 2,\ 2 \le z \le 6 - r^2$ **(b)** 8π

27. $\frac{405\pi}{2} \approx 636.17$ **29.** $\frac{2}{3}$ **31.** 243π

33. $\int_{0}^{2\pi} \int_{0}^{1} \int_{0}^{4} f(r\cos\theta,\ r\sin\theta,\ z)\, r\, dz\, dr\, d\theta$

35. $\int_{0}^{\pi} \int_{0}^{1} \int_{0}^{r^2} f(r\cos\theta,\ r\sin\theta,\ z)\, r\, dz\, dr\, d\theta$

37. $z = \frac{H}{R}r;\ V = \frac{\pi R^2 H}{3}$ **39.** 16π

41. $V =$
$$2\int_{0}^{2\pi} \left(\int_{b}^{a} \left(\int_{0}^{\sqrt{a^2-r^2}} (r)\, dz\right) dr\right) d\theta = \frac{4}{3}\pi\left(a^2 - b^2\right)^{3/2} = \frac{4}{3}\pi r h^3$$

43. $V = \int_{0}^{2\pi} \left(\int_{0}^{\pi/3} \left(\int_{\sec(\phi)}^{2} \left(\rho^2 \sin(\phi)\right) d\rho\right) d\phi\right) d\theta = \frac{5\pi}{3}$

45. $-\frac{\pi}{16}$ **47.** $\frac{8\pi}{15}$ **49.** $\frac{8\pi}{5}$ **51.** $\frac{5\pi}{8}$ **53.** π **55.** $\frac{4\pi a^3}{3}$

57. (b) $J = \int_{0}^{2\pi} \left(\int_{0}^{\infty} \left(e^{-r^2} r\right) dr\right) d\theta = \pi$

Section 15.5 Preliminary Questions

1. 5 kg/m³   **2. (a)**

3. The probability that $0 \le X \le 1$ and $0 \le Y \le 1$; the probability that $0 \le X + Y \le 1$

Section 15.5 Exercises

1. $\frac{2}{3}$ **3.** $4\left(1 - e^{-100}\right) \times 10^{-6}$ C $\approx 4 \times 10^{-6}$ C

5. $10{,}000 - 18{,}000e^{-4/5} \approx 1{,}912$

7. $25\pi \left(3 \times 10^{-8} \text{ C}\right) \approx 2.356 \times 10^{-6}$ C

9. $\approx 2.593 \times 10^{10}$ kg **11.** $\left(0, \frac{2}{5}\right)$ **13.** $\left(\frac{4R}{3\pi}, \frac{4R}{3\pi}\right)$

15. $(0.555, 0)$ **17.** $\left(0, 0, \frac{3R}{8}\right)$ **19.** $\left(0, 0, \frac{9}{8}\right)$

21. $\left(0, 0, \frac{13}{2\left(17-6\sqrt{6}\right)}\right)$ **23.** $(2, 1)$ **25.** $\left(\frac{1}{6}, \frac{1}{6}\right)$ **27.** $\frac{16}{15\pi}$

29. (a) $\frac{M}{4ab}$ **(b)** $I_x = \frac{Mb^2}{3}$; $I_0 = \frac{M(a^2 + b^2)}{3}$ **(c)** $\frac{b}{\sqrt{3}}$

31. $I_0 = 8{,}000$ kg $\cdot$ m^2; $I_x = 4{,}000$ kg $\cdot$ m^2

33. $\frac{9}{2}$ **35.** $\frac{243}{20}$ **37.** $\left(\frac{a}{2}, \frac{2b}{5}\right)$ **39.** $\frac{a^2b^4}{60}$

41. $I_x = \frac{MR^2}{4}$; kinetic energy required is $\frac{25MR^2}{2}$ J

47. (a) $I = 182.5$ g $\cdot$ cm^2 **(b)** $\omega \approx 126.92$ rad/s

49. $\frac{13}{72}$ **51.** $\frac{1}{64}$ **53.** $C = 15$; probability is $\frac{5}{8}$.

55. (a) $C = 4$ **(b)** $\frac{1}{48\pi} + \frac{1}{32} \approx 0.038$

Section 15.6 Preliminary Questions

1. (b)

2. (a) $G(1, 0) = (2, 0)$ **(b)** $G(1, 1) = (1, 3)$

(c) $G(2, 1) = (3, 3)$

3. Area $(G(R)) = 36$ **4.** Area $(G(R)) = 0.06$

Section 15.6 Exercises

1. (a) Image of the u-axis is the line $y = \frac{1}{2}x$; image of the v-axis is the y-axis.

(b) The parallelogram with vertices $(0, 0)$, $(10, 5)$ $(10, 2)$, $(0, 7)$

(c) The segment joining the points $(2, 3)$ and $(10, 8)$

(d) The triangle with vertices $(0, 1)$, $(2, 1)$, and $(2, 2)$

3. G is not one-to-one; G is one-to-one on the domain $\{(u, v) : u \geq 0\}$, and G is one-to-one on the domain $\{(u, v) : u \leq 0\}$.

(a) The positive x-axis including the origin and the y-axis, respectively

(b) The rectangle $[0, 1] \times [-1, 1]$

(c) The curve $y = \sqrt{x}$ for $0 \leq x \leq 1$

(d)

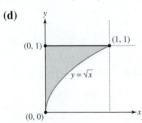

5. $y = 3x - c$ **7.** $y = \frac{17}{6}x$ **11.** $\text{Jac}(G) = 1$

13. $\text{Jac}(G) = -10$ **15.** $\text{Jac}(G) = 1$ **17.** $\text{Jac}(G) = 4$

19. $G(u, v) = (4u + 2v, u + 3v)$ **21.** $\frac{2329}{12} \approx 194.08$

23. (a) Area $(G(R)) = 105$ **(b)** Area $(G(R)) = 126$

25. $\text{Jac}(G) = \frac{2u}{v}$; for $R = [1, 4] \times [1, 4]$, area $(G(R)) = 15 \ln 4$

27.

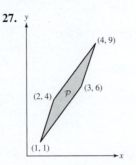

$$G(u, v) = (1 + 2u + v, \ 1 + 5u + 3v)$$

29. 82 **31.** 80 **33.** $\frac{56}{45}$ **35.** $\frac{\pi(e^{36} - 1)}{6}$

37.

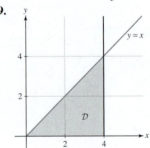

$$\iint_D y^{-1} \, dx \, dy = 1$$

39. $\iint_D e^{xy} \, dA = (e^{20} - e^{10}) \ln 2$

41. (b) $-\frac{1}{x+y}$ **(c)** $I = 9$ **45.** $\frac{\pi^2}{8}$

Chapter 15 Review

1. (a) $S_{2,3} = 240$ **(b)** $S_{2,3} = 510$ **(c)** 520

3. $S_{4,4} = 2.9375$ **5.** $\frac{32}{3}$ **7.** $\frac{\sqrt{3}-1}{2}$

9.

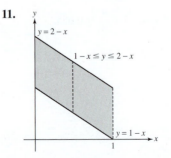

$$\iint_D \cos y \, dA = 1 - \cos 4$$

11.

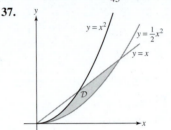

$$\iint_D e^{x+2y} \, dA = \frac{1}{2}e(e + 1)(e - 1)^2$$

13.

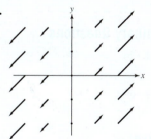

$$\iint_D ye^{1+x}\, dA = 0.5(e^2 - 2e^{1.5} + e)$$

15. $\int_0^9 \int_{-\sqrt{9-y}}^{\sqrt{9-y}} f(x,\ y)\, dx\, dy$ **17.** $\frac{1}{24}$ **19.** $18(\sqrt{2} - 1)$

21. $1 - \cos 1$ **23.** 6π **25.** $\pi/2$ **27.** 10 **29.** $\frac{\pi}{4} + \frac{2}{3}$ **31.** π

33. $\frac{1}{4}$ **35.** $\int_0^{\pi/2} \int_0^1 \int_0^r zr\, dz\, dr\, d\theta = \pi/16$ **37.** $\frac{2\pi(-1 + e^8)}{3e^8}$

41. $\frac{256\pi}{15} \approx 53.62$ **43.** 1280π **45.** $\left(-\frac{1}{4}R,\ 0,\ \frac{5}{8}H\right)$

47. $\left(-\frac{2}{11\pi}R,\ -\frac{2}{11\pi}R(2 - \sqrt{3}),\ \frac{1}{2}H\right)$.   **49.** $\left(0,\ 0,\ \frac{2}{3}\right)$

51. $\left(\frac{8}{15},\ \frac{16}{15\pi},\ \frac{16}{15\pi}\right)$ **53.** $\frac{19}{33}$ **55.** $\frac{4}{7}$

57. $G(u,\ v) = (3u + v,\ -u + 4v)$; $\text{Area}\,(G(R)) = 156$

59. $\text{Area}(D) \approx \frac{1}{5}$

61. (a)

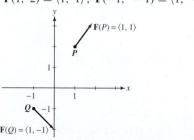

(d) $\frac{3}{4}(e^2 - \sqrt{e})$

Chapter 16

Section 16.1 Preliminary Questions

1. (b)   **2.**

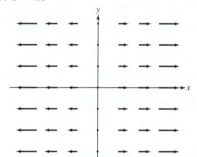

3. $\mathbf{F} = \langle 0,\ -z,\ y\rangle$ **4.** $f_1(x,\ y,\ z) = xyz + 1$

Section 16.1 Exercises

1. $\mathbf{F}(1,\ 2) = \langle 1,\ 1\rangle$, $\mathbf{F}(-1,\ -1) = \langle 1,\ -1\rangle$

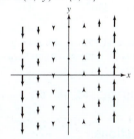

3. $\mathbf{F}(P) = \langle 0,\ 1,\ 0\rangle$, $\mathbf{F}(Q) = \langle 2,\ 0,\ 2\rangle$

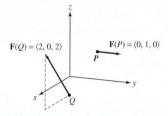

5. $\mathbf{F} = \langle 1,\ 0\rangle$

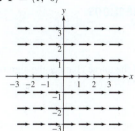

7. $\mathbf{F} = x\mathbf{i}$

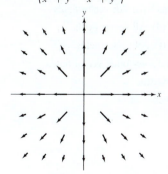

9. $\mathbf{F}(x,\ y) = \langle 0,\ x\rangle$

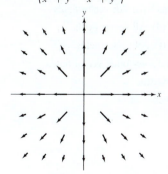

11. $\mathbf{F} = \left\langle \frac{x}{x^2 + y^2},\ \frac{y}{x^2 + y^2}\right\rangle$

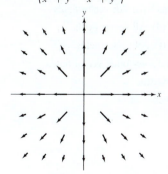

13. Plot (D)   **15.** Plot (B)   **17.** Plot (C)   **19.** Plot (B)

21. $(0, y)$ **23.** $\text{div}(\mathbf{F}) = y + z$, $\text{curl}(\mathbf{F}) = \langle y, 3x^2, -x\rangle$

25. $\text{div}(\mathbf{F}) = 1 - 4xz - x + 2x^2z$, $\text{curl}(\mathbf{F}) = \langle -1, 2x^2 - 2xz^2, -y\rangle$

27. $\text{div}(\mathbf{F}) = 0$, $\text{curl}(\mathbf{F}) = \langle 1 - 3z^2, 1 - 2x, 1 + 2y\rangle$

29. $\text{div}(\mathbf{F}) = 0$, $\text{curl}(\mathbf{F}) = \langle 0, \sin x, \cos x - e^y\rangle$

39. $f(x, y) = \frac{1}{2}x^2 + \frac{1}{2}y^2 + K$

41. $f(x, y) = e^{xy} + K$

43. $f(x, y, z) = xyz^2 + K$ **45.** $f(x, y, z) = \sin(xyz) + K$

47. $f(x, y, z) = ax + by + cz + K$

49. **(b)** $e_p(1, 1) = \langle -2, -1 \rangle / \sqrt{5}$;

(c) $f(x, y, z) = \sqrt{(x-a)^2 + (y-b)^2}$

51. (A)

Section 16.2 Preliminary Questions

1. 50   **2.** (a), (c), (d), (e)

3. **(a)** True

(b) False. Reversing the orientation of the curve changes the sign of the vector line integral.

4. **(a)** 0   **(b)** -5

Section 16.2 Exercises

1. **(a)** $f(\mathbf{r}(t)) = 6t + 4t^2; ds = 2\sqrt{11}\, dt$

(b) $\int_0^1 \left(6t + 4t^2\right) 2\sqrt{11}\, dt = \frac{26\sqrt{11}}{3}$

3. **(a)** $\mathbf{F}(\mathbf{r}(t)) = \left\langle t^{-2}, t^2 \right\rangle; d\mathbf{r} = \left\langle 1, -t^{-2} \right\rangle dt$

(b) $\int_1^2 \left(t^{-1} - 1\right) dt = -\frac{1}{2}$

5. $\sqrt{2}\left(\pi + \frac{\pi^3}{3}\right)$ **7.** $\pi^2/2$ **9.** 2.8   **11.** $\frac{128\sqrt{29}}{3} \approx 229.8$

13. $\frac{\sqrt{3}}{2}(e - 1) \approx 1.488$ **15.** $\frac{2}{3}\left((e^2 + 5)^{3/2} - 2^{3/2}\right)$

17. 39; the distance between $(8, -6, 24)$ and $(20, -15, 60)$

19. $\frac{16}{3}$ **21.** 0   **23.** $2(e^2 - e^{-2}) - (e - e^{-1}) \approx 12.157$ **25.** $\frac{10}{9}$

27. $-\frac{8}{3}$ **29.** $\frac{13}{2}$ **31.** $\frac{\pi}{2}$ **33.** 339.5587   **35.** $2 - e - \frac{1}{e}$

37. **(a)** -8 **(b)** -11 **(c)** -16

39. $\approx 7.6; \approx 4$ **41.** (A) Zero, (B) Negative, (C) Zero   **43.** 64π g

45. $\approx 10.4 \times 10^{-6}$ C   **47.** $\approx 22{,}743.10$ volts   **49.** $\approx -10{,}097$ volts

51. 1   **53.** $\frac{27}{28}$ **55.** **(a)** ABC **(b)** CBA

59. $\frac{1}{3}\left((4\pi^2 + 1)^{3/2} - 1\right) \approx 85.5 \times 10^{-6}$ C   **65.** 18   **67.** $e - 1$

Section 16.3 Preliminary Questions

1. Closed

2. **(a)** Conservative vector fields   **(b)** All vector fields

(c) Conservative vector fields   **(d)** All vector fields

(e) Conservative vector fields   **(f)** All vector fields

(g) Conservative vector fields and *some* other vector fields

3. **(a)** Always true   **(b)** Always true

(c) True under additional hypotheses on D

4. **(a)** 4   **(b)** -4

Section 16.3 Exercises

1. 0   **3.** $-\frac{9}{4}$ **5.** $32e - 1$ **7.** $f(x, y, z) = zx + y$

9. $f(x, y, z) = y^2 x + e^z y$

11. The vector field is not conservative.

13. $f(x, y, z) = z \tan x + zy$ **15.** $f(x, y, z) = x^2 y + 5x - 4zy$

17. 16   **19.** 1   **21.** 6   **23.** $\frac{2}{3}$; 0   **25.** 6.2×10^9 J

27. **(a)** $f(x, y, z) = -gz$ **(b)** ≈ 82.8 m/s

29. (A) 2π, (B) 2π, (C) 0, (D) -2π, (E) 4π

Section 16.4 Preliminary Questions

1. 50

2. A distortion factor that indicates how much the area of R_{ij} is altered under the map G.

3. Area$(\mathcal{S}) \approx 0.0006$ **4.** $\iint_{\mathcal{S}} f(x, y, z)\, dS \approx 0.6$

5. Area$(\mathcal{S}) = 20$ **6.** $\left(\frac{2}{3}, \frac{2}{3}, \frac{1}{3}\right)$

Section 16.4 Exercises

1. **(a)** v   **(b)** iii   **(c)** i   **(d)** iv   **(e)** ii

3. **(a)** $\mathbf{T}_u = \langle 2, 1, 3 \rangle$, $\mathbf{T}_v = \langle 0, -1, 1 \rangle$, $\mathbf{n}(u, v) = \langle 4, -2, -2 \rangle$

(b) Area$(\mathcal{S}) = 4\sqrt{6}$ **(c)** $\iint_{\mathcal{S}} f(x, y, z)\, dS = \frac{32\sqrt{6}}{3}$

5. **(a)** $\mathbf{T}_x = \langle 1, 0, y \rangle$, $\mathbf{T}_y = \langle 0, 1, x \rangle$, $\mathbf{N}(x, y) = \langle -y, -x, 1 \rangle$

(b) $\frac{(2\sqrt{2} - 1)\pi}{6}$ **(c)** $\frac{\sqrt{2} + 1}{15}$

7. $\mathbf{T}_u = \langle 2, 1, 3 \rangle$, $\mathbf{T}_v = \langle 1, -4, 0 \rangle$, $\mathbf{N}(u, v) = 3\langle 4, 1, -3 \rangle$, $4x + y - 3z = 0$

9. $\mathbf{T}_\theta = \langle -\sin\theta\sin\phi, \cos\theta\sin\phi, 0 \rangle$, $\mathbf{T}_\phi = \langle \cos\theta\cos\phi, \sin\theta\cos\phi, -\sin\phi \rangle$, $\mathbf{N}(u, v) = -\cos\theta\sin^2\phi\mathbf{i} - \sin\theta\sin^2\phi\mathbf{j} - \sin\phi\cos\phi\mathbf{k}$, $y + z = \sqrt{2}$

11. Area$(\mathcal{S}) \approx 0.2078$ **13.** $\frac{\sqrt{2}}{5}$ **15.** $\frac{37\sqrt{37} - 1}{4} \approx 56.02$ **17.** $\frac{\pi}{6}$

19. $4\pi(1 - e^{-4})$ **21.** $\frac{\sqrt{3}}{6}$ **23.** $\frac{7\pi}{3}$ **25.** $\frac{5\sqrt{10}}{27} - \frac{1}{54}$

27. Area$(\mathcal{S}) = 16$ **29.** $3e^3 - 6e^2 + 3e + 1 \approx 25.08$

31. Area$(\mathcal{S}) = 4\pi R^2$

33. **(a)** Area$(\mathcal{S}) \approx 1.0780$ **(b)** ≈ 0.09814

35. Area$(\mathcal{S}) = \frac{5\sqrt{29}}{4} \approx 6.73$ **37.** Area$(\mathcal{S}) = \pi$ **39.** 48π

43. Area$(\mathcal{S}) = \frac{\pi}{6}\left(17\sqrt{17} - 1\right) \approx 36.18$ **47.** $4\pi^2 ab$

49. $f(r) = -\frac{Gm}{2Rr}\left(\sqrt{R^2 + r^2} - |R - r|\right)$

Section 16.5 Preliminary Questions

1. (b)   **2.** (c)   **3.** (a)   **4.** (b)

5. **(a)** 0   **(b)** π **(c)** π

6. $\approx 0.05\sqrt{2} \approx 0.0707$ **7.** 0

Section 16.5 Exercises

1. **(a)** $\mathbf{N} = \langle 2v, -4uv, 1 \rangle$, $\mathbf{F} \cdot \mathbf{N} = 2v^3 + u$

(b) $\frac{4}{\sqrt{69}}$ **(c)** 265

3. 4   **5.** -4 **7.** $\frac{27}{12}(3\pi + 4)$ **9.** $\frac{693}{5}$ **11.** $\frac{11}{12}$ **13.** $\frac{9\pi}{4}$

15. $(e - 1)^2$ **17.** 270

19. **(a)** $18\pi e^{-3}$ **(b)** $\frac{\pi}{2}e^{-1}$

21. $\left(2 - \frac{6}{\sqrt{13}}\right)\pi k$ **23.** $\frac{2\pi}{3}$ m^3/s   **25.** 4π **27.** $\frac{16\pi}{3}$

29. **(a)** 1   **(b)** 1

33. $\Phi(t) = -1.56 \times 10^{-5}e^{-0.1t}$ T-m^2; voltage drop $= -1.56 \times 10^{-6}e^{-0.1t}$ V

35. The flow is greatest at $z = 3$.

Chapter 16 Review

1. (a) $\langle -15, 8 \rangle$ **(b)** $\langle 4, 8 \rangle$ **(c)** $\langle 9, 1 \rangle$

3.

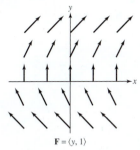

$\mathbf{F} = \langle y, 1 \rangle$

5. $\mathbf{F}(x, y) = \langle 2x, -1 \rangle$

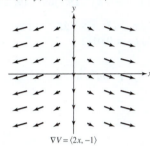

$\nabla V = \langle 2x, -1 \rangle$

7. $\operatorname{div}(\mathbf{F}) = 2x + 2y + 2z$, $\operatorname{curl}(\mathbf{F}) = \langle 0, 0, 0 \rangle$

9. $\operatorname{div}(\mathbf{F}) = 3x^2 y + y^2$, $\operatorname{curl}(\mathbf{F}) = \langle 2yz - 2xz, 0, z^2 - x^3 \rangle$

11. $\operatorname{div}(\mathbf{F}) = -1$, $\operatorname{curl}(\mathbf{F}) = \langle 0, 0, -1 \rangle$

13. $\operatorname{div}(\mathbf{F}) = 4x^2 e^{-x^2 - y^2 - z^2} + 4y^2 e^{-x^2 - y^2 - z^2} + 4z^2 e^{-x^2 - y^2 - z^2} - 6e^{-x^2 - y^2 - z^2}$; $\operatorname{curl}(\mathbf{F}) = \langle 0, 0, 0 \rangle$

15. $\operatorname{curl}(\mathbf{F}) = \langle -2z, 0, 2y \rangle$ **19.** $f(x, y) = x^4 y^5$

21. $\mathbf{F}$ is conservative, $f(x, y, z) = 2x + 4y + e^z$

23. $f(x, y) = \frac{x^4 y^4}{4}$

25. $\mathbf{F}$ is conservative, $f(x, y, z) = y \ln\left(x^2 + z\right)$

27. $\mathbf{F} = \langle 1 + by, 1 + bx \rangle$ **29.** -2

31. $\sqrt{2}(\sin 3 - 2\cos 3 + \sin 1 + 2\cos 1 + 4) \approx 11.375$

33. (a) 2 **(b)** 0 **37.** $\frac{1}{3}$ **39.** $4 \ln\left(1 + (\ln 2)^4 + e^2\right) \approx 8.616$

41. $\frac{52}{29} \approx 1.79$ **43.** $3e^{3/2} - \frac{15}{2} \approx 5.945$

45. Area $(S) =$
$\int_{-1}^{1} \int_{-1}^{1} \sqrt{125u^2 - 100uv + 425v^2 + 81}\, du\, dv \approx 62.911$

47. $\frac{4}{9}\left(2\sqrt{2} - 1\right)$

49. (a) Zero since $f(x, y, z) = y^3$ is odd and symmetric about the xz-plane

(b) Positive since $f(x, y, z) = z^3$ is non-negative

(c) Zero since $f(-x, y, z) = -xyz = -f(x, y, z)$ is symmetric about the yz-plane

(d) Negative since $f(x, y, z) = z^2 - 2$ is negative

51. (a) $\mathbf{N} = \left\langle \frac{2}{\sqrt{5}}, \frac{2\sqrt{3}}{\sqrt{5}}, 2 \right\rangle$

(b) Area $(G(R)) = \frac{6}{\sqrt{5}} \cdot 0.1 \cdot 0.05 \approx 0.0134$

53. $\iint_S \mathbf{F} \cdot dS =$
$\int_0^1 \left(\int_0^1 (-4(2u - v) - 2(2v + 5) + 7(-u - 3v))\, dv \right) du = -28$

55. $\frac{1}{4} - \frac{\sinh 1}{3}$ **57.** $\frac{\sqrt{6}\pi}{9}$ **59.** 27π **61.** $\frac{9\sqrt{3}}{4}$

Chapter 17

Section 17.1 Preliminary Questions

1. $\mathbf{F} = \left\langle -e^y, x^2 \right\rangle$

2.

c

3. Yes **4.** (a), (c)

Section 17.1 Exercises

1. $\oint_C xy\, dx + y\, dy =$

$\int_0^{2\pi} (\cos\theta \sin\theta (-\sin\theta) + \sin\theta \cos\theta)\, d\theta = 0 =$

$\int_{-1}^{1} \int_{-\sqrt{1-x^2}}^{\sqrt{1-x^2}} (0 - x)\, dy\, dx = \int\int_{\mathcal{D}} \left(\frac{\partial}{\partial x} y - \frac{\partial}{\partial y} (xy) \right) dA$

3. 0 **5.** $-\frac{\pi}{4}$ **7.** $\frac{1}{6}$ **9.** $\frac{(e^2 - 1)(e^4 - 5)}{2}$

11. (a) $V(x, y) = x^2 e^y$ **13.** $I = 34$ **15.** $\frac{1}{2}$ **17.** $\frac{8}{3}$

19. (c) $A = \frac{3}{2}$ **23.** $9 + \frac{15\pi}{2}$ **25.** 214π

27. (A) Zero **(B)** Positive **(C)** Negative **(D)** Zero

29. -0.10 **31.** $R = \sqrt{\frac{2}{3}}$ **33.** Triangle (A), 3; Polygon (B), 12

35. -4 **37.** 5π **39.** $\frac{9\pi}{4}$ **41.** 29.2 buffalo per minute

Section 17.2 Preliminary Questions

1.

(A)   (B)

2. (a)

3. A vector field $\mathbf{A}$ such that $\mathbf{F} = \operatorname{curl}(\mathbf{A})$ is a vector potential for $\mathbf{F}$

4. (b)

5. $\mathbf{F}$ must be the curl of some other vector field $\mathbf{A}$ (see Theorem 2).

Section 17.2 Exercises

1. $\iint_C \mathbf{F} \cdot ds = \iint_S \operatorname{curl}(\mathbf{F}) \cdot d\mathbf{S} = \pi$

3. $\iint_C \mathbf{F} \cdot ds = \iint_S \operatorname{curl}(\mathbf{F}) \cdot d\mathbf{S} = e^{-1} - 1$

5. $\left\langle -3z^2 e^{z^3}, 2ze^{z^2} + z\sin(xz), 2 \right\rangle$; 2π

7. $\langle -2, 3, 5 \rangle$, 20π **9.** 0 **11.** -45π **13.** 0

15. (a)   **(b)** 140π

17. (a) $\mathbf{A} = \left\langle 0, 0, e^y - e^{x^2} \right\rangle$ **(c)** $\iint_S \mathbf{F} \cdot d\mathbf{S} = \frac{\pi}{2}$

19. (a) $\iint_S \mathbf{B} \cdot d\mathbf{S} = r^2 B\pi$ **(b)** $\int_{\partial S} \mathbf{A} \cdot ds = 0$

21. $\int\int_S B\, dS = 18b$ **23.** $c = 2a$ and b is arbitrary.

27. $\iint_S \mathbf{F} \cdot d\mathbf{S} = 25$

Section 17.3 Preliminary Questions

1. $\iint_S \mathbf{F} \cdot d\mathbf{S} = 0$

2. Since the integrand is positive for all $(x, y, z) \neq (0, 0, 0)$, the triple integral, hence also the flux, is positive.

3. (a), (b), (d), (f) are meaningful; (b) and (d) are automatically zero.

4. (c) **5.** $\mathrm{div}(\mathbf{F}) = 1$ and flux $= \int \mathrm{div}(\mathbf{F})\, dV =$ volume

Section 17.3 Exercises

1. $\iint_S \mathbf{F} \cdot d\mathbf{S} = \iiint_{\mathcal{R}} \mathrm{div}(\mathbf{F})\, dV = \iiint_{\mathcal{R}} 0\, dV = 0$

3. $\iint_S \mathbf{F} \cdot d\mathbf{S} = \iiint_{\mathcal{R}} \mathrm{div}(\mathbf{F})\, dV = 4\pi$ **5.** $\frac{4\pi}{15}$ **7.** 60π

9. $\frac{3{,}616}{105}$ **11.** $\frac{32\pi}{5}$ **13.** 64π **15.** 81π **17.** π **19.** $\frac{13}{3}$

21. $\frac{4\pi}{3}$ **23.** $\frac{16\pi}{3} + \frac{9\sqrt{3}}{2} \approx 24.549$ **25.** ≈ 1.57 m^3/s

27. (b) 0 (c) 0

(d) Since $\mathbf{E}$ is not defined at the origin, which is inside the ball W, we cannot use the Divergence Theorem.

29. $(-4) \cdot \left[\frac{256\pi}{3} - 1\right] \approx -1068.33$

31. $\mathrm{Div}(f\mathbf{F}) = f\,\mathrm{Div}(\mathbf{F}) + \mathbf{F} \cdot \nabla f$

35. (d) Flux through a closed surface is 0.

Chapter 17 Review

1. 0 **3.** -30 **5.** $\frac{3}{5}$

7. (a) 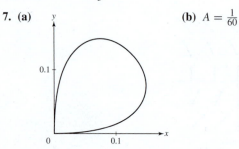 (b) $A = \frac{1}{60}$

9. $\frac{1}{3}$ **11.** $\frac{1}{3}$ **13.** $\{(x, y) \mid y = x \text{ or } y = -x\}$ **15.** 36π **19.** 2π

21. $\mathbf{A} = \langle yz, 0, 0 \rangle$ and the flux is 8π. **23.** $\frac{296}{3}$ **25.** -128π

27. Volume$(W) = 2$ **29.** $4 \cdot 0.0009\pi \approx 0.0113$

31. $2x - y + 4z = 0$ **33.** (b) $\frac{\pi}{2}$

35. (c) 0 (d) $\int_{C_1} \mathbf{F} \cdot d\mathbf{s} = -4$, $\int_{C_2} \mathbf{F} \cdot d\mathbf{s} = 4$

41. $V = \frac{4\pi}{3} abc$

REFERENCES

The online source MacTutor History of Mathematics Archive **www-history.mcs .st-and.ac.uk** has been a valuable source of historical information.

Section 10.1

(EX 68) Adapted from G. Klambauer, *Aspects of Calculus*, Springer-Verlag, New York, 1986, p. 393.

Section 10.2

(EX 48) Adapted from *Calculus Problems for a New Century*, Robert Fraga, ed., Mathematical Association of America, Washington, DC, 1993, p. 137.

(EX 49) Adapted from *Calculus Problems for a New Century*, Robert Fraga, ed., Mathematical Association of America, Washington, DC, 1993, p. 138.

(EX 61) Adapted from George Andrews, "The Geometric Series in Calculus," *American Mathematical Monthly* 105, 1:36-40 (1998).

(EX 64) Adapted from Larry E. Knop, "Cantor's Disappearing Table," *The College Mathematics Journal* 16, 5:398-399 (1985).

Section 10.4

(EX 33) Adapted from *Calculus Problems for a New Century*, Robert Fraga, ed., Mathematical Association of America, Washington, DC, 1993, p. 145.

Section 11.2

(EX 41) Adapted from Richard Courant and Fritz John, *Differential and Integral Calculus*, Wiley-Interscience, New York, 1965.

Section 11.3

(EX 58) Adapted from *Calculus Problems for a New Century*, Robert Fraga, ed., Mathematical Association of America, Washington, DC, 1993.

Section 12.4

(EX 65) Adapted from Ethan Berkove and Rich Marchand, "The Long Arm of Calculus," *The College Mathematics Journal* 29, 5:376-386 (November 1998).

Section 13.3

(EX 21) Adapted from *Calculus Problems for a New Century*, Robert Fraga, ed., Mathematical Association of America, Washington, DC, 1993.

Section 13.4

(EX 68) Damien Gatinel, Thanh Hoang-Xuan, and Dimitri T. Azar, "Determination of Corneal Asphericity After Myopia Surgery with the Excimer Laser: A Mathematical Model," *Investigative Opthalmology and Visual Science* 42: 1736-1742 (2001).

Section 13.5

(EX 57, 60) Adapted from notes to the course "Dynamics and Vibrations" at Brown University, http://www.engin.brown.edu/ courses/en4/.

Section 14.8

(EX 46) Adapted from C. Henry Edwards, "Ladders, Moats, and Lagrange Multipliers," *Mathematica Journal* 4, Issue 1 (Winter 1994).

Section 15.3

(FIGURE 10 COMPUTATION) The computation is based on Jeffrey Nunemacher, "The Largest Unit Ball in Any Euclidean Space," in *A Century of Calculus*, Part II, Mathematical Association of America, Washington DC, 1992.

Section 15.6

(CONCEPTUAL INSIGHT) See R. Courant and F. John, *Introduction to Calculus and Analysis*, Springer-Verlag, New York, 1989, p. 534.

Section 16.2

(FIGURE 9) Inspired by Tevian Dray and Corinne A. Manogue, "The Murder Mystery Method for Determining Whether a Vector Field Is Conservative," *The College Mathematics Journal*, May 2003.

Section 16.3

(EX 21) Adapted from *Calculus Problems for a New Century*, Robert Fraga, ed., Mathematical Association of America, Washington, DC, 1993.

Appendix D

(PROOF OF THEOREM 6) A proof without this simplifying assumption can be found in R. Courant and F. John, *Introduction to Calculus and Analysis, Vol. 1*, Springer-Verlag, New York, 1989.

Section 7.6

(EX 1–5, 9) Problems suggested by Brian Bradie, Christopher Newport University.

Section 7.7

(EX 84) Problem suggested by Chris Bishop, SUNY Stony Brook.

Section 7.9

See R. Courant and F. John, *Introduction to Calculus and Analysis, Vol. 1*, Springer-Verlag, New York, 1989.

Section 8.1

(EX 54) Adapted from G. Klambauer, *Aspects of Calculus*, Springer-Verlag, New York, 1986, Ch 6.

Section 9.1

(EX 57) Adapted from E. Batschelet, *Introduction to Mathematics for Life Scientists*, Springer-Verlag, New York, 1979.

(EX 59) Adapted from *Calculus Problems for a New Century*, Robert Fraga, ed., Mathematical Association of America, Washington, DC, 1993.

(EX 60, 65) Adapted from M. Tenenbaum and H. Pollard, *Ordinary Differential Equations*, Dover, New York, 1985.

Section 10.1

(EX 68) Adapted from G. Klambauer, *Aspects of Calculus*, Springer-Verlag, New York, 1986, p. 393.

Section 10.2

(EX 48) Adapted from *Calculus Problems for a New Century*, Robert Fraga, ed., Mathematical Association of America, Washington, DC, 1993, p. 137.

(EX 49) Adapted from *Calculus Problems for a New Century*, Robert Fraga, ed., Mathematical Association of America, Washington, DC, 1993, p. 138.

(EX 61) Adapted from George Andrews, "The Geometric Series in Calculus," *American Mathematical Monthly* 105, 1:36-40 (1998).

(EX 64) Adapted from Larry E. Knop, "Cantor's Disappearing Table," *The College Mathematics Journal* 16, 5:398-399 (1985).

Section 10.4

(EX 33) Adapted from *Calculus Problems for a New Century*, Robert Fraga, ed., Mathematical Association of America, Washington, DC, 1993, p. 145.

Section 11.2

(EX 41) Adapted from Richard Courant and Fritz John, *Differential and Integral Calculus*, Wiley-Interscience, New York, 1965.

Section 11.3

(EX 58) Adapted from *Calculus Problems for a New Century*, Robert Fraga, ed., Mathematical Association of America, Washington, DC, 1993.

Section 12.4

(EX 65) Adapted from Ethan Berkove and Rich Marchand, "The Long Arm of Calculus," *The College Mathematics Journal* 29, 5:376-386 (November 1998).

Section 13.3

(EX 21) Adapted from *Calculus Problems for a New Century*, Robert Fraga, ed., Mathematical Association of America, Washington, DC, 1993.

Section 13.4

(EX 68) Damien Gatinel, Thanh Hoang-Xuan, and Dimitri T. Azar, "Determination of Corneal Asphericity After Myopia Surgery with the Excimer Laser: A Mathematical Model," *Investigative Opthalmology and Visual Science* 42: 1736-1742 (2001).

Section 13.5

(EX 57, 60) Adapted from notes to the course "Dynamics and Vibrations" at Brown University, http://www.engin.brown.edu/courses/en4/.

Section 14.8

(EX 46) Adapted from C. Henry Edwards, "Ladders, Moats, and Lagrange Multipliers," *Mathematica Journal* 4, Issue 1 (Winter 1994).

Section 15.3

(FIGURE 10 COMPUTATION) The computation is based on Jeffrey Nunemacher, "The Largest Unit Ball in Any Euclidean Space," in *A Century of Calculus*, Part II, Mathematical Association of America, Washington DC, 1992.

Section 15.6

(CONCEPTUAL INSIGHT) See R. Courant and F. John, *Introduction to Calculus and Analysis*, Springer-Verlag, New York, 1989, p. 534.

Section 16.2

(FIGURE 9) Inspired by Tevian Dray and Corinne A. Manogue, "The Murder Mystery Method for Determining Whether a Vector Field Is Conservative," *The College Mathematics Journal*, May 2003.

Section 16.3

(EX 21) Adapted from *Calculus Problems for a New Century*, Robert Fraga, ed., Mathematical Association of America, Washington, DC, 1993.

Appendix D

(PROOF OF THEOREM 6) A proof without this simplifying assumption can be found in R. Courant and F. John, *Introduction to Calculus and Analysis, Vol. 1*, Springer-Verlag, New York, 1989.

INDEX

ELEMENTARY FUNCTIONS

Power Functions $f(x) = x^a$

$f(x) = x^n$, n a positive integer

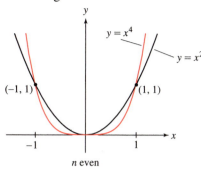

n even

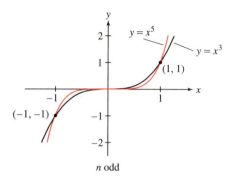

n odd

Asymptotic behavior of a polynomial function
of even degree and positive leading coefficient

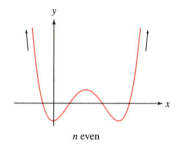

n even

Asymptotic behavior of a polynomial function
of odd degree and positive leading coefficient

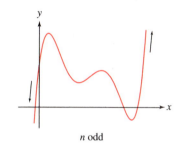

n odd

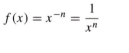
$f(x) = x^{-n} = \dfrac{1}{x^n}$

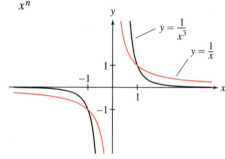

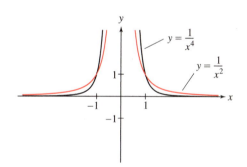

Inverse Trigonometric Functions

$\arcsin x = \sin^{-1} x = \theta$

$\Leftrightarrow \quad \sin \theta = x, \quad -\dfrac{\pi}{2} \le \theta \le \dfrac{\pi}{2}$

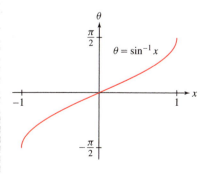

$\arccos x = \cos^{-1} x = \theta$

$\Leftrightarrow \quad \cos \theta = x, \quad 0 \le \theta \le \pi$

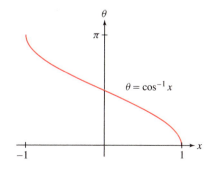

$\arctan x = \tan^{-1} x = \theta$

$\Leftrightarrow \quad \tan \theta = x, \quad -\dfrac{\pi}{2} < \theta < \dfrac{\pi}{2}$

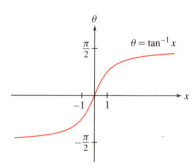

Exponential and Logarithmic Functions

$$\boxed{\log_a x = y \quad \Leftrightarrow \quad a^y = x}$$

$$\boxed{\ln x = y \quad \Leftrightarrow \quad e^y = x}$$

$$\log_a(xy) = \log_a x + \log_a y$$

$$\log_a(a^x) = x \qquad a^{\log_a x} = x$$

$$\ln(e^x) = x \qquad e^{\ln x} = x$$

$$\log_a\left(\frac{x}{y}\right) = \log_a x - \log_a y$$

$$\log_a 1 = 0 \qquad \log_a a = 1$$

$$\ln 1 = 0 \qquad \ln e = 1$$

$$\log_a(x^r) = r \log_a x$$

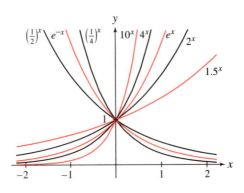

$$\lim_{x \to \infty} a^x = \infty, \quad a > 1$$

$$\lim_{x \to \infty} a^x = 0, \quad 0 < a < 1$$

$$\lim_{x \to -\infty} a^x = 0, \quad a > 1$$

$$\lim_{x \to -\infty} a^x = \infty, \quad 0 < a < 1$$

$$\lim_{x \to 0^+} \log_a x = -\infty$$

$$\lim_{x \to \infty} \log_a x = \infty$$

Hyperbolic Functions

$$\sinh x = \frac{e^x - e^{-x}}{2} \qquad \operatorname{csch} x = \frac{1}{\sinh x}$$

$$\cosh x = \frac{e^x + e^{-x}}{2} \qquad \operatorname{sech} x = \frac{1}{\cosh x}$$

$$\tanh x = \frac{\sinh x}{\cosh x} \qquad \coth x = \frac{\cosh x}{\sinh x}$$

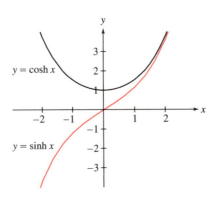

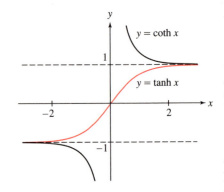

$$\sinh(x + y) = \sinh x \cosh y + \cosh x \sinh y$$

$$\cosh(x + y) = \cosh x \cosh y + \sinh x \sinh y$$

$$\sinh 2x = 2 \sinh x \cosh x$$

$$\cosh 2x = \cosh^2 x + \sinh^2 x$$

Inverse Hyperbolic Functions

$$y = \sinh^{-1} x \quad \Leftrightarrow \quad \sinh y = x$$

$$y = \cosh^{-1} x \quad \Leftrightarrow \quad \cosh y = x \text{ and } y \geq 0$$

$$y = \tanh^{-1} x \quad \Leftrightarrow \quad \tanh y = x$$

$$\sinh^{-1} x = \ln\left(x + \sqrt{x^2 + 1}\right)$$

$$\cosh^{-1} x = \ln\left(x + \sqrt{x^2 - 1}\right) \quad x > 1$$

$$\tanh^{-1} x = \frac{1}{2} \ln\left(\frac{1 + x}{1 - x}\right) \quad -1 < x < 1$$

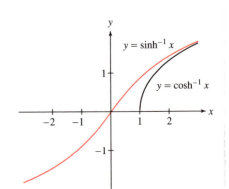

DIFFERENTIATION

Differentiation Rules

1. $\dfrac{d}{dx}(c) = 0$

2. $\dfrac{d}{dx}x = 1$

3. $\dfrac{d}{dx}(x^n) = nx^{n-1}$ (Power Rule)

4. $\dfrac{d}{dx}[cf(x)] = cf'(x)$

5. $\dfrac{d}{dx}[f(x) + g(x)] = f'(x) + g'(x)$

6. $\dfrac{d}{dx}[f(x)g(x)] = f(x)g'(x) + g(x)f'(x)$ (Product Rule)

7. $\dfrac{d}{dx}\left[\dfrac{f(x)}{g(x)}\right] = \dfrac{g(x)f'(x) - f(x)g'(x)}{[g(x)]^2}$ (Quotient Rule)

8. $\dfrac{d}{dx}f(g(x)) = f'(g(x))g'(x)$ (Chain Rule)

9. $\dfrac{d}{dx}f(x)^n = nf(x)^{n-1}f'(x)$ (General Power Rule)

10. $\dfrac{d}{dx}f(kx + b) = kf'(kx + b)$

11. $g'(x) = \dfrac{1}{f'(g(x))}$ where $g(x)$ is the inverse $f^{-1}(x)$

12. $\dfrac{d}{dx}\ln f(x) = \dfrac{f'(x)}{f(x)}$

Trigonometric Functions

13. $\dfrac{d}{dx}\sin x = \cos x$

14. $\dfrac{d}{dx}\cos x = -\sin x$

15. $\dfrac{d}{dx}\tan x = \sec^2 x$

16. $\dfrac{d}{dx}\csc x = -\csc x \cot x$

17. $\dfrac{d}{dx}\sec x = \sec x \tan x$

18. $\dfrac{d}{dx}\cot x = -\csc^2 x$

Inverse Trigonometric Functions

19. $\dfrac{d}{dx}(\sin^{-1} x) = \dfrac{1}{\sqrt{1 - x^2}}$

20. $\dfrac{d}{dx}(\cos^{-1} x) = -\dfrac{1}{\sqrt{1 - x^2}}$

21. $\dfrac{d}{dx}(\tan^{-1} x) = \dfrac{1}{1 + x^2}$

22. $\dfrac{d}{dx}(\csc^{-1} x) = -\dfrac{1}{|x|\sqrt{x^2 - 1}}$

23. $\dfrac{d}{dx}(\sec^{-1} x) = \dfrac{1}{|x|\sqrt{x^2 - 1}}$

24. $\dfrac{d}{dx}(\cot^{-1} x) = -\dfrac{1}{1 + x^2}$

Exponential and Logarithmic Functions

25. $\dfrac{d}{dx}(e^x) = e^x$

26. $\dfrac{d}{dx}(a^x) = (\ln a)a^x$

27. $\dfrac{d}{dx}\ln|x| = \dfrac{1}{x}$

28. $\dfrac{d}{dx}(\log_a x) = \dfrac{1}{(\ln a)x}$

Hyperbolic Functions

29. $\dfrac{d}{dx}(\sinh x) = \cosh x$

30. $\dfrac{d}{dx}(\cosh x) = \sinh x$

31. $\dfrac{d}{dx}(\tanh x) = \operatorname{sech}^2 x$

32. $\dfrac{d}{dx}(\operatorname{csch} x) = -\operatorname{csch} x \coth x$

33. $\dfrac{d}{dx}(\operatorname{sech} x) = -\operatorname{sech} x \tanh x$

34. $\dfrac{d}{dx}(\coth x) = -\operatorname{csch}^2 x$

Inverse Hyperbolic Functions

35. $\dfrac{d}{dx}(\sinh^{-1} x) = \dfrac{1}{\sqrt{1 + x^2}}$

36. $\dfrac{d}{dx}(\cosh^{-1} x) = \dfrac{1}{\sqrt{x^2 - 1}}$

37. $\dfrac{d}{dx}(\tanh^{-1} x) = \dfrac{1}{1 - x^2}$

38. $\dfrac{d}{dx}(\operatorname{csch}^{-1} x) = -\dfrac{1}{|x|\sqrt{x^2 + 1}}$

39. $\dfrac{d}{dx}(\operatorname{sech}^{-1} x) = -\dfrac{1}{x\sqrt{1 - x^2}}$

40. $\dfrac{d}{dx}(\coth^{-1} x) = \dfrac{1}{1 - x^2}$

INTEGRATION

Substitution

If an integrand has the form $f(u(x))u'(x)$, then rewrite the entire integral in terms of u and its differential $du = u'(x)\,dx$:

$$\int f(u(x))u'(x)\,dx = \int f(u)\,du$$

Integration by Parts Formula

$$\int u(x)v'(x)\,dx = u(x)v(x) - \int u'(x)v(x)\,dx$$

TABLE OF INTEGRALS

Basic Forms

1. $\displaystyle\int u^n\,du = \frac{u^{n+1}}{n+1} + C, \quad n \neq -1$

2. $\displaystyle\int \frac{du}{u} = \ln|u| + C$

3. $\displaystyle\int e^u\,du = e^u + C$

4. $\displaystyle\int a^u\,du = \frac{a^u}{\ln a} + C$

5. $\displaystyle\int \sin u\,du = -\cos u + C$

6. $\displaystyle\int \cos u\,du = \sin u + C$

7. $\displaystyle\int \sec^2 u\,du = \tan u + C$

8. $\displaystyle\int \csc^2 u\,du = -\cot u + C$

9. $\displaystyle\int \sec u \tan u\,du = \sec u + C$

10. $\displaystyle\int \csc u \cot u\,du = -\csc u + C$

11. $\displaystyle\int \tan u\,du = \ln|\sec u| + C$

12. $\displaystyle\int \cot u\,du = \ln|\sin u| + C$

13. $\displaystyle\int \sec u\,du = \ln|\sec u + \tan u| + C$

14. $\displaystyle\int \csc u\,du = \ln|\csc u - \cot u| + C$

15. $\displaystyle\int \frac{du}{\sqrt{a^2 - u^2}} = \sin^{-1}\frac{u}{a} + C$

16. $\displaystyle\int \frac{du}{a^2 + u^2} = \frac{1}{a}\tan^{-1}\frac{u}{a} + C$

Exponential and Logarithmic Forms

17. $\displaystyle\int u e^{au}\,du = \frac{1}{a^2}(au - 1)e^{au} + C$

18. $\displaystyle\int u^n e^{au}\,du = \frac{1}{a}u^n e^{au} - \frac{n}{a}\int u^{n-1}e^{au}\,du$

19. $\displaystyle\int e^{au}\sin bu\,du = \frac{e^{au}}{a^2 + b^2}(a\sin bu - b\cos bu) + C$

20. $\displaystyle\int e^{au}\cos bu\,du = \frac{e^{au}}{a^2 + b^2}(a\cos bu + b\sin bu) + C$

21. $\displaystyle\int \ln u\,du = u\ln u - u + C$

22. $\displaystyle\int u^n \ln u\,du = \frac{u^{n+1}}{(n+1)^2}[(n+1)\ln u - 1] + C$

23. $\displaystyle\int \frac{1}{u\ln u}\,du = \ln|\ln u| + C$

Hyperbolic Forms

24. $\displaystyle\int \sinh u\,du = \cosh u + C$

25. $\displaystyle\int \cosh u\,du = \sinh u + C$

26. $\displaystyle\int \tanh u\,du = \ln\cosh u + C$

27. $\displaystyle\int \coth u\,du = \ln|\sinh u| + C$

28. $\displaystyle\int \operatorname{sech} u\,du = \tan^{-1}|\sinh u| + C$

29. $\displaystyle\int \operatorname{csch} u\,du = \ln\left|\tanh \frac{1}{2}u\right| + C$

30. $\displaystyle\int \operatorname{sech}^2 u\,du = \tanh u + C$

31. $\displaystyle\int \operatorname{csch}^2 u\,du = -\coth u + C$

32. $\displaystyle\int \operatorname{sech} u \tanh u\,du = -\operatorname{sech} u + C$

33. $\displaystyle\int \operatorname{csch} u \coth u\,du = -\operatorname{csch} u + C$

Trigonometric Forms

34. $\displaystyle\int \sin^2 u\,du = \frac{1}{2}u - \frac{1}{4}\sin 2u + C$

35. $\displaystyle\int \cos^2 u\,du = \frac{1}{2}u + \frac{1}{4}\sin 2u + C$

36. $\displaystyle\int \tan^2 u\,du = \tan u - u + C$

37. $\displaystyle\int \cot^2 u\,du = -\cot u - u + C$

38. $\displaystyle\int \sin^3 u\,du = -\frac{1}{3}(2 + \sin^2 u)\cos u + C$

39. $\displaystyle\int \cos^3 u\,du = \frac{1}{3}(2 + \cos^2 u)\sin u + C$

40. $\displaystyle\int \tan^3 u\,du = \frac{1}{2}\tan^2 u + \ln|\cos u| + C$

41. $\displaystyle\int \cot^3 u\,du = -\frac{1}{2}\cot^2 u - \ln|\sin u| + C$

42. $\displaystyle\int \sec^3 u\,du = \frac{1}{2}\sec u \tan u + \frac{1}{2}\ln|\sec u + \tan u| + C$

43. $\displaystyle\int \csc^3 u\, du = -\frac{1}{n}\csc u \cot u + \frac{1}{n}\ln|\csc u - \cot u| + C$

44. $\displaystyle\int \sin^n u\, du = -\frac{1}{n}\sin^{n-1}u \cos u + \frac{n-1}{n}\int \sin^{n-2}u\, du$

45. $\displaystyle\int \cos^n u\, du = \frac{1}{n}\cos^{n-1}u \sin u + \frac{n-1}{n}\int \cos^{n-2}u\, du$

46. $\displaystyle\int \tan^n u\, du = \frac{1}{n-1}\tan^{n-1}u - \int \tan^{n-2}u\, du$

47. $\displaystyle\int \cot^n u\, du = \frac{-1}{n-1}\cot^{n-1}u - \int \cot^{n-2}u\, du$

48. $\displaystyle\int \sec^n u\, du = \frac{1}{n-1}\tan u \sec^{n-2}u + \frac{n-2}{n-1}\int \sec^{n-2}u\, du$

49. $\displaystyle\int \csc^n u\, du = \frac{-1}{n-1}\cot u \csc^{n-2}u + \frac{n-2}{n-1}\int \csc^{n-2}u\, du$

50. $\displaystyle\int \sin au \sin bu\, du = \frac{\sin(a-b)u}{2(a-b)} - \frac{\sin(a+b)u}{2(a+b)} + C$

51. $\displaystyle\int \cos au \cos bu\, du = \frac{\sin(a-b)u}{2(a-b)} + \frac{\sin(a+b)u}{2(a+b)} + C$

52. $\displaystyle\int \sin au \cos bu\, du = -\frac{\cos(a-b)u}{2(a-b)} - \frac{\cos(a+b)u}{2(a+b)} + C$

53. $\displaystyle\int u \sin u\, du = \sin u - u \cos u + C$

54. $\displaystyle\int u \cos u\, du = \cos u + u \sin u + C$

55. $\displaystyle\int u^n \sin u\, du = -u^n \cos u + n\int u^{n-1}\cos u\, du$

56. $\displaystyle\int u^n \cos u\, du = u^n \sin u - n\int u^{n-1}\sin u\, du$

57. $\displaystyle\int \sin^n u \cos^m u\, du$

$\displaystyle = -\frac{\sin^{n-1}u \cos^{m+1}u}{n+m} + \frac{n-1}{n+m}\int \sin^{n-2}u \cos^m u\, du$

$\displaystyle = \frac{\sin^{n+1}u \cos^{m-1}u}{n+m} + \frac{m-1}{n+m}\int \sin^n u \cos^{m-2}u\, du$

Inverse Trigonometric Forms

58. $\displaystyle\int \sin^{-1} u\, du = u\sin^{-1}u + \sqrt{1-u^2} + C$

59. $\displaystyle\int \cos^{-1} u\, du = u\cos^{-1}u - \sqrt{1-u^2} + C$

60. $\displaystyle\int \tan^{-1} u\, du = u\tan^{-1}u - \frac{1}{2}\ln(1+u^2) + C$

61. $\displaystyle\int u\sin^{-1} u\, du = \frac{2u^2-1}{4}\sin^{-1}u + \frac{u\sqrt{1-u^2}}{4} + C$

62. $\displaystyle\int u\cos^{-1} u\, du = \frac{2u^2-1}{4}\cos^{-1}u - \frac{u\sqrt{1-u^2}}{4} + C$

63. $\displaystyle\int u\tan^{-1} u\, du = \frac{u^2+1}{2}\tan^{-1}u - \frac{u}{2} + C$

64. $\displaystyle\int u^n \sin^{-1} u\, du = \frac{1}{n+1}\left[u^{n+1}\sin^{-1}u - \int \frac{u^{n+1}\,du}{\sqrt{1-u^2}}\right],\quad n\neq -1$

65. $\displaystyle\int u^n \cos^{-1} u\, du = \frac{1}{n+1}\left[u^{n+1}\cos^{-1}u + \int \frac{u^{n+1}\,du}{\sqrt{1-u^2}}\right],\quad n\neq -1$

66. $\displaystyle\int u^n \tan^{-1} u\, du = \frac{1}{n+1}\left[u^{n+1}\tan^{-1}u - \int \frac{u^{n+1}\,du}{1+u^2}\right],\quad n\neq -1$

Forms Involving $\sqrt{a^2 - u^2},\, a > 0$

67. $\displaystyle\int \sqrt{a^2-u^2}\, du = \frac{u}{2}\sqrt{a^2-u^2} + \frac{a^2}{2}\sin^{-1}\frac{u}{a} + C$

68. $\displaystyle\int u^2\sqrt{a^2-u^2}\, du = \frac{u}{8}(2u^2-a^2)\sqrt{a^2-u^2} + \frac{a^4}{8}\sin^{-1}\frac{u}{a} + C$

69. $\displaystyle\int \frac{\sqrt{a^2-u^2}}{u}\, du = \sqrt{a^2-u^2} - a\ln\left|\frac{a+\sqrt{a^2-u^2}}{u}\right| + C$

70. $\displaystyle\int \frac{\sqrt{a^2-u^2}}{u^2}\, du = -\frac{1}{u}\sqrt{a^2-u^2} - \sin^{-1}\frac{u}{a} + C$

71. $\displaystyle\int \frac{u^2\,du}{\sqrt{a^2-u^2}} = -\frac{u}{2}\sqrt{a^2-u^2} + \frac{a^2}{2}\sin^{-1}\frac{u}{a} + C$

72. $\displaystyle\int \frac{du}{u\sqrt{a^2-u^2}} = -\frac{1}{a}\ln\left|\frac{a+\sqrt{a^2-u^2}}{u}\right| + C$

73. $\displaystyle\int \frac{du}{u^2\sqrt{a^2-u^2}} = -\frac{1}{a^2 u}\sqrt{a^2-u^2} + C$

74. $\displaystyle\int (a^2-u^2)^{3/2}\, du = -\frac{u}{8}(2u^2-5a^2)\sqrt{a^2-u^2} + \frac{3a^4}{8}\sin^{-1}\frac{u}{a} + C$

75. $\displaystyle\int \frac{du}{(a^2-u^2)^{3/2}} = \frac{u}{a^2\sqrt{a^2-u^2}} + C$

Forms Involving $\sqrt{u^2 - a^2},\, a > 0$

76. $\displaystyle\int \sqrt{u^2-a^2}\, du = \frac{u}{2}\sqrt{u^2-a^2} - \frac{a^2}{2}\ln\left|u+\sqrt{u^2-a^2}\right| + C$

77. $\displaystyle\int u^2\sqrt{u^2-a^2}\, du$

$\displaystyle = \frac{u}{8}(2u^2-a^2)\sqrt{u^2-a^2} - \frac{a^4}{8}\ln\left|u+\sqrt{u^2-a^2}\right| + C$

78. $\displaystyle\int \frac{\sqrt{u^2-a^2}}{u}\, du = \sqrt{u^2-a^2} - a\cos^{-1}\frac{a}{|u|} + C$

79. $\displaystyle\int \frac{\sqrt{u^2-a^2}}{u^2}\, du = -\frac{\sqrt{u^2-a^2}}{u} + \ln\left|u+\sqrt{u^2-a^2}\right| + C$

80. $\displaystyle\int \frac{du}{\sqrt{u^2-a^2}} = \ln\left|u+\sqrt{u^2-a^2}\right| + C$

81. $\displaystyle\int \frac{u^2\,du}{\sqrt{u^2-a^2}} = \frac{u}{2}\sqrt{u^2-a^2} + \frac{a^2}{2}\ln\left|u+\sqrt{u^2-a^2}\right| + C$

82. $\displaystyle\int \frac{du}{u^2\sqrt{u^2-a^2}} = \frac{\sqrt{u^2-a^2}}{a^2 u} + C$

83. $\displaystyle\int \frac{du}{(u^2-a^2)^{3/2}} = -\frac{u}{a^2\sqrt{u^2-a^2}} + C$

Forms Involving $\sqrt{a^2 + u^2},\, a > 0$

84. $\displaystyle\int \sqrt{a^2+u^2}\, du = \frac{u}{2}\sqrt{a^2+u^2} + \frac{a^2}{2}\ln\left(u+\sqrt{a^2+u^2}\right) + C$

85. $\displaystyle\int u^2\sqrt{a^2+u^2}\, du$

$\displaystyle = \frac{u}{8}(a^2+2u^2)\sqrt{a^2+u^2} - \frac{a^4}{8}\ln\left(u+\sqrt{a^2+u^2}\right) + C$

86. $\displaystyle\int \frac{\sqrt{a^2+u^2}}{u}\, du = \sqrt{a^2+u^2} - a\ln\left|\frac{a+\sqrt{a^2+u^2}}{u}\right| + C$

87. $\displaystyle\int \frac{\sqrt{a^2+u^2}}{u^2}\, du = -\frac{\sqrt{a^2+u^2}}{u} + \ln\left(u+\sqrt{a^2+u^2}\right) + C$

88. $\int \dfrac{du}{\sqrt{a^2+u^2}} = \ln\left(u+\sqrt{a^2+u^2}\right) + C$

89. $\int \dfrac{u^2\,du}{\sqrt{a^2+u^2}} = \dfrac{u}{2}\sqrt{a^2+u^2} - \dfrac{a^2}{2}\ln\left(u+\sqrt{a^2+u^2}\right) + C$

90. $\int \dfrac{du}{u\sqrt{a^2+u^2}} = -\dfrac{1}{a}\ln\left|\dfrac{\sqrt{a^2+u^2}+a}{u}\right| + C$

91. $\int \dfrac{du}{u^2\sqrt{a^2+u^2}} = -\dfrac{\sqrt{a^2+u^2}}{a^2 u} + C$

92. $\int \dfrac{du}{(a^2+u^2)^{3/2}} = \dfrac{u}{a^2\sqrt{a^2+u^2}} + C$

Forms Involving $a + bu$

93. $\int \dfrac{u\,du}{a+bu} = \dfrac{1}{b^2}\left(a+bu - a\ln|a+bu|\right) + C$

94. $\int \dfrac{u^2\,du}{a+bu} = \dfrac{1}{2b^3}\left[(a+bu)^2 - 4a(a+bu) + 2a^2\ln|a+bu|\right] + C$

95. $\int \dfrac{du}{u(a+bu)} = \dfrac{1}{a}\ln\left|\dfrac{u}{a+bu}\right| + C$

96. $\int \dfrac{du}{u^2(a+bu)} = -\dfrac{1}{au} + \dfrac{b}{a^2}\ln\left|\dfrac{a+bu}{u}\right| + C$

97. $\int \dfrac{u\,du}{(a+bu)^2} = \dfrac{a}{b^2(a+bu)} + \dfrac{1}{b^2}\ln|a+bu| + C$

98. $\int \dfrac{du}{u(a+bu)^2} = \dfrac{1}{a(a+bu)} - \dfrac{1}{a^2}\ln\left|\dfrac{a+bu}{u}\right| + C$

99. $\int \dfrac{u^2\,du}{(a+bu)^2} = \dfrac{1}{b^3}\left(a+bu - \dfrac{a^2}{a+bu} - 2a\ln|a+bu|\right) + C$

100. $\int u\sqrt{a+bu}\,du = \dfrac{2}{15b^2}(3bu-2a)(a+bu)^{3/2} + C$

101. $\int u^n\sqrt{a+bu}\,du$
$$= \dfrac{2}{b(2n+3)}\left[u^n(a+bu)^{3/2} - na\int u^{n-1}\sqrt{a+bu}\,du\right]$$

102. $\int \dfrac{u\,du}{\sqrt{a+bu}} = \dfrac{2}{3b^2}(bu-2a)\sqrt{a+bu} + C$

103. $\int \dfrac{u^n\,du}{\sqrt{a+bu}} = \dfrac{2u^n\sqrt{a+bu}}{b(2n+1)} - \dfrac{2na}{b(2n+1)}\int \dfrac{u^{n-1}\,du}{\sqrt{a+bu}}$

104. $\int \dfrac{du}{u\sqrt{a+bu}} = \dfrac{1}{\sqrt{a}}\ln\left|\dfrac{\sqrt{a+bu}-\sqrt{a}}{\sqrt{a+bu}+\sqrt{a}}\right| + C, \quad \text{if } a > 0$

$$= \dfrac{2}{\sqrt{-a}}\tan^{-1}\sqrt{\dfrac{a+bu}{-a}} + C, \quad \text{if } a < 0$$

105. $\int \dfrac{du}{u^n\sqrt{a+bu}} = -\dfrac{\sqrt{a+bu}}{a(n-1)u^{n-1}} - \dfrac{b(2n-3)}{2a(n-1)}\int \dfrac{du}{u^{n-1}\sqrt{a+bu}}$

106. $\int \dfrac{\sqrt{a+bu}}{u}\,du = 2\sqrt{a+bu} + a\int \dfrac{du}{u\sqrt{a+bu}}$

107. $\int \dfrac{\sqrt{a+bu}}{u^2}\,du = -\dfrac{\sqrt{a+bu}}{u} + \dfrac{b}{2}\int \dfrac{du}{u\sqrt{a+bu}}$

Forms Involving $\sqrt{2au - u^2}$, $a > 0$

108. $\int \sqrt{2au-u^2}\,du = \dfrac{u-a}{2}\sqrt{2au-u^2} + \dfrac{a^2}{2}\cos^{-1}\left(\dfrac{a-u}{a}\right) + C$

109. $\int u\sqrt{2au-u^2}\,du$
$$= \dfrac{2u^2-au-3a^2}{6}\sqrt{2au-u^2} + \dfrac{a^3}{2}\cos^{-1}\left(\dfrac{a-u}{a}\right) + C$$

110. $\int \dfrac{du}{\sqrt{2au-u^2}} = \cos^{-1}\left(\dfrac{a-u}{a}\right) + C$

111. $\int \dfrac{du}{u\sqrt{2au-u^2}} = -\dfrac{\sqrt{2au-u^2}}{au} + C$

ESSENTIAL THEOREMS

Intermediate Value Theorem

If $f(x)$ is continuous on a closed interval $[a, b]$ and $f(a) \neq f(b)$, then for every value M between $f(a)$ and $f(b)$, there exists at least one value $c \in (a, b)$ such that $f(c) = M$.

Mean Value Theorem

If $f(x)$ is continuous on a closed interval $[a, b]$ and differentiable on (a, b), then there exists at least one value $c \in (a, b)$ such that

$$f'(c) = \dfrac{f(b) - f(a)}{b - a}$$

Extreme Values on a Closed Interval

If $f(x)$ is continuous on a closed interval $[a, b]$, then $f(x)$ attains both a minimum and a maximum value on $[a, b]$. Furthermore, if $c \in [a, b]$ and $f(c)$ is an extreme value (min or max), then c is either a critical point of $f(x)$ in (a, b) or one of the endpoints a or b.

The Fundamental Theorem of Calculus, Part I

Assume that $f(x)$ is continuous on $[a, b]$ and let $F(x)$ be an antiderivative of $f(x)$ on $[a, b]$. Then

$$\int_a^b f(x)\,dx = F(b) - F(a)$$

Fundamental Theorem of Calculus, Part II

Assume that $f(x)$ is a continuous function on $[a, b]$. Then the area function $A(x) = \displaystyle\int_a^x f(t)\,dt$ is an antiderivative of $f(x)$, that is,

$$A'(x) = f(x) \quad \text{or equivalently} \quad \dfrac{d}{dx}\int_a^x f(t)\,dt = f(x)$$

Furthermore, $A(x)$ satisfies the initial condition $A(a) = 0$.